Selected Keys of the Graphing Calculator

Magnifies or reduces a portion of the curve being viewed and can "square" the graph to reduce distortion.

Used to determine certain important values associated with a graph.

Controls the values that are used when creating a table.

Determines the portion of the curve(s) shown and the scale of the graph.

Used to display the coordinates of points on a curve.

Used to enter the equation(s) that is to be graphed.

Used to display x- and y-values in a table.

Controls whether graphs are drawn sequentially or simultaneously and if the window is split.

Used to graph equations that were entered using the Y= key.

Activates the secondary functions printed above many keys.

Used to move the cursor and adjust contrast.

Used to delete previously entered characters.

Used to fit curves to data.

Used to write the variable, x.

Used to access a previously named function or equation.

These keys are similar to those found on a scientific calculator (see inside back cover).

Used to raise a base to a power.

Used as a negative sign.

Intermediate Algebra
GRAPHS AND MODELS

SECOND EDITION

Intermediate Algebra
GRAPHS AND MODELS

SECOND EDITION

Marvin L. Bittinger
Indiana University Purdue University Indianapolis

David J. Ellenbogen
Community College of Vermont

Barbara L. Johnson
Indiana University Purdue University Indianapolis

PEARSON

Addison
Wesley

Boston   San Francisco   New York
London   Toronto   Sydney   Tokyo   Singapore   Madrid
Mexico City   Munich   Paris   Cape Town   Hong Kong   Montreal

Publisher	Greg Tobin
Editor in Chief	Maureen O'Connor
Acquisitions Editor	Jennifer Crum
Project Editor	Suzanne Alley
Editorial Assistant	Katie Nopper
Managing Editor	Ron Hampton
Production Supervisor	Kathleen A. Manley
Editorial and Production Services	Martha K. Morong/Quadrata, Inc.
Art Editor and Photo Researcher	Geri Davis/The Davis Group, Inc.
Compositor	Beacon Publishing Services
Media Producer	Lynne Blaszak
Software Development	Mary Dougherty and Marty Wright
Marketing Manager	Dona Kenly
Marketing Coordinator	Lindsay Skay
Prepress Supervisor	Caroline Fell
Manufacturing Buyer	Evelyn Beaton
Design Supervisor	Dennis Schaefer
Text Designer	Geri Davis/The Davis Group, Inc.
Cover Designer	KoDesign Studio
Cover Photograph	Larry Lee Photography/CORBIS

For permission to use copyrighted material, grateful acknowledgment is made to the copyright holders on page A-52, which is hereby made part of this copyright page.

Library of Congress Cataloging-in-Publication Data

Bittinger, Marvin L.
    Intermediate algebra: graphs and models / Marvin L. Bittinger,
    David J. Ellenbogen, Barbara L. Johnson.—2nd ed.
      p. cm.
    ISBN 0-321-12709-9 (SE: alk. paper)—ISBN 0-321-16864-X (AIE: alk. paper)
    1. Algebra.   I. Ellenbogen, David.   II. Johnson, Barbara L. (Barbara Loreen),
    1962-III.
    Title.

QA154.3.B5824   2003
512.9—dc21

                2003048213

1 2 3 4 5 6 7 8 9 10—RRDW—07 06 05 04 03

Contents

Preface

Appropriate for a one-term course in intermediate algebra, *Intermediate Algebra: Graphs and Models*, Second Edition, is intended for those students who have completed a first course in algebra. This text is more interactive than most other intermediate algebra texts. Our goal is to enhance the learning process through the use of technology by encouraging students to visualize the mathematics and by providing as much support as possible to help students in their study of algebra. This text is part of a series that includes the following texts:

- *Elementary Algebra: Graphs and Models*
- *Elementary and Intermediate Algebra: Graphs and Models*, Second Edition

Content Features

Problem Solving

One distinguishing feature of our approach is our treatment of and emphasis on problem solving. We use problem solving and applications to motivate the material wherever possible, and we include real-life applications and problem-solving techniques throughout the text. Problem solving not only encourages students to think about how mathematics can be used, it helps to prepare them for more advanced material in future courses.

- In Chapter 1, we introduce the five-step process for solving problems: (1) *Familiarize*, (2) *Translate*, (3) *Carry out*, (4) *Check*, and (5) *State* the answer. These steps are then used consistently throughout the text whenever we encounter a problem-solving situation. Repeated use of this problem-solving strategy gives students a sense that they have a starting point for any type of problem they encounter, and frees them to focus on the mathematics necessary to successfully translate the problem situation. We often use estimation and carefully checked guesses to help with the *Familiarize* and *Check* steps (see pp. 54, 111, 209, and 445–446).

Algebraic/Graphical Side-by-Sides

Algebraic/graphical side-by-sides give students a direct comparison between these two problem-solving methods. They show the connection between algebraic and graphical or visual solutions and demonstrate that there is more than one way to obtain a result. This approach also illustrates the comparative efficiency and accuracy of the two methods. (See pp. 382 and 620.)

Applications

Interesting applications of mathematics help motivate both students and instructors. Solving applied problems gives students the opportunity to see their conceptual understanding put to use in a real way. In the Second Edition of *Intermediate Algebra: Graphs and Models*, not only have we increased the total number of applications and real-data problems overall, but we have increased the number of source lines to better highlight the real-world data. As in the past, art is integrated into the applications and exercises to aid the student in visualizing the mathematics. (See pp. 110, 260, 386, and 546.)

Interactive Discoveries

Interactive Discoveries invite students to develop analytical and reasoning skills while taking an active role in the learning process. These discoveries can be used as lecture launchers to introduce new topics at the beginning of a class and quickly guide students through a concept, or as out-of-class concept discoveries. (See pp. 27 and 125.)

Pedagogical Features

New! **Connecting the Concepts.** To help students understand the big picture, Connecting the Concepts subsections within each chapter relate the concept at hand to previously learned and upcoming concepts. Because students may occasionally lose sight of the forest because of the trees, this feature will help them keep better track of their bearings as they encounter new material. (See pp. 66, 183, 198, and 342.)

New! **Study Tips.** Most plentiful in the first three chapters when students are still establishing their study habits, Study Tips are found in the margins and interspersed throughout the first seven chapters. Our Study Tips range from how to approach assignments, to reminders of the various study aids that are available, to strategies for preparing for a final exam. (See pp. 18, 81, and 516.)

New! **Calculator References.** Throughout each chapter, calculator references (indicated by the icon) act as an in-text manual to introduce and explain various graphing-calculator techniques. The references are placed so the student can learn the technique just prior to practicing it. (See pp. 9–10 and 424.)

New! ***Aha!* Exercises.** Designated by Aha!, these exercises can be solved quickly if the student has the proper insight. The Aha! designation is used the first time a new insight can be used on a particular type of exercise and indicates to the student that there is a simpler way to complete the exercise that requires less lengthy computation. It's then up to the student to find the simpler approach and, in subsequent exercises, to determine if and when that particular insight can be used again. Occasionally an *Aha!* exercise is easily answered by looking at the preceding odd-numbered exercise. Our hope is that the *Aha!* exercises will discourage rote learning and reward students who look before they leap into a problem. (See pp. 11, 155, 188, 344, and 363.)

Synthesis Exercises. Located within the review exercises, these exercises offer opportunities for students to synthesize skills and concepts from earlier sections with the present material, and often provide students with deeper insights into the current topic. Synthesis exercises are generally more challenging than those in the main body of the exercise set. (See pp. 119, 133, 289, and 321.)

Thinking and Writing Exercises. In this edition, nearly every set of exercises includes at least four thinking and writing exercises. Two of these are more basic and appear just before the Skill Maintenance exercises. The other thinking and writing exercises are more challenging and appear as Synthesis exercises. All are marked with τ^W and require answers that are one or more complete sentences. This type of problem has been found to aid in student comprehension, critical thinking, and conceptualization. Because some instructors may collect answers to writing exercises, and because more than one answer may be correct, answers to thinking and writing exercises are not listed at the back of the text. (See pp. 70, 159, 191, and 420.)

Collaborative Corners. In today's professional world, teamwork is essential. We continue to provide optional Collaborative Corner features throughout the text that require students to work in groups to explore and solve mathematical problems. There are one to three Collaborative Corners per chapter, each one appearing after the appropriate exercise set. (See pp. 72, 200–201, 390, and 476.)

Cumulative Review. After Chapters 3, 6, 9, and 10, we have included a Cumulative Review, which reviews skills and concepts from preceding chapters of the text. (See pp. 248, 482, 733, and 780.)

What's New in the Second Edition?

We have rewritten many key topics in response to user and reviewer feedback and have made significant improvements in design, art, pedagogy, and an expanded supplements package. Detailed information about the content changes is available in the form of a conversion guide. Please ask your local Addison-Wesley sales consultant for more information. Following is a list of the major changes in this edition.

New Design
You will see that the page dimension for this edition is larger, which allows for an open look and a typeface that is easier to read. In addition, we continue to pay close attention to the pedagogical use of color to make sure that it is used to present concepts in the clearest possible manner.

Content Changes

A variety of content changes have been made throughout the text. Some of the more significant changes are listed below.

- Chapter 2 now includes a brief introduction to interpolation and extrapolation. The concept of slope is now closely linked with the idea of rate of change, beginning in Section 2.4.
- The topic of variation is moved from Section 8.6 into Section 6.8. As a result, Chapter 8 is shortened to 9 sections.
- Chapter 7 is rewritten so that Section 7.3 is now strictly on multiplying radical expressions. Section 7.4 is now strictly on division of radical expressions. Section 7.5 is now devoted to expressions with two or more radical terms.
- Chapter 9 now begins with Composite and Inverse Functions (formerly Section 9.2) and then moves to Exponential Functions (formerly Section 9.1). This results in a better flow of topics and facilitates coverage of Composite and Inverse Functions as a stand-alone topic if desired.
- Conic sections are now included in Appendixes A and B. The coverage of circles, the distance formula, and the midpoint formula is moved from Section 7.7 to Appendix A to consolidate related topics.
- Previously used in synthesis exercises only, factoring sums and differences of cubes is now covered in Section 5.7 and used in later sections such as manipulations with rational expressions.

Supplements for the Instructor

New! Annotated Instructor's Edition
(ISBN 0-321-16864-X)

The *Annotated Instructor's Edition* includes all the answers to the exercise sets, usually right on the page where the exercises appear, and Teaching Tips in the margins that give insights and classroom discussion suggestions that will be especially useful for new instructors. These handy answers and ready Teaching Tips will help both new and experienced instructors save classroom preparation time.

Instructor's Solutions Manual
(ISBN 0-321-16862-3)

The *Instructor's Solutions Manual* by Judith A. Penna contains worked-out solutions to all exercises in the exercise sets, including the thinking and writing exercises. It also includes a sample test with answers for each chapter and answers to the exercises in the appendixes. The sample tests are also included in the *Student's Solutions Manual*.

Printed Test Bank/Instructor's Resource Guide
(ISBN 0-321-16861-5)

This supplement contains the following:

- Extra practice problems.
- Black-line masters of grids and number lines for transparency masters or test preparation.
- A videotape index.

The test bank portion contains the following:

- Six free-response test forms for each chapter, following the format of and with the same level of difficulty as the tests in the main text.
- Two multiple-choice test forms for each chapter.
- Eight alternative forms of the final examination, three with questions organized by type, three with questions organized by chapter, and two with multiple-choice questions.

MyMathLab

MyMathLab.com is a complete, online course for Addison-Wesley mathematics textbooks that integrates interactive, multimedia instruction correlated to the textbook content. MyMathLab can be easily customized to suit the needs of students and instructors and provides a comprehensive and efficient online course-management system that allows for diagnosis, assessment, and tracking of students' progress.

MyMathLab features:

- Fully interactive multimedia textbooks are built in CourseCompass, a version of Blackboard™ designed specifically for Addison-Wesley.
- Chapter and section folders from the textbook contain a wide range of instructional content: videos, software tools, audio clips, animations, and electronic supplements.
- Hyperlinks take you directly to online testing, diagnosis, tutorials, and gradebooks in MathXL—Addison-Wesley's tutorial and testing system for mathematics and statistics.
- Instructors can create, copy, edit, assign, and track all tests for their course as well as track student tutorial and testing performance.
- With push-button ease, instructors can remove, hide, or annotate Addison-Wesley preloaded content, add their own course documents, or change the order in which material is presented.
- Using the communication tools found in MyMathLab, instructors can hold online office hours, host a discussion board, create communication groups within their class, send e-mails, and maintain a course calendar.
- Print supplements are available online, side by side with their textbooks.

For more information, visit our Web site at www.mymathlab.com or contact your Addison-Wesley sales representative for a live demonstration.

TestGen with QuizMaster
(ISBN 0-321-16903-4)

TestGen enables instructors to build, edit, print, and administer tests using a computerized bank of questions developed to cover all the objectives of the text. Instructors can modify test-bank questions or add new questions by using the built-in question editor, which allows users to create graphs, import graphics, insert math notation, and insert variable numbers or text. Tests can be printed or administered online via the Web or other network. TestGen comes packaged with QuizMaster, which allows students to take tests on a local area network. The software is available on a dual-platform Windows/Macintosh CD-ROM.

MathXL

MathXL is an online testing, homework, and tutorial system that uses algorithmically generated exercises correlated to your textbook. Instructors can assign tests and homework provided by Addison-Wesley or create and customize their own tests and homework assignments. Instructors can also track their students' results and tutorial work in an online gradebook. Students can take chapter tests and receive personalized study plans that will diagnose weaknesses and link them to areas they need to study and retest. Students can also work unlimited practice problems and receive tutorial instruction for areas in which they need improvement. MathXL can be packaged with new copies of *Intermediate Algebra: Graphs and Models*, Second Edition. Please contact your Addison-Wesley representative for details.

Supplements for the Student

Graphing Calculator Manual
(ISBN 0-321-16863-1)

The *Graphing Calculator Manual* by Judith A. Penna, with the assistance of Daphne Bell, contains keystroke level instruction for the Texas Instruments TI-83/83+, TI-86, and TI-89.

Bundled free with every copy of the text, the *Graphing Calculator Manual* uses actual examples and exercises from *Intermediate Algebra: Graphs and Models*, Second Edition, to help teach students to use their graphing calculator. The order of topics in the *Graphing Calculator Manual* mirrors that of the text, providing a just-in-time mode of instruction.

Student's Solutions Manual
(ISBN 0-321-16870-4)

The *Student's Solutions Manual* by Judith A. Penna contains completely worked-out solutions with step-by-step annotations for all the odd-numbered exercises in the exercise sets in the text, with the exception of the thinking and writing exercises. It also includes a self-test with answers for each chapter and a final examination.

MyMathLab

MyMathLab.com is a complete, online course for Addison-Wesley mathematics textbooks that integrates interactive, multimedia instruction correlated to the textbook content. MyMathLab can be easily customized to suit the needs of students and instructors and provides a comprehensive and efficient online course-management system that allows for diagnosis, assessment, and tracking of students' progress.

MyMathLab features:

- Fully interactive multimedia textbooks are built in CourseCompass, a version of Blackboard™ designed specifically for Addison-Wesley.
- Chapter and section folders from the textbook contain a wide range of instructional content: videos, software tools, audio clips, animations, and electronic supplements.
- Hyperlinks take you directly to online testing, diagnosis, tutorials, and gradebooks in MathXL—Addison-Wesley's tutorial and testing system for mathematics and statistics.
- Print supplements are available online, side by side with their textbooks.

For more information, visit our Web site at www.mymathlab.com.

InterAct Math Tutorial Software
(ISBN 0-321-16906-9)

Available on CD-ROM, this interactive tutorial software provides algorithmically generated practice exercises that are correlated at the objective level to the odd-numbered exercises in the text. Every exercise in the program is accompanied by an example and a guided solution designed to involve students in the solution process. Selected problems also include a video clip to help students visualize concepts. The software tracks student activity and scores and can generate printed summaries of students' progress.

MathXL

MathXL is an online testing, homework, and tutorial system that uses algorithmically generated exercises correlated to your textbook. Students can take chapter tests and receive personalized study plans that will diagnose weaknesses and link them to areas they need to study and retest. Students can also work unlimited practice problems and receive tutorial instruction for areas in which they need improvement.

Addison-Wesley Math Tutor Center

The Addison-Wesley Math Tutor Center is staffed by qualified college mathematics instructors who tutor students on examples and exercises from the textbook. Tutoring is provided via toll-free telephone, toll-free fax, e-mail, and the Internet. Interactive web-based technology allows students and tutors to view and listen to live instruction in real-time over the Internet! The Math Tutor Center is accessed through a registration number that can be packaged with a new textbook or purchased separately. (*Note*: MyMathLab students obtain access to the Tutor Center through their MyMathLab access code.)

Videotape Series
(ISBN 0-321-16902-6)

This series of videotapes, created specifically for *Intermediate Algebra: Graphs and Models*, Second Edition, features an engaging team of lecturers who provide comprehensive lessons on every objective in the text. Many of the video segments are staged as an office hour during which a student poses questions and works through problems with the instructor.

Digital Video Tutor
(ISBN 0-321-16907-7)

This supplement provides the entire set of videotapes for the text in digital format on CD-ROM, making it easy and convenient for students to watch video segments from a computer, either at home or on campus. Available for purchase with the text at minimal cost, the Digital Video Tutor is ideal for distance learning and supplemental instruction.

Acknowledgments

No book can be produced without a team of professionals who take pride in their work and are willing to put in long hours. Laurie A. Hurley deserves special thanks for her careful accuracy checks, well-thought-out suggestions, and uncanny eye for detail. We are extremely grateful for Judy Penna's careful reading of the text and outstanding work in organizing and preparing the printed supplements. Thanks to Mark Stevenson for authoring the *Printed Test Bank*. Dawn Mulheron, Julie Stephenson, and Daphne Bell of Motlow State Community College provided enormous help, often in the face of great time pressure, as accuracy checkers. We are also indebted to Chris Burditt and Jann MacInnes for their many fine ideas that appear in our Collaborative Corners and Janet Wyatt for her recommendations for Teaching Tips featured in the *Annotated Instructor's Edition*.

Martha Morong, of Quadrata, Inc., provided editorial and production services of the highest quality imaginable—she is simply a joy to work with. Geri Davis, of the Davis Group, Inc., performed superb work as designer, art editor, and photo researcher, and always with a disposition that can brighten an otherwise gray day. Network Graphics generated the graphs, charts, and many of the illustrations. Not only are the people at Network reliable, but they clearly take pride in their work. The many hand-drawn illustrations appear thanks to Jim Bryant, a gifted artist with true mathematical sensibilities.

Our team at Addison-Wesley deserves special thanks. Editorial Assistant Katie Nopper coordinated all the reviews and managed many of the day-to-day details—always in a pleasant and reliable manner. Project Editor Suzanne Alley expertly provided information and a steadying influence along with gentle prodding at just the right moments. Senior Acquisitions Editor Jenny Crum provided many fine suggestions along with unflagging support. Senior Production Supervisor Kathy Manley exhibited patience when others would have shown frustration. Designer Dennis Schaefer's willingness to listen and then creatively respond resulted in a book that is beautiful to look at. Senior Marketing Manager Dona Kenly and Senior Marketing Coordinator Lindsay Skay skillfully kept us in touch with the needs of faculty; Associate Producer Lynne Blaszak provided us with the technological guidance so necessary for our many supplements and our fine video series. To all of these people we owe a real debt of gratitude.

Reviewers

Marwan Abu-Sawwa, *Florida Community College at Jacksonville*
Peggy Clifton, *Redlands Community College*
Helen Denat, *Berkshire Community College*
Steven Drucker, *Santa Rosa Junior College*
Al Giambrone, *Sinclair Community College*
Alice Hemingway, *Western Nebraska Community College*
Kandace Kling, *Portland Community College, Sylvania*
Donna Mills, *Frederick Community College*
Dan Munton, *Santa Rosa Junior College*
Sharon O'Donnell, *Chicago State University*
Jeannette O'Rourke, *Middlesex County College*
Jeff Parent, *Oakland Community College, Royal Oak*
Shirley Pereira, *Grossmont College*
Don Ransford, *Edison Community College*
Adam Rubin, *Harper College*
Alberta Sautter, *Southeast Community College, Beatrice*
Trudy Streilein, *Northern Virginia Community College*
Kevin Yokoyama, *College of the Redwoods*
Bella Zamansky, *University of Cincinnati*
Linda Zimmerman, *New Mexico State University*

M.L.B.
D.J.E.
B.L.J.

Feature Walkthrough

Chapter Openers

Each chapter opens with a real-data application, including a data table (numerical representation) and a graphical representation of the data. Data tables and graphs are used frequently throughout the body of the text to show the relevance of the material as well as to appeal to visual learners.

Functions, Linear Equations, and Models

2

2.1 Functions
2.2 Solving Linear Equations
2.3 Applications and Formulas
2.4 Linear Functions: Slope, Graphs, and Models
2.5 Another Look at Linear Graphs
2.6 Introduction to Curve Fitting: Point–Slope Form
2.7 Domains and the Algebra of Functions

SUMMARY AND REVIEW

TEST

A *function* is a certain kind of relationship between sets. Functions are very important in mathematics in general, and in problem solving in particular. In this chapter, you will learn what we mean by a function and then begin to use functions to solve problems.

A function that can be described by a linear equation is called a *linear function*. We will study graphs of linear equations in detail and will use linear equations and functions to model and solve applications.

APPLICATION

PAPER RECYCLING. The following table shows the amount of paper recovered in the United States for various years. Estimate the amount of paper that will be recovered in 2003.

YEAR	PAPER RECOVERED (IN MILLIONS OF TONS)
1988	26.2
1990	29.1
1992	34.0
1994	39.7
1996	43.1
1998	45.1
2000	49.4

Source: www.afandpa.org

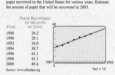

The graph shows that a linear function will fit the data and can be used to make a prediction.

This problem appears as Example 5 in Section 2.6.

---

24   CHAPTER 1   BASICS OF ALGEBRA AND GRAPHING

Since $b - c = b + (-c)$, it is also true by the distributive law $a(b - c) = ab - ac$.

EXAMPLE 4   Obtain an expression equivalent to $5x(y + 4)$ by multiplying.

Solution   We use the distributive law to get

$$5x(y + 4) = 5xy + 5x \cdot 4 \qquad \text{Using the distributive law}$$
$$= 5xy + 5 \cdot 4 \cdot x \qquad \text{Using the commutative law of multiplication}$$
$$= 5xy + 20x. \qquad \text{Simplifying}$$

The expressions $5x(y + 4)$ and $5xy + 20x$ are equivalent. They same number for any replacements of x and y.

When we do the opposite of what we did in Example 4, we say **factoring** an expression. This allows us to rewrite a sum as a produc

TEACHING TIP

Be aware that students may omit the 1 in factorizations such as $10t + 5 = 5(2t + 1)$.

EXAMPLE 5   Obtain an expression equivalent to $3x - 6$ by facto

Solution   We use the distributive law to get

$$3x - 6 = 3 \cdot x - 3 \cdot 2 = 3(x - 2).$$

In Example 5, since the product of 3 and $x - 2$ is $3x - 6$, we and $x - 2$ are **factors** of $3x - 6$. Thus "factor" can act as a noun o

Combining Like Terms

In an expression like $8a^3 + 17 + 4/b + (-6a^3b)$, the parts that are by addition signs are called *terms*. A **term** is a number, a variable, numbers and/or variables, or a quotient of numbers and/or variables. 17, 4/b, and $-6a^3b$ are terms in $8a^3 + 17 + 4/b + (-6a^3b)$. W have variable factors that are exactly the same, we refer to those ter or **similar**, **terms**. Thus, $3x^2y$ and $-7x^2y$ are similar terms, but $3x^2$ are not. We can often simplify expressions by **combining**, or **colle terms**.

EXAMPLE 6   Combine like terms: $3a + 5a^2 + 4a + a^2$.

Solution

$$3a + 5a^2 + 4a + a^2 = 3a + 4a + 5a^2 + a^2 \qquad \text{Using the commutat}$$
$$= (3 + 4)a + (5 + 1)a^2 \qquad \text{Using the distributi law. Note } a^2 = 1a^2.$$
$$= 7a + 6a^2$$

TEACHING TIP

Some students may have difficulty seeing that $a^2 = 1 \cdot a^2$.

Sometimes we must use the distributive law to remove grouping symbols before combining like terms. Remember to remove the innermost grouping symbols first.

Annotated Instructor's Edition

The *Annotated Instructor's Edition* includes all the answers to the exercise sets, usually right on the page where the exercises appear, and Teaching Tips in the margins that give insights and classroom discussion suggestions that will be especially useful for new instructors. These handy answers and ready Teaching Tips will help both new and experienced instructors save preparation time.

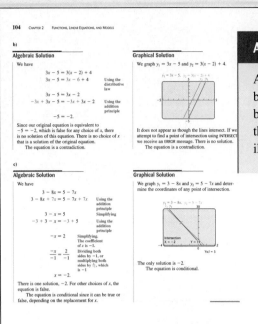

Algebraic/Graphical Side-by-Sides

Algebraic/Graphical Side-by-Sides give students a direct comparison between these two problem-solving methods. They show the connection between algebraic and graphical or visual solutions and demonstrate that there is more than one way to obtain a result. This approach also illustrates the comparative efficiency and accuracy of the two methods.

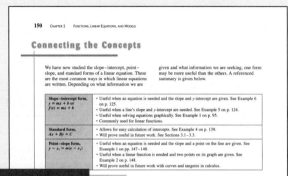

Connecting the Concepts

To help students understand the big picture, Connecting the Concepts subsections relate the concept at hand to previously learned and upcoming concepts. Because students occasionally lose sight of the forest because of the trees, this feature helps students keep their bearings as they encounter new material.

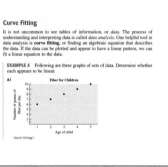

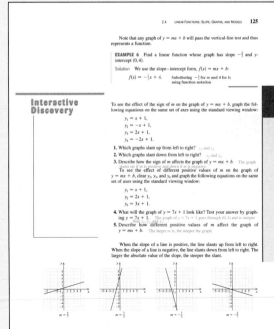

Interactive Discoveries

Interactive Discoveries invite students to develop analytical and reasoning skills while taking an active role in the learning process. These discoveries can be used as lecture launchers to introduce new topics at the start of a class and quickly guide students through a concept, or as out-of-class concept discoveries.

Real-Data Applications

Real-data applications, often denoted in examples and exercises with a source line, serve to motivate students by showing them how the mathematics they are studying in their text relates to their everyday lives. The number of application problems has been increased significantly in this edition.

Graphs as Models

Graphs can represent a large amount of information in a concise way. They also help to visualize problem situations and to show relationships between two quantities. When one of those quantities depends on the other, we generally represent the dependent variable on the vertical axis.

EXAMPLE 6 Sales of Cellular Phones. The following graph shows the number of cellular phones sold worldwide for the years 1996 to 2001. (*Source:* Gartner Dataquest)

a) How many cellular phones were sold in 2001?

b) In what year was the number of cellular phones sold the greatest?

c) Between what years did the number of cellular phones sold decrease?

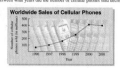

Worldwide Sales of Cellular Phones

TEACHING TIP
Reinforce that although a value read from a graph is an estimate, such a value is often all that is required.

Solution

a) First, we note that "year" is on the horizontal axis and "Sales of Cellular Phones" is on the vertical axis. We also note that the number of cellular phones sold is given in millions. After locating 2001 on the horizontal axis, we move up to the graph and horizontally to the corresponding value on the vertical axis. We see that in 2001, about 400 million phones were sold.

b) We find the highest point on the graph and then move down to the horizontal axis. The number of cellular phones sold was greatest in 2000.

c) As we move from 1996 to 2000 along the horizontal axis, the graph rises, which indicates that cellular phone sales were increasing from one year to the next. The graph falls as we move from 2000 to 2001, so we know that there was a decrease in cellular phone sales between 2000 and 2001.

Coordinates and Points

Coordinates of ordered pairs are entered as *data* in lists, using the STAT menu. To enter or change data, press STAT and then choose the EDIT option. The lists of numbers will appear as three columns on the screen. If there are already numbers in the lists, clear them by moving the cursor to the title of the list (L₁, L₂, and so on) and pressing CLEAR ENTER.

EXAMPLE 8 Graph: $y = x^2 - 5$.

Solution We select numbers for x and find the corresponding values for y. For example, if we choose -2 for x, we get $y = (-2)^2 - 5 = 4 - 5 = -1$. The table lists several ordered pairs.

x	y
0	-5
-1	-4
1	-4
-2	-1
2	-1
-3	4
3	4

TEACHING TIP
To help students see that a graph "pictures" all solutions of an equation, ask them to select a point on the graph and check to see if it is a solution of the equation.

Next, we plot the points. The more points plotted of the graph becomes. Since the value of $x^2 - 5$ gr away from the origin, the graph rises steeply on eith

---

The line that *best* describes the data may not actually go through any of the given points. There are different methods for finding an equation of a line that fits a set of data. These methods generally consider all the points, not just two, when fitting an equation to data. The most commonly used method is *linear regression*.

The development of the method of linear regression belongs to a later mathematics course, but most graphing calculators offer regression as a way of fitting a line or curve to a set of data.

Linear Regression

Fitting a curve to data is done using the STAT CALC menu. Enter the values of the independent variable as L₁ and the values of the dependent variable as L₂. To fit a line to the data, choose the LinReg option of the STAT CALC menu and press ENTER. (If the data are entered in lists other than L₁ and L₂, the names of the list must follow the LinReg command.) The "best-fit" equation will be given in slope–intercept form $y = ax + b$. The values of r^2 and r that may appear on the screen give an indication of how well the regression line fits the data. When r^2 is close to 1, the regression line is a good fit for the data. We call r the *coefficient of correlation*. The values of r^2 and r will appear only if the *diagnostics* are turned on.

Curve Fitting

Curve fitting is a visual theme introduced in Chapter 2 and provided as an optional topic with quadratic, cubic, quartic, exponential, and logarithmic functions. Primarily with, but sometimes without, the graphing calculator, students are able to visualize trends and make estimations and predictions based on real-world data. The content is optional and can be skipped without any loss of continuity.

---

x	$f(x)$
-1	3
0	3
2	3

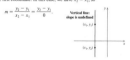

We see from Example 1 the following:

The graph of any constant function of the form $f(x) = b$ or $y = b$ is a horizontal line that crosses the y-axis at $(0, b)$.

Suppose that two different points are on a vertical line. They then have the same first coordinate. In this case, we have $x_2 = x_1$, so

$$m = \frac{y_2 - y_1}{x_2 - x_1} = \frac{y_2 - y_1}{0}.$$

Study Tip

If you are finding it difficult to master a particular topic or concept, talk about it with a classmate. Verbalizing your questions about the material might help clarify it for you. If your classmate is also having difficulty, it is possible that a majority of your classmates are confused and you can ask your instructor to explain the concept again.

Since we cannot divide by 0, this is undefined. Note that when we say that $(y_2 - y_1)/0$ is undefined, it means that we have agreed to not attach any meaning to that expression.

Slope of a Vertical Line

The slope of a vertical line is undefined.

EXAMPLE 2 Graph: $x = -2$.

Solution With y missing, no matter which value of y is chosen, x must be -2. Thus the pairs $(-2, 3)$, $(-2, 0)$, and $(-2, -4)$ all satisfy the equation. The graph is a line parallel to the y-axis. Note that since y is missing, this equation cannot be written in slope–intercept form.

Study Tips

To help students develop good study habits throughout this course, Study Tips are interspersed throughout the first seven chapters in the margins. These tips range from how to approach assignments, to reminders of the various study aids that are available, to strategies for preparing for a final exam.

Five-Step Problem-Solving Process

Bittinger's five-step problem-solving process (Familiarize, Translate, Carry Out, Check, State) is introduced early in the text and used consistently throughout the text whenever students encounter an application problem. This provides students with a consistent framework for solving application problems.

Use of Color

Specific colors have been designated for the graphing calculator screens and in other technical art. Lines are both labeled and color coded. This consistent use of color makes the presentation of graphs more attractive and easier to read than traditional graphing calculator screens.

Annotated Examples

Learning is carefully guided with numerous color-coded art pieces and step-by-step annotations alongside examples. Substitutions and annotations are highlighted in red so students can see exactly what is happening in each step.

1.6

The Five-Step Strategy ■ Translating to Algebraic Expressions and Equations ■
Models ■ Graphs as Models

Mathematical Models and Problem Solving

We now begin to study and practice the "art" of problem solving. Although we are interested mainly in using algebra to solve problems, much of what we say here applies to all methods of problem solving.

What do we mean by a *problem*? Perhaps you have already used algebra to solve some "real-world" problems. What procedure did you use? Was there anything in your approach that could be used to solve problems of a more general nature? These are some questions that we will answer in this section.

In this text, we do not restrict the use of the word "problem" to computational situations involving arithmetic or algebra, such as $589 + 437 = a$ or $3x + 5x = 9$. We mean instead some question to which we wish to find an answer. Perhaps this can best be illustrated with some sample problems:

1. Can I afford to rent a bigger apartment?
2. If I exercise twice a week and eat 3000 calories a day, will I lose weight?
3. Do I have enough time to take 4 courses while working 20 hours a week?
4. My fishing boat travels 12 km/h in still water. How long will it take me to cruise 25 km upstream if the river's current is 3 km/h?

Although these problems are all different, there is a general strategy that can be applied to all of them.

The Five-Step Strategy

Since you have already studied some algebra, you have had some experience with problem solving. The following steps constitute a strategy that you may already have used and a good strategy for problem solving in general.

Five Steps for Problem Solving with Algebra
1. *Familiarize* yourself with the problem.
2. *Translate* to mathematical language.
3. *Carry out* some mathematical manipulation.
4. *Check* your possible answer in the original problem.
5. *State* the answer clearly.

Of the five steps, probably the most important is the first: becoming familiar with the problem situation. Here are some ways in which this can be done.

Solution To determine the slope of each line, we solve for y to find slope–intercept form:

$$3x - y = 7$$
$$-y = -3x + 7 \qquad \text{Adding } -3x \text{ to both sides}$$
$$y = 3x - 7; \qquad \text{Multiplying both sides by } -1$$
$$x + 3y = 1$$
$$3y = -x + 1 \qquad \text{Adding } -x \text{ to both sides}$$
$$y = -\tfrac{1}{3}x + \tfrac{1}{3}. \qquad \text{Multiplying both sides by } \tfrac{1}{3}$$

The slopes of the lines are 3 and $-\tfrac{1}{3}$. Since $3 \cdot \left(-\tfrac{1}{3}\right) = -1$, the lines are perpendicular.

To check, we graph $y_1 = 3x - 7$ and $y_2 = -\tfrac{1}{3}x + \tfrac{1}{3}$. The graphs are shown in a standard viewing window on the left below. Note that the lines do not appear to be perpendicular. If we press ZOOM 5, the lines are graphed in a squared viewing window as shown in the graph on the right below, and they do appear perpendicular. We can perform a better visual check by laying a corner of a piece of paper on the screen. If the lines are perpendicular, they should exactly fit the corner.

EXAMPLE 9 Consider the function given by $f(x) = \tfrac{2}{3}x - 8$.

a) Write an equation for a linear function g with a graph parallel to the graph of f and a y-intercept of $\left(0, \tfrac{3}{4}\right)$.

b) Write an equation for a linear function h with a graph perpendicular to the graph of f and a y-intercept of $\left(0, \tfrac{3}{4}\right)$.

Solution

a) Since g is linear, it can be written in the form $g(x) = mx + b$. To find g, we must determine its slope and y-intercept. The slope of the line given by $f(x) = \tfrac{2}{3}x - 8$ is $\tfrac{2}{3}$. Therefore, the slope of a parallel line is $\tfrac{2}{3}$. Since we are given that the y-intercept of the graph of g is $\left(0, \tfrac{3}{4}\right)$, we have $g(x) = \tfrac{2}{3}x + \tfrac{3}{4}$.

b) Since the slope of the graph of f is $\tfrac{2}{3}$, the slope of a line perpendicular to the graph is $-\tfrac{3}{2}$. Thus, $m = -\tfrac{3}{2}$ and $b = \tfrac{3}{4}$, so $h(x) = -\tfrac{3}{2}x + \tfrac{3}{4}$.

PROBLEM SOLVING

Graphing calculators are especially useful when equations contain fractions or decimals or when the coordinates of the intersection are not integers.

EXAMPLE 5 Solve graphically:

$$3.45x + 4.21y = 8.39,$$
$$7.12x - 5.43y = 6.18.$$

Solution First, we solve for y in each equation:

$$3.45x + 4.21y = 8.39$$
$$4.21y = 8.39 - 3.45x \qquad \text{Subtracting } 3.45x \text{ from both sides}$$
$$y = (8.39 - 3.45x)/4.21; \qquad \text{Dividing both sides by } 4.21$$

$$7.12x - 5.43y = 6.18$$
$$-5.43y = 6.18 - 7.12x \qquad \text{Subtracting } 7.12x \text{ from both sides}$$
$$y = (6.18 - 7.12x)/(-5.43). \qquad \text{Dividing both sides by } -5.43$$

It is not necessary to simplify further. We have the system

$$y = (8.39 - 3.45x)/4.21,$$
$$y = (6.18 - 7.12x)/(-5.43).$$

Next, we enter both equations and graph using the same viewing window. By using the INTERSECT feature in the CALC menu, we see that, to the nearest hundredth, the solution is (1.47, 0.79).

Models

Sometimes two or more sets of data can be modeled by a system of linear equations.

EXAMPLE 6 Travel. The numbers of U.S. travelers to Canada and to Europe are listed in the following table.

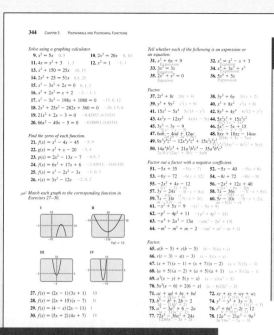

Solve using a graphing calculator.

9. $x^2 = 5x$ 0, 5
10. $2x^2 = 20x$ 0, 10
11. $4x = x^2 + 3$ 1, 3
12. $x^2 = 1$ −1, 1
13. $x^2 + 150 = 25x$ 10, 15
14. $2x^2 + 25 = 51x$ 0.5, 25
15. $x^3 − 3x^2 + 2x = 0$ 0, 1, 2
16. $x^3 + 2x^2 = x + 2$ −2, −1, 1
17. $x^3 − 3x^2 − 198x + 1080 = 0$ −15, 6, 12
18. $2x^3 + 25x^2 − 282x + 360 = 0$ −20, 1.5, 6
19. $21x^2 + 2x − 3 = 0$ −0.42857, 0.33333
20. $66x^2 − 49x − 5 = 0$ −0.09091, 0.83333

Find the zeros of each function.

21. $f(x) = x^2 − 4x − 45$ −5, 9
22. $g(x) = x^2 + x − 20$ −5, 4
23. $p(x) = 2x^2 − 13x − 7$ −0.5, 7
24. $f(x) = 6x^2 + 17x + 6$ −2.42013, −0.41320
25. $f(x) = x^3 − 2x^2 − 3x$ −1, 0, 3
26. $r(x) = 3x^3 − 12x$ −2, 0, 2

Match each graph to the corresponding function in Exercises 27–30.

I   II   III   IV

27. $f(x) = (2x − 1)(3x + 1)$ III
28. $f(x) = (2x + 15)(x − 7)$ II
29. $f(x) = (4 − x)(2x − 11)$ I
30. $f(x) = (5x + 2)(4x + 7)$ IV

Tell whether each of the following is an expression or an equation.

31. $x^2 + 6x + 9$ Expression
32. $x^2_1 = x^2 − x + 3$ Equation
33. $3x^2 = 3x$ Equation
34. $x^3 + 3x^2 + x^2$ Expression
35. $2x^2 + x^2 = 0$ Equation
36. $5x^4 + 5x$ Expression

Factor.

37. $2t^2 + 8t$ $2t(t + 4)$
38. $3y^2 + 6y$ $3y(y + 2)$
39. $y^3 + 9y^2$ $y^2(y + 9)$
40. $x^3 + 8x^2$ $x^2(x + 8)$
41. $15x^2 − 5x^4$ $5x^2(3 − x^2)$
42. $8y^2 + 4y^4$ $4y^2(2 + y^2)$
43. $4x^2y − 12xy^2$ $4xy(x − 3y)$
44. $5x^2y^3 + 15x^3y^2$
45. $3y^2 − 3y − 9$
46. $5x^2 − 5x + 15$
47. $6ab − 4ad + 12ac$
48. $8xy + 10xz − 14xw$
49. $9x^3y^6z^2 − 12x^4y^4z^4 + 15x^2y^3z^5$
50. $14a^4b^3c^3 + 21a^3b^7c^4 − 35a^2b^6c^5$

Factor out a factor with a negative coefficient.

51. $−5x + 35$ $−5(x − 7)$
52. $−5x − 40$ $−5(x + 8)$
53. $−6y − 72$ $−6(y + 12)$
54. $−8t + 72$ $−8(t − 9)$
55. $−2x^2 + 4x − 12$
56. $−2x^2 + 12x + 40$
57. $3y^2 − 24x$
58. $7x − 56y$
59. $7a − 14t$
60. $5r − 10s$
61. $−x^3 + 5x^2 − 9$
62. $−p^3 − 4p^2 + 11$
63. $−a^4 + 2a^3 − 13a$
64. $−m^3 − m^2 + m − 2$

Factor.

65. $a(b − 5) + c(b − 5)$
66. $r(t − 3) − s(t − 3)$
67. $(x + 7)(x − 1) + (x + 7)(x − 2)$
68. $(a + 5)(a − 2) + (a + 5)(a + 1)$
69. $a^2(x − y) + 5(y − x)$
70. $5x^2(x − 6) + 2(6 − x)$
71. $ac + ad + bc + bd$
72. $xy + xz + wy + wz$
73. $b^3 − b^2 + 2b − 2$
74. $y^3 + y^2 + 3y − 3$
75. $a^3 − 3a^2 + 6 − 2a$
76. $t^3 + 6t^2 − 2t − 12$
77. $72x^3 − 36x^2 + 24x$
78. $12a^4 − 21a^3 − 9a^2$

Aha!

In an effort to discourage rote learning and reward students who think a problem through before solving it, these exercises can be solved quickly if the student has the proper insight. The *Aha!* designation is used the first time a new insight can be used on a particular type of exercise and indicates to the student that there is a way to complete the exercise with more concise computation. It's then up to the student to find the simpler approach and, in subsequent exercises, to determine if and when that particular insight might be used again.

Exercise Sets

Exercise sets include both graphing calculator and non-graphing calculator exercises. In some cases, detailed instruction lines indicate the approach that seems best, and others let students choose the most efficient approach, thereby encouraging critical thinking.

In this chapter, we consider only the real-number zeros of functions. We can solve, or find the roots of, the equation $f(x) = 0$ by finding the zeros of the function f.

Finding Zeros of Functions

Many graphing calculators have a ZERO or ROOT option in the CALC menu that will calculate an x-intercept of a graph.

To find the zero of a function, enter the function and choose a viewing window that shows the intersection of the graph of the function with the x-axis. Then choose the ZERO option from the CALC menu. The calculator first asks us for a Left Bound. This is an x-value less than, or to the left of, a zero. After visually determining the approximate location of a zero, or x-intercept of the graph, use the left and right arrow keys to move the cursor to a location on the graph left of the zero, or enter an x-value to the left of the zero. Then press ENTER. Next, enter a Right Bound in a similar way, and then a Guess. The calculator will return the coordinates of the x-intercept between the left and right bounds.

Calculator References

Throughout each chapter, calculator references act as an in-text manual to introduce and explain various graphing-calculator techniques. The references are placed so that the student can learn the technique just prior to practicing it.

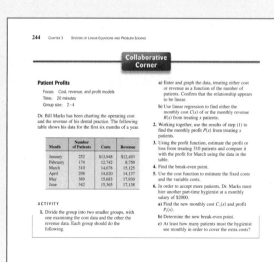

Collaborative Corner

Patient Profits

Focus: Cost, revenue, and profit models
Time: 20 minutes
Group size: 2–4

Dr. Bill Marks has been charting the operating cost and the revenue of his dental practice. The following table shows his data for the first six months of a year.

Month	Number of Patients	Costs	Revenue
January	252	$13,948	$12,493
February	174	12,742	8,750
March	310	14,678	15,125
April	298	14,620	14,137
May	369	15,683	17,930
June	342	15,365	17,138

ACTIVITY

1. Divide the group into two smaller groups, with one examining the cost data and the other the revenue data. Each group should do the following.

a) Enter and graph the data, treating either cost or revenue as a function of the number of patients. Confirm that the relationship appears to be linear.

b) Use linear regression to find either the monthly cost $C(x)$ or the monthly revenue $R(x)$ from treating x patients.

2. Working together, use the results of step (1) to find the monthly profit $P(x)$ from treating x patients.

3. Using the profit function, estimate the profit or loss from treating 310 patients and compare it with the profit for March using the data in the table.

4. Find the break-even point.

5. Use the cost function to estimate the fixed costs and the variable costs.

6. In order to accept more patients, Dr. Marks must hire another part-time hygienist at a monthly salary of $2000.

a) Find the new monthly cost $C_1(x)$ and profit $P_1(x)$.

b) Determine the new break-even point.

c) At least how many patients must the hygienist see monthly in order to cover the extra costs?

Collaborative Corners

Collaborative Corner exercises appear approximately three times per chapter, after the section exercise set. Designed for group work, these optional activities provide instructors and students with the opportunity to incorporate group learning and/or class discussions.

END-OF-CHAPTER MATERIAL

At the end of each chapter, students can practice all they have learned as well as tie the current chapter material to material covered in earlier chapters.

Chapter Summary and Review

The Chapter Summary and Review provides an extensive set of review exercises along with a list of important properties and formulas covered in that chapter. This feature provides an excellent preparation for chapter tests and the final examination.

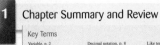

1 Chapter Summary and Review

Key Terms

Variable, p. 2
Constant, p. 2
Algebraic expression, p. 2
Exponential notation, p. 3
Exponent, or power, p. 3
Base, p. 3
Substituting, p. 3
Evaluating the expression, p. 3
Equation, p. 5
Solution, p. 5
Inequality, p. 5
Natural numbers, p. 6
Whole numbers, p. 6
Integers, p. 6
Roster notation, p. 7
Set-builder notation, p. 7
Element, p. 7
Rational numbers, p. 7
Fraction notation, p. 8

Decimal notation, p. 8
Irrational numbers, p. 8
Real numbers, p. 8
Subset, p. 9
Cursor, p. 10
Contrast, p. 10
Home screen, p. 10
Absolute value, p. 13
Opposites, p. 14
Additive inverses, p. 14
Reciprocals, p. 17
Multiplicative inverses, p. 17
Indeterminate, p. 18
Menu, p. 19
Submenus, p. 19
Equivalent expressions, p. 22
Factoring, p. 24
Term, p. 24

Like terms, p. 24
Combining like terms, p. 24
Scientific notation, p. 38
Axes, p. 43
x, y-coordinate system, p. 43
Cartesian coordinate system, p. 43
Ordered pairs, p. 44
Origin, p. 44
Coordinates, p. 44
Quadrants, p. 45
Viewing window, p. 46
Standard viewing window, p. 46
Graph, p. 47
Dependent variable, p. 49
Independent variable, p. 49
Linear equation, p. 49
Nonlinear equation, p. 50
Mathematical model, p. 60

Important Properties and Formulas

Area of a rectangle:	$A = lw$
Area of a square:	$A = s^2$
Area of a parallelogram:	$A = bh$
Area of a trapezoid:	$A = \frac{h}{2}(b_1 + b_2)$
Area of a triangle:	$A = \frac{1}{2}bh$
Area of a circle:	$A = \pi r^2$
Circumference of a circle:	$C = \pi d$
Volume of a cube:	$V = s^3$
Volume of a right circular cylinder:	$V = \pi r^2 h$
Perimeter of a square:	$P = 4s$
Distance traveled:	$d = rt$
Simple interest:	$I = Prt$

Key Terms

A comprehensive list of Key Terms from the chapter is provided along with the corresponding page numbers to serve as a reference point for students.

Important Properties and Formulas

Following the Key Terms in most chapters and summarized in one place just before the review exercises, a list of Important Properties and Formulas offers students a quick review before beginning the review exercises.

Review Exercises

The following review exercises are for practice. Answers are at the back of the book. If you need to, restudy the section indicated alongside the answer.

Evaluate each expression using the values provided.
1. $3x - (4 - y)$, for $x = 10$ and $y = 2$ [1.1] 28
2. $7x^2 - 5y + zx$, for $x = -2.78$, $y = 1.5$, and $z = 3.2$ [1.1] 60.514459
3. Name the set consisting of the first six even natural numbers using both roster notation and set-builder notation. [1.1] {2, 4, 6, 8, 10, 12}; {x | x is an even integer between 1 and 13}
4. Find the area of a triangular sign that has a base of 60 cm and a height of 70 cm. [1.1] 2100 cm²

Tell whether each number is a solution of the given equation or inequality.
5. $10 - 3x = 1$; (a) 7; (b) 3 [1.1] (a) No; (b) yes

Simplify. Do not use negative exponents in the answer.
32. $3^{-4} \cdot 3^7$ [1.4] 3^3, or 27   33. $(5a^2)^3$ [1.4] 125a^6
34. $(-2a^{-1}b^2)^{-3}$ [1.4] $-\frac{a^3}{8b^6}$ 35. $\left(\frac{x^3y^3}{z^4}\right)^{-2}$ [1.4] $\frac{z^8}{x^6y^6}$
36. $\left(\frac{2a^{-3}b}{4a^2b^{-1}}\right)^4$ [1.4] $\frac{b^8}{16a^{20}}$

Simplify.
37. $\frac{7(5 - 2 \cdot 3) - 3^2}{4^2 - 3^2}$ [1.2] $-\frac{16}{7}$
38. $1 - (2 - 5)^2 + 5 \div 10 \cdot 4^2$ [1.2] 0
39. Convert 0.000000103 to scientific notation. [1.4] 1.03×10^{-7}
40. One parsec (a unit that is used in astronomy) is 30,860,000,000,000 km. Write scientific notation for this number. [1.4] 3.086×10^{13}

Review Exercises

At the end of each chapter, students are provided with an extensive set of Review Exercises that provides practice for the key chapter material. These exercises include a wide variety of exercises, including synthesis exercises.

Chapter Test 1

1. Evaluate $a^3 - 5b + b \div ac$ for $a = -2$, $b = 6$, and $c = 3$. [1.1] -47
2. The base of a triangular stamp measures 3 cm and its height 2.5 cm. Find the area of the stamp.
[1.1] 3.75 cm²

3. Tell whether each number is a solution of the equation $14 - 5x = 4$.
a) 1 [1.1] No
b) 2 [1.1] Yes
c) 0 [1.1] No

Perform the indicated operation.
4. $-25 + (-16)$ [1.2] -41 5. $-10.5 + 6.8$ [1.2] -3.7
6. $6\frac{2}{3} + (-8.32)$ [1.2]   7. $29.5 - 43.7$ [1.2] -14.2
8. $-17.8 - 25.4$ [1.2] -43.2 9. $-6.4(5.3)$ [1.2] -33.92
10. $-\frac{3}{7} - (-\frac{3}{4})$ [1.2] $-\frac{3}{4}$ 11. $-\frac{5}{6}$
12. $\frac{-40.2}{6}$ [1.2] 6   13. $\frac{5}{6} \div$
14. Simplify: $5 + (1 - 3)^2 - 7 \div 2^2$.
18. [1.4] $\frac{72}{x^{10}y^5}$

Determine whether the ordered pair is a solution.
26. $(0, -5)$; $x + 4y = -20$ [1.5] Yes
27. $(1, -4)$; $-2p + 5q = 18$ [1.5] No
Graph.
28. $y = -5x + 4$ 29. $y = -2x^2 + 3$
30. Create a table of solutions of the equation
$$y = 10 - x^2$$
for integer values of x from -3 to 3. Then graph.
31. Translate to an algebraic expression: [1.6] Let m and n represent the numbers.
Three more than the product of two numbers.
32. Translate to an equation:
Greg's scores on five tests are 94, 80, 76, 91, and 75. What must Greg score on the sixth test so that his average will be 85?

Gas Mileage.   The following graph shows the gas mileage of a truck traveling at different speeds.

Chapter Test

Following the Review Exercises, a sample Chapter Test allows students to review and test comprehension of chapter skills prior to taking an instructor's exam.

1-3 Cumulative Review

Solve.
1. $-14.3 + 29.17 = x$ [2.2] 14.87
2. $x + 9.4 = -12.6$ [2.2] -22
3. $3.9(-11) = x$ [2.2] -42.9
4. $-2.4x = -48$ [2.2] 20
5. $4x + 7 = -14$ [2.2]
6. $-3 + 5x = 2x + 15$
7. $3n - (4n - 2) = 7$
8. $6y - 5(3y - 4) = 10$

9. $14 + 2c = -3(c + 4) - 6$ [2.2] $-\frac{32}{5}$
10. $5x - [4 - 2(6x - 1)] = 12$ [2.2]

Simplify. Do not leave negative exponents in your answers.
11. $x^4 \cdot x^{-6} \cdot x^{13}$ [1.4] x^{11}
12. $(4x^{-2}y^3)(-10x^{-4}y^{-2})$ [1.4] $-\frac{40}{x^6}$

Cumulative Review

Following every three chapters, students will find a cumulative review, which covers skills and concepts from all preceding chapters of the text.

Basics of Algebra and Graphing

The principal theme of this text is problem solving in algebra. Both symbolic algebra and graphs are used to develop models and solve applications. In this chapter, we introduce the basics of algebra, graphing, and the graphing calculator. We then use algebra and graphs to develop the idea of a mathematical model.

APPLICATION

ENDANGERED SPECIES. The following table lists the numbers of U.S. species of mammals considered endangered during various years (*Source*: U.S. Fish and Wildlife Service). Use the data to draw a line graph. The number of species of endangered mammals can be *modeled* by graphing the data.

Year	Number of Species of Endangered Mammals
1980	32
1985	43
1990	53
1993	56
1997	57
1999	61
2001	64

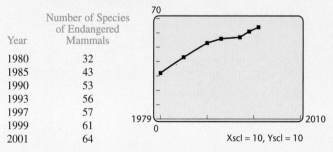

This problem appears as Exercise 49 in Exercise Set 1.6.

1.1

Algebraic Expressions and Their Use ■ Evaluating Algebraic Expressions ■ Equations and Inequalities ■ Sets of Numbers ■ Introduction to the Graphing Calculator

Some Basics of Algebra

The primary difference between algebra and arithmetic is the use of *variables*. In this section, we will see how variables can be used to represent various situations. We will also examine the different types of numbers that will be represented by variables throughout this text.

Algebraic Expressions and Their Use

We are all familiar with expressions like

$$95 + 21, \quad 57 \times 34, \quad 9 - 4, \quad \text{and} \quad \frac{35}{71}.$$

In algebra, we use these as well as expressions like

$$x + 21, \quad l \cdot w, \quad 9 - s, \quad \text{and} \quad \frac{d}{t}.$$

When a letter is used to stand for various numbers, it is called a **variable**. If a letter represents one particular number, it is called a **constant**. Let $d =$ the number of hours it takes the moon to orbit the earth. Then d is a constant. If $a =$ the age of a baby chick, in minutes, then a is a variable since a changes as time passes.

An **algebraic expression** consists of variables, numbers, and operation signs. All of the expressions above are examples of algebraic expressions. When an equals sign is placed between two expressions, an **equation** is formed.

Algebraic expressions and equations arise frequently in problem-solving situations. Suppose, for example, that we want to determine by how much the gas mileage of the most fuel-efficient cars increased between 1999 and 2002.

TEACHING TIP

Begin emphasizing the difference between an algebraic expression and an algebraic equation. You might give several examples or compare them to an English phrase and sentence.

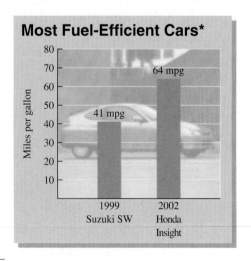

Most Fuel-Efficient Cars*

64 mpg

41 mpg

Miles per gallon

1999
Suzuki SW

2002
Honda
Insight

*Source: U.S. Department of Energy; figures represent a combination of highway and city driving.

By using x to represent the increase in mileage, we can form an equation:

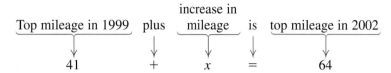

To find a **solution,** we can subtract 41 on both sides of the equation:

$x = 64 - 41$

$x = 23.$

The mileage increased by 23 miles per gallon (mpg) from 1999 to 2002.

Algebraic expressions can indicate operations other than addition, subtraction, multiplication, and division. Another type of notation used in some algebraic expressions is *exponential notation*. Many different kinds of numbers can be used as *exponents*. Here we establish the meaning of a^n when n is a counting number, $1, 2, 3, \ldots$.

Exponential Notation

The expression a^n, in which n is a counting number, means

$$\underbrace{a \cdot a \cdot a \cdot \cdots \cdot a \cdot a}_{n \text{ factors.}}$$

In a^n, a is called the *base* and n is the *exponent,* or *power.* When no exponent appears, it is assumed to be 1. Thus, $a^1 = a$.

The expression a^n is read "a raised to the nth power" or simply "a to the nth." We often read s^2 as "s-squared" and x^3 as "x-cubed." This terminology comes from the fact that the area of a square of side s is $s \cdot s = s^2$ and the volume of a cube of side x is $x \cdot x \cdot x = x^3$.

Area = s^2 s Volume = x^3

Evaluating Algebraic Expressions

When we replace a variable with a number, we say that we are **substituting** for the variable. The calculation that follows the substitution is called **evaluating the expression**.

Geometric formulas are often evaluated. In the following example, we use the formula for the area A of a triangle with a base of length b and a height of length h:

$A = \frac{1}{2} \cdot b \cdot h.$

This is an important formula that is worth remembering.

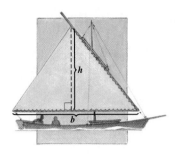

EXAMPLE 1   The base of a triangular sail is 8 m and the height is 6.3 m. Find the area of the sail.

Solution   We substitute 8 for b and 6.3 for h and multiply:

$$\tfrac{1}{2} \cdot b \cdot h = \tfrac{1}{2} \cdot 8 \cdot 6.3 \qquad \textbf{We use color to highlight the substitution.}$$
$$= 25.2 \text{ square meters (sq m or } m^2\text{).}$$

Exponential notation tells us that 5^2 means $5 \cdot 5$, or 25, but what does $1 + 2 \cdot 5^2$ mean? If we add 1 and 2 and multiply by 25, we get 75. If we multiply 2 times 5^2, or 25, and add 1, we get 51. A third possibility is to square $2 \cdot 5$ to get 100 and then add 1 to get 101. The following convention indicates that only the second of these approaches is correct: We square 5, then multiply, and then add.

Rules for Order of Operations

1. Simplify within any grouping symbols, such as parentheses and brackets.
2. Simplify all exponential expressions.
3. Perform all multiplication and division working from left to right.
4. Perform all addition and subtraction working from left to right.

EXAMPLE 2   Evaluate $5 + 2(a - 1)^2$ for $a = 4$.

Solution

$$
\begin{aligned}
5 + 2(a - 1)^2 &= 5 + 2(4 - 1)^2 &&\textbf{Substituting} \\
&= 5 + 2(3)^2 &&\textbf{Working within parentheses first} \\
&= 5 + 2(9) &&\textbf{Simplifying } 3^2 \\
&= 5 + 18 &&\textbf{Multiplying} \\
&= 23 &&\textbf{Adding}
\end{aligned}
$$

Step (3) in the rules for order of operations tells us to divide before we multiply when division appears first, reading left to right. Similarly, if subtraction appears before addition, we subtract before we add.

TEACHING TIP

Emphasize that $6 \div 2y^2$ is not the same as $6 \div (2y^2)$.

EXAMPLE 3   Evaluate $9 - x^3 + 6 \div 2y^2$ for $x = 2$ and $y = 5$.

Solution

$$
\begin{aligned}
9 - x^3 + 6 \div 2y^2 &= 9 - 2^3 + 6 \div 2(5)^2 &&\textbf{Substituting} \\
&= 9 - 8 + 6 \div 2 \cdot 25 &&\textbf{Simplifying } 2^3 \textbf{ and } 5^2 \\
&= 9 - 8 + 3 \cdot 25 &&\textbf{Dividing} \\
&= 9 - 8 + 75 &&\textbf{Multiplying} \\
&= 1 + 75 &&\textbf{Subtracting} \\
&= 76 &&\textbf{Adding}
\end{aligned}
$$

Equations and Inequalities

An equation can be true or false. For example, the equation

$$13 + 3 = 16$$

is true, and the equation

$$4 + 25 = 30$$

is false. The equation

$$x + 12 = 35$$

is neither true nor false. However, when the algebraic expressions on each side of the equation are evaluated for a value for x, the equation becomes true or false, depending on the replacement for x. A replacement that makes the equation true is called a *solution* of the equation.

EXAMPLE 4   Tell whether each number is a solution of the equation $x + 12 = 35$: **(a)** 15; **(b)** 23.

Solution

a) We replace x with 15 and evaluate the expressions on both sides of the equals sign.

$$\begin{array}{c}
x + 12 = 35 \\
\hline
15 + 12 \ ? \ 35 \\
27 \ \mid \ 35 \qquad \text{FALSE}
\end{array}$$

Since $27 = 35$ is false, 15 *is not* a solution of the equation.

b) We replace x with 23 and evaluate the expressions on both sides of the equals sign.

$$\begin{array}{c}
x + 12 = 35 \\
\hline
23 + 12 \ ? \ 35 \\
35 \ \mid \ 35 \qquad \text{TRUE}
\end{array}$$

Since $35 = 35$ is true, 23 *is* a solution of the equation.

The symbols $<$ (is less than), $>$ (is greater than), $\leq$ (is less than or equal to), and $\geq$ (is greater than or equal to) are inequality symbols. An **inequality** is formed when an inequality symbol is placed between two algebraic expressions. Like equations, inequalities can be true, false, or neither true nor false. A *solution* of an inequality is a replacement that makes the inequality true.

EXAMPLE 5   Tell whether each number is a solution of the inequality $3y + 2 \leq 11$: **(a)** 0; **(b)** 5; **(c)** 3.

Solution

a) We replace y with 0.

$$3y + 2 \leq 11$$

$$
\begin{array}{c|c}
3 \cdot 0 + 2 \ ? \ 11 & \\
0 + 2 & \\
2 & 11 \quad \text{TRUE}
\end{array}
$$

Inserting the inequality symbol, we have $2 \leq 11$, read "2 is less than or equal to 11." Since 2 is less than 11, this inequality is true, and 2 *is* a solution of the inequality.

b) We replace y with 5.

$$3y + 2 \leq 11$$

$$
\begin{array}{c|c}
3 \cdot 5 + 2 \ ? \ 11 & \\
15 + 2 & \\
17 & 11 \quad \text{FALSE}
\end{array}
$$

Since the statement "17 is less than or equal to 11" is false, 5 *is not* a solution of the inequality.

c) We replace y with 3.

$$3y + 2 \leq 11$$

$$
\begin{array}{c|c}
3 \cdot 3 + 2 \ ? \ 11 & \\
9 + 2 & \\
11 & 11 \quad \text{TRUE}
\end{array}
$$

Since the statement "11 is less than or equal to 11" is true (11 is equal to 11), 3 *is* a solution of the inequality.

Sets of Numbers

When evaluating algebraic expressions, and in problem solving in general, we often must examine the *type* of numbers used. For example, if a formula is used to determine an optimal class size, any fractional results must be rounded up or down since it is impossible to have a fractional part of a student. Three frequently used sets of numbers are listed below.

Natural Numbers, Whole Numbers, and Integers

Natural Numbers (or Counting Numbers)   Those numbers used for counting: $\{1, 2, 3, \ldots\}$

Whole Numbers   The set of natural numbers with 0 included: $\{0, 1, 2, 3, \ldots\}$

Integers   The set of all whole numbers and their opposites:

$$\{\ldots, -4, -3, -2, -1, 0, 1, 2, 3, 4, \ldots\}$$

The dots are called ellipses and indicate that the pattern continues without end.

Study Tip

The integers correspond to the points on a number line as follows:

To fill in the numbers between these points, we must describe two more sets of numbers. This requires us to first discuss set notation.

The set containing the numbers -2, 1, and 3 can be written $\{-2, 1, 3\}$. This way of writing a set is known as **roster notation**. Roster notation was used for the sets listed above. A second type of set notation, **set-builder notation**, specifies conditions under which a number is in the set. The following example of set-builder notation is read as shown:

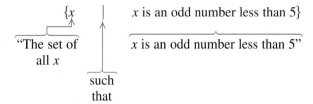

Set-builder notation is generally used when it is difficult to list a set using roster notation.

EXAMPLE 6   Using both roster notation and set-builder notation, represent the set consisting of the first 15 even natural numbers.

Solution

Using roster notation: $\{2, 4, 6, 8, 10, 12, 14, 16, 18, 20, 22, 24, 26, 28, 30\}$

Using set-builder notation: $\{n \mid n$ is an even number between 1 and 31$\}$

The symbol $\in$ is used to indicate that an element belongs to a set. Thus if $A = \{2, 4, 6, 8\}$, we can write $4 \in A$ to indicate that 4 *is an element of A*. We can also write $5 \notin A$ to indicate that 5 *is not an element of A*.

EXAMPLE 7   Classify the statement $8 \in \{x \mid x$ is an integer$\}$ as true or false.

Solution   Since 8 *is* an integer, the statement is true. In other words, since 8 is an integer, it belongs to the set of all integers.

With set-builder notation, we can describe the set of all *rational numbers*.

Rational Numbers

Numbers that can be expressed as an integer divided by a nonzero integer are called *rational numbers*:

$$\left\{ \frac{p}{q} \;\middle|\; p \text{ is an integer, } q \text{ is an integer, and } q \neq 0 \right\}.$$

Rational numbers can be written using fraction or decimal notation. *Fraction notation* uses symbolism like the following:

$$\frac{5}{8}, \quad \frac{12}{-7}, \quad \frac{-17}{15}, \quad -\frac{9}{7}, \quad \frac{39}{1}, \quad \frac{0}{6}.$$

In *decimal notation,* rational numbers either *terminate* (end) or *repeat* (have a repeating block of digits).

EXAMPLE 8   When written in decimal form, does each of the following numbers terminate or repeat? **(a)** $\frac{5}{8}$; **(b)** $\frac{6}{11}$.

Solution

a) Since $\frac{5}{8}$ means $5 \div 8$, we perform long division to find that $\frac{5}{8} = 0.625$, a decimal that ends. Thus, $\frac{5}{8}$ can be written as a terminating decimal.

b) Using long division, we find that $6 \div 11 = 0.5454\ldots$, so we can write $\frac{6}{11}$ as a repeating decimal. Repeating decimal notation can be abbreviated by writing a bar over the repeating block of digits—in this case, $0.\overline{54}$.

Many numbers, like π, $\sqrt{2}$, and $-\sqrt{15}$, can be only approximated by rational numbers. For example, $\sqrt{2}$ is the number for which $\sqrt{2} \cdot \sqrt{2} = 2$. A calculator's representation of $\sqrt{2}$ as 1.414213562 is an approximation since $(1.414213562)^2$ is not exactly 2.

To see that $\sqrt{2}$ is a "real" point on the number line, note that when a right triangle has two legs of length 1, the remaining side has length $\sqrt{2}$. Thus we can "measure" $\sqrt{2}$ and locate it precisely on a number line.

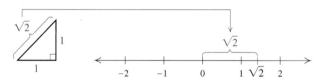

Numbers like π, $\sqrt{2}$, and $-\sqrt{15}$ are said to be **irrational**. Decimal notation for irrational numbers neither terminates nor repeats.

The set of all rational numbers, combined with the set of all irrational numbers, gives us the set of all **real numbers**.

Real Numbers

Numbers that are either rational or irrational are called *real numbers.* The set of all real numbers is often represented as $\mathbb{R}$:

$$\mathbb{R} = \{x \,|\, x \text{ is rational or } x \text{ is irrational}\}.$$

Every point on the number line represents some real number and every real number is represented by some point on the number line.

TEACHING TIP

An example like $\frac{39}{1}$ can be used to show that all integers are rational numbers.

The following figure shows the relationships among various kinds of numbers, along with examples of how real numbers can be sorted.

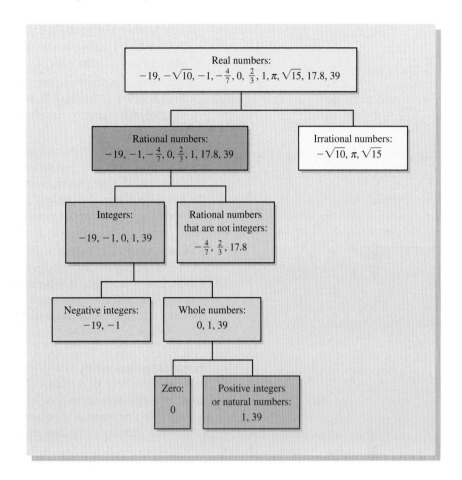

When all members of one set are found in a second set, the first set is a **subset** of the second set. Thus if $A = \{2, 4, 6\}$ and $B = \{1, 2, 4, 5, 6\}$, we write $A \subseteq B$ to indicate that A *is a subset of B*. Similarly, if $\mathbb{N}$ represents the set of all natural numbers and $\mathbb{Z}$ the set of all integers, we can write $\mathbb{N} \subseteq \mathbb{Z}$. Additional statements can be made using other sets in the diagram above.

Study Tip

Be sure that you understand and can use mathematical principles such as the rules for order of operations before you rely on a graphing calculator to do the operations. When you are learning the rules, a calculator can be useful to check your answers.

Introduction to the Graphing Calculator

Graphing calculators and computers equipped with graphing software can be valuable aids in understanding and applying algebra. Features and keystrokes vary among the many brands and models of graphing calculators available. In this text, we use features that are common to most graphing calculators. Specific keystrokes and instructions for certain calculators are included in the Graphing Calculator Manual that accompanies this book. For other procedures, you should consult your instructor or the user's manual for your particular calculator.

There are two important things to keep in mind as you proceed through this text.

1. You will not learn to use a graphing calculator by simply reading about it; you must in fact *use* the calculator. Press the keys on your own calculator as you read the text, do the calculator exercises in the exercise set, and experiment with new options.

2. Your user's manual contains more information about your calculator than appears in this text. If you need additional explanation and examples, be sure to consult the manual.

Keypad A diagram of the keypad of a graphing calculator appears at the front of this text. The organization and labeling of the keys differ for different calculators. Note that there are options written above keys as well as on the keys. To access the options shown above the keys, press 2nd or ALPHA and then the key below the desired option.

Screen After you have turned the calculator on, you should see a blinking rectangle, or **cursor**, at the top left corner of the screen. If you do not see anything, try adjusting the **contrast**. On many calculators, this is done by pressing 2nd and the up or down arrow keys. To perform computations, you should be in the **home screen**. Pressing 2nd QUIT will return you to the home screen.

Performing Operations To perform addition, subtraction, multiplication, and division using a graphing calculator, type in the expression as it is written. The entire expression will appear on the screen, and you can check your typing. The operation of division is usually shown by the symbol /. If the expression appears to be correct, press ENTER. At that time, the calculator will evaluate the expression and display the result on the screen.

Catalog A graphing calculator's catalog lists all the functions of the calculator in alphabetical order. On many calculators, CATALOG is the second option associated with the 0 key. To copy an item from the catalog to the screen, press 2nd CATALOG and then scroll through the list using the up and down arrow keys until the desired item is indicated. To move through the list more quickly, press the key associated with the first letter of the item. The indicator will move to the first item beginning with that letter. When the desired item is indicated, press ENTER.

Exponents On most graphing calculators, you enter exponents using the ∧ key before the exponent. For the exponent 2, you can often use an x^2 key.

Order of Operations Graphing calculators generally follow the rules for order of operations, so expressions can be entered as they are written. Extra parentheses may need to be added to make the meaning clear. Grouping symbols such as brackets or braces are entered as parentheses.

EXAMPLE 9   Evaluate $2(y - 3)^2 + 7$ for $y = 5$.

Solution   We replace y with 5 and enter the expression.

```
2(5−3)²+7
                          15
■
```

The result is 15.

1.1

FOR EXTRA HELP

Digital Video
Tutor CD 1
Videotape 1

Student's
Solutions
Manual

Tutor
Center
AW Math
Tutor Center

InterAct Math

Math XL
MathXL

MyMathLab

Exercise Set

To the student and the instructor: **Throughout this text,** *selected exercises are marked with the icon* Aha!*. These "Aha!" exercises can be answered quite easily if the student pauses to inspect the exercise rather than proceed mechanically. This is done to discourage rote memorization. Some "Aha!" exercises are left unmarked to encourage students to* always *pause before working a problem.*

Evaluate each expression for the values provided.

1. $7x + y$, for $x = 3$ and $y = 4$ 25

2. $6a - b$, for $a = 5$ and $b = 3$ 27

3. $2c \div 3b$, for $b = 4$ and $c = 6$ 16

4. $3z \div 2y$, for $y = 1$ and $z = 6$ 9

5. $25 + r^2 - s$, for $r = 3$ and $s = 7$ 27

6. $n^3 + 2 - p$, for $n = 2$ and $p = 5$ 5

Aha!   **7.** $3n^2p - 3pn^2$, for $n = 5$ and $p = 9$ 0

8. $2a^3b - 2b^2$, for $a = 3$ and $b = 7$ 280

9. $5x \div (2 + x - y)$, for $x = 6$ and $y = 2$ 5

10. $3(m + 2n) \div m$, for $m = 7$ and $n = 0$ 3

11. $[10 - (a - b)]^2$, for $a = 7$ and $b = 2$ 25

12. $[17 - (x + y)]^2$, for $x = 4$ and $y = 1$ 144

13. $m + [n(3 + n)]^2$, for $m = 9$ and $n = 2$ 109

14. $a^2 - [3(a - b)]^2$, for $a = 7$ and $b = 5$ 13

In Exercises 15–18, find the area of a triangular window with the given base and height.

15. Base $= 5$ ft, height $= 7$ ft   17.5 sq ft

16. Base $= 2.9$ m, height $= 2.1$ m   3.045 sq m

17. Base $= 7$ m, height $= 3.2$ m   11.2 sq m

18. Base $= 3.6$ ft, height $= 4$ ft   7.2 sq ft

Tell whether each number is a solution of the given equation or inequality.

19. $12 - y = 5$; **(a)** 7; **(b)** 5; **(c)** 12   (a) Yes; (b) no; (c) no

20. $2x + 4 = 14$; **(a)** 3; **(b)** 7; **(c)** 5   (a) No; (b) no; (c) yes

21. $5 - x \le 2$; **(a)** 0; **(b)** 4; **(c)** 3   (a) No; (b) yes; (c) yes

22. $3y - 5 > 10$; **(a)** 3; **(b)** 5; **(c)** 7   (a) No; (b) no; (c) yes

23. $3m - 8 < 13$; **(a)** 6; **(b)** 7; **(c)** 9 (a) Yes; (b) no; (c) no

24. $15 - 3n = 6$; **(a)** 3; **(b)** 2; **(c)** 5 (a) Yes; (b) no; (c) no

Use roster notation to write each set.

25. The set of all vowels in the alphabet {a, e, i, o, u} or {a, e, i, o, u, y}

26. The set of all days of the week {Sunday, Monday, Tuesday, Wednesday, Thursday, Friday, Saturday}

27. The set of all odd natural numbers {1, 3, 5, 7, ...}

28. The set of all even natural numbers {2, 4, 6, 8, ...}

29. The set of all natural numbers that are multiples of 5 {5, 10, 15, 20, ...}

30. The set of all natural numbers that are multiples of 10 {10, 20, 30, 40, ...}

Use set-builder notation to write each set.

31. The set of all odd numbers between 10 and 20 $\{x \mid x$ is an odd number between 10 and 20$\}$

32. The set of all multiples of 4 between 22 and 35 $\{x \mid x$ is a multiple of 4 between 22 and 35$\}$

33. {0, 1, 2, 3, 4} $\{x \mid x$ is a whole number less than 5$\}$

34. {−3, −2, −1, 0, 1, 2} $\{x \mid x$ is an integer greater than −4 and less than 3$\}$

35. The set of all multiples of 5 between 7 and 79 $\{n \mid n$ is a multiple of 5 between 7 and 79$\}$

36. The set of all even numbers between 9 and 99 $\{x \mid x$ is an even number between 9 and 99$\}$

Classify each statement as true or false. The following sets are used:

$\mathbb{N}$ = the set of natural numbers;
$\mathbb{W}$ = the set of whole numbers;
$\mathbb{Z}$ = the set of integers;
$\mathbb{Q}$ = the set of rational numbers;
$\mathbb{H}$ = the set of irrational numbers;
$\mathbb{R}$ = the set of real numbers.

37. $9 \in \mathbb{N}$ True

38. $5.1 \in \mathbb{N}$ False

39. $\mathbb{N} \subseteq \mathbb{W}$ True

40. $\mathbb{W} \subseteq \mathbb{Z}$ True

41. $\sqrt{8} \in \mathbb{Q}$ False

42. $\frac{2}{3} \in \mathbb{H}$ False

43. $\mathbb{H} \subseteq \mathbb{R}$ True

44. $\sqrt{10} \in \mathbb{R}$ True

45. $4.3 \notin \mathbb{Z}$ True

46. $\mathbb{Z} \not\subseteq \mathbb{N}$ True

47. $\mathbb{Q} \subseteq \mathbb{R}$ True

48. $\mathbb{Q} \subseteq \mathbb{Z}$ False

Evaluate each of the following using a graphing calculator.

49. $13 - (y - 4)^3 + 10$, for $y = 6$ 15

50. $(t + 4)^2 - 12 \div (19 - 17) + 68$, for $t = 5$ 143

51. $3.86 + 2.7(2.1x + 1.7)$, for $x = 5.82$ 41.4494

52. $1.5(3.982 - a) + 2.3^2$, for $a = 2.19$ 7.978

53. $3(m + 2n) \div m$, for $m = 1.6$ and $n = 5.9$ 25.125

54. $1.5 + (2x - y)^2$, for $x = 3.25$ and $y = 1.7$ 24.54

55. $\frac{1}{2}(x + 5/z)^2$, for $x = 141$ and $z = 0.2$ 13,778

56. $a - \frac{3}{4}(2a - b)$, for $a = 116$ and $b = 207$ 97.25

To the student and the instructor: **The icon** $\mathbb{TW}$ **is used to denote thinking and writing exercises. These exercises are meant to be answered with one or more English sentences. Because many writing exercises have a variety of correct answers, these solutions are not listed in the answers at the back of the book.**

$\mathbb{TW}$ **57.** What is the difference between rational numbers and integers?

$\mathbb{TW}$ **58.** Werner insists that $15 - 4 + 1 \div 2 \cdot 3$ is 2. What error is he making?

Synthesis

To the student and the instructor: Synthesis exercises are designed to challenge students to extend the concepts or skills studied in each section. Many synthesis exercises require the assimilation of skills and concepts from several sections.

$\mathbb{TW}$ **59.** Is the following true or false, and why?

$$\{2, 4, 6\} \subseteq \{2, 4, 6\}$$

$\mathbb{TW}$ **60.** On a quiz, Francesca answers $6 \in \mathbb{Z}$ while Jacob writes $\{6\} \in \mathbb{Z}$. Jacob's answer does not receive full credit while Francesca's does. Why?

Use roster notation to write each set.

61. The set of all whole numbers that are not natural numbers   {0}

62. The set of all integers that are not whole numbers   {−1, −2, −3, ...}

63. $\{x \mid x = 5n, n$ is a natural number$\}$ {5, 10, 15, 20, ...}

64. $\{x \mid x = 3n, n$ is a natural number$\}$ {3, 6, 9, 12, ...}

65. $\{x \mid x = 2n + 1, n$ is a whole number$\}$ {1, 3, 5, 7, ...}

66. $\{x \mid x = 2n, n$ is an integer$\}$ {..., −4, −2, 0, 2, 4, ...}

67. Draw a right triangle that could be used to measure $\sqrt{13}$ units.

1.2

Absolute Value ■ Order ■ Addition, Subtraction, and Opposites ■ Multiplication, Division, and Reciprocals

Operations with Real Numbers

In this section, we review how real numbers are added, subtracted, multiplied, and divided. First, however, it is important to discuss absolute value.

Absolute Value

It is convenient to have a notation that represents a number's distance from zero on the number line.

> ### Absolute Value
>
> The notation $|a|$, read "the absolute value of a," represents the number of units that a is from zero.

EXAMPLE 1   Find the absolute value: **(a)** $|-3|$; **(b)** $|2.5|$; **(c)** $|0|$.

Solution

a) $|-3| = 3$ -3 is 3 units from 0.

b) $|2.5| = 2.5$ 2.5 is 2.5 units from 0.

c) $|0| = 0$ 0 is 0 units from itself.

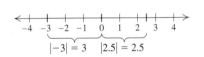

Note that absolute value is never negative.

Order

We use inequality symbols to indicate how two real numbers compare with each other. For any two numbers on the number line, the one to the left is said to be less than, or smaller than, the one to the right. In the figure below, we have $-6 < -1$ (since -6 is to the left of -1) and $|-6| > |-1|$ (since 6 is to the right of 1).

EXAMPLE 2   Determine whether each inequality is a true statement: **(a)** $-7 < -2$; **(b)** $1 > -4$; **(c)** $-3 \geq -2$.

Solution

a) $-7 < -2$ is *true* because -7 is to the left of -2 on a number line.

b) $1 > -4$ is *true* because 1 is to the right of -4.

c) $-3 \geq -2$ is *false* because -3 is to the left of -2.

Addition, Subtraction, and Opposites

We are now ready to review the addition of real numbers.

Addition of Two Real Numbers

1. *Positive numbers*: Add the numbers. The result is positive.

2. *Negative numbers*: Add absolute values. Make the answer negative.

3. *A negative and a positive number*: If the numbers have the same absolute value, the answer is 0. Otherwise, subtract the smaller absolute value from the larger one:

   a) If the positive number is further from 0, make the answer positive.

   b) If the negative number is further from 0, make the answer negative.

4. *One number is zero*: The sum is the other number.

EXAMPLE 3 Add: **(a)** $-9 + (-5)$; **(b)** $-3.2 + 9.7$; **(c)** $-\frac{3}{4} + \frac{1}{3}$.

Solution

a) $-9 + (-5)$ — We add the absolute values, getting 14. The answer is *negative*, -14.

b) $-3.2 + 9.7$ — The absolute values are 3.2 and 9.7. Subtract 3.2 from 9.7 to get 6.5. The positive number is further from 0, so the answer is *positive*, 6.5.

c) $-\frac{3}{4} + \frac{1}{3} = -\frac{9}{12} + \frac{4}{12}$ — The absolute values are $\frac{9}{12}$ and $\frac{4}{12}$. Subtract to get $\frac{5}{12}$. The negative number is further from 0, so the answer is *negative*, $-\frac{5}{12}$.

When numbers like 7 and -7 are added, the result is 0. Such numbers are called **opposites**, or **additive inverses**, of one another.

The Law of Opposites

For any two numbers a and $-a$,

$$a + (-a) = 0.$$

(When opposites are added, their sum is 0.)

EXAMPLE 4 Find the opposite: **(a)** -17.5; **(b)** $\frac{4}{5}$; **(c)** 0.

Solution

a) The opposite of -17.5 is 17.5 because $-17.5 + 17.5 = 0$.

b) The opposite of $\frac{4}{5}$ is $-\frac{4}{5}$ because $\frac{4}{5} + \left(-\frac{4}{5}\right) = 0$.

c) The opposite of 0 is 0 because $0 + 0 = 0$.

To name the opposite, we use the symbol "−" and read the symbolism $-a$ as "the opposite of a."

> **CAUTION!** $-a$ does not necessarily denote a negative number. In particular, when a represents a *negative* number, $-a$ is *positive*.

EXAMPLE 5   Find $-x$ for the following: **(a)** $x = -2$; **(b)** $x = \frac{3}{4}$.

Solution

a) If $x = -2$, then $-x = -(-2) = 2$.   **The opposite of -2 is 2.**

b) If $x = \frac{3}{4}$, then $-x = -\frac{3}{4}$.   **The opposite of $\frac{3}{4}$ is $-\frac{3}{4}$.**

Using the notation of opposites, we can formally define absolute value.

> **Absolute Value**
>
> $$|x| = \begin{cases} x & \text{if } x \geq 0, \\ -x & \text{if } x < 0 \end{cases}$$
>
> (When x is nonnegative, the absolute value of x is x. When x is negative, the absolute value of x is the opposite of x. Thus, $|x|$ is never negative.)

A negative number is said to have a negative "sign" and a positive number a positive "sign." To subtract, we can add an opposite. Thus we sometimes say that we "change the sign of the number being subtracted and then add."

EXAMPLE 6   Subtract: **(a)** $5 - 9$; **(b)** $-1.2 - (-3.7)$; **(c)** $-\frac{4}{5} - \frac{2}{3}$.

Solution

a) $5 - 9 = 5 + (-9)$ **Change the sign and add.**

$\qquad = -4$

b) $-1.2 - (-3.7) = -1.2 + 3.7$ **Instead of *subtracting* -3.7, we *add* 3.7.**

$\qquad = 2.5$

c) $-\frac{4}{5} - \frac{2}{3} = -\frac{4}{5} + \left(-\frac{2}{3}\right)$

$\qquad = -\frac{12}{15} + \left(-\frac{10}{15}\right)$ **Finding a common denominator**

$\qquad = -\frac{22}{15}$

Multiplication, Division, and Reciprocals

In this text, we direct graphing-calculator exploration of mathematical concepts with Interactive Discovery features like the one that follows. Such explorations are a part of the development of the material presented and should be performed as you read the text.

When real numbers are multiplied, the sign of the product depends on the signs of the factors.

Interactive Discovery

Consider the following patterns of products. If the patterns are to continue, what should the missing products be? Use a calculator to check your answers. How can we determine the sign of a product from the signs of the factors?

1.
$$3 \cdot 2 = 6$$
$$3 \cdot 1 = 3$$
$$3 \cdot 0 = 0$$
$$3 \cdot (-1) = ?$$
$$3 \cdot (-2) = ? \quad -3, -6$$

2.
$$(-5) \cdot 2 = -10$$
$$(-5) \cdot 1 = -5$$
$$(-5) \cdot 0 = 0$$
$$(-5) \cdot (-1) = ?$$
$$(-5) \cdot (-2) = ? \quad 5, 10$$

You may have noticed that when one factor is positive and one is negative, the product is negative. When both factors are positive or both are negative, the product is positive.

Division is defined in terms of multiplication. For example, $10 \div (-2) = -5$ because $(-5)(-2) = 10$. Thus the rules for division are the same as those for multiplication.

TEACHING TIP

You can also examine repeated addition or repeated subtraction, beginning at 0, to develop the rules for multiplication—for example, $3 \cdot (-2) = 0 + (-2) + (-2) + (-2) = -6$
and
$(-2)(-5) = 0 - (-5) - (-5) = 10$.

Multiplication or Division of Two Real Numbers

1. To multiply or divide two numbers with *unlike signs,* multiply or divide their absolute values. The answer is *negative.*

2. To multiply or divide two numbers that have the *same sign,* multiply or divide their absolute values. The answer is *positive.*

EXAMPLE 7   Multiply or divide: **(a)** $\left(-\frac{2}{3}\right)\left(-\frac{3}{8}\right)$; **(b)** $20 \div (-4)$; **(c)** $\dfrac{-45}{-15}$.

Solution

a) $\left(-\frac{2}{3}\right)\left(-\frac{3}{8}\right) = \frac{6}{24} = \frac{1}{4}$ Multiply absolute values. The answer is positive.

b) $20 \div (-4) = -5$ Divide absolute values. The answer is negative.

c) $\dfrac{-45}{-15} = 3$ Divide absolute values. The answer is positive.

Note that since
$$\frac{-8}{2} = \frac{8}{-2} = -\frac{8}{2} = -4,$$
we have the following generalization.

The Sign of a Fraction

For any number a and any nonzero number b,

$$\frac{-a}{b} = \frac{a}{-b} = -\frac{a}{b}.$$

Recall that

$$\frac{a}{b} = \frac{a}{1} \cdot \frac{1}{b} = a \cdot \frac{1}{b}.$$

That is, if we prefer, we can multiply by $1/b$ rather than divide by b. Provided that b is not 0, the numbers b and $1/b$ are called **reciprocals,** or **multiplicative inverses**, of each other.

The Law of Reciprocals

For any two numbers a and $1/a$ ($a \neq 0$),

$$a \cdot \frac{1}{a} = 1.$$

(When reciprocals are multiplied, their product is 1.)

EXAMPLE 8 Find the reciprocal: **(a)** $\frac{7}{8}$; **(b)** $-\frac{3}{4}$; **(c)** -8.

Solution

a) The reciprocal of $\frac{7}{8}$ is $\frac{8}{7}$ because $\frac{7}{8} \cdot \frac{8}{7} = 1$.

b) The reciprocal of $-\frac{3}{4}$ is $-\frac{4}{3}$.

c) The reciprocal of -8 is $\frac{1}{-8}$, or $-\frac{1}{8}$.

To divide, we can multiply by a reciprocal. We sometimes say that we "invert and multiply."

EXAMPLE 9 Divide: **(a)** $-\frac{1}{4} \div \frac{3}{5}$; **(b)** $-\frac{6}{7} \div (-10)$.

Solution

a) $-\frac{1}{4} \div \frac{3}{5} = -\frac{1}{4} \cdot \frac{5}{3}$ "Inverting" $\frac{3}{5}$ and changing division to multiplication

$= -\frac{5}{12}$

b) $-\frac{6}{7} \div (-10) = -\frac{6}{7} \cdot \left(-\frac{1}{10}\right) = \frac{6}{70}$, or $\frac{3}{35}$

Thus far, we have never divided by 0 or, equivalently, had a denominator of 0. There is a reason for this. Suppose 5 were divided by 0. The answer would have to be a number that, when multiplied by 0, gave 5. But any number times 0 is 0. Thus we cannot divide 5 or any other nonzero number by 0.

What if we divide 0 by 0? In this case, our solution would need to be some number that, when multiplied by 0, gave 0. But then *any* number would work

TEACHING TIP

To show students why we invert and multiply, show that $\dfrac{a}{b} \div \dfrac{c}{d}$ can be written as $\dfrac{\frac{a}{b}}{\frac{c}{d}}$ or $\dfrac{a}{b} \cdot \dfrac{d}{c}$, and simplify.

as a solution to $0 \div 0$. This could lead to contradictions like $7 = 4$ so we agree to exclude division of 0 by 0 also.

> ### Division by Zero
> We never divide by 0. If asked to divide a nonzero number by 0, we say that the answer is *undefined*. If asked to divide 0 by 0, we say that the answer is *indeterminate*.

The rules for order of operations discussed in Section 1.1 apply to *all* real numbers, regardless of their signs.

EXAMPLE 10   Simplify: **(a)** $(-5)^2$; **(b)** -5^2.

Solution   An exponent is written immediately after the base. Thus, in $(-5)^2$, the base is (-5), and in -5^2, the base is 5.

a) $(-5)^2 = 25$ **Squaring -5**

b) $-5^2 = -25$ **Squaring 5 and then taking the opposite**

Note that $(-5)^2 \neq -5^2$.

EXAMPLE 11   Simplify: $7 - 5^2 + 6 \div 2(-5)^2$.

Solution

$$7 - 5^2 + 6 \div 2(-5)^2 = 7 - 25 + 6 \div 2 \cdot 25 \quad \text{Simplifying } 5^2 \text{ and } (-5)^2$$
$$= 7 - 25 + 3 \cdot 25 \quad \text{Dividing}$$
$$= 7 - 25 + 75 \quad \text{Multiplying}$$
$$= -18 + 75 \quad \text{Subtracting}$$
$$= 57 \quad \text{Adding}$$

Besides parentheses, brackets, and braces, groupings may be indicated by a fraction bar, absolute-value symbol, or radical sign $(\sqrt{})$.

EXAMPLE 12   Calculate: $\dfrac{12|7 - 9| + 4 \cdot 5}{(-3)^4 + 2^3}$.

Solution   We simplify the numerator and the denominator before we divide the results:

$$\frac{12|7 - 9| + 4 \cdot 5}{(-3)^4 + 2^3} = \frac{12|-2| + 20}{81 + 8}$$
$$= \frac{12(2) + 20}{89}$$
$$= \frac{44}{89}. \quad \text{Multiplying and adding. This shows } 44 \div 89.$$

Study Tip

Take the time to include all the steps when working your homework problems. Doing so will help you organize your thinking and avoid computational errors. It will also give you complete, step-by-step solutions of the exercises that will make sense when studying for quizzes and tests.

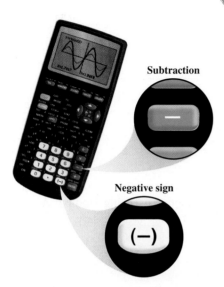

Subtraction

Negative sign

Entering Expressions

Graphing calculators have different keys for writing negatives and subtracting. The key labeled $\boxed{(-)}$ is used to create a negative sign, whereas the one labeled $\boxed{-}$ is used for subtraction.

Some grouping symbols, such as a fraction bar and an absolute-value symbol, may require parentheses to indicate the grouping.

Menus The absolute-value option is accessed using a **menu**, a list of options that appears when a key is pressed. For example, pressing $\boxed{\text{MATH}}$ results in a screen like the one on the left below. Four menu titles, or **submenus**, are listed across the top of the screen. In the screen on the left, the MATH submenu is highlighted and the options in that menu are listed. We refer to this submenu as MATH MATH, which means we first press $\boxed{\text{MATH}}$ and then choose the MATH submenu.

```
MATH  NUM  CPX  PRB
1:► Frac
2:► Dec
3: 3
4: 3√ (
5: x√
6: fMin(
7↓fMax(
```

```
MATH  NUM  CPX  PRB
1: abs(
2: round(
3: iPart(
4: fPart(
5: int(
6: min(
7↓max(
```

We use the left and right arrow keys to highlight the desired menu. In the screen shown on the right above, the NUM menu is highlighted. The options in the highlighted menu appear on the screen. Note in both figures that the menus contain more options than can fit on a screen, as indicated by the arrows in entry 7. The remaining options will appear as the down arrow is pressed.

To copy an item on the home screen, we highlight its number using the up or down arrow keys and press $\boxed{\text{ENTER}}$, or simply press the number of the item.

EXAMPLE 13 Calculate: $\dfrac{14 - 3|-16 + 38|}{4|-2^4 - 3^2|}$.

Solution There is no fraction bar on the calculator, so the expression must be rewritten using parentheses:

$$(14 - 3|-16 + 38|) \div (4|-2^4 - 3^2|).$$

For most graphing calculators, $|x|$ is written abs(x) and is accessed through a menu. We press $\boxed{\text{MATH}}$ $\boxed{\triangleright}$ $\boxed{1}$ to copy ABS on the screen. The expressions within the absolute-value symbols must be enclosed in parentheses. Some calculators will supply the left parenthesis; for these, you must close the expression with a right parenthesis.

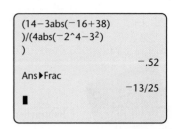

Many graphing calculators can convert decimal notation to fraction notation, as shown in the screen on the right above. We press ⃞MATH ⃞1 to copy ▶ FRAC on the screen. The answer is -0.52, or $-\frac{13}{25}$.

Example 13 demonstrates that being able to use a calculator does not make it any less important to understand the mathematics being studied. A solid understanding of the rules for order of operations is necessary in order to locate parentheses properly in an expression.

1.2

Exercise Set

FOR EXTRA HELP

Digital Video Tutor CD 1 Videotape 1 | Student's Solutions Manual | AW Math Tutor Center | InterAct Math | MathXL | MyMathLab

Find each absolute value.

1. $|-9|$ 9

2. $|-7|$ 7

3. $|6|$ 6

4. $|47|$ 47

5. $|-6.2|$ 6.2

6. $|-7.9|$ 7.9

7. $|0|$ 0

8. $\left|3\frac{3}{4}\right|$ $3\frac{3}{4}$

9. $\left|1\frac{7}{8}\right|$ $1\frac{7}{8}$

10. $|7.24|$ 7.24

11. $|-4.21|$ 4.21

12. $|-5.309|$ 5.309

Determine whether each inequality is a true statement.

13. $-6 \leq -2$ True

14. $-1 \leq -5$ False

15. $-9 > 1$ False

16. $-3 < 0$ True

17. $0 \geq -5$ True

18. $-6 \leq -6$ True

19. $-8 < -3$ True

20. $2 \geq -15$ True

21. $-4 \geq -4$ True

22. $-7 < -7$ False

23. $-5 > -5$ False

24. $-2 > -12$ True

Add.

25. $4 + 7$ 11

26. $8 + 3$ 11

27. $-4 + (-7)$ -11

28. $-8 + (-3)$ -11

29. $-3.9 + 2.7$ -1.2

30. $-1.9 + 7.3$ 5.4

31. $\frac{2}{7} + \left(-\frac{3}{5}\right)$ $-\frac{11}{35}$

32. $\frac{3}{8} + \left(-\frac{2}{5}\right)$ $-\frac{1}{40}$

33. $-4.9 + (-3.6)$ -8.5

34. $-2.1 + (-7.5)$ -9.6

35. $-\frac{1}{9} + \frac{2}{3}$ $\frac{5}{9}$

36. $-\frac{1}{2} + \frac{4}{5}$ $\frac{3}{10}$

37. $0 + (-4.5)$ -4.5

38. $-3.19 + 0$ -3.19

39. $-7.24 + 7.24$ 0

40. $-9.46 + 9.46$ 0

41. $15.9 + (-22.3)$ -6.4

42. $21.7 + (-28.3)$ -6.6

Find the opposite, or additive inverse.

43. 3.14 -3.14

44. 5.43 -5.43

45. $-4\frac{1}{3}$ $4\frac{1}{3}$

46. $2\frac{3}{5}$ $-2\frac{3}{5}$

47. 0 0

48. $-2\frac{3}{4}$ $2\frac{3}{4}$

Find $-x$ for each of the following.

49. $x = 7$ -7

50. $x = 3$ -3

51. $x = -2.7$ 2.7

52. $x = -1.9$ 1.9

53. $x = 1.79$ -1.79

54. $x = 3.14$ -3.14

55. $x = 0$ 0

56. $x = -1$ 1

Subtract.

57. $9 - 2$ 7

58. $10 - 3$ 7

59. $2 - 9$ -7

60. $3 - 10$ -7

61. $-6 - (-10)$ 4

62. $-3 - (-9)$ 6

63. $-5 - 14$ -19

64. $-7 - 8$ -15

65. $2.7 - 5.8$ -3.1

66. $3.7 - 3.7$ 0

67. $-\frac{3}{5} - \frac{1}{2}$ $-\frac{11}{10}$

68. $-\frac{2}{3} - \frac{1}{5}$ $-\frac{13}{15}$

Aha! **69.** $-3.9 - (-3.9)$ 0

70. $-5.4 - (-4.3)$ -1.1

71. $0 - (-7.9)$ 7.9

72. $0 - 5.3$ -5.3

Multiply.

73. $(-5)6$ -30

74. $(-4)7$ -28

75. $(-3)(-8)$ 24

76. $(-7)(-8)$ 56

77. $(4.2)(-5)$ -21

78. $(3.5)(-8)$ -28

79. $\frac{3}{7}(-1)$ $-\frac{3}{7}$

80. $-1 \cdot \frac{2}{5}$ $-\frac{2}{5}$

81. $(-17.45) \cdot 0$ 0

82. 15.2×0 0

83. $(-3.2) \times (-1.7)$ 5.44

84. $(1.9) \cdot (4.3)$ 8.17

Divide, if possible. If an expression is undefined, state so.

85. $\frac{-10}{-2}$ 5

86. $\frac{-15}{-3}$ 5

87. $\frac{-100}{20}$ -5

88. $\frac{-50}{5}$ -10

89. $\frac{73}{-1}$ -73

90. $\frac{-62}{1}$ -62

91. $\frac{0}{-7}$ 0

92. $\frac{0}{-11}$ 0

93. $\frac{-6}{0}$ Undefined

94. $\frac{-15}{0}$ Undefined

Find the reciprocal, or multiplicative inverse.

95. 4 $\frac{1}{4}$

96. 3 $\frac{1}{3}$

97. -9 $-\frac{1}{9}$

98. -5 $-\frac{1}{5}$

99. $\frac{2}{3}$ $\frac{3}{2}$

100. $\frac{4}{7}$ $\frac{7}{4}$

101. $-\frac{3}{11}$ $-\frac{11}{3}$

102. $-\frac{7}{3}$ $-\frac{3}{7}$

Divide.

103. $\frac{2}{3} \div \frac{4}{5}$ $\frac{5}{6}$

104. $\frac{2}{7} \div \frac{6}{5}$ $\frac{5}{21}$

105. $-\frac{3}{5} \div \frac{1}{2}$ $-\frac{6}{5}$

106. $\left(-\frac{4}{7}\right) \div \frac{1}{3}$ $-\frac{12}{7}$

107. $\left(-\frac{2}{9}\right) \div (-8)$ $\frac{1}{36}$

108. $\left(-\frac{2}{11}\right) \div (-6)$ $\frac{1}{33}$

Aha! **109.** $-\frac{12}{7} \div \left(-\frac{12}{7}\right)$ 1

110. $\left(-\frac{2}{7}\right) \div (-1)$ $\frac{2}{7}$

Calculate using the rules for order of operations.

111. $7 - (8 - 3 \cdot 2^3)$ 23

112. $19 - (4 + 2 \cdot 3^2)$ -3

113. $\frac{5 \cdot 2 - 4^2}{27 - 2^4}$ $-\frac{6}{11}$

114. $\frac{7 \cdot 3 - 5^2}{9 + 4 \cdot 2}$ $-\frac{4}{17}$

115. $\frac{3^4 - (5 - 3)^4}{8 - 2^3}$ Undefined

116. $\frac{4^3 - (7 - 4)^2}{3^2 - 7}$ $\frac{55}{2}$

117. $\frac{(2 - 3)^3 - 5|2 - 4|}{7 - 2 \cdot 5^2}$ $\frac{11}{43}$

118. $\frac{8 \div 4 \cdot 6|4^2 - 5^2|}{9 - 4 + 11 - 4^2}$ Undefined

119. $|2^2 - 7|^3 + 1$ 28

120. $|-2 - 3| \cdot 4^2 - 1$ 79

121. $28 - (-5)^2 + 15 \div (-3) \cdot 2$ -7

122. $43 - (-9 + 2)^2 + 18 \div 6 \cdot (-2)$ -12

123. $12 - \sqrt{11 - (3 + 4)} \div [-5 - (-6)]^2$ 10

124. $15 - 1 + \sqrt{5^2 - (3 + 1)^2}(-1)$ 11

Aha! **125.** $-12.86 - 5.2(-1.7 - 3.8)^2 \cdot 0$ -12.86

126. $\dfrac{0 \cdot [(-1.2)^2 + (9.2 - 1.78)^3]}{-2.5 - (1.7 + 0.9)}$ 0

Match each algebraic expression with the correct series of keystrokes. Check your answer by calculating the expression both by hand and by using the calculator.

127. $\dfrac{5(3 - 7) + 4^3}{(-2 - 3)^2}$ (a)

128. $(5(3 - 7) + 4)^3 \div (-2) - 3^2$ (c)

129. $5(3 - 7) + 4^3 \div (-2 - 3)^2$ (d)

130. $\dfrac{5(3 - 7) + 4^3}{-2 - 3^2}$ (b)

a) $\boxed{(}\,\boxed{5}\,\boxed{(}\,\boxed{3}\,\boxed{-}\,\boxed{7}\,\boxed{)}\,\boxed{+}\,\boxed{4}\,\boxed{\wedge}\,\boxed{3}\,\boxed{)}\,\boxed{\div}$
$\boxed{(}\,\boxed{(}\,\boxed{(-)}\,\boxed{2}\,\boxed{-}\,\boxed{3}\,\boxed{)}\,\boxed{x^2}\,\boxed{)}\,\boxed{\text{ENTER}}$

b) $\boxed{(}\,\boxed{5}\,\boxed{(}\,\boxed{3}\,\boxed{-}\,\boxed{7}\,\boxed{)}\,\boxed{+}\,\boxed{4}\,\boxed{\wedge}\,\boxed{3}\,\boxed{)}\,\boxed{\div}$
$\boxed{(}\,\boxed{(-)}\,\boxed{2}\,\boxed{-}\,\boxed{3}\,\boxed{x^2}\,\boxed{)}\,\boxed{\text{ENTER}}$

c) $\boxed{(}\,\boxed{5}\,\boxed{(}\,\boxed{3}\,\boxed{-}\,\boxed{7}\,\boxed{)}\,\boxed{+}\,\boxed{4}\,\boxed{)}\,\boxed{\wedge}\,\boxed{3}\,\boxed{\div}$
$\boxed{(-)}\,\boxed{2}\,\boxed{-}\,\boxed{3}\,\boxed{x^2}\,\boxed{\text{ENTER}}$

d) $\boxed{5}\,\boxed{(}\,\boxed{3}\,\boxed{-}\,\boxed{7}\,\boxed{)}\,\boxed{+}\,\boxed{4}\,\boxed{\wedge}\,\boxed{3}\,\boxed{\div}$
$\boxed{(}\,\boxed{(-)}\,\boxed{2}\,\boxed{-}\,\boxed{3}\,\boxed{)}\,\boxed{x^2}\,\boxed{\text{ENTER}}$

TW **131.** Describe in your own words a method for determining the sign of the sum of a positive number and a negative number.

TW **132.** Explain in your own words the difference between the opposite of a number and the reciprocal of a number.

Skill Maintenance

To the student and the instructor: Exercises included for Skill Maintenance review skills previously studied in the text. Usually these exercises provide preparation for the next section of the text. The numbers in brackets immediately following the directions or exercise indicate the section in which the skill was introduced. The answers to all Skill Maintenance exercises appear at the back of the book. If a Skill Maintenance exercise gives you difficulty, review the material in the indicated section of the text.

Evaluate. [1.1]

133. $2(x + 5)$ and $2x + 10$, for $x = 3$ $16; 16$

134. $2a - 3$ and $a - 3 + a$, for $a = 7$ $11; 11$

Synthesis

TW **135.** Explain in your own words why 7/0 is undefined.

TW **136.** Explain in your own words why 0 has an opposite but not a reciprocal.

Insert one pair of parentheses to convert each false statement into a true statement.

137. $8 - 5^3 + 9 = 36$ $(8 - 5)^3 + 9 = 36$

138. $2 \cdot 7 + 3^2 \cdot 5 = 104$ $2 \cdot (7 + 3^2 \cdot 5) = 104$

139. $5 \cdot 2^3 \div 3 - 4^4 = 40$ $5 \cdot 2^3 \div (3 - 4)^4 = 40$

140. $2 - 7 \cdot 2^2 + 9 = -11$ $(2 - 7) \cdot 2^2 + 9 = -11$

141. Find the greatest value of a for which $|a| \geq 6.2$ and $a < 0$. -6.2

1.3

The Commutative, Associative, and Distributive Laws ■ Combining Like Terms ■ Checking by Evaluating

Equivalent Algebraic Expressions

In algebra, we are often interested in manipulating expressions without changing their value. In this section, we look at some properties of real numbers that allow us to write *equivalent expressions*.

Equivalent Expressions

Two expressions that have the same value for all possible replacements are called *equivalent expressions*.

The Commutative, Associative, and Distributive Laws

When a pair of real numbers are added or multiplied, the order in which the numbers are written does not affect the result.

The Commutative Laws

For any real numbers a and b,

$a + b = b + a$; $a \cdot b = b \cdot a$.
(for Addition)   (for Multiplication)

The commutative laws provide one way of writing equivalent expressions.

EXAMPLE 1 Use a commutative law to write an expression equivalent to $7x + 9$.

Solution   Using the commutative law of addition, we have

$7x + 9 = 9 + 7x.$

We can also use the commutative law of multiplication to write

$$7x + 9 = x7 + 9.$$

The expressions $7x + 9$, $9 + 7x$, and $x7 + 9$ are all equivalent. They name the same number for any replacement of x.

The *associative laws* also enable us to form equivalent expressions. We use the associative laws to change the *grouping* of numbers being added or multiplied.

The Associative Laws

For any real numbers a, b, and c,

$a + (b + c) = (a + b) + c;$ $a \cdot (b \cdot c) = (a \cdot b) \cdot c.$
(for Addition)   (for Multiplication)

EXAMPLE 2   Write an expression equivalent to $(3x + 7y) + 9z$, using the associative law of addition.

Solution   We have

$$(3x + 7y) + 9z = 3x + (7y + 9z).$$

The expressions $(3x + 7y) + 9z$ and $3x + (7y + 9z)$ are equivalent. They name the same number for any replacements of x, y, and z.

EXAMPLE 3   Use the commutative and associative laws to write an expression equivalent to

$$\frac{5}{x} \cdot (yz).$$

Solution   Answers may vary. We use the associative law first and then the commutative law.

$$\frac{5}{x} \cdot (yz) = \left(\frac{5}{x} \cdot y\right) \cdot z \quad \text{Using the associative law of multiplication}$$

$$= \left(y \cdot \frac{5}{x}\right) \cdot z \quad \text{Using the commutative law of multiplication inside the parentheses}$$

The *distributive law* that follows provides another way of forming equivalent expressions. In essence, the distributive law allows us to rewrite the *product* of a and $b + c$ as the *sum* of ab and ac.

The Distributive Law

For any numbers a, b, and c,

$$a(b + c) = ab + ac.$$

Since $b - c = b + (-c)$, it is also true by the distributive law that $a(b - c) = ab - ac$.

EXAMPLE 4 Obtain an expression equivalent to $5x(y + 4)$ by multiplying.

Solution We use the distributive law to get

$$5x(y + 4) = 5xy + 5x \cdot 4 \qquad \text{Using the distributive law}$$
$$= 5xy + 5 \cdot 4 \cdot x \qquad \text{Using the commutative law of multiplication}$$
$$= 5xy + 20x. \qquad \text{Simplifying}$$

The expressions $5x(y + 4)$ and $5xy + 20x$ are equivalent. They name the same number for any replacements of x and y.

When we do the opposite of what we did in Example 4, we say that we are **factoring** an expression. This allows us to rewrite a sum as a product.

TEACHING TIP

Be aware that students may omit the 1 in factorizations such as $10x + 5 = 5(2x + 1)$.

EXAMPLE 5 Obtain an expression equivalent to $3x - 6$ by factoring.

Solution We use the distributive law to get

$$3x - 6 = 3 \cdot x - 3 \cdot 2 = 3(x - 2).$$

In Example 5, since the product of 3 and $x - 2$ is $3x - 6$, we say that 3 and $x - 2$ are **factors** of $3x - 6$. Thus "factor" can act as a noun or as a verb.

Combining Like Terms

In an expression like $8a^5 + 17 + 4/b + (-6a^3b)$, the parts that are separated by addition signs are called *terms*. A **term** is a number, a variable, a product of numbers and/or variables, or a quotient of numbers and/or variables. Thus, $8a^5$, 17, $4/b$, and $-6a^3b$ are terms in $8a^5 + 17 + 4/b + (-6a^3b)$. When terms have variable factors that are exactly the same, we refer to those terms as **like**, or **similar**, **terms**. Thus, $3x^2y$ and $-7x^2y$ are similar terms, but $3x^2y$ and $4xy^2$ are not. We can often simplify expressions by **combining**, or **collecting, like terms**.

EXAMPLE 6 Combine like terms: $3a + 5a^2 + 4a + a^2$.

Solution

$$3a + 5a^2 + 4a + a^2 = 3a + 4a + 5a^2 + a^2 \qquad \text{Using the commutative law}$$
$$= (3 + 4)a + (5 + 1)a^2 \qquad \text{Using the distributive law. Note that } a^2 = 1a^2.$$
$$= 7a + 6a^2$$

TEACHING TIP

Some students may have difficulty seeing that $a^2 = 1 \cdot a^2$.

Sometimes we must use the distributive law to remove grouping symbols before combining like terms. Remember to remove the innermost grouping symbols first.

EXAMPLE 7   Simplify: $3x + 2[4 + 5(x + 2y)]$.

Solution

$$3x + 2[4 + 5(x + 2y)] = 3x + 2[4 + 5x + 10y] \quad \text{Using the distributive law}$$

$$= 3x + 8 + 10x + 20y \quad \text{Using the distributive law}$$

$$= 13x + 8 + 20y \quad \text{Combining like terms}$$

The product of a number and -1 is its opposite, or additive inverse—for example,

$$-1 \cdot 8 = -8 \qquad \text{(the opposite of 8).}$$

Thus we have $-8 = -1 \cdot 8$, and in general, $-x = -1 \cdot x$. We can use this fact along with the distributive law when parentheses are preceded by a negative sign or subtraction.

EXAMPLE 8   Simplify $-(a - b)$, using multiplication by -1.

Solution   We have

$$-(a - b) = -1 \cdot (a - b) \quad \text{Replacing } - \text{ with multiplication by } -1$$

$$= -1 \cdot a - (-1) \cdot b \quad \text{Using the distributive law}$$

$$= -a - (-b) \quad \text{Replacing } -1 \cdot a \text{ with } -a \text{ and } (-1) \cdot b \text{ with } -b$$

$$= -a + b, \text{ or } b - a. \quad \text{Try to go directly to this step.}$$

The expressions $-(a - b)$ and $b - a$ are equivalent. They name the same number for all replacements of a and b.

TEACHING TIP

Some students may not immediately recognize that $-a + b = b - a$.

Example 8 illustrates a useful shortcut worth remembering:

$a - b$ and $b - a$ are opposites of each other.

EXAMPLE 9   Simplify: $9x - 5y - (5x + y - 7)$.

Solution

$$9x - 5y - (5x + y - 7) = 9x - 5y - 5x - y + 7 \quad \text{Taking the opposite of } 5x + y - 7$$

$$= 4x - 6y + 7 \quad \text{Combining like terms}$$

EXAMPLE 10   Simplify by combining like terms:

$$3x^2 - 2\{3[x - 5x^2 - 4(x + 3) + 7] - x\}.$$

Solution

$$3x^2 - 2\{3[x - 5x^2 - 4(x + 3) + 7] - x\}$$

$$= 3x^2 - 2\{3[x - 5x^2 - 4x - 12 + 7] - x\}$$ 　Multiplying to remove the innermost parentheses using the distributive law

$$= 3x^2 - 2\{3[-5x^2 - 3x - 5] - x\}$$ 　Combining like terms inside the brackets

$$= 3x^2 - 2\{-15x^2 - 9x - 15 - x\}$$ 　Using the distributive law to remove the brackets

$$= 3x^2 - 2\{-15x^2 - 10x - 15\}$$ 　Combining like terms

$$= 3x^2 + 30x^2 + 20x + 30$$ 　Using the distributive law

$$= 33x^2 + 20x + 30$$ 　Combining like terms

Editing Expressions

It is possible to correct an error or change a value in an expression by using the *arrow, insert,* and *delete* keys.

As the expression is being typed, use the arrow keys to move the cursor to the character you want to change. Pressing the INSERT key (the 2nd function associated with the DELETE key) changes the calculator between the INSERT and OVERWRITE modes. The OVERWRITE mode is often indicated by a rectangular cursor and the INSERT mode by an underscore cursor. The following table shows how to make changes.

Insert a character in front of the cursor.	In the INSERT mode, press the character you wish to insert.
Replace the character under the cursor.	In the OVERWRITE mode, press the character you want as the replacement.
Delete the character under the cursor.	Press DEL .

After you have evaluated an expression by pressing ENTER , the expression can be recalled to the screen by pressing 2nd ENTRY . (ENTRY is the 2nd function associated with the ENTER key.) Then it can be edited as described above. Pressing ENTER will then evaluate the edited expression.

EXAMPLE 11   Evaluate $13 - 5[x + 6(4 - x) - 18]$ for $x = 1.3$ and for $x = -5$.

Solution   To evaluate the expression for $x = 1.3$, we type $13 - 5(1.3 + 6(4 - 1.3) - 18)$ and press ENTER . Then, to evaluate the expression for $x = -5$, we press 2nd ENTRY and replace each occurrence of 1.3 with -5. Since 1.3 consists of three characters and -5

two characters, we delete one character of 1.3 and overwrite the other two with -5. Remember to use the $\boxed{(-)}$ key to enter the negative sign. Then we press $\boxed{\text{ENTER}}$.

```
13-5(1.3+6(4-1.3
)-18)
                    15.5
13-5(-5+6(4--5)-
18)
                    -142
```

We see that $13 - 5[x + 6(4 - x) - 18]$ is 15.5 when $x = 1.3$ and -142 when $x = -5$.

Checking by Evaluating

Recall that equivalent expressions have the same value for all possible replacements. Thus one way to check whether two expressions are equivalent is to evaluate both expressions using the same replacements for the variables.

Interactive Discovery

1. Evaluate both expressions in the following table for the given values for x, and determine whether the expressions appear to be equivalent.

x	$3x + 5x^2 + 4x + x^2$	$7x + 5x^2$	Equivalent?
0	0	0	
5	185	160	No
-3	33	24	

2. Next, evaluate both expressions in the following table for the given values for x, and determine whether these expressions appear to be equivalent.

x	$2x - 3[4 - 6(x - 1)]$	$20x - 30$	Equivalent?
0	-30	-30	
5	70	70	Yes
-3	-90	-90	

Can two expressions that are *not* equivalent have the same value for *some* possible replacements?

Checking by Evaluating

To check whether two expressions are equivalent, evaluate both expressions using the same replacements for the variables.

1. If the values are different, the expressions *are not* equivalent.
2. If the values are the same, the expressions *may be* equivalent.
3. Evaluating expressions for several different replacements makes the check more certain.

CAUTION!   Checking by evaluating can be very useful when a quick, partial check is needed. However, showing that two expressions have the same value for one replacement *does not prove* that they are equivalent.

EXAMPLE 12   Check Example 9 by evaluating the expressions.

Solution   We evaluate the given expression, $9x - 5y - (5x + y - 7)$, and the simplified expression, $4x - 6y + 7$. We choose the replacement values of -1 for x and 2 for y.

$$9x - 5y - (5x + y - 7) = 9(-1) - 5(2) - (5(-1) + 2 - 7)$$

                      Replacing x with -1 and y with 2

$$= -9 - 10 - (-5 + 2 - 7)$$ **Multiplying**

$$= -9 - 10 - (-10)$$ **Simplifying within the parentheses**

$$= -9 - 10 + 10$$ **Removing parentheses**

$$= -9$$ **Adding and subtracting from left to right**

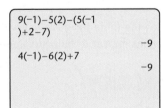

$$4x - 6y + 7 = 4(-1) - 6(2) + 7$$ **Replacing x with -1 and y with 2**

$$= -4 - 12 + 7$$ **Multiplying**

$$= -9$$ **Adding and subtracting from left to right**

The computations can also be performed using a graphing calculator, as shown in the figure at left. Since $-9 = -9$, we have a partial check.

1.3

Exercise Set

Write an equivalent expression using a commutative law. Answers may vary.

$xy + 6; 6 + yx$

1. $4a + 7b$
  $7b + 4a;\ a4 + b7;\ 4a + b7$
2. $6 + xy$

3. $(7x)y$
  $y(7x);\ (x7)y$
4. $-9(ab)$
  $(ab)(-9);\ -9(ba)$

Write an equivalent expression using an associative law.

5. $(3x)y$ $3(xy)$
6. $-7(ab)$ $(-7a)b$

7. $x + (2y + 5)$
  $(x + 2y) + 5$
8. $(3y + 4) + 10$
  $3y + (4 + 10)$

Write an equivalent expression using the distributive law.

9. $3(a + 7)$ $3a + 21$
10. $8(x + 1)$ $8x + 8$

11. $4(x - y)$ $4x - 4y$
12. $9(a - b)$ $9a - 9b$

13. $-5(2a + 3b)$
  $-10a - 15b$
14. $-2(3c + 5d)$
  $-6c - 10d$

15. $9a(b - c + d)$
  $9ab - 9ac + 9ad$
16. $5x(y - z + w)$
  $5xy - 5xz + 5xw$

Find an equivalent expression by factoring.

17. $5x + 25$ $5(x + 5)$
18. $7a + 7b$ $7(a + b)$

19. $3p - 9$ $3(p - 3)$
20. $15x - 3$ $3(5x - 1)$

21. $7x - 21y + 14z$
  $7(x - 3y + 2z)$
22. $6y - 9x - 3w$
  $3(2y - 3x - w)$

Aha! **23.** $255 - 34b$
  $17(15 - 2b)$
24. $132a + 33$
  $33(4a + 1)$

25. $xy + x$
  $x(y + 1)$
26. $ab + b$
  $b(a + 1)$

List the terms of each expression.

27. $4x - 5y + 3$ $4x, -5y, 3$
28. $2a + 7b - 13$
  $2a, 7b, -13$

29. $x^2 - 6x - 7$
  $x^2, -6x, -7$
30. $-3y^2 + 6y + 10$
  $-3y^2, 6y, 10$

Simplify by combining like terms. Use the distributive law as needed.

31. $3x + 7x$ $10x$
32. $9x + 3x$ $12x$

33. $7rt - 9rt$ $-2rt$
34. $3ab + 7ab$ $10ab$

35. $9t^2 + t^2$ $10t^2$
36. $7a^2 + a^2$ $8a^2$

37. $12a - a$ $11a$
38. $15x - x$ $14x$

39. $n - 8n$ $-7n$
40. $x - 6x$ $-5x$

41. $5x - 3x + 8x$ $10x$
42. $3x - 11x + 2x$ $-6x$

43. $4x - 2x^2 + 3x$ $7x - 2x^2$
44. $9a - 5a^2 + 4a$
  $13a - 5a^2$

45. $6a + 7a^2 - a + 4a^2$ $5a + 11a^2$

46. $9x + 2x^3 + 5x - 6x^2$ $14x + 2x^3 - 6x^2$

47. $4x - 7 + 18x + 25$ $22x + 18$
48. $13p + 5 - 4p + 7$ $9p + 12$
49. $-7t^2 + 3t + 5t^3 - t^3 + 2t^2 - t$ $-5t^2 + 2t + 4t^3$
50. $-9n + 8n^2 + n^3 - 2n^2 - 3n + 4n^3$
  $-12n + 6n^2 + 5n^3$
51. $7a - (2a + 5)$ $5a - 5$
52. $x - (5x + 9)$ $-4x - 9$
53. $m - (3m - 1)$ $-2m + 1$
54. $5a - (4a - 3)$ $a + 3$
55. $3d - 7 - (5 - 2d)$ $5d - 12$
56. $8x - 9 - (7 - 5x)$ $13x - 16$
57. $-2(x + 3) - 5(x - 4)$ $-7x + 14$
58. $-9(y + 7) - 6(y - 3)$ $-15y - 45$
59. $4x - 7(2x - 3)$ $-10x + 21$
60. $9y - 4(5y - 6)$ $-11y + 24$
61. $9a - [7 - 5(7a - 3)]$ $44a - 22$
62. $12b - [9 - 7(5b - 6)]$ $47b - 51$
63. $5\{-2a + 3[4 - 2(3a + 5)]\}$ $-100a - 90$
64. $7\{-7x + 8[5 - 3(4x + 6)]\}$ $-721x - 728$
65. $2y + \{7[3(2y - 5) - (8y + 7)] + 9\}$ $-12y - 145$
66. $7b - \{6[4(3b - 7) - (9b + 10)] + 11\}$
  $-11b + 217$

Evaluate both expressions in each pair for $x = 1.3$, $x = -3$, and $x = 0$. Then determine whether the expressions appear to be equivalent.

67. $2x - 3(x + 5)$, $-16.3; -12; -15$
  $-x + 15$ $13.7; 18; 15;$ not equivalent

68. $2(x + 5) - 5(x + 2)$, $-3.9; 9; 0$
  $7x$ $9.1; -21; 0;$ not equivalent

69. $4(x + 3)$, $17.2; 0; 12$
  $4x + 12$ $17.2; 0; 12;$ equivalent

70. $7x + 1$, $10.1; -20; 1$
  $7(x + 1)$ $16.1; -14; 7;$ not equivalent

TW **71.** What is the difference between the associative law of multiplication and the distributive law?

TW **72.** Write a sentence in which the word "factor" appears once as a verb and once as a noun.

Skill Maintenance

Find each absolute value. [1.2]

73. $|-3|$ 3

74. $|3.59|$ 3.59

75. Find the opposite of -35. [1.2] 35

76. Find the reciprocal of -35. [1.2] $-\frac{1}{35}$

Synthesis

TW **77.** Explain how the distributive and commutative laws can be used to rewrite $3x + 6y + 4x + 2y$ as $7x + 8y$.

TW **78.** Lee simplifies the expression $a(b + c)$ by omitting the parentheses and writing $ab + c$. Are these expressions equivalent? Why or why not?

Simplify.

79. $11(a - 3) + 12a - \{6[4(3b - 7) - (9b + 10)] + 11\}$
$23a - 18b + 184$

80. $-3[9(x - 4) + 5x] - 8\{3[5(3y + 4)] - 12\}$ $-42x - 360y - 276$

81. $z - \{2z + [3z - (4z + 5z) - 6z] + 7z\} - 8z$ $-4z$

82. $x + [f - (f + x)] + [x - f] + 3x$ $4x - f$

83. $x - \{x + 1 - [x + 2 - (x - 3 - \{x + 4 - [x - 5 + (x - 6)]\})]\}$ $-x + 19$

84. Use the commutative, associative, and distributive laws to show that $5(a + bc)$ is equivalent to $c(b5) + a5$. Use only one law in each step of your work.

TW **85.** Are subtraction and division commutative? Why or why not?

TW **86.** Are subtraction and multiplication associative? Why or why not?

84. $5(a + bc)$
$= 5a + 5(bc)$ Distributive
$= 5(bc) + 5a$ Commutative, addition
$= (bc)5 + a5$ Commutative, multiplication
$= (cb)5 + a5$ Commutative, multiplication
$= c(b5) + a5$ Associative, multiplication

1.4

The Product and Quotient Rules ▪ The Zero Exponent ▪ Negative Integers as Exponents ▪ Raising Powers to Powers ▪ Raising a Product or a Quotient to a Power ▪ Scientific Notation

Exponential and Scientific Notation

In Section 1.1, we discussed how whole-number exponents are used. We now develop rules for manipulating exponents and define what zero and negative integers will mean as exponents.

The Product and Quotient Rules

Interactive Discovery

Calculate the value of each of the following expressions. Then match each numbered expression with the lettered expression having the same value.

1. $3^4 \cdot 3^5$ (b)

2. $3^2 \cdot 3^5$ (c)

3. $3^5 \cdot 3^6$ (a)

4. $3^4 \cdot 3$ (d)

a. 3^{11}

b. 3^9

c. 3^7

d. 3^5

How are the exponents of the equivalent expressions related?

The result in the Interactive Discovery above is generalized in the *product rule*.

Multiplying with Like Bases: The Product Rule

For any number a and any positive integers m and n,

$$a^m \cdot a^n = a^{m+n}.$$

(When multiplying powers, if the bases are the same, keep the base and add the exponents.)

EXAMPLE 1   Multiply and simplify: **(a)** $m^5 \cdot m^7$; **(b)** $(5a^2b^3)(3a^4b^5)$.

Solution

a) $m^5 \cdot m^7 = m^{5+7} = m^{12}$ **Multiplying powers by adding exponents**

b) $(5a^2b^3)(3a^4b^5) = 5 \cdot 3 \cdot a^2 \cdot a^4 \cdot b^3 \cdot b^5$ **Using the associative and commutative laws**

$$= 15a^{2+4}b^{3+5}$$ **Multiplying; using the product rule**

$$= 15a^6b^8$$

CAUTION! $5^8 \cdot 5^6 = 5^{14}$; $5^8 \cdot 5^6 \neq 25^{14}$.

Next, we simplify a quotient.

Interactive Discovery

Calculate the value of each of the following expressions. Then match each numbered expression with the lettered expression having the same value.

1. $\dfrac{3^{11}}{3^{10}}$ (d)   **2.** $\dfrac{3^7}{3^2}$ (a)   **a.** 3^5 **b.** 3^4

3. $\dfrac{3^5}{3}$ (b)   **4.** $\dfrac{3^8}{3^2}$ (c)   **c.** 3^6 **d.** 3^1

How are the exponents of the equivalent expressions related?

The generalization of the result in the Interactive Discovery above is the *quotient rule*.

Dividing with Like Bases: The Quotient Rule

For any nonzero number a and any positive integers m and n, $m > n$,

$$\frac{a^m}{a^n} = a^{m-n}.$$

(When dividing powers, if the bases are the same, keep the base and subtract the exponent of the denominator from the exponent of the numerator.)

EXAMPLE 2   Divide and simplify: **(a)** $\dfrac{r^9}{r^3}$; **(b)** $\dfrac{10x^{11}y^5}{2x^4y^3}$.

Solution

a) $\dfrac{r^9}{r^3} = r^{9-3} = r^6$ Using the quotient rule

b) $\dfrac{10x^{11}y^5}{2x^4y^3} = 5 \cdot x^{11-4} \cdot y^{5-3}$ Dividing; using the quotient rule

$= 5x^7y^2$

CAUTION! $\dfrac{7^8}{7^2} = 7^6$; $\dfrac{7^8}{7^2} \neq 7^4$.

The Zero Exponent

Suppose now that the bases in the numerator and the denominator are both raised to the same power. On the one hand, any (nonzero) expression divided by itself is equal to 1. For example,

$$\frac{t^5}{t^5} = 1 \quad \text{and} \quad \frac{6^4}{6^4} = 1.$$

On the other hand, if we continue subtracting exponents when dividing powers with the same base, we have

$$\frac{t^5}{t^5} = t^{5-5} = t^0 \quad \text{and} \quad \frac{6^4}{6^4} = 6^{4-4} = 6^0.$$

This suggests that t^5/t^5 equals both 1 *and* t^0. It also suggests that $6^4/6^4$ equals both 1 *and* 6^0. This leads to the following definition.

The Zero Exponent

For any nonzero real number a,

$$a^0 = 1.$$

(Any nonzero number raised to the zero power is 1. 0^0 is undefined.)

EXAMPLE 3   Evaluate each of the following for $x = 2.9$: **(a)** x^0; **(b)** $-x^0$; **(c)** $(-x)^0$.

Solution

a) $x^0 = 2.9^0 = 1$ Using the definition of 0 as an exponent

b) $-x^0 = -2.9^0 = -1$ The exponent 0 pertains only to the 2.9.

c) $(-x)^0 = (-2.9)^0 = 1$ The base here is -2.9.

Parts (b) and (c) of Example 3 illustrate an important result:

Since $-a^n$ means $-1 \cdot a^n$, in general, $-a^n \neq (-a)^n$.

Negative Integers as Exponents

Later in this text we will explain what numbers like $\frac{2}{9}$ or $\sqrt{2}$ mean as exponents. Until then, integer exponents will suffice.

To develop a definition for negative integer exponents, we simplify $5^3/5^7$ two ways. First we proceed as in arithmetic:

$$\frac{5^3}{5^7} = \frac{5 \cdot 5 \cdot 5}{5 \cdot 5 \cdot 5 \cdot 5 \cdot 5 \cdot 5 \cdot 5} = \frac{5 \cdot 5 \cdot 5 \cdot 1}{5 \cdot 5 \cdot 5 \cdot 5 \cdot 5 \cdot 5 \cdot 5}$$

$$= \frac{5 \cdot 5 \cdot 5}{5 \cdot 5 \cdot 5} \cdot \frac{1}{5 \cdot 5 \cdot 5 \cdot 5}$$

$$= \frac{1}{5^4}.$$

Were we to apply the quotient rule, we would have

$$\frac{5^3}{5^7} = 5^{3-7} = 5^{-4}.$$

These two expressions for $5^3/5^7$ suggest that

$$5^{-4} = \frac{1}{5^4}.$$

This leads to the definition of negative exponents.

Negative Exponents

For any real number a that is nonzero and any integer n,

$$a^{-n} = \frac{1}{a^n}.$$

(The numbers a^{-n} and a^n are reciprocals of each other.)

The definitions above preserve the following pattern:

$$4^3 = 4 \cdot 4 \cdot 4,$$
$$4^2 = 4 \cdot 4, \qquad \text{Dividing both sides by 4}$$
$$4^1 = 4, \qquad \text{Dividing both sides by 4}$$
$$4^0 = 1, \qquad \text{Dividing both sides by 4}$$
$$4^{-1} = \frac{1}{4}, \qquad \text{Dividing both sides by 4}$$
$$4^{-2} = \frac{1}{4 \cdot 4} = \frac{1}{4^2}. \qquad \text{Dividing both sides by 4}$$

CAUTION!   A negative exponent does not, in itself, indicate that an expression is negative. As shown above,

$$4^{-2} \neq 4(-2).$$

EXAMPLE 4 Express using positive exponents and, if possible, simplify.

a) 3^{-2} **b)** $5x^{-4}y^3$ **c)** $\dfrac{1}{7^{-2}}$

Solution

a) $3^{-2} = \dfrac{1}{3^2} = \dfrac{1}{9}$

b) $5x^{-4}y^3 = 5\left(\dfrac{1}{x^4}\right)y^3 = \dfrac{5y^3}{x^4}$

c) Since $\dfrac{1}{a^n} = a^{-n}$, we have

$$\dfrac{1}{7^{-2}} = 7^{-(-2)} = 7^2, \text{ or } 49.$$ **Remember: n can be a negative integer.**

The result from part (c) above can be generalized.

Factors and Negative Exponents

For any nonzero real numbers a and b and any integers m and n,

$$\dfrac{1}{a^{-n}} = \dfrac{a^n}{1} \quad \text{and} \quad \dfrac{a^{-n}}{b^{-m}} = \dfrac{b^m}{a^n}.$$

(A factor can be moved to the other side of the fraction bar if the sign of the exponent is changed.)

EXAMPLE 5 Write an equivalent expression without negative exponents:

$$\dfrac{-5^2 x^{-2} y^{-5}}{z^{-4} w^{-3}}.$$

Solution We can move a factor to the other side of the fraction bar if we change the sign of its exponent:

$$\dfrac{-5^2 x^{-2} y^{-5}}{z^{-4} w^{-3}} = \dfrac{-25 z^4 w^3}{x^2 y^5}.$$ **Note that -5^2 does not have a negative exponent, so it is not moved.**

The product and quotient rules apply for all integer exponents.

EXAMPLE 6 Simplify: **(a)** $7^{-3} \cdot 7^8$; **(b)** $\dfrac{b^{-5}}{b^{-4}}$.

Solution

a) $7^{-3} \cdot 7^8 = 7^{-3+8}$ Using the product rule

$= 7^5$

Using the product rule

b) $\dfrac{b^{-5}}{b^{-4}} = b^{-5-(-4)} = b^{-1}$

$= \dfrac{1}{b}$ Writing the answer in an equivalent form, without a negative exponent

Example 6(b) can also be simplified as follows:

$$\frac{b^{-5}}{b^{-4}} = \frac{b^4}{b^5} = b^{4-5} = b^{-1} = \frac{1}{b}.$$

CAUTION!   Only *factors* with negative exponents can be moved to the other side of the fraction bar—for example,

$$\frac{a^{-3}b^2}{c^{-4}d} = \frac{c^4 b^2}{a^3 d}, \quad \text{but} \quad \frac{a^{-3}+b^2}{c^{-4}+d} \neq \frac{c^4+b^2}{a^3+d}.$$

Raising Powers to Powers

Next, consider an expression like $(3^4)^2$:

$(3^4)^2 = (3^4)(3^4)$ We are raising 3^4 to the second power.

$= (3 \cdot 3 \cdot 3 \cdot 3)(3 \cdot 3 \cdot 3 \cdot 3)$

$= 3 \cdot 3 \cdot 3 \cdot 3 \cdot 3 \cdot 3 \cdot 3 \cdot 3$ Using the associative law

$= 3^8.$

Note that in this case, we could have multiplied the exponents:

$(3^4)^2 = 3^{4 \cdot 2} = 3^8.$

Likewise, $(y^8)^3 = (y^8)(y^8)(y^8) = y^{24}$. Once again, we get the same result if we multiply the exponents:

$(y^8)^3 = y^{8 \cdot 3} = y^{24}.$

Raising a Power to a Power: The Power Rule

For any real number a and any integers m and n for which a^m and a^{mn} exist,

$$(a^m)^n = a^{mn}.$$

(To raise a power to a power, multiply the exponents.)

EXAMPLE 7   Simplify: **(a)** $(3^5)^4$; **(b)** $(y^{-5})^7$; **(c)** $(a^{-3})^{-7}$.

Solution

a) $(3^5)^4 = 3^{5 \cdot 4} = 3^{20}$

b) $(y^{-5})^7 = y^{-5 \cdot 7} = y^{-35} = \dfrac{1}{y^{35}}$

c) $(a^{-3})^{-7} = a^{(-3)(-7)} = a^{21}$

Raising a Product or a Quotient to a Power

When an expression inside parentheses is raised to a power, the inside expression is the base. Let's compare $2a^3$ and $(2a)^3$.

$$2a^3 = 2 \cdot a \cdot a \cdot a; \qquad (2a)^3 = (2a)(2a)(2a)$$
$$= 2 \cdot 2 \cdot 2 \cdot a \cdot a \cdot a$$
$$= 2^3 a^3 = 8a^3$$

We see that $2a^3$ and $(2a)^3$ are *not* equivalent. Note also that to simplify $(2a)^3$, we can raise each factor to the power 3. This leads to the following rule.

Raising a Product to a Power

For any integer n, and any real numbers a and b for which $(ab)^n$ exists,

$$(ab)^n = a^n b^n.$$

(To raise a product to a power, raise each factor to that power.)

EXAMPLE 8 Simplify: **(a)** $(-2x)^3$; **(b)** $(-3x^5 y^{-1})^{-4}$.

Solution

a) $(-2x)^3 = (-2)^3 \cdot x^3$ **Raising each factor to the third power**
$$= -8x^3$$

b) $(-3x^5 y^{-1})^{-4} = (-3)^{-4}(x^5)^{-4}(y^{-1})^{-4}$ **Raising each factor to the negative fourth power**

$$= \frac{1}{(-3)^4} \cdot x^{-20} y^4$$ **Multiplying powers; writing $(-3)^4$ as $\frac{1}{(-3)^4}$**

$$= \frac{y^4}{81x^{20}}$$ **Note that $x^{-20} = \frac{1}{x^{20}}$.**

TEACHING TIP

Some students may want to multiply the exponent and the coefficient, too, to get $12x^{-20}y^4$ for Example 8(b). Writing out the first step may help them avoid this mistake.

There is a similar rule for raising a quotient to a power.

Raising a Quotient to a Power

For any integer n, and any real numbers a and b for which a/b, a^n, and b^n exist,

$$\left(\frac{a}{b}\right)^n = \frac{a^n}{b^n}.$$

(To raise a quotient to a power, raise both the numerator and the denominator to that power.)

EXAMPLE 9 Simplify: **(a)** $\left(\dfrac{x^2}{2}\right)^4$; **(b)** $\left(\dfrac{y^2 z^3}{5}\right)^{-3}$.

Solution

a) $\left(\dfrac{x^2}{2}\right)^4 = \dfrac{(x^2)^4}{2^4} = \dfrac{x^8}{16}$ ← $2 \cdot 4 = 8$
← $2^4 = 16$

b) $\left(\dfrac{y^2z^3}{5}\right)^{-3} = \dfrac{(y^2z^3)^{-3}}{5^{-3}}$

$= \dfrac{5^3}{(y^2z^3)^3}$ Moving factors to the other side of the fraction bar and changing each -3 to 3

$= \dfrac{125}{y^6z^9}$

The rule for raising a quotient to a power allows us to derive a useful result for manipulating negative exponents:

$$\left(\frac{a}{b}\right)^{-n} = \frac{a^{-n}}{b^{-n}} = \frac{b^n}{a^n} = \left(\frac{b}{a}\right)^n.$$

Using this result, we can simplify Example 9(b) as follows:

$\left(\dfrac{y^2z^3}{5}\right)^{-3} = \left(\dfrac{5}{y^2z^3}\right)^3$ Taking the reciprocal of the base and changing the exponent's sign

$= \dfrac{5^3}{(y^2z^3)^3} = \dfrac{125}{y^6z^9}.$

Definitions and Properties of Exponents

The following summary assumes that no denominators are 0 and that 0^0 is not considered. For any integers m and n,

1 as an exponent:	$a^1 = a$
0 as an exponent:	$a^0 = 1$
Negative exponents:	$a^{-n} = \dfrac{1}{a^n}; \dfrac{1}{a^{-n}} = a^n$
	$\dfrac{a^{-n}}{b^{-m}} = \dfrac{b^m}{a^n}$
	$\left(\dfrac{a}{b}\right)^{-n} = \left(\dfrac{b}{a}\right)^n$
The Product Rule:	$a^m \cdot a^n = a^{m+n}$
The Quotient Rule:	$\dfrac{a^m}{a^n} = a^{m-n}$
The Power Rule:	$(a^m)^n = a^{mn}$
Raising a product to a power:	$(ab)^n = a^n b^n$
Raising a quotient to a power:	$\left(\dfrac{a}{b}\right)^n = \dfrac{a^n}{b^n}$

Scientific Notation

TEACHING TIP

You may want to use a calculator to illustrate uses of scientific notation.

Very large and very small numbers that occur in science and other fields are often written in **scientific notation**, using exponents.

> ### Scientific Notation
>
> *Scientific notation* for a number is an expression of the form $N \times 10^m$, where N is in decimal notation, $1 \leq N < 10$, and m is an integer.

Note that $10^b/10^b = 10^b \cdot 10^{-b} = 1$. To convert a number to scientific notation, we can multiply by 1, writing 1 in the form $10^b/10^b$ or $10^b \cdot 10^{-b}$.

EXAMPLE 10   Population Projections.   It has been estimated that in 2010, the world population will be 6,832,000,000 (*Source: The Statistical Abstract of the United States,* 2000). Write scientific notation for this number.

Solution   To write 6,832,000,000 as 6.832×10^m for some integer m, we must move the decimal point in 6,832,000,000 to the left 9 places. This can be accomplished by dividing—and then multiplying—by 10^9:

$$6{,}832{,}000{,}000 = \frac{6{,}832{,}000{,}000}{10^9} \cdot 10^9 \qquad \textbf{Multiplying by 1: } \frac{10^9}{10^9} = 1$$

$$= 6.832 \times 10^9. \qquad \textbf{This is scientific notation.}$$

EXAMPLE 11   Write scientific notation for the mass of a grain of sand:

0.0648 gram (g).

Solution   To write 0.0648 as 6.48×10^m for some integer m, we must move the decimal point 2 places to the right. To do this, we multiply—and then divide—by 10^2:

$$0.0648 = \frac{0.0648 \times 10^2}{10^2} \qquad \textbf{Multiplying by 1: } \frac{10^2}{10^2} = 1$$

$$= \frac{6.48}{10^2}$$

$$= 6.48 \times 10^{-2} \text{ g}. \qquad \textbf{Writing scientific notation}$$

Try to make conversions to scientific notation mentally if possible. In doing so, remember that negative powers of 10 are used for small numbers and positive powers of 10 are used for large numbers.

EXAMPLE 12 Convert mentally to scientific notation: **(a)** 82,500,000; **(b)** 0.0000091.

Solution

a) $82{,}500{,}000 = 8.25 \times 10^7$ *Check*: **Multiplying 8.25 by 10^7 moves the decimal point 7 places to the right.**

b) $0.0000091 = 9.1 \times 10^{-6}$ *Check*: **Multiplying 9.1 by 10^{-6} moves the decimal point 6 places to the left.**

EXAMPLE 13 Convert mentally to decimal notation: **(a)** 4.371×10^7; **(b)** 1.73×10^{-5}.

Solution

a) $4.371 \times 10^7 = 43{,}710{,}000$ **Moving the decimal point 7 places to the right**

b) $1.73 \times 10^{-5} = 0.0000173$ **Moving the decimal point 5 places to the left**

We can use the associative and commutative laws to multiply and divide using scientific notation.

EXAMPLE 14 Multiply and write scientific notation for the answer: $(7.2 \times 10^5)(4.3 \times 10^9)$.

Solution We have

$(7.2 \times 10^5)(4.3 \times 10^9) = (7.2 \times 4.3)(10^5 \times 10^9)$ **Using the commutative and associative laws**

$= 30.96 \times 10^{14}.$ **Adding exponents; simplifying**

The result is not yet written in scientific notation (because $30.96 > 10$), so we convert 30.96 to scientific notation and simplify:

$$30.96 \times 10^{14} = (3.096 \times 10^1) \times 10^{14} = 3.096 \times 10^{15}.$$

EXAMPLE 15 Divide and write scientific notation for the answer:

$$\frac{3.48 \times 10^{-7}}{4.64 \times 10^6}.$$

Solution

$\frac{3.48 \times 10^{-7}}{4.64 \times 10^6} = \frac{3.48}{4.64} \times \frac{10^{-7}}{10^6}$ **Separating factors**

$= 0.75 \times 10^{-13}$ **Subtracting exponents; simplifying**

$= (7.5 \times 10^{-1}) \times 10^{-13}$ **Converting 0.75 to scientific notation**

$= 7.5 \times 10^{-14}$ **Adding exponents**

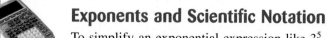

Exponents and Scientific Notation

To simplify an exponential expression like 3^5, we can use a calculator's exponentiation key, usually labeled $\boxed{\wedge}$. If the exponent is 2, we can also use the $\boxed{x^2}$ key. If it is -1, we can use the $\boxed{x^{-1}}$ key. If the exponent is a single number, not an expression, we do not need to enclose it in parentheses.

   Graphing calculators will accept entries using scientific notation, and will normally write very large or very small numbers using scientific notation. The $\boxed{\text{EE}}$ key, the 2nd function associated with the $\boxed{.}$ key, is used to enter scientific notation. On the calculator screen, a notation like E22 represents $\times 10^{22}$.

EXAMPLE 16   Calculate 3^5, $(-4.7)^2$, and $(-8)^{-1}$.

Solution   To calculate 3^5, we press $\boxed{3}$ $\boxed{\wedge}$ $\boxed{5}$ $\boxed{\text{ENTER}}$.
   To calculate $(-4.7)^2$, we press $\boxed{(}$ $\boxed{(-)}$ $\boxed{4}$ $\boxed{.}$ $\boxed{7}$ $\boxed{)}$ $\boxed{\wedge}$ $\boxed{2}$ $\boxed{\text{ENTER}}$ or $\boxed{(}$ $\boxed{(-)}$ $\boxed{4}$ $\boxed{.}$ $\boxed{7}$ $\boxed{)}$ $\boxed{x^2}$ $\boxed{\text{ENTER}}$.
   To calculate $(-8)^{-1}$, we press $\boxed{(}$ $\boxed{(-)}$ $\boxed{8}$ $\boxed{)}$ $\boxed{x^{-1}}$ $\boxed{\text{ENTER}}$ or $\boxed{(}$ $\boxed{(-)}$ $\boxed{8}$ $\boxed{)}$ $\boxed{\wedge}$ $\boxed{(-)}$ $\boxed{1}$ $\boxed{\text{ENTER}}$.

```
3^5
                    243
(-4.7)^2
                  22.09
(-4.7)2
                  22.09
```

```
(-8)-1
                   -.125
(8)^-1
                   -.125
```

We see that $3^5 = 243$, $(-4.7)^2 = 22.09$, and $(-8)^{-1} = -0.125$.

EXAMPLE 17   Calculate $(7.5 \times 10^8)(1.2 \times 10^{-14})$.

Solution   We press $\boxed{7}$ $\boxed{.}$ $\boxed{5}$ $\boxed{\text{2nd}}$ $\boxed{\text{EE}}$ $\boxed{8}$ $\boxed{\times}$ $\boxed{1}$ $\boxed{.}$ $\boxed{2}$ $\boxed{\text{2nd}}$ $\boxed{\text{EE}}$ $\boxed{(-)}$ $\boxed{1}$ $\boxed{4}$ $\boxed{\text{ENTER}}$.

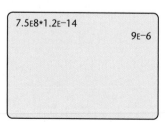

```
7.5E8*1.2E-14
                    9E-6
```

The result shown is then read as 9×10^{-6}.

1.4

Exercise Set

Multiply and simplify. Leave the answer in exponential notation.

1. $5^6 \cdot 5^3$ 5^9

2. $6^3 \cdot 6^5$ 6^8

3. $t^0 \cdot t^8$ t^8

4. $x^0 \cdot x^5$ x^5

5. $6x^5 \cdot 3x^2$ $18x^7$

6. $4a^3 \cdot 2a^7$ $8a^{10}$

7. $(-3m^4)(-7m^9)$ $21m^{13}$

8. $(-2a^5)(7a^4)$ $-14a^9$

9. $(x^3 y^4)(x^7 y^6 z^0)$ $x^{10} y^{10}$

10. $(m^6 n^5)(m^4 n^7 p^0)$ $m^{10} n^{12}$

Divide and simplify.

11. $\dfrac{a^9}{a^3}$ a^6

12. $\dfrac{x^{12}}{x^3}$ x^9

13. $\dfrac{12t^7}{4t^2}$ $3t^5$

14. $\dfrac{20a^{20}}{5a^4}$ $4a^{16}$

15. $\dfrac{m^7 n^9}{m^2 n^5}$ $m^5 n^4$

16. $\dfrac{m^{12} n^9}{m^4 n^6}$ $m^8 n^3$

17. $\dfrac{28 x^{10} y^9 z^8}{-7 x^2 y^3 z^2}$ $-4x^8 y^6 z^6$

18. $\dfrac{18 x^8 y^6 z^7}{-3 x^2 y^3 z}$ $-6x^6 y^3 z^6$

Evaluate each of the following for $x = -2$.

19. $-x^0$ -1

20. $(-x)^0$ 1

21. $(4x)^0$ 1

22. $4x^0$ 4

Simplify.

23. $(-3)^4$ 81

24. $(-2)^6$ 64

25. -3^4 -81

26. -2^6 -64

27. $(-4)^{-2}$ $\frac{1}{16}$

28. $(-5)^{-2}$ $\frac{1}{25}$

29. -4^{-2} $-\frac{1}{16}$

30. -5^{-2} $-\frac{1}{25}$

31. -1^{-4} -1

Write an equivalent expression without negative exponents and, if possible, simplify.

32. a^{-3} $\frac{1}{a^3}$

33. n^{-6} $\frac{1}{n^6}$

34. $\dfrac{1}{5^{-3}}$ 5^3, or 125 Aha!

35. $\dfrac{1}{2^{-6}}$ 2^6, or 64

36. $4x^{-7}$ $\frac{4}{x^7}$

37. $2a^2 b^{-5}$ $\frac{2a^2}{b^5}$

38. $\dfrac{y^{-5}}{x^2}$ $\frac{1}{x^2 y^5}$

39. $\dfrac{y^4 z^{-3}}{x^{-2}}$ $\frac{x^2 y^4}{z^3}$

40. $\dfrac{x^{-2} y^7}{z^{-4}}$ $\frac{y^7 z^4}{x^2}$

Write an equivalent expression with negative exponents.

41. $\dfrac{1}{3^4}$ 3^{-4}

42. $\dfrac{1}{(-8)^6}$ $(-8)^{-6}$

43. x^5 $\frac{1}{x^{-5}}$ Aha!

44. n^3 $\frac{1}{n^{-3}}$

45. $6x^2$ $\frac{1}{6^{-1} x^{-2}}$

46. $(-4y)^5$ $\frac{6}{x^{-2}}$, or $\frac{1}{(-4y)^{-5}}$

47. $\dfrac{1}{(5y)^3}$ $(5y)^{-3}$

48. $\dfrac{1}{3y^4}$ $\frac{y^{-4}}{3}$, or $3^{-1} y^{-4}$

Simplify. Should negative exponents appear in the answer, write a second answer using only positive exponents.

49. $7^{-1} \cdot 7^{-6}$ 7^{-7}, or $\frac{1}{7^7}$

50. $3^{-4} \cdot 3^{-8}$ 3^{-12}, or $\frac{1}{3^{12}}$

51. $b^2 \cdot b^{-5}$ b^{-3}, or $\frac{1}{b^3}$

52. $a^4 \cdot a^{-3}$ a

53. $a^{-3} \cdot a^4 \cdot a^2$ a^3

54. $x^{-8} \cdot x^5 \cdot x^3$ 1

55. $(5a^{-2} b^{-3})(2a^{-4} b)$ $\square$

56. $(3a^{-5} b^{-7})(2ab^{-2})$ $\square$

57. $\dfrac{10^{-3}}{10^6}$ 10^{-9}, or $\frac{1}{10^9}$

58. $\dfrac{12^{-4}}{12^8}$ 12^{-12}, or $\frac{1}{12^{12}}$

59. $\dfrac{2^{-7}}{2^{-5}}$ 2^{-2}, or $\frac{1}{2^2}$, or $\frac{1}{4}$

60. $\dfrac{9^{-4}}{9^{-6}}$ 9^2, or 81

61. $\dfrac{8^{-7}}{8^{-2}}$ 8^{-5}, or $\frac{1}{8^5}$

62. $\dfrac{10^{-4}}{10^{-6}}$ 10^2, or 100

63. $\dfrac{-5 x^{-2} y^4 z^7}{30 x^{-5} y^6 z^{-3}}$ $\square$

64. $\dfrac{9 a^6 b^{-4} c^7}{27 a^{-4} b^5 c^9}$ $\square$

65. $(x^4)^3$ x^{12}

66. $(a^3)^2$ a^6

67. $(9^3)^{-4}$ 9^{-12}, or $\frac{1}{9^{12}}$

68. $(8^4)^{-3}$ 8^{-12}, or $\frac{1}{8^{12}}$

69. $(7^{-8})^{-5}$ 7^{40}

70. $(6^{-4})^{-3}$ 6^{12}

71. $(a^3 b)^4$ $a^{12} b^4$

72. $(x^3 y)^5$ $x^{15} y^5$

73. $5(x^2 y^2)^{-7}$ $\square$

74. $7(a^3 b^4)^{-5}$ $\square$

75. $(a^{-5} b^2)^3 (a^4 b^{-1})^2$ $a^{-7} b^4$, or $\frac{b^4}{a^7}$

76. $(x^2 y^{-3})^3 (x^{-4} y)^2$ $x^{-2} y^{-7}$, or $\frac{1}{x^2 y^7}$

77. $(5x^{-3} y^2)^{-4} (5x^{-3} y^2)^4$ 1

78. $(2a^{-1} b^3)^{-2} (2a^{-1} b^3)^{-2}$ $2^{-4} a^4 b^{-12}$, or $\frac{a^4}{16 b^{12}}$

79. $\dfrac{(3x^3 y^4)^3}{6xy^3}$ $\frac{9}{2} x^8 y^9$

80. $\dfrac{(5a^3 b)^2}{10a^2 b}$ $\frac{5}{2} a^4 b$

81. $\left(\dfrac{-4x^4 y^{-2}}{5x^{-1} y^4}\right)^{-4}$ $\frac{5^4 y^{24}}{4^4 x^{20}}$

82. $\left(\dfrac{2x^3 y^{-2}}{3y^{-3}}\right)^3$ $\frac{2^3 x^9 y^3}{3^3}$

83. $\left(\dfrac{4a^3 b^{-9}}{2a^{-2} b^5}\right)^0$ 1

84. $\left(\dfrac{5x^0 y^{-7}}{2x^{-2} y^4}\right)^0$ 1

$\square$ Answers to Exercises 55, 56, 63, 64, 73, and 74 can be found on p. A-53.

85. $\left(\dfrac{6a^3b^{-4}}{5a^{-2}b^2}\right)^{-2}$ $\dfrac{5^2b^{12}}{6^2a^{10}}$

86. $\left(\dfrac{3m^4n^2}{5m^8n^{-5}}\right)^{-4}$ $\dfrac{5^4m^{16}}{3^4n^{28}}$

87. $\left(\dfrac{8x^3y^{-10}}{6x^{-5}y^{-6}}\right)^3$ $\dfrac{8^3x^{24}}{6^3y^{12}}$, or $\dfrac{64x^{24}}{27y^{12}}$

88. $\left(\dfrac{6x^{-5}y^{-3}}{4x^{-7}y^{-1}}\right)^{-2}$ $\dfrac{4^2y^4}{6^2x^4}$, or $\dfrac{4y^4}{3x^4}$

Convert to scientific notation.

89. 47,000,000,000 4.7×10^{10}

90. 2,600,000,000,000 2.6×10^{12}

91. 0.000000016 1.6×10^{-8}

92. 0.000000263 2.63×10^{-7}

93. 407,000,000,000 4.07×10^{11}

94. 3,090,000,000,000 $\boxdot$

95. 0.000000603 6.03×10^{-7}

96. 0.00000000802 8.02×10^{-9}

Convert to decimal notation.

97. 4×10^{-4} 0.0004

98. 5×10^{-5} 0.00005

99. 6.73×10^8 673,000,000

100. 9.24×10^7 92,400,000

101. 8.923×10^{-10} $\boxdot$

102. 7.034×10^{-2} 0.07034

103. 9.03×10^{10} 90,300,000,000

104. 9.001×10^{10} 90,010,000,000

Simplify and write scientific notation for the answer.

105. $(2.3 \times 10^6)(4.2 \times 10^{-11})$ 9.66×10^{-5}

106. $(6.5 \times 10^3)(5.2 \times 10^{-8})$ 3.38×10^{-4}

107. $(2.34 \times 10^{-8})(5.7 \times 10^{-4})$ 1.3338×10^{-11}

108. $(3.26 \times 10^{-6})(8.2 \times 10^{-6})$ 2.6732×10^{-11}

109. $(2.506 \times 10^{-7})(1.408 \times 10^{10})$ 3.528448×10^3

110. $(1.6158 \times 10^6)(9.075 \times 10^{-8})$ 1.4663385×10^{-1}

111. $\dfrac{5.1 \times 10^6}{3.4 \times 10^3}$ 1.5×10^3

112. $\dfrac{8.5 \times 10^8}{3.4 \times 10^5}$ 2.5×10^3

113. $\dfrac{7.5 \times 10^{-9}}{2.5 \times 10^{-4}}$ 3×10^{-5}

114. $\dfrac{12.6 \times 10^8}{4.2 \times 10^{-3}}$ 3×10^{11}

115. $\dfrac{1.23 \times 10^8}{6.87 \times 10^{-13}}$ $\boxdot$

116. $\dfrac{4.95 \times 10^{-3}}{1.64 \times 10^{10}}$ $\boxdot$

117. $\dfrac{780{,}000{,}000 \times 0.00071}{0.000005}$ 1.1076×10^{11}

118. $\dfrac{830{,}000{,}000 \times 0.12}{3{,}100{,}000}$ Approximately 3.213×10

119. $5.9 \times 10^{23} + 2.4 \times 10^{23}$ 8.3×10^{23}

120. $1.8 \times 10^{-34} + 5.4 \times 10^{-34}$ 7.2×10^{-34}

TW **121.** Explain why $(-1)^n = 1$ for any even number n.

TW **122.** List two advantages of using scientific notation.
Answers may vary.

Skill Maintenance

123. Subtract: $-\frac{5}{6} - \left(-\frac{3}{4}\right)$. [1.2] $-\frac{1}{12}$

124. Multiply: $(-7.2)(-4.3)$. [1.2]   30.96

125. Multiply: $-2(4x - 6y)$. [1.3] $-8x + 12y$

126. Factor: $8x - 10$. [1.3] $2(4x - 5)$

Synthesis

TW **127.** Explain why $(-17)^{-8}$ is positive.

TW **128.** Is the following true or false, and why?
$$5^{-6} > 4^{-9}$$

TW **129.** Some numbers exceed the limits of the calculator. Enter 1.3×10^{-1000} and 1.3×10^{1000} and explain the results.

Simplify. Assume that all variables represent nonzero integers.

130. $\dfrac{12a^{x-2}}{3a^{2x+2}}$ $4a^{-x-4}$

131. $\dfrac{-12x^{a+1}}{4x^{2-a}}$ $-3x^{2a-1}$

132. $(3^{a+2})^a$ 3^{a^2+2a}

133. $(12^{3-a})^{2b}$ 12^{6b-2ab}

134. $\dfrac{4x^{2a+3}y^{2b-1}}{2x^{a+1}y^{b+1}}$ $2x^{a+2}y^{b-2}$

135. $\dfrac{25x^{a+b}y^{b-a}}{-5x^{a-b}y^{b+a}}$ $-5x^{2b}y^{-2a}$

136. Compare $8 \cdot 10^{-90}$ and $9 \cdot 10^{-91}$. Which is the larger value? How much larger? Write scientific notation for the difference. $\boxdot$

137. Write the reciprocal of 8.00×10^{-23} in scientific notation. 1.25×10^{22}

138. Evaluate: $(4096)^{0.05}(4096)^{0.2}$.   8

139. What is the ones digit in 513^{128}?   1

140. A grain of sand is placed on the first square of a chessboard, two grains on the second square, four grains on the third, eight on the fourth, and so on. Without a calculator, use scientific notation to approximate the number of grains of sand required for the 64th square. (*Hint*: Use the fact that $2^{10} \approx 10^3$.) 8×10^{18} grains

$\boxdot$ Answers to Exercises 94, 101, 115, 116, and 136 can be found on p. A-53.

Collaborative Corner

Paired Problem Solving

Focus: Problem solving, scientific notation, and unit conversion

Time: 15–25 minutes

Group size: 3

ACTIVITY

Given that the earth's average distance from the sun is 1.5×10^{11} meters, determine the earth's orbital speed around the sun in miles per hour. Assume a circular orbit and use the following guidelines.

1. Two students should attempt to solve this problem while the third group member silently observes and writes notes describing the efforts of the other two.

2. After 10–15 minutes, all observers should share their observations with the class as a whole, answering these three questions:

   a) What successful strategies were used?

   b) What unsuccessful strategies were used?

   c) What recommendations do the observers have to make for students working in pairs to solve a problem?

1.5

Points and Ordered Pairs ■ Quadrants and Scale ■ Graphs of Equations ■ Nonlinear Equations

Graphs

It has often been said that a picture is worth a thousand words. As we turn our attention to the study of graphs, we discover that in mathematics this is quite literally the case. Graphs are a compact means of displaying information and provide a visual approach to problem solving.

Points and Ordered Pairs

Whereas on a number line, each point corresponds to a number, on a plane, each point corresponds to a pair of numbers. The idea of using two perpendicular number lines, called **axes** (pronounced ak-sēz; singular, **axis**) to identify points in a plane is commonly attributed to the great French mathematician and philosopher René Descartes (1596–1650). In honor of Descartes, this representation is also called the **Cartesian coordinate system**. The variable x is normally represented on the horizontal axis and the variable y on the vertical axis, so we will also refer to the *x, y*-**coordinate system**.

To identify a point that appears on the x, y-coordinate system, we write a pair of numbers in the form (x, y), where x is the number directly above or below the point on the horizontal axis and y is the number to the left or right of the point on the vertical axis (see the dashed lines in the figure below). Thus, $(2, 3)$ and $(3, 2)$ are different points. Because the order in which the numbers are listed is important, these are called **ordered pairs**.

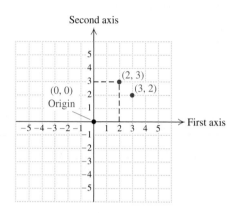

EXAMPLE 1 Plot the points $(-4, 3)$, $(-5, -3)$, $(0, 4)$, and $(2.5, 0)$.

Solution   To plot $(-4, 3)$, we note that the first number, -4, tells us the distance in the first, or horizontal, direction. We go 4 units *left* of the origin. The second number tells us the distance in the second, or vertical, direction. We go 3 units *up*. The point $(-4, 3)$ is then marked, or "plotted."

The points $(-5, -3)$, $(0, 4)$, and $(2.5, 0)$ are also plotted below.

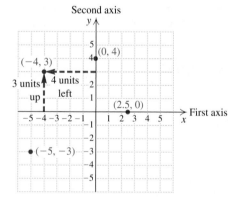

The numbers in an ordered pair are called **coordinates**. In $(-4, 3)$, the *first coordinate* is -4 and the *second coordinate** is 3. The point with coordinates $(0, 0)$ is called the **origin**.

---

*The first coordinate is sometimes called the **abscissa** and the second coordinate the **ordinate**.

Quadrants and Scale

The axes divide the plane into four regions called **quadrants**, as shown here.

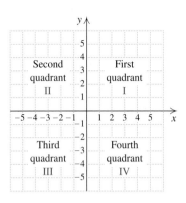

In region I (the *first* quadrant), both coordinates of a point are positive. In region II (the *second* quadrant), the first coordinate is negative and the second coordinate is positive. In the third quadrant, both coordinates are negative, and in the fourth quadrant, the first coordinate is positive and the second coordinate is negative.

Points with one or more 0's as coordinates, such as $(0, -6)$, $(4, 0)$, and $(0, 0)$, are on axes and *not* in quadrants.

The coordinate plane extends without end in all directions. We draw only part of it when we plot points. Although it is standard to show portions of all four quadrants, as in the graphs above, it may be more practical to show a different portion of the plane in order to display information more clearly. Sometimes a different *scale* is selected for each axis.

TEACHING TIP

You may wish to ask students to explain why a graph typically includes the origin and axes.

EXAMPLE 2   Plot the points $(10, 44)$, $(95, 120)$, $(55, 130)$, and $(70, 15)$.

Solution   The smallest first coordinate is 10 and the largest is 95. The smallest second coordinate is 15 and the largest is 130. Thus we need show only the first quadrant. We must show at least 95 units of the first axis and at least 130 units of the second axis. Because it would be impractical to label the axes with all the natural numbers, we will label only every tenth unit. We say that we use a scale of 10.

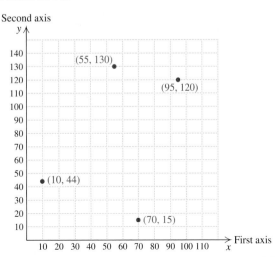

The first axis is commonly referred to as the x-axis and the second as the y-axis. This is true when graphing on a calculator.

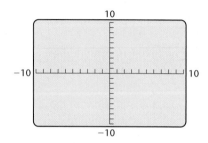

Windows

On a graphing calculator, the rectangular portion of the screen in which a graph appears is called the **viewing window**. Windows are described by four numbers of the form [L, R, B, T], representing the **L**eft and **R**ight endpoints of the x-axis and the **B**ottom and **T**op endpoints of the y-axis. The **standard viewing window** is the window determined by the settings $[-10, 10, -10, 10]$.

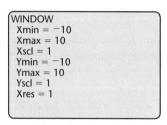

```
WINDOW
 Xmin = -10
 Xmax = 10
 Xscl = 1
 Ymin = -10
 Ymax = 10
 Yscl = 1
 Xres = 1
```

Pressing $\boxed{\text{WINDOW}}$, we set the left and right endpoints of the horizontal axis by setting Xmin and Xmax. Similarly, the values of Ymin and Ymax determine the bottom and top endpoints of the vertical axis shown. The scales for the axes are set using Xscl and Yscl. Xres indicates the pixel resolution, which we generally set as Xres = 1. In this text, the window dimensions are written outside the graphs.

Graphs of Equations

If an equation has two variables, its solutions are pairs of numbers. When such a solution is written as an ordered pair, the first number listed in the pair generally replaces the variable that occurs first alphabetically.

EXAMPLE 3   Determine whether the pairs $(4, 2)$, $(-1, -4)$, and $(2, 5)$ are solutions of the equation $y = 3x - 1$.

Solution   To determine whether each pair is a solution, we replace x with the first coordinate and y with the second coordinate. When the replacements make the equation true, we say that the ordered pair is a solution.

$y = 3x - 1$	$y = 3x - 1$	$y = 3x - 1$
2 ? $3(4) - 1$	-4 ? $3(-1) - 1$	5 ? $3(2) - 1$
$12 - 1$	$-3 - 1$	$6 - 1$
$2 \mid 11$	$-4 \mid -4$	$5 \mid 5$

Since $2 = 11$ is false, the pair $(4, 2)$ *is not* a solution.

Since $-4 = -4$ is *true*, the pair $(-1, -4)$ *is a* solution.

Since $5 = 5$ is *true*, the pair $(2, 5)$ *is a* solution.

In fact, there is an infinite number of solutions of $y = 3x - 1$. We can use a graph as a convenient way of representing these solutions. Thus to **graph** an equation means to make a drawing that represents all of its solutions.

EXAMPLE 4   Graph the equation $y = x$.

Solution   We label the horizontal axis as the x-axis and the vertical axis as the y-axis.

Next, we find some ordered pairs that are solutions of the equation. We can choose a value for x, substitute in the equation, and find the corresponding value for y. These numbers then form one ordered pair, which represents one solution of the equation. Here are a few pairs that satisfy the equation $y = x$:

$$(0, 0), \quad (1, 1), \quad (5, 5), \quad (-1, -1), \quad (-6, -6).$$

Plotting these points, we can see that if we were to plot a million solutions, the dots would merge into a solid line. Observing the pattern, we can draw the line with a ruler. The line is the graph of the equation $y = x$. We label the line $y = x$.

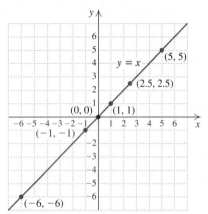

Note that the coordinates of *any* point on the line—for example, $(2.5, 2.5)$—satisfy the equation $y = x$. Note too that the line continues indefinitely in both directions—only part of it is shown.

Entering and Graphing Equations

Equations are entered using the equation-editor screen, often accessed by pressing $\boxed{\text{Y=}}$. The first part of each equation, "Y=," is supplied by the calculator, including a subscript that identifies the equation. An equation can be cleared by positioning the cursor on the equation and pressing $\boxed{\text{CLEAR}}$. On many calculators, a symbol before the Y indicates the graph style, and a highlighted = indicates that the equation selected is to be graphed. An equation can be selected or deselected by positioning the cursor on the = and pressing $\boxed{\text{ENTER}}$. Selected equations are then graphed by pressing $\boxed{\text{GRAPH}}$. *Note*: If the window is not set appropriately, the graph may not appear on the screen at all.

Study Tip

A number of examples throughout this text are worked using two methods. The steps are shown side by side, as in Example 6 on this page. You should read both methods and compare the steps and the results.

EXAMPLE 5   Graph $y = 2x$ using a graphing calculator.

Solution   After pressing Y= and clearing any other equations present, we enter $y = 2x$. The standard $[-10, 10, -10, 10]$ window is a good choice for this graph. We check the window dimensions and then graph the equation.

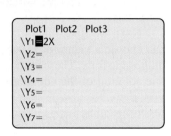

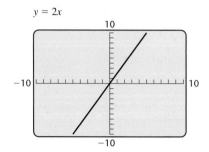

The equation in the next example is graphed both by hand and by using a graphing calculator. Let's compare the processes and the results.

EXAMPLE 6   Graph the equation $y = -\frac{1}{2}x$.

Solution

By Hand

We find some ordered pairs that are solutions. This time we list the pairs in a table. To find an ordered pair, we can choose *any* number for x and then determine y. By choosing even numbers for x, we can avoid fractions when calculating y. For example, if we choose 4 for x, we get $y = \left(-\frac{1}{2}\right)(4)$, or -2. If x is -6, we get $y = \left(-\frac{1}{2}\right)(-6)$, or 3. We find several ordered pairs, plot them, and draw the line.

x	y	(x, y)
4	-2	$(4, -2)$
-6	3	$(-6, 3)$
0	0	$(0, 0)$
2	-1	$(2, -1)$

Choose any x.

Compute y.

Form the pair.

Plot the points and draw the line.

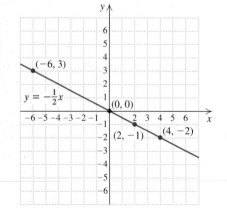

With a Graphing Calculator

To enter the equation, we access the equation editor and clear any equations present. We then enter the equation as $y = -(1/2)x$. Remember to use the $(-)$ key for the negative sign. Also note that some calculators require parentheses around a fraction coefficient. The standard viewing window is a good choice for this graph.

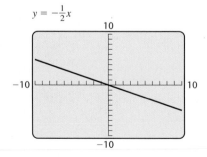

We need not create a table of values in order to graph an equation using a graphing calculator. However, such tables are useful in many situations. To list ordered pairs that are solutions, we can use the TABLE feature of a graphing calculator.

Tables

A TABLE feature lists ordered pairs that are solutions of an equation. Since the value of *y depends* on the choice of the value for *x*, we say that *y* is the **dependent variable** and *x* is the **independent variable**.

After entering an equation on the equation-editor screen, we can view a table of solutions. A table is set up by pressing 2nd TBLSET. (TBLSET is the 2nd feature associated with the WINDOW key.)

If we want to choose the values for the independent variable, we set Indpnt to Ask, as shown on the left below.

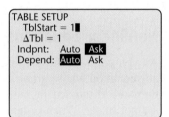
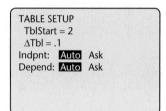

If Indpnt is set to Auto, the calculator will provide values for *x*, beginning with the value specified as TblStart and continuing by adding the value of ΔTbl to the preceding value for *x*. The figure on the right above shows a beginning value of 2 and an increment, or ΔTbl, of 0.1.

To view the values in a table, we press 2nd TABLE. (TABLE is the 2nd feature associated with the GRAPH key.) The up and down arrow keys allow us to scroll up and down the table.

EXAMPLE 7   Create a table of ordered pairs that are solutions of the equation $y = -\frac{1}{2}x$. Use integer values of *x* beginning at -3.

Solution   We first enter the equation $y = -(1/2)x$ and then set up the table, as shown on the left below. Then we press 2nd TABLE to view the table. We see that some solutions of the equation are $(-3, 1.5)$, $(-2, 1)$, $(-1, 0.5)$, and so on.

Connecting the Concepts

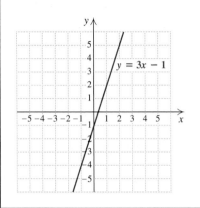

INTERPRETING GRAPHS: SOLUTIONS

A solution of an equation like $5 = 3x - 1$ is a number such as 2. The graph of the solution of this equation is a point on a number line.

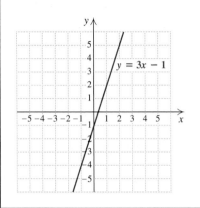

A solution of an equation like $y = 3x - 1$ is an ordered pair such as $(2, 5)$. The graph of the solutions of this equation is a line in a coordinate plane, as shown in the figure at left.

Nonlinear Equations

As you can see, the graphs in Examples 4–6 are straight lines. We refer to any equation whose graph is a straight line as a **linear equation**. Linear equations are discussed in more detail in Chapter 2. Many equations, however, are not linear. When ordered pairs that are solutions of such an equation are plotted, the pattern formed is not a straight line. Let's look at some of these **nonlinear equations**.

EXAMPLE 8 Graph: $y = x^2 - 5$.

Solution   We select numbers for x and find the corresponding values for y. For example, if we choose -2 for x, we get $y = (-2)^2 - 5 = 4 - 5 = -1$. The table lists several ordered pairs.

x	y
0	−5
−1	−4
1	−4
−2	−1
2	−1
−3	4
3	4

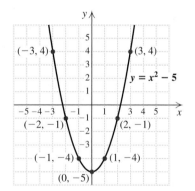

TEACHING TIP

To help students see that a graph "pictures" all solutions of an equation, ask them to select a point on the graph and check to see if it is a solution of the equation.

Next, we plot the points. The more points plotted, the clearer the shape of the graph becomes. Since the value of $x^2 - 5$ grows rapidly as x moves away from the origin, the graph rises steeply on either side of the y-axis.

Curves similar to the one in Example 8 are studied in detail in Chapter 8.

EXAMPLE 9   Use a graphing calculator to create a table of solutions of $y = |x|$ for integer values of x beginning at -3. Then graph the equation. Compare the values in the table with the graph.

Solution   We first enter the equation on the equation-editor screen as $y = abs(x)$. (On many calculators, abs(is in the MATH NUM menu.) We then create a table of values by setting Indpnt to Auto, letting TblStart $= -3$ and ΔTbl $= 1$. We use a standard viewing window for the graph.

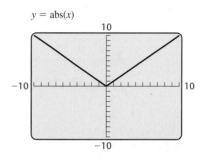

X	Y₁
-3	3
-2	2
-1	1
0	0
1	1
2	2
3	3

X = -3

$y = abs(x)$

From the table, we see that the absolute value of a positive number is the same as the absolute value of its opposite. Thus, for example, the x-values 3 and -3 both are paired with the y-value 3. Note that the graph is V-shaped and centered at the origin.

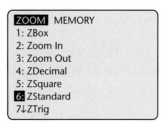

Choosing a Viewing Window

A standard viewing window is not always the best window to use. Choosing an appropriate viewing window for a graph can be challenging. There is generally no one "correct" window; the choice can vary according to personal preference and can also be dictated by the portion of the graph that you need to see. It often involves trial and error, but as you learn more about graphs of equations, you will be able to determine an appropriate viewing window more directly. For now, start with a standard window, and change the dimensions if needed. The ZOOM menu can help in setting window dimensions; for example, choosing ZStandard from the menu will graph the selected equations using the standard viewing window.

ZOOM MEMORY
1: ZBox
2: Zoom In
3: Zoom Out
4: ZDecimal
5: ZSquare
6: ZStandard
7↓ZTrig

1.5

Exercise Set

Give the coordinates of each point.

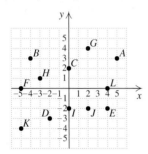

1. *A, B, C, D, E,* and *F* $(5,3), (-4,3), (0,2), (-2,-3),$
$(4,-2),$ and $(-5,0)$

2. *G, H, I, J, K,* and *L* $(2,4), (-3,1), (0,-2), (2,-2),$
$(-5,-4),$ and $(4,0)$

Plot the points. Label each point with the indicated letter.

3. $A(3,0), B(4,2), C(5,4), D(6,6), E(3,-4), F(3,-3),$
$G(3,-2), H(3,-1)$ ⊡

4. $A(1,1), B(2,3), C(3,5), D(4,7), E(-2,1), F(-2,2),$
$G(-2,3), H(-2,4), J(-2,5), K(-2,6)$ ⊡

5. Plot the points $M(2,3), N(5,-3),$ and $P(-2,-3).$
Draw $\overline{MN}, \overline{NP},$ and $\overline{MP}.$ ($\overline{MN}$ means the line segment from M to $N.$) What kind of geometric figure is formed? What is its area?   ⊡

6. Plot the points $Q(-4,3), R(5,3), S(2,-1),$ and
$T(-7,-1).$ Draw $\overline{QR}, \overline{RS}, \overline{ST},$ and $\overline{TQ}.$ What kind of figure is formed? What is its area?   ⊡

Name the quadrant in which each point is located.

7. $(-4,-9)$ III

8. $(2,17)$ I

9. $(-6,1)$ II

10. $(4,-8)$ IV

11. $\left(3,\frac{1}{2}\right)$ I

12. $(-1,-7)$ III

13. $(6.9,-2)$ IV

14. $(-4,31)$ II

Determine whether each ordered pair is a solution of the given equation.

15. $(1,-1); y = 3x - 4$ Yes **16.** $(2,5); y = 4x - 3$ Yes

17. $(2,4); 5s - t = 8$ No   **18.** $(1,3); 4p + q = 5$ No

19. $(3,5); 4x - y = 7$ Yes **20.** $(2,7); 5x - y = 3$ Yes

21. $\left(0,\frac{3}{5}\right); 6a + 5b = 3$ Yes **22.** $\left(0,\frac{3}{2}\right); 3f + 4g = 6$ Yes

23. $(2,-1); 4r + 3s = 5$ Yes **24.** $(2,-4); 5w + 2z = 2$ Yes

25. $(5,3); x - 3y = -4$ Yes **26.** $(1,2); 2x - 5y = -6$ No

27. $(-1,3); y = 3x^2$ Yes   **28.** $(2,4); 2r^2 - s = 5$ No

29. $(2,3); 5s^2 - t = 7$ No   **30.** $(2,3); y = x^3 - 5$ Yes

Graph by hand.

31. $y = x + 1$ ⊡

32. $y = x + 3$ ⊡

33. $y = -x$ ⊡

34. $y = 3x$ ⊡

35. $y = -2x + 3$ ⊡

36. $y = -3x + 1$ ⊡

Aha! **37.** $y + 2x = 3$ ⊡

38. $y + 3x = 1$ ⊡

39. $y = \frac{2}{3}x + 1$ ⊡

40. $y = \frac{1}{3}x + 2$ ⊡

Using a graphing calculator, create a table of solutions for integer values of x beginning at −3. Then graph.

41. $y = -\frac{3}{2}x + 1$ ⊡

42. $y = -\frac{2}{3}x - 2$ ⊡

43. $y = \frac{3}{4}x + 1$ ⊡

44. $y = \frac{3}{4}x + 2$ ⊡

45. $y = -x^2$ ⊡

46. $y = x^2$ ⊡

47. $y = x^2 - 3$ ⊡

48. $y = x^2 + 2$ ⊡

49. $y = |x| + 2$ ⊡

50. $y = -|x|$ ⊡

51. $y = 3 - x^2$ ⊡

52. $y = 4 - x^2$ ⊡

53. $y = x^3$ ⊡

54. $y = x^3 - 2$ ⊡

Graph each equation using both viewing windows indicated. Determine which window best shows the shape of the graph and where it crosses the x- and y-axes.

55. $y = x - 15$ (b)
   a) $[-10, 10, -10, 10],$ Xscl $= 1,$ Yscl $= 1$
   b) $[-20, 20, -20, 20],$ Xscl $= 5,$ Yscl $= 5$

56. $y = -3x + 30$ (b)
   a) $[-10, 10, -10, 10],$ Xscl $= 1,$ Yscl $= 1$
   b) $[-20, 20, -20, 40],$ Xscl $= 5,$ Yscl $= 5$

57. $y = 5x^2 - 8$ (a)
   a) $[-10, 10, -10, 10],$ Xscl $= 1,$ Yscl $= 1$
   b) $[-3, 3, -3, 3],$ Xscl $= 1,$ Yscl $= 1$

⊡ Answers to Exercises 3–6 and 31–54 can be found on pp. A-53 and A-54.

58. $y = \frac{1}{10}x^2 + \frac{1}{3}$ (a)

    a) $[-10, 10, -10, 10]$, Xscl $= 1$, Yscl $= 1$

    b) $[-0.5, 0.5, -0.5, 0.5]$, Xscl $= 0.1$, Yscl $= 0.1$

59. $y = 4x^3 - 12$ (b)

    a) $[-10, 10, -10, 10]$, Xscl $= 1$, Yscl $= 1$

    b) $[-5, 5, -20, 10]$, Xscl $= 1$, Yscl $= 5$

60. $y = |4x^3 - 12|$ (b)

    a) $[-10, 10, -10, 10]$, Xscl $= 1$, Yscl $= 1$

    b) $[-3, 3, 0, 20]$, Xscl $= 1$, Yscl $= 5$

61. Determine which of the equations in the odd-numbered exercises 31–59 are linear.   31–43 and 55

62. Determine which of the equations in the even-numbered exercises 32–60 are linear.   32–44 and 56

TW **63.** What can be said about the location of two points that have the same first coordinates and second coordinates that are opposites of each other?

TW **64.** Examine Example 8 and explain why it is unwise to draw a graph after plotting just two points.

Skill Maintenance

Tell whether the number is a solution of the given equation or inequality. [1.1]

65. $3x - 5 = 10$; 5 Yes

66. $4y \geq 18$; 6 Yes

67. $3n - 7 < 5$; 4 No

68. $2t + 3 = 5$; 2 No

Evaluate.

69. $5s - 3t$, for $s = 2$ and $t = 4$ [1.2] -2

70. $(2m + n)^2$, for $m = 3$ and $n = 1$ [1.1] 49

Aha! **71.** $(5 - x)^4(x + 2)^3$, for $x = -2$ [1.2] 0

72. $2x^2 + 4x - 9$, for $x = -3$ [1.2] -3

Synthesis

TW **73.** Without making a drawing, how can you tell that the graph of $y = x - 30$ passes through three quadrants?

TW **74.** At what point will the line passing through $(a, -1)$ and $(a, 5)$ intersect the line that passes through $(-3, b)$ and $(2, b)$? Why?

TW **75.** Graph $y = 6x$, $y = 3x$, $y = \frac{1}{2}x$, $y = -6x$, $y = -3x$, and $y = -\frac{1}{2}x$ using the same set of axes or viewing window, and compare the slants of the lines. Describe the pattern that relates the slant of the line to the multiplier of x.

TW **76.** Using the same set of axes or viewing window, graph $y = 2x$, $y = 2x - 3$, and $y = 2x + 3$. Describe the pattern relating each line to the number that is added to $2x$.

77. Which of the following equations have $\left(-\frac{1}{3}, \frac{1}{4}\right)$ as a solution?   (a), (d)

    a) $-\frac{3}{2}x - 3y = -\frac{1}{4}$

    b) $8y - 15x = \frac{7}{2}$

    c) $0.16y = -0.09x + 0.1$

    d) $2(-y + 2) - \frac{1}{4}(3x - 1) = 4$

78. If $(2, -3)$ and $(-5, 4)$ are the endpoints of a diagonal of a square, what are the coordinates of the other two vertices? What is the area of the square?
$(2, 4), (-5, -3)$; 49 square units

79. If $(-10, -2)$, $(-3, 4)$, and $(6, 4)$ are the coordinates of three consecutive vertices of a parallelogram, what are the coordinates of the fourth vertex?
$(-1, -2)$, $(-19, -2)$, or $(13, 10)$

80. One value of y for the equation $y = 3.2x - 5$ is -11.4. Use the TABLE feature of a graphing calculator to determine the x-value that is paired with -11.4. -2

81. Graph each of the following equations and determine which appear to be linear. Try to find a way to tell if an equation is linear without graphing it.

    a) $y = 3x + 2$ Equations (a), (c), and (d)

    b) $y = \frac{1}{2}x^2 - 5$ appear to be linear.

    c) $y = 8$

    d) $y = 4 - \frac{1}{5}x$

    e) $y = |3 - x|$

    f) $y = 4x^3$

82. The graph of $y = 0.5x^2 - 15x + 64$ crosses the x-axis twice. Determine a viewing window that shows both intersections of $y = 0.5x^2 - 15x + 64$ and the x-axis.
Answers may vary. $[0, 30, -50, 10]$ with Xscl $= 10$ and Yscl $= 10$ is a good choice.

1.6

Mathematical Models and Problem Solving

We now begin to study and practice the "art" of problem solving. Although we are interested mainly in using algebra to solve problems, much of what we say here applies to all methods of problem solving.

What do we mean by a *problem*? Perhaps you have already used algebra to solve some "real-world" problems. What procedure did you use? Was there anything in your approach that could be used to solve problems of a more general nature? These are some questions that we will answer in this section.

In this text, we do not restrict the use of the word "problem" to computational situations involving arithmetic or algebra, such as $589 + 437 = a$ or $3x + 5x = 9$. We mean instead some question to which we wish to find an answer. Perhaps this can best be illustrated with some sample problems:

1. Can I afford to rent a bigger apartment?
2. If I exercise twice a week and eat 3000 calories a day, will I lose weight?
3. Do I have enough time to take 4 courses while working 20 hours a week?
4. My fishing boat travels 12 km/h in still water. How long will it take me to cruise 25 km upstream if the river's current is 3 km/h?

Although these problems are all different, there is a general strategy that can be applied to all of them.

The Five-Step Strategy

Since you have already studied some algebra, you have had some experience with problem solving. The following steps constitute a strategy that you may already have used and a good strategy for problem solving in general.

Five Steps for Problem Solving with Algebra

1. *Familiarize* yourself with the problem.
2. *Translate* to mathematical language.
3. *Carry out* some mathematical manipulation.
4. *Check* your possible answer in the original problem.
5. *State* the answer clearly.

Of the five steps, probably the most important is the first: becoming familiar with the problem situation. Here are some ways in which this can be done.

The First Step in Problem Solving with Algebra

To familiarize yourself with the problem:

1. If the problem is written, read it carefully.

2. Reread the problem, perhaps aloud. Verbalize the problem to yourself.

3. List the information given and restate the question being asked. Select a variable or variables to represent any unknown(s) and clearly state what each variable represents. Be descriptive! For example, let t = the flight time, in hours; let p = Paul's weight, in kilograms; and so on.

4. Find additional information. Look up formulas or definitions with which you are not familiar. Geometric formulas appear on the inside back cover of this text; important words appear in the index. Consult an expert in the field or a reference librarian.

5. Create a table, using variables, in which both known and unknown information is listed. Look for possible patterns.

6. Make and label a drawing.

7. Estimate or guess an answer and check to see whether it is correct.

EXAMPLE 1   How might you familiarize yourself with the situation of Problem 1: "Can I afford to rent a bigger apartment?"

Solution   Clearly more information is needed to solve this problem. You might:

a) Estimate the rent of some apartments in which you are interested.

b) Examine what your savings are and how your income is budgeted.

c) Determine how much rent you can afford.

When enough information is known, it might be wise to make a chart or table to help you reach an answer.   ▬▬▬▬

EXAMPLE 2   How might you familiarize yourself with Problem 4: "How long will it take the boat to cruise 25 km upstream?"

Solution   First read the question *very* carefully. This may even involve speaking aloud. You may need to reread the problem several times to fully understand what information is given and what information is required. A sketch or table is often helpful.

boat   current

Distance to be Traveled	25 km
Speed of Boat in Still Water	12 km/h
Speed of Current	3 km/h
Speed of Boat Upstream	?
Time Required	?

To continue the familiarization process, we should determine, possibly with the aid of outside references, what relationships exist among the various quantities in the problem. With some effort it can be learned that the current's speed should be subtracted from the boat's speed in still water to determine the boat's speed going upstream. We also need to either find or recall an extremely important formula:

Distance = Speed × Time.

We rewrite part of the table, letting $t =$ the number of hours required for the boat to cruise 25 km upstream.

Distance to be Traveled	25 km
Speed of Boat Upstream	$12 - 3 = 9$ km/h
Time Required	t

At this point we might try a guess. Suppose the boat traveled upstream for 2 hr. The boat would have then traveled

$$9\,\frac{\text{km}}{\text{hr}} \times 2\,\text{hr} = 18\,\text{km}. \qquad \text{Note that } \frac{\text{km}}{\text{hr}} \cdot \text{hr} = \text{km}.$$

$$\text{Speed} \times \text{Time} = \text{Distance}$$

Since $18 \neq 25$, our guess is wrong. Still, examining how we checked our guess sheds light on how to translate the problem to an equation. Note that a better guess, when multiplied by 9, would yield a number closer to 25.

The second step in problem solving is to translate the situation to mathematical language. In algebra, this often means forming an equation.

The Second Step in Problem Solving with Algebra

Translate the problem to mathematical language. In some cases, translation can be done by writing an algebraic expression, but most problems in this text are best solved by translating to an equation.

In the third step of our process, we work with the results of the first two steps. Often this will require us to use the algebra that we have studied.

The Third Step in Problem Solving with Algebra

Carry out some mathematical manipulation. If you have translated to an equation, this means to solve the equation.

To complete the problem-solving process, we should always **check** our solution and then **state** the solution in a clear and precise manner. To check, we make sure that our answer is reasonable, that all the conditions of the original problem have been satisfied, and that appropriate units are used such as "km/h" or "seconds." If our answer checks, we write a complete English sentence stating the solution. The five steps are listed again below. Try to apply them regularly in your work.

Five Steps for Problem Solving with Algebra

1. *Familiarize* yourself with the problem.
2. *Translate* to mathematical language.
3. *Carry out* some mathematical manipulation.
4. *Check* your possible answer in the original problem.
5. *State* the answer clearly.

Translating to Algebraic Expressions and Equations

One way to model a problem situation is to use an equation. In order to translate problems to equations, we must first be able to translate phrases to algebraic expressions. To do this, we need to know which words correspond to which symbols. The following table lists some examples.

Key Words

Addition	Subtraction	Multiplication	Division
add	subtract	multiply	divide
sum of	difference of	product of	divided by
plus	minus	times	quotient of
increased by	decreased by	twice	ratio
more than	less than	of	per

When the value of a number is unknown, we represent it with a variable.

TEACHING TIP

Emphasize the proper order when representing "less than." It may help to translate "more than" in the same order. For example, "five less than a number" translates to $x - 5$ and "five more than a number" translates to $x + 5$.

Phrase	Algebraic Expression
Five *more than* some number	$n + 5$
Half *of* a number	$\frac{1}{2}t$ or $\frac{t}{2}$
Five *more than* three *times* some number	$3p + 5$
The *difference of* two numbers	$x - y$
Six *less than* the *product of* two numbers	$rs - 6$
Seventy-six percent *of* some number	$0.76z$ or $\frac{76}{100}z$

Note that an expression like rs represents a product and can also be written as $r \cdot s$, $r \times s$, or $(r)(s)$.

EXAMPLE 3   Translate to an algebraic expression:

Five less than forty-three percent of the quotient of two numbers.

Solution   We let r and s represent the two numbers.

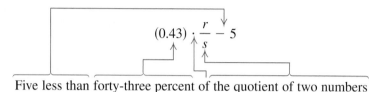

$$(0.43) \cdot \frac{r}{s} - 5$$

Five less than forty-three percent of the quotient of two numbers

Sometimes a problem can be translated to an equation. It is often helpful to reword the problem using phrases that can be translated directly to algebraic expressions. The key word "is" translates to an equals sign, $=$. Remember to familiarize yourself with each problem situation before attempting to translate it.

EXAMPLE 4   Translate the following problem to an equation. Do not solve.

There are 737 million acres of forested land in the United States. This is one-third of the United States. How many acres are in the United States? (*Source*: envirosystemsinc.com)

Solution

1. **Familiarize.**   Read the problem carefully. The question asked is, "How many acres are in the United States?," so we let

   $x =$ the number of acres in the United States.

   The problem gives us two facts about forested land. The number of acres of forested land is 737 million, and the forested land is one-third of the United States, or $\frac{1}{3}x$.

2. **Translate.**   We reword the problem and translate.

Rewording:	The number of acres of forested land	is	one-third of the United States.
Translating:	737,000,000	$=$	$\frac{1}{3}x$

   The translation is the equation

   $$737{,}000{,}000 = \frac{1}{3}x,$$

   where x is the number of acres in the United States.

EXAMPLE 5   Allie had a $10 gift certificate toward a meal at the Point Steakhouse. She was charged 5% tax on the cost of the meal before the certificate was used. In addition to the gift certificate, Allie spent $3.44. What was the cost of the meal before tax and the gift certificate?

Solution

1. **Familiarize.**   Read the problem carefully. The question asks for the original cost of the meal, so we let

x = the cost of the meal before tax and the gift certificate.

It is often helpful to guess an answer and check the guess. The steps followed in the check often lead to a successful translation of the problem. Let's guess that the meal originally cost $12, and fill in a table or chart with the corresponding amounts.

Original Cost	$12
Tax	
Cost of Meal plus Tax	
Gift Certificate	$10
Amount Paid	

Original Cost	$12
Tax	$ 0.60
Cost of Meal plus Tax	$12.60
Gift Certificate	$10
Amount Paid	$ 2.60

The tax was 5% of the original cost of the meal, or 5% of $12, or $0.60, so the cost of the meal plus tax was $12.60. The amount of the gift certificate, $10, was subtracted from the cost of the meal plus tax, leaving $2.60. Since $2.60 ≠ $3.44, we know that our guess was not correct. However, we did learn that we must calculate the tax on the original cost, add it to the original cost, and then subtract the amount of the gift certificate in order to find the amount paid.

2. **Translate.**   We reword the problem and translate.

Rewording: The amount paid   is   the original cost   plus   tax   minus   the gift certificate.

Translating: 3.44   = x + $0.05x$ −   10

The translation is the equation

$$3.44 = x + 0.05x - 10$$

or, if we combine like terms,

$$3.44 = 1.05x - 10.$$

Models

When we translate a problem into mathematical language, we say that we *model* the problem. A **mathematical model** is a representation, using mathematics, of a real-world situation. Following are some examples of models.

1. REAL-WORLD SITUATION

Redwood trees.   The girth, or distance around the trunk, of a redwood tree is 48 ft. What is the diameter of the trunk?

MATHEMATICAL MODEL: FORMULA

A *formula* can be a mathematical model. We use the formula for the circumference of, or distance around, a circle:

$$C = \pi d, \quad \text{where } d \text{ is the diameter of the trunk, in feet.}$$

Since $C = 48$, we have the model

$$48 = \pi d.$$

2. REAL-WORLD SITUATION

Endangered species.   In 1941, there were just 15 whooping cranes in existence. Through recovery efforts, there were 418 birds in 2000. By how much did the whooping-crane population increase? (*Source: Endangered Species Bulletin,* May/June 2000, Volume XXV, No. 3, p. 15)

MATHEMATICAL MODEL: EQUATION OR INEQUALITY

An *equation* or *inequality* can be used to model a situation. We can translate this problem to the equation

$$15 + x = 418,$$

where x is the increase in the whooping-crane population.

3. REAL-WORLD SITUATION

Scooter-related injuries.   After a new version of a foot-propelled scooter became popular in 2000, the number of scooter-related injuries increased dramatically. The number requiring emergency care for various months is shown in the following table. Use the given data to estimate the number of injuries in August 2000. (*Source*: *MMWR Weekly,* December 15, 2000)

Month	Number of Scooter-Related Injuries
May 2000	600
July 2000	2100
September 2000	8600
October 2000	7400

MATHEMATICAL MODEL: GRAPH

We can visualize this situation by plotting the points (May, 600), (July, 2100), (September, 8600), and (October, 7400). Because the number of injuries varies from 600 to 8600, we use a scale of 1000 on the vertical axis. We connect the points to form a line graph. An estimate of the number of injuries in August can then be read from the graph.

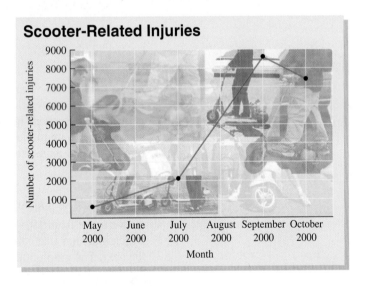

Throughout the remainder of this text, we will be learning how to develop and use such models.

Graphs as Models

Graphs can represent a large amount of information in a concise way. They also help to visualize problem situations and to show relationships between two quantities. When one of those quantities depends on the other, we generally represent the dependent variable on the vertical axis.

> **EXAMPLE 6   Sales of Cellular Phones.**   The following graph shows the number of cellular phones sold worldwide for the years 1996 to 2001. (*Source*: Gartner Dataquest)
>
> **a)** How many cellular phones were sold in 2001?
>
> **b)** In what year was the number of cellular phones sold the greatest?
>
> **c)** Between what years did the number of cellular phones sold decrease?

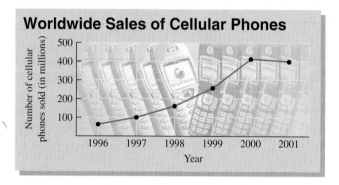

Worldwide Sales of Cellular Phones

TEACHING TIP

Reinforce that although a value read from a graph is an estimate, such a value is often all that is required.

Solution

a) First, we note that "year" is on the horizontal axis and "Sales of Cellular Phones" is on the vertical axis. We also note that the number of cellular phones sold is given in millions. After locating 2001 on the horizontal axis, we move up to the graph and horizontally to the corresponding value on the vertical axis. We see that in 2001, about 400 million phones were sold.

b) We find the highest point on the graph and then move down to the horizontal axis. The number of cellular phones sold was greatest in 2000.

c) As we move from 1996 to 2000 along the horizontal axis, the graph rises, which indicates that cellular phone sales were increasing from one year to the next. The graph falls as we move from 2000 to 2001, so we know that there was a decrease in cellular phone sales between 2000 and 2001.

Coordinates and Points

Coordinates of ordered pairs are entered as *data* in lists, using the STAT menu. To enter or change data, press $\boxed{\text{STAT}}$ and then choose the EDIT option. The lists of numbers will appear as three columns on the screen. If there are already numbers in the lists, clear them by moving the cursor to the title of the list (L1, L2, and so on) and pressing $\boxed{\text{CLEAR}}$ $\boxed{\text{ENTER}}$.

To enter a number in a list, move the cursor to the correct position, type in the number, and press ENTER. Enter the first coordinates of the ordered pairs as one list and the second coordinates as another list. The coordinates of each point should be at the same position on both lists.

To plot the points, first turn on the STAT PLOT feature. Press 2nd STAT PLOT. (STAT PLOT is the 2nd feature associated with the Y= key.) The graphing calculator allows several different sets of points to be displayed at once. See the screen on the left below. Choose the Plot you wish to define; if you are simply plotting one set of points, choose Plot1 by pressing 1. Then turn Plot1 on by positioning the cursor over On and pressing ENTER, as shown on the right below.

The remaining items on the screen define the plot. Use the down arrow key to move to the next item.

The points entered can be used to make a line graph or a bar graph as well as to graph points. The screen on the right below shows six available types of graphs. To plot points, choose the first type of graph shown, a scatter diagram or scatterplot. The second option in the list is a line graph, in which the points are connected. The third type is a bar graph. The last three types will not be discussed in this course.

The next item on the screen, Xlist, should be the list in which the first coordinates were entered, probably L_1, and Ylist the list in which the second coordinates were entered, probably L_2. To enter the name of a list, press 2nd L_1, 2nd L_2, and so on. (L_1 through L_6 are the 2nd features associated with the number keys 1 through 6.)

The last choice on the screen is the type of mark used to plot the points. Different marks can be used to distinguish among several sets of data.

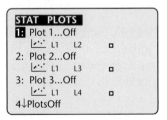
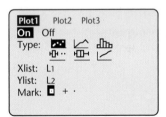

When the STAT PLOT feature has been set correctly, choose window dimensions that will allow all the points to be seen and press GRAPH. After graphing, pressing TRACE displays the coordinates of the point indicated by the cursor. The left and right arrow keys move the cursor along a graph or from point to point. Pressing 2nd CALC and choosing the VALUE option allows a user to enter a particular x-value and find the associated y-value. If you wish to use VALUE or to enter an x-value using TRACE, the window must include the x-value.

When you no longer wish to plot a set of data, turn off STAT PLOT by pressing 2nd STAT PLOT, choosing the appropriate Plot, and highlighting Off. To turn off all the plots, press 4 to choose the PlotsOff option and press ENTER.

Year	Number of Weekly Newspapers
1960	8174
1970	7612
1980	7954
1990	7606
2000	7689

CAUTION!   The graphing calculator will attempt to graph any equations selected in the Y= screen and any points described by an active Plot. Be sure to clear or deselect any unwanted equations and turn off any unwanted Plots before graphing.

EXAMPLE 7   Weekly Newspapers.   The table at left shows the number of weekly newspapers in the United States for various years from 1960 to 2000. Use the data to draw a line graph. (*Source*: Newspaper Association of America, *Facts About Newspapers*, 2001.)

Solution   We first enter the years in list L1 and the number of newspapers in L2, as shown on the left below. We turn the Plot feature On and choose the second type of graph, the line graph, as shown on the right below. The Xlist should be L1, and the Ylist should be L2.

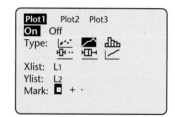

We set the viewing window so that all the points will be shown. We use a viewing window of [1950, 2010, 7500, 8500], with Xscl = 10 and Yscl = 100. Many graphing calculators have a ZOOMSTAT option that will automatically select an appropriate window and graph the points.

After clearing or deselecting any equations present in the equation-editor screen, we press GRAPH . The points are plotted and connected to form a line graph, as shown in the screen on the left below. Note that there are no axes shown on the screen, since the window dimensions chosen did not include 0. Only a portion of the first quadrant is shown.

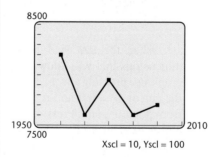

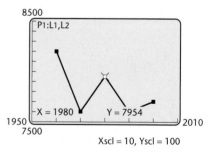

To read a value visually from the graph, we must know the portion of the axes shown and the scale used. We can also press TRACE and the left and right arrow keys to show the coordinates of the point. The screen shown on the right above displays the coordinates of the point indicated by the cursor, (1980, 7954).

EXAMPLE 8   Model Rockets.   Suppose that a model rocket is launched upward with an initial velocity of 96 ft/sec. Its height h, in feet, after t seconds is given by

$$h = -16t^2 + 96t.$$

a) For how long will the rocket climb?

b) How high will the rocket go?

c) After how long will the rocket reach the ground?

Solution   Before graphing the equation, we make sure that STAT PLOT is turned off. We then replace h with y and t with x, and graph the equation, using the viewing window [0, 10, 0, 200], with Yscl = 10.

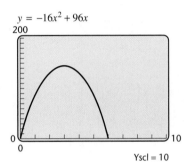
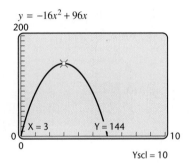

a) The vertical axis shows the height of the rocket, in feet, and the horizontal axis the time, in seconds. We press ⟨TRACE⟩ and use the left and right arrow keys or enter an estimated x-value to move the cursor along the graph to the greatest y-value. The greatest y-value occurs when x is 3, as shown in the graph on the right above. Later in this text, we will consider methods to determine more precisely where such a *maximum* value occurs. The rocket will climb for about 3 sec.

b) In part (a), we found that the highest point on the graph is (3, 144). Thus the rocket will climb about 144 ft.

c) The rocket is at the ground when the height y is 0. Using TRACE, we see that y is 0 when x is about 6. The rocket will reach the ground about 6 sec after it has been launched.

Connecting the Concepts

INTERPRETING GRAPHS: READING GRAPHS

When a graph is used to model a problem, information about that problem can be read directly from the graph. Listed here are a few kinds of information that a graph contains.

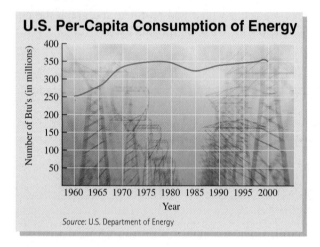

U.S. Per-Capita Consumption of Energy

Source: U.S. Department of Energy

1. *Labels and units.* When reading a graph, look first at any labels. The title of the graph and the labels on the axes tell what information the graph represents. The scales used on each axis indicate the numbers represented by a point. If an equation is listed, the graph represents an equation rather than a set of data.

   The title of the graph shown here indicates that it represents the per capita consumption of energy in the United States. The horizontal axis is labeled "Year," indicating that the graph shows the years from 1960 to 2000. The vertical axis is labeled "Number of Btu's (in millions)" and is

scaled in units of 50. Thus a second coordinate of 300 represents 300,000,000 Btu's.

2. *Relationships.* The coordinates of a point on a graph indicate a relationship between the quantities represented on the horizontal and vertical axes.

   The graph shown here gives the number of Btu's, in millions, that each person used, on average, in a given year. Thus the point (1960, 250) on the graph indicates that the U.S. per capita consumption of energy in 1960 was 250,000,000 Btu's. Note that a value read from a graph is often an approximation.

3. *Trends.* As the graph continues from left to right, the quantity represented on the horizontal axis increases. The quantity represented on the vertical axis may increase, decrease, or remain constant. If the graph rises from left to right, the quantity on the vertical axis is increasing; if the graph falls from left to right, the quantity is decreasing; if the graph is level, the quantity is not changing, so it is constant.

   The graph shown here indicates that U.S. per capita energy consumption increased from 1960 to 1979, decreased from 1979 to 1985, and then increased again from 1985 to 1999. From 1999 to 2000, there was a slight decrease.

4. *Maximums and minimums.* A "high" point or a "low" point on a graph indicates a maximum or minimum value for the quantity on the vertical axis. The highest point on the graph shown here is approximately (1999, 360). Thus the maximum U.S. per capita consumption of energy for the years 1960 to 2000 was 360,000,000 Btu's in 1999.

1.6

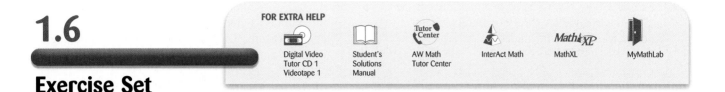

FOR EXTRA HELP

Digital Video
Tutor CD 1
Videotape 1

Student's
Solutions
Manual

Tutor
Center
AW Math
Tutor Center

InterAct Math

Math XL
MathXL

MyMathLab

Exercise Set

Use mathematical symbols to translate each phrase.

1. Three less than some number   Let *n* represent the
number; $n - 3$

2. Four more than some number   Let *n* represent the
number; $4 + n$, or $n + 4$

3. Twelve times a number   Let *t* represent the number; $12t$

4. Twice a number   Let *x* represent the number; $2x$

5. Sixty-five percent of some number   Let *x* represent
the number; $0.65x$, or $\frac{65}{100}x$

6. Thirty-nine percent of some number   Let *x* represent
the number; $0.39x$, or $\frac{39}{100}x$

7. Ten more than twice a number
Let *y* represent the number; $2y + 10$

8. Six less than half of a number   Let *y* represent
the number; $\frac{1}{2}y - 6$

9. Eight more than ten percent of some number   Let *x*
Let *s* represent the number; $0.1s + 8$, or $\frac{10}{100}s + 8$

10. Five less than six percent of some number
represent the number; $0.06s - 5$, or $\frac{6}{100}s - 5$

11. One less than the difference of two numbers
Let *m* and *n* represent the numbers; $m - n - 1$

12. Two more than the product of two numbers
Let *m* and *n* represent the numbers; $mn + 2$

13. Ninety miles per every four gallons of gas
$90 \div 4$, or $\frac{90}{4}$

14. One hundred words per every sixty seconds
$100 \div 60$, or $\frac{100}{60}$

*For each problem, familiarize yourself with the situation.
Then translate to mathematical language. You need not
actually solve the problem; just carry out the first two
steps of the five-step strategy.*

15. The sum of two numbers is 65. One of the numbers
is 7 more than the other. What are the numbers?   Let *s*
and $x + 7$ represent the numbers; $x + (x + 7) = 65$

16. The sum of two numbers is 83. One of the numbers
is 11 more than the other. What are the numbers?
Let *x* and $x + 11$ represent the numbers; $x + (x + 11) = 83$

17. The number 128 is 0.4 of what number?
Let *x* represent the number; $128 = 0.4x$

18. The number 456 is $\frac{1}{3}$ of what number?   Let *n* represent
the number; $456 = \frac{1}{3}n$

19. The quotient of two numbers is 12.3. If the divisor is
4, find the other number.   Let *a* represent the number;
$\frac{a}{4} = 12.3$

20. One number is less than another by 65. The sum of
the numbers is 92. What is the smaller number?
Let *x* represent the larger number; then $x - 65$ represents
the smaller number; $x + (x - 65) = 92$

21. The length of a rectangle is twice its width and its
perimeter is 21 m. Find the dimensions of the
rectangle.   Let *w* represent the rectangle's width;
$21 = 2(2w) + 2w$

22. The width of a rectangle is one-third its length and
its perimeter is 32 m. Find the dimensions of the
rectangle.   Let *y* represent the length; then $\frac{1}{3}y$ represents
the width; $2y + 2\left(\frac{1}{3}y\right) = 32$

23. A reclining chair is on sale for $490. This is 35%
off the original price. What was the original price?
Let *p* represent the original price; $490 = p - 0.35p$

24. Approximately 30% of school-aged children live in
apartment communities (*Source*: National Multi
Housing Council). There are 1250 students at
Southside Middle School. How many would be
expected to live in apartment communities?   Let *c*
represent the number of children in apartments; $c = 0.30(1250)$

25. The length of an American bullfrog is 64% of the
length of a Goliath frog. (*Source*: *Visual Encyclo-
pedia*. London: Dorling Kindersley Limited, 2001).
If a Goliath frog is 350 mm long, how long is an
American bullfrog?   Let *a* represent the length of an
American bullfrog; $a = 0.64(350)$

26. Easy Chair Bookstore sells books for 80% more than
it pays for them. A gardening book sells for $22.00.
How much did the bookstore pay for the book?
Let *b* represent the price the store paid; $22 = b + 0.8b$

27. Rhonda bicycled 25 mi at a rate of 15 mph. For how
long did she ride?   Let *n* represent the number of hours
Rhonda rode; $25 = 15n$

28. Brock ran 6 mi in 45 min. Logan ran 9 mi at the
same rate. For how long did Logan run?   Let *t*
represent the number of minutes Logan ran; $9 = \frac{6}{45}t$

29. *Angles in a Triangle.* The degree measures of the angles in a triangle are three consecutive integers. Find the measure of the angles. Let x represent the measure of the smallest angle; $x + (x + 1) + (x + 2) = 180$

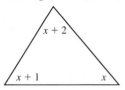

30. *Pricing.* The Sound Connection prices TDK 100-min blank audiotapes by raising the wholesale price 50% and adding 25 cents. What must a tape's wholesale price be if the tape is to sell for $1.99? Let w represent the wholesale price; $w + 0.5w + 0.25 = 1.99$

31. *Pricing.* Becker Lumber gives contractors a 10% discount on all orders. After the discount, a contractor's order cost $279. What was the original cost of the order? Let c represent the original cost of the order, in dollars; $c - 0.1c = 279$

32. *Pricing.* Miller Oil offers a 5% discount to customers who pay promptly for an oil delivery. The Blancos promptly paid $142.50 for their December oil bill. What would the cost have been had they not promptly paid? Let b represent the original amount of the bill, in dollars; $b - 0.05b = 142.50$

33. *Cruising Altitude.* A Boeing 747 has been instructed to climb from its present altitude of 8000 ft to a cruising altitude of 29,000 ft. If the plane ascends at a rate of 3500 ft/min, how long will it take to reach the cruising altitude? Let t represent the number of minutes spent climbing; $3500t = 29,000 - 8000$

34. A piece of wire 10 m long is to be cut into two pieces, one of them $\frac{2}{3}$ as long as the other. How should the wire be cut? Let x represent the longer length; $x + \frac{2}{3}x = 10$

⊡ Answers to Exercises 40–42 can be found on p. A-54.

35. *Angles in a Triangle.* One angle of a triangle is three times as great as a second angle. The third angle measures 12° less than twice the second angle. Find the measures of the angles. Let x represent the measure of the second angle, in degrees; $3x + x + (2x - 12) = 180$

36. *Angles in a Triangle.* One angle of a triangle is four times as great as a second angle. The third angle measures 5° more than twice the second angle. Find the measures of the angles. Let x represent the measure of the second angle, in degrees; $4x + x + (2x + 5) = 180$

37. Find two consecutive even integers such that two times the first plus three times the second is 76. Let x represent the first even number; $2x + 3(x + 2) = 76$

38. Find three consecutive odd integers such that the sum of the first, twice the second, and three times the third is 70. Let n represent the first odd number; $n + 2(n + 2) + 3(n + 4) = 70$

39. A steel rod 90 cm long is to be cut into two pieces, each to be bent to make an equilateral triangle. The length of a side of one triangle is to be twice the length of a side of the other. How should the rod be cut? Let s represent the length, in centimeters, of a side of the smaller triangle; $3s + 3 \cdot 2s = 90$

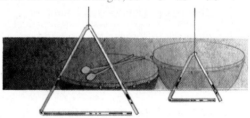

40. A piece of wire 100 cm long is to be cut into two pieces, and those pieces are each to be bent to make a square. The area of one square is to be 144 cm² greater than that of the other. How should the wire be cut? (*Remember*: Do not solve.) ⊡

41. *Test Scores.* Deirdre's scores on five tests are 93, 89, 72, 80, and 96. What must the score be on her next test so that the average will be 88? ⊡

42. *Pricing.* Whitney's Appliances is having a sale on 13 TV sets. They are displayed in order of increasing price from left to right. The price of each set differs by $20 from either set next to it. For the price of the set at the extreme right, a customer can buy both the second and seventh sets. What is the price of the least expensive set? ⊡

Heart Attacks and Cholesterol.   For Exercises 43 and 44, use the following graph, which shows the annual heart attack rate per 10,000 men on the basis of blood cholesterol level. *

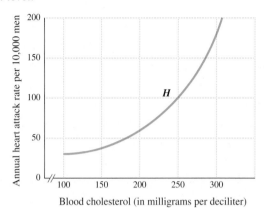

Blood cholesterol (in milligrams per deciliter)

43. Approximate the annual heart attack rate for those men whose blood cholesterol level is 225 mg/dl.
75 heart attacks per 10,000 men

44. Approximate the annual heart attack rate for those men whose blood cholesterol level is 275 mg/dl.
125 heart attacks per 10,000 men

Voting Attitudes.   For Exercises 45 and 46, use this graph, which shows the percentage of people responding yes to the question, "If your (political) party nominated a generally well-qualified person for president who happened to be a woman, would you vote for that person?" (Source: The New York Times, August 13, 2000)

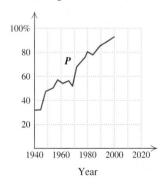

Year

---

*Copyright 1989, CSPI. Adapted from *Nutrition Action Healthletter* (1875 Connecticut Avenue, N.W., Suite 300, Washington, DC 20009-5728. $24 for 10 issues).

45. Approximate the percentage of Americans willing to vote for a woman for president in 1960. That is, find $P(1960)$.   56%

46. Approximate the percentage of Americans willing to vote for a woman for president in 2000. That is, find $P(2000)$.   92%

Blood Alcohol Level.   The following table can be used to predict the number of drinks required for a person of a specified weight to be considered legally intoxicated (blood alcohol level of 0.08 or above). One 12-oz glass of beer, a 5-oz glass of wine, or a cocktail containing 1 oz of a distilled liquor all count as one drink. Assume that all drinks are consumed within one hour.

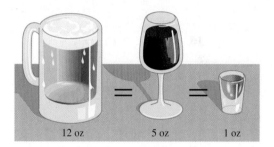

12 oz   5 oz   1 oz

Input, Body Weight (in pounds)	Output, Number of Drinks
100	2.5
160	4
180	4.5
200	5

47. Use the data in the table above to draw a graph and to estimate the number of drinks that a 140-lb person would have to consume to be considered intoxicated.

48. Use the graph from Exercise 47 to estimate the number of drinks a 120-lb person would have to consume to be considered intoxicated.   3 drinks

47. 3.5 drinks

Body weight (in pounds)

49. *Endangered Species.* The following table lists the numbers of U.S. species of mammals considered endangered species during various years (*Source*: U.S. Fish and Wildlife Service). Use the data to draw a line graph with a graphing calculator.

Year	Number of Species of Endangered Mammals
1980	32
1985	43
1990	53
1993	56
1997	57
1999	61
2001	64

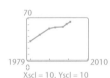

Xscl = 10, Yscl = 10

50. *Child Expenditure.* The following table lists annual expenditures on a child in 1999 by families with an average income of $50,000 (*Source*: Department of Agriculture, Center for Nutrition Policy and Promotion, *Expenditures on Children by Families, 1999 Annual Report*). Use the data to draw a line graph with a graphing calculator.

Age of Child	Annual Expenditure
1	$8450
4	8660
7	8700
10	8650
13	9390
16	9530

Yscl = 500

U.S. Farms. *The number of U.S. farms f, in millions, can be approximated by the equation*

$$f = -\frac{1}{100}t + 2.5,$$

where t is the number of years after 1970. For example, t = 0 corresponds to 1970, t = 1 corresponds to 1971, and so on. (*Source: Based on information from* The New York Times Almanac, 2002).

51. Graph the equation and use the graph to estimate how many farms there were in the United States in 1985.
2,350,000

52. Use the graph from Exercise 51 to approximate in what year there will be 2.0 million farms in the United States.
2020

Hotel Rooms. *The average hotel or motel room rate in the United States can be approximated by the equation*

$$R = 0.3t^2 - 0.1t + 57.84,$$

where t is the number of years after 1990 (*Source: Based on information from the* Statistical Abstract of the United States, 2000, *U.S. Bureau of the Census*).

53. Graph the equation and use the graph to estimate in what year the average room rate was $60.   1993

54. Use the graph from Exercise 53 to estimate the average room rate in 2002. $99.84

Researchers at Yale University have suggested that the following graphs* may represent three different aspects of love.

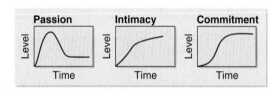

TW **55.** In what unit would you measure time if the horizontal length of each graph were ten units? Why?

TW **56.** Do you agree with the researchers that these graphs should be shaped as they are? Why or why not?

Skill Maintenance

Simplify. [1.4]

57. $(-2)^4$ 16   **58.** -2^4 -16 **59.** $z^{-2} \cdot z^6$ z^4

*From "A Triangular Theory of Love," by R. J. Sternberg, 1986, *Psychological Review,* **93**(2), 119–135. Copyright 1986 by the American Psychological Association, Inc. Reprinted by permission.

60. $\dfrac{z^{-2}}{z^6}$ z^{-8}, or $\dfrac{1}{z^8}$ **61.** $2(x^2y^3)^4$

$2x^8y^{12}$

62. $(2x^2y^3)^4$

$16x^8y^{12}$

Synthesis

TW **63.** How can a guess or estimate help prepare you for the *Translate* step in problem solving?

TW **64.** Describe at least two benefits of using a graph to model a situation.

65. Match each sentence with the most appropriate of the four graphs shown.

a) Roberta worked part time until September, full time until December, and overtime until Christmas.   III

b) Clyde worked full time until September, half time until December, and full time until Christmas.   II

c) Clarissa worked overtime until September, full time until December, and overtime until Christmas.   I

d) Doug worked part time until September, half time until December, and full time until Christmas.   IV

I

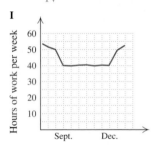

II

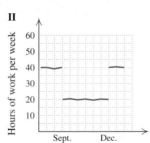

III

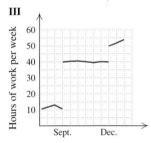

IV

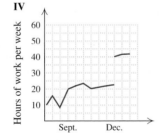

66. Match each sentence with the most appropriate of the four graphs shown below.

a) Carpooling to work, Terry spent 10 min on local streets, then 20 min cruising on the freeway, and then 5 min on local streets to his office.   IV

b) For her commute to work, Sharon drove 10 min to the train station, rode the express for 20 min, and then walked for 5 min to her office.   III

c) For his commute to school, Roger walked 10 min to the bus stop, rode the express for 20 min, and then walked for 5 min to his class.   I

d) Coming home from school, Kristy waited 10 min for the school bus, rode the bus for 20 min, and then walked 5 min to her house.   II

I

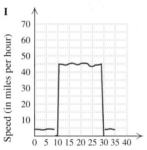

Time from the start (in minutes)

II

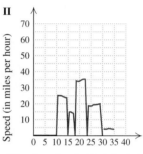

Time from the start (in minutes)

III

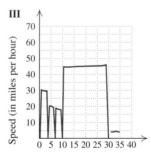

Time from the start (in minutes)

IV

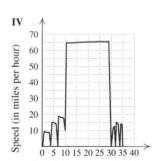

Time from the start (in minutes)

Collaborative Corner

No Room for Emergencies

Focus: Models
Time: 20 minutes
Group size: 3

The number of emergency room visits throughout the United States is increasing, due in part to a greater number of people seeking emergency treatment for illnesses that require routine medical care. At the same time, the number of hospitals is decreasing. Use the information in the following table to perform the listed activities. (*Sources*: U.S. Bureau of the Census; *Statistical Abstract of the United States,* 2000; *The New York Times Almanac,* 2002; Centers for Disease Control; www.cdc.gov)

Year	Number of Emergency Room Visits	Number of Hospitals	Number of Emergency Room Visits per Hospital	U.S. Population	Number of Emergency Room Visits per Person
1995	95,000,000	6291		263,082,000	
1996	94,000,000	6201		265,502,000	
1997	95,000,000	6097		268,048,000	
1998	100,000,000	6021		270,509,000	
1999	103,000,000	5890		272,945,000	
2000	108,000,000	5800		281,422,000	

ACTIVITY

1. Calculate, as a group, the number of emergency room visits per hospital and the number of emergency room visits per person for the years 1995 to 2000. Record that information in the table in the appropriate columns.

2. Each group member should choose one of the following columns:

   Number of emergency room visits;

   Number of emergency room visits per hospital;

   Number of emergency room visits per person.

He or she should create a line graph, using the figures in that column on the vertical axis and the corresponding years on the horizontal axis.

3. On the basis of the line graph created, each group member should determine whether the following statement appears to be true:

   "The number of emergency room visits in the United States is increasing."

4. Compare the graphs.

5. Working together as a group, write a sentence or paragraph that describes what each graph indicates.

Chapter Summary and Review

Key Terms

Variable, p. 2
Constant, p. 2
Algebraic expression, p. 2
Exponential notation, p. 3
Exponent, or power, p. 3
Base, p. 3
Substituting, p. 3
Evaluating the expression, p. 3
Equation, p. 5
Solution, p. 5
Inequality, p. 5
Natural numbers, p. 6
Whole numbers, p. 6
Integers, p. 6
Roster notation, p. 7
Set-builder notation, p. 7
Element, p. 7
Rational numbers, p. 7
Fraction notation, p. 8

Decimal notation, p. 8
Irrational numbers, p. 8
Real numbers, p. 8
Subset, p. 9
Cursor, p. 10
Contrast, p. 10
Home screen, p. 10
Absolute value, p. 13
Opposites, p. 14
Additive inverses, p. 14
Reciprocals, p. 17
Multiplicative inverses, p. 17
Indeterminate, p. 18
Menu, p. 19
Submenus, p. 19
Equivalent expressions, p. 22
Factoring, p. 24
Term, p. 24

Like terms, p. 24
Combining like terms, p. 24
Scientific notation, p. 38
Axes, p. 43
x, y-coordinate system, p. 43
Cartesian coordinate system, p. 43
Ordered pairs, p. 44
Origin, p. 44
Coordinates, p. 44
Quadrants, p. 45
Viewing window, p. 46
Standard viewing window, p. 46
Graph, p. 47
Dependent variable, p. 49
Independent variable, p. 49
Linear equation, p. 50
Nonlinear equation, p. 50
Mathematical model, p. 60

Important Properties and Formulas

Area of a rectangle:	$A = lw$
Area of a square:	$A = s^2$
Area of a parallelogram:	$A = bh$
Area of a trapezoid:	$A = \dfrac{h}{2}(b_1 + b_2)$
Area of a triangle:	$A = \frac{1}{2}bh$
Area of a circle:	$A = \pi r^2$
Circumference of a circle:	$C = \pi d$
Volume of a cube:	$V = s^3$
Volume of a right circular cylinder:	$V = \pi r^2 h$
Perimeter of a square:	$P = 4s$
Distance traveled:	$d = rt$
Simple interest:	$I = Prt$

Addition of Two Real Numbers

1. *Positive numbers*: Add the numbers. The result is positive.

2. *Negative numbers*: Add absolute values. Make the answer negative.

3. *A negative and a positive number*: If the numbers have the same absolute value, the answer is 0. Otherwise, subtract the smaller absolute value from the larger one:

   a) If the positive number is further from 0, make the answer positive.

   b) If the negative number is further from 0, make the answer negative.

4. *One number is zero*: The sum is the other number.

Multiplication of Two Real Numbers

1. To multiply two numbers with *unlike signs,* multiply their absolute values. The answer is *negative.*

2. To multiply two numbers with the *same sign,* multiply their absolute values. The answer is *positive.*

Division of Two Real Numbers

1. To divide two numbers with *unlike signs,* divide their absolute values. The answer is *negative.*

2. To divide two numbers with the *same sign,* divide their absolute values. The answer is *positive.*

The law of opposites: $a + (-a) = 0$

The law of reciprocals: $a \cdot \dfrac{1}{a} = 1, a \neq 0$

Absolute value: $|x| = \begin{cases} x, & \text{if } x \geq 0, \\ -x, & \text{if } x < 0 \end{cases}$

For any number a and any nonzero number b,

$$\frac{-a}{b} = \frac{a}{-b} = -\frac{a}{b}.$$

Rules for Order of Operations

1. Simplify within any grouping symbols.

2. Simplify all exponential expressions.

3. Perform all multiplication and division, working from left to right.

4. Perform all addition and subtraction, working from left to right.

Commutative laws: $a + b = b + a,$
$ab = ba$

Associative laws: $a + (b + c) = (a + b) + c,$
$a(bc) = (ab)c$

Distributive law: $a(b + c) = ab + ac$

Definitions and Rules for Exponents

For any integers m and n (assuming 0 is not raised to a nonpositive power):

Zero as an exponent:

$$a^0 = 1$$

Negative integers as exponents:

$$a^{-n} = \frac{1}{a^n}; \frac{1}{a^{-n}} = a^n;$$

$$\frac{a^{-n}}{b^{-m}} = \frac{b^m}{a^n}; \left(\frac{a}{b}\right)^{-n} = \left(\frac{b}{a}\right)^n$$

Multiplying with like bases:

$$a^m \cdot a^n = a^{m+n} \quad \text{(Product Rule)}$$

Dividing with like bases:

$$\frac{a^m}{a^n} = a^{m-n}; a \neq 0 \quad \text{(Quotient Rule)}$$

Raising a product to a power:

$$(ab)^n = a^n b^n$$

Raising a power to a power:

$$(a^m)^n = a^{mn} \quad \text{(Power Rule)}$$

Raising a quotient to a power:

$$\left(\frac{a}{b}\right)^n = \frac{a^n}{b^n}; b \neq 0$$

Scientific notation for a number is an expression of the type $N \times 10^m$, where $1 \leq N < 10$, N is in decimal notation, and m is an integer.

Review Exercises

The following review exercises are for practice. Answers are at the back of the book. If you need to, restudy the section indicated alongside the answer.

Evaluate each expression using the values provided.

1. $3x - (4 - y)$, for $x = 10$ and $y = 2$ [1.1] 28

2. $7x^2 - 5y \div zx$, for $x = -2.78$, $y = 1.5$, and $z = 3.2$
[1.1] 60.614425

3. Name the set consisting of the first six even natural numbers using both roster notation and set-builder notation.   [1.1] $\{2, 4, 6, 8, 10, 12\}$; $\{x \,|\, x$ is an even integer between 1 and 13$\}$

4. Find the area of a triangular sign that has a base of 60 cm and a height of 70 cm.   [1.1] 2100 cm²

Tell whether each number is a solution of the given equation or inequality.

5. $10 - 3x = 1$; **(a)** 7; **(b)** 3   [1.1] **(a)** No; **(b)** yes

6. $5a + 2 \le 7$; **(a)** 0; **(b)** 1   [1.1] **(a)** Yes; **(b)** yes

Find the absolute value.

7. $|-7.3|$ [1.2] 7.3 **8.** $|4.09|$ [1.2] 4.09 **9.** $|0|$ [1.2] 0

Perform the indicated operation.

10. $-8.4 + (-3.7)$ [1.2] -12.1 **11.** $\left(-\frac{4}{5}\right) + \left(\frac{1}{7}\right)$ [1.2] $-\frac{23}{35}$

12. $\left(-\frac{1}{3}\right) + \frac{4}{5}$ [1.2] $\frac{7}{15}$ **13.** $-7.9 - 3.6$ [1.2] -11.5

14. $-\frac{2}{3} - \left(-\frac{1}{2}\right)$ [1.2] $-\frac{1}{6}$ **15.** $12.5 - 17.9$ [1.2] -5.4

16. $(-5.1)(-3)$ [1.2] 15.3 **17.** $\left(-\frac{2}{3}\right)\left(\frac{5}{8}\right)$ [1.2] $-\frac{5}{12}$

18. $\frac{72.8}{-8}$ [1.2] -9.1 **19.** $-7 \div \frac{4}{3}$ [1.2] $-\frac{21}{4}$

20. Find $-a$ if $a = -4.01$.   [1.2] 4.01

Use a commutative law to write an equivalent expression.

21. $5 + a$ [1.3] $a + 5$ **22.** $3y$ [1.3] $y3$ **23.** $5x + y$ [1.3] $y + 5x$, or $x5 + y$, or $y + x5$

Use an associative law to write an equivalent expression.

24. $(4 + a) + b$ [1.3] $4 + (a + b)$ **25.** $(xy)7$ [1.3] $x(y7)$

26. Obtain an expression that is equivalent to $7mn + 14m$ by factoring.   [1.3] $7m(n + 2)$

27. Combine like terms: $5x^3 - 8x^2 + x^3 + 2$.   [1.3] $6x^3 - 8x^2 + 2$

28. Simplify: $7x - 4[2x + 3(5 - 4x)]$.   [1.3] $47x - 60$

29. Multiply and simplify: $(5a^2b^7)(-2a^3b)$.   [1.4] $-10a^5b^8$

30. Divide and simplify: $\frac{12x^3y^8}{3x^2y^2}$.   [1.4] $4xy^6$

31. Evaluate a^0, a^2, and $-a^2$ for $a = -5.3$.   [1.4] 1; 28.09; -28.09

Simplify. Do not use negative exponents in the answer.

32. $3^{-4} \cdot 3^7$ [1.4] 3^3, or 27 **33.** $(5a^2)^3$ [1.4] $125a^6$

34. $(-2a^{-3}b^2)^{-3}$ [1.4] $-\frac{a^9}{8b^6}$ **35.** $\left(\frac{x^2y^3}{z^4}\right)^{-2}$ [1.4] $\frac{z^8}{x^4y^6}$

36. $\left(\frac{2a^{-2}b}{4a^3b^{-3}}\right)^4$ [1.4] $\frac{b^{16}}{16a^{20}}$

Simplify.

37. $\frac{7(5 - 2 \cdot 3) - 3^2}{4^2 - 3^2}$ [1.2] $-\frac{16}{7}$

38. $1 - (2 - 5)^2 + 5 \div 10 \cdot 4^2$ [1.2] 0

39. Convert 0.000000103 to scientific notation.   [1.4] 1.03×10^{-7}

40. One *parsec* (a unit that is used in astronomy) is 30,860,000,000,000 km. Write scientific notation for this number.   [1.4] 3.086×10^{13}

Simplify and write scientific notation for each answer.

41. $(8.7 \times 10^{-9}) \times (4.3 \times 10^{15})$ [1.4] 3.741×10^7

42. $\frac{1.2 \times 10^{-12}}{1.5 \times 10^{-7}}$ [1.4] 8×10^{-6}

Determine whether the ordered pair is a solution.

43. $(3, 7)$; $4p - q = 5$ [1.5] Yes

44. $(-2, 4)$; $x - 2y = 12$ [1.5] No

45. $\left(0, -\frac{1}{2}\right)$; $3a - 4b = 2$ [1.5] Yes

46. $(8, -2)$; $3c + 2d = 28$ [1.5] No

Graph.

47. $y = -3x + 2$ ⊡ **48.** $y = -x^2 + 1$ ⊡

49. $y = 3 - |x|$ ⊡ **50.** $y = 6$ ⊡

51. Using a graphing calculator, complete the following table for the equation   [1.5] 3, 1, -1, -3, -1, 1, 3
$$y = 2|x| - 3.$$
Then graph.

X	Y1
-3	
-2	
-1	
0	
1	
2	
3	
X = -3	

⊡ Answers to Exercises 47–50 can be found on p. A-54.

Translate to an equation.

52. 13 less than twice a number is 21. [1.6] Let *n* represent the number; $2n - 13 = 21$

53. A number is 17 less than another number. The sum of the numbers is 115. Find the smaller number. [1.6] Let *n* represent the larger number; $n + (n - 17) = 115$

54. One angle of a triangle measures three times the second angle. The third angle measures twice the second angle. Find the measures of the angles. [1.6] Let *x* represent the measure of the second angle; $3x + x + 2x = 180$

55. *Air Quality.* The following graph shows the number of days with unhealthy air (an air quality index of over 100) for Dallas, Texas, over several years (*Source*: Environmental Protection Agency, *National Air Quality and Emissions Trends Report 1999*).

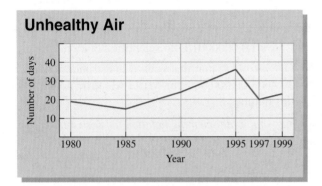

a) How many unhealthy days did Dallas have in 1985? [1.6] About 15 unhealthy days

b) In what year were there the most unhealthy days in Dallas? [1.6] 1995

c) What was the amount of decrease in the number of unhealthy days in Dallas between 1995 and 1997? [1.6] About 15

56. *Residential Remodeling.* The following table lists the amounts spent in the United States on residential remodeling (*Sources*: U.S. Bureau of the Census; *Statistical Abstract of the United States, 2000*). Use the data to draw a line graph with a graphing calculator. [1.6]

Year	Remodeling Expenses (in billions)
1980	$ 46.3
1985	80.3
1990	106.8
1991	97.5
1992	103.7
1993	108.3
1994	115.0
1995	111.7
1996	114.9
1997	118.6
1998	120.7

Synthesis

TW **57.** Explain the difference between a solution of an equation like $y = 2x + 1$ and a solution of an equation like $3x + 5 = 2$.

TW **58.** Explain why it is necessary to have a standard set of rules for order of operations.

59. Evaluate $a + b(c - a^2)^0 + (abc)^{-1}$ for $a = 2$, $b = -3$, and $c = -4$. [1.2], [1.4] $-\frac{23}{24}$

60. Simplify:
$$\frac{(3^{-2})^a \cdot (3^b)^{-2a}}{(3^{-2})^b \cdot (9^{-b})^{-3a}}.$$
[1.4] $3^{-2a+2b-8ab}$

61. Use the commutative law for addition once and the distributive law twice to show that
$$a2 + cb + cd + ad = a(d + 2) + c(b + d).$$

62. Find an irrational number between $\frac{1}{2}$ and $\frac{3}{4}$.
[1.2] $0.56556555655556\ldots$; answers may vary

61. [1.3] $a2 + cb + cd + ad = ad + a2 + cb + cd = a(d + 2) + c(b + d)$

Chapter Test 1

1. Evaluate $a^3 - 5b + b \div ac$ for $a = -2$, $b = 6$, and $c = 3$.   [1.2] -47

2. The base of a triangular stamp measures 3 cm and its height 2.5 cm. Find the area of the stamp.

[1.1]
3.75 cm²

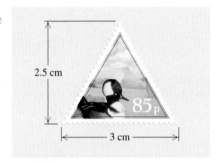

2.5 cm

3 cm

85p

3. Tell whether each number is a solution of the equation $14 - 5x = 4$.

 a) 1   [1.1] No
 b) 2   [1.1] Yes
 c) 0   [1.1] No

Perform the indicated operation.

4. $-25 + (-16)$ [1.2] -41 **5.** $-10.5 + 6.8$ [1.2] -3.7

6. $6.21 + (-8.32)$ [1.2] -2.11 **7.** $29.5 - 43.7$ [1.2] -14.2

8. $-17.8 - 25.4$ [1.2] -43.2 **9.** $-6.4(5.3)$ [1.2] -33.92

10. $-\frac{7}{3} - \left(-\frac{3}{4}\right)$ [1.2] $-\frac{19}{12}$ **11.** $-\frac{2}{7}\left(-\frac{5}{14}\right)$ [1.2] $\frac{5}{49}$

12. $\frac{-42.6}{-7.1}$ [1.2] 6   **13.** $\frac{2}{5} \div \left(-\frac{3}{10}\right)$ [1.2] $-\frac{4}{3}$

14. Simplify: $5 + (1 - 3)^2 - 7 \div 2^2 \cdot 6$.   [1.2] $-\frac{3}{2}$

15. Use a commutative law to write an expression equivalent to $7x + y$.   [1.3] $y + 7x$, or $x7 + y$, or $y + x7$

16. Combine like terms: $4y - 10 - 7y - 19$.   [1.3] $-3y - 29$

17. Simplify: $9x - 3(2x - 5) - 7$.   [1.3] $3x + 8$

Simplify. Do not use negative exponents in the answer.

18. $(12x^{-4}y^{-7})(-6x^{-6}y)$

19. -3^{-2} [1.4] $-\frac{1}{9}$

20. $(-6x^2y^{-4})^{-2}$ [1.4] $\frac{y^8}{36x^4}$ **21.** $\left(\frac{2x^3y^{-6}}{-4y^{-2}}\right)^2$ [1.4] $\frac{x^6}{4y^8}$

22. $(5x^3y)^0$ [1.4] 1

Simplify and write scientific notation for the answer.

23. $(9.05 \times 10^{-3})(2.22 \times 10^{-5})$ [1.4] 2.0091×10^{-7}

24. $\frac{5.6 \times 10^7}{2.8 \times 10^{-3}}$ [1.4] 2.0×10^{10} **25.** $\frac{1.0614 \times 10^{-5}}{3.48 \times 10^{-10}}$ [1.4] 3.05×10^4

Determine whether the ordered pair is a solution.

26. $(0, -5)$; $x + 4y = -20$ [1.5] Yes

27. $(1, -4)$; $-2p + 5q = 18$ [1.5] No

Graph.

28. $y = -5x + 4$ ⊡   **29.** $y = -2x^2 + 3$ ⊡

30. Create a table of solutions of the equation
$$y = 10 - x^2$$
for integer values of x from -3 to 3. Then graph.   ⊡

31. Translate to an algebraic expression: [1.6] Let m and n represent the numbers; $mn + 3$, or $3 + mn$
Three more than the product of two numbers.

32. Translate to an equation: ⊡

Greg's scores on five tests are 94, 80, 76, 91, and 75. What must Greg score on the sixth test so that his average will be 85?

Gas Mileage.   The following graph shows the gas mileage of a truck traveling at different speeds.

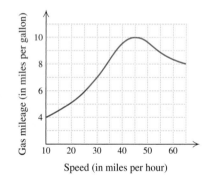

33. At what speed is the gas mileage highest?   [1.6] 45 mph

34. What is the gas mileage when the truck is traveling at 30 mph?   [1.6] 7 mpg

Synthesis

Simplify.

35. $(4x^{3a}y^{b+1})^{2c}$ [1.4] $16^c x^{6ac} y^{2bc+2c}$

36. $\frac{-27a^{x+1}}{3a^{x-2}}$ [1.4] $-9a^3$

37. $\frac{(-16x^{x-1}y^{y-2})(2x^{x+1}y^{y+1})}{(-7x^{x+2}y^{y+2})(8x^{x-2}y^{y-1})}$ [1.4] $\frac{4}{7y^2}$

18. [1.4] $-\frac{72}{x^{10}y^6}$

⊡ Answers to Exercises 28–30 and 32 can be found on p. A-54.

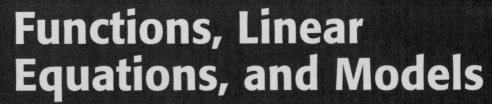

Functions, Linear Equations, and Models

2.1 Functions

2.2 Solving Linear Equations

2.3 Applications and Formulas

2.4 Linear Functions: Slope, Graphs, and Models

2.5 Another Look at Linear Graphs

2.6 Introduction to Curve Fitting: Point – Slope Form

2.7 Domains and the Algebra of Functions

SUMMARY AND REVIEW

TEST

A *function* is a certain kind of relationship between sets. Functions are very important in mathematics in general, and in problem solving in particular. In this chapter, you will learn what we mean by a function and then begin to use functions to solve problems.

A function that can be described by a linear equation is called a *linear function.* We will study graphs of linear equations in detail and will use linear equations and functions to model and solve applications.

APPLICATION

PAPER RECYCLING. The following table shows the amount of paper recovered in the United States for various years. Estimate the amount of paper that will be recovered in 2003. The graph shows that a linear function will fit the data and can be used to make a prediction.

Year	Paper Recovered (in millions of tons)
1988	26.2
1990	29.1
1992	34.0
1994	39.7
1996	43.1
1998	45.1
2000	49.4

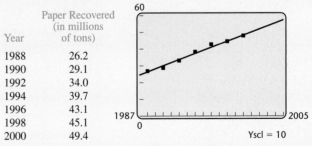

Source: www.afandpa.org

This problem appears as Example 5 in Section 2.6.

2.1

Functions and Graphs ■ Function Notation and Equations ■ Applications: Interpolation and Extrapolation

Functions

We now develop the idea of a *function*—one of the most important concepts in mathematics. A function is a special kind of correspondence between two sets. For example,

To each person in a class	there corresponds	his or her biological mother.
To each item in a shop	there corresponds	its price.
To each real number	there corresponds	the cube of that number.

In each example, the first set is called the **domain**. The second set is called the **range**. For any member of the domain, there is *just one* member of the range to which it corresponds. This kind of correspondence is called a **function**.

Note in the list above that although two members of a class may have the same biological mother, and two items in a shop may have the same price, each correspondence is still a function since every member of the domain is paired with exactly one member of the range.

Example 1   Determine whether each correspondence is a function.

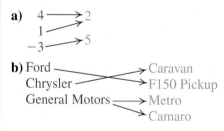

Solution

a) The correspondence *is* a function because each member of the domain corresponds to *just one* member of the range.

b) The correspondence *is not* a function because a member of the domain (General Motors) corresponds to more than one member of the range.

Function

A *function* is a correspondence between a first set, called the *domain,* and a second set, called the *range,* such that each member of the domain corresponds to *exactly one* member of the range.

Example 2   Determine whether each correspondence is a function.

Domain	Correspondence	Range
a) An elevator full of people	Each person's weight	A set of positive numbers
b) $\{-2, 0, 1, 2\}$	Each number's square	$\{0, 1, 4\}$
c) The books in a college bookstore	Each book's author	A set of people

Solution

a) The correspondence *is* a function, because each person has *only one* weight.

b) The correspondence *is* a function, because every number has *only one* square.

c) The correspondence *is not* a function, because some books have *more than one* author.

Functions and Graphs

The functions in Examples 1(a) and 2(b) can be expressed as sets of ordered pairs. Example 1(a) can be written $\{(-3, 5), (1, 2), (4, 2)\}$ and Example 2(b) can be written $\{(-2, 4), (0, 0), (1, 1), (2, 4)\}$. We can graph these functions as follows.

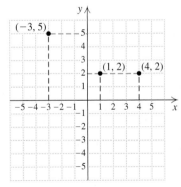

The function $\{(-3, 5), (1, 2), (4, 2)\}$
Domain is $\{-3, 1, 4\}$
Range is $\{5, 2\}$

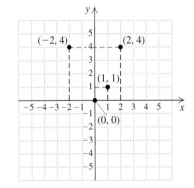

The function $\{(-2, 4), (0, 0), (1, 1), (2, 4)\}$
Domain is $\{-2, 0, 1, 2\}$
Range is $\{4, 0, 1\}$

When a function is given as a set of ordered pairs, the domain is the set of all first coordinates and the range is the set of all second coordinates. Functions are generally represented by lower- or upper-case letters.

Example 3   Find the domain and the range of the function f shown here.

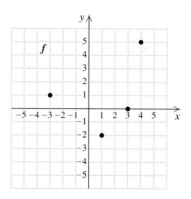

Solution   Here f can be written $\{(-3, 1), (1, -2), (3, 0), (4, 5)\}$. The domain is the set of all first coordinates, $\{-3, 1, 3, 4\}$, and the range is the set of all second coordinates, $\{1, -2, 0, 5\}$. We can also find the domain and the range directly from the graph, without first listing all pairs.

Example 4   For the function f shown here, determine each of the following.

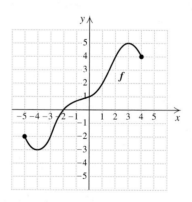

a) What member of the range is paired with 2
b) The domain of f
c) What member of the domain is paired with -3
d) The range of f

Solution

a) To determine what member of the range is paired with 2, we locate 2 on the horizontal axis (this is where the domain is located). Next, we find the point on the graph of f for which 2 is the first coordinate. From that point, we can look to the vertical axis to find the corresponding y-coordinate, 4. The "input" 2 has the "output" 4.

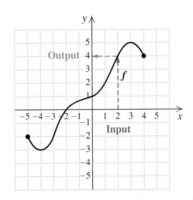 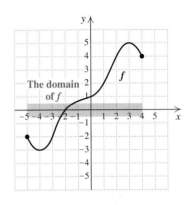

TEACHING TIP

It may help students to think of identifying the domain from left to right and the range from bottom to top.

b) The domain of the function is the set of all x-values that are in the graph (see the figure on the right above). Because there are no breaks in the graph of f, these extend continuously from -5 to 4 and can be viewed as the curve's shadow, or *projection,* on the x-axis. Thus the domain is $\{x \mid -5 \le x \le 4\}$.

c) To determine what member of the domain is paired with -3, we locate -3 on the vertical axis. (This is where the range is located; see below.) From there we look left and right to the graph of f to find any points for which -3 is the second coordinate. One such point exists, $(-4, -3)$. We note that -4 is the only element of the domain paired with -3.

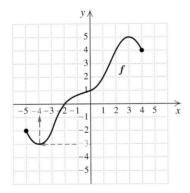 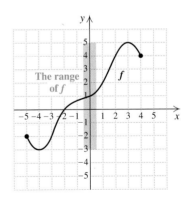

d) The range of the function is the set of all y-values that are in the graph. (See the graph on the right above.) These extend continuously from -3 to 5, and can be viewed as the curve's projection on the y-axis. Thus the range is $\{y \mid -3 \le y \le 5\}$.

Note that if a graph contains two or more points with the same first coordinate, that graph cannot represent a function (otherwise one member of the domain would correspond to more than one member of the range). This observation is the basis of the *vertical-line test.*

The Vertical-Line Test

If it is possible for a vertical line to cross a graph more than once, then the graph is not the graph of a function.

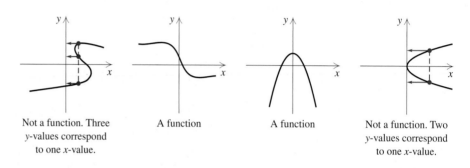

| Not a function. Three *y*-values correspond to one *x*-value. | A function | A function | Not a function. Two *y*-values correspond to one *x*-value. |

Graphs that do not represent functions still do represent *relations.*

Relation

A *relation* is a correspondence between a first set, called the *domain*, and a second set, called the *range*, such that each member of the domain corresponds to *at least one* member of the range.

Thus, although the correspondences and graphs above are not all functions, they *are* all relations.

Function Notation and Equations

We often think of an element of the domain of a function as an *input* and its corresponding element of the range as an *output.* In Example 3, the function f is written

$$\{(-3, 1), (1, -2), (3, 0), (4, 5)\}.$$

Thus, for an input of -3, the corresponding output is 1, and for an input of 3, the corresponding output is 0.

We use *function notation* to indicate what output corresponds to a given input. For the function f defined above, we write

$$f(-3) = 1, \qquad f(1) = -2, \qquad f(3) = 0, \quad \text{and} \quad f(4) = 5.$$

The notation $f(x)$ is read "f of x," "f at x," or "the value of f at x."

Note that $f(x)$ *does not mean f times x.*

Most functions are described by equations. For example, $f(x) = 2x + 3$ describes the function that takes an input x, multiplies it by 2, and then adds 3.

$$f(x) \;=\; \underset{\text{Double}}{\overset{\overset{\text{Input}}{\downarrow}}{2x}} \;+\; \underbrace{3}_{\text{Add 3}}$$

To calculate the output $f(4)$, we take the input 4, double it, and add 3 to get 11. That is, we substitute 4 into the formula for $f(x)$:

$$f(4) = 2 \cdot 4 + 3$$
$$= 11. \longleftarrow \text{Output}$$

To understand function notation, it can help to imagine a "function machine." Think of putting an input into the machine. For the function $f(x) = 2x + 3$, the machine will double the input and then add 3. The result will be the output.

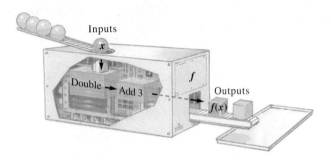

Sometimes, in place of $f(x) = 2x + 3$, we write $y = 2x + 3$, where it is understood that the value of y, the *dependent variable,* depends on our choice of x, the *independent variable.* To understand why $f(x)$ notation is so useful, consider two equivalent statements:

a) If $f(x) = 2x + 3$, then $f(4) = 11$.

b) If $y = 2x + 3$, then the value of y is 11 when x is 4.

The notation used in part (a) is far more concise.

EXAMPLE 5   Find the indicated function value.

a) $f(5)$, for $f(x) = 3x + 2$

b) $g(-2)$, for $g(r) = 5r^2 + 3r$

c) $h(4)$, for $h(x) = 7$

d) $F(a + 1)$, for $F(x) = 3x + 2$

e) $F(a) + 1$, for $F(x) = 3x + 2$

Solution Finding function values is much like evaluating an algebraic expression.

a) $f(5) = 3 \cdot 5 + 2 = 17$

b) $g(-2) = 5(-2)^2 + 3(-2)$

$= 5 \cdot 4 - 6 = 14$

c) For the function given by $h(x) = 7$, all inputs share the same output, 7. Therefore, $h(4) = 7$. The function h is an example of a *constant function*.

d) $F(a + 1) = 3(a + 1) + 2$ **The input is $a + 1$.**

$= 3a + 3 + 2 = 3a + 5$

e) $F(a) + 1 = [3(a) + 2] + 1$ **The input is a.**

$= [3a + 2] + 1 = 3a + 3$

Note that whether we write $f(x) = 3x + 2$, or $f(t) = 3t + 2$, or $f(\square) = 3\square + 2$, we still have $f(5) = 17$. Thus the independent variable can be thought of as a *dummy variable*. The letter chosen for the dummy variable is not as important as the algebraic manipulations to which it is subjected.

Function Notation

The values of a function that is described by an equation can be found directly using a graphing calculator. For example, if

$$f(x) = x^2 - 6x + 7$$

is entered as

$$y_1 = x^2 - 6x + 7,$$

then $f(2)$ can be calculated by evaluating $Y_1(2)$. After entering the function, move to the home screen by pressing $\boxed{\text{2nd}}$ $\boxed{\text{QUIT}}$. The notation "Y_1" can be found by pressing $\boxed{\text{VARS}}$, choosing the Y-VARS submenu, selecting the FUNCTION option, and then selecting Y_1. After Y_1 appears on the home screen, press $\boxed{(}\,\boxed{2}\,\boxed{)}$ $\boxed{\text{ENTER}}$ to evaluate $Y_1(2)$.

EXAMPLE 6   For $f(a) = 2a^2 - 3a + 1$, find $f(3)$ and $f(-5.1)$.

Solution   We first enter the function into the graphing calculator, replacing a with x and the notation $f(a)$ with Y_1. The equation $Y_1 = 2x^2 - 3x + 1$ represents the same function, with the understanding that the Y_1-values are the outputs and the x-values are the inputs; $Y_1(x)$ is equivalent to $f(a)$.

To find $f(3)$, we enter $Y_1(3)$ and find that $f(3) = 10$.

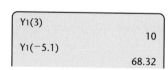

To find $f(-5.1)$, we can press $\boxed{\text{2nd}}$ $\boxed{\text{ENTRY}}$ and edit the previous entry. We see that $f(-5.1) = 68.32$.

We can also find function values from the graph of the function.

EXAMPLE 7   Find $g(2)$ for $g(x) = 2x - 5$.

Solution   We enter and graph the function, using the standard viewing window. Most graphing calculators have a VALUE option in the CALC menu that will calculate a function's value given a value of x. For most calculators, that x-value must be shown in the viewing window; that is, Xmin $\leq x \leq$ Xmax. In this case, when $x = 2$, $y = -1$, so $g(2) = -1$.

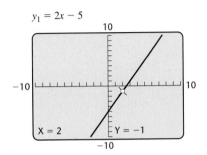

Applications: Interpolation and Extrapolation

Function notation is often used in formulas. For example, to emphasize that the area A of a circle is a function of its radius r, instead of

$$A = \pi r^2,$$

we can write

$$A(r) = \pi r^2.$$

When a function is given as a graph in a problem-solving situation, we are often asked to determine certain quantities on the basis of the graph. Later in this text, we will develop models that can be used for calculations. For now we simply use the graph to estimate the coordinates of an unknown point by using other points with known coordinates. When the unknown point is *between* the known points, this process is called **interpolation**. If the unknown point extends *beyond* the known points, the process is called **extrapolation**.

EXAMPLE 8   Registered Nurses.   The number of registered nurses working full-time increased by over 700,000 from 1980 to 2000. According to the National Sample Survey of Registered Nurses, conducted every four years, 850,000 registered nurses worked full-time in 1980. The figure grew to 975,000 in 1984, 1,100,000 in 1988, 1,225,000 in 1992, 1,500,000 in 1996, and 1,600,000 in 2000. Estimate the number of full-time registered nurses in 1990 (a year in which the survey was not conducted) and in 2004.

Solution

1. and **2. Familiarize and Translate.**   The given information enables us to plot and connect six points. We let the horizontal axis represent the year and the vertical axis the number of registered nurses employed full-time that year. We label the function itself N.

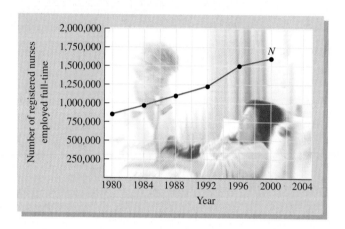

3. Carry out.   To estimate the number of registered nurses working full-time in 1990, we locate the point directly above 1990 on the horizontal axis. We then estimate the second coordinate by moving horizontally from that point to the y-axis. Although our result is not exact, we see that $N(1990) \approx 1{,}175{,}000$.

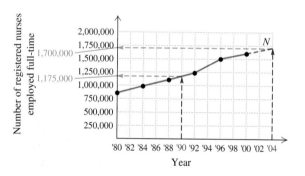

To estimate the number of registered nurses working full-time in 2004, we extend the graph and extrapolate. It appears that $N(2004) \approx 1{,}700{,}000$.

4. Check.   A precise check requires consulting an outside information source. Since 1,175,000 is between 1,100,000 and 1,225,000, and since 1,700,000 is greater than 1,600,000, our estimates seem plausible.

5. State.   In 1990, about 1,175,000 registered nurses were employed full-time. By 2004, that number can be predicted to be about 1,700,000.

Connecting the Concepts

DATA INTERPRETATION: PREDICTIONS

In business and in many other fields, decisions are often made on the basis of a prediction of the future. Although extrapolation from past data is an important tool in making predictions, other factors should be considered as well.

Look, for example, at the following graph of global sales of cellular phones from 1996 to 2000 (*Source*: Gartner Dataquest).

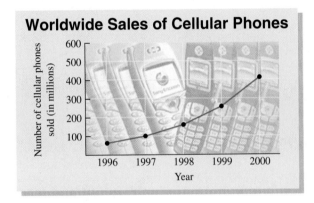

On the basis of past data, we might predict sales of cellular phones to be well over 500 million in 2001. However, it turns out that fewer were sold in 2001 than in 2000.

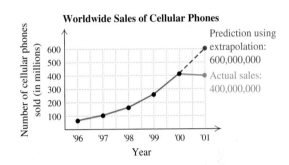

In this case, factors such as the economy, population size, and saturation of the cellular-phone network contributed to the actual numbers. Although the past can be an excellent predictor of the future, any prediction is to some extent a guess.

2.1

Exercise Set

FOR EXTRA HELP

Digital Video Tutor CD 2 Videotape 2 | Student's Solutions Manual | AW Math Tutor Center | InterAct Math | MathXL | MyMathLab

Determine whether each correspondence is a function.

1. 3 ⟶ a
5 ⟶ b
7   c
9   d
     e   No

2. 1 ⟶ a
2 ⟶ b
3   c
4 ⟶ d
5   Yes

3.

Girl's Age (in months)	Average Daily Weight Gain (in grams)
2 ⟶	21.8
9 ⟶	11.7
16 ⟶	8.5
23 ⟶	7.0

Source: American Family Physician, December 1993, p. 1435.   Yes

4.

Boy's Age (in months)	Average Daily Weight Gain (in grams)
2	→ 24.3
9	→ 11.7
16	→ 8.2
23	→ 7.0

Source: *American Family Physician*, December 1993, p. 1435.   Yes

5.

Basketball Player	Height
Kobe Bryant	6'1"
Vince Carter	
Shaquille O'Neal	6'7"
David Robinson	
John Stockton	7'1" Yes

6.

Olympics Site	Year
Lake Placid	1980
Calgary	2002
Squaw Valley	1960
Salt Lake City	1988
	1932 No

Determine whether each of the following is a function. Identify any relations that are not functions.

Domain	Correspondence	Range
7. A yard full of Christmas trees	Each tree's price	A set of prices
	Function	
8. The swordfish stored in a boat	Each fish's weight	A set of weights
	Function	
9. The members of a rock band	An instrument the person can play	A set of instruments
	A relation but not a function	
10. The students in a math class	Each person's seat number	A set of numbers
	Function	
11. A set of numbers	Square each number and then add 4.	A set of numbers
	Function	
12. A set of shapes	The area of each shape	A set of numbers
	Function	

For each graph of a function, determine **(a)** $f(1)$; **(b)** *the domain;* **(c)** *any x-values for which* $f(x) = 2$; *and* **(d)** *the range.*

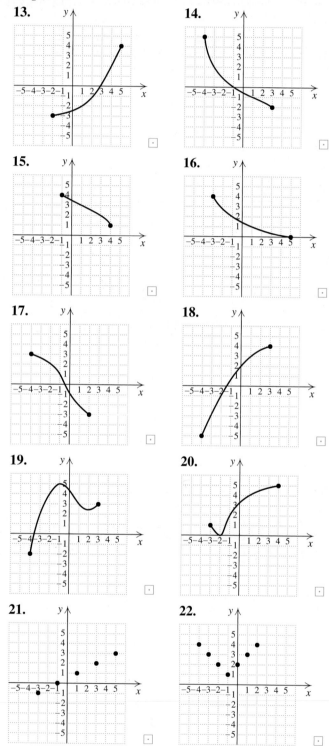

13.

14.

15.

16.

17.

18.

19.

20.

21.

22.

⊡ Answers to Exercises 13–22 can be found on pp. A-54 and A-55.

23.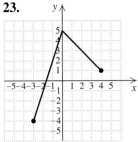
⊡

24.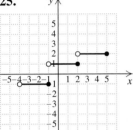
⊡

25.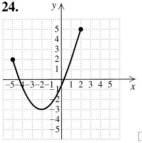
⊡

26.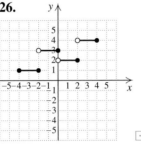
⊡

Determine whether each of the following is the graph of a function.

27.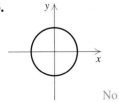
Yes

28.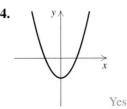
No

29.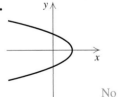
Yes

30.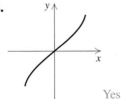
No

31.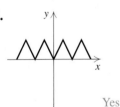
No

32.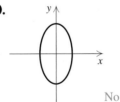
Yes

33.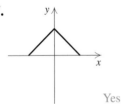
No

34.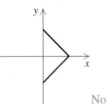
Yes

⊡ Answers to Exercises 23–26 can be found on p. A-55.

Find the function values.

35. $g(x) = 2x + 3$

　a) $g(0)$　3　　**b)** $g(-4)$　-5　　**c)** $g(-7)$　-11

　d) $g(8)$　19　　**e)** $g(a + 2)$　　**f)** $g(a) + 2$
　　　　　　　　　　　　$2a + 7$　　　　$2a + 5$

36. $h(x) = 3x - 2$

　a) $h(4)$　10　　**b)** $h(8)$　22　　**c)** $h(-3)$　-11

　d) $h(-4)$　-14　**e)** $h(a - 1)$　　**f)** $h(a) - 1$
　　　　　　　　　　　　$3a - 5$　　　　$3a - 3$

37. $f(n) = 5n^2 + 4n$

　a) $f(0)$　0　　**b)** $f(-1)$　1　　**c)** $f(3)$　57

Aha! **d)** $f(t)$　$5t^2 + 4t$　**e)** $f(2a)$　　**f)** $2 \cdot f(a)$
　　　　　　　　　　　　$20a^2 + 8a$　　$10a^2 + 8a$

38. $g(n) = 3n^2 - 2n$

　a) $g(0)$　0　　**b)** $g(-1)$　5　　**c)** $g(3)$　21

　d) $g(t)$　$3t^2 - 2t$　**e)** $g(2a)$　　**f)** $2 \cdot g(a)$
　　　　　　　　　　　　$12a^2 - 4a$　　$6a^2 - 4a$

39. $f(x) = \dfrac{x - 3}{2x - 5}$

　a) $f(0)$　$\frac{3}{5}$　　**b)** $f(4)$　$\frac{1}{3}$　　**c)** $f(-1)$　$\frac{4}{7}$

　d) $f(3)$　0　　**e)** $f(x + 2)$　$\dfrac{x - 1}{2x - 1}$

40. $s(x) = \dfrac{3x - 4}{2x + 5}$

　a) $s(10)$　$\frac{26}{25}$　**b)** $s(2)$　$\frac{2}{9}$　　**c)** $s\left(\frac{1}{2}\right)$　$-\dfrac{5}{12}$

　d) $s(-1)$　$-\frac{7}{3}$　**e)** $s(x + 3)$　$\dfrac{3x + 5}{2x + 11}$

The function A described by $A(s) = s^2 \dfrac{\sqrt{3}}{4}$ gives the area of an equilateral triangle with side s.

41. Find the area when a side measures 4 cm.
　$4\sqrt{3}\,\text{cm}^2 \approx 6.93\,\text{cm}^2$

42. Find the area when a side measures 6 in.
　$9\sqrt{3}\,\text{in}^2 \approx 15.59\,\text{in}^2$

The function V described by $V(r) = 4\pi r^2$ gives the surface area of a sphere with radius r.

43. Find the area when the radius is 3 in.
　$36\pi\,\text{in}^2 \approx 113.10\,\text{in}^2$

44. Find the area when the radius is 5 cm.
　$100\pi\,\text{cm}^2 \approx 314.16\,\text{cm}^2$

Chemistry. The function F described by

$$F(C) = \tfrac{9}{5}C + 32$$

gives the Fahrenheit temperature corresponding to the Celsius temperature C.

45. Find the Fahrenheit temperature equivalent
　to $-10°C$.　14°F

46. Find the Fahrenheit temperature equivalent to 5°C.
41°F

Archaeology.   The function H described by

$$H(x) = 2.75x + 71.48$$

can be used to predict the height, in centimeters, of a woman whose humerus (the bone from the elbow to the shoulder) is x cm long. Predict the height of a woman whose humerus is the length given.

47. 32 cm   159.48 cm

48. 35 cm   167.73 cm

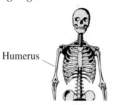

Humerus

Incidence of AIDS.   The following table indicates the number of cases of AIDS reported in the United States in each of several years (Source: U.S. Centers for Disease Control and Prevention).

Input, Year	Output, Number of Cases Reported
1990	43,164
1994	77,103
1996	68,459
1999	46,400

49. Use the data in the table above to draw a graph and to estimate the number of cases of AIDS reported in 2002.   ⊡

50. Use the graph from Exercise 49 to estimate the number of cases of AIDS reported in 1997.
About 61,000 cases

Population Growth.   The town of Eagledale recorded the following dates and populations.

Input, Year	Output, Population (in tens of thousands)
1995	5.8
1997	6
1999	7
2001	7.5

51. About 65,000

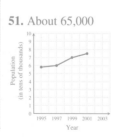

51. Use the data in the table above to draw a graph of the population as a function of time. Then estimate what the population was in 1998.

⊡ Answers to Exercises 49 and 53 can be found on p. A-55.

52. Use the graph from Exercise 51 to predict Eagledale's population in 2003.   About 80,000

53. *Retailing.*   Shoreside Gifts is experiencing constant growth. They recorded a total of $250,000 in sales in 1996 and $285,000 in 2001. Use a graph that displays the store's total sales as a function of time to predict total sales for 2005.   ⊡

54. Use the graph constructed in Exercise 53 to estimate what the total sales were in 1999.   About $271,000

ᵀᵂ **55.** Which would you trust more and why: estimates made using interpolation or those made using extrapolation?

ᵀᵂ **56.** Suppose a function *f* is defined by

$$f(x) = x^2 - 5x + 7.$$

Describe in your own words the operations performed on the input to produce the output.

Skill Maintenance

Simplify by combining like terms. [1.3]

57. $3x - 5 - 9x + 15$
$-6x + 10$

58. $x - (9x + 3)$
$-8x - 3$

59. $2x + 4 - 6(5 - 7x)$
$44x - 26$

60. $9y - 3(2y + 7)$
$3y - 21$

61. $3 - 2[5(x - 7) + 1]$ $-10x + 71$

62. $2x - \{3[4 - 2(1 - x)] + 7x\}$ $-11x - 6$

Simplify. [1.2]

63. $\dfrac{10 - 3^2}{9 - 2 \cdot 3}$ $\dfrac{1}{3}$

64. $\dfrac{2^4 - 10}{6 - 4 \cdot 3}$ -1

Synthesis

ᵀᵂ **65.** For the function given by $n(z) = ab + wz$, what is the independent variable? How can you tell?

ᵀᵂ **66.** Explain in your own words why every function is a relation, but not every relation is a function.

For Exercises 67 and 68, let $f(x) = 3x^2 - 1$ and $g(x) = 2x + 5$.

67. Find $f(g(-4))$ and $g(f(-4))$.   26; 99

68. Find $f(g(-1))$ and $g(f(-1))$.   26; 9

The function $V(r) = \frac{4}{3}\pi r^3$ gives the volume of a sphere with radius r. Use a table of values or a graph to find the radius of a sphere with the given volume.

69. 50 cm³   About 2.29 cm   **70.** 1.2 in³   About 0.659 in.

Pregnancy.   For Exercises 71–74, use the following graph of a woman's "stress test." This graph shows the size of a pregnant woman's contractions as a function of time.

71. How large is the largest contraction that occurred during the test?   About 22 mm

72. At what time during the test did the largest contraction occur?   About 2 min, 50 sec

TW 73. On the basis of the information provided, how large a contraction would you expect 60 seconds after the end of the test? Why?

74. What is the frequency of the largest contraction?   1 every 3 min

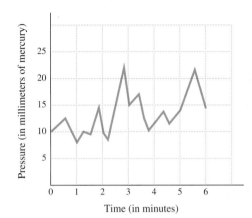

75. The *greatest integer function* $f(x) = [\![x]\!]$ is defined as follows: $[\![x]\!]$ is the greatest integer that is less than or equal to x. For example, if $x = 3.74$, then $[\![x]\!] = 3$; and if $x = -0.98$, then $[\![x]\!] = -1$. Graph the greatest integer function for $-5 \le x \le 5$. (The notation $f(x) = \text{INT}[x]$ is used in many graphing calculators and computer programs.)

76. Suppose that a function g is such that $g(-1) = -7$ and $g(3) = 8$. Find a formula for g if $g(x)$ is of the form $g(x) = mx + b$, where m and b are constants.

77. *Energy Expenditure.*   On the basis of the information given below, what burns more energy: walking $4\frac{1}{2}$ mph for two hours or bicycling 14 mph for one hour?   Bicycling 14 mph for 1 hr

Approximate Energy Expenditure by a 150-Pound Person in Various Activities

Activity	Calories per Hour
Walking, $2\frac{1}{2}$ mph	210
Bicycling, $5\frac{1}{2}$ mph	210
Walking, $3\frac{3}{4}$ mph	300
Bicycling, 13 mph	660

Source: Based on material prepared by Robert E. Johnson, M.D., Ph.D., and colleagues, University of Illinois.

75.

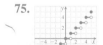

76. $g(x) = \frac{15}{4}x - \frac{13}{4}$

Collaborative Corner

Calculating License Fees

Focus: Functions
TIme: 15 – 20 minutes
Group size: 3 – 4

The California Department of Motor Vehicles calculates automobile registration fees (VLF) according to the schedule shown below.

ACTIVITY

1. Determine the original sale price of the oldest vehicle owned by a member of your group. If necessary, use the price and age of a family member's vehicle. Be sure to note the year in which the car was purchased.

2. Use the schedule below to calculate the vehicle license fee (VLF) for the vehicle in part (1) above for each year from the year of purchase to the present. To speed your work, each group member can find the fee for a few different years.

3. Graph the results from part (2). On the x-axis, plot years beginning with the year of purchase, and on the y-axis, plot $V(x)$, the VLF as a function of year.

4. What is the lowest VLF that the owner of this car will ever have to pay, according to this schedule? Compare your group's answer with other groups' answers.

5. Does your group feel that California's method for calculating registration fees is fair? Why or why not? How could it be improved?

6. Try, as a group, to find an algebraic form for the function $y = V(x)$.

7. *Optional out-of-class extension*: Create a program for a graphing calculator that accepts two inputs (initial value of the vehicle and year of purchase) and produces $V(x)$ as the output.

DMV
A Public Service Agency

VEHICLE LICENSE FEE INFORMATION

The 2% **Vehicle License Fee (VLF)** is in lieu of a personal property tax on vehicles. Most VLF revenue is returned to City and County Local Governments (see reverse side). The license fee charged is based upon the sale price or vehicle value when initially registered in California. The vehicle value is adjusted for any subsequent sale or transfer, that occurred 8/19/91 or later, excluding sales or transfers between specified relatives.

The VLF is calculated by rounding the sale price to the nearest **odd** hundred dollar. That amount is reduced by a percentage utilizing an eleven year schedule (shown to the right), and 2% of that amount is the fee charged. See the accompanying example for a vehicle purchased last year for $9,199. This would be the second registration year following that purchase.

WHERE DO YOUR DMV FEES GO? SEE REVERSE SIDE.

DMV77 8(REV.8/95)   95 30123

PERCENTAGE SCHEDULE
Rev. & Tax. Code Sec. 10753.2
(Trailer coaches have a different schedule)

1st Year	100%	7th Year	40%
2nd Year	90%	8th Year	30%
3rd Year	80%	9th Year	25%
4th Year	70%	10th year	20%
5th Year	60%	11th Year	
6th Year	50%	onward	15%

VLF CALCULATION EXAMPLE

Purchase Price:	$9,199
Rounded to:	$9,100
Times the Percentage:	90%
Equals Fee Basis of:	$8,190
Times 2% Equals:	$163.80
Rounded to:	$164

2.2

Solving Equations Graphically ■ Equivalent Equations ■ The Addition and Multiplication Principles ■ Types of Equations

Solving Linear Equations

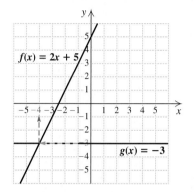

$f(x) = 2x + 5$

$g(x) = -3$

To *solve* an equation means to find all the replacements for the variable that make the equation true. In this section, we solve *linear equations* using graphical and algebraic methods.

Solving Equations Graphically

To see how solutions of equations are related to graphs, consider the graphs of the functions given by $f(x) = 2x + 5$ and $g(x) = -3$.

At the point where the graphs intersect, $f(x) = g(x)$. Thus, for that particular x-value, we have $2x + 5 = -3$. In other words, the solution of $2x + 5 = -3$ is the x-coordinate of the point of intersection of the graphs of $f(x) = 2x + 5$ and $g(x) = -3$. Careful inspection suggests that -4 is that x-value. To check, note that $f(-4) = 2(-4) + 5 = -3$.

EXAMPLE 1 Solve graphically: $\frac{1}{2}x + 3 = 2$.

Solution   To find the x-value for which $\frac{1}{2}x + 3$ will equal 2, we graph $f(x) = \frac{1}{2}x + 3$ and $g(x) = 2$ on the same set of axes. Since the intersection appears to be $(-2, 2)$, the solution is apparently -2.

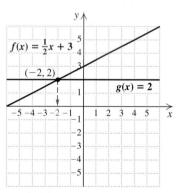

$f(x) = \frac{1}{2}x + 3$

$(-2, 2)$

$g(x) = 2$

Check:
$$\frac{1}{2}x + 3 = 2$$

$$\frac{\frac{1}{2}(-2) + 3 \;?\; 2}{\begin{array}{c} -1 + 3 \\ 2 \end{array} \Big|\; 2} \quad \text{TRUE}$$

The solution is -2.

CAUTION!   The x-coordinate of the point of intersection, not the y-coordinate, indicates the solution of the equation.

EXAMPLE 2   Solve graphically: $-\frac{3}{4}x + 6 = 2x - 1$.

Solution   We graph $f(x) = -\frac{3}{4}x + 6$ and $g(x) = 2x - 1$ on the same set of axes. It appears that the lines intersect at $(2.5, 4)$.

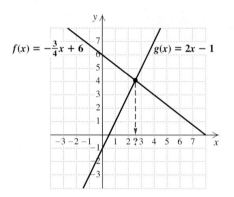

Check:

$$-\frac{3}{4}x + 6 = 2x - 1$$

$$
\begin{array}{c|c}
-\frac{3}{4}(2.5) + 6 \;\; ? & 2(2.5) - 1 \\
-1.875 + 6 & 5 - 1 \\
4.125 & 4 \qquad\qquad \text{FALSE}
\end{array}
$$

Our check shows that 2.5 is *not* the solution, although it may not be off by much. To find the exact solution, we need either a more precise way of determining the coordinates of the point of intersection or a different approach altogether.

CAUTION!   When using a hand-drawn graph to solve an equation, it is important to use graph paper and to work as neatly as possible. Use a straightedge when drawing lines and be sure to erase any mistakes completely.

Point of Intersection

A graphing calculator can be used to determine the point of intersection of graphs. Often it provides a more accurate solution than graphs drawn by hand. We can trace along the graph of one of the functions to find the coordinates of the point of intersection, enlarging the graph using a ZOOM feature if necessary. Most graphing calculators can find the point of intersection directly using an INTERSECT feature, found in the CALC menu.

To find the point of intersection of two graphs, press 2nd CALC and choose the INTERSECT option. Because more than two equations may be graphed, the questions FIRST CURVE? and SECOND CURVE? are used to identify the graphs with which we are concerned. Position the cursor on each graph, in turn, and press ENTER, using the up and down arrow keys if necessary.

Since graphs may intersect at more than one point, we must visually identify the point of intersection in which we are interested and indicate the general location of that point. Enter a guess either by moving the cursor near the point of intersection and pressing ENTER or by typing a guess and pressing ENTER. The calculator then returns the coordinates of the point of intersection.

EXAMPLE 3   Solve using a graphing calculator: $-\frac{3}{4}x + 6 = 2x - 1$.

Solution   In Example 2, we saw that the graphs of $f(x) = -\frac{3}{4}x + 6$ and $g(x) = 2x - 1$ intersect near the point $(2.5, 4)$. To determine more precisely the point of intersection, we graph $y_1 = -\frac{3}{4}x + 6$ and $y_2 = 2x - 1$ using the same viewing window. After choosing the INTERSECT feature, we indicate which two graphs, or *curves*, we are considering. Then we enter a guess. The coordinates of the point of intersection then appear at the bottom of the screen.

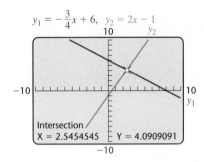

It appears from the screen above that the solution is 2.5454545. To check, we evaluate both sides of the equation $-\frac{3}{4}x + 6 = 2x - 1$ for this value of x. The first coordinate of the point of intersection is stored as X, so we evaluate Y₁(X) and Y₂(X) from the home screen, as shown on the left below.

Although the check shows that 2.5454545 is the solution, it is actually an approximation of the solution. In some cases, the calculator can give us the exact answer. Since the x-coordinate of the point of intersection is stored as X, converting X to fraction notation will give an exact solution. From the screen on the right above, we see that $\frac{28}{11}$ is the solution of the equation.

In many cases, we cannot find an exact solution using a graphing calculator. Instead, we need an algebraic approach.

EXAMPLE 4   Cost Projections. Cleartone Communications charges $50 for a cellular phone and $40 per month for calls made under its Call Anywhere plan. Formulate and graph a mathematical model for the cost. Then use the model to estimate the time required for the total cost to reach $250.

Solution

1. **Familiarize.**   The problem describes a situation in which a monthly fee is charged after an initial purchase has been made. After 1 month of service, the total cost will be $50 + $40 = $90. After 2 months, the total cost will be $50 + $40 · 2 = $130. This can be generalized in a model if we let $C(t)$ represent the total cost, in dollars, for t months of service.

2. **Translate.**   We reword and translate as follows:

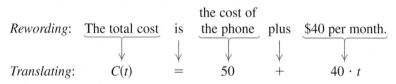

where $t \geq 0$ (since there cannot be a negative number of months).

3. **Carry out.**   To estimate the time required for the total cost to reach $250, we are estimating the solution of

$$40t + 50 = 250. \quad \textbf{Replacing } C(t) \textbf{ with 250}$$

We do this by graphing $C(t) = 40t + 50$ and $y = 250$ and looking for the point of intersection. On a graphing calculator, we let $y_1 = 40x + 50$ and $y_2 = 250$, and adjust the window dimensions to include the point of intersection. Using either a hand-drawn graph or one generated by a graphing calculator, we find a point of intersection at $(5, 250)$.

Thus we estimate that it takes 5 months for the total cost to reach $250.

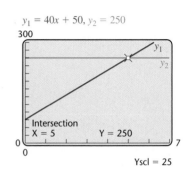

4. Check.   We evaluate:

$$C(5) = 40 \cdot 5 + 50$$
$$= 200 + 50$$
$$= 250.$$

Our estimate turns out to be precise.

5. State.   It takes 5 months for the total cost to reach $250.

Equivalent Equations

Algebraic methods of solving equations use principles of algebra to write *equivalent equations* from which solutions can be more readily found.

> ### Equivalent Equations
> Two equations are *equivalent* if they have the same solution(s).

Consider the equations $2x = 6$ and $10x = 30$. The equation $2x = 6$ is true only when x is 3. Similarly, $10x = 30$ is true only when x is 3. Since both equations have the same solution, they are equivalent.

The equations $x + 4 = 9$ and $x = 5$ are also equivalent. Each equation has only one solution, the number 5. Thus the equations are equivalent.

Now consider the equations $3x = 4x$ and $3/x = 4/x$. When x is replaced with 0, neither $3/x$ nor $4/x$ is defined, so 0 is *not* a solution of $3/x = 4/x$. Since 0 *is* a solution of $3x = 4x$, the equations are not equivalent.

The Addition and Multiplication Principles

Suppose that a and b represent the same number and that some number c is added to a. If c is also added to b, we will get two equal sums, since a and b are the same number. The same is true if we multiply both a and b by c. In this manner, we can produce equivalent equations.

> ## The Addition and Multiplication Principles for Equations
>
> For any real numbers a, b, and c:
>
> **a)** $a = b$ is equivalent to $a + c = b + c$;
>
> **b)** $a = b$ is equivalent to $a \cdot c = b \cdot c$, provided $c \neq 0$.

EXAMPLE 5   Solve: $y - 4.7 = 13.9$.

Solution

$$y - 4.7 = 13.9$$

$$y - 4.7 + 4.7 = 13.9 + 4.7 \qquad \text{Using the addition principle;}$$
$$\text{adding 4.7}$$

$$y + 0 = 13.9 + 4.7 \qquad \text{The law of opposites}$$

$$y = 18.6$$

Check:

$$\frac{y - 4.7 = 13.9}{18.6 - 4.7 \ ? \ 13.9} \qquad \text{Substituting 18.6 for } y$$
$$13.9 \ | \ 13.9 \qquad \text{TRUE}$$

The solution is 18.6. ▬▬▬▬▬

In Example 5, why did we add 4.7 to both sides? Because 4.7 is the opposite of -4.7 and we wanted the variable y alone on one side of the equation. Adding 4.7 gave us $y + 0$, or just y, on the left side. This led to the equivalent equation, $y = 18.6$, from which the solution, 18.6, is immediately apparent.

EXAMPLE 6   Solve: $\frac{2}{5}x = \frac{9}{10}$.

Solution   We have

$$\frac{2}{5}x = \frac{9}{10}$$

$$\frac{5}{2} \cdot \frac{2}{5}x = \frac{5}{2} \cdot \frac{9}{10} \qquad \text{Using the multiplication principle, we multiply}$$
$$\text{both sides by } \tfrac{5}{2}, \text{ the reciprocal of } \tfrac{2}{5}.$$

$$1x = \frac{45}{20} \qquad \text{The law of reciprocals}$$

$$x = \frac{9}{4} \qquad \text{Simplifying}$$

The check is left to the student. The solution is $\frac{9}{4}$. ▬▬▬▬▬

In Example 6, why did we multiply by $\frac{5}{2}$? Because $\frac{5}{2}$ is the reciprocal of $\frac{2}{5}$ and we wanted x alone on one side of the equation. When we multiplied by $\frac{5}{2}$, we got $1x$, or just x, on the left side. This led to the equivalent equation $x = \frac{9}{4}$, from which the solution, $\frac{9}{4}$, is clear.

There is no need for a subtraction or division principle since subtraction can be regarded as adding opposites and division can be regarded as multiplying by reciprocals.

The addition and multiplication principles can be used together to solve equations. We combine all the terms containing the variable on one side of the equation and then use the multiplication principle.

EXAMPLE 7   Solve: $-\frac{3}{4}x + 6 = 2x - 1$.

Solution

$$-\frac{3}{4}x + 6 = 2x - 1$$

$$-\frac{3}{4}x + 6 - 6 = 2x - 1 - 6$$ Using the addition principle: adding the opposite of 6, which is -6, to both sides or subtracting 6 from both sides

$$-\frac{3}{4}x = 2x - 7$$ **Simplifying**

$$-\frac{3}{4}x - 2x = 2x - 7 - 2x$$ Using the addition principle: subtracting $2x$ from both sides

$$-\frac{3}{4}x - \frac{8}{4}x = -7$$ **Simplifying. All the terms containing x are on the left side of the equation.**

$$-\frac{11}{4}x = -7$$ **Combining like terms**

$$-\frac{4}{11}\left(-\frac{11}{4}x\right) = -\frac{4}{11}(-7)$$ Using the multiplication principle: multiplying both sides by $-\frac{4}{11}$, the reciprocal of $-\frac{11}{4}$

$$x = \frac{28}{11}$$ **The law of reciprocals; simplifying**

Check:

$$\frac{-\frac{3}{4}x + 6 = 2x - 1}{-\frac{3}{4}\left(\frac{28}{11}\right) + 6 \;?\; 2\left(\frac{28}{11}\right) - 1}$$

$$-\frac{21}{11} + \frac{66}{11} \;\Big|\; \frac{56}{11} - \frac{11}{11}$$

$$\frac{45}{11} \;\Big|\; \frac{45}{11}$$ TRUE

The solution is $\frac{28}{11}$.

Compare the solutions of Examples 2, 3, and 7. The algebraic approach gave the exact solution, the graphical solution found by hand was an approximation, and the graphical solution found using a calculator was an approximation that we were able to convert to an exact solution. One method is not always "better" than another. Often in applications, an approximate solution is sufficient. Also, graphical methods provide a visualization of the problem that algebraic methods do not.

Connecting the Concepts

It is important to distinguish between forming *equivalent expressions* and writing *equivalent equations*.

In Chapter 1, we used the commutative, associative, and distributive laws to write equivalent expressions that take on the same value when the variables are replaced with numbers. This is how we "simplify" an expression.

In Examples 5–7 of this section, we used the addition and multiplication principles to write a series of equivalent equations that led to an equation for which the solution is clear. Because the equations were equivalent, the solution of the last equation was also a solution of the original equation.

Equivalent Equations	Equivalent Expressions
$y - 4.7 = 13.9$	$3x + 2[4 + 5(x + 2y)]$
$y - 4.7 + 4.7 = 13.9 + 4.7$	$= 3x + 2[4 + 5x + 10y]$
$y + 0 = 13.9 + 4.7$	$= 3x + 8 + 10x + 20y$
$y = 18.6$	$= 13x + 8 + 20y$

Often, as in Example 8, which follows, we merge these ideas by forming an equivalent equation by replacing part of an equation with an equivalent expression.

EXAMPLE 8   Solve: $5x - 2(x - 5) = 7x - 2$.

Solution

$$5x - 2(x - 5) = 7x - 2$$

$$5x - 2x + 10 = 7x - 2 \qquad \text{Using the distributive law to write an equivalent expression for } -2(x - 5)$$

$$3x + 10 = 7x - 2 \qquad \text{Combining like terms}$$

$$3x + 10 - 3x = 7x - 2 - 3x \qquad \text{Using the addition principle: adding } -3x, \text{ the opposite of } 3x, \text{ to both sides}$$

$$10 = 4x - 2 \qquad \text{Combining like terms}$$

$$10 + 2 = 4x - 2 + 2 \qquad \text{Using the addition principle}$$

$$12 = 4x \qquad \text{Simplifying}$$

$$\tfrac{1}{4} \cdot 12 = \tfrac{1}{4} \cdot 4x \qquad \text{Using the multiplication principle: multiplying both sides by } \tfrac{1}{4}, \text{ the reciprocal of 4}$$

$$3 = x \qquad \text{The law of reciprocals; simplifying. Note that } x \text{ can be on the right side.}$$

TEACHING TIP

Remind students to check solutions
in the original equation.

Check:

$$\begin{array}{c|c} \multicolumn{2}{c}{5x - 2(x - 5) = 7x - 2} \\ \hline 5 \cdot 3 - 2(3 - 5) \; ? \; 7 \cdot 3 - 2 \\ 15 - 2(-2) \mid 21 - 2 \\ 15 + 4 \mid 19 \\ 19 \mid 19 \qquad \text{TRUE} \end{array}$$

The solution is 3.

Types of Equations

The equations that we have solved in this section are called *linear equations*. A **linear equation** in one variable—say, x—is an equation equivalent to one of the form $ax = b$ with a and b constants and $a \neq 0$.

Don't rush to solve equations in your head. Work neatly, keeping in mind that the number of steps in a solution is less important than producing a simpler, yet equivalent, equation in each step.

Every equation falls into one of three categories. An **identity** is an equation that is true for all replacements that can be used on both sides of the equation (for example, $x + 3 = 2 + x + 1$). A **contradiction** is an equation, like $n + 5 = n + 7$, that is *never* true. A **conditional equation**, like $2x + 5 = 17$, is sometimes true and sometimes false, depending on what the replacement of x is. Most of the equations examined in this text are conditional.

EXAMPLE 9   Solve each of the following equations and classify the equation as an identity, a contradiction, or a conditional equation.

a) $2x + 7 = 7(x + 1) - 5x$

b) $3x - 5 = 3(x - 2) + 4$

c) $3 - 8x = 5 - 7x$

Solution   We solve each equation both algebraically and graphically to compare the methods and to provide a check.

a)

Algebraic Solution

We have

$2x + 7 = 7(x + 1) - 5x$

$2x + 7 = 7x + 7 - 5x$ **Using the distributive law**

$2x + 7 = 2x + 7.$ **Combining like terms**

The equation $2x + 7 = 2x + 7$ is true regardless of the replacement for x, so all real numbers are solutions.

   The equation is an identity.

Graphical Solution

We graph $y_1 = 2x + 7$ and $y_2 = 7(x + 1) - 5x$. The graphs are the same line.

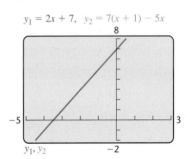

$y_1 = 2x + 7, \quad y_2 = 7(x + 1) - 5x$

We can also see this by looking at the table of values for Y1 and Y2.

X	Y₁	Y₂
0	7	7
1	9	9
2	11	11
3	13	13
4	15	15
5	17	17
6	19	19
X = 0		

Since $y_1 = y_2$ for any choice of x, all real numbers are solutions.

   The equation is an identity.

b)

Algebraic Solution

We have

$$3x - 5 = 3(x - 2) + 4$$

$$3x - 5 = 3x - 6 + 4 \qquad \text{Using the distributive law}$$

$$3x - 5 = 3x - 2$$

$$-3x + 3x - 5 = -3x + 3x - 2 \qquad \text{Using the addition principle}$$

$$-5 = -2.$$

Since our original equation is equivalent to $-5 = -2$, which is false for any choice of x, there is no solution of this equation. There is no choice of x that is a solution of the original equation.

    The equation is a contradiction.

Graphical Solution

We graph $y_1 = 3x - 5$ and $y_2 = 3(x - 2) + 4$.

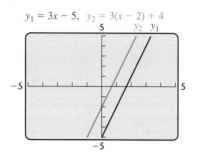

It does not appear as though the lines intersect. If we attempt to find a point of intersection using INTERSECT, we receive an ERROR message. There is no solution.

    The equation is a contradiction.

c)

Algebraic Solution

We have

$$3 - 8x = 5 - 7x$$

$$3 - 8x + 7x = 5 - 7x + 7x \qquad \text{Using the addition principle}$$

$$3 - x = 5 \qquad \text{Simplifying}$$

$$-3 + 3 - x = -3 + 5 \qquad \text{Using the addition principle}$$

$$-x = 2 \qquad \text{Simplifying. The coefficient of } x \text{ is } -1.$$

$$\frac{-x}{-1} = \frac{2}{-1} \qquad \text{Dividing both sides by } -1, \text{ or multiplying both sides by } \frac{1}{-1}, \text{ which is } -1$$

$$x = -2.$$

There is one solution, -2. For other choices of x, the equation is false.

    The equation is conditional since it can be true or false, depending on the replacement for x.

Graphical Solution

We graph $y_1 = 3 - 8x$ and $y_2 = 5 - 7x$ and determine the coordinates of any point of intersection.

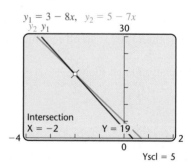

The only solution is -2.

    The equation is conditional.

TEACHING TIP

Some students will want to write
{0} for the empty set. You may want
to explain the difference between
{0} and { }.

We will sometimes refer to the set of solutions, or **solution set**, of a particular equation. Thus the solution set for Example 9(c) is {−2}. The solution set for Example 9(a) is simply ℝ, the set of all real numbers, and the solution set for Example 9(b) is the **empty set**, denoted ∅ or { }. As its name suggests, the empty set is the set containing no elements.

2.2

Exercise Set

FOR EXTRA HELP

Digital Video Tutor CD 2 Videotape 2 · Student's Solutions Manual · Tutor Center AW Math Tutor Center · InterAct Math · MathXL · MyMathLab

Estimate the solution of each equation from the associated graph.

1. $2x - 1 = -5$ −2

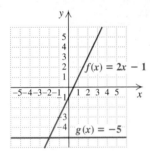

2. $\frac{1}{2}x + 1 = 3$ 4

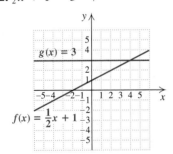

3. $2x + 3 = x - 1$ −4

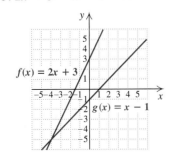

4. $2 - x = 3x - 2$ 1

5. $\frac{1}{2}x + 3 = x - 1$ 8

$y_1 = (1/2)x + 3, \ y_2 = x - 1$

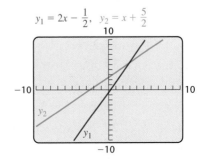

6. $2x - \frac{1}{2} = x + \frac{5}{2}$ 3

$y_1 = 2x - \frac{1}{2}, \ y_2 = x + \frac{5}{2}$

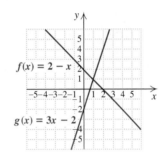

7. $f(x) = g(x)$ 0

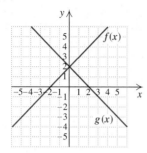

8. $f(x) = g(x)$ 2

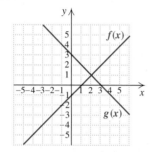

9. Estimate the value of x for which $y_1 = y_2$.   5

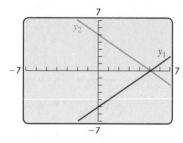

10. Estimate the value of x for which $y_1 = y_2$.   1

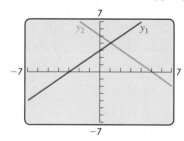

Solve graphically.

11. $x - 3 = 4$ 7

12. $x + 4 = 6$ 2

13. $2x + 1 = 7$ 3

14. $3x - 5 = 1$ 2

15. $\frac{1}{3}x - 2 = 1$ 9

16. $\frac{1}{2}x + 3 = -1$ −8

17. $x + 3 = 5 - x$ 1

18. $x - 7 = 3x - 3$ −2

19. $5 - \frac{1}{2}x = x - 4$ 6

20. $3 - x = \frac{1}{2}x - 3$ 4

21. $2x - 1 = -x + 3$
$1\frac{1}{3}$

22. $-3x + 4 = 3x - 4$
$1\frac{1}{3}$

Use a graph to estimate the solution in each of the following.

23. *Telephone Charges.*   Skytone Calling charges $50 for a telephone and $25 per month under its economy plan. Estimate the time required for the total cost to reach $150.   4 months

24. *Cellular Phone Charges.*   The Cellular Connection charges $60 for a cellular phone and $40 per month under its economy plan. Estimate the time required for the total cost to reach $260.   5 months

25. *Parking Fees.*   Karla's Parking charges $3.00 to park plus 50¢ for each 15-min unit of time. Estimate how long someone can park for $7.50.*
2 hr, 15 min

26. *Cost of a Road Call.*   Dave's Foreign Auto Village charges $35 for a road call plus $10 for each 15-min unit of time. Estimate the time required for a road call that cost $75.   1 hr

27. *Cost of a FedEx Delivery.*   In 2001, for Standard delivery, within 150 mi, of packages weighing from 6 to 16 lb, FedEx charged $18.75 for the first 6 lb plus $0.75 for each additional pound. Estimate the weight of a package that cost $24 to ship.
13 lb

28. *Copying Costs.*   A local Mailboxes Etc.® store charges $2.25 for binding plus 5¢ per page for each spiralbound copy of a town report. Estimate the length of a spiralbound report that cost $3.50.
25 pages

29. Weekly pay at Bikes for Hikes is $219 plus a 3.5% sales commission. If a salesperson's pay was $370.03, what did that salesperson's sales total?
$4315.14

30. It costs Bert's Shirts $38 plus $2.35 a shirt to print tee shirts for a day camp. Camp Weehawken paid Bert's $623.15 for shirts. How many shirts were printed?   249 shirts

---

*More precise, nonlinear models of Exercises 25 and 26 appear in Exercises 103 and 104, respectively.

Determine whether the two equations in each pair are equivalent.

31. $3x = 15$ and $5x = 25$ Yes

32. $6x = 24$ and $15x = 60$ Yes

33. $t + 5 = 11$ and $3t = 18$ Yes

34. $t - 3 = 7$ and $3t = 24$ No

35. $12 - x = 3$ and $2x = 20$ No

36. $3x - 4 = 8$ and $3x = 12$ Yes

37. $5x = 2x$ and $\dfrac{4}{x} = 3$ No

38. $6 = 2x$ and $5 = \dfrac{2}{3 - x}$ No

Solve. Be sure to check.

39. $x - 2.9 = 13.4$ 16.3

40. $y + 4.3 = 11.2$ 6.9

41. $8t = 72$ 9

42. $9t = 63$ 7

43. $4x - 12 = 60$ 18

44. $4x - 6 = 70$ 19

45. $3n + 5 = 29$ 8

46. $7t + 11 = 74$ 9

47. $3x + 5x = 56$ 7

48. $3x + 7x = 120$ 12

49. $9y - 7y = 42$ 21

50. $8t - 3t = 65$ 13

51. $-6y - 10y = -32$ 2

52. $-9y - 5y = 28$ -2

53. $2(x + 6) = 8x$ 2

54. $3(y + 5) = 8y$ 3

55. $70 = 10(3t - 2)$ 3

56. $27 = 9(5y - 2)$ 1

57. $180(n - 2) = 900$ 7

58. $210(x - 3) = 840$ 7

59. $5y - (2y - 10) = 25$ 5

60. $8x - (3x - 5) = 40$ 7

61. $7y - 1 = 23 - 5y$ 2

62. $14t + 20 = 8t - 22$ -7

63. $\frac{1}{5} + \frac{3}{10}x = \frac{4}{5}$ 2

64. $-\frac{5}{2}x + \frac{1}{2} = -18$ $\frac{37}{5}$

65. $0.9y - 0.7 = 4.2$ $\frac{49}{9}$

66. $0.8t - 0.3t = 6.5$ 13

67. $4.23x - 17.898 = -1.65x - 42.454$ -4.17619

68. $-0.00458y + 1.7787 = 13.002y - 1.005$ 0.21402

69. $7r - 2 + 5r = 6r + 6 - 4r$ $\frac{4}{5}$

70. $9m - 15 - 2m = 6m - 1 - m$ 7

71. $\frac{1}{4}(16y + 8) - 17 = -\frac{1}{2}(8y - 16)$ $\frac{23}{8}$

72. $\frac{1}{3}(6t + 48) - 20 = -\frac{1}{4}(12t - 72)$ $\frac{22}{5}$

73. $5 + 2(x - 3) = 2[5 - 4(x + 2)]$ $-\frac{1}{2}$

74. $3[2 - 4(x - 1)] = 3 - 4(x + 2)$ $\frac{23}{8}$

Find each solution set. Then classify each equation as a conditional equation, an identity, or a contradiction.

75. $5x + 7 - 3x = 2x$ $\varnothing$; contradiction

76. $3t + 5 + t = 5 + 4t$ $\mathbb{R}$; identity

77. $1 + 9x = 3(4x + 1) - 2$ $\{0\}$; conditional

78. $4 + 7x = 7(x + 1)$ $\varnothing$; contradiction

Aha! **79.** $-9t + 2 = -9t - 7(6 \div 2(49) + 8)$
$\varnothing$; contradiction

80. $-8t + 2 = 2 - 8t - 5(8 \div 4(1 + 3^4))$
$\varnothing$; contradiction

81. $2\{9 - 3[-2x - 4]\} = 12x + 42$ $\mathbb{R}$; identity

82. $3\{7 - 2[7x - 4]\} = -40x + 45$ $\{0\}$; conditional

TW **83.** Explain the difference between equivalent expressions and equivalent equations.

TW **84.** When an equation is solved graphically, why is it important that the solution be checked algebraically?

Skill Maintenance

85. Write the set consisting of the positive integers less than 10, using both roster notation and set-builder notation. [1.1] $\{1, 2, 3, 4, 5, 6, 7, 8, 9\}$; $\{x \,|\, x$ is a positive integer less than 10$\}$

86. Write the set consisting of the negative integers greater than -9, using both roster notation and set-builder notation. [1.1] $\{-8, -7, -6, -5, -4, -3, -2, -1\}$; $\{x \,|\, x$ is a negative integer greater than $-9\}$

Translate to an algebraic expression. [1.6]

87. Three less than the product of two numbers
Let m and n represent the numbers; $mn - 3$

88. Ten more than the difference of two numbers   Let x and y represent the numbers; $(x - y) + 10$, or $10 + (x - y)$

89. Nine more than twice a number
Let n represent the number; $9 + 2n$, or $2n + 9$

90. Forty-two percent of half of a number
Let t represent the number; $0.42\left(\frac{t}{2}\right)$

Synthesis

TW **91.** Karl solves the equation $x + 5 = 2x + 8$ graphically and states that the solution is $(-3, 2)$. Is this correct? Why or why not?

TW **92.** As the first step in solving
$$2x + 5 = -3,$$
a student multiplies both sides by $\frac{1}{2}$. Is this incorrect? Why or why not?

Solve and check.

93. $4x - \{3x - [2x - (5x - (7x - 1))]\} = 4x + 7$ 8

94. $3x - \{5x - [7x - (4x - (3x + 1))]\} = 3x + 5$ 4

95. $17 - 3\{5 + 2[x - 2]\} + 4\{x - 3(x + 7)\}$
$= 9\{x + 3[2 + 3(4 - x)]\}$ $\frac{224}{29}$

96. $23 - 2\{4 + 3[x - 1]\} + 5\{x - 2(x + 3)\}$
$= 7\{x - 2[5 - (2x + 3)]\}$ $\frac{19}{46}$

Solve graphically. Be sure to check.

1.6666667, 3; or $\frac{5}{3}$, 3

97. $2x = |x + 1|$ 1

98. $x - 1 = |4 - 2x|$

99. $\frac{1}{2}x = 3 - |x|$ −6, 2

100. $2 - |x| = 1 - 3x$

−0.25, or −$\frac{1}{4}$

101. $x^2 = x + 2$ −1, 2

102. $x^2 = x$ 0, 1

103. (Refer to Exercise 25.) It costs as much to park at Karla's for 16 min as it does for 29 min. Thus the linear graph drawn in the solution of Exercise 25 is not a precise representation of the situation. Draw a graph with a series of "steps" that more accurately reflects the situation.

103.

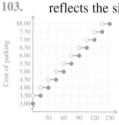

104. (Refer to Exercise 26.) A 32-min road call with Dave's costs the same as a 44-min road call. Thus the linear graph drawn in the solution of Exercise 26 is not a precise representation of the situation. Draw a graph with a series of "steps" that more accurately reflects the situation.

TW **105.** Create an equation for which it is preferable to use the multiplication principle *before* using the addition principle. Explain why it is best to solve the equation in this manner.

Problem Solving ■ Formulas

Applications and Formulas

104.

In Chapter 1, we introduced the five-step problem-solving approach used in this text. In Section 1.6, we discussed the first two steps: *familiarizing* ourselves with the problem situation and *translating* the problem situation to mathematical language. We now use our equation-solving skills to continue with the last three steps of the process.

Five Steps for Problem Solving with Algebra

1. *Familiarize* yourself with the problem.
2. *Translate* to mathematical language.
3. *Carry out* some mathematical manipulation.
4. *Check* your possible answer in the original problem.
5. *State* the answer clearly.

Problem Solving

At this point, our study of algebra has just begun. Thus we have few algebraic tools with which to work problems. As the number of tools in our algebraic "toolbox" increases, so will the difficulty of the problems we can solve. For now our problems may not seem difficult; however, to gain practice with the problem-solving process, you should try to use all five steps. Later some steps may be shortened or combined.

EXAMPLE 1 Purchasing. Elka pays $1187.20 for a computer. If the price paid includes a 6% sales tax, what is the price of the computer itself?

Solution

1. Familiarize. First, we familiarize ourselves with the problem. Note that tax is calculated from, and then added to, the computer's price. Let's guess that the computer's price is $1000. To check the guess, we calculate the amount of tax, $(0.06)($1000) = 60, and add it to $1000:

$$\$1000 + (0.06)(\$1000) = \$1000 + \$60$$
$$= \$1060. \quad \textbf{\$1060} \neq \textbf{\$1187.20}$$

Our guess was too low, but the manner in which we checked the guess will guide us in the next step. We let

$C =$ the computer's price, in dollars.

2. **Translate.**   Our guess leads us to the following translation:

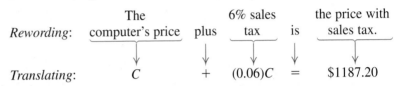

Rewording: The computer's price   plus   6% sales tax   is   the price with sales tax.

Translating: C $+$ $(0.06)C$ $=$ $\$1187.20$

3. **Carry out.**   Next, we carry out some mathematical manipulation:

$$C + (0.06)C = 1187.20$$
$$1.06C = 1187.20 \qquad \textbf{Combining like terms}$$
$$\frac{1}{1.06} \cdot 1.06C = \frac{1}{1.06} \cdot 1187.20 \qquad \textbf{Using the multiplication principle}$$
$$C = 1120.$$

4. **Check.**   To check the answer in the original problem, note that the tax on a computer costing $1120 would be $(0.06)($1120) = 67.20. When this is added to $1120, we have

$$\$1120 + \$67.20, \text{ or } \$1187.20.$$

We can also check by solving the equation graphically, as shown at left. Thus, $1120 checks in the original problem.

5. **State.**   We clearly state the answer: The computer itself costs $1120.

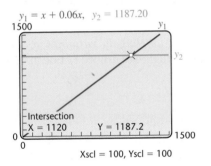

$y_1 = x + 0.06x, \quad y_2 = 1187.20$

EXAMPLE 2   Home Maintenance.   In an effort to make their home more energy-efficient, Alma and Drew purchased 200 in. of 3M Press-In-Place window glazing. This will be just enough to outline their two square skylights. If the length of the sides of the larger skylight is $1\frac{1}{2}$ times the length of the sides of the smaller one, how should the glazing be cut?

Solution

1. **Familiarize.**   Note that the *perimeter* of (distance around) each square is four times the length of a side. Furthermore, if s is used to represent the length of a side of the smaller square, then $\left(1\frac{1}{2}\right)s$ will represent the length of a side of the larger square. We make a drawing (see the following page) and note that the two perimeters must add up to 200 in.

   Perimeter of a square $= 4 \cdot$ length of a side

2. Translate.   Rewording the problem can help us translate:

	The perimeter of one square	plus	the perimeter of the other	is	200 in.
Rewording:	↓	↓	↓	↓	↓
Translating:	$4s$	$+$	$4\left(1\frac{1}{2}s\right)$	$=$	200

3. Carry out.   We solve the equation:

$$4s + 4\left(1\tfrac{1}{2}s\right) = 200$$

$$4s + 6s = 200 \qquad \text{Simplifying}$$

$$10s = 200 \qquad \text{Combining like terms}$$

$$s = \frac{1}{10} \cdot 200 \qquad \text{Multiplying both sides by } \tfrac{1}{10}$$

$$s = 20. \qquad \text{Simplifying}$$

4. Check.   If 20 is the length of the smaller side, then $\left(1\frac{1}{2}\right)(20) = 30$ is the length of the larger side. The two perimeters would then be

$$4 \cdot 20 \text{ in.} = 80 \text{ in.} \quad \text{and} \quad 4 \cdot 30 \text{ in.} = 120 \text{ in.}$$

Since 80 in. + 120 in. = 200 in., our answer checks.

5. State.   The glazing should be cut into two pieces, one 80 in. long and the other 120 in. long.

We cannot stress too much the importance of labeling the variables in your problem. In Example 2, solving for s was not enough: We needed to find $4s$ and $4\left(1\frac{1}{2}s\right)$ to determine the numbers we were after.

Formulas

A *formula* is an equation that uses letters to represent a relationship between two or more quantities. Some important geometric formulas are $A = \pi r^2$ (for the area A of a circle of radius r), $C = \pi d$ (for the circumference C of a circle of diameter d), and $A = b \cdot h$ (for the area A of a parallelogram of height h and base length b).* A more complete list of geometric formulas appears on the inside back cover.

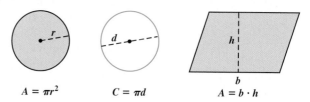

$$A = \pi r^2 \qquad C = \pi d \qquad A = b \cdot h$$

Sometimes problem solving will require the use of scientific notation.

---

*The Greek letter π, read "pi," is *approximately* 3.14159265358979323846264. Often 3.14 or 22/7 is used to approximate π when a calculator with a π key is unavailable.

EXAMPLE 3 Telecommunications. A fiber-optic wire will be used for 375 km of transmission line. The wire has a diameter of 1.2 cm. What is the volume of wire needed for the line?

Solution

1. Familiarize. Making a drawing, we see that we have a cylinder (a very *long* one). Its length is 375 km and the circular base has a diameter of 1.2 cm.

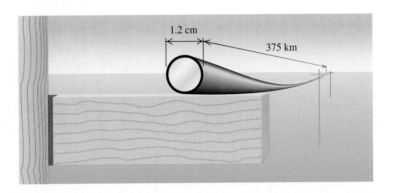

Recall that the formula for the volume of a cylinder is

$$V = \pi r^2 h,$$

where r is the radius and h is the height (in this case, the length of the wire).

2. Translate. We will use the volume formula, but it is important to make the units consistent. Let's express everything in meters:

Length: 375 km = 375,000 m,   or 3.75×10^5 m;
Diameter: 1.2 cm = 0.012 m,   or 1.2×10^{-2} m.

The radius, which we will need in the formula, is half the diameter:

Radius: 0.6×10^{-2} m,   or 6×10^{-3} m.

We now substitute into the above formula:

$$V = \pi(6 \times 10^{-3})^2(3.75 \times 10^5).$$

3. Carry out. We do the calculations:

$$V = \pi \times (6 \times 10^{-3})^2(3.75 \times 10^5)$$
$$= \pi \times 6^2 \times 10^{-6} \times 3.75 \times 10^5 \qquad \textbf{Using the properties of exponents}$$
$$= (\pi \times 6^2 \times 3.75) \times (10^{-6} \times 10^5)$$
$$= 135\pi \times 10^{-1} \qquad \textbf{Multiplying}$$
$$\approx 42.41. \qquad \textbf{Using the } \boxed{\pi} \textbf{ key on a calculator and rounding}$$

4. Check. We recheck the translation and calculations. The answer checks.

5. State. The volume of the wire is about 42.41 m³ (cubic meters).

TEACHING TIP

If students struggle with solving formulas, you may wish to substitute numbers first. For example, in $A = lw$, if the area is 200 ft^2 and the width is 10 ft, what is the length?

Suppose we remember the area and the width of a rectangular room and want to find the length. To do so, we could "solve" the formula $A = l \cdot w$ (Area = Length · Width) for l, using the same principles that we use for solving equations.

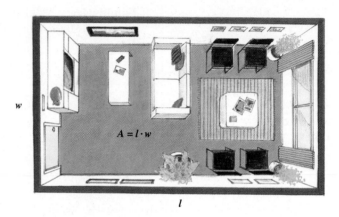

w

$A = l \cdot w$

l

EXAMPLE 4   **Area of a Rectangle.**   Solve the formula $A = l \cdot w$ for l.

Solution

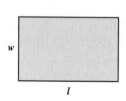

w

l

$$A = l \cdot w \qquad \text{We want this letter alone.}$$

$$\frac{A}{w} = \frac{l \cdot w}{w} \qquad \begin{array}{l}\text{Dividing both sides by } w,\\ \text{or multiplying both sides}\\ \text{by } 1/w\end{array}$$

$$\frac{A}{w} = l \cdot \frac{w}{w}$$

$$\frac{A}{w} = l \qquad \begin{array}{l}\text{Simplifying by removing a}\\ \text{factor equal to 1: } \dfrac{w}{w} = 1\end{array}$$

Thus to find the length of a rectangular room, we can divide the area of the room by its width. Were we to do this calculation for a variety of rectangular rooms, the formula $l = A/w$ would be more convenient than repeatedly substituting into $A = l \cdot w$ and solving for l each time.

EXAMPLE 5   **Simple Interest.**   The formula $I = Prt$ is used to determine the simple interest I earned when a principal of P dollars is invested for t years at an interest rate r. Solve this formula for t.

Solution

$$I = Prt \qquad \text{We want this letter alone.}$$

$$\frac{I}{Pr} = \frac{Prt}{Pr} \qquad \text{Dividing both sides by } Pr, \text{ or multiplying both sides by } \frac{1}{Pr}$$

$$\frac{I}{Pr} = t \qquad \text{Simplifying by removing a factor equal to 1: } \frac{Pr}{Pr} = 1$$

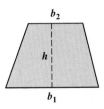

EXAMPLE 6   Area of a Trapezoid.   A trapezoid is a geometric shape with four sides, exactly two of which, the bases, are parallel to each other. The formula for calculating the area A of a trapezoid with bases b_1 and b_2 (read "b sub one" and "b sub two") and height h is given by

$$A = \frac{h}{2}(b_1 + b_2),$$

A derivation of this formula is outlined in Exercise 107 of this section.

where the *subscripts* 1 and 2 distinguish one base from the other. Solve for b_1.

Solution   There are several ways to "remove" the parentheses. We could distribute $h/2$, but an easier approach is to multiply both sides by the reciprocal of $h/2$.

$$A = \frac{h}{2}(b_1 + b_2)$$

$$\frac{2}{h} \cdot A = \frac{2}{h} \cdot \frac{h}{2}(b_1 + b_2) \qquad \text{Multiplying both sides by } \frac{2}{h}$$

$$\left(\text{or dividing by } \frac{h}{2} \right)$$

$$\frac{2A}{h} = b_1 + b_2 \qquad \text{Simplifying. The right side is "cleared" of fractions.}$$

$$\frac{2A}{h} - b_2 = b_1 \qquad \text{Adding } -b_2 \text{ to both sides}$$

The similarities between solving formulas and solving equations can be seen below. In (a), we solve as we did before; in (b), we do not carry out all calculations; and in (c), we cannot carry out all calculations because the numbers are unknown. The same steps are used each time.

a)
$$9 = \frac{3}{2}(x + 5)$$
$$\frac{2}{3} \cdot 9 = \frac{2}{3} \cdot \frac{3}{2}(x + 5)$$
$$6 = x + 5$$
$$1 = x$$

b)
$$9 = \frac{3}{2}(x + 5)$$
$$\frac{2}{3} \cdot 9 = \frac{2}{3} \cdot \frac{3}{2}(x + 5)$$
$$\frac{2 \cdot 9}{3} = x + 5$$
$$\frac{2 \cdot 9}{3} - 5 = x$$

c)
$$A = \frac{h}{2}(b_1 + b_2)$$
$$\frac{2}{h} \cdot A = \frac{2}{h} \cdot \frac{h}{2}(b_1 + b_2)$$
$$\frac{2A}{h} = b_1 + b_2$$
$$\frac{2A}{h} - b_2 = b_1$$

EXAMPLE 7   Simple Interest.   The formula $A = P + Prt$ gives the amount A that a principal of P dollars will be worth in t years when invested at simple interest rate r. Solve the formula for P.

Solution

$$A = P + Prt \qquad \text{We want this letter alone.}$$

$$A = P(1 + rt) \qquad \begin{array}{l}\text{Factoring (using the distributive law)}\\ \text{to combine like terms}\end{array}$$

$$\frac{A}{1 + rt} = \frac{P(1 + rt)}{1 + rt} \qquad \begin{array}{l}\text{Dividing both sides by } 1 + rt, \text{ or}\\[4pt] \text{multiplying both sides by } \dfrac{1}{1 + rt}\end{array}$$

$$\frac{A}{1 + rt} = P \qquad \text{Simplifying}$$

This last equation can be used to determine how much should be invested at interest rate r in order to have A dollars t years later.

Note in Example 7 that the factoring enabled us to write P once rather than twice. This is comparable to combining like terms when solving an equation like $16 = x + 7x$.

Most graphing calculators graph only functions and thus require us to solve for the dependent variable. An equation written in another form must be solved for y before it can be entered into a graphing calculator.

EXAMPLE 8   Graph: $3x - 4y = 2y + 7$.

Solution   In order to graph $3x - 4y = 2y + 7$ using a graphing calculator, we must first solve for y:

$$3x - 4y = 2y + 7 \qquad \text{We want this letter alone.}$$

$$3x - 4y - 2y = 7 \qquad \text{Adding } -2y \text{ to both sides}$$

$$3x - 6y = 7 \qquad \text{Combining like terms}$$

$$-6y = -3x + 7 \qquad \text{Adding } -3x \text{ to both sides}$$

$$y = \frac{-3x + 7}{-6}. \qquad \text{Dividing both sides by } -6$$

Now we can enter the equation and graph it, as shown at left. Since $y = (-3x + 7)/(-6)$ is equivalent to $3x - 4y = 2y + 7$, their graphs are the same.

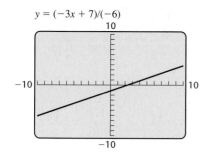

$y = (-3x + 7)/(-6)$

You may find the following summary useful.

To Solve a Formula for a Specified Letter

1. Get all terms with the letter being solved for on one side of the equation and all other terms on the other side, using the addition principle. To do this may require removing parentheses.
   - To remove parentheses, either divide both sides by the multiplier in front of the parentheses or use the distributive law.
2. When all terms with the specified letter are on the same side, factor (if necessary) to combine like terms.
3. Solve for the letter in question by dividing both sides by the multiplier of that letter.

2.3

Exercise Set

FOR EXTRA HELP

 Digital Video Tutor CD 2 Videotape 2   Student's Solutions Manual   Tutor Center AW Math Tutor Center   InterAct Math   Math XP MathXL   MyMathLab

Solve using all five problem-solving steps.

1. The sum of two numbers is 65. One of the numbers is 7 more than the other. What are the numbers? 29, 36

2. The sum of two numbers is 83. One of the numbers is 11 more than the other. What are the numbers? 36, 47

3. The product of two numbers is 72. If one number is 48, find the other number. $\frac{3}{2}$

4. The number 596 is $\frac{2}{3}$ of what number?   894

5. The length of a rectangle is twice its width and its perimeter is 15 ft. Find the dimensions of the rectangle.   Length: 5 ft; width: $\frac{5}{2}$ ft

6. The width of a rectangle is $\frac{1}{4}$ its length and its perimeter is 50 m. Find the dimensions of the rectangle.   Length: 20 m; width: 5 m

7. *Boating.*   The Delta Queen is a paddleboat that tours the Mississippi River near New Orleans, Louisiana. It is not uncommon for the Delta Queen to run 7 mph in still water and for the Mississippi to flow at a rate of 3 mph (*Source*: Delta Queen information). At these rates, how long will it take the boat to cruise 2 mi upstream? $\frac{1}{2}$ hr

8. *Swimming.*   Fran swims at a rate of 5 km/h in still water. The Lazy River flows at a rate of 2.3 km/h. How long will it take Fran to swim 1.8 km upstream? $\frac{2}{3}$ hr

9. *Moving Sidewalks.*   The moving sidewalk in O'Hare Airport is 300 ft long and moves at a rate of 5 ft/sec. If Alida walks at a rate of 4 ft/sec, how long will it take her to walk the length of the moving sidewalk? $\frac{100}{3}$ sec

4 ft/sec

5 ft/sec

10. *Aviation.*   A Cessna airplane traveling 390 km/h in still air encounters a 65-km/h headwind. How long will it take the plane to travel 725 km into the wind? $\frac{29}{13}$ hr

11. **Angle Measures.**   Two angles in a triangle have the same measure. The third angle is twice as great as the others. Find the measures of the angles.

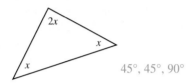

    45°, 45°, 90°

12. *Angle Measures.*   One angle of a triangle is one-half as great as the second angle. The third angle measures 5° more than twice the second angle. Find the measures of the angles.   25°, 50°, 105°

13. *Power Usage.*   On a hot day, Twin City Electric set a record for power usage of 4998 megawatts of electricity. This was 2% higher than the previous record. What was the previous power usage record? 4900 megawatts

14. *Record Precipitation.*   One rainy May, Jamesport received 8.17 in. of rain. This was 5% less than the record. What was the record amount of rain for May? 8.6 in.

15. Kim and Kelly bought a $\frac{1}{2}$-lb slab of fudge. They split it so that Kim got $\frac{2}{3}$ as much fudge as Kelly. How much fudge did Kim get? $\frac{1}{5}$ lb

16. *Tank Capacity.*   On Monday, Karen filled the gas tank when it was $\frac{3}{4}$ empty. On Friday, she filled it when it was $\frac{1}{2}$ empty. Her total purchase of gasoline on both days was $18\frac{3}{4}$ gal. How many gallons of gasoline does the tank hold? 15 gal

17. **Herbal Supplements.**   The sales of herbal supplements in 2002 were $4.2 million. This was a 110% increase over sales of herbal supplements in 1997. (*Source*: FIND/SVP) What was the amount of sales of herbal supplements in 1997? $2 million

18. *Government Workers.*   The number of government employees in Marion County, Indiana, decreased 15% from 1990 to 2000 (*Source*: *Indianapolis Star* 5/16/02). If 45,900 were employed in 2000, how many were employed in 1990? 54,000 employees

19. *Photography.*   A photo shop charges $6.30 to develop a roll of film and $0.22 per exposure to print it. LaKenya is charged $14.66 to develop and print a roll of film. How many prints did she receive? 38 prints

20. *Water Usage.*   The Rosado family has budgeted $15.00 per month for home water usage. The monthly service charge is $7.20 and the volume charge is $1.04 for each 100 cubic feet of water. What is the most water the Rosados can use and stay within their budget? 750 cubic feet

21. *Interest.*   Appliances Now allows customers to delay payment on a purchase for one year. It charges 12% simple interest on the purchase price. Tom bought a compact disc player and agreed to pay $504 a year later. What was the price of the compact disc player? $450

22. *Discount.*   Elridge Furniture discounts furniture 8% to customers paying cash. Jinney paid $1007.40 cash for a roll-top desk. What was the original price of the desk? $1095

Write the answers for Exercises 23–32 using scientific notation.

23. *Astronomy.*   Venus has a nearly circular orbit of the sun. If the average distance from the sun to Venus is 1.08×10^8 km, how far does Venus travel in one orbit? 6.79×10^8 km

24. *Office Supplies.*   A ream of copier paper weighs 2.25 kg. How much does a sheet of copier paper weigh? 4.5 g

25. *Printing and Engraving.*   A ton of $5 bills is worth $4,540,000. How many pounds does a $5 bill weigh? 2.2×10^{-3} lb

26. *Astronomy.*   The average distance of Earth from the sun is about 9.3×10^7 mi. About how far does Earth travel in a yearly orbit about the sun? (Assume a circular orbit.) 5.84×10^8 mi

27. *Astronomy.*   The brightest star in the night sky, Sirius, is about 4.704×10^{13} mi from Earth. The distance that light travels in one year, or one light year, is 5.88×10^{12} mi. How many light years is it from Earth to Sirius?   8 light years

28. *Astronomy.*   The diameter of the Milky Way galaxy is approximately 5.88×10^{17} mi. The distance that light travels in one year, or one light year, is 5.88×10^{12} mi. How many light years is it from one end of the galaxy to the other? 1×10^5 light years

29. *Biology.*   An average of 4.55×10^{11} bacteria live in each pound of U.S. mud (*Source*: *Harper's Magazine*, April 1996, p. 13). There are 60 drops in one teaspoon and 6 teaspoons in an ounce. How many bacteria live in a drop of U.S. mud? 7.90×10^7 bacteria

30. *Astronomy.*   If a star 5.9×10^{14} mi from Earth were to explode today, its light would not reach us for 100 yr. How far does light travel in 13 weeks? 1.475×10^{12} mi

31. *Astronomy.*   The diameter of Jupiter is about 1.43×10^5 km. A day on Jupiter lasts about 10 hr. At what speed is Jupiter's equator spinning? 4.49×10^4 km/h

32. *Finance.*   A *mil* is one thousandth of a dollar. The taxation rate in a certain school district is 5.0 mils for every dollar of assessed valuation. The assessed valuation for the district is 13.4 million dollars. How much tax revenue will be raised? $\$6.7 \times 10^4$

Solve.

33. $d = rt$, for r (a distance formula) $r = \dfrac{d}{t}$

34. $d = rt$, for t $t = \dfrac{d}{r}$

35. $F = ma$, for a (a physics formula) $a = \dfrac{F}{m}$

36. $A = lw$, for w (an area formula) $w = \dfrac{A}{l}$

37. $W = EI$, for I (an electricity formula) $I = \dfrac{W}{E}$

38. $W = EI$, for E $E = \dfrac{W}{I}$

39. $V = lwh$, for h (a volume formula) $h = \dfrac{V}{lw}$

40. $I = Prt$, for r (a formula for interest) $r = \dfrac{I}{Pt}$

41. $L = \dfrac{k}{d^2}$, for k
(a formula for intensity of sound or light) $k = Ld^2$

42. $F = \dfrac{mv^2}{r}$, for m (a physics formula) $m = \dfrac{Fr}{v^2}$

43. $G = w + 150n$, for n
(a formula for the gross weight of a bus) $n = \dfrac{G - w}{150}$

44. $P = b + 0.5t$, for t (a formula for parking prices) ⊡

45. $2w + 2h + l = p$, for l $l = p - 2w - 2h$
(a formula used when shipping boxes)

46. $2w + 2h + l = p$, for w $w = \dfrac{p - 2h - l}{2}$

47. $Ax + By = C$, for y (a formula for graphing lines) ⊡

48. $P = 2l + 2w$, for l (a perimeter formula) ⊡

49. $C = \frac{5}{9}(F - 32)$, for F (a temperature formula) ⊡

50. $T = \frac{3}{10}(I - 12{,}000)$, for I (a tax formula) ⊡

51. $A = \dfrac{h}{2}(b_1 + b_2)$, for b_2 (an area formula) $b_2 = \dfrac{2A}{h} - b_1$

52. $A = \dfrac{h}{2}(b_1 + b_2)$, for h (an area formula) $h = \dfrac{2A}{b_1 + b_2}$

53. $v = \dfrac{d_2 - d_1}{t}$, for t (a physics formula) $t = \dfrac{d_2 - d_1}{v}$

(*Hint*: Multiply by t to "clear" fractions.)

54. $v = \dfrac{s_2 - s_1}{m}$, for m $m = \dfrac{s_2 - s_1}{v}$

55. $v = \dfrac{d_2 - d_1}{t}$, for d_1 $d_1 = d_2 - vt$

56. $v = \dfrac{s_2 - s_1}{m}$, for s_1 $s_1 = s_2 - vm$

57. $r = m + mnp$, for m $m = \dfrac{r}{1 + np}$

58. $p = x - xyz$, for x $x = \dfrac{p}{1 - yz}$

59. $y = ab - ac^2$, for a $a = \dfrac{y}{b - c^2}$

60. $d = mn - mp^3$, for m $m = \dfrac{d}{n - p^3}$

Solve for y.

61. $3x + 6y = 9$ $y = \dfrac{3 - x}{2}$

62. $4x - 7y = 6$ $y = \dfrac{6 - 4x}{-7}$

63. $x = y - 7$ $y = x + 7$

64. $2 = 4x - y$ $y = 4x - 2$

65. $y - 3(x + 2) = 4 + 2y$ $y = -3x - 10$

66. $3x - 2(x + y) = y - x$ $y = \frac{2}{3}x$

67. $4y + x^2 = x + 1$ $y = \dfrac{-x^2 + x + 1}{4}$

68. $2x^2 - y + 3x = 0$ $y = 2x^2 + 3x$

69. *Investing.*   Janos has $\$2600$ to invest for 6 months. If he needs the money to earn $\$156$ in that time, at what rate of simple interest must Janos invest?   12%

70. *Banking.*   Yvonne plans to buy a one-year certificate of deposit (CD) that earns 7% simple interest. If she needs the CD to earn $\$110$, how much should Yvonne invest? $\$1571.43$

⊡ Answers to Exercises 44 and 47–50 can be found on p. A-55.

71. *Geometry.* The area of a parallelogram is 78 cm². The base of the figure is 13 cm. What is the height? 6 cm

72. *Geometry.* The area of a parallelogram is 72 cm². The height of the figure is 6 cm. How long is the base? 12 cm

Projected Birth Weight. *Ultrasonic images of 29-week-old fetuses can be used to predict weight. One model, developed by Thurnau,* is P = 9.337da − 299; a second model, developed by Weiner,[†] is P = 94.593c + 34.227a − 2134.616. For both formulas, P is the estimated fetal weight, in grams; d the diameter of the fetal head, in centimeters; c the circumference of the fetal head, in centimeters; and a the circumference of the fetal abdomen, in centimeters.*

73. Use Thurnau's model to estimate the diameter of a fetus' head at 29 weeks when the estimated weight is 1614 g and the circumference of the fetal abdomen is 24.1 cm. 8.5 cm

74. Use Weiner's model to estimate the circumference of a fetus' head at 29 weeks when the estimated weight is 1277 g and the circumference of the fetal abdomen is 23.4 cm. 27.6 cm

75. *Gardening.* A garden is being constructed in the shape of a trapezoid. The dimensions are as shown in the figure. The unknown dimension is to be such that the area of the garden is 90 ft². Find that unknown dimension. 9 ft

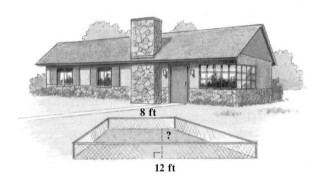

8 ft
?
12 ft

76. *Fencing.* A rectangular garden is being constructed, and 76 ft of fencing is available. The width of the garden is to be 13 ft. What should the length be, in order to use just 76 ft of fence? 25 ft

Aha! **77.** *Investing.* Bok Lum Chan is going to invest $1000 at simple interest at 9%. How long will it take for the investment to be worth $1090? 1 yr

78. *Investing.* Trey is going to invest $950 at simple interest at 7%. How long will it take for his investment to be worth $1349? 6 yr

Waiting Time. *In an effort to minimize waiting time for patients at a doctor's office without increasing a physician's idle time, Michael Goiten of Massachusetts General Hospital has developed a model. Goiten suggests that the interval time I, in minutes, between scheduled appointments be related to the total number of minutes T that a physician spends with patients in a day and the number of scheduled appointments N according to the formula I = 1.08(T/N). (Source: New England Journal of Medicine, 30 August 1990: 604−608)*

79. Dr. Cruz determines that she has a total of 8 hr a day to see patients. If she insists on an interval time of 15 min, according to Goiten's model, how many appointments should she make in one day? 34 appointments

80. Dr. Pitman insists on an interval time of 20 min and must be able to schedule 25 appointments a day. According to Goiten's model, how many hours a day should the doctor be prepared to spend with patients? 463 min, or 7.7 hr

TW **81.** Why is it important to check the solution from Step (3) (*Carry out*) in the original wording of the problem being solved?

TW **82.** Is every rectangle a trapezoid? Why or why not?

Skill Maintenance

Graph by hand. [1.5]

83. $y = -x$ ▫

84. $y = -x + 4$ ▫

85. $x + y = 5$ ▫

86. $y = 2x$ ▫

Graph using a graphing calculator. [1.5]

87. $y = |x + 1|$ ▫

88. $y = -2x^2$ ▫

Synthesis

TW **89.** Predictions made using the models of Exercises 73 and 74 are often off by as much as 10%. Does this mean the models should be discarded? Why or why not?

▫ Answers to Exercises 83–88 can be found on p. A-55.

---

*Thurnau, G. R., R. K. Tamura, R. E. Sabbagha, et al. *Am. J. Obstet Gynecol* 1983; **145**:557.
[†]Weiner, C. P., R. E. Sabbagha, N. Vaisrub, et al. *Obstet Gynecol* 1985; **65**:812.

TW **90.** Both of the models used in Exercises 73 and 74 have *P* alone on one side of the equation. Why?

For Exercises 91–94, use the formulas

$$D = \frac{m}{V} \quad \text{and} \quad V = \pi r^2 h,$$

where D is the density, m the mass, V the volume, r the length of the radius, and h the height of a circular cylinder.

91. The density of platinum is 21.5 g/cm³. If the ring shown in the figure below is crafted out of platinum, how much will it weigh?   About 10.9 g

92. A collector suspects that a silver coin is not solid silver. The density of silver is 10.5 g/cm³ and the coin is 0.2 cm thick with a radius of 2 cm. If the coin is really silver, how much should it weigh?
About 26 g

93. The density of a penny is 8.93 g/cm³. The mass of a roll of pennies is 177.6 g. If the diameter of a penny is 1.85 cm, how tall is a roll of pennies?
About 7.4 cm

94. The density of copper is 8.93 g/cm³. How long must a copper wire be if it is 1 cm thick and has a mass of 4280 g?   About 610 cm

95. *Test Scores.*   Tico's scores on four tests are 83, 91, 78, and 81. How many points above the average must Tico score on the next test in order to raise his average 2 points?   10 points above the average

96. *Geometry.*   The height and sides of a triangle are four consecutive integers. The height is the first integer, and the base is the fourth integer. The perimeter of the triangle is 42 in. Find the area of the triangle.   90 in²

97. *Home Prices.*   Panduski's real estate prices increased 6% from 1999 to 2000 and 2% from 2000 to 2001. From 2001 to 2002, prices dropped 1%. If a house sold for $117,743 in 2002, what was its worth in 1999? (Round to the nearest dollar.)
$110,000

98. *Adjusted Wages.*   Blanche's salary is reduced *n*% during a period of financial difficulty. By what number should her salary be multiplied in order to bring it back to where it was before the reduction?
$\frac{100}{100-n}$

Solve.

99. $s = v_i t + \frac{1}{2}at^2$, for *a*

100. $A = 4lw + w^2$, for *l*

101. $\frac{P_1 V_1}{T_1} = \frac{P_2 V_2}{T_2}$, for T_2

102. $\frac{P_1 V_1}{T_1} = \frac{P_2 V_2}{T_2}$, for T_1

103. $\frac{b}{a-b} = c$, for *b* $b = \frac{ac}{1+c}$

104. $m = \frac{(d/e)}{(e/f)}$, for *d* $d = \frac{me^2}{f}$

105. $\frac{a}{a+b} = c$, for *a* $a = \frac{bc}{1-c}$

Aha! **106.** $s + \frac{s+t}{s-t} = \frac{1}{t} + \frac{s+t}{s-t}$, for *t* $t = \frac{1}{s}$

TW **107.** To derive the formula for the area of a trapezoid, consider the area of two trapezoids, one of which is upside down.

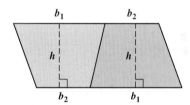

Explain why the total area of the two trapezoids is given by $h(b_1 + b_2)$. Then explain why the area of a trapezoid is given by $\frac{h}{2}(b_1 + b_2)$.

TW **108.** Write a problem for a classmate to solve. Devise the problem so that the solution is "The material should be cut into two pieces, one 30 cm long and the other 45 cm long."

TW **109.** Write a problem for a classmate to solve. Devise the problem so that the solution is "The first angle is 40°, the second angle is 50°, and the third angle is 90°."

Graph.

110. $2x^3 - y + 7 = 4y$

111. $y - 2(x^2 + y) = 0$

112. $2x^3 - 3y = 7(x - y)$

113. $|x+2| - 3y = 4y$

110. 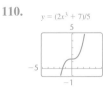   **111.**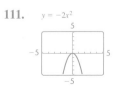

112. 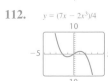   **113.**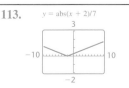

99. $a = \frac{2s - 2v_i t}{t^2}$ **100.** $l = \frac{A - w^2}{4w}$

101. $T_2 = \frac{T_1 P_2 V_2}{P_1 V_1}$ **102.** $T_1 = \frac{P_1 V_1 T_2}{P_2 T_2}$

Collaborative Corner

Who Pays What?

Focus: Problem solving
Time: 15 minutes
Group size: 5

Suppose that two of the five members in each group are celebrating birthdays and the entire group goes out to lunch. Suppose further that each member whose birthday it is gets treated to his or her lunch by the other *four* members. Finally, suppose that all meals cost the same amount and that the total bill is $40.00.*

---

*This activity was inspired by "The Birthday-Lunch Problem," *Mathematics Teaching in the Middle School,* vol. 2, no. 1, September–October 1996, pp. 40–42.

ACTIVITY

1. Determine, as a group, how much each group member should pay for the lunch described above. Then explain how this determination was made.

2. Compare the results and methods used for part (1) with those of the other groups in the class.

2.4

Slope–Intercept Form of an Equation ■ Applications

Linear Functions: Slope, Graphs, and Models

TEACHING TIP

You may want to point out that $f(x) = mx + b$ and $y = mx + b$ are often used interchangeably.

Different functions have different graphs. In this section, we examine functions and real-life models with graphs that are straight lines. Such functions and their graphs are called *linear* and can be written in the form $f(x) = mx + b$, where m and b are constants.

Slope–Intercept Form of an Equation

We can learn much about functions by examining their graphs. Also, we can tell whether a function is linear by examining the equation that describes the function.

Interactive Discovery

Graph each of the following functions using a standard viewing window:

$$f(x) = 2x,$$
$$g(x) = -2x,$$
$$h(x) = 0.3x.$$

1. Do the graphs appear to be linear?   Yes
2. At what point do they cross the x-axis?   The origin, or $(0, 0)$
3. At what point do they cross the y-axis?   The origin, or $(0, 0)$

The pattern you may have observed is true in general.

A function given by an equation of the form $f(x) = mx$ is a *linear function*. Its graph is a straight line passing through the origin.

What happens to the graph of $f(x) = mx$ if we add a number b to get an equation of the form $f(x) = mx + b$?

Interactive Discovery

1. The value of y_2 is 5 more than that of y_1 for the same value of x. The value of y_3 is 7 less than that of y_1.
2. The graph of y_4 will look like the graph of y_1, shifted down 3.2 units.
3. The graph is shifted up or down, depending on the sign of b.

Graph the following equations on the same set of axes using a standard viewing window:

$$y_1 = 0.5x, \qquad y_2 = 0.5x + 5, \qquad y_3 = 0.5x - 7.$$

1. By examining the graphs and creating a table of values, explain how the values of y_2 and y_3 differ from those of y_1.
2. Predict what the graph of $y_4 = 0.5x + (-3.2)$ will look like. Test your description by graphing y_4 on the same set of axes as $y_1 = 0.5x$.
3. Describe what happens to the graph of $y_1 = 0.5x$ when a number b is added to $0.5x$.

The pattern you may have observed is also true for other values of m.

EXAMPLE 1   Graph on the same set of axes $y = 2x$ and $y = 2x + 3$.

Solution   We first make a table of solutions of both equations.

	y	y
x	$y = 2x$	$y = 2x + 3$
0	0	3
1	2	5
-1	-2	1
2	4	7
-2	-4	-1
3	6	9

Study Tip

Make an effort to do your homework as soon as possible after each class. Make this part of your routine, choosing a time and a place where you can focus with a minimum of distractions.

We then plot these points. Drawing a blue line for $y = 2x + 3$ and a red line for $y = 2x$, we observe that the graph of $y = 2x + 3$ is simply the graph of $y = 2x$ shifted, or *translated*, 3 units up. The lines are parallel.

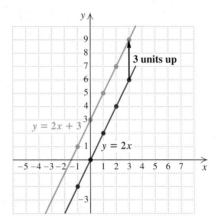

Note that the graph of $y = 2x + 3$ passes through the point $(0, 3)$. In general, we have the following.

The graph of $y = mx + b$, $b \neq 0$, is a line parallel to $y = mx$, passing through the point $(0, b)$.

The point $(0, b)$ is called the **y-intercept**. Often we refer to the number b as the y-intercept.

TEACHING TIP

Consider mentioning that for any point on the y-axis, we have $x = 0$.

EXAMPLE 2 For each equation, find the y-intercept.

a) $y = -5x + 4$ **b)** $f(x) = 5.3x - 12$

Solution

a) The y-intercept is $(0, 4)$, or simply 4.

b) The y-intercept is $(0, -12)$, or simply -12.

Interactive Discovery

Graph the following equations on the same set of axes using a standard viewing window:

$$y_1 = 3x + 1,$$
$$y_2 = -2x + 1,$$
$$y_3 = 0.6x + 1.$$

1. Do the lines have the same y-intercept?   Yes

2. Do the lines have the same slant?   No

Deselect y_2 and y_3 and graph the following equations on the same set of axes using a standard viewing window:

$$y_1 = 3x + 1,$$
$$y_4 = 3x - 3,$$
$$y_5 = 3x + 2.5.$$

3. Do the lines have the same y-intercept?   No

4. Do the lines have the same slant?   Yes

5. On the basis of the graphs of the five equations given here, does it appear that the slant of a line $y = mx + b$ depends on the number m or the number b? m

The number m, in the equation $y = mx + b$, is responsible for the slant of a line. Note that, for the graphs in Example 1, both equations have $m = 2$, and the slant of the red line seems to match the slant of the blue line. We call the slant of a line its *slope* and can visualize the slope as a geometric ratio using the following definition.

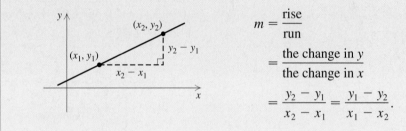

Slope

The *slope* of the line passing through (x_1, y_1) and (x_2, y_2) is given by

$$m = \frac{\text{rise}}{\text{run}}$$
$$= \frac{\text{the change in } y}{\text{the change in } x}$$
$$= \frac{y_2 - y_1}{x_2 - x_1} = \frac{y_1 - y_2}{x_1 - x_2}.$$

In the definition above, (x_1, y_1) and (x_2, y_2)—read "x sub-one, y sub-one and x sub-two, y sub-two"—represent two different points on a line. It does not matter which point is considered (x_1, y_1) and which is considered (x_2, y_2) so long as coordinates are subtracted in the same order in both the numerator and the denominator.

The letter m is traditionally used for slope. This usage has its roots in the French verb *monter*, to climb.

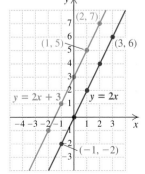

EXAMPLE 3   Find the slope of the lines drawn in Example 1.

Solution   To find the slope of a line, we can use the coordinates of any two points on that line. We use $(1, 5)$ and $(2, 7)$ to find the slope of the blue line in Example 1:

$$\text{Slope} = \frac{\text{rise}}{\text{run}} = \frac{\text{change in } y}{\text{change in } x} = \frac{y_2 - y_1}{x_2 - x_1} = \frac{7 - 5}{2 - 1} = 2.$$

To find the slope of the red line in Example 1, we use $(-1, -2)$ and $(3, 6)$:

$$\text{Slope} = \frac{\text{rise}}{\text{run}} = \frac{\text{change in } y}{\text{change in } x} = \frac{6 - (-2)}{3 - (-1)} = \frac{8}{4} = 2.$$

In Example 3, we found that the lines given by $y = 2x + 3$ and $y = 2x$ both have a slope of 2. This supports (but does not prove) the following:

The slope of any line written in the form $y = mx + b$ is m.

A proof of this result is outlined in Exercise 99 on p. 135.

Note in the definition of slope that coordinates can be subtracted in either order, so long as the order is the same in the numerator and the denominator. Thus, in Example 3, the slope of the blue line can also be found by

$$\frac{y_1 - y_2}{x_1 - x_2} = \frac{5 - 7}{1 - 2} = \frac{-2}{-1} = 2.$$

EXAMPLE 4 Determine the slope of the line given by $y = \frac{2}{3}x + 4$, and graph the line.

Solution Here $m = \frac{2}{3}$, so the slope is $\frac{2}{3}$. This means that from *any* point on the graph, we can locate a second point by simply going *up* 2 units (the change in y) and to the right 3 units (the change in x). Where do we start? Because the y-intercept, $(0, 4)$, is known to be on the graph, we calculate that $(0 + 3, 4 + 2)$, or $(3, 6)$, is also on the graph. Knowing two points, we can draw the graph.

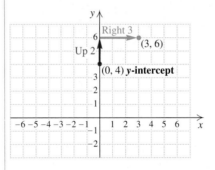
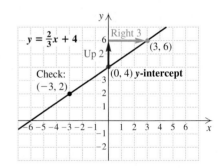

Important: To check the graph, we use some other value for x—say, -3—substitute in the equation, and determine y (in this case, 2). We plot that point and see that it *is* on the line. Were it not, we would know that some error had been made.

The Slope–Intercept Equation

Any equation $y = mx + b$ has a graph that is a straight line. It goes through the y-*intercept* $(0, b)$ and has slope m. Any equation of the form $y = mx + b$ is said to be a *slope–intercept equation*.

EXAMPLE 5 Determine the slope and the y-intercept of the line given by $y = -\frac{1}{3}x + 2$.

Solution The equation $y = -\frac{1}{3}x + 2$ is written in the form $y = mx + b$, with $m = -\frac{1}{3}$ and $b = 2$. Thus the slope is $-\frac{1}{3}$ and the y-intercept is $(0, 2)$.

Note that any graph of $y = mx + b$ will pass the vertical-line test and thus represents a function.

EXAMPLE 6 Find a linear function whose graph has slope $-\frac{2}{3}$ and y-intercept $(0,4)$.

Solution   We use the slope–intercept form, $f(x) = mx + b$:

$$f(x) = -\frac{2}{3}x + 4. \qquad \textbf{Substituting } -\frac{2}{3} \textbf{ for } m \textbf{ and 4 for } b;$$
$$\textbf{using function notation}$$

Interactive Discovery

To see the effect of the sign of m on the graph of $y = mx + b$, graph the following equations on the same set of axes using the standard viewing window:

$$y_1 = x + 1,$$
$$y_2 = -x + 1,$$
$$y_3 = 2x + 1,$$
$$y_4 = -2x + 1.$$

1. Which graphs slant up from left to right? y_1 and y_3
2. Which graphs slant down from left to right? y_2 and y_4
3. Describe how the sign of m affects the graph of $y = mx + b$.   The graph slants up if m is positive and down if m is negative.
   To see the effect of different positive values of m on the graph of $y = mx + b$, clear y_2, y_3, and y_4 and graph the following equations on the same set of axes using the standard viewing window:

$$y_1 = x + 1,$$
$$y_2 = 2x + 1,$$
$$y_3 = 3x + 1.$$

4. What will the graph of $y = 7x + 1$ look like? Test your answer by graphing $y = 7x + 1$.   The graph of $y = 7x + 1$ goes through $(0, 1)$ and is steeper than y_1, y_2, and y_3.
5. Describe how different positive values of m affect the graph of $y = mx + b$.   The larger m is, the steeper the graph.

When the slope of a line is positive, the line slants up from left to right. When the slope of a line is negative, the line slants down from left to right. The larger the absolute value of the slope, the steeper the slant.

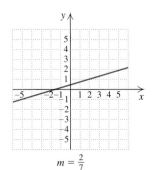

$$m = \frac{2}{7}$$

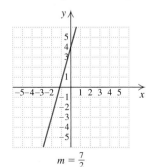

$$m = \frac{7}{2}$$

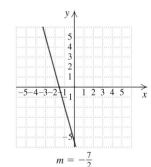

$$m = -\frac{7}{2}$$

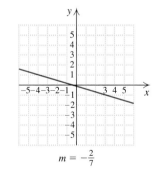

$$m = -\frac{2}{7}$$

EXAMPLE 7   Graph: $f(x) = -\frac{1}{2}x + 5$.

Solution   The y-intercept is $(0, 5)$. The slope is $-\frac{1}{2}$, or $\frac{-1}{2}$. From the y-intercept, we go *down* 1 unit and *to the right* 2 units. That gives us the point $(2, 4)$. We can now draw the graph.

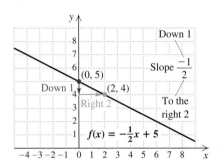

As a new type of check, we rename the slope and find another point:

$$-\frac{1}{2} = \frac{1}{-2}.$$

Thus we can go *up* 1 unit and then *to the left* 2 units. This gives the point $(-2, 6)$. Since $(-2, 6)$ is on the line, we have a check.

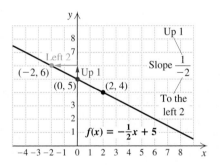

Note that $m = -\frac{1}{2}$ is negative; thus the graph slants downward from left to right.

Often the easiest way to graph an equation is to rewrite it in slope–intercept form and then proceed as in Examples 4 and 7 above.

EXAMPLE 8   Determine the slope and the y-intercept for the equation $5x - 4y = 8$. Then graph.

Solution   We convert to a slope–intercept equation:

$$5x - 4y = 8$$
$$-4y = -5x + 8 \qquad \text{Adding } -5x \text{ to both sides}$$
$$y = -\tfrac{1}{4}(-5x + 8) \qquad \text{Multiplying both sides by } -\tfrac{1}{4}$$
$$y = \tfrac{5}{4}x - 2. \qquad \text{Using the distributive law}$$

Because we have an equation of the form $y = mx + b$, we know that the slope is $\frac{5}{4}$ and the y-intercept is $(0, -2)$. We plot $(0, -2)$, and from there we go *up* 5 units and *to the right* 4 units, and plot a second point at $(4, 3)$. We then draw the graph.

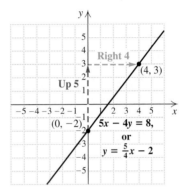

Note that $m = \frac{5}{4}$ is positive; thus the graph slants upward from left to right.

As a check, we calculate the coordinates of another point on the line. For $x = 2$, we have

$$5 \cdot 2 - 4y = 8$$
$$10 - 4y = 8$$
$$-4y = -2$$
$$y = \frac{1}{2}.$$

Thus, $\left(2, \frac{1}{2}\right)$ should appear on the graph. Since it *does* appear to be on the line, we have a check.

Applications

TEACHING TIP

You might introduce terms such as *grade* and *pitch* at this time.

Because slope is a ratio that indicates how a change in the vertical direction corresponds to a change in the horizontal direction, it has many real-world applications. Foremost is the use of slope to represent a *rate of change*.

EXAMPLE 9 Online Travel Plans. More and more Americans are making their travel arrangements using the Internet. Use the graph below to find the rate at which this number is growing (*Source*: Travel Industry Association of America). Note that the jagged "break" on the vertical axis is used to avoid including a large portion of unused grid.

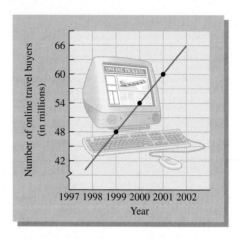

Solution Since the graph is linear, we can use any pair of points to determine the rate of change. We choose the points (1999, 48 million) and (2001, 60 million), which gives us

$$\text{Rate of change} = \frac{60 \text{ million} - 48 \text{ million}}{2001 - 1999}$$

$$= \frac{12 \text{ million}}{2 \text{ years}} = \frac{6 \text{ million}}{1 \text{ year}} = 6 \text{ million per year.}$$

The number of Americans making online travel arrangements is growing at the rate of approximately 6 million users per year.

EXAMPLE 10 Cellular Phone Use. In 1999, there were approximately 86 million cellular phone customers in the United States. By 2001, the figure had grown to 110 million. (*Source*: based on data from the Cellular Telecommunications Industry Association) At what rate is the number of cellular phone customers changing?

Solution The rate at which the number of customers is changing is given by

$$\text{Rate of change} = \frac{\text{change in number of customers}}{\text{change in time}}$$

$$= \frac{110 \text{ million customers} - 86 \text{ million customers}}{2001 - 1999}$$

$$= \frac{24 \text{ million customers}}{2 \text{ years}}$$

$$= 12 \text{ million customers per year.}$$

Between 1999 and 2001, the number of cellular phone customers grew at a rate of approximately 12 million customers per year.

EXAMPLE 11   Running Speed.   Stephanie runs 10 km during each workout. For the first 7 km, her pace is twice as fast as it is for the last 3 km. Which of the following graphs best describes Stephanie's workout?

A.
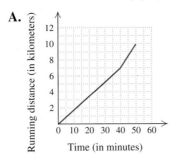

B.
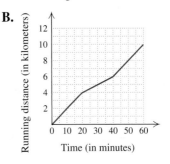

C.
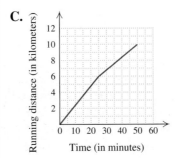

D.
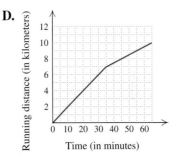

Solution   The slopes in graph A increase as we move to the right. This would indicate that Stephanie ran faster for the *last* part of her workout. Thus graph A is not the correct one.

The slopes in graph B indicate that Stephanie slowed down in the middle of her run and then resumed her original speed. Thus graph B does not correctly model the situation either.

According to graph C, Stephanie slowed down not at the 7-km mark, but at the 6-km mark. Thus graph C is also incorrect.

Graph D indicates that Stephanie ran the first 7 km in 35 min, a rate of 0.2 km/min. It also indicates that she ran the final 3 km in 30 min, a rate of 0.1 km/min. This means that Stephanie's rate was twice as fast for the first 7 km, so graph D provides a correct description of her workout.

Linear functions arise continually in today's world. As with the rate problems appearing in Examples 9–11, it is critical to use proper units in all answers.

EXAMPLE 12   Salvage Value.   Tyline Electric uses the function

$$S(t) = -700t + 3500$$

to determine the *salvage value* $S(t)$, in dollars, of a color photocopier t years after its purchase.

a) What do the numbers -700 and 3500 signify?

b) How long will it take the copier to *depreciate* completely?

c) What is the domain of S?

Solution Drawing, or at least visualizing, a graph can be useful here.

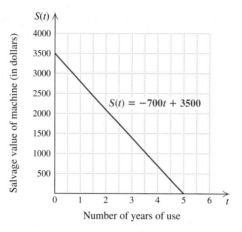

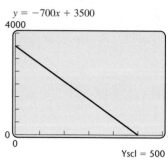

a) At time $t = 0$, we have $S(0) = -700 \cdot 0 + 3500 = 3500$. Thus the number 3500 signifies the original cost of the copier, in dollars.

This function is written in slope–intercept form. Since the output is measured in dollars and the input in years, the number -700 signifies that the value of the copier is declining at a rate of $700 per year.

b) The copier will have depreciated completely when its value drops to 0. To learn when this occurs, we determine when $S(t) = 0$:

$$S(t) = 0 \qquad \text{A graph is not always available.}$$
$$-700t + 3500 = 0 \qquad \text{Substituting } -700t + 3500 \text{ for } S(t)$$
$$-700t = -3500 \qquad \text{Subtracting 3500 from both sides}$$
$$t = 5. \qquad \text{Dividing both sides by } -700$$

The copier will have depreciated completely in 5 yr.

c) Neither the number of years of service nor the salvage value can be negative. In part (b) we found that after 5 yr the salvage value will have dropped to 0. Thus the domain of S is $\{t \mid 0 \le t \le 5\}$. Either graph above serves as a visual check of this result.

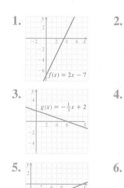

2.4

Exercise Set

Graph.

1. $f(x) = 2x - 7$

2. $g(x) = 3x - 7$

3. $g(x) = -\frac{1}{3}x + 2$

4. $f(x) = -\frac{1}{2}x - 5$

5. $h(x) = \frac{2}{5}x - 4$

6. $h(x) = \frac{4}{5}x + 2$

Determine the y-intercept.

7. $y = 5x + 7$ $(0, 7)$

8. $y = 4x - 9$ $(0, -9)$

9. $f(x) = -2x - 6$ $(0, -6)$

10. $g(x) = -5x + 7$ $(0, 7)$

11. $y = -\frac{3}{8}x - 4.5$ $(0, -4.5)$

12. $y = \frac{15}{7}x + 2.2$ $(0, 2.2)$

13. $g(x) = 2.9x - 9$ $(0, -9)$ **14.** $f(x) = -3.1x + 5$
$(0, 5)$
15. $y = 37x + 204$ $(0, 204)$ **16.** $y = -52x + 700$
$(0, 700)$

For each pair of points, find the slope of the line containing them.

17. $(6, 9)$ and $(4, 5)$ 2

18. $(8, 7)$ and $(2, -1)$ $\frac{4}{3}$

19. $(3, 8)$ and $(9, -4)$ -2

20. $(17, -12)$ and $(-9, -15)$ $\frac{3}{26}$

21. $(-16.3, 12.4)$ and $(-5.2, 8.7)$ $-\frac{1}{3}$

22. $(12.4, -5.8)$ and $(-14.5, -15.6)$ $\frac{98}{269}$

23. $(-9.7, 43.6)$ and $(4.5, 43.6)$ 0

24. $(-2.8, -3.1)$ and $(-1.8, -2.6)$ $\frac{1}{2}$

Aha! **25.** Use the slope and the y-intercept of each line to match each equation with the correct graph.

a) $y = 3x - 5$ II
b) $y = 0.7x + 1$ IV
c) $y = -0.25x - 3$ III
d) $y = -4x + 2$ I

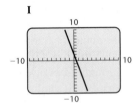
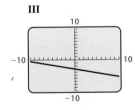

26. Use the slope and the y-intercept of each line to match each equation with the correct graph.

a) $y = \frac{1}{2}x - 5$ II
b) $y = 2x + 3$ IV
c) $y = -3x + 1$ I
d) $y = -\frac{3}{4}x - 2$ III

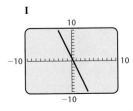

$\boxdot$ Answers to Exercises 27–46 can be found on pp. A-55 and A-56.

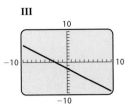
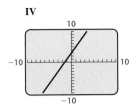

Determine the slope and the y-intercept. Then draw a graph by hand. Be sure to check as in Example 4 or Example 7.

27. $y = \frac{5}{2}x + 3$ $\boxdot$ **28.** $y = \frac{2}{5}x + 4$ $\boxdot$

29. $f(x) = -\frac{5}{2}x + 1$ $\boxdot$ **30.** $f(x) = -\frac{2}{5}x + 3$ $\boxdot$

31. $2x - y = 5$ $\boxdot$ **32.** $2x + y = 4$ $\boxdot$

33. $f(x) = \frac{1}{3}x + 2$ $\boxdot$ **34.** $f(x) = -3x + 6$ $\boxdot$

35. $7y + 2x = 7$ $\boxdot$ **36.** $4y + 20 = x$ $\boxdot$

Aha! **37.** $f(x) = -0.25x$ $\boxdot$ **38.** $f(x) = 1.5x - 3$ $\boxdot$

39. $4x - 5y = 10$ $\boxdot$ **40.** $5x + 4y = 4$ $\boxdot$

41. $f(x) = \frac{5}{4}x - 2$ $\boxdot$ **42.** $f(x) = \frac{4}{3}x + 2$ $\boxdot$

43. $12 - 4f(x) = 3x$ $\boxdot$ **44.** $15 + 5f(x) = -2x$ $\boxdot$

Aha! **45.** $g(x) = 2.5$ $\boxdot$ **46.** $g(x) = \frac{3}{4}x$ $\boxdot$

Find a linear function whose graph has the given slope and y-intercept.

47. Slope $\frac{2}{3}$, y-intercept $(0, -9)$ $f(x) = \frac{2}{3}x - 9$

48. Slope $-\frac{3}{4}$, y-intercept $(0, 12)$ $f(x) = -\frac{3}{4}x + 12$

49. Slope -5, y-intercept $(0, 2)$ $f(x) = -5x + 2$

50. Slope 2, y-intercept $(0, -1)$ $f(x) = 2x - 1$

51. Slope $-\frac{7}{9}$, y-intercept $(0, 5)$ $f(x) = -\frac{7}{9}x + 5$

52. Slope $-\frac{4}{11}$, y-intercept $(0, 9)$ $f(x) = -\frac{4}{11}x + 9$

53. Slope 5, y-intercept $\left(0, \frac{1}{2}\right)$ $f(x) = 5x + \frac{1}{2}$

54. Slope 6, y-intercept $\left(0, \frac{2}{3}\right)$ $f(x) = 6x + \frac{2}{3}$

For each graph, find the rate of change. Remember to use appropriate units. See Example 9.

55.
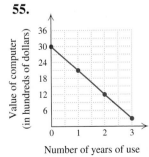

56.
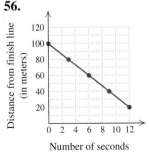

55. The value is decreasing at a rate of $900 per year.
56. The distance from the finish line is decreasing at a rate of $6\frac{2}{3}$ meters per second.

57. ⊡

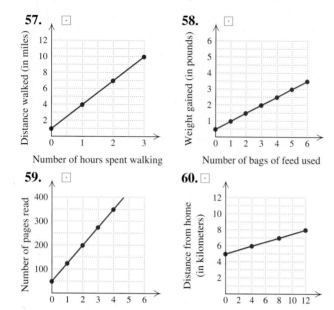

y-axis: Distance walked (in miles) — 2, 4, 6, 8, 10, 12
x-axis: Number of hours spent walking — 0, 1, 2, 3

58. ⊡

y-axis: Weight gained (in pounds) — 1, 2, 3, 4, 5, 6
x-axis: Number of bags of feed used — 0, 1, 2, 3, 4, 5, 6

59. ⊡

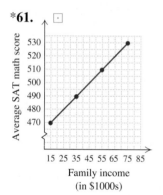

y-axis: Number of pages read — 100, 200, 300, 400
x-axis: Number of days spent reading — 0, 1, 2, 3, 4, 5, 6

60. ⊡

y-axis: Distance from home (in kilometers) — 2, 4, 6, 8, 10, 12
x-axis: Number of minutes spent running — 0, 2, 4, 6, 8, 10, 12

*61. ⊡

y-axis: Average SAT math score — 470, 480, 490, 500, 510, 520, 530
x-axis: Family income (in \$1000s) — 15 25 35 45 55 65 75 85

*62. ⊡

y-axis: Average SAT verbal score — 480, 490, 500, 510, 520, 530, 540
x-axis: Family income (in \$1000s) — 15 25 35 45 55 65 75 85

63. *Skiing Rate.*   A cross-country skier reaches the 3-km mark of a race in 15 min and the 12-km mark 45 min later. Find the speed of the skier.   12 km/h

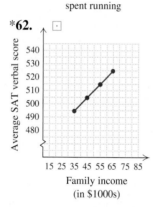

12 km
3 km

64.   *Running Rate.*   An Olympic marathoner passes the 5-mi point of a race after 30 min and reaches the 25-mi point 2 hr later. Find the speed of the marathoner.   10 mph

*Based on data from the College Board Online.
⊡ Answers to Exercises 57–62 can be found on p. A-56.

65. *Rate of Computer Hits.*   At the beginning of 1999, SciMor.com had already received 80,000 hits at their website. At the beginning of 2001, that number had climbed to 430,000. Calculate the rate at which the number of hits is increasing.   175,000 hits/yr

66. *Work Rate.*   As a painter begins work, one-fourth of a house has already been painted. Eight hours later, the house is two-thirds done. Calculate the painter's work rate. $\frac{5}{96}$ of the house per hour

67. *Rate of Descent.*   A plane descends to sea level from 12,000 ft after being airborne for $1\frac{1}{2}$ hr. The entire flight time is 2 hr and 10 min. Determine the average rate of descent of the plane.   300 ft/min

68. *Growth in Overseas Travel.*   In 1988, the number of U.S. visitors overseas was about 11.6 million. In 1995, the number grew to 18.7 million. (*Source: Statistical Abstract of the United States*) Determine the rate at which the number of U.S. visitors overseas was growing. $\frac{71}{70}$ million/yr

69. *Nursing.*   Match each sentence with the most appropriate of the four graphs shown.
   a) The rate at which fluids were given intravenously was doubled after 3 hr.   II
   b) The rate at which fluids were given intravenously was gradually reduced to 0.   IV
   c) The rate at which fluids were given intravenously remained constant for 5 hr.   I
   d) The rate at which fluids were given intravenously was gradually increased.   III

I

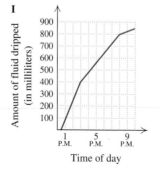

y-axis: Amount of fluid dripped (in milliliters) — 100, 200, 300, 400, 500, 600, 700, 800, 900
x-axis: Time of day — 1 P.M., 5 P.M., 9 P.M.

II

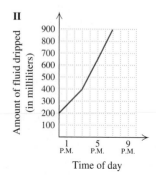

y-axis: Amount of fluid dripped (in milliliters) — 100, 200, 300, 400, 500, 600, 700, 800, 900
x-axis: Time of day — 1 P.M., 5 P.M., 9 P.M.

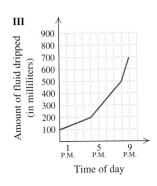

III

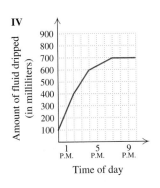

IV

70. *Market Research.*   Match each sentence with the most appropriate graph below.

a) After January 1, daily sales continued to rise, but at a slower rate.   III

b) After January 1, sales decreased faster than they ever grew.   IV

c) The rate of growth in daily sales doubled after January 1.   I

d) After January 1, daily sales decreased at half the rate that they grew in December.   II

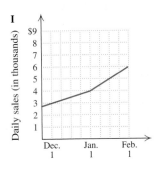

I

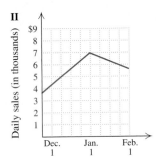

II

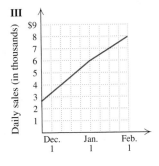

III

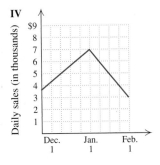

IV

In Exercises 71–84, each model is of the form
$f(x) = mx + b$. *In each case, determine what m and b signify.*

71. *Catering.*   When catering a party for x people, Jennette's Catering uses the formula $C(x) = 25x + 75$, where $C(x)$ is the cost of the party, in dollars.   25 signifies that the cost per person is $25; 75 signifies that the setup cost for the party is $75.

72. *Weekly Pay.*   Each salesperson at Knobby's Furniture is paid $P(x)$ dollars, where $P(x) = 0.05x + 200$ and x is the value of the salesperson's sales for the week.   ⊡

73. *Hair Growth.*   After Tina gets a "buzz cut," the length $L(t)$ of her hair, in inches, is given by $L(t) = \frac{1}{2}t + 1$, where t is the number of months after she gets the haircut. $\frac{1}{2}$ signifies that Tina's hair grows $\frac{1}{2}$ in. per month; 1 signifies that her hair is 1 in. long when cut.

74. *Landscaping.*   After being cut, the length $G(t)$ of the lawn, in inches, at Great Harrington Community College is given by $G(t) = \frac{1}{8}t + 2$, where t is the number of days since the lawn was cut.   ⊡

75. *Life Expectancy of American Women.*   The life expectancy of American women t years after 1950 is given by $A(t) = \frac{3}{20}t + 72$.   ⊡

76. *Natural Gas Demand.*   The demand, in quadrillions of joules, for natural gas is approximated by $D(t) = \frac{1}{5}t + 20$, where t is the number of years after 1960.   ⊡

77. *Sales of Cotton Goods.*   The function given by $f(t) = 2.6t + 17.8$ can be used to estimate the yearly sales of cotton goods, in billions of dollars, t years after 1975.   ⊡

78. *Cost of a Movie Ticket.*   The average price $P(t)$, in dollars, of a movie ticket is given by $P(t) = 0.1522t + 4.29$, where t is the number of years since 1990.   ⊡

79. *Cost of a Taxi Ride.*   The cost, in dollars, of a taxi ride in Pelham is given by $C(d) = 0.75d + 2$, where d is the number of miles traveled.   ⊡

80. *Cost of Renting a Truck.*   The cost, in dollars, of a one-day truck rental is given by $C(d) = 0.3d + 20$, where d is the number of miles driven.   ⊡

⊡ Answers to Exercises 72 and 74–80 can be found on p. A-56.

81. *Salvage Value.*   Green Glass Recycling uses the function given by $F(t) = -5000t + 90,000$ to determine the salvage value $F(t)$, in dollars, of a waste removal truck t years after it has been put into use.

a) What do the numbers -5000 and $90,000$ signify? ⊡
b) How long will it take the truck to depreciate completely?   18 yr
c) What is the domain of F? $\{t\,|\,0 \le t \le 18\}$

82. *Salvage Value.*   Consolidated Shirt Works uses the function given by $V(t) = -2000t + 15,000$ to determine the salvage value $V(t)$, in dollars, of a color separator t years after it has been put into use.

a) What do the numbers -2000 and $15,000$ signify? ⊡
b) How long will it take the machine to depreciate completely?   7.5 yr
c) What is the domain of V? $\{t\,|\,0 \le t \le 7.5\}$

83. *Trade-in Value.*   The trade-in value of a Homelite snowblower can be determined using the function given by $v(n) = -150n + 900$. Here $v(n)$ is the trade-in value, in dollars, after n winters of use.

a) What do the numbers -150 and 900 signify? ⊡
b) When will the trade-in value of the snowblower be \$300?   After 4 winters of use
c) What is the domain of v? $\{0, 1, 2, 3, 4, 5, 6\}$

84. *Trade-in Value.*   The trade-in value of a John Deere riding lawnmower can be determined using the function given by $T(x) = -300x + 2400$. Here $T(x)$ is the trade-in value, in dollars, after x summers of use.

a) What do the numbers -300 and 2400 signify? ⊡
b) When will the value of the mower be \$1200? ⊡
c) What is the domain of T? $\{0, 1, 2, 3, 4, 5, 6, 7, 8\}$

TW **85.** A student makes a mistake when using a graphing calculator to draw $4x + 5y = 12$ and the following screen appears. Use algebra to show that a mistake has been made. What do you think the mistake was?

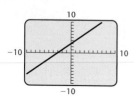

TW **86.** A student makes a mistake when using a graphing calculator to draw $5x - 2y = 3$ and the following screen appears. Use algebra to show that a mistake has been made. What do you think the mistake was?

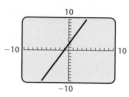

Skill Maintenance

Simplify.

87. $9\{2x - 3[5x + 2(-3x + y^0 - 2)]\}$ [1.3], [1.4] $45x + 54$

88. $(-5a^2b^3)^3$ [1.4] $-125a^6b^9$

89. $(13m^2n^3)(-2m^5n)$ [1.4] $-26m^7n^4$

90. $2x - 3[y - 2(x + 3) + 10]$ [1.3] $8x - 3y - 12$

Aha! **91.** $(4x^2y^3)(-3x^5y)^0$ [1.4] $4x^2y^3$

92. $(2x^{-5}y^{-7})^{-2}$ [1.4] $\dfrac{x^{10}y^{14}}{4}$

Synthesis

TW **93.** *Economics.*   In 2000, the federal debt could be modeled using $D(t) = mt + 6000$, where $D(t)$ is in billions of dollars t years after 2000, and m is a constant. If you were president of the United States, would you want m to be positive or negative? Why?

TW **94.** *Economics.*   Examine the function given in Exercise 93. What units of measure must be used for m? Why?

TW **95.** Hope Diswerkz claims that her firm's profits continue to go up, but the rate of increase is going down.

a) Sketch a graph that might represent her firm's profits as a function of time.
b) Explain why the graph can go up while the rate of increase goes down.

TW **96.** Belly Up, Inc., is losing \$1.5 million per year while Spinning Wheels, Inc., is losing \$170 an hour. Which firm would you rather own and why?

In Exercises 97 and 98, assume that r, p, and s are constants and that x and y are variables. Determine the slope and the y-intercept.

97. $rx + py = s$ Slope: $-\dfrac{r}{p}$; y-intercept: $\left(0, \dfrac{s}{p}\right)$

98. $rx + py = s - ry$ ⊡

⊡ Answers to Exercises 81(a), 82(a), 83(a), 84(a), 84(b), and 98 can be found on pp. A-56 and A-57.

99. Let (x_1, y_1) and (x_2, y_2) be two distinct points on the graph of $y = mx + b$. Use the fact that both pairs are solutions of the equation to prove that m is the slope of the line given by $y = mx + b$. (*Hint*: Use the slope formula.)

Given that $f(x) = mx + b$, classify each of the following as true or false.

100. $f(c + d) = f(c) + f(d)$ False

101. $f(cd) = f(c)f(d)$ False

102. $f(kx) = kf(x)$ False

103. $f(c - d) = f(c) - f(d)$ False

104. Find k such that the line containing $(-3, k)$ and $(4, 8)$ is parallel to the line containing $(5, 3)$ and $(1, -6)$. $-\frac{31}{4}$

105. Match each sentence with the most appropriate graph below.

   a) Annie drove 2 mi to a lake, swam 1 mi, and then drove 3 mi to a store.   III

   b) During a preseason workout, Rico biked 2 mi, ran for 1 mi, and then walked 3 mi.   IV

   c) James bicycled 2 mi to a park, hiked 1 mi over the notch, and then took a 3-mi bus ride back to the park.   I

   d) After hiking 2 mi, Marcy ran for 1 mi before catching a bus for the 3-mi ride into town.   II

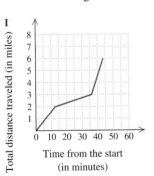

I

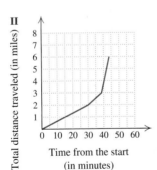

II

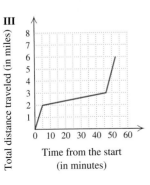

III

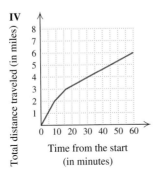

IV

106. Find the slope of the line that contains the given pair of points.

   a) $(5b, -6c), (b, -c)$ $-\frac{5c}{4b}$

   b) $(b, d), (b, d + e)$ Undefined

   c) $(c + f, a + d), (c - f, -a - d)$ $\frac{a + d}{f}$

107. *Cost of a Speeding Ticket.* The penalty schedule shown below is used to determine the cost of a speeding ticket in certain states. Use this schedule to graph the cost of a speeding ticket as a function of the number of miles per hour over the limit that a driver is going.

STATE POLICE
SPEEDING VIOLATION
FINES

1 – 10 mph over limit: $5.00/mph plus $17.50 surcharge
11 – 20 mph over limit: $6.00/mph plus $17.50 surcharge
21 – 30 mph over limit: $7.00/mph plus $17.50 surcharge
31+   mph over limit: $8.00/mph plus $17.50 surcharge

Officer will enter mph over limit in line 5a

108. Graph the equations

$$y_1 = 1.4x + 2, \qquad y_2 = 0.6x + 2,$$
$$y_3 = 1.4x + 5, \quad \text{and} \quad y_4 = 0.6x + 5$$

using a graphing calculator. If possible, use the SIMULTANEOUS mode so that you cannot tell which equation is being graphed first. Then decide which line corresponds to each equation.

99. Since (x_1, y_1) and (x_2, y_2) are two points on the graph of $y = mx + b$, then $y_1 = mx_1 + b$ and $y_2 = mx_2 + b$. Using the definition of slope, we have

$$\text{Slope} = \frac{y_2 - y_1}{x_2 - x_1}$$
$$= \frac{(mx_2 + b) - (mx_1 + b)}{x_2 - x_1}$$
$$= \frac{m(x_2 - x_1)}{x_2 - x_1}$$
$$= m.$$

107.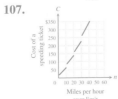

2.5

Zero Slope and Lines with Undefined Slope ■ Graphing Using Intercepts ■
Parallel and Perpendicular Lines ■ Recognizing Linear Equations

Another Look at Linear Graphs

In Section 2.4, we graphed linear equations using slopes and y-intercepts. We now graph lines that have slope 0 or that have an undefined slope. We also graph lines using both x- and y-intercepts and learn how to recognize whether the graphs of two linear equations will be parallel or perpendicular.

Zero Slope and Lines with Undefined Slope

If two different points have the same second coordinate, what is the slope of the line joining them? In this case, we have $y_2 = y_1$, so

$$m = \frac{y_2 - y_1}{x_2 - x_1} = \frac{0}{x_2 - x_1} = 0.$$

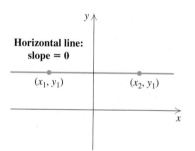

Horizontal line: slope = 0

(x_1, y_1) (x_2, y_1)

Slope of a Horizontal Line
Every horizontal line has a slope of 0.

EXAMPLE 1 Graph: $f(x) = 3$.

Solution Recall from Section 2.1 that a function of this type is called a *constant function*. Writing slope–intercept form,

$$f(x) = 0 \cdot x + 3,$$

we see that the y-intercept is $(0, 3)$ and the slope is 0. Thus we can graph f by plotting the point $(0, 3)$ and, from there, determining a slope of 0. Because $0 = 0/2$ (any nonzero number could be used in place of 2), we can draw the graph by going up 0 units and to the right 2 units. As a check, we also find some ordered pairs. Note that for any choice of x-value, $f(x)$ must be 3.

x	f(x)
−1	3
0	3
2	3

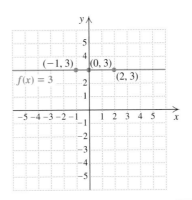

We see from Example 1 the following:

The graph of any constant function of the form $f(x) = b$ or $y = b$ is a horizontal line that crosses the y-axis at $(0, b)$.

Suppose that two different points are on a vertical line. They then have the same first coordinate. In this case, we have $x_2 = x_1$, so

$$m = \frac{y_2 - y_1}{x_2 - x_1} = \frac{y_2 - y_1}{0}.$$

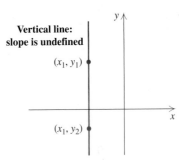

**Vertical line:
slope is undefined**

Study Tip

If you are finding it difficult to master a particular topic or concept, talk about it with a classmate. Verbalizing your questions about the material might help clarify it for you. If your classmate is also having difficulty, it is possible that a majority of your classmates are confused and you can ask your instructor to explain the concept again.

Since we cannot divide by 0, this is undefined. Note that when we say that $(y_2 - y_1)/0$ is undefined, it means that we have agreed to not attach any meaning to that expression.

Slope of a Vertical Line
The slope of a vertical line is undefined.

EXAMPLE 2   Graph: $x = -2$.

Solution   With y missing, no matter which value of y is chosen, x must be -2. Thus the pairs $(-2, 3)$, $(-2, 0)$, and $(-2, -4)$ all satisfy the equation. The graph is a line parallel to the y-axis. Note that since y is missing, this equation cannot be written in slope–intercept form.

TEACHING TIP

You may want to ask students *why* a vertical line cannot be written in slope–intercept form.

x	y
-2	3
-2	0
-2	-4

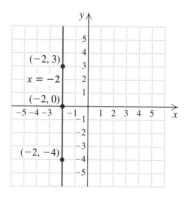

Example 2 shows the following:

The graph of any equation of the form $x = a$ is a vertical line that crosses the x-axis at $(a, 0)$.

EXAMPLE 3   Find the slope of each given line. If the slope is undefined, state this.

a) $3y + 2 = 14$

b) $2x = 10$

Solution

a) We solve for y:

$$3y + 2 = 14$$
$$3y = 12 \qquad \text{Subtracting 2 from both sides}$$
$$y = 4. \qquad \text{Dividing both sides by 3}$$

The graph of $y = 4$ is a horizontal line. Since $3y + 2 = 14$ is equivalent to $y = 4$, the slope of the line $3y + 2 = 14$ is 0.

b) When y does not appear, we solve for x:

$$2x = 10$$
$$x = 5. \qquad \text{Dividing both sides by 2}$$

The graph of $x = 5$ is a vertical line. Since $2x = 10$ is equivalent to $x = 5$, the slope of the line $2x = 10$ is undefined.

Graphing Using Intercepts

Any line that is not horizontal or vertical will cross both the x- and y-axes. We have already seen that the point at which a line crosses the y-axis is called the *y-intercept*. Similarly, the point at which a line crosses the x-axis is called the *x-intercept*. Any time the x- and y-intercepts are not both $(0, 0)$, they provide enough information to graph a linear equation. Recall that to find the y-intercept, we replace x with 0 and solve for y. To find the x-intercept, we replace y with 0 and solve for x.

> ### To Determine Intercepts
>
> The x-intercept is $(a, 0)$. To find a, let $y = 0$ and solve the original equation for x.
>
> The y-intercept is $(0, b)$. To find b, let $x = 0$ and solve the original equation for y.

EXAMPLE 4   Graph the equation $3x + 2y = 12$ by using intercepts.

Solution   *To find the y-intercept, we let x = 0 and solve for y:*

$$3 \cdot 0 + 2y = 12$$
$$2y = 12$$
$$y = 6.$$

The y-intercept is $(0, 6)$.

To find the x-intercept, we let y = 0 and solve for x:

$$3x + 2 \cdot 0 = 12$$
$$3x = 12$$
$$x = 4.$$

The x-intercept is $(4, 0)$.

We plot the two intercepts and draw the line. A third point could be calculated and used as a check.

TEACHING TIP

Students often confuse the x- and y-intercepts. Encourage students to visualize the y-intercept on the y-axis and the x-intercept on the x-axis to help determine when to let $x = 0$ and when to let $y = 0$.

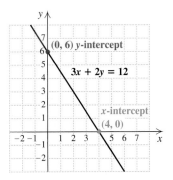

EXAMPLE 5   Graph $f(x) = 2x + 5$ by using intercepts.

Solution   Because the function is in slope–intercept form, we know that the y-intercept is $(0, 5)$. To find the x-intercept, we replace $f(x)$ with 0 and solve for x:

$$0 = 2x + 5$$
$$-5 = 2x$$
$$-\tfrac{5}{2} = x.$$

The x-intercept is $\left(-\tfrac{5}{2}, 0\right)$.

We plot the intercepts $(0, 5)$ and $\left(-\frac{5}{2}, 0\right)$ and draw the line. As a check, we can calculate the slope:

$$m = \frac{5 - 0}{0 - \left(-\frac{5}{2}\right)}$$

$$= \frac{5}{\frac{5}{2}}$$

$$= 5 \cdot \frac{2}{5}$$

$$= 2.$$

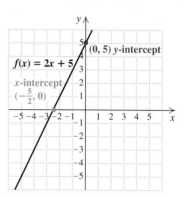

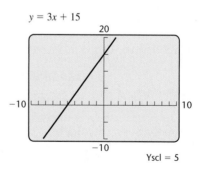

TEACHING TIP

You may wish to remind students that unless a line goes through the origin, the x-intercept and the y-intercept will be two different points. Some students will drop the zeros and combine the x- and y-intercepts to form one ordered pair.

The slope is 2, as expected.

The intercepts of the graph of a function can help us determine an appropriate viewing window on a graphing calculator.

EXAMPLE 6 Determine a viewing window that shows the intercepts of the graph of the function $f(x) = 3x + 15$.

Solution In this case the function is in slope–intercept form, so we know that the y-intercept is $(0, 15)$. To find the x-intercept, we replace $f(x)$ with 0 and solve for x:

$$0 = 3x + 15$$
$$-15 = 3x \qquad \text{Subtracting 15 from both sides}$$
$$-5 = x.$$

The x-intercept is $(-5, 0)$.

A standard viewing window will not show the y-intercept, $(0, 15)$. Thus we adjust the Ymax value and choose a viewing window of $[-10, 10, -10, 20]$, with Yscl $= 5$. Other choices of window dimensions are also possible.

Parallel and Perpendicular Lines

If two lines are vertical, they are parallel. How can we tell whether nonvertical lines are parallel?

Interactive Discovery

1. The lines appear to be parallel.
2. The lines appear to be parallel.
3. The lines in Exercise 1 still appear parallel; the lines in Exercise 2 are not parallel.

1. Graph $y = 3x - 7$ and $y - 3x = 8$ on a graphing calculator using a standard viewing window and determine whether the lines appear to be parallel.

2. Again using a standard viewing window, graph $7x + 10y = 50$ and $4x + 5y = -25$. Do these lines appear to be parallel?

3. Now graph each pair of equations given above using the viewing window $[-100, 100, -100, 100]$, with Xscl $= 10$ and Yscl $= 10$. Do the lines still appear to be parallel?

Graphs of nonparallel lines may appear parallel in certain viewing windows. In order to determine whether two lines are parallel, we can look at their slopes.

Slope and Parallel Lines

Two lines are parallel if they have the same slope.

EXAMPLE 7   Determine whether each of the given pairs of lines is parallel.

a) $y = 3x - 7$ and $y - 3x = 8$
b) $7x + 10y = 50$ and $4x + 5y = -25$

Solution

a) The slope of the line $y = 3x - 7$ is 3. To find the slope of the line $y - 3x = 8$, we first solve for y:

$$y - 3x = 8$$
$$y = 3x + 8. \qquad \text{Adding } 3x \text{ to both sides}$$

The slope of the line $y - 3x = 8$ is also 3, so the lines are parallel.

b) We find the slope of each line:

$$7x + 10y = 50$$
$$10y = -7x + 50 \qquad \text{Adding } -7x \text{ to both sides}$$
$$y = -\tfrac{7}{10}x + 5; \qquad \text{Multiplying both sides by } \tfrac{1}{10}$$
$$\text{The slope is } -\tfrac{7}{10}.$$

$$4x + 5y = -25$$
$$5y = -4x - 25 \qquad \text{Adding } -4x \text{ to both sides}$$
$$y = -\tfrac{4}{5}x - 5. \qquad \text{Multiplying both sides by } \tfrac{1}{5}$$
$$\text{The slope is } -\tfrac{4}{5}.$$

Since the slopes are not the same, the lines are not parallel.

If one line is vertical and another is horizontal, they are perpendicular. There are other instances in which two lines are perpendicular.

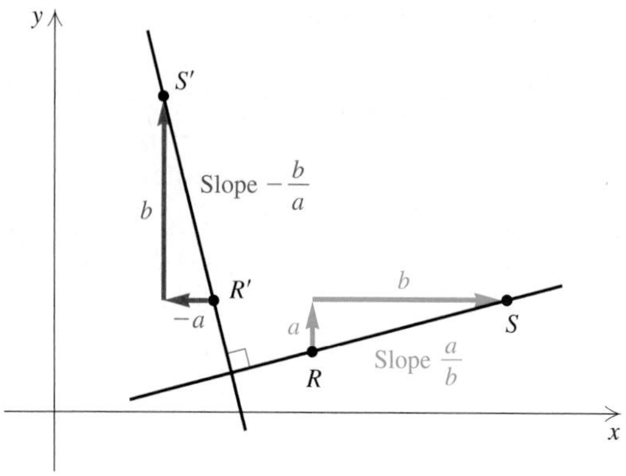

You may want to use sketches to illustrate that if two lines are perpendicular, then if one rises, the other falls, and if one is "steep," the other is not.

Consider a line $\overleftrightarrow{RS}$, as shown in the graph above, with slope a/b. Then think of rotating the figure 90° to get a line $\overleftrightarrow{R'S'}$ perpendicular to $\overleftrightarrow{RS}$. For the new line, the rise and the run are interchanged, but the run is now negative. Thus the slope of the new line is $-b/a$. Let's multiply the slopes:

$$\frac{a}{b}\left(-\frac{b}{a}\right) = -1.$$

This can help us determine which lines are perpendicular.

Slope and Perpendicular Lines

Two lines are perpendicular if the product of their slopes is -1. (If one line has slope m, the slope of a line perpendicular to it is $-1/m$. That is, we take the reciprocal and change the sign.) Lines are also perpendicular if one is vertical and the other is horizontal.

Squaring a Viewing Window

If the units on the x-axis are a different length than those on the y-axis, two lines that are perpendicular may not appear to be so when graphed. Finding a viewing window with units the same length on both axes is called *squaring* the viewing window.

Windows can be squared by choosing the ZSQUARE option in the ZOOM menu. They can be squared manually by choosing the portions of the axes shown in the correct proportion. For example, if the ratio of the length of a viewing window to its height is 3:2, a window like $[-9, 9, -6, 6]$ will be squared.

EXAMPLE 8 Determine whether the lines given by the equations $3x - y = 7$ and $x + 3y = 1$ are perpendicular, and check by graphing.

Solution   To determine the slope of each line, we solve for y to find slope–intercept form:

$$3x - y = 7$$
$$-y = -3x + 7 \qquad \text{Adding } -3x \text{ to both sides}$$
$$y = 3x - 7; \qquad \text{Multiplying both sides by } -1$$

$$x + 3y = 1$$
$$3y = -x + 1 \qquad \text{Adding } -x \text{ to both sides}$$
$$y = -\tfrac{1}{3}x + \tfrac{1}{3}. \qquad \text{Multiplying both sides by } \tfrac{1}{3}$$

The slopes of the lines are 3 and $-\frac{1}{3}$. Since $3 \cdot \left(-\frac{1}{3}\right) = -1$, the lines are perpendicular.

To check, we graph $y_1 = 3x - 7$ and $y_2 = -\frac{1}{3}x + \frac{1}{3}$. The graphs are shown in a standard viewing window on the left below. Note that the lines do not appear to be perpendicular. If we press [ZOOM] [5], the lines are graphed in a squared viewing window as shown in the graph on the right below, and they do appear perpendicular. We can perform a better visual check by laying a corner of a piece of paper on the screen. If the lines are perpendicular, they should exactly fit the corner.

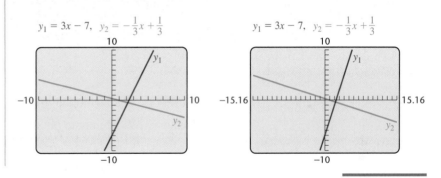

EXAMPLE 9   Consider the function given by $f(x) = \frac{2}{3}x - 8$.

a) Write an equation for a linear function g with a graph parallel to the graph of f and a y-intercept of $\left(0, \frac{3}{4}\right)$.

b) Write an equation for a linear function h with a graph perpendicular to the graph of f and a y-intercept of $\left(0, \frac{3}{4}\right)$.

Solution

a) Since g is linear, it can be written in the form $g(x) = mx + b$. To find g, we must determine its slope and y-intercept. The slope of the line given by $f(x) = \frac{2}{3}x - 8$ is $\frac{2}{3}$. Therefore, the slope of a parallel line is $\frac{2}{3}$. Since we are given that the y-intercept of the graph of g is $\left(0, \frac{3}{4}\right)$, we have $g(x) = \frac{2}{3}x + \frac{3}{4}$.

b) Since the slope of the graph of f is $\frac{2}{3}$, the slope of a line perpendicular to the graph is $-\frac{3}{2}$. Thus, $m = -\frac{3}{2}$ and $b = \frac{3}{4}$, so $h(x) = -\frac{3}{2}x + \frac{3}{4}$.

Recognizing Linear Equations

Is every equation of the form $Ax + By = C$ linear? To find out, suppose that A and B are nonzero and solve for y:

$$Ax + By = C$$

$$By = -Ax + C \qquad \text{Adding } -Ax \text{ to both sides}$$

$$y = -\frac{A}{B}x + \frac{C}{B}. \qquad \text{Dividing both sides by } B$$

Since the last equation is a slope–intercept equation, we see that $Ax + By = C$ is a linear equation when $A \neq 0$ and $B \neq 0$.

But what if A or B (but not both) is 0? If A is 0, then $By = C$ and $y = C/B$. If B is 0, then $Ax = C$ and $x = C/A$. In the first case, the graph is a horizontal line; in the second case, the line is vertical. In either case, $Ax + By = C$ is a linear equation when A or B (but not both) is 0. We have now justified the following result.

The Standard Form of a Linear Equation

Any equation of the form $Ax + By = C$, where A, B, and C are real numbers and A and B are not both 0, is linear.

Any equation of the form $Ax + By = C$ is said to be a linear equation in *standard form*.

EXAMPLE 10 Determine whether the equation $y = x^2 - 5$ is linear.

Solution   We attempt to put the equation in standard form:

$$y = x^2 - 5$$

$$-x^2 + y = -5. \qquad \text{Adding } -x^2 \text{ to both sides}$$

This last equation is not linear because it has an x^2-term.

We can see this as well from the graph of the equation.

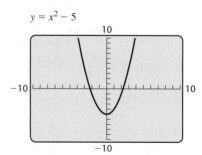

Only linear equations have graphs that are straight lines. Also, only linear graphs have a constant slope. Were you to try to calculate the slope between several pairs of points in Example 10, you would find that the slopes vary.

2.5

Exercise Set

FOR EXTRA HELP

Digital Video
Tutor CD 2
Videotape 3

Student's
Solutions
Manual

Tutor
Center

AW Math
Tutor Center

InterAct Math

Math*XP*

MathXL

MyMathLab

For each equation, find the slope. If the slope is undefined, state this.

1. $y - 7 = 5$ 0

2. $x + 1 = 7$ Undefined

3. $3x = 6$ Undefined

4. $y - 3 = 5$ 0

5. $4y = 20$ 0

6. $19 = -6y$ 0

7. $9 + x = 12$ Undefined

8. $2x = 18$ Undefined

9. $2x - 4 = 3$ Undefined

10. $5y - 1 = 16$ 0

11. $5y - 4 = 35$ 0

12. $2x - 17 = 3$ Undefined

13. $3y + x = 3y + 2$ Undefined

14. $x - 4y = 12 - 4y$ Undefined

15. $5x - 2 = 2x - 7$ Undefined

16. $5y + 3 = y + 9$ 0

Aha! **17.** $y = -\frac{2}{3}x + 5$ $-\frac{2}{3}$

18. $y = -\frac{3}{2}x + 4$ $-\frac{3}{2}$

Graph by hand.

19. $y = 4$ ⊡

20. $x = -1$ ⊡

21. $x = 2$ ⊡

22. $y = 5$ ⊡

23. $4 \cdot f(x) = 20$ ⊡

24. $6 \cdot g(x) = 12$ ⊡

25. $3x = -15$ ⊡

26. $2x = 10$ ⊡

27. $4 \cdot g(x) + 3x = 12 + 3x$ ⊡

28. $3 - f(x) = 2$ ⊡

Find the intercepts. Then graph by using the intercepts and a third point as a check.

29. $x + y = 5$ ⊡

30. $x + y = 4$ ⊡

31. $y = 2x + 8$ ⊡

32. $y = 3x + 9$ ⊡

33. $3x + 5y = 15$ ⊡

34. $5x - 4y = 20$ ⊡

35. $2x - 3y = 18$ ⊡

36. $3x + 2y = 12$ ⊡

37. $7x = 3y - 21$ ⊡

38. $5y = -15 + 3x$ ⊡

39. $f(x) = 3x - 8$ ⊡

40. $g(x) = 2x - 9$ ⊡

41. $1.4y - 3.5x = -9.8$ ⊡

42. $3.6x - 2.1y = 22.68$ ⊡

43. $5x + 2g(x) = 7$ ⊡

44. $3x - 4f(x) = 11$ ⊡

For each function, determine which of the given viewing windows will show both intercepts.

45. $f(x) = 20 - 4x$ (c)

   a) $[-10, 10, -10, 10]$ **b)** $[-5, 10, -5, 10]$
   c) $[-10, 10, -10, 30]$ **d)** $[-10, 10, -30, 10]$

 Answers to Exercises 19–44 can be found on p. A-57.

46. $g(x) = 3x + 7$ (a)

   a) $[-10, 10, -10, 10]$ **b)** $[-1, 15, -1, 15]$
   c) $[-15, 5, -15, 5]$ **d)** $[-10, 10, -30, 0]$

47. $p(x) = -35x + 7000$ (d)

   a) $[-10, 10, -10, 10]$
   b) $[-35, 0, 0, 7000]$
   c) $[-1000, 1000, -1000, 1000]$
   d) $[0, 500, 0, 10,000]$

48. $r(x) = 0.2 - 0.01x$ (b)

   a) $[-10, 10, -10, 10]$ **b)** $[-5, 30, -1, 1]$
   c) $[-1, 1, -5, 30]$ **d)** $[0, 0.01, 0, 0.2]$

Without graphing, tell whether the graphs of each pair of equations are parallel.

49. $x + 8 = y$,   Yes
$y - x = -5$

50. $2x - 3 = y$,   Yes
$y - 2x = 9$

51. $y + 9 = 3x$,   Yes
$3x - y = -2$

52. $y + 8 = -6x$,   No
$-2x + y = 5$

53. $f(x) = 3x + 9$,   No
$2y = -6x - 2$

54. $f(x) = -7x - 9$,
$-3y = 21x + 7$ Yes

Without graphing, tell whether the graphs of each pair of equations are perpendicular.

55. $f(x) = 4x - 3$,   Yes
$4y = 7 - x$

56. $2x - 5y = -3$,   No
$2x + 5y = 4$

57. $x + 2y = 7$,   No
$2x + 4y = 4$

58. $y = -x + 7$,   Yes
$f(x) = x + 3$

Write an equation for a linear function parallel to the given line with the given y-intercept.

59. $y = 3x - 2$; $(0, 9)$ $f(x) = 3x + 9$

60. $y = -5x + 7$; $(0, -2)$ $f(x) = -5x - 2$

61. $2x + y = 3$; $(0, -5)$ $f(x) = -2x - 5$

62. $3x = y + 10$; $(0, 1)$ $f(x) = 3x + 1$

63. $2x + 5y = 8$; $\left(0, -\frac{1}{3}\right)$ $f(x) = -\frac{2}{5}x - \frac{1}{3}$

64. $3x - 6y = 4$; $\left(0, \frac{4}{5}\right)$ $f(x) = \frac{1}{2}x + \frac{4}{5}$

Aha! **65.** $3y = 12$; $(0, -5)$ $f(x) = -5$

66. $5 = 10y$; $(0, 12)$ $f(x) = 12$

Write an equation for a linear function perpendicular to the given line with the given y-intercept.

67. $y = x - 3$; $(0, 4)$
$f(x) = -x + 4$

68. $y = 2x - 7$; $(0, -3)$
$f(x) = -\frac{1}{2}x - 3$

69. $2x + 3y = 6$; $(0, -4)$
$f(x) = \frac{3}{2}x - 4$

70. $4x + 2y = 8$; $(0, 8)$
$f(x) = \frac{1}{2}x + 8$

71. $5x - y = 13$; $\left(0, \frac{1}{5}\right)$
$f(x) = -\frac{1}{5}x + \frac{1}{5}$

72. $2x - 5y = 7$; $\left(0, -\frac{1}{8}\right)$
$f(x) = -\frac{5}{2}x - \frac{1}{8}$

Determine whether each equation is linear. Find the slope of any nonvertical lines.

73. $5x - 3y = 15$
Linear; $\frac{5}{3}$

74. $3x + 5y + 15 = 0$
Linear; $-\frac{3}{5}$

75. $16 + 4y = 10$
Linear; 0

76. $3x - 12 = 0$
Linear; line is vertical

77. $3g(x) = 6x^2$
Not linear

78. $2x + 4f(x) = 8$
Linear; $-\frac{1}{2}$

79. $3y = 7(2x - 4)$
Linear; $\frac{14}{3}$

80. $2(5 - 3x) = 5y$
Linear; $-\frac{6}{5}$

81. $g(x) - \dfrac{1}{x} = 0$ Not linear

82. $f(x) + \dfrac{1}{x} = 0$
Not linear

83. $\dfrac{f(x)}{5} = x^2$ Not linear

84. $\dfrac{g(x)}{2} = 3 + x$
Linear; 2

85. *Meteorology.*   Wind chill is a measure of how cold the wind makes you feel. Below are some measurements of wind chill for a 15-mph breeze. How can you tell from the data that a linear function will give an approximate fit?

Temperature	15-mph Wind Chill
30°	19°
25°	13°
20°	6°
15°	0°
10°	−7°
5°	−13°
0°	−19°

Sources: National Oceanic & Atmospheric Administration; www.nws.noaa.gov

86. Explain why examining the graphs of two lines is not the best way to tell whether the lines are parallel.

Skill Maintenance

Simplify.

87. $-\frac{3}{7} \cdot \frac{7}{3}$ [1.2] −1

88. $\frac{5}{4}\left(-\frac{4}{5}\right)$ [1.2] −1

89. $3(2x - y + 7)$ [1.3]
$6x - 3y + 21$

90. $-2(x + 5y - 1)$ [1.3]
$-2x - 10y + 2$

91. $-5[x - (-3)]$ [1.3]
$-5x - 15$

92. $-2[x - (-4)]$ [1.3]
$-2x - 8$

93. $\frac{2}{3}\left[x - \left(-\frac{1}{2}\right)\right] - 1$ [1.3]
$\frac{2}{3}x - \frac{2}{3}$

94. $-\frac{3}{2}\left(x - \frac{2}{5}\right) - 3$
[1.3] $-\frac{3}{2}x - \frac{12}{5}$

Synthesis

95. Jim tries to avoid fractions as often as possible. Under what conditions will graphing using intercepts allow him to avoid fractions? Why?

96. Under what condition(s) will the *x*- and *y*-intercepts of a line coincide? What would the equation for such a line look like?

97. Give an equation, in standard form, for the line whose *x*-intercept is 5 and whose *y*-intercept is −4.
$4x - 5y = 20$

98. Find the *x*-intercept of $y = mx + b$, assuming that $m \neq 0$. $\left(-\frac{b}{m}, 0\right)$

In Exercises 99–102, assume that r, p, and s are nonzero constants and that x and y are variables. Determine whether each equation is linear.

99. $rx + 3y = p - s$
Linear

100. $py = sx - ry + 2$
Linear

101. $r^2x = py + 5$ Linear

102. $x/r - py = 17$
Linear

103. Suppose that two linear equations have the same *y*-intercept but that equation A has an *x*-intercept that is half the *x*-intercept of equation B. How do the slopes compare?   The slope of equation B is $\frac{1}{2}$ the slope of equation A.

Consider the linear equation

$$ax + 3y = 5x - by + 8.$$

104. Find *a* and *b* if the graph is a horizontal line passing through $(0, 2)$. $a = 5, b = 1$

105. Find *a* and *b* if the graph is a vertical line passing through $(4, 0)$. $a = 7, b = -3$

106. Since a vertical line is not the graph of a function, many graphing calculators cannot graph equations of the form $x = a$. Some graphing calculators can draw vertical lines using the DRAW menu. Use the VERTICAL option of the DRAW menu to graph each of the following equations.
   a) $x = 3.6$
   b) $x = -1.52$
   c) $3x - 5 = 7x + 2$
   d) $2(x - 5) = x + 10$

107. A table of values can be used to determine whether two lines are parallel. If the difference between the *y*-values for two lines is the same for all *x*-values, the lines are parallel. Use the TABLE feature of a graphing calculator to determine whether each of the following pairs of lines is parallel.
   a) $2.3x - 3.4y = 9.8$,   Yes
      $1.84x = 2.72y - 17.4$
   b) $2.56y + 3.2x - 7.2 = 0$,   No
      $5.12y + 6.3x = 14.3$

2.6

Point–Slope Equations ■ Curve Fitting ■ Linear Regression

Introduction to Curve Fitting: Point–Slope Form

Specifying a line's slope and one point through which the line passes enables us to draw the line. In this section, we study how this same information can be used to produce an *equation* of the line.

Point–Slope Equations

Suppose that a line of slope m passes through the point (x_1, y_1). For any other point (x, y) to lie on this line, we must have

$$\frac{y - y_1}{x - x_1} = m.$$

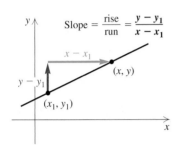

$$\text{Slope} = \frac{\text{rise}}{\text{run}} = \frac{y - y_1}{x - x_1}$$

It is tempting to use this last equation as an equation of the line of slope m that passes through (x_1, y_1). The only problem with this form is that when x and y are replaced with x_1 and y_1, we have $\frac{0}{0} = m$, a false equation. To avoid this difficulty, we multiply both sides by $x - x_1$ and simplify:

$$(x - x_1)\frac{y - y_1}{x - x_1} = m(x - x_1) \qquad \textbf{Multiplying both sides by } x - x_1$$

$$y - y_1 = m(x - x_1). \qquad \textbf{Removing a factor equal to 1:}$$
$$\frac{x - x_1}{x - x_1} = 1$$

This is the *point–slope* form of a linear equation.

Point–Slope Equation

The *point–slope equation* of a line with slope m, passing through (x_1, y_1), is

$$y - y_1 = m(x - x_1).$$

EXAMPLE 1 Find and graph an equation of the line passing through $(3, 4)$ with slope $-\frac{1}{2}$.

Solution   We substitute in the point–slope equation:

$$y - y_1 = m(x - x_1)$$
$$y - 4 = -\tfrac{1}{2}(x - 3). \qquad \textbf{Substituting}$$

To graph this point–slope equation, we count off a slope of $-\frac{1}{2}$, starting at $(3, 4)$. Then we draw the line.

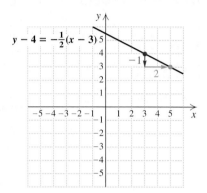

$$y - 4 = -\frac{1}{2}(x - 3)$$

EXAMPLE 2   Find a linear function that has a graph passing through the points $(-1, -5)$ and $(3, -2)$.

Solution   We first determine the slope of the line and then use the point–slope equation. Note that

$$m = \frac{-5 - (-2)}{-1 - 3} = \frac{-3}{-4} = \frac{3}{4}.$$

Since the line passes through $(3, -2)$, we have

$$y - (-2) = \tfrac{3}{4}(x - 3) \qquad \textbf{Substituting into the point–slope equation}$$
$$y + 2 = \tfrac{3}{4}x - \tfrac{9}{4}. \qquad \textbf{Using the distributive law}$$

Before using function notation, we isolate y:

$$y = \tfrac{3}{4}x - \tfrac{9}{4} - 2 \qquad \textbf{Subtracting 2 from both sides}$$
$$y = \tfrac{3}{4}x - \tfrac{17}{4} \qquad\qquad -\tfrac{9}{4} - \tfrac{8}{4} = -\tfrac{17}{4}$$
$$f(x) = \tfrac{3}{4}x - \tfrac{17}{4}. \qquad \textbf{Using function notation}$$

You can check that substituting $(-1, -5)$ instead of $(3, -2)$ in the point–slope equation will yield the same expression for $f(x)$.

EXAMPLE 3   Tattoo Removal.   In 1996, an estimated 275,000 Americans visited a doctor for tattoo removal. That figure was expected to grow to 410,000 in 2000. (*Source*: Mike Meyers, staff writer, Star-Tribune Newspaper of the Twin Cities Minneapolis–St. Paul, copyright 2000) Assuming constant growth since 1995, how many people will visit a doctor for tattoo removal in 2005?

Solution

1. Familiarize.   Constant growth indicates a constant rate of change, so a linear relationship can be assumed. We let n represent the number of people, in thousands, who visit a doctor for tattoo removal and t the number of years since 1995. Since we will write the number of tattoo removals as a function of the number of years since 1995, we write t on the horizontal axis and n on the vertical axis and form the pairs $(1, 275)$ and $(5, 410)$. After choosing suitable scales on the two axes, we draw the graph.

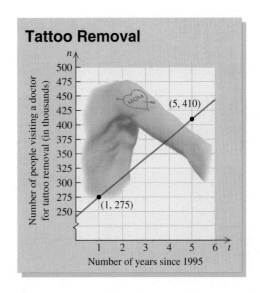

2. Translate.   To find an equation relating n and t, we first find the slope of the line. This corresponds to the *growth rate*:

$$m = \frac{410 \text{ thousand people } - 275 \text{ thousand people}}{5 \text{ years } - 1 \text{ year}}$$

$$= \frac{135 \text{ thousand people}}{4 \text{ years}}$$

$$= 33.75 \text{ thousand people per year.}$$

Next, we use the point–slope equation and solve for n:

$n - 275 = 33.75(t - 1)$	**Writing point–slope form**
$n - 275 = 33.75t - 33.75$	**Using the distributive law**
$n = 33.75t + 241.25.$	**Adding 275 to both sides**

3. Carry out.   Using function notation, we have

$$n(t) = 33.75t + 241.25.$$

To estimate the number of people who will visit a doctor for tattoo removal in 2005, we find

$n(10) = 33.75 \cdot 10 + 241.25$	**2005 is 10 years from 1995.**
$= 578.75.$	**This represents 578,750 people.**

4. Check.   To check, we can repeat our calculations. We could also extend the graph to see if $(10, 578.75)$ appears to be on the line.

5. State.   Assuming constant growth, there will be about 578,750 people visiting a doctor for tattoo removal in 2005.

Connecting the Concepts

We have now studied the slope–intercept, point–slope, and standard forms of a linear equation. These are the most common ways in which linear equations are written. Depending on what information we are given and what information we are seeking, one form may be more useful than the others. A referenced summary is given below.

Slope–intercept form, $y = mx + b$ or $f(x) = mx + b$	• Useful when an equation is needed and the slope and y-intercept are given. See Example 6 on p. 125. • Useful when a line's slope and y-intercept are needed. See Example 5 on p. 124. • Useful when solving equations graphically. See Example 1 on p. 95. • Commonly used for linear functions.
Standard form, $Ax + By = C$	• Allows for easy calculation of intercepts. See Example 4 on p. 139. • Will prove useful in future work. See Sections 3.1–3.3.
Point–slope form, $y - y_1 = m(x - x_1)$	• Useful when an equation is needed and the slope and a point on the line are given. See Example 1 on pp. 147–148. • Useful when a linear function is needed and two points on its graph are given. See Example 2 on p. 148. • Will prove useful in future work with curves and tangents in calculus.

Curve Fitting

It is not uncommon to see tables of information, or *data*. The process of understanding and interpreting data is called *data analysis*. One helpful tool in data analysis is **curve fitting**, or finding an algebraic equation that describes the data. If the data can be plotted and appear to have a linear pattern, we can fit a linear equation to the data.

EXAMPLE 4 Following are three graphs of sets of data. Determine whether each appears to be linear.

a)

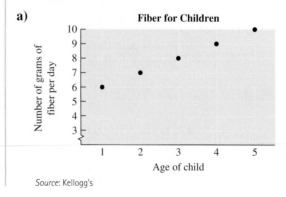

Source: Kellogg's

b)

Source: U.S. Department of Transportation, Office of Consumer Affairs, *Air Travel Consumer Report*

c)

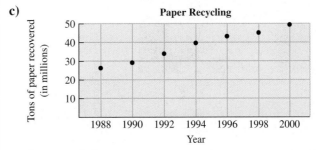

Source: www.afandpa.org

Solution   In order for data to be linear, the points must lie, at least approximately, on a straight line. The rate of change is constant for a linear function, so the change in the quantity on the vertical axis should be about the same for each unit on the horizontal axis.

a) The points lie on a straight line, so the data are linear. The rate of change is constant: 1 gram of fiber for each year of age of a child.

b) Note that the number of complaints is nearly the same for the years 1990 through 1995 and then it begins to increase. The rate of change is not constant, and the points do not lie on a straight line. The data are not linear.

c) The points lie approximately on a straight line. The data appear to be linear.

If data appear to be linear, we can fit a linear function to the data using the point–slope equation.

EXAMPLE 5   Paper Recycling.   The amount of paper recovered in the United States for various years is shown in the following table and graph.

a) Fit a linear function to the data.

b) Graph the function and use it to estimate the amount of paper that will be recovered in 2003.

Year	Amount of Paper Recovered (in millions of tons)
1988	26.2
1990	29.1
1992	34.0
1994	39.7
1996	43.1
1998	45.1
2000	49.4

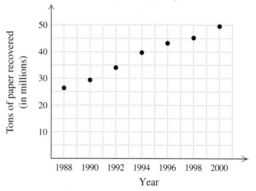

Paper Recycling

Source: www.afandpa.org

Solution

a) To make the numbers easier to work with, we will redefine the year t as the number of years since 1988. Thus, 1988 corresponds to $t = 0$, 1990 corresponds to $t = 2$, and so on. We let P represent the amount of paper recovered, in millions of tons. Thus we are looking for an equation of the form $P(t) = mx + b$.

We choose two points and determine the slope of the line containing those points. Since the data do not lie exactly on a straight line, the function found will depend on the choice of the points. We choose the points $(1990, 29.1)$ and $(1998, 45.1)$, or $(2, 29.1)$ and $(10, 45.1)$, with the first coordinate now representing the number of years since 1988. The slope of the line containing these points is

$$m = \frac{45.1 - 29.1}{10 - 2} = \frac{16}{8} = 2.$$

Then, using the point $(2, 29.1)$ and the slope 2, we have

$$P(t) - 29.1 = 2(t - 2) \qquad \textbf{Substituting into the point–slope equation}$$
$$P(t) - 29.1 = 2t - 4 \qquad \textbf{Multiplying}$$
$$P(t) = 2t + 25.1. \qquad \textbf{Solving for } P(t)$$

b) If we graph $P(t) = 2t + 25.1$ with the data, we see that the line only approximates the data. Not all the points lie on the line.

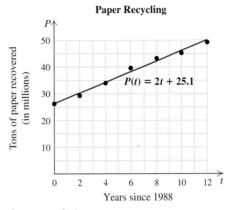

Paper Recycling

Source: www.afandpa.org

To estimate the amount of paper that will be recovered in 2003, we find $P(15)$:

$$P(15) = 2 \cdot 15 + 25.1 = 30 + 25.1 = 55.1 \qquad \textbf{2003} - \textbf{1988} = \textbf{15}$$

About 55.1 million tons of paper will be recovered in 2003.

The line that *best* describes the data may not actually go through any of the given points. There are different methods for finding an equation of a line that fits a set of data. These methods generally consider all the points, not just two, when fitting an equation to data. The most commonly used method is *linear regression*.

The development of the method of linear regression belongs to a later mathematics course, but most graphing calculators offer regression as a way of fitting a line or curve to a set of data.

Linear Regression

Fitting a curve to data is done using the STAT CALC menu. Enter the values of the independent variable as L1 and the values of the dependent variable as L2. To fit a line to the data, choose the LinReg option of the STAT CALC menu and press ENTER . (If the data are entered in lists other than L1 and L2, the names of the list must follow the LinReg command.) The "best-fit" equation will be given in slope–intercept form $y = ax + b$. The values of r^2 and r that may appear on the screen give an indication of how well the regression line fits the data. When r^2 is close to 1, the regression line is a good fit for the data. We call r the *coefficient of correlation*. The values of r^2 and r will appear only if the *diagnostics* are turned on.

In order to graph the equation found by regression, we need to copy it to the equation-editor screen. One way to do this is to list the function name after the linear regression command but before pressing [ENTER], as shown on the following screen.

LinReg(ax+ b) Y₁

EXAMPLE 6   Use linear regression to fit a line to the data on paper recycling given in Example 5. Graph the line with the data and estimate the amount of paper recovered in 2003.

Solution   We first enter the data, with the number of years since 1988 as L1 and the number of millions of tons of paper recycled as L2.

L1	L2	L3	1
0	26.2	------	
2	29.1		
4	34		
6	39.7		
8	43.1		
10	45.1		
12	49.4		
L1(1)=0			

Next, we turn Plot1 on, if necessary, and clear any equations listed in the [Y=] screen. Since years vary from 0 to 12 and the amount of paper varies from 26.2 million tons to 49.4 million tons, we set a viewing window of $[0, 15, 0, 60]$, with Yscl = 10.

To calculate the equation, we choose the LinReg option in the STAT CALC menu, then select Y1 from the VARS Y-VARS function menu, and finally press [ENTER].

The screen on the left below indicates that the equation is

$$y = 1.976785714x + 26.225.$$

The screen in the middle shows this equation entered as Y1, and the screen on the right shows the points plotted and the line graphed.

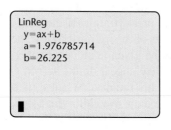

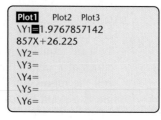

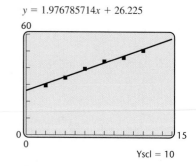

$y = 1.976785714x + 26.225$

Yscl = 10

1.
$y - 3 = -2(x - 2)$

2.
$y - 4 = 5(x - 7)$

3.
$y - 7 = 3(x - 4)$

4.
$y - 3 = 2(x - 7)$

5.
$y - (-4) = \frac{1}{2}(x - (-2))$,
or $y + 4 = \frac{1}{2}(x + 2)$

6.
$y - (-7) = x - (-5)$, or
$y + 7 = x + 5$

7.
$y - 0 = -1(x - 8)$, or
$y = -(x - 8)$

8.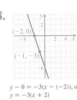
$y - 0 = -3(x - (-2))$, or
$y = -3(x + 2)$

9.
$y - 8 = \frac{2}{5}(x - (-3))$,
or $y - 8 = \frac{2}{5}(x + 3)$

10.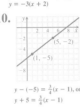
$y - (-5) = \frac{3}{4}(x - 1)$, or
$y + 5 = \frac{3}{4}(x - 1)$

Using the VALUE option of the CALC menu for $x = 15$ or entering $Y_1(15)$ on the home screen, we estimate the number of tons of paper recovered in 2003 to be 55.9 million tons.

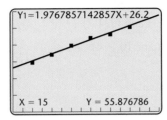

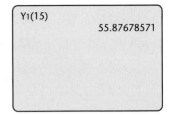

We can compare the equations and predictions found in Examples 5 and 6. For ease in comparison, we round the coefficients of the linear regression equation.

Method	Function	Amount of Paper Recovered in 2003
Point–slope equation	$P(t) = 2t + 25.1$	55.1 million tons
Linear regression	$y = 1.98x + 26.2$	55.9 million tons

Although linear regression provides an equation that more accurately represents *all* the data, in many cases a point–slope equation is perfectly adequate, especially if you are working by hand.

21. $f(x) = -\frac{3}{5}x - \frac{23}{5}$ **22.** $f(x) = -\frac{1}{5}x + \frac{3}{5}$ **25.** $f(x) = \frac{2}{7}x - 5$
26. $f(x) = \frac{1}{4}x + 3$

2.6

Exercise Set

FOR EXTRA HELP

Digital Video Tutor CD 2 Videotape 3 | Student's Solutions Manual | Tutor Center AW Math Tutor Center | InterAct Math | MathXL | MyMathLab

Find an equation in point–slope form of the line having the specified slope and containing the point indicated. Then graph the line.

1. $m = -2, (2, 3)$

2. $m = 5, (7, 4)$

3. $m = 3, (4, 7)$

4. $m = 2, (7, 3)$

5. $m = \frac{1}{2}, (-2, -4)$

6. $m = 1, (-5, -7)$

7. $m = -1, (8, 0)$

8. $m = -3, (-2, 0)$

9. $m = \frac{2}{5}, (-3, 8)$

10. $m = \frac{3}{4}, (1, -5)$

For each point–slope equation listed, state the slope and a point on the graph.

11. $y - 4 = \frac{2}{7}(x - 1)$ $\frac{2}{7}; (1, 4)$ **12.** $y - 3 = 9(x - 2)$ $9; (2, 3)$

$-5; (7, -2)$
13. $y + 2 = -5(x - 7)$

$-\frac{2}{9}; (-5, 1)$
14. $y - 1 = -\frac{2}{9}(x + 5)$

15. $y - 1 = -\frac{5}{3}(x + 2)$ $-\frac{5}{3}; (-2, 1)$
16. $y + 7 = -4(x - 9)$ $-4; (9, -7)$

Aha! **17.** $y = \frac{4}{7}x$ $\frac{4}{7}; (0, 0)$
18. $y = 3x$ $3; (0, 0)$

Find an equation of the line having the specified slope and containing the indicated point. Write your final answer as a linear function in slope–intercept form.

19. $m = 4, (2, -3)$
$f(x) = 4x - 11$

20. $m = -4, (-1, 5)$
$f(x) = -4x + 1$

21. $m = -\frac{3}{5}, (4, -7)$

22. $m = -\frac{1}{5}, (-2, 1)$

23. $m = -0.6, (-3, -4)$
$f(x) = -0.6x - 5.8$

24. $m = 2.3, (4, -5)$
$f(x) = 2.3x - 14.2$

Aha! **25.** $m = \frac{2}{7}, (0, -5)$
26. $m = \frac{1}{4}, (0, 3)$

Find an equation of the line containing each pair of points. Write your final answer as a linear function in slope–intercept form.

27. (1, 4) and (5, 6)

$f(x) = -\frac{5}{2}x + 11$

28. (2, 6) and (4, 1)

$f(x) = \frac{1}{2}x + \frac{7}{2}$

29. (2.5, −3) and (6.5, 3)

$f(x) = 1.5x - 6.75$

30. (2, −1.3) and (7, 1.7)

$f(x) = 0.6x - 2.5$

Aha! **31.** (1, 3) and (0, −2)

$f(x) = 5x - 2$

32. (−3, 0) and (0, −4)

$f(x) = -\frac{4}{3}x - 4$

33. (−2, −3) and (−4, −6) $f(x) = \frac{3}{2}x$

34. (−4, −7) and (−2, −1) $f(x) = 3x + 5$

In Exercises 35–44, assume that a constant rate of change exists for each model formed.

35. *Records in the 400-meter Run.*   In 1930, the record for the 400-m run was 46.8 sec. In 1970, it was 43.8 sec. Let $R(t)$ represent the record in the 400-m run and t the number of years since 1930.

a) Find a linear function that fits the data.

b) Use the function of part (a) to estimate the record in 2003; in 2006.   41.325 sec; 41.1 sec

c) When will the record be 40 sec?   2021

36. *Records in the 1500-meter Run.*   In 1930, the record for the 1500-m run was 3.85 min. In 1950, it was 3.70 min. Let $R(t)$ represent the record in the 1500-m run and t the number of years since 1930.

a) Find a linear function that fits the data.

b) Use the function of part (a) to estimate the record in 2002; in 2006.   3.31 min; 3.28 min

c) When will the record be 3.1 min?   2030

37. *PAC Contributions.*   In 1992, Political Action Committees (PACs) contributed $178.6 million to congressional candidates. In 2000, the figure rose to $243.1 million. (*Source: Congressional Research Service and Federal Election Commission*) Let $A(t)$ represent the amount of PAC contributions, in millions, and t the number of years since 1992.

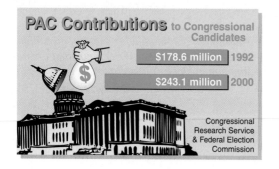

38. *Consumer Demand.*   Suppose that 6.5 million lb of coffee are sold when the price is $8 per pound, and 4.0 million lb are sold when it is $9 per pound.

$A(t) = 8.0625t + 178.6$

a) Find a linear function that fits the data.

b) Use the function of part (a) to estimate the amount of PAC contributions in 2006. $291.475 million

a) Find a linear function that expresses the amount of coffee sold as a function of the price per pound. $A(p) = -2.5p + 26.5$

b) Use the function of part (a) to estimate how much consumers would be willing to buy at a price of $6 per pound.   11.5 million lb

39. *Recycling.*   In 1993, Americans recycled 43.8 million tons of solid waste. In 1998, the figure grew to 62.2 million tons. (*Source: Statistical Abstract of the United States, 2000*) Let $N(t)$ represent the number of tons recycled, in millions, and t the number of years since 1993.

$N(t) = 3.68t + 43.8$

a) Find a linear function that fits the data.

b) Use the function of part (a) to estimate the amount recycled in 2005.   87.96 million tons

40. *Seller's Supply.*   Suppose that suppliers are willing to sell 5.0 million lb of coffee at a price of $8 per pound and 7.0 million lb at $9 per pound.

a) Find a linear function that expresses the amount suppliers are willing to sell as a function of the price per pound. $A(p) = 2p - 11$

b) Use the function of part (a) to estimate how much suppliers would be willing to sell at a price of $6 per pound.   1 million lb

Aha! **41.** *Life Expectancy of Females in the United States.*   In 1990, the life expectancy of females was 78.8 yr. In 2000, it was 79.8 yr. (*Source: Vital Statistics of the United States*) Let $E(t)$ represent life expectancy and t the number of years since 1990.

$E(t) = 0.1t + 78.8$

a) Find a linear function that fits the data.

b) Use the function of part (a) to estimate the life expectancy of females in 2008.   80.6 yr

42. *Life Expectancy of Males in the United States.*   In 1990, the life expectancy of males was 71.8 yr. In 2000, it was 74.4 yr. Let $E(t)$ represent life expectancy and t the number of years since 1990.

a) Find a linear function that fits the data.

b) Use the function of part (a) to estimate the life expectancy of males in 2007.   76.22 yr

35. (a) $R(t) = -0.075t + 46.8$ **36. (a)** $R(t) = -0.0075t + 3.85$ **42. (a)** $E(t) = 0.26t + 71.8$

43. *National Park Land.*   In 1994, the National Park system consisted of about 74.9 million acres. By 1998, the figure had grown to 77.7 million acres. (*Source*: *Statistical Abstract of the United States*, 2000) Let $A(t)$ represent the amount of land in the National Park system, in millions of acres, t years after 1994.

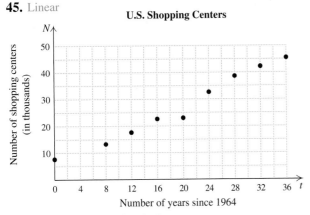

$A(t) = 0.7t + 74.9$

a) Find a linear function that fits the data.

b) Use the function of part (a) to estimate the amount of land in the National Park system in 2006.
83.3 million acres

44. *Pressure at Sea Depth.*   The pressure 100 ft beneath the ocean's surface is approximately 4 atm (atmospheres), whereas at a depth of 200 ft, the pressure is about 7 atm.

a) Find a linear function that expresses pressure as a function of depth. $P(d) = 0.03d + 1$

b) Use the function of part (a) to determine the pressure at a depth of 690 ft.   21.7 atm

Determine whether the data in each graph appear to be linear.

45. Linear

U.S. Shopping Centers

Source: International Council of Shopping Centers

46.

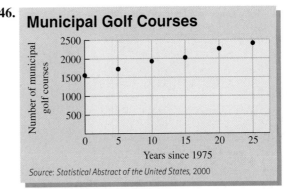

Municipal Golf Courses

Source: *Statistical Abstract of the United States*, 2000

Linear

47.

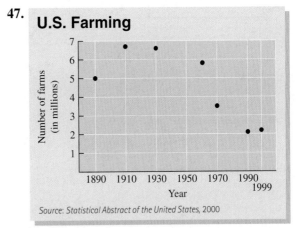

U.S. Farming

Source: *Statistical Abstract of the United States*, 2000

Not linear

48.

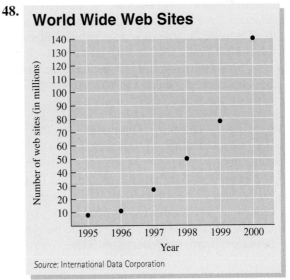

World Wide Web Sites

Source: International Data Corporation

Not linear

49.

Media Usage by Eighteen-Year-Olds

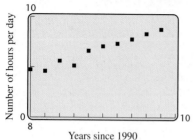

Source: Veronis, Suhler & Associates

Linear

50.

U.S Worker Leisure Time

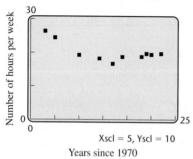

Source: The Wall Street Journal Almanac, 1998

Not linear

51. *Shopping Centers.*   The number of shopping centers S in the United States has greatly increased since 1964, as shown by the graph in Exercise 45.

a) Use the points $(8, 13)$ and $(24, 33)$ to find a linear function $S(t)$ that expresses the number of shopping centers, in thousands, in the United States t years after 1964. $S(t) = 1.25t + 3$

b) Use the function of part (a) to estimate the number of shopping centers in 2010.   60,500 shopping centers

52. *Golf Courses.*   The number of municipal golf courses G is increasing, as shown by the graph in Exercise 46.

a) Use the points $(5, 1704)$ and $(15, 2012)$ to find a linear function $G(t)$ that expresses the number of municipal golf courses t years after 1975. $G(t) = 30.8t + 1550$

b) Use the function of part (a) to estimate the number of municipal golf courses in 2006.   About 2505 golf courses

53. *Life Expectancy of Females in the United States.*   The following table shows the life expectancy of women who were born in the United States in selected years.

Life Expectancy of Women

Year	Life Expectancy (in years)
1950	71.1
1960	73.1
1970	74.4
1980	77.5
1990	78.8
2000	79.8

Sources: Vital Statistics of the United States; *The New York Times Almanac*, 2002

a) Use linear regression to find a linear function that can be used to predict the life expectancy W of a woman as a function of the year in which she was born. (Let $x =$ the number of years since 1900.) $W(x) = 0.182x + 62.13333333$

b) Predict the life expectancy of a woman in 2008 and compare your answer with the answer to Exercise 41.   81.8 yr; this estimate is higher

54. *Life Expectancy of Males in the United States.*   The following table shows the life expectancy of males born in the United States in selected years.

Life Expectancy of Men

Year	Life Expectancy (in years)
1950	65.6
1960	66.6
1970	67.1
1980	70.0
1990	71.8
2000	74.4

Sources: Vital Statistics of the United States; *The New York Times Almanac*, 2002

a) Use linear regression to find a linear function that can be used to predict the life expectancy M of a man as a function of the year in which he was born. (Let $x =$ the number of years since 1900.)

b) Predict the life expectancy of a man in 2007 and compare your answer with the answer to Exercise 42.   74.96 yr; this estimate is lower

54. (a) $M(x) = 0.1785714286x + 55.85714286$

55. *Cost of a Cellular Phone.* As the number of people using cellular phones has increased, the average local monthly bill has declined. The following table shows the average monthly bill for cellular phone subscribers in the United States for various years.

Monthly Cellular Phone Bills

Year	Average Local Monthly Bill
1987	$96.83
1989	89.30
1991	72.74
1993	61.48
1995	51.00
1997	42.78
1999	41.24

Sources: The Wall Street Almanac, 1998; Statistical Abstract of the United States, 2000

a) Use linear regression to find a linear function that can be used to predict the average local monthly cellular phone bill C as a function of the year. (Let x = the number of years since 1987.)
b) Estimate the average local monthly bill for a cellular phone in 2003. $14.78

56. *Child-Rearing Costs.* The following table shows the estimated annual expenditures on a child in 1999 by a family with an annual income of $50,000.

Annual Family Expenditure per Child

Age of Child	Annual Expenditure
1	$8450
4	8660
7	8700
13	9390
16	9530

Source: Department of Agriculture, Center for Nutrition Policy and Promotion, Expenditures on Children by Families, 1999 Annual Report

a) Use linear regression to find a linear function that can be used to predict the annual expenditure E on a child as a function of the age x of the child.
b) Estimate the annual expenditure on a 10-year-old child in 1999. $9083

55. (a) $C(x) = -5.027678571x + 95.21892857$
56. (a) $E(x) = 75.93023256x + 8323.372093$

TW **57.** Rosewood Graphics recently promised its employees 6% raises each year for the next five years. Amy currently earns $30,000 a year. Can she use a linear function to predict her salary for the next five years? Why or why not?

TW **58.** Examine the model of the monthly cellular phone bill given in Exercise 55. Will the estimate for a monthly bill in 2006 be reasonable? What might you conclude will happen with average monthly cellular phone cost?

Skill Maintenance

Simplify. [1.3]
59. $(3x^2 + 5x) + (2x - 4)$ $3x^2 + 7x - 4$
60. $(5t^2 - 2t) - (4t + 3)$ $5t^2 - 6t - 3$

Evaluate.

61. $\dfrac{2t - 6}{4t + 1}$, for $t = 3$ [1.1] 0
62. $(3t - 1)(4t + 20)$, for $t = -5$ [1.2] 0
63. $2x - 5y$, for $x = 3$ and $y = -1$ [1.2] 11
64. $3x - 7y$, for $x = -2$ and $y = 4$ [1.2] -34

Synthesis

TW **65.** In 1986, Political Action Committees contributed $132.7 million to congressional candidates. Does this information make your answer to Exercise 37(b) seem too low or too high? Why?

TW **66.** On the basis of your answers to Exercises 41 and 42, would you predict that at some point in the future the life expectancy of males will exceed that of females? Why or why not?

For Exercises 67–72, assume that a linear equation models each situation.

67. *Depreciation of a Computer.* After 6 mos of use, the value of Pearl's computer had dropped to $900. After 8 mos, the value had gone down to $750. How much did the computer cost originally? $1350

68. *Temperature Conversion.* Water freezes at 32° Fahrenheit and at 0° Celsius. Water boils at 212°F and at 100°C. What Celsius temperature corresponds to a room temperature of 70°F? 21.1°C

69. *Cellular Phone Charges.* The total cost of Mel's cellular phone was $230 after 5 mos of service and $390 after 9 mos. What costs had Mel already incurred when his service just began? $30

70. *Operating Expenses.*   The total cost for operating Ming's Wings was $7500 after 4 mos and $9250 after 7 mos. Predict the total cost after 10 mos. $11,000

71. *Medical Insurance.*   In 1993, health insurance companies collected $124.7 billion in premiums and paid out $103.6 billion in benefits. In 1996, they collected $137.1 billion in premiums and paid out $113.8 billion in benefits. Estimate the percentage of premiums paid out in benefits in 2005.   82.8%

72. On the basis of the information given in Exercises 38 and 40, determine at what price the supply will equal the demand. $8.33 per pound

Write an equation of the line containing the specified point and parallel to the indicated line.

73. $(3, 7)$, $x + 2y = 6$ $y = -\frac{1}{2}x + \frac{17}{2}$

74. $(-1, 4)$, $3x - y = 7$ $y = 3x + 7$

Write an equation of the line containing the specified point and perpendicular to the indicated line.

75. $(2, 5)$, $2x + y = -3$ $y = \frac{1}{2}x + 4$

76. $(4, 0)$, $x - 3y = 0$ $y = -3x + 12$

77. Specify the domain of your answer to Exercise 38(a). $\{p \mid 0 < p \le 10.6\}$

78. Specify the domain of your answer to Exercise 40(a). $\{p \mid p > 5.5\}$

79. For a linear function g, $g(3) = -5$ and $g(7) = -1$.
   a) Find an equation for g. $g(x) = x - 8$
   b) Find $g(-2)$. -10
   c) Find a such that $g(a) = 75$.   83

2.7

The Sum, Difference, Product, or Quotient of Two Functions ▪ Domains and Graphs

Domains and the Algebra of Functions

We now examine four ways in which functions can be combined, followed by a reexamination of domains.

The Sum, Difference, Product, or Quotient of Two Functions

Suppose that a is in the domain of two functions, f and g. The input a is paired with $f(a)$ by f and with $g(a)$ by g. The outputs can then be added to get $f(a) + g(a)$.

Interactive Discovery

Enter $y_1 = x + 4$ and $y_2 = x^2 + 1$. Next, enter $y_3 = y_1 + y_2$.

1. Set up a table that shows the values of the three functions. If y_1 represents $f(x)$ and y_2 represents $g(x)$, what does y_3 represent? $f(x) + g(x)$

2. Graph y_1, y_2, and y_3 using the same viewing window. How could you draw the graph of y_3 given the graphs of y_1 and y_2?   Add the y-values of y_1 and y_2 for each x-value.
Adding the expressions for y_1 and y_2 algebraically, we get

$$(x + 4) + (x^2 + 1) = x^2 + x + 5.$$

3. Enter $y_4 = x^2 + x + 5$ and compare the values of y_3 and y_4 using a table or a graph. How are these functions related? $y_3 = y_4$

We see that if $f(x) = x + 4$ and $g(x) = x^2 + 1$, then $f(x) + g(x) = x^2 + x + 5$, which can be regarded as a "new" function, written $(f + g)(x)$.

The Algebra of Functions

If f and g are functions and x is in the domain of both functions, then:

1. $(f + g)(x) = f(x) + g(x)$;
2. $(f - g)(x) = f(x) - g(x)$;
3. $(f \cdot g)(x) = f(x) \cdot g(x)$;
4. $(f/g)(x) = f(x)/g(x)$, provided $g(x) \neq 0$.

EXAMPLE 1   For $f(x) = x^2 - x$ and $g(x) = x + 2$, find the following.

a) $(f + g)(3)$

b) $(f - g)(x)$ and $(f - g)(-1)$

c) $(f/g)(x)$ and $(f/g)(-4)$

d) $(f \cdot g)(3)$

Solution

a) Since $f(3) = 3^2 - 3 = 6$ and $g(3) = 3 + 2 = 5$, we have

$$(f + g)(3) = f(3) + g(3)$$
$$= 6 + 5 \qquad \textbf{Substituting}$$
$$= 11.$$

Alternatively, we could first find $(f + g)(x)$:

$$(f + g)(x) = f(x) + g(x)$$
$$= x^2 - x + x + 2$$
$$= x^2 + 2. \qquad \textbf{Combining like terms}$$

Thus,

$$(f + g)(3) = 3^2 + 2 = 11. \qquad \textbf{Our results match.}$$

b) We have

$$(f - g)(x) = f(x) - g(x)$$
$$= x^2 - x - (x + 2) \qquad \textbf{Substituting}$$
$$= x^2 - 2x - 2. \qquad \textbf{Removing parentheses and combining like terms}$$

Thus,

$$(f - g)(-1) = (-1)^2 - 2(-1) - 2 \qquad \textbf{Using } (f - g)(x) \textbf{ is faster than using } f(x) - g(x).$$
$$= 1. \qquad \textbf{Simplifying}$$

c) We have

$$(f/g)(x) = f(x)/g(x)$$
$$= \frac{x^2 - x}{x + 2}. \qquad \textbf{We assume that } x \neq -2.$$

Thus,

$$(f/g)(-4) = \frac{(-4)^2 - (-4)}{-4 + 2} \qquad \textbf{Substituting}$$

$$= \frac{20}{-2} = -10.$$

d) Using our work in part (a), we have

$$(f \cdot g)(3) = f(3) \cdot g(3)$$
$$= 6 \cdot 5$$
$$= 30.$$

It is also possible to compute $(f \cdot g)(3)$ by first multiplying $x^2 - x$ and $x + 2$ using methods we will discuss in Chapter 5.

Domains and Graphs

Although applications involving products and quotients of functions rarely appear in newspapers, situations involving sums or differences of functions often do appear in print. For example, the following graphs are similar to those published by the California Department of Education to promote breakfast programs in which students eat a balanced meal of fruit or juice, toast or cereal, and 2% or whole milk. The combination of carbohydrate, protein, and fat gives a sustained release of energy, delaying the onset of hunger for several hours.

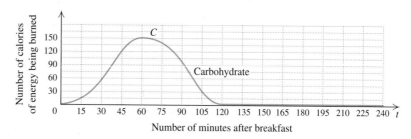

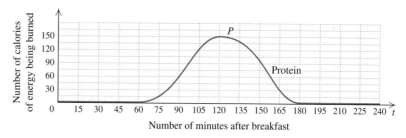

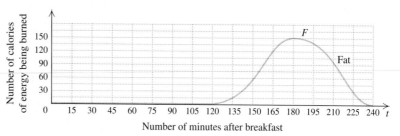

When the three graphs are superimposed, and the calorie expenditures added, it becomes clear that a balanced meal results in a steady, sustained supply of energy.

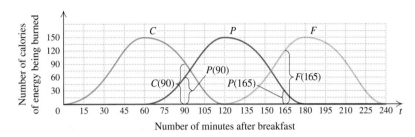

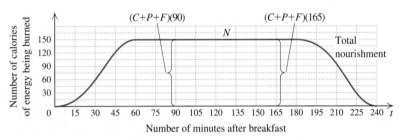

Note that for $t > 120$, we have $C(t) = 0$; for $t < 60$ or $t > 180$, we have $P(t) = 0$; and for $t < 120$, we have $F(t) = 0$. For any point on this last graph, we have

$$N(t) = (C + P + F)(t) = C(t) + P(t) + F(t).$$

To find $(f + g)(a)$, $(f - g)(a)$, $(f \cdot g)(a)$, or $(f/g)(a)$, we must first be able to find $f(a)$ and $g(a)$. Thus we need to ensure that a is in the domain of both f and g.

When a function is described by an equation, the domain is often unspecified. In such cases, the domain is the set of all numbers for which function values can be calculated.

EXAMPLE 2   For each equation, determine the domain of f.

a) $f(x) = |x|$

b) $f(x) = \dfrac{7}{2x - 6}$

Solution

a) We ask ourselves, "Is there any number x for which we cannot compute $|x|$?" Since we can find the absolute value of *any* number, the answer is no. Thus the domain of f is $\mathbb{R}$, the set of all real numbers.

b) Is there any number x for which $\dfrac{7}{2x - 6}$ cannot be computed? Since $\dfrac{7}{2x - 6}$ cannot be computed when $2x - 6$ is 0, the answer is yes. To determine what x-value causes the denominator to be 0, we set up and solve an equation:

$$2x - 6 = 0 \qquad \text{Setting the denominator equal to 0}$$
$$2x = 6 \qquad \text{Adding 6 to both sides}$$
$$x = 3. \qquad \text{Dividing both sides by 2}$$

Thus, 3 is *not* in the domain of f, whereas all other real numbers are. The domain of f is $\{x \mid x$ is a real number *and* $x \neq 3\}$.

A table can be used to determine whether a number is in a function's domain.

EXAMPLE 3 If $f(x) = \dfrac{7}{2x - 6}$, use a table to determine whether 0, 5, and 3 are in the domain of the function.

Solution We enter $y = 7/(2x - 6)$ and set up a table. Since we want to supply the values of x, we set Indpnt to Ask in the Table Setup before viewing the table. To find the function values, we enter x-values of 0, 5, and 3.

From the table, we see that $f(0) \approx -1.167$ and $f(5) = 1.75$. The ERROR entry indicates that $f(3)$ is not defined. This tells us that 0 and 5 are in the domain of f and 3 is not.

X	Y1	
0	−1.167	
5	1.75	
3	ERROR	
Y1=ERROR		

CAUTION! A graphing calculator is useful for checking whether a particular value of x is in the domain of a function. However, algebraic methods are needed in order to determine the entire domain, since it is impossible to check every real number with a calculator.

Interactive Discovery

Let $f(x) = \dfrac{5}{x}$ and $g(x) = \dfrac{2x - 6}{x + 1}$. Enter $y_1 = f(x)$, $y_2 = g(x)$, and $y_3 = y_1 + y_2$. Create a table of values for the functions with TblStart = −3, ΔTbl = 1, and Indpnt set to Auto.

1. What number is not in the domain of f? 0

2. What number is not in the domain of g? −1

3. What numbers are not in the domain of $f + g$? −1, 0

Now enter $y_4 = y_1 - y_2$, $y_5 = y_1 \cdot y_2$, and $y_6 = y_1/y_2$. Create a table of values for the functions.

4. Does the domain of $f - g$ appear to be the same as the domain of $f + g$?
Yes

5. Does the domain of $f \cdot g$ appear to be the same as the domain of $f + g$?
Yes

6. Does the domain of f/g appear to be the same as the domain of $f + g$?
No; 3 is not in the domain of f/g.

In the Interactive Discovery above, because division by 0 is not defined, we have

the domain of $f = \{x \mid x$ is a real number *and* $x \neq 0\}$

and

the domain of $g = \{x \mid x$ is a real number *and* $x \neq -1\}$.

In order to find $f(a) + g(a)$, $f(a) - g(a)$, or $f(a) \cdot g(a)$, we must know that a is in *both* of the above domains. Thus,

the domain of $f + g$ = the domain of $f - g$ = the domain of $f \cdot g$
$$= \{x \mid x \text{ is a real number } and \ x \neq 0 \ and \ x \neq -1\}.$$

The domain of f/g also excludes the number 3, because $g(3) = 0$:

the domain of $f/g = \{x \mid x$ is a real number *and*
$x \neq 0$ *and* $x \neq -1$ *and* $x \neq 3\}$.

Determining the Domain

The domain of $f + g$, $f - g$, or $f \cdot g$ is the set of all values common to the domains of f and g.

The domain of f/g is the set of all values common to the domains of f and g, excluding any values for which $g(x)$ is 0.

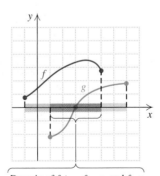

Domain of $f + g$, $f - g$, and $f \cdot g$

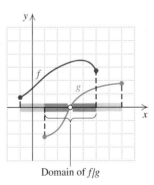

Domain of f/g

EXAMPLE 4   Given $f(x) = 1/x$ and $g(x) = 2x - 7$, find the domains of $f + g$, $f - g$, $f \cdot g$, and f/g.

Solution   The domain of f is $\{x \mid x$ is a real number *and* $x \neq 0\}$. The domain of g is $\mathbb{R}$. The domains of $f + g$, $f - g$, and $f \cdot g$ are the set of all elements common to both the domain of f and the domain of g. We have

$$\text{the domain of } f + g = \text{the domain of } f - g = \text{the domain of } f \cdot g$$
$$= \{x \,|\, x \text{ is a real number } and \; x \neq 0\}.$$

The domain of f/g is $\{x \,|\, x$ is a real number $and \; x \neq 0\}$, with the additional restriction that $g(x) \neq 0$. To determine what x-values would make $g(x) = 0$, we solve:

$$2x - 7 = 0 \qquad \textbf{Replacing } g(x) \textbf{ with } 2x - 7$$
$$2x = 7$$
$$x = \tfrac{7}{2}.$$

Since $g(x) = 0$ for $x = \tfrac{7}{2}$,

$$\text{the domain of } f/g = \left\{x \,|\, x \text{ is a real number } and \; x \neq 0 \; and \; x \neq \tfrac{7}{2}\right\}.$$

Division by 0 is not the only condition that can force restrictions on the domain of a function. In Chapter 7, we will examine functions similar to that given by $f(x) = \sqrt{x}$, for which the concern is taking the square root of a negative number.

2.7

Exercise Set

Let $f(x) = -3x + 1$ and $g(x) = x^2 + 2$. Find the following.

1. $f(2) + g(2)$ 1
2. $f(-1) + g(-1)$ 7
3. $f(5) - g(5)$ −41
4. $f(4) - g(4)$ −29
5. $f(-1) \cdot g(-1)$ 12
6. $f(-2) \cdot g(-2)$ 42
7. $f(-4)/g(-4)$ $\frac{13}{18}$
8. $f(3)/g(3)$ $-\frac{8}{11}$
9. $g(1) - f(1)$ 5
10. $g(2)/f(2)$ $-\frac{6}{5}$
11. $(f + g)(x)$ $x^2 - 3x + 3$
12. $(g - f)(x)$ $x^2 + 3x + 1$

Let $F(x) = x^2 - 2$ and $G(x) = 5 - x$. Find the following.

13. $(F + G)(x)$ $x^2 - x + 3$
14. $(F + G)(a)$ $a^2 - a + 3$
15. $(F + G)(-4)$ 23
16. $(F + G)(-5)$ 33
17. $(F - G)(3)$ 5
18. $(F - G)(2)$ −1
19. $(F \cdot G)(-3)$ 56
20. $(F \cdot G)(-4)$ 126
21. $(F/G)(x)$ $\frac{x^2 - 2}{5 - x}, x \neq 5$
22. $(G - F)(x)$ $-x^2 - x + 7$
23. $(F/G)(-2)$ $\frac{2}{7}$
24. $(F/G)(-1)$ $-\frac{1}{6}$

The following graph shows the number of women, in millions, who had a child in the last year. Here $W(t)$ represents the number of women under 30 who gave birth in year t, $R(t)$ the number of women 30 and older who gave birth in year t, and $N(t)$ the total number of women who gave birth in year t.

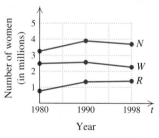

Source: Statistical Abstract of the United States, 2000

25. Use estimates of $R(1980)$ and $W(1980)$ to estimate $N(1980)$. $0.75 + 2.5 = 3.25$

26. Use estimates of $R(1990)$ and $W(1990)$ to estimate $N(1990)$. $1.3 + 2.6 = 3.9$

27. Which group of women was responsible for the drop in the number of births from 1990 to 1998?   Women under 30

28. Which group of women was responsible for the rise in the number of births from 1980 to 1990?   Women 30 and older

Often function addition is represented by stacking the individual functions directly on top of each other. The graph below indicates how the three major airports servicing New York City have been utilized. The braces indicate the values of the individual functions.

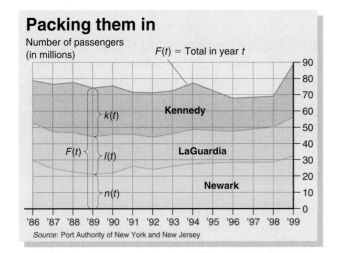

Packing them in

Number of passengers (in millions)

$F(t)$ = Total in year t

Source: Port Authority of New York and New Jersey

29. Estimate $(n + l)$ ('98). What does it represent?   ⊡

30. Estimate $(k + l)$ ('98). What does it represent?   ⊡

31. Estimate $(k - l)$ ('94). What does it represent?   ⊡

32. Estimate $(k - n)$ ('94). What does it represent?   ⊡

33. Estimate $(n + l + k)$ ('99). What does it represent?   ⊡

34. Estimate $(n + l + k)$ ('98). What does it represent?   ⊡

35. Find the domain of f.

a) $f(x) = \dfrac{5}{x - 3}$ ⊡

b) $f(x) = \dfrac{7}{6 - x}$ ⊡

c) $f(x) = 2x + 1$ ⊡

d) $f(x) = x^2 + 3$ ⊡

e) $f(x) = \dfrac{3}{2x - 5}$ ⊡

f) $f(x) = |3x - 4|$ ⊡

36. Find the domain of g.

a) $g(x) = \dfrac{3}{x - 1}$ ⊡

b) $g(x) = |5 - x|$ ⊡

c) $g(x) = \dfrac{9}{x + 3}$ ⊡

d) $g(x) = \dfrac{4}{3x + 4}$ ⊡

e) $g(x) = x^3 - 1$ ⊡

f) $g(x) = 7x - 8$ ⊡

⊡ Answers to Exercises 29–36 can be found on p. A-57.

For each pair of functions f and g, determine the domain of the sum, difference, and product of the two functions.

37. $f(x) = x^2$, $g(x) = 7x - 4$ ℝ

38. $f(x) = 5x - 1$, $g(x) = 2x^2$ ℝ

39. $f(x) = \dfrac{1}{x - 3}$, $g(x) = 4x^3$ $\{x \mid x$ is a real number *and* $x \neq 3\}$

40. $f(x) = 3x^2$, $g(x) = \dfrac{1}{x - 9}$ $\{x \mid x$ is a real number *and* $x \neq 9\}$

41. $f(x) = \dfrac{2}{x}$, $g(x) = x^2 - 4$ $\{x \mid x$ is a real number *and* $x \neq 0\}$

42. $f(x) = x^3 + 1$, $g(x) = \dfrac{5}{x}$ $\{x \mid x$ is a real number *and* $x \neq 0\}$

43. $f(x) = x + \dfrac{2}{x - 1}$, $g(x) = 3x^3$ $\{x \mid x$ is a real number *and* $x \neq 1\}$

44. $f(x) = 9 - x^2$, $g(x) = \dfrac{3}{x - 6} + 2x$ $\{x \mid x$ is a real number *and* $x \neq 6\}$

45. $f(x) = \dfrac{3}{x - 2}$, $g(x) = \dfrac{5}{4 - x}$ $\{x \mid x$ is a real number *and* $x \neq 2$ *and* $x \neq 4\}$

46. $f(x) = \dfrac{5}{x - 3}$, $g(x) = \dfrac{1}{x - 2}$ $\{x \mid x$ is a real number *and* $x \neq 3$ *and* $x \neq 2\}$

For each pair of functions f and g, determine the domain of f/g.

47. $f(x) = x^4$, $g(x) = x - 3$ $\{x \mid x$ is a real number *and* $x \neq 3\}$

48. $f(x) = 2x^3$, $g(x) = 5 - x$ $\{x \mid x$ is a real number *and* $x \neq 5\}$

49. $f(x) = 3x - 2$, $g(x) = 2x - 8$ $\{x \mid x$ is a real number *and* $x \neq 4\}$

50. $f(x) = 5 + x$, $g(x) = 6 - 2x$ $\{x \mid x$ is a real number *and* $x \neq 3\}$

51. $f(x) = \dfrac{3}{x - 4}$, $g(x) = 5 - x$ $\{x \mid x$ is a real number *and* $x \neq 4$ *and* $x \neq 5\}$

52. $f(x) = \dfrac{1}{2 - x}$, $g(x) = 7 - x$ $\{x \mid x$ is a real number *and* $x \neq 2$ *and* $x \neq 7\}$

53. $f(x) = \dfrac{2x}{x + 1}$, $g(x) = 2x + 5$ $\{x \mid x$ is a real number *and* $x \neq -1$ *and* $x \neq -\frac{5}{2}\}$

54. $f(x) = \dfrac{7x}{x - 2}$, $g(x) = 3x + 7$ $\{x \mid x$ is a real number *and* $x \neq 2$ *and* $x \neq -\frac{7}{3}\}$

For Exercises 55–62, consider the functions F and G as shown.

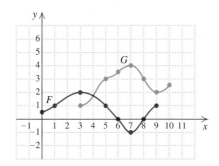

60. $\{x \mid 3 \le x \le 9\}$; $\{x \mid 3 \le x \le 9\}$; $\{x \mid 3 \le x \le 9$ *and $x \ne 6$ and $x \ne 8\}$*

55. Determine $(F + G)(5)$ and $(F + G)(7)$.   4; 3

56. Determine $(F \cdot G)(6)$ and $(F \cdot G)(9)$.   0; 2

57. Determine $(G - F)(7)$ and $(G - F)(3)$.   5; −1

58. Determine $(F/G)(3)$ and $(F/G)(7)$.   2; $-\frac{1}{4}$

59. Find the domains of F, G, $F + G$, and F/G.
$\{x \mid 0 \le x \le 9\}$; $\{x \mid 3 \le x \le 10\}$; $\{x \mid 3 \le x \le 9\}$; $\{x \mid 3 \le x \le 9\}$

60. Find the domains of $F - G$, $F \cdot G$, and G/F.

61. Graph $F + G$.   ⊡   **62.** Graph $G - F$.   ⊡

In the following graph, $W(t)$ represents the number of gallons of whole milk, $L(t)$ the number of gallons of lowfat milk, and $S(t)$ the number of gallons of skim milk consumed by the average American in year t.*

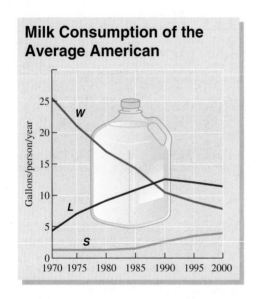

Milk Consumption of the Average American

TW **63.** Explain in words what $(W - S)(t)$ represents and what it would mean to have $(W - S)(t) < 0$.

TW **64.** Consider $(W + L + S)(t)$ and explain why you feel that total milk consumption per person did or did not change over the years 1970–2000.

Skill Maintenance

Evaluate each expression using the values provided. [1.1]

65. $3x^2 - 5y$, for $x = 10$ and $y = 6$ 270

66. $3x + 10 \div 5y$, for $x = 4$ and $y = 2$ 16

---

*Sources: Copyright 1990, CSPI. Adapted from *Nutrition Action Healthletter* (1875 Connecticut Avenue, N.W., Suite 300, Washington, DC 20009-5728. $24.00 for 10 issues); USDA Agricultural Fact Book 2000, USDA Economic Research Service

⊡ Answers to Exercises 61, 62, 79, and 81 can be found on p. A-57.

Write scientific notation. [1.4] 4.506×10^{12}

67. 0.00000703 7.03×10^{-6} **68.** $4,506,000,000,000$

Write in decimal notation. [1.4]
                                                            0.0006037
69. 4.3×10^8 $430,000,000$ **70.** 6.037×10^{-4}

Translate each of the following. Do not solve. [1.6]

71. The sum of two consecutive integers is 145.   Let x
represent the first integer; $x + (x + 1) = 145$
72. The difference between a number and its opposite
is 20.   Let n represent the number; $n - (-n) = 20$

Synthesis

TW **73.** If $f(x) = c$, where c is some positive constant,
describe how the graphs of $y = g(x)$ and
$y = (f + g)(x)$ will differ.

TW **74.** Examine the graphs following Example 1 and explain how they might be modified to represent the
absorption of 200 mg of Advil® taken four times
a day. $\{x \mid x$ is a real number *and*
                          $x \ne -\frac{5}{2}$ and $x \ne -3$ and
75. Find the domain of f/g, if $x \ne 1$ and $x \ne -1\}$

$$f(x) = \frac{3x}{2x + 5} \quad \text{and} \quad g(x) = \frac{x^4 - 1}{3x + 9}.$$

76. Find the domain of F/G, if

$$F(x) = \frac{1}{x - 4} \quad \text{and} \quad G(x) = \frac{x^2 - 4}{x - 3}. \quad \{x \mid x \text{ is a}$$
real number *and $x \ne 4$ and $x \ne 3$ and $x \ne 2$ and $x \ne -2\}$*
77. Sketch the graph of two functions f and g such that
the domain of f/g is   Answers may vary.

$$\{x \mid -2 \le x \le 3 \text{ and } x \ne 1\}.$$

78. Find the domain of m/n, if

$$m(x) = 3x \text{ for } -1 < x < 5$$

and

$$n(x) = 2x - 3. \quad \begin{array}{l} \{x \mid x \text{ is a real number } and \\ \quad -1 < x < 5 \text{ and } x \ne \frac{3}{2}\} \end{array}$$

79. Find the domains of $f + g$, $f - g$, $f \cdot g$, and f/g, if

$$f = \{(-2, 1), (-1, 2), (0, 3), (1, 4), (2, 5)\}$$

and

$$g = \{(-4, 4), (-3, 3), (-2, 4), (-1, 0), (0, 5), (1, 6)\}.$$

80. For f and g as defined in Exercise 79, find   ⊡
$(f + g)(-2)$, $(f \cdot g)(0)$, and $(f/g)(1)$.   5; 15; $\frac{2}{3}$

81. Write equations for two functions f and g such that
the domain of $f + g$ is

$$\{x \mid x \text{ is a real number and } x \ne -2 \text{ and } x \ne 5\}.$$
                                                            ⊡
82. Using the window $[-5, 5, -1, 9]$, graph $y_1 = 5$,
$y_2 = x + 2$, and $y_3 = \sqrt{x}$. Then predict what shape
the graphs of $y_1 + y_2$, $y_1 + y_3$, and $y_2 + y_3$ will take.
Use a graph to check each prediction.

83. Let $y_1 = 2.5x + 1.5$, $y_2 = x - 3$, and $y_3 = y_1/y_2$.
Depending on whether the CONNECTED or DOT mode
is used, the graph of y_3 appears as follows.

CONNECTED MODE   DOT MODE

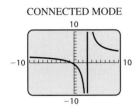 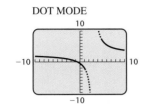

Use algebra to determine which graph more accurately represents y_3.

84. Use the TABLE feature on a graphing calculator to
check your answers to Exercises 37, 43, 45, and 51.

85. Use the graphs of f and g, shown below, to match
each of $(f + g)(x)$, $(f - g)(x)$, $(f \cdot g)(x)$, and
$(f/g)(x)$ with its graph.

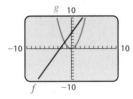

a) $(f + g)(x)$ IV   **b)** $(f - g)(x)$ I
c) $(f \cdot g)(x)$ II   **d)** $(f/g)(x)$ III

I

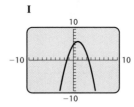

II

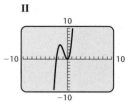

III

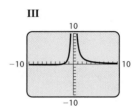

IV

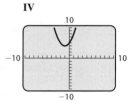

83. The domain of $y_3 = \dfrac{2.5x + 1.5}{x - 3}$ is $\{x \mid x$ is a real number
and $x \neq 3\}$. The CONNECTED mode graph contains the line
$x = 3$, whereas the DOT mode graph contains no points having
3 as the first coordinate. Thus the DOT mode graph represents
y_3 more accurately.

Collaborative Corner

Time On Your Hands

   Focus: The algebra of functions
   Time: 10–15 minutes
   Group size: 2–3

The graph and data at right chart the average
retirement age $R(x)$ and life expectancy $E(x)$ of
U.S. citizens in year x.

ACTIVITY

1. Working as a team, perform the appropriate
calculations and then graph $E - R$.

2. What does $(E - R)(x)$ represent? In what fields
of study or business might the function $E - R$
prove useful?

3. Should E and R really be calculated separately
for men and women? Why or why not?

4. What advice would you give to someone
considering early retirement?

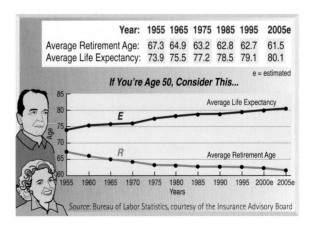

Year:	1955	1965	1975	1985	1995	2005e
Average Retirement Age:	67.3	64.9	63.2	62.8	62.7	61.5
Average Life Expectancy:	73.9	75.5	77.2	78.5	79.1	80.1

e = estimated

If You're Age 50, Consider This...

Source: Bureau of Labor Statistics, courtesy of the Insurance Advisory Board

2 Chapter Summary and Review

Key Terms

Domain, p. 80
Range, p. 80
Function, p. 80
Projection, p. 83
Relation, p. 84
Input, p. 84
Output, p. 84
Dependent variable, p. 85
Independent variable, p. 85
Constant function, p. 86
Dummy variable, p. 86

Interpolation, p. 87
Extrapolation, p. 87
Equivalent equations, p. 99
Identity, p. 103
Contradiction, p. 103
Conditional equation, p. 103
Solution set, p. 105
Empty set, p. 105
Formula, p. 110
Linear function, p. 121
y-intercept, p. 122

Slope, p. 123
Rise, p. 123
Run, p. 123
Rate of change, p. 127
Salvage value, p. 129
Depreciate, p. 129
Zero slope, p. 136
Undefined slope, p. 137
x-intercept, p. 138
Curve fitting, p. 150
Linear regression, p. 153

Important Properties and Formulas

The Vertical-Line Test

A graph represents a function if it is not possible to draw a vertical line that intersects the graph more than once.

___

The addition principle for equations:

$$a = b \quad \text{is equivalent to} \quad a + c = b + c.$$

The multiplication principle for equations:

$$\text{For } c \neq 0, a = b \text{ is equivalent to } a \cdot c = b \cdot c.$$

___

Five Steps for Problem Solving with Algebra

1. *Familiarize* yourself with the problem situation.
2. *Translate* to mathematical language.
3. *Carry out* some mathematical manipulation.
4. *Check* your possible answer in the original problem.
5. *State* the answer clearly.

___

To solve a formula for a given letter, identify the letter, and:

1. Multiply on both sides to clear fractions or decimals, if that is needed.
2. Combine like terms on each side where convenient.
3. Get all terms with the letter being solved for on one side of the equation and all other terms on the other side, using the addition principle.
4. Combine like terms again, if necessary. This may require factoring.
5. Solve for the letter in question, using the multiplication principle.

___

$$\text{Slope} = m = \frac{\text{rise}}{\text{run}} = \frac{\text{change in } y}{\text{change in } x} = \frac{y_2 - y_1}{x_2 - x_1}$$

___

Every horizontal line has a slope of 0.
The slope of a vertical line is undefined.

The x-intercept is $(a, 0)$. To find a, let $y = 0$ and solve the original equation for x.

The y-intercept is $(0, b)$. To find b, let $x = 0$ and solve the original equation for y.

The slope–intercept equation of a line is

$$y = mx + b.$$

The point–slope equation of a line is

$$y - y_1 = m(x - x_1).$$

The standard form of a linear equation is

$$Ax + By = C.$$

Parallel lines: The slopes are equal.
Perpendicular lines: The product of the slopes is -1.

The Algebra of Functions

1. $(f + g)(x) = f(x) + g(x)$
2. $(f - g)(x) = f(x) - g(x)$
3. $(f \cdot g)(x) = f(x) \cdot g(x)$
4. $(f/g)(x) = f(x)/g(x)$, provided $g(x) \neq 0$

Review Exercises

1. For the following graph of f, determine **(a)** $f(2)$; **(b)** the domain of f; **(c)** any x-values for which $f(x) = 2$; and **(d)** the range of f.

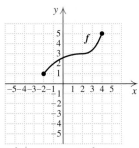

[2.1] **(a)** 3; **(b)** $\{x \mid -2 \le x \le 4\}$; **(c)** -1; **(d)** $\{y \mid 1 \le y \le 5\}$

2. Using the graph below, estimate the solution of $f(x) = g(x)$.   [2.2] 2

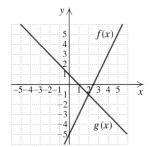

3. The function $A(t) = 0.233t + 5.87$ can be used to estimate the median age of cars in the United States t years after 1990 (*Source*: The Polk Co.). (In this context, a median age of 3 yr means that half the cars are more than 3 yr old and half are less.) Predict the median age of cars in 2010; that is, find $A(20)$.   [2.1] 10.53 yr

Solve each equation graphically and check.

4. $\frac{2}{3}x - 5 = 3$ [2.2] 12

5. $3x - 7 = 2(x + 1)$ [2.2] 9

Solve. If the solution set is $\varnothing$ or $\mathbb{R}$, classify the equation as a contradiction or as an identity.

6. $x - 4.9 = 1.7$ [2.2] 6.6

7. $\frac{2}{3}a = 9$ [2.2] $\frac{27}{2}$

8. $-9x + 4(2x - 3) = 5(2x - 3) + 7$ [2.2] $-\frac{4}{11}$

9. $3(x - 4) + 2 = x + 2(x - 5)$ [2.2] $\mathbb{R}$; identity

10. $5t - (9 - t) = 4t + 2(3 + t)$ [2.2] $\varnothing$; contradiction

11. $1.7x - 4.03 = -12.4(x - 0.2)$ [2.2] 0.46170

Solve.

12. A number is 17 less than another number. The sum of the two numbers is 115. Find the smaller number. [2.3] 49

13. *Landscaping.* A mature tree can be moved using a tree spade. Standard tree spades range from 20 in. to 92 in. in diameter. The diameter of the largest tree that can be moved successfully is one-tenth the diameter of the spade. (*Source: Popular Mechanics*, April 1998: 102) How large a tree spade would be needed to move a tree that is $5\frac{1}{2}$ in. in diameter?

[2.3] 55 in.

14. One angle of a triangle measures three times the second angle. The third angle measures twice the second angle. Find the measures of the angles. [2.3] 90°, 30°, 60°

15. A sheet of plastic has a thickness of 0.00015 mm. The sheet is 1.2 m by 79 m. Use scientific notation to find the volume of the sheet. ⊡

16. Solve for y: $6x - 5y = 3$.   [2.3] $y = \dfrac{3 - 6x}{-5}$, or $\dfrac{6x - 3}{5}$

17. Solve for m: $P = m/S$.   [2.3] $m = PS$

18. Solve for x: $c = mx - rx$.   [2.3] $x = \dfrac{c}{m - r}$

Find the slope and the y-intercept.

19. $g(x) = -4x - 9$ [2.4] Slope: -4; y-intercept: $(0, -9)$

20. $-6y + 2x = 7$ [2.4] Slope: $\frac{1}{3}$; y-intercept: $\left(0, -\frac{7}{6}\right)$

The following table shows the annual soft-drink production in the United States, in number of 12-oz cans per person (Sources: National Soft Drink Association; Beverage World).

Input, Year	Output, Number of 12-oz Cans per Person
1977	360
1987	480
1997	580

21. Use the data in the table to draw a graph and to estimate the annual soft-drink production in 1990. [2.1] About 510 cans per person

22. Use the graph from Exercise 21 to estimate the annual soft-drink production in 2005. [2.1] About 670 cans per person

Find the slope of each line. If the slope is undefined, state this.

23. Containing the points $(4, 5)$ and $(-3, 1)$ [2.4] $\frac{4}{7}$

24. Containing $(-16.4, 2.8)$ and $(-16.4, 3.5)$ [2.5] Undefined

25. Find the rate of change for the graph below. Use appropriate units. [2.4] $2500 of personal income per year since high school

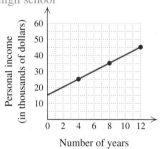

Number of years since high school

26. By March 1, 2000, U.S. builders had begun construction of 265,000 homes. By July 1, 2000, there were 795,000 begun. (*Source*: U.S. Department of Commerce) Calculate the rate at which new homes were being started. [2.4] 132,500 homes per month

27. The average cost of tuition at a state university t years after 1997 can be estimated by $C(t) = 645t + 9800$. What do the numbers 645 and 9800 signify? [2.4] 645 signifies that tuition is increasing at a rate of $645 a year; 9800 represents tuition costs in 1997

28. Find an equation for a linear function whose graph has slope $\frac{2}{7}$ and y-intercept $(0, -6)$. [2.4] $f(x) = \frac{2}{7}x - 6$

Graph by hand.

29. $f(x) = 5$ ⊡

30. $3 - x = 9$ ⊡

31. $-2x + 4y = 8$ ⊡

32. $x + 7y = 14$ ⊡

33. Determine an appropriate viewing window for the graph of $f(x) = -\frac{1}{7}x + 14$. Answers may vary.   ⊡

Determine whether each pair of lines is parallel, perpendicular, or neither.

34. $y + 5 = -x$,
$x - y = 2$
[2.5] Perpendicular

35. $3x - 5 = 7y$,
$7y - 3x = 7$
[2.5] Parallel

Determine whether each of these is a linear equation.

36. $2x - 7 = 0$ [2.5] Linear

37. $3x - 8f(x) = 7$ [2.5] Linear

38. $2a + 7b^2 = 3$ [2.5] Not linear

39. $2p - \dfrac{7}{q} = 1$ [2.5] Not linear

40. Find an equation in point–slope form of the line with slope -2 and containing $(-3, 4)$.   ⊡

41. Using function notation, write an equation for the line containing $(2, 5)$ and $(-4, -3)$. [2.6] $f(x) = \frac{4}{3}x + \frac{7}{3}$

⊡ Answers to Exercises 15, 29–33, and 40 can be found on pp. A-57 and A-58.

42. To join the Family Fitness Center, it costs $75 plus $15 a month. Use a graph to estimate the time required for the total cost to reach $180. [2.5] 7 months

43. In 1955, the U.S. minimum wage was $0.75, and in 1997, it was $5.15. Let W = the minimum wage, in dollars, t years after 1955.

a) Find an equation for a linear function that fits the data. [2.6] $W(t) = \frac{11}{105}t + 0.75$

b) Use the function of part (a) to predict the minimum wage in 2005. [2.6] $5.99

Answering Machines. *The following table shows the sales of telephone answering machines for the years 1990–2001.*

Sales of Answering Machines

Year	Sales (in millions)
1990	13.6
1991	15.4
1992	14.6
1993	16.3
1995	17.5
2000	23.0
2001	24.6

Source: Consumer Electronics Manufacturers Association

44. [2.6]

Linear

Yscl = 5

44. Graph the data and determine whether the relationship appears to be linear.

45. Let t = the number of years after 1990, and use the points $(2, 14.6)$ and $(10, 23.0)$ to find a linear function A, for which $A(t)$ expresses the sales, in millions, of answering machines t years after 1990. [2.6] $A(t) = 1.05t + 12.5$

46. Use linear regression to find a linear function A that can be used to estimate the number of answering machines sold, in millions, as a function of x years since 1980. [2.6] $A(x) = 0.9688442211x + 13.4281407$

Find the domain of f.

47. $f(x) = \dfrac{5}{x + 3}$ [2.7] $\{x \mid x$ is a real number and $x \neq -3\}$

48. $f(x) = \dfrac{x + 3}{5}$ [2.7] $\mathbb{R}$

Let $g(x) = 3x - 6$ and $h(x) = x^2 + 1$. Find the following.

49. $g(0)$ [2.1] -6

50. $h(-5)$ [2.1] 26

51. $(g \cdot h)(4)$ [2.7] 102

52. $(g - h)(-2)$ [2.7] -17

53. $(g/h)(-1)$ [2.7] $-\frac{9}{2}$

54. $g(a + b)$ [2.1] $3a + 3b - 6$

55. The domains of $g + h$ and $g \cdot h$ [2.7] $\mathbb{R}$

56. The domain of h/g [2.7] $\{x \mid x$ is a real number *and* $x \neq 2\}$

Synthesis

ᵀᵂ **57.** Explain why every function is a relation, but not every relation is a function.

ᵀᵂ **58.** Explain why the slope of a vertical line is undefined whereas the slope of a horizontal line is 0.

59. Find the y-intercept of the function given by
$$f(x) + 3 = 0.17x^2 + (5 - 2x)^x - 7.$$
[1.4], [2.5] -9

60. Determine the value of a such that the lines
$$3x - 4y = 12 \quad \text{and} \quad ax + 6y = -9$$
are parallel. [2.5] $-\frac{9}{2}$

61. Homespun Jellies charges $2.49 for each jar of preserves. Shipping charges are $3.75 for handling, plus $0.60 per jar. Find a linear function for determining the cost of shipping x jars of preserves. [2.6] $f(x) = 3.09x + 3.75$

62. Match each sentence with the most appropriate graph below.

a) Joni walks for 10 min to the train station, rides the train for 15 min, and then walks 5 min to the office. [2.4] III

b) During a workout, Phil bikes for 10 min, runs for 15 min, and then walks for 5 min. [2.4] IV

c) Sam pilots his motorboat for 10 min to the middle of the lake, fishes for 15 min, and then motors for another 5 min to another spot. [2.4] I

d) Patti waits 10 min for her train, rides the train for 15 min, and then runs for 5 min to her job. [2.4] II

I

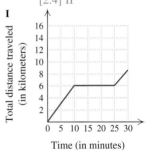

II

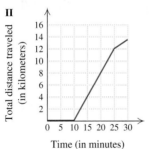

III

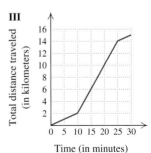

IV

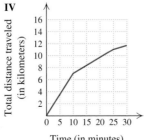

Chapter Test 2

1. For the following graph of f, determine **(a)** $f(-2)$; **(b)** the domain of f; **(c)** any x-value for which $f(x) = \frac{1}{2}$; and **(d)** the range of f. [2.1] **(a)** 1; **(b)** $\{x \mid -3 \le x \le 4\}$; **(c)** 3; **(d)** $\{y \mid -1 \le y \le 2\}$

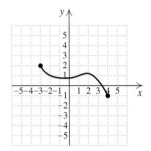

2. The function $S(t) = 1.2t + 21.4$ can be used to estimate the total U.S. sales of books, in billions of dollars, t years after 1992. [2.1] $40.6 billion
a) Predict the total U.S. sales of books in 2008.
b) What do the numbers 1.2 and 21.4 signify? ⊡

3. There were 43.3 million international visitors to the United States in 1995, and 48.5 million in 1999 (*Source*: Tourism Industries, International Trade Administration, Department of Commerce). Draw a graph and estimate the number of international visitors in 1997. [2.1] 46 million

4. Solve graphically and check:
$$1.2x - 5 = 3.6 + x. \quad \text{[2.2] 43}$$

Solve. If the solution set is ℝ or ∅, classify the equation as an identity or a contradiction.

5. $13x - 7 = 41x + 49$ [2.2] -2

6. $8t - (5 - 2t) = 5(2t - 1)$ [2.2] ℝ; identity

Solve.

7. Find three consecutive even integers such that the sum of the first, two times the second, and three times the third is 124. [2.3] 18, 20, 22

8. The changes in the salary of a vice president of a corporation for three consecutive years are, respectively, a 10% increase, a 15% increase, and a 5% decrease. What is the percent of total change for those three years? [2.3] 20.175% increase

⊡ Answers to Exercises 2(b) and 18–20 can be found on p. A-58.

9. *Astronomy.* The average distance from the planet Venus to the sun is 6.7×10^7 mi. About how far does Venus travel in one orbit around the sun? (Assume a circular orbit.) [2.3] About 4.2×10^8 mi

10. The surface area of a rectangular solid of length l, width w, and height h is given by $S = 2lh + 2lw + 2wh$. Solve for l.

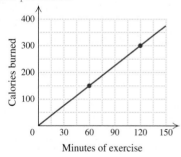

[2.3]
$$l = \frac{S - 2wh}{2h + 2w}$$

11. Solve for y: $3x^2 + 2y = x - 7$. [2.3] $y = \dfrac{x - 7 - 3x^2}{2}$

Find the slope and the y-intercept.

12. $f(x) = -\frac{3}{5}x + 12$ [2.4] Slope: $-\frac{3}{5}$; y-intercept: $(0, 12)$

13. $-5y - 2x = 7$ [2.4] Slope: $-\frac{2}{5}$; y-intercept: $\left(0, -\frac{7}{5}\right)$

Find the slope of the line containing the following points. If the slope is undefined, state this.

14. $(-2, -2)$ and $(6, 3)$ [2.4] $\frac{5}{8}$

15. $(-3.1, 5.2)$ and $(-4.4, 5.2)$ [2.4] 0

16. Find the rate of change for the graph below. Use appropriate units. [2.4] 75 calories per 30 minutes, or 2.5 calories per minute

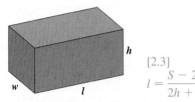

17. Find a linear function whose graph has slope -5 and y-intercept $(0, -1)$. [2.4] $f(x) = -5x - 1$

Graph by hand.

18. $f(x) = 5x - 2$ ⊡   **19.** $3 - x = 5 + x$ ⊡

20. $-2x + 5y = 12$ ⊡

21. Determine whether the standard viewing window shows the *x*- and *y*-intercepts of the graph of $f(x) = 2x + 9$.   [2.5] Yes

Determine without graphing whether each pair of lines is parallel, perpendicular, or neither.

22. $4y + 2 = 3x$,
$-3x + 4y = -12$
[2.5] Parallel

23. $y = -2x + 5$,
$2y - x = 6$
[2.5] Perpendicular

24. Which of the following are linear equations?
a) $8x - 7 = 0$
b) $4b - 9a^2 = 2$ [2.5] (a), (c)
c) $2x - 5y = 3$

25. Find an equation in point–slope form of the line with slope 4 and containing $(-2, -4)$.

26. Use function notation to write an equation for the line containing $(3, -1)$ and $(4, -2)$.

27. If you rent a van for one day and drive it 250 mi, the cost is \$100. If you drive it 300 mi, the cost is \$115. Let $C(m) =$ the cost, in dollars, of driving *m* miles.
a) Find a linear function that fits the data.
b) Use the function to determine how much it will cost to rent the van for one day and drive it 500 mi.   [2.6] \$175

28. *Accidental Deaths.*   The number of accidental deaths in the United States per 100,000 population is dropping, as shown in the following table.

Year	Number of Accidental Deaths per 100,000
1910	84.4
1920	71.2
1930	80.5
1940	73.4
1950	60.3
1960	52.1
1970	56.2
1980	46.5
1990	36.9
1999	35.5

Source: U.S. Department of Health and Human Services, National Center for Health Statistics

a) Use linear regression to find a linear function *A* that can be used to predict the number of accidental deaths per 100,000 population as a function of the number of years *x* since 1900.
b) Predict the number of accidental deaths per 100,000 population in 2010.
[2.6] 29.1 accidental deaths

29. Find the domain of $g(x) = \dfrac{x - 6}{2x + 1}$.

30. Find the following, given that $g(x) = -3x - 4$ and $h(x) = x^2 + 1$.
a) $h(-2)$ [2.1] 5
b) $(g \cdot h)(3)$ [2.7] -130
c) The domain of h/g [2.7] $\{x \mid x$ is a real number $and\ x \neq -\frac{4}{3}\}$

Synthesis

31. The function $f(t) = 5 + 15t$ can be used to determine a bicycle racer's location, in miles from the starting line, measured *t* hours after passing the 5-mi mark.   [2.1], [2.4] 30 mi
a) How far from the start will the racer be 1 hr and 40 min after passing the 5-mi mark?
b) Assuming a constant rate, how fast is the racer traveling?   [2.1], [2.4] 15 mph

Find an equation of the line.

32. Containing $(-3, 2)$ and parallel to the line $2x - 5y = 8$ [2.5], [2.6] $y = \frac{2}{5}x + \frac{16}{5}$

33. Containing $(-3, 2)$ and perpendicular to the line $2x - 5y = 8$ [2.5], [2.6] $y = -\frac{5}{2}x - \frac{11}{2}$

34. The graph of the function $f(x) = mx + b$ contains the points $(r, 3)$ and $(7, s)$. Express *s* in terms of *r* if the graph is parallel to the line $3x - 2y = 7$.
[2.6] $s = -\frac{3}{2}r + \frac{27}{2}$, or $\frac{27 - 3r}{2}$

35. Given that $f(x) = 5x^2 + 1$ and $g(x) = 4x - 3$, find an expression for $h(x)$ so that the domain of $f/g/h$ is $\{x \mid x$ is a real number $and\ x \neq \frac{3}{4}$ $and\ x \neq \frac{2}{7}\}$.
Answers may vary.   [2.7] $h(x) = 7x - 2$

---

25. [2.6] $y - (-4) = 4(x - (-2))$, or $y + 4 = 4(x + 2)$
26. [2.6] $f(x) = -x + 2$
27. [2.6] **(a)** $C(m) = 0.3m + 25$
28. [2.6] **(a)** $A(x) = -0.554571187x + 90.14595817$
29. [2.7] $\{x \mid x$ is a real number $and\ x \neq -\frac{1}{2}\}$

Systems of Linear Equations and Problem Solving

3

The most difficult part of problem solving is nearly always translating the problem situation to mathematical language. Once a problem has been translated, the solution is generally straightforward. In this chapter, we study *systems of equations* and how to solve them using graphing, substitution, elimination, and matrices. Systems of equations often provide the easiest way to translate into mathematics real-world situations from fields such as psychology, sociology, business, education, engineering, and science.

APPLICATION

TRAVEL. The numbers of U.S. travelers to Canada and to Europe are shown in the following table and graph. Estimate the year in which the numbers of travelers to Canada and to Europe will be the same. A system of linear equations can be used to solve this problem.

Year	Travelers to Canada (in millions)	Travelers to Europe (in millions)
1992	11.8	7.1
1994	12.5	8.2
1996	12.9	8.7
1998	14.9	11.1
2000	15.1	13.4

Source: Office of Travel and Tourism Industries

Yscl = 2

This problem appears as Example 6 in Section 3.1.

Translating ■ Identifying Solutions ■ Solving Systems Graphically ■ Models

Systems of Equations in Two Variables

Translating

Problems involving two unknown quantities are often solved most easily if we can first translate the situation to two equations in two unknowns.

EXAMPLE 1 Real Estate. Translate the following problem situation to mathematical language, using two equations.

> In 1996, the Simon Property Group and the DeBartolo Realty Corporation merged to form the largest real estate company in the United States, owning 183 shopping centers in 32 states. Prior to merging, Simon owned twice as many properties as DeBartolo. How many properties did each company own before the merger?

DeBartolo shareholders would receive 0.68 share of Simon common stock for each share of DeBartolo common stock. Simon also would agree to repay $1.5 billion in DeBartolo debt. At Tuesday's closing price of $ 23.625 a share for common stock, the transaction is valued at roughly $3 billion.

Executives say the proposed company, Simon DeBartolo Group, would be the largest real estate company in the United States, worth $7.5 billion. **Not included in the deal:** DeBartolo's ownership stake in the San Francisco 49ers, or the Indiana Pacers, owned separately by the Simon

Solution

1. **Familiarize.** We have already seen problems in which we need to look up certain formulas or the meaning of certain words. Here we need only observe that the words shopping centers and properties are being used interchangeably.

    Often problems contain information that has no bearing on the situation being discussed. In this case, the number 32 is irrelevant to the question being asked. Instead we focus on the number of properties owned and the phrase "twice as many." Rather than guess and check, let's proceed to the next step, using x for the number of properties originally owned by Simon and y for the number of properties originally owned by DeBartolo.

2. **Translate.** There are two statements to translate. First we look at the total number of properties involved:

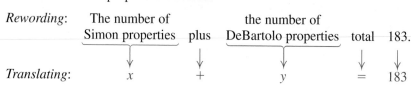

Rewording: The number of Simon properties plus the number of DeBartolo properties total 183.

Translating: x $+$ y $=$ 183

The second statement compares the number of properties that each company held before merging:

Rewording: The number of   twice the number of
               Simon properties   was   DeBartolo properties.

Translating: x $=$ $2 \cdot y$

We have now translated the problem to a pair, or **system, of equations**:

$$x + y = 183,$$
$$x = 2y.$$

We will complete the solution of this problem in Section 3.3.

Problems like Example 1 *can* be solved using one variable; however, as problems become complicated, you will find that using more than one variable (and more than one equation) is often the preferable approach.

EXAMPLE 2   Purchasing.   Recently the Champlain Valley Community Music Center purchased 120 stamps for $33.90. If the stamps were a combination of 23¢ postcard stamps and 37¢ first-class stamps, how many of each type were bought?

Solution

1. **Familiarize.**   To familiarize ourselves with this problem, let's guess that the music center bought 60 stamps at 23¢ each and 60 stamps at 37¢ each. The total cost would then be

$$60 \cdot \$0.23 + 60 \cdot \$0.37 = \$13.80 + \$22.20, \text{ or } \$36.00.$$

Since $\$36.00 \neq \$33.90$, our guess is incorrect. Rather than guess again, let's see how algebra can be used to translate the problem.

2. **Translate.**   We let $p =$ the number of postcard stamps and $f =$ the number of first-class stamps. The information can be organized in a table, which will help with the translating.

Type of Stamp	Postcard	First-class	Total
Number Sold	p	f	120
Price	$0.23	$0.37	
Amount	$0.23p	$0.37f	$33.90

$p + f = 120$

$0.23p + 0.37f = 33.90$

The first row of the table and the first sentence of the problem indicate that a total of 120 stamps were bought:

$$p + f = 120.$$

TEACHING TIP

You may want to encourage students to use tables to organize information. A properly organized table can lead directly to a translation.

Since each postcard stamp cost \$0.23 and p stamps were bought, $0.23p$ represents the amount paid, in dollars, for the postcard stamps. Similarly, $0.37f$ represents the amount paid, in dollars, for the first-class stamps. This leads to a second equation:

$$0.23p + 0.37f = 33.90. \qquad \text{Note that all units are in dollars.}$$

Multiplying both sides by 100, we can clear the decimals. This gives the following system of equations as the translation:

$$p + f = 120, \qquad \text{We complete the solution of}$$
$$23p + 37f = 3390. \qquad \text{this problem in Section 3.3.}$$

Identifying Solutions

A *solution* of a system of equations in two variables is an ordered pair of numbers that makes *both* equations true.

TEACHING TIP

Some students may need to be reminded that a solution of a system is a solution of *all* equations in the system.

EXAMPLE 3   Determine whether $(-4, 7)$ is a solution of the system

$$x + y = 3,$$
$$5x - y = -27.$$

Solution   We use alphabetical order of the variables. Thus we replace x with -4 and y with 7:

$$\begin{array}{c|c} x + y = 3 \\ \hline -4 + 7 \ ? \ 3 \\ 3 \ | \ 3 \qquad \text{TRUE} \end{array} \qquad \begin{array}{c|c} 5x - y = -27 \\ \hline 5(-4) - 7 \ ? \ -27 \\ -20 - 7 \\ -27 \ | \ -27 \qquad \text{TRUE} \end{array}$$

The pair $(-4, 7)$ makes both equations true, so it is a solution of the system. We can also describe the solution by writing $x = -4$ and $y = 7$. Set notation can also be used to list the solution set $\{(-4, 7)\}$.

Solving Systems Graphically

Recall that the graph of an equation is a drawing that represents its solution set. If we graph the equations in Example 3, we find that $(-4, 7)$ is the only point common to both lines. Thus one way to solve a system of two equations is to graph both equations and identify any points of intersection. The coordinates of each point of intersection represent a solution of that system.

$$x + y = 3,$$
$$5x - y = -27$$

Most pairs of lines have exactly one point in common. We will soon see, however, that this is not always the case.

EXAMPLE 4   Solve each system graphically.

a) $y - x = 1$,
   $y + x = 3$

b) $y = -3x + 5$,
   $y = -3x - 2$

c) $3y - 2x = 6$,
   $-12y + 8x = -24$

Solution

a) We graph each equation using any method studied in Chapter 2. All ordered pairs from line L_1 in the graph below are solutions of the first equation. All ordered pairs from line L_2 are solutions of the second equation. The point of intersection has coordinates that make *both* equations true. Apparently, $(1, 2)$ is the solution. Graphing is not always accurate, so solving by graphing may yield approximate answers. Our check below shows that $(1, 2)$ is indeed the solution.

$y - x = 1$,
$y + x = 3$

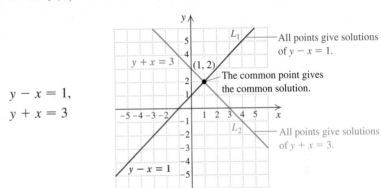

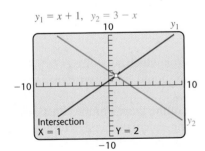

Check:

We can also solve the system using the INTERSECT feature in the CALC menu of a graphing calculator, as shown in the figure at left.

b) We graph the equations. The lines have the same slope, -3, and different y-intercepts, so they are parallel. There is no point at which they cross, so the system has no solution.

$y = -3x + 5$,
$y = -3x - 2$

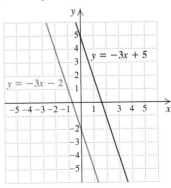

$y_1 = -3x + 5, \quad y_2 = -3x - 2$

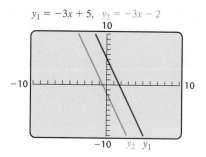

TEACHING TIP

Some students may think *all* ordered pairs are solutions. You may want to emphasize that the system is true for an infinite number of ordered pairs, but those pairs must be of a certain form.

What happens when we try to solve this system using a graphing calculator? We enter and graph both equations as shown at left, noting that the graphs appear to be parallel. When we attempt to find the intersection using the INTERSECT feature, we get an error message.

c) We graph the equations and find that the same line is drawn twice. Thus any solution of one equation is a solution of the other. Each equation has an infinite number of solutions, so the system itself has an infinite number of solutions. We check one solution, $(0, 2)$, which is the y-intercept of each equation.

$$3y - 2x = 6,$$
$$-12y + 8x = -24$$

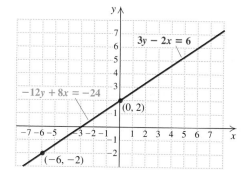

Check:

$3y - 2x = 6$	
$3(2) - 2(0)$? 6	
$6 - 0$	
6	6 TRUE

$-12y + 8x = -24$	
$-12(2) + 8(0)$? -24	
$-24 + 0$	
-24	-24 TRUE

You can check that $(-6, -2)$ is another solution of both equations. In fact, any pair that is a solution of one equation is a solution of the other equation as well. Thus the solution set is

$$\{(x, y) \mid 3y - 2x = 6\}$$

or, in words, "the set of all pairs (x, y) for which $3y - 2x = 6$." Since the two equations are equivalent, we could have written instead $\{(x, y) \mid -12y + 8x = -24\}$.

If we attempt to find the intersection of the graphs of the equations in this system using INTERSECT, the graphing calculator will return as the intersection whatever point we choose as the guess. This is a point of intersection, as is any other point on the graph of the lines.

When we graph a system of two linear equations in two variables, one of the following three outcomes will occur.

1. The lines have one point in common, and that point is the only solution of the system (see Example 4a). Any system that has at least one solution is said to be **consistent**.

2. The lines are parallel, with no point in common, and the system has no solution (see Example 4b). This type of system is called **inconsistent**.

3. The lines coincide, sharing the same graph. Because every solution of one equation is a solution of the other, the system has an infinite number of solutions (see Example 4c). Since it has a solution, this type of system is *consistent*.

When one equation in a system can be obtained by multiplying both sides of another equation by a constant, the two equations are said to be **dependent**. Thus the equations in Example 4(c) are dependent, but those in Examples 4(a) and 4(b) are **independent**. For systems of three or more equations, the definitions of dependent and independent must be slightly modified.

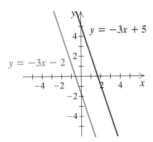

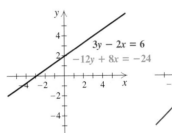

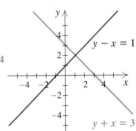

Graphs are parallel.
The system is *inconsistent* because there is no solution. Since the equations are not equivalent, they are *independent*.

Equations have the same graph.
The system is *consistent* and has an infinite number of solutions. The equations are *dependent* since they are equivalent.

Graphs intersect at one point.
The system is *consistent* and has one solution. Since neither equation is a multiple of the other, they are *independent*.

Connecting the Concepts

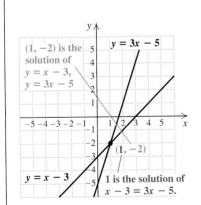

INTERPRETING GRAPHS: POINT OF INTERSECTION

Two graphs cross at a point of intersection. What the coordinates of that point represent depends on the type of problem being solved.

In Section 2.2, we solved linear equations by graphing two functions and looking for points of intersection. For example, to solve

$$x - 3 = 3x - 5,$$

we graph $f(x) = x - 3$ and $g(x) = 3x - 5$. The x-coordinate of the point of intersection, 1, gives the solution of the equation.

In this section, we are solving systems of equations. Now, the solution is the ordered pair given by the coordinates of the point of intersection. For example, to solve the system

$$y = x - 3,$$
$$y = 3x - 5,$$

we graph $y = x - 3$ and $y = 3x - 5$. The solution of the system is the ordered pair $(1, -2)$.

Graphing calculators are especially useful when equations contain fractions or decimals or when the coordinates of the intersection are not integers.

EXAMPLE 5   Solve graphically:

$$3.45x + 4.21y = 8.39,$$
$$7.12x - 5.43y = 6.18.$$

Solution   First, we solve for y in each equation:

$$3.45x + 4.21y = 8.39$$
$$4.21y = 8.39 - 3.45x \qquad \text{**Subtracting 3.45x from both sides**}$$
$$y = (8.39 - 3.45x)/4.21; \qquad \text{**Dividing both sides by 4.21**}$$

$$7.12x - 5.43y = 6.18$$
$$-5.43y = 6.18 - 7.12x \qquad \text{**Subtracting 7.12x from both sides**}$$
$$y = (6.18 - 7.12x)/(-5.43). \qquad \text{**Dividing both sides by -5.43**}$$

It is not necessary to simplify further. We have the system

$$y = (8.39 - 3.45x)/4.21,$$
$$y = (6.18 - 7.12x)/(-5.43).$$

Next, we enter both equations and graph using the same viewing window. By using the INTERSECT feature in the CALC menu, we see that, to the nearest hundredth, the solution is $(1.47, 0.79)$.

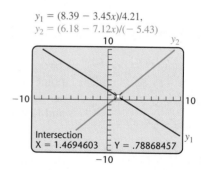

$y_1 = (8.39 - 3.45x)/4.21,$
$y_2 = (6.18 - 7.12x)/(-5.43)$

Intersection
X = 1.4694603   Y = .78868457

Models

Sometimes two or more sets of data can be modeled by a system of linear equations.

EXAMPLE 6   Travel.   The numbers of U.S. travelers to Canada and to Europe are listed in the following table.

Year	U.S. Travelers to Canada (in millions)	U.S. Travelers to Europe (in millions)
1992	11.8	7.1
1994	12.5	8.2
1996	12.9	8.7
1998	14.9	11.1
2000	15.1	13.4

Source: Office of Travel and Tourism Industries

a) Use the points $(2, 11.8)$ and $(10, 15.1)$ to find a linear equation that can be used to estimate the number of travelers f to Canada, in millions, x years after 1990.

b) Use the points $(2, 7.1)$ and $(10, 13.4)$ to find a linear equation that can be used to estimate the number of travelers g to Europe, in millions, x years after 1990.

c) Use the equations found in parts (a) and (b) to estimate the year in which the number of U.S. travelers to Europe will be the same as the number of U.S. travelers to Canada.

d) Use linear regression to find two linear equations that can be used to estimate the number of U.S. travelers to Canada and Europe, in millions, x years after 1990.

e) Use the equations found in part (d) to estimate the year in which the number of U.S. travelers to Europe will be the same as the number of U.S. travelers to Canada.

Solution

a) We first enter and graph the data using a graphing calculator, entering the number of years after 1990 as L1, the number of travelers to Canada, in millions, as L2, and the number of travelers to Europe, in millions, as L3. In the following graph, the points marked with boxes represent the number of travelers to Canada, and the points marked with plus signs represent the number of travelers to Europe. Both sets of data are approximately linear, and we can see that the number of travelers to Europe is increasing more rapidly. It appears likely that the numbers will be equal in the future, if the trends continue.

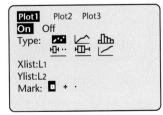

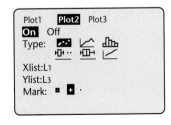

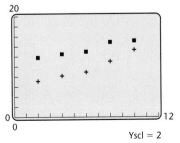

We use the points $(2, 11.8)$ and $(10, 15.1)$ to find a linear equation that describes travel to Canada. The slope of the line is given by

$$m = \frac{15.1 - 11.8}{10 - 2} = \frac{3.3}{8} = 0.4125.$$

Using the variables x and f and the point–slope equation, we have

$$f - 11.8 = 0.4125(x - 2) \qquad \text{Using the point } (2, 11.8) \text{ and the slope } 0.4125$$

$$f - 11.8 = 0.4125x - 0.825$$

$$f = 0.4125x + 10.975.$$

b) To find a linear equation that describes travel to Europe, we use the points $(2, 7.1)$ and $(10, 13.4)$. The slope of the line is given by

$$m = \frac{13.4 - 7.1}{10 - 2} = \frac{6.3}{8} = 0.7875.$$

Using the variables x and g and the point–slope equation, we have

$$g - 7.1 = 0.7875(x - 2) \qquad \text{Using the point } (2, 7.1) \text{ and the slope } 0.7875$$

$$g - 7.1 = 0.7875x - 1.575$$

$$g = 0.7875x + 5.525.$$

c) When the number of travelers is the same, $f = g$. If we let y represent the total number of travelers, we can estimate the year in which the numbers of travelers to Canada and to Europe will be the same by solving the system

$$y = 0.4125x + 10.975,$$

$$y = 0.7875x + 5.525.$$

We will use a graphing calculator to solve the system. We enter $y_1 = 0.4125x + 10.975$ and $y_2 = 0.7875x + 5.525$, graph the equations, and find the coordinates of the point of intersection. The following graph also shows the data for reference. Note that the window size was extended to include the point of intersection.

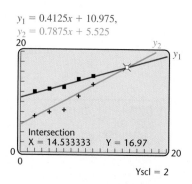

The point of intersection, $(14.53, 16.97)$ represents the solution of the system of equations. Since x represents the number of years after 1990, we estimate that the number of U.S. travelers to Europe will be the same as the number of travelers to Canada about 15 years after 1990, or in 2005.

d) Since the number of travelers to Canada was entered as L2, we first find the linear regression line when L1 is the independent variable and L2 is the dependent variable and store this equation as Y1. Since L1 and L2 are the default lists, we need not specify them, as shown on the left below.

To find the second regression line, we must specify that L1 is the independent variable and L3 the dependent variable; this equation we store as Y2, as shown in the middle below.

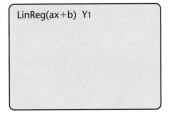

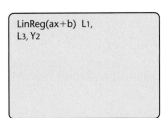

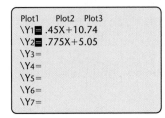

The screen on the right above gives us the two equations:

$$y_1 = 0.45x + 10.74 \quad \text{and} \quad y_2 = 0.775x + 5.05.$$

e) We graph the equations along with the data and determine the coordinates of the point of intersection—approximately $(17.51, 18.62)$, as shown in the following graph. Using these equations, we can estimate that the number of U.S. travelers to Europe will be the same as the number traveling to Canada about 18 years after 1990, or in 2008.

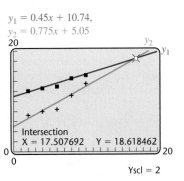

Graphing is helpful when solving systems because it allows us to "see" the solution. It can also be used on systems of nonlinear equations, and in many applications, it provides a satisfactory answer. However, graphing often lacks precision, especially when fractional or decimal solutions are involved. In Section 3.2, we will develop two algebraic methods of solving systems. Both methods produce exact answers.

3.1

Exercise Set

Determine whether the ordered pair is a solution of the given system of equations. Remember to use alphabetical order of variables.

1. $(1,2)$; $4x - y = 2$,
$\quad 10x - 3y = 4$ Yes

2. $(-1,-2)$; $2x + y = -4$,
$\quad\quad\quad\quad x - y = 1$ Yes

3. $(2,5)$; $y = 3x - 1$,
$\quad\quad 2x + y = 4$ No

4. $(-1,-2)$; $x + 3y = -7$,
$\quad\quad\quad\quad 3x - 2y = 12$ No

5. $(1,5)$; $x + y = 6$,
$\quad\quad\; y = 2x + 3$ Yes

6. $(5,2)$; $a + b = 7$,
$\quad\quad\; 2a - 8 = b$ Yes

Aha!   **7.** $(3,1)$; $3x + 4y = 13$,
$\quad\quad\; 6x + 8y = 26$ Yes

8. $(4,-2)$; $-3x - 2y = -8$,
$\quad\quad\quad\quad 8 = 3x + 2y$ Yes

Solve each system graphically. Be sure to check your solution. If a system has an infinite number of solutions, use set-builder notation to write the solution set. If a system has no solution, state this.

9. $x - y = 3$,
$\quad x + y = 5$ $(4,1)$

10. $x + y = 4$,
$\quad\; x - y = 2$ $(3,1)$

11. $3x + y = 5$,
$\quad\;\; x - 2y = 4$ $(2,-1)$

12. $2x - y = 4$,
$\quad\;\; 5x - y = 13$ $(3,2)$

13. $4y = x + 8$,
$\quad\;\; 3x - 2y = 6$ $(4,3)$

14. $4x - y = 9$,
$\quad\quad x - 3y = 16$ $(1,-5)$

15. $x = y - 1$,
$\quad\; 2x = 3y$ $(-3,-2)$

16. $a = 1 + b$,
$\quad\; b = 5 - 2a$ $(2,1)$

Aha!   **17.** $x = -3$,
$\quad\;\; y = 2$ $(-3,2)$

18. $x = 4$,
$\quad\; y = -5$ $(4,-5)$

19. $y = -5.43x + 10.89$,
$\quad\; y = 6.29x - 7.04$
Approximately $(1.53, 2.58)$

20. $y = 123.52x + 89.32$,
$\quad\; y = -89.22x + 33.76$
Approximately $(-0.26, 57.06)$

21. $t + 2s = -1$,
$\quad\quad s = t + 10$ $(3,-7)$

22. $b + 2a = 2$,
$\quad\quad a = -3 - b$ $(5,-8)$

23. $2b + a = 11$,
$\quad\;\; a - b = 5$ $(7,2)$

24. $y = -\frac{1}{3}x - 1$,
$\quad\; 4x - 3y = 18$ $(3,-2)$

25. $y = -\frac{1}{4}x + 1$,
$\quad\; 2y = x - 4$ $(4,0)$

26. $6x - 2y = 2$,
$\quad\; 9x - 3y = 1$
No solution

27. $2.18x + 7.81y = 13.78$,
$\quad\; 5.79x - 3.45y = 8.94$
Approximately $(2.23, 1.14)$

28. $2.6x - 1.1y = 4$,
$\quad\; 1.32y = 3.12x - 5.04$
No solution

29. $y - x = 5$,
$\quad\; 2x - 2y = 10$ No solution

30. $y = -x - 1$,
$\quad\; 4x - 3y = 24$ $(3,-4)$

31. $57y - 45x = 33$,
$\quad\; 95y - 22 = 30x$ Approximately $(-0.73, 0)$, or $\left(-\frac{11}{15}, 0\right)$

32. $-9.25x - 12.94y = -3.88$,
$\quad\; 21.83x + 16.33y = 13.69$ Approximately $(0.87, -0.32)$

33. $y = 3 - x$,
$\quad\; 2x + 2y = 6$ $\{(x,y) \mid y = 3 - x\}$

34. $y - x = 3$,
$\quad\; 4x + y = -2$ $(-1,2)$

35. $1.9x = 4.8y + 1.7$,
$\quad\; 12.92x + 23.8 = 32.64y$ No solution

36. $2x - 3y = 6$,
$\quad\; 3y - 2x = -6$ $\{(x,y) \mid 2x - 3y = 6\}$

37. For the systems in the odd-numbered exercises 9–35, which are consistent?
All but Exercises 29 and 35

38. For the systems in the even-numbered exercises 10–36, which are consistent?
All but Exercises 26 and 28

39. For the systems in the odd-numbered exercises 9–35, which contain dependent equations?
Exercise 33

40. For the systems in the even-numbered exercises 10–36, which contain dependent equations?   Exercise 36

Translate each problem situation to a system of equations. Do not attempt to solve, but save for later use.

41. The difference between two numbers is 11. Twice the smaller plus three times the larger is 123. What are the numbers?   Let x represent the larger number and y the smaller number; $x - y = 11, 3x + 2y = 123$

42. The sum of two numbers is -42. The first number minus the second number is 52. What are the numbers?   Let x represent the first number and y the second number; $x + y = -42, x - y = 52$

43. *Retail Sales.*   Paint Town sold 45 paintbrushes, one kind at $8.50 each and another at $9.75 each. In all, $398.75 was taken in for the brushes. How many of each kind were sold?   ⊡

44. *Retail Sales.*   Mountainside Fleece sold 40 neck-warmers. Polarfleece neckwarmers sold for $9.90 each and wool ones sold for $12.75 each. In all, $421.65 was taken in for the neckwarmers. How many of each type were sold?   ⊡

45. *Geometry.*   Two angles are supplementary.* One angle is 3° less than twice the other. Find the measures of the angles. Let x and y represent the angles; $x + y = 180, x = 2y - 3$

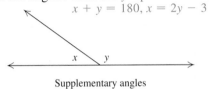

Supplementary angles

46. *Geometry.*   Two angles are complementary.† The sum of the measures of the first angle and half the second angle is 64°. Find the measures of the angles. Let x represent the first angle and y the second angle; $x + y = 90, x + \frac{1}{2}y = 64$

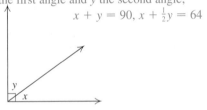

Complementary angles

47. *Basketball Scoring.*   Wilt Chamberlain once scored 100 points, setting a record for points scored in an NBA game. Chamberlain took only two-point shots and (one-point) free throws and made a total of 64 shots. How many shots of each type did he make? ⊡

48. *Fundraising.*   The St. Mark's Community Barbecue served 250 dinners. A child's plate cost $3.50 and an adult's plate cost $7.00. A total of $1347.50 was collected. How many of each type of plate was served? ⊡

49. *Sales of Pharmaceuticals.*   In 2002, the Diabetic Express charged $23.97 for a vial of Humulin insulin and $34.39 for a vial of Novolin insulin. If a total of $1406.90 was collected for 50 vials, how many vials of each type were sold?   ⊡

50. *Court Dimensions.*   The perimeter of a standard basketball court is 288 ft. The length is 44 ft longer than the width. Find the dimensions.   ⊡

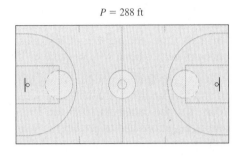

51. *Court Dimensions.*   The perimeter of a standard tennis court used for doubles is 228 ft. The width is 42 ft less than the length. Find the dimensions.   ⊡

52. *Basketball Scoring.*   The Fenton College Cougars made 40 field goals in a recent basketball game, some 2-pointers and the rest 3-pointers. Altogether the 40 baskets counted for 89 points. How many of each type of field goal was made?   ⊡

53. *Hockey Rankings.*   Hockey teams receive 2 points for a win and 1 point for a tie. The Wildcats once won a championship with 60 points. They won 9 more games than they tied. How many wins and how many ties did the Wildcats have?   ⊡

54. *Radio Airplay.*   Roscoe must play 12 commercials during his 1-hr radio show. Each commercial is either 30 sec or 60 sec long. If the total commercial time during that hour is 10 min, how many commercials of each type does Roscoe play?   ⊡

55. *Nontoxic Floor Wax.*   A nontoxic floor wax can be made from lemon juice and food-grade linseed oil. The amount of oil should be twice the amount of lemon juice. How much of each ingredient is needed to make 32 oz of floor wax? (The mix should be spread with a rag and buffed when dry.)   ⊡

56. *Lumber Production.*   Denison Lumber can convert logs into either lumber or plywood. In a given day, the mill turns out 42 pallets of plywood and lumber. It makes a profit of $25 on a pallet of lumber and $40 on a pallet of plywood. How many pallets of each type must be produced and sold in order to make a profit of $1245?   ⊡

57. *Video Rentals.*   J.P.'s Video rents general-interest films for $3.00 each and children's films for $1.50

*The sum of the measures of two supplementary angles is 180°.
†The sum of the measures of two complementary angles is 90°.

⊡ Answers to Exercises 43, 44, and 47–56 can be found on p. A-58.

each. In one day, a total of $213 was taken in from the rental of 77 videos. How many of each type of video was rented? ⊡

58. *Airplane Seating.* An airplane has a total of 152 seats. The number of coach-class seats is 5 more than six times the number of first-class seats. How many of each type of seat are there on the plane? ⊡

59. *Work Force.* The following table shows the percent of men and women in the civilian labor force for various years. Use the data from 1980 and 1998 to write a system of equations and to predict the year in which the percent of women in the work force will be the same as the percent of men in the work force. ⊡

Year	Percent of Women in Work Force	Percent of Men in Work Force
1970	43.3	79.7
1980	51.5	77.4
1990	57.5	76.4
1995	58.9	75.0
1998	59.8	74.9
1999	60.0	74.7

Source: *Statistical Abstract of the United States*, 2000

60. *Automobile Production.* The following table shows the amounts of production for both passenger cars and trucks and buses in the United States for various years. Use the data from 1980 and 1996 to write a system of equations and to estimate the year in which the number of cars produced equaled the number of trucks and buses produced. ⊡

Year	Passenger Car Production (in thousands)	Truck and Bus Production (in thousands)
1980	6376	1634
1990	6078	3706
1992	5664	4038
1993	5981	4917
1994	6614	5649
1995	6351	5635
1996	6083	5716

Source: American Automobile Manufacturers Association

61. *Recycling.* In the United States, the amount of paper being recycled is slowly catching up to the amount of paper waste being generated, as shown in

the following table. Use the data from 1994 and 1998 to write a system of equations and to predict the year in which the amount recycled will equal the amount generated. ⊡

Year	Amount of Paper Waste Generated (in millions of tons)	Amount of Paper Recycled (in millions of tons)
1990	72.7	27.8
1992	74.3	33.0
1994	80.8	36.5
1996	79.7	41.6
1998	84.1	41.6

Source: *Statistical Abstract of the United States*, 2000

62. *Dining Out.* The number of meals per person annually purchased in a restaurant and the number annually purchased for takeout are given in the following table for various years. Use the data from 1990 and 1993 to write a system of equations and to estimate the year in which the number of meals eaten on-premise equaled the number eaten off-premise. ⊡

Year	Number of Meals (per person annually) Eaten On-Premise	Number of Meals (per person annually) Eaten Off-Premise
1990	64	55
1991	64	56
1992	63	57
1993	62	59

Source: *The Wall Street Journal Almanac*, 1998

63. *Beverage Consumption.* The following table shows the number of gallons of milk and soft drinks consumed per capita in the United States for years from 1950 to 2000. Use the data from 1950 and 2000 to write a system of equations and to estimate the year in which the per capita consumption of both milk and soft drinks was the same. ⊡

Year	Per Capita Milk Consumption (in gallons)	Per Capita Soft Drink Consumption (in gallons)
1950	38	10
1960	33	12
1970	31	23
1980	28	35
1990	26	47
2000	24	53

Source: USDA Economic Research Service

⊡ Answers to Exercises 57–63 can be found on p. A-58.

64. *Ice Cream Consumption.*   The following table shows the number of pounds of regular and lowfat ice cream consumed per capita in the United States for various years. Use the data from 1980 and 1998 to predict the year in which the consumption of lowfat ice cream will equal that of regular ice cream.   ⊡

Year	Per Capita Regular Ice Cream Consumption (in pounds)	Per Capita Lowfat Ice Cream Consumption (in pounds)
1980	17.5	7.1
1985	18.1	6.9
1990	15.8	7.7
1995	15.7	7.5
1998	16.6	8.3

Source: U.S. Department of Agriculture, Economic Research Service

65. Use linear regression to fit a line to each set of data in Exercise 59. Use the equations found by regression to predict the year in which the percentage of women in the work force will be the same as the percentage of men in the work force.   ⊡

66. Use linear regression to fit a line to each set of data in Exercise 60. Use the equations found by regression to estimate the year in which the number of cars produced equaled the number of trucks and buses produced.   ⊡

67. Use linear regression to fit a line to each set of data in Exercise 61. Use the equations found by regression to predict the year in which the amount recycled will equal the amount generated.   ⊡

68. Use linear regression to fit a line to each set of data in Exercise 64. Use the equations found by regression to predict the year in which the consumption of lowfat ice cream will equal that of regular ice cream.   ⊡

TW **69.** Examine Exercises 61 and 67. Is it possible for the amount of paper recycled to exceed the amount produced? Why or why not?

TW **70.** Explain the difference between solving a linear equation graphically and solving a system of equations graphically.

Skill Maintenance

Solve. [2.2]

71. $2(4x - 3) - 7x = 9$ 15

⊡ Answers to Exercises 64–68 can be found on p. A-58.

72. $6y - 3(5 - 2y) = 4$ $\frac{19}{12}$

73. $4x - 5x = 8x - 9 + 11x$ $\frac{9}{20}$

74. $8x - 2(5 - x) = 7x + 3$ $\frac{13}{3}$

Solve. [2.3]

75. $3x + 4y = 7$, for y $y = -\frac{3}{4}x + \frac{7}{4}$

76. $2x - 5y = 9$, for y $y = \frac{2}{5}x - \frac{9}{5}$

Synthesis

Technology in U.S. Schools.   For Exercises 77–80, consider the following graph.

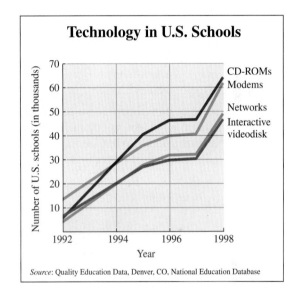

Source: Quality Education Data, Denver, CO, National Education Database

TW **77.** Is it accurate to state that there have always been more schools with CD-ROMs than with networks? Why or why not?

TW **78.** Is it accurate to state that there have always been more schools with networks than with interactive videodisks? Why or why not?

79. During which year did the number of schools with CD-ROMs first exceed the number of schools with modems?   1994

80. During which year did the number of schools owning an interactive videodisk increase the most? At what rate did it increase?   1997; about 15,000 schools per year

81. For each of the following conditions, write a system of equations.   Answers may vary.

   a) $(5, 1)$ is a solution. $x + y = 6$, $x - y = 4$

   b) There is no solution. $x + y = 1$, $2x + 2y = 3$

   c) There is an infinite number of solutions.
   $x + y = 1$, $2x + 2y = 2$

82. A system of linear equations has $(1, -1)$ and $(-2, 3)$ as solutions. Determine: *Answers may vary.*
a) a third point that is a solution, and $(4, -5)$
b) how many solutions there are. *Infinitely many*

83. The solution of the following system is $(4, -5)$. Find A and B.

$$Ax - 6y = 13,$$
$$x - By = -8. \quad A = -\tfrac{17}{4}, B = -\tfrac{12}{5}$$

Translate to a system of equations. Do not solve.

84. *Ages.* Burl is twice as old as his son. Ten years ago, Burl was three times as old as his son. How old are they now? *Let x represent Burl's age now and y his son's age now; $x = 2y, x - 10 = 3(y - 10)$*

85. *Work Experience.* Lou and Juanita are mathematics professors at a state university. Together, they have 46 years of service. Two years ago, Lou had taught 2.5 times as many years as Juanita. How long has each taught at the university? ⊡

86. *Design.* A piece of posterboard has a perimeter of 156 in. If you cut 6 in. off the width, the length becomes four times the width. What are the dimensions of the original piece of posterboard? ⊡

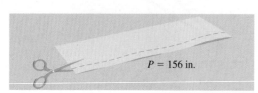

$P = 156$ in.

⊡ Answers to Exercises 85–87 can be found on p. A-58.

87. *Nontoxic Scouring Powder.* A nontoxic scouring powder is made up of 4 parts baking soda and 1 part vinegar. How much of each ingredient is needed for a 16-oz mixture? ⊡

Solve graphically.

88. $y = |x|,$
$x + 4y = 15 \quad (-5, 5), (3, 3)$

89. $x - y = 0,$
$y = x^2 \quad (0, 0), (1, 1)$

In Exercises 90–93, match each system with the appropriate graph from the selections given.

a) 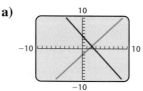   **b)**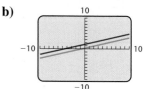

c) 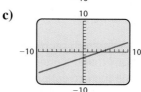   **d)**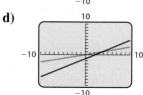

90. $x = 4y,$
$3x - 5y = 7 \quad (d)$

91. $2x - 8 = 4y,$
$x - 2y = 4 \quad (c)$

92. $8x + 5y = 20,$
$4x - 3y = 6 \quad (a)$

93. $x = 3y - 4,$
$2x + 1 = 6y \quad (b)$

3.2

The Substitution Method ▪ The Elimination Method ▪ Comparing Methods

Solving by Substitution or Elimination

The Substitution Method

One algebraic (nongraphical) method for solving systems of equations, the *substitution method*, relies on having a variable isolated.

EXAMPLE 1 Solve the system

$$x + y = 4, \quad (1) \quad \text{For easy reference, we have numbered the equations.}$$
$$x = y + 1. \quad (2)$$

Solution Equation (2) says that x and $y + 1$ name the same number. Thus we can substitute $y + 1$ for x in equation (1):

$$x + y = 4 \quad \text{Equation (1)}$$
$$(y + 1) + y = 4. \quad \text{Substituting } y + 1 \text{ for } x$$

TEACHING TIP

You may wish to point out that the value of the second variable can be most easily found using the equation that defined it in the substitution step.

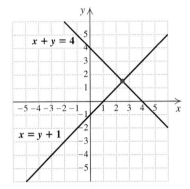

A visualization of Example 1.
Note that the coordinates of the intersection are not obvious.

We solve this last equation, using methods learned earlier:

$$(y + 1) + y = 4$$
$$2y + 1 = 4 \qquad \text{Removing parentheses and combining like terms}$$
$$2y = 3 \qquad \text{Subtracting 1 from both sides}$$
$$y = \tfrac{3}{2}. \qquad \text{Dividing by 2}$$

We now return to the original pair of equations and substitute $\tfrac{3}{2}$ for y in either equation so that we can solve for x. For this problem, calculations are slightly easier if we use equation (2):

$$x = y + 1 \qquad \text{Equation (2)}$$
$$= \tfrac{3}{2} + 1 \qquad \text{Substituting } \tfrac{3}{2} \text{ for } y$$
$$= \tfrac{3}{2} + \tfrac{2}{2} = \tfrac{5}{2}.$$

We obtain the ordered pair $\left(\tfrac{5}{2}, \tfrac{3}{2}\right)$. A check ensures that it is a solution:

Check:

$$\begin{array}{c|c}
x + y = 4 & x = y + 1 \\
\hline
\tfrac{5}{2} + \tfrac{3}{2} \ ? \ 4 & \tfrac{5}{2} \ ? \ \tfrac{3}{2} + 1 \\
\tfrac{8}{2} & \tfrac{3}{2} + \tfrac{2}{2} \\
4 \ \big| \ 4 \quad \text{TRUE} & \tfrac{5}{2} \ \big| \ \tfrac{5}{2} \quad \text{TRUE}
\end{array}$$

Since $\left(\tfrac{5}{2}, \tfrac{3}{2}\right)$ checks, it is the solution.

The exact solution to Example 1 is difficult to find by graphing by hand because it involves fractions. Despite this, the graph shown does serve as a check and provides a visualization of the problem.

If neither equation in a system has a variable alone on one side, we first isolate a variable in one equation and then substitute.

EXAMPLE 2   Solve the system

$$2x + y = 6, \qquad (1)$$
$$3x + 4y = 4. \qquad (2)$$

Solution   First, we select an equation and solve for one variable. To isolate y, we can subtract $2x$ from both sides of equation (1):

$$2x + y = 6 \qquad (1)$$
$$y = 6 - 2x. \qquad (3) \qquad \text{Subtracting } 2x \text{ from both sides}$$

Next, we proceed as in Example 1, by substituting:

$$3x + 4(6 - 2x) = 4 \qquad \text{Substituting } 6 - 2x \text{ for } y \text{ in equation (2).}$$
$$\text{Use parentheses!}$$
$$3x + 24 - 8x = 4 \qquad \text{Distributing to remove parentheses}$$
$$3x - 8x = 4 - 24 \qquad \text{Subtracting 24 from both sides}$$
$$-5x = -20 \qquad \text{Combining like terms}$$
$$x = 4. \qquad \text{Dividing both sides by } -5$$

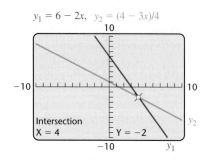

$y_1 = 6 - 2x, \quad y_2 = (4 - 3x)/4$

Intersection
X = 4   Y = −2

Next, we substitute 4 for x in either equation (1), (2), or (3). It is easiest to use equation (3) because it has already been solved for y:

$$y = 6 - 2x$$
$$= 6 - 2(4)$$
$$= 6 - 8 = -2.$$

The pair $(4, -2)$ appears to be the solution. We check in equations (1) and (2).

Check:

$$\frac{2x + y = 6}{2(4) + (-2) \; ? \; 6}$$
$$8 - 2 \quad\quad$$
$$6 \; | \; 6 \quad \text{TRUE}$$

$$\frac{3x + 4y = 4}{3(4) + 4(-2) \; ? \; 4}$$
$$12 - 8 \quad\quad$$
$$4 \; | \; 4 \quad \text{TRUE}$$

We can also check by solving using a graphing calculator, as shown at left. Since $(4, -2)$ checks, it is the solution.

Some systems have no solution, as we saw graphically in Section 3.1. How do we recognize such systems if we are solving by an algebraic method?

EXAMPLE 3 Solve the system

$$y = -3x + 5, \quad (1)$$
$$y = -3x - 2. \quad (2)$$

Solution We solved this system graphically in Example 4(b) of Section 3.1, and found that the lines are parallel and the system has no solution. Let's now try to solve the system by substitution. Proceeding as in Example 1, we substitute $-3x - 2$ for y in the first equation:

$$y = -3x + 5$$
$$-3x - 2 = -3x + 5 \quad \text{Substituting } -3x - 2 \text{ for } y \text{ in equation (1)}$$
$$-2 = 5. \quad \text{Adding } 3x \text{ to both sides; } -2 = 5 \text{ is a contradiction.}$$

When we add $3x$ to get the x-terms on one side, the x-terms drop out and we end up with a contradiction—that is, an equation that is always false. When solving algebraically yields a contradiction, the system has no solution; it is inconsistent.

The Elimination Method

The *elimination method* for solving systems of equations makes use of the *addition principle*: If $a = b$, then $a + c = b + c$. Consider the following system:

$$2x - 3y = 0, \quad (1)$$
$$-4x + 3y = -1. \quad (2)$$

To see why the elimination method works well with this system, note the $-3y$ in one equation and the $3y$ in the other. These terms are opposites, or additive inverses. If we add all terms on the left side of the equations, $-3y$ and $3y$ add to 0, and in effect, the variable y is "eliminated."

To use the addition principle for equations, note that according to equation (2), $-4x + 3y$ and -1 are the same number. Thus we can work vertically and add $-4x + 3y$ to the left side of equation (1) and -1 to the right side:

$$\begin{array}{ll} 2x - 3y = 0 & \text{(1)} \\ \underline{-4x + 3y = -1} & \text{(2)} \\ -2x + 0y = -1. & \textbf{Adding} \end{array}$$

This eliminates the variable y. We can now solve for x:

$$-2x = -1$$
$$x = \tfrac{1}{2}.$$

Next, we substitute $\tfrac{1}{2}$ for x in equation (1) and solve for y:

$$2 \cdot \tfrac{1}{2} - 3y = 0 \qquad \textbf{Substituting. We also could have used equation (2).}$$
$$1 - 3y = 0$$
$$-3y = -1, \text{ so } y = \tfrac{1}{3}.$$

Check:

$$\begin{array}{c|c} 2x - 3y = 0 \\ \hline 2\left(\tfrac{1}{2}\right) - 3\left(\tfrac{1}{3}\right) \;?\; 0 \\ 1 - 1 & \\ 0 & 0 \qquad \text{TRUE} \end{array} \qquad \begin{array}{c|c} -4x + 3y = -1 \\ \hline -4\left(\tfrac{1}{2}\right) + 3\left(\tfrac{1}{3}\right) \;?\; -1 \\ -2 + 1 & \\ -1 & -1 \qquad \text{TRUE} \end{array}$$

Since $\left(\tfrac{1}{2}, \tfrac{1}{3}\right)$ checks, it is the solution.

To eliminate a variable, we must sometimes multiply before adding.

EXAMPLE 4   Solve the system

$$5x + 4y = 22, \qquad \text{(1)}$$
$$-3x + 8y = 18. \qquad \text{(2)}$$

Solution   If we add the left sides of the two equations, we will not eliminate a variable. However, if the $4y$ in equation (1) were changed to $-8y$, we would. To do this, we multiply both sides of equation (1) by -2:

$$\begin{array}{ll} -10x - 8y = -44 & \textbf{Multiplying both sides of equation (1) by } -2 \\ \underline{-3x + 8y = 18} & \\ -13x + 0 = -26 & \textbf{Adding} \\ x = 2. & \textbf{Solving for } x \end{array}$$

Then

$$\begin{array}{ll} -3 \cdot 2 + 8y = 18 & \textbf{Substituting 2 for } x \textbf{ in equation (2)} \\ -6 + 8y = 18 & \\ \left.\begin{array}{l} 8y = 24 \\ y = 3. \end{array}\right\} & \textbf{Solving for } y \end{array}$$

We obtain $(2, 3)$, or $x = 2$, $y = 3$. We can check either by substituting in the original equations or by solving graphically, as shown in the figure at left. The solution is $(2, 3)$.

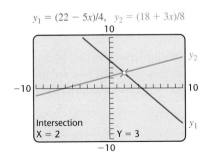

$y_1 = (22 - 5x)/4, \; y_2 = (18 + 3x)/8$

Sometimes we must multiply twice in order to make two terms become opposites.

EXAMPLE 5   Solve the system

$$2x + 3y = 17, \quad (1)$$
$$5x + 7y = 29. \quad (2)$$

Solution   We multiply in order to be able to eliminate the x-terms.

$$2x + 3y = 17, \longrightarrow \text{Multiplying both sides by 5} \longrightarrow 10x + 15y = 85$$

$$5x + 7y = 29 \longrightarrow \text{Multiplying both sides by } -2 \longrightarrow \underline{-10x - 14y = -58}$$

$$0 + y = 27 \qquad \text{Adding}$$
$$y = 27.$$

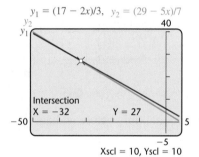

$y_1 = (17 - 2x)/3, \quad y_2 = (29 - 5x)/7$

Intersection
X = −32   Y = 27

Xscl = 10, Yscl = 10

Next, we substitute to find x:

$$2x + 3 \cdot 27 = 17 \qquad \textbf{Substituting 27 for } y \textbf{ in equation (1)}$$
$$2x + 81 = 17$$
$$\left. \begin{array}{r} 2x = -64 \\ x = -32. \end{array} \right\} \quad \textbf{Solving for } x$$

To check, we use a graphing calculator, as shown in the figure at left. We obtain $(-32, 27)$, or $x = -32$, $y = 27$, as the solution.

EXAMPLE 6   Solve the system

$$3y - 2x = 6, \qquad (1)$$
$$-12y + 8x = -24, \qquad (2)$$

Solution   We graphed this system in Example 4(c) of Section 3.1, and found that the lines coincide and the system has an infinite number of solutions. Suppose we were to solve this system using the elimination method:

$$12y - 8x = 24 \qquad \textbf{Multiplying both sides of equation (1) by 4}$$
$$\underline{-12y + 8x = -24}$$
$$0 = 0. \qquad \textbf{We obtain an identity; } 0 = 0 \textbf{ is always true.}$$

Note that both variables have been eliminated and what remains is an identity—that is, an equation that is always true. Any pair that is a solution of equation (1) is also a solution of equation (2). The equations are dependent and the solution set is infinite:

$$\{(x, y) \mid 3y - 2x = 6\}.$$

Special Cases

When solving a system of two linear equations in two variables:

1. If an identity is obtained, such as $0 = 0$, then the system has an infinite number of solutions. The equations are dependent and, since a solution exists, the system is consistent.*

2. If a contradiction is obtained, such as $0 = 7$, then the system has no solution. The system is inconsistent.

Should decimals or fractions appear, it often helps to *clear* them before solving.

EXAMPLE 7   Solve the system

$$0.2x + 0.3y = 1.7,$$
$$\tfrac{1}{7}x + \tfrac{1}{5}y = \tfrac{29}{35}.$$

Solution   We have

$0.2x + 0.3y = 1.7,$ ⟶ **Multiplying both sides by 10** ⟶ $2x + 3y = 17$

$\tfrac{1}{7}x + \tfrac{1}{5}y = \tfrac{29}{35}$ ⟶ **Multiplying both sides by 35** ⟶ $5x + 7y = 29.$

We multiplied both sides of the first equation by 10 to clear the decimals. Multiplication by 35, the least common denominator, clears the fractions in the second equation. The problem is now identical to Example 5. The solution is $(-32, 27)$, or $x = -32$, $y = 27$.

Comparing Methods

Often it is helpful to use both an algebraic and a graphical method to solve a system of equations. In Examples 2, 4, and 5, we solved each system algebraically and checked by solving graphically. A graph helps us visualize a solution that can be found exactly using algebra. On the other hand, algebra can help us determine an appropriate viewing window for a graph. Note that the graph of the system in Example 5 is shown in the viewing window $[-50, 5, -5, 40]$. If that system were graphed using a standard viewing window, the lines would appear to be parallel. When graphing a system of equations, we generally choose a viewing window that shows any points of intersection.

The following table is a summary that compares the graphical, substitution, and elimination methods for solving systems of equations.

---

*Consistent systems and dependent equations are discussed in greater detail in Section 3.4.

Connecting the Concepts

We now have three different methods for solving systems of equations. Each method has certain strengths and weaknesses, as outlined below.

Method	Strengths	Weaknesses
Graphical	Solutions are displayed graphically. Works with any system that can be graphed.	Inexact when solutions involve numbers that are not integers. Solution may not appear on the part of the graph drawn.
Substitution	Yields exact solutions. Easy to use when a variable is alone on one side.	Introduces extensive computations with fractions when solving more complicated systems. Solutions are not displayed graphically.
Elimination	Yields exact solutions. Easy to use when fractions or decimals appear in the system. The preferred method for systems of 3 or more equations in 3 or more variables (see Section 3.4).	Solutions are not displayed graphically.

3.2

Exercise Set

FOR EXTRA HELP

Digital Video Tutor CD 3 Videotape 4 • Student's Solutions Manual • Tutor Center AW Math Tutor Center • InterAct Math • MathXL • MyMathLab

For Exercises 1–48, if a system has an infinite number of solutions, use set-builder notation to write the solution set. If a system has no solution, state this.

Solve using the substitution method.

1. $y = 5 - 4x,$
$2x - 3y = 13$ $(2, -3)$

2. $x = 8 - 4y,$
$3x + 5y = 3$ $(-4, 3)$

3. $2y + x = 9,$
$x = 3y - 3$ $\left(\frac{21}{5}, \frac{12}{5}\right)$

4. $9x - 2y = 3,$
$3x - 6 = y$ $(-3, -15)$

5. $3s - 4t = 14,$
$5s + t = 8$ $(2, -2)$

6. $m - 2n = 16,$
$4m + n = 1$ $(2, -7)$

7. $4x - 2y = 6,$
$2x - 3 = y$ $\{(x, y) \mid 2x - 3 = y\}$

8. $t = 4 - 2s,$
$t + 2s = 6$ No solution

9. $-5s + t = 11,$
$4s + 12t = 4$ $(-2, 1)$

10. $5x + 6y = 14,$
$-3y + x = 7$ $(4, -1)$

11. $2x + 2y = 2,$
$3x - y = 1$ $\left(\frac{1}{2}, \frac{1}{2}\right)$

12. $4p - 2q = 16,$
$5p + 7q = 1$ $(3, -2)$

13. $3a - b = 7,$
$2a + 2b = 5$ $\left(\frac{19}{8}, \frac{1}{8}\right)$

14. $5x + 3y = 4,$
$x - 4y = 3$ $\left(\frac{25}{23}, -\frac{11}{23}\right)$

15. $2x - 3 = y,$
$y - 2x = 1$ No solution

16. $a - 2b = 3,$
$3a = 6b + 9$ $\{(a, b) \mid a - 2b = 3\}$

Solve using the elimination method.

17. $x + 3y = 7,$
$-x + 4y = 7$ $(1, 2)$

18. $x + y = 9,$
$2x - y = -3$ $(2, 7)$

19. $2x + y = 6,$
$x - y = 3$ $(3, 0)$

20. $x - 2y = 6,$
$-x + 3y = -4$ $(10, 2)$

21. $9x + 3y = -3,$
$2x - 3y = -8$ $(-1, 2)$

22. $6x - 3y = 18,$
$6x + 3y = -12$ $\left(\frac{1}{2}, -5\right)$

23. $5x + 3y = 19,$
$2x - 5y = 11$ $\left(\frac{128}{31}, -\frac{17}{31}\right)$

24. $3x + 2y = 3,$
$9x - 8y = -2$ $\left(\frac{10}{21}, \frac{11}{14}\right)$

25. $5r - 3s = 24,$
$3r + 5s = 28$ $(6, 2)$

26. $5x - 7y = -16,$
$2x + 8y = 26$ $(1, 3)$

27. $6s + 9t = 12,$
$4s + 6t = 5$ No solution

28. $10a + 6b = 8,$
$5a + 3b = 2$ No solution

29. $\frac{1}{2}x - \frac{1}{6}y = 3,$
$\frac{2}{5}x + \frac{1}{2}y = 2$ $\left(\frac{110}{19}, -\frac{12}{19}\right)$

30. $\frac{1}{3}x + \frac{1}{5}y = 7,$
$\frac{1}{6}x - \frac{2}{5}y = -4$ $(12, 15)$

31. $\dfrac{x}{2} + \dfrac{y}{3} = \dfrac{7}{6},$
$\dfrac{2x}{3} + \dfrac{3y}{4} = \dfrac{5}{4}$ $(3, -1)$

32. $\dfrac{2x}{3} + \dfrac{3y}{4} = \dfrac{11}{12},$
$\dfrac{x}{3} + \dfrac{7y}{18} = \dfrac{1}{2}$ $(-2, 3)$

Aha! **33.** $12x - 6y = -15,$
$-4x + 2y = 5$ $\{(x, y) | -4x + 2y = 5\}$

34. $8s + 12t = 16,$
$6s + 9t = 12$ $\{(s, t) | 6s + 9t = 12\}$

35. $0.2a + 0.3b = 1,$
$0.3a - 0.2b = 4$ $\left(\frac{140}{13}, -\frac{50}{13}\right)$

36. $-0.4x + 0.7y = 1.3,$
$0.7x - 0.3y = 0.5$ $(2, 3)$

Solve using any appropriate method.

37. $a - 2b = 16,$
$b + 3 = 3a$ $(-2, -9)$

38. $5x - 9y = 7,$
$7y - 3x = -5$ $\left(\frac{1}{2}, -\frac{1}{2}\right)$

39. $10x + y = 306,$
$10y + x = 90$ $(30, 6)$

40. $3(a - b) = 15,$
$4a = b + 1$ $\left(-\frac{4}{3}, -\frac{19}{3}\right)$

41. $3y = x - 2,$
$x = 2 + 3y$ $\{(x, y) | x = 2 + 3y\}$

42. $x + 2y = 8,$
$x = 4 - 2y$ No solution

43. $3s - 7t = 5,$
$7t - 3s = 8$ No solution

44. $2s - 13t = 120,$
$-14s + 91t = -840$ $\{(s, t) | 2s - 13t = 120\}$

45. $0.05x + 0.25y = 22,$
$0.15x + 0.05y = 24$ $(140, 60)$

46. $1.3x - 0.2y = 12,$
$0.4x + 17y = 89$ $(10, 5)$

47. $13a - 7b = 9,$
$2a - 8b = 6$ $\left(\frac{1}{3}, -\frac{2}{3}\right)$

48. $3a - 12b = 9,$
$14a - 11b = 5$ $\left(-\frac{13}{45}, -\frac{37}{45}\right)$

In Exercises 49–52, determine which of the given viewing windows below shows the point of intersection of the graphs of the equations in the given system. Check by graphing.

a) $[-5, 5, -5, 5]$
b) $[25, 50, 0, 10]$
c) $[0, 20, 0, 10]$
d) $[100, 200, 0, 100]$

49. The system of Exercise 45 (d)

50. The system of Exercise 38 (a)

51. The system of Exercise 39 (b)

52. The system of Exercise 46 (c)

TW 53. Describe a procedure that can be used to write an inconsistent system of equations.

TW 54. Describe a procedure that can be used to write a system that has an infinite number of solutions.

Skill Maintenance

55. The fare for a taxi ride from Johnson Street to Elm Street is $5.20. If the rate of the taxi is $1.00 for the first $\frac{1}{2}$ mi and 30¢ for each additional $\frac{1}{4}$ mi, how far is it from Johnson Street to Elm Street? [2.3] 4 mi

56. A student's average after 4 tests is 78.5. What score is needed on the fifth test in order to raise the average to 80? [2.3] 86

57. *Home Remodeling.* In a recent year, Americans spent $35 billion to remodel bathrooms and kitchens. Twice as much was spent on kitchens as on bathrooms. (*Source: Indianapolis Star*) How much was spent on each? [2.3] $11\frac{2}{3}$ billion on bathrooms; $23\frac{1}{3}$ billion on kitchens

58. A 480-m wire is cut into three pieces. The second piece is three times as long as the first. The third is four times as long as the second. How long is each piece? [2.3] 30 m, 90 m, 360 m

59. *Car Rentals.* Badger Rent-A-Car rents a compact car at a daily rate of $34.95 plus 10¢ per mile. A businessperson is allotted $80 for car rental. How many miles can she travel on the $80 budget? [2.3] 450.5 mi

60. *Car Rentals.* Badger rents midsized cars at a rate of $43.95 plus 10¢ per mile. A tourist has a car-rental budget of $90. How many miles can he travel on the $90? [2.3] 460.5 mi

Synthesis

TW **61.** A student solving the system

$$17x + 19y = 102,$$
$$136x + 152y = 826$$

graphs both equations on a graphing calculator and gets the following screen.

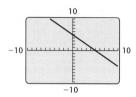

The student then (incorrectly) concludes that the equations are dependent and the solution set is infinite. How can algebra be used to convince the student that a mistake has been made?

TW **62.** Some systems are more easily solved by substitution and some are more easily solved by elimination. Write guidelines that could be used to help someone determine which method to use.

63. If $(1, 2)$ and $(-3, 4)$ are two solutions of $f(x) = mx + b$, find m and b. $m = -\frac{1}{2}, b = \frac{5}{2}$

64. If $(0, -3)$ and $\left(-\frac{3}{2}, 6\right)$ are two solutions of $px - qy = -1$, find p and q. $p = 2, q = -\frac{1}{3}$

65. Determine a and b for which $(-4, -3)$ is a solution of the system

$$ax + by = -26,$$
$$bx - ay = 7. \qquad a = 5, b = 2$$

66. Solve for x and y in terms of a and b:

$$5x + 2y = a,$$
$$x - y = b. \qquad \left(\dfrac{a + 2b}{7}, \dfrac{a - 5b}{7}\right)$$

Solve.

67. $\dfrac{x + y}{2} - \dfrac{x - y}{5} = 1,$

$\dfrac{x - y}{2} + \dfrac{x + y}{6} = -2 \qquad \left(-\dfrac{32}{17}, \dfrac{38}{17}\right)$

68. $3.5x - 2.1y = 106.2,$ Approximately
$4.1x + 16.7y = -106.28 \qquad (23.118879, -12.039964)$

Each of the following is a system of nonlinear equations. However, each is reducible to linear, since an appropriate substitution (say, u for 1/x and v for 1/y) yields a linear system. Make such a substitution, solve for the new variables, and then solve for the original variables. $\left(-\dfrac{1}{4}, -\dfrac{1}{2}\right)$

69. $\dfrac{2}{x} + \dfrac{1}{y} = 0,$

$\dfrac{5}{x} + \dfrac{2}{y} = -5 \qquad \left(-\dfrac{1}{5}, \dfrac{1}{10}\right)$

70. $\dfrac{1}{x} - \dfrac{3}{y} = 2,$

$\dfrac{6}{x} + \dfrac{5}{y} = -34$

Collaborative Corner

How Many Two's? How Many Three's?

   Focus: Systems of linear equations

   Time: 20 minutes

   Group size: 3

The box score at the top of the following page, from a basketball game in 2002 between the Los Angeles Lakers and the San Antonio Spurs, contains information on how many field goals and free throws each player attempted and made. For example, the line "Duncan 15-26 9-9 40" means that San Antonio's Tim Duncan made 15 field goals out of 26 attempts and 9 free throws out of 9 attempts, for a total of 40 points. (Each free throw is worth 1 point and each field goal is worth either 2 or 3 points, depending on how far from the basket it was shot.)

ACTIVITY

1. Work as a group to develop a system of two equations in two unknowns that can be used to determine how many 2-pointers and how many 3-pointers were made by the Lakers' Kobe Bryant.

2. Each group member should solve the system from part (1) in a different way: one person algebraically, one person by making a table and methodically checking all combinations of 2- and 3-pointers, and one person by guesswork. Compare answers when this has been completed.

3. Determine, as a group, how many 2- and 3-pointers the Los Angeles Lakers made as a team.

■**Lakers 88, Spurs 81:** The Los Angeles Lakers outscored the San Antonio Spurs 13–5 over the final 6:11 Monday night for an 88–81 victory in San Antonio. The Lakers won their 17th straight in a rough game at the Alamodome, where San Antonio had lost just eight in the regular season.
(*Source*: espn.go.com/nba)

LA Lakers (88)
Fox 5-12 0-0 11, Grant 2-4 2-4 6, O'Neal 8-21 3-6 19, Bryant 11-24 5-6 28, Fisher 6-9 0-0 16, Horry 2-5 0-2 6, Lue 0-1 0-0 0, Shaw 1-3 0-0 2, Madsen 0-0 0-0 0, Totals 35-79 10-18 88.

San Antonio (81)
Duncan 15-26 9-9 40, Ferry 1-6 0-0 3, Robinson 3-9 1-2 7, Daniels 5-15 12-12 24, Porter 2-8 1-2 5, Rose 1-2 0-0 2, S. Elliott 0-1 0-0 0, Johnson 0-4 0-0 0, Walker 0-0 0-0 0. Totals 27-71 23-25 81.

LA Lakers	21	17	28	22	—	88
San Antonio	22	24	21	14	—	81

3.3

Total-Value and Mixture Problems ■ Motion Problems

Solving Applications: Systems of Two Equations

You are in a much better position to solve problems now that systems of equations can be used. Using systems often makes the translating step easier.

EXAMPLE 1   Real Estate.   In 1996, the Simon Property Group and the DeBartolo Realty Corporation merged to form the largest real estate company in the United States, owning 183 shopping centers in 32 states. Prior to merging, Simon owned twice as many properties as DeBartolo. How many properties did each company own before the merger?

Solution   The *Familiarize* and *Translate* steps were outlined in Example 1 of Section 3.1. The resulting system of equations is

$$x + y = 183,$$
$$x = 2y,$$

where x is the number of properties originally owned by Simon and y is the number of properties originally owned by DeBartolo.

3. Carry out.   We solve the system of equations both algebraically and graphically.

Algebraic Solution

Since one equation already has a variable isolated, let's use the substitution method:

$$x + y = 183$$
$$2y + y = 183 \qquad \text{Substituting } 2y \text{ for } x$$
$$3y = 183 \qquad \text{Combining like terms}$$
$$y = 61.$$

Returning to the second equation, we substitute 61 for y and compute x:

$$x = 2y = 2 \cdot 61 = 122.$$

We have $x = 122$, $y = 61$.

Graphical Solution

We graph $y_1 = 183 - x$ and $y_2 = \frac{1}{2}x$ and find the point of intersection of the graphs. Writing $y_1 = 183 - x$ in slope–intercept form—that is, $y_1 = -x + 183$—we see that the y-intercept is $(0, 183)$. Only positive values of x and y make sense in this problem, so we choose a viewing window of $[0, 200, 0, 200]$, with Xscl $= 10$ and Yscl $= 10$.

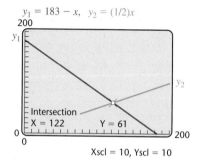

We have a solution of $(122, 61)$.

4. Check.   Apparently Simon owned 122 properties and DeBartolo 61. The sum of 122 and 61 is 183, so the total number of properties is correct. Since 122 is twice 61, the numbers check.

5. State.   Prior to merging, Simon owned 122 properties and DeBartolo owned 61.

Total-Value and Mixture Problems

EXAMPLE 2 Purchasing. Recently the Champlain Valley Community Music Center purchased 120 stamps for $33.90. If the stamps were a combination of 23¢ postcard stamps and 37¢ first-class stamps, how many of each type were bought?

Solution   The *Familiarize* and *Translate* steps were completed in Example 2 of Section 3.1.

3. Carry out.   We are to solve the system of equations

$$p + f = 120, \qquad (1)$$
$$23p + 37f = 3390, \qquad (2) \qquad \text{Working in cents rather than dollars}$$

where p is the number of postcard stamps bought and f is the number of first-class stamps bought.

Algebraic Solution

Because both equations are in the form
$Ax + By = C$, let's use the elimination method to
solve the system. We can eliminate p by multiplying
both sides of equation (1) by -23 and adding them
to the corresponding sides of equation (2):

$$-23p - 23f = -2760 \quad \text{Multiplying both sides of equation (1) by } -23$$

$$\underline{23p + 37f = 3390}$$
$$14f = 630 \quad \text{Adding}$$
$$f = 45. \quad \text{Solving for } f$$

To find p, we substitute 45 for f in equation (1) and
then solve for p:

$$p + f = 120 \quad \text{Equation (1)}$$
$$p + 45 = 120 \quad \text{Substituting 45 for } f$$
$$p = 75. \quad \text{Solving for } p$$

We obtain $(45, 75)$, or $f = 45$, $p = 75$.

Graphical Solution

We replace f with x and p with y and solve for y:

$$y + x = 120 \quad \text{Solving for } y \text{ in equation (1)}$$
$$y = 120 - x,$$

and

$$23y + 37x = 3390 \quad \text{Solving for } y \text{ in equation (2)}$$
$$23y = 3390 - 37x$$
$$y = (3390 - 37x)/23.$$

Since the number of each kind of stamp is between
0 and 120, an appropriate viewing window is
$[0, 120, 0, 120]$.

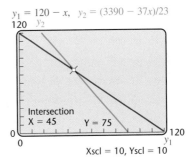

The point of intersection is $(45, 75)$. Since f was replaced with x and p with y, we have $f = 45$, $p = 75$.

4. **Check.** We check in the original problem. Recall that f is the number of
first-class stamps and p the number of postcard stamps.

Number of stamps: $f + p = 45 + 75 = 120$
Cost of first-class stamps: $\$0.37f = 0.37 \times 45 = \$16.65$
Cost of postcard stamps: $\$0.23p = 0.23 \times 75 = \underline{\$17.25}$
Total $= \$33.90$

The numbers check.

5. **State.**   The music center bought 45 first-class stamps and 75 postcard
stamps.

Example 2 involved two types of items (first-class stamps and postcard
stamps), the quantity of each type bought, and the total value of the items. We
refer to this type of problem as a *total-value problem*.

EXAMPLE 3   Blending Teas.   Tara's Tea Terrace sells loose Black tea for 95¢ an ounce and Lapsang Souchong for $1.43 an ounce. Tara wants to make a 1-lb mixture of the two types, called Imperial Blend, that sells for $1.10 an ounce. How much tea of each type should Tara use?

Solution

1. **Familiarize.**   This problem is similar to Example 2. Rather than postcard stamps and first-class stamps, we have ounces of Black tea and ounces of Lapsang Souchong. Instead of a different price for each type of stamp, we have a different price per ounce for each type of tea. Finally, rather than knowing the total cost of the stamps, we know the weight and the price per ounce of the Imperial Blend. It is important to note that we can find the total value of the blend by multiplying 16 ounces (1 lb) times $1.10 per ounce. Although we could make and check a guess, we proceed to let $b =$ the number of ounces of Black tea and $l =$ the number of ounces of Lapsang Souchong.

2. **Translate.**   Since a 16-oz batch is being made, we must have

$$b + l = 16.$$

To find a second equation, we use the fact that the total value of the 16-oz blend must match the combined value of the separate ingredients. Note that we convert $1.43 and $1.10 to cents in order to use the same unit for all prices.

Rewording: The value of the Black tea   plus   the value of the Lapsang Souchong   is   the value of the Imperial Blend.

Translating: $b \cdot 95$ $+$ $l \cdot 143$ $=$ $16 \cdot 110$

These equations can also be obtained from a table.

	Black Tea	Lapsang Souchong	Imperial Blend	
Number of Ounces	b	l	16	$b + l = 16$
Price per Ounce	95¢	143¢	110¢	
Value of Tea	$95b$	$143l$	$16 \cdot 110$, or 1760¢	$95b + 143l = 1760$

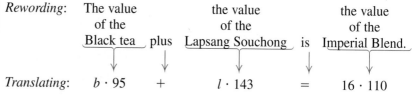

We have translated to a system of equations:

$$b + l = 16, \qquad (1)$$
$$95b + 143l = 1760. \qquad (2)$$

3. Carry out.   We solve the system both algebraically and graphically.

Algebraic Solution

We can solve using the substitution method. When equation (1) is solved for b, we have $b = 16 - l$. Substituting $16 - l$ for b in equation (2), we find l:

$95(16 - l) + 143l = 1760$	Substituting
$1520 - 95l + 143l = 1760$	Using the distributive law
$48l = 240$	Combining like terms; subtracting 1520 from both sides
$l = 5.$	Dividing both sides by 48

We have $l = 5$ and, from equation (1) above, $b + l = 16$. Thus, $b = 11$.

Graphical Solution

We replace b with x and l with y and solve for y, giving us the system of equations

$$y = 16 - x,$$
$$y = (1760 - 95x)/143.$$

We know from the problem that the number of ounces of each kind of tea is between 0 and 16, so we choose the viewing window $[0, 16, 0, 16]$, graph the equations, and find the point of intersection.

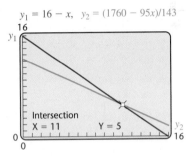

$$y_1 = 16 - x, \quad y_2 = (1760 - 95x)/143$$

The point of intersection is $(11, 5)$. Since b was replaced with x and l with y, we have $b = 11, l = 5$.

4. Check.   If 11 oz of Black tea and 5 oz of Lapsang Souchong are combined, a 16-oz, or 1-lb, blend will result. The value of 11 oz of Black tea is 11($0.95), or $10.45. The value of 5 oz of Lapsang Souchong is 5($1.43), or $7.15, so the combined value of the blend is $10.45 + $7.15 = $17.60. A 16-oz batch, priced at $1.10 an ounce, would also be worth $17.60, so our answer checks.

5. State.   The Imperial Blend should be made by combining 11 oz of Black tea with 5 oz of Lapsang Souchong.

EXAMPLE 4   Student Loans.   Dawn's student loans totaled $9600. Part was a Perkins loan made at 5% simple interest and the rest was a Federal Education Loan made at 8% simple interest. After one year, Dawn's loans accumulated $633 in interest. What was the original amount of each loan?

Solution

1. Familiarize.   We begin with a guess. If $7000 was borrowed at 5% and $2600 was borrowed at 8%, the two loans would total $9600. The interest would then be 0.05($7000), or $350, and 0.08($2600), or $208, for a total of $558 in interest. Because $558 < 633$, we now know that more than $2600 was borrowed at the higher interest rate. Our guess is wrong, but checking it familiarized us with the problem.

2. **Translate.**   We let $p =$ the amount of the Perkins loan and $f =$ the amount of the Federal Education Loan. We then organize a table in which each column comes from the formula for simple interest:

$$Principal \cdot Rate \cdot Time = Interest.$$

	Perkins Loan	Federal Loan	Total
Principal	p	f	$9600	→ $p + f = 9600$
Rate of Interest	5%, or 0.05	8%, or 0.08	
Time	1 yr	1 yr	
Interest	$0.05p$	$0.08f$	$633	→ $0.05p + 0.08f = 633$

The total amount borrowed is found in the first row of the table:

$$p + f = 9600.$$

A second equation, representing the accumulated interest, can be found in the last row:

$0.05p + 0.08f = 633$,   or $5p + 8f = 63,300$.   **Clearing decimals**

3. **Carry out.**   The system can be solved by elimination:

$$p + f = 9600, \quad \longrightarrow \text{Multiplying both} \longrightarrow \quad -5p - 5f = -48,000$$
$$5p + 8f = 63,300 \quad \text{sides by } -5 \qquad \underline{\quad 5p + 8f = 63,300}$$
$$3f = 15,300$$
$$p + f = 9600 \longleftarrow f = 5100$$
$$p + 5100 = 9600$$
$$p = 4500.$$

We find that $p = 4500$ and $f = 5100$.

4. **Check.**   The total amount borrowed is $4500 + $5100, or $9600. The interest on $4500 at 5% for 1 yr is 0.05($4500), or $225. The interest on $5100 at 8% for 1 yr is 0.08($5100), or $408. The total amount of interest is $225 + $408, or $633, so the numbers check. We could also check using a graphing calculator.

5. **State.**   The Perkins loan was for $4500 and the Federal Education Loan was for $5100.

Before proceeding to Example 5, briefly scan Examples 2–4 for similarities. Note that in each case, one of the equations in the system is a simple sum while the other equation represents a sum of products. Example 5 continues this pattern with what is commonly called a *mixture problem*.

Problem-Solving Tip

When solving a problem, see if it is patterned or modeled after a problem that you have already solved.

Study Tip

Do not be surprised if your success rate drops some as you work through the exercises in this section. *This is normal.* Your success rate will increase as you gain experience with these types of problems and use some of the study tips already listed.

EXAMPLE 5 Mixing Fertilizers. Yardbird Gardening, Inc., carries two brands of fertilizer containing nitrogen and water. "Gently Green" is 5% nitrogen and "Sun Saver" is 15% nitrogen. Yardbird Gardening needs to combine the two types of solutions in order to make 100 L of a solution that is 12% nitrogen. How much of each brand should be used?

Solution

1. Familiarize. We make a drawing and then make a guess to gain familiarity with the problem.

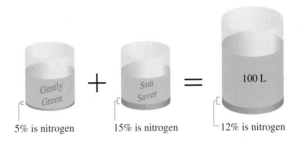

Suppose that 40 L of Gently Green and 60 L of Sun Saver are mixed. The resulting mixture will be the right size, 100 L, but will it be the right strength? To find out, note that 40 L of Gently Green would contribute $0.05(40) = 2$ L of nitrogen to the mixture while 60 L of Sun Saver would contribute $0.15(60) = 9$ L of nitrogen to the mixture. Altogether, 40 L of Gently Green and 60 L of Sun Saver would make 100 L of a mixture that has $2 + 9 = 11$ L of nitrogen. Since this would mean that the final mixture is only 11% nitrogen, our guess of 40 L and 60 L is incorrect. Still, the process of checking our guess has familiarized us with the problem.

2. Translate. We let g = the number of liters of Gently Green and s = the number of liters of Sun Saver. The information can be organized in a table.

	Gently Green	Sun Saver	Mixture	
Number of Liters	g	s	100	$\rightarrow g + s = 100$
Percent of Nitrogen	5%, or 0.05	15%, or 0.15	12%, or 0.12	
Amount of Nitrogen	$0.05g$	$0.15s$	0.12×100, or 12 liters	$\rightarrow 0.05g + 0.15s = 12$

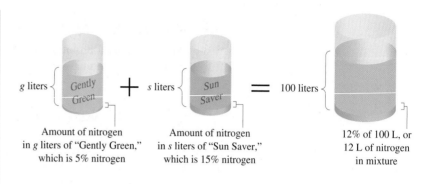

Amount of nitrogen in g liters of "Gently Green," which is 5% nitrogen

Amount of nitrogen in s liters of "Sun Saver," which is 15% nitrogen

12% of 100 L, or 12 L of nitrogen in mixture

If we add g and s in the first row of the table, we get one equation. It represents the total amount of mixture: $g + s = 100$.

If we add the amounts of nitrogen listed in the third row of the table, we get a second equation. This equation represents the amount of nitrogen in the mixture: $0.05g + 0.15s = 12$.

After clearing decimals, we have translated the problem to the system

$$g + s = 100, \quad (1)$$
$$5g + 15s = 1200. \quad (2)$$

3. Carry out. We use the elimination method to solve the system:

$$-5g - 5s = -500 \qquad \textbf{Multiplying both sides of equation (1) by } -5$$
$$\underline{5g + 15s = 1200}$$
$$10s = 700 \qquad \textbf{Adding}$$
$$s = 70; \qquad \textbf{Solving for } s$$

$$g + 70 = 100 \qquad \textbf{Substituting into equation (1)}$$
$$g = 30. \qquad \textbf{Solving for } g$$

4. Check. Remember, g is the number of liters of Gently Green and s is the number of liters of Sun Saver.

Total amount of mixture:
$$g + s = 30 + 70 = 100$$

Total amount of nitrogen:
$$5\% \text{ of } 30 + 15\% \text{ of } 70 = 1.5 + 10.5 = 12$$

Percentage of nitrogen in mixture:
$$\frac{\text{Total amount of nitrogen}}{\text{Total amount of mixture}} = \frac{12}{100} = 12\%$$

The numbers check in the original problem. We can also check graphically, as shown at left, where x replaces g and y replaces s.

5. State. Yardbird Gardening should mix 30 L of Gently Green with 70 L of Sun Saver.

$y_1 = 100 - x, \quad y_2 = (1200 - 5x)/15$

100

y_1

y_2

Intersection
X = 30   Y = 70

0

0   100

Xscl = 10, Yscl = 10

Motion Problems

When a problem deals with distance, speed (rate), and time, recall the following.

Distance, Rate, and Time Equations

If r represents rate, t represents time, and d represents distance, then

$$d = rt, \qquad r = \frac{d}{t}, \qquad \text{and} \qquad t = \frac{d}{r}.$$

Be sure to remember at least one of these equations. The others can be obtained by using algebraic manipulations as needed.

EXAMPLE 6 Train Travel. A Vermont Railways freight train, loaded with logs, leaves Boston, heading to Washington D.C., at a speed of 60 km/h. Two hours later, an Amtrak® Metroliner leaves Boston, bound for Washington D.C., on a parallel track at 90 km/h. At what point will the Metroliner catch up to the freight train?

Solution

1. **Familiarize.** Let's make a guess—say, 180 km—and check to see if it is correct. The freight train, traveling 60 km/h, would reach a point 180 km from Boston in $\frac{180}{60} = 3$ hr. The Metroliner, traveling 90 km/h, would cover 180 km in $\frac{180}{90} = 2$ hr. Since 3 hr is *not* two hours more than 2 hr, our guess of 180 km is incorrect. Although our guess is wrong, we see that the time that the trains are running and the point at which they meet are both unknown. We let $t =$ the number of hours that the freight train is running before they meet and $d =$ the distance at which the trains meet. Since the freight train has a 2-hr head start, the Metroliner runs for $t - 2$ hours before catching up to the freight train, at which point both trains have traveled the same distance.

60 km/h
d kilometers
t hours

90 km/h
d kilometers
t − 2 hours

Trains meet here − − −

2. **Translate.** We can organize the information in a chart. The formula *Distance = Rate · Time* guides our choice of rows and columns.

	Distance	Rate	Time	
Freight Train	d	60	t	$\longrightarrow d = 60t$
Metroliner	d	90	$t - 2$	$\longrightarrow d = 90(t - 2)$

Using *Distance = Rate · Time* twice, we get two equations:

$$d = 60t, \qquad (1)$$
$$d = 90(t - 2). \qquad (2)$$

3. Carry out.   We solve the system both algebraically and graphically.

Algebraic Solution

We solve the system using the substitution method:

$60t = 90(t - 2)$ **Substituting 60t for d in equation (2)**

$60t = 90t - 180$ **Using the distributive law**

$-30t = -180$

$t = 6.$

The time for the freight train is 6 hr, which means that the time for the Metroliner is $6 - 2$, or 4 hr. Remember that it is distance, not time, that the problem asked for. Thus for $t = 6$, we have $d = 60 \cdot 6 = 360$ km.

Graphical Solution

The variables used in the equations are d and t. Note that in both equations, d is given in terms of t. If we replace d with y and t with x, we can enter the equations directly. Since $y =$ distance and $x =$ time, we use a viewing window of $[0, 10, 0, 500]$. We graph $y_1 = 60x$ and $y_2 = 90(x - 2)$ and find the point of intersection.

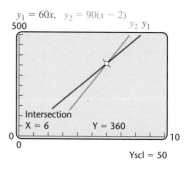

The point of intersection is $(6, 360)$. The problem asks for the distance that the trains travel. Recalling that $y =$ distance, we have the distance that both trains travel is 360 km.

4. Check.   At 60 km/h, the freight train will travel $60 \cdot 6$, or 360 km, in 6 hr. At 90 km/h, the Metroliner will travel $90 \cdot (6 - 2) = 360$ km in 4 hr. The numbers check.

5. State.   The freight train will catch up to the Metroliner at a point 360 km from Boston.

EXAMPLE 7   Jet Travel.   An F16 jet flies 4 hr west with a 60-mph tailwind. Returning *against* the wind takes 5 hr. Find the speed of the jet with no wind. (Thanks to the Vermont Air National Guard for realistic wind and plane speed information.)

TEACHING TIP

You might ask students to think of another situation that can cause speed to increase or decrease. Lead them, if possible, to the idea of the current of a river or stream.

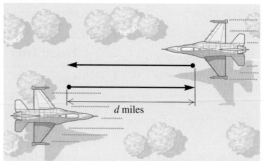

Solution

1. **Familiarize.**   We imagine the situation and make a drawing. Note that the wind *speeds up* the jet on the outbound flight, but *slows down* the jet on the return flight. Since the distances traveled each way must be the same, we can check a guess of the jet's speed with no wind. Suppose the speed of the jet with no wind is 400 mph. The jet would then fly $400 + 60 = 460$ mph with the wind and $400 - 60 = 340$ mph into the wind. In 4 hr, the jet would travel $460 \cdot 4 = 1840$ mi with the wind and $340 \cdot 5 = 1700$ mi against the wind. Since $1840 \neq 1700$, our guess of 400 mph is incorrect. Rather than guess again, let's have $r =$ the speed, in miles per hour, of the jet in still air. Then $r + 60 =$ the jet's speed with the wind and $r - 60 =$ the jet's speed against the wind. We also let $d =$ the distance traveled, in miles.

2. **Translate.**   The information can be organized in a chart. The distances traveled are the same, so we use *Distance = Rate* (or *Speed*) $\cdot$ *Time*. Each row of the chart gives an equation.

	Distance	Rate	Time	
With Wind	d	$r + 60$	4	$\longrightarrow d = (r + 60)4$
Against Wind	d	$r - 60$	5	$\longrightarrow d = (r - 60)5$

The two equations constitute a system:

$$d = (r + 60)4, \qquad (1)$$
$$d = (r - 60)5. \qquad (2)$$

3. **Carry out.**   We solve the system using substitution:

$(r - 60)5 = (r + 60)4$ **Substituting $(r - 60)5$ for d in equation (1)**

$5r - 300 = 4r + 240$ **Using the distributive law**

$r = 540.$ **Solving for r**

4. **Check.**   When $r = 540$, the speed with the wind is $540 + 60 = 600$ mph, and the speed against the wind is $540 - 60 = 480$ mph. The distance with the wind, $600 \cdot 4 = 2400$ mi, matches the distance into the wind, $480 \cdot 5 = 2400$ mi, so we have a check.

5. **State.**   The speed of the jet with no wind is 540 mph.

Tips for Solving Motion Problems

1. Draw a diagram using an arrow or arrows to represent distance and the direction of each object in motion.

2. Organize the information in a chart.

3. Look for times, distances, or rates that are the same. These often can lead to an equation.

4. Translating to a system of equations allows for the use of two variables.

5. Always make sure that you have answered the question asked.

3.3

Exercise Set

1.–18. *For Exercises 1–18, solve Exercises 41–58 from pp. 188–190.* ⊡

19. *Sales.*   Staples® recently sold a box of Flair® felt-tip pens for $12 and a four-pack of Sanford® Uni-ball® pens for $8. At the start of a recent fall semester, a combination of 40 boxes and four-packs of these pens was sold for a total of $372. How many of each type were purchased?   Boxes: 13; four-packs: 27

20. *Sales.*   Staples recently sold a wirebound graph-paper notebook for $2.50 and a college-ruled note-book made of recycled paper for $2.30. At the start of a recent spring semester, a combination of 50 of these notebooks was sold for a total of $118.60. How many of each type were sold?   Graph-paper: 18; college-ruled: 32

21. *Blending Coffees.*   The Coffee Counter charges $9.00 per pound for Kenyan French Roast coffee and $8.00 per pound for Sumatran coffee. How much of each type should be used to make a 20-lb blend that sells for $8.40 per pound?   Kenyan: 8 lb; Sumatran: 12 lb

22. *Mixed Nuts.*   The Nutty Professor sells cashews for $6.75 per pound and Brazil nuts for $5.00 per pound. How much of each type should be used to make a 50-lb mixture that sells for $5.70 per pound?   Cashews: 20 lb; Brazil nuts: 30 lb

Aha! **23.** *Catering.*   Casella's Catering is planning a wedding reception. The bride and groom would like to serve a nut mixture containing 25% peanuts. Casella has available mixtures that are either 40% or 10% peanuts. How much of each type should be mixed to get a 10-lb mixture that is 25% peanuts?   5 lb of each

24. *Livestock Feed.*   Soybean meal is 16% protein and corn meal is 9% protein. How many pounds of each should be mixed to get a 350-lb mixture that is 12% protein?   Soybean meal: 150 lb; corn meal: 200 lb

25. *Ink Remover.*   Etch Clean Graphics uses one cleanser that is 25% acid and a second that is 50% acid. How many liters of each should be mixed to get 10 L of a solution that is 40% acid?   25%-acid: 4 L; 50%-acid: 6 L

⊡ Answers to Exercises 1–18 can be found on p. A-58.

26. *Blending Granola.*   Deep Thought Granola is 25% nuts and dried fruit. Oat Dream Granola is 10% nuts and dried fruit. How much of Deep Thought and how much of Oat Dream should be mixed to form a 20-lb batch of granola that is 19% nuts and dried fruit?   Deep Thought: 12 lb; Oat Dream: 8 lb

27. *Student Loans.*   Lomasi's two student loans totaled $12,000. One of her loans was at 6% simple interest and the other at 9%. After one year, Lomasi owed $855 in interest. What was the amount of each loan? $7500 at 6%; $4500 at 9%

28. *Investments.*   An executive nearing retirement made two investments totaling $15,000. In one year, these investments yielded $1432 in simple interest. Part of the money was invested at 9% and the rest at 10%. How much was invested at each rate? $6800 at 9%; $8200 at 10%

29. *Automotive Maintenance.*   "Arctic Antifreeze" is 18% alcohol and "Frost-No-More" is 10% alcohol. How many liters of each should be mixed to get 20 L of a mixture that is 15% alcohol?   Arctic Antifreeze: 12.5 L; Frost-No-More: 7.5 L

30. *Food Science.* The following bar graph shows the milk fat percentages in three dairy products. How many pounds each of whole milk and cream should be mixed to form 200 lb of milk for cream cheese? Whole milk: $169\frac{3}{13}$ lb; cream: $30\frac{10}{13}$ lb

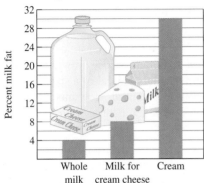

31. *Real Estate.* The perimeter of an oceanfront lot is 190 m. The width is one-fourth of the length. Find the dimensions. Length: 76 m; width: 19 m

32. *Architecture.* The rectangular ground floor of the John Hancock building has a perimeter of 860 ft. The length is 100 ft more than the width. Find the length and the width. Length: 265 ft; width: 165 ft

33. *Making Change.* Cecilia makes a $9.25 purchase at the bookstore with a $20 bill. The store has no bills and gives her change in quarters and fifty-cent pieces. There are 30 coins in all. How many of each kind are there? Quarters: 17; fifty-cent pieces: 13

34. *Teller Work.* Ashford goes to a bank and gets change for a $50 bill consisting of all $5 bills and $1 bills. There are 22 bills in all. How many of each kind are there? $5 bills: 7; $1 bills: 15

35. *Train Travel.* A train leaves Danville Junction and travels north at a speed of 75 km/h. Two hours later, an express train leaves on a parallel track and travels north at 125 km/h. How far from the station will they meet? 375 km

36. *Car Travel.* Two cars leave Salt Lake City, traveling in opposite directions. One car travels at a speed of 80 km/h and the other at 96 km/h. In how many hours will they be 528 km apart? 3 hr

37. *Canoeing.* Alvin paddled for 4 hr with a 6-km/h current to reach a campsite. The return trip against the same current took 10 hr. Find the speed of Alvin's canoe in still water. 14 km/h

38. *Boating* Mia's motorboat took 3 hr to make a trip downstream with a 6-mph current. The return trip against the same current took 5 hr. Find the speed of the boat in still water. 24 mph

39. *Point of No Return.* A plane flying the 3458-mi trip from New York City to London has a 50-mph tailwind. The flight's *point of no return* is the point at which the flight time required to return to New York is the same as the time required to continue to London. If the speed of the plane in still air is 360 mph, how far is New York from the point of no return? About 1489 mi

40. *Point of No Return.* A plane is flying the 2553-mi trip from Los Angeles to Honolulu into a 60-mph headwind. If the speed of the plane in still air is 310 mph, how far from Los Angeles is the plane's point of no return? (See Exercise 39). About 1524 mi

Tʷ **41.** Write at least three study tips of your own for someone beginning this exercise set.

Tʷ **42.** In what ways are Examples 3 and 4 similar? In what sense are their systems of equations similar?

Skill Maintenance

Evaluate.

43. $2x - 3y + 12$, for $x = 5$ and $y = 2$ [1.1] 16

44. $7x - 4y + 9$, for $x = 2$ and $y = 3$ [1.1] 11

45. $5a - 7b + 3c$, for $a = -2$, $b = 3$, and $c = 1$ [1.1], [1.2] -28

46. $3a - 8b - 2c$, for $a = -4$, $b = -1$, and $c = 3$ [1.1], [1.2] -10

47. $4 - 2y + 3z$, for $y = \frac{1}{3}$ and $z = \frac{1}{4}$ [1.1] $\frac{49}{12}$

48. $3 - 5y + 4z$, for $y = \frac{1}{2}$ and $z = \frac{1}{5}$ [1.1] $\frac{13}{10}$

Synthesis

TW **49.** Suppose that in Example 3 you are asked only for the amount of Black tea needed for the Imperial Blend. Would the method of solving the problem change? Why or why not?

TW **50.** Write a problem similar to Example 2 for a classmate to solve. Design the problem so that the solution is "The florist sold 14 hanging plants and 9 flats of petunias."

51.–54. *For Exercises 51–54, solve Exercises 84–87 from Exercise Set 3.1.* ☐

55. *Retail.* Some of the world's best and most expensive coffee is Hawaii's Kona coffee. In order for coffee to be labeled "Kona Blend," it must contain at least 30% Kona beans. Bean Town Roasters has 40 lb of Mexican coffee. How much Kona coffee must they add if they wish to market it as Kona Blend? $\frac{120}{7}$ lb

56. *Automotive Maintenance.* The radiator in Michelle's car contains 6.3 L of antifreeze and water. This mixture is 30% antifreeze. How much of this mixture should she drain and replace with pure antifreeze so that there will be a mixture of 50% antifreeze? 1.8 L

57. *Exercise.* Natalie jogs and walks to school each day. She averages 4 km/h walking and 8 km/h jogging. From home to school is 6 km and Natalie makes the trip in 1 hr. How far does she jog in a trip? 4 km

58. *Book Sales.* A limited edition of a book published by a historical society was offered for sale to members. The cost was one book for $12 or two books for $20 (maximum of two per member). The society sold 880 books, for a total of $9840. How many members ordered two books? 180 members

59. The tens digit of a two-digit positive integer is 2 more than three times the units digit. If the digits are interchanged, the new number is 13 less than half the given number. Find the given integer. (*Hint*: Let x = the tens-place digit and y = the units-place digit; then $10x + y$ is the number.) 82

60. *Train Travel.* A train leaves Union Station for Central Station, 216 km away, at 9 A.M. One hour later, a train leaves Central Station for Union Station. They meet at noon. If the second train had started at 9 A.M. and the first train at 10:30 A.M., they would still have met at noon. Find the speed of each train. First train: 36 km/h; second train: 54 km/h

61. *Wood Stains.* Williams' Custom Flooring has 0.5 gal of stain that is 20% brown and 80% neutral. A customer orders 1.5 gal of a stain that is 60% brown and 40% neutral. How much pure brown stain and how much neutral stain should be added to the original 0.5 gal in order to fill the order?* Brown: 0.8 gal; neutral: 0.2 gal

62. *Fuel Economy.* Grady's station wagon gets 18 miles per gallon (mpg) in city driving and 24 mpg in highway driving. The car is driven 465 mi on 23 gal of gasoline. How many miles were driven in the city and how many were driven on the highway? City: 261 mi; highway: 204 mi

63. *Biochemistry.* Industrial biochemists routinely use a machine to mix a buffer of 10% acetone by adding 100% acetone to water. One day, instead of adding 5 L of acetone to create a vat of buffer, a machine added 10 L. How much additional water was needed to lower the concentration to 10%? 45 L

64. *Gender.* Phil and Phyllis are siblings. Phyllis has twice as many brothers as she has sisters. Phil has the same number of brothers as sisters. How many girls and how many boys are in the family? 3 girls, 4 boys

65. See Exercise 61 above. Let x = the amount of pure brown stain added to the original 0.5 gal. Find a function $P(x)$ that can be used to determine the percentage of brown stain in the 1.5-gal mixture. On a graphing calculator, draw the graph of P and use INTERSECT to confirm the answer to Exercise 61.

$$P(x) = \frac{0.1 + x}{1.5}$$

(This expresses the percent as a decimal quantity.)

*This problem was suggested by Professor Chris Burditt of Yountville, California.

☐ Answers to Exercises 51–54 can be found on p. A-58.

3.4

Identifying Solutions ■ Solving Systems in Three Variables ■ Dependency, Inconsistency, and Geometric Considerations

Systems of Equations in Three Variables

Some problems translate directly to two equations. Others call for a translation to three or more equations. Here we learn how to solve systems of three linear equations. Later, we will use such systems in problem-solving situations.

Identifying Solutions

A **linear equation in three variables** is an equation equivalent to one in the form $Ax + By + Cz = D$, where A, B, C, and D are real numbers. We refer to the form $Ax + By + Cz = D$ as *standard form* for a linear equation in three variables.

A solution of a system of three equations in three variables is an ordered triple (x, y, z) that makes *all three* equations true.

EXAMPLE 1   Determine whether $\left(\frac{3}{2}, -4, 3\right)$ is a solution of the system

$$4x - 2y - 3z = 5,$$
$$-8x - y + z = -5,$$
$$2x + y + 2z = 5.$$

Solution   We substitute $\left(\frac{3}{2}, -4, 3\right)$ into the three equations, using alphabetical order.

By Hand

We have the following:

$$\frac{4x - 2y - 3z = 5}{4 \cdot \tfrac{3}{2} - 2(-4) - 3 \cdot 3 \ ? \ 5}$$
$$6 + 8 - 9 \ \Big|$$
$$5 \ \Big| \ 5 \quad \text{TRUE}$$

$$\frac{-8x - y + z = -5}{-8 \cdot \tfrac{3}{2} - (-4) + 3 \ ? \ -5}$$
$$-12 + 4 + 3 \ \Big|$$
$$-5 \ \Big| \ -5 \quad \text{TRUE}$$

$$\frac{2x + y + 2z = 5}{2 \cdot \tfrac{3}{2} + (-4) + 2 \cdot 3 \ ? \ 5}$$
$$3 - 4 + 6 \ \Big|$$
$$5 \ \Big| \ 5 \quad \text{TRUE}$$

Using a Graphing Calculator

We store $\frac{3}{2}$ as X, -4 as Y, and 3 as Z, using the ALPHA key to enter Y and Z.

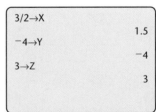

Now we enter the expression on the left side of each equation and press ENTER. If the value is the same as the right side of the equation, the ordered triple makes the equation true.

```
4X−2Y−3Z
                    5
−8X−Y+Z
                   −5
2X+Y+2Z
                    5
```

The triple makes all three equations true, so it is a solution.

Solving Systems in Three Variables

Graphical methods for solving linear equations in three variables are problematic, because a three-dimensional coordinate system is required and the graph of a linear equation in three variables is a plane. The substitution method *can* be used but becomes very cumbersome unless one or more of the equations has only two variables. Fortunately, the elimination method allows us to manipulate a system of three equations in three variables so that a simpler system of two equations in two variables is formed. Once that simpler system has been solved, we can substitute into one of the three original equations and solve for the third variable.

EXAMPLE 2   Solve the following system of equations:

$$\begin{aligned} x + y + z &= 4, &(1)\\ x - 2y - z &= 1, &(2)\\ 2x - y - 2z &= -1. &(3) \end{aligned}$$

Solution   We select *any* two of the three equations and work to get one equation in two variables. Let's add equations (1) and (2):

$$\begin{array}{ll} x + y + z = 4 & (1)\\ \underline{x - 2y - z = 1} & (2)\\ 2x - y = 5. & (4) \qquad \text{Adding to eliminate } z \end{array}$$

Next, we select a different pair of equations and eliminate the *same variable* we did above. Let's use equations (1) and (3) to again eliminate z. Be careful here! A common error is to eliminate a different variable in this step.

$$\begin{array}{l} x + y + z = 4,\\ 2x - y - 2z = -1 \end{array} \xrightarrow[\text{equation (1) by 2}]{\textbf{Multiplying both sides of}} \begin{array}{l} 2x + 2y + 2z = 8\\ \underline{2x - y - 2z = -1}\\ 4x + y = 7 \quad (5) \end{array}$$

Now we solve the resulting system of equations (4) and (5). That solution will give us two of the numbers in the solution of the original system.

$$\begin{array}{ll} 2x - y = 5 & (4)\\ \underline{4x + y = 7} & (5)\\ 6x = 12 & \text{Adding}\\ x = 2 \end{array}$$

Note that we now have two equations in two variables. Had we not eliminated the same variable in both of the above steps, this would not be the case.

We can use either equation (4) or (5) to find y. We choose equation (5):

$$\begin{aligned} 4x + y &= 7 &&(5)\\ 4 \cdot 2 + y &= 7 &&\textbf{Substituting 2 for } x \textbf{ in equation (5)}\\ 8 + y &= 7\\ y &= -1. \end{aligned}$$

We now have $x = 2$ and $y = -1$. To find the value for z, we use any of the original three equations and substitute to find the third number, z. Let's use equation (1) and substitute our two numbers in it:

$$x + y + z = 4 \qquad (1)$$
$$2 + (-1) + z = 4 \qquad \textbf{Substituting 2 for } x \textbf{ and } -1 \textbf{ for } y$$
$$1 + z = 4$$
$$z = 3.$$

We have obtained the triple $(2, -1, 3)$. It should check in *all three* equations:

$$\frac{x + y + z = 4}{2 + (-1) + 3 \ ? \ 4} \qquad\qquad \frac{x - 2y - z = 1}{2 - 2(-1) - 3 \ ? \ 1}$$
$$\qquad 4 \ | \ 4 \quad \text{TRUE} \qquad\qquad\qquad 1 \ | \ 1 \quad \text{TRUE}$$

$$\frac{2x - y - 2z = -1}{2 \cdot 2 - (-1) - 2 \cdot 3 \ ? \ -1}$$
$$-1 \ | \ -1 \quad \text{TRUE}$$

The solution is $(2, -1, 3)$.

Solving Systems of Three Linear Equations

To use the elimination method to solve systems of three linear equations:

1. Write all equations in the standard form $Ax + By + Cz = D$.

2. Clear any decimals or fractions.

3. Choose a variable to eliminate. Then select two of the three equations and work to get one equation in two variables.

4. Next, use a different pair of equations and eliminate the same variable that you did in step (3).

5. Solve the system of equations that resulted from steps (3) and (4).

6. Substitute the solution from step (5) into one of the original three equations and solve for the third variable. Then check.

EXAMPLE 3   Solve the system

$$4x - 2y - 3z = 5, \qquad (1)$$
$$-8x - y + z = -5, \qquad (2)$$
$$2x + y + 2z = 5. \qquad (3)$$

Solution

1., 2. The equations are already in standard form with no fractions or decimals.

3. Next, select a variable to eliminate. We decide on y because the y-terms are opposites of each other in equations (2) and (3). We add:

$$
\begin{array}{rl}
-8x - y + z = -5 & \quad (2) \\
\underline{2x + y + 2z = 5} & \quad (3) \\
-6x + 3z = 0. & \quad (4) \qquad \text{Adding}
\end{array}
$$

4. We use another pair of equations to create a second equation in x and z. That is, we eliminate the same variable, y, as in step (3). We use equations (1) and (3):

$$
\begin{array}{l}
4x - 2y - 3z = 5, \\
2x + y + 2z = 5
\end{array}
\xrightarrow[\textbf{equation (3) by 2}]{\textbf{Multiplying both sides of}}
\begin{array}{rl}
4x - 2y - 3z = 5 \\
\underline{4x + 2y + 4z = 10} \\
8x + z = 15. \quad (5)
\end{array}
$$

5. Now we solve the resulting system of equations (4) and (5). That allows us to find two parts of the ordered triple.

$$
\begin{array}{l}
-6x + 3z = 0, \\
8x + z = 15
\end{array}
\xrightarrow[\textbf{equation (5) by } -3]{\textbf{Multiplying both sides of}}
\begin{array}{rl}
-6x + 3z = 0 \\
\underline{-24x - 3z = -45} \\
-30x = -45 \\
x = \frac{-45}{-30} = \frac{3}{2}
\end{array}
$$

We use equation (5) to find z:

$$
\begin{array}{rl}
8x + z = 15 \\
8 \cdot \frac{3}{2} + z = 15 & \quad \textbf{Substituting } \tfrac{3}{2} \textbf{ for } x \\
12 + z = 15 \\
z = 3.
\end{array}
$$

6. Finally, we use any of the original equations and substitute to find the third number, y. We choose equation (3):

$$
\begin{array}{rl}
2x + y + 2z = 5 & \quad (3) \\
2 \cdot \frac{3}{2} + y + 2 \cdot 3 = 5 & \quad \textbf{Substituting } \tfrac{3}{2} \textbf{ for } x \textbf{ and 3 for } z \\
3 + y + 6 = 5 \\
y + 9 = 5 \\
y = -4.
\end{array}
$$

The solution is $\left(\frac{3}{2}, -4, 3\right)$. The check was performed as Example 1.

TEACHING TIP

You may want to remind students to list the values in the ordered triple in alphabetical order of the variables.

Sometimes, certain variables are missing at the outset.

EXAMPLE 4   Solve the system

$$
\begin{array}{rl}
x + y + z = 180, & \quad (1) \\
x - z = -70, & \quad (2) \\
2y - z = 0. & \quad (3)
\end{array}
$$

Solution

1., 2. The equations appear in standard form with no fractions or decimals.

3., 4. Note that there is no x-term in equation (3) and no y-term in equation (2). It will save some steps if we eliminate one of these variables; we choose to eliminate y. Since, at the outset, we already have one equation with no y, we need only one more equation with y eliminated. We use equations (1) and (3):

$$
\begin{array}{ll}
x + y + z = 180, & \xrightarrow{\substack{\textbf{Multiplying both sides of}\\ \textbf{equation (1) by } -2}} \quad -2x - 2y - 2z = -360 \\
2y - z = \quad 0 & 2y - z = 0 \\
& \overline{-2x - 3z = -360.} \quad (4)
\end{array}
$$

5., 6. Now we solve the resulting system of equations (2) and (4):

$$
\begin{array}{ll}
x - z = -70, & \xrightarrow{\substack{\textbf{Multiplying both sides of}\\ \textbf{equation (2) by } 2}} \quad 2x - 2z = -140 \\
-2x - 3z = -360 & -2x - 3z = -360 \\
& \overline{ -5z = -500} \\
& z = 100.
\end{array}
$$

Continuing as in Examples 2 and 3, we get the solution $(30, 50, 100)$. The check is left to the student.

Dependency, Inconsistency, and Geometric Considerations

Each equation in Examples 2, 3, and 4 has a graph that is a plane in three dimensions. The solutions are points common to the planes of each system. Since three planes can have an infinite number of points in common or no points at all in common, we need to generalize the concept of *consistency*.

One solution: planes intersecting in exactly one point. System is consistent.

The planes intersect along a common line. An infinite number of points are common to the three planes. System is consistent.

Three parallel planes. There is no common point of intersection. System is inconsistent.

Planes intersect two at a time, but there is no point common to all three. System is inconsistent.

Consistency

A system of equations that has at least one solution is said to be **consistent**.

A system of equations that has no solution is said to be **inconsistent**.

EXAMPLE 5 Solve:

$$y + 3z = 4, \qquad (1)$$
$$-x - y + 2z = 0, \qquad (2)$$
$$x + 2y + z = 1. \qquad (3)$$

Solution The variable x is missing in equation (1). By adding equations (2) and (3), we can find a second equation in which x is missing:

$$\begin{array}{ll} -x - y + 2z = 0 & (2) \\ \underline{x + 2y + z = 1} & (3) \\ y + 3z = 1. & (4) \qquad \textbf{Adding} \end{array}$$

Equations (1) and (4) form a system in y and z. We solve as before:

$$y + 3z = 4, \xrightarrow{\textbf{Multiplying both sides of equation (1) by } -1} -y - 3z = -4$$
$$y + 3z = 1 \qquad\qquad\qquad\qquad \underline{y + 3z = 1}$$
$$\textbf{This is a contradiction.} \longrightarrow 0 = -3. \qquad \textbf{Adding}$$

Since we end up with a *false* equation, or contradiction, we know that the system has no solution. It is *inconsistent*.

The notion of *dependency* from Section 3.1 can also be extended.

EXAMPLE 6 Solve:

$$2x + y + z = 3, \qquad (1)$$
$$x - 2y - z = 1, \qquad (2)$$
$$3x + 4y + 3z = 5. \qquad (3)$$

Solution Our plan is to first use equations (1) and (2) to eliminate z. Then we will select another pair of equations and again eliminate z:

$$\begin{array}{l} 2x + y + z = 3 \\ \underline{x - 2y - z = 1} \\ 3x - y = 4. \qquad (4) \end{array}$$

Next, we use equations (2) and (3) to eliminate z again:

$$x - 2y - z = 1, \xrightarrow{\textbf{Multiplying both sides of equation (2) by } 3} 3x - 6y - 3z = 3$$
$$3x + 4y + 3z = 5 \qquad\qquad\qquad\qquad \underline{3x + 4y + 3z = 5}$$
$$6x - 2y = 8. \qquad (5)$$

We now try to solve the resulting system of equations (4) and (5):

$$3x - y = 4, \xrightarrow{\textbf{Multiplying both sides of equation (4) by } -2} -6x + 2y = -8$$
$$6x - 2y = 8 \qquad\qquad\qquad\qquad \underline{6x - 2y = 8}$$
$$0 = 0. \qquad (6)$$

Equation (6), which is an identity, indicates that equations (1), (2), and (3) are *dependent*. This means that the original system of three equations is equivalent to a system of two equations. One way to see this is to observe that two times equation (1), minus equation (2), is equation (3). Thus removing equation (3) from the system does not affect the solution of the system.* In writing an answer to this problem, we simply state that "the equations are dependent."

Recall that when dependent equations appeared in Section 3.1, the solution sets were always infinite in size and were written in set-builder notation. There, all systems of dependent equations were *consistent*. This is not always the case for systems of three or more equations. The following figures illustrate some possibilities geometrically.

The planes intersect along a common line. The equations are dependent and the system is consistent. There is an infinite number of solutions.

The planes coincide. The equations are dependent and the system is consistent. There is an infinite number of solutions.

Two planes coincide. The third plane is parallel. The equations are dependent and the system is inconsistent. There is no solution.

3.4

Exercise Set

FOR EXTRA HELP

Digital Video Tutor CD 3 Videotape 4 | Student's Solutions Manual | Tutor Center AW Math Tutor Center | InterAct Math | MathXL | MyMathLab

1. Determine whether $(2, -1, -2)$ is a solution of the system

$$x + y - 2z = 5,$$
$$2x - y - z = 7,$$
$$-x - 2y + 3z = 6. \quad \text{No}$$

2. Determine whether $(1, -2, 3)$ is a solution of the system

$$x + y + z = 2,$$
$$x - 2y - z = 2,$$
$$3x + 2y + z = 2. \quad \text{Yes}$$

Solve each system. If a system's equations are dependent or if there is no solution, state this.

3.
$$x + y + z = 6,$$
$$2x - y + 3z = 9,$$
$$-x + 2y + 2z = 9$$
$(1, 2, 3)$

4.
$$2x - y + z = 10,$$
$$4x + 2y - 3z = 10,$$
$$x - 3y + 2z = 8$$
$(4, 0, 2)$

5.
$$2x - y - 3z = -1,$$
$$2x - y + z = -9,$$
$$x + 2y - 4z = 17$$
$(-1, 5, -2)$

6.
$$x - y + z = 6,$$
$$2x + 3y + 2z = 2,$$
$$3x + 5y + 4z = 4$$
$(2, -2, 2)$

7.
$$2x - 3y + z = 5,$$
$$x + 3y + 8z = 22,$$
$$3x - y + 2z = 12$$
$(3, 1, 2)$

8.
$$6x - 4y + 5z = 31,$$
$$5x + 2y + 2z = 13,$$
$$x + y + z = 2$$
$(3, -2, 1)$

---

*A set of equations is dependent if at least one equation can be expressed as a sum of multiples of other equations in that set.

9. $3a - 2b + 7c = 13,$
$a + 8b - 6c = -47,$
$7a - 9b - 9c = -3$
$(-3, -4, 2)$

10. $x + y + z = 0,$
$2x + 3y + 2z = -3,$
$-x + 2y - 3z = -1$
$(7, -3, -4)$

11. $2x + 3y + z = 17,$
$x - 3y + 2z = -8,$
$5x - 2y + 3z = 5$
$(2, 4, 1)$

12. $2x + y - 3z = -4,$
$4x - 2y + z = 9,$
$3x + 5y - 2z = 5$
$(2, 1, 3)$

13. $2x + y + z = -2,$
$2x - y + 3z = 6,$
$3x - 5y + 4z = 7$
$(-3, 0, 4)$

14. $2x + y + 2z = 11,$
$3x + 2y + 2z = 8,$
$x + 4y + 3z = 0$
$(2, -5, 6)$

15. $x - y + z = 4,$
$5x + 2y - 3z = 2,$
$4x + 3y - 4z = -2$
The equations are dependent.

16. $-2x + 8y + 2z = 4,$
$x + 6y + 3z = 4,$
$3x - 2y + z = 0$
The equations are dependent.

17. $a + 2b + c = 1,$
$7a + 3b - c = -2,$
$a + 5b + 3c = 2$
$(3, -5, 8)$

18. $4x - y - z = 4,$
$2x + y + z = -1,$
$6x - 3y - 2z = 3$
$\left(\frac{1}{2}, 4, -6\right)$

19. $5x + 3y + \frac{1}{2}z = \frac{7}{2},$
$0.5x - 0.9y - 0.2z = 0.3,$
$3x - 2.4y + 0.4z = -1$ $\left(\frac{3}{5}, \frac{2}{3}, -3\right)$

20. $r + \frac{3}{2}s + 6t = 2,$
$2r - 3s + 3t = 0.5,$
$r + s + t = 1$ $\left(\frac{1}{2}, \frac{1}{3}, \frac{1}{6}\right)$

21. $3p + 2r = 11,$
$q - 7r = 4,$
$p - 6q = 1$ $\left(4, \frac{1}{2}, -\frac{1}{2}\right)$

22. $4a + 9b = 8,$
$8a + 6c = -1,$
$6b + 6c = -1$ $\left(\frac{1}{2}, \frac{2}{3}, -\frac{5}{6}\right)$

23. $x + y + z = 105,$
$10y - z = 11,$
$2x - 3y = 7$ $(17, 9, 79)$

24. $x + y + z = 57,$
$-2x + y = 3,$
$x - z = 6$ $(15, 33, 9)$

25. $2a - 3b = 2,$
$7a + 4c = \frac{3}{4},$
$2c - 3b = 1$ $\left(\frac{1}{4}, -\frac{1}{2}, -\frac{1}{4}\right)$

26. $a - 3c = 6,$
$b + 2c = 2,$
$7a - 3b - 5c = 14$ $(3, 4, -1)$

Aha! **27.** $x + y + z = 182,$
$y = 2 + 3x,$
$z = 80 + x$
$(20, 62, 100)$

28. $l + m = 7,$
$3m + 2n = 9,$
$4l + n = 5$
$(2, 5, -3)$

29. $x + y = 0,$
$x + z = 1,$
$2x + y + z = 2$
No solution

30. $x + z = 0,$
$x + y + 2z = 3,$
$y + z = 2$
No solution

31. $y + z = 1,$
$x + y + z = 1,$
$x + 2y + 2z = 2$
The equations are dependent.

32. $x + y + z = 1,$
$-x + 2y + z = 2,$
$2x - y = -1$
The equations are dependent.

TW **33.** Abbie recommends that a frustrated classmate double- and triple-check each step of work when attempting to solve a system of three equations. Is this good advice? Why or why not?

TW **34.** Describe a method for writing an inconsistent system of three equations in three variables.

Skill Maintenance

Translate each sentence to mathematics. [1.6]

35. One number is twice another.   Let x and y represent the numbers; $x = 2y$

36. The sum of two numbers is three times the first number.   Let x and y represent the numbers; $x + y = 3x$

37. The sum of three consecutive numbers is 45.

38. One number plus twice another number is 17.   Let x and y represent the numbers; $x + 2y = 17$

39. The sum of two numbers is five times a third number.   Let x, y, and z represent the numbers; $x + y = 5z$

40. The product of two numbers is twice their sum.   Let x and y represent the numbers; $xy = 2(x + y)$

Synthesis

37. Let x represent the first number; $x + (x + 1) + (x + 2) = 45$

TW **41.** Is it possible for a system of three equations to have exactly two ordered triples in its solution set? Why or why not?

TW **42.** Describe a procedure that could be used to solve a system of four equations in four variables.

Solve.

43. $\dfrac{x + 2}{3} - \dfrac{y + 4}{2} + \dfrac{z + 1}{6} = 0,$
$\dfrac{x - 4}{3} + \dfrac{y + 1}{4} - \dfrac{z - 2}{2} = -1,$
$\dfrac{x + 1}{2} + \dfrac{y}{2} + \dfrac{z - 1}{4} = \dfrac{3}{4}$ $(1, -1, 2)$

44. $w + x + y + z = 2,$
$w + 2x + 2y + 4z = 1,$
$w - x + y + z = 6,$
$w - 3x - y + z = 2$ $(1, -2, 4, -1)$

45. $w + x - y + z = 0,$
$w - 2x - 2y - z = -5,$
$w - 3x - y + z = 4,$
$2w - x - y + 3z = 7$ $(-3, -1, 0, 4)$

For Exercises 46 and 47, let u represent $1/x$, v represent $1/y$, and w represent $1/z$. Solve for u, v, and w, and then solve for x, y, and z.

46. $\dfrac{2}{x} - \dfrac{1}{y} - \dfrac{3}{z} = -1,$

$\dfrac{2}{x} - \dfrac{1}{y} + \dfrac{1}{z} = -9,$

$\dfrac{1}{x} + \dfrac{2}{y} - \dfrac{4}{z} = 17$

$\left(-1, \frac{1}{5}, -\frac{1}{2}\right)$

47. $\dfrac{2}{x} + \dfrac{2}{y} - \dfrac{3}{z} = 3,$

$\dfrac{1}{x} - \dfrac{2}{y} - \dfrac{3}{z} = 9,$

$\dfrac{7}{x} - \dfrac{2}{y} + \dfrac{9}{z} = -39$

$\left(-\frac{1}{2}, -1, -\frac{1}{3}\right)$

Determine k so that each system is dependent.

48. $x - 3y + 2z = 1,$
$2x + y - z = 3,$
$9x - 6y + 3z = k$
12

49. $5x - 6y + kz = -5,$
$x + 3y - 2z = 2,$
$2x - y + 4z = -1$
14

In each case, three solutions of an equation in x, y, and z are given. Find the equation.

50. $Ax + By + Cz = 12;$
$\left(1, \frac{3}{4}, 3\right), \left(\frac{4}{3}, 1, 2\right),$ and $(2, 1, 1)$ $3x + 4y + 2z = 12$

51. $z = b - mx - ny;$
$(1, 1, 2), (3, 2, -6),$ and $\left(\frac{3}{2}, 1, 1\right)$ $z = 8 - 2x - 4y$

52. Write an inconsistent system of equations that contains dependent equations.   Answers may vary.
$x + y + z = 1, 2x + 2y + 2z = 2, x + y + z = 3$

Collaborative Corner

Finding the Preferred Approach

Focus: Systems of three linear equations

Time: 10–15 minutes

Group size: 3

Consider the six steps outlined on p. 217 along with the following system:

$$2x + 4y = 3 - 5z,$$
$$0.3x = 0.2y + 0.7z + 1.4,$$
$$0.04x + 0.03y = 0.07 + 0.04z.$$

ACTIVITY

1. Working independently, each group member should solve the system above. One person

should begin by eliminating x, one should first eliminate y, and one should first eliminate z. Write neatly so that others can follow your steps.

2. Once all group members have solved the system, compare your answers. If the answers do not check, exchange notebooks and check each other's work. If a mistake is detected, allow the person who made the mistake to make the repair.

3. Decide as a group which of the three approaches above (if any) ranks as easiest and which (if any) ranks as most difficult. Then compare your rankings with the other groups in the class.

3.5

Applications of Three Equations in Three Unknowns

Solving Applications: Systems of Three Equations

Solving systems of three or more equations is important in many applications. Such systems arise in the natural and social sciences, business, and engineering. In mathematics, purely numerical applications also arise.

EXAMPLE 1   The sum of three numbers is 4. The first number minus twice the second, minus the third is 1. Twice the first number minus the second, minus twice the third is -1. Find the numbers.

Solution

1. **Familiarize.**   There are three statements involving the same three numbers. Let's label these numbers x, y, and z.

2. **Translate.**   We can translate directly as follows.

<u>The sum of the three numbers</u>   <u>is</u> <u>4.</u>

$$x + y + z \qquad\qquad = 4$$

<u>The first number</u> <u>minus</u> <u>twice the second</u> <u>minus</u> <u>the third</u> <u>is</u> <u>1.</u>

$$x \qquad - \qquad 2y \qquad - \qquad z \quad = 1$$

<u>Twice the first number</u> <u>minus</u> <u>the second</u> <u>minus</u> <u>twice the third</u> <u>is</u> <u>-1.</u>

$$2x \qquad - \qquad y \qquad - \qquad 2z \quad = -1$$

We now have a system of three equations:

$$
\begin{aligned}
x + y + z &= 4, \\
x - 2y - z &= 1, \\
2x - y - 2z &= -1.
\end{aligned}
$$

3. **Carry out.**   We need to solve the system of equations. Note that we found the solution, $(2, -1, 3)$, in Example 2 of Section 3.4.

4. **Check.**   The first statement of the problem says that the sum of the three numbers is 4. That checks, because $2 + (-1) + 3 = 4$. The second statement says that the first number minus twice the second, minus the third is 1: $2 - 2(-1) - 3 = 1$. That checks. The check of the third statement is left to the student.

5. **State.**   The three numbers are 2, -1, and 3. ━━━━━

EXAMPLE 2   Architecture.   In a triangular cross section of a roof, the largest angle is 70° greater than the smallest angle. The largest angle is twice as large as the remaining angle. Find the measure of each angle.

Solution

1. Familiarize. The first thing we do is make a drawing, or a sketch.

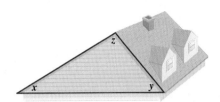

Since we don't know the size of any angle, we use x, y, and z to represent the three measures, from smallest to largest. Recall that the measures of the angles in any triangle add up to $180°$.

2. Translate. This geometric fact about triangles gives us one equation:

$$x + y + z = 180.$$

Two of the statements can be translated almost directly.

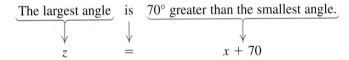

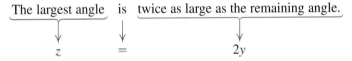

We now have a system of three equations:

$$\begin{array}{lll} x + y + z = 180, & \quad & x + y + z = 180, \\ x + 70 = z, & \text{or} \quad x \quad - z = -70, & \textbf{Rewriting in} \\ 2y = z; & \quad 2y - z = 0. & \textbf{standard form} \end{array}$$

3. Carry out. The system was solved in Example 4 of Section 3.4. The solution is $(30, 50, 100)$.

4. Check. The sum of the numbers is 180, so that checks. The measure of the largest angle, $100°$, is $70°$ greater than the measure of the smallest angle, $30°$, so that checks. The measure of the largest angle is also twice the measure of the remaining angle, $50°$. Thus we have a check.

5. State. The angles in the triangle measure $30°$, $50°$, and $100°$.

EXAMPLE 3   Cholesterol Levels. Americans have become very conscious of their cholesterol levels. Recent studies indicate that a child's intake of cholesterol should be no more than 300 mg per day. By eating 1 egg, 1 cupcake, and 1 slice of pizza, a child consumes 302 mg of cholesterol. If the child eats 2 cupcakes and 3 slices of pizza, he or she takes in 65 mg of cholesterol. By eating 2 eggs and 1 cupcake, a child consumes 567 mg of cholesterol. How much cholesterol is in each item?

Solution

1. **Familiarize.**   After we have read the problem a few times, it becomes clear that an egg contains considerably more cholesterol than the other foods. Let's guess that one egg contains 200 mg of cholesterol and one cupcake contains 50 mg. Because of the third sentence in the problem, it would follow that a slice of pizza contains 52 mg of cholesterol since $200 + 50 + 52 = 302$.

   To see if our guess satisfies the other statements in the problem, we find the amount of cholesterol that 2 cupcakes and 3 slices of pizza would contain: $2 \cdot 50 + 3 \cdot 52 = 256$. Since this does not match the 65 mg listed in the fourth sentence of the problem, our guess was incorrect. Rather than guess again, we examine how we checked our guess and let g, c, and $s =$ the number of milligrams of cholesterol in an egg, a cupcake, and a slice of pizza, respectively.

2. **Translate.**   By rewording some of the sentences in the problem, we can translate it into three equations.

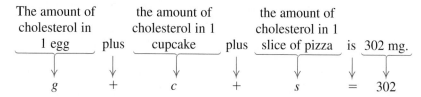

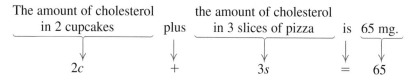

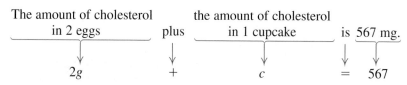

   We now have a system of three equations:

   $$g + c + s = 302,$$
   $$2c + 3s = 65,$$
   $$2g + c = 567.$$

3. **Carry out.**   We solve and get $g = 274$, $c = 19$, $s = 9$.

4. **Check.**   The sum of 274, 19, and 9 is 302 so the total cholesterol in 1 egg, 1 cupcake, and 1 slice of pizza checks. Two cupcakes and three slices of pizza would contain $2 \cdot 19 + 3 \cdot 9 = 65$ mg, while two eggs and one cupcake would contain $2 \cdot 274 + 19 = 567$ mg of cholesterol. The answer checks.

5. **State.**   An egg contains 274 mg of cholesterol, a cupcake contains 19 mg of cholesterol, and a slice of pizza contains 9 mg of cholesterol.

3.5

Exercise Set

Solve.

1. The sum of three numbers is 57. The second is 3 more than the first. The third is 6 more than the first. Find the numbers.   16, 19, 22

2. The sum of three numbers is 5. The first number minus the second plus the third is 1. The first minus the third is 3 more than the second. Find the numbers.   4, 2, −1

3. The sum of three numbers is 26. Twice the first minus the second is 2 less than the third. The third is the second minus three times the first. Find the numbers.   8, 21, −3

4. The sum of three numbers is 105. The third is 11 less than ten times the second. Twice the first is 7 more than three times the second. Find the numbers.   17, 9, 79

5. *Geometry.*   In triangle *ABC*, the measure of angle *B* is three times that of angle *A*. The measure of angle *C* is 20° more than that of angle *A*. Find the angle measures.   32°, 96°, 52°

6. *Geometry.*   In triangle *ABC*, the measure of angle *B* is twice the measure of angle *A*. The measure of angle *C* is 80° more than that of angle *A*. Find the angle measures.   25°, 50°, 105°

7. *Automobile Pricing.*   A recent basic model of a particular automobile had a price of $12,685. The basic model with the added features of automatic transmission and power door locks was $14,070. The basic model with air conditioning (AC) and power door locks was $13,580. The basic model with AC and automatic transmission was $13,925. What was the individual cost of each of the three options?   Automatic transmission: $865; power door locks: $520; air conditioning: $375

8. *Lens Production.*   When Sight-Rite's three polishing machines, A, B, and C, are all working, 5700 lenses can be polished in one week. When only A and B are working, 3400 lenses can be polished in one week. When only B and C are working, 4200 lenses can be polished in one week. How many lenses can be polished in a week by each machine?   A: 1500; B: 1900; C: 2300

9. *Welding Rates.*   Elrod, Dot, and Wendy can weld 74 linear feet per hour when working together. Elrod and Dot together can weld 44 linear feet per hour, while Elrod and Wendy can weld 50 linear feet per hour. How many linear feet per hour can each weld alone?   Elrod: 20; Dot: 24; Wendy: 30

Aha!

10. *Telemarketing.*   Sven, Tillie, and Isaiah can process 740 telephone orders per day. Sven and Tillie together can process 470 orders, while Tillie and Isaiah together can process 520 orders per day. How many orders can each person process alone?   Sven: 220; Tillie: 250; Isaiah: 270

11. *Restaurant Management.*   In a recent year, Kyle worked at Dunkin® Donuts, where a 10-oz cup of coffee cost $1.05, a 14-oz cup cost $1.35, and a 20-oz cup cost $1.65. During one busy period, Kyle served 34 cups of coffee, emptying five 96-oz pots while collecting a total of $45. How many cups of each size did Kyle fill?   10-oz cups: 11; 14-oz cups: 15; 20-oz cups: 8

10 oz $1.05   14 oz $1.35   20 oz $1.65

12. *Advertising.*   In a recent year, companies spent a total of $84.8 billion on newspaper, television, and radio ads. The total amount spent on television and radio ads was only $2.6 billion more than the amount spent on newspaper ads alone. The amount spent on newspaper ads was $5.1 billion more than what was spent on television ads. How much was spent on each form of advertising?   Newspaper: $41.1 billion; television: $36 billion; radio: $7.7 billion

13. *Investments.*   A business class divided an imaginary investment of $80,000 among three mutual funds.

The first fund grew by 10%, the second by 6%, and the third by 15%. Total earnings were $8850. The earnings from the first fund were $750 more than the earnings from the third. How much was invested in each fund? First fund: $45,000; second fund: $10,000; third fund: $25,000

14. *Restaurant Management.* McDonald's® recently sold small soft drinks for 87¢, medium soft drinks for $1.08, and large soft drinks for $1.54. During a lunch-time rush, Chris sold 40 soft drinks for a total of $43.40. The number of small and large drinks, combined, was 10 fewer than the number of medium drinks. How many drinks of each size were sold? Small: 10; medium: 25; large: 5

| small | medium | large |
| $0.87 | $1.08 | $1.54 |

15. *Nutrition.* A dietician in a hospital prepares meals under the guidance of a physician. Suppose that for a particular patient a physician prescribes a meal to have 800 calories, 55 g of protein, and 220 mg of vitamin C. The dietician prepares a meal of roast beef, baked potatoes, and broccoli according to the data in the following table.

	Calories	Protein (in grams)	Vitamin C (in milligrams)
Roast Beef, 3 oz	300	20	0
Baked Potato	100	5	20
Broccoli, 156 g	50	5	100

How many servings of each food are needed in order to satisfy the doctor's orders? Roast beef: 2; baked potato: 1; broccoli: 2

16. *Nutrition.* Repeat Exercise 15 but replace the broccoli with asparagus, for which a 180-g serving contains 50 calories, 5 g of protein, and 44 mg of vitamin C. Which meal would you prefer eating? Roast beef: $1\frac{1}{8}$; baked potatoes: $2\frac{3}{4}$; asparagus: $3\frac{3}{4}$

17. *Crying Rate.* The sum of the average number of times a man, a woman, and a one-year-old child cry each month is 71.7. A one-year-old cries 46.4 more times than a man. The average number of times a one-year-old cries per month is 28.3 more times than the average number of times combined that a man and a woman cry. What is the average number of times per month that each cries? Man: 3.6; woman: 18.1; one-year-old child: 50

18. *Obstetrics.* In the United States, the highest incidence of fraternal twin births occurs among Asian-Americans, then African-Americans, and then Caucasians. Out of every 15,400 births, the total number of fraternal twin births for all three is 739, where there are 185 more for Asian-Americans than African-Americans and 231 more for Asian-Americans than Caucasians. How many births of fraternal twins are there for each group out of every 15,400 births? Asian-American: 385; African-American: 200; Caucasian: 154

19. *Basketball Scoring.* The New York Knicks recently scored a total of 92 points on a combination of 2-point field goals, 3-point field goals, and 1-point foul shots. Altogether, the Knicks made 50 baskets and 19 more 2-pointers than foul shots. How many shots of each kind were made? Two-point field goals: 32; three-point field goals: 5; foul shots: 13

20. *History.* Find the year in which the first U.S. transcontinental railroad was completed. The following are some facts about the number. The sum of the digits in the year is 24. The ones digit is 1 more than the hundreds digit. Both the tens and the ones digits are multiples of 3. 1869

TW **21.** Problems like Exercises 11 and 12 could be classified as total-value problems. How do these problems differ from the total-value problems of Section 3.3?

TW **22.** Write a problem for a classmate to solve. Design the problem so that it translates to a system of three equations in three variables.

Skill Maintenance

Simplify.

23. $5(-3) + 7$ [1.2] -8 **24.** $-4(-6) + 9$ [1.2] 33

25. $-6(8) + (-7)$ [1.2] $\overset{-55}{}$

26. $7(-9) + (-8)$ [1.2] $\overset{-71}{}$

27. $-7(2x - 3y + 5z)$

[1.3] $-14x + 21y - 35z$

28. $-6(4a + 7b - 9c)$

[1.3] $-24a - 42b + 54c$

29. $-4(2a + 5b) + 3a + 20b$ [1.3] $-5a$

30. $3(2x - 7y) + 5x + 21y$ [1.3] $11x$

Synthesis

TW **31.** Consider Exercise 19. Suppose there were no foul shots made. Would there still be a solution? Why or why not?

TW **32.** Consider Exercise 11. Suppose Kyle collected $46. Could the problem still be solved? Why or why not?

33. Find a three-digit positive integer such that the sum of all three digits is 14, the tens digit is 2 more than the ones digit, and if the digits are reversed, the number is unchanged.   464

34. *Ages.*   Tammy's age is the sum of the ages of Carmen and Dennis. Carmen's age is 2 more than the sum of the ages of Dennis and Mark. Dennis's age is four times Mark's age. The sum of all four ages is 42. How old is Tammy?   20 yr

35. *Ticket Revenue.*   The Pops concert audience of 100 people consists of adults, students, and children.

The ticket prices are $10 for adults, $3 for students, and 50¢ for children. The total amount of money taken in is $100. How many adults, students, and children are in attendance? Does there seem to be some information missing? Do some more careful reasoning.   Adults: 5; students: 1; children: 94

36. *Sharing Raffle Tickets.*   Hal gives Tom as many raffle tickets as Tom first had and Gary as many as Gary first had. In like manner, Tom then gives Hal and Gary as many tickets as each then has. Similarly, Gary gives Hal and Tom as many tickets as each then has. If each finally has 40 tickets, with how many tickets does Tom begin?   35 tickets

37. Find the sum of the angle measures at the tips of the star in this figure.   180°

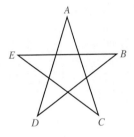

3.6

Matrices and Systems ■ Row-Equivalent Operations

Elimination Using Matrices

In solving systems of equations, we perform computations with the constants. The variables play no important role until the end. Thus we can simplify writing a system by omitting the variables. For example, the system

$$3x + 4y = 5,$$
$$x - 2y = 1$$

simplifies to

$$\begin{matrix} 3 & 4 & 5 \\ 1 & -2 & 1 \end{matrix}$$

if we do not write the variables, the operation of addition, and the equals signs.

Matrices and Systems

In the example above, we have written a rectangular array of numbers. Such an array is called a **matrix** (plural, **matrices**). We ordinarily write brackets around matrices. The following are matrices:

$$\begin{bmatrix} -3 & 1 \\ 0 & 5 \end{bmatrix}, \quad \begin{bmatrix} 2 & 0 & -1 & 3 \\ -5 & 2 & 7 & -1 \\ 4 & 5 & 3 & 0 \end{bmatrix}, \quad \begin{bmatrix} 2 & 3 \\ 7 & 15 \\ -2 & 23 \\ 4 & 1 \end{bmatrix}$$

The individual numbers are called *elements* or *entries*.

The **rows** of a matrix are horizontal, and the **columns** are vertical.

$$A = \begin{bmatrix} 5 & -2 & -2 \\ 1 & 0 & 1 \\ 4 & -3 & 2 \end{bmatrix} \begin{matrix} \leftarrow \text{row 1} \\ \leftarrow \text{row 2} \\ \leftarrow \text{row 3} \end{matrix}$$

$$\text{column 1} \quad \text{column 2} \quad \text{column 3}$$

Let's see how matrices can be used to solve a system.

EXAMPLE 1 Solve the system

$$5x - 4y = -1,$$
$$-2x + 3y = 2.$$

As an aid for understanding, we list the corresponding system in the margin.

$$5x - 4y = -1,$$
$$-2x + 3y = 2$$

Solution We write a matrix using only coefficients and constants, listing x-coefficients in the first column and y-coefficients in the second. Note that in each matrix a dashed line separates the coefficients from the constants:

$$\begin{bmatrix} 5 & -4 & \vdots & -1 \\ -2 & 3 & \vdots & 2 \end{bmatrix}.$$

Our goal is to transform

$$\begin{bmatrix} 5 & -4 & \vdots & -1 \\ -2 & 3 & \vdots & 2 \end{bmatrix} \quad \text{into the form} \quad \begin{bmatrix} a & b & \vdots & c \\ 0 & d & \vdots & e \end{bmatrix}.$$

The variables x and y can then be reinserted to form equations from which we can complete the solution.

We do calculations that are similar to those that we would do if we wrote the entire equations. The first step is to multiply and/or interchange the rows so that each number in the first column below the first number is a multiple of that number. Here that means multiplying Row 2 by 5. This corresponds to multiplying both sides of the second equation by 5.

$$5x - 4y = -1,$$
$$-10x + 15y = 10$$

$$\begin{bmatrix} 5 & -4 & \vdots & -1 \\ -10 & 15 & \vdots & 10 \end{bmatrix} \quad \textbf{New Row 2 = 5(Row 2 from above)}$$

Next, we multiply the first row by 2, add this to Row 2, and write that result as the "new" Row 2. This corresponds to multiplying the first equation by 2 and adding the result to the second equation in order to eliminate a variable. Write out these computations as necessary—we perform them mentally.

$$5x - 4y = -1,$$
$$7y = 8$$

$$\begin{bmatrix} 5 & -4 & \vdots & -1 \\ 0 & 7 & \vdots & 8 \end{bmatrix} \quad \begin{matrix} 2(5 \quad -4 \ \vdots \ -1) = (10 \quad -8 \ \vdots \ -2) \textbf{ and} \\ (10 \quad -8 \ \vdots \ -2) + (-10 \quad 15 \ \vdots \ 10) = (0 \quad 7 \ \vdots \ 8) \\ \textbf{New Row 2 = 2(Row 1) + (Row 2)} \end{matrix}$$

If we now reinsert the variables, we have

$$5x - 4y = -1, \quad (1)$$
$$7y = 8. \quad (2)$$

We can now proceed as before, solving equation (2) for y:

$$7y = 8 \quad (2)$$
$$y = \tfrac{8}{7}.$$

TEACHING TIP

You may wish to point out that at the start of our work, we should make sure that all equations are in standard form.

$2x - y + 4z = -3,$
$x \quad\;\; - 4z = 5,$
$6x - y + 2z = 10$

$ax + by + cz = d,$
$\quad\;\; ey + fz = g,$
$\quad\quad\quad\; hz = i$

$x \quad\;\; - 4z = 5,$
$2x - y + 4z = -3,$
$6x - y + 2z = 10$

$x \quad\;\; - 4z = 5,$
$\quad -y + 12z = -13,$
$6x - y + 2z = 10$

$x \quad\;\; - 4z = 5,$
$\quad -y + 12z = -13,$
$\quad -y + 26z = -20$

Next, we substitute $\frac{8}{7}$ for y in equation (1):

$$5x - 4y = -1 \quad (1)$$
$$5x - 4 \cdot \tfrac{8}{7} = -1 \quad \text{Substituting } \tfrac{8}{7} \text{ for } y \text{ in equation (1)}$$
$$x = \tfrac{5}{7}. \quad \text{Solving for } x$$

The solution is $\left(\frac{5}{7}, \frac{8}{7}\right)$. The check is left to the student.

All the systems of equations shown in the margin by Example 1 are **equivalent systems of equations**; that is, they all have the same solution.

EXAMPLE 2 Solve the system

$$2x - y + 4z = -3,$$
$$x \quad\;\; - 4z = 5,$$
$$6x - y + 2z = 10.$$

Solution We first write a matrix, using only the constants. Where there are missing terms, we must write 0's:

$$\begin{bmatrix} 2 & -1 & 4 & -3 \\ 1 & 0 & -4 & 5 \\ 6 & -1 & 2 & 10 \end{bmatrix}.$$

Our goal is to transform the matrix to one of the form

$$\begin{bmatrix} a & b & c & d \\ 0 & e & f & g \\ 0 & 0 & h & i \end{bmatrix}.$$

A matrix of this form can be rewritten as a system of equations that is equivalent to the original system, and from which a solution can be easily found.

The first step is to multiply and/or interchange the rows so that each number in the first column is a multiple of the first number in the first row. In this case, we do so by interchanging Rows 1 and 2:

$$\begin{bmatrix} 1 & 0 & -4 & 5 \\ 2 & -1 & 4 & -3 \\ 6 & -1 & 2 & 10 \end{bmatrix} \quad \text{This corresponds to interchanging the first two equations.}$$

Next, we multiply the first row by -2, add it to the second row, and replace Row 2 with the result:

$$\begin{bmatrix} 1 & 0 & -4 & 5 \\ 0 & -1 & 12 & -13 \\ 6 & -1 & 2 & 10 \end{bmatrix}. \quad \begin{array}{l} -2(1 \;\; 0 \;\; -4 \;\; 5) = (-2 \;\; 0 \;\; 8 \;\; -10) \text{ and} \\ (-2 \;\; 0 \;\; 8 \;\; -10) + (2 \;\; -1 \;\; 4 \;\; -3) = \\ (0 \;\; -1 \;\; 12 \;\; -13) \end{array}$$

Now we multiply the first row by -6, add it to the third row, and replace Row 3 with the result:

$$\begin{bmatrix} 1 & 0 & -4 & 5 \\ 0 & -1 & 12 & -13 \\ 0 & -1 & 26 & -20 \end{bmatrix}. \quad \begin{array}{l} -6(1 \;\; 0 \;\; -4 \;\; 5) = (-6 \;\; 0 \;\; 24 \;\; -30) \text{ and} \\ (-6 \;\; 0 \;\; 24 \;\; -30) + (6 \;\; -1 \;\; 2 \;\; 10) = \\ (0 \;\; -1 \;\; 26 \;\; -20) \end{array}$$

$$x \quad - \quad 4z = 5,$$
$$-y + 12z = -13,$$
$$14z = -7$$

Next, we multiply Row 2 by -1, add it to the third row, and replace Row 3 with the result:

$$\begin{bmatrix} 1 & 0 & -4 & \vdots & 5 \\ 0 & -1 & 12 & \vdots & -13 \\ 0 & 0 & 14 & \vdots & -7 \end{bmatrix}.$$

$-1(0 \quad -1 \quad 12 \;\vdots\; -13) = (0 \quad 1 \quad -12 \;\vdots\; 13)$
and $(0 \quad 1 \quad -12 \;\vdots\; 13) + (0 \quad -1 \quad 26 \;\vdots\; -20) = (0 \quad 0 \quad 14 \;\vdots\; -7)$

Reinserting the variables gives us

$$x \quad - \quad 4z = 5,$$
$$- y + 12z = -13,$$
$$14z = -7.$$

We now solve this last equation for z and get $z = -\frac{1}{2}$. Next, we substitute $-\frac{1}{2}$ for z in the preceding equation and solve for y: $-y + 12\left(-\frac{1}{2}\right) = -13$, so $y = 7$. Since there is no y-term in the first equation of this last system, we need only substitute $-\frac{1}{2}$ for z to solve for x: $x - 4\left(-\frac{1}{2}\right) = 5$, so $x = 3$. The solution is $\left(3, 7, -\frac{1}{2}\right)$. The check is left to the student.

The operations used in the preceding example correspond to those used to produce equivalent systems of equations. We call the matrices **row-equivalent** and the operations that produce them **row-equivalent operations**.

Row-Equivalent Operations

TEACHING TIP

You may want to explain that row-equivalent matrices are not *equal* but the solutions of the corresponding equations are the same.

Row-Equivalent Operations

Each of the following row-equivalent operations produces a row-equivalent matrix:

a) Interchanging any two rows.

b) Multiplying all elements of a row by a nonzero constant.

c) Replacing a row with the sum of that row and a multiple of another row.

The best overall method for solving systems of equations is by row-equivalent matrices; even computers are programmed to use them. Matrices are part of a branch of mathematics known as linear algebra. They are also studied in many courses in finite mathematics.

When we solved the systems in Examples 1 and 2, we used row-equivalent operations to write an equivalent system of equations that we could solve without using elimination. We can continue to use row-equivalent operations to write a row-equivalent matrix in **reduced row-echelon form,** from which the solution of the system can often be read directly.

> **Reduced Row-Echelon Form**
>
> A matrix is in reduced row-echelon form if:
>
> **1.** All rows consisting entirely of zeros are at the bottom of the matrix.
> **2.** The first nonzero number in any nonzero row is 1, called a leading 1.
> **3.** The leading 1 in any row is farther to the left than the leading 1 in any lower row.
> **4.** Each column that contains a leading 1 has zeros everywhere else.

EXAMPLE 3   Solve the system

$$5x - 4y = -1,$$
$$-2x + 3y = 2$$

by writing a reduced row-echelon matrix.

Solution   We began this solution in Example 1, and used row-equivalent operations to write the row-equivalent matrix:

$$\begin{bmatrix} 5 & -4 & \vdots & -1 \\ 0 & 7 & \vdots & 8 \end{bmatrix}.$$

$$5x - 4y = -1,$$
$$7y = 8$$

Before we can write reduced row-echelon form, the first nonzero number in any nonzero row must be 1. We multiply Row 1 by $\frac{1}{5}$ and Row 2 by $\frac{1}{7}$:

$$\begin{bmatrix} 1 & -\frac{4}{5} & \vdots & -\frac{1}{5} \\ 0 & 1 & \vdots & \frac{8}{7} \end{bmatrix}.$$ **New Row 1 = $\frac{1}{5}$ (Row 1 from above)**
New Row 2 = $\frac{1}{7}$ (Row 1 from above)

$$x - \frac{4}{5}y = -\frac{1}{5},$$
$$y = \frac{8}{7}$$

There are no zero rows, there is a leading 1 in each row, and the leading 1 in Row 1 is farther to the left than the leading 1 in Row 2. Thus conditions (1)–(3) for reduced row-echelon form have been met.

To satisfy condition (4), we must obtain a 0 in Row 1, Column 2, above the leading 1 in Row 2. We multiply Row 2 by $\frac{4}{5}$ and add it to Row 1:

$$\begin{bmatrix} 1 & 0 & \vdots & \frac{5}{7} \\ 0 & 1 & \vdots & \frac{8}{7} \end{bmatrix}.$$ $\frac{4}{5}\begin{pmatrix}0 & 1 & \vdots & \frac{8}{7}\end{pmatrix} = \begin{pmatrix}0 & \frac{4}{5} & \vdots & \frac{32}{35}\end{pmatrix}$ and
$\begin{pmatrix}0 & \frac{4}{5} & \vdots & \frac{32}{35}\end{pmatrix} + \begin{pmatrix}1 & -\frac{4}{5} & \vdots & -\frac{1}{5}\end{pmatrix} = \begin{pmatrix}1 & 0 & \vdots & \frac{5}{7}\end{pmatrix}$
New Row 1 = $\frac{4}{5}$(Row 2) + Row 1

$$x = \frac{5}{7},$$
$$y = \frac{8}{7}$$

This matrix is in reduced row-echelon form. If we reinsert the variables, we have

$$x = \frac{5}{7},$$
$$y = \frac{8}{7}.$$

The solution, $\left(\frac{5}{7}, \frac{8}{7}\right)$, can be read directly from the last column of the matrix.

Finding reduced row-echelon form often involves extensive calculations with fractions or decimals. Thus it is common to use a computer or a graphing calculator to store and manipulate matrices.

Matrix Operations

On many graphing calculators, matrix operations are accessed by pressing [MATRX]. The MATRX menu has three submenus: NAMES, MATH, and EDIT.

The EDIT menu allows matrices to be entered. For example, to enter

$$A = \begin{bmatrix} 1 & 2 & 5 \\ -3 & 0 & 4 \end{bmatrix},$$

choose option [A] from the MATRX EDIT menu. Enter the **dimensions**, or number of rows and columns, of the matrix first, listing the number of rows before the number of columns. Thus the dimensions of A are 2×3, read "2 by 3." Then enter each element of A by pressing the number and [ENTER]. The notation $2, 3 = 4$ indicates that the 3rd entry of the 2nd row is 4. Matrices are generally entered row by row rather than column by column.

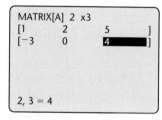

After entering the matrix, exit the matrix editor by pressing [2nd] [QUIT]. To access a matrix once it has been entered, use the MATRX NAMES submenu as shown above on the right. The brackets around A indicate that it is a matrix.

The submenu MATRX MATH lists operations that can be performed on matrices.

EXAMPLE 4   Solve the following system using a graphing calculator:

$$\begin{aligned} 2x + 5y - 8z &= 7, \\ 3x + 4y - 3z &= 8, \\ 5y - 2x &= 9. \end{aligned}$$

Solution   Before writing a matrix to represent this system, we rewrite the third equation in the form $ax + by + cz = d$:

$$\begin{aligned} 2x + 5y - 8z &= 7, \\ 3x + 4y - 3z &= 8, \\ -2x + 5y \quad\;\; &= 9. \end{aligned}$$

3.6 ELIMINATION USING MATRICES **235**

The matrix that represents this system is thus

$$\begin{bmatrix} 2 & 5 & -8 & 7 \\ 3 & 4 & -3 & 8 \\ -2 & 5 & 0 & 9 \end{bmatrix}.$$

We enter the matrix as A using the MATRX EDIT menu, noting that its dimensions are 3×4. Once we have entered each element of the matrix, we return to the home screen to perform operations. The contents of matrix A can be displayed on the screen, using the MATRX NAMES menu.

```
[A]
           [[2 5 -8 7]
            [3 4 -3 8]
            [-2 5 0 9]]
```

Each of the row-equivalent operations can be performed using the MATRX MATH menu. On many calculators, it is also possible to go directly to the reduced row-echelon form. To find this form, from the home screen, we choose the rref option from the MATRX MATH menu and then choose [A] from the MATRX NAMES menu. Finally, press MATH 1 ENTER to write the entries in the reduced row-echelon form using fraction notation.

```
rref ( [A] )▶Frac
       [[1 0 0 1/2]
        [0 1 0 2]
        [0 0 1 1/2]]
```

The reduced row-echelon form shown on the screen above is equivalent to the system of equations

$$x = \tfrac{1}{2},$$
$$y = 2,$$
$$z = \tfrac{1}{2}.$$

The solution of the system is thus $\left(\tfrac{1}{2}, 2, \tfrac{1}{2}\right)$, which can be read directly from the last column of the reduced row-echelon matrix.

3.6

Exercise Set

13. $\left(\frac{93}{34}, -\frac{7}{34}, \frac{67}{34}\right)$ **14.** $\left(\frac{23}{5}, \frac{5}{2}, -2\right)$

Solve using matrices.

1. $9x - 2y = 5,$
$3x - 3y = 11$ $\left(-\frac{1}{3}, -4\right)$

2. $4x + y = 7,$
$5x - 3y = 13$ $(2, -1)$

3. $x + 4y = 8,$
$3x + 5y = 3$ $(-4, 3)$

4. $x + 4y = 5,$
$-3x + 2y = 13$
$(-3, 2)$

5. $6x - 2y = 4,$
$7x + y = 13$ $\left(\frac{3}{2}, \frac{5}{2}\right)$

6. $3x + 4y = 7,$
$-5x + 2y = 10$
$\left(-1, \frac{5}{2}\right)$

7. $3x + 2y + 2z = 3,$
$x + 2y - z = 5,$
$2x - 4y + z = 0$ $\left(2, \frac{1}{2}, -2\right)$

8. $4x - y - 3z = 19,$
$8x + y - z = 11,$
$2x + y + 2z = -7$
$\left(\frac{3}{2}, -4, -3\right)$

9. $p - 2q - 3r = 3,$
$2p - q - 2r = 4,$
$4p + 5q + 6r = 4$
$(2, -2, 1)$

10. $x + 2y - 3z = 9,$
$2x - y + 2z = -8,$
$3x - y - 4z = 3$
$(-1, 2, -2)$

11. $3p + 2r = 11,$
$q - 7r = 4,$
$p - 6q = 1$ $\left(4, \frac{1}{2}, -\frac{1}{2}\right)$

12. $4a + 9b = 8,$
$8a + 6c = -1,$
$6b + 6c = -1$
$\left(\frac{1}{2}, \frac{2}{3}, -\frac{5}{6}\right)$

13. $3x + y = 8,$
$4x + 5y - 3z = 4,$
$7x + 2y - 9z = 1$

14. $5x - 8y = 3,$
$2y + 8z = -11,$
$5x + 7z = 9$

15. $-0.01x + 0.7y = -0.9,$
$0.5x - 0.3y + 0.18z = 0.01,$ Approximately
$50x + 6y - 75z = 12$ $(-0.53, -1.29, -0.62)$

16. $4x + 2y + 5z = 1,$
$-3x + 7y + 2z = -9,$
$7x - 5y + z = 3$ $\left(-\frac{167}{68}, -\frac{227}{68}, \frac{7}{2}\right)$

17. $2x + 2y - 2z - 2w = -10,$
$w + y + z + x = -5,$
$x - y + 4z + 3w = -2,$
$w - 2y + 2z + 3x = -6$ $(1, -3, -2, -1)$

18. $-w - 3y + z + 2x = -8,$
$x + y - z - w = -4,$
$w + y + z + x = 22,$
$x - y - z - w = -14$ $(7, 4, 5, 6)$

Solve using matrices.

19. *Coin Value.*   A collection of 43 coins consists of dimes and quarters. The total value is $7.60. How many dimes and how many quarters are there?
Dimes: 21; quarters: 22

20. *Coin Value.*   A collection of 34 coins consists of dimes and nickels. The total value is $1.90. How many dimes and how many nickels are there?
Dimes: 4; nickels: 30

21. *Mixed Granola.*   Grace sells two kinds of granola. One is worth $4.05 per pound and the other is worth $2.70 per pound. She wants to blend the two granolas to get a 15-lb mixture worth $3.15 per pound. How much of each kind of granola should be used? $4.05-granola: 5 lb; $2.70-granola: 10 lb

22. *Trail Mix.*   Phil mixes nuts worth $1.60 per pound with oats worth $1.40 per pound to get 20 lb of trail mix worth $1.54 per pound. How many pounds of nuts and how many pounds of oats should be used?
Nuts: 14 lb; oats: 6 lb

23. *Investments.*   Elena receives $212 per year in simple interest from three investments totaling $2500. Part is invested at 7%, part at 8%, and part at 9%. There is $1100 more invested at 9% than at 8%. Find the amount invested at each rate. $400 at 7%; $500 at 8%; $1600 at 9%

24. *Investments.*   Miguel receives $306 per year in simple interest from three investments totaling $3200. Part is invested at 8%, part at 9%, and part at 10%. There is $1900 more invested at 10% than at 9%. Find the amount invested at each rate. $500 at 8%; $400 at 9%; $2300 at 10%

ᵀᵂ **25.** Explain how you can recognize dependent equations when solving with matrices.

ᵀᵂ **26.** Explain how you can recognize an inconsistent system when solving with matrices.

Skill Maintenance

Simplify. [1.2]

27. $5(-3) - (-7)4$ 13

28. $8(-5) - (-2)9$ -22

29. $-2(5 \cdot 3 - 4 \cdot 6) - 3(2 \cdot 7 - 15) + 4(3 \cdot 8 - 5 \cdot 4)$ 37

30. $6(2 \cdot 7 - 3(-4)) - 4(3(-8) - 10) + 5(4 \cdot 3 - (-2)7)$ 422

Synthesis

TW **31.** If the matrices

$$\begin{bmatrix} a_1 & b_1 & c_1 \\ d_1 & e_1 & f_1 \end{bmatrix} \quad \text{and} \quad \begin{bmatrix} a_2 & b_2 & c_2 \\ d_2 & e_2 & f_2 \end{bmatrix}$$

share the same solution, does it follow that the corresponding entries are all equal to each other ($a_1 = a_2$, $b_1 = b_2$, etc.)? Why or why not?

TW **32.** Explain how the row-equivalent operations make use of the addition, multiplication, and distributive properties.

33. The sum of the digits in a four-digit number is 10. Twice the sum of the thousands digit and the tens digit is 1 less than the sum of the other two digits. The tens digit is twice the thousands digit. The ones digit equals the sum of the thousands digit and the hundreds digit. Find the four-digit number.   1324

34. Solve for x and y:
$$ax + by = c,$$
$$dx + ey = f. \quad x = \frac{ce - bf}{ae - bd}, \, y = \frac{af - cd}{ae - bd}$$

Break-Even Analysis ■ Supply and Demand

Business and Economics Applications

Break-Even Analysis

When a company manufactures x units of a product, it spends money. This is **total cost** and can be thought of as a function C, where $C(x)$ is the total cost of producing x units. When the company sells x units of the product, it takes in money. This is **total revenue** and can be thought of as a function R, where $R(x)$ is the total revenue from the sale of x units. **Total profit** is the money taken in less the money spent, or total revenue minus total cost. Total profit from the production and sale of x units is a function P given by

$$\textbf{Profit} = \textbf{Revenue} - \textbf{Cost,} \quad \text{or} \quad P(x) = R(x) - C(x).$$

If $R(x)$ is greater than $C(x)$, the company makes money. If $C(x)$ is greater than $R(x)$, the company has a loss. When $R(x) = C(x)$, the company breaks even.

There are two kinds of costs. First, there are costs like rent, insurance, machinery, and so on. These costs, which must be paid whether a product is produced or not, are called *fixed costs*. When a product is being produced, there are costs for labor, materials, marketing, and so on. These are called *variable costs*, because they vary according to the amount being produced. The sum of the fixed cost and the variable cost gives the *total cost* of producing a product.

CAUTION! Do not confuse "cost" with "price." When we discuss the *cost* of an item, we are referring to what it costs to produce the item. The *price* of an item is what a consumer pays to purchase the item and is used when calculating revenue.

EXAMPLE 1   Manufacturing Radios.   Ergs, Inc., is planning to make a new kind of radio. Fixed costs will be $90,000, and it will cost $15 to produce each radio (variable costs). Each radio sells for $26.

a) Find the total cost $C(x)$ of producing x radios.

b) Find the total revenue $R(x)$ from the sale of x radios.

c) Find the total profit $P(x)$ from the production and sale of x radios.

d) What profit or loss will the company realize from the production and sale of 3000 radios? of 14,000 radios?

e) Graph the total-cost, total-revenue, and total-profit functions using the same set of axes. Determine the break-even point.

Solution

a) Total cost is given by

$$C(x) = \text{(Fixed costs)} \ \text{plus} \ \text{(Variable costs)},$$

or $C(x) = \quad 90{,}000 \quad + \quad 15x,$

where x is the number of radios produced.

b) Total revenue is given by

$$R(x) = 26x. \qquad \text{\$26 times the number of radios sold. We assume that every radio produced is sold.}$$

c) Total profit is given by

$$P(x) = R(x) - C(x) \qquad \text{Profit is revenue minus cost.}$$
$$= 26x - (90{,}000 + 15x)$$
$$= 11x - 90{,}000.$$

d) Profits will be

$$P(3000) = 11 \cdot 3000 - 90{,}000 = -\$57{,}000$$

when 3000 radios are produced and sold, and

$$P(14{,}000) = 11 \cdot 14{,}000 - 90{,}000 = \$64{,}000$$

when 14,000 radios are produced and sold. These values can also be found using a graphing calculator, as shown in the figure at left. Thus the company loses money if only 3000 radios are sold, but makes money if 14,000 are sold.

$y_1 = 11x - 90000$

Y₁(3000)	
	−57000
Y₁(14000)	
	64000

e) The graphs of each of the three functions are shown below:

$R(x) = 26x,$ **This represents the revenue function.**

$C(x) = 90,000 + 15x,$ **This represents the cost function.**

$P(x) = 11x - 90,000.$ **This represents the profit function.**

$R(x)$, $C(x)$, and $P(x)$ are all in dollars.

The revenue function has a graph that goes through the origin and has a slope of 26. The cost function has an intercept on the $-axis of 90,000 and has a slope of 15. The profit function has an intercept on the $-axis of $-90,000$ and has a slope of 11. It is shown by the dashed line. The red dashed line shows a "negative" profit, which is a loss. (That is what is known as "being in the red.") The black dashed line shows a "positive" profit, or gain. (That is what is known as "being in the black.")

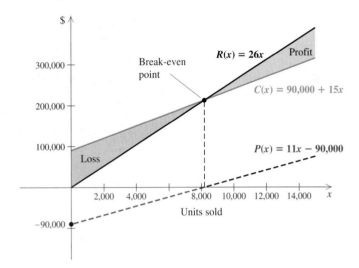

Profits occur where the revenue is greater than the cost. Losses occur where the revenue is less than the cost. The **break-even point** occurs where the graphs of R and C cross. To find this point, we solve a system:

$R(x) = 26x,$

$C(x) = 90,000 + 15x.$

Since both revenue and cost are in *dollars* and they are equal at the break-even point, the system can be rewritten as

$d = 26x,$ (1)

$d = 90,000 + 15x$ (2)

and solved using substitution:

$26x = 90,000 + 15x$ **Substituting 26x for d in equation (2)**

$11x = 90,000$

$x \approx 8181.8.$

The system can also be solved using a graphing calculator, as shown in the figure at left.

TEACHING TIP

Be aware that students may need help setting window dimensions when solving problems graphically, since the units may be large.

$y_1 = 26x, \quad y_2 = 90,000 + 15x$

300,000 $y_1 \quad y_2$

0   14,000

Intersection

X = 8181.8182   Y = 212727.27

$-100,000$

Xscl = 2000, Yscl = 100,000

The firm will break even if it produces and sells about 8182 radios (8181 will yield a tiny loss and 8182 a tiny gain), and takes in a total of $R(8182) = 26 \cdot 8182 = \$212,732$ in revenue. Note that the x-coordinate of the break-even point can also be found by solving $P(x) = 0$. The break-even point is (8182 radios, \$212,732).

Supply and Demand

As the price of coffee varies, the amount sold varies. The table and graph below show that consumers will demand less as the price goes up.

Demand Function, *D*

Price, p, per Kilogram	Quantity, $D(p)$ (in millions of kilograms)
$ 8.00	25
9.00	20
10.00	15
11.00	10
12.00	5

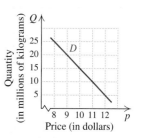

As the price of coffee varies, the amount available varies. The table and graph below show that sellers will supply less as the price goes down.

Supply Function, *S*

Price, p, per Kilogram	Quantity, $S(p)$ (in millions of kilograms)
$ 9.00	5
9.50	10
10.00	15
10.50	20
11.00	25

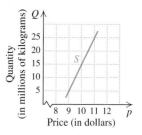

Let's look at the above graphs together. We see that as price increases, demand decreases. As price increases, supply increases. The point of intersection is called the **equilibrium point**. At that price, the amount that the seller will supply is the same amount that the consumer will buy. The situation is analogous to a buyer and a seller negotiating the price of an item. The equilibrium point is the price and quantity that they finally agree on.

Any ordered pair of coordinates from the graph is (price, quantity), because the horizontal axis is the price axis and the vertical axis is the quantity axis. If D is a demand function and S is a supply function, then the equilibrium point is where demand equals supply:

$$D(p) = S(p).$$

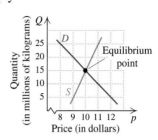

EXAMPLE 2   Find the equilibrium point for the demand and supply functions given:

$$D(p) = 1000 - 60p, \quad (1)$$
$$S(p) = 200 + 4p. \quad (2)$$

Solution   Since both demand and supply are *quantities* and they are equal at the equilibrium point, we rewrite the system as

$$q = 1000 - 60p, \quad (1)$$
$$q = 200 + 4p. \quad (2)$$

We substitute $200 + 4p$ for q in equation (1) and solve:

$$200 + 4p = 1000 - 60p \quad \text{Substituting } 200 + 4p \text{ for } q \text{ in equation (1)}$$
$$200 + 64p = 1000 \qquad \qquad \text{Adding } 60p \text{ to both sides}$$
$$64p = 800 \qquad \qquad \text{Adding } -200 \text{ to both sides}$$
$$p = \frac{800}{64} = 12.5.$$

Thus the equilibrium price is $12.50 per unit.

To find the equilibrium quantity, we substitute $12.50 into either $D(p)$ or $S(p)$. We use $S(p)$:

$$S(12.5) = 200 + 4(12.5) = 200 + 50 = 250.$$

Thus the equilibrium quantity is 250 units, and the equilibrium point is ($12.50, 250). The graph shown at left confirms the solution.

$y_1 = 1000 - 60x, \quad y_2 = 200 + 4x$

Intersection
X = 12.5   Y = 250

Xscl = 2, Yscl = 50

3.7

Exercise Set

FOR EXTRA HELP

Digital Video Tutor CD 3 Videotape 4 | Student's Solutions Manual | AW Math Tutor Center | InterAct Math | MathXL | MyMathLab

For each of the following pairs of total-cost and total-revenue functions, find **(a)** *the total-profit function and* **(b)** *the break-even point.*

1. $C(x) = 45x + 300{,}000;$ (a) $P(x) = 20x - 300{,}000;$
$R(x) = 65x$ (b) 15,000 units

2. $C(x) = 25x + 270{,}000;$ (a) $P(x) = 45x - 270{,}000;$
$R(x) = 70x$ (b) 6000 units

3. $C(x) = 10x + 120{,}000;$ (a) $P(x) = 50x - 120{,}000;$
$R(x) = 60x$ (b) 2400 units

4. $C(x) = 30x + 49,500;$ **(a)** $P(x) = 55x - 49,500;$
$R(x) = 85x$ **(b)** 900 units

5. $C(x) = 40x + 22,500;$ **(a)** $P(x) = 45x - 22,500;$
$R(x) = 85x$ **(b)** 500 units

6. $C(x) = 20x + 10,000;$ **(a)** $P(x) = 80x - 10,000;$
$R(x) = 100x$ **(b)** 125 units

7. $C(x) = 22x + 16,000;$ **(a)** $P(x) = 18x - 16,000;$
$R(x) = 40x$ **(b)** 889 units

8. $C(x) = 15x + 75,000;$ **(a)** $P(x) = 40x - 75,000;$
$R(x) = 55x$ **(b)** 1875 units

Aha! **9.** $C(x) = 75x + 100,000;$ **(a)** $P(x) = 50x - 100,000;$
$R(x) = 125x$ **(b)** 2000 units

10. $C(x) = 20x + 120,000;$ **(a)** $P(x) = 30x - 120,000;$
$R(x) = 50x$ **(b)** 4000 units

Find the equilibrium point for each of the following pairs of demand and supply functions.

11. $D(p) = 1000 - 10p,$
$S(p) = 230 + p$
($70, 300)

12. $D(p) = 2000 - 60p,$
$S(p) = 460 + 94p$
($10, 1400)

13. $D(p) = 760 - 13p,$
$S(p) = 430 + 2p$
($22, 474)

14. $D(p) = 800 - 43p,$
$S(p) = 210 + 16p$
($10, 370)

15. $D(p) = 7500 - 25p,$
$S(p) = 6000 + 5p$
($50, 6250)

16. $D(p) = 8800 - 30p,$
$S(p) = 7000 + 15p$
($40, 7600)

17. $D(p) = 1600 - 53p,$
$S(p) = 320 + 75p$
($10, 1070)

18. $D(p) = 5500 - 40p,$
$S(p) = 1000 + 85p$
($36, 4060)

Solve.

19. *Computer Manufacturing.* Biz.com Electronics is planning to introduce a new line of computers. The fixed costs for production are $125,300. The variable costs for producing each computer are $450. The revenue from each computer is $800. Find the following.
$C(x) = 125,300 + 450x$
a) The total cost $C(x)$ of producing x computers
b) The total revenue $R(x)$ from the sale of x computers $R(x) = 800x$
c) The total profit $P(x)$ from the production and sale of x computers $P(x) = 350x - 125,300$
d) The profit or loss from the production and sale of 100 computers; of 400 computers ⊡
e) The break-even point   (358 computers, $286,400)

20. *Manufacturing Lamps.* City Lights, Inc., is planning to manufacture a new type of lamp. The fixed costs for production are $22,500. The variable costs for producing each lamp are estimated to be $40. The revenue from each lamp is to be $85. Find the following.

$C(x) = 22,500 + 40x$
a) The total cost $C(x)$ of producing x lamps
b) The total revenue $R(x)$ from the sale of x lamps ⊡
c) The total profit $P(x)$ from the production and sale of x lamps $P(x) = 45x - 22,500$
d) The profit or loss from the production and sale of 3000 lamps; of 400 lamps ⊡
e) The break-even point   (500 lamps, $42,500)

21. *Manufacturing Caps.* Martina's Custom Printing is planning on adding painter's caps to its product line. For the first year, the fixed costs for setting up production are $16,404. The variable costs for producing a dozen caps are $6.00. The revenue on each dozen caps will be $18.00. Find the following.
a) The total cost $C(x)$ of producing x dozen caps ⊡
b) The total revenue $R(x)$ from the sale of x dozen caps $R(x) = 18x$
c) The total profit $P(x)$ from the production and sale of x dozen caps $P(x) = 12x - 16,404$
d) The profit or loss from the production and sale of 3000 dozen caps; of 1000 dozen caps ⊡
e) The break-even point   (1367 dozen caps, $24,606)

22. *Sport Coat Production.* Sarducci's is planning a new line of sport coats. For the first year, the fixed costs for setting up production are $10,000. The variable costs for producing each coat are $30. The revenue from each coat is to be $80. Find the following.
$C(x) = 10,000 + 30x$
a) The total cost $C(x)$ of producing x coats
b) The total revenue $R(x)$ from the sale of x coats ⊡
c) The total profit $P(x)$ from the production and sale of x coats $P(x) = 50x - 10,000$
d) The profit or loss from the production and sale of 2000 coats; of 50 coats $90,000 profit, $7500 loss
e) The break-even point   (200 coats, $16,000)

23. *Dog Food Production.* Great Foods will soon begin producing a new line of puppy food. The marketing department predicts that the demand function will be $D(p) = -14.97p + 987.35$ and the supply function will be $S(p) = 98.55p - 5.13$.

a) To the nearest cent, what price per unit should be charged in order to have equilibrium between supply and demand? $8.74
b) The production of the puppy food involves $5265 in fixed costs and $2.10 per unit in variable costs. If the price per unit is the value you found in part (a), how many units must be sold in order to break even?   793 units

⊡ Answers to Exercises 19(d), 20(b), 20(d), 21(a), 21(d), and 22(b) can be found on p. A-58.

24. *Computer Production.*   Number Solutions Computers is planning a new line of computers, each of which will sell for $970. For the first year, the fixed costs in setting up production are $1,235,580 and the variable costs for each computer are $697.

a) What is the break-even point? (Round to the nearest whole number.)   (4526 units, $4,390,220)

b) The marketing department at Number Solutions is not sure that $970 is the best price. Their demand function for the new computers is given by $D(p) = -304.5p + 374,580$ and their supply function is given by $S(p) = 788.7p - 576,504$. What price p would result in equilibrium between supply and demand? $870

TW **25.** In Example 1, the slope of the line representing Revenue is the sum of the slopes of the other two lines. This is not a coincidence. Explain why.

TW **26.** Variable costs and fixed costs are often compared to the slope and the y-intercept, respectively, of an equation for a line. Explain why you feel this analogy is or is not valid.

Skill Maintenance

Solve. [2.2]

27. $3x - 9 = 27$ 12   **28.** $4x - 7 = 53$ 15

29. $4x - 5 = 7x - 13$ $\frac{8}{3}$ **30.** $2x + 9 = 8x - 15$ 4

31. $7 - 2(x - 8) = 14$ $\frac{9}{2}$ **32.** $6 - 4(3x - 2) = 10$
$\frac{1}{3}$

Synthesis

TW **33.** Ian claims that since his fixed costs are $1000, he need sell only 20 birdbaths at $50 each in order to break even. Does this sound plausible? Why or why not?

TW **34.** In this section, we examined supply and demand functions for coffee. Does it seem realistic to you for the graph of D to have a constant slope? Why or why not?

35. *Yo-yo Production.*   Bing Boing Hobbies is willing to produce 100 yo-yo's at $2.00 each and 500 yo-yo's at $8.00 each. Research indicates that the public will buy 500 yo-yo's at $1.00 each and 100 yo-yo's at $9.00 each. Find the equilibrium point.
($5, 300 yo-yo's)

36. *Loudspeaker Production.*   Fidelity Speakers, Inc., has fixed costs of $15,400 and variable costs of $100 for each pair of speakers produced. If the speakers sell for $250 a pair, how many pairs of speakers must be produced (and sold) in order to have enough profit to cover the fixed costs of two additional facilities? Assume that all fixed costs are identical.   308 pairs

37. *Peanut Butter.*   The following table lists the data for supply and demand of an 18-oz jar of peanut butter at various prices.

Price	Supply (in millions)	Demand (in millions)
$1.59	23.4	22.5
1.29	19.2	24.8
1.69	26.8	22.2
1.19	18.4	29.7
1.99	30.7	19.3

a) Use linear regression to find the supply function $S(p)$ for suppliers of peanut butter at price p.

b) Use linear regression to find the demand function $D(p)$ for consumers of peanut butter at price p.

c) Find the equilibrium point.

38. *Funnel Cakes.*   Each year, the Harvey County Fair sets prices for concession vendors. The following table lists the data for supply and demand of a funnel cake at different prices.

Price	Supply (in thousands)	Demand (in thousands)
$1.50	3.6	5.4
1.75	4.8	5.2
2.00	6.2	5.0
2.50	7.2	4.2
2.25	7.1	4.0

a) Use linear regression to find the supply function $S(p)$ for suppliers of funnel cakes at price p.

b) Use linear regression to find the demand function $D(p)$ for consumers of funnel cakes at price p.

c) Find the equilibrium point.

37. (a) $S(p) = 15.97p - 1.05$; **(b)** $D(p) = -11.26p + 41.16$; **(c)** ($1.55, 23.7 million jars)   **38. (a)** $S(p) = 3.8p - 1.82$; **(b)** $D(p) = -1.44p + 7.64$; **(c)** ($1.81, 5.0 thousand)

Collaborative Corner

Patient Profits

Focus: Cost, revenue, and profit models

Time: 20 minutes

Group size: 2–4

Dr. Bill Marks has been charting the operating cost and the revenue of his dental practice. The following table shows his data for the first six months of a year.

Month	Number of Patients	Costs	Revenue
January	252	$13,948	$12,493
February	174	12,742	8,750
March	310	14,678	15,125
April	298	14,620	14,137
May	369	15,683	17,930
June	342	15,365	17,138

ACTIVITY

1. Divide the group into two smaller groups, with one examining the cost data and the other the revenue data. Each group should do the following.

a) Enter and graph the data, treating either cost or revenue as a function of the number of patients. Confirm that the relationship appears to be linear.

b) Use linear regression to find either the monthly cost $C(x)$ of or the monthly revenue $R(x)$ from treating x patients.

2. Working together, use the results of step (1) to find the monthly profit $P(x)$ from treating x patients.

3. Using the profit function, estimate the profit or loss from treating 310 patients and compare it with the profit for March using the data in the table.

4. Find the break-even point.

5. Use the cost function to estimate the fixed costs and the variable costs.

6. In order to accept more patients, Dr. Marks must hire another part-time hygienist at a monthly salary of $2000.

a) Find the new monthly cost $C_1(x)$ and profit $P_1(x)$.

b) Determine the new break-even point.

c) At least how many patients must the hygienist see monthly in order to cover the extra costs?

Chapter Summary and Review

Key Terms

System of equations, p. 179
Solution of a system, p. 180
Consistent, p. 182
Inconsistent, p. 182
Dependent, p. 183
Independent, p. 183
Substitution method, p. 192
Elimination method, p. 194
Total-value problem, p. 203
Mixture problem, p. 206
Motion problem, p. 208

Linear equation in three
   variables, p. 215
Matrix (matrices), p. 229
Elements, p. 229
Entries, p. 229
Rows, p. 230
Columns, p. 230
Equivalent systems, p. 231
Row-equivalent, p. 232
Reduced row-echelon form,
   p. 232

Dimensions, p. 234
Total cost, p. 237
Total revenue, p. 237
Total profit, p. 237
Fixed costs, p. 237
Variable costs, p. 237
Break-even point, p. 239
Demand function, p. 240
Supply function, p. 240
Equilibrium point, p. 240

Important Properties and Formulas

When solving a system of two linear equations in two variables:

1. If an identity is obtained, such as $0 = 0$, then the system has an infinite number of solutions. The equations are dependent and, since a solution exists, the system is consistent.

2. If a contradiction is obtained, such as $0 = 7$, then the system has no solution. The system is inconsistent.

To use the elimination method to solve systems of three linear equations:

1. Write all equations in the standard form $Ax + By + Cz = D$.

2. Clear any decimals or fractions.

3. Choose a variable to eliminate. Then select two of the three equations and work to get one equation in two variables.

4. Next, use a different pair of equations and eliminate the *same* variable that you did in step (3).

5. Solve the system of equations that resulted from steps (3) and (4).

6. Substitute the solution from step (5) into one of the original three equations and solve for the third variable. Then check.

Row-Equivalent Operations

Each of the following row-equivalent operations produces a row-equivalent matrix:

a) Interchanging any two rows.

b) Multiplying all elements of a row by a nonzero constant.

c) Replacing a row by the sum of that row and a multiple of another row.

Review Exercises

For Exercises 1–9, if a system has an infinite number of solutions, use set-builder notation to write the solution set. If a system has no solution, state this.

Solve graphically.

1. $3x + 2y = -4,$
$y = 3x + 7$ [3.1] $(-2, 1)$

2. $2x + 3y = 12,$
$4x - y = 10$
[3.1] $(3, 2)$

Solve using the substitution method.

3. $9x - 6y = 2,$
$x = 4y + 5$ [3.2] $\left(-\frac{11}{15}, -\frac{43}{30}\right)$

4. $y = x + 2,$
$y - x = 8$ [3.2] No solution

5. $x - 3y = -2,$
$7y - 4x = 6$ [3.2] $\left(-\frac{4}{5}, \frac{2}{5}\right)$

Solve using the elimination method.

6. $8x - 2y = 10,$
$-4y - 3x = -17$ [3.2] $\left(\frac{37}{19}, \frac{53}{19}\right)$

7. $4x - 7y = 18,$
$9x + 14y = 40$ [3.2] $\left(\frac{76}{17}, -\frac{2}{119}\right)$

8. $3x - 5y = -4,$
$5x - 3y = 4$
[3.2] $(2, 2)$

9. $1.5x - 3 = -2y,$
$3x + 4y = 6$
[3.2] $\{(x, y) \mid 3x + 4y = 6\}$

Solve.

10. Glynn bought two equally-priced DVD's and one videocassette for $72. If he had purchased one DVD and two videocassettes at the same price, he would have spent $15 less. What is the price of a DVD? What is the price of a videocassette?   [3.3] DVD: $29; videocassette: $14

11. A freight train leaves Chicago at midnight traveling south at a speed of 44 mph. One hour later, a passenger train, going 55 mph, travels south from Chicago on a parallel track. How many hours will the passenger train travel before it overtakes the freight train?   [3.3] 4 hr

12. Yolanda wants 14 L of fruit punch that is 10% juice. At the store, she finds punch that is 15% juice and punch that is 8% juice. How much of each should she purchase?   [3.3] 8% juice: 10 L; 15% juice: 4 L

Solve. If a system's equations are dependent or if there is no solution, state this.

13. $x + 4y + 3z = 2,$
$2x + y + z = 10,$
$-x + y + 2z = 8$ [3.4] $(4, -8, 10)$

14. $4x + 2y - 6z = 34,$
$2x + y + 3z = 3,$
$6x + 3y - 3z = 37$ [3.4] The equations are dependent.

15. $2x - 5y - 2z = -4,$
$7x + 2y - 5z = -6,$
$-2x + 3y + 2z = 4$ [3.4] $(2, 0, 4)$

16. $-5x + 5y = -6,$
$2x - 2y = 4$ [3.2] No solution

17. $3x + y = 2,$
$x + 3y + z = 0,$
$x + z = 2$ [3.4] $\left(\frac{8}{9}, -\frac{2}{3}, \frac{10}{9}\right)$

Solve.

18. In triangle *ABC*, the measure of angle *A* is four times the measure of angle *C*, and the measure of angle *B* is 45° more than the measure of angle *C*. What are the measures of the angles of the triangle?   [3.5] *A*: 90°; *B*: 67.5°; *C*: 22.5°

19. Find the three-digit number in which the sum of the digits is 11, the tens digit is 3 less than the sum of the hundreds and ones digits, and the ones digit is 5 less than the hundreds digit.   [3.5] 641

20. Lynn has $159 in her purse, consisting of $20, $5, and $1 bills. The number of $20 bills is the same as the total number of $1 and $5 bills. If she has 14 bills in her purse, how many of each denomination does she have?   [3.5] $20 bills: 7; $5 bills: 3; $1 bills: 4

Solve using matrices. Show your work.

21. $3x + 4y = -13,$
$5x + 6y = 8$
[3.6] $\left(55, -\frac{89}{2}\right)$

22. $3x - y + z = -1,$
$2x + 3y + z = 4,$
$5x + 4y + 2z = 5$
[3.6] $(-1, 1, 3)$

23. $11x + 4y = 7z + 3,$
$5y + 2 = 5z,$
$8x - 3z = 11$ [3.6] $\left(-\frac{32}{15}, -\frac{439}{45}, -\frac{421}{45}\right)$

24. Find the equilibrium point for the demand and supply functions

$$S(p) = 60 + 7p$$

and

$$D(p) = 120 - 13p.$$ [3.7] ($3, 81)

25. Robbyn is beginning to produce organic honey. For the first year, the fixed costs for setting up production are $18,000. The variable costs for producing each pint of honey are $1.50. The revenue from each pint of honey is $6. Find the following.

a) The total cost $C(x)$ of producing x pints of honey   [3.7] $C(x) = 1.5x + 18{,}000$

b) The total revenue $R(x)$ from the sale of x pints of honey   [3.7] $R(x) = 6x$

c) The total profit $P(x)$ from the production and sale of x pints of honey   [3.7] $P(x) = 4.5x - 18{,}000$

d) The profit or loss from the production and sale of 1500 pints of honey; of 5000 pints of honey

e) The break-even point   [3.7] (4000 pints of honey, $24,000)

25. (d) [3.7] $11,250 loss; $4500 profit

Synthesis

TW **26.** How would you go about solving a problem that involves four variables?

TW **27.** Explain how a system of equations can be both dependent and inconsistent.

28. Robbyn is quitting a job that pays $27,000 a year to make honey (see Exercise 25). How many pints of honey must she produce and sell in order to make the same amount that she made in the job she left?   [3.7] 10,000 pints

29. Solve graphically:
$$y = x + 2,$$
$$y = x^2 + 2. \quad [3.1] \ (0, 2), (1, 3)$$

30. The graph of $f(x) = ax^2 + bx + c$ contains the points $(-2, 3)$, $(1, 1)$, and $(0, 3)$. Find a, b, and c and give a formula for the function.   [3.5] $a = -\frac{2}{3}$, $b = -\frac{4}{3}$, $c = 3$; $f(x) = -\frac{2}{3}x^2 - \frac{4}{3}x + 3$

Chapter Test **3**

Solve graphically.

1. $2x + y = 8,$
$y - x = 2$ [3.1] (2, 4)

2. $16x - 7y = 25,$
$8x + 3y = 19$ [3.1] (2, 1)

Solve, if possible, using the substitution method.

3. $x + 3y = -8,$
$4x - 3y = 23$ [3.2] $\left(3, -\frac{11}{3}\right)$

4. $2x + 4y = -6,$
$y = 3x - 9$ [3.2] $\left(\frac{15}{7}, -\frac{18}{7}\right)$

Solve, if possible, using the elimination method.

5. $4x - 6y = 3,$
$6x - 4y = -3$ [3.2] $\left(-\frac{3}{2}, -\frac{3}{2}\right)$

6. $4y + 2x = 18,$
$3x + 6y = 26$ [3.2] No solution

7. The perimeter of a rectangle is 96. The length of the rectangle is 6 less than twice the width. Find the dimensions of the rectangle.   [3.3] Length: 30 units; width: 18 units

8. Between her home mortgage (loan), car loan, and credit card bill (loan), Rema is $75,300 in debt. Rema's credit card bill accumulates 1.5% interest, her car loan 1% interest, and her mortgage 0.6% interest each month. After one month, her total accumulated interest is $460.50. The interest on Rema's credit card bill was $4.50 more than the interest on her car loan. Find the amount of each loan.   [3.5] Mortgage: $74,000; car loan: $600; credit card bill: $700

Solve. If a system's equations are dependent or if there is no solution, state this.   [3.4] $\left(2, -\frac{1}{2}, -1\right)$

9. $-3x + y - 2z = 8,$
$-x + 2y - z = 5,$
$2x + y + z = -3$ [3.4] The equations are dependent.

10. $6x + 2y - 4z = 15,$
$-3x - 4y + 2z = -6,$
$4x - 6y + 3z = 8$

11. $2x + 2y = 0,$
$4x + 4z = 4,$
$2x + y + z = 2$ [3.4] No solution

12. $3x + 3z = 0,$
$2x + 2y = 2,$
$3y + 3z = 3$ [3.4] (0, 1, 0)

Solve using matrices.

13. $7x - 8y = 10,$
$9x + 5y = -2$ [3.6] $\left(\frac{34}{107}, -\frac{104}{107}\right)$

14. $x + 3y - 3z = 12,$
$3x - y + 4z = 0,$
$-x + 2y - z = 1$ [3.6] (3, 1, -2)

15. $4y - z = 8,$
$15x - 10 = 6z,$
$x + 20y - 8 = 6z$ [3.6] $\left(\frac{202}{9}, \frac{281}{18}, \frac{490}{9}\right)$

16. An electrician, a carpenter, and a plumber are hired to work on a house. The electrician earns $21 per hour, the carpenter $19.50 per hour, and the plumber $24 per hour. The first day on the job, they worked a total of 21.5 hr and earned a total of $469.50. If the plumber worked 2 more hours than the carpenter did, how many hours did the electrician work?   [3.5] 3.5 hr

17. Find the equilibrium point for the demand and supply functions   [3.7] ($3, 55)
$$D(p) = 79 - 8p \quad \text{and} \quad S(p) = 37 + 6p.$$

18. Complete Communications, Inc., is producing a new family radio service model. For the first year, the fixed costs for setting up production are $40,000. The variable costs for producing each radio are $25. The revenue from each radio is $70. Find the following. [3.7] $C(x) = 25x + 40,000$
a) The total cost $C(x)$ of producing x radios
b) The total revenue $R(x)$ from the sale of x radios
c) The total profit $P(x)$ from the production and sale of x radios [3.7] $P(x) = 45x - 40,000$
d) The profit or loss from the production and sale of 300 radios; of 900 radios [3.7] $26,500 loss, $500 profit
e) The break-even point [3.7] (889 radios, $62,230)

18. (b) [3.7] $R(x) = 70x$

Synthesis

19. The graph of the function $f(x) = mx + b$ contains the points $(-1, 3)$ and $(-2, -4)$. Find m and b. [2.4], [3.3] $m = 7, b = 10$

20. At a county fair, an adult's ticket sold for $5.50, a senior citizen's ticket for $4.00, and a child's ticket for $1.50. On opening day, the number of adults' and senior citizens' tickets sold was 30 more than the number of children's tickets sold. The number of adults' tickets sold was 6 more than four times the number of senior citizens' tickets sold. Total receipts from the ticket sales were $11,219.50. How many of each type of ticket were sold? [3.5] Adult: 1346; senior citizen: 335; child: 1651

1-3 Cumulative Review

Solve.

1. $-14.3 + 29.17 = x$ [2.2] 14.87

2. $x + 9.4 = -12.6$ [2.2] -22

3. $3.9(-11) = x$ [2.2] -42.9

4. $-2.4x = -48$ [2.2] 20

5. $4x + 7 = -14$ [2.2] $-\frac{21}{4}$

6. $-3 + 5x = 2x + 15$ [2.2] 6

7. $3n - (4n - 2) = 7$ [2.2] -5

8. $6y - 5(3y - 4) = 10$ [2.2] $\frac{10}{9}$

9. $14 + 2c = -3(c + 4) - 6$ [2.2] $-\frac{32}{5}$

10. $5x - [4 - 2(6x - 1)] = 12$ [2.2] $\frac{18}{17}$

Simplify. Do not leave negative exponents in your answers.

11. $x^4 \cdot x^{-6} \cdot x^{13}$ [1.4] x^{11}

12. $(4x^{-3}y^2)(-10x^4y^{-7})$ [1.4] $-\dfrac{40x}{y^5}$

13. $(6x^2y^3)^2(-2x^0y^4)^3$ [1.4] $-288x^4y^{18}$

14. $\dfrac{y^4}{y^{-6}}$ [1.4] y^{10}

15. $\dfrac{-10a^7b^{-11}}{25a^{-4}b^{22}}$ [1.4] $-\dfrac{2a^{11}}{5b^{33}}$ **16.** $\left(\dfrac{3x^4y^{-2}}{4x^{-5}}\right)^4$ [1.4] $\dfrac{81x^{36}}{256y^8}$

17. $(1.95 \times 10^{-3})(5.73 \times 10^8)$ [1.4] 1.11735×10^6

18. $\dfrac{2.42 \times 10^5}{6.05 \times 10^{-2}}$ [1.4] 4×10^6

19. Solve $A = \frac{1}{2}h(b + t)$ for b. [2.3] $b = \dfrac{2A}{h} - t$, or $\dfrac{2A - ht}{h}$

20. Determine whether $(-3, 4)$ is a solution of $5a - 2b = -23$. [1.5] Yes

Graph.

21. $y = -2x + 3$ ⊡

22. $y = x^2 - 1$ ⊡

23. $4x + 16 = 0$ ⊡

24. $-3x + 2y = 6$ ⊡

25. Find the slope and the y-intercept of the line with equation $-4y + 9x = 12$. [2.4] Slope: $\frac{9}{4}$; y-intercept: $(0, -3)$

26. Find the slope, if it exists, of the line containing the points $(2, 7)$ and $(-1, 3)$. [2.4] $\frac{4}{3}$

27. Find an equation of the line with slope -3 and containing the point $(2, -11)$. [2.6] $y = -3x - 5$

28. Find an equation of the line containing the points $(-6, 3)$ and $(4, 2)$. [2.6] $y = -\frac{1}{10}x + \frac{12}{5}$

29. Determine whether the lines are parallel or perpendicular:

$$2x = 4y + 7,$$
$$x - 2y = 5. \quad \text{[2.5] Parallel}$$

30. Find an equation of the line perpendicular to $2x - 3y = 7$ with y-intercept $(0, -3)$.
[2.5] $y = -\frac{3}{2}x - 3$

31. For the graph of f shown, determine the domain, the range, $f(-3)$, and any value of x for which $f(x) = 5$.
[2.1] $\{-5, -3, -1, 1, 3\}$; $\{-3, -2, 1, 4, 5\}$; -2; 3

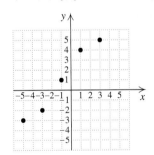

32. Determine the domain of the function given by

$$f(x) = \frac{7}{2x - 1}. \quad \text{[2.7]} \{x \,|\, x \text{ is a real number } and \ x \neq \frac{1}{2}\}$$

⊡ Answers to Exercises 21–24, 46, and 47 can be found on p. A-59.

Given $g(x) = 4x - 3$ and $h(x) = -2x^2 + 1$, find the following function values.

33. $h(4)$ [2.1] -31

34. $-g(0)$ [2.1] 3

35. $(g \cdot h)(-1)$ [2.7] 7

36. $g(a) - h(2a)$
[2.7] $8a^2 + 4a - 4$

Solve.

37. $3x + y = 4,$
$6x - y = 5$ [3.2] $(1, 1)$

38. $4x + 4y = 4,$
$5x - 3y = -19$ [3.2] $(-2, 3)$

39. $6x - 10y = -22,$
$-11x - 15y = 27$ [3.2] $\left(-3, \frac{2}{5}\right)$

40. $x + y + z = -5,$
$2x + 3y - 2z = 8,$
$x - y + 4z = -21$ [3.4] $(-3, 2, -4)$

41. $2x + 5y - 3z = -11,$
$-5x + 3y - 2z = -7,$
$3x - 2y + 5z = 12$ [3.4] $(0, -1, 2)$

42. $0.5x + 1.2y = -1.2,$
$4.8x - 6.4y = 2$ [3.6] $\left(-\frac{33}{56}, -\frac{169}{224}\right)$

43. $16x + 25y = 14 + z,$
$x + 2y = 1,$
$6y + 12z = 9x + 7$ [3.6] $\left(\frac{11}{15}, \frac{2}{15}, \frac{16}{15}\right)$

44. The sum of two numbers is 26. Three times the smaller plus twice the larger is 60. Find the numbers.
[3.3] 8, 18

45. In 1997, there were 652 endangered or threatened species of U.S. animals and plants that had recovery plans. By August 2002, there were 976 species with recovery plans. (*Source*: U.S. Fish and Wildlife Service, Department of the Interior) Find the rate at which recovery plans were being formed.
[2.4] 64.8 recovery plans per year

46. The number of U.S. pleasure trips, in millions, t years after 1994 can be estimated by $P(t) = 9t + 616$ (*Source*: Travel Industry Association of America). What do the numbers 9 and 616 signify? ⊡

47. In 1989, there were 6.6 million U.S. aircraft departures, and in 2001, there were 8.8 million departures (*Source*: Air Transport Association of America). Let $A(t) =$ the number of departures, in millions, t years after 1989.

a) Find an equation for a linear function that fits the data. ⊡

b) Use the function of part (a) to predict the number of departures in 2010. ⊡

48. A bag of Chewies candy contains 32% green candy. A bag of Crunchies candy contains 12% green candy. How much of each type should be mixed in order to obtain 20 ounces that is 25% green candy? [3.3] Chewies: 13 oz; Crunchies: 7 oz

49. Find three consecutive odd numbers such that the sum of four times the first number and five times the third number is 47. [2.3] 3, 5, 7

50. Belinda's scores on four tests are 83, 92, 100, and 85. What must the score be on the fifth test so that the average will be 90? [2.3] 90

51. The perimeter of a rectangle is 32 cm. If five times the width equals three times the length, what are the dimensions of the rectangle? [3.3] Length: 10 cm; width: 6 cm

52. There are 4 more nickels than dimes in a bank. The total amount of money in the bank is $2.45. How many of each type of coin are in the bank? [3.3] Nickels: 19; dimes: 15

53. One month Lori and Jon spent $680 for electricity, rent, and telephone. The electric bill was $\frac{1}{4}$ of the rent and the rent was $400 more than the phone bill. How much was the electric bill? [3.5] $120

54. A hockey team played 64 games one season. It won 15 more games than it tied and lost 10 more games than it won. How many games did it win? lose? tie? [3.5] Wins: 23; losses: 33; ties: 8

55. Reggie, Jenna, and Achmed are counting calories. For lunch one day, Reggie ate two cookies and a banana, for a total of 260 calories. Jenna had a cup of yogurt and a banana, for a total of 245 calories. Achmed ate a cookie, a cup of yogurt, and two bananas, for a total of 415 calories. How many calories are in each item? [3.5] Cookie: 90; banana: 80; yogurt: 165

56. *Stress.* The percent of teens of different ages who experience emotional stress is shown in the following table.

Age	13	15	17	19
Percent of Teens Who Experience Stress	15%	18%	19%	20%

Source: The Wall Street Journal, May 30, 2002

a) Use the data to draw a line graph.
b) Use linear regression to find a linear function that can be used to estimate the percent of teens of age x who are experiencing emotional stress.
c) Use the function from part (b) to estimate the percent of teens of age 16 who are experiencing emotional stress. [2.1] 18%

Synthesis

57. Simplify: $(6x^{a+2}y^{b+2})(-2x^{a-2}y^{y+1})$. [1.4] $-12x^{2a}y^{b+y+3}$

58. An automotive dealer discovers that when $1000 is spent on radio advertising, weekly sales increase by $101,000. When $1250 is spent on radio advertising, weekly sales increase by $126,000. Assuming that sales increase according to a linear equation, by what amount would sales increase when $1500 is spent on radio advertising? [2.6] $151,000

59. Given that $f(x) = mx + b$ and that $f(5) = -3$ when $f(-4) = 2$, find m and b. [2.6], [3.3] $m = -\frac{5}{9}, b = -\frac{2}{9}$

56. (a) [1.6] **(b)** [2.6] $f(x) = 0.8x + 5.2$
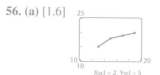
Xscl = 2, Yscl = 5

Inequalities and Problem Solving

4.1 Inequalities and Applications

4.2 Intersections, Unions, and Compound Inequalities

4.3 Absolute-Value Equations and Inequalities

4.4 Inequalities in Two Variables

SUMMARY AND REVIEW

TEST

I nequalities are mathematical sentences containing symbols such as $<$ (is less than). Principles similar to those used for solving equations enable us to solve inequalities and the problems that translate to inequalities. In this chapter, we develop procedures for solving a variety of inequalities and systems of inequalities.

APPLICATION

WINTER OLYMPIC GAMES. The number of nations participating in the Winter Olympic Games has been increasing over the years, as shown in the following table. Use the data from 1984 and 1994 to predict the years in which more than 100 nations will participate. A graph of the data shows that a linear function can be used to model the problem.

Year	Number of Nations in Winter Olympic Games
1980	37
1984	49
1988	57
1992	64
1994	67
1998	72
2002	77

Source: Olympic.org

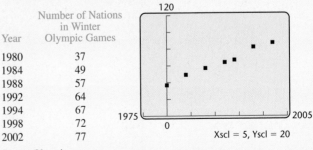

This problem appears as Example 8 in Section 4.1.

4.1

Inequalities and Applications

Solving Inequalities ■ Interval Notation ■ Graphical Solutions ■ The Addition Principle for Inequalities ■ The Multiplication Principle for Inequalities ■ Using the Principles Together ■ Problem Solving

Solving Inequalities

We can extend our equation-solving skills to the solving of inequalities. An **inequality** is any sentence containing $<$, $>$, $\leq$, $\geq$, or $\neq$ (see Section 1.1)—for example,

$$-2 < a, \quad x > 4, \quad x + 3 \leq 6, \quad 7y \geq 10y - 4, \quad \text{and} \quad 5x \neq 10.$$

Any replacement for the variable that makes an inequality true is called a **solution**. The set of all solutions is called the **solution set**. When all solutions of an inequality are found, we say that we have **solved** the inequality.

EXAMPLE 1 Determine whether the given number is a solution of the inequality.

a) $x + 3 < 6;$ 5   **b)** $2x - 3 > -5;$ 1

Solution

a) We substitute to get $5 + 3 < 6$, or $8 < 6$, a false sentence. Thus, 5 *is not* a solution.

b) We substitute to get $2 \cdot 1 - 3 > -5$, or $-1 > -5$, a true sentence. Thus 1 *is* a solution.

The *graph* of an inequality is a drawing that represents its solutions. An inequality in one variable can be graphed on a number line. Inequalities in two variables can be graphed on a coordinate plane, and are considered later in this chapter.

EXAMPLE 2 Graph $x < 4$ on a number line.

Solution   The solutions are all real numbers less than 4, so we shade all numbers less than 4. Since 4 is not a solution, we use an open dot at 4.

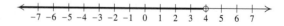

We can write the solution set using *set-builder notation* (see Section 1.1):

$$\{x \mid x < 4\}.$$

This is read

"The set of all x such that x is less than 4."

Interval Notation

Another way to write solutions of an inequality in one variable is to use **interval notation**. Interval notation uses parentheses, ( ), and brackets, [ ].

If a and b are real numbers such that $a < b$, we define the **open interval (a, b)** as the set of all numbers x for which $a < x < b$. Thus,

$$(a, b) = \{x \mid a < x < b\}. \qquad \text{Parentheses are used to exclude endpoints.}$$

Its graph excludes the endpoints:

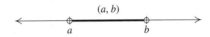

> **CAUTION!**   Do not confuse the *interval* (a, b) with the *ordered pair* (a, b). The context in which the notation appears usually makes the meaning clear.

The **closed interval $[a, b]$** is defined as the set of all numbers x for which $a \leq x \leq b$. Thus,

$$[a, b] = \{x \mid a \leq x \leq b\}. \qquad \text{Brackets are used to include endpoints.}$$

Its graph includes the endpoints*:

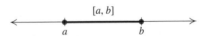

There are two kinds of **half-open intervals**, defined as follows:

1. $(a, b] = \{x \mid a < x \leq b\}$. This is open on the left. Its graph is as follows:

2. $[a, b) = \{x \mid a \leq x < b\}$. This is open on the right. Its graph is as follows:

---

*Some books use the representations ⟵———⟶ and ⟵———⟶ instead of, respec-
$\qquad\qquad\qquad\qquad\qquad\quad a\quad b \qquad\qquad\quad a\quad b$
tively, ———— and ————.
$\qquad a\quad b\qquad\quad a\quad b$

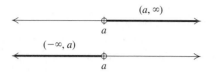

We use the symbols ∞ and $-\infty$ to represent positive and negative infinity, respectively (see the figure at left). The symbol ∞ is used when there is no upper limit to the set of numbers, and $-\infty$ is used when there is no lower limit. Thus the notation (a, ∞) represents the set of all real numbers greater than a, and $(-\infty, a)$ represents the set of all real numbers less than a.

The notations $[a, \infty)$ and $(-\infty, a]$ are used when we want to include the endpoint a.

EXAMPLE 3 Graph $y \geq -2$ on a number line and write the solution set using both set-builder and interval notations.

Solution Using set-builder notation, we write the solution set as $\{y \mid y \geq -2\}$.

Using interval notation, we write the solution set as $[-2, \infty)$.

To graph the solution, we shade all numbers to the right of -2 and use a solid dot to indicate that -2 is also a solution.

TEACHING TIP

You may want to discuss why the infinity symbol is used with an open interval.

Graphical Solutions

Solving inequalities graphically involves first finding a point of intersection.

Connecting the Concepts

INTERPRETING GRAPHS: RELATIONSHIPS BETWEEN GRAPHS

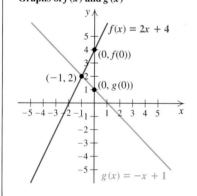

Graphs of $f(x)$ and $g(x)$

Consider the graphs of the functions $f(x) = 2x + 4$ and $g(x) = -x + 1$.

At the point of intersection, $(-1, 2)$, $f(x) = g(x)$, or $2x + 4 = -x + 1$. Thus we can say that $2x + 4 = -x + 1$ when $x = -1$.

There are no x-values other than -1 for which $f(x) = g(x)$. For all other x-values, either $f(x) > g(x)$ or $f(x) < g(x)$. For example, $f(0) = 2 \cdot 0 + 4 = 4$ and $g(0) = -0 + 1 = 1$. Thus, $f(0) > g(0)$. On the graph, the point $(0, f(0))$ lies above the point $(0, g(0))$. In fact, $f(x) > g(x)$ for all x-values greater than -1 and $f(x) < g(x)$ for all x-values less than -1. The relationship between f and g changes at the point of intersection.

Since $f(x) < g(x)$ for $x < -1$, we also know that $2x + 4 < -x + 1$ for $x < -1$. In this way, a graph can be used to solve an inequality.

Graph of solution set of $f(x) = g(x)$

Graph of solution set of $f(x) < g(x)$

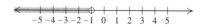

EXAMPLE 4   Solve graphically: $16 - 7x \geq 10x - 4$.

Solution   We let $y_1 = 16 - 7x$ and $y_2 = 10x - 4$, and graph y_1 and y_2 in the window $[-5, 5, -5, 15]$.

From the graph, we see that $y_1 = y_2$ at the point of intersection. To the left of the point of intersection, $y_1 > y_2$. You can check this by calculating the values of both y_1 and y_2 for an x-value to the left of the point of intersection—say, for $x = 0.5$. Since $12.5 > 1$, we know that $y_1 > y_2$ when $x = 0.5$. Similarly, we see that $y_1 < y_2$ to the right of the point of intersection.

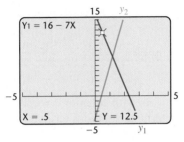 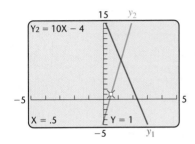

The solution set will be all x-values to the left of the point of intersection, as well as the x-coordinate of the point of intersection. Using INTERSECT, we find that the x-coordinate of the point of intersection is approximately 1.1764706. Thus the solution set is approximately $(-\infty, 1.1764706]$, or converting to fraction notation, $\left(-\infty, \frac{20}{17}\right]$.

On many graphing calculators, the interval that is the solution set can be indicated by using the VARS and TEST keys. To find the solution of Example 4, we also enter and graph $y_3 = y_1 \geq y_2$ (the "$\geq$" symbol can be found in the TEST menu). Where this is true, the value of y_3 will be 1, and where it is false, the value will be 0. The solution set is thus displayed as an interval, shown by a horizontal line 1 unit above the x-axis. The endpoint of the interval corresponds to the intersection of the graphs of the equations. Using DOT mode for y_3 will result in a more accurate graph.

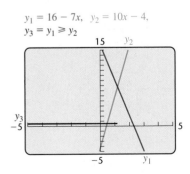

Algebraic methods for solving inequalities make use of addition and multiplication principles similar to those used to solve equations.

The Addition Principle for Inequalities

Two inequalities are *equivalent* if they have the same solution set. For example, the inequalities $x > 4$ and $4 < x$ are equivalent. Just as the addition principle for equations produces equivalent equations, the addition principle for inequalities produces equivalent inequalities.

> ## The Addition Principle for Inequalities
>
> For any real numbers a, b, and c:
>
> $$a < b \text{ is equivalent to } a + c < b + c;$$
> $$a > b \text{ is equivalent to } a + c > b + c.$$
>
> Similar statements hold for $\leq$ and $\geq$.

As with equations, we try to get the variable alone on one side in order to determine solutions easily.

EXAMPLE 5   Solve and graph: **(a)** $x + 5 > 1$; **(b)** $4x - 1 \geq 5x - 2$.

Solution

a)
$$x + 5 > 1$$
$$x + 5 + (-5) > 1 + (-5) \qquad \text{\textbf{Using the addition principle}}$$
$$x > -4 \qquad\qquad\qquad\qquad \text{\textbf{to add −5 to both sides}}$$

When an inequality—like this last one—has an infinite number of solutions, we cannot possibly check them all. Instead, we can perform a partial check by substituting one member of the solution set (here we use -1) into the original inequality:

$$\frac{x + 5 > 1}{-1 + 5 \;?\; 1}$$
$$4 \;\mid\; 1 \qquad \text{TRUE}$$

Since $4 > 1$ is true, we have our check. The solution set is $\{x \mid x > -4\}$, or $(-4, \infty)$. The graph is as follows:

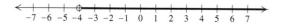

b)
$$4x - 1 \geq 5x - 2$$
$$4x - 1 + 2 \geq 5x - 2 + 2 \qquad \text{\textbf{Adding 2 to both sides}}$$
$$4x + 1 \geq 5x \qquad\qquad\qquad \text{\textbf{Simplifying}}$$
$$4x + 1 - 4x \geq 5x - 4x \qquad \text{\textbf{Adding −4x to both sides}}$$
$$1 \geq x \qquad\qquad\qquad\qquad \text{\textbf{Simplifying}}$$

We know that $1 \geq x$ has the same meaning as $x \leq 1$. You can check that any number less than or equal to 1 is a solution. As another check, we can solve graphically, as shown in the figure at left.

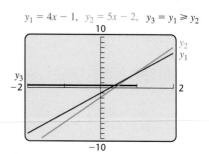

$y_1 = 4x - 1, \quad y_2 = 5x - 2, \quad y_3 = y_1 \geq y_2$

The solution set is $\{x \mid 1 \geq x\}$ or, more commonly, $\{x \mid x \leq 1\}$. Using interval notation, we write the solution set as $(-\infty, 1]$. The graph is as follows:

The Multiplication Principle for Inequalities

The multiplication principle for inequalities differs from the multiplication principle for equations.

Consider this true inequality:

$$4 < 9.$$

If we multiply both sides of $4 < 9$ by 2, we get another true inequality:

$$4 \cdot 2 < 9 \cdot 2, \quad \text{or} \quad 8 < 18.$$

If we multiply both sides of $4 < 9$ by -2, we get a false inequality:

FALSE $\longrightarrow 4(-2) < 9(-2), \quad \text{or} \quad -8 < -18. \longleftarrow$ FALSE

This is because multiplication (or division) by a negative number changes the sign of the number being multiplied (or divided). When the signs of both numbers in an inequality are changed, the position of the numbers on the number line with respect to each other is reversed.

$-8 > -18. \longleftarrow$ TRUE

The < symbol has been reversed!

The Multiplication Principle for Inequalities

For any real numbers a and b, and for any *positive* number c,

$a < b$ is equivalent to $ac < bc$;

$a > b$ is equivalent to $ac > bc$.

For any real numbers a and b, and for any *negative* number c,

$a < b$ is equivalent to $ac > bc$;

$a > b$ is equivalent to $ac < bc$.

Similar statements hold for $\leq$ and $\geq$.

TEACHING TIP

Many students reverse the inequality symbol after simplifying but not in the multiplication or division step. Point out that the direction must be reversed in the same step that the multiplication or division by a negative number is done. Each step must be an equivalent inequality.

Since division by c is the same as multiplication by $1/c$, there is no need for a separate division principle.

> **CAUTION!** Remember that whenever we multiply or divide both sides of an inequality by a negative number, we must reverse the inequality symbol.

EXAMPLE 6 Solve and graph: **(a)** $3a < \frac{3}{4}$; **(b)** $-5x \geq -80$.

Solution

a) $3a < \frac{3}{4}$

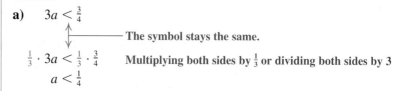

$\frac{1}{3} \cdot 3a < \frac{1}{3} \cdot \frac{3}{4}$ Multiplying both sides by $\frac{1}{3}$ or dividing both sides by 3

$a < \frac{1}{4}$

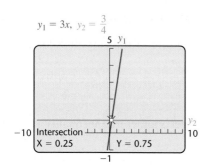

Any number less than $\frac{1}{4}$ is a solution. The graph shown at left serves as a check. The solution set is $\left\{a \mid a < \frac{1}{4}\right\}$, or $\left(-\infty, \frac{1}{4}\right)$. The graph is as follows:

b) $-5x \geq -80$

The symbol must be reversed.

$\dfrac{-5x}{-5} \leq \dfrac{-80}{-5}$ Dividing both sides by -5 or multiplying both sides by $-\frac{1}{5}$

$x \leq 16$

We get the same solution by solving graphically, as shown in the figure at left.

The solution set is $\{x \mid x \leq 16\}$, or $(-\infty, 16]$. The graph is as follows:

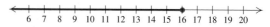

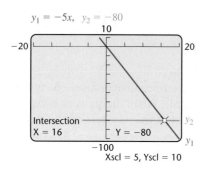

Using the Principles Together

EXAMPLE 7 Solve.

a) $16 - 7y \geq 10y - 4$ **b)** $-3(x + 8) - 5x > 4x - 9$

Solution We use the addition and multiplication principles together in solving inequalities in much the same way as in solving equations.

a) $16 - 7y \geq 10y - 4$

$-16 + 16 - 7y \geq -16 + 10y - 4$ Adding -16 to both sides

$-7y \geq 10y - 20$

$-10y + (-7y) \geq -10y + 10y - 20$ Adding $-10y$ to both sides

$-17y \geq -20$

The symbol must be reversed.

$-\frac{1}{17} \cdot (-17y) \leq -\frac{1}{17} \cdot (-20)$ Multiplying both sides by $-\frac{1}{17}$ or dividing both sides by -17

$y \leq \frac{20}{17}$

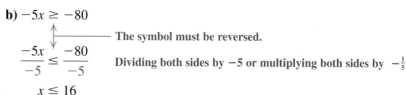

The solution set is $\left\{y \mid y \leq \frac{20}{17}\right\}$, or $\left(-\infty, \frac{20}{17}\right]$.

b) $-3(x + 8) - 5x > 4x - 9$

$\quad -3x - 24 - 5x > 4x - 9$ **Using the distributive law**

$\quad\quad\quad -24 - 8x > 4x - 9$

$\quad -24 - 8x + 8x > 4x - 9 + 8x$ **Adding $8x$ to both sides**

$\quad\quad\quad\quad -24 > 12x - 9$

$\quad\quad -24 + 9 > 12x - 9 + 9$ **Adding 9 to both sides**

$\quad\quad\quad\quad\quad -15 > 12x$

                       The symbol stays the same.

$\quad\quad\quad\quad\quad -\frac{5}{4} > x$ **Dividing by 12 and simplifying**

The solution set is $\left\{x \mid -\frac{5}{4} > x\right\}$, or $\left\{x \mid x < -\frac{5}{4}\right\}$, or $\left(-\infty, -\frac{5}{4}\right)$.

Problem Solving

Many problem-solving situations translate to inequalities. In addition to "is less than" and "is more than," other phrases are commonly used.

Important Words	Sample Sentence	Translation
is at least	Max is at least 5 years old.	$m \geq 5$
is at most	At most 6 people could fit in the elevator.	$n \leq 6$
cannot exceed	Total weight in the elevator cannot exceed 2000 pounds.	$w \leq 2000$
must exceed	The speed must exceed 15 mph.	$s > 15$
is between	Heather's income is between $23,000 and $35,000.	$23{,}000 < h < 35{,}000$
no more than	Bing weighs no more than 90 pounds.	$w \leq 90$
no less than	Saul would accept no less than $4000 for the piano.	$p \geq 4000$

EXAMPLE 8 Winter Olympic Games. The number of nations participating in the Winter Olympic Games has been increasing over the years, as shown in the following table. Use the data from 1984 and 1994 to predict the years in which more than 100 nations will participate.

Year	Number of Nations Participating in Winter Olympic Games
1980	37
1984	49
1988	57
1992	64
1994	67
1998	72
2002	77

Solution

1. **Familiarize.** We let x represent the number of years after 1980 and y the number of nations participating in the Winter Olympic Games, and graph the data to determine whether the relationship appears to be linear.

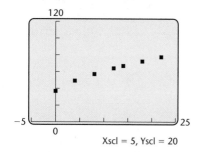

Since it does appear linear, we will find a linear function that can approximate the data. We will then use that function to predict the years in which more than 100 nations will participate in the Winter Olympic Games.

2. **Translate.** We use the data from 1984 and 1994 and the point–slope equation to find a linear function that approximates the data. The slope of the line containing the points $(4, 49)$ and $(14, 67)$ is

$$m = \frac{67 - 49}{14 - 4} = \frac{18}{10} = 1.8.$$

Using the point–slope equation, we have

$$
\begin{aligned}
y - y_1 &= m(x - x_1) \\
y - 49 &= 1.8(x - 4) && \text{Using } m = 1.8 \text{ and } (x_1, y_1) = (4, 49) \\
y - 49 &= 1.8x - 7.2 && \text{Using the distributive law} \\
y &= 1.8x + 41.8. && \text{Simplifying}
\end{aligned}
$$

Letting W represent the number of nations participating in the Winter Olympic Games x years after 1980, we have the linear model

$$W(x) = 1.8x + 41.8.$$

To predict the years in which more than 100 nations will participate in the Winter Olympic Games, we solve the inequality $W(x) > 100$, or

$$1.8x + 41.8 > 100.$$

3. Carry out.   We solve the inequality:

$$1.8x + 41.8 > 100$$
$$1.8x > 58.2 \qquad \textbf{Subtracting 41.8 from both sides}$$
$$x > 32.3. \qquad \textbf{Dividing both sides by 1.8 and rounding}$$

4. Check.   A partial check is to substitute a value for x greater than 32.3:

$$W(33) = 1.8(33) + 41.8 = 101.2.$$

Since $W(33)$ is greater than 100, we have a partial check of the answer.

5. State.   There will be more than 100 nations participating in the Winter Olympic Games that are held more than 32.3 years after 1980, or held in a year after 2012.

   In Example 8, we used the data from 1984 and 1994 to find a linear function that fits the data. Choosing two different points would result in a different function. To fit a linear function using all the data, we use linear regression.

EXAMPLE 9   Use the data from Example 8 and linear regression to find a linear function that can be used to estimate the number of nations y that will participate in the Winter Olympic Games x years after 1980. Then use the function to predict the years in which more than 100 nations will participate.

Solution   We enter the data as in Example 8, and find the linear regression equation, as shown in the figure on the right below. We then graph $y_1 = 1.768312102x + 40.72452229$ and $y_2 = 100$. The x-coordinate of the point of intersection of the graphs of y_1 and y_2 gives the number of years after 1980 when 100 nations will participate.

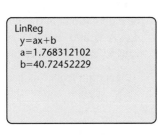

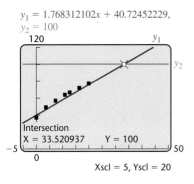

Since 100 nations will participate in the Winter Olympics approximately 33.5 years after 1980, we predict that more than 100 nations will participate in Winter Olympic Games held after 2013.

EXAMPLE 10   Earnings Plans.   On a new job, Rose can be paid in one of two ways:

Plan A: A salary of $600 per month, plus a commission of 4% of sales;

Plan B: A salary of $800 per month, plus a commission of 6% of sales in excess of $10,000.

For what amount of monthly sales is plan A better than plan B, if we assume that sales are always more than $10,000?

Solution

1. Familiarize.   Listing the given information in a table will be helpful.

Plan A: Monthly Income	Plan B: Monthly Income
$600 salary 4% of sales *Total*: $600 + 4% of sales	$800 salary 6% of sales over $10,000 *Total*: $800 + 6% of sales over $10,000

Next, suppose that Rose sold a certain amount—say, $12,000—in one month. Which plan would be better? Under plan A, she would earn $600 plus 4% of $12,000, or

$600 + 0.04(12,000) = \$1080.$

Since with plan B commissions are paid only on sales in excess of $10,000, Rose would earn $800 plus 6% of ($12,000 − $10,000), or

$800 + 0.06(2000) = \$920.$

This shows that for monthly sales of $12,000, plan A is better. Similar calculations will show that for sales of $30,000 a month, plan B is better. To determine *all* values for which plan A earns more money, we must solve an inequality that is based on the calculations above.

2. Translate.   We let S = the amount of monthly sales, in dollars. Examining the calculations in the *Familiarize* step, we see that monthly income from plan A is $600 + 0.04S$ and from plan B is $800 + 0.06(S − 10,000)$. We want to find all values of S for which

$$\underbrace{\text{Income from plan A}} \quad \underbrace{\text{is greater than}} \quad \underbrace{\text{income from plan B}}$$

$$600 + 0.04S \quad > \quad 800 + 0.06(S − 10,000).$$

3. Carry out.   We solve the problem both algebraically and graphically.

Algebraic Solution

We solve the inequality:

$600 + 0.04S > 800 + 0.06(S - 10{,}000)$

$600 + 0.04S > 800 + 0.06S - 600$ **Using the distributive law**

$600 + 0.04S > 200 + 0.06S$ **Combining like terms**

$400 > 0.02S$ **Subtracting 200 and 0.04S from both sides**

$20{,}000 > S$, or $S < 20{,}000$.   **Dividing both sides by 0.02**

Graphical Solution

We graph the equations $y_1 = 600 + 0.04x$ and $y_2 = 800 + 0.06(x - 10000)$.

$y_1 = 600 + 0.04x$,
$y_2 = 800 + 0.06(x - 10000)$

Intersection X = 20000   Y = 1400

Xscl = 10000, Yscl = 100

We see that $y_1 > y_2$ to the left of the point of intersection, $(20{,}000, 1400)$. Thus $y_1 > y_2$ for $x < 20{,}000$.

5. $\{y \mid y < 6\}, (-\infty, 6)$

6. $\{x \mid x > 4\}, (4, \infty)$

7. $\{x \mid x \ge -4\}, [-4, \infty)$

8. $\{t \mid t \le 6\}, (-\infty, 6]$

9. $\{t \mid t > -3\}, (-3, \infty)$

10. $\{y \mid y < -3\}, (-\infty, -3)$

11. $\{x \mid x \le -7\}, (-\infty, -7]$

12. $\{x \mid x \ge -6\}, [-6, \infty)$

4. Check. For $S = 20{,}000$, the income from plan A is

$600 + 4\% \cdot 20{,}000$, or \$1400.

The income from plan B is

$800 + 6\% \cdot (20{,}000 - 10{,}000)$, or \$1400.

This confirms that for sales totaling \$20,000, Rose's pay is the same under either plan.

In the *Familiarize* step, we saw that for sales of \$12,000, plan A pays more. Since $12{,}000 < 20{,}000$, this is a partial check. Since we cannot check all possible values of S, we will stop here.

5. State. For monthly sales of less than \$20,000, plan A is better.

4.1

FOR EXTRA HELP

Digital Video Tutor CD 4 Videotape 5 | Student's Solutions Manual | Tutor Center AW Math Tutor Center | InterAct Math | MathXL | MyMathLab

Exercise Set

Determine whether the given numbers are solutions of the inequality.

1. $x - 1 \ge 7$; $-4, 0, 8, 13$ No, no, yes, yes

2. $3x + 5 \le -10$; $-5, -10, 0, 27$ Yes, yes, no, no

3. $t - 6 > 2t - 1$; $0, -8, -9, -3$ No, yes, yes, no

4. $5y - 9 < 3 - y$; $2, -3, 0, 3$ No, yes, yes, no

Graph each inequality, and write the solution set using both set-builder and interval notation.

5. $y < 6$ **6.** $x > 4$

7. $x \ge -4$ **8.** $t \le 6$

9. $t > -3$ **10.** $y < -3$

11. $x \le -7$ **12.** $x \ge -6$

Solve. Then graph the solution set.

13. $x + 9 > 4$ ⊡

14. $x + 5 > 2$ ⊡

15. $a + 7 \le -13$ ⊡

16. $a + 9 \le -12$ ⊡

17. $x - 5 \le 7$ ⊡

18. $t + 14 \ge 9$ ⊡

19. $y - 9 > -18$ ⊡

20. $y - 8 > -14$ ⊡

21. $y - 20 \le -6$ ⊡

22. $x - 11 \le -2$ ⊡

23. $9t < -81$ ⊡

24. $8x \ge 24$ ⊡

25. $0.5x < 25$ ⊡

26. $0.3x < -18$ ⊡

27. $-8y \le 3.2$ ⊡

28. $-9x \ge -8.1$ ⊡

29. $-\frac{5}{6}y \le -\frac{3}{4}$ ⊡

30. $-\frac{3}{4}x \ge -\frac{5}{8}$ ⊡

31. $5y + 13 > 28$ ⊡

32. $2x + 7 < 19$ ⊡

33. $-9x + 3x \ge -24$ ⊡

34. $5y + 2y \le -21$ ⊡

35. Let $f(x) = 8x - 9$ and $g(x) = 3x - 11$. Find all values of x for which $f(x) < g(x)$. ⊡

36. Let $f(x) = 2x - 7$ and $g(x) = 5x - 9$. Find all values of x for which $f(x) < g(x)$. ⊡

37. Let $f(x) = 0.4x + 5$ and $g(x) = 1.2x - 4$. Find all values of x for which $g(x) \ge f(x)$. ⊡

38. Let $f(x) = \frac{3}{8} + 2x$ and $g(x) = 3x - \frac{1}{8}$. Find all values of x for which $g(x) \ge f(x)$. ⊡

Solve each inequality using the given graph.

39. $f(x) \ge g(x)$
$\{x \mid x \ge 2\}$, or $[2, \infty)$

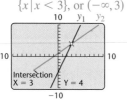

40. $f(x) < g(x)$
$\{x \mid x < -1\}$, or $(-\infty, -1)$

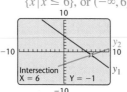

41. $y_1 < y_2$
$\{x \mid x < 3\}$, or $(-\infty, 3)$

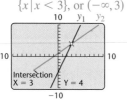

42. $y_1 \ge y_2$
$\{x \mid x \le 6\}$, or $(-\infty, 6]$

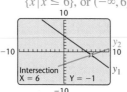

⊡ Answers to Exercises 13–38 can be found on p. A-59.

43. The graphs of $y_1 = -\frac{1}{2}x + 5$, $y_2 = x - 1$, and $y_3 = 2x - 3$ are as shown below. Solve each inequality, referring only to the figure.

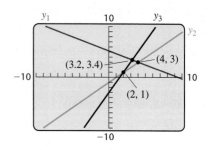

a) $-\frac{1}{2}x + 5 > x - 1$ $\{x \mid x < 4\}$, or $(-\infty, 4)$

b) $x - 1 \le 2x - 3$ $\{x \mid x \ge 2\}$, or $[2, \infty)$

c) $2x - 3 \ge -\frac{1}{2}x + 5$ $\{x \mid x \ge 3.2\}$, or $[3.2, \infty)$

44. The graphs of $f(x) = 2x + 1$, $g(x) = -\frac{1}{2}x + 3$, and $h(x) = x - 1$ are as shown below. Solve each inequality, referring only to the figure.

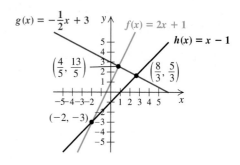

a) $2x + 1 \le x - 1$ $\{x \mid x \le -2\}$, or $(-\infty, -2]$

b) $x - 1 > -\frac{1}{2}x + 3$ $\{x \mid x > \frac{8}{3}\}$, or $\left(\frac{8}{3}, \infty\right)$

c) $-\frac{1}{2}x + 3 < 2x + 1$ $\{x \mid x > \frac{4}{5}\}$, or $\left(\frac{4}{5}, \infty\right)$

Solve.   **50.** $\{x \mid x \le -\frac{23}{2}\}$, or $\left(-\infty, -\frac{23}{2}\right]$

45. $4(3y - 2) \ge 9(2y + 5)$ $\{y \mid y \le -\frac{53}{6}\}$, or $\left(-\infty, -\frac{53}{6}\right]$

46. $4m + 7 \ge 14(m - 3)$ $\{m \mid m \le \frac{49}{10}\}$, or $\left(-\infty, \frac{49}{10}\right]$

47. $5(t - 3) + 4t < 2(7 + 2t)$ $\{t \mid t < \frac{29}{5}\}$, or $\left(-\infty, \frac{29}{5}\right)$

48. $2(4 + 2x) > 2x + 3(2 - 5x)$ $\{x \mid x > -\frac{2}{17}\}$, or $\left(-\frac{2}{17}, \infty\right)$

49. $5[3m - (m + 4)] > -2(m - 4)$ $\{m \mid m > \frac{7}{3}\}$, or $\left(\frac{7}{3}, \infty\right)$

50. $8x - 3(3x + 2) - 5 \ge 3(x + 4) - 2x$

51. $19 - (2x + 3) \le 2(x + 3) + x$ $\{x \mid x \ge 2\}$, or $[2, \infty)$

52. $13 - (2c + 2) \ge 2(c + 2) + 3c$ $\{c \mid c \le 1\}$, or $(-\infty, 1]$

53. $\frac{1}{4}(8y + 4) - 17 < -\frac{1}{2}(4y - 8)$ $\{y \mid y < 5\}$, or $(-\infty, 5)$

54. $\frac{1}{3}(6x + 24) - 20 > -\frac{1}{4}(12x - 72)$ $\{x \mid x > 6\}$, or $(6, \infty)$

55. $2[8 - 4(3 - x)] - 2 \geq 8[2(4x - 3) + 7] - 50$ ⊡

56. $5[3(7 - t) - 4(8 + 2t)] - 20 \leq -6[2(6 + 3t) - 4]$ ⊡

Solve.

57. *Truck Rentals.*   Campus Entertainment rents a truck for $45 plus 20¢ per mile. A budget of $75 has been set for the rental. For what mileages will they not exceed the budget?   Mileages less than or equal to 150 mi

58. *Truck Rentals.*   Metro Concerts can rent a truck for either $55 with unlimited mileage or $29 plus 40¢ per mile. For what mileages would the unlimited mileage plan save money?   Mileages greater than 65 mi

59. *Insurance Claims.*   After a serious automobile accident, most insurance companies will replace the damaged car with a new one if repair costs exceed 80% of the NADA, or "blue-book," value of the car. Miguel's car recently sustained $9200 worth of damage but was not replaced. What was the blue-book value of his car? $11,500 or more

60. *Phone Rates.*   A long-distance telephone call using Down East Calling costs 10 cents for the first minute and 8 cents for each additional minute. The same call, placed on Long Call Systems, costs 15 cents for the first minute and 6 cents for each additional minute. For what length phone calls is Down East Calling less expensive?   Calls shorter than $3\frac{1}{2}$ min

Phone Rates.   In Vermont, Verizon charges customers $13.55 for monthly service plus 2.2¢ per minute for local phone calls between 9 A.M. and 9 P.M. weekdays. The charge for off-peak local calls is 0.5¢ per minute. Calls are free after the total monthly charges reach $39.40.

61. Assume that only peak local calls were made. For how long must a customer speak on the phone if the $39.40 maximum charge is to apply?   For 1175 min or more

62. Assume that only off-peak calls were made. For how long must a customer speak on the phone if the $39.40 maximum charge is to apply?   For 5170 min or more

63. *Checking-Account Rates.*   The Hudson Bank offers two checking-account plans. Their Anywhere plan charges 20¢ per check whereas their Acu-checking plan costs $2 per month plus 12¢ per check. For what numbers of checks per month will the Acu-checking plan cost less?   More than 25 checks per month

64. *Moving Costs.*   Musclebound Movers charges $85 plus $40 an hour to move households across town. Champion Moving charges $60 an hour for cross-town moves. For what lengths of time is Champion more expensive?   More than 4.25 hr

65. *Wages.*   Toni can be paid in one of two ways:

Plan A: A salary of $400 per month, plus a commission of 8% of gross sales;

Plan B: A salary of $610 per month, plus a commission of 5% of gross sales.

For what amount of gross sales should Toni select plan A?   Gross sales greater than $7000

66. *Wages.*   Branford can be paid for his masonry work in one of two ways:

Plan A: $300 plus $9.00 per hour;

Plan B: Straight $12.50 per hour.

Suppose that the job takes *n* hours. For what values of *n* is plan B better for Branford?   Values of n greater than $85\frac{5}{7}$ hr

67. *Insurance Benefits.*   Bayside Insurance offers two plans. Under plan A, Giselle would pay the first $50 of her medical bills and 20% of all bills after that. Under plan B, Giselle would pay the first $250 of bills, but only 10% of the rest. For what amount of medical bills will plan B save Giselle money? (Assume that her bills will exceed $250.)   More than $1850

68. *Wedding Costs.*   The Arnold Inn offers two plans for wedding parties. Under plan A, the inn charges $30 for each person in attendance. Under plan B, the inn charges $1300 plus $20 for each person in excess of the first 25 who attend. For what size parties will plan B cost less? (Assume that more than 25 guests will attend.)   Parties of more than 80

⊡ Answers to Exercises 55 and 56 can be found on p. A-59.

70. (a) Fahrenheit temperatures less than 1945.4°

69. *Show Business.*   Slobberbone receives $750 plus 15% of receipts over $750 for playing a club date. If a club charges a $6 cover charge, how many people must attend in order for the band to receive at least $1200?   At least 625 people

70. *Temperature Conversion.*   The function

$$C(F) = \tfrac{5}{9}(F - 32)$$

can be used to find the Celsius temperature $C(F)$ that corresponds to $F°$ Fahrenheit.

a) Gold is solid at Celsius temperatures less than 1063°C. Find the Fahrenheit temperatures for which gold is solid.

b) Silver is solid at Celsius temperatures less than 960.8°C. Find the Fahrenheit temperatures for which silver is solid.   Fahrenheit temperatures less than 1761.44°

71. *Manufacturing.*   Ergs, Inc., is planning to make a new kind of radio. Fixed costs will be $90,000, and variable costs will be $15 for the production of each radio. The total-cost function for x radios is

$$C(x) = 90,000 + 15x.$$

The company makes $26 in revenue for each radio sold. The total-revenue function for x radios is

$$R(x) = 26x.$$

(See Section 3.7.)

a) When $R(x) < C(x)$, the company loses money. Find the values of x for which the company loses money. $\{x \mid x < 8181\tfrac{9}{11}\}$, or $\{x \mid x \le 8181\}$

b) When $R(x) > C(x)$, the company makes a profit. Find the values of x for which the company makes a profit. $\{x \mid x > 8181\tfrac{9}{11}\}$, or $\{x \mid x \ge 8182\}$

72. *Publishing.*   The demand and supply functions for a locally produced poetry book are approximated by

$$D(p) = 2000 - 60p \quad \text{and}$$
$$S(p) = 460 + 94p,$$

where p is the price in dollars (see Section 3.7).

a) Find those values of p for which demand exceeds supply. $\{p \mid p < 10\}$

b) Find those values of p for which demand is less than supply. $\{p \mid p > 10\}$

73. *Trampoline Injuries.*   The following table shows the number of trampoline injuries in the United States for various years. Use the data from 1992 and 1997 to find a linear function that can be used to estimate the number t of trampoline injuries x years after 1990. Then predict those years in which the number of trampoline injuries will exceed 100,000.

Year	Number of Injuries
1990	29,600
1991	35,500
1992	39,000
1993	40,500
1994	58,500
1995	58,400
1997	82,722

Source: National Safety Council, Itasca, IL, *Injury Facts, 1999 Edition*

$t(x) = 8744.4x + 21,511.2$; years after 1998

74. *Construction.*   The following table shows the number of square feet, in millions, of new educational buildings that were constructed during various years. Use the data from 1993 and 1997 to find a linear function that can be used to estimate the number of square feet b of new educational buildings constructed x years after 1990. Then predict those years for which there will be more than 400 million square feet of new educational buildings built. $b(x) = 9.75x + 135.75$; years after 2017

Year	Square Feet of Floor Space of New Educational Buildings (in millions)
1990	152
1993	165
1995	186
1997	204
1999	262

Source: *Statistical Abstract of the United States*, 2000

75. *Sports Participation.* The following table shows the number of people who participated in swimming and exercising with equipment for various years. Use the data from 1990 and 1998 to find two linear functions that can be used to estimate the number participating in swimming and in exercising with equipment *x* years after 1985. Then predict those years for which the number of people who exercise with equipment will be greater than the number of people who swim.

Year	Number of People Who Swim (in millions)	Number of People Who Exercise With Equipment (in millions)
1985	73.3	32.1
1990	67.5	35.3
1995	61.5	44.4
1996	60.2	47.8
1998	58.2	46.1

Sources: The Wall Street Journal Almanac, 1998; Statistical Abstract of the United States, 2000

75. Swimming: $f(x) = -1.1625x + 73.3125$; equipment: $g(x) = 1.35x + 28.55$; years after 2002
76. Public: $f(x) = 0.12625x - 0.0605$; private: $g(x) = 0.015375x + 2.02325$; years after 1998

76. *Nursery School.* The following table shows the number of children enrolled in public and private nursery schools in the United States for various years. Use the data from 1990 and 1998 to find two linear functions that can be used to estimate the number of children enrolled in public nursery schools and the number enrolled in private nursery schools *x* years after 1980. Then predict those years for which the number of children enrolled in public nursery schools will be greater than the number enrolled in private nursery schools.

Year	Public Nursery School Enrollment (in millions)	Private Nursery School Enrollment (in millions)
1980	0.628	1.353
1985	0.846	1.631
1990	1.202	2.177
1995	1.950	2.381
1998	2.212	2.300

Source: U.S. Bureau of the Census

77. *High Jump.* The following table shows the world record for the men's high jump for various years. Use linear regression to find a linear function that can be used to predict the high-jump world record *x* years after 1900. Then predict those years for which the world record will exceed 2.5 m.
$f(x) = 0.0056862962x + 1.891866531$; years after 2006

Year	High Jump (in meters)	Athlete
1912	2.00	George L. Horine (USA)
1924	2.04	Harold M. Osborn (USA)
1934	2.06	Walter Marty (USA)
1941	2.11	Lester Steers (USA)
1957	2.162	Yuriy Styepanov (USSR)
1970	2.29	Ni Zhinquin (China)
1980	2.36	Gerd Wessig (Germany)
1993	2.45	Javier Sotomayor (Cuba)

Source: McWhirter, Norris, Book of Historical Records. Sterling Publishing, New York, NY, 2000, p. 255

78. *Mile Run.* The following table shows the world record for the mile run for various years. Use linear regression to find a linear function that can be used to predict the mile run world record *x* years after 1954. Then predict those years for which the world record will be less than 3.5 min.
$f(x) = -0.0058087885x + 3.962605066$; years after 2033

Year	Time for Mile (in minutes)	Athlete
1954	3.99	Sir Roger Bannister (Great Britain)
1962	3.9017	Peter Snell (New Zealand)
1975	3.8233	John Walker (New Zealand)
1981	3.7888	Sebastian Coe (Great Britain)
1993	3.7398	Noureddine Morceli (Algeria)
1999	3.7188	Hican El Guerrouj (Morocco)

Source: McWhirter, Norris, Book of Historical Records. Sterling Publishing, New York, NY, 2000, p. 253

79. Use linear regression to solve Exercise 75. ▫

80. Use linear regression to solve Exercise 76. ▫

TW **81.** Explain in your own words why the inequality symbol must be reversed when both sides of an inequality are multiplied by a negative number.

TW **82.** Why isn't roster notation used to write solutions of inequalities?

▫ Answers to Exercises 79 and 80 can be found on p. A-59.

Skill Maintenance

Find the domain of f. [2.7] $\{x \mid x \text{ is a real number } and \ x \neq -3\}$

83. $f(x) = \dfrac{3}{x - 2}$

$\{x \mid x \text{ is a real number } and \ x \neq 2\}$

84. $f(x) = \dfrac{x - 5}{4x + 12}$

85. $f(x) = \dfrac{5x}{7 - 2x}$

$\{x \mid x \text{ is a real number } and \ x \neq \frac{7}{2}\}$

86. $f(x) = \dfrac{x + 3}{9 - 4x}$

$\{x \mid x \text{ is a real number } and \ x \neq \frac{9}{4}\}$

Simplify. [1.3]

87. $9x - 2(x - 5)$

$7x + 10$

88. $8x + 7(2x - 1)$

$22x - 7$

Synthesis

ᵀᵂ **89.** A Presto photocopier costs \$510 and an Exact Image photocopier costs \$590. Write a problem that involves the cost of the copiers, the cost per page of photocopies, and the number of copies for which the Presto machine is the more expensive machine to own.

ᵀᵂ **90.** Explain how the addition principle can be used to avoid ever needing to multiply or divide both sides of an inequality by a negative number.

Solve. Assume that a, b, c, d, and m are positive constants.

91. $3ax + 2x \geq 5ax - 4$; assume $a > 1$

$\left\{x \mid x \leq \dfrac{2}{a - 1}\right\}$

92. $6by - 4y \leq 7by + 10$

$\left\{y \mid y \geq -\dfrac{10}{b + 4}\right\}$

93. $a(by - 2) \geq b(2y + 5)$; assume $a > 2$ ⊡

94. $c(6x - 4) < d(3 + 2x)$; assume $3c > d$ ⊡

95. $c(2 - 5x) + dx > m(4 + 2x)$; assume $5c + 2m < d$

96. $a(3 - 4x) + cx < d(5x + 2)$; assume $c > 4a + 5d$ ⊡

Determine whether each statement is true or false. If false, give an example that shows this.

97. For any real numbers a, b, c, and d, if $a < b$ and $c < d$, then $a - c < b - d$.   False; $2 < 3$ and $4 < 5$, but $2 - 4 = 3 - 5$.

98. For all real numbers x and y, if $x < y$, then $x^2 < y^2$.   False; $-3 < -2$, but $9 > 4$.

ᵀᵂ **99.** Are the inequalities

$$x < 3 \quad \text{and} \quad x + \frac{1}{x} < 3 + \frac{1}{x}$$

equivalent? Why or why not?

ᵀᵂ **100.** Are the inequalities

$$x < 3 \quad \text{and} \quad 0 \cdot x < 0 \cdot 3$$

equivalent? Why or why not?

Solve. Then graph.

101. $x + 5 \leq 5 + x$ $\mathbb{R}$

102. $x + 8 < 3 + x$ $\varnothing$

103. $x^2 > 0$ $\{x \mid x \text{ is a real number } and \ x \neq 0\}$

⊡ Answers to Exercises 93–96 can be found on p. A-59.

4.2

Intersections of Sets and Conjunctions of Sentences ■ Unions of Sets and Disjunctions of Sentences ■ Interval Notation and Domains

Intersections, Unions, and Compound Inequalities

We now consider **compound inequalities**—that is, sentences like "$-2 < x$ *and* $x < 1$" or "$x < -3$ *or* $x > 3$" that are formed using the word *and* or the word *or*.

Intersections of Sets and Conjunctions of Sentences

The **intersection** of two sets A and B is the set of all elements that are common to both A and B. We denote the intersection of sets A and B as

$$A \cap B.$$

The intersection of two sets is often pictured as shown here.

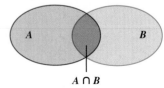

$A \cap B$

EXAMPLE 1   Find the intersection: $\{1, 2, 3, 4, 5\} \cap \{-2, -1, 0, 1, 2, 3\}$.

Solution   The numbers 1, 2, and 3 are common to both sets, so the intersection is $\{1, 2, 3\}$.

When two or more sentences are joined by the word *and* to make a compound sentence, the new sentence is called a **conjunction** of the sentences. The following is a conjunction of inequalities:

$$-2 < x \quad \text{and} \quad x < 1.$$

A number is a solution of a conjunction if it is a solution of both of the separate parts. For example, -1 is a solution because it is a solution of $-2 < x$ as well as $x < 1$.

Below we show the graph of $-2 < x$, followed by the graph of $x < 1$, and finally the graph of the conjunction $-2 < x$ *and* $x < 1$. *Note that the solution set of a conjunction is the intersection of the solution sets of the individual sentences.*

$\{x \mid -2 < x\}$ $(-2, \infty)$

$\{x \mid x < 1\}$ $(-\infty, 1)$

$\{x \mid -2 < x\} \cap \{x \mid x < 1\}$ $(-2, 1)$
$= \{x \mid -2 < x \text{ and } x < 1\}$

Mathematical Use of the Word "and"

The word "and" corresponds to "intersection" and to the symbol "∩". Any solution of a conjunction must make each part of the conjunction true.

Because there are numbers that are both greater than -2 and less than 1, the conjunction $-2 < x$ *and* $x < 1$ can be abbreviated by $-2 < x < 1$. Thus the interval $(-2, 1)$ can be represented as $\{x \mid -2 < x < 1\}$, the set of all numbers that are *simultaneously* greater than -2 *and* less than 1. Note that for $a < b$,

$a < x$ **and** $x < b$ **can be abbreviated** $a < x < b$;

and, equivalently,

$b > x$ **and** $x > a$ **can be abbreviated** $b > x > a$.

EXAMPLE 2   Solve and graph the solution set: $-1 \leq 2x + 5 < 13$.

Solution   This inequality is an abbreviation for the conjunction

$$-1 \leq 2x + 5 \quad \text{and} \quad 2x + 5 < 13.$$

Algebraic Solution

The word *and* corresponds to set *intersection*. To solve the conjunction, we solve each of the two inequalities separately and then find the intersection of the solution sets:

$$-1 \le 2x + 5 \quad and \quad 2x + 5 < 13$$

$$-6 \le 2x \qquad and \qquad 2x < 8 \qquad \text{Subtracting 5 from both sides of each inequality}$$

$$-3 \le x \qquad and \qquad x < 4. \qquad \text{Dividing both sides of each inequality by 2}$$

We now abbreviate the answer:

$$-3 \le x < 4.$$

The solution set is $\{x \mid -3 \le x < 4\}$, or, in interval notation, $[-3, 4)$. The graph is the intersection of the two separate solution sets.

$\{x \mid -3 \le x\}$

$\{x \mid x < 4\}$

$\{x \mid -3 \le x\} \cap \{x \mid x < 4\}$
$= \{x \mid -3 \le x < 4\}$

Graphical Solution

We graph the equations $y_1 = -1$, $y_2 = 2x + 5$, and $y_3 = 13$, and determine those x-values for which $y_1 \le y_2$ *and* $y_2 < y_3$.

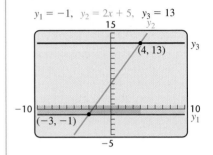

$y_1 = -1, \ y_2 = 2x + 5, \ y_3 = 13$

Using INTERSECT, we find that the graphs of y_1 and y_2 intersect at the point $(-3, -1)$ and the graphs of y_2 and y_3 intersect at the point $(4, 13)$. From the graph, we see that $y_1 < y_2$ for x-values greater than -3, as indicated by the purple and blue shading on the x-axis. We also see that $y_2 < y_3$ for x-values less than 4, as shown by the purple and red shading on the x-axis. The solution set, indicated by the purple shading, is the intersection of these sets, as well as the number -3. It includes all x-values for which the line $y_2 = 2x + 5$ is both on or above the line $y_1 = -1$ *and* below the line $y_3 = 13$. This can be written $\{x \mid -3 \le x < 4\}$, or $[-3, 4)$.

The steps in Example 2 are sometimes combined as follows:

$$-1 \le 2x + 5 < 13$$
$$-1 - 5 \le 2x + 5 - 5 < 13 - 5$$
$$-6 \le 2x < 8$$
$$-3 \le x < 4.$$

Such an approach saves some writing and will prove useful in Section 4.3.

Connecting the Concepts

Graphs of $y = 2x + 1$ and $y = 4 - x$

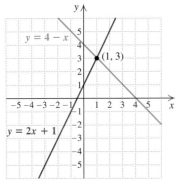

Graph of solution set of $2x + 1 = 4 - x$

$\{1\}$

Graph of solution set of
$y = 2x + 1$,
$y = 4 - x$

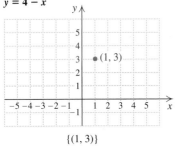

$\{(1, 3)\}$

Graph of solution set of $2x + 1 < 4 - x$

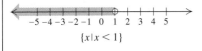

$\{x \mid x < 1\}$

INTERPRETING GRAPHS:
GRAPHS OF SOLUTION SETS

When solving an equation or an inequality graphically, it is important to distinguish the graphs used in the solution from the graph of the solution set. For example, we can use the graphs of the equations

$$y = 2x + 1$$

and

$$y = 4 - x,$$

shown at left, to solve several types of equations and inequalities.

In Chapter 2, we solved linear equations by finding the x-coordinate of a point of intersection. To solve the equation

$$2x + 1 = 4 - x,$$

we graph $y = 2x + 1$ and $y = 4 - x$ as above. The point of intersection is $(1, 3)$, and the solution set of the equation is $\{1\}$. The graphs of the equations were used to find the solution; the graph of the solution set is simply a point on the number line.

In Chapter 3, we solved systems of linear equations by finding the coordinates of a point of intersection. To solve the system

$$y = 2x + 1,$$
$$y = 4 - x,$$

we graph the equations as above. The solution set of the system is $\{(1, 3)\}$. The graphs of the equations were used to find the solution; the graph of the solution set is a point on the x, y-plane.

In this chapter, we use the graphs of equations to solve inequalities. To solve the inequality

$$2x + 1 < 4 - x,$$

we graph $y = 2x + 1$ and $y = 4 - x$ as above. Since the graph of $y = 2x + 1$ lies below the graph of $y = 4 - x$ to the left of the point of intersection, the solution set is $\{x \mid x < 1\}$. The graphs of the equations were used to find the solution; the graph of the solution set is an interval on a number line.

EXAMPLE 3   Solve and graph the solution set: $2x - 5 \geq -3$ *and* $5x + 2 \geq 17$.

Solution   We first solve each inequality separately:

$$
\begin{array}{ccc}
2x - 5 \geq -3 & \textit{and} & 5x + 2 \geq 17 \\
2x \geq 2 & \textit{and} & 5x \geq 15 \\
x \geq 1 & \textit{and} & x \geq 3.
\end{array}
$$

Next, we find the intersection of the two separate solution sets.

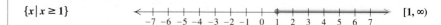

$\{x \mid x \geq 1\}$ $[1, \infty)$

$\{x \mid x \geq 3\}$ $[3, \infty)$

$\{x \mid x \geq 1\} \cap \{x \mid x \geq 3\}$
$= \{x \mid x \geq 3\}$ $[3, \infty)$

The numbers common to both sets are greater than or equal to 3. Thus the solution set is $\{x \mid x \geq 3\}$, or, in interval notation, $[3, \infty)$. You should check that any number in $[3, \infty)$ satisfies the conjunction whereas numbers outside $[3, \infty)$ do not. The graph shown at left serves as another check.

Sometimes there are no numbers that make both parts of a conjunction true.

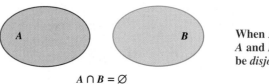

$A \cap B = \varnothing$

When $A \cap B = \varnothing$, A and B are said to be *disjoint*.

$y_1 = 2x - 5, \ y_2 = 5x + 2,$
$y_3 = -3, \ y_4 = 17$

EXAMPLE 4 Solve and graph the solution set: $2x - 3 > 1 \ and \ 3x - 1 < 2$.

Solution We solve each inequality separately:

$$2x - 3 > 1 \quad and \quad 3x - 1 < 2$$
$$2x > 4 \quad and \quad 3x < 3$$
$$x > 2 \quad and \quad x < 1.$$

The solution set is the intersection of the individual inequalities.

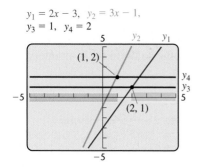

$\{x \mid x > 2\}$ $(2, \infty)$

$\{x \mid x < 1\}$ $(-\infty, 1)$

$\{x \mid x > 2\} \cap \{x \mid x < 1\}$
$= \{x \mid x > 2 \ and \ x < 1\} = \varnothing$ $\varnothing$

Since no number is both greater than 2 and less than 1, the solution set is the empty set, $\varnothing$. The graph shown at left confirms that the solution set is empty.

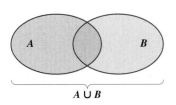

$A \cup B$

Unions of Sets and Disjunctions of Sentences

The **union** of two sets A and B is the collection of elements belonging to A and/or B. We denote the union of A and B by

$A \cup B.$

The union of two sets is often pictured as shown at left.

EXAMPLE 5   Find the union: $\{2, 3, 4\} \cup \{3, 5, 7\}$.

Solution   The numbers in either or both sets are 2, 3, 4, 5, and 7, so the union is $\{2, 3, 4, 5, 7\}$. ─────────

TEACHING TIP

You might note that a member of both sets is not listed twice in the set representing the union.

When two or more sentences are joined by the word *or* to make a compound sentence, the new sentence is called a **disjunction** of the sentences. Here is an example:

$x < -3$ *or* $x > 3.$

A number is a solution of a disjunction if it is a solution of either of the separate parts. For example, -5 is a solution of this disjunction since -5 is a solution of $x < -3$. Below we show the graph of $x < -3$, followed by the graph of $x > 3$, and finally the graph of the disjunction $x < -3$ *or* $x > 3$. *Note that the solution set of a disjunction is the union of the solution sets of the individual sentences.*

$\{x \mid x < -3\}$ (graph: $-6\ -5\ -4\ -3\ -2\ -1\ 0\ 1\ 2\ 3\ 4\ 5\ 6$) $(-\infty, -3)$

$\{x \mid x > 3\}$ (graph: $-6\ -5\ -4\ -3\ -2\ -1\ 0\ 1\ 2\ 3\ 4\ 5\ 6$) $(3, \infty)$

$\{x \mid x < -3\} \cup \{x \mid x > 3\}$ (graph: $-6\ -5\ -4\ -3\ -2\ -1\ 0\ 1\ 2\ 3\ 4\ 5\ 6$) $(-\infty, -3) \cup (3, \infty)$
$= \{x \mid x < -3 \ or \ x > 3\}$

The solution set of $x < -3$ *or* $x > 3$ is $\{x \mid x < -3 \ or \ x > 3\}$, or, in interval notation, $(-\infty, -3) \cup (3, \infty)$. There is no simpler way to write the solution.

Mathematical Use of the Word "or"

The word "or" corresponds to "union" and to the symbol "$\cup$". For a number to be a solution of a disjunction, it must be in *at least one* of the solution sets of the individual sentences.

EXAMPLE 6   Solve: $7 + 2x < -1$ *or* $13 - 5x \leq 3$.

Solution   We solve each inequality separately, retaining the word *or*:

$$7 + 2x < -1 \quad or \quad 13 - 5x \leq 3$$
$$2x < -8 \quad or \quad -5x \leq -10$$

Reverse the symbol.

$$x < -4 \quad or \quad x \geq 2.$$

To find the solution set of the disjunction, we consider the individual graphs. We graph $x < -4$ and then $x \geq 2$. Then we take the union of the graphs.

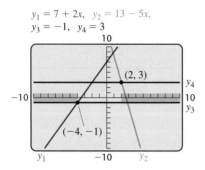

$y_1 = 7 + 2x, \ y_2 = 13 - 5x,$
$y_3 = -1, \ y_4 = 3$

$\{x \mid x < -4\}$ $(-\infty, -4)$

$\{x \mid x \geq 2\}$ $[2, \infty)$

$\{x \mid x < -4\} \cup \{x \mid x \geq 2\}$
$= \{x \mid x < -4 \ or \ x \geq 2\}$ $(-\infty, -4) \cup [2, \infty)$

The solution set is $\{x \mid x < -4 \ or \ x \geq 2\}$, or $(-\infty, -4) \cup [2, \infty)$. This is confirmed by the graph shown at left.

CAUTION!   A compound inequality like

$$x < -4 \quad or \quad x \geq 2,$$

as in Example 6, *cannot* be expressed as $2 \leq x < -4$ because to do so would be to say that x is *simultaneously* less than -4 and greater than or equal to 2. No number is both less than -4 *and* greater than 2, but many are less than -4 *or* greater than 2.

TEACHING TIP

Be aware that some students will also try to abbreviate the inequality as $-4 < x \geq 2$.

EXAMPLE 7   Solve: $-2x - 5 < -2$ *or* $x - 3 < -10$.

Solution   We solve the inequalities separately, retaining the word *or*:

$$-2x - 5 < -2 \quad or \quad x - 3 < -10$$
$$-2x < 3 \quad or \quad x < -7$$

Reverse the symbol.

$$x > -\frac{3}{2} \quad or \quad x < -7.$$

Keep the word "or."

The solution set is $\left\{x \mid x < -7 \ or \ x > -\frac{3}{2}\right\}$, or $(-\infty, -7) \cup \left(-\frac{3}{2}, \infty\right)$.

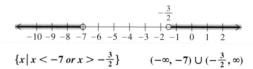

$\{x \mid x < -7 \ or \ x > -\frac{3}{2}\}$ $(-\infty, -7) \cup (-\frac{3}{2}, \infty)$

EXAMPLE 8   Solve: $3x - 11 < 4$ *or* $4x + 9 \geq 1$.

Solution   We solve the individual inequalities separately, retaining the word *or*:

$$3x - 11 < 4 \quad or \quad 4x + 9 \geq 1$$
$$3x < 15 \quad or \qquad 4x \geq -8$$
$$x < 5 \quad or \qquad x \geq -2.$$

Keep the word "or."

To find the solution set, we first look at the individual graphs.

$\{x \mid x < 5\}$

$(-\infty, 5)$

$\{x \mid x \geq -2\}$

$[-2, \infty)$

$\{x \mid x < 5\} \cup \{x \mid x \geq -2\}$
$= \{x \mid x < 5 \text{ or } x \geq -2\}$

$(-\infty, \infty) = \mathbb{R}$

Since *all* numbers are less than 5 or greater than or equal to -2, the two sets fill the entire number line. Thus the solution set is $\mathbb{R}$, the set of all real numbers.

Interval Notation and Domains

In Section 2.7, we saw that if $g(x) = (5x - 2)/(3x - 7)$, then the domain of g is $\{x \mid x$ is a real number *and* $x \neq \frac{7}{3}\}$. We can now represent such a set using interval notation:

$$\{x \mid x \text{ is a real number } and \ x \neq \tfrac{7}{3}\} = \left(-\infty, \tfrac{7}{3}\right) \cup \left(\tfrac{7}{3}, \infty\right).$$

EXAMPLE 9   Use interval notation to write the domain of f if $f(x) = \sqrt{x + 2}$.

Solution   The expression $\sqrt{x + 2}$ is not a real number when $x + 2$ is negative. Thus the domain of f is the set of all x-values for which $x + 2 \geq 0$. Since $x + 2 \geq 0$ is equivalent to $x \geq -2$, we have

$$\text{Domain of } f = \{x \mid x \geq -2\} = [-2, \infty).$$

TEACHING TIP

Emphasize that it is not x that must be positive, but $x + 2$.

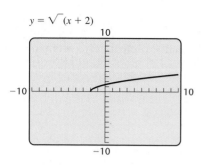

$y = \sqrt{(x+2)}$

We can check that $[-2, \infty)$ is the domain of f either by using a table of values or by viewing the graph of $f(x)$. By tracing the curve, we can confirm that no y-value is given for x-values less than -2. The domain is $[-2, \infty)$.

Interactive Discovery

1. Domain of $f = (-\infty, 3]$; domain of $g = [-1, \infty)$
2. Domain of $f + g$ = domain of $f - g$ = domain of $f \cdot g = [-1, 3]$
3. By finding the intersection of the domains of f and g

Consider the functions $f(x) = \sqrt{3 - x}$ and $g(x) = \sqrt{x + 1}$. Enter these in a graphing calculator as $y_1 = \sqrt{(3 - x)}$ and $y_2 = \sqrt{(x + 1)}$.

1. Determine algebraically the domain of f and the domain of g. Then graph y_1 and y_2, and trace each curve to verify the domains.

2. Graph $y_1 + y_2$, $y_1 - y_2$, and $y_1 \cdot y_2$. Use the graphs to determine the domains of $f + g$, $f - g$, and $f \cdot g$.

3. How can the domains of the sum, the difference, and the product of f and g be found algebraically?

Domain of the Sum, Difference, or Product of Functions

The domain of the sum, the difference, or the product of the functions f and g is the intersection of the domains of f and g.

EXAMPLE 10 Find the domain of $f + g$ if $f(x) = \sqrt{2x - 5}$ and $g(x) = \sqrt{x + 1}$.

Solution We first find the domain of f and the domain of g. The domain of f is the set of all x-values for which $2x - 5 \geq 0$, or $\{x \mid x \geq \frac{5}{2}\}$, or $[\frac{5}{2}, \infty)$. Similarly, the domain of g is $\{x \mid x \geq -1\}$, or $[-1, \infty)$. The intersection of the domains is $\{x \mid x \geq \frac{5}{2}\}$, or $[\frac{5}{2}, \infty)$. We can confirm this at least approximately by tracing the graph of $f + g$.

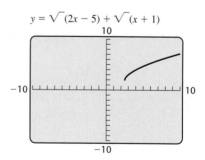

$y = \sqrt{(2x - 5)} + \sqrt{(x + 1)}$

Thus the

Domain of $f + g = \{x \mid x \geq \frac{5}{2}\}$, or $[\frac{5}{2}, \infty)$.

4.2

Exercise Set

Find each indicated intersection or union.

1. $\{7, 9, 11\} \cap \{9, 11, 13\}$ $\{9, 11\}$

2. $\{2, 4, 8\} \cup \{8, 9, 10\}$ $\{2, 4, 8, 9, 10\}$

3. $\{1, 5, 10, 15\} \cup \{5, 15, 20\}$ $\{1, 5, 10, 15, 20\}$

4. $\{2, 5, 9, 13\} \cap \{5, 8, 10\}$ $\{5\}$

5. $\{a, b, c, d, e, f\} \cap \{b, d, f\}$ $\{b, d, f\}$

6. $\{a, b, c\} \cup \{a, c\}$ $\{a, b, c\}$

7. $\{r, s, t\} \cup \{r, u, t, s, v\}$ $\{r, s, t, u, v\}$

8. $\{m, n, o, p\} \cap \{m, o, p\}$ $\{m, o, p\}$

9. $\{3, 6, 9, 12\} \cap \{5, 10, 15\}$ $\varnothing$

10. $\{1, 5, 9\} \cup \{4, 6, 8\}$ $\{1, 4, 5, 6, 8, 9\}$

11. $\{3, 5, 7\} \cup \varnothing$ $\{3, 5, 7\}$

12. $\{3, 5, 7\} \cap \varnothing$ $\varnothing$

Graph and write interval notation for each compound inequality.

13. $3 < x < 8$ ⊡

14. $0 \le y \le 4$ ⊡

15. $-6 \le y \le -2$ ⊡

16. $-9 \le x < -5$ ⊡

17. $x < -2 \; or \; x > 3$ ⊡

18. $x < -5 \; or \; x > 1$ ⊡

19. $x \le -1 \; or \; x > 5$ ⊡

20. $x \le -5 \; or \; x > 2$ ⊡

21. $-4 \le -x < 2$ ⊡

22. $x > -7 \; and \; x < -2$ ⊡

23. $x > -2 \; and \; x < 4$ ⊡

24. $3 > -x \ge -1$ ⊡

25. $5 > a \; or \; a > 7$ ⊡

26. $t \ge 2 \; or \; -3 > t$ ⊡

27. $x \ge 5 \; or \; -x \ge 4$ ⊡

28. $-x < 3 \; or \; x < -6$ ⊡

29. $4 > y \; and \; y \ge -6$ ⊡

30. $6 > -x \ge 0$ ⊡

31. $x < 7 \; and \; x \ge 3$ ⊡

32. $x \ge -3 \; and \; x < 3$ ⊡

Aha! **33.** $t < 2 \; or \; t < 5$ ⊡

34. $t > 4 \; or \; t > -1$ ⊡

35. $x > -1 \; or \; x \le 3$ ⊡

36. $4 > x \; or \; x \ge -3$ ⊡

37. $x \ge 5 \; and \; x > 7$ ⊡

38. $x \le -4 \; and \; x < 1$ ⊡

Solve and graph each solution set.

39. $-1 < t + 2 < 7$ ⊡

40. $-3 < t + 1 \le 5$ ⊡

41. $2 < x + 3 \; and \; x + 1 \le 5$ ⊡

42. $-1 < x + 2 \; and \; x - 4 < 3$ ⊡

43. $-7 \le 2a - 3 \; and \; 3a + 1 < 7$ ⊡

44. $-4 \le 3n + 2 \; and \; 2n - 3 \le 5$ ⊡

Aha! **45.** $x + 7 \le -2 \; or \; x + 7 \ge -3$ ⊡

46. $x + 5 < -3 \; or \; x + 5 \ge 4$ ⊡

47. $2 \le f(x) \le 8$, where $f(x) = 3x - 1$ ⊡

48. $7 \ge g(x) \ge -2$, where $g(x) = 3x - 5$ ⊡

49. $-21 \le f(x) < 0$, where $f(x) = -2x - 7$ ⊡

50. $4 > g(t) \ge 2$, where $g(t) = -3t - 8$ ⊡

51. $f(x) \le 2 \; or \; f(x) \ge 8$, where $f(x) = 3x - 1$ ⊡

52. $g(x) \le -2 \; or \; g(x) \ge 10$, where $g(x) = 3x - 5$ ⊡

53. $f(x) < -1 \; or \; f(x) > 1$, where $f(x) = 2x - 7$ ⊡

54. $g(x) < -7 \; or \; g(x) > 7$, where $g(x) = 3x + 5$ ⊡

55. $6 > 2a - 1 \; or \; -4 \le -3a + 2$ ⊡

56. $3a - 7 > -10 \; or \; 5a + 2 \le 22$ ⊡

57. $a + 4 < -1 \; and \; 3a - 5 < 7$ ⊡

58. $1 - a < -2 \; and \; 2a + 1 > 9$ ⊡

59. $3x + 2 < 2 \; or \; 4 - 2x < 14$ ⊡

60. $2x - 1 > 5 \; or \; 3 - 2x \ge 7$ ⊡

61. $2t - 7 \le 5 \; or \; 5 - 2t > 3$ ⊡

62. $5 - 3a \le 8 \; or \; 2a + 1 > 7$ ⊡

⊡ Answers to Exercises 13–62 can be found on pp. A-59 and A-60.

63. Use the graph of $f(x) = 2x - 5$ shown below to solve $-7 < 2x - 5 < 7$. $(-1, 6)$

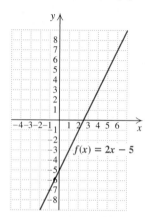

64. Use the graph of $g(x) = 4 - x$ shown below to solve $4 - x < -2$ or $4 - x > 7$. $(-\infty, -3) \cup (6, \infty)$

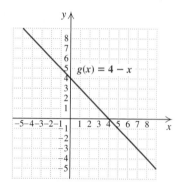

For f(x) as given, use interval notation to write the domain of f.

65. $f(x) = \dfrac{9}{x + 7}$ $(-\infty, -7) \cup (-7, \infty)$

66. $f(x) = \dfrac{2}{x + 3}$ $(-\infty, -3) \cup (-3, \infty)$

67. $f(x) = \sqrt{x - 6}$ $[6, \infty)$

68. $f(x) = \sqrt{x - 2}$ $[2, \infty)$

69. $f(x) = \dfrac{x + 3}{2x - 5}$ $\left(-\infty, \frac{5}{2}\right) \cup \left(\frac{5}{2}, \infty\right)$

70. $f(x) = \dfrac{x - 1}{3x + 4}$ $\left(-\infty, -\frac{4}{3}\right) \cup \left(-\frac{4}{3}, \infty\right)$

71. $f(x) = \sqrt{2x + 8}$ $[-4, \infty)$

72. $f(x) = \sqrt{8 - 4x}$ $(-\infty, 2]$

73. $f(x) = \sqrt{8 - 2x}$ $(-\infty, 4]$

74. $f(x) = \sqrt{10 - 2x}$ $(-\infty, 5]$

For f(x) and g(x) as given, use interval notation to write the domains of f + g, f − g, and f · g.

75. $f(x) = \sqrt{x + 3}$, $g(x) = \sqrt{4 - x}$ $[-3, 4]$

76. $f(x) = \sqrt{2x + 1}$, $g(x) = \sqrt{3 - 5x}$ $\left[-\frac{1}{2}, \frac{3}{5}\right]$

77. $f(x) = \sqrt{4x - 3}$, $g(x) = \sqrt{2x + 9}$ $\left[\frac{3}{4}, \infty\right)$

78. $f(x) = \sqrt{5 - x}$, $g(x) = \sqrt{1 - x}$ $(-\infty, 1]$

79. $f(x) = \dfrac{x}{x - 5}$, $g(x) = \sqrt{3x + 2}$ $\left[-\frac{2}{3}, 5\right) \cup (5, \infty)$

80. $f(x) = \sqrt{4 - 5x}$, $g(x) = \dfrac{3x}{2x + 1}$ $\left(-\infty, -\frac{1}{2}\right) \cup \left(-\frac{1}{2}, \frac{4}{5}\right]$

TW **81.** Why can the conjunction $2 < x$ and $x < 5$ be rewritten as $2 < x < 5$, but the disjunction $2 < x$ or $x < 5$ cannot be rewritten as $2 < x < 5$?

TW **82.** Can the solution set of a disjunction be empty? Why or why not?

Skill Maintenance

Graph.

83. $y = 5$ [2.5] ▢

84. $y = -2$ [2.5] ▢

85. $f(x) = |x|$ [2.1] ▢

86. $g(x) = x - 1$ [2.4] ▢

Solve each system graphically. [3.1]

87. $y = x - 3$,
$y = 5$
$(8, 5)$

88. $y = x + 2$,
$y = -3$
$(-5, -3)$

Synthesis

TW **89.** What can you conclude about a, b, c, and d, if $[a, b] \cup [c, d] = [a, d]$? Why?

TW **90.** What can you conclude about a, b, c, and d, if $[a, b] \cap [c, d] = [a, b]$? Why?

91. *Minimizing Tolls.* A $3.00 toll is charged to cross the bridge from Sanibel Island to mainland Florida. A six-month pass, costing $15.00, reduces the toll to $0.50. A one-year pass, costing $150, allows for free crossings. How many crossings per year does it take, on average, for the two six-month passes to be the most economical choice? Assume a constant number of trips per month. Between 12 and 240 trips

92. *Pressure at Sea Depth.* The function

$$P(d) = 1 + \frac{d}{33} \qquad 0 \text{ ft} \le d \le 198 \text{ ft}$$

gives the pressure, in atmospheres (atm), at a depth of d feet in the sea. For what depths d is the pressure at least 1 atm and at most 7 atm?

▢ Answers to Exercises 83–86 can be found on p. A-60.

93. *Converting Dress Sizes.* The function

$$f(x) = 2(x + 10)$$

can be used to convert dress sizes x in the United States to dress sizes $f(x)$ in Italy. For what dress sizes in the United States will dress sizes in Italy be between 32 and 46? Sizes between 6 and 13

94. *Solid-Waste Generation.* The function

$$w(t) = 0.01t + 4.3$$

can be used to estimate the number of pounds of solid waste, $w(t)$, produced daily, on average, by each person in the United States, t years after 1991. For what years will waste production range from 4.5 to 4.75 lb per person per day? From 2011 through 2036

95. *Temperatures of Liquids.* The formula

$$C = \tfrac{5}{9}(F - 32)$$

can be used to convert Fahrenheit temperatures F to Celsius temperatures C. $1945.4° \le F < 4820°$

a) Gold is liquid for Celsius temperatures C such that $1063° \le C < 2660°$. Find a comparable inequality for Fahrenheit temperatures.

b) Silver is liquid for Celsius temperatures C such that $960.8° \le C < 2180°$. Find a comparable inequality for Fahrenheit temperatures. $1761.44° \le F < 3956°$

96. *Records in the Women's 100-m Dash.* As of 2002, the record for the women's 100-m dash was 10.49 sec, set by Florence Griffith Joyner in 1988. The function

$$R(t) = -0.0433t + 10.49$$

can be used to predict the world record in the women's 100-m dash t years after 1988. Predict (in terms of an inequality) those years for which the world record was between 11.5 and 10.8 sec. (Measure from the middle of 1988.) $1965 \le y \le 1981$

Solve and graph.

97. $4a - 2 \le a + 1 \le 3a + 4$

98. $4m - 8 > 6m + 5$ *or* $5m - 8 < -2$

99. $x - 10 < 5x + 6 \le x + 10$

100. $3x < 4 - 5x < 5 + 3x$

Determine whether each sentence is true or false for all real numbers a, b, and c.

101. If $-b < -a$, then $a < b$. True

102. If $a \le c$ and $c \le b$, then $b > a$. False

103. If $a < c$ and $b < c$, then $a < b$. False

104. If $-a < c$ and $-c > b$, then $a > b$. True

For $f(x)$ as given, use interval notation to write the domain of f. $\left[-\tfrac{5}{2}, 1\right) \cup (1, \infty)$ $(-\infty, -7) \cup \left(-7, \tfrac{3}{4}\right]$

105. $f(x) = \dfrac{\sqrt{5 + 2x}}{x - 1}$ **106.** $f(x) = \dfrac{\sqrt{3 - 4x}}{x + 7}$

107. On many graphing calculators, the TEST key provides access to inequality symbols, while the LOGIC option of that same key accesses the conjunction *and* and the disjunction *or*. Thus, if $y_1 = x > -2$ and $y_2 = x < 4$, Exercise 23 can be checked by forming the expression $y_3 = y_1$ *and* y_2. The interval(s) in the solution set are shown as a horizontal line 1 unit above the x-axis. (Be careful to "deselect" y_1 and y_2 so that only y_3 is drawn.) Use this approach to check your answers to Exercises 29 and 60.

97. $\left\{a \mid -\tfrac{3}{2} \le a \le 1\right\}$, or $\left[-\tfrac{3}{2}, 1\right]$;

98. $\left\{m \mid m < \tfrac{6}{5}\right\}$, or $\left(-\infty, \tfrac{6}{5}\right)$

99. $\{x \mid -4 < x \le 1\}$, or $(-4, 1]$;

100. $\left\{x \mid -\tfrac{1}{8} < x < \tfrac{1}{2}\right\}$, or $\left(-\tfrac{1}{8}, \tfrac{1}{2}\right)$

Collaborative Corner

Saving on Shipping Costs

Focus: Compound inequalities and
         solution sets

Time: 20–30 minutes

Group size: 2–3

At present (2002), the U.S. Postal Service charges
23 cents per ounce plus an additional 14-cent
delivery fee (1 oz or less costs 37 cents; more than
1 oz, but not more than 2 oz, costs 60 cents; and so
on). Rapid Delivery charges $1.05 per pound plus an
additional $2.50 delivery fee (up to 16 oz costs
$3.55; more than 16 oz, but less than or equal to
32 oz, costs $4.60; and so on). Let x be the weight, in
ounces, of an item being mailed.*

_____

*Based on an article by Michael Contino in *Mathematics Teacher*,
May 1995.

ACTIVITY

One group member should determine the function p,
where $p(x)$ represents the cost, in dollars, of mailing
x ounces at a post office. Another group member
should determine the function r, where $r(x)$
represents the cost, in dollars, of mailing x ounces
with Rapid Delivery. The third group member should
graph p and r on the same set of axes. Finally,
working together, use the graph to determine those
weights for which the Postal Service is less expensive.
Express your answer using both set-builder and
interval notation.

 4.3   Equations with Absolute Value ■ Inequalities with Absolute Value

Absolute-Value Equations and Inequalities

Equations with Absolute Value

Recall from Section 1.2 the definition of absolute value.

Absolute Value

The absolute value of x, denoted $|x|$, is defined as

$$|x| = \begin{cases} x, & \text{if } x \geq 0, \\ -x, & \text{if } x < 0. \end{cases}$$

(When x is nonnegative, the absolute value of x is x. When x is negative,
the absolute value of x is the opposite of x.)

Interactive Discovery

1. Graph $f(x) = |x|$ and $g(x) = 4$. Use the graph to solve $|x| = 4$. $\{-4, 4\}$

2. What are the solutions of an equation $|x| = a$ if a is positive? $-a, a$

3. Graph $f(x) = |x|$ and $h(x) = 0$. Use the graph to solve $|x| = 0$. $\{0\}$

4. Graph $f(x) = |x|$ and $r(x) = -7$. Use the graph to solve $|x| = -7$. $\varnothing$

5. What are the solutions of an equation $|x| = a$ if a is negative?   No solution

In general, the following is true:

If a is positive, the equation $|x| = a$ has two solutions, a and $-a$.

If a is 0, the equation $|x| = a$ has one solution, 0.

If a is negative, the equation $|x| = a$ has no solution.

Other equations involving absolute value can be solved graphically.

EXAMPLE 1   Solve: $|x - 3| = 5$.

Solution   We let $y_1 = \text{abs}(x - 3)$ and $y_2 = 5$. The solution set of the equation consists of the x-coordinates of any points of intersection of the graphs of y_1 and y_2. We see that the graphs intersect at two points. We use INTERSECT twice to find the coordinates of the points of intersection. Each time, we enter a GUESS that is close to the point we are looking for.

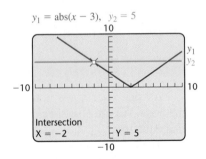
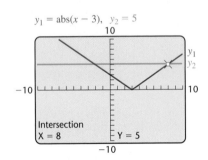

The graphs intersect at $(-2, 5)$ and $(8, 5)$. Thus the solutions are -2 and 8. To check, we substitute each in the original equation.

Check:

For -2:

$$\frac{|x - 3| = 5}{|-2 - 3| \ ? \ 5}$$
$$\frac{|-5|}{5 \ | \ 5} \quad \text{TRUE}$$

For 8:

$$\frac{|x - 3| = 5}{|8 - 3| \ ? \ 5}$$
$$\frac{|5|}{5 \ | \ 5} \quad \text{TRUE}$$

The solution set is $\{-2, 8\}$.

We can solve equations involving absolute value algebraically using the following principle.

Study Tip

The absolute-value principle will be used with a variety of replacements for X. Make sure that the principle, as stated here, makes sense before going further.

The Absolute-Value Principle for Equations

For any positive number p and any algebraic expression X:

a) The solutions of $|X| = p$ are those numbers that satisfy
$$X = -p \quad or \quad X = p.$$

b) The equation $|X| = 0$ is equivalent to the equation $X = 0$.

c) The equation $|X| = -p$ has no solution.

EXAMPLE 2 Find the solution set: **(a)** $|2x + 5| = 13$; **(b)** $|4 - 7x| = -8$.

a)

Algebraic Solution

We use the absolute-value principle, replacing X with $2x + 5$ and p with 13:

$$|X| = p$$
$$|2x + 5| = 13$$
$$2x + 5 = -13 \quad or \quad 2x + 5 = 13$$
$$2x = -18 \quad or \quad 2x = 8$$
$$x = -9 \quad or \quad x = 4.$$

The solutions are -9 and 4.

Graphical Solution

We graph $y_1 = \text{abs}(2x + 5)$ and $y_2 = 13$ and use INTERSECT to find any points of intersection, moving the cursor close to the desired point when prompted for a GUESS. The graphs intersect at $(-9, 13)$ and $(4, 13)$. The x-coordinates, -9 and 4, are the solutions.

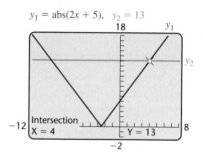

$y_1 = \text{abs}(2x + 5), \quad y_2 = 13$

As one check, we note that we found the same solution using both methods. We can also check by substituting the possible solutions in the original equation.

The solution set is $\{-9, 4\}$.

b)

Algebraic Solution

The absolute-value principle reminds us that absolute value is always nonnegative. Thus the equation $|4 - 7x| = -8$ has no solution. The solution set is $\varnothing$.

Graphical Solution

We graph $y_1 = \text{abs}(4 - 7x)$ and $y_2 = -8$. There are no points of intersection, so the solution set is $\varnothing$.

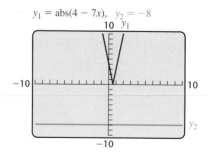

$y_1 = \text{abs}(4 - 7x), \quad y_2 = -8$

To use the absolute-value principle, we must be sure that the absolute-value expression is alone on one side of the equation.

EXAMPLE 3   Given that $f(x) = 2|x + 3| + 1$, find all x for which $f(x) = 15$.

Algebraic Solution

Since we are looking for $f(x) = 15$, we substitute:

$$f(x) = 15$$

$2|x + 3| + 1 = 15$ **Replacing $f(x)$ with $2|x + 3| + 1$**

$2|x + 3| = 14$ **Subtracting 1 from both sides**

$|x + 3| = 7$ **Dividing both sides by 2**

$x + 3 = -7$ *or* $x + 3 = 7$ **Replacing X with $x + 3$ and p with 7 in the absolute-value principle**

$x = -10$ *or* $x = 4$.

The solutions are -10 and 4.

Graphical Solution

We graph $y_1 = 2 \text{ abs}(x + 3) + 1$ and $y_2 = 15$ and determine the coordinates of any points of intersection. We make certain that the viewing window is large enough to show all such points. Choosing an appropriate viewing window may involve some trial and error.

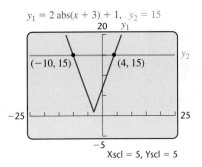

$y_1 = 2 \text{ abs}(x + 3) + 1, \ y_2 = 15$

Xscl = 5, Yscl = 5

The solutions are the x-coordinates of the points of intersection, -10 and 4.

To check, note that we found the same solution using both methods. We can also check that $f(-10) = f(4) = 15$. The solution set is $\{-10, 4\}$.

We can think of a number's absolute value as its distance from 0 on a number line. For example, we can interpret the equation

$$|x| = 4$$

to mean that the number x is 4 units from 0 on a number line.

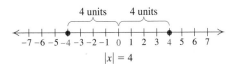

$|x| = 4$

There are two such numbers, 4 and -4. Thus the solution set is $\{-4, 4\}$.

EXAMPLE 4 Solve: $|x - 2| = 3$.

Solution   Because this equation is of the form $|a - b| = c$, it can be solved two ways.

Method 1.   We interpret $|x - 2| = 3$ as stating that the number $x - 2$ is 3 units from zero. Using the absolute-value principle, we replace X with $x - 2$ and p with 3:

$$|X| = p$$
$$|x - 2| = 3$$
$$x - 2 = -3 \quad or \quad x - 2 = 3 \qquad \textbf{Using the absolute-value principle}$$
$$x = -1 \quad or \qquad x = 5.$$

Method 2.   This approach is helpful in calculus. The expressions $|a - b|$ and $|b - a|$ can be used to represent the *distance between a and b* on the number line. For example, the distance between 7 and 8 is given by $|8 - 7|$ or $|7 - 8|$. From this viewpoint, the equation $|x - 2| = 3$ states that the distance between x and 2 is 3 units. We draw a number line and locate all numbers that are 3 units from 2.

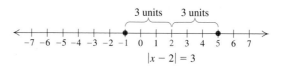

$$|x - 2| = 3$$

The solutions of $|x - 2| = 3$ are -1 and 5.

Check: The check consists of observing that both methods give the same solutions. The graphical solution shown at left serves as another check. The solution set is $\{-1, 5\}$.

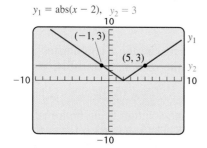

$y_1 = \text{abs}(x - 2), \quad y_2 = 3$

Sometimes an equation has two absolute-value expressions. Consider $|a| = |b|$. This means that a and b are the same distance from zero.

If a and b are the same distance from zero, then either they are the same number or they are opposites.

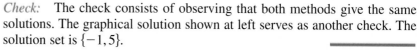

The Principle of Equal Absolute Values

For any algebraic expressions X and Y:

If $|X| = |Y|$,   then $X = Y$ or $X = -Y$.

EXAMPLE 5 Solve: $|2x - 3| = |x + 5|$.

Algebraic Solution

Either $2x - 3 = x + 5$ (they are the same number) or $2x - 3 = -(x + 5)$ (they are opposites). We solve each equation separately:

This assumes these numbers are the same.

This assumes these numbers are opposites.

$$
\begin{array}{lll}
2x - 3 = x + 5 & or & 2x - 3 = -(x + 5) \\
x - 3 = 5 & or & 2x - 3 = -x - 5 \\
x = 8 & or & 3x - 3 = -5 \\
& & 3x = -2 \\
& & x = -\frac{2}{3}.
\end{array}
$$

The solutions are 8 and $-\frac{2}{3}$. The solution set is $\left\{-\frac{2}{3}, 8\right\}$.

Graphical Solution

We graph $y_1 = \text{abs}(2x - 3)$ and $y_2 = \text{abs}(x + 5)$ and look for any points of intersection. The x-coordinates of the points of intersection are -0.6666667 and 8.

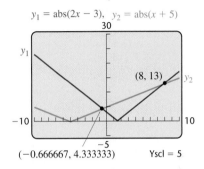

$y_1 = \text{abs}(2x - 3), \quad y_2 = \text{abs}(x + 5)$

$(8, 13)$

$(-0.666667, 4.333333)$ Yscl = 5

The solution set is $\{-0.6666667, 8\}$. Converted to fraction notation, the solution set is $\left\{-\frac{2}{3}, 8\right\}$.

Inequalities with Absolute Value

Our methods for solving equations with absolute value can be adapted for solving inequalities. Inequalities of this sort arise regularly in more advanced courses.

Interactive Discovery

Graph $y_1 = \text{abs}(x)$ and $y_2 = 4$. The solution set of the inequality $|x| < 4$ consists of all x-values for which the graph of $y_1 = \text{abs}(x)$ is below the horizontal line $y_2 = 4$. Similarly, the solution set of the inequality $|x| > 4$ is the set of all x-values for which (x, y_1) is above the line $y_2 = 4$.

Use the graph to write the solution set of each of the following equations using interval notation.

1. $|x| < 4$ $(-4, 4)$

2. $|x| > 4$ $(-\infty, -4) \cup (4, \infty)$

3. $|x| \leq 4$ $[-4, 4]$

4. $|x| \geq 4$ $(-\infty, -4] \cup [4, \infty)$

5. What is the solution set of $|x| > -3$? $\mathbb{R}$

6. What is the solution set of $|x| \leq -3$? $\varnothing$

We can also visualize solutions of inequalities involving absolute value using the number line. The solutions of $|x| < 4$ are all numbers for which the *distance from* 0 *is less than* 4. Looking at the number line, we can see that numbers like $-3, -2, -\frac{1}{2}, 0, \frac{1}{2}, 2,$ and 3 are all solutions. In fact, the solution set is $\{x \mid -4 < x < 4\}$, or, in interval notation, $(-4, 4)$. The graph is as follows:

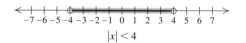

$$|x| < 4$$

The solutions of $|x| \geq 4$ are all numbers for which the *distance from* 0 *is greater than or equal to* 4—in other words, those numbers x such that $x \leq -4$ or $4 \leq x$. The solution set is $\{x \mid x \leq -4 \ or \ x \geq 4\}$. In interval notation, the solution is $(-\infty, -4] \cup [4, \infty)$. The graph is as follows:

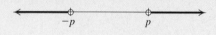

$$|x| \geq 4$$

The following is a general principle for solving problems in which absolute-value symbols appear.

Principles for Solving Absolute-Value Problems

For any positive number p and any expression X:

a) The solutions of $|X| = p$ are those numbers that satisfy $X = -p$ or $X = p$.

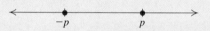

b) The solutions of $|X| < p$ are those numbers that satisfy $-p < X < p$.

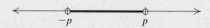

c) The solutions of $|X| > p$ are those numbers that satisfy $X < -p$ or $p < X$.

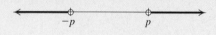

Of course, if p is negative, any value of X will satisfy the inequality $|X| > p$ because absolute value is never negative. By the same reasoning, $|X| < p$ has no solution when p is not positive. Thus, $|2x - 7| > -3$ is true for any real number x, and $|2x - 7| < -3$ has no solution.

Note that an inequality of the form $|X| < p$ corresponds to a *conjunction*, whereas an inequality of the form $|X| > p$ corresponds to a *disjunction*.

EXAMPLE 6 Solve: $|3x - 2| < 4$.

Algebraic Solution

We use part (b) of the principles listed above. In this case, X is $3x - 2$ and p is 4.

$$|X| < p$$

$|3x - 2| < 4$ Replacing X with $3x - 2$ and p with 4

$-4 < 3x - 2 < 4$ The number $3x - 2$ must be within 4 units of zero.

$-2 < 3x < 6$ Adding 2

$-\frac{2}{3} < x < 2$.   Multiplying by $\frac{1}{3}$

The solution set is $\left\{ x \mid -\frac{2}{3} < x < 2 \right\}$, or, in interval notation, $\left(-\frac{2}{3}, 2 \right)$. The graph is as follows.

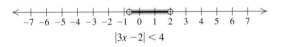
$|3x - 2| < 4$

Graphical Solution

We graph $y_1 = \text{abs}(3x - 2)$ and $y_2 = 4$.

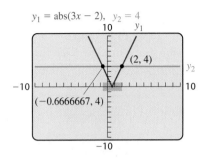

$y_1 = \text{abs}(3x - 2), \; y_2 = 4$

The solution set consists of the x-values for which $y_1 < y_2$. We can see that $y_1 < y_2$ between the points $(-0.6666667, 4)$ and $(2, 4)$. The solution set of the inequality is thus the interval $(-0.6666667, 2)$, as indicated by the shading on the x-axis. Converted to fraction notation, the solution is $\left(-\frac{2}{3}, 2 \right)$.

EXAMPLE 7 Given that $f(x) = |4x + 2|$, find all x for which $f(x) \geq 6$.

Algebraic Solution

We have

$$f(x) \geq 6,$$

or $|4x + 2| \geq 6$.   **Substituting**

To solve, we use part (c) of the principles listed above. In this case, X is $4x + 2$ and p is 6.

$$|X| \geq p$$

$|4x + 2| \geq 6$ Replacing X with $4x + 2$ and p with 6

$4x + 2 \leq -6 \quad or \quad 6 \leq 4x + 2$

The number $4x + 2$ must be at least 6 units from zero.

$4x \leq -8 \quad or \quad 4 \leq 4x$ Adding -2

$x \leq -2 \quad or \quad 1 \leq x$ Multiplying by $\frac{1}{4}$

The solution set is $\{ x \mid x \leq -2 \; or \; x \geq 1 \}$, or, in interval notation, $(-\infty, -2] \cup [1, \infty)$. The graph is as follows.

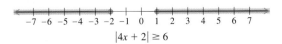
$|4x + 2| \geq 6$

Graphical Solution

To find all values of x for which $|4x + 2| \geq 6$, we graph $y_1 = \text{abs}(4x + 2)$ and $y_2 = 6$ and determine the points of intersection of the graphs.

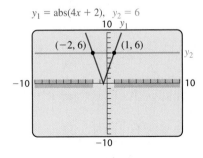

$y_1 = \text{abs}(4x + 2), \; y_2 = 6$

The solution set consists of all x-values for which the graph of $y_1 = |4x + 2|$ is *on or above* the graph of $y_2 = 6$. The graph of y_1 lies *on* the graph of y_2 when $x = -2$ or $x = 1$. The graph of y_1 is *above* the graph of y_2 when $x < -2$ or $x > 1$. Thus the solution set is $(-\infty, -2] \cup [1, \infty)$, as indicated by the shading on the x-axis.

4.3

Exercise Set

33. $\left\{\frac{3}{2}, \frac{7}{2}\right\}$ **34.** $\left\{-\frac{1}{3}, 3\right\}$ **35.** $\left\{-\frac{4}{3}, 4\right\}$ **36.** $\left\{-\frac{3}{2}, \frac{17}{2}\right\}$

Use the following graph to solve Exercises 1–6.

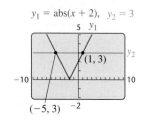

$y_1 = \text{abs}(x + 2),\ \ y_2 = 3$

1. $|x + 2| = 3$ $\{-5, 1\}$

2. $|x + 2| \le 3$ $[-5, 1]$ *Aha!*

3. $|x + 2| < 3$ $(-5, 1)$

4. $|x + 2| > 3$ $(-\infty, -5) \cup (1, \infty)$

5. $|x + 2| \ge 3$ $(-\infty, -5] \cup [1, \infty)$

6. $|x + 2| = -1$ $\varnothing$

Solve.

7. $|x| = 4$ $\{-4, 4\}$

8. $|x| = 9$ $\{-9, 9\}$

Aha! **9.** $|x| = -5$ $\varnothing$

10. $|x| = -3$ $\varnothing$

11. $|y| = 7.3$ $\{-7.3, 7.3\}$

12. $|p| = 0$ $\{0\}$

13. $|m| = 0$ $\{0\}$

14. $|t| = 5.5$ $\{-5.5, 5.5\}$

15. $|5x + 2| = 7$ $\left\{-\frac{9}{5}, 1\right\}$

16. $|2x - 3| = 4$ $\left\{-\frac{1}{2}, \frac{7}{2}\right\}$

17. $|7x - 2| = -9$ $\varnothing$

18. $|3x - 10| = -8$ $\varnothing$

19. $|x - 3| = 8$ $\{-5, 11\}$

20. $|x - 2| = 6$ $\{-4, 8\}$ *Aha!*

21. $|x - 6| = 1$ $\{5, 7\}$

22. $|x - 5| = 3$ $\{2, 8\}$

23. $|x - 4| = 5$ $\{-1, 9\}$

24. $|x - 7| = 9$ $\{-2, 16\}$

25. $|2y| - 5 = 13$ $\{-9, 9\}$

26. $|5x| - 3 = 37$ $\{-8, 8\}$

27. $7|z| + 2 = 16$ $\{-2, 2\}$

28. $5|q| - 2 = 9$ $\left\{-\frac{11}{5}, \frac{11}{5}\right\}$

29. $\left|\dfrac{4 - 5x}{6}\right| = 3$ $\left\{-\frac{14}{5}, \frac{22}{5}\right\}$

30. $\left|\dfrac{2x - 1}{3}\right| = 5$ $\{-7, 8\}$ *Aha!*

31. $|t - 7| + 1 = 4$ $\{4, 10\}$

32. $|m + 5| + 9 = 16$ $\{-12, 2\}$

33. $3|2x - 5| - 7 = -1$

34. $5 - 2|3x - 4| = -5$

35. Let $f(x) = |3x - 4|$. Find all x for which $f(x) = 8$.

36. Let $f(x) = |2x - 7|$. Find all x for which $f(x) = 10$.

37. Let $f(x) = |x| - 2$. Find all x for which $f(x) = 6.3$. $\{-8.3, 8.3\}$

38. Let $f(x) = |x| + 7$. Find all x for which $f(x) = 18$. $\{-11, 11\}$

39. Let $f(x) = \left|\dfrac{3x - 2}{5}\right|$. Find all x for which $f(x) = 2$. $\left\{-\frac{8}{3}, 4\right\}$

40. Let $f(x) = \left|\dfrac{1 - 2x}{3}\right|$. Find all x for which $f(x) = 1$. $\{-1, 2\}$

Solve.

41. $|x + 4| = |2x - 7|$ $\{1, 11\}$

42. $|3x + 5| = |x - 6|$ $\left\{-\frac{11}{2}, \frac{1}{4}\right\}$

43. $|x - 9| = |x + 6|$ $\left\{\frac{3}{2}\right\}$

44. $|x + 4| = |x - 3|$ $\left\{-\frac{1}{2}\right\}$

45. $|5t + 7| = |4t + 3|$ ⊡

46. $|3a - 1| = |2a + 4|$ ⊡

47. $|n - 3| = |3 - n|$ $\mathbb{R}$

48. $|y - 2| = |2 - y|$ $\mathbb{R}$

49. $|7 - a| = |a + 5|$ $\{1\}$

50. $|6 - t| = |t + 7|$ $\left\{-\frac{1}{2}\right\}$

51. $\left|\frac{1}{2}x - 5\right| = \left|\frac{1}{4}x + 3\right|$ $\left\{32, \frac{8}{3}\right\}$

52. $\left|2 - \frac{2}{3}x\right| = \left|4 + \frac{7}{8}x\right|$ $\left\{-\frac{48}{37}, -\frac{144}{5}\right\}$

Solve and graph.

53. $|a| \le 7$ ⊡

54. $|x| < 2$ ⊡

55. $|x| > 8$ ⊡

56. $|a| \ge 3$ ⊡

57. $|t| > 0$ ⊡

58. $|t| \ge 1.7$ ⊡

59. $|x - 3| < 5$ ⊡

60. $|x - 1| < 3$ ⊡

61. $|x + 2| \le 6$ ⊡

62. $|x + 4| \le 1$ ⊡

63. $|x - 3| + 2 > 7$ ⊡

64. $|x - 4| + 5 > 2$ ⊡

65. $|2y - 7| > -5$ ⊡

66. $|3y - 4| > 8$ ⊡

67. $|3a - 4| + 2 \ge 8$ ⊡

68. $|2a - 5| + 1 \ge 9$ ⊡

69. $|y - 3| < 12$ ⊡

70. $|p - 2| < 3$ ⊡

71. $9 - |x + 4| \le 5$ ⊡

72. $12 - |x - 5| \le 9$

73. $|4 - 3y| > 8$ ⊡

74. $|7 - 2y| < -6$ $\varnothing$

75. $|3 - 4x| < -5$ $\varnothing$

76. $7 + |4a - 5| \le 26$ ⊡

77. $\left|\dfrac{2 - 5x}{4}\right| \ge \dfrac{2}{3}$ ⊡

78. $\left|\dfrac{1 + 3x}{5}\right| > \dfrac{7}{8}$ ⊡

79. $|m + 5| + 9 \le 16$ ⊡

80. $|t - 7| + 3 \ge 4$ ⊡

81. $25 - 2|a + 3| > 19$ ⊡

82. $30 - 4|a + 2| > 12$ ⊡

83. Let $f(x) = |2x - 3|$. Find all x for which $f(x) \le 4$.   ⊡

84. Let $f(x) = |5x + 2|$. Find all x for which $f(x) \le 3$.   ⊡

85. Let $f(x) = 2 + |3x - 4|$. Find all x for which $f(x) \ge 13$.   ⊡

86. Let $f(x) = |2 - 9x|$. Find all x for which $f(x) \ge 25$.   ⊡

⊡ Answers to Exercises 45, 46, 53–73, and 76–86 can be found on p. A-60.

87. Let $f(x) = 7 + |2x - 1|$. Find all x for which $f(x) < 16$.

88. Let $f(x) = 5 + |3x + 2|$. Find all x for which $f(x) < 19$.

^{TW} **89.** Explain in your own words why -7 is not a solution of $|x| < 5$.

^{TW} **90.** Explain in your own words why $[6, \infty)$ is only part of the solution of $|x| \geq 6$.

Skill Maintenance

Solve using substitution or elimination. [3.2]

91. $2x - 3y = 7,$
$3x + 2y = -10$ $\left(-\frac{16}{13}, -\frac{41}{13}\right)$

92. $3x - 5y = 9,$
$4x - 3y = 1$ $(-2, -3)$

93. $x = -2 + 3y,$
$x - 2y = 2$ $(10, 4)$

94. $y = 3 - 4x,$
$2x - y = -9$
$(-1, 7)$

Solve graphically. [3.1]

95. $x + 2y = 9,$
$3x - y = -1$ $(1, 4)$

96. $2x + y = 7,$
$-3x - 2y = 10$
$(24, -41)$

Synthesis

^{TW} **97.** Is it possible for an equation in x of the form $|ax + b| = c$ to have exactly one solution? Why or why not?

^{TW} **98.** Isabel is using the following graph to solve $|x - 3| < 4$. How can you tell that a mistake has been made?

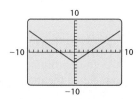

99. From the definition of absolute value, $|x| = x$ only when $x \geq 0$. Solve $|3t - 5| = 3t - 5$ using this same reasoning. $\left\{t \mid t \geq \frac{5}{3}\right\}$, or $\left[\frac{5}{3}, \infty\right)$

Solve.

100. $|x + 2| > x$ $\mathbb{R}$, or $(-\infty, \infty)$

101. $2 \leq |x - 1| \leq 5$

102. $|5t - 3| = 2t + 4$ $\left\{x \mid -4 \leq x \leq -1 \text{ or } 3 \leq x \leq 6\right\}$ or $[-4, -1] \cup [3, 6]$

103. $t - 2 \leq |t - 3|$

Find an equivalent inequality with absolute value.

104. $-3 < x < 3$ $|x| < 3$

105. $-5 \leq y \leq 5$ $|y| \leq 5$

106. $x \leq -6 \text{ or } 6 \leq x$ $|x| \geq 6$

107. $x < -4 \text{ or } 4 < x$ $|x| > 4$

108. $x < -8 \text{ or } 2 < x$

109. $-5 < x < 1$ $|x + 2| < 3$

110. x is less than 2 units from 7. $|x + 3| > 5$

111. x is less than 1 unit from 5. $|x - 7| < 2$, or $|7 - x| < 2$
$|x - 5| < 1$, or $|5 - x| < 1$

Write an absolute-value inequality for which the interval shown is the solution.

112. $|x - 3| \leq 4$

113. $|x - 2| < 6$

114. $|x + 4| < 3$

115. $|x - 7| \leq 5$

116. *Motion of a Spring.*   A weighted spring is bouncing up and down so that its distance d above the ground satisfies the inequality $|d - 6 \text{ ft}| \leq \frac{1}{2}$ ft (see the figure below). Find all possible distances d.

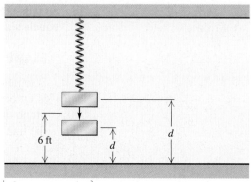

$\left\{d \mid 5\frac{1}{2} \text{ ft} \leq d \leq 6\frac{1}{2} \text{ ft}\right\}$, or $\left[5\frac{1}{2}, 6\frac{1}{2}\right]$

---

87. $\{x \mid -4 < x < 5\}$, or $(-4, 5)$

88. $\left\{x \mid -\frac{16}{3} < x < 4\right\}$, or $\left(-\frac{16}{3}, 4\right)$

102. $\left\{-\frac{1}{7}, \frac{7}{3}\right\}$ **103.** $\left\{x \mid x \leq \frac{5}{2}\right\}$, or $\left(-\infty, \frac{5}{2}\right]$

4.4

Graphs of Linear Inequalities ▪ Systems of Linear Inequalities

Inequalities in Two Variables

In Section 4.1, we graphed inequalities in one variable on a number line. Now we graph inequalities in two variables on a plane.

Graphs of Linear Inequalities

When the equals sign in a linear equation is replaced with an inequality sign, a **linear inequality** is formed. Solutions of linear inequalities are ordered pairs.

EXAMPLE 1 Determine whether $(-3, 2)$ and $(6, -7)$ are solutions of the inequality $5x - 4y > 13$.

Solution   Below, on the left, we replace x with -3 and y with 2. On the right, we replace x with 6 and y with -7.

$$
\begin{array}{l}
\underline{5x - 4y > 13} \\
5(-3) - 4 \cdot 2 \; \overset{?}{\vert} \; 13 \\
\quad -15 - 8 \; \vert \\
\quad\quad -23 \; \vert \; 13 \qquad \text{FALSE}
\end{array}
\qquad\qquad
\begin{array}{l}
\underline{5x - 4y > 13} \\
5(6) - 4(-7) \; \overset{?}{\vert} \; 13 \\
\quad 30 + 28 \; \vert \\
\quad\quad 58 \; \vert \; 13 \qquad \text{TRUE}
\end{array}
$$

Since $-23 > 13$ is false, $(-3, 2)$ is not a solution.

Since $58 > 13$ is true, $(6, -7)$ is a solution.

The graph of a linear equation is a straight line. The graph of a linear inequality is a *half-plane*, bordered by the graph of the *related equation*. To find an inequality's related equation, we simply replace the inequality sign with an equals sign.

EXAMPLE 2 Graph: $y \le x$.

Solution   We first graph the related equation $y = x$. Every solution of $y = x$ is an ordered pair, like $(3, 3)$, in which both coordinates are the same. The graph of $y = x$ is shown on the left below. Since the inequality symbol is $\le$, the line is drawn solid and is part of the graph of $y \le x$.

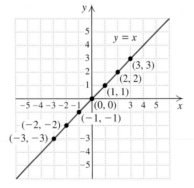
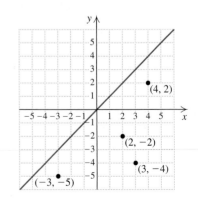

Study Tip

It is never too soon to begin reviewing for the final examination. Take a few minutes each week to review important problems, formulas, and properties. There is also at least one Connecting the Concepts in each chapter. Spend time reviewing the information in this special feature.

Note that in the graph on the right at the bottom of the preceding page each ordered pair on the half-plane below $y = x$ contains a y-coordinate that is less than the x-coordinate. All these pairs represent solutions of $y \leq x$. We check one pair, $(4, 2)$, as follows:

$$\frac{y \leq x}{2 \mid 4} \quad \text{TRUE}$$

It turns out that *any* point on the same side of $y = x$ as $(4, 2)$ is also a solution. Thus, if one point in a half-plane is a solution, then *all* points in that half-plane are solutions. We complete the drawing of the solution set by shading the half-plane below $y = x$. The complete solution set consists of the shaded half-plane and the line. Note too that for any inequality of the form $y \leq f(x)$ or $y < f(x)$, we shade *below* the graph of $y = f(x)$.

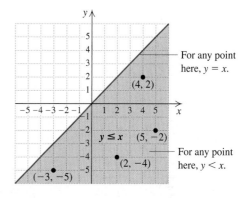

EXAMPLE 3   Graph: $8x + 3y > 24$.

Solution   First, we sketch the line $8x + 3y = 24$. Since the inequality sign is $>$, points on this line do not represent solutions of the inequality, so the line is drawn dashed. Points representing solutions of $8x + 3y > 24$ are in either the half-plane above the line or the half-plane below the line. To determine which, we select a point that is not on the line and determine whether it is a solution of $8x + 3y > 24$. We try $(-3, 4)$ as a *test point*:

$$\frac{8x + 3y > 24}{\begin{array}{c|c} 8(-3) + 3 \cdot 4 \; ? \; 24 & \\ -24 + 12 & \\ -12 & 24 \end{array}} \quad \text{FALSE}$$

Since $-12 > 24$ is *false*, $(-3, 4)$ is not a solution. Thus no point in the half-plane containing $(-3, 4)$ is a solution. The points in the other half-plane *are* solutions, so we shade that half-plane and obtain the graph shown at right.

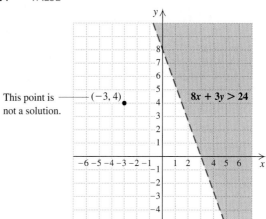

Steps for Graphing Linear Inequalities

1. Replace the inequality sign with an equals sign and graph this related equation. If the inequality symbol is $<$ or $>$, draw the line dashed. If the inequality symbol is $\leq$ or $\geq$, draw the line solid.

2. The graph consists of a half-plane on one side of the line and, if the line is solid, the line as well.

    a) If the inequality is of the form $y < f(x)$ or $y \leq f(x)$, shade *below* the line. If the inequality is of the form $y > f(x)$ or $y \geq f(x)$, shade *above* the line.

    b) If y is not isolated, either use algebra to isolate y and proceed as in part (a) or select a point not on the line. If the point represents a solution of the inequality, shade the half-plane containing the point. If it does not, shade the other half-plane.

Most graphing calculators can draw the solution set of an inequality in two variables. The procedures used vary widely. Some calculators allow you to choose to shade above or below a boundary line at the time that you enter the equation.

EXAMPLE 4   Use a graphing calculator to graph $8x + 3y > 24$ (see Example 3).

Solution   First, we solve for y:

$$8x + 3y > 24$$
$$3y > -8x + 24$$
$$y > -\tfrac{8}{3}x + 8.$$

Next, we enter the related equation $y_1 = (-8/3)x + 8$ and select the part of the graph to shade. On many calculators, we do this by moving the cursor to the GRAPHSTYLE icon just to the left of y_1. Since the inequality states that y is *greater than* $-\tfrac{8}{3}x + 8$, we shade the half-plane *above* the line $y = -\tfrac{8}{3}x + 8$. We press ENTER until ◥ appears. Now we can view the graph of the inequality.

TEACHING TIP

Point out that on nearly all graphing calculators, the boundary line will always appear solid.

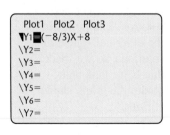

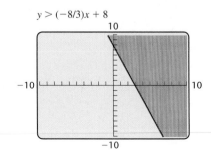

$y > (-8/3)x + 8$

EXAMPLE 5   Graph: $6x - 2y < 12$.

Solution   We graph both by hand and using a graphing calculator.

By Hand

We first graph the related equation, $6x - 2y = 12$, as a dashed line. This line passes through the points $(2, 0)$, $(0, -6)$, and $(3, 3)$, and serves as the boundary of the solution set of the inequality. Since y is not isolated, we determine which half-plane to shade by testing a point *not* on the line. The pair $(0, 0)$ is easy to substitute:

$$\frac{6x - 2y < 12}{\begin{array}{c|c} 6 \cdot 0 - 2 \cdot 0 \ ? \ 12 \\ 0 - 0 \\ 0 \ | \ 12 \quad \text{TRUE} \end{array}}$$

Since the inequality $0 < 12$ is *true*, the point $(0, 0)$ is a solution, as are all points in the half-plane containing $(0, 0)$. The graph is shown below.

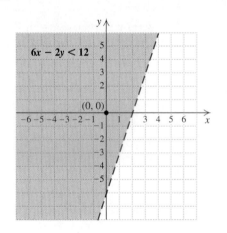

Using a Graphing Calculator

We solve the inequality for y:

$$6x - 2y < 12$$
$$-2y < 12 - 6x$$
$$y > -6 + 3x \text{ or } y > 3x - 6.$$

We enter the related equation, $y = 3x - 6$, into the graphing calculator and shade the half-plane above the line.

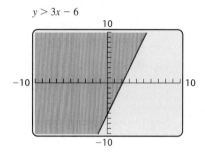

Note the calculator does not draw a dashed line, so the graph of $y > 3x - 6$ appears to be the same as the graph of $y \geq 3x - 6$.

Although a graphing calculator can be a very useful tool, some equations and inequalities are actually quicker to graph by hand. It is important to be able to quickly sketch basic linear equations and inequalities by hand. Also, when we are using a calculator to graph equations, knowing the basic shape of the graph can serve as a check that the equation was entered correctly.

EXAMPLE 6   Graph $x > -3$ on a plane.

Solution   There is a missing variable in this inequality. If we graph the inequality on a line, its graph is as follows:

However, we can also write this inequality as $x + 0y > -3$ and graph it on a plane. We can use the same technique as in the examples above. First, we graph the related equation $x = -3$ in the plane, using a dashed line. Then we test some point, say, $(2, 5)$:

$$\frac{x + 0y > -3}{2 + 0 \cdot 5 \; ? \; -3}$$
$$2 \mid -3 \qquad \text{TRUE}$$

Since $(2, 5)$ is a solution, all points in the half-plane containing $(2, 5)$ are solutions. We shade that half-plane. Another approach is to simply note that the solutions of $x > -3$ are all pairs with first coordinates greater than -3.

Although many graphing calculators can graph this inequality from the DRAW menu, it is simpler to graph it by hand.

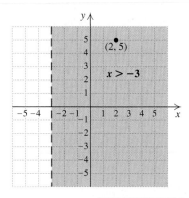

EXAMPLE 7 Graph $y \le 4$ on a plane.

Solution   The inequality is of the form $y \le f(x)$, so we shade below the solid line representing solutions of $y = f(x)$, or in this case, $y = 4$.

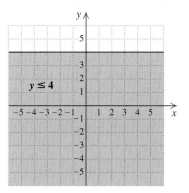

This inequality can also be graphed by drawing $y = 4$ and testing a point above or below the line. The student should check that this results in a graph identical to the one above.

Systems of Linear Inequalities

To graph a system of equations, we graph the individual equations and then find the intersection of the individual graphs. We do the same thing for a system of inequalities, that is, we graph each inequality and find the intersection of the individual graphs.

EXAMPLE 8 Graph the system

$$x + y \le 4,$$
$$x - y < 4.$$

Solution   To graph $x + y \le 4$, we graph $x + y = 4$ using a solid line. Since the test point $(0, 0)$ *is* a solution and $(0, 0)$ is below the line, we shade the half-plane below the graph red. The arrows near the ends of the line are another way of indicating the half-plane containing solutions.

Next, we graph $x - y < 4$. We graph $x - y = 4$ using a dashed line and consider $(0, 0)$ as a test point. Again, $(0, 0)$ is a solution, so we shade that side of the line blue. The solution set of the system is the region that is shaded purple (both red and blue) and part of the line $x + y = 4$.

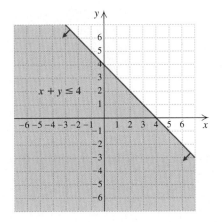
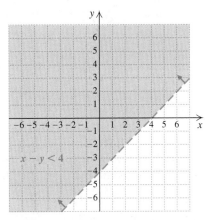
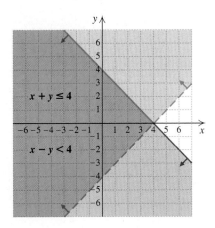

On a graphing calculator, the intersection of the graphs of inequalities is shown by using different shading patterns.

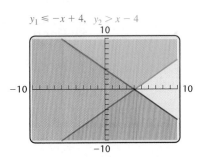

EXAMPLE 9   Graph: $-2 < x \le 3$.

Solution   This is a system of inequalities:

$$-2 < x,$$
$$x \le 3.$$

We graph the equation $-2 = x$, and see that the graph of the first inequality is the half-plane to the right of the line $-2 = x$. It is shaded red.

We graph the second inequality, starting with the line $x = 3$, and find that its graph is the line and also the half-plane to its left. It is shaded blue.

The solution set of the system is the region that is the intersection of the individual graphs. Since it is shaded both blue and red, it appears to be purple. All points in this region have x-coordinates that are greater than -2 but do not exceed 3.

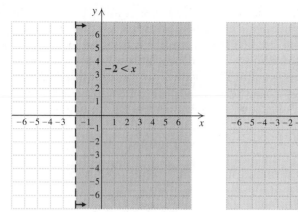 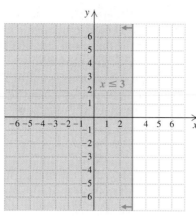 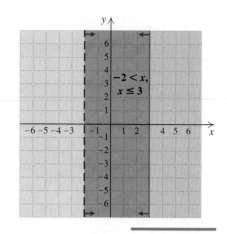

A system of inequalities may have a graph that consists of a polygon and its interior. The corners of such a graph are called *vertices* (singular, *vertex*).

EXAMPLE 10   Graph the system of inequalities. Find the coordinates of any vertices formed.

$$6x - 2y \le 12, \quad (1)$$
$$y - 3 \le 0, \quad (2)$$
$$x + y \ge 0. \quad (3)$$

Solution   We graph both by hand and using a graphing calculator

By Hand

We graph the lines

$$6x - 2y = 12,$$
$$y - 3 = 0,$$
and $$x + y = 0$$

using solid lines. The regions for each inequality are indicated by the arrows near the ends of the lines. We note where the regions overlap and shade the region of solutions purple.

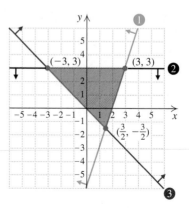

To find the vertices, we solve three different systems of equations. The system of related equations from inequalities (1) and (2) is

$$6x - 2y = 12,$$
$$y - 3 = 0.$$

Solving, we obtain the vertex $(3, 3)$.

The system of related equations from inequalities (1) and (3) is

$$6x - 2y = 12,$$
$$x + y = 0.$$

Solving, we obtain the vertex $\left(\frac{3}{2}, -\frac{3}{2}\right)$.

The system of related equations from inequalities (2) and (3) is

$$y - 3 = 0,$$
$$x + y = 0.$$

Solving, we obtain the vertex $(-3, 3)$.

Using a Graphing Calculator

First, we solve each inequality for y and obtain the equivalent system of inequalities

$$y_1 \geq (12 - 6x)/(-2),$$
$$y_2 \leq 3,$$
$$y_3 \geq -x.$$

We enter the related equations

$$y_1 = (12 - 6x)/(-2),$$
$$y_2 = 3,$$

and $\quad y_3 = -x$

and shade the half-planes *above* y_1, *below* y_2, and *above* y_3. The intersection of the half-planes is the graph of the system.

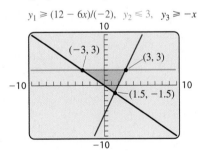

$y_1 \geq (12 - 6x)/(-2), \quad y_2 \leq 3, \quad y_3 \geq -x$

The vertices are the points of intersection of the graphs of the related equations y_1 and y_2, y_1 and y_3, and y_2 and y_3. We can find them by using INTERSECT and choosing the appropriate curves. The vertices are $(3, 3)$, $(1.5, -1.5)$, and $(-3, 3)$.

Graphs of systems of inequalities and the coordinates of vertices are used to solve a variety of problems using a branch of mathematics called *linear programming*.

Connecting the Concepts

We have now solved a variety of equations, inequalities, systems of equations, and systems of inequalities. In each case, there are different ways to represent the solution. Below is a list of the different types of problems we have solved, along with illustrations of each type.

Type	Example	Solution	Graph
Linear equations in one variable	$2x - 8 = 3(x + 5)$	A number	
Linear inequalities in one variable	$-3x + 5 > 2$	A set of numbers; an interval	
Linear equations in two variables	$2x + y = 7$	A set of ordered pairs	
Linear inequalities in two variables	$x + y \geq 4$	A set of ordered pairs	
System of equations in two variables	$x + y = 3,$ $5x - y = -27$	An ordered pair or a (possibly empty) set of ordered pairs	
System of inequalities in two variables	$6x - 2y \leq 12,$ $y - 3 \leq 0,$ $x + y \geq 0$	A set of ordered pairs	

Keeping in mind how these solutions vary and what their graphs look like will help you as you progress further in this book and in mathematics in general.

4.4

Exercise Set

FOR EXTRA HELP

| Digital Video Tutor CD 4 Videotape 5 | Student's Solutions Manual | Tutor Center AW Math Tutor Center | InterAct Math | MathXL | MyMathLab |

Determine whether each ordered pair is a solution of the given inequality.

1. $(-4, 2)$; $2x + 3y < -1$ Yes

2. $(3, -6)$; $4x + 2y \leq -2$ No

3. $(8, 14)$; $2y - 3x \geq 9$ No

4. $(7, 20)$; $3x - y > -1$ Yes

Graph on a plane.

5. $y > \frac{1}{2}x$ ▢

6. $y > 2x$ ▢

7. $y \geq x - 3$ ▢

8. $y < x + 3$ ▢

9. $y \leq x + 4$ ▢

10. $y > x - 2$ ▢

11. $x - y \leq 5$ ▢

12. $x + y < 4$ ▢

13. $2x + 3y < 6$ ▢

14. $3x + 4y \leq 12$ ▢

15. $2y - x \leq 4$ ▢

16. $2y - 3x > 6$ ▢

17. $2x - 2y \geq 8 + 2y$ ▢

18. $3x - 2 \leq 5x + y$ ▢

19. $y \geq 2$ ▢

20. $x < -5$ ▢

21. $x \leq 7$ ▢

22. $y > -3$ ▢

23. $-2 < y < 6$ ▢

24. $-4 < y < -1$ ▢

25. $-4 \leq x \leq 5$ ▢

26. $-3 \leq y \leq 4$ ▢

27. $0 \leq y \leq 3$ ▢

28. $0 \leq x \leq 6$ ▢

Graph using a graphing calculator.

29. $y > x + 3.5$ ▢

30. $7y \leq 2x + 5$ ▢

31. $8x - 2y < 11$ ▢

32. $11x + 13y + 4 \geq 0$ ▢

Graph each system.

33. $y > x,$
$y < -x + 2$ ▢

34. $y < x,$
$y > -x + 1$ ▢

35. $y \geq x,$
$y \geq 2x - 4$ ▢

36. $y \geq x,$
$y \leq -x + 4$ ▢

37. $y \leq -3,$
$x \geq -1$ ▢

38. $y \geq -3,$
$x \geq 1$ ▢

39. $x > -4,$
$y < -2x + 3$ ▢

40. $x < 3,$
$y > -3x + 2$ ▢

41. $y \leq 3,$
$y \geq -x + 2$ ▢

42. $y \geq -2,$
$y \geq x + 3$ ▢

▢ Answers to Exercises 5–54 can be found on p. A-61.

43. $x + y \leq 6,$
$x - y \leq 4$ ▢

44. $x + y < 1,$
$x - y < 2$ ▢

45. $y + 3x > 0,$
$y + 3x < 2$ ▢

46. $y - 2x \geq 1,$
$y - 2x \leq 3$ ▢

Graph each system of inequalities. Find the coordinates of any vertices formed.

47. $y \leq 2x - 1,$
$y \geq -2x + 1,$
$x \leq 3$ ▢

48. $2y - x \leq 2,$
$y - 3x \geq -4,$
$y \geq -1$ ▢

49. $x + 2y \leq 12,$
$2x + y \leq 12,$
$x \geq 0,$
$y \geq 0$ ▢

50. $x - y \leq 2,$
$x + 2y \geq 8,$
$y \leq 4$ ▢

51. $8x + 5y \leq 40,$
$x + 2y \leq 8,$
$x \geq 0,$
$y \geq 0$ ▢

52. $4y - 3x \geq -12,$
$4y + 3x \geq -36,$
$y \leq 0,$
$x \leq 0$ ▢

53. $y - x \geq 1,$
$y - x \leq 3,$
$2 \leq x \leq 5$ ▢

54. $3x + 4y \geq 12,$
$5x + 6y \leq 30,$
$1 \leq x \leq 3$ ▢

ᵀᵂ **55.** In Example 8, is the point $(4, 0)$ part of the solution set? Why or why not?

ᵀᵂ **56.** When graphing linear inequalities, Ron makes a habit of always shading above the line when the symbol $\geq$ is used. Is this wise? Why or why not?

Skill Maintenance

57. *Catering.* Sandy's Catering needs to provide 10 lb of mixed nuts for a wedding reception. Peanuts cost $2.50 per pound and fancy nuts cost $7 per pound. If $40 has been allocated for nuts, how many pounds of each type should be mixed? [3.3]   Peanuts: $6\frac{2}{3}$ lb; fancy nuts: $3\frac{1}{3}$ lb

58. *Household Waste.* The Hendersons generate two and a half times as much trash as their neighbors, the Savickis. Together, the two households produce 14 bags of trash each month. How much trash does each household produce? [3.3]   Hendersons: 10 bags; Savickis: 4 bags

59. *Paid Admissions.* There were 203 tickets sold for a volleyball game. For activity-card holders the price was $1.25, and for noncard holders the price was $2. The total amount of money collected was $310. How many of each type of ticket were sold? [3.3]
Activity-card holders: 128; noncard holders: 75

60. *Paid Admissions.* There were 200 tickets sold for an alumni basketball game. Tickets for students were $2 each and for adults were $3 each. The total amount collected was $530. How many of each type of ticket were sold? [3.3] Students: 70; adults: 130

61. *Landscaping.* Grass seed is being spread on a triangular traffic island. If the grass seed can cover an area of 200 ft² and the island's base is 16 ft long, how tall a triangle can the seed fill? [2.3] 25 ft

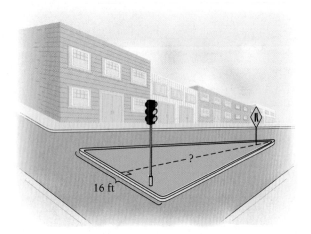

62. *Interest Rate.* What rate of interest is required in order for a principal of $320 to earn $17.60 in half a year? [2.3] 11%

Synthesis

TW **63.** Explain how a system of linear inequalities could have a solution set containing exactly one pair.

TW **64.** Do all systems of linear inequalities have solutions? Why or why not?

Graph.
65. $x + y > 8,$
$x + y \le -2$

66. $x + y \ge 1,$
$-x + y \ge 2,$
$x \le 4,$
$y \ge 0,$
$y \le 4,$
$x \le 2$

☐ Answers to Exercises 68–72 can be found on pp. A-61 and A-62.

67. $x - 2y \le 0,$
$-2x + y \le 2,$
$x \le 2,$
$y \le 2,$
$x + y \le 4$

68. Write four systems of four inequalities that describe a 2-unit by 2-unit square that has $(0,0)$ as one of the vertices. ☐

69. *Luggage Size.* Unless an additional fee is paid, most major airlines will not check any luggage that is more than 62 in. long. The U.S. Postal Service will ship a package only if the sum of the package's length and girth (distance around its midsection) does not exceed 108 in. Concert Productions is ordering several 62-in. long trunks that will be both mailed and checked as luggage. Using w and h for width and height (in inches), respectively, write and graph an inequality that represents all acceptable combinations of width and height. ☐

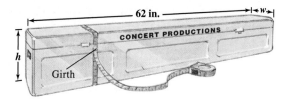

70. *Hockey Wins and Losses.* The Skating Stars figure that they need at least 60 points for the season in order to make the playoffs. A win is worth 2 points and a tie is worth 1 point. Graph a system of inequalities that describes the situation. (*Hint*: Let w = the number of wins and t = the number of ties.) ☐

71. *Elevators.* Many elevators have a capacity of 1 metric ton (1000 kg). Suppose that c children, each weighing 35 kg, and a adults, each 75 kg, are on an elevator. Graph a system of inequalities that indicates when the elevator is overloaded. ☐

72. *Widths of a Basketball Floor.* Sizes of basketball floors vary due to building sizes and other constraints such as cost. The length L is to be at most 94 ft and the width W is to be at most 50 ft. Graph a system of inequalities that describes the possible dimensions of a basketball floor. ☐

Write a system of inequalities for each region shown.

73.

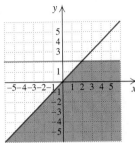

$y \leq x,$
$y \leq 2$

74.

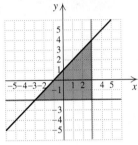

$y \leq x + 1,$
$x \leq 3,$
$y \geq -2$

75.

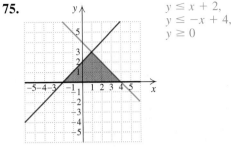

$y \leq x + 2,$
$y \leq -x + 4,$
$y \geq 0$

76.

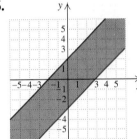

$y \leq x + 2,$
$y \geq x - 3$

Collaborative Corner

The Rule of 85

Focus: Linear inequalities

Time: 20–30 minutes

Group size: 3

Under a proposed "Rule of 85," full-time faculty in the California State Teachers Retirement System (kindergarten through community college) who are *a* years old with *y* years of service would have the option of retirement if $a + y \geq 85$.

ACTIVITY

1. Decide, as a group, the age range of full-time teachers. Express this age range as an inequality involving *a*.

2. Decide, as a group, the number of years someone could teach full-time before retiring. Express this answer as a compound inequality involving *y*.

3. Using the Rule of 85 and the answers to parts (1) and (2) above, write a system of inequalities. Then, using a scale of 5 yr per square, graph the system. To facilitate comparisons with graphs from other groups, plot *a* on the horizontal axis and *y* on the vertical axis.

4. Compare the graphs from all groups. Try to reach consensus on the graph that most clearly illustrates what the status would be of someone who would have the option of retirement under the Rule of 85.

5. If your instructor is agreeable to the idea, attempt to represent him or her with a point on your graph.

4 Chapter Summary and Review

Key Terms

Important Properties and Formulas

The Addition Principle for Inequalities

For any real numbers a, b, and c:

$a < b$ is equivalent to $a + c < b + c$;
$a > b$ is equivalent to $a + c > b + c$.

Similar statements hold for $\leq$ and $\geq$.

The Multiplication Principle for Inequalities

For any real numbers a and b, and for any positive number c,

$a < b$ is equivalent to $ac < bc$;
$a > b$ is equivalent to $ac > bc$.

For any real numbers a and b, and for any *negative* number c,

$a < b$ is equivalent to $ac > bc$;
$a > b$ is equivalent to $ac < bc$.

Similar statements hold for $\leq$ and $\geq$.

Set intersection:

$$A \cap B = \{x \mid x \text{ is in } A \text{ and } x \text{ is in } B\}$$

Set union:

$$A \cup B = \{x \mid x \text{ is in } A \text{ or in } B, \text{ or both}\}$$

Intersection corresponds to "and"; union corresponds to "or."

Absolute Value

$|x| = x$ if $x \geq 0$; $|x| = -x$ if $x < 0$.

The Absolute-Value Principles for Equations and Inequalities

For any positive number p and any algebraic expression X:

a) The solutions of $|X| = p$ are those numbers that satisfy $X = -p$ *or* $X = p$.

b) The solutions of $|X| < p$ are those numbers that satisfy $-p < X < p$.

c) The solutions of $|X| > p$ are those numbers that satisfy $X < -p$ *or* $p < X$.

If $|X| = 0$, then $X = 0$. If p is negative, then $|X| = p$ and $|X| < p$ have no solution, and any value of X will satisfy $|X| > p$.

The Principle of Equal Absolute Values

For any algebraic expressions X and Y:

If $|X| = |Y|$, then $X = Y$ or $X = -Y$.

Steps for Graphing Linear Inequalities

1. Graph the related equation. Draw the line *dashed* if the inequality symbol is $<$ or $>$ and *solid* if the inequality symbol is $\leq$ or $\geq$.

2. The graph consists of a half-plane on one side of the line and, if the line is solid, the line as well.

a) Shade *below* the line if the inequality is of the form $y < f(x)$ or $y \leq f(x)$. Shade *above* the line if the inequality is of the form $y > f(x)$ or $y \geq f(x)$.

b) If y is not isolated, either isolate y and proceed as in part (a) or select a test point not on the line. If the point represents a solution of the inequality, shade the half-plane containing the point. If it does not, shade the other half-plane.

36. [4.4]

37. [4.4]

38. [4.4]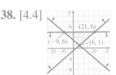

Review Exercises

Graph each inequality and write the solution set using both set-builder and interval notation.

1. $x \leq -2$ ⊡

2. $a + 7 \leq -14$ ⊡

3. $y - 5 \geq -12$ ⊡

4. $4y > -15$ ⊡

5. $-0.3y < 9$ ⊡

6. $-6x - 5 < 4$ ⊡

7. $-\frac{1}{2}x - \frac{1}{4} > \frac{1}{2} - \frac{1}{4}x$ ⊡

8. $0.3y - 7 < 2.6y + 15$ ⊡

9. $-2(x - 5) \geq 6(x + 7) - 12$ ⊡

10. Let $f(x) = 3x - 5$ and $g(x) = 11 - x$. Find all values of x for which $f(x) \leq g(x)$.   [4.1] $\{x \mid x \leq 4\}$, or $(-\infty, 4]$

Solve.

11. Jessica can choose between two summer jobs. She can work as a checker in a discount store for $8.40 an hour, or she can mow lawns for $12.00 an hour. In order to mow lawns, she must buy a $450 lawn-mower. How many hours of labor will it take Jessica to make more money mowing lawns?   [4.1] More than 125 hr

12. Clay is going to invest $4500, part at 6% and the rest at 7%. What is the most he can invest at 6% and still be guaranteed $300 in interest each year?   [4.1] $1500

13. Find the intersection:

$$\{1, 2, 5, 6, 9\} \cap \{1, 3, 5, 9\}.$$ [4.2] $\{1, 5, 9\}$

14. Find the union:

$$\{1, 2, 5, 6, 9\} \cup \{1, 3, 5, 9\}.$$ [4.2] $\{1, 2, 3, 5, 6, 9\}$

Graph and write interval notation.

15. $x \leq 3 \text{ and } x > -5$ [4.2] ───; $(-5, 3]$

16. $x \leq 3 \text{ or } x > -5$ [4.2] ───; $(-\infty, \infty)$

Solve and graph each solution set.

17. $-4 < x + 3 \leq 5$ ⊡

18. $-15 < -4x - 5 < 0$ ⊡

19. $3x < -9 \text{ or } -5x < -5$ ⊡

20. $2x + 5 < -17 \text{ or } -4x + 10 \leq 34$ ⊡

21. $2x + 7 \leq -5 \text{ or } x + 7 \geq 15$ ⊡

22. $f(x) < -5 \text{ or } f(x) > 5$, where $f(x) = 3 - 5x$ ⊡

For $f(x)$ as given, use interval notation to write the domain of f.

23. $f(x) = \dfrac{x}{x - 3}$ [4.2] $(-\infty, 3) \cup (3, \infty)$

24. $f(x) = \sqrt{x + 3}$ [4.2] $[-3, \infty)$

25. $f(x) = \sqrt{8 - 3x}$ [4.2] $\left(-\infty, \frac{8}{3}\right]$

Solve.

26. $|x| = 4$ [4.3] $\{-4, 4\}$

27. $|t| \geq 3.5$ ⊡

28. $|x - 2| = 7$ [4.3] $\{-5, 9\}$

29. $|2x + 5| < 12$ ⊡

30. $|3x - 4| \geq 15$ ⊡

31. $|2x + 5| = |x - 9|$ ⊡

32. $|5n + 6| = -8$ [4.3] ∅

33. $\left|\dfrac{x + 4}{8}\right| \leq 1$ [4.3] $\{x \mid -12 \leq x \leq 4\}$, or $[-12, 4]$

34. $2|x - 5| - 7 > 3$ [4.3] $\{x \mid x < 0 \text{ or } x > 10\}$, or $(-\infty, 0) \cup (10, \infty)$

35. Let $f(x) = |3x - 5|$. Find all x for which $f(x) < 0$.   [4.3] ∅

36. Graph $x - 2y \geq 6$ on a plane.

Graph each system of inequalities. Find the coordinates of any vertices formed.

37. $x + 3y > -1$,
  $x + 3y < 4$

38. $x - 3y \leq 3$,
  $x + 3y \geq 9$,
  $y \leq 6$

⊡ Answers to Exercises 1–9, 17–22, 27, and 29–31 can be found on p. A-62.

Synthesis

TW **39.** Explain in your own words why $|X| = p$ has two solutions when p is positive and no solution when p is negative.

TW **40.** Explain why the graph of the solution of a system of linear inequalities is the intersection, not the union, of the individual graphs.

1. [4.1] $\{x \mid x < 12\}$, or $(-\infty, 12)$ ⟵━━┼━━━◦━⟶
                                          0   12

3. [4.1] $\{y \mid y \le -2\}$, or $(-\infty, -2]$ ⟵━◦━┼━⟶
                                          -2 0

5. [4.1] $\left\{x \mid x > \frac{5}{2}\right\}$, or $\left(\frac{5}{2}, \infty\right)$ ⟵━┼━◦━⟶
                                          0   $\frac{5}{2}$

[4.3] $\left\{x \mid -\frac{8}{3} \le x \le -2\right\}$, or $\left[-\frac{8}{3}, -2\right]$

41. Solve: $|2x + 5| \le |x + 3|$.

42. Classify as true or false: If $x < 3$, then $x^2 < 9$. If false, give an example showing why. [4.1] False; $-4 < 3$ is true, but $(-4)^2 < 9$ is false.

43. Just-For-Fun manufactures marbles with a 1.1-cm diameter and a ± 0.03-cm manufacturing tolerance, or allowable variation in diameter. Write the tolerance as an inequality with absolute value.
[4.3] $|d - 1.1| \le 0.03$

2. [4.1] $\{y \mid y > -50\}$, or $(-50, \infty)$ ⟵◦━┼━⟶
                                          -50 0

4. [4.1] $\left\{a \mid a \le \frac{11}{5}\right\}$, or $\left(-\infty, \frac{11}{5}\right]$ ⟵━━┼━━━•━⟶
                                          0    $\frac{11}{5}$

6. [4.1] $\left\{x \mid x \le \frac{7}{4}\right\}$, or $\left(-\infty, \frac{7}{4}\right]$ ⟵━━┼━━•━⟶
                                          0   $\frac{7}{4}$

Chapter Test 4

Graph each inequality and write the solution set using both set-builder and interval notation.

1. $x - 2 < 10$

2. $-0.6y < 30$

3. $-4y - 3 \ge 5$

4. $3a - 5 \le -2a + 6$

5. $4(5 - x) < 2x + 5$

6. $-8(2x + 3) + 6(4 - 5x) \ge 2(1 - 7x) - 4(4 + 6x)$

7. Let $f(x) = -5x - 1$ and $g(x) = -9x + 3$. Find all values of x for which $f(x) > g(x)$. [4.1] $\{x \mid x > 1\}$, or $(1, \infty)$

8. Lia can rent a van for either $40 with unlimited mileage or $30 with 100 free miles and an extra charge of 15¢ for each mile over 100. For what numbers of miles traveled would the unlimited mileage plan save Lia money? [4.1] More than $166\frac{2}{3}$ mi

9. A refrigeration repair company charges $40 for the first half-hour of work and $30 for each additional hour. Blue Mountain Camp has budgeted $100 to repair its walk-in cooler. For what lengths of a service call will the budget not be exceeded?
[4.1] Less than or equal to 2.5 hr

10. Find the intersection: [4.2] $\{3, 5\}$

$$\{1, 3, 5, 7, 9\} \cap \{3, 5, 11, 13\}.$$

11. Find the union: [4.2] $\{1, 3, 5, 7, 9, 11, 13\}$

$$\{1, 3, 5, 7, 9\} \cup \{3, 5, 11, 13\}.$$

12. Write the domain of f using interval notation if $f(x) = \sqrt{7 - x}$. [4.2] $(-\infty, 7]$

Solve and graph each solution set.

13. $-3 < x - 2 < 4$ ▫

14. $-11 \le -5t - 2 < 0$ ▫

15. $3x - 2 < 7 \; or \; x - 2 > 4$
[4.2] $\{x \mid x < 3 \; or \; x > 6\}$, or $(-\infty, 3) \cup (6, \infty)$ ⟵━◦━┼━◦━⟶
                                                                          0 3   6

16. $-3x > 12 \; or \; 4x > -10$ ▫

17. $-\frac{1}{3} \le \frac{1}{6}x - 1 < \frac{1}{4}$ ▫

18. $|x| = 9$ ▫

19. $|a| > 3$ ▫

20. $|4x - 1| < 4.5$ ▫

21. $|-5t - 3| \ge 10$ ▫

22. $|2 - 5x| = -10$ [4.3] $\varnothing$

23. $g(x) < -3 \; or \; g(x) > 3$, where $g(x) = 4 - 2x$ ▫

24. Let $f(x) = |x + 10|$ and $g(x) = |x - 12|$. Find all values of x for which $f(x) = g(x)$. [4.3] $\{1\}$

Graph the system of inequalities. Find the coordinates of any vertices formed. **25.** [4.4]

25. $x + y \ge 3$,
$\quad x - y \ge 5$

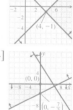

26. $2y - x \ge -7$,
$\quad 2y + 3x \le 15$, **26.** [4.4]
$\quad y \le 0$,
$\quad x \le 0$

Synthesis

Solve. Write the solution set using interval notation.

27. $|2x - 5| \le 7 \; and \; |x - 2| \ge 2$ [4.3] $[-1, 0] \cup [4, 6]$

28. $7x < 8 - 3x < 6 + 7x$ [4.2] $\left(\frac{1}{5}, \frac{4}{5}\right)$

29. Write an absolute-value inequality for which the interval shown is the solution. [4.3] $|x + 13| \le 2$

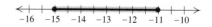

▫ Answers to Exercises 13, 14, 16–21, and 23 can be found on p. A-62.

Polynomials and Polynomial Functions

5

- **5.1** Introduction to Polynomials and Polynomial Functions
- **5.2** Multiplication of Polynomials
- **5.3** Polynomial Equations and Factoring
- **5.4** Equations Containing Trinomials of the Type $x^2 + bx + c$
- **5.5** Equations Containing Trinomials of the Type $ax^2 + bx + c$, $a \neq 1$
- **5.6** Equations Containing Perfect-Square Trinomials and Differences of Squares
- **5.7** Equations Containing Sums or Differences of Cubes
- **5.8** Applications of Polynomial Equations

SUMMARY AND REVIEW

TEST

A polynomial is a type of algebraic expression that contains one or more terms. We define polynomials more specifically in this chapter, and examine polynomial functions both algebraically and graphically.

APPLICATION

BACHELOR'S DEGREES. The number of bachelor's degrees earned in the biological and life sciences for various years is shown in the following table and graph. Fit a polynomial function of degree 3 (cubic) to the data, and use the function to predict the number of bachelor's degrees that will be earned in the biological and life sciences in 2005.

Year	Bachelor's Degrees in Biological/Life Science
1971	35,743
1976	54,275
1980	46,370
1986	38,524
1990	37,204
1994	51,383
2000	63,532

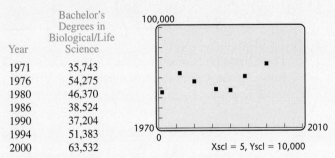

Source: U.S. National Center for Education Statistics, *Digest of Education Statistics*, annual

This problem appears as Example 4 in Section 5.8.

5.1

Algebraic Expressions and Polynomials ■ Polynomial Functions ■ Adding Polynomials ■ Opposites and Subtraction

Introduction to Polynomials and Polynomial Functions

We now begin the study of *polynomials* and *polynomial functions*. Earlier in this text, we examined several other types of functions.

Connecting the Concepts

FAMILIES OF GRAPHS OF FUNCTIONS

Often, the shape of a graph of a function can be predicted by examining the equation describing the function.

We examined *linear functions* and their graphs in Chapter 2. Functions f that can be expressed in the form $f(x) = mx + b$ have graphs that are straight lines. Their slope, or rate of change, is constant. We can think of linear functions as a "family" of functions, with the function $f(x) = x$ being the simplest such function.

We have also studied some nonlinear functions. In Chapter 4, we solved equations and inequalities containing absolute-value symbols by examining graphs of absolute-value functions. An *absolute-value function* of the form $f(x) = |ax + b|$, where $a \neq 0$, has a graph similar to that of $f(x) = |x|$.

Another nonlinear function that we considered in Chapter 4 when examining domains is a square-root function. A *square-root function* of the form $f(x) = \sqrt{ax + b}$, where $a \neq 0$, has a graph similar to the graph of $f(x) = \sqrt{x}$.

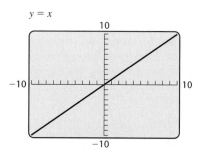

$y = x$

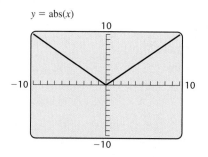

$y = \text{abs}(x)$

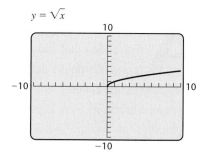

$y = \sqrt{x}$

These basic functions, along with others that we will develop throughout the text, form a **library of functions**. Knowing the general shape of the graphs of different types of functions will help in analyzing both data and graphs.

In this chapter, we look at polynomial functions. The graphs of polynomial functions have certain common characteristics.

Answers may include: The graph of a polynomial function has no sharp corners; there are no holes or breaks; the domain is the set of all real numbers.

The following table lists some polynomial functions and some nonpolynomial functions. Graph each function and compare the graphs.

Polynomial Functions	Nonpolynomial Functions
$f(x) = x^2 + 3x + 5$ $f(x) = 4$ $f(x) = -0.5x^4 + 5x - 2.3$	$f(x) = \lvert x - 4 \rvert$ $f(x) = 1 + \sqrt{2x - 5}$ $f(x) = \dfrac{x - 7}{2x}$

What are some characteristics of graphs of polynomial functions?

You may have noticed the following:

The graph of a polynomial function is "smooth," that is, there are no sharp corners.

The graph of a polynomial function is continuous, that is, there are no holes or breaks.

The domain of a polynomial function, unless otherwise specified, is all real numbers.

There are other characteristics of polynomial functions, but before discussing them, we need to establish some terminology.

Algebraic Expressions and Polynomials

In Chapter 1, we introduced algebraic expressions like

$$5x^2, \quad \frac{3}{x^2 + 5}, \quad 9a^3b^4, \quad 3x^{-2}, \quad 6x^2 + 3x + 1, \quad -9, \quad \text{and} \quad 5 - 2x.$$

Of the expressions listed, $5x^2$, $9a^3b^4$, $3x^{-2}$, and -9 are examples of *terms*. A **term** is simply a number or a product of a number and a variable or variables raised to a power.

When all variables in a term are raised to whole-number powers, the term is a **monomial** (pronounced mä-nō-mē-əl). Of the terms listed above, $5x^2$, $9a^3b^4$, and -9 are monomials. The **degree** of a monomial is the sum of the exponents of the variables. Thus, $5x^2$ has degree 2 and $9a^3b^4$ has degree 7. Nonzero constant terms, like -9, can be written $-9x^0$ and therefore have degree 0. The term 0 itself is said to have no degree.

The number 5 is said to be the **coefficient** of $5x^2$. Thus the coefficient of $9a^3b^4$ is 9 and the coefficient of $-2x$ is -2. The coefficient of a constant term is just that constant.

A **polynomial** is a monomial or a sum of monomials. Of the expressions listed, $5x^2$, $9a^3b^4$, $6x^2 + 3x + 1$, -9, and $5 - 2x$ are polynomials. In fact, with the exception of $9a^3b^4$, these are all polynomials *in one variable*. The expres-

TEACHING TIP

This section contains many new terms. Encourage students to use the new vocabulary. You may want to have students verbalize examples in class.

sion $9a^3b^4$ is a *polynomial in two variables*. Note that $5 - 2x$ is the sum of 5 and $-2x$. Thus, 5 and $-2x$ are the terms in the polynomial $5 - 2x$.

The **leading term** of a polynomial is the term of highest degree. Its coefficient is called the **leading coefficient**. The **degree of a polynomial** is the same as the degree of its leading term.

EXAMPLE 1 For each polynomial given, find the degree of each term, the degree of the polynomial, the leading term, and the leading coefficient.

a) $2x^3 + 8x^2 - 17x - 3$
b) $6x^2 + 8x^2y^3 - 17xy - 24xy^2z^4 + 2y + 3$

Solution

a) $2x^3 + 8x^2 - 17x - 3$ b) $6x^2 + 8x^2y^3 - 17xy - 24xy^2z^4 + 2y + 3$

Term	$2x^3$	$8x^2$	$-17x$	-3	$6x^2$	$8x^2y^3$	$-17xy$	$-24xy^2z^4$	$2y$	3
Degree	3	2	1	0	2	5	2	7	1	0
Leading Term	$2x^3$				$-24xy^2z^4$					
Leading Coefficient	2				-24					
Degree of Polynomial	3				7					

A polynomial of degree 0 or 1 is called **linear**. A polynomial in one variable is said to be **quadratic** if it is of degree 2 and **cubic** if it is of degree 3.

The following are some names for certain kinds of polynomials.

Type	Definition	Examples
Monomial	A polynomial of one term	$4, -3p, 5x^2, -7a^2b^3, 0, xyz$
Binomial	A polynomial of two terms	$2x + 7, a - 3b, 5x^2 + 7y^3$
Trinomial	A polynomial of three terms	$x^2 - 7x + 12, 4a^2 + 2ab + b^2$

We generally arrange polynomials in one variable so that the exponents *decrease* from left to right. This is called **descending order**. Some polynomials may be written with exponents *increasing* from left to right, which is **ascending order**. Generally, if an exercise is written in one kind of order, the answer is written in that same order.

EXAMPLE 2 Arrange in ascending order: $12 + 2x^3 - 7x + x^2$.

Solution

$$12 + 2x^3 - 7x + x^2 = 12 - 7x + x^2 + 2x^3$$

Polynomials in several variables can be arranged with respect to the powers of one of the variables.

EXAMPLE 3   Arrange in descending powers of x:

$$y^4 + 2 - 5x^2 + 3x^3y + 7xy^2.$$

Solution

$$y^4 + 2 - 5x^2 + 3x^3y + 7xy^2 = 3x^3y - 5x^2 + 7xy^2 + y^4 + 2$$

Polynomial Functions

A *polynomial function* is a function in which outputs are determined by evaluating a polynomial. For example, the function P given by

$$P(x) = 5x^7 + 3x^5 - 4x^2 - 5$$

is an example of a polynomial function. To evaluate a polynomial function, we substitute a number for the variable just as in Chapter 2. In this text, we limit ourselves to polynomial functions in one variable.

EXAMPLE 4   Find $P(-5)$ for the polynomial function given by $P(x) = -x^2 + 4x - 1$.

Solution   We evaluate the function using several methods discussed in Chapter 2.

Using algebraic substitution. We substitute -5 for x and carry out the operations using the rules for order of operations:

$$\begin{aligned} P(-5) &= -(-5)^2 + 4(-5) - 1 \\ &= -25 - 20 - 1 \quad \text{We square the input before} \\ &\qquad\qquad\qquad\quad \text{taking its opposite.} \\ &= -46. \end{aligned}$$

Using function notation on a graphing calculator. We let $y_1 = -x^2 + 4x - 1$. We enter the function into the graphing calculator and evaluate $Y1(-5)$ using the Y-VARS menu.

```
Y₁(−5)
                    −46
```

Using a table. If $y_1 = -x^2 + 4x - 1$, we can find the value of y_1 for $x = -5$ by setting Indpnt to Ask in the TABLE SETUP.

X	Y₁	
−5	−46	

X =

Using the graph of the function. We graph $y_1 = -x^2 + 4x - 1$ and find the value of y_1 for $x = -5$ using VALUE in the CALC menu.

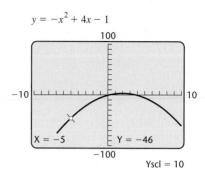

$$y = -x^2 + 4x - 1$$

X = -5   Y = -46

Yscl = 10

No matter which method we use, we have $P(-5) = -46$.

EXAMPLE 5 Veterinary Medicine. Gentamicin is an antibiotic frequently used by veterinarians. The concentration, in micrograms per milliliter (mcg/mL), of Gentamicin in a horse's bloodstream t hours after injection can be approximated by the polynomial function

$$C(t) = -0.005t^4 + 0.003t^3 + 0.35t^2 + 0.5t.$$

(*Source*: Michele Tulis, DVM, telephone interview)

a) What is the concentration 2 hr after injection?

b) Use the graph below to estimate $C(4)$.

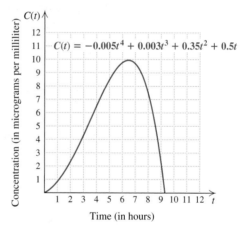

$$C(t) = -0.005t^4 + 0.003t^3 + 0.35t^2 + 0.5t$$

Concentration (in micrograms per milliliter)

Time (in hours)

Solution

a) We evaluate the function for $t = 2$:

$$
\begin{aligned}
C(2) &= -0.005(2)^4 + 0.003(2)^3 + 0.35(2)^2 + 0.5(2) \\
&= -0.005(16) + 0.003(8) + 0.35(4) + 0.5(2) \\
&= -0.08 + 0.024 + 1.4 + 1 \\
&= -0.08 + 2.424 \\
&= 2.344.
\end{aligned}
$$

We carry out the calculation using the rules for order of operations.

The concentration after 2 hr is 2.344 mcg/mL.

b) To estimate $C(4)$, the concentration after 4 hr, we locate 4 on the horizontal axis. From there we move vertically to the graph of the function and then horizontally to the $C(t)$-axis, as shown below. This locates a value of about 6.5. Thus,

$$C(4) \approx 6.5.$$

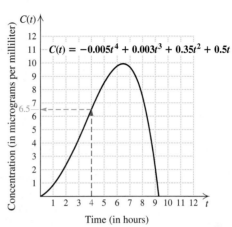

Recall from Section 2.7 that the domain of a function, when not specified, is the set of all real numbers for which the function is defined. If no restrictions are given, a polynomial function is defined for all real numbers. In other words,

The domain of a polynomial function is $(-\infty, \infty)$.

Note in Example 5 that the domain of the function C is restricted by the context of the problem. Since Gentamicin cannot have a negative concentration, values of t that make C negative are not in the domain of the function for this problem. Thus the domain of C, as estimated from its graph, is approximately $[0, 9.3]$.

For a polynomial function with an unrestricted domain, the range may be $(-\infty, \infty)$, or there may be a maximum or a minimum value of the function. We can estimate the range of a function from its graph. In Example 5, the values of C vary from a minimum of 0 to a maximum of approximately 10. Thus we estimate that the range of C is about $[0, 10]$.

EXAMPLE 6   Estimate the range of each of the following functions from its graph.

a)

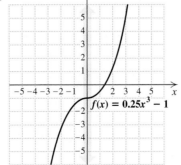

b)

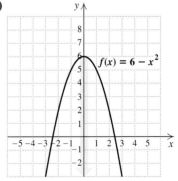

c)

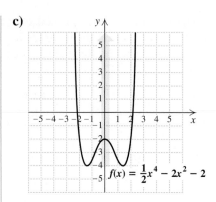

$f(x) = \frac{1}{2}x^4 - 2x^2 - 2$

Solution   The domain of each function represented in the graphs is $(-\infty, \infty)$.

a) Since there is no maximum or minimum value indicated on the graph of the first function, we estimate its range to be $(-\infty, \infty)$. The range is indicated by the shading on the y-axis.

b) The second function has a maximum value of about 6, and no minimum is indicated, so its range is about $(-\infty, 6]$.

c) The third function has a minimum value of about -4 and no maximum value, so its range is $[-4, \infty)$.

When we are estimating the range of a polynomial function from its graph, care must be taken to show enough of the graph to see its behavior. As you learn more about graphs of polynomial functions, you will be able to more readily choose appropriate scales or viewing windows. In this chapter, we will supply the graph or an appropriate viewing window for the graph of a polynomial function.

EXAMPLE 7   Graph each of the following functions in the standard viewing window and estimate the range of the function.

a) $f(x) = x^3 - 4x^2 + 5$ **b)** $g(x) = x^4 - 4x^2 + 5$

Solution

a) The graph of $y = x^3 - 4x^2 + 5$ is shown on the left below. It appears that the graph extends downward and upward indefinitely, so the range of the function is $(-\infty, \infty)$.

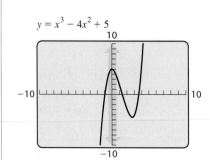

$y = x^3 - 4x^2 + 5$

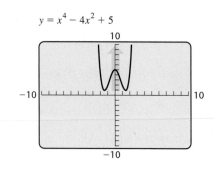

$y = x^4 - 4x^2 + 5$

b) The graph of $y = x^4 - 4x^2 + 5$ is shown on the right at the bottom of the preceding page. The graph extends upward indefinitely but does not go below a y-value of 1. (This can be confirmed by tracing along the graph.) Since there is a minimum function value of 1 and no maximum function value, the range of the function is $[1, \infty)$.

Adding Polynomials

Recall from Section 1.3 that when two terms have the same variable(s) raised to the same power(s), they are **similar**, or **like**, **terms** and can be "combined" or "collected."

EXAMPLE 8   Combine like terms.

a) $3x^2 - 4y + 2x^2$
b) $9x^3 + 5x - 4x^2 - 2x^3 + 5x^2$
c) $3x^2y + 5xy^2 - 3x^2y - xy^2$

Solution

a) $3x^2 - 4y + 2x^2 = 3x^2 + 2x^2 - 4y$ **Rearranging terms using the commutative law for addition**

$\qquad\qquad\qquad\quad = (3 + 2)x^2 - 4y$ **Using the distributive law**

$\qquad\qquad\qquad\quad = 5x^2 - 4y$

b) $9x^3 + 5x - 4x^2 - 2x^3 + 5x^2 = 7x^3 + x^2 + 5x$ **We usually perform the middle steps mentally and write just the answer.**

c) $3x^2y + 5xy^2 - 3x^2y - xy^2 = 4xy^2$

The sum of two polynomials can be found by writing a plus sign between them and then combining like terms.

EXAMPLE 9   Add: $(-3x^3 + 2x - 4) + (4x^3 + 3x^2 + 2)$.

Solution

$$(-3x^3 + 2x - 4) + (4x^3 + 3x^2 + 2) = x^3 + 3x^2 + 2x - 2$$

Checking

A graphing calculator can be used to check whether two algebraic expressions given by y_1 and y_2 are equivalent. Some of the methods listed below have been described before; here they are listed together as a summary.

1. *Comparing graphs.* Graph both y_1 and y_2 on the same set of axes using the SEQUENTIAL mode. If the expressions are equivalent, the graphs will be identical. Using the PATH graphstyle for y_2 allows us to see if the graph of y_2 is being traced over the graph of y_1.

2. *Subtracting expressions.* Let $y_3 = y_1 - y_2$. If the expressions are equivalent, the graph of y_3 will be $y = 0$, or the x-axis. Using the PATH graphstyle allows us to see if the graph of y_3 is being traced over the x-axis.

3. *Comparing values.* Viewing a table allows us to compare values of y_1 and y_2. If the expressions are equivalent, the values will be the same for any given x-value. This check becomes more certain as we use more x-values.

4. *Comparing both graphs and values.* Many graphing calculators will split the viewing screen and show both a graph and a table. This choice is usually made using the MODE key. In the bottom line of the screen on the left below, the FULL mode indicates a full-screen table or graph. In the HORIZ mode, the screen is split in half horizontally, and in the G-T mode, the screen is split vertically. In the screen in the middle below, a graph and a table of values are shown for $y = x^2 - 1$ using the HORIZ mode. The screen on the right shows the same graph and table in the G-T mode.

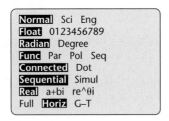 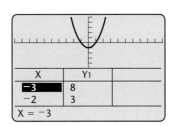 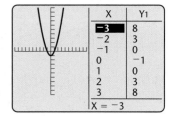

Interactive Discovery

1. The graphs of y_1 and y_2 should be the same.

$y_1 = (-3x^3 + 2x - 4) + (4x^3 + 3x^2 + 2)$,
$y_2 = x^3 + 3x^2 + 2x - 2$

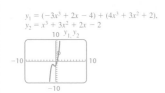

2. The graph of y_3 should be the x-axis.

$y_3 = y_2 - y_1$

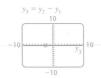

In Example 9, we added polynomials to obtain the identity

$$(-3x^3 + 2x - 4) + (4x^3 + 3x^2 + 2) = x^3 + 3x^2 + 2x - 2.$$

Let $y_1 = (-3x^3 + 2x - 4) + (4x^3 + 3x^2 + 2)$

and $y_2 = x^3 + 3x^2 + 2x - 2.$

1. If the addition is correct, how should the graphs of y_1 and y_2 compare? Check the addition by graphing y_1 and y_2.

2. Now let $y_3 = y_1 - y_2$. If our addition is correct, what will be the graph of y_3? Check Example 9 by graphing y_3.

3. As a third check, compare the values of y_1 and y_2 using a table.

Now check the addition

$$(-2x^4 + 3x^2 + 5) + (6x^4 - 2x - 8) = 4x^4 + 3x^2 - 2x.$$

Let $y_1 = (-2x^4 + 3x^2 + 5) + (6x^4 - 2x - 8)$

and

$y_2 = 4x^4 + 3x^2 - 2x.$

3.

X	Y₁	Y₂
−3	8	8
−2	2	2
−1	2	2
0		
1	4	4
2	22	22
3	58	58

X = −3

4. The graphs appear to coincide.

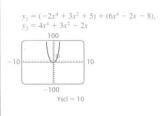

$y_1 = (-2x^4 + 3x^2 + 5) + (6x^4 - 2x - 8),$
$y_2 = 4x^4 + 3x^2 - 2x$

Yscl = 10

5. The table shows that the values of y_1 and y_2 are different for the same x-value.

X	Y₁	Y₂
−3	354	357
−2	77	80

X = −3

6. The graph of y_3 is not the x-axis.

X	Y₃
−3	−3
−2	−3
−1	−3
0	−3
1	−3
2	−3
3	−3

X = −3

It may be easier to see that the difference is not zero than that the graphs do not coincide.

X	Y₁	Y₂
0	−7	−7
1	7	7
2	53	53
3	149	149
4	313	313
5	563	563
6	917	917

X = 0

4. Graph y_1 and y_2 using a viewing window of $[-10, 10, -100, 100]$. Do the graphs appear to coincide?

5. Use a screen split horizontally to compare the graphs with a table of values. Are the values of y_1 and y_2 the same for any given x-value?

6. Graph $y_3 = y_1 - y_2$. Is the graph of y_3 the x-axis? Use a screen split vertically to compare a table of values with the graph. What advantages does this check have over graphing y_1 and y_2?

7. Is the addition $(-2x^4 + 3x^2 + 5) + (6x^4 - 2x - 8) = 4x^4 + 3x^2 - 2x$ correct?   No

Graphs and tables of values provide good checks of the results of algebraic manipulation. When checking by graphing, remember that two different graphs might not appear different in some viewing windows. It is also possible for two different functions to have the same value for several entries in a table. However, when enough values of two polynomial functions are the same, a table does provide a foolproof check.

> If the values of two nth-degree polynomial functions are the same for $n + 1$ or more values, then the polynomials are equivalent.

Using columns for addition is sometimes helpful. To do so, we write the polynomials one under the other, aligning like terms and leaving spaces for missing terms.

EXAMPLE 10   Add: $4x^3 + 4x - 5$ and $-x^3 + 7x^2 - 2$. Check using a table of values.

Solution   We have

$$
\begin{array}{r}
4x^3 + 4x - 5 \\
-x^3 + 7x^2 - 2 \\
\hline
3x^3 + 7x^2 + 4x - 7
\end{array}
$$

To check, we let

$$y_1 = (4x^3 + 4x - 5) + (-x^3 + 7x^2 - 2)$$

and $y_2 = 3x^3 + 7x^2 + 4x - 7.$

Because the polynomials are of degree 3, it suffices to show that $y_1 = y_2$ for four different x-values. This can be accomplished quickly by creating a table of values, as shown at left. The answer checks.

Note that if a graphing calculator were not available, it would suffice to show, manually, that $y_1 = y_2$ for four different x-values.

Opposites and Subtraction

If the sum of two polynomials is 0, the polynomials are *opposites,* or *additive inverses,* of each other. For example,

$$(3x^2 - 5x + 2) + (-3x^2 + 5x - 2) = 0,$$

so the opposite of $(3x^2 - 5x + 2)$ must be $(-3x^2 + 5x - 2)$. We can say the same thing using algebraic symbolism, as follows:

The opposite of $(3x^2 - 5x + 2)$ is $(-3x^2 + 5x - 2)$.

$$- \quad (3x^2 - 5x + 2) \quad = \quad -3x^2 + 5x - 2$$

To form the opposite of a polynomial, we can think of distributing the "−" sign, or multiplying each term of the polynomial by −1, and removing the parentheses. The effect is to change the sign of each term in the polynomial.

TEACHING TIP

To reinforce the importance of parentheses, you may want to discuss the difference between $9x - (2x^2 + 3x - 1)$ and $9x - 2x^2 + 3x - 1$.

The Opposite of a Polynomial

The *opposite* of a polynomial P can be written as $-P$ or, equivalently, by replacing each term with its opposite.

EXAMPLE 11 Write two equivalent expressions for the opposite of
$$7xy^2 - 6xy - 4y + 3.$$

Solution

a) The opposite of $7xy^2 - 6xy - 4y + 3$ can be written as
$$-(7xy^2 - 6xy - 4y + 3). \qquad \text{Writing the opposite of } P \text{ as } -P$$

b) The opposite of $7xy^2 - 6xy - 4y + 3$ can also be written as
$$-7xy^2 + 6xy + 4y - 3. \qquad \text{Multiplying each term by } -1 \text{ and removing parentheses}$$

To subtract a polynomial, we add its opposite.

EXAMPLE 12 Subtract: $(-5x^2 + 4) - (2x^2 + 3x - 1)$.

Solution We have
$$(-5x^2 + 4) - (2x^2 + 3x - 1)$$
$$= (-5x^2 + 4) + (-2x^2 - 3x + 1) \qquad \text{Adding the opposite of the polynomial being subtracted}$$
$$= -7x^2 - 3x + 5.$$

We can check the result using graphs or a table of values.

With practice, you will find that you can skip more steps, by mentally taking the opposite of each term and then combining like terms. Eventually, all you will write is the answer.

To use columns for subtraction, we mentally change the signs of the terms being subtracted.

EXAMPLE 13 Subtract:

$$(4x^2y - 6x^3y^2 + x^2y^2) - (4x^2y + x^3y^2 + 3x^2y^3 - 8x^2y^2).$$

Solution

Write: (Subtract)

$$\begin{array}{r} 4x^2y - 6x^3y^2 \qquad\quad + x^2y^2 \\ -(4x^2y + \ x^3y^2 + 3x^2y^3 - 8x^2y^2) \\ \hline \end{array}$$

Think: (Add)

$$\begin{array}{r} 4x^2y - 6x^3y^2 \qquad\quad + x^2y^2 \\ -4x^2y - \ x^3y^2 - 3x^2y^3 + 8x^2y^2 \\ \hline -7x^3y^2 - 3x^2y^3 + 9x^2y^2 \end{array}$$

Take the opposite of each term mentally and add.

5.1

Exercise Set

Determine the degree of each term and the degree of the polynomial.

1. $-6x^5 - 8x^3 + x^2 + 3x - 4$ 5, 3, 2, 1, 0; 5

2. $t^3 - 5t^2 + 2t + 7$ 3, 2, 1, 0; 3

3. $y^3 + 2y^7 + x^2y^4 - 8$ 3, 7, 6, 0; 7

4. $-2u^2 + 3v^5 - u^3v^4 - 7$ 2, 5, 7, 0; 7

5. $a^5 + 4a^2b^4 + 6ab + 4a - 3$ 5, 6, 2, 1, 0; 6

6. $8p^6 + 2p^4t^4 - 7p^3t + 5p^2 - 14$ 6, 8, 4, 2, 0; 8

Arrange in descending order. Then find the leading term and the leading coefficient.

7. $19 - 4y^3 + 7y - 6y^2$ $-4y^3 - 6y^2 + 7y + 19;\ -4y^3;\ -4$

8. $3 - 5y + 6y^2 + 11y^3 - 18y^4$ $-18y^4 + 11y^3 + 6y^2 - 5y + 3;\ -18y^4;\ -18$

9. $5x^2 + 3x^7 - x + 12$ $3x^7 + 5x^2 - x + 12;\ 3x^7;\ 3$

10. $9 - 3x - 10x^4 + 7x^2$ $-10x^4 + 7x^2 - 3x + 9;\ -10x^4;\ -10$

11. $a + 5a^3 - a^7 - 19a^2 + 8a^5$ $-a^7 + 8a^5 + 5a^3 - 19a^2 + a;\ -a^7;\ -1$

12. $a^3 - 7 + 11a^4 + a^9 - 5a^2$ $a^9 + 11a^4 + a^3 - 5a^2 - 7;\ a^9;\ 1$

Arrange in ascending powers of x.

13. $6x - 9 + 3x^4 - 5x^2$ $-9 + 6x - 5x^2 + 3x^4$

14. $-3x^4 + 4x - x^3 + 9$ $9 + 4x - x^3 - 3x^4$

15. $7x^3y + 3xy^3 + x^2y^2 - 5x^4$ $3xy^3 + x^2y^2 + 7x^3y - 5x^4$

16. $5x^2y^2 - 9xy + 8x^3y^2 - 5x^4$ $-9xy + 5x^2y^2 + 8x^3y^2 - 5x^4$

17. $4ax - 7ab + 4x^6 - 7ax^2$ $-7ab + 4ax - 7ax^2 + 4x^6$

18. $5xy^8 - 3ax^5 + 4ax^3 - 12a + 5x^5$ $-12a + 5xy^8 + 4ax^3 + 5x^5 - 3ax^5$

Find the specified function values.

19. Find $P(4)$ and $P(0)$: $P(x) = 3x^2 - 2x + 7$.   47; 7

20. Find $Q(3)$ and $Q(-1)$: $Q(x) = -4x^3 + 7x^2 - 6$. $-51;\ 5$

21. Find $P(-2)$ and $P(\frac{1}{3})$: $P(y) = 8y^3 - 12y - 5$. $-45;\ -8\frac{19}{27}$

22. Find $Q(-3)$ and $Q(0)$:   282; -9
$$Q(y) = -8y^3 + 7y^2 - 4y - 9.$$

Evaluate each polynomial for x = 4.

23. $-7x + 5$ -23

24. $4x - 13$ 3

25. $x^3 - 5x^2 + x$ -12

26. $7 - x + 3x^2$ 51

Evaluate each polynomial function for x = −1.

27. $f(x) = -5x^3 + 3x^2 - 4x - 3$ 9

28. $g(x) = -4x^3 + 2x^2 + 5x - 7$ −6

Electing Officers. *For a club consisting of n people, the number of ways in which a president, vice president, and treasurer can be elected can be determined using the function given by*

$$p(n) = n^3 - 3n^2 + 2n.$$

29. The Southside Rugby Club has 20 members. In how many ways can they elect a president, vice president, and treasurer?   6840

30. The Stage Right drama club has 12 members. In how many ways can a president, vice president, and treasurer be elected?   1320

Falling Distance. *The distance s(t), in feet, traveled by a body falling freely from rest in t seconds is approximated by the function given by*

$$s(t) = 16t^2.$$

31. A paintbrush falls from a scaffold and takes 3 sec to hit the ground. How high is the scaffold?   144 ft

$s(t) = \mathbf{16}t^2$

32. A stone is dropped from the Briar Cliff lookout and takes 5 sec to hit the ground. How high is the cliff?   400 ft

Total Revenue. *An electronics firm is marketing a new kind of DVD player. The firm determines that when it sells x DVD players, its total revenue is*

$$R(x) = 280x - 0.4x^2 \text{ dollars}.$$

33. What is the total revenue from the sale of 75 DVD players? $18,750

34. What is the total revenue from the sale of 100 DVD players? $24,000

Total Cost. *The electronics firm determines that the total cost, in dollars, of producing x DVD players is given by*

$$C(x) = 5000 + 0.6x^2.$$

35. What is the total cost of producing 75 DVD players? $8375

36. What is the total cost of producing 100 DVD players? $11,000

Daily Accidents. *The number of daily accidents (the average number of accidents per day) involving drivers of age a is approximated by the polynomial function*

$$P(a) = 0.4a^2 - 40a + 1039.$$

37. Find the number of daily accidents involving a 20-year-old driver.   399 accidents

38. Find the number of daily accidents involving a 25-year-old driver.   289 accidents

NASCAR Attendance. *Attendance at NASCAR auto races has grown rapidly over the past 10 years. Attendance A, in millions, can be approximated by the polynomial function given by*

$$A(x) = 0.0024x^3 - 0.005x^2 + 0.31x + 3,$$

where x is the number of years since 1989. Use the following graph for Exercises 39–42.

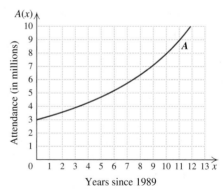

Source: NASCAR

39. Estimate the attendance at NASCAR races in 1995.   5.2 million

40. Estimate the attendance at NASCAR races in 1999.   8 million

41. Approximate $A(8)$.   6.4 million

42. Approximate $A(12)$.   10.1 million

Medicine. *Ibuprofen is a medication used to relieve pain. The polynomial function*

$$M(t) = 0.5t^4 + 3.45t^3 - 96.65t^2 + 347.7t,$$
$$0 \le t \le 6$$

can be used to estimate the number of milligrams of ibuprofen in the bloodstream t hours after 400 mg of the medication has been swallowed (Source: Based on data from Dr. P. Carey, Burlington, VT). Use the following graph for Exercises 43–46.

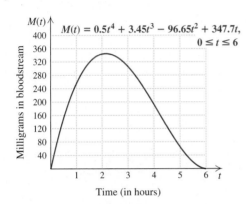

$M(t) = 0.5t^4 + 3.45t^3 - 96.65t^2 + 347.7t,$
$0 \le t \le 6$

43. Use the graph above to estimate the number of milligrams of ibuprofen in the bloodstream 2 hr after 400 mg has been swallowed. About 340 mg

44. Use the graph above to estimate the number of milligrams of ibuprofen in the bloodstream 4 hr after 400 mg has been swallowed. About 185 mg

45. Approximate the range of M. $[0, 345]$

46. Approximate the domain of M. $[0, 6]$

Surface Area of a Right Circular Cylinder. *The surface area of a right circular cylinder is given by the polynomial*

$$2\pi rh + 2\pi r^2,$$

where h is the height, r is the radius of the base, and h and r are given in the same units.

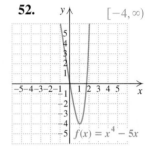

47. A 16-oz beverage can has height 6.3 in. and radius 1.2 in. Find the surface area of the can. (Use a calculator with a $\boxed{\pi}$ key or use 3.141592654 for π.) 56.5 in²

48. A 12-oz beverage can has height 4.7 in. and radius 1.2 in. Find the surface area of the can. (Use a calculator with a $\boxed{\pi}$ key or use 3.141592654 for π.) 44.5 in²

Estimate the range of each function from its graph.

49. $(-\infty, 3]$

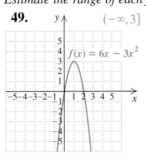

$f(x) = 6x - 3x^2$

50. $(-\infty, \infty)$

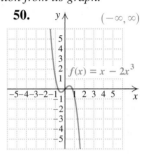

$f(x) = x - 2x^3$

51. $(-\infty, \infty)$

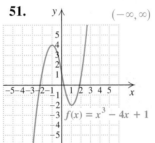

$f(x) = x^3 - 4x + 1$

52. $[-4, \infty)$

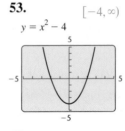

$f(x) = x^4 - 5x$

53. $[-4, \infty)$

$y = x^2 - 4$

54. $(-\infty, \infty)$

$y = 0.3x^3$

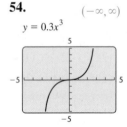

55. $[-65, \infty)$

$y = x^4 - 5x^3 + x - 2$

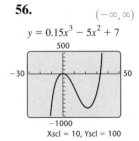

Yscl = 10

56. $(-\infty, \infty)$

$y = 0.15x^3 - 5x^2 + 7$

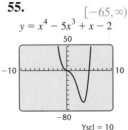

Xscl = 10, Yscl = 100

Use a graphing calculator to graph each polynomial function in the indicated viewing window, and estimate its range.

57. $f(x) = x^2 + 2x + 1, [-10, 10, -10, 10]$ $[0, \infty)$

58. $p(x) = x^2 + x - 6, [-10, 10, -10, 10]$ $[-6.25, \infty)$

59. $q(x) = -2x^2 + 5, [-10, 10, -10, 10]$ $(-\infty, 5]$

60. $g(x) = 1 - x^2, [-10, 10, -10, 10]$ $(-\infty, 1]$

61. $p(x) = -2x^3 + x + 5, [-10, 10, -10, 10]$ $(-\infty, \infty)$

62. $f(x) = -x^4 + 2x^3 - 10$, $[-5, 5, -30, 10]$,
Yscl = 5 $\quad (-\infty, -8.3]$

63. $g(x) = x^4 + 2x^3 - 5$, $[-5, 5, -10, 10]$ $\quad [-6.7, \infty)$

64. $q(x) = x^5 - 2x^4$, $[-5, 5, -5, 5]$ $\quad (-\infty, \infty)$

Combine like terms.

65. $5a + 6 - 4 + 2a^3 - 6a + 2$ $\quad 2a^3 - a + 4$

66. $6x + 13 - 8 - 7x + 5x^2 + 10$ $\quad 5x^2 - x + 15$

67. $3a^2b + 4b^2 - 9a^2b - 7b^2$ $\quad -6a^2b - 3b^2$

68. $5x^2y^2 + 4x^3 - 8x^2y^2 - 12x^3$ $\quad -8x^3 - 3x^2y^2$

69. $9x^2 - 3xy + 12y^2 + x^2 - y^2 + 5xy + 4y^2$
$\quad 10x^2 + 2xy + 15y^2$

70. $a^2 - 2ab + b^2 + 9a^2 + 5ab - 4b^2 + a^2$
$\quad 11a^2 + 3ab - 3b^2$

Add.

71. $(8a + 6b - 3c) + (4a - 2b + 2c)$ $\quad 12a + 4b - c$

72. $(7x - 5y + 3z) + (9x + 12y - 8z)$ $\quad 16x + 7y - 5z$

73. $(a^2 - 3b^2 + 4c^2) + (-5a^2 + 2b^2 - c^2)$
$\quad -4a^2 - b^2 + 3c^2$

74. $(x^2 - 5y^2 - 9z^2) + (-6x^2 + 9y^2 - 2z^2)$
$\quad -5x^2 + 4y^2 - 11z^2$

75. $(x^2 + 2x - 3xy - 7) + (-3x^2 - x + 2xy + 6)$
$\quad -2x^2 + x - xy - 1$

76. $(3a^2 - 2b + ab + 6) + (-a^2 + 5b - 5ab - 2)$
$\quad 2a^2 + 3b - 4ab + 4$

77. $(8x^2y - 3xy^2 + 4xy) + (-2x^2y - xy^2 + xy)$
$\quad 6x^2y - 4xy^2 + 5xy$

78. $(9ab - 3ac + 5bc) + (13ab - 15ac - 8bc)$
$\quad 22ab - 18ac - 3bc$

79. $(2r^2 + 12r - 11) + (6r^2 - 2r + 4) + (r^2 - r - 2)$
$\quad 9r^2 + 9r - 9$

80. $(5x^2 + 19x - 23) + (-7x^2 - 11x + 12) + (-x^2 - 9x + 8)$ $\quad -3x^2 - x - 3$

81. $\left(\frac{1}{8}xy - \frac{3}{5}x^3y^2 + 4.3y^3\right) + \left(-\frac{1}{3}xy - \frac{3}{4}x^3y^2 - 2.9y^3\right)$

82. $\left(\frac{2}{3}xy + \frac{5}{6}xy^2 + 5.1x^2y\right) + \left(-\frac{4}{5}xy + \frac{3}{4}xy^2 - 3.4x^2y\right)$

Write two equivalent expressions for the opposite, or additive inverse, of each polynomial.

83. $5x^3 - 7x^2 + 3x - 9$ $\quad -(5x^3 - 7x^2 + 3x - 9)$,
$\quad -5x^3 + 7x^2 - 3x + 9$

84. $-8y^4 - 18y^3 + 4y - 7$ $\quad -(-8y^4 - 18y^3 + 4y - 7)$,
$\quad 8y^4 + 18y^3 - 4y + 7$

85. $-12y^5 + 4ay^4 - 7by^2$ $\quad -(-12y^5 + 4ay^4 - 7by^2)$,
$\quad 12y^5 - 4ay^4 + 7by^2$

86. $7ax^3y^2 - 8by^4 - 7abx - 12ay$
$\quad -(7ax^3y^2 - 8by^4 - 7abx - 12ay)$,
$\quad -7ax^3y^2 + 8by^4 + 7abx + 12ay$

Subtract.

87. $(7x - 5) - (-3x + 4)$ $\quad 10x - 9$

88. $(8y + 2) - (-6y - 5)$ $\quad 14y + 7$

89. $(-3x^2 + 2x + 9) - (x^2 + 5x - 4)$ $\quad -4x^2 - 3x + 13$

90. $(-9y^2 + 4y + 8) - (4y^2 + 2y - 3)$ $\quad -13y^2 + 2y + 11$

91. $(6a - 2b + c) - (3a + 2b - 2c)$ $\quad 3a - 4b + 3c$

92. $(7x - 4y + z) - (4x + 6y - 3z)$ $\quad 3x - 10y + 4z$

93. $(3x^2 - 2x - x^3) - (5x^2 - 8x - x^3)$ $\quad -2x^2 + 6x$

81. $-\frac{5}{24}xy - \frac{27}{20}x^3y^2 + 1.4y^3$ **82.** $-\frac{2}{15}xy + \frac{19}{12}xy^2 + 1.7x^2y$

94. $(8y^2 - 3y - 4y^3) - (3y^2 - 9y - 7y^3)$
$\quad 5y^2 + 6y + 3y^3$

95. $(5a^2 + 4ab - 3b^2) - (9a^2 - 4ab + 2b^2)$
$\quad -4a^2 + 8ab - 5b^2$

96. $(7y^2 - 14yz - 8z^2) - (12y^2 - 8yz + 4z^2)$
$\quad -5y^2 - 6yz - 12z^2$

97. $(6ab - 4a^2b + 6ab^2) - (3ab^2 - 10ab - 12a^2b)$
$\quad 8a^2b + 16ab + 3ab^2$

98. $(10xy - 4x^2y^2 - 3y^3) - (-9x^2y^2 + 4y^3 - 7xy)$
$\quad 17xy + 5x^2y^2 - 7y^3$

99. $\left(\frac{5}{8}x^4 - \frac{1}{4}x^2 - \frac{1}{2}\right) - \left(-\frac{3}{8}x^4 + \frac{3}{4}x^2 + \frac{1}{2}\right)$ $\quad x^4 - x^2 - 1$

100. $\left(\frac{5}{6}y^4 - \frac{1}{2}y^2 - 7.8y\right) - \left(-\frac{3}{8}y^4 + \frac{3}{4}y^2 + 3.4y\right)$

Total Profit. Total profit is defined as total revenue minus total cost. In Exercises 101 and 102, $R(x)$ and $C(x)$ are the revenue and cost, respectively, from the sale of x futons.

101. If $R(x) = 280x - 0.4x^2$ and $C(x) = 5000 + 0.6x^2$, find the profit from the sale of 70 futons. $\quad$ \$9700

102. If $R(x) = 280x - 0.7x^2$ and $C(x) = 8000 + 0.5x^2$, find the profit from the sale of 100 futons. $\quad$ \$8000

In Exercises 103–106, tell which of the following statements are true for the given polynomials.

a) *The graphs of y_1 and y_2 coincide.*
b) *The graphs of y_1 and y_2 differ.*
c) $y_1 - y_2 = 0$
d) $y_1 - y_2 \neq 0$

103. $y_1 = (5x^3 + 2x + 3) + (3x^3 - 1)$,
$y_2 = 8x^3 + 2x + 2$ $\quad$ (a) and (c)

104. $y_1 = (7x^2 - 9x + 3) - (x^3 - 9x + 3)$,
$y_2 = -x^3 + 7x^2$ $\quad$ (a) and (c)

105. $y_1 = (x^2 + 8x + 1) - (-x^2 + 3x + 5)$,
$y_2 = 5x - 4$ $\quad$ (b) and (d)

106. $y_1 = (x^4 + x^2 + 4) + (2x^3 - 3x^2 - 7)$,
$y_2 = 3x^4 - 2x^2 - 3$ $\quad$ (b) and (d)

TW 107. Is the sum of two binomials always a binomial? Why or why not?

TW 108. Ani claims that she can add any two polynomials but finds subtraction difficult. What advice would you offer her?

Skill Maintenance

Simplify.

109. $2(x + 3) + 5(x + 2)$ [1.3] $\quad 7x + 16$

110. $7(a + 2) + 3(a + 15)$ [1.3] $\quad 10a + 59$

111. $a(a - 1) + 4(a - 1)$ [1.3] $\quad a^2 + 3a - 4$

112. $x(x - 3) + 2(x - 1)$ [1.3] $\quad x^2 - x - 2$

113. $x^5 \cdot x^4$ [1.4] $\quad x^9$

114. $a^2 \cdot a^6$ [1.4] $\quad a^8$

100. $\frac{29}{24}y^4 - \frac{5}{4}y^2 - 11.2y$

Synthesis

TW **115.** Write a problem in which revenue and cost functions are given and a profit function, $P(x)$, is required. Devise the problem so that $P(0) < 0$ and $P(100) > 0$.

TW **116.** A student who is trying to graph

$$p(x) = 0.05x^4 - x^2 + 5$$

gets the following screen. How can the student tell at a glance that a mistake has been made?

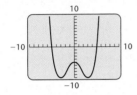

For $P(x)$ and $Q(x)$ as given, find the following.

$$P(x) = 13x^5 - 22x^4 - 36x^3 + 40x^2 - 16x + 75,$$
$$Q(x) = 42x^5 - 37x^4 + 50x^3 - 28x^2 + 34x + 100$$

117. $2[P(x)] + Q(x)$

118. $3[P(x)] - Q(x)$

119. $2[Q(x)] - 3[P(x)]$

120. $4[P(x)] + 3[Q(x)]$

121. *Volume of a Display.*   The number of spheres in a triangular pyramid with x layers is given by the function

$$N(x) = \tfrac{1}{6}x^3 + \tfrac{1}{2}x^2 + \tfrac{1}{3}x.$$

The volume of a sphere of radius r is given by the function

$$V(r) = \tfrac{4}{3}\pi r^3,$$

where π can be approximated as 3.14.

Chocolate Heaven has a window display of truffles piled in a triangular pyramid formation 5 layers deep. (See the figure at the top of the next column.) If the diameter of each truffle is 3 cm, find the volume of chocolate in the display.
494.55 cm³

122. If one large truffle were to have the same volume as the display of truffles in Exercise 121, what would be its diameter?   About 9.8 cm

123. Find a polynomial function that gives the outside surface area of a box like this one, with an open top and dimensions as shown. $5x^2 - 8x$

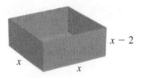

124. Develop a formula for the surface area of a right circular cylinder in which h is the height, in *centimeters*, and r is the radius, in *meters*. (See Exercises 47 and 48.) $200\pi rh + 20{,}000\pi r^2$ cm², or $0.02\pi rh + 2\pi r^2$ m²

Perform the indicated operation. Assume that the exponents are natural numbers.

125. $(2x^{2a} + 4x^a + 3) + (6x^{2a} + 3x^a + 4)$
$8x^{2a} + 7x^a + 7$

126. $(3x^{6a} - 5x^{5a} + 4x^{3a} + 8) -$
$(2x^{6a} + 4x^{4a} + 3x^{3a} + 2x^{2a})$
$x^{6a} - 5x^{5a} - 4x^{4a} + x^{3a} - 2x^{2a} + 8$

127. $(2x^{5b} + 4x^{4b} + 3x^{3b} + 8) -$
$(x^{5b} + 2x^{3b} + 6x^{2b} + 9x^b + 8)$
$x^{5b} + 4x^{4b} + x^{3b} - 6x^{2b} - 9x^b$

117. $68x^5 - 81x^4 - 22x^3 + 52x^2 + 2x + 250$ **118.** $-3x^5 - 29x^4 - 158x^3 + 148x^2 - 82x + 125$
119. $45x^5 - 8x^4 + 208x^3 - 176x^2 + 116x - 25$ **120.** $178x^5 - 199x^4 + 6x^3 + 76x^2 + 38x + 600$

Collaborative Corner

How Many Handshakes?

Focus: Polynomial functions
Time: 20 minutes
Group size: 5

ACTIVITY

1. All group members should shake hands with each other. Without "double counting," determine how many handshakes occurred.

2. Complete the table in the next column.

3. Join another group to determine the number of handshakes for a group of size 10.

4. Try to find a function of the form $H(n) = an^2 + bn$, for which $H(n)$ is the number of different handshakes that are possible in a group of n people. Make sure that $H(n)$

Group Size	Number of Handshakes
1	
2	
3	
4	
5	

produces all of the values in the table above. (*Hint*: Use the table to twice select n and $H(n)$. Then solve the resulting system of equations for a and b.)

5.2

Multiplication of Polynomials

Multiplying Monomials ■ Multiplying Monomials and Binomials ■ Multiplying Any Two Polynomials ■ The Product of Two Binomials: FOIL ■ Squares of Binomials ■ Products of Sums and Differences ■ Function Notation

Just like numbers, polynomials can be multiplied. The product of two polynomials $P(x)$ and $Q(x)$ is a polynomial $R(x)$ that gives the same value as $P(x) \cdot Q(x)$ for any replacement of x.

Multiplying Monomials

To multiply monomials, we first multiply their coefficients. Then we multiply the variables using the rules for exponents and the commutative and associative laws. With practice, we can work mentally, writing only the answer.

EXAMPLE 1 Multiply and simplify.

a) $(-8x^4y^7)(5x^3y^2)$
b) $(-2x^2yz^5)(-6x^5y^{10}z^2)$

Solution

a) $(-8x^4y^7)(5x^3y^2) = -8 \cdot 5 \cdot x^4 \cdot x^3 \cdot y^7 \cdot y^2$ **Using the associative and commutative laws**

$$= -40x^{4+3}y^{7+2}$$ **Multiplying coefficients; adding exponents**

$$= -40x^7y^9$$

b) $(-2x^2yz^5)(-6x^5y^{10}z^2) = (-2)(-6) \cdot x^2 \cdot x^5 \cdot y \cdot y^{10} \cdot z^5 \cdot z^2$

$$= 12x^7y^{11}z^7$$ **Multiplying coefficients; adding exponents**

Multiplying Monomials and Binomials

The distributive law is the basis for multiplying polynomials other than monomials. We first multiply a monomial and a binomial.

EXAMPLE 2   Multiply: **(a)** $2x(3x - 5)$; **(b)** $3a^2b(a^2 - b^2)$.

Solution

a) $2x(3x - 5) = 2x \cdot 3x - 2x \cdot 5$ **Using the distributive law**

$$= 6x^2 - 10x$$ **Multiplying monomials**

b) $3a^2b(a^2 - b^2) = 3a^2b \cdot a^2 - 3a^2b \cdot b^2$ **Using the distributive law**

$$= 3a^4b - 3a^2b^3$$

We can use graphs or tables to check the multiplication of polynomials in one variable. For example, we can check the product found in Example 2(a) by letting

$$y_1 = 2x(3x - 5)$$

and $y_2 = 6x^2 - 10x.$

The following table of values for y_1 and y_2 shows that the answer found is correct.

X	Y₁	Y₂
0	0	0
1	−4	−4
2	4	4
3	24	24
4	56	56
5	100	100
6	156	156
X = 0		

The distributive law is also used when multiplying two binomials. In this case, however, we begin by distributing a *binomial* rather than a monomial. With practice, some of the following steps can be combined.

EXAMPLE 3 Multiply: $(y^3 - 5)(2y^3 + 4)$.

Solution

$$(y^3 - 5)(2y^3 + 4) = (y^3 - 5)2y^3 + (y^3 - 5)4 \qquad \text{"Distributing"}$$
$$\text{the } y^3 - 5$$

$$= 2y^3(y^3 - 5) + 4(y^3 - 5) \qquad \textbf{Using the commutative law for multiplication. Try to do this step mentally.}$$

$$= 2y^3 \cdot y^3 - 2y^3 \cdot 5 + 4 \cdot y^3 - 4 \cdot 5 \qquad \textbf{Using the distributive law (twice)}$$

$$= 2y^6 - 10y^3 + 4y^3 - 20 \qquad \textbf{Multiplying the monomials}$$

$$= 2y^6 - 6y^3 - 20 \qquad \textbf{Combining like terms}$$

TEACHING TIP

Lead students to realize that there are four separate multiplications of terms.

Multiplying Any Two Polynomials

Repeated use of the distributive law enables us to multiply *any* two polynomials, regardless of how many terms are in each.

EXAMPLE 4 Multiply: $(p + 2)(p^4 - 2p^3 + 3)$.

Solution By the distributive law, we have

$$(p + 2)(p^4 - 2p^3 + 3)$$
$$= (p + 2)(p^4) - (p + 2)(2p^3) + (p + 2)(3)$$
$$= p^4(p + 2) - 2p^3(p + 2) + 3(p + 2) \qquad \textbf{Using a commutative law}$$
$$= p^4 \cdot p + p^4 \cdot 2 - 2p^3 \cdot p - 2p^3 \cdot 2 + 3 \cdot p + 3 \cdot 2$$
$$= p^5 + 2p^4 - 2p^4 - 4p^3 + 3p + 6$$
$$= p^5 - 4p^3 + 3p + 6. \qquad \textbf{Combining like terms}$$

In order for these polynomials to be equivalent, they must have the same value for at least 6 x-values, because the degree is 5. The table at left shows that this is true.

$y_1 = (x + 2)(x^4 - 2x^3 + 3)$,
$y_2 = x^5 - 4x^3 + 3x + 6$

X	Y₁	Y₂
0	6	6
1	6	6
2	12	12
3	150	150
4	786	786
5	2646	2646
6	6936	6936

X = 0

The Product of Two Polynomials

The *product* of two polynomials $P(x)$ and $Q(x)$ is found by multiplying each term of $P(x)$ by every term of $Q(x)$ and then combining like terms.

It is also possible to stack the polynomials, multiplying each term at the top by every term below, keeping like terms in columns, and leaving spaces for missing terms. Then we add just as we do in long multiplication with numbers.

EXAMPLE 5   Multiply: $(5x^3 + x - 4)(-2x^2 + 3x + 6)$.

Solution   We have

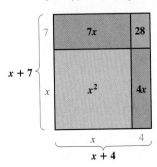

$y_1 = (5x^3 + x - 4)(-2x^2 + 3x + 6)$,
$y_2 = -10x^5 + 15x^4 + 28x^3 + 11x^2 - 6x - 24$,
$y_3 = y_1 - y_2$

$$
\begin{array}{r}
5x^3 + x - 4 \\
-2x^2 + 3x + 6 \\
\hline
30x^3 + 6x - 24 \\
15x^4 + 3x^2 - 12x \\
-10x^5 - 2x^3 + 8x^2 \\
\hline
-10x^5 + 15x^4 + 28x^3 + 11x^2 - 6x - 24.
\end{array}
$$

Multiplying by 6
Multiplying by 3x
Multiplying by $-2x^2$
Adding

We check by showing that the difference of the original expression and the resulting product is 0, as shown in the graph and table at left.

EXAMPLE 6   Multiply $4x^4y - 7x^2y + 3y$ by $2y - 3x^2y$.

Solution

$$
\begin{array}{r}
4x^4y - 7x^2y + 3y \\
-3x^2y + 2y \\
\hline
8x^4y^2 - 14x^2y^2 + 6y^2 \\
-12x^6y^2 + 21x^4y^2 - 9x^2y^2 \\
\hline
-12x^6y^2 + 29x^4y^2 - 23x^2y^2 + 6y^2
\end{array}
$$

Writing descending powers of x
Multiplying by 2y
Multiplying by $-3x^2y$
Adding

The Product of Two Binomials: FOIL

We now consider what are called *special products*. These products of polynomials occur often and can be simplified using shortcuts that we now develop.

To find a faster special-product rule for the product of two binomials, consider $(x + 7)(x + 4)$. We multiply each term of $(x + 7)$ by each term of $(x + 4)$:

$$(x + 7)(x + 4) = x \cdot x + x \cdot 4 + 7 \cdot x + 7 \cdot 4.$$

This multiplication illustrates a pattern that occurs whenever two binomials are multiplied:

A visualization of
$(x + 7)(x + 4)$ using areas

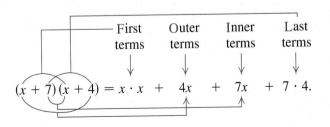

We use the mnemonic device FOIL to remember this method for multiplying.

The FOIL Method

To multiply two binomials $A + B$ and $C + D$, multiply the First terms
AC, the Outer terms AD, the Inner Terms BC, and then the Last terms
BD. Then combine like terms, if possible.

$$(A + B)(C + D) = AC + AD + BC + BD$$

1. Multiply First terms: AC.
2. Multiply Outer terms: AD.
3. Multiply Inner terms: BC.
4. Multiply Last terms: BD.
$$\downarrow$$
$$\text{FOIL}$$

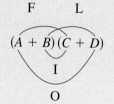

EXAMPLE 7   Multiply.

a) $(x + 5)(x - 8)$ **b)** $(2x + 3y)(x - 4y)$ **c)** $(5xy + 2x)(x^2 + 2xy^2)$

Solution

$$\quad\quad\quad\quad\quad\quad\text{F}\quad\text{O}\quad\text{I}\quad\text{L}$$
a) $(x + 5)(x - 8) = x^2 - 8x + 5x - 40$
$$= x^2 - 3x - 40 \quad\quad \textbf{Combining like terms}$$

b) $(2x + 3y)(x - 4y) = 2x^2 - 8xy + 3xy - 12y^2 \quad\quad \textbf{Using FOIL}$
$$= 2x^2 - 5xy - 12y^2 \quad\quad \textbf{Combining like terms}$$

c) $(5xy + 2x)(x^2 + 2xy^2) = 5x^3y + 10x^2y^3 + 2x^3 + 4x^2y^2$

$$\textbf{There are no like terms to combine.}$$

Squares of Binomials

**Interactive
Discovery**

Determine which of the following are identities.

1. $(x + 3)^2 = x^2 + 9$ Not an identity
2. $(x + 3)^2 = x^2 + 6x + 9$ Identity
3. $(x - 3)^2 = x^2 - 6x - 9$ Not an identity
4. $(x - 3)^2 = x^2 - 9$ Not an identity
5. $(x - 3)^2 = x^2 - 6x + 9$ Identity

A fast method for squaring binomials can be developed using FOIL:

$$(A + B)^2 = (A + B)(A + B)$$
$$= A^2 + AB + AB + B^2 \quad\quad \textbf{Note that } AB \textbf{ occurs twice.}$$
$$= A^2 + 2AB + B^2;$$

$$(A - B)^2 = (A - B)(A - B)$$
$$= A^2 - AB - AB + B^2 \qquad \text{Note that } -AB \text{ occurs twice.}$$
$$= A^2 - 2AB + B^2.$$

A visualization of $(A + B)^2$ **using areas**

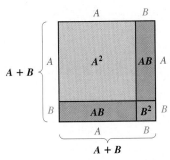

Squaring a Binomial

$$(A + B)^2 = A^2 + 2AB + B^2;$$
$$(A - B)^2 = A^2 - 2AB + B^2$$

The square of a binomial is the square of the first term, plus twice the product of the two terms, plus the square of the last term.

It can help to remember the words of the rules and say them while multiplying.

EXAMPLE 8 Multiply: **(a)** $(y - 5)^2$; **(b)** $(2x + 3y)^2$; **(c)** $\left(\frac{1}{2}x - 3y^4\right)^2$.

Solution

$$(A - B)^2 = A^2 - 2 \cdot A \cdot B + B^2$$

a) $(y - 5)^2 = y^2 - 2 \cdot y \cdot 5 + 5^2$ Note that $-2 \cdot y \cdot 5$ is twice the product of y and -5.
$$= y^2 - 10y + 25$$

b) $(2x + 3y)^2 = (2x)^2 + 2 \cdot 2x \cdot 3y + (3y)^2$
$$= 4x^2 + 12xy + 9y^2 \qquad \text{Raising a product to a power}$$

c) $\left(\frac{1}{2}x - 3y^4\right)^2 = \left(\frac{1}{2}x\right)^2 - 2 \cdot \frac{1}{2}x \cdot 3y^4 + (3y^4)^2$

$$2 \cdot \frac{1}{2}x \cdot (-3y^4) = -2 \cdot \frac{1}{2}x \cdot 3y^4$$

$$= \frac{1}{4}x^2 - 3xy^4 + 9y^8 \qquad \text{Raising a product to a power; multiplying exponents}$$

CAUTION! Note that $(y - 5)^2 \neq y^2 - 5^2$. (To see this, replace y with 6 and note that $(6 - 5)^2 = 1^2 = 1$ and $6^2 - 5^2 = 36 - 25 = 11$.) More generally.

$$(A + B)^2 \neq A^2 + B^2 \quad \text{and} \quad (A - B)^2 \neq A^2 - B^2.$$

Products of Sums and Differences

Another pattern emerges when we are multiplying a sum and a difference of the same two terms.

Interactive Discovery

Determine which of the following are identities.

1. $(x + 3)(x - 3) = x^2 - 9$ Identity
2. $(x + 3)(x - 3) = x^2 + 9$ Not an identity
3. $(x + 3)(x - 3) = x^2 - 6x + 9$ Not an identity
4. $(x + 3)(x - 3) = x^2 + 6x + 9$ Not an identity

The pattern you may have observed is true in general. Note the following:

$$(A + B)(A - B) = A^2 - AB + AB - B^2$$
$$= A^2 - B^2. \qquad -AB + AB = 0$$

Study Tip

Spending extra time studying the special products in this section will help you in Sections 5.4–5.6.

The Product of a Sum and a Difference

$$(A + B)(A - B) = A^2 - B^2$$

This is called a *difference of two squares*.

The product of the sum and difference of the same two terms is the square of the first term minus the square of the second term.

EXAMPLE 9   Multiply.

a) $(y + 5)(y - 5)$
b) $(2xy^2 + 3x)(2xy^2 - 3x)$
c) $(0.2t - 1.4m)(0.2t + 1.4m)$
d) $\left(\frac{2}{3}n - m^3\right)\left(\frac{2}{3}n + m^3\right)$

Solution

$$(A + B)(A - B) = A^2 - B^2$$

a) $(y + 5)(y - 5) = y^2 - 5^2$ **Replacing A with y and B with 5**
$$= y^2 - 25$$ **Try to do problems like this mentally.**

b) $(2xy^2 + 3x)(2xy^2 - 3x) = (2xy^2)^2 - (3x)^2$
$$= 4x^2y^4 - 9x^2$$ **Raising a product to a power**

c) $(0.2t - 1.4m)(0.2t + 1.4m) = (0.2t)^2 - (1.4m)^2$
$$= 0.04t^2 - 1.96m^2$$

d) $\left(\frac{2}{3}n - m^3\right)\left(\frac{2}{3}n + m^3\right) = \left(\frac{2}{3}n\right)^2 - (m^3)^2$
$$= \frac{4}{9}n^2 - m^6$$

EXAMPLE 10   Multiply using the special products developed in this section.

a) $(5y + 4 + 3x)(5y + 4 - 3x)$

b) $(3xy^2 + 4y)(-3xy^2 + 4y)$

c) $(a - 5b)(a + 5b)(a^2 - 25b^2)$

Solution

a) $(5y + 4 + 3x)(5y + 4 - 3x) = (5y + 4)^2 - (3x)^2$ **Try to be alert for situations like this.**

$$= 25y^2 + 40y + 16 - 9x^2$$

We can also multiply $(5y + 4 + 3x)(5y + 4 - 3x)$ using columns, but not as quickly.

b) $(3xy^2 + 4y)(-3xy^2 + 4y) = (4y + 3xy^2)(4y - 3xy^2)$ **Rewriting**

$$= (4y)^2 - (3xy^2)^2$$

$$= 16y^2 - 9x^2y^4$$

c) $(a - 5b)(a + 5b)(a^2 - 25b^2) = (a^2 - 25b^2)(a^2 - 25b^2)$

$$= (a^2 - 25b^2)^2$$

$$= (a^2)^2 - 2(a^2)(25b^2) + (25b^2)^2$$

Squaring a binomial

$$= a^4 - 50a^2b^2 + 625b^4$$

Function Notation

Let's stop for a moment and look back at what we have done in this section. We have shown, for example, that

$$(x - 2)(x + 2) = x^2 - 4,$$

that is, $x^2 - 4$ and $(x - 2)(x + 2)$ are equivalent expressions.

From the viewpoint of functions, if

$$f(x) = x^2 - 4 \quad \text{and} \quad g(x) = (x - 2)(x + 2),$$

then for any given input x, the outputs $f(x)$ and $g(x)$ are identical. Thus the graphs of these functions are identical and we say that f and g represent the same function. Functions like these are graphed in detail in Chapter 8.

x	$f(x)$	$g(x)$
3	5	5
2	0	0
1	-3	-3
0	-4	-4
-1	-3	-3
-2	0	0
-3	5	5

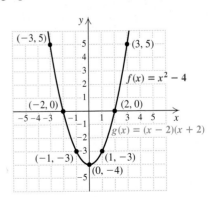

Our work with multiplying can be used when evaluating functions.

EXAMPLE 11   Given $f(x) = x^2 - 4x + 5$, find and simplify each of the following.

a) $f(a + 3)$

b) $f(a + h) - f(a)$

Solution

a) To find $f(a + 3)$, we replace x with $a + 3$. Then we simplify:

$$f(a + 3) = (a + 3)^2 - 4(a + 3) + 5$$
$$= a^2 + 6a + 9 - 4a - 12 + 5$$
$$= a^2 + 2a + 2.$$

b) To find $f(a + h)$ and $f(a)$, we replace x with $a + h$ and a, respectively.

$$f(a + h) - f(a) = [(a + h)^2 - 4(a + h) + 5] - [a^2 - 4a + 5]$$
$$= a^2 + 2ah + h^2 - 4a - 4h + 5 - a^2 + 4a - 5$$
$$= 2ah + h^2 - 4h$$

5.2

Exercise Set

FOR EXTRA HELP

Digital Video Tutor CD 4 Videotape 6 · Student's Solutions Manual · AW Math Tutor Center · InterAct Math · MathXL · MyMathLab

Multiply.

1. $8a^2 \cdot 4a$ $32a^3$

2. $-5x^3 \cdot 2x$ $-10x^4$

3. $5x(-4x^2y)$ $-20x^3y$

4. $-3ab^2(2a^2b^2)$ $-6a^3b^4$

5. $(2x^3y^2)(-5x^2y^4)$ $-10x^5y^6$

6. $(7a^2bc^4)(-8ab^3c^2)$ $-56a^3b^4c^6$

7. $7x(3 - x)$ $21x - 7x^2$

8. $3a(a^2 - 4a)$ $3a^3 - 12a^2$

9. $5cd(4c^2d - 5cd^2)$ $20c^3d^2 - 25c^2d^3$

10. $a^2(2a^2 - 5a^3)$ $2a^4 - 5a^5$

11. $(2x + 5)(3x - 4)$ $6x^2 + 7x - 20$

12. $(2a + 3b)(4a - b)$ $8a^2 + 10ab - 3b^2$

13. $(m + 2n)(m - 3n)$ $m^2 - mn - 6n^2$

14. $(m - 5)(m + 5)$ $m^2 - 25$

15. $(3y + 8x)(y - 7x)$ $3y^2 - 13xy - 56x^2$

16. $(x + y)(x - 2y)$ $x^2 - xy - 2y^2$

17. $(a^2 - 2b^2)(a^2 - 3b^2)$ $a^4 - 5a^2b^2 + 6b^4$

18. $(2m^2 - n^2)(3m^2 - 5n^2)$ $6m^4 - 13m^2n^2 + 5n^4$

19. $(x - 4)(x^2 + 4x + 16)$ $x^3 - 64$

20. $(y + 3)(y^2 - 3y + 9)$ $y^3 + 27$

21. $(x + y)(x^2 - xy + y^2)$ $x^3 + y^3$

22. $(a - b)(a^2 + ab + b^2)$ $a^3 - b^3$

23. $(a^2 + a - 1)(a^2 + 4a - 5)$ $a^4 + 5a^3 - 2a^2 - 9a + 5$

24. $(x^2 - 2x + 1)(x^2 + x + 2)$ $x^4 - x^3 + x^2 - 3x + 2$

25. $(3a^2b - 2ab + 3b^2)(ab - 2b + a)$
$3a^3b^2 - 8a^2b^2 + 3a^3b + 7ab^2 - 2a^2b + 3ab^3 - 6b^3$

26. $(2x^2 + y^2 - 2xy)(x^2 - 2y^2 - xy)$
$-4x^3y + 3xy^3 - x^2y^2 - 2y^4 + 2x^4$

27. $\left(x - \frac{1}{2}\right)\left(x - \frac{1}{4}\right)$ $x^2 - \frac{3}{4}x + \frac{1}{8}$

28. $\left(b - \frac{1}{3}\right)\left(b - \frac{1}{3}\right)$ $b^2 - \frac{2}{3}b + \frac{1}{9}$

29. $(1.2x - 3y)(2.5x + 5y)$ $3x^2 - 1.5xy - 15y^2$

30. $(40a - 0.24b)(0.3a + 10b)$ $12a^2 + 399.928ab - 2.4b^2$

31. Let $P(x) = 3x^2 - 5$ and $Q(x) = 4x^2 - 7x + 1$.
Find $P(x) \cdot Q(x)$. $12x^4 - 21x^3 - 17x^2 + 35x - 5$

32. Let $P(x) = x^2 - x + 1$ and $Q(x) = x^3 + x^2 + 5$.
Find $P(x) \cdot Q(x)$. $x^5 + 6x^2 - 5x + 5$

Multiply.

33. $(a + 4)(a + 5)$
$a^2 + 9a + 20$

34. $(x + 3)(x + 2)$
$x^2 + 5x + 6$

35. $(y - 8)(y + 3)$
$y^2 - 5y - 24$

36. $(y - 1)(y + 5)$
$y^2 + 4y - 5$

37. $(x + 5)^2$
$x^2 + 10x + 25$

38. $(y - 7)^2$
$y^2 - 14y + 49$

39. $(x - 2y)^2$
$x^2 - 4xy + 4y^2$

40. $(2s + 3t)^2$
$4s^2 + 12st + 9t^2$

41. $(2x + 9)(x + 2)$
$2x^2 + 13x + 18$

42. $(3b + 2)(2b - 5)$
$6b^2 - 11b - 10$

43. $(10a - 0.12b)^2$
$100a^2 - 2.4ab + 0.0144b^2$

44. $(10p^2 + 2.3q)^2$
$100p^4 + 46p^2q + 5.29q^2$

45. $(2x - 3y)(2x + y)$
$4x^2 - 4xy - 3y^2$

46. $(2a - 3b)(2a - b)$
$4a^2 - 8ab + 3b^2$

47. $(2x^3 - 3y^2)^2$
$4x^6 - 12x^3y^2 + 9y^4$

48. $(3s^2 + 4t^3)^2$
$9s^4 + 24s^2t^3 + 16t^6$

49. $(a^2b^2 + 1)^2$
$a^4b^4 + 2a^2b^2 + 1$

50. $(x^2y - xy^2)^2$
$x^4y^2 - 2x^3y^3 + x^2y^4$

51. Let $P(x) = 4x - 1$. Find $P(x) \cdot P(x)$. $16x^2 - 8x + 1$

52. Let $Q(x) = 3x^2 + 1$. Find $Q(x) \cdot Q(x)$. $9x^4 + 6x^2 + 1$

53. Let $F(x) = 2x - \frac{1}{3}$. Find $[F(x)]^2$. $4x^2 - \frac{4}{3}x + \frac{1}{9}$

54. Let $G(x) = 5x - \frac{1}{2}$. Find $[G(x)]^2$. $25x^2 - 5x + \frac{1}{4}$

Multiply.

55. $(c + 2)(c - 2)$ $c^2 - 4$

56. $(x - 3)(x + 3)$ $x^2 - 9$

57. $(4x + 1)(4x - 1)$
$16x^2 - 1$

58. $(3 - 2x)(3 + 2x)$
$9 - 4x^2$

59. $(3m - 2n)(3m + 2n)$
$9m^2 - 4n^2$

60. $(3x + 5y)(3x - 5y)$
$9x^2 - 25y^2$

61. $(x^3 + yz)(x^3 - yz)$ $x^6 - y^2z^2$

62. $(4a^3 + 5ab)(4a^3 - 5ab)$ $16a^6 - 25a^2b^2$

63. $(-mn + m^2)(mn + m^2)$ $-m^2n^2 + m^4$, or $m^4 - m^2n^2$

64. $(-3b + a^2)(3b + a^2)$ $-9b^2 + a^4$, or $a^4 - 9b^2$

65. $(x + 1)(x - 1)(x^2 + 1)$ $x^4 - 1$

66. $(y - 2)(y + 2)(y^2 + 4)$ $y^4 - 16$

67. $(a - b)(a + b)(a^2 - b^2)$ $a^4 - 2a^2b^2 + b^4$

68. $(2x - y)(2x + y)(4x^2 - y^2)$ $16x^4 - 8x^2y^2 + y^4$

Aha! **69.** $(a + b + 1)(a + b - 1)$ $a^2 + 2ab + b^2 - 1$

70. $(m + n + 2)(m + n - 2)$ $m^2 + 2mn + n^2 - 4$

71. $(2x + 3y + 4)(2x + 3y - 4)$ $4x^2 + 12xy + 9y^2 - 16$

72. $(3a - 2b + c)(3a - 2b - c)$ $9a^2 - 12ab + 4b^2 - c^2$

73. *Compounding Interest.* Suppose that P dollars is invested in a savings account at interest rate i, compounded annually, for 2 yr. The amount A in the account after 2 yr is given by
$$A = P(1 + i)^2.$$ $A = P + 2Pi + Pi^2$

Find an equivalent expression for A.

74. *Compounding Interest.* Suppose that P dollars is invested in a savings account at interest rate i,

compounded semiannually, for 1 yr. The amount A in the account after 1 yr is given by
$$A = P\left(1 + \frac{i}{2}\right)^2.$$ $A = P + Pi + \frac{Pi^2}{4}$

Find an equivalent expression for A.

75. Given $f(x) = x^2 + 5$, find and simplify.

a) $f(t - 1)$ $t^2 - 2t + 6$

b) $f(a + h) - f(a)$ $2ah + h^2$

c) $f(a) - f(a - h)$ $2ah - h^2$

76. Given $f(x) = x^2 + 7$, find and simplify.

a) $f(p + 1)$ $p^2 + 2p + 8$

b) $f(a + h) - f(a)$ $2ah + h^2$

c) $f(a) - f(a - h)$ $2ah - h^2$

TW **77.** Find two binomials whose product is $x^2 - 25$ and explain how you decided on those two binomials.

TW **78.** Find two binomials whose product is $x^2 - 6x + 9$ and explain how you decided on those two binomials.

Skill Maintenance

Solve. [2.3]

79. $ab + ac = d$, for a $a = \dfrac{d}{b + c}$

80. $xy + yz = w$, for y $y = \dfrac{w}{x + z}$

81. $mn + m = p$, for m $m = \dfrac{p}{n + 1}$

82. $rs + s = t$, for s $s = \dfrac{t}{r + 1}$

83. *Value of Coins.* There are 50 dimes in a roll of dimes, 40 nickels in a roll of nickels, and 40 quarters in a roll of quarters. Kacie has 13 rolls of coins, which have a total value of $89. There are three more rolls of dimes than nickels. How many of each type of roll does she have? [3.5]   Dimes: 5; nickels: 2; quarters: 6

84. *Wages.* Takako worked a total of 17 days last month at her father's restaurant. She earned $50 a day during the week and $60 a day during the weekend. Last month Takako earned $940. How many weekdays did she work? [3.3]   8 weekdays

Synthesis

TW **85.** We have seen that $(a - b)(a + b) = a^2 - b^2$. Explain how this result can be used to develop a fast way of multiplying $95 \cdot 105$.

TW **86.** A student incorrectly claims that since $2x^2 \cdot 2x^2 = 4x^4$, it follows that $5x^5 \cdot 5x^5 = 25x^{25}$. What mistake is the student making?

Multiply. Assume that variables in exponents represent natural numbers.

87. $[(-x^a y^b)^4]^a$ $x^{4a^2} y^{4ab}$

88. $(z^{n^2})^{n^3}(z^{4n^3})^{n^2}$ z^{5n^5}

89. $(a^x b^{2y})(\frac{1}{2} a^{3x} b)^2$ $\frac{1}{4} a^{7x} b^{2y+2}$

90. $(a^x b^y)^{w+z}$ $a^{xw+xz} b^{yw+yz}$

91. $y^3 z^n (y^{3n} z^3 - 4yz^{2n})$ $y^{3n+3} z^{n+3} - 4y^4 z^{3n}$

92. $[(a+b)(a-b)][5-(a+b)][5+(a+b)]$

93. $(a - b + c - d)(a + b + c + d)$ $-a^4 - 2a^3 b + 25a^2 + 2ab^3 - 25b^2 + b^4$ $a^2 + 2ac + c^2 - b^2 - 2bd - d^2$

Aha!

94. $(\frac{2}{3}x + \frac{1}{3}y + 1)(\frac{2}{3}x - \frac{1}{3}y - 1)$ $\frac{4}{9}x^2 - \frac{1}{9}y^2 - \frac{2}{3}y - 1$

95. $(4x^2 + 2xy + y^2)(4x^2 - 2xy + y^2)$ $16x^4 + 4x^2 y^2 + y^4$

96. $(x^2 - 3x + 5)(x^2 + 3x + 5)$ $x^4 + x^2 + 25$

97. $(x^a + y^b)(x^a - y^b)(x^{2a} + y^{2b})$ $x^{4a} - y^{4b}$

98. $(x - 1)(x^2 + x + 1)(x^3 + 1)$ $x^6 - 1$

99. $(x^{a-b})^{a+b}$ $x^{a^2-b^2}$

100. $(M^{x+y})^{x+y}$ $M^{x^2+2xy+y^2}$

Aha! **101.** $(x - a)(x - b)(x - c) \cdots (x - z)$ 0

102. Draw rectangles similar to those on p. 325 to show that $(x + 2)(x + 5) = x^2 + 7x + 10$. ☐

103. Use a graphing calculator to determine whether each of the following is an identity.   (b) and (c) are identities.
   a) $(x - 1)^2 = x^2 - 1$
   b) $(x - 2)(x + 3) = x^2 + x - 6$
   c) $(x - 1)^3 = x^3 - 3x^2 + 3x - 1$
   d) $(x + 1)^4 = x^4 + 1$
   e) $(x + 1)^4 = x^4 + 4x^3 + 8x^2 + 4x + 1$

Answer to Exercise 102 can be found on p. A-62.

Collaborative Corner

Algebra and Number Tricks

Focus: Polynomial multiplication
Time: 15–20 minutes
Group size: 2

Consider the following dialogue:

Jinny: Cal, let me do a number trick with you. Think of a number between 1 and 7. I'll have you perform some manipulations to this number, you'll tell me the result, and I'll tell you your number.

Cal: Okay. I've thought of a number.

Jinny: Good. Write it down so I can't see it, double it, and then subtract x from the result.

Cal: Hey, this is algebra!

Jinny: I know. Now square your binomial and subtract x^2.

Cal: How did you know I had an x^2? I *thought* this was rigged!

Jinny: It is. Now, divide by 4 and tell me either your constant term or your x-term. I'll tell you the other term and the number you chose.

Cal: Okay. The constant term is 16.

Jinny: Then the other term is $-4x$ and the number you chose is 4.

Cal: You're right! How did you do it?

ACTIVITY

1. Each group member should follow Jinny's instructions. Then determine how Jinny determined Cal's number and the other term.

2. Suppose that, at the end, Cal told Jinny the x-term. How would Jinny have determined Cal's number and the other term?

3. Would Jinny's "trick" work with *any* real number? Why do you think she specified numbers between 1 and 7?

4. Each group member should create a new number "trick" and perform it on the other group member. Be sure to include a variable so that both members can gain practice with polynomials.

5.3

Graphical Solutions ▪ The Principle of Zero Products ▪ Terms with Common Factors ▪ Factoring by Grouping ▪ Factoring and Equations

Polynomial Equations and Factoring

Whenever two polynomials are set equal to each other, the result is a **polynomial equation**. In this section, we learn how to solve such equations both graphically and algebraically by *factoring*.

Graphical Solutions

EXAMPLE 1   Solve: $x^2 = 6x$.

Solution   We can find real-number solutions of a polynomial equation by finding the points of intersection of two graphs, much as we did in Section 2.2. Alternatively, we can rewrite the equation so that one side is 0 and then find the *x*-intercepts of one graph.

Points of intersection.   To solve $x^2 = 6x$ using the first method, we graph $y_1 = x^2$ and $y_2 = 6x$ and find the coordinates of any points of intersection.

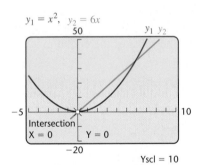

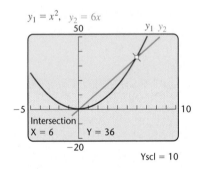

There are two points of intersection of the graphs, so there are two solutions of the equation. The *x*-coordinates of the points of intersection are 0 and 6, so the solutions of the equation are 0 and 6.

x-intercepts.   Using the second method, we rewrite the equation so that one side is 0:

$$x^2 = 6x$$
$$x^2 - 6x = 0. \qquad \textbf{Adding } -6x \textbf{ to both sides}$$

To solve $x^2 - 6x = 0$, we graph the function $f(x) = x^2 - 6x$ and look for values of x for which $f(x) = 0$, or the x-intercepts of the graph.

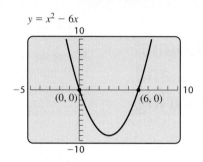

$y = x^2 - 6x$

The solutions of the equation are the x-coordinates of the x-intercepts of the graph, 0 and 6. Both 0 and 6 check in the original equation.

In Example 1, the values 0 and 6 are called **zeros** of the function $f(x) = x^2 - 6x$. They are also referred to as **roots** of the equation $f(x) = 0$.

Zeros and Roots

The x-values for which a function $f(x)$ is 0 are called the *zeros* of the function.

The x-values for which an equation such as $f(x) = 0$ is true are called the *roots* of the equation.

In this chapter, we consider only the real-number zeros of functions. We can solve, or find the roots of, the equation $f(x) = 0$ by finding the zeros of the function f.

Finding Zeros of Functions

Many graphing calculators have a ZERO or ROOT option in the CALC menu that will calculate an x-intercept of a graph.

To find the zero of a function, enter the function and choose a viewing window that shows the intersection of the graph of the function with the x-axis. Then choose the ZERO option from the CALC menu. The calculator first asks us for a Left Bound. This is an x-value less than, or to the left of, a zero. After visually determining the approximate location of a zero, or x-intercept of the graph, use the left and right arrow keys to move the cursor to a location on the graph left of the zero, or enter an x-value to the left of the zero. Then press $\boxed{\text{ENTER}}$. Next, enter a Right Bound in a similar way, and then a Guess. The calculator will return the coordinates of the x-intercept between the left and right bounds.

EXAMPLE 2 Find the zeros of the function given by

$$f(x) = x^3 - 3x^2 - 4x + 12.$$

Solution   First, we graph the equation $y = x^3 - 3x^2 - 4x + 12$, choosing a viewing window that shows the x-intercepts of the graph. It may require trial and error to choose an appropriate viewing window. We might try the standard viewing window first, as shown on the left below. It appears that there are three zeros, but we cannot see the shape of the graph for x-values between -2 and 1; Ymax should be increased. We might next try a viewing window of $[-10, 10, -100, 100]$, with Yscl $= 10$, as shown in the middle below. Here we get a better idea of the overall shape of the graph, but we cannot see all three x-intercepts clearly.

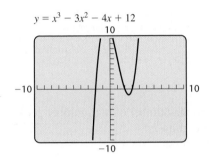

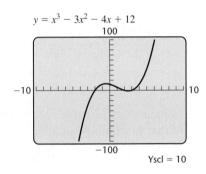

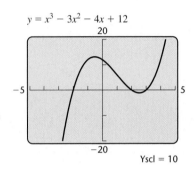

We magnify the portion of the graph close to the origin by making the window dimensions smaller. The window shown on the right above, $[-5, 5, -20, 20]$, is a good choice for viewing the zeros of this function. There are other good choices as well. The zeros of the function seem to be about -2, 2, and 3.

To find the zero that appears to be about -2, we first choose the ZERO option from the CALC menu. We then choose a Left Bound to the left of -2 on the x-axis. Next, we choose a Right Bound to the right of -2 on the x-axis. For a Guess, we choose an x-value close to -2. We see that -2 is indeed a zero of the function f.

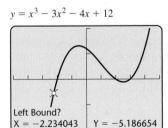

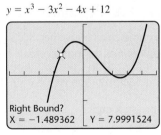

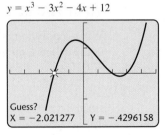

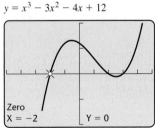

Using the same procedure for each of the other two zeros, we find that the zeros of the function $f(x) = x^3 - 3x^2 - 4x + 12$ are -2, 2, and 3.

We saw in the discussion after Example 1 that $f(x) = x^2 - 6x$ has 2 zeros. We also saw in Example 2 that $f(x) = x^3 - 3x^2 - 4x + 12$ has 3 zeros. We can tell something about the number of zeros of a polynomial function from its degree.

Interactive Discovery

Each of the following is a second-degree polynomial (quadratic) function. Use a graph to determine the number of real-number zeros of each function.

1. $f(x) = x^2 - 2x + 1$ 1

2. $g(x) = x^2 + 9$ 0

3. $h(x) = 2x^2 + 7x - 4$ 2

Each of the following is a third-degree polynomial (cubic) function. Use a graph to determine the *number* of real-number zeros of each function.

4. $f(x) = 2x^3 - 5x^2 - 3x$ 3

5. $g(x) = x^3 - 7x^2 + 16x - 12$ 2

6. $h(x) = x^3 - 4x^2 + 5x - 20$ 1

7. Compare the degree of each function with the number of real-number zeros of that function. What conclusion can you draw?
The number of real-number zeros is less than or equal to the degree.

A second-degree polynomial function will have 0, 1, or 2 real-number zeros. A third-degree polynomial function will have 1, 2, or 3 real-number zeros. This result can be generalized, although we will not prove it here.

> An nth-degree polynomial function will have at most n zeros.

"Seeing" all the zeros of a polynomial function when solving an equation graphically depends on a good choice of viewing window. Many polynomial equations can be solved algebraically. One principle used in solving polynomial equations is the *principle of zero products*.

The Principle of Zero Products

When a polynomial is written as a product, we say that it is *factored*. The product is called a *factorization* of the polynomial. For example,

$$3x(x + 4) \quad \text{is a factorization of} \quad 3x^2 + 12x$$

since $3x(x + 4)$ is a product and

$$3x(x + 4) = 3x^2 + 12x.$$

The polynomials $3x$ and $x + 4$ are *factors* of $3x^2 + 12x$. The zeros of a polynomial function are related to the factorization of the polynomial.

Interactive Discovery

Consider the function f given by $f(x) = 3x^2 + 12x$. If $g(x) = 3x$ and $h(x) = x + 4$, then $f(x) = g(x) \cdot h(x)$.

1. Graph $f(x)$ and find the zeros. $-4, 0$

2. Using the same window, graph $g(x)$ and $h(x)$ and find the zero of each function. $g(x)$: 0; $h(x)$: -4

3. How are the zeros of $f(x)$ and the zeros of $g(x)$ and $h(x)$ related?
The zeros of f are the zeros of g and h.

We see that if a polynomial function f is given by $f(x) = g(x) \cdot h(x)$, then $f(x) = 0$ at the same x-values for which $g(x) = 0$ or $h(x) = 0$.

The zeros of a polynomial function are the zeros of the functions described by the factors of the polynomial.

This result leads us to the following principle, which is used to solve factored polynomials algebraically.

The Principle of Zero Products

For any real numbers a and b:

If $ab = 0$, then $a = 0$ or $b = 0$.
If $a = 0$ or $b = 0$, then $ab = 0$.

The principle of zero products is based on a multiplication property of 0. When we multiply two or more numbers, the product is 0 if one of the factors is 0. Conversely, if a product is 0, then at least one of the factors must be 0. Note that in order to solve a polynomial equation using the principle of zero products, we must have 0 on one side of the equation and the other side must be factored.

EXAMPLE 3   Solve: $(x - 3)(x + 2) = 0$.

Solution   The principle of zero products says that in order for $(x - 3)(x + 2)$ to be 0, at least one factor must be 0. Thus,

$$x - 3 = 0 \quad or \quad x + 2 = 0. \qquad \text{Using the principle of zero products}$$

Each of these linear equations is then solved separately:

$$x = 3 \quad or \quad x = -2.$$

We check both algebraically and graphically, as follows.

Algebraic Solution

For 3:

$$\frac{(x-3)(x+2)=0}{(3-3)(3+2)\ ?\ 0}$$
$$0(5)$$

$$0\ |\ 0 \quad \text{TRUE}$$

For -2:

$$\frac{(x-3)(x+2)=0}{(-2-3)(-2+2)\ ?\ 0}$$
$$(-5)(0)$$

$$0\ |\ 0 \quad \text{TRUE}$$

Graphical Solution

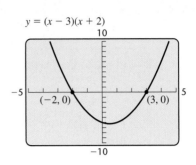

The solutions are -2 and 3.

TEACHING TIP

Be aware that some students will attempt to solve Example 4 by dividing both sides of the equation by a. Lead them to see why they cannot do this.

EXAMPLE 4 Given that $f(x) = x(3x+2)$, find all the values of a for which $f(a) = 0$.

Solution We are looking for all numbers a for which $f(a) = 0$. Since $f(a) = a(3a+2)$, we must have

$$a(3a+2)=0 \qquad \text{Setting } f(a) \text{ equal to } 0$$
$$a=0 \quad or \quad 3a+2=0 \qquad \textbf{Using the principle of zero products}$$
$$a=0 \quad or \qquad a=-\tfrac{2}{3}.$$

We check by evaluating $f(0)$ and $f\left(-\tfrac{2}{3}\right)$. One way to check with a graphing calculator is to enter $y_1 = x(3x+2)$ and calculate $Y_1(0)$ and $Y_1(-2/3)$.

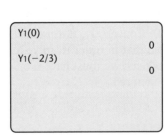

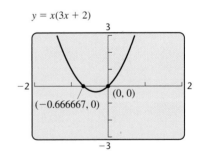

We can also graph $f(x) = x(3x+2)$ and observe that its zeros are $-\tfrac{2}{3}$ and 0. Thus, to have $f(a) = 0$, we must have $a = 0$ or $a = -\tfrac{2}{3}$.

If the polynomial in an equation of the form $p(x) = 0$ is not in factored form, we must factor it before we can use the principle of zero products to solve the equation. To **factor** an expression means to write an equivalent expression that is a product.

Terms with Common Factors

When factoring, we first look for factors common to every term in an expression and then use the distributive law.

EXAMPLE 5   Factor out a common factor: $4x^2 - 8$.

Solution

$$4x^2 - 8 = 4 \cdot x^2 - 4 \cdot 2 \qquad \text{Noting that 4 is a common factor}$$
$$= 4(x^2 - 2) \qquad \text{Using the distributive law}$$

In some cases, there is more than one common factor. In Example 5, both 2 and 4 are common factors. For the polynomial $5x^4 + 20x^3$, some common factors are 5, x^3, and $5x^3$. If there is more than one common factor, we factor out the *largest*, or *greatest, common factor*, that is, the common factor with the largest coefficient and the highest degree. In $5x^4 + 20x^3$, the largest common factor is $5x^3$.

EXAMPLE 6   Factor out a common factor.

a) $5x^4 + 20x^3$ **b)** $12x^2y - 20x^3y$ **c)** $10p^6q^2 - 4p^5q^3 - 2p^4q^4$

Solution

a) $5x^4 + 20x^3 = 5x^3(x + 4)$ Try to write your answer directly. Multiply mentally to check your answer.

b) $12x^2y - 20x^3y = 4x^2y(3 - 5x)$
   Check: $4x^2y \cdot 3 = 12x^2y$ and $4x^2y(-5x) = -20x^3y$, so $4x^2y(3 - 5x) = 12x^2y - 20x^3y$.

c) $10p^6q^2 - 4p^5q^3 - 2p^4q^4 = 2p^4q^2(5p^2 - 2pq - q^2)$
   The check is left to the student.

The polynomials in Examples 5 and 6 cannot be factored further unless (in the cases of Examples 5 and 6c) square roots or (in the case of Example 6b) fractions are used. In both examples, we have **factored completely** over the set of integers. The factors used are said to be **prime polynomials** over the set of integers.

When a factor contains more than one term, it is usually desirable for the leading coefficient to be positive. To achieve this may require factoring out a common factor with a negative coefficient.

EXAMPLE 7   Factor out a common factor with a negative coefficient.

a) $-4x - 24$ **b)** $-2x^3 + 6x^2 - 10x$

Solution

a) $-4x - 24 = -4(x + 6)$
b) $-2x^3 + 6x^2 - 10x = -2x(x^2 - 3x + 5)$

EXAMPLE 8 Height of a Thrown Object.   Suppose that a baseball is thrown upward with an initial velocity of 64 ft/sec and an initial height of 0 ft. Its height in feet, $h(t)$, after t seconds is given by

$$h(t) = -16t^2 + 64t.$$

Find an equivalent expression for $h(t)$ by factoring out a common factor with a negative coefficient.

Solution   We factor out $-16t$ as follows:

$$h(t) = -16t^2 + 64t = -16t(t - 4). \qquad \textbf{Check: } -16t \cdot t = -16t^2 \text{ and}$$
$$-16t(-4) = 64t.$$

Note that we can obtain function values using either expression for $h(t)$, since factoring forms equivalent expressions. For example,

$$h(1) = -16 \cdot 1^2 + 64 \cdot 1 = 48$$
$$\text{and} \quad h(1) = -16 \cdot 1(1 - 4) = 48. \qquad \textbf{Using the factorization}$$

$y_1 = -16x^2 + 64x,$
$y_2 = (-16x)(x - 4)$

X	Y1	Y2
0	0	0
1	48	48
2	64	64
3	48	48
4	0	0
5	−80	−80
6	−192	−192

X = 0

We can evaluate the expressions $-16t^2 + 64t$ and $-16t(t - 4)$ in Example 8 using any value for t. The results should always match. Thus a quick partial check of any factorization is to evaluate the factorization and the original polynomial for one or two convenient replacements. The check in Example 8 becomes foolproof if three replacements are used. Recall that, in general, an nth-degree factorization is correct if it checks for $n + 1$ different replacements. The table shown at left confirms that the factorization is correct.

Factoring by Grouping

The largest common factor is sometimes a binomial.

EXAMPLE 9   Factor: $(a - b)(x + 5) + (a - b)(x - y^2)$.

Solution   Here the largest common factor is the binomial $a - b$:

$$(a - b)(x + 5) + (a - b)(x - y^2) = (a - b)[(x + 5) + (x - y^2)]$$
$$= (a - b)[2x + 5 - y^2].$$

Often, in order to identify a common binomial factor in a polynomial with four terms, we must regroup into two groups of two terms each.

EXAMPLE 10   Factor.

a) $y^3 + 3y^2 + 4y + 12$ **b)** $4x^3 - 15 + 20x^2 - 3x$

Solution

a) $y^3 + 3y^2 + 4y + 12 = (y^3 + 3y^2) + (4y + 12)$ Each grouping has a common factor.

$\qquad\qquad = y^2(y + 3) + 4(y + 3)$ Factoring out a common factor from each binomial

$\qquad\qquad = (y + 3)(y^2 + 4)$ Factoring out $y + 3$

b) When we try grouping $4x^3 - 15 + 20x^2 - 3x$ as

$$(4x^3 - 15) + (20x^2 - 3x),$$

we are unable to factor $4x^3 - 15$. When this happens, we can rearrange the polynomial and try a different grouping:

$4x^3 - 15 + 20x^2 - 3x = 4x^3 + 20x^2 - 3x - 15$ Using the commutative law to rearrange the terms

$\qquad\qquad = 4x^2(x + 5) - 3(x + 5)$

$\qquad\qquad = (x + 5)(4x^2 - 3).$

In Example 8 of Section 1.3 (see p. 25), we saw that

$$b - a, \qquad -(a - b), \quad \text{and} \quad -1(a - b)$$

are equivalent. Remembering this can help anytime we wish to reverse subtraction (see the third step below).

EXAMPLE 11   Factor: $ax - bx + by - ay$.

Solution   We have

$ax - bx + by - ay = (ax - bx) + (by - ay)$ Grouping

$\qquad\qquad = x(a - b) + y(b - a)$ Factoring each binomial

$\qquad\qquad = x(a - b) + y(-1)(a - b)$ Factoring out -1 to reverse $b - a$

$\qquad\qquad = x(a - b) - y(a - b)$ Simplifying

$\qquad\qquad = (a - b)(x - y).$ Factoring out $a - b$

We can always check our factoring by multiplying:

Check: $(a - b)(x - y) = ax - ay - bx + by = ax - bx + by - ay.$

Some polynomials with four terms, like $x^3 + x^2 + 3x - 3$, are prime. Not only is there no common monomial factor, but no matter how we group terms, there is no common binomial factor:

$$x^3 + x^2 + 3x - 3 = x^2(x + 1) + 3(x - 1);$$ No common factor

$$x^3 + 3x + x^2 - 3 = x(x^2 + 3) + (x^2 - 3);$$ No common factor

$$x^3 - 3 + x^2 + 3x = (x^3 - 3) + x(x + 3).$$ No common factor

Factoring and Equations

Factoring can help us solve polynomial equations.

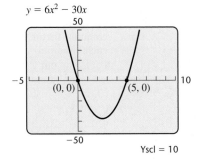

$y = 6x^2 - 30x$

EXAMPLE 12   Solve: $6x^2 = 30x$.

Solution   We can use the principle of zero products if there is a 0 on one side of the equation and the other side is in factored form:

$$6x^2 = 30x$$
$$6x^2 - 30x = 0 \qquad \text{Subtracting } 30x. \text{ One side is now 0.}$$
$$6x(x - 5) = 0 \qquad \text{Factoring}$$
$$6x = 0 \quad or \quad x - 5 = 0 \qquad \text{Using the principle of zero products}$$
$$x = 0 \quad or \qquad x = 5.$$

We check by substitution or graphically, as shown in the figure at left. The solutions are 0 and 5.

Connecting the Concepts

When we factor an expression such as $6x^2 - 30x$, we are finding an equivalent expression:

$$6x^2 - 30x = 6x(x - 5).$$

We are *not* "solving" the expression. Remember that we can *factor expressions* and *solve equations*. In the process of solving an equation, we may factor an expression within the equation:

$$6x^2 - 30x = 0 \qquad \text{This is an equation that we can attempt to solve.}$$
$$6x(x - 5) = 0. \qquad \text{We factor the expression } 6x^2 - 30x.$$

Do not make the mistake of attempting to solve an expression such as $6x^2 - 30x$.

The principle of zero products can be used to show that an nth-degree polynomial function can have at most n zeros. In Example 12, we wrote a quadratic polynomial as a product of two linear factors. Each linear factor corresponded to one zero of the polynomial function. In general, a polynomial function of degree n can have at most n linear factors, so a polynomial function of degree n can have at most n zeros. Thus, when solving an nth-degree polynomial equation, we need not look for more than n zeros.

To Use the Principle of Zero Products

1. Obtain a 0 on one side of the equation using the addition principle.

2. Factor the nonzero side of the equation.

3. Set each factor that is not a constant equal to 0.

4. Solve the resulting equations.

5.3

Exercise Set

In Exercises 1 and 2, use the graph to solve $f(x) = 0$.

1. $-3, 5$

2. $-5, -2$

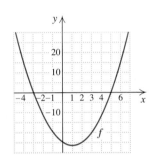

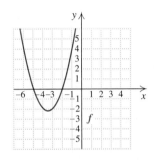

In Exercises 3 and 4, use the graph to find the zeros of the function f.

3. $-2, 0$

4. $-1, 3$

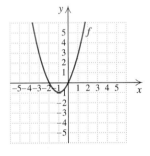

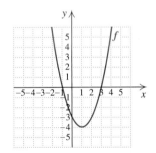

5. Use the following graph to solve $x^2 + 2x = 3$.
$-3, 1$

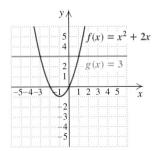

6. Use the following graph to solve $x^2 = 4$. $-2, 2$

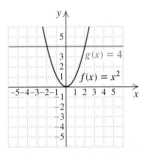

7. Use the following graph to solve $x^2 + 2x - 8 = 0$.
$-4, 2$

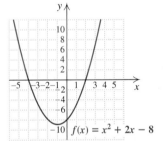

8. Use the following graph to find the zeros of the function given by $f(x) = x^2 - 2x + 1$. 1

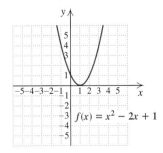

Solve using a graphing calculator.

9. $x^2 = 5x$ 0, 5

10. $2x^2 = 20x$ 0, 10

11. $4x = x^2 + 3$ 1, 3

12. $x^2 = 1$ −1, 1

13. $x^2 + 150 = 25x$ 10, 15

14. $2x^2 + 25 = 51x$ 0.5, 25

15. $x^3 - 3x^2 + 2x = 0$ 0, 1, 2

16. $x^3 + 2x^2 = x + 2$ −2, −1, 1

17. $x^3 - 3x^2 - 198x + 1080 = 0$ −15, 6, 12

18. $2x^3 + 25x^2 - 282x + 360 = 0$ −20, 1.5, 6

19. $21x^2 + 2x - 3 = 0$ −0.42857, 0.33333

20. $66x^2 - 49x - 5 = 0$ −0.09091, 0.83333

Find the zeros of each function.

21. $f(x) = x^2 - 4x - 45$ −5, 9

22. $g(x) = x^2 + x - 20$ −5, 4

23. $p(x) = 2x^2 - 13x - 7$ −0.5, 7

24. $f(x) = 6x^2 + 17x + 6$ −2.42013, −0.41320

25. $f(x) = x^3 - 2x^2 - 3x$ −1, 0, 3

26. $r(x) = 3x^3 - 12x$ −2, 0, 2

Aha! *Match each graph to the corresponding function in Exercises 27–30.*

I

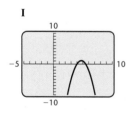

II

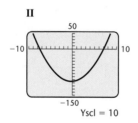
Yscl = 10

III

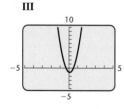

IV

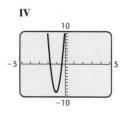

27. $f(x) = (2x - 1)(3x + 1)$ III

28. $f(x) = (2x + 15)(x - 7)$ II

29. $f(x) = (4 - x)(2x - 11)$ I

30. $f(x) = (5x + 2)(4x + 7)$ IV

Tell whether each of the following is an expression or an equation.

31. $x^2 + 6x + 9$ Expression

32. $x^3 = x^2 - x + 3$ Equation

33. $3x^2 = 3x$ Equation

34. $x^4 + 3x^3 + x^2$ Expression

35. $2x^3 + x^2 = 0$ Equation

36. $5x^4 + 5x$ Expression

Factor.

37. $2t^2 + 8t$ $2t(t + 4)$

38. $3y^2 + 6y$ $3y(y + 2)$

39. $y^3 + 9y^2$ $y^2(y + 9)$

40. $x^3 + 8x^2$ $x^2(x + 8)$

41. $15x^2 - 5x^4$ $5x^2(3 - x^2)$

42. $8y^2 + 4y^4$ $4y^2(2 + y^2)$

43. $4x^2y - 12xy^2$ $4xy(x - 3y)$

44. $5x^2y^3 + 15x^3y^2$ $5x^2y^2(y + 3x)$

45. $3y^2 - 3y - 9$ $3(y^2 - y - 3)$

46. $5x^2 - 5x + 15$ $5(x^2 - x + 3)$

47. $6ab - 4ad + 12ac$ $2a(3b - 2d + 6c)$

48. $8xy + 10xz - 14xw$ $2x(4y + 5z - 7w)$

49. $9x^3y^6z^2 - 12x^4y^4z^4 + 15x^2y^5z^3$ $3x^2y^4z^2(3xy^2 - 4x^2z^2 + 5yz)$

50. $14a^4b^3c^5 + 21a^3b^5c^4 - 35a^4b^4c^3$ $7a^3b^3c^3(2ac^2 + 3b^2c - 5ab)$

Factor out a factor with a negative coefficient.

51. $-5x + 35$ $-5(x - 7)$

52. $-5x - 40$ $-5(x + 8)$

53. $-6y - 72$ $-6(y + 12)$

54. $-8t + 72$ $-8(t - 9)$

55. $-2x^2 + 4x - 12$ $-2(x^2 - 2x + 6)$

56. $-2x^2 + 12x + 40$ $-2(x^2 - 6x - 20)$

57. $3y - 24x$ $-3(-y + 8x)$, or $-3(8x - y)$

58. $7x - 56y$ $-7(-x + 8y)$, or $-7(8y - x)$

59. $7s - 14t$ $-7(-s + 2t)$, or $-7(2t - s)$

60. $5r - 10s$ $-5(-r + 2s)$, or $-5(2s - r)$

61. $-x^2 + 5x - 9$ $-(x^2 - 5x + 9)$

62. $-p^3 - 4p^2 + 11$ $-(p^3 + 4p^2 - 11)$

63. $-a^4 + 2a^3 - 13a$ $-a(a^3 - 2a^2 + 13)$

64. $-m^3 - m^2 + m - 2$ $-(m^3 + m^2 - m + 2)$

Factor.

65. $a(b - 5) + c(b - 5)$ $(b - 5)(a + c)$

66. $r(t - 3) - s(t - 3)$ $(t - 3)(r - s)$

67. $(x + 7)(x - 1) + (x + 7)(x - 2)$ $(x + 7)(2x - 3)$

68. $(a + 5)(a - 2) + (a + 5)(a + 1)$ $(a + 5)(2a - 1)$

69. $a^2(x - y) + 5(y - x)$ $(x - y)(a^2 - 5)$

70. $5x^2(x - 6) + 2(6 - x)$ $(x - 6)(5x^2 - 2)$

71. $ac + ad + bc + bd$ $(c + d)(a + b)$

72. $xy + xz + wy + wz$ $(y + z)(x + w)$

73. $b^3 - b^2 + 2b - 2$ $(b - 1)(b^2 + 2)$

74. $y^3 - y^2 + 3y - 3$ $(y - 1)(y^2 + 3)$

75. $a^3 - 3a^2 + 6 - 2a$ $(a - 3)(a^2 - 2)$

76. $t^3 + 6t^2 - 2t - 12$ $(t + 6)(t^2 - 2)$

77. $72x^3 - 36x^2 + 24x$ $12x(6x^2 - 3x + 2)$

78. $12a^4 - 21a^3 - 9a^2$ $3a^2(4a^2 - 7a - 3)$

79. $x^6 - x^5 - x^3 + x^4$ $x^3(x - 1)(x^2 + 1)$

80. $y^4 - y^3 - y + y^2$ $y(y - 1)(y^2 + 1)$

81. $2y^4 + 6y^2 + 5y^2 + 15$ $(y^2 + 3)(2y^2 + 5)$

82. $2xy - x^2y - 6 + 3x$ $(2 - x)(xy - 3)$

83. *Height of a Baseball.*   A baseball is popped up with an upward velocity of 72 ft/sec. Its height in feet, $h(t)$, after t seconds is given by

$$h(t) = -16t^2 + 72t.$$

a) Find an equivalent expression for $h(t)$ by factoring out a common factor with a negative coefficient. $h(t) = -8t(2t - 9)$

b) Perform a partial check of part (a) by evaluating both expressions for $h(t)$ at $t = 1$. $h(1) = 56$ ft

84. *Height of a Rocket.*   A model rocket is launched upward with an initial velocity of 96 ft/sec. Its height in feet, $h(t)$, after t seconds is given by

$$h(t) = -16t^2 + 96t.$$

a) Find an equivalent expression for $h(t)$ by factoring out a common factor with a negative coefficient. $h(t) = -16t(t - 6)$

b) Check your factoring by evaluating both expressions for $h(t)$ at $t = 1$. $h(1) = 80$ ft

85. *Airline Routes.*   When an airline links n cities so that from any one city it is possible to fly directly to each of the other cities, the total number of direct routes is given by

$$R(n) = n^2 - n.$$

Find an equivalent expression for $R(n)$ by factoring out a common factor. $R(n) = n(n - 1)$

86. *Surface Area of a Silo.*   A silo is a structure that is shaped like a right circular cylinder with a half sphere on top. The surface area of a silo of height h and radius r (including the area of the base) is given by the polynomial $2\pi rh + \pi r^2$. Find an equivalent expression by factoring out a common factor. $\pi r(2h + r)$

87. *Total Profit.*   When x hundred CD players are sold, Rolics Electronics collects a profit of $P(x)$, where

$$P(x) = x^2 - 3x,   P(x) = x(x - 3)$$

and $P(x)$ is in thousands of dollars. Find an equivalent expression by factoring out a common factor.

88. *Total Profit.*   After t weeks of production, Claw Foot, Inc., is making a profit of $P(t) = t^2 - 5t$ from sales of their surfboards. Find an equivalent expression by factoring out a common factor.
$P(t) = t(t - 5)$

89. *Total Revenue.*   Urban Sounds is marketing a new MP3 player. The firm determines that when it sells x units, the total revenue R is given by the polynomial function

$$R(x) = 280x - 0.4x^2 \text{ dollars.}$$

Find an equivalent expression for $R(x)$ by factoring out $0.4x$. $R(x) = 0.4x(700 - x)$

90. *Total Cost.*   Urban Sounds determines that the total cost C of producing x MP3 players is given by the polynomial function

$$C(x) = 0.18x + 0.6x^2.$$

Find an equivalent expression for $C(x)$ by factoring out $0.6x$. $C(x) = 0.6x(0.3 + x)$

91. *Counting Spheres in a Pile.*   The number N of spheres in a triangular pile like the one shown here is a polynomial function given by

$$N(x) = \tfrac{1}{6}x^3 + \tfrac{1}{2}x^2 + \tfrac{1}{3}x,$$

where x is the number of layers and $N(x)$ is the number of spheres. Find an equivalent expression for $N(x)$ by factoring out $\tfrac{1}{6}$. $N(x) = \tfrac{1}{6}(x^3 + 3x^2 + 2x)$

92. *Number of Games in a League.*   If there are n teams in a league and each team plays every other team once, we can find the total number of games played by using the polynomial function $f(n) = \tfrac{1}{2}n^2 - \tfrac{1}{2}n$. Find an equivalent expression by factoring out $\tfrac{1}{2}$. $f(n) = \tfrac{1}{2}(n^2 - n)$

93. *High-fives.* When a team of n players all give each other high-fives, a total of $H(n)$ hand slaps occurs, where

$$H(n) = \tfrac{1}{2}n^2 - \tfrac{1}{2}n. \quad H(n) = \tfrac{1}{2}n(n-1)$$

Find an equivalent expression by factoring out $\tfrac{1}{2}n$.

94. *Number of Diagonals.* The number of diagonals of a polygon having n sides is given by the polynomial function

$$P(n) = \tfrac{1}{2}n^2 - \tfrac{3}{2}n.$$

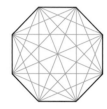

Find an equivalent expression for $P(n)$ by factoring out $\tfrac{1}{2}$. $\quad P(n) = \tfrac{1}{2}(n^2 - 3n)$

Solve using the principle of zero products.

95. $x(x+1) = 0 \quad -1, 0$

96. $5x(x-2) = 0 \quad 0, 2$

97. $x^2 - 3x = 0 \quad 0, 3$

98. $2x^2 + 8x = 0 \quad -4, 0$

99. $-5x^2 = 15x \quad -3, 0$

100. $2x - 4x^2 = 0 \quad 0, \tfrac{1}{2}$

101. $12x^4 + 4x^3 = 0 \quad -\tfrac{1}{3}, 0$

102. $21x^3 = 7x^2 \quad 0, \tfrac{1}{3}$

TW **103.** Under what conditions would it be easier to evaluate a polynomial *after* it has been factored?

TW **104.** Gabrielle claims that the zeros of the function given by $f(x) = x^4 - 3x^2 + 7x + 20$ are $-1, 1, 2, 4,$ and 5. How can you tell, without performing any calculations, that she cannot be correct?

Skill Maintenance

Simplify. [1.2]

105. $2(-3) + 4(-5) \quad -26$

106. $-7(-2) + 5(-3) \quad -1$

107. $4(-6) - 3(2) \quad -30$

108. $5(-3) + 2(-2) \quad -19$

109. *Geometry.* The perimeter of a triangle is 174. The lengths of the three sides are consecutive even numbers. What are the lengths of the sides of the triangle? [2.3] $\quad 56, 58, 60$

110. *Manufacturing.* In a factory, there are three machines A, B, and C. When all three are running, they produce 222 suitcases per day. If A and B work but C does not, they produce 159 suitcases per day. If B and C work but A does not, they produce 147 suitcases. What is the daily production of each machine? [3.5] A: 75; B: 84; C: 63

Synthesis

TW **111.** Is it true that if the coefficients and exponents of a polynomial are all prime numbers, then the polynomial itself is prime? Why or why not?

TW **112.** Following Example 8, we stated that checking the factorization of a second-degree polynomial by making a single replacement is only a *partial* check. Write an *incorrect* factorization and explain how evaluating both the polynomial and the factorization might not catch the mistake.

113. Use the results of Exercise 7 to factor $x^2 + 2x - 8$. $(x + 4)(x - 2)$

114. Use the results of Exercise 8 to factor $x^2 - 2x + 1$. $(x - 1)(x - 1)$

Complete each of the following. $\quad x^5y^4 + x^4y^6 = x^3y(x^2y^3 + xy^5)$

115. $x^5y^4 + \underline{\quad} = x^3y(\underline{\quad} + xy^5)$

116. $a^3b^7 - \underline{\quad} = \underline{\quad}(ab^4 - c^2)$
$\quad a^3b^7 - a^2b^3c^2 = a^2b^3(ab^4 - c^2)$

Factor.

117. $rx^2 - rx + 5r + sx^2 - sx + 5s \quad (x^2 - x + 5)(r + s)$

118. $3a^2 + 6a + 30 + 7a^2b + 14ab + 70b$
$\quad (a^2 + 2a + 10)(3 + 7b)$

119. $a^4x^4 + a^4x^2 + 5a^4 + a^2x^4 + a^2x^2 + 5a^2 + 5x^4 + 5x^2 + 25 \quad (x^4 + x^2 + 5)(a^4 + a^2 + 5)$
(*Hint*: Use three groups of three.)

Factor out the smallest power of x in each of the following.

120. $x^{1/2} + 5x^{3/2} \quad x^{1/2}(1 + 5x)$

121. $x^{1/3} - 7x^{4/3}$
$\quad x^{1/3}(1 - 7x)$

122. $x^{3/4} + x^{1/2} - x^{1/4}$
$\quad x^{1/4}(x^{1/2} + x^{1/4} - 1)$

123. $x^{1/3} - 5x^{1/2} + 3x^{3/4}$
$\quad x^{1/3}(1 - 5x^{1/6} + 3x^{5/12})$

Factor. Assume that all exponents are natural numbers.

124. $2x^{3a} + 8x^a + 4x^{2a} \quad 2x^a(x^{2a} + 4 + 2x^a)$

125. $3a^{n+1} + 6a^n - 15a^{n+2} \quad 3a^n(a + 2 - 5a^2)$

126. $4x^{a+b} + 7x^{a-b} \quad x^a(4x^b + 7x^{-b})$

127. $7y^{2a+b} - 5y^{a+b} + 3y^{a+2b} \quad y^{a+b}(7y^a - 5 + 3y^b)$

5.4

Factoring Trinomials of the Type $x^2 + bx + c$ ■ Equations Containing Trinomials ■ Zeros and Factoring

Equations Containing Trinomials of the Type $x^2 + bx + c$

In this section, we expand our list of the types of polynomials that we can factor so that we can solve a wider variety of polynomial equations. We begin by factoring trinomials of the type $x^2 + bx + c$ and then solve equations containing such trinomials. We then see how to use zeros of a function to factor a polynomial.

Factoring Trinomials of the Type $x^2 + bx + c$

When trying to factor trinomials of the type $x^2 + bx + c$, we can use a trial-and-error procedure.

Constant Term Positive

Recall the FOIL method of multiplying two binomials:

$$
\begin{array}{cccc}
 & \text{F} & \text{O} \quad \text{I} & \text{L} \\
(x + 3)(x + 5) = & x^2 + & \underline{5x + 3x} & + 15 \\
 & \downarrow & \downarrow & \downarrow \\
= & x^2 + & 8x & + 15.
\end{array}
$$

Because the leading coefficient in each binomial is 1, the leading coefficient in the product is also 1. To factor $x^2 + 8x + 15$, we think of FOIL: The first term, x^2, is the product of the First terms of two binomial factors, so the first term in each binomial must be x. The challenge is to find two numbers p and q such that

$$
\begin{aligned}
x^2 + 8x + 15 &= (x + p)(x + q) \\
&= x^2 + qx + px + pq.
\end{aligned}
$$

Note that the Outer and Inner products, qx and px, can be written as $(p + q)x$. The Last product, pq, will be a constant. Thus the numbers p and q must be selected so that their product is 15 and their sum is 8. In this case, we know from above that these numbers are 3 and 5. The factorization is

$$(x + 3)(x + 5), \quad \text{or} \quad (x + 5)(x + 3). \qquad \textbf{Using a commutative law}$$

In general, to factor $x^2 + (p + q)x + pq$, we use FOIL in reverse:

$$x^2 + (p + q)x + pq = (x + p)(x + q).$$

EXAMPLE 1 Factor: $x^2 + 9x + 8$.

Solution We think of FOIL in reverse. The first term of each factor is x. We are looking for numbers p and q such that

$$x^2 + 9x + 8 = (x + p)(x + q) = x^2 + (p + q)x + pq.$$

Thus we search for factors of 8 whose sum is 9.

Pair of Factors	Sum of Factors
2, 4	6
1, 8	9

The numbers we need are 1 and 8.

The factorization is thus $(x + 1)(x + 8)$. The student should check by multiplying to confirm that the product is the original trinomial.

When factoring trinomials with a leading coefficient of 1, it suffices to consider all pairs of factors along with their sums, as we did above. At times, however, you may be tempted to form factors without calculating any sums. It is essential that you check any attempt made in this manner! For example, if we attempt the factorization

$$x^2 + 9x + 8 \stackrel{?}{=} (x + 2)(x + 4),$$

a check reveals that

$$(x + 2)(x + 4) = x^2 + 6x + 8 \neq x^2 + 9x + 8.$$

This type of trial-and-error procedure becomes easier to use with time. As you gain experience, you will find that many trials can be performed mentally.

EXAMPLE 2 Factor: $t^2 - 9t + 20$.

Solution Since the constant term is positive and the coefficient of the middle term is negative, we look for a factorization of 20 in which both factors are negative. Their sum must be -9.

$y_1 = x^2 - 9x + 20,$
$y_2 = (x - 4)(x - 5)$

X	Y1	Y2
0	20	20
1	12	12
2	6	6
3	2	2
4	0	0
5	0	0
6	2	2

X = 0

Pair of Factors	Sum of Factors
$-1, -20$	-21
$-2, -10$	-12
$-4, -5$	-9

The numbers we need are -4 and -5.

The factorization is $(t - 4)(t - 5)$. We check by comparing values using a table, as shown at left.

Constant Term Negative

When the constant term of a trinomial is negative, we look for one negative factor and one positive factor. The sum of the factors must still be the coefficient of the middle term.

EXAMPLE 3   Factor: $x^3 - x^2 - 30x$.

Solution   *Always* look first for a common factor! This time there is one, x. We factor it out:

$$x^3 - x^2 - 30x = x(x^2 - x - 30).$$

Now we consider $x^2 - x - 30$. We need a factorization of -30 in which one factor is positive, the other factor is negative, and the sum of the factors is -1. Since the sum is to be negative, the negative factor must have the greater absolute value. Thus we need consider only the following pairs of factors.

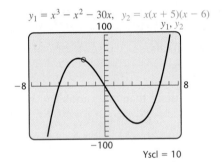
$y_1 = x^3 - x^2 - 30x$, $y_2 = x(x + 5)(x - 6)$
Yscl = 10

Pair of Factors	Sum of Factors
1, −30	−29
3, −10	−7
5, −6	−1 ← — The numbers we need are 5 and −6.

The factorization of $x^2 - x - 30$ is $(x + 5)(x - 6)$. *Don't forget to include the factor that was factored out earlier!* In this case, the factorization of the original trinomial is $x(x + 5)(x - 6)$. We check by graphing $y_1 = x^3 - x^2 - 30x$ and $y_2 = x(x + 5)(x - 6)$, as shown at left.

EXAMPLE 4   Factor: $2x^2 + 34x - 220$.

Solution   *Always* look first for a common factor! This time we can factor out 2:

$$2x^2 + 34x - 220 = 2(x^2 + 17x - 110).$$

We next look for a factorization of -110 in which one factor is positive, the other factor is negative, and the sum of the factors is 17. Since the sum is to be positive, we examine only pairs of factors in which the positive term has the larger absolute value.

Pair of Factors	Sum of Factors
−1, 110	109
−2, 55	53
−5, 22	17 ← — The numbers we need are −5 and 22.

The factorization of $x^2 + 17x - 110$ is $(x - 5)(x + 22)$. The factorization of the original trinomial, $2x^2 + 34x - 220$, is $2(x - 5)(x + 22)$.

Some polynomials are not factorable using integers.

EXAMPLE 5 Factor: $x^2 - x - 7$.

Solution There are no factors of -7 whose sum is -1. This trinomial is *not* factorable into binomials with integer coefficients. Although $x^2 - x - 7$ can be factored using more advanced techniques, for our purposes the polynomial is **prime**.

> **Tips for Factoring $x^2 + bx + c$**
>
> **1.** If necessary, rewrite the trinomial in descending order. Search for factors of c that add up to b. Remember the following:
>
> - If c is positive, the signs of the factors are the same as the sign of b.
> - If c is negative, one factor is positive and the other is negative.
> - If the sum of the two factors is the opposite of b, changing the signs of both factors will give the desired factors whose sum is b.
>
> **2.** Check the result by multiplying the binomials.

These tips also apply when a trinomial has more than one variable.

EXAMPLE 6 Factor: $x^2 - 2xy - 48y^2$.

Solution We look for numbers p and q such that

$$x^2 - 2xy - 48y^2 = (x + py)(x + qy).$$ **The x's and y's can be written in the binomials in advance.**

Our thinking is much the same as if we were factoring $x^2 - 2x - 48$. We look for factors of -48 whose sum is -2. Those factors are 6 and -8. Thus,

$$x^2 - 2xy - 48y^2 = (x + 6y)(x - 8y).$$

The check is left to the student.

Equations Containing Trinomials

We can now use our new factoring skills to solve some polynomial equations.

EXAMPLE 7 Solve: $x^2 + 9x + 8 = 0$.

Solution We solve using two methods.

CAUTION! This is an equation that we can try to solve. Do not try to solve an expression!

Algebraic Solution

We use the principle of zero products:

$x^2 + 9x + 8 = 0$

$(x + 1)(x + 8) = 0$ **Using the factorization from Example 1**

$x + 1 = 0$ *or* $x + 8 = 0$ **Using the principle of zero products**

$x = -1$ *or* $x = -8.$ **Solving each equation for x**

The solutions are -1 and -8.

Graphical Solution

The real-number solutions of $x^2 + 9x + 8 = 0$ are the first coordinates of the x-intercepts of the graph of $f(x) = x^2 + 9x + 8$.

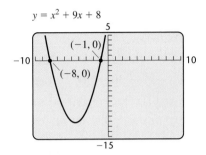

We find that -1 and -8 are solutions.

Check:

For -1:

$$\frac{x^2 + 9x + 8 = 0}{\begin{array}{c} (-1)^2 + 9(-1) + 8 \ ? \ 0 \\ 1 - 9 + 8 \\ 0 \mid 0 \quad \text{TRUE} \end{array}}$$

For -8:

$$\frac{x^2 + 9x + 8 = 0}{\begin{array}{c} (-8)^2 + 9(-8) + 8 \ ? \ 0 \\ 64 - 72 + 8 \\ 0 \mid 0 \quad \text{TRUE} \end{array}}$$

The solutions are -1 and -8.

EXAMPLE 8 Solve: $(t - 10)(t + 1) = -24$.

Solution   We solve using both the principle of zero products and a graphing calculator.

Algebraic Solution

Note that the left side of the equation is factored. It may be tempting to set both factors equal to -24 and solve, but this is *not* correct. The principle of zero products requires 0 on one side of the equation. We begin by multiplying the left side.

$(t - 10)(t + 1) = -24$

$t^2 - 9t - 10 = -24$ **Multiplying**

$t^2 - 9t + 14 = 0$ **Adding 24 to both sides to get 0 on one side**

$(t - 2)(t - 7) = 0$ **Factoring**

$t - 2 = 0$ *or* $t - 7 = 0$ **Using the principle of zero products**

$t = 2$ *or* $t = 7$

The solutions are 2 and 7.

Graphical Solution

There is no need to multiply. We let $y_1 = (x - 10)(x + 1)$ and $y_2 = -24$ and look for any points of intersection of the graphs. The x-coordinates of these points are the solutions of the equation.

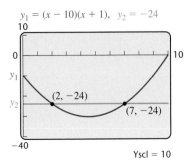

We find that 2 and 7 are solutions.

EXAMPLE 9   Solve: $x^3 = x^2 + 30x$.

Solution   We solve using two methods.

Algebraic Solution

In order to solve $x^3 = x^2 + 30x$ using the principle of zero products, we must rewrite the equation with 0 on one side.

We have

$$x^3 = x^2 + 30x$$

$$x^3 - x^2 - 30x = 0 \qquad \text{Getting 0 on one side}$$

$$x(x - 6)(x + 5) = 0 \qquad \text{Using the factorization from Example 3}$$

$$x = 0 \quad or \quad x - 6 = 0 \quad or \quad x + 5 = 0 \qquad \text{Using the principle of zero products}$$

$$x = 0 \quad or \qquad x = 6 \quad or \qquad x = -5.$$

The solutions are 0, 6, and -5.

Graphical Solution

We let $y_1 = x^3$ and $y_2 = x^2 + 30x$ and look for points of intersection of the graphs.

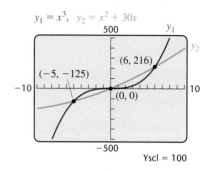

The viewing window $[-10, 10, -500, 500]$ shows three points of intersection. Since the polynomial equation is of degree 3, we know there will be no more than 3 solutions. The x-coordinates of the points of intersection are -5, 0, and 6.

We check by substituting -5, 0, and 6 into the original equation. The solutions are -5, 0, and 6.

Zeros and Factoring

Note in Example 7 the relationship between the factors of $x^2 + 9x + 8$ and the solutions of the equation $x^2 + 9x + 8 = 0$. The graphs of $f(x) = x^2 + 9x + 8$, $g(x) = x + 1$, and $h(x) = x + 8$ are shown below. Note that -8 is a zero of both $f(x)$ and $h(x)$ and -1 is a zero of both $f(x)$ and $g(x)$. We can use the principle of zero products "in reverse" to factor a polynomial and to write a function with given zeros.

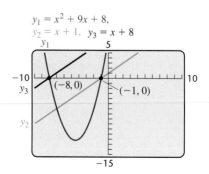

EXAMPLE 10   Factor: $x^2 - 13x - 608$.

Solution   The factorization of the trinomial $x^2 - 13x - 608$ will be in the form

$$x^2 - 13x - 608 = (x + p)(x + q).$$

We could use trial and error to find p and q, but there are many factors of 608. Thus we factor the trinomial using the principle of zero products.

Note that the roots of the equation

$$x^2 - 13x - 608 = 0$$

are the zeros of the function $f(x) = x^2 - 13x - 608$. We graph the function and find the zeros.

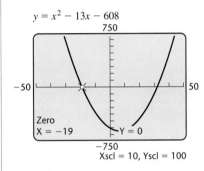

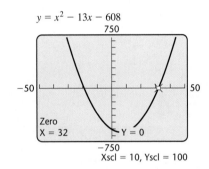

The zeros are -19 and 32. Each zero is a root of $x^2 - 13x - 608 = 0$. We also know that -19 is a zero of the linear function $g(x) = x + 19$ and that 32 is a zero of the linear function $h(x) = x - 32$. This suggests that $x + 19$ and $x - 32$ are factors of $x^2 - 13x - 608$. Using the principle of zero products in reverse, we now have

$$x^2 - 13x - 608 = (x + 19)(x - 32).$$

Multiplication indicates that the factorization is correct.

We can also use the principle of zero products to write a function whose zeros are given.

1. $(x + 2)(x + 6)$ **2.** $(x + 1)(x + 5)$
3. $(t + 3)(t + 5)$ **4.** $(y + 3)(y + 9)$
5. $(x - 9)(x + 3)$ **6.** $(t - 5)(t + 3)$
7. $2(n - 5)(n - 5)$, or $2(n - 5)^2$
8. $2(a - 4)(a - 4)$, or $2(a - 4)^2$
9. $a(a + 8)(a - 9)$ **10.** $x(x + 9)(x - 6)$
11. $(x + 9)(x + 5)$ **12.** $(y + 8)(y + 4)$
13. $(x + 5)(x - 2)$ **14.** $(x + 3)(x - 2)$
15. $3(x + 2)(x + 3)$ **16.** $5(y + 1)(y + 7)$
17. $-(x - 8)(x + 7)$, or $(-x + 8)(x + 7)$, or $(8 - x)(7 + x)$ **18.** $(8 - y)(4 + y)$, or $-(y - 8)(y + 4)$, or $(-y + 8)(y + 4)$
19. $-y(y - 8)(y + 4)$, or $y(-y + 8)(y + 4)$, or $y(8 - y)(4 + y)$ **20.** $-x(x - 8)(x + 7)$, or $x(8 - x)(7 + x)$, or $x(-x + 8)(x + 7)$
21. $x^2(x + 16)(x - 5)$
22. $y^2(y + 12)(y - 7)$ **23.** Prime
24. Prime   **25.** $(p - 8q)(p + 3q)$
26. $(x + 3y)(x + 9y)$
27. $(y + 4z)(y + 4z)$, or $(y + 4z)^2$
28. $(x - 7y)(x - 7y)$, or $(x - 7y)^2$
29. $p^2(p + 1)(p + 79)$
30. $x^2(x + 49)(x + 1)$

EXAMPLE 11 Write a polynomial function $f(x)$ whose zeros are -1, 0, and 3.

Solution   Each zero of the polynomial function f is a zero of a linear factor of the polynomial. We write a linear function for each given zero:

$$-1 \text{ is a zero of } g(x) = x + 1;$$
$$0 \text{ is a zero of } h(x) = x;$$
and $3 \text{ is a zero of } k(x) = x - 3.$

Thus a polynomial function with zeros -1, 0, and 3 is

$$f(x) = (x + 1) \cdot x \cdot (x - 3);$$

multiplying gives us

$$f(x) = x^3 - 2x^2 - 3x.$$

5.4

Exercise Set

Factor.

1. $x^2 + 8x + 12$ **2.** $x^2 + 6x + 5$

3. $t^2 + 8t + 15$ **4.** $y^2 + 12y + 27$

5. $x^2 - 27 - 6x$ **6.** $t^2 - 15 - 2t$

7. $2n^2 - 20n + 50$ **8.** $2a^2 - 16a + 32$

9. $a^3 - a^2 - 72a$ **10.** $x^3 + 3x^2 - 54x$

11. $14x + x^2 + 45$ **12.** $12y + y^2 + 32$

13. $3x + x^2 - 10$ **14.** $x + x^2 - 6$

15. $3x^2 + 15x + 18$ **16.** $5y^2 + 40y + 35$

17. $56 + x - x^2$ **18.** $32 + 4y - y^2$

19. $32y + 4y^2 - y^3$ **20.** $56x + x^2 - x^3$

21. $x^4 + 11x^3 - 80x^2$ **22.** $y^4 + 5y^3 - 84y^2$

23. $x^2 + 12x + 13$ **24.** $x^2 - 3x + 7$

25. $p^2 - 5pq - 24q^2$ **26.** $x^2 + 12xy + 27y^2$

27. $y^2 + 8yz + 16z^2$ **28.** $x^2 - 14xy + 49y^2$

29. $p^4 + 80p^3 + 79p^2$ **30.** $x^4 + 50x^3 + 49x^2$

31. Use the results of Exercise 1 to solve $x^2 + 8x + 12 = 0$. $-6, -2$

32. Use the results of Exercise 2 to solve $x^2 + 6x + 5 = 0$. $-5, -1$

33. Use the results of Exercise 7 to solve $2n^2 + 50 = 20n$. 5

34. Use the results of Exercise 18 to solve $32 + 4y = y^2$. $-4, 8$

35. Use the results of Exercise 9 to solve $a^3 - a^2 = 72a$. $-8, 0, 9$

36. Use the results of Exercise 20 to solve $56x + x^2 = x^3$. $-7, 0, 8$

In Exercises 37–40, use the graph to solve the given equation. Check by substituting into the equation.

Aha! **37.** $x^2 + 4x - 5 = 0$ $-5, 1$ **38.** $x^2 - x - 6 = 0$ $-2, 3$

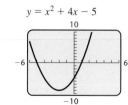

$y = x^2 + 4x - 5$

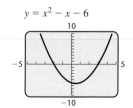

$y = x^2 - x - 6$

39. $x^2 + x - 6 = 0$ $-3, 2$ **40.** $x^2 + 8x + 15 = 0$ $-5, -3$

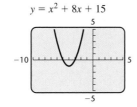

Find the zeros of each function.

41. $f(x) = x^2 - 4x - 45$ $-5, 9$

42. $f(x) = x^2 + x - 20$ $-5, 4$

43. $r(x) = x^3 - 2x^2 - 3x$ $-1, 0, 3$

44. $g(x) = 3x^2 + 21x + 30$ $-5, -2$

Solve.

45. $x^2 + 4x = 45$ $-9, 5$ **46.** $t^2 - 3t = 28$ $-4, 7$

47. $x^2 - 9x = 0$ $0, 9$ **48.** $a^2 + 18a = 0$ $-18, 0$

49. $a^3 - 3a^2 = 40a$ $-5, 0, 8$ **50.** $x^3 - 2x^2 = 63x$ $-7, 0, 9$

51. $(x - 3)(x + 2) = 14$ $-4, 5$ **52.** $(z + 4)(z - 2) = -5$ $-3, 1$

53. $35 - x^2 = 2x$ $-7, 5$ **54.** $40 - x^2 + 3x = 0$ $-5, 8$

In Exercises 55 and 56, use the graph to factor the given polynomial.

55. $x^2 + 10x - 264$ $(x + 22)(x - 12)$

56. $x^2 + 16x - 336$ $(x + 28)(x - 12)$

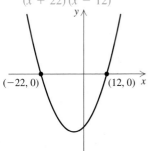

$f(x) = x^2 + 10x - 264$

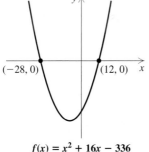

$f(x) = x^2 + 16x - 336$

In Exercises 57–60, use a graph to help factor each polynomial.

57. $x^2 + 40x + 384$ $(x + 24)(x + 16)$

58. $x^2 - 13x - 300$ $(x + 12)(x - 25)$

59. $x^2 + 26x - 2432$ $(x + 64)(x - 38)$

60. $x^2 - 46x + 504$ $(x - 18)(x - 28)$

Write a polynomial function that has the given zeros. Answers may vary.

61. $-1, 2$ ⊡   **62.** $2, 5$ ⊡

63. $-7, -10$ ⊡   **64.** $8, -3$ ⊡

65. $0, 1, 2$ ⊡   **66.** $-3, 0, 5$ ⊡

TW **67.** Sandra says that she will never miss a point of intersection when solving graphically because she

⊡ Answers to Exercises 61–66 and 77 can be found on p. A-62.

always uses a $[-100, 100, -100, 100]$ window. Is she correct? Why or why not?

TW **68.** How can one conclude that $x^2 - 59x + 6$ is a prime polynomial without performing any trials?

Skill Maintenance

Factor. [5.3]

69. $10x^3 - 35x^2 + 5x$ $5x(2x^2 - 7x + 1)$

70. $12t^3 - 40t^2 - 8t$ $4t(3t^2 - 10t - 2)$

Simplify.

71. $(5a^4)^3$ [1.4] $125a^{12}$ **72.** $(-2x^2)^5$ [1.4] $-32x^{10}$

73. If $g(x) = -5x^2 - 7x$, find $g(-3)$. [2.1] -24

74. *Height of a Rocket.* A model rocket is launched upward with an initial velocity of 96 ft/sec from a height of 880 ft. Its height in feet, $h(t)$, after t seconds is given by

$$h(t) = -16t^2 + 96t + 880.$$

What is its height after 0 sec, 1 sec, 3 sec, 8 sec, and 10 sec? [2.1]   880 ft; 960 ft; 1024 ft; 624 ft; 240 ft

Synthesis

TW **75.** Explain how the following graph of

$$y = x^2 + 3x - 2 - (x - 2)(x + 1)$$

can be used to show that

$$x^2 + 3x - 2 \neq (x - 2)(x + 1).$$

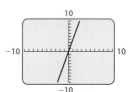

TW **76.** Explain why Example 11 reads "Write *a* polynomial function $f(x)$" and not "Write *the* polynomial function $f(x)$."

77. Use the following graph of $f(x) = x^2 - 2x - 3$ to solve $x^2 - 2x - 3 = 0$ and to solve $x^2 - 2x - 3 < 5$.   ⊡

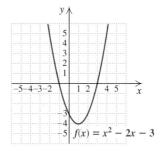

78. Use the following graph of $g(x) = -x^2 - 2x + 3$ to solve $-x^2 - 2x + 3 = 0$ and to solve $-x^2 - 2x + 3 \geq -5$.

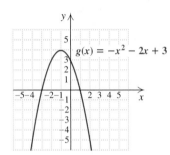

79. Find a polynomial function f for which $f(2) = 0$, $f(-1) = 0$, $f(3) = 0$, and $f(0) = 30$.

80. Find a polynomial function g for which $g(-3) = 0$, $g(1) = 0$, $g(5) = 0$, and $g(0) = 45$.

In Exercises 81–84, use a graphing calculator to find any solutions that exist accurate to two decimal places.

81. $-x^2 + 13.80x = 47.61$ 6.90

82. $-x^2 + 3.63x + 34.34 = x^2$ $-3.33, 5.15$

83. $x^3 - 3.48x^2 + x = 3.48$ 3.48

84. $x^2 + 4.68 = 1.2x$ No real solution

Factor. Assume that variables in exponents represent positive integers.

85. $2a^4b^6 - 3a^2b^3 - 20ab^2$ $ab^2(2a^3b^4 - 3ab - 20)$

86. $5x^8y^6 + 35x^4y^3 + 60$ $5(x^4y^3 + 4)(x^4y^3 + 3)$

87. $x^2 - \frac{4}{25} + \frac{3}{5}x$ $(x + \frac{4}{5})(x - \frac{1}{5})$

88. $y^2 - \frac{8}{49} + \frac{2}{7}y$ $(y + \frac{4}{7})(y - \frac{2}{7})$

89. $y^2 + 0.4y - 0.05$ $(y - 0.1)(y + 0.5)$

90. $x^{2a} + 5x^a - 24$ $(x^a + 8)(x^a - 3)$

91. $x^2 + ax + bx + ab$ $(x + a)(x + b)$

92. $bdx^2 + adx + bcx + ac$ $(bx + a)(dx + c)$

93. $a^2p^{2a} + a^2p^a - 2a^2$ $a^2(p^a + 2)(p^a - 1)$

Aha! **94.** $(x + 3)^2 - 2(x + 3) - 35$ $(x - 4)(x + 8)$

95. Find all integers m for which $x^2 + mx + 75$ can be factored. $76, -76, 28, -28, 20, -20$

96. Find all integers q for which $x^2 + qx - 32$ can be factored. $31, -31, 14, -14, 4, -4$

97. One factor of $x^2 - 345x - 7300$ is $x + 20$. Find the other factor. $x - 365$

78. $\{-3, 1\}; [-4, 2], \text{ or } \{x \mid -4 \leq x \leq 2\}$ **79.** $f(x) = 5x^3 - 20x^2 + 5x + 30;$ answers may vary
80. $g(x) = 3x^3 - 9x^2 - 39x + 45;$ answers may vary

5.5

Factoring Trinomials of the Type $ax^2 + bx + c, a \neq 1$ ■ Equations and Functions

Equations Containing Trinomials of the Type $ax^2 + bx + c, a \neq 1$

Factoring Trinomials of the Type $ax^2 + bx + c$, $a \neq 1$

Now we look at trinomials in which the leading coefficient is not 1. We consider two factoring methods. Use the one that works best for you or what your instructor chooses for you.

Method 1: The FOIL Method

We first consider the **FOIL method** for factoring trinomials of the type

$$ax^2 + bx + c, \text{ where } a \neq 1.$$

Consider the following multiplication.

$$
\begin{array}{c}
\ \ \ \mathrm{F}\qquad\ \mathrm{O}\qquad\ \mathrm{I}\qquad\ \mathrm{L}\\
\ \ \ \downarrow\qquad\ \downarrow\qquad\ \downarrow\qquad\ \downarrow\\
(3x + 2)(4x + 5) = 12x^2 + \underbrace{15x + 8x}\ + 10\\
\ \ \downarrow\qquad\quad\downarrow\qquad\quad\downarrow\\
= 12x^2 + \quad 23x \quad + 10
\end{array}
$$

To factor $12x^2 + 23x + 10$, we must reverse what we just did. We look for two binomials whose product is this trinomial. The product of the First terms must be $12x^2$. The product of the Outer terms plus the product of the Inner terms must be $23x$. The product of the Last terms must be 10. We know from the preceding discussion that the factorization is

$$(3x + 2)(4x + 5).$$

In general, however, finding such an answer involves trial and error. We use the following method.

To Factor $ax^2 + bx + c$ Using FOIL

1. Factor out the largest common factor, if one exists. Here we assume none does.

2. Find two **First** terms whose product is ax^2:

$$(\blacksquare x + \ \ )(\blacksquare x + \ \ ) = ax^2 + bx + c.$$
$$\text{FOIL}$$

3. Find two **Last** terms whose product is c:

$$(\ x + \blacksquare)(\ x + \blacksquare) = ax^2 + bx + c.$$
$$\text{FOIL}$$

4. Repeat steps (2) and (3) until a combination is found for which the sum of the **Outer** and **Inner** products is bx:

$$(\blacksquare x + \blacksquare)(\blacksquare x + \blacksquare) = ax^2 + bx + c.$$
$$\text{I}\qquad\qquad\qquad\text{FOIL}$$
$$\text{O}$$

EXAMPLE 1   Factor: $3x^2 + 10x - 8$.

Solution

1. First, observe that there is no common factor (other than 1 or -1).

2. Next, factor the first term, $3x^2$. The only possibility for factors is $3x \cdot x$. Thus, if a factorization exists, it must be of the form

$$(3x + \blacksquare)(x + \blacksquare).$$

We need to find the right numbers for the blanks.

3. Note that the constant term, -8, can be factored as $(-8)(1)$, $8(-1)$, $(-2)4$, and $2(-4)$, as well as $(1)(-8)$, $(-1)8$, $4(-2)$, and $(-4)2$.

4. Find a pair of factors for which the sum of the products (the "outer" and "inner" parts of FOIL) is the middle term, $10x$. Each possibility should be checked by multiplying:

$$(3x - 8)(x + 1) = 3x^2 - 5x - 8. \qquad \text{O} + \text{I} = 3x + (-8x) = -5x$$

This gives a middle term with a negative coefficient. Since a positive coefficient is needed, a second possibility must be tried:

$$(3x + 8)(x - 1) = 3x^2 + 5x - 8. \qquad \text{O} + \text{I} = -3x + 8x = 5x$$

Note that changing the signs of the two constant terms changes only the sign of the middle term. We try again:

$$(3x - 2)(x + 4) = 3x^2 + 10x - 8. \qquad \text{**This is what we wanted.**}$$

Thus the desired factorization is $(3x - 2)(x + 4)$.

EXAMPLE 2   Factor: $6x^6 - 19x^5 + 10x^4$.

Solution

1. First, factor out the common factor x^4:

$$x^4(6x^2 - 19x + 10).$$

2. Note that $6x^2 = 6x \cdot x$ and $6x^2 = 3x \cdot 2x$. Thus, $6x^2 - 19x + 10$ may factor into

$$(3x + \blacksquare)(2x + \blacksquare) \quad \text{or} \quad (6x + \blacksquare)(x + \blacksquare).$$

3. We factor the last term, 10. The possibilities are $10 \cdot 1, (-10)(-1), 5 \cdot 2$, and $(-5)(-2)$, as well as $1 \cdot 10, (-1)(-10), 2 \cdot 5$, and $(-2)(-5)$.

4. There are 8 possibilities for *each* factorization in step (2). We need factors for which the sum of the products (the "outer" and "inner" parts of FOIL) is the middle term, $-19x$. Since the x-coefficient is negative, we consider pairs of negative factors. Each possible factorization must be checked by multiplying:

$$(3x - 10)(2x - 1) = 6x^2 - 23x + 10.$$

We try again:

$$(3x - 5)(2x - 2) = 6x^2 - 16x + 10.$$

Actually this last attempt could have been rejected by simply noting that $2x - 2$ has a common factor, 2. Since the *largest* common factor was removed in step (1), no other common factors can exist. We try again, reversing the -5 and -2:

$$(3x - 2)(2x - 5) = 6x^2 - 19x + 10. \qquad \text{**This is what we wanted.**}$$

The factorization of $6x^2 - 19x + 10$ is $(3x - 2)(2x - 5)$. But do not forget the common factor! We must include it to get the complete factorization of the original trinomial:

$$6x^6 - 19x^5 + 10x^4 = x^4(3x - 2)(2x - 5).$$

The graphs of the original polynomial and the factorization coincide, as shown in the figure at left.

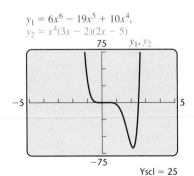

$y_1 = 6x^6 - 19x^5 + 10x^4$,
$y_2 = x^4(3x - 2)(2x - 5)$

Yscl = 25

Tips for Factoring $ax^2 + bx + c$ with FOIL

1. If the largest common factor has been factored out of the original trinomial, then no binomial factor can have a common factor (other than 1 or -1).

2. If a and c are both positive, then the signs in the factors will be the same as the sign of b.

3. When a possible factoring produces the opposite of the desired middle term, reverse the signs of the constants in the factors.

4. Be systematic about your trials. Keep track of those possibilities that you have tried and those that you have not.

Keep in mind that this method of factoring involves trial and error. With practice, you will find yourself making fewer and better guesses.

Method 2: The ac-Method

The second method for factoring trinomials of the type $ax^2 + bx + c, a \neq 1$, is known as the **ac-method**. It involves not only trial and error and FOIL but also factoring by grouping. We know that

$$x^2 + 7x + 10 = x^2 + 2x + 5x + 10$$
$$= x(x + 2) + 5(x + 2)$$
$$= (x + 2)(x + 5),$$

but what if the leading coefficient is not 1? Consider $6x^2 + 23x + 20$. The method is similar to what we just did with $x^2 + 7x + 10$, but we need two more steps.* First, multiply the leading coefficient, 6, and the constant, 20, to get 120. Then find a factorization of 120 in which the sum of the factors is the coefficient of the middle term: 23. The middle term is then split into a sum or difference using these factors.

TEACHING TIP

Some students have difficulty listing all the factors of a number. You may want to review with them an organized method for listing all factors.

$6x^2 + 23x + 20$

(1) Multiply 6 and 20: $6 \cdot 20 = 120$.

(2) Factor 120: $120 = 8 \cdot 15$, and $8 + 15 = 23$.

(3) Split the middle term: $23x = 8x + 15x$.

(4) Factor by grouping.

We factor by grouping as follows:

$$6x^2 + 23x + 20 = 6x^2 + 8x + 15x + 20$$
$$= 2x(3x + 4) + 5(3x + 4) \left.\rule{0pt}{30pt}\right\} \quad \text{Factoring by}$$
$$= (3x + 4)(2x + 5). \quad \quad \text{grouping}$$

---

*The rationale behind these steps is outlined in Exercise 93.

> **To Factor $ax^2 + bx + c$ Using the ac-Method**
>
> **1.** Make sure that any common factors have been factored out.
>
> **2.** Multiply the leading coefficient a and the constant c.
>
> **3.** Try to factor the product ac so that the sum of the factors is b. That is, find integers p and q so that $pq = ac$ and $p + q = b$.
>
> **4.** Split the middle term. That is, write bx as $px + qx$.
>
> **5.** Factor by grouping.

EXAMPLE 3 Factor: $3x^2 + 10x - 8$.

Solution

1. First, look for a common factor. There is none (other than 1 or -1).

2. Multiply the leading coefficient and the constant, 3 and -8.

$$3(-8) = -24.$$

3. Try to factor -24 so that the sum of the factors is 10:

$$-24 = 12(-2) \quad \text{and} \quad 12 + (-2) = 10.$$

4. Split $10x$ using the results of step (3):

$$10x = 12x - 2x.$$

5. Finally, factor by grouping:

$$
\begin{aligned}
3x^2 + 10x - 8 &= 3x^2 + 12x - 2x - 8 \\
&= 3x(x + 4) - 2(x + 4) \\
&= (x + 4)(3x - 2).
\end{aligned}
\left.\vphantom{\begin{aligned} \\ \\ \end{aligned}}\right\} \quad
\begin{aligned}
&\textbf{Factoring by} \\
&\textbf{grouping}
\end{aligned}
$$

We check the solution in two ways.

Algebraic Check

$$
\begin{aligned}
(x + 4)(3x - 2) &= 3x^2 - 2x + 12x - 8 \\
&= 3x^2 + 10x - 8
\end{aligned}
$$

Check by Evaluating

We use a table to check.

$y_1 = 3x^2 + 10x - 8,$
$y_2 = (x + 4)(3x - 2)$

X	Y₁	Y₂
0	-8	-8
1	5	5
2	24	24
3	49	49
4	80	80
5	117	117
6	160	160

X = 0

Equations and Functions

We now use our new factoring skill to solve a polynomial equation. We factor a polynomial *expression* and use the principle of zero products to solve the *equation*.

EXAMPLE 4   Solve: $6x^6 - 19x^5 + 10x^4 = 0$.

Solution   We note at the outset that the polynomial is of degree 6, so there will be at most 6 solutions of the equation.

Algebraic Solution

We solve as follows:

$$6x^6 - 19x^5 + 10x^4 = 0$$
$$x^4(3x - 2)(2x - 5) = 0$$

                                 Factoring as in Example 2

$$x^4 = 0 \quad or \quad 3x - 2 = 0 \quad or \quad 2x - 5 = 0$$

                        Using the principle of zero products

$$x = 0 \quad or \quad x = \tfrac{2}{3} \quad or \quad x = \tfrac{5}{2}.$$

          Solving for x; $x^4 = x \cdot x \cdot x \cdot x = 0$ when $x = 0$

The solutions are $0, \tfrac{2}{3}$, and $\tfrac{5}{2}$.

Graphical Solution

We find the x-intercepts of the function

$$f(x) = 6x^6 - 19x^5 + 10x^4$$

using the ZERO option of the CALC menu.

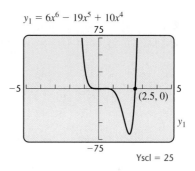

Using the viewing window $[-5, 5, -75, 75]$, we see that 2.5 is a zero of the function. From this window, we cannot tell how many times the graph of the function intersects the x-axis between -1 and 1. We magnify that portion of the x-axis.

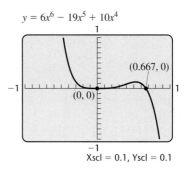

The x-coordinates of the x-intercepts are 0, 0.667, and 2.5.

Since $\tfrac{5}{2} = 2.5$ and $\tfrac{2}{3} \approx 0.667$, we obtain the same solutions using both methods. The solutions are $0, \tfrac{2}{3}$, and $\tfrac{5}{2}$.

Example 4 illustrates two disadvantages of solving equations graphically: (1) It can be easy to "miss" solutions in some viewing windows and (2) solutions may be given as approximations.

Our work with factoring can help us when we are working with functions.

EXAMPLE 5   Given that $f(x) = 3x^2 - 4x$, find all values of a for which $f(a) = 4$.

Solution   We want all numbers a for which $f(a) = 4$. Since $f(a) = 3a^2 - 4a$, we must have

$$3a^2 - 4a = 4 \qquad \text{Setting } f(a) \text{ equal to 4}$$
$$3a^2 - 4a - 4 = 0 \qquad \text{Getting 0 on one side}$$
$$(3a + 2)(a - 2) = 0 \qquad \text{Factoring}$$
$$3a + 2 = 0 \quad or \quad a - 2 = 0$$
$$a = -\tfrac{2}{3} \quad or \qquad a = 2.$$

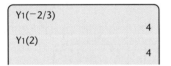

To check using a graphing calculator, we enter $y_1 = 3x^2 - 4x$ and calculate $Y1(-2/3)$ and $Y1(2)$. To have $f(a) = 4$, we must have $a = -\tfrac{2}{3}$ or $a = 2$.

EXAMPLE 6   Find the domain of F if $F(x) = \dfrac{x - 2}{x^2 + 2x - 15}$.

Solution   The domain of F is the set of all values for which

$$\frac{x - 2}{x^2 + 2x - 15}$$

is a real number. Since division by 0 is undefined, $F(x)$ cannot be calculated for any x-value for which the denominator, $x^2 + 2x - 15$, is 0. To make sure these values are *excluded*, we solve:

$$x^2 + 2x - 15 = 0 \qquad \text{Setting the denominator equal to 0}$$
$$(x - 3)(x + 5) = 0 \qquad \text{Factoring}$$
$$x - 3 = 0 \quad or \quad x + 5 = 0$$
$$x = 3 \quad or \qquad x = -5. \qquad \text{These are the values to \textit{exclude}.}$$

TEACHING TIP

Be aware that some students may think the restriction is $x \neq 0$. Reinforce the difficulty of dividing by 0.

X	Y1
3	ERROR
−5	ERROR
0	.13333
.5	.10909
8	.09231
127	.00764
−3	.41667
X = 3	

We can check to see whether either 3 or -5 is in the domain of F using a graphing calculator, as shown at left. We enter $y = (x - 2)/(x^2 + 2x - 15)$ and set up a table with Indpnt set to Ask. When the x-values of 3 and -5 are supplied, the graphing calculator should return an error message, indicating that those numbers are not in the domain of the function. There will be a function value given for any other value of x.

The domain of F is $\{x \mid x \text{ is a real number } \textit{and } x \neq -5 \textit{ and } x \neq 3\}$.

5.5

Exercise Set

Factor.

1. $6x^2 - 5x - 25$
   $(3x + 5)(2x - 5)$
2. $3x^2 - 16x - 12$
   $(3x + 2)(x - 6)$
3. $10y^3 - 12y - 7y^2$
   $y(2y - 3)(5y + 4)$
4. $6x^3 - 15x - x^2$
   $x(3x - 5)(2x + 3)$
5. $24a^2 - 14a + 2$
   $2(4a - 1)(3a - 1)$
6. $3a^2 - 10a + 8$
   $(3a - 4)(a - 2)$
7. $35y^2 + 34y + 8$
   $(7y + 4)(5y + 2)$
8. $9a^2 + 18a + 8$
   $(3a + 2)(3a + 4)$
9. $4t + 10t^2 - 6$
   $2(5t - 3)(t + 1)$
10. $8x + 30x^2 - 6$
    $2(5x + 3)(3x - 1)$
11. $8x^2 - 16 - 28x$
    $4(x - 4)(2x + 1)$
12. $18x^2 - 24 - 6x$
    $6(3x - 4)(x + 1)$
13. $14x^4 - 19x^3 - 3x^2$
    $x^2(2x - 3)(7x + 1)$
14. $70x^4 - 68x^3 + 16x^2$
    $2x^2(5x - 2)(7x - 4)$
15. $12a^2 - 4a - 16$
    $4(3a - 4)(a + 1)$
16. $12a^2 - 14a - 20$
    $2(6a + 5)(a - 2)$
17. $9x^2 + 15x + 4$
    $(3x + 4)(3x + 1)$
18. $6y^2 - y - 2$
    $(3y - 2)(2y + 1)$

Aha! 19. $4x^2 + 15x + 9$
    $(4x + 3)(x + 3)$
20. $2y^2 + 7y + 6$
    $(2y + 3)(y + 2)$
21. $-8t^2 - 8t + 30$
    $-2(2t - 3)(2t + 5)$
22. $-36a^2 + 21a - 3$
    $-3(4a - 1)(3a - 1)$
23. $8 - 6z - 9z^2$
    $(4 + 3z)(2 - 3z)$
24. $3 + 35a - 12a^2$
    $(3 - a)(1 + 12a)$
25. $18xy^3 + 3xy^2 - 10xy$
    $xy(6y + 5)(3y - 2)$
26. $3x^3y^2 - 5x^2y^2 - 2xy^2$
    $xy^2(3x + 1)(x - 2)$
27. $24x^2 - 2 - 47x$
    $(x - 2)(24x + 1)$
28. $15z^2 - 10 - 47z$
    $(5z + 1)(3z - 10)$
29. $63x^3 + 111x^2 + 36x$
    $3x(7x + 3)(3x + 4)$
30. $50t^3 + 115t^2 + 60t$
    $5t(5t + 4)(2t + 3)$
31. $48x^4 + 4x^3 - 30x^2$
    $2x^2(4x - 3)(6x + 5)$
32. $40y^4 + 4y^2 - 12$
    $4(5y^2 + 3)(2y^2 - 1)$
33. $12a^2 - 17ab + 6b^2$
    $(4a - 3b)(3a - 2b)$
34. $20p^2 - 23pq + 6q^2$
    $(4p - 3q)(5p - 2q)$
35. $2x^2 + xy - 6y^2$
    $(2x - 3y)(x + 2y)$
36. $8m^2 - 6mn - 9n^2$
    $(4m + 3n)(2m - 3n)$
37. $6x^2 - 29xy + 28y^2$
    $(3x - 4y)(2x - 7y)$
38. $10p^2 + 7pq - 12q^2$
    $(5p - 4q)(2p + 3q)$
39. $9x^2 - 30xy + 25y^2$
    $(3x - 5y)(3x - 5y),$ or $(3x - 5y)^2$
40. $4p^2 + 12pq + 9q^2$
    $(2p + 3q)(2p + 3q),$ or $(2p + 3q)^2$
41. $9x^2y^2 + 5xy - 4$
    $(9xy - 4)(xy + 1)$
42. $7a^2b^2 + 13ab + 6$
    $(7ab + 6)(ab + 1)$
43. Use the results of Exercise 1 to solve
    $6x^2 - 5x - 25 = 0.$ $-\frac{5}{3}, \frac{5}{2}$
44. Use the results of Exercise 2 to solve
    $3x^2 - 16x - 12 = 0.$ $-\frac{2}{3}, 6$
45. Use the results of Exercise 23 to solve
    $9z^2 + 6z = 8.$ $-\frac{4}{3}, \frac{2}{3}$
46. Use the results of Exercise 24 to solve
    $3 + 35a = 12a^2.$ $-\frac{1}{12}, 3$
47. Use the results of Exercise 29 to solve
    $63x^3 + 111x^2 + 36x = 0.$ $-\frac{4}{3}, -\frac{3}{7}, 0$

48. Use the results of Exercise 30 to solve
    $50t^3 + 115t^2 + 60t = 0.$ $-\frac{3}{2}, -\frac{4}{5}, 0$

Solve.

49. $3x^2 - 8x + 4 = 0$ $\frac{2}{3}, 2$
50. $9x^2 - 15x + 4 = 0$ $\frac{1}{3}, \frac{4}{3}$
51. $4t^3 + 11t^2 + 6t = 0$ $-2, -\frac{3}{4}, 0$
52. $8n^3 + 10n^2 + 3n = 0$
53. $6x^2 = 13x + 5$ $-\frac{1}{3}, \frac{5}{2}$
54. $40x^2 + 43x = 6$ $-\frac{6}{5}, \frac{1}{8}$
55. $x(5 + 12x) = 28$ $-\frac{7}{4}, \frac{4}{3}$
56. $a(1 + 21a) = 10$ $-\frac{5}{7}, \frac{2}{3}$
57. Find the zeros of the function given by
    $f(x) = 2x^2 - 13x - 7.$ $-\frac{1}{2}, 7$
58. Find the zeros of the function given by
    $g(x) = 6x^2 + 13x + 6.$ $-\frac{3}{2}, -\frac{2}{3}$
59. Let $f(x) = x^2 + 12x + 40.$ Find a such that
    $f(a) = 8.$ $-8, -4$
60. Let $f(x) = x^2 + 14x + 50.$ Find a such that
    $f(a) = 5.$ $-9, -5$
61. Let $g(x) = 2x^2 + 5x.$ Find a such that $g(a) = 12.$ $-4, \frac{3}{2}$
62. Let $g(x) = 2x^2 - 15x.$ Find a such that $g(a) = -7.$ $\frac{1}{2}, 7$
63. Let $h(x) = 12x + x^2.$ Find a such that $h(a) = -27.$ $-9, -3$
64. Let $h(x) = 4x - x^2.$ Find a such that $h(a) = -32.$ $-4, 8$

Find the domain of the function f given by each of the following.

65. $f(x) = \dfrac{3}{x^2 - 4x - 5}$ ⊡
66. $f(x) = \dfrac{2}{x^2 - 7x + 6}$
67. $f(x) = \dfrac{x - 5}{9x - 18x^2}$ ⊡
68. $f(x) = \dfrac{1 + x}{3x - 15x^2}$ ⊡
69. $f(x) = \dfrac{7}{5x^3 - 35x^2 + 50x}$ $\{x \mid x$ is a real number *and* $x \neq 0$ and $x \neq 2$ and $x \neq 5\}$
70. $f(x) = \dfrac{3}{2x^3 - 2x^2 - 12x}$ $\{x \mid x$ is a real number *and* $x \neq -2$ and $x \neq 0$ and $x \neq 3\}$

TW 71. Emily has factored a polynomial as $(a - b)(x - y),$ while Jorge has factored the same polynomial as $(b - a)(y - x).$ Who is correct, and why?

⊡ Answers to Exercises 65–68 can be found on p. A-62.

TW **72.** Austin says that the domain of the function

$$F(x) = \frac{x + 3}{3x^2 - x - 2}$$

is $\left\{-\frac{2}{3}, 1\right\}$. Is he correct? Why or why not?

Skill Maintenance

73. The width of a rectangle is 7 ft less than its length. If the width is increased by 2 ft, the perimeter is then 66 ft. What is the area of the original rectangle? [3.3] 228 ft²

74. If $f(x) = 7 - x^2$, find $f(-3)$. [2.1] -2

75. Find the slope and the y-intercept of the line given by $4x - 3y = 8$. [2.4] Slope: $\frac{4}{3}$; y-intercept: $\left(0, -\frac{8}{3}\right)$

Solve. [4.3]

76. $|x| = 27$ $\{-27, 27\}$ **77.** $|5x - 6| \le 39$

78. $|5x - 6| > 39$

Synthesis

TW **79.** Suppose that $(rx + p)(sx - q) = ax^2 - bx + c$ is true. Explain how this can be used to factor $ax^2 + bx + c$.

TW **80.** Describe in your own words an approach that can be used to factor any "nonprime" trinomial of the form $ax^2 + bx + c$.

77. $\left\{x \,\middle|\, -\frac{33}{5} \le x \le 9\right\}$, or $\left[-\frac{33}{5}, 9\right]$ **78.** $\left\{x \,\middle|\, x < -\frac{33}{5} \text{ or } x > 9\right\}$, or $\left(-\infty, -\frac{33}{5}\right) \cup (9, \infty)$

Use a graph to help factor each polynomial.

81. $4x^2 + 120x + 675$ **82.** $4x^2 + 164x + 1197$
  $(2x + 15)(2x + 45)$ $(2x + 63)(2x + 19)$
83. $3x^3 + 150x^2 - 3672x$ **84.** $5x^4 + 20x^3 - 1600x^2$
  $3x(x + 68)(x - 18)$ $5x^2(x + 20)(x - 16)$

Solve.

85. $(8x + 11)(12x^2 - 5x - 2) = 0$ $-\frac{11}{8}, -\frac{1}{4}, \frac{2}{3}$

86. $(x + 1)^3 = (x - 1)^3 + 26$ $-2, 2$

87. $(x - 2)^3 = x^3 - 2$ 1

Factor. Assume that variables in exponents represent positive integers.

88. $6(x - 7)^2 + 13(x - 7) - 5$ $(2x - 9)(3x - 22)$

89. $2a^4b^6 - 3a^2b^3 - 20$ $(2a^2b^3 + 5)(a^2b^3 - 4)$

90. $5x^8y^6 + 35x^4y^3 + 60$ $5(x^4y^3 + 4)(x^4y^3 + 3)$

91. $4x^{2a} - 4x^a - 3$ $(2x^a - 3)(2x^a + 1)$

92. $2ar^2 + 4asr + as^2 - asr$ $a(2r + s)(r + s)$

93. To better understand factoring $ax^2 + bx + c$ by the ac-method, suppose that

$$ax^2 + bx + c = (mx + r)(nx + s).$$

Show that if $P = ms$ and $Q = rn$, then $P + Q = b$ and $PQ = ac$.

Since $ax^2 + bx + c = (mx + r)(nx + s)$, from FOIL we know that $a = mn$, $c = rs$, and $b = ms + rn$. If $P = ms$ and $Q = rn$, then $b = P + Q$. Since $ac = mnrs = msrn$, we have $ac = PQ$.

5.6

Perfect-Square Trinomials ■ Differences of Squares ■ More Factoring by Grouping ■ Solving Equations

Equations Containing Perfect-Square Trinomials and Differences of Squares

We now introduce a faster way to factor trinomials that are squares of binomials. A method for factoring differences of squares is also developed. These factoring methods enable us to solve more types of polynomial equations using the principle of zero products.

Perfect-Square Trinomials

Consider the trinomial

$$x^2 + 6x + 9.$$

To factor it, we can proceed as in Section 5.4 and look for factors of 9 that add to 6. These factors are 3 and 3 and the factorization is

$$x^2 + 6x + 9 = (x + 3)(x + 3) = (x + 3)^2.$$

TEACHING TIP

To check students' understanding of perfect-square trinomials, you can ask the class to give examples.

Note that the result is the square of a binomial. Because of this, we call $x^2 + 6x + 9$ a **perfect-square trinomial**. Although trial and error can be used to factor a perfect-square trinomial, a faster procedure can be used if we recognize when a trinomial is a perfect square.

> ### Determining a Perfect-Square Trinomial
> - Two of the terms must be squares, such as A^2 and B^2.
> - There must be no minus sign before A^2 or B^2.
> - The remaining term is twice the product of A and B, $2AB$, or its opposite, $-2AB$.

EXAMPLE 1   Determine whether each polynomial is a perfect-square trinomial.

a) $x^2 + 10x + 25$
b) $4x + 16 + 3x^2$
c) $100y^2 + 81 - 180y$

Solution

a) • Two of the terms in $x^2 + 10x + 25$ are squares: x^2 and 25.
 • There is no minus sign before either x^2 or 25.
 • The remaining term, $10x$, is twice the product of the square roots, x and 5. Thus, $x^2 + 10x + 25$ *is* a perfect-square trinomial.

b) In $4x + 16 + 3x^2$, only one term, 16, is a square ($3x^2$ is not a square because 3 is not a perfect-square integer and $4x$ is not a square because x is not a square).

Thus, $4x + 16 + 3x^2$ *is not* a perfect-square trinomial.

c) It can help to first write the polynomial in descending order:

$$100y^2 - 180y + 81.$$

 • Two of the terms, $100y^2$ and 81, are squares.
 • There is no minus sign before either $100y^2$ or 81.
 • If the product of the square roots, $10y$ and 9, is doubled, we get the opposite of the remaining term: $2(10y)(9) = 180y$ (the opposite of $-180y$).

Thus, $100y^2 + 81 - 180y$ *is* a perfect-square trinomial.

To factor a perfect-square trinomial, we reuse the patterns that we learned in Section 5.2.

> ### Factoring a Perfect-Square Trinomial
> $A^2 + 2AB + B^2 = (A + B)^2$;
> $A^2 - 2AB + B^2 = (A - B)^2$

EXAMPLE 2 Factor.

a) $x^2 - 10x + 25$

b) $16y^2 + 49 + 56y$

c) $-20xy + 4y^2 + 25x^2$

Caution! **These are expressions that we may be able to factor. They are *not* equations and thus we do not try to solve them.**

Solution

a) $x^2 - 10x + 25 = (x - 5)^2$ We find the square terms and write the square roots with a minus sign between them.

Note the sign!

b) $16y^2 + 49 + 56y = 16y^2 + 56y + 49$ Using a commutative law

$= (4y + 7)^2$ We find the square terms and write the square roots with a plus sign between them.

c) $-20xy + 4y^2 + 25x^2 = 4y^2 - 20xy + 25x^2$ Writing descending order with respect to y

$= (2y - 5x)^2$

This square can also be expressed as

$25x^2 - 20xy + 4y^2 = (5x - 2y)^2.$

As always, any factorization can be checked by multiplying:

$(5x - 2y)^2 = (5x - 2y)(5x - 2y) = 25x^2 - 20xy + 4y^2.$

When factoring, always look first for a factor common to all the terms.

EXAMPLE 3 Factor: **(a)** $2x^2 - 12xy + 18y^2$; **(b)** $-4y^2 - 144y^8 + 48y^5$.

Solution

a) We first look for a common factor. This time, there is a common factor, 2.

$2x^2 - 12xy + 18y^2 = 2(x^2 - 6xy + 9y^2)$ Factoring out the 2

$= 2(x - 3y)^2$ Factoring the perfect-square trinomial

b) $-4y^2 - 144y^8 + 48y^5 = -4y^2(1 + 36y^6 - 12y^3)$ Factoring out the common factor

$= -4y^2(36y^6 - 12y^3 + 1)$ Changing order. Note that $(y^3)^2 = y^6$.

$= -4y^2(6y^3 - 1)^2$ Factoring the perfect-square trinomial

Differences of Squares

When an expression like $x^2 - 9$ is recognized as a difference of two squares, we can reverse another pattern first seen in Section 5.2.

Factoring a Difference of Two Squares

$$A^2 - B^2 = (A + B)(A - B)$$

To factor a difference of two squares, write the product of the sum and the difference of the quantities being squared.

EXAMPLE 4   Factor: **(a)** $x^2 - 9$; **(b)** $25y^6 - 49x^2$.

Solution

a) $x^2 - 9 = x^2 - 3^2 = (x + 3)(x - 3)$

$$A^2 \quad - \quad B^2 = (A + B)(A - B)$$
$$\downarrow \qquad \downarrow \qquad \downarrow \quad \downarrow \quad \downarrow \quad \downarrow \qquad \downarrow$$

b) $25y^6 - 49x^2 = (5y^3)^2 - (7x)^2 = (5y^3 + 7x)(5y^3 - 7x)$

As always, the first step in factoring is to look for common factors.

EXAMPLE 5   Factor: **(a)** $5 - 5x^2y^6$; **(b)** $16x^4y - 81y$.

Solution

a) $5 - 5x^2y^6 = 5(1 - x^2y^6)$ **Factoring out the common factor**

$ = 5[1^2 - (xy^3)^2]$ **Rewriting x^2y^6 as a quantity squared**

$ = 5(1 + xy^3)(1 - xy^3)$ **Factoring the difference of squares**

b) $16x^4y - 81y = y(16x^4 - 81)$ **Factoring out the common factor**

$ = y[(4x^2)^2 - 9^2]$

$ = y(4x^2 + 9)(4x^2 - 9)$ **Factoring the difference of squares**

$ = y(4x^2 + 9)(2x + 3)(2x - 3)$ **Factoring $4x^2 - 9$, which is *also* a difference of squares**

In Example 5(b), it is tempting to try to factor $4x^2 + 9$. Note that it is a sum of two squares. Apart from possibly removing a common factor, do not try to factor a sum of squares using real numbers. Note also in Example 5(b) that $4x^2 - 9$ *could* be factored further. Whenever a factor itself can be factored, do so. We say that we have *factored completely* when none of the factors can be factored further.

More Factoring by Grouping

Sometimes, when factoring a polynomial with four terms, we may be able to factor further.

TEACHING TIP

Remind students that anytime a new polynomial factor appears, they should check to see if that polynomial can be factored further.

EXAMPLE 6 Factor: $x^3 + 3x^2 - 4x - 12$.

Solution

$$
\begin{aligned}
x^3 + 3x^2 - 4x - 12 &= x^2(x + 3) - 4(x + 3) && \text{\textbf{Factoring by grouping}} \\
&= (x + 3)(x^2 - 4) && \text{\textbf{Factoring out } } x + 3 \\
&= (x + 3)(x + 2)(x - 2) && \text{\textbf{Factoring } } x^2 - 4
\end{aligned}
$$

A difference of squares can have four or more terms. For example, one of the squares may be a trinomial. In this case, a type of grouping can be used.

EXAMPLE 7 Factor: **(a)** $x^2 + 6x + 9 - y^2$; **(b)** $a^2 - b^2 + 8b - 16$.

Solution

a) $x^2 + 6x + 9 - y^2 = (x^2 + 6x + 9) - y^2$ **Grouping as a perfect-square trinomial minus y^2 to show a difference of squares**

$$
\begin{aligned}
&= (x + 3)^2 - y^2 \\
&= (x + 3 + y)(x + 3 - y)
\end{aligned}
$$

b) Grouping $a^2 - b^2 + 8b - 16$ into two groups of two terms does not yield a common binomial factor, so we look for a perfect-square trinomial. In this case, the perfect-square trinomial is being subtracted from a^2:

$$
\begin{aligned}
a^2 - b^2 + 8b - 16 &= a^2 - (b^2 - 8b + 16) && \text{\textbf{Factoring out } } -1 \text{ \textbf{and rewriting as subtraction}} \\
&= a^2 - (b - 4)^2 && \text{\textbf{Factoring the perfect-square trinomial}} \\
&= (a + (b - 4))(a - (b - 4)) && \text{\textbf{Factoring a difference of squares}} \\
&= (a + b - 4)(a - b + 4) && \text{\textbf{Removing parentheses}}
\end{aligned}
$$

Solving Equations

We can now solve polynomial equations involving differences of squares and perfect-square trinomials.

EXAMPLE 8 Solve: $x^3 + 3x^2 = 4x + 12$.

Solution   We first note that the equation is a third-degree polynomial equation. Thus it will have 3 or fewer solutions.

Algebraic Solution

We have

$$x^3 + 3x^2 = 4x + 12$$

$$x^3 + 3x^2 - 4x - 12 = 0 \qquad \text{Getting 0 on one side}$$

$$(x + 3)(x + 2)(x - 2) = 0$$

Factoring; using the results of Example 6

$$x + 3 = 0 \quad or \quad x + 2 = 0 \quad or \quad x - 2 = 0$$

Using the principle of zero products

$$x = -3 \quad or \qquad x = -2 \quad or \qquad x = 2.$$

The solutions are -3, -2, and 2.

Graphical Solution

We let $f(x) = x^3 + 3x^2$ and $g(x) = 4x + 12$ and look for any points of intersection of the graphs of f and g.

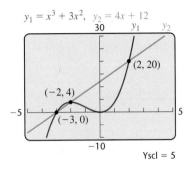

$y_1 = x^3 + 3x^2, \quad y_2 = 4x + 12$

(2, 20)

(−2, 4)

(−3, 0)

Yscl = 5

The x-coordinates of the points of intersection are -3, -2, and 2. These are the solutions of the equation.

As a partial check, note that both methods yield the same solutions. Substituting -3, -2, and 2 into the original equation, we see that all three numbers check. The solutions are -3, -2, and 2.

EXAMPLE 9 Find the zeros of the function given by $f(x) = x^3 + x^2 - x - 1$.

Solution This function is a third-degree polynomial function, so there will be at most 3 zeros. To find the zeros of the function, we find the roots of the equation $x^3 + x^2 - x - 1 = 0$.

Algebraic Solution

We first factor the polynomial:

$$x^3 + x^2 - x - 1 = 0$$

$$x^2(x + 1) - 1(x + 1) = 0 \qquad \text{Factoring by grouping}$$

$$(x + 1)(x^2 - 1) = 0$$

$$(x + 1)(x + 1)(x - 1) = 0. \qquad \text{Factoring a difference of squares}$$

Then we use the principle of zero products:

$$x + 1 = 0 \quad or \quad x + 1 = 0 \quad or \quad x - 1 = 0$$

$$x = -1 \quad or \qquad x = -1 \quad or \qquad x = 1.$$

The solutions are -1 and 1.

Graphical Solution

We graph $f(x) = x^3 + x^2 - x - 1$ and find the zeros of the function.

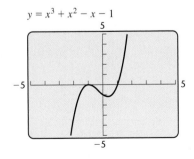

$y = x^3 + x^2 - x - 1$

The zeros are -1 and 1.

The numbers -1 and 1 check. The zeros are -1 and 1.

In Example 9, the factor $(x + 1)$ appeared twice in the factorization. When a factor appears two or more times, we say that we have a **repeated root**. If the factor occurs twice, we say that we have a **double root**, or a **root of multiplicity two**.

In this chapter, we have emphasized the relationship between factoring and equation solving. There are many polynomial equations, however, that cannot be solved by factoring. Nonetheless, we can find approximations of any real-number solutions using a graphing calculator.

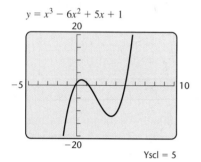

$y = x^3 - 6x^2 + 5x + 1$

X	Y₁	
−.166	9E−5	
1.217	9.5E−4	
4.949	.00328	

X = −.166

EXAMPLE 10   Solve: $x^3 - 6x^2 + 5x + 1 = 0$.

Solution   The polynomial $x^3 - 6x^2 + 5x + 1$ is prime; it cannot be factored using rational coefficients. We solve the equation by graphing $f(x) = x^3 - 6x^2 + 5x + 1$ and finding the zeros of the function. Since the polynomial is of degree 3, we need not look for more than 3 zeros.

There are three x-intercepts of the graph. We use the ZERO feature to find each root, or zero. Since the solutions are irrational, the calculator will give approximations. Rounding each to the nearest thousandth, we have the solutions -0.166, 1.217, and 4.949. We check the solutions using a table (see the figure at left). Note that the function values are close to 0 for each approximate solution.

We have considered both algebraic and graphical methods of solving polynomial equations. It is important to understand and be able to use both methods. Some of the advantages and disadvantages of each method are given in the following table.

	Advantages	Disadvantages
Algebraic Method	• Can find exact answers • Works well when the polynomial is in factored form or can be readily factored • Can be used to find solutions that are not real numbers (see Chapter 8)	• Cannot be used if the polynomial is not factorable • Can be difficult to use if factorization is not readily apparent
Graphical Method	• Does not require the polynomial to be factored • Can visualize solutions	• Easy to miss solutions if an appropriate viewing window is not chosen • Gives approximations of solutions • For most graphing calculators, only real-number solutions can be found.

5.6

Exercise Set

FOR EXTRA HELP

Digital Video
Tutor CD 5
Videotape 7

Student's
Solutions
Manual

Tutor Center

AW Math
Tutor Center

InterAct Math

Math XL
MathXL

MyMathLab

Factor.

1. $x^2 + 8x + 16$ $(x + 4)^2$

2. $t^2 + 6t + 9$ $(t + 3)^2$

3. $a^2 + 16a + 64$ $(a + 8)^2$

4. $a^2 - 14a + 49$ $(a - 7)^2$

5. $2a^2 + 8a + 8$ $2(a + 2)^2$

6. $4a^2 - 16a + 16$ $4(a - 2)^2$

7. $y^2 + 36 - 12y$ $(y - 6)^2$

8. $y^2 + 36 + 12y$ $(y + 6)^2$

9. $24a^2 + a^3 + 144a$ $a(a + 12)^2$

10. $-18y^2 + y^3 + 81y$ $y(y - 9)^2$

11. $32x^2 + 48x + 18$ $2(4x + 3)^2$

12. $2x^2 - 40x + 200$ $2(x - 10)^2$

13. $64 + 25a^2 - 80a$ $(5a - 8)^2$

14. $1 - 8d + 16d^2$ $(1 - 4d)^2$

15. $0.25x^2 + 0.30x + 0.09$ $(0.5x + 0.3)^2$

16. $0.04x^2 - 0.28x + 0.49$ $(0.2x - 0.7)^2$

17. $p^2 - 2pq + q^2$ $(p - q)^2$

18. $m^2 + 2mn + n^2$ $(m + n)^2$

19. $a^3 - 10a^2 + 25a$ $a(a - 5)^2$

20. $y^3 + 8y^2 + 16y$ $y(y + 4)^2$

21. $25a^2 - 30ab + 9b^2$ $(5a - 3b)^2$

22. $49p^2 - 84pq + 36q^2$ $(7p - 6q)^2$

23. $4t^2 - 8tr + 4r^2$ $4(t - r)^2$

24. $5a^2 - 10ab + 5b^2$ $5(a - b)^2$

25. $x^2 - 16$ $(x + 4)(x - 4)$

26. $y^2 - 100$ $(y + 10)(y - 10)$

27. $p^2 - 49$ $(p + 7)(p - 7)$

28. $m^2 - 64$ $(m + 8)(m - 8)$

29. $a^2b^2 - 81$ $(ab + 9)(ab - 9)$

30. $p^2q^2 - 25$ $(pq + 5)(pq - 5)$

31. $6x^2 - 6y^2$ $6(x + y)(x - y)$

32. $8x^2 - 8y^2$ $8(x + y)(x - y)$

33. $7xy^4 - 7xz^4$ $7x(y^2 + z^2)(y + z)(y - z)$

34. $25ab^4 - 25az^4$ $25a(b^2 + z^2)(b + z)(b - z)$

35. $4a^3 - 49a$ $a(2a + 7)(2a - 7)$

36. $9x^4 - 25x^2$ $x^2(3x + 5)(3x - 5)$

37. $3x^8 - 3y^8$ $3(x^4 + y^4)(x^2 + y^2)(x + y)(x - y)$

38. $9a^4 - a^2b^2$ $a^2(3a + b)(3a - b)$

39. $9a^4 - 25a^2b^4$ $a^2(3a + 5b^2)(3a - 5b^2)$

40. $16x^6 - 121x^2y^4$ $x^2(4x^2 + 11y^2)(4x^2 - 11y^2)$

41. $\frac{1}{25} - x^2$ $\left(\frac{1}{5} + x\right)\left(\frac{1}{5} - x\right)$

42. $\frac{1}{16} - y^2$ $\left(\frac{1}{4} + y\right)\left(\frac{1}{4} - y\right)$

43. $(a + b)^2 - 9$ $(a + b + 3)(a + b - 3)$

44. $(p + q)^2 - 25$ $(p + q + 5)(p + q - 5)$

45. $x^2 - 6x + 9 - y^2$ $(x - 3 + y)(x - 3 - y)$

46. $a^2 - 8a + 16 - b^2$ $(a - 4 + b)(a - 4 - b)$

47. $m^2 - 2mn + n^2 - 25$ $(m - n + 5)(m - n - 5)$

48. $x^2 + 2xy + y^2 - 9$ $(x + y + 3)(x + y - 3)$

49. $36 - (x + y)^2$ $(6 + x + y)(6 - x - y)$

50. $49 - (a + b)^2$ $(7 + a + b)(7 - a - b)$

51. $r^2 - 2r + 1 - 4s^2$ $(r - 1 + 2s)(r - 1 - 2s)$

52. $c^2 + 4cd + 4d^2 - 9p^2$ $(c + 2d + 3p)(c + 2d - 3p)$

Aha! **53.** $16 - a^2 - 2ab - b^2$ $(4 + a + b)(4 - a - b)$

54. $9 - x^2 - 2xy - y^2$ $(3 + x + y)(3 - x - y)$

55. $m^3 - 7m^2 - 4m + 28$ $(m - 7)(m + 2)(m - 2)$

56. $x^3 + 8x^2 - x - 8$ $(x + 8)(x + 1)(x - 1)$

57. $a^3 - ab^2 - 2a^2 + 2b^2$ $(a + b)(a - b)(a - 2)$

58. $p^2q - 25q + 3p^2 - 75$ $(p + 5)(p - 5)(q + 3)$

59. Use the results of Exercise 1 to solve $x^2 + 8x + 16 = 0$. -4

60. Use the results of Exercise 4 to solve $a^2 - 14a + 49 = 0$. 7

61. Use the results of Exercise 25 to solve $x^2 = 16$. $-4, 4$

62. Use the results of Exercise 26 to solve $y^2 = 100$. $-10, 10$

Solve. Round any irrational solutions to the nearest thousandth.

63. $a^2 + 1 = 2a$ 1

64. $r^2 + 16 = 8r$ 4

65. $2x^2 - 24x + 72 = 0$ 6

66. $-t^2 - 16t - 64 = 0$ -8

67. $x^2 - 9 = 0$ $-3, 3$

68. $r^2 - 64 = 0$ $-8, 8$

69. $a^2 = \frac{1}{25}$ $-\frac{1}{5}, \frac{1}{5}$

70. $x^2 = \frac{1}{100}$ $-\frac{1}{10}, \frac{1}{10}$

71. $8x^3 + 1 = 4x^2 + 2x$ $-\frac{1}{2}, \frac{1}{2}$

72. $27x^3 + 18x^2 = 12x + 8$ $-\frac{2}{3}, \frac{2}{3}$

73. $x^3 + 3 = 3x^2 + x$ $-1, 1, 3$

74. $x^3 + x^2 = 16x + 16$ $-4, -1, 4$

75. $x^2 - 3x - 7 = 0$ $-1.541, 4.541$

76. $x^2 - 5x + 1 = 0$ $0.209, 4.791$

77. $2x^2 + 8x + 1 = 0$ $-3.871, -0.129$

78. $3x^2 + x - 1 = 0$ $-0.768, 0.434$

79. $x^3 + 3x^2 + x - 1 = 0$ $-2.414, -1, 0.414$

80. $x^3 + x^2 + x - 1 = 0$ 0.544

81. Let $f(x) = x^2 - 12x$. Find a such that $f(a) = -36$. 6

82. Let $g(x) = x^2$. Find a such that $g(a) = 144$. $-12, 12$

Find the zeros of each function.

83. $f(x) = x^2 - 16$ $-4, 4$

84. $f(x) = x^2 - 8x + 16$ 4

85. $f(x) = 2x^2 + 4x + 2$ -1

86. $f(x) = 3x^2 - 27$ $-3, 3$

87. $f(x) = x^3 - 2x^2 - x + 2$ $-1, 1, 2$

88. $f(x) = x^3 + x^2 - 4x - 4$ $-2, -1, 2$

TW **89.** Are the product and power rules for exponents (see Section 1.4) important when factoring differences of squares? Why or why not?

90. Describe a procedure that could be used to find a polynomial with four terms that can be factored as a difference of two squares.

Skill Maintenance

Simplify.

91. $(2a^4b^5)^3$ [1.4] $8a^{12}b^{15}$

92. $(5x^2y^4)^3$ [1.4] $125x^6y^{12}$

93. $(x + y)^3$ [5.2] $x^3 + 3x^2y + 3xy^2 + y^3$

94. $(a + 1)^3$ [5.2] $a^3 + 3a^2 + 3a + 1$

Solve.

95. $x - y + z = 6,$
$2x + y - z = 0,$ $(2, -1, 3)$
$x + 2y + z = 3$ [3.4]

96. $|5 - 7x| \geq 9$ [4.3]
$\{x \mid x \leq -\frac{4}{7} \text{ or } x \geq 2\}$, or
$(-\infty, -\frac{4}{7}] \cup [2, \infty)$

97. $|5 - 7x| \leq 9$ [4.3]
$\{x \mid -\frac{4}{7} \leq x \leq 2\}$, or $[-\frac{4}{7}, 2]$

98. $5 - 7x > -9 + 12x$
[4.1] $\{x \mid x < \frac{14}{19}\}$, or
$(-\infty, \frac{14}{19})$

Synthesis

TW **99.** Without finding the entire factorization, determine the number of factors of $x^{256} - 1$. Explain how you arrived at your answer.

TW **100.** Under what conditions can a sum of two squares be factored?

Factor completely. Assume that variables in exponents represent positive integers.

101. $-\frac{8}{27}r^2 - \frac{10}{9}rs - \frac{1}{6}s^2 + \frac{2}{3}rs$ $-\frac{1}{54}(4r + 3s)^2$

102. $\frac{1}{36}x^8 + \frac{2}{9}x^4 + \frac{4}{9}$ $(\frac{1}{6}x^4 + \frac{2}{3})^2$, or $\frac{1}{36}(x^4 + 4)^2$

103. $0.09x^8 + 0.48x^4 + 0.64$ $(0.3x^4 + 0.8)^2$, or $\frac{1}{100}(3x^4 + 8)^2$

104. $a^2 + 2ab + b^2 - c^2 + 6c - 9$
$(a + b + c - 3)(a + b - c + 3)$

105. $r^2 - 8r - 25 - s^2 - 10s + 16$
$(r + s + 1)(r - s - 9)$

106. $x^{2a} - y^2$ $(x^a + y)(x^a - y)$

107. $x^{4a} - y^{2b}$ $(x^{2a} + y^b)(x^{2a} - y^b)$

TW **108.** $4y^{4a} + 20y^{2a} + 20y^{2a} + 100$ $4(y^{2a} + 5)^2$

109. $25y^{2a} - (x^{2b} - 2x^b + 1)$
$(5y^a + x^b - 1)(5y^a - x^b + 1)$

110. $8(a - 3)^2 - 64(a - 3) + 128$ $8(a - 7)^2$

111. $3(x + 1)^2 + 12(x + 1) + 12$ $3(x + 3)^2$

112. $5c^{100} - 80d^{100}$ $5(c^{50} + 4d^{50})(c^{25} + 2d^{25})(c^{25} - 2d^{25})$

113. $9x^{2n} - 6x^n + 1$ $(3x^n - 1)^2$

114. $c^{2w+1} + 2c^{w+1} + c$ $c(c^w + 1)^2$

115. If $P(x) = x^2$, use factoring to simplify
$$P(a + h) - P(a).\quad h(2a + h)$$

116. If $P(x) = x^4$, use factoring to simplify
$$P(a + h) - P(a).\quad h(2a + h)(2a^2 + 2ah + h^2)$$

117. *Volume of Carpeting.* The volume of a carpet that is rolled up can be estimated by the polynomial $\pi R^2h - \pi r^2h$.

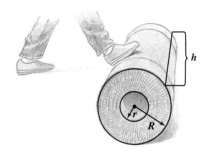

a) Factor the polynomial. $\pi h(R + r)(R - r)$

b) Use both the original and the factored forms to find the volume of a roll for which $R = 50$ cm, $r = 10$ cm, and $h = 4$ m. Use 3.14 for π. 3,014,400 cm³

118. Use a graphing calculator to show that
$$(x^2 - 3x + 2)^4 = x^8 + 81x^4 + 16$$
is *not* an identity.

Enter $y_1 = (x^2 - 3x + 2)^4$ and $y_2 = x^8 + 81x^4 + 16$ and look at the table of values. Note that $y_1 \neq y_2$.

5.7

Factoring Sums or Differences of Cubes ▪ Solving Equations

Equations Containing Sums or Differences of Cubes

Factoring Sums or Differences of Cubes

We have seen that a difference of two squares can be factored but (unless a common factor exists) a *sum* of two squares cannot be factored. The situation is different with cubes: The difference *or sum* of two cubes can always be factored. To see this, consider the following products:

$$(A + B)(A^2 - AB + B^2) = A(A^2 - AB + B^2) + B(A^2 - AB + B^2)$$
$$= A^3 - A^2B + AB^2 + A^2B - AB^2 + B^3$$
$$= A^3 + B^3 \quad \textbf{Combining like terms}$$

and

$$(A - B)(A^2 + AB + B^2) = A(A^2 + AB + B^2) - B(A^2 + AB + B^2)$$
$$= A^3 + A^2B + AB^2 - A^2B - AB^2 - B^3$$
$$= A^3 - B^3. \quad \textbf{Combining like terms}$$

These products allow us to factor a sum or a difference of two cubes.

TEACHING TIP

Consider having students describe the patterns in the signs of the terms in the factorizations.

> **Factoring a Sum or Difference of Two Cubes**
> $A^3 + B^3 = (A + B)(A^2 - AB + B^2);$
> $A^3 - B^3 = (A - B)(A^2 + AB + B^2)$

When factoring a sum or a difference of cubes, it can be helpful to remember the following:

$$1^3 = 1,$$
$$2^3 = 8,$$
$$3^3 = 27,$$
$$4^3 = 64,$$
$$5^3 = 125,$$
$$6^3 = 216,$$
$$x^3 = x^3,$$
$$(x^2)^3 = x^6,$$
$$(x^3)^3 = x^9,$$

and so on. We say that 2 is the *cube root* of 8, that 3 is the cube root of 27, and so on.

EXAMPLE 1 Factor: $x^3 - 27$.

Solution   We have

$$x^3 - 27 = x^3 - 3^3.$$

In one set of parentheses, we write the first cube root, x, minus the second cube root, 3:

$$(x - 3)(\qquad).$$

To get the other factor, we think of $x - 3$ and do the following:

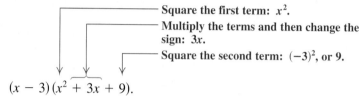

Square the first term: x^2.
Multiply the terms and then change the sign: $3x$.
Square the second term: $(-3)^2$, or 9.

$$(x - 3)(x^2 + 3x + 9).$$

Thus, $x^3 - 27 = (x - 3)(x^2 + 3x + 9)$.

In Example 1, note that $x^2 + 3x + 9$ cannot be factored further. As a general rule, unless A and B share a common factor, the trinomials $A^2 + AB + B^2$ and $A^2 - AB + B^2$ are prime.

EXAMPLE 2   Factor.

a) $125x^3 + y^3$ **b)** $m^6 + 64$

c) $128y^7 - 250x^6y$ **d)** $r^6 - s^6$

Solution

a) We have

$$125x^3 + y^3 = (5x)^3 + y^3.$$

In one set of parentheses, we write the cube root of the first term, $5x$, plus the cube root of the second term, y:

$$(5x + y)(\qquad).$$

To get the other factor, we think of $5x + y$ and do the following:

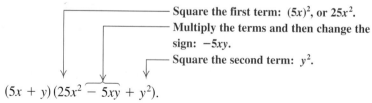

Square the first term: $(5x)^2$, or $25x^2$.
Multiply the terms and then change the sign: $-5xy$.
Square the second term: y^2.

$$(5x + y)(25x^2 - 5xy + y^2).$$

Thus, $125x^3 + y^3 = (5x + y)(25x^2 - 5xy + y^2)$.

b) We have

$$m^6 + 64 = (m^2)^3 + 4^3.$$ **Rewriting as quantities cubed**

Next, we reuse the pattern used in part (a) above:

$$A^3 + B^3 = (A + B)(A^2 - A \cdot B + B^2)$$

$$(m^2)^3 + 4^3 = (m^2 + 4)((m^2)^2 - m^2 \cdot 4 + 4^2)$$
$$= (m^2 + 4)(m^4 - 4m^2 + 16).$$

c) We have

$$128y^7 - 250x^6y = 2y(64y^6 - 125x^6) \quad \text{Remember: } \textit{Always} \textbf{ look for a common factor.}$$
$$= 2y[(4y^2)^3 - (5x^2)^3] \quad \textbf{Rewriting as quantities cubed}$$

To factor $(4y^2)^3 - (5x^2)^3$, it is essential to remember the pattern used in Example 1:

$$A^3 - B^3 = (A - B)(A^2 + A \cdot B + B^2)$$

$$(4y^2)^3 - (5x^2)^3 = (4y^2 - 5x^2)((4y^2)^2 + 4y^2 \cdot 5x^2 + (5x^2)^2)$$
$$= (4y^2 - 5x^2)(16y^4 + 20x^2y^2 + 25x^4).$$

Thus,

$$128y^7 - 250x^6y = 2y(4y^2 - 5x^2)(16y^4 + 20x^2y^2 + 25x^4).$$

d) We have

$$r^6 - s^6 = (r^3)^2 - (s^3)^2$$
$$= (r^3 + s^3)(r^3 - s^3) \quad \textbf{Factoring a difference of two } \textit{squares}$$
$$= (r + s)(r^2 - rs + s^2)(r - s)(r^2 + rs + s^2).$$
$$\textbf{Factoring the sum and difference of two cubes}$$

In Example 2(d), suppose we first factored $r^6 - s^6$ as a difference of two cubes:

$$(r^2)^3 - (s^2)^3 = (r^2 - s^2)(r^4 + r^2s^2 + s^4)$$
$$= (r + s)(r - s)(r^4 + r^2s^2 + s^4).$$

In this case, we might have missed some factors; $r^4 + r^2s^2 + s^4$ can be factored as $(r^2 - rs + s^2)(r^2 + rs + s^2)$, but we probably would never have suspected that such a factorization exists. Given a choice, it is generally better to factor as a difference of squares before factoring as a sum or difference of cubes.

Try to remember the following:

Useful Factoring Facts

Sum of cubes:	$A^3 + B^3 = (A + B)(A^2 - AB + B^2)$;
Difference of cubes:	$A^3 - B^3 = (A - B)(A^2 + AB + B^2)$;
Difference of squares:	$A^2 - B^2 = (A + B)(A - B)$;
Sum of squares:	$A^2 + B^2$ cannot be factored using real numbers if the largest common factor has been removed.

Solving Equations

We can now solve equations involving sums or differences of cubes.

EXAMPLE 3   Solve: $x^3 = 27$.

Solution   We rewrite the equation to get 0 on one side and factor:

$$x^3 = 27$$
$$x^3 - 27 = 0 \qquad \text{Subtracting 27 from both sides}$$
$$(x - 3)(x^2 + 3x + 9) = 0. \qquad \text{Using the factorization from Example 1}$$

The second-degree factor, $x^2 + 3x + 9$, does not factor using real coefficients. When we apply the principle of zero products, we have

$$x - 3 = 0 \quad or \quad x^2 + 3x + 9 = 0. \qquad \text{Using the principle of zero products}$$
$$x = 3 \qquad \text{We cannot factor } x^2 + 3x + 9.$$

We have one solution, 3. In Chapter 8, we will learn how to solve equations like $x^2 + 3x + 9 = 0$. For now, we can solve the equation $x^3 = 27$ graphically, by looking for any points of intersection of the graphs of $y_1 = x^3$ and $y_2 = 27$.

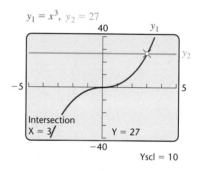

It appears that the only real-number solution is the one found algebraically. The solution of $x^3 = 27$ is 3.

5.7

Exercise Set

Factor completely.

1. $t^3 + 27$
$(t + 3)(t^2 - 3t + 9)$

2. $x^3 + 64$
$(x + 4)(x^2 - 4x + 16)$

3. $x^3 - 8$
$(x - 2)(x^2 + 2x + 4)$

4. $z^3 - 1$
$(z - 1)(z^2 + z + 1)$

5. $m^3 - 64$
$(m - 4)(m^2 + 4m + 16)$

6. $x^3 - 27$
$(x - 3)(x^2 + 3x + 9)$

7. $8a^3 + 1$
$(2a + 1)(4a^2 - 2a + 1)$

8. $27x^3 + 1$
$(3x + 1)(9x^2 - 3x + 1)$

9. $27 - 8t^3$
$(3 - 2t)(9 + 6t + 4t^2)$

10. $64 - 125x^3$
$(4 - 5x)(16 + 20x + 25x^2)$

$(2x + 3)(4x^2 - 6x + 9)$ $(3y + 4)(9y^2 - 12y + 16)$

11. $8x^3 + 27$

12. $27y^3 + 64$

13. $y^3 - z^3$

14. $x^3 - y^3$

15. $x^3 + \frac{1}{27}$

16. $a^3 + \frac{1}{8}$

17. $2y^3 - 128$

18. $8t^3 - 8$

$2(y - 4)(y^2 + 4y + 16)$ $8(t - 1)(t^2 + t + 1)$

19. $8a^3 + 1000$

20. $54x^3 + 2$

$8(a + 5)(a^2 - 5a + 25)$ $2(3x + 1)(9x^2 - 3x + 1)$

21. $rs^3 + 64r$

22. $ab^3 + 125a$

$r(s + 4)(s^2 - 4s + 16)$ $a(b + 5)(b^2 - 5b + 25)$

23. $2y^3 - 54z^3$

24. $5x^3 - 40z^3$

$2(y - 3z)(y^2 + 3yz + 9z^2)$ $5(x - 2z)(x^2 + 2xz + 4z^2)$

25. $y^3 + 0.125$

26. $x^3 + 0.001$

$(y + 0.5)(y^2 - 0.5y + 0.25)$ $(x + 0.1)(x^2 - 0.1x + 0.01)$

27. $125c^6 - 8d^6$

28. $64x^6 - 8t^6$

$8(2x^2 - t^2)(4x^4 + 2x^2t^2 + t^4)$

29. $3z^5 - 3z^2$

30. $2y^4 - 128y$

$3z^2(z - 1)(z^2 + z + 1)$ $2y(y - 4)(y^2 + 4y + 16)$

31. $t^6 + 1$

32. $z^6 - 1$

$(t^2 + 1)(t^4 - t^2 + 1)$

33. $p^6 - q^6$

34. $t^6 + 64y^6$

35. $a^9 + b^{12}c^{15}$

36. $x^{12} - y^3z^{12}$

Solve.

37. $x^3 + 1 = 0$ -1

38. $t^3 - 8 = 0$ 2

39. $8x^3 = 27$ $\frac{3}{2}$

40. $64x^3 + 27 = 0$ $-\frac{3}{4}$

41. $2t^3 - 2000 = 0$ 10

42. $375 = 24x^3$ $\frac{5}{2}$

TW 43. How could you use factoring to convince someone that $x^3 + y^3 \neq (x + y)^3$?

TW 44. Is the following statement true or false and why? If A^3 and B^3 have a common factor, then A and B have a common factor.

Skill Maintenance

45. *Height of a Baseball.* A baseball is thrown upward with an initial velocity of 80 ft/sec from a 224-ft-high cliff. Its height in feet, $h(t)$, after t seconds is given by

$$h(t) = -16t^2 + 80t + 224.$$

What is the height of the ball after 0 sec, 1 sec, 3 sec, 4 sec, and 6 sec? [2.1]   224 ft; 288 ft; 320 ft; 288 ft; 128 ft

In Exercises 46–48, find an equation for a linear equation whose graph has the given characteristics.

46. Slope: $-\frac{3}{4}$; y-intercept: $(0, 5)$ [2.4] $f(x) = -\frac{3}{4}x - 5$

$f(x) = 3x - 8$

47. Slope: 3; contains the point $(1, -5)$ [2.5]

48. Contains the points $(3, -2)$ and $(4, 0)$ [2.5]

$f(x) = 2x - 8$

Solve. [2.2]

49. $3x - 5 = 0$ $\frac{5}{3}$

50. $2x + 7 = 0$ $-\frac{7}{2}$

Synthesis

TW 51. Explain how the geometric model below can be used to verify the formula for factoring $a^3 - b^3$.

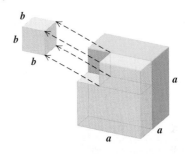

TW 52. Explain how the formula for factoring a *difference* of two cubes can be used to factor $x^3 + 8$.

Factor.

53. $x^{6a} - y^{3b}$

54. $2x^{3a} + 16y^{3b}$

Aha! 55. $(x + 5)^3 + (x - 5)^3$

56. $\frac{1}{16}x^{3a} + \frac{1}{2}y^{6a}z^{9b}$

57. $5x^3y^6 - \frac{5}{8}$

58. $x^3 - (x + y)^3$

59. $x^{6a} - (x^{2a} + 1)^3$

$-(3x^{4a} + 3x^{2a} + 1)$

60. $(x^{2a} - 1)^3 - x^{6a}$

$-(3x^{4a} - 3x^{2a} + 1)$

61. $t^4 - 8t^3 - t + 8$ $(t - 8)(t - 1)(t^2 + t + 1)$

62. If $P(x) = x^3$, use factoring to simplify

$$P(a + h) - P(a).$$ $h(3a^2 + 3ah + h^2)$

63. If $Q(x) = x^6$, use factoring to simplify

$$Q(a + h) - Q(a).$$

$h(2a + h)(a^2 + ah + h^2)(3a^2 + 3ah + h^2)$

64. Using one viewing window, graph the following.

a) $f(x) = x^3$

b) $g(x) = x^3 - 8$

c) $h(x) = (x - 2)^3$

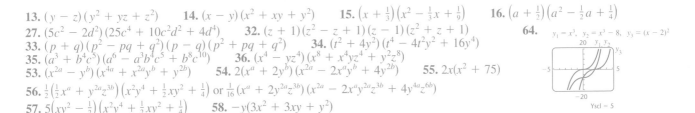

13. $(y - z)(y^2 + yz + z^2)$ **14.** $(x - y)(x^2 + xy + y^2)$ **15.** $\left(x + \frac{1}{3}\right)\left(x^2 - \frac{1}{3}x + \frac{1}{9}\right)$ **16.** $\left(a + \frac{1}{2}\right)\left(a^2 - \frac{1}{2}a + \frac{1}{4}\right)$

27. $(5c^2 - 2d^2)(25c^4 + 10c^2d^2 + 4d^4)$ **32.** $(z + 1)(z^2 - z + 1)(z - 1)(z^2 + z + 1)$

33. $(p + q)(p^2 - pq + q^2)(p - q)(p^2 + pq + q^2)$ **34.** $(t^2 + 4y^2)(t^4 - 4t^2y^2 + 16y^4)$

35. $(a^3 + b^4c^5)(a^6 - a^3b^4c^5 + b^8c^{10})$ **36.** $(x^4 - yz^4)(x^8 + x^4yz^4 + y^2z^8)$

53. $(x^{2a} - y^b)(x^{4a} + x^{2a}y^b + y^{2b})$ **54.** $2(x^a + 2y^b)(x^{2a} - 2x^ay^b + 4y^{2b})$ **55.** $2x(x^2 + 75)$

56. $\frac{1}{2}\left(\frac{1}{2}x^a + y^{2a}z^{3b}\right)\left(x^2y^4 + \frac{1}{2}xy^2 + \frac{1}{4}\right)$ or $\frac{1}{16}(x^a + 2y^{2a}z^{3b})(x^{2a} - 2x^ay^{2a}z^{3b} + 4y^{4a}z^{6b})$

57. $5\left(xy^2 - \frac{1}{2}\right)\left(x^2y^4 + \frac{1}{2}xy^2 + \frac{1}{4}\right)$ **58.** $-y(3x^2 + 3xy + y^2)$

64.

$y_1 = x^3$, $y_2 = x^3 - 8$, $y_3 = (x - 2)^3$

Yscl = 5

Applications of Polynomial Equations

Problem Solving ■ Fitting Polynomial Functions to Data

Polynomial equations and functions occur frequently in problem solving and in data analysis.

Problem Solving

Some problems can be translated to polynomial equations that we can now solve. The problem-solving process is the same as that used for other kinds of problems.

EXAMPLE 1 Prize Tee Shirts. During intermission at sporting events, it has become common for team mascots to use a powerful slingshot to launch tightly rolled tee shirts into the stands. The height $h(t)$, in feet, of an airborne tee shirt t seconds after being launched can be approximated by

$$h(t) = -15t^2 + 75t + 10.$$

After peaking, a rolled-up tee shirt is caught by a fan 70 ft above ground level. How long was the tee shirt in the air?

Solution

1. **Familiarize.** We make a drawing and label it, using the information provided (see the figure). If we wanted to, we could evaluate $h(t)$ for a few values of t. Note that t cannot be negative, since it represents time from launch.

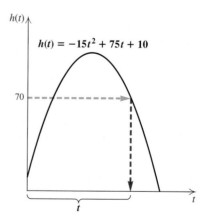

2. **Translate.** The relevant function has been provided. Since we are asked to determine how long it will take for the shirt to reach someone 70 ft above ground level, we are interested in the value of t for which $h(t) = 70$:

$$-15t^2 + 75t + 10 = 70.$$

3. Carry out.   We solve both algebraically and graphically, as follows.

Algebraic Solution

We solve by factoring:

$$-15t^2 + 75t + 10 = 70$$

$$-15t^2 + 75t - 60 = 0 \qquad \text{Subtracting 70 from both sides}$$

$$\left.\begin{array}{l} -15(t^2 - 5t + 4) = 0 \\ -15(t - 4)(t - 1) = 0 \end{array}\right\} \qquad \text{Factoring}$$

$$t - 4 = 0 \quad or \quad t - 1 = 0$$

$$t = 4 \quad or \qquad t = 1.$$

The solutions appear to be 4 and 1.

Graphical Solution

We graph $y_1 = -15x^2 + 75x + 10$ and $y_2 = 70$ and look for any points of intersection of the graphs.

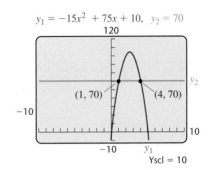

Using the INTERSECT option of the CALC menu, we see that the graphs intersect at $(1, 70)$ and $(4, 70)$. Thus the solutions of the equation are 1 and 4.

4. Check.   We have

$$h(4) = -15 \cdot 4^2 + 75 \cdot 4 + 10 = -240 + 300 + 10 = 70 \text{ ft};$$

$$h(1) = -15 \cdot 1^2 + 75 \cdot 1 + 10 = -15 + 75 + 10 = 70 \text{ ft}.$$

Both 4 and 1 check. However, the problem states that the tee shirt is caught after peaking. Thus we reject 1 since that would indicate when the height of the tee shirt was 70 ft on the way *up*.

5. State.   The tee shirt was in the air for 4 sec before being caught 70 ft above ground level.

The following problem involves the **Pythagorean theorem**, which relates the lengths of the sides of a right triangle. A **right triangle** has a 90°, or right, angle, which is denoted in the triangle by the symbol ⌐ or ⌐. The longest side, opposite the 90° angle, is called the **hypotenuse**. The other sides, called **legs**, form the two sides of the right angle.

The Pythagorean Theorem

In any right triangle, if a and b are the lengths of the legs and c is the length of the hypotenuse, then

$$a^2 + b^2 = c^2.$$

The symbol ⌐ denotes a 90° angle.

TEACHING TIP

You may wish to discuss algebraic representations of "consecutive integers," "consecutive even integers," and "consecutive odd integers."

EXAMPLE 2   Carpentry.   In order to build a deck at a right angle to their house, Lucinda and Felipe decide to place a stake in the ground a precise distance from the back wall of their house. This stake will combine with two marks on the house to form a right triangle. From a course in geometry, Lucinda remembers that there are three consecutive integers that can work as sides of a right triangle. Find the measurements of that triangle.

Solution

1. Familiarize.   Recall that x, $x + 1$, and $x + 2$ can be used to represent three unknown consecutive integers. Since $x + 2$ is the largest number, it must represent the hypotenuse. The legs serve as the sides of the right angle, so one leg must be formed by the marks on the house. We make a drawing in which

$x =$ the distance between the marks on the house,

$x + 1 =$ the length of the other leg,

and

$x + 2 =$ the length of the hypotenuse.

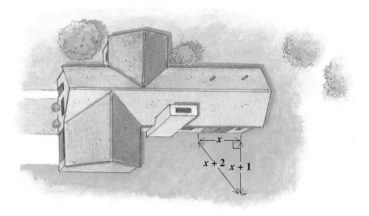

2. Translate.   Applying the Pythagorean theorem, we translate as follows:

$$a^2 + b^2 = c^2$$
$$x^2 + (x + 1)^2 = (x + 2)^2.$$

3. Carry out.   We solve both algebraically and graphically.

Algebraic Solution

We solve the equation as follows:

$$x^2 + (x + 1)^2 = (x + 2)^2$$

$x^2 + (x^2 + 2x + 1) = x^2 + 4x + 4$ **Squaring the binomials**

$2x^2 + 2x + 1 = x^2 + 4x + 4$ **Combining like terms**

$x^2 - 2x - 3 = 0$ **Subtracting $x^2 + 4x + 4$ from both sides**

$(x - 3)(x + 1) = 0$ **Factoring**

$x - 3 = 0$ or $x + 1 = 0$

Using the principle of zero products

$x = 3$ or $x = -1$.

Graphical Solution

We graph $y_1 = x^2 + (x + 1)^2$ and $y_2 = (x + 2)^2$ and look for the points of intersection of the graphs. A standard $[-10, 10, -10, 10]$ viewing window shows one point of intersection.

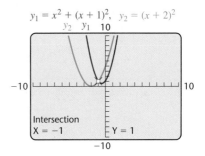

To see if the graphs intersect at another point, we use a larger viewing window. The window $[-10, 10, -10, 50]$ shows that there is indeed another point of intersection.

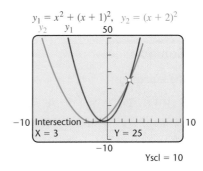

The points of intersection are $(-1, 1)$ and $(3, 25)$, so the solutions are -1 and 3.

TEACHING TIP

You may want to have students discuss how, or even try, to locate a point precisely 4 yd from one mark and 5 yd from a second mark.

4. **Check.**   The integer -1 cannot be a length of a side because it is negative. For $x = 3$, we have $x + 1 = 4$, and $x + 2 = 5$. Since $3^2 + 4^2 = 5^2$, the lengths 3, 4, and 5 determine a right triangle. Thus, 3, 4, and 5 check.

5. **State.**   Lucinda and Felipe should use a triangle with sides having a ratio of $3:4:5$. Thus, if the marks on the house are 3 yd apart, they should locate the stake at the point in the yard that is precisely 4 yd from one mark and 5 yd from the other mark.

EXAMPLE 3   Display of a Sports Card.   A valuable sports card is 4 cm wide and 5 cm long. The card is to be sandwiched by two pieces of Lucite, each of which is $5\frac{1}{2}$ times the area of the card. Determine the dimensions of the Lucite that will ensure a uniform border around the card.

Solution

1. **Familiarize.** We make a drawing and label it, using x to represent the width of the border, in centimeters. Since the border extends uniformly around the entire card, the length of the Lucite must be $5 + 2x$ and the width must be $4 + 2x$.

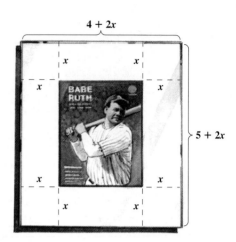

2. **Translate.** We rephrase the information given and translate as follows:

$$\underbrace{\text{Area of Lucite}} \quad \underset{\downarrow}{\text{is}} \quad \underbrace{5\tfrac{1}{2} \text{ times}} \quad \underbrace{\text{area of card.}}$$

$$(5 + 2x)(4 + 2x) \quad = \quad 5\tfrac{1}{2} \cdot \quad \quad 5 \cdot 4$$

3. **Carry out.** We solve both algebraically and graphically, as follows.

Algebraic Solution

We solve the equation:

$$(5 + 2x)(4 + 2x) = 5\tfrac{1}{2} \cdot 5 \cdot 4$$

$20 + 10x + 8x + 4x^2 = 110$ **Multiplying**

$4x^2 + 18x - 90 = 0$ **Finding standard form**

$2x^2 + 9x - 45 = 0$ **Multiplying by $\tfrac{1}{2}$ on both sides**

$(2x + 15)(x - 3) = 0$ **Factoring**

$2x + 15 = 0$ *or* $x - 3 = 0$

              Principle of zero products

$x = -7\tfrac{1}{2}$ *or* $x = 3$.

Graphical Solution

We graph $y_1 = (5 + 2x)(4 + 2x)$ and $y_2 = 5.5(5)(4)$ and look for the points of intersection of the graphs. Since we have $y_2 = 110$, we use a viewing window with Ymax > 110.

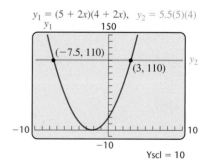

Using INTERSECT, we find the points of intersection $(-7.5, 110)$ and $(3, 110)$. The solutions are -7.5 and 3.

4. **Check.** We check 3 in the original problem. (Note that $-7\tfrac{1}{2}$ is not a solution because measurements cannot be negative.) If the border is 3 cm wide, the Lucite will have a length of $5 + 2 \cdot 3$, or 11 cm, and a width of $4 + 2 \cdot 3$, or 10 cm. The area of the Lucite is thus $11 \cdot 10$, or 110 cm^2.

Since the area of the card is 20 cm² and 110 cm² is $5\frac{1}{2}$ times 20 cm², the number 3 checks.

5. State.   Each piece of Lucite should be 11 cm long and 10 cm wide.

Fitting Polynomial Functions to Data

In Chapters 2 and 3, we modeled data using linear functions, or polynomial functions of degree 1. Some data that are not linear can be modeled using polynomial functions with higher degrees.

EXAMPLE 4   Bachelor's Degrees.   The number of bachelor's degrees earned in the biological and life sciences for various years is shown in the following table and graph.

a) Fit a polynomial function of degree 3 (cubic) to the data.

b) Use the function found in part (a) to estimate the number of bachelor's degrees earned in the biological and life sciences in 1999 and to predict the number that will be earned in 2005.

c) In what year or years were approximately 45,000 bachelor's degrees earned in the biological and life sciences?

Year	Number of Bachelor's Degrees Earned in Biological/Life Science
1971	35,743
1976	54,275
1980	46,370
1986	38,524
1990	37,204
1994	51,383
2000	63,532

U.S. National Center for Education Statistics, *Digest of Education Statistics,* annual

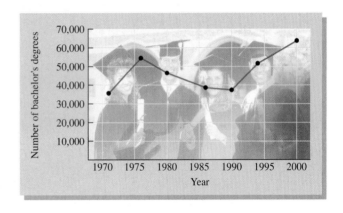

Solution

a) We enter the data in a graphing calculator. To avoid large numbers, we let $x = $ the number of years since 1970 and y the number of bachelor's degrees, in thousands, earned x years after 1970. The years are entered as L1 and the number of degrees as L2.

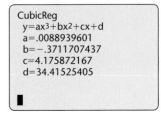

From the graph, we can see that the data do not lie on a straight line, nor do they approximate a straight line. Thus the data are not linear. We use regression to fit a cubic function, or a polynomial of degree 3, to the data. To do so, we choose option 6: CubicReg from the STAT CALC menu. Pressing STAT ▷ 6 VARS ▷ 1 1 ENTER will calculate the regression equation and copy it to Y1. Rounding the coefficients to the nearest thousandth, we find that the cubic function that fits the data is $f(x) = 0.009x^3 - 0.371x^2 + 4.176x + 34.415$.

b) To estimate or predict the number of degrees for years not given, we will use the graph of the equation. To see a scatterplot of the data as well as the graph of the equation, we first check that STAT PLOT is turned on for PLOT1. We need to choose window settings that will show the given years as well as the requested projections. For this situation, we use a window setting of $[0, 40, 0, 100]$, with Xscl = 10 and Yscl = 10. We press GRAPH and use the VALUE option of the CALC menu to find the function value when $x = 29$ (for 1999) and when $x = 35$ (for 2005).

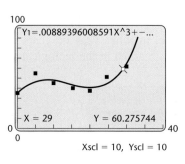

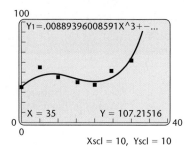

On the basis of this model, we estimate that 60,276 bachelor's degrees were earned in 1999 and that 107,215 will be earned in 2005.

c) To find the year or years in which there were 45,000 bachelor's degrees earned in the biological and life sciences, we solve the equation $f(x) = 45$, where $f(x)$ is the cubic function found in part (a). To solve graphically, we graph $y_2 = 45$ on the same graph as y_1 above, and find all points of intersection of the graphs.

From the graph, we see that there are three points of intersection. Using the INTERSECT option of the CALC menu, we find that the coordinates of the points are $(3.5714729, 45)$, $(13.526583, 45)$, and $(24.634844, 45)$. Rounding, this gives us x-values of 4, 14, and 25, which correspond to the years 1974, 1984, and 1995. Thus we estimate that 45,000 bachelor's degrees were earned in the biological and life sciences in 1974, 1984, and 1995.

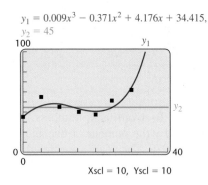

Connecting the Concepts

ESTIMATION AND PREDICTION

When we use data to predict values not given, we assume that the trends shown by the data are consistent for values not shown. In general, estimating values between given data points (also called *interpolation*) is more reliable than predicting values beyond the last data point (called *extrapolation*). However, both estimation and prediction depend on the assumptions we make.

In Example 4, we estimated that 60,276 biological/life science degrees were earned in 1999. We assumed that the number of degrees earned increased for years between 1994 and 2000. However, research would show us that the number of bachelor's degrees earned in those disciplines in 1999 was in fact 65,868, more than were earned in 2000.

Similarly, our prediction of 107,215 biological/life science degrees earned in 2005

depends on the increasing pattern continuing. However, before the year 2005, we cannot know whether that assumption is valid.

Many personal, business, and government decisions are made based on predictions of the future. Mathematics and curve fitting can be very helpful in making those predictions, but we must keep in mind the underlying assumptions and the knowledge that a prediction is not a fact.

In Example 4, the data did not approximate the graph of a linear function (studied in Chapter 2) or the graph of a quadratic function (studied in Chapter 8). A cubic function more closely approximates the data. The best type of model to use for a situation is often difficult to determine.

5.8

Exercise Set

FOR EXTRA HELP

Digital Video Tutor CD 5 Videotape 7

Student's Solutions Manual

Tutor Center
AW Math Tutor Center

InterAct Math

Math XP
MathXL

MyMathLab

Solve.

1. The square of a number plus the number is 132. What is the number?   −12, 11

2. The square of a number plus the number is 156. What is the number?   −13, 12

3. A photo is 5 cm longer than it is wide. Find the length and the width if the area is 84 cm².
Length: 12 cm; width: 7 cm

4. An envelope is 4 cm longer than it is wide. The area is 96 cm². Find the length and the width.
Length: 12 cm; width: 8 cm

5. *Geometry.*   If each of the sides of a square is lengthened by 4 m, the area becomes 49 m². Find the length of a side of the original square.   3 m

6. *Geometry.*   If each of the sides of a square is lengthened by 6 cm, the area becomes 144 cm². Find the length of a side of the original square.   6 cm

7. *Framing a Picture.*   A picture frame measures 12 cm by 20 cm, and 84 cm² of picture shows. Find the width of the frame.   3 cm

8. *Framing a Picture.*   A picture frame measures 14 cm by 20 cm, and 160 cm² of picture shows. Find the width of the frame.   2 cm

9. *Landscaping.*   A rectangular lawn measures 60 ft by 80 ft. Part of the lawn is torn up to install a sidewalk of uniform width around it. The area of the new lawn is 2400 ft². How wide is the sidewalk?   10 ft

10. *Landscaping.*   A rectangular garden is 30 ft by 40 ft. Part of the garden is removed in order to install a walkway of uniform width around it. The area of the new garden is one-half the area of the old garden. How wide is the walkway?   5 ft

11. Three consecutive even integers are such that the square of the third is 76 more than the square of the second. Find the three integers.   16, 18, 20

12. Three consecutive even integers are such that the square of the first plus the square of the third is 136. Find the three integers.   −10, −8, −6 or 6, 8, 10

13. *Tent Design.*   The triangular entrance to a tent is 2 ft taller than it is wide. The area of the entrance is 12 ft². Find the height and the base.
Height: 6 ft; base: 4 ft

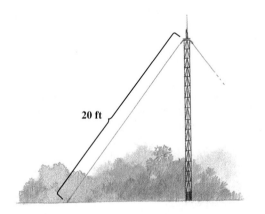

Area = 12 ft²

14. *Antenna Wires.*   A wire is stretched from the ground to the top of an antenna tower, as shown. The wire is 20 ft long. The height of the tower is 4 ft greater than the distance d from the tower's base to the bottom of the wire. Find the distance d and the height of the tower.   Distance d: 12 ft; tower height: 16 ft

20 ft

15. *Sailing.*   A triangular sail is 9 m taller than it is wide. The area is 56 m². Find the height and the base of the sail.
Height: 16 m; base: 7 m

Area = 56 m²

16. *Parking Lot Design.*   A rectangular parking lot is 50 ft longer than it is wide. Determine the dimensions of the parking lot if it measures 250 ft diagonally.   Length: 200 ft; width: 150 ft

17. *Ladder Location.*   The foot of an extension ladder is 9 ft from a wall. The height that the ladder reaches on the wall and the length of the ladder are consecutive integers. How long is the ladder?   41 ft

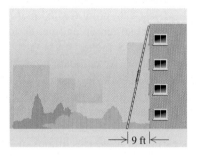

→|9 ft|←

18. *Ladder Location.*   The foot of an extension ladder is 10 ft from a wall. The ladder is 2 ft longer than the height that it reaches on the wall. How far up the wall does the ladder reach?   24 ft

19. *Garden Design.*   Ignacio is planning a garden that is 25 m longer than it is wide. The garden will have an area of 7500 m². What will its dimensions be?
Length: 100 m; width: 75 m

20. *Garden Design.*   A flower bed is to be 3 m longer than it is wide. The flower bed will have an area of 108 m². What will its dimensions be?
Length: 12 m; width: 9 m

21. *Cabinet Making.*   Dovetail Woodworking determines that the revenue R, in thousands of dollars, from the sale of x sets of cabinets is given by $R(x) = 2x^2 + x$. If the cost C, in thousands of dollars, of producing x sets of cabinets is given by $C(x) = x^2 - 2x + 10$, how many sets must be produced and sold in order for the company to break even?   2 sets

22. *Camcorder Production.*   Suppose that the cost of making x video cameras is $C(x) = \frac{1}{9}x^2 + 2x + 1$, where $C(x)$ is in thousands of dollars. If the revenue from the sale of x video cameras is given by $R(x) = \frac{5}{36}x^2 + 2x$, where $R(x)$ is in thousands of dollars, how many cameras must be sold in order for the firm to break even?   6 video cameras

23. *Prize Tee Shirts.*   Using the model in Example 1, determine how long a tee shirt has been airborne if it is caught on the way *up* by a fan 100 ft above ground level.   2 sec

24. *Prize Tee Shirts.*   Using the model in Example 1, determine how long a tee shirt has been airborne if it is caught on the way *down* by a fan 10 ft above ground level.   5 sec

25. *Fireworks Displays.*   Fireworks are typically launched from a mortar with an upward velocity (initial speed) of about 64 ft/sec. The height $h(t)$, in feet, of a "weeping willow" display, t seconds after having been launched from an 80-ft-high rooftop, is given by

$$h(t) = -16t^2 + 64t + 80.$$

After how long will the cardboard shell from the fireworks reach the ground?   5 sec

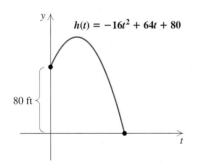

$h(t) = -16t^2 + 64t + 80$

80 ft

26. *Safety Flares.*   Suppose that a flare is launched upward with an initial velocity of 80 ft/sec from a height of 224 ft. Its height in feet, $h(t)$, after t seconds is given by

$$h(t) = -16t^2 + 80t + 224.$$

After how long will the flare reach the ground?   7 sec

27. *Residential Electricity Cost.*   The following table lists the average residential retail cost of electricity in the United States for various years.

a) Use regression to find a quartic (degree 4) polynomial function E that can be used to estimate the average retail cost of electricity, in cents per kilowatt-hour, x years after 1960.   ⊡

b) Estimate the average retail cost of electricity in 1990 and predict the cost in 2004.   ⊡

c) In what year or years between 1960 and 2002 was the average retail cost of electricity 9¢ per kilowatt-hour?   1964, 1977, 1992

Year	Average Retail Cost of Electricity (in cents per kilowatt-hour)
1960	11.6
1966	9.4
1972	7.5
1978	8.9
1982	10.4
1992	8.8
1998	7.9
2002	8.0

Sources: Energy Information Administration; U.S. Department of Health and Human Services

28. *Farms.*   The following table shows the number of U.S. farms for various years.

a) Use regression to find a quartic (degree 4) polynomial function f that can be used to estimate the number of U.S. farms, in millions, x years after 1850.   ⊡

b) Estimate the number of farms in 1990 and predict the number of farms in 2005.   ⊡

c) In what year or years between 1850 and 2000 were there 3 million farms in the United States?   1876, 1974

Year	Number of U.S. Farms (in millions)
1850	1.449
1890	4.565
1920	6.454
1950	5.388
1980	2.440
2000	2.172

Source: *The New York Times Almanac*, 2002

⊡ Answers to Exercises 27(a) and (b) and 28(a) and (b) can be found on p. A-62.

29. *Bachelor's Degrees.* The following table shows the number of bachelor's degrees earned in engineering in the United States for various years.

   a) Use regression to find a quadratic (degree 2) polynomial function d that can be used to estimate the number of bachelor's degrees earned in engineering x years after 1970. ⊡

   b) Estimate the number of bachelor's degrees in engineering earned in 1999 and predict the number that will be earned in 2005. ⊡

   c) In what year or years between 1971 and 2000 were 60,000 engineering bachelor's degrees earned? 1976, 1998

Year	Number of Bachelor's Degrees Earned in Engineering
1971	44,898
1980	68,893
1986	76,225
1994	62,220
2000	58,427

Source: U.S. National Center for Education Statistics, *Digest of Education Statistics*, annual

30. *Teen Birth Rate.* The following table shows the U.S. birth rate for women ages 15 to 19.

   a) Use regression to find a cubic (degree 3) polynomial function r that can be used to estimate the teen birth rate, in number of births per 1000 women, x years after 1970. ⊡

   b) Estimate the teen birth rate in 1993 and predict the rate in 2005. 61.3; 12.8

   c) In what year or years between 1970 and 1999 was the teen birth rate 60 births per 1000 women? 1972, 1989, 1995

Year	U.S. Birth Rate for Women Ages 15 to 19 (in number of births per 1000 women)
1970	69
1975	55.5
1981	51
1991	62
1999	49.6

Source: U.S. Centers for Disease Control and Prevention

31. *Recycling.* The following table lists the percent of plastic soft drink bottles that were recycled for various years.

   a) Use regression to find a quadratic (degree 2) polynomial function b that can be used to estimate the percent of bottles recycled x years after 1990. ⊡

   b) Estimate the percent of bottles recycled in 1992 and in 2000. 40%; 7%

   c) In what year after 1991 were only 30% of bottles recycled? 1998

Year	Percent of Soft Drink Bottles That Were Recycled
1991	34
1994	47
1996	40
1997	38

Source: National Soft Drink Association

32. *Newspapers.* The following table lists the number of weekly newspapers in the United States for various years.

   a) Use regression to find a quartic (degree 4) polynomial function n that can be used to estimate the number of weekly newspapers x years after 1960.

   b) Estimate the number of weekly newspapers in 1985 and predict the number in 2005. 7860; 8985 ⊡

   c) In what year or years between 1960 and 2000 were there 8000 weekly newspapers published? 1961

Year	Number of Weekly Newspapers
1960	8174
1970	7612
1980	7954
1990	7606
2000	7689

Source: Newspaper Association of America, *Facts About Newspapers*, 2001

TW 33. Tyler disregards any negative solutions that he finds when solving applied problems. Is this approach correct? Why or why not?

TW 34. Write a chart of the population of two imaginary cities. Devise the numbers in such a way that one city has linear growth and the other has nonlinear growth.

⊡ Answers to Exercises 29(a) and (b), 30(a), 31(a), and 32(a) can be found on p. A-62.

Skill Maintenance

Simplify. [1.2]

35. $\dfrac{5 - 10 \cdot 3}{-4 + 11 \cdot 4}$ $-\frac{5}{8}$

36. $\dfrac{2 \cdot 3 - 5 \cdot 2}{7 - 3^2}$ 2

37. *Driving.*   At noon, two cars start from the same location going in opposite directions at different speeds. After 7 hr, they are 651 mi apart. If one car is traveling 15 mph slower than the other car, what are their respective speeds? [3.3]   Faster car: 54 mph; slower car: 39 mph

38. *Television Sales.*   At the beginning of the month, J.C.'s Appliances had 150 televisions in stock. During the month, they sold 45% of their conventional televisions and 60% of their surround-sound televisions. If they sold a total of 78 televisions, how many of each type did they sell? [3.3]   Conventional: 36; surround-sound: 42

Solve.

39. $2x - 14 + 9x > -8x + 16 + 10x$ [4.1] $\left\{x \mid x > \frac{10}{3}\right\}$, or $\left(\frac{10}{3}, \infty\right)$

40. $x + y = 0,$
$z - y = -2,$
$x - z = 6$ [3.4] $(2, -2, -4)$

Synthesis

TW **41.** Explain how you might decide what kind of function would fit a particular set of data.

TW **42.** Compare the data in Exercise 29 with that in Example 4. Describe the trends occurring between 1990 and 2000. How might this affect decisions made by college administrations?

43. *Box Construction.*   A rectangular piece of tin is twice as long as it is wide. Squares 2 cm on a side are cut out of each corner, and the ends are turned up to make a box whose volume is 480 cm^3. What are the dimensions of the piece of tin?
Length: 28 cm; width: 14 cm

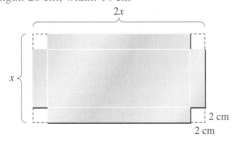

44. *Highway Fatalities.*   The function given by $n(t) = 0.015t^2 - 0.9t + 28$ can be used to estimate the number of deaths $n(t)$ per 100,000 drivers, for drivers t years above age 20 (*Source*: based on information in the *Statistical Abstract of the United States*, 2000). What age group of drivers has a fatality rate of 14.5 deaths per 100,000 drivers?
50-yr-olds

45. *Navigation.*   A tugboat and a freighter leave the same port at the same time at right angles. The freighter travels 7 km/h slower than the tugboat. After 4 hr, they are 68 km apart. Find the speed of each boat.   Tugboat: 15 km/h; freighter: 8 km/h

46. *Skydiving.*   During the first 13 sec of a jump, a skydiver falls approximately $11.12t^2$ feet in t seconds. A small heavy object (with less wind resistance) falls about $15.4t^2$ feet in t seconds. Suppose that a skydiver jumps from 30,000 ft, and 1 sec later a camera falls out of the airplane. How long will it take the camera to catch up to the skydiver?   About 5.7 sec

Collaborative Corner

Diet and Health

Focus: Polynomial functions and data analysis

Time: 20 minutes

Group size: 3

A country's average life expectancy and infant mortality rates are related to the diet of its citizens. The following table shows daily caloric intake, life expectancy, and infant mortality rates for several countries.

Country	Daily Caloric Intake, c	Life Expectancy, L	Infant Mortality Rate, M (in number of deaths per 1000 births)
Argentina	2880	72.3	26.1
Bolivia	2100	62.0	60.2
Canada	3482	80.0	5.5
Dominican Republic	2359	70.4	40.8
Germany	3443	76.7	22.2
Haiti	1707	50.2	98.4
Mexico	3181	75.0	20.7
United States	3671	76.3	6.2

Sources: The Universal Almanac; Statistical Abstract of the United States

ACTIVITY

1. Consider life expectancy as a function of daily caloric intake.

   a) Use regression to fit a linear function, a quadratic function, and a cubic function to the data, with each group member modeling one type of function.

   b) Use each of the three functions to estimate the life expectancy for a country with a daily caloric intake of 1500 calories; of 4500 calories.

   c) Decide as a group which of the three functions best models the situation.

   d) Is there an "optimal" daily caloric intake; that is, one for which the life expectancy is greatest?

2. Consider infant mortality rate as a function of daily caloric intake.

   a) Use regression to fit a linear function, a quadratic function, and a cubic function to the data, with each group member modeling one type of function.

   b) Use each of the three functions to estimate the infant mortality rate for a country with a daily caloric intake of 1500 calories; of 4500 calories.

   c) Decide as a group which of the three functions best models the situation.

   d) Is there an "optimal" daily caloric intake; that is, one for which the infant mortality rate is lowest?

5 Chapter Summary and Review

Key Terms

Linear function, p. 306
Absolute-value function, p. 306
Square-root function, p. 306
Library of functions, p. 306
Term, p. 307
Monomial, p. 307
Degree, p. 307
Coefficient, p. 307
Polynomial, p. 307
Leading term, p. 308
Leading coefficient, p. 308
Degree of a polynomial, p. 308
Linear, p. 308
Quadratic, p. 308
Cubic, p. 308

Binomial, p. 308
Trinomial, p. 308
Descending order, p. 308
Ascending order, p. 308
Polynomial function, p. 309
Similar, or like, terms, p. 313
Opposites, or additive inverses, p. 316
Special products, p. 325
FOIL, p. 325
Square of a binomial, p. 327
Difference of two squares, p. 328
Polynomial equation, p. 333
Zero, p. 334
Root, p. 334

Factor, p. 338
Factored completely, p. 339
Prime polynomial, p. 339
Factoring by grouping, p. 340
Perfect-square trinomial, p. 365
Repeated root, p. 370
Double root, p. 370
Root of multiplicity two, p. 370
Pythagorean theorem, p. 379
Right triangle, p. 379
Hypotenuse, p. 379
Leg, p. 379

Important Properties and Formulas

Factoring Formulas

$A^2 + 2AB + B^2 = (A + B)^2;$

$A^2 - 2AB + B^2 = (A - B)^2;$

$A^2 - B^2 = (A + B)(A - B);$

$A^3 + B^3 = (A + B)(A^2 - AB + B^2);$

$A^3 - B^3 = (A - B)(A^2 + AB + B^2)$

To Factor $ax^2 + bx + c$ Using FOIL

1. Factor out the largest common factor, if one exists. Here we assume none does.

2. Find two **First** terms whose product is ax^2:

$$(\quad x + \quad)(\quad x + \quad) = ax^2 + bx + c.$$
FOIL

3. Find two **Last** terms whose product is c:

$$(\quad x + \quad)(\quad x + \quad) = ax^2 + bx + c.$$
FOIL

4. Repeat steps (2) and (3) until a combination is found for which the sum of the **Outer** and **Inner** products is bx:

$$(\quad x + \quad)(\quad x + \quad) = ax^2 + bx + c.$$
I
O
FOIL

To Factor $ax^2 + bx + c$ Using the ac-method

1. Make sure that any common factors have been factored out.
2. Multiply the leading coefficient a and the constant c.
3. Try to factor the product ac so that the sum of the factors is b. That is, find integers p and q so that $pq = ac$ and $p + q = b$.
4. Split the middle term. That is, write bx as $px + qx$.
5. Factor by grouping.

To Factor a Polynomial

a) Always factor out the largest common factor.
b) Once any common factor has been factored out, look at the number of terms.

   Two terms: Try factoring as a difference of squares first. Next, try factoring as a sum or a difference of cubes. Do *not* try to factor a *sum* of squares.

Three terms: Try factoring as a perfect-square trinomial. Next, try trial and error, using the FOIL method or the ac-method.

Four or more terms: Try factoring by grouping and factoring out a common binomial factor. Next, try grouping into a difference of squares, one of which is a trinomial.

c) Always *factor completely*. If a factor with more than one term can itself be factored further, do so.
d) Check the factorization by multiplying.

The Principle of Zero Products

For any real numbers a and b:

If $ab = 0$, then $a = 0$ or $b = 0$.
If $a = 0$ or $b = 0$, then $ab = 0$.

Review Exercises

1. Given the polynomial
   $$2xy^6 - 7x^8y^3 + 2x^3 - 3,$$
   determine the degree of each term and the degree of the polynomial.   [5.1] 7, 11, 3, 0; 11

2. Given the polynomial
   $$4x - 5x^3 + 2x^2 - 7,$$
   arrange in descending order and determine the leading term and the leading coefficient.
   [5.1] $-5x^3 + 2x^2 + 4x - 7$; $-5x^3$; -5

3. Arrange in ascending powers of x:
   $$3x^6y - 7x^8y^3 + 2x^3 - 3x^2.$$
   [5.1] $-3x^2 + 2x^3 + 3x^6y - 7x^8y^3$

4. Find $P(0)$ and $P(-1)$:
   $$P(x) = x^3 - x^2 + 4x.$$ [5.1] 0; -6

5. Evaluate the polynomial function for $x = -2$:
   $$P(x) = 4 - 2x - x^2.$$ [5.1] 4

Combine like terms.

6. $6 - 4a + a^2 - 2a^3 - 10 + a$ [5.1] $-2a^3 + a^2 - 3a - 4$

7. $4x^2y - 3xy^2 - 5x^2y + xy^2$ [5.1] $-x^2y - 2xy^2$

Add. [5.1] $-x^3 + 2x^2 + 5x + 2$

8. $(-6x^3 - 4x^2 + 3x + 1) + (5x^3 + 2x + 6x^2 + 1)$

9. $(3x^4 + 3x^3 - 8x + 9) + (-6x^4 + 4x + 7 + 3x)$ [5.1] $-3x^4 + 3x^3 - x + 16$

10. $(-9xy^2 - xy + 6x^2y) + (-5x^2y - xy + 4xy^2)$ [5.1] $-5xy^2 - 2xy + x^2y$

Subtract.

11. $(3x - 5) - (-6x + 2)$ [5.1] $9x - 7$

12. $(4a - b + 3c) - (6a - 7b - 4c)$ [5.1] $-2a + 6b + 7c$

13. $(8x^2 - 4xy + y^2) - (2x^2 + 3xy - 2y^2)$ [5.1] $6x^2 - 7xy + 3y^2$

Multiply.

14. $(3x^2y)(-6xy^3)$ [5.2] $-18x^3y^4$

15. $(x^4 - 2x^2 + 3)(x^4 + x^2 - 1)$ [5.2] $x^8 - x^6 + 5x^2 - 3$

16. $(4ab + 3c)(2ab - c)$ [5.2] $8a^2b^2 + 2abc - 3c^2$

17. $(2x + 5y)(2x - 5y)$ [5.2] $4x^2 - 25y^2$

18. $(2x - 5y)^2$ [5.2] $4x^2 - 20xy + 25y^2$

19. $(x + 3)(2x - 1)$ [5.2] $2x^2 + 5x - 3$

20. $(x^2 + 4y^3)^2$ [5.2] $x^4 + 8x^2y^3 + 16y^6$

21. $(x - 5)(x^2 + 5x + 25)$ [5.2] $x^3 - 125$

22. $\left(x - \frac{1}{3}\right)\left(x - \frac{1}{6}\right)$ [5.2] $x^2 - \frac{1}{2}x + \frac{1}{18}$

Factor.

23. $6x^2 + 5x$ [5.3] $x(6x + 5)$

24. $9y^4 - 3y^2$ [5.3] $3y^2(3y^2 - 1)$

25. $15x^4 - 18x^3 + 21x^2 - 9x$
[5.3] $3x(5x^3 - 6x^2 + 7x - 3)$

26. $a^2 - 12a + 27$ [5.4] $(a - 9)(a - 3)$

27. $3m^2 + 14m + 8$ **28.** $25x^2 + 20x + 4$
[5.5] $(3m + 2)(m + 4)$ [5.6] $(5x + 2)^2$

29. $4y^2 - 16$ **30.** $5x^2 + x^3 - 14x$
[5.6] $4(y + 2)(y - 2)$ [5.4] $x(x - 2)(x + 7)$

31. $ax + 2bx - ay - 2by$ **32.** $3y^3 + 6y^2 - 5y - 10$
[5.3] $(a + 2b)(x - y)$ [5.3] $(y + 2)(3y^2 - 5)$

33. $a^4 - 81$ **34.** $4x^4 + 4x^2 + 20$
[5.6] $(a^2 + 9)(a + 3)(a - 3)$ [5.3] $4(x^4 + x^2 + 5)$

35. $27x^3 - 8$ **36.** $0.064b^3 - 0.125c^3$ ⊡
[5.7] $(3x - 2)(9x^2 + 6x + 4)$

37. $y^5 + y$ [5.3] $y(y^4 + 1)$ **38.** $2z^8 - 16z^6$
                                [5.3] $2z^6(z^2 - 8)$

39. $54x^6y - 2y$ **40.** $36x^2 - 120x + 100$
[5.7] $2y(3x^2 - 1)(9x^4 + 3x^2 + 1)$ [5.6] $4(3x - 5)^2$

41. $6t^2 + 17pt + 5p^2$ **42.** $x^3 + 2x^2 - 9x - 18$
[5.5] $(3t + p)(2t + 5p)$ [5.6] $(x + 3)(x - 3)(x + 2)$

43. $a^2 - 2ab + b^2 - 4t^2$ [5.6] $(a - b + 2t)(a - b - 2t)$

44. The following graph is that of a polynomial function $p(x)$. Use the graph to find the zeros of p.
[5.3] $-2, 1, 5$

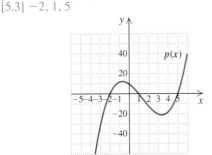

45. Find the zeros of the function given by
$$f(x) = x^2 - 11x + 28.$$ [5.4] $4, 7$

Solve. Where appropriate, round answers to the nearest thousandth.   [5.6] 10   [5.5] $\frac{2}{3}, \frac{3}{2}$

46. $x^2 - 20x = -100$ **47.** $6b^2 - 13b + 6 = 0$

48. $8t^2 = 14t$ [5.3] $0, \frac{7}{4}$ **49.** $r^2 = 16$ [5.6] $-4, 4$

50. $a^3 = 4a^2 + 21a$ **51.** $(x - 1)(x - 4) = 10$
[5.4] $-3, 0, 7$ [5.4] $-1, 6$

52. $x^3 - 5x^2 - 16x + 80 = 0$ [5.6] $-4, 4, 5$

53. $x^2 + 180 = 27x$ [5.4] $12, 15$

54. Let $f(x) = x^2 - 7x - 40$. Find a such that $f(a) = 4$.   [5.4] $-4, 11$

55. Find the domain of the function f given by
$$f(x) = \frac{x - 3}{3x^2 + 19x - 14}.$$ ⊡

56. The area of a square is 5 more than four times the length of a side. What is the length of a side of the square?   [5.8] 5

57. The sum of the squares of three consecutive odd numbers is 83. Find the numbers.
[5.8] 3, 5, 7; $-7, -5, -3$

58. A photograph is 3 in. longer than it is wide. When a 2-in. border is placed around the photograph, the total area of the photograph and the border is 108 in². Find the dimensions of the photograph.
[5.8] Length: 8 in.; width: 5 in.

59. Tim is designing a rectangular garden with a width of 8 ft. The path that leads diagonally across the garden is 2 ft longer than the length of the garden. How long is the path?   [5.8] 17 ft

8 ft

60. The following table lists the U.S. generation of electricity from petroleum for various years.

   a) Use regression to find a cubic (degree 3) polynomial function p that can be used to estimate the amount of electricity generated from petroleum x years after 1960.   ⊡

   b) Predict the amount of electricity generated from petroleum in 2004.   166 billion kilowatt-hours

   c) In what year or years between 1960 and 2000 was there 100 billion kilowatt-hours of electricity generated from petroleum?   1962, 1991, 2000

Year	U.S. Generation of Electricity from Petroleum (in billions of kilowatt-hours)
1960	48
1970	184
1975	289
1980	246
1985	100
1990	124
1995	75
2000	109

Source: U.S. Department of Energy, *Annual Energy Review (2000)*

⊡ Answers to Exercises 36, 55, and 60(a) can be found on p. A-63.

Synthesis

TW **61.** Explain how to find the roots of a polynomial function from its graph.

TW **62.** Explain in your own words why there must be a 0 on one side of an equation before you can use the principle of zero products. $[5.2] \, z^{5n^5}$

Factor.

$[5.7] \, -2(3x^2 + 1)$

63. $128x^6 - 2y^6$ **64.** $(x - 1)^3 - (x + 1)^3$
$[5.7] \, 2(2x - y)(4x^2 + 2xy + y^2)(2x + y)(4x^2 - 2xy + y^2)$

Multiply.

65. $[a - (b - 1)][(b - 1)^2 + a(b - 1) + a^2]$
$[5.2], [5.7] \, a^3 - b^3 + 3b^2 - 3b + 1$

66. $(z^{n^2})^n^3 (z^{4n^3})^{n^2}$

67. Solve: $(x + 1)^3 = x^2(x + 1)$. $[5.6] \, -1, -\frac{1}{2}$

Chapter Test 5

Given the polynomial $3xy^3 - 4x^2y + 5x^5y^4 - 2x^4y$.

1. Determine the degree of the polynomial. $[5.1] \, 9$

2. Arrange in descending powers of x.
$[5.1] \, 5x^5y^4 - 2x^4y - 4x^2y + 3xy^3$

3. Determine the leading term of the polynomial $8a - 2 + a^2 - 4a^3$. $[5.1] \, -4a^3$

4. Given $P(x) = 2x^3 + 3x^2 - x + 4$, find $P(0)$ and $P(-2)$. $[5.1] \, 4; 2$

5. Given $P(x) = x^2 - 5x$, find and simplify
$$P(a + h) - P(a).\quad [5.2] \, 2ah + h^2 - 5h$$

6. Combine like terms:
$$5xy - 2xy^2 - 2xy + 5xy^2.\quad [5.1] \, 3xy + 3xy^2$$

Add. $[5.1] \, -3x^3 + 3x^2 - 6y - 7y^2$

7. $(-6x^3 + 3x^2 - 4y) + (3x^3 - 2y - 7y^2)$

8. $(5m^3 - 4m^2n - 6mn^2 - 3n^3) +$
$(9mn^2 - 4n^3 + 2m^3 + 6m^2n)$
$[5.1] \, 7m^3 + 2m^2n + 3mn^2 - 7n^3$

Subtract.

9. $(9a - 4b) - (3a + 4b)$ $[5.1] \, 6a - 8b$

10. $(6y^2 - 2y - 5y^3) - (4y^2 - 7y - 6y^3)$
$[5.1] \, 2y^2 + 5y + y^3$

Multiply.

11. $(-4x^2y)(-16xy^2)$ $[5.2] \, 64x^3y^3$

12. $(6a - 5b)(2a + b)$ $[5.2] \, 12a^2 - 4ab - 5b^2$

13. $(x - y)(x^2 - xy - y^2)$ $[5.2] \, x^3 - 2x^2y + y^3$

14. $(2x^3 + 5)^2$ $[5.2] \, 4x^6 + 20x^3 + 25$

15. $(4y - 9)^2$ $[5.2] \, 16y^2 - 72y + 81$

16. $(x - 2y)(x + 2y)$ $[5.2] \, x^2 - 4y^2$

Factor.

17. $15x^2 - 5x^4$ $[5.3] \, 5x^2(3 - x^2)$

18. $y^3 + 5y^2 - 4y - 20$ $[5.6] \, (y + 5)(y + 2)(y - 2)$

19. $p^2 - 12p - 28$ $[5.4] \, (p - 14)(p + 2)$

20. $12m^2 + 20m + 3$ $[5.5] \, (6m + 1)(2m + 3)$

21. $9y^2 - 25$ $[5.6] \, (3y + 5)(3y - 5)$

22. $3r^3 - 3$ $[5.7] \, 3(r - 1)(r^2 + r + 1)$

23. $9x^2 + 25 - 30x$ $[5.6] \, (3x - 5)^2$

24. $x^8 - y^8$ $[5.6] \, (x^4 + y^4)(x^2 + y^2)(x + y)(x - y)$

25. $y^2 + 8y + 16 - 100t^2$ $[5.6] \, (y + 4 + 10t)(y + 4 - 10t)$

26. $20a^2 - 5b^2$ $[5.6] \, 5(2a - b)(2a + b)$

27. $24x^2 - 46x + 10$ $[5.5] \, 2(4x - 1)(3x - 5)$

28. $16a^7b + 54ab^7$ $[5.7] \, 2ab(2a^2 + 3b^2)(4a^4 - 6a^2b^2 + 9b^4)$

29. $4y^4x + 36yx^2 + 8y^2x^3 - 16xy$

30. The graph shown below is that of the polynomial function $p(x)$. Use the graph to determine the zeros of p. $[5.3] \, -3, -1, 2, 4$

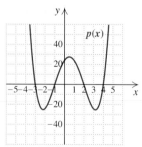

29. $[5.3] \, 4xy(y^3 + 9x + 2x^2y - 4)$

31. Find the zeros of the function given by
$f(x) = 2x^2 - 11x - 40.$ [5.5] $-\frac{5}{2}, 8$

Solve. Where appropriate, round solutions to the nearest thousandth.

32. $x^2 - 18 = 3x$ [5.4] $-3, 6$ **33.** $5t^2 = 125$ [5.6] $-5, 5$

34. $2x^2 + 21 = -17x$ [5.5] $-7, -\frac{3}{2}$ **35.** $9x^2 + 3x = 0$ [5.3] $-\frac{1}{3}, 0$

36. $x^2 + 81 = 18x$ [5.6] 9 **37.** $x^2(x + 1) = 8x$ [5.4] $-3.372, 0, 2.372$

38. Let $f(x) = 3x^2 - 15x + 11$. Find a such that $f(a) = 11$.   [5.5] $0, 5$

39. Find the domain of the function f given by
$$f(x) = \frac{3 - x}{x^2 + 2x + 1}.$$

40. A photograph is 3 cm longer than it is wide. Its area is 40 cm². Find its length and its width.

41. To celebrate a town's centennial, fireworks are launched over a lake off a dam 36 ft above the water. The height of a display, t seconds after it has been launched, is given by
$$h(t) = -16t^2 + 64t + 36.$$
After how long will the shell from the fireworks reach the water?   [5.8] $4\frac{1}{2}$ sec

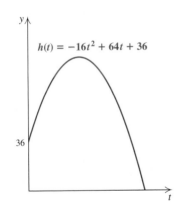

42. *Birth Rate.*   The following table lists the average number of live births to women who are x years old.

Age	Average Number of Live Births per 1000 Women
16	34
18.5	86.5
22	111.1
27	113.9
32	84.5
37	35.4
42	6.8

Source: U.S. Centers for Disease Control and Prevention

a) Use regression to find a cubic (degree 3) polynomial function b that can be used to estimate the average number of live births per 1000 women who are x years old.

b) Estimate the average number of live births per 1000 20-yr-old women.

c) For what age or ages are there approximately 100 live births per 1000 women?

Synthesis

43. a) Multiply: $(x^2 + x + 1)(x^3 - x^2 + 1)$.   [5.2] $x^5 + x + 1$

 b) Factor: $x^5 + x + 1$.   [5.2], [5.7] $(x^2 + x + 1)(x^3 - x^2 + 1)$

44. Factor: $6x^{2n} - 7x^n - 20$.   [5.5] $(3x^n + 4)(2x^n - 5)$

39. [5.5] $\{x \mid x$ is a real number *and* $x \neq -1\}$
40. [5.8] Length: 8 cm; width: 5 cm
42. [5.8] **(a)** $0.03135826x^3 - 3.21956840x^2 + 10117516232x - 886.92962439$; **(b)** 100; **(c)** 20, 30

Rational Expressions, Equations, and Functions

Like fractions in arithmetic, a rational expression is an expression that indicates division. In this chapter, we add, subtract, multiply, and divide rational expressions and use them in equations and functions. We then use rational expressions to solve problems that we could not have solved before.

APPLICATION

ULTRAVIOLET INDEX. The ultraviolet, or UV, index is a measure issued daily by the National Weather Service that indicates the strength of the sun's rays in a particular locale. The following table shows, for people whose skin is highly sensitive, how many minutes of exposure to the sun is safe (will not cause sunburn) on days with various UV ratings (*Source*: *Indianapolis Star*, 7/18/02). Find an equation of variation and estimate the safe exposure time when the UV rating is 5. The graph shows that we can use an *equation of inverse variation* to estimate the safe exposure time.

UV Index	Safe Exposure Time (in minutes)
2	30
4	15
6	10
8	8
10	6

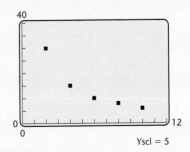

This problem appears as Example 10 in Section 6.8.

6.1

Rational Functions ▪ Multiplying ▪ Simplifying Rational Expressions ▪ Dividing and Simplifying

Rational Expressions and Functions: Multiplying and Dividing

An expression that consists of a polynomial divided by a nonzero polynomial is called a **rational expression**. The following are examples of rational expressions:

$$\frac{3}{4}, \quad \frac{x}{y}, \quad \frac{9}{a+b}, \quad \frac{x^2 + 7xy - 4}{x^3 - y^3}, \quad \frac{1 + z^3}{1 - z^6}.$$

Rational Functions

Like polynomials, certain rational expressions are used to describe functions. Such functions are called **rational functions**.

Following is the graph of the rational function

$$f(x) = \frac{1}{x}.$$

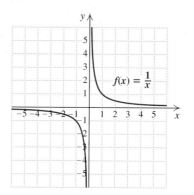

Graphs of rational functions vary widely in shape, but some general statements can be drawn from the graph shown above.

1. Graphs of rational functions may not be continuous—that is, there may be a break in the graph. The graph of $f(x) = 1/x$ consists of two unconnected parts, with a break at $x = 0$.

2. The domain of a rational function may not include all real numbers. The domain of $f(x) = 1/x$ does not include 0.

Although we will not consider graphs of rational functions in detail in this course, you should add the graph of $f(x) = 1/x$ to your library of functions.

EXAMPLE 1   The function given by

$$H(t) = \frac{t^2 + 5t}{2t + 5}$$

gives the time, in hours, for two machines, working together, to complete a job that the first machine could do alone in t hours and the other machine could do in $t + 5$ hours. How long will the two machines, working together, require for the job if the first machine alone would take **(a)** 1 hour? **(b)** 5 hours?

Solution

a) $H(1) = \dfrac{1^2 + 5 \cdot 1}{2 \cdot 1 + 5} = \dfrac{1 + 5}{2 + 5} = \dfrac{6}{7}$ hr

b) $H(5) = \dfrac{5^2 + 5 \cdot 5}{2 \cdot 5 + 5} = \dfrac{25 + 25}{10 + 5} = \dfrac{50}{15} = \dfrac{10}{3}$ hr, or $3\frac{1}{3}$ hr

In Section 5.5, we found that the domain of a rational function must exclude any numbers for which the denominator is 0. For a function like H above, the denominator is 0 when t is $-\frac{5}{2}$, so the domain of H is $\left(-\infty, -\frac{5}{2}\right) \cup \left(-\frac{5}{2}, \infty\right)$.
    How is the domain of a rational function related to its graph?

Interactive Discovery

Consider the function given by

$$f(x) = \frac{2}{x - 3}.$$

1. What number is not in the domain of the function?   3
2. Let $y = 2/(x - 3)$. Graph the function and trace along the graph, starting at $x = 0$ and moving to the right. What happens to the function values as x gets closer to 3?   The function values get very small.
3. Now trace along the graph, starting at $x = 6$ and moving to the left. What happens to the function values as x gets closer to 3?   The function values get very large.
4. What happens when you attempt to find the value of the function for $x = 3$?   A table and Y1(3) return error messages. The VALUE option of the CALC menu returns nothing for Y when X = 3.

The pattern you may have observed in the Interactive Discovery is true in general.

If a is not in the domain of f, the graph of f does not cross the line $x = a$.

Consider the graph of the function in Example 1,

$$H(t) = \frac{t^2 + 5t}{2t + 5}.$$

We consider all values for t, although negative values would not be considered for time. Note that the graph consists of two unconnected "branches." Since $-\frac{5}{2}$ is not in the domain of H, a vertical line drawn at $-\frac{5}{2}$ does not touch the graph of H. The line $x = -\frac{5}{2}$ is called a *vertical asymptote*.

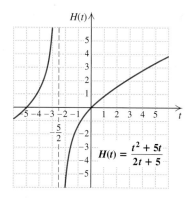

Now consider the graph of

$$H(t) = \frac{t^2 + 5t}{2t + 5}$$

as drawn using a graphing calculator. The vertical line that appears on the screen on the left below is not part of the graph, nor should it be considered a vertical asymptote. Since a graphing calculator graphs an equation by plotting points and connecting them, the vertical line is actually connecting a point just to the left of the line $x = -\frac{5}{2}$ with a point just to the right of the line $x = -\frac{5}{2}$. Such a vertical line may or may not appear for values not in the domain of a function.

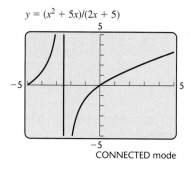

CONNECTED mode

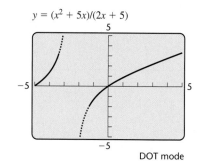

DOT mode

If we tell the calculator not to connect the points that it plots, the vertical line will not appear. Changing from CONNECTED mode to DOT mode gives us the graph on the right above.

Multiplying

The calculations that are performed with rational expressions resemble those performed in arithmetic.

> ### Products of Rational Expressions
>
> To multiply two rational expressions, multiply numerators and multiply denominators:
>
> $$\frac{A}{B} \cdot \frac{C}{D} = \frac{AC}{BD}, \quad \text{where } B \neq 0, D \neq 0.$$

EXAMPLE 2   Multiply: $\dfrac{x+1}{y-3} \cdot \dfrac{x^2}{y+1}$.

Solution

$$\frac{x+1}{y-3} \cdot \frac{x^2}{y+1} = \frac{(x+1)x^2}{(y-3)(y+1)}$$

Multiplying the numerators and multiplying the denominators. The product is a single rational expression.

Recall from arithmetic that multiplication by 1 can be used to find equivalent expressions:

$$\frac{3}{5} = \frac{3}{5} \cdot \frac{2}{2} \qquad \text{Multiplying by } \tfrac{2}{2}, \text{ which is 1}$$

$$= \frac{6}{10}. \qquad \tfrac{3}{5} \text{ and } \tfrac{6}{10} \text{ represent the same number.}$$

Similarly, multiplication by 1 can be used to find equivalent rational expressions:

$$\left.\begin{aligned} \frac{x-5}{x+2} &= \frac{x-5}{x+2} \cdot \frac{x+3}{x+3} \\[4pt] &= \frac{(x-5)(x+3)}{(x+2)(x+3)}. \end{aligned}\right\}$$

Multiplying by $\dfrac{x+3}{x+3}$, which, provided $x \neq -3$, is 1

The expressions

$$\frac{x-5}{x+2} \quad \text{and} \quad \frac{(x-5)(x+3)}{(x+2)(x+3)}$$

are equivalent: So long as x is replaced with a number other than -2 or -3, both expressions represent the same number. For example, if $x = 4$, then

$$\frac{x-5}{x+2} = \frac{4-5}{4+2} = \frac{-1}{6}$$

and

$$\frac{(x-5)(x+3)}{(x+2)(x+3)} = \frac{(4-5)(4+3)}{(4+2)(4+3)} = \frac{-1 \cdot 7}{6 \cdot 7} = \frac{-7}{42} = \frac{-1}{6}.$$

EXAMPLE 3 Multiply to find an equivalent expression:

$$\frac{4-x}{y-x}\cdot\frac{-1}{-1}.$$

Solution We have

$$\frac{4-x}{y-x}\cdot\frac{-1}{-1}=\frac{-4+x}{-y+x}=\frac{x-4}{x-y}.$$

Multiplying by $-1/-1$ is the same as multiplying by 1, so

$$\frac{4-x}{y-x} \text{ is equivalent to } \frac{x-4}{x-y}.$$

TEACHING TIP

It is important that students understand that they are multiplying by a form of 1, not by -1.

Simplifying Rational Expressions

As in arithmetic, rational expressions are *simplified* by "removing" a factor equal to 1. This reverses the process shown above:

$$\frac{6}{10}=\frac{3\cdot 2}{5\cdot 2}=\frac{3}{5}\cdot\frac{2}{2}=\frac{3}{5}.$$ **We "removed" the factor that equals 1: $\frac{2}{2}=1$.**

Similarly,

$$\frac{(x-5)(x+3)}{(x+2)(x+3)}=\frac{x-5}{x+2}\cdot\frac{x+3}{x+3}=\frac{x-5}{x+2}.$$ **We "removed" the factor that equals 1: $\frac{x+3}{x+3}=1$.**

EXAMPLE 4 Simplify: $\frac{7a^2+21a}{14a}$.

Solution We first factor the numerator and the denominator, looking for the largest factor common to both. Once the greatest common factor is found, we use it to write 1 and simplify:

$$\frac{7a^2+21a}{14a}=\frac{7a(a+3)}{7\cdot 2\cdot a}$$ **Factoring. The greatest common factor is $7a$.**

$$=\frac{7a}{7a}\cdot\frac{a+3}{2}$$ **Rewriting as a product of two rational expressions**

$$=1\cdot\frac{a+3}{2}$$ **$\frac{7a}{7a}=1$; try to do this step mentally.**

$$=\frac{a+3}{2}.$$ **Removing the factor 1**

As a check, we let $y_1=(7x^2+21x)/(14x)$ and $y_2=(x+3)/2$. Let's compare values in a table. Note that the y-values are the same for any given x-value except 0. We exclude 0 because although $(x+3)/2$ is defined when $x=0$, $(7x^2+21x)/(14x)$ is not. Both expressions represent the same value when x is replaced with a number that can be used in *either* expression, so they are equivalent.

X	Y1	Y2
−3	0	0
−2	.5	.5
−1	1	1
0	ERROR	1.5
1	2	2
2	2.5	2.5
3	3	3

X = −3

A rational expression is said to be **simplified** when no factors equal to 1 can be removed. If the largest common factor is not found, simplifying may require two or more steps. For example, suppose we remove 7/7 instead of $(7a)/(7a)$ in Example 4. We would then have

$$\frac{7a^2 + 21a}{14a} = \frac{7(a^2 + 3a)}{7 \cdot 2a}$$

$$= \frac{a^2 + 3a}{2a}.$$

Removing a factor equal to 1: $\frac{7}{7} = 1$

Here, since 7 is not the *greatest* common factor, we need to simplify further:

$$\frac{a^2 + 3a}{2a} = \frac{a(a + 3)}{a \cdot 2}$$

$$= \frac{a + 3}{2}.$$

Removing another factor equal to 1: $a/a = 1$.
The rational expression is now simplified.

In Example 4, we found that

$$\frac{7x^2 + 21x}{14x} = \frac{x + 3}{2}.$$

These expressions are equivalent; their values are equal for every value of x except $x = 0$, for which the expression on the left is not defined. Special care must be taken, however, when simplifying a rational expression within a function. If

$$f(x) = \frac{7x^2 + 21x}{14x} \qquad \text{and} \qquad g(x) = \frac{x + 3}{2},$$

we cannot say that $f = g$, because the domains of the functions are not the same. If it is necessary to simplify a rational expression that defines a function, the domain must be carefully specified.

EXAMPLE 5   Given the function $f(x) = \dfrac{x^2 - 4}{2x^2 - 3x - 2}$.

a) Find the domain of f.

b) Simplify the rational expression.

c) Find the domain of the function given by the simplified rational expression.

Solution

a) To find the domain of f, we set the denominator equal to 0 and solve:

$$2x^2 - 3x - 2 = 0$$

$$(2x + 1)(x - 2) = 0 \qquad \text{Factoring the denominator}$$

$$2x + 1 = 0 \quad or \quad x - 2 = 0 \qquad \text{Using the principle of zero products}$$

$$x = -\tfrac{1}{2} \quad or \qquad x = 2.$$

The numbers $-\frac{1}{2}$ and 2 are not in the domain of the function. Thus the domain of f is $\left\{x \mid x \text{ is a real number } \textit{and } x \neq -\frac{1}{2} \textit{ and } x \neq 2\right\}$.

b) We have

$$\frac{x^2 - 4}{2x^2 - 3x - 2} = \frac{(x - 2)(x + 2)}{(2x + 1)(x - 2)} \qquad \text{Factoring the numerator and the denominator}$$

$$= \frac{x - 2}{x - 2} \cdot \frac{x + 2}{2x + 1} \qquad \text{Factoring the rational expression;}$$
$$\frac{x - 2}{x - 2} = 1$$

$$= \frac{x + 2}{2x + 1}. \qquad \text{Removing a factor equal to 1}$$

The simplified form of the expression is $\dfrac{x + 2}{2x + 1}$.

c) The function given by the simplified form of the rational expression is

$$g(x) = \frac{x + 2}{2x + 1}.$$

The function is undefined when

$$2x + 1 = 0$$

or $\qquad x = -\frac{1}{2}$.

The domain of g is $\left\{x \mid x \text{ is a real number } \textit{and } x \neq -\frac{1}{2}\right\}$.

In Example 5, we cannot say that

$$f(x) = \frac{x + 2}{2x + 1}$$

unless we specify that 2 is not in the domain of f.

Interactive Discovery

Let

$$f(x) = \frac{x^2 - 4}{2x^2 - 3x - 2} \quad \text{and} \quad g(x) = \frac{x + 2}{2x + 1}.$$

As we saw in Example 5, the rational expressions describing these functions are equivalent but the domains of the functions are different.

1. Graph $f(x)$. How many vertical asymptotes does the graph appear to have? What is the vertical asymptote?   One vertical asymptote; $x = -\frac{1}{2}$

2. Graph $g(x)$ and determine the vertical asymptote. $x = -\frac{1}{2}$

If a function $f(x)$ is described by a *simplified* rational expression, and a is a number that makes the denominator 0, then $x = a$ is a vertical asymptote of the graph of $f(x)$.

EXAMPLE 6   Determine the vertical asymptotes of the graph of

$$f(x) = \frac{9x^2 + 6x - 3}{12x^2 - 12}.$$

Solution   We first simplify the rational expression describing the function:

$$\frac{9x^2 + 6x - 3}{12x^2 - 12} = \frac{3(x + 1)(3x - 1)}{3 \cdot 4(x + 1)(x - 1)} \qquad \text{**Factoring the numerator and the denominator**}$$

$$= \frac{3(x + 1)}{3(x + 1)} \cdot \frac{3x - 1}{4(x - 1)} \qquad \text{**Factoring the rational expression**}$$

$$= \frac{3x - 1}{4(x - 1)}. \qquad \text{**Removing a factor equal to 1**}$$

The denominator of the simplified expression, $4(x - 1)$, is 0 when $x = 1$. Thus, $x = 1$ is a vertical asymptote of the graph. Although the domain of the function also excludes -1, there is no asymptote at $x = -1$, only a "hole." This hole should be indicated on a hand-drawn graph but it may not be obvious on a graphing-calculator screen. A table will show that -1 is not in the domain of this function.

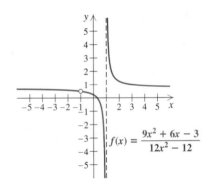

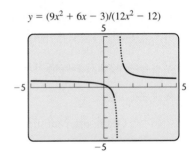

$$y = (9x^2 + 6x - 3)/(12x^2 - 12)$$

X	Y1
−3	.625
−2	.58333
−1	ERROR
0	.25
1	ERROR
2	1.25
3	1

X = −3

Canceling

Canceling is a shortcut that you may have used for removing a factor equal to 1 when working with fractions. With caution, we mention it here as a possible way to speed up your work. Canceling is one way to remove factors equal to 1 in products. It *cannot* be done in sums or when adding expressions together. If your instructor permits canceling (not all do), it is essential that it be done with care and understanding. The simplification in Example 6 might have been done faster as follows:

$$\frac{9x^2 + 6x - 3}{12x^2 - 12} = \frac{3\cancel{(x + 1)}(3x - 1)}{3 \cdot 4\cancel{(x + 1)}(x - 1)} \qquad \text{**When a factor that equals 1 is noted, it is "canceled" as shown.**}$$

$$= \frac{3x - 1}{4(x - 1)}. \qquad \text{**Removing a factor equal to 1:**} \quad \frac{3(x + 1)}{3(x + 1)} = 1$$

Interactive Discovery

Graph each pair of functions and determine whether they are equivalent.

1. $f(x) = \dfrac{x + 3}{x}$; $g(x) = 3$ No

2. $f(x) = \dfrac{4x + 3}{2}$; $g(x) = 2x + 3$ No

3. $f(x) = \dfrac{5}{5 + x}$; $g(x) = \dfrac{1}{x}$ No

Note that the pairs of functions in the preceding Interactive Discovery are not equivalent.

> **CAUTION!** Canceling is often performed incorrectly:
>
> $$\dfrac{\cancel{x} + 3}{\cancel{x}} = 3, \quad \dfrac{\cancel{4}x + 3}{\cancel{2}} = 2x + 3, \quad \dfrac{\cancel{5}}{\cancel{5} + x} = \dfrac{1}{x}$$
>
> $\quad$ Incorrect! $\qquad$ Incorrect! $\qquad$ Incorrect!
>
> To check that these are not equivalent, substitute a number for x.
>
> In each of these situations, the expressions canceled are *not* factors that equal 1. Factors are parts of products. For example, in $x \cdot 3$, x and 3 are factors, but in $x + 3$, x and 3 are *not* factors, but terms. If you can't factor, you can't cancel! If in doubt, don't cancel!

Often, after multiplying two rational expressions, it is possible to simplify the result.

EXAMPLE 7 Multiply. (Write the product as a single rational expression.) Then simplify by removing a factor equal to 1.

a) $\dfrac{x + 2}{x - 3} \cdot \dfrac{x^2 - 4}{x^2 + x - 2}$ $\qquad\qquad$ **b)** $\dfrac{1 - a^3}{a^2} \cdot \dfrac{a^5}{a^2 - 1}$

Solution

a) $\dfrac{x + 2}{x - 3} \cdot \dfrac{x^2 - 4}{x^2 + x - 2} = \dfrac{(x + 2)(x^2 - 4)}{(x - 3)(x^2 + x - 2)}$ Multiplying the numerators and also the denominators

$= \dfrac{(x + 2)(x - 2)(x + 2)}{(x - 3)(x + 2)(x - 1)}$ Factoring the numerator and the denominator and finding common factors

$= \dfrac{(x + 2)\cancel{(x + 2)}(x - 2)}{(x - 3)\cancel{(x + 2)}(x - 1)}$ Removing a factor equal to 1: $\dfrac{x + 2}{x + 2} = 1$

$= \dfrac{(x + 2)(x - 2)}{(x - 3)(x - 1)}$ Simplifying

TEACHING TIP

Some students may try to "solve" these rational expressions. You may wish to remind students of the differences between an expression and an equation.

b) $\dfrac{1 - a^3}{a^2} \cdot \dfrac{a^5}{a^2 - 1}$

$= \dfrac{(1 - a^3)a^5}{a^2(a^2 - 1)}$

$= \dfrac{(1 - a)(1 + a + a^2)a^5}{a^2(a - 1)(a + 1)}$ Factoring a difference of cubes and a difference of squares

$= \dfrac{-1(a - 1)(1 + a + a^2)a^5}{a^2(a - 1)(a + 1)}$ *Important!* **Factoring out −1 reverses the subtraction.**

$= \dfrac{\cancel{(a - 1)}a^2 \cdot a^3(-1)(1 + a + a^2)}{\cancel{(a - 1)}a^2(a + 1)}$ Rewriting a^5 as $a^2 \cdot a^3$; removing a factor equal to 1: $\dfrac{(a - 1)a^2}{(a - 1)a^2} = 1$

$= \dfrac{-a^3(1 + a + a^2)}{a + 1}$ Simplifying

As in Example 6, there is no need for us to multiply out the numerator or the denominator of the final result.

Dividing and Simplifying

Two expressions are reciprocals of each other if their product is 1. As in arithmetic, to find the reciprocal of a rational expression, we interchange numerator and denominator.

The reciprocal of $\dfrac{x}{x^2 + 3}$ is $\dfrac{x^2 + 3}{x}$.

The reciprocal of $y - 8$ is $\dfrac{1}{y - 8}$.

Quotients of Rational Expressions

For any rational expressions A/B and C/D, with $B, C, D \neq 0$,

$$\frac{A}{B} \div \frac{C}{D} = \frac{A}{B} \cdot \frac{D}{C}.$$

(To divide two rational expressions, multiply by the reciprocal of the divisor. We often say that we "*invert* and multiply.")

EXAMPLE 8   Divide. Simplify by removing a factor equal to 1 if possible.

a) $\dfrac{x - 2}{x + 1} \div \dfrac{x + 5}{x - 3}$

b) $\dfrac{a^2 - 1}{a - 1} \div \dfrac{a^2 - 2a + 1}{a + 1}$

Study Tip

The procedures covered in this chapter are by their nature rather long. As is often the case in mathematics, it may help to write out each step as you do the problems. If you have difficulty, consider starting over with a new sheet of paper. Don't squeeze your work into a small amount of space. When using lined paper, consider using two spaces at a time, writing the fraction bar on a line of the paper.

Solution

a) $\dfrac{x-2}{x+1} \div \dfrac{x+5}{x-3} = \dfrac{x-2}{x+1} \cdot \dfrac{x-3}{x+5}$ **Multiplying by the reciprocal of the divisor**

$= \dfrac{(x-2)(x-3)}{(x+1)(x+5)}$ **Multiplying the numerators and the denominators**

b) $\dfrac{a^2-1}{a-1} \div \dfrac{a^2-2a+1}{a+1} = \dfrac{a^2-1}{a-1} \cdot \dfrac{a+1}{a^2-2a+1}$ **Multiplying by the reciprocal of the divisor**

$= \dfrac{(a^2-1)(a+1)}{(a-1)(a^2-2a+1)}$ **Multiplying the numerators and the denominators**

$= \dfrac{(a+1)(a-1)(a+1)}{(a-1)(a-1)(a-1)}$ **Factoring the numerator and the denominator**

$= \dfrac{(a+1)\cancel{(a-1)}(a+1)}{(a-1)\cancel{(a-1)}(a-1)}$ **Removing a factor equal to 1:** $\dfrac{a-1}{a-1}=1$

$= \dfrac{(a+1)(a+1)}{(a-1)(a-1)}$ **Simplifying**

6.1

Exercise Set

Photo Developing.   *Rik usually takes 3 hr more than Pearl does to process a day's orders at Liberty Place Photo. If Pearl takes t hr to process a day's orders, the function given by*

$$H(t) = \dfrac{t^2 + 3t}{2t + 3}$$

can be used to determine how long it would take if they worked together.

1. How long will it take them, working together, to complete a day's orders if Pearl can process the orders alone in 5 hr? $\frac{40}{13}$ hr, or $3\frac{1}{13}$ hr

2. How long will it take them, working together, to complete a day's orders if Pearl can process the orders alone in 7 hr? $\frac{70}{17}$ hr, or $4\frac{2}{17}$ hr

For each rational function, find the function values indicated, provided the value exists.

3. $v(t) = \dfrac{4t^2 - 5t + 2}{t + 3}$; $v(0), v(-2), v(7)$ $\frac{2}{3};\ 28;\ \frac{163}{10}$

4. $f(x) = \dfrac{5x^2 + 4x - 12}{6 - x}$; $f(0), f(-1), f(3)$ $-2;\ -\frac{11}{7};\ 15$

5. $g(x) = \dfrac{2x^3 - 9}{x^2 - 4x + 4}$; $g(0), g(2), g(-1)$ $-\frac{9}{4};$ does not exist; $-\frac{11}{9}$

6. $r(t) = \dfrac{t^2 - 5t + 4}{t^2 - 9}$; $r(1), r(2), r(-3)$ $0;\ \frac{2}{5};$ does not exist

Multiply to obtain a single equivalent expression. Do not simplify. Assume that all denominators are nonzero.

7. $\dfrac{4x}{4x} \cdot \dfrac{x - 3}{x + 2}$

8. $\dfrac{3 - a^2}{a - 7} \cdot \dfrac{-1}{-1}$

9. $\dfrac{t - 2}{t + 3} \cdot \dfrac{-1}{-1}$ $\dfrac{(t-2)(-1)}{(t+3)(-1)}$

10. $\dfrac{x - 4}{x + 5} \cdot \dfrac{x - 5}{x - 5}$ $\dfrac{(x-4)(x-5)}{(x+5)(x-5)}$

Simplify by removing a factor equal to 1.

11. $\dfrac{15x}{5x^2}$ $\frac{3}{x}$

12. $\dfrac{7a^3}{21a}$

13. $\dfrac{18t^3}{27t^7}$ $\dfrac{2}{3t^4}$

14. $\dfrac{8y^5}{4y^9}$ $\dfrac{2}{y^4}$

15. $\dfrac{2a - 10}{2}$ $a - 5$

16. $\dfrac{3a + 12}{3}$ $a + 4$

17. $\dfrac{15}{25a - 30}$ $\dfrac{3}{5a - 6}$

18. $\dfrac{21}{6x - 9}$ $\dfrac{7}{2x - 3}$

19. $\dfrac{3x - 12}{3x + 15}$ $\dfrac{x - 4}{x + 5}$

20. $\dfrac{4y - 20}{4y + 12}$ $\dfrac{y - 5}{y + 3}$

21. $\dfrac{5x + 20}{x^2 + 4x}$ $\dfrac{5}{x}$

22. $\dfrac{3x + 21}{x^2 + 7x}$ $\dfrac{3}{x}$

23. $\dfrac{3a - 1}{2 - 6a}$ $-\dfrac{1}{2}$

24. $\dfrac{6 - 5a}{10a - 12}$ $-\dfrac{1}{2}$

25. $\dfrac{8t - 16}{t^2 - 4}$ $\dfrac{8}{t + 2}$

26. $\dfrac{t^2 - 9}{5t + 15}$ $\dfrac{t - 3}{5}$

27. $\dfrac{2t - 1}{1 - 4t^2}$ $-\dfrac{1}{1 + 2t}$

28. $\dfrac{3a - 2}{4 - 9a^2}$ $-\dfrac{1}{2 + 3a}$

29. $\dfrac{12 - 6x}{5x - 10}$ $-\dfrac{6}{5}$

30. $\dfrac{21 - 7x}{3x - 9}$ $-\dfrac{7}{3}$

31. $\dfrac{a^2 - 25}{a^2 + 10a + 25}$ $\dfrac{a - 5}{a + 5}$

32. $\dfrac{a^2 - 16}{a^2 - 8a + 16}$ $\dfrac{a + 4}{a - 4}$

33. $\dfrac{x^2 + 9x + 8}{x^2 - 3x - 4}$ $\dfrac{x + 8}{x - 4}$

34. $\dfrac{t^2 - 8t - 9}{t^2 + 5t + 4}$ $\dfrac{t - 9}{t + 4}$

35. $\dfrac{16 - t^2}{t^2 - 8t + 16}$ $\dfrac{4 + t}{4 - t}$

36. $\dfrac{25 - p^2}{p^2 + 10p + 25}$ $\dfrac{5 - p}{5 + p}$

37. $\dfrac{x^3 - 1}{x^2 - 1}$ $\dfrac{x^2 + x + 1}{x + 1}$

38. $\dfrac{a^3 + 8}{a^2 - 4}$ $\dfrac{a^2 - 2a + 4}{a - 2}$

39. $\dfrac{3y^3 + 24}{y^2 - 2y + 4}$ $3(y + 2)$

40. $\dfrac{x^3 - 27}{5x^2 + 15x + 45}$ $\dfrac{x - 3}{5}$

Determine the vertical asymptotes of the graph of each function.

41. $f(x) = \dfrac{3x - 12}{3x + 15}$
  $x = -5$

42. $f(x) = \dfrac{4x - 20}{4x + 12}$
  $x = -3$

43. $g(x) = \dfrac{12 - 6x}{5x - 10}$
  No vertical asymptotes

44. $r(x) = \dfrac{21 - 7x}{3x - 9}$
  No vertical asymptotes

45. $t(x) = \dfrac{x^3 + 3x^2}{x^2 + 6x + 9}$
  $x = -3$

46. $g(x) = \dfrac{x^2 - 4}{2x^2 - 5x + 2}$
  $x = \frac{1}{2}$

47. $f(x) = \dfrac{x^2 - x - 6}{x^2 - 6x + 8}$
  $x = 2, x = 4$

48. $f(x) = \dfrac{x^2 + 2x + 1}{x^2 - 2x + 1}$
  $x = 1$

In Exercises 49–54, match each function with one of the following graphs.

a)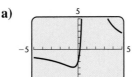

b)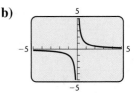

c)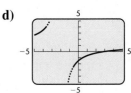

d)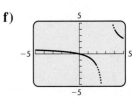

e)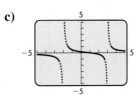

f)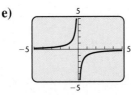

49. $h(x) = \dfrac{1}{x}$ (b)

50. $q(x) = -\dfrac{1}{x}$ (e)

51. $f(x) = \dfrac{x}{x - 3}$ (f)

52. $g(x) = \dfrac{x - 3}{x + 2}$ (d)

53. $r(x) = \dfrac{4x - 2}{x^2 - 2x + 1}$ (a)

54. $t(x) = \dfrac{x - 1}{x^2 - x - 6}$ (c)

Multiply and, if possible, simplify.

55. $\dfrac{5a^3}{3b} \cdot \dfrac{7b^3}{10a^7}$ $\dfrac{7b^2}{6a^4}$

56. $\dfrac{25a}{9b^8} \cdot \dfrac{3b^5}{5a^2}$ $\dfrac{5}{3ab^3}$

57. $\dfrac{8x - 16}{5x} \cdot \dfrac{x^3}{5x - 10}$ $\dfrac{8x^2}{25}$

58. $\dfrac{5t^3}{4t - 8} \cdot \dfrac{6t - 12}{10t}$ $\dfrac{3t^2}{4}$

59. $\dfrac{y^2 - 16}{4y + 12} \cdot \dfrac{y + 3}{y - 4}$ $\dfrac{y + 4}{4}$

60. $\dfrac{m^2 - n^2}{4m + 4n} \cdot \dfrac{m + n}{m - n}$ $\dfrac{m + n}{4}$

61. $\dfrac{x^2 - 16}{x^2} \cdot \dfrac{x^2 - 4x}{x^2 - x - 12}$ $\dfrac{(x + 4)(x - 4)}{x(x + 3)}$

62. $\dfrac{y^2 + 10y + 25}{y^2 - 9} \cdot \dfrac{y^2 + 3y}{y + 5}$ $\dfrac{y(y + 5)}{y - 3}$

63. $\dfrac{7a - 14}{4 - a^2} \cdot \dfrac{5a^2 + 6a + 1}{35a + 7}$ $\dfrac{a + 1}{2 + a}$

64. $\dfrac{a^2 - 1}{2 - 5a} \cdot \dfrac{15a - 6}{a^2 + 5a - 6}$ $-\dfrac{3(a + 1)}{a + 6}$

65. $\dfrac{t^3 - 4t}{t - t^4} \cdot \dfrac{t^4 - t}{4t - t^3}$ 1
  Aha!

66. $\dfrac{x^2 - 6x + 9}{12 - 4x} \cdot \dfrac{x^6 - 9x^4}{x^3 - 3x^2}$ $\quad \dfrac{x^2(x+3)(x-3)}{-4}$

67. $\dfrac{x^2 - 2x - 35}{2x^3 - 3x^2} \cdot \dfrac{4x^3 - 9x}{7x - 49}$ $\quad \dfrac{(x+5)(2x+3)}{7x}$

68. $\dfrac{y^2 - 10y + 9}{y^2 - 1} \cdot \dfrac{1 - y^2}{y^2 - 5y - 36}$ $\quad \dfrac{1-y}{y+4}$

69. $\dfrac{c^3 + 8}{c^5 - 4c^3} \cdot \dfrac{c^6 - 4c^5 + 4c^4}{c^2 - 2c + 4}$ $\quad c(c-2)$

70. $\dfrac{x^3 - 27}{x^4 - 9x^2} \cdot \dfrac{x^5 - 6x^4 + 9x^3}{x^2 + 3x + 9}$ $\quad \dfrac{x(x-3)^2}{x+3}$

71. $\dfrac{a^3 - b^3}{3a^2 + 9ab + 6b^2} \cdot \dfrac{a^2 + 2ab + b^2}{a^2 - b^2}$ $\quad \dfrac{a^2 + ab + b^2}{3(a+2b)}$

72. $\dfrac{x^3 + y^3}{x^2 + 2xy - 3y^2} \cdot \dfrac{x^2 - y^2}{3x^2 + 6xy + 3y^2}$ $\quad \dfrac{x^2 - xy + y^2}{3(x+3y)}$

73. $\dfrac{4x^2 - 9y^2}{8x^3 - 27y^3} \cdot \dfrac{4x^2 + 6xy + 9y^2}{4x^2 + 12xy + 9y^2}$ $\quad \dfrac{1}{2x+3y}$

74. $\dfrac{3x^2 - 3y^2}{27x^3 - 8y^3} \cdot \dfrac{6x^2 + 5xy - 6y^2}{6x^2 + 12xy + 6y^2}$ $\quad \dfrac{(x-y)(2x+3y)}{2(x+y)(9x^2 + 6xy + 4y^2)}$

Divide and, if possible, simplify.

75. $\dfrac{9x^5}{8y^2} \div \dfrac{3x}{16y^9}$ $\quad 6x^4y^7$

76. $\dfrac{16a^7}{3b^5} \div \dfrac{8a^3}{6b}$ $\quad \dfrac{4a^4}{b^4}$

77. $\dfrac{5x + 10}{x^8} \div \dfrac{x + 2}{x^3}$ $\quad \dfrac{5}{x^5}$

78. $\dfrac{3y + 15}{y^7} \div \dfrac{y + 5}{y^2}$ $\quad \dfrac{3}{y^5}$

79. $\dfrac{x^2 - 4}{x^3} \div \dfrac{x^5 - 2x^4}{x + 4}$ $\quad \dfrac{(x+2)(x+4)}{x^7}$

80. $\dfrac{y^2 - 9}{y^2} \div \dfrac{y^5 + 3y^4}{y + 2}$ $\quad \dfrac{(y-3)(y+2)}{y^6}$

81. $\dfrac{25x^2 - 4}{x^2 - 9} \div \dfrac{2 - 5x}{x + 3}$ $\quad -\dfrac{5x+2}{x-3}$

82. $\dfrac{4a^2 - 1}{a^2 - 4} \div \dfrac{2a - 1}{2 - a}$ $\quad \dfrac{-2a - 1}{a+2}$

83. $\dfrac{5y - 5x}{15y^3} \div \dfrac{x^2 - y^2}{3x + 3y}$ $\quad -\dfrac{1}{y^3}$

84. $\dfrac{x^2 - y^2}{4x + 4y} \div \dfrac{3y - 3x}{12x^2}$ $\quad -x^2$

85. $\dfrac{x^2 - 16}{x^2 - 10x + 25} \div \dfrac{3x - 12}{x^2 - 3x - 10}$ $\quad \dfrac{(x+4)(x+2)}{3(x-5)}$

86. $\dfrac{y^2 - 36}{y^2 - 8y + 16} \div \dfrac{3y - 18}{y^2 - y - 12}$ $\quad \dfrac{(y+6)(y+3)}{3(y-4)}$

87. $\dfrac{y^3 + 3y}{y^2 - 9} \div \dfrac{y^2 + 5y - 14}{y^2 + 4y - 21}$ $\quad \dfrac{y(y^2+3)}{(y+3)(y-2)}$

88. $\dfrac{a^3 + 4a}{a^2 - 16} \div \dfrac{a^2 + 8a + 15}{a^2 + a - 20}$ $\quad \dfrac{a(a^2+4)}{(a+4)(a+3)}$

89. $\dfrac{x^3 - 64}{x^3 + 64} \div \dfrac{x^2 - 16}{x^2 - 4x + 16}$ $\quad \dfrac{x^2 + 4x + 16}{(x+4)^2}$

90. $\dfrac{8y^3 - 27}{64y^3 - 1} \div \dfrac{4y^2 - 9}{16y^2 + 4y + 1}$ $\quad \dfrac{4y^2 + 6y + 9}{(4y-1)(2y+3)}$

91. $\dfrac{8a^3 + b^3}{2a^2 + 3ab + b^2} \div \dfrac{8a^2 - 4ab + 2b^2}{4a^2 + 4ab + b^2}$ $\quad \dfrac{(2a+b)^2}{2(a+b)}$

92. $\dfrac{x^3 + 8y^3}{2x^2 + 5xy + 2y^2} \div \dfrac{x^3 - 2x^2y + 4xy^2}{8x^2 - 2y^2}$ $\quad \dfrac{2(2x-y)}{x}$

TW **93.** To check Example 4, Kara lets
$$y_1 = \frac{7x^2 + 21x}{14x} \quad \text{and} \quad y_2 = \frac{x+3}{2}.$$
Since the graphs of y_1 and y_2 appear to be identical, Kara believes that the domains of the functions described by y_1 and y_2 are the same, $\mathbb{R}$. How could you convince Kara otherwise?

TW **94.** Nancy *incorrectly* simplifies $\dfrac{x+2}{x}$ as
$$\frac{x+2}{x} = \frac{\cancel{x}+2}{\cancel{x}} = 1 + 2 = 3.$$
She insists this is correct because it checks when x is replaced with 1. Explain her misconception and give an example that could convince her that the simplification is incorrect.

Skill Maintenance

Simplify.

95. $\dfrac{3}{10} - \dfrac{8}{15}$ [1.2] $\quad -\dfrac{7}{30}$

96. $\dfrac{3}{8} - \dfrac{7}{10}$ [1.2] $\quad -\dfrac{13}{40}$

97. $\dfrac{2}{3} \cdot \dfrac{5}{7} - \dfrac{5}{7} \cdot \dfrac{1}{6}$ [1.2] $\quad \dfrac{5}{14}$

98. $\dfrac{4}{7} \cdot \dfrac{1}{5} - \dfrac{3}{10} \cdot \dfrac{2}{7}$ [1.2] $\quad \dfrac{1}{35}$

99. $(8x^3 - 5x^2 + 6x + 2) - (4x^3 + 2x^2 - 3x + 7)$ [5.1] $\quad 4x^3 - 7x^2 + 9x - 5$

100. $(6t^4 + 9t^3 - t^2 + 4t) - (8t^4 - 2t^3 - 6t + 3)$
  [5.1] $-2t^4 + 11t^3 - t^2 + 10t - 3$

Synthesis

TW **101.** Tony *incorrectly* argues that since

$$\frac{a^2 - 4}{a - 2} = \frac{a^2}{a} + \frac{-4}{-2} = a + 2,$$

is correct, it follows that

$$\frac{x^2 + 9}{x + 1} = \frac{x^2}{x} + \frac{9}{1} = x + 9.$$

Explain his misconception.

TW **102.** Explain why the graphs of $f(x) = 5x$ and
$g(x) = \dfrac{5x^2}{x}$ differ.

103. Let

$$g(x) = \frac{2x + 3}{4x - 1}.$$

Determine each of the following.

a) $g(x + h)$
b) $g(2x - 2) \cdot g(x)$
c) $g\left(\frac{1}{2}x + 1\right) \cdot g(x)$

104. Graph the function given by

$$f(x) = \frac{x^2 - 9}{x - 3}.$$

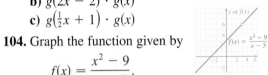

(*Hint*: Determine the domain of f and simplify.)

Perform the indicated operations and simplify.

105. $\left[\dfrac{r^2 - 4s^2}{r + 2s} \div (r + 2s)\right] \cdot \dfrac{2s}{r - 2s} \quad \dfrac{2s}{r + 2s}$

106. $\left[\dfrac{d^2 - d}{d^2 - 6d + 8} \cdot \dfrac{d - 2}{d^2 + 5d}\right] \div \dfrac{5d}{d^2 - 9d + 20}$

Aha! **107.** $\left[\dfrac{6t^2 - 26t + 30}{8t^2 - 15t - 21} \cdot \dfrac{5t^2 - 9t - 15}{6t^2 - 14t - 20}\right] \div$
  $\dfrac{6t^2 - 26t + 30}{8t^2 - 15t - 21} \qquad \dfrac{5t^2 - 9t - 15}{6t^2 - 14t - 20}$

Simplify.

108. $\dfrac{x(x + 1) - 2(x + 3)}{(x + 3)(x + 1)(x + 2)} \quad \dfrac{x - 3}{(x + 3)(x + 1)}$

109. $\dfrac{m^2 - t^2}{m^2 + t^2 + m + t + 2mt} \quad \dfrac{m - t}{m + t + 1}$

110. $\dfrac{a^3 - 2a^2 + 2a - 4}{a^3 - 2a^2 - 3a + 6} \quad \dfrac{a^2 + 2}{a^2 - 3}$

111. $\dfrac{x^3 + x^2 - y^3 - y^2}{x^2 - 2xy + y^2} \quad \dfrac{x^2 + xy + y^2 + x + y}{x - y}$

112. $\dfrac{u^6 + v^6 + 2u^3v^3}{u^3 - v^3 + u^2v - uv^2} \quad \dfrac{(u^2 - uv + v^2)^2}{u - v}$

113. $\dfrac{x^5 - x^3 + x^2 - 1 - (x^3 - 1)(x + 1)^2}{(x^2 - 1)^2} \quad -\dfrac{2x}{x - 1}$

114. Let

$$f(x) = \frac{4}{x^2 - 1} \quad \text{and} \quad g(x) = \frac{4x^2 + 8x + 4}{x^3 - 1}.$$

Find each of the following.

a) $(f \cdot g)(x)$
b) $(f/g)(x)$
c) $(g/f)(x)$

Determine the domain and the range of each function from its graph.

115.

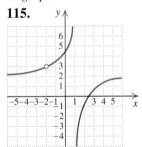

116.

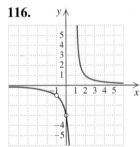

117.

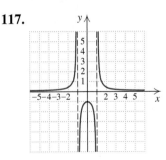

---

103. **(a)** $\dfrac{2x + 2h + 3}{4x + 4h - 1}$; **(b)** $\dfrac{2x + 3}{8x - 9}$; **(c)** $\dfrac{x + 5}{4x - 1}$ **106.** $\dfrac{(d - 1)(d - 5)}{5d(d + 5)}$ **114. (a)** $\dfrac{16(x + 1)}{(x - 1)^2(x^2 + x + 1)}$; **(b)** $\dfrac{x^2 + x + 1}{(x + 1)^3}$;

(c) $\dfrac{(x + 1)^3}{x^2 + x + 1}$ **115.** Domain: $(-\infty, -2) \cup (-2, 1) \cup (1, \infty)$; range: $(-\infty, 2) \cup (2, 3) \cup (3, \infty)$
116. Domain: $(-\infty, -1) \cup (-1, 0) \cup (0, 1) \cup (1, \infty)$; range: $(-\infty, -3) \cup (-3, -1) \cup (-1, 0) \cup (0, \infty)$
117. Domain: $(-\infty, -1) \cup (-1, 1) \cup (1, \infty)$; range: $(-\infty, -1] \cup (0, \infty)$

6.2

When Denominators Are the Same ▪ When Denominators Are Different

Rational Expressions and Functions: Adding and Subtracting

Rational expressions are added in much the same way as the fractions of arithmetic.

When Denominators Are the Same

Addition and Subtraction with Like Denominators

To add or subtract when denominators are the same, add or subtract the numerators and keep the same denominator.

$$\frac{A}{C} + \frac{B}{C} = \frac{A+B}{C} \quad \text{and} \quad \frac{A}{C} - \frac{B}{C} = \frac{A-B}{C}, \quad \text{where } C \neq 0.$$

X	Y₁	Y₂
−2	−2.5	−2.5
−1	−6	−6
0	ERROR	ERROR
1	8	8
2	4.5	4.5
3	3.3333	3.3333
4	2.75	2.75

X = −2

EXAMPLE 1   Add: $\dfrac{3+x}{x} + \dfrac{4}{x}$.

Solution   We have

$$\frac{3+x}{x} + \frac{4}{x} = \frac{3+x+4}{x} = \frac{x+7}{x}.$$ Because x is not a factor of the numerator and the denominator, the result cannot be simplified.

To check, we let $y_1 = (3+x)/x + 4/x$ and $y_2 = (7+x)/x$. The table at left shows that $y_1 = y_2$ for all x not equal to 0.

EXAMPLE 2   Add: $\dfrac{4x^2 - 5xy}{x^2 - y^2} + \dfrac{2xy - y^2}{x^2 - y^2}$.

Solution

$$\frac{4x^2 - 5xy}{x^2 - y^2} + \frac{2xy - y^2}{x^2 - y^2} = \frac{4x^2 - 3xy - y^2}{x^2 - y^2}$$ Adding the numerators and combining like terms. The denominator is unchanged.

$$= \frac{(x-y)(4x+y)}{(x-y)(x+y)}$$ Factoring the numerator and the denominator and looking for common factors

$$= \frac{\cancel{(x-y)}(4x+y)}{\cancel{(x-y)}(x+y)}$$ Removing a factor equal to 1: $\dfrac{x-y}{x-y} = 1$

$$= \frac{4x+y}{x+y}$$ Simplifying

Recall that a fraction bar is a grouping symbol. The next example shows that when a numerator is subtracted, care must be taken to subtract, or change the sign of, *each* term in that polynomial.

EXAMPLE 3   If

$$f(x) = \frac{4x + 5}{x + 3} - \frac{x - 2}{x + 3},$$

find a simplified form of $f(x)$.

Solution

$$f(x) = \frac{4x + 5}{x + 3} - \frac{x - 2}{x + 3}$$

$$= \frac{4x + 5 - (x - 2)}{x + 3} \qquad \text{The parentheses remind us to subtract } both \text{ terms.}$$

$$= \frac{4x + 5 - x + 2}{x + 3}$$

$$= \frac{3x + 7}{x + 3}$$

When Denominators Are Different

In order to add rational expressions such as

$$\frac{7}{12xy^2} + \frac{8}{15x^3y} \quad \text{or} \quad \frac{x}{x^2 - y^2} + \frac{y}{x^2 - 4xy + 3y^2},$$

we must first find common denominators. As in arithmetic, our work is easier when we use the *least common multiple* (LCM) of the denominators.

Least Common Multiple

To find the least common multiple (LCM) of two or more expressions, find the prime factorization of each expression and form a product that contains each factor the greatest number of times that it occurs in any one prime factorization.

EXAMPLE 4   Find the least common multiple of each pair of polynomials.

a) $21x$ and $3x^2$ **b)** $x^2 + x - 12$ and $x^2 - 16$

Solution

a) We write the prime factorizations of $21x$ and $3x^2$:

$$21x = 3 \cdot 7 \cdot x \quad \text{and} \quad 3x^2 = 3 \cdot x \cdot x.$$

The factors 3, 7, and x must appear in the LCM if $21x$ is to be a factor of the LCM. The other polynomial, $3x^2$, is not a factor of $3 \cdot 7 \cdot x$ because the prime factors of $3x^2$—namely, 3, x, and x—do not all appear in $3 \cdot 7 \cdot x$.

However, if $3 \cdot 7 \cdot x$ is multiplied by another factor of x, a product is formed that contains both $21x$ and $3x^2$ as factors:

21x is a factor.

$$\text{LCM} = 3 \cdot 7 \cdot x \cdot x = 21x^2.$$

$3x^2$ is a factor.

Note that each factor (3, 7, and x) is used the greatest number of times that it occurs as a factor of either $21x$ or $3x^2$. The LCM is $3 \cdot 7 \cdot x \cdot x$, or $21x^2$.

b) We factor both expressions:

$$x^2 + x - 12 = (x - 3)(x + 4),$$
$$x^2 - 16 = (x + 4)(x - 4).$$

The LCM must contain each polynomial as a factor. By multiplying the factors of $x^2 + x - 12$ by $x - 4$, we form a product that contains both $x^2 + x - 12$ and $x^2 - 16$ as factors:

$x^2 + x - 12$ is a factor.

$$\text{LCM} = (x - 3)(x + 4)(x - 4). \quad \textbf{There is no need to multiply this out.}$$

$x^2 - 16$ is a factor.

Before adding or subtracting rational expressions with unlike denominators, we determine the *least common denominator*, or LCD, by finding the LCM of the denominators. Each rational expression is then multiplied by a form of 1, as needed, to form an equivalent expression that has the LCD.

EXAMPLE 5   Add: $\dfrac{2}{21x} + \dfrac{5}{3x^2}$.

TEACHING TIP

Be aware that some students may want to simplify the rational expressions containing common denominators before they add or subtract them.

Solution   In Example 4(a), we found that the LCD is $3 \cdot 7 \cdot x \cdot x$, or $21x^2$. We now multiply each rational expression by 1, using expressions for 1 that give us the LCD in each expression. To determine what to use, ask "$21x$ times what is $21x^2$?" and "$3x^2$ times what is $21x^2$?" The answers are x and 7, respectively, so we multiply by x/x and $7/7$:

$$\frac{2}{21x} \cdot \frac{x}{x} + \frac{5}{3x^2} \cdot \frac{7}{7} = \frac{2x}{21x^2} + \frac{35}{21x^2} \qquad \text{We now have a common denominator.}$$

$$= \frac{2x + 35}{21x^2}. \qquad \text{This expression cannot be simplified.}$$

EXAMPLE 6   Add: $\dfrac{x^2}{x^2 + 2xy + y^2} + \dfrac{2x - 2y}{x^2 - y^2}$.

Solution   Before we look for an LCD, each denominator is factored:

$$\frac{x^2}{x^2 + 2xy + y^2} + \frac{2x - 2y}{x^2 - y^2} = \frac{x^2}{(x + y)(x + y)} + \frac{2x - 2y}{(x + y)(x - y)}.$$

Although the numerators need not always be factored, we do so if it enables us to simplify. In this case, the rightmost rational expression can be simplified:

$$\frac{x^2}{x^2 + 2xy + y^2} + \frac{2x - 2y}{x^2 - y^2} = \frac{x^2}{(x + y)(x + y)} + \frac{2(x - y)}{(x + y)(x - y)} \quad \text{Factoring}$$

$$= \frac{x^2}{(x + y)(x + y)} + \frac{2}{x + y}. \quad \begin{array}{l} \text{Removing a factor} \\ \text{equal to 1:} \\ \dfrac{x - y}{x - y} = 1 \end{array}$$

Note that the LCM of $(x + y)(x + y)$ and $(x + y)$ is $(x + y)(x + y)$. To get the LCD in the second expression, we multiply by 1, using $(x + y)/(x + y)$. Then we add and, if possible, simplify.

$$\frac{x^2}{(x + y)(x + y)} + \frac{2}{x + y} = \frac{x^2}{(x + y)(x + y)} + \frac{2}{x + y} \cdot \frac{x + y}{x + y}$$

$$= \frac{x^2}{(x + y)(x + y)} + \frac{2x + 2y}{(x + y)(x + y)} \quad \begin{array}{l} \text{We now} \\ \text{have the} \\ \text{LCD.} \end{array}$$

$$= \frac{x^2 + 2x + 2y}{(x + y)(x + y)} \quad \begin{array}{l} \text{Since the numerator} \\ \text{cannot be factored, we} \\ \text{cannot simplify further.} \end{array}$$

EXAMPLE 7   Subtract: $\dfrac{2y + 1}{y^2 - 7y + 6} - \dfrac{y + 3}{y^2 - 5y - 6}.$

Solution

$$\frac{2y + 1}{y^2 - 7y + 6} - \frac{y + 3}{y^2 - 5y - 6}$$

$$= \frac{2y + 1}{(y - 6)(y - 1)} - \frac{y + 3}{(y - 6)(y + 1)} \quad \begin{array}{l} \text{The LCD is} \\ (y - 6)(y - 1)(y + 1). \end{array}$$

$$= \frac{2y + 1}{(y - 6)(y - 1)} \cdot \frac{y + 1}{y + 1} - \frac{y + 3}{(y - 6)(y + 1)} \cdot \frac{y - 1}{y - 1}$$

$$\underset{\text{↑ Multiplying by 1 to get the ↑}}{}$$
$$\text{LCD in each expression}$$

$$= \frac{(2y + 1)(y + 1) - (y + 3)(y - 1)}{(y - 6)(y - 1)(y + 1)}$$

$$= \frac{2y^2 + 3y + 1 - (y^2 + 2y - 3)}{(y - 6)(y - 1)(y + 1)} \quad \text{The parentheses are important.}$$

$$= \frac{2y^2 + 3y + 1 - y^2 - 2y + 3}{(y - 6)(y - 1)(y + 1)}$$

$$= \frac{y^2 + y + 4}{(y - 6)(y - 1)(y + 1)} \quad \begin{array}{l} \text{We leave the denominator} \\ \text{in factored form.} \end{array}$$

TEACHING TIP

You might remind students that the expression for 1 is formed by looking at the LCD and determining what factors are missing.

EXAMPLE 8 Add: $\dfrac{3}{8a} + \dfrac{1}{-8a}$.

Solution

$$\dfrac{3}{8a} + \dfrac{1}{-8a} = \dfrac{3}{8a} + \dfrac{-1}{-1} \cdot \dfrac{1}{-8a}$$

> When denominators are opposites, we multiply one rational expression by $-1/-1$ to get the LCD.

$$= \dfrac{3}{8a} + \dfrac{-1}{8a}$$

$$= \dfrac{2}{8a}$$

$$= \dfrac{2 \cdot 1}{2 \cdot 4a} \qquad \text{Simplifying by removing a factor equal to 1: } \dfrac{2}{2} = 1$$

$$= \dfrac{1}{4a}$$

EXAMPLE 9 Subtract: $\dfrac{5x}{x-2y} - \dfrac{3y-7}{2y-x}$.

Solution

$$\dfrac{5x}{x-2y} - \dfrac{3y-7}{2y-x} = \dfrac{5x}{x-2y} - \dfrac{-1}{-1} \cdot \dfrac{3y-7}{2y-x} \qquad \text{Note that } x-2y \text{ and } 2y-x \text{ are opposites.}$$

$$= \dfrac{5x}{x-2y} - \dfrac{7-3y}{x-2y} \qquad \begin{array}{l} \text{Performing the multiplication.} \\ \textit{Note: } -1(2y-x) = -2y+x \\ \qquad\qquad\qquad = x - 2y. \end{array}$$

$$= \dfrac{5x - (7-3y)}{x-2y}$$

$$= \dfrac{5x - 7 + 3y}{x-2y} \qquad \begin{array}{l} \text{Subtracting. The parentheses are important.} \end{array}$$

In Example 9, you may have noticed that when $3y - 7$ is multiplied by -1 and subtracted, the result is $-7 + 3y$, which is equivalent to the original $3y - 7$. Thus, instead of multiplying the numerator by -1 and then subtracting, we could have simply *added* $3y - 7$ to $5x$, as in the following:

$$\dfrac{5x}{x-2y} - \dfrac{3y-7}{2y-x} = \dfrac{5x}{x-2y} + (-1) \cdot \dfrac{3y-7}{2y-x} \qquad \begin{array}{l} \text{Rewriting subtraction as addition} \end{array}$$

$$= \dfrac{5x}{x-2y} + \dfrac{1}{-1} \cdot \dfrac{3y-7}{2y-x} \qquad \text{Writing } -1 \text{ as } \dfrac{1}{-1}$$

$$= \dfrac{5x}{x-2y} + \dfrac{3y-7}{x-2y} \qquad \begin{array}{l} \text{The opposite of } 2y-x \\ \text{is } x-2y. \end{array}$$

$$= \dfrac{5x + 3y - 7}{x-2y}. \qquad \begin{array}{l} \text{This checks with the answer to Example 9.} \end{array}$$

EXAMPLE 10   Perform the indicated operations and simplify:

$$\frac{2x}{x^2-4}+\frac{5}{2-x}-\frac{1}{2+x}.$$

Solution   We have

$$\frac{2x}{x^2-4}+\frac{5}{2-x}-\frac{1}{2+x}=\frac{2x}{(x-2)(x+2)}+\frac{5}{2-x}-\frac{1}{2+x} \quad \text{Factoring}$$

$$=\frac{2x}{(x-2)(x+2)}+\frac{-1}{-1}\cdot\frac{5}{(2-x)}-\frac{1}{x+2} \quad \begin{array}{l}\text{Multiplying by } \dfrac{-1}{-1} \\ \text{since } 2-x \text{ is the} \\ \text{opposite of } x-2\end{array}$$

$$=\frac{2x}{(x-2)(x+2)}+\frac{-5}{x-2}-\frac{1}{x+2} \quad \begin{array}{l}\text{The LCD is} \\ (x-2)(x+2).\end{array}$$

$$=\frac{2x}{(x-2)(x+2)}+\frac{-5}{x-2}\cdot\frac{x+2}{x+2}-\frac{1}{x+2}\cdot\frac{x-2}{x-2} \quad \begin{array}{l}\text{Multiplying} \\ \text{by 1 to get} \\ \text{the LCD}\end{array}$$

$$=\frac{2x-5(x+2)-(x-2)}{(x-2)(x+2)}$$

$$=\frac{2x-5x-10-x+2}{(x-2)(x+2)}$$

$$=\frac{-4x-8}{(x-2)(x+2)}$$

$$=\frac{-4(x+2)}{(x-2)(x+2)}$$

$$=\frac{-4\cancel{(x+2)}}{(x-2)\cancel{(x+2)}} \quad \text{Removing a factor equal to 1: } \frac{x+2}{x+2}=1$$

$$=\frac{-4}{x-2}, \text{ or } -\frac{4}{x-2}.$$

Another correct answer is $\dfrac{4}{2-x}$. It is found by writing $-\dfrac{4}{x-2}$ as $\dfrac{4}{-(x-2)}$ and then using the distributive law to remove parentheses.

Our work in Example 10 indicates that if

$$f(x)=\frac{2x}{x^2-4}+\frac{5}{2-x}-\frac{1}{2+x}$$

and

$$g(x)=\frac{-4}{x-2},$$

then, for $x\neq-2$ and $x\neq2$, we have $f=g$. Note that whereas the domain of f includes all real numbers except -2 or 2, the domain of g excludes only 2. This is illustrated in the graphs below. Note that the graphs of f and g coincide, except for $x=-2$. Methods for drawing such graphs by hand are discussed in more advanced courses. The graphs are for visualization only.

Study Tip

Because there are many steps to these problems, we recommend that you check each step as you go.

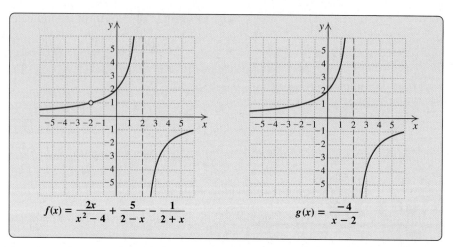

A computer-generated visualization of Example 10

Whenever rational expressions are simplified, a quick partial check is to evaluate both the original and the simplified expressions for a convenient choice of x. For instance, to check Example 10, if $x = 1$, we have

$$f(1) = \frac{2 \cdot 1}{1^2 - 4} + \frac{5}{2 - 1} - \frac{1}{2 + 1}$$

$$= \frac{2}{-3} + \frac{5}{1} - \frac{1}{3} = 5 - \frac{3}{3} = 4$$

and

$$g(1) = \frac{-4}{1 - 2} = \frac{-4}{-1} = 4.$$

Since both functions include the pair $(1, 4)$, our algebra was *probably* correct. Although this is only a partial check (on rare occasions, an incorrect answer might "check"), because it is so easy to perform, it is nonetheless very useful. Further evaluation provides a more definitive check.

6.2

Exercise Set

FOR EXTRA HELP

Digital Video Tutor CD 5 Videotape 8 | Student's Solutions Manual | Tutor Center AW Math Tutor Center | InterAct Math | MathXL | MyMathLab

Perform the indicated operations. Simplify when possible.

1. $\dfrac{5}{3a} + \dfrac{7}{3a}$ $\quad \dfrac{4}{a}$

2. $\dfrac{3}{2y} + \dfrac{5}{2y}$ $\quad \dfrac{4}{y}$

3. $\dfrac{1}{4a^2b} - \dfrac{5}{4a^2b}$ $\quad -\dfrac{1}{a^2b}$

4. $\dfrac{5}{3m^2n^2} - \dfrac{4}{3m^2n^2}$ $\quad \dfrac{1}{3m^2n^2}$

5. $\dfrac{a - 5b}{a + b} + \dfrac{a + 7b}{a + b}$ $\quad 2$

6. $\dfrac{x - 3y}{x + y} + \dfrac{x + 5y}{x + y}$ $\quad 2$

7. $\dfrac{4y + 2}{y - 2} - \dfrac{y - 3}{y - 2}$ $\quad \dfrac{3y + 5}{y - 2}$

8. $\dfrac{3t + 2}{t - 4} - \dfrac{t - 2}{t - 4}$ $\quad \dfrac{2t + 4}{t - 4}$

9. $\dfrac{3x-4}{x^2-5x+4} + \dfrac{3-2x}{x^2-5x+4}$ $\dfrac{1}{x-4}$

10. $\dfrac{5x-4}{x^2-6x-7} + \dfrac{5-4x}{x^2-6x-7}$ $\dfrac{1}{x-7}$

11. $\dfrac{3a-2}{a^2-25} - \dfrac{4a-7}{a^2-25}$ $\dfrac{-1}{a+5}$

12. $\dfrac{2a-5}{a^2-9} - \dfrac{3a-8}{a^2-9}$ $\dfrac{-1}{a+3}$

13. $\dfrac{a^2}{a-b} + \dfrac{b^2}{b-a}$ $a+b$

14. $\dfrac{s^2}{r-s} + \dfrac{r^2}{s-r}$ $-(s+r)$

15. $\dfrac{7}{x} - \dfrac{8}{-x}$ $\dfrac{15}{x}$

16. $\dfrac{2}{a} - \dfrac{5}{-a}$ $\dfrac{7}{a}$

17. $\dfrac{x-7}{x^2-16} - \dfrac{x-1}{16-x^2}$ $\dfrac{2}{x+4}$

18. $\dfrac{y-4}{y^2-25} - \dfrac{9-2y}{25-y^2}$ ⊡

19. $\dfrac{t^2+3}{t^4-16} + \dfrac{7}{16-t^4}$ $\dfrac{1}{t^2+4}$

20. $\dfrac{y^2-5}{y^4-81} + \dfrac{4}{81-y^4}$ ⊡

21. $\dfrac{m-3n}{m^3-n^3} - \dfrac{2n}{n^3-m^3}$ ⊡

22. $\dfrac{r-6s}{r^3-s^3} - \dfrac{5s}{s^3-r^3}$ ⊡

23. $\dfrac{a+2}{a-4} + \dfrac{a-2}{a+3}$ ⊡

24. $\dfrac{a+3}{a-5} + \dfrac{a-2}{a+4}$ ⊡

25. $4 + \dfrac{x-3}{x+1}$ $\dfrac{5x+1}{x+1}$

26. $3 + \dfrac{y+2}{y-5}$ $\dfrac{4y-13}{y-5}$

27. $\dfrac{4xy}{x^2-y^2} + \dfrac{x-y}{x+y}$ $\dfrac{x+y}{x-y}$

28. $\dfrac{5ab}{a^2-b^2} + \dfrac{a+b}{a-b}$ ⊡

29. $\dfrac{8}{2x^2-7x+5} + \dfrac{3x+2}{2x^2-x-10}$ $\dfrac{3x^2+7x+14}{(2x-5)(x-1)(x+2)}$

30. $\dfrac{7}{3y^2+y-4} + \dfrac{9y+2}{3y^2-2y-8}$ $\dfrac{3y-4}{(y-1)(y-2)}$

31. $\dfrac{4}{x+1} + \dfrac{x+2}{x^2-1} + \dfrac{3}{x-1}$ $\dfrac{8x+1}{(x+1)(x-1)}$

32. $\dfrac{-2}{y+2} + \dfrac{5}{y-2} + \dfrac{y+3}{y^2-4}$ $\dfrac{4y+17}{(y+2)(y-2)}$

33. $\dfrac{x+6}{5x+10} - \dfrac{x-2}{4x+8}$ $\dfrac{-x+34}{20(x+2)}$

34. $\dfrac{a+3}{5a+25} - \dfrac{a-1}{3a+15}$ $\dfrac{-2a+14}{15(a+5)}$

35. $\dfrac{5ab}{a^2-b^2} - \dfrac{a-b}{a+b}$ $\dfrac{-a^2+7ab-b^2}{(a-b)(a+b)}$

36. $\dfrac{6xy}{x^2-y^2} - \dfrac{x+y}{x-y}$ $\dfrac{-x^2+4xy-y^2}{(x+y)(x-y)}$

37. $\dfrac{x}{x^2+9x+20} - \dfrac{4}{x^2+7x+12}$ $\dfrac{x-5}{(x+5)(x+3)}$

38. $\dfrac{x}{x^2+11x+30} - \dfrac{5}{x^2+9x+20}$ $\dfrac{x-6}{(x+6)(x+4)}$

39. $\dfrac{3y}{y^2-7y+10} - \dfrac{2y}{y^2-8y+15}$ $\dfrac{y}{(y-2)(y-3)}$

40. $\dfrac{5x}{x^2-6x+8} - \dfrac{3x}{x^2-x-12}$ $\dfrac{2x^2+21x}{(x-4)(x-2)(x+3)}$

41. $\dfrac{2x+1}{x-y} + \dfrac{5x^2-5xy}{x^2-2xy+y^2}$ $\dfrac{7x+1}{x-y}$

42. $\dfrac{2-3a}{a-b} + \dfrac{3a^2+3ab}{a^2-b^2}$ $\dfrac{2}{a-b}$

43. $\dfrac{3y+2}{y^2+5y-24} + \dfrac{7}{y^2+4y-32}$ $\dfrac{3y^2-3y-29}{(y-3)(y+8)(y-4)}$

44. $\dfrac{3x+2}{x^2-7x+10} + \dfrac{2x}{x^2-8x+15}$ $\dfrac{5x^2-11x-6}{(x-5)(x-2)(x-3)}$

45. $\dfrac{a-3}{a^2-16} - \dfrac{3a-2}{a^2+2a-24}$ $\dfrac{-2a^2-7a-10}{(a+4)(a-4)(a+6)}$

46. $\dfrac{t+4}{t^2-9} - \dfrac{3t-1}{t^2+2t-3}$ $\dfrac{-2t^2+13t-7}{(t+3)(t-3)(t-1)}$

47. $\dfrac{2}{a^2-5a+4} + \dfrac{-2}{a^2-4}$ $\dfrac{10a-16}{(a-4)(a-1)(a-2)(a+2)}$

48. $\dfrac{3}{a^2-7a+6} + \dfrac{-3}{a^2-9}$ $\dfrac{21a-45}{(a-6)(a-1)(a+3)(a-3)}$

49. $5 + \dfrac{t}{t+2} - \dfrac{8}{t^2-4}$ $\dfrac{2(3t-7)}{t-2}$

50. $2 + \dfrac{t}{t-3} - \dfrac{18}{t^2-9}$ $\dfrac{3(t+4)}{t+3}$

51. $\dfrac{2y-6}{y^2-9} - \dfrac{y}{y-1} + \dfrac{y^2+2}{y^2+2y-3}$ $\dfrac{-y}{(y+3)(y-1)}$

52. $\dfrac{x-1}{x^2-1} - \dfrac{x}{x-2} + \dfrac{x^2+2}{x^2-x-2}$ 0

Aha! **53.** $\dfrac{5y}{1-4y^2} - \dfrac{2y}{2y+1} + \dfrac{5y}{4y^2-1}$ $-\dfrac{2y}{2y+1}$

54. $\dfrac{4x}{x^2-1} + \dfrac{3x}{1-x} - \dfrac{4}{x-1}$ $\dfrac{-3x^2-3x-4}{(x+1)(x-1)}$

55. $\dfrac{2}{x^2-5x+6} - \dfrac{4}{x^2-2x-3} + \dfrac{2}{x^2+4x+3}$ ⊡

⊡ Answers to Exercises 18, 20–24, 28, and 55 can be found on p. A-63.

56. $\dfrac{1}{t^2 + 5t + 6} - \dfrac{2}{t^2 + 3t + 2} - \dfrac{1}{t^2 + 5t + 6}$ $\dfrac{-2}{t^2 + 3t + 2}$

TW **57.** Janine found that the sum of two rational expressions was $(3 - x)/(x - 5)$. The answer given at the back of the book is $(x - 3)/(5 - x)$. Is Janine's answer incorrect? Why or why not?

TW **58.** When two rational expressions are added or subtracted, should the numerator of the result be factored? Why or why not?

Skill Maintenance

Simplify. Use only positive exponents in your answer. [1.4]

59. $\dfrac{15x^{-7}y^{12}z^4}{35x^{-2}y^6z^{-3}}$ $\dfrac{3y^6z^7}{7x^5}$

60. $\dfrac{21a^{-4}b^6c^8}{27a^{-2}b^{-5}c}$ $\dfrac{7b^{11}c^7}{9a^2}$

61. $\dfrac{34s^9t^{-40}r^{30}}{10s^{-3}t^{20}r^{-10}}$ $\dfrac{17s^{12}r^{40}}{5t^{60}}$

62. Find an equation for the line that passes through the point $(0, 3)$ and is perpendicular to the line $f(x) = -\frac{4}{5}x + 7$. [2.5] $y = \frac{5}{4}x + 3$

63. *Value of Coins.* There are 50 dimes in a roll of dimes, 40 nickels in a roll of nickels, and 40 quarters in a roll of quarters. Robert has a total of 12 rolls of coins with a total value of $70.00. If he has 3 more rolls of nickels than dimes, how many of each roll of coins does he have? [3.5]   Dimes: 2 rolls; nickels: 5 rolls; quarters: 5 rolls

64. *Audiotapes.* Anna wants to buy tapes for her work at the campus radio station. She needs some 30-min tapes and some 60-min tapes. If she buys 12 tapes with a total recording time of 10 hr, how many tapes of each length did she buy? [3.3]   30-min tapes: 4; 60-min tapes: 8

Synthesis

TW **65.** Many students make the mistake of always multiplying denominators when looking for a common denominator. Use Example 7 to explain why this approach can yield results that are more difficult to simplify.

TW **66.** Is the sum of two rational expressions always a rational expression? Why or why not?

67. *Prescription Drugs.* After visiting her doctor, Corinna went to the pharmacy for a two-week supply of Zyrtec®, a 20-day supply of Albuterol®, and a 30-day supply of Pepcid®. Corinna refills each prescription as soon as her supply runs out. How long will it be until she can refill all three prescriptions on the same day?   420 days

68. *Astronomy.* The earth, Jupiter, Saturn, and Uranus all revolve around the sun. The earth takes 1 yr, Jupiter 12 yr, Saturn 30 yr, and Uranus 84 yr. How frequently do these four planets line up with each other?   Every 420 yr

69. *Music.* To duplicate a common African polyrhythm, a drummer needs to play sextuplets (6 beats per measure) on a tom-tom while simultaneously playing quarter notes (4 beats per measure) on a bass drum. Into how many equally-sized parts must a measure be divided, in order to precisely execute this rhythm?   12 parts

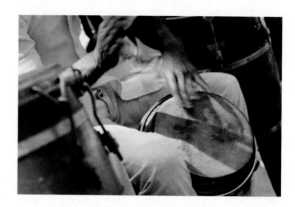

70. *Home Appliances.* Refrigerators last an average of 20 yr, clothes washers about 14 yr, and dishwashers about 10 yr (*Source*: U.S. Department of Energy). In 1980, Westgate College bought new refrigerators for its dormitories. In 1990, the college bought new clothes washers and dishwashers. Predict the year in which the college will need to replace all three types of appliances at once.   2060

Find the LCM. $x^4(x^2 + 1)(x + 1)(x - 1)(x^2 + x + 1)(x^2 - x + 1)$

71. $x^8 - x^4,\ x^5 - x^2,\ x^5 - x^3,\ x^5 + x^2$

72. $2a^3 + 2a^2b + 2ab^2,\ a^6 - b^6,\ 2b^2 + ab - 3a^2,$ $2a^2b + 4ab^2 + 2b^3$

$2ab(a^2 + ab + b^2)(a + b)^2(a^2 - ab + b^2)(a - b)(-2b - 3a)$

73. The LCM of two expressions is $8a^4b^7$. One of the expressions is $2a^3b^7$. List all the possibilities for the other expression. $8a^4$, $8a^4b$, $8a^4b^2$, $8a^4b^3$, $8a^4b^4$, $8a^4b^5$, $8a^4b^6$, $8a^4b^7$

74. Determine the domain and the range of the function graphed below.   Domain: $\{x \,|\, x$ is a real number *and* $x \neq -2$ *and* $x \neq 1\}$; range: $\{y \,|\, y$ is a real number *and* $y \neq 2$ *and* $y \neq 3\}$

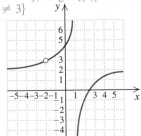

If

$$f(x) = \frac{x^3}{x^2 - 4} \quad and \quad g(x) = \frac{x^2}{x^2 + 3x - 10},$$

find each of the following.

75. $(f + g)(x)$

76. $(f - g)(x)$

77. $(f \cdot g)(x)$

78. $(f/g)(x)$

79. The domain of $f + g$

80. The domain of f/g

Perform the indicated operations and simplify.

81. $5(x - 3)^{-1} + 4(x + 3)^{-1} - 2(x + 3)^{-2}$

82. $4(y - 1)(2y - 5)^{-1} + 5(2y + 3)(5 - 2y)^{-1} + (y - 4)(2y - 5)^{-1}$ $\dfrac{5y + 23}{5 - 2y}$

83. $\dfrac{x + 4}{6x^2 - 20x} \cdot \left(\dfrac{x}{x^2 - x - 20} + \dfrac{2}{x + 4} \right)$ $\dfrac{1}{2x(x - 5)}$

84. $\dfrac{x^2 - 7x + 12}{x^2 - x - 29/3} \cdot \left(\dfrac{3x + 2}{x^2 + 5x - 24} + \dfrac{7}{x^2 + 4x - 32} \right)$ $\dfrac{3}{x + 8}$

85. $\dfrac{8t^5}{2t^2 - 10t + 12} \div \left(\dfrac{2t}{t^2 - 8t + 15} - \dfrac{3t}{t^2 - 7t + 10} \right)$ $-4t^4$

86. $\dfrac{9t^3}{3t^3 - 12t^2 + 9t} \div \left(\dfrac{t + 4}{t^2 - 9} - \dfrac{3t - 1}{t^2 + 2t - 3} \right)$ $\dfrac{3t^2(t + 3)}{-2t^2 + 13t - 7}$

Determine the domain and estimate the range of each function.

87. $f(x) = 2 + \dfrac{x - 3}{x + 1}$ Domain: $(-\infty, -1) \cup (-1, \infty)$; range: $(-\infty, 3) \cup (3, \infty)$

88. $g(x) = \dfrac{2}{(x + 1)^2} + 5$ Domain: $(-\infty, -1) \cup (-1, \infty)$; range: $(5, \infty)$

89. $r(x) = \dfrac{1}{x^2} + \dfrac{1}{(x - 1)^2}$ Domain: $(-\infty, 0) \cup (0, 1) \cup (1, \infty)$; range: $(0, \infty)$

---

75. $\dfrac{x^4 + 6x^3 + 2x^2}{(x + 2)(x - 2)(x + 5)}$ **76.** $\dfrac{x^4 + 4x^3 - 2x^2}{(x + 2)(x - 2)(x + 5)}$ **77.** $\dfrac{x^5}{(x^2 - 4)(x^2 + 3x - 10)}$ **78.** $\dfrac{x(x + 5)}{x + 2}$ (*Note:* $x \neq 0$,

$x \neq -5$, and $x \neq 2$ are additional restrictions, since $g(0) = 0$, -5 is not in the domain of g, and 2 is not in the domain of either f

or g.)   **79.** $\{x \,|\, x$ is a real number *and* $x \neq -5$ *and* $x \neq -2$ *and* $x \neq 2\}$ **80.** $\{x \,|\, x$ is a real number *and* $x \neq -5$ *and* $x \neq -2$

and $x \neq 0$ *and* $x \neq 2\}$ **81.** $\dfrac{9x^2 + 28x + 15}{(x - 3)(x + 3)^2}$

6.3

Multiplying by 1 ■ Dividing Two Rational Expressions

Complex Rational Expressions

A **complex rational expression** is a rational expression that contains rational expressions within its numerator and/or its denominator. Here are some examples:

$$\frac{x + \dfrac{5}{x}}{4x}, \qquad \frac{\dfrac{x - y}{x + y}}{\dfrac{2x - y}{3x + y}}, \qquad \frac{\dfrac{7x}{3} - \dfrac{4}{x}}{\dfrac{5x}{6} + \dfrac{8}{3}}, \qquad \frac{\dfrac{r}{6} + \dfrac{r^2}{144}}{\dfrac{r}{12}}.$$ The rational expressions within each complex rational expression are red.

Complex rational expressions arise in a variety of real-world applications. For example, the complex rational expression on the far right above is used when calculating the size of certain loan payments.

In this section, we simplify complex rational expressions—that is, we write equivalent expressions that do *not* contain rational expressions within a numerator or a denominator. For example, the first expression listed above,

$$\frac{x + \dfrac{5}{x}}{4x},$$ $\dfrac{5}{x}$ **is a rational expression in the numerator.**

is equivalent to

$$\frac{x^2 + 5}{4x^2}.$$ **This rational expression is no longer "complex."**

Two methods are used to simplify complex rational expressions.

Method 1: Multiplying by 1

One method of simplifying a complex rational expression is to multiply the entire expression by 1. To write 1, we use the LCD of the rational expressions within the complex rational expression.

EXAMPLE 1 Simplify:

$$\frac{\dfrac{1}{a^3b} + \dfrac{1}{b}}{\dfrac{1}{a^2b^2} - \dfrac{1}{b^2}}.$$

Solution   The denominators within the complex rational expression are a^3b, b, a^2b^2, and b^2. Thus the LCD is a^3b^2. We multiply by 1, using $(a^3b^2)/(a^3b^2)$:

$$\frac{\dfrac{1}{a^3b} + \dfrac{1}{b}}{\dfrac{1}{a^2b^2} - \dfrac{1}{b^2}} = \frac{\dfrac{1}{a^3b} + \dfrac{1}{b}}{\dfrac{1}{a^2b^2} - \dfrac{1}{b^2}} \cdot \frac{a^3b^2}{a^3b^2}$$ **Multiplying by 1, using the LCD**

$$= \frac{\left(\dfrac{1}{a^3b} + \dfrac{1}{b}\right)a^3b^2}{\left(\dfrac{1}{a^2b^2} - \dfrac{1}{b^2}\right)a^3b^2}$$ **Multiplying the numerator and the denominator. Remember to use parentheses.**

$$= \frac{\dfrac{1}{a^3b} \cdot a^3b^2 + \dfrac{1}{b} \cdot a^3b^2}{\dfrac{1}{a^2b^2} \cdot a^3b^2 - \dfrac{1}{b^2} \cdot a^3b^2}$$ **Using the distributive law to carry out the multiplications**

$$= \frac{\dfrac{\cancel{a^3b}}{\cancel{a^3b}} \cdot b + \dfrac{\cancel{b}}{\cancel{b}} \cdot a^3b}{\dfrac{\cancel{a^2b^2}}{\cancel{a^2b^2}} \cdot a - \dfrac{\cancel{b^2}}{\cancel{b^2}} \cdot a^3}$$ **Removing factors that equal 1. Study this carefully.**

TEACHING TIP

Students should realize that, if done correctly, the expression will no longer be "complex" after multiplying.

$$= \frac{b + a^3 b}{a - a^3} \qquad \text{Simplifying}$$

$$= \frac{b(1 + a^3)}{a(1 - a^2)} \qquad \text{Factoring}$$

$$= \frac{b(1 + a)(1 - a + a^2)}{a(1 + a)(1 - a)} \qquad \begin{array}{l}\text{Factoring further and identi-}\\\text{fying a factor that equals 1}\end{array}$$

$$= \frac{b(1 - a + a^2)}{a(1 - a)}. \qquad \text{Simplifying}$$

Using Multiplication by 1 to Simplify a Complex Rational Expression

1. Find the LCD of all rational expressions *within* the complex rational expression.

2. Multiply the complex rational expression by 1, writing 1 as the LCD divided by itself.

3. Distribute and simplify so that the numerator and the denominator of the complex rational expression are polynomials.

4. Factor and, if possible, simplify.

Note that use of the LCD when multiplying by a form of 1 clears the numerator and the denominator of the complex rational expression of all rational expressions.

EXAMPLE 2   Simplify:

$$\frac{\dfrac{3}{2x - 2} - \dfrac{1}{x + 1}}{\dfrac{1}{x - 1} + \dfrac{x}{x^2 - 1}}.$$

Solution   In this case, to find the LCD, we have to factor first:

$$\frac{\dfrac{3}{2x - 2} - \dfrac{1}{x + 1}}{\dfrac{1}{x - 1} + \dfrac{x}{x^2 - 1}} = \frac{\dfrac{3}{2(x - 1)} - \dfrac{1}{x + 1}}{\dfrac{1}{x - 1} + \dfrac{x}{(x - 1)(x + 1)}} \qquad \begin{array}{l}\text{The LCD is}\\2(x - 1)(x + 1).\end{array}$$

$$= \frac{\dfrac{3}{2(x - 1)} - \dfrac{1}{x + 1}}{\dfrac{1}{x - 1} + \dfrac{x}{(x - 1)(x + 1)}} \cdot \frac{2(x - 1)(x + 1)}{2(x - 1)(x + 1)} \qquad \begin{array}{l}\text{Multiplying by 1,}\\\text{using the LCD}\end{array}$$

$$= \frac{\dfrac{3}{2(x-1)} \cdot 2(x-1)(x+1) - \dfrac{1}{x+1} \cdot 2(x-1)(x+1)}{\dfrac{1}{x-1} \cdot 2(x-1)(x+1) + \dfrac{x}{(x-1)(x+1)} \cdot 2(x-1)(x+1)} \longleftarrow$$

Using the distributive law

$$= \frac{\dfrac{2(x-1)}{2(x-1)} \cdot 3(x+1) - \dfrac{x+1}{x+1} \cdot 2(x-1)}{\dfrac{x-1}{x-1} \cdot 2(x+1) + \dfrac{(x-1)(x+1)}{(x-1)(x+1)} \cdot 2x}$$

Removing factors that equal 1

$$= \frac{3(x+1) - 2(x-1)}{2(x+1) + 2x}$$
Simplifying

$$= \frac{3x+3-2x+2}{2x+2+2x}$$
Using the distributive law

$$= \frac{x+5}{4x+2}.$$
Combining like terms

Entering Rational Expressions

We can check the simplification of rational expressions in one variable using a graphing calculator. For most calculators, parentheses replace the fraction bars in rational expressions. Care must be taken to place the parentheses properly. Thus we use parentheses around the entire numerator and parentheses around the entire denominator of a complex rational expression. If necessary, we also use parentheses around numerators and denominators of rational expressions within the complex rational expression. Following are some examples of complex rational expressions rewritten with parentheses.

Complex Rational Expression	Expression Rewritten with Parentheses
$\dfrac{x + \dfrac{5}{x}}{4x}$	$(x + 5/x)/(4x)$
$\dfrac{\dfrac{x+1}{x}}{\dfrac{x-1}{x^2}}$	$((x+1)/x)/((x-1)/x^2)$
$\dfrac{\dfrac{3}{2x-2} - \dfrac{1}{x+1}}{\dfrac{1}{x-1} + \dfrac{x}{x^2-1}}$	$(3/(2x-2) - 1/(x+1))/(1/(x-1) + x/(x^2-1))$

The following tips may help in placing parentheses properly.

1. Close each set of parentheses: For every left parenthesis, there should be a right parenthesis.
2. Remember the rules for order of operations when deciding how to place parentheses.
3. When in doubt, place parentheses around every numerator and around every denominator. Extra parentheses will make the expression look more complicated, but will not change its value.

Method 2: Dividing Two Rational Expressions

Another method for simplifying complex rational expressions involves first adding or subtracting, as necessary, to obtain one rational expression in the numerator and one rational expression in the denominator. The problem is thereby simplified to one involving the division of two rational expressions.

EXAMPLE 3   Simplify:

$$\frac{\dfrac{3}{x} - \dfrac{2}{x^2}}{\dfrac{3}{x-2} + \dfrac{1}{x^2}}.$$

TEACHING TIP

Consider pointing out that for this method, we work with the numerator and the denominator separately.

Solution

$$\frac{\dfrac{3}{x} - \dfrac{2}{x^2}}{\dfrac{3}{x-2} + \dfrac{1}{x^2}} = \frac{\dfrac{3}{x}\cdot\dfrac{x}{x} - \dfrac{2}{x^2}}{\dfrac{3}{x-2}\cdot\dfrac{x^2}{x^2} + \dfrac{1}{x^2}\cdot\dfrac{x-2}{x-2}}$$

Multiplying $3/x$ by 1 to obtain x^2 as a common denominator

Multiplying by 1, twice, to obtain $x^2(x-2)$ as a common denominator

$$= \frac{\dfrac{3x}{x^2} - \dfrac{2}{x^2}}{\dfrac{3x^2}{(x-2)x^2} + \dfrac{x-2}{x^2(x-2)}}$$

There is now a common denominator in the numerator and a common denominator in the denominator of the complex rational expression.

$$= \frac{\dfrac{3x-2}{x^2}}{\dfrac{3x^2+x-2}{(x-2)x^2}}$$

Subtracting in the numerator and adding in the denominator. We now have one rational expression divided by another rational expression.

$$= \frac{3x-2}{x^2} \div \frac{3x^2+x-2}{(x-2)x^2}$$

Rewriting with a division symbol

$$= \frac{3x-2}{x^2} \cdot \frac{(x-2)x^2}{3x^2+x-2}$$

To divide, multiply by the reciprocal of the divisor.

$$= \frac{(3x-2)(x-2)x^2}{x^2(3x-2)(x+1)}$$

Factoring and removing a factor equal to 1: $\dfrac{x^2(3x-2)}{x^2(3x-2)} = 1$

$$= \frac{x-2}{x+1}$$

$y_1 = (3/x - 2/x^2)/(3/(x - 2) + 1/x^2),$
$y_2 = (x - 2)/(x + 1)$

X	Y1	Y2
−3	2.5	2.5
−2	4	4
−1	ERROR	ERROR
0	ERROR	−2
1	−.5	−.5
2	ERROR	0
3	.25	.25

X = −3

To check, we let $y_1 = (3/x - 2/x^2)/(3/(x - 2) + 1/x^2)$ and $y_2 = (x - 2)/(x + 1)$. A table of values, like that shown at left, indicates that the values of y_1 and y_2 are the same for all x-values for which both expressions are defined.

Using Division to Simplify a Complex Rational Expression

1. Add or subtract, as necessary, to get one rational expression in the numerator.

2. Add or subtract, as necessary, to get one rational expression in the denominator.

3. Perform the indicated division (invert the divisor and multiply).

4. Simplify, if possible, by removing any factors that equal 1.

EXAMPLE 4 Simplify:

$$\frac{1 + \dfrac{2}{x}}{1 - \dfrac{4}{x^2}}.$$

Solution We have

$$\frac{1 + \dfrac{2}{x}}{1 - \dfrac{4}{x^2}} = \frac{\dfrac{x}{x} + \dfrac{2}{x}}{\dfrac{x^2}{x^2} - \dfrac{4}{x^2}} \quad \begin{array}{l} \text{Finding a common denominator} \\[2em] \text{Finding a common denominator} \end{array}$$

$$= \frac{\dfrac{x + 2}{x}}{\dfrac{x^2 - 4}{x^2}} \quad \begin{array}{l} \text{Adding in the numerator} \\[1.5em] \text{Subtracting in the denominator} \end{array}$$

$$= \frac{x + 2}{x} \cdot \frac{x^2}{x^2 - 4} \quad \begin{array}{l} \text{Multiplying by the reciprocal of the divisor} \end{array}$$

$$= \frac{(x + 2) \cdot x^2}{x(x + 2)(x - 2)} \quad \begin{array}{l} \text{Factoring. Remember to simplify when possible.} \end{array}$$

$$= \frac{\cancel{(x + 2)}\cancel{x} \cdot x}{\cancel{x}\cancel{(x + 2)}(x - 2)} \quad \begin{array}{l} \text{Removing a factor equal to 1:} \\ \dfrac{(x + 2)x}{(x + 2)x} = 1 \end{array}$$

$$= \frac{x}{x - 2}. \quad \text{Simplifying}$$

As a quick partial check, we select a convenient value for x—say, 1:

$$\frac{1 + \dfrac{2}{1}}{1 - \dfrac{4}{1^2}} = \frac{1 + 2}{1 - 4} = \frac{3}{-3} = -1 \quad \begin{array}{l} \textbf{We evaluated the original} \\ \textbf{expression for } x = 1. \end{array}$$

TEACHING TIP

Some students may need guidance with the use of parentheses when entering complex rational expressions into a graphing calculator.

and

$$\frac{1}{1-2} = \frac{1}{-1} = -1.$$ **We evaluated the simplified expression for $x = 1$.**

Since both expressions yield the same result, our simplification is probably correct. More evaluation would provide a more definitive check and can be done using the TABLE feature of a graphing calculator.

If negative exponents occur, we first find an equivalent expression using positive exponents and then proceed as in the preceding examples.

EXAMPLE 5   Simplify:

$$\frac{a^{-1} + b^{-1}}{a^{-3} + b^{-3}}.$$

TEACHING TIP

You may wish to review negative exponents. Consider using an example like $(x + 1)^{-1} \div (x^2 - 1)^{-1}$.

Solution

$$\frac{a^{-1} + b^{-1}}{a^{-3} + b^{-3}} = \frac{\dfrac{1}{a} + \dfrac{1}{b}}{\dfrac{1}{a^3} + \dfrac{1}{b^3}}$$ **Rewriting with positive exponents. We continue, using method 2.**

$$= \frac{\dfrac{1}{a} \cdot \dfrac{b}{b} + \dfrac{1}{b} \cdot \dfrac{a}{a}}{\dfrac{1}{a^3} \cdot \dfrac{b^3}{b^3} + \dfrac{1}{b^3} \cdot \dfrac{a^3}{a^3}}$$ **Finding a common denominator**

Finding a common denominator

$$= \frac{\dfrac{b}{ab} + \dfrac{a}{ab}}{\dfrac{b^3}{a^3b^3} + \dfrac{a^3}{a^3b^3}}$$

$$= \frac{\dfrac{b + a}{ab}}{\dfrac{b^3 + a^3}{a^3b^3}}$$ **Adding in the numerator**

Adding in the denominator

$$= \frac{b + a}{ab} \cdot \frac{a^3b^3}{b^3 + a^3}$$ **Multiplying by the reciprocal of the divisor**

$$= \frac{(b + a) \cdot ab \cdot a^2b^2}{ab(b + a)(b^2 - ab + a^2)}$$ **Factoring and looking for common factors**

$$= \frac{\cancel{(b + a)} \cdot \cancel{ab} \cdot a^2b^2}{\cancel{ab}\cancel{(b + a)}(b^2 - ab + a^2)}$$ **Removing a factor equal to 1:** $\dfrac{(b + a)ab}{(b + a)ab} = 1$

$$= \frac{a^2b^2}{b^2 - ab + a^2}$$

There is no one method that is best to use. For expressions like

$$\frac{\dfrac{3x+1}{x-5}}{\dfrac{2-x}{x+3}} \quad \text{or} \quad \frac{\dfrac{3}{x}-\dfrac{2}{x}}{\dfrac{1}{x+1}+\dfrac{5}{x+1}},$$

the second method is probably easier to use since it is little or no work to write the expression as a quotient of two rational expressions.

On the other hand, expressions like

$$\frac{\dfrac{3}{a^2b}-\dfrac{4}{bc^3}}{\dfrac{1}{b^3c}+\dfrac{2}{ac^4}} \quad \text{or} \quad \frac{\dfrac{5}{a^2-b^2}+\dfrac{2}{a^2+2ab+b^2}}{\dfrac{1}{a-b}+\dfrac{4}{a+b}}$$

require fewer steps if we use the first method. Either method can be used with any complex rational expression.

6.3

Exercise Set

Simplify. If possible, use a second method or evaluation as a check.

1. $\dfrac{7+\dfrac{1}{a}}{\dfrac{1}{a}-3}$ $\dfrac{7a+1}{1-3a}$

2. $\dfrac{\dfrac{1}{y}+2}{\dfrac{1}{y}-3}$ $\dfrac{1+2y}{1-3y}$

3. $\dfrac{x-x^{-1}}{x+x^{-1}}$ $\dfrac{x^2-1}{x^2+1}$

4. $\dfrac{y+y^{-1}}{y-y^{-1}}$ $\dfrac{y^2+1}{y^2-1}$

5. $\dfrac{\dfrac{6}{x}+\dfrac{7}{y}}{\dfrac{7}{x}-\dfrac{6}{y}}$ $\dfrac{6y+7x}{7y-6x}$

6. $\dfrac{\dfrac{5}{z}+\dfrac{2}{y}}{\dfrac{4}{z}-\dfrac{1}{y}}$ $\dfrac{2z+5y}{4y-z}$

7. $\dfrac{\dfrac{x^2-y^2}{xy}}{\dfrac{x-y}{y}}$ $\dfrac{x+y}{x}$

8. $\dfrac{\dfrac{a^2-b^2}{ab}}{\dfrac{a-b}{b}}$ $\dfrac{a+b}{a}$

9. $\dfrac{\dfrac{3x}{y}-x}{2y-\dfrac{y}{x}}$ $\dfrac{x^2(3-y)}{y^2(2x-1)}$

10. $\dfrac{1-\dfrac{2}{3x}}{x-\dfrac{4}{9x}}$ $\dfrac{3}{3x+2}$

11. $\dfrac{\dfrac{a^{-1}+b^{-1}}{a^2-b^2}}{ab}$ $\dfrac{1}{a-b}$

12. $\dfrac{\dfrac{x^{-1}+y^{-1}}{x^2-y^2}}{xy}$ $\dfrac{1}{x-y}$

Aha! **13.** $\dfrac{8 + \dfrac{8}{d}}{1 + \dfrac{1}{d}}$ 8

14. $\dfrac{\dfrac{x}{5y^3} + \dfrac{3}{10y}}{\dfrac{3}{10y} + \dfrac{x}{5y^3}}$ 1

31. $\dfrac{\dfrac{y^2}{y^2 - 9} - \dfrac{y}{y + 3}}{\dfrac{y}{y^2 - 9} - \dfrac{1}{y - 3}}$ $-y$

32. $\dfrac{\dfrac{y^2}{y^2 - 25} - \dfrac{y}{y - 5}}{\dfrac{y}{y^2 - 25} - \dfrac{1}{y + 5}}$ $-y$

15. $\dfrac{\dfrac{1}{x + h} - \dfrac{1}{x}}{h}$ $-\dfrac{1}{x(x + h)}$

16. $\dfrac{\dfrac{1}{a - h} - \dfrac{1}{a}}{h}$ $\dfrac{1}{a(a - h)}$

33. $\dfrac{\dfrac{a}{a + 3} + \dfrac{4}{5a}}{\dfrac{a}{2a + 6} + \dfrac{3}{a}}$ $\dfrac{2(5a^2 + 4a + 12)}{5(a^2 + 6a + 18)}$

34. $\dfrac{\dfrac{a}{a + 2} + \dfrac{5}{a}}{\dfrac{a}{2a + 4} + \dfrac{1}{3a}}$ $\dfrac{6a^2 + 30a + 60}{3a^2 + 2a + 4}$

17. $\dfrac{\dfrac{x^2 - x - 12}{x^2 - 2x - 15}}{\dfrac{x^2 + 8x + 12}{x^2 - 5x - 14}}$ $\dfrac{(x - 4)(x - 7)}{(x - 5)(x + 6)}$

18. $\dfrac{\dfrac{a^2 - 4}{a^2 + 3a + 2}}{\dfrac{a^2 - 5a - 6}{a^2 - 6a - 7}}$ $\dfrac{(a - 2)(a - 7)}{(a + 1)(a - 6)}$

35. $\dfrac{c + \dfrac{8}{c^2}}{1 + \dfrac{2}{c}}$ $\dfrac{c^2 - 2c + 4}{c}$

36. $\dfrac{x^{-1} + y^{-1}}{x^{-3} + y^{-3}}$ $\dfrac{x^2 y^2}{y^2 - xy + x^2}$

19. $\dfrac{\dfrac{1}{x - 2} + \dfrac{3}{x - 1}}{\dfrac{2}{x - 1} + \dfrac{5}{x - 2}}$ $\dfrac{4x - 7}{7x - 9}$

20. $\dfrac{\dfrac{2}{y - 3} + \dfrac{1}{y + 1}}{\dfrac{3}{y + 1} + \dfrac{4}{y - 3}}$ $\dfrac{3y - 1}{7y - 5}$

37. $\dfrac{x^2 + xy + y^2}{\dfrac{x^2}{y} - \dfrac{y^2}{x}}$ $\dfrac{xy}{x - y}$

38. $\dfrac{\dfrac{a^2}{b} + \dfrac{b^2}{a}}{a^2 - ab + b^2}$ $\dfrac{a + b}{ab}$

21. $\dfrac{a(a + 3)^{-1} - 2(a - 1)^{-1}}{a(a + 3)^{-1} - (a - 1)^{-1}}$ $\dfrac{a^2 - 3a - 6}{a^2 - 2a - 3}$

22. $\dfrac{a(a + 2)^{-1} - 3(a - 3)^{-1}}{a(a + 2)^{-1} - (a - 3)^{-1}}$ $\dfrac{a^2 - 6a - 6}{a^2 - 4a - 2}$

39. $\dfrac{\dfrac{1}{x^2 - 3x + 2} + \dfrac{1}{x^2 - 4}}{\dfrac{1}{x^2 + 4x + 4} + \dfrac{1}{x^2 - 4}}$ $\dfrac{(2x + 1)(x + 2)}{2x(x - 1)}$

23. $\dfrac{\dfrac{x}{x^2 + 3x - 4} - \dfrac{1}{x^2 + 3x - 4}}{\dfrac{x}{x^2 + 6x + 8} + \dfrac{3}{x^2 + 6x + 8}}$ $\dfrac{x + 2}{x + 3}$

40. $\dfrac{\dfrac{1}{x^2 + 3x + 2} + \dfrac{1}{x^2 - 1}}{\dfrac{1}{x^2 - 1} + \dfrac{1}{x^2 - 4x + 3}}$ $\dfrac{(2x + 1)(x - 3)}{2(x + 2)(x - 1)}$

24. $\dfrac{\dfrac{x}{x^2 + 5x - 6} + \dfrac{6}{x^2 + 5x - 6}}{\dfrac{x}{x^2 - 5x + 4} - \dfrac{2}{x^2 - 5x + 4}}$ $\dfrac{x - 4}{x - 2}$

41. $\dfrac{\dfrac{3}{a^2 - 4a + 3} + \dfrac{3}{a^2 - 5a + 6}}{\dfrac{3}{a^2 - 3a + 2} + \dfrac{3}{a^2 + 3a - 10}}$ $\dfrac{(2a - 3)(a + 5)}{2(a - 3)(a + 2)}$

25. $\dfrac{\dfrac{2}{a^2 - 1} + \dfrac{1}{a + 1}}{\dfrac{3}{a^2 - 1} + \dfrac{2}{a - 1}}$ $\dfrac{a + 1}{2a + 5}$

26. $\dfrac{\dfrac{3}{a^2 - 9} + \dfrac{2}{a + 3}}{\dfrac{4}{a^2 - 9} + \dfrac{1}{a + 3}}$ $\dfrac{2a - 3}{a + 1}$

42. $\dfrac{\dfrac{1}{a^2 + 7a + 10} - \dfrac{2}{a^2 - 7a + 12}}{\dfrac{2}{a^2 - a - 6} - \dfrac{1}{a^2 + a - 20}}$ $\dfrac{-a^2 - 21a - 8}{a^2 + 3a - 34}$

27. $\dfrac{\dfrac{5}{x^2 - 4} - \dfrac{3}{x - 2}}{\dfrac{4}{x^2 - 4} - \dfrac{2}{x + 2}}$ ⊡

28. $\dfrac{\dfrac{4}{x^2 - 1} - \dfrac{3}{x + 1}}{\dfrac{5}{x^2 - 1} - \dfrac{2}{x - 1}}$ ⊡

Aha! **43.** $\dfrac{\dfrac{y}{y^2 - 4} - \dfrac{2y}{y^2 + y - 6}}{\dfrac{2y}{y^2 + y - 6} - \dfrac{y}{y^2 - 4}}$ -1

29. $\dfrac{\dfrac{y}{y^2 - 4} + \dfrac{5}{4 - y^2}}{\dfrac{y^2}{y^2 - 4} + \dfrac{25}{4 - y^2}}$ $\dfrac{1}{y + 5}$

30. $\dfrac{\dfrac{y}{y^2 - 1} + \dfrac{3}{1 - y^2}}{\dfrac{y^2}{y^2 - 1} + \dfrac{9}{1 - y^2}}$ $\dfrac{1}{y + 3}$

44. $\dfrac{\dfrac{y}{y^2 - 1} - \dfrac{3y}{y^2 + 5y + 4}}{\dfrac{3y}{y^2 - 1} - \dfrac{y}{y^2 - 4y + 3}}$ $\dfrac{-2y^2 + 13y - 21}{2(y^2 - y - 20)}$, or $\dfrac{-(y - 3)(2y - 7)}{2(y + 4)(y - 5)}$

⊡ Answers to Exercises 27 and 28 can be found on p. A-63.

TW **45.** Michael *incorrectly* simplifies

$$\frac{a + b^{-1}}{a + c^{-1}} \quad \text{as} \quad \frac{a + c}{a + b}.$$

What mistake is he making and how could you convince him that this is incorrect?

TW **46.** To simplify a complex rational expression in which the sum of two fractions is divided by the difference of the same two fractions, which method is easier? Why?

Skill Maintenance

Solve. [2.2]

47. $2(3x - 1) + 5(4x - 3) = 3(2x + 1)$ 1

48. $5(2x + 3) - 3(4x + 1) = 2(3x - 5)$ $\frac{11}{4}$

49. Solve for y: $\dfrac{t}{s + y} = r$. [2.3] $y = \dfrac{t - rs}{r}$

50. Solve: $|2x - 3| = 7$. [4.3] $\{-2, 5\}$

51. *Framing.*   Andrea has two rectangular frames. The first frame is 3 cm shorter, and 4 cm narrower, than the second frame. If the perimeter of the second frame is 1 cm less than twice the perimeter of the first, what is the perimeter of each frame? [3.3]
First frame: 15 cm; second frame: 29 cm

52. *Earnings.*   Antonio received $28 in tips on Monday, $22 in tips on Tuesday, and $36 in tips on Wednesday. How much will Antonio need to receive in tips on Thursday if his average for the four days is to be $30? [2.3] $34

Synthesis

TW **53.** Use algebra to determine the domain of the function given by

$$f(x) = \frac{\dfrac{1}{x - 2}}{\dfrac{x}{x - 2} - \dfrac{5}{x - 2}}.$$

Then explain how a graphing calculator could be used to check your answer.

TW **54.** In arithmetic, we are taught that

$$\frac{a}{b} \div \frac{c}{d} = \frac{a}{b} \cdot \frac{d}{c}$$

(to divide by a fraction, we invert and multiply). Use method 1 to explain *why* we do this.

Simplify.

55. $\dfrac{5x^{-2} + 10x^{-1}y^{-1} + 5y^{-2}}{3x^{-2} - 3y^{-2}}$ $\dfrac{5(y + x)}{3(y - x)}$

56. $(a^2 - ab + b^2)^{-1}(a^2b^{-1} + b^2a^{-1}) \times$ $\dfrac{b - a}{ab}$
$\qquad (a^{-2} - b^{-2})(a^{-2} + 2a^{-1}b^{-1} + b^{-2})^{-1}$

57. *Astronomy.*   When two galaxies are moving in opposite directions at velocities v_1 and v_2, an observer in one of the galaxies would see the other galaxy receding at speed

$$\frac{v_1 + v_2}{1 + \dfrac{v_1v_2}{c^2}},$$

where c is the speed of light. Determine the observed speed if v_1 and v_2 are both one-fourth the speed of light. $\dfrac{8c}{17}$

Find and simplify

$$\frac{f(x + h) - f(x)}{h}$$

for each rational function f in Exercises 58–61.

58. $f(x) = \dfrac{2}{x^2}$ $\dfrac{-2(2x + h)}{x^2(x + h)^2}$ **59.** $f(x) = \dfrac{3}{x}$ $\dfrac{-3}{x(x + h)}$

60. $f(x) = \dfrac{x}{1 - x}$ $\dfrac{1}{(1 - x - h)(1 - x)}$ **61.** $f(x) = \dfrac{2x}{1 + x}$ $\dfrac{2}{(1 + x + h)(1 + x)}$

62. If

$$F(x) = \frac{3 + \dfrac{1}{x}}{2 - \dfrac{8}{x^2}}, \quad \{x \mid x \text{ is a real number } and \; x \neq 0 \text{ and } x \neq -2 \text{ and } x \neq 2\}$$

find the domain of F.

63. If

$$G(x) = \frac{x - \dfrac{1}{x^2 - 1}}{\dfrac{1}{9} - \dfrac{1}{x^2 - 16}}, \quad \{x \mid x \text{ is a real number } and \; x \neq \pm 1 \text{ and } x \neq \pm 4 \text{ and } \; x \neq \pm 5\}$$

find the domain of G.

64. Find the reciprocal of y if $\dfrac{x^2}{x^4 + x^3 + x^2 + x + 1}$

$$y = x^2 + x + 1 + \frac{1}{x} + \frac{1}{x^2}.$$

65. For $f(x) = \dfrac{2}{2 + x}$, find $f(f(a))$. $\dfrac{2 + a}{3 + a}$

66. For $g(x) = \dfrac{x + 3}{x - 1}$, find $g(g(a))$.   *a*

67. Let

$$f(x) = \left[\dfrac{\dfrac{x + 3}{x - 3} + 1}{\dfrac{x + 3}{x - 3} - 1}\right]^4.$$

Find a simplified form of $f(x)$ and specify the domain of f. $\dfrac{x^4}{81}$; $\{x \mid x$ is a real number *and* $x \neq 3\}$, or $(-\infty, 3) \cup (3, \infty)$

68. *Financial Planning.*   Alexis wishes to invest a portion of each month's pay in an account that pays 7.5% interest. If he wants to have $30,000 in the account after 10 yr, the amount invested each month is given by

$$\dfrac{30{,}000 \cdot \dfrac{0.075}{12}}{\left(1 + \dfrac{0.075}{12}\right)^{120} - 1}.$$

Find the amount of Alexis' monthly investment. $168.61

Collaborative Corner

Which Method Is Best?

Focus: Complex rational expressions
Time: 10–15 minutes
Group size: 2–3

ACTIVITY

Consider the steps in Examples 2 and 3 for simplifying a complex rational expression by each of the two methods. Then, work as a group to simplify

$$\dfrac{\dfrac{5}{x + 1} - \dfrac{1}{x}}{\dfrac{2}{x^2} + \dfrac{4}{x}}$$

subject to the following conditions.

1. The group should predict which method will more easily simplify this expression.

2. Using the method selected in part (1), one group member should perform the first step in the simplification and then pass the problem on to another member of the group. That person then checks the work, performs the next step, and passes the problem on to another group member. If a mistake is found, the problem should be passed to the person who made the mistake for repair. This process continues until, eventually, the simplification is complete.

3. At the same time that part (2) is being performed, another group member should perform the first step of the solution using the method not selected in part (1). He or she should then pass the problem to another group member and so on, just as in part (2).

4. What method *was* easier? Why? Compare your responses with those of other groups.

6.4

Solving Rational Equations

Rational Equations

Connecting the Concepts

It is not unusual to learn a skill that enables us to write equivalent expressions and then to put that skill to work solving a new type of equation. That is precisely what we are now doing: Sections 6.1–6.3 have been devoted to writing equivalent expressions in which rational expressions appeared. There we used least common denominators to add or subtract. Here in Section 6.4, we return to the task of solving equations, but this time the equations contain rational expressions and the LCD is used as part of the multiplication principle.

Study Tip

Make sure that all the examples make sense to you before attempting any exercises. This will prove to be a most efficient use of time.

Solving Rational Equations

In Sections 6.1–6.3, we learned how to *simplify expressions*. We now learn to *solve* a new type of *equation*. A **rational equation** is an equation that contains one or more rational expressions. Here are some examples:

$$\frac{2}{3} - \frac{5}{6} = \frac{1}{t}, \qquad \frac{a-1}{a-5} = \frac{4}{a^2 - 25}, \qquad x^3 + \frac{6}{x} = 5.$$

As you will see in Section 6.5, equations of this type occur frequently in applications. To solve rational equations, recall that one way to *clear fractions* from an equation is to multiply both sides of the equation by the LCD.

To Solve a Rational Equation

Multiply both sides of the equation by the LCD. This is called *clearing fractions* and produces an equation similar to those we have already solved.

Recall that division by 0 is undefined. Note, too, that variables usually appear in at least one denominator of a rational equation. Thus certain numbers can often be ruled out as possible solutions before we even attempt to solve a given rational equation.

EXAMPLE 1   Solve: $\dfrac{x+4}{3x} + \dfrac{x+8}{5x} = 2$.

Algebraic Solution

Because the left side of this equation is undefined when x is 0, we state at the outset that $x \ne 0$. Next, we multiply both sides of the equation by the LCD, $3 \cdot 5 \cdot x$, or $15x$:

$$15x\left(\frac{x+4}{3x} + \frac{x+8}{5x}\right) = 15x \cdot 2 \qquad \text{Multiplying by the LCD to clear fractions}$$

$$15x \cdot \frac{x+4}{3x} + 15x \cdot \frac{x+8}{5x} = 15x \cdot 2 \qquad \text{Using the distributive law}$$

$$\frac{5 \cdot 3x \cdot (x+4)}{3x} + \frac{3 \cdot 5x \cdot (x+8)}{5x} = 30x \qquad \text{Locating factors equal to 1}$$

$$5(x+4) + 3(x+8) = 30x \qquad \text{Removing factors equal to 1:} \quad \frac{3x}{3x} = 1; \frac{5x}{5x} = 1$$

$$5x + 20 + 3x + 24 = 30x \qquad \text{Using the distributive law}$$

$$8x + 44 = 30x$$

$$44 = 22x$$

$$2 = x. \qquad \text{This should check since } x \ne 0.$$

Graphical Solution

We let $y_1 = (x+4)/(3x) + (x+8)/(5x)$ and $y_2 = 2$. The solutions of the equation will be the x-coordinates of any points of intersection.

$y_1 = (x+4)/(3x) + (x+8)/(5x), \quad y_2 = 2$

It appears from the graph that there is one point of intersection, $(2, 2)$. The solution is 2.

Check:

$$\frac{x+4}{3x} + \frac{x+8}{5x} = 2$$

$$\frac{2+4}{3 \cdot 2} + \frac{2+8}{5 \cdot 2} \;?\; 2$$

$$\frac{6}{6} + \frac{10}{10}$$

$$2 \;\bigg|\; 2 \quad \text{TRUE}$$

The number 2 is the solution.

Note that when we clear fractions, all denominators "disappear." This leaves an equation without rational expressions, which we know how to solve.

EXAMPLE 2 Solve: $\dfrac{x-1}{x-5} = \dfrac{4}{x-5}$.

Algebraic Solution

To ensure that neither denominator is 0, we state at the outset the restriction that $x \neq 5$. Then we proceed as before, multiplying both sides by the LCD, $x - 5$:

$$(x-5) \cdot \frac{x-1}{x-5} = (x-5) \cdot \frac{4}{x-5}$$

$$x - 1 = 4$$

$$x = 5. \qquad \textbf{But recall that } x \neq 5.$$

In this case, it is important to remember that, because of the restriction above, 5 cannot be a solution. A check confirms the necessity of that restriction.

Check:

$$\frac{x-1}{x-5} = \frac{4}{x-5}$$

$$\begin{array}{c|c} \dfrac{5-1}{5-5} & \dfrac{4}{5-5} \\[2mm] \hline \dfrac{4}{0} & \dfrac{4}{0} \end{array} \quad \textbf{Division by 0 is undefined.}$$

Graphical Solution

We graph $y_1 = (x-1)/(x-5)$ and $y_2 = 4/(x-5)$ and look for any points of intersection of the graphs.

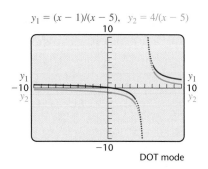

$y_1 = (x-1)/(x-5), \quad y_2 = 4/(x-5)$

DOT mode

It appears that the graphs may intersect around $x = 4$ or $x = 6$. If we use INTERSECT, the calculator returns an error message. This in itself does not tell us that there is no point of intersection, because we do not see all the graph in the viewing window. In order to state positively that there is no solution, we must solve the equation with algebra.

This equation has no solution.

To see why 5 is not a solution of Example 2, note that the multiplication principle for equations requires that we multiply both sides by a *nonzero* number. When both sides of an equation are multiplied by an expression containing variables, it is possible that certain replacements will make that expression equal to 0. Thus it is safe to say that *if* a solution of

$$\frac{x-1}{x-5} = \frac{4}{x-5}$$

exists, then it is also a solution of $x - 1 = 4$. We *cannot* conclude that every solution of $x - 1 = 4$ is a solution of the original equation.

> **CAUTION!** When solving rational equations, do not forget to list any restrictions as part of the first step.

Example 2 illustrates a disadvantage of graphical methods of solving equations. It can be difficult to tell whether two graphs actually intersect. Also, it is easy to "miss" seeing a solution. This is especially true for rational equations, where there are usually two or more branches of the graph. Solving by graphing does make a good check, however, of an algebraic solution.

EXAMPLE 3   Solve: $\dfrac{x^2}{x-3} = \dfrac{9}{x-3}$.

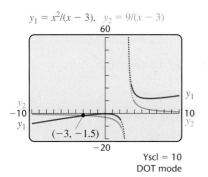

$y_1 = x^2/(x-3),\ \ y_2 = 9/(x-3)$

$(-3, -1.5)$

Yscl = 10
DOT mode

Solution   Note that $x \neq 3$. We multiply both sides by the LCD, $x - 3$:

$$(x-3) \cdot \dfrac{x^2}{x-3} = (x-3) \cdot \dfrac{9}{x-3}$$

$$x^2 = 9 \qquad\qquad \textbf{Simplifying}$$

$$x^2 - 9 = 0 \qquad\qquad \textbf{Getting 0 on one side}$$

$$(x-3)(x+3) = 0 \qquad\qquad \textbf{Factoring}$$

$$x = 3 \quad or \quad x = -3. \qquad \textbf{Using the principle of zero products}$$

Although 3 is a solution of $x^2 = 9$, it must be rejected as a solution of the rational equation. You should perform a check to confirm that -3 *is* a solution despite the fact that 3 is not. This check can also be done graphically, as shown at left.

EXAMPLE 4   Solve: $\dfrac{2}{x+5} + \dfrac{1}{x-5} = \dfrac{16}{x^2-25}$.

Solution   To find all restrictions and to assist in finding the LCD, we factor:

$$\dfrac{2}{x+5} + \dfrac{1}{x-5} = \dfrac{16}{(x+5)(x-5)}. \qquad \textbf{Factoring } x^2 - 25$$

Note that $x \neq -5$ and $x \neq 5$. We multiply by the LCD, $(x+5)(x-5)$, and then use the distributive law:

$$(x+5)(x-5)\left(\dfrac{2}{x+5} + \dfrac{1}{x-5}\right) = (x+5)(x-5) \cdot \dfrac{16}{(x+5)(x-5)}$$

$$(x+5)(x-5)\dfrac{2}{x+5} + (x+5)(x-5)\dfrac{1}{x-5}$$

$$= (x+5)(x-5) \cdot \dfrac{16}{(x+5)(x-5)}$$

$$2(x-5) + (x+5) = 16$$

$$2x - 10 + x + 5 = 16$$

$$3x - 5 = 16$$

$$3x = 21$$

$$x = 7.$$

A check will confirm that the solution is 7.

EXAMPLE 5 Let $f(x) = x + \dfrac{6}{x}$. Find all values of a for which $f(a) = 5$.

Solution Since $f(a) = a + \dfrac{6}{a}$, the problem asks that we find all values of a for which

$$a + \frac{6}{a} = 5.$$

First note that $a \neq 0$. To solve for a, we multiply both sides of the equation by the LCD, a:

$$a\left(a + \frac{6}{a}\right) = 5 \cdot a \qquad \text{Multiplying both sides by } a. \\ \text{Parentheses are important.}$$

$$a \cdot a + a \cdot \frac{6}{a} = 5a \qquad \text{Using the distributive law}$$

$$a^2 + 6 = 5a \qquad \text{Simplifying}$$

$$a^2 - 5a + 6 = 0 \qquad \text{Getting 0 on one side}$$

$$(a - 3)(a - 2) = 0 \qquad \text{Factoring}$$

$$a = 3 \quad or \quad a = 2. \qquad \text{Using the principle of zero products}$$

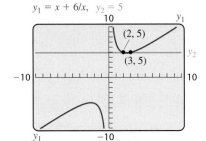

$y_1 = x + 6/x, \quad y_2 = 5$

Check: $\quad f(3) = 3 + \dfrac{6}{3} = 3 + 2 = 5;$

$$f(2) = 2 + \frac{6}{2} = 2 + 3 = 5.$$

The solutions are 2 and 3.

The graph at left shows the points of intersection of the graph of $f(x)$ and the line $y = 5$. It also shows that we have two solutions of $f(x) = 5$, 2 and 3. For $a = 2$ or $a = 3$, we have $f(a) = 5$.

6.4

Exercise Set

Solve.

1. $\dfrac{4}{5} + \dfrac{1}{3} = \dfrac{x}{9}$ $\quad \dfrac{51}{5}$

2. $\dfrac{7}{8} + \dfrac{2}{5} = \dfrac{x}{20}$ $\quad \dfrac{51}{2}$

5. $\dfrac{1}{3} - \dfrac{1}{x} = \dfrac{5}{6}$ $\quad -2$

6. $\dfrac{5}{8} - \dfrac{1}{a} = \dfrac{2}{5}$ $\quad \dfrac{40}{9}$

3. $\dfrac{x}{3} - \dfrac{x}{4} = 12$ $\quad 144$

4. $\dfrac{y}{5} - \dfrac{y}{3} = 15$ $\quad -\dfrac{225}{2}$

7. $\dfrac{1}{2} - \dfrac{2}{7} = \dfrac{3}{2x}$ $\quad 7$

8. $\dfrac{2}{3} - \dfrac{1}{5} = \dfrac{7}{3x}$ $\quad 5$

9. $\dfrac{12}{15} - \dfrac{1}{3x} = \dfrac{4}{5}$
No solution

10. $\dfrac{2}{6} + \dfrac{1}{2x} = \dfrac{1}{3}$
No solution

11. $\dfrac{4}{3y} - \dfrac{3}{y} = \dfrac{10}{3}$ $-\dfrac{1}{2}$

12. $y + \dfrac{4}{y} = -5$ $-4, -1$

13. $\dfrac{x-2}{x-4} = \dfrac{2}{x-4}$
No solution

14. $\dfrac{y-1}{y-3} = \dfrac{2}{y-3}$
No solution

15. $\dfrac{5}{4t} = \dfrac{7}{5t-2}$ $-\dfrac{10}{3}$

16. $\dfrac{3}{x-2} = \dfrac{5}{x+4}$ 11

Aha! **17.** $\dfrac{x^2+4}{x-1} = \dfrac{5}{x-1}$ -1

18. $\dfrac{x^2-1}{x+2} = \dfrac{3}{x+2}$ 2

19. $\dfrac{6}{a+1} = \dfrac{a}{a-1}$ $2, 3$

20. $\dfrac{4}{a-7} = \dfrac{-2a}{a+3}$ $2, 3$

21. $\dfrac{60}{t-5} - \dfrac{18}{t} = \dfrac{40}{t}$ -145

22. $\dfrac{50}{t-2} - \dfrac{16}{t} = \dfrac{30}{t}$ -23

23. $\dfrac{3}{x} + \dfrac{x}{x+2} = \dfrac{4}{x^2+2x}$ -1

24. $\dfrac{x}{x+1} + \dfrac{5}{x} = \dfrac{1}{x^2+x}$ -4

In Exercises 25–30, a rational function f is given. Find all values of a for which f(a) is the indicated value.

25. $f(x) = 2x - \dfrac{15}{x}$; $f(a) = 7$ $-\dfrac{3}{2}, 5$

26. $f(x) = 2x - \dfrac{6}{x}$; $f(a) = 1$ $-\dfrac{3}{2}, 2$

27. $f(x) = \dfrac{x-5}{x+1}$; $f(a) = \dfrac{3}{5}$ 14

28. $f(x) = \dfrac{x-3}{x+2}$; $f(a) = \dfrac{1}{5}$ $\dfrac{17}{4}$

29. $f(x) = \dfrac{12}{x} - \dfrac{12}{2x}$; $f(a) = 8$ $\dfrac{3}{4}$

30. $f(x) = \dfrac{6}{x} - \dfrac{6}{2x}$; $f(a) = 5$ $\dfrac{3}{5}$

Solve.

31. $\dfrac{5}{x+2} - \dfrac{3}{x-2} = \dfrac{2x}{4-x^2}$ 4

32. $\dfrac{y+3}{y+2} - \dfrac{y}{y^2-4} = \dfrac{y}{y-2}$ -3

33. $\dfrac{2}{a+4} + \dfrac{2a-1}{a^2+2a-8} = \dfrac{1}{a-2}$ 3

34. $\dfrac{3}{x^2-6x+9} + \dfrac{x-2}{3x-9} = \dfrac{x}{2x-6}$ $-6, 5$

35. $\dfrac{2}{x+3} - \dfrac{3x+5}{x^2+4x+3} = \dfrac{5}{x+1}$ No solution

36. $\dfrac{3-2y}{y+1} - \dfrac{10}{y^2-1} = \dfrac{2y+3}{1-y}$ No solution

37. $\dfrac{x-1}{x^2-2x-3} + \dfrac{x+2}{x^2-9} = \dfrac{2x+5}{x^2+4x+3}$ $-\dfrac{7}{3}$

38. $\dfrac{2x+1}{x^2-3x-10} + \dfrac{x-1}{x^2-4} = \dfrac{3x-1}{x^2-7x+10}$ $\dfrac{5}{14}$

39. $\dfrac{3}{x^2-x-12} + \dfrac{1}{x^2+x-6} = \dfrac{4}{x^2+3x-10}$ $\dfrac{1}{7}$

40. $\dfrac{3}{x^2-2x-3} - \dfrac{1}{x^2-1} = \dfrac{2}{x^2-8x+7}$ $\dfrac{3}{5}$

^{TW} **41.** Explain how one can easily produce rational equations for which no solution exists. (*Hint*: Examine Example 2.)

^{TW} **42.** Below are unlabeled graphs of $f(x) = x + 2$ and $g(x) = (x^2 - 4)/(x - 2)$. How could you determine which graph represents *f* and which graph represents *g*?

Skill Maintenance

43. *Test Questions.* There are 70 questions on a test. The questions are either multiple-choice, true–false, or fill-in. There are twice as many true–false as fill-in and 5 fewer multiple-choice than true–false. How many of each type of question are there on the test? [3.5]
Multiple-choice: 25; true–false: 30; fill-in: 15

44. *Dinner Prices.* The Danville Volunteer Fire Department served 250 dinners. A child's dinner cost $3 and an adult's dinner cost $8. The total amount of money collected was $1410. How many of each type of dinner was served? [3.3]
Child's: 118; adult's: 132

45. *Agriculture.* The perimeter of a rectangular corn field is 628 m. The length of the field is 6 m greater than the width. Find the area of the field. [2.3] 24,640 m²

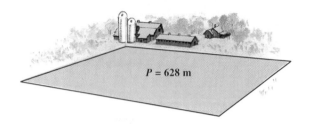

P = 628 m

46. Find two consecutive positive even numbers whose product is 288. [5.8] 16 and 18

47. Determine whether each of the following systems is consistent or inconsistent.
 a) $2x - 3y = 4,$
 $4x - 6y = 7$ [3.1] Inconsistent
 b) $x - 3y = 2,$
 $2x - 3y = 1$ [3.1] Consistent

48. Solve: $|x - 2| > 3$. [4.3] $\{x \mid x < -1 \text{ or } x > 5\}$, or $(-\infty, -1) \cup (5, \infty)$

Synthesis

TW **49.** Karin and Kyle are working together to solve the equation

$$\frac{x}{x^2 - x - 6} = 2x - 1.$$

Karin obtains the graph below using the DOT mode of a graphing calculator, while Kyle obtains the graph at the top of the next column using the CONNECTED mode. Which approach is the better method of showing the solutions? Why?

$y_1 = x/(x^2 - x - 6),\ y_2 = 2x - 1$

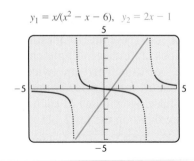

$y_1 = x/(x^2 - x - 6),\ y_2 = 2x - 1$

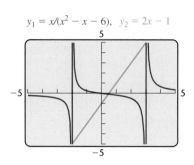

TW **50.** Is the following statement true or false: "For any real numbers a, b, and c, if $ac = bc$, then $a = b$"? Explain why you answered as you did.

For each pair of functions f and g, find all values of a for which f(a) = g(a).

51. $f(x) = \dfrac{x - \dfrac{2}{3}}{x + \dfrac{1}{2}},\ g(x) = \dfrac{x + \dfrac{2}{3}}{x - \dfrac{3}{2}}$ $\dfrac{1}{5}$

52. $f(x) = \dfrac{2 - \dfrac{x}{4}}{2},\ g(x) = \dfrac{\dfrac{x}{4} - 2}{\dfrac{x}{2} + 2}$ $-8, 8$

53. $f(x) = \dfrac{x + 3}{x + 2} - \dfrac{x + 4}{x + 3},\ g(x) = \dfrac{x + 5}{x + 4} - \dfrac{x + 6}{x + 5}$ $\dfrac{7}{2}$

54. $f(x) = \dfrac{1}{1 + x} + \dfrac{x}{1 - x},\ g(x) = \dfrac{1}{1 - x} - \dfrac{x}{1 + x}$
 $\{a \mid a \text{ is a real number } and\ a \neq -1\ and\ a \neq 1\}$

55. $f(x) = \dfrac{0.793}{x} + 18.15,\ g(x) = \dfrac{6.034}{x} - 43.17$ 0.0854697

56. $f(x) = \dfrac{2.315}{x} - \dfrac{12.6}{17.4},\ g(x) = \dfrac{6.71}{x} + 0.763$ −2.955341202

Recall that identities are true for any possible replacement of the variable(s). Determine whether each of the following equations is an identity.

57. $\dfrac{x^2 + 6x - 16}{x - 2} = x + 8,\ x \neq 2$ Yes

58. $\dfrac{x^3 + 8}{x^2 - 4} = \dfrac{x^2 - 2x + 4}{x - 2},\ x \neq -2, x \neq 2$ Yes

Problems Involving Work ▪ Problems Involving Motion

Solving Applications Using Rational Equations

Now that we are able to solve rational equations, it is possible to solve problems that we could not have handled before. The five problem-solving steps remain the same.

Problems Involving Work

EXAMPLE 1   Sue can mow a lawn in 4 hr. Lenny can mow the same lawn in 5 hr. How long would it take both of them, working together, to mow the lawn?

Solution

1. Familiarize.   We familiarize ourselves with the problem by considering two *incorrect* ways of translating the problem to mathematical language.

a) One *incorrect* way to translate the problem is to add the two times:

$$4 \text{ hr} + 5 \text{ hr} = 9 \text{ hr}.$$

Think about this. Sue can do the job *alone* in 4 hr. If Sue and Lenny work together, whatever time it takes them must be *less* than 4 hr.

b) Another *incorrect* approach is to assume that each person mows half the lawn. Were this the case,

Sue would mow $\frac{1}{2}$ the lawn in $\frac{1}{2}(4 \text{ hr})$, or 2 hr

and

Lenny would mow $\frac{1}{2}$ the lawn in $\frac{1}{2}(5 \text{ hr})$, or $2\frac{1}{2}$ hr.

But time would be wasted since Sue would finish $\frac{1}{2}$ hr before Lenny.
Were Sue to help Lenny after completing her half, the entire job would
take between 2 and $2\frac{1}{2}$ hr. This information provides a partial check on
any answer we get—the answer should be between 2 and $2\frac{1}{2}$ hr.

Let's consider how much of the job each person completes in 1 hr, 2 hr,
3 hr, and so on. Since Sue takes 4 hr to mow the entire lawn, in 1 hr she
mows $\frac{1}{4}$ of the lawn. Since Lenny takes 5 hr to mow the entire lawn, in 1 hr
he mows $\frac{1}{5}$ of the lawn. Thus Sue works at a rate of $\frac{1}{4}$ lawn per hr, and
Lenny works at a rate of $\frac{1}{5}$ lawn per hr. (Note that we are assuming that
each person works at a constant rate.)

Working together, Sue and Lenny mow $\frac{1}{4} + \frac{1}{5}$ of the lawn in 1 hr, so
their rate—as a team—is $\frac{5}{20} + \frac{4}{20} = \frac{9}{20}$ lawn per hr.

In 2 hr, Sue mows $\frac{1}{4} \cdot 2$ of the lawn and Lenny mows $\frac{1}{5} \cdot 2$ of the lawn.
Working together, they mow

$$\frac{1}{4} \cdot 2 + \frac{1}{5} \cdot 2, \text{ or } \frac{9}{10} \text{ of the lawn in 2 hr.} \qquad \text{Note that } \frac{9}{20} \cdot 2 = \frac{9}{10}.$$

Continuing this pattern, we can form a table.

	Fraction of the Lawn Mowed		
Time	**By Sue**	**By Lenny**	**Together**
1 hr	$\frac{1}{4}$	$\frac{1}{5}$	$\frac{1}{4} + \frac{1}{5}$, or $\frac{9}{20}$
2 hr	$\frac{1}{4} \cdot 2$	$\frac{1}{5} \cdot 2$	$\left(\frac{1}{4} + \frac{1}{5}\right)2$, or $\frac{9}{20} \cdot 2$, or $\frac{9}{10}$
3 hr	$\frac{1}{4} \cdot 3$	$\frac{1}{5} \cdot 3$	$\left(\frac{1}{4} + \frac{1}{5}\right)3$, or $\frac{9}{20} \cdot 3$, or $\frac{27}{20}$
t hr	$\frac{1}{4} \cdot t$	$\frac{1}{5} \cdot t$	$\left(\frac{1}{4} + \frac{1}{5}\right)t$, or $\frac{9}{20} \cdot t$

This is too
little; $\frac{9}{10} < 1$.

This is too
much; $\frac{27}{20} > 1$.

Sue and Lenny complete the job when the fraction of the lawn mowed
is 1, representing one entire lawn. From the table, we note that the
number of hours t required for Sue and Lenny to mow exactly one lawn is
between 2 hr and 3 hr.

2. Translate.   From the table, we see that t must be some number for which

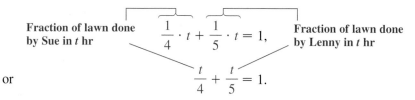

Fraction of lawn done
by Sue in t hr

$$\frac{1}{4} \cdot t + \frac{1}{5} \cdot t = 1,$$

Fraction of lawn done
by Lenny in t hr

or

$$\frac{t}{4} + \frac{t}{5} = 1.$$

3. Carry out.   We solve the equation

$$\frac{t}{4} + \frac{t}{5} = 1$$

both algebraically and graphically.

Algebraic Solution

We solve the equation:

$$\frac{t}{4} + \frac{t}{5} = 1$$

$$20\left(\frac{t}{4} + \frac{t}{5}\right) = 20 \cdot 1 \qquad \text{Multiplying by the LCD}$$

$$\frac{20t}{4} + \frac{20t}{5} = 20 \qquad \text{Distributing the 20}$$

$$5t + 4t = 20 \qquad \text{Simplifying}$$

$$9t = 20$$

$$t = \frac{20}{9}, \text{ or } 2\frac{2}{9}.$$

Graphical Solution

We graph $y_1 = x/4 + x/5$ and $y_2 = 1$.

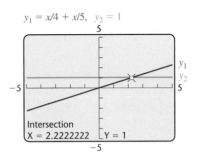

$y_1 = x/4 + x/5, \quad y_2 = 1$

Intersection
X = 2.2222222   Y = 1

Since the graphs intersect at the point (2.2222222, 1), the solution is about 2.22.

4. Check. In $\frac{20}{9}$ hr, Sue mows $\frac{1}{4} \cdot \frac{20}{9}$, or $\frac{5}{9}$, of the lawn and Lenny mows $\frac{1}{5} \cdot \frac{20}{9}$, or $\frac{4}{9}$, of the lawn. Together, they mow $\frac{5}{9} + \frac{4}{9}$, or 1 lawn. The fact that our solution is between 2 and $2\frac{1}{2}$ hr (see step 1 above) is also a check.

5. State. It will take $2\frac{2}{9}$ hr for Sue and Lenny, working together, to mow the lawn.

EXAMPLE 2 It takes Ruth 9 hr more than Willie to repaint a car. Working together, they can do the job in 20 hr. How long would it take, each working alone, to repaint a car?

Solution

1. Familiarize. Unlike Example 1, this problem does not provide us with the times required by the individuals to do the job alone. Let's have $w =$ the number of hours it would take Willie working alone and $w + 9 =$ the number of hours it would take Ruth working alone.

2. Translate. Using the same reasoning as in Example 1, we see that Willie completes $\dfrac{1}{w}$ of the job in 1 hr and Ruth completes $\dfrac{1}{w + 9}$ of the job in 1 hr. In 2 hr, Willie completes $\dfrac{1}{w} \cdot 2$ of the job and Ruth completes

$\dfrac{1}{w+9} \cdot 2$ of the job. We are told that, working together, Willie and Ruth can complete the entire job in 20 hr. This gives the following:

Fraction of job done by Willie in 20 hr $\quad \dfrac{1}{w} \cdot 20 + \dfrac{1}{w+9} \cdot 20 = 1 \quad$ Fraction of job done by Ruth in 20 hr

or $\qquad \dfrac{20}{w} + \dfrac{20}{w+9} = 1.$

3. Carry out. We solve the equation

$$\frac{20}{w} + \frac{20}{w+9} = 1$$

both algebraically and graphically.

Algebraic Solution

We have

$$\frac{20}{w} + \frac{20}{w+9} = 1$$

$$w(w+9)\left(\frac{20}{w} + \frac{20}{w+9}\right) = w(w+9)1 \qquad \text{Multiplying by the LCD}$$

$$(w+9)20 + w \cdot 20 = w(w+9) \qquad \begin{array}{l}\text{Distributing and}\\ \text{simplifying}\end{array}$$

$$40w + 180 = w^2 + 9w$$

$$0 = w^2 - 31w - 180 \qquad \begin{array}{l}\text{Obtaining 0 on}\\ \text{one side}\end{array}$$

$$0 = (w-36)(w+5) \qquad \text{Factoring}$$

$$w - 36 = 0 \quad or \quad w + 5 = 0 \qquad \begin{array}{l}\text{Principle of zero}\\ \text{products}\end{array}$$

$$w = 36 \quad or \qquad w = -5.$$

Graphical Solution

We let $y_1 = 20/x + 20/(x+9)$ and $y_2 = 1$. We graph y_1 and y_2 and find the points of intersection. If we use the window $[-8, 8, -10, 10]$, we see that there is a point of intersection near the x-value -5. It also appears that the graphs may intersect at a point off the screen to the right. Thus we adjust the window to $[-8, 50, 0, 5]$ and see that there is indeed another point of intersection.

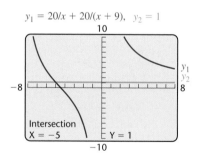

$y_1 = 20/x + 20/(x+9), \; y_2 = 1$

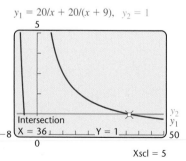

$y_1 = 20/x + 20/(x+9), \; y_2 = 1$

We have $x = 36$ or $x = -5$.

4. **Check.** Since negative time has no meaning in the problem, -5 is not a solution to the original problem. The number 36 checks since, if Willie takes 36 hr alone and Ruth takes $36 + 9 = 45$ hr alone, in 20 hr they would have completed

$$\frac{20}{36} + \frac{20}{45} = \frac{5}{9} + \frac{4}{9} = 1 \text{ paint job.}$$

5. **State.** It would take Willie 36 hr to repaint a car alone, and Ruth 45 hr.

The equations used in Examples 1 and 2 can be generalized as follows.

Modeling Work Problems

If

$a =$ the time needed for A to complete the work alone,

$b =$ the time needed for B to complete the work alone, and

$t =$ the time needed for A and B to complete the work together,

then

$$\frac{t}{a} + \frac{t}{b} = 1.$$

The following are equivalent equations that can also be used:

$$\frac{1}{a} \cdot t + \frac{1}{b} \cdot t = 1 \quad \text{and} \quad \frac{1}{a} + \frac{1}{b} = \frac{1}{t}.$$

Problems Involving Motion

Problems dealing with distance, rate (or speed), and time are called **motion problems**. To translate them, we use either the basic motion formula, $d = rt$, or the formulas $r = d/t$ or $t = d/r$, which can be derived from $d = rt$.

EXAMPLE 3 A racer is bicycling 15 km/h faster than a person on a mountain bike. In the time it takes the racer to travel 80 km, the person on the mountain bike has gone 50 km. Find the speed of each bicyclist.

Solution

1. **Familiarize.** Let's guess that the person on the mountain bike is going 10 km/h. The racer would then be traveling $10 + 15$, or 25 km/h. At 25 km/h, the racer will travel 80 km in $\frac{80}{25} = 3.2$ hr. Going 10 km/h, the mountain bike will cover 50 km in $\frac{50}{10} = 5$ hr. Since $3.2 \neq 5$, our guess was wrong, but we can see that if $r =$ the rate, in kilometers per hour, of the slower bike, then the rate of the racer $= r + 15$.

Making a drawing and constructing a table can be helpful.

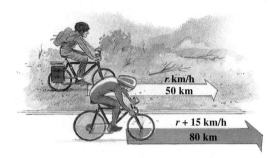

	Distance	Speed	Time
Mountain Bike	50	r	t
Racing Bike	80	$r + 15$	t

2. Translate. By looking at how we checked our guess, we see that in the **Time** column of the table, the t's can be replaced, using the formula *Time = Distance/Rate*, as follows.

	Distance	Speed	Time
Mountain Bike	50	r	$50/r$
Racing Bike	80	$r + 15$	$80/(r + 15)$

Since we are told that the times must be the same, we can write an equation:

$$\frac{50}{r} = \frac{80}{r + 15}.$$

3. Carry out. We solve the equation

$$\frac{50}{r} = \frac{80}{r + 15}$$

both algebraically and graphically.

Algebraic Solution

We have

$$\frac{50}{r} = \frac{80}{r + 15}$$

$$r(r + 15)\frac{50}{r} = r(r + 15)\frac{80}{r + 15} \qquad \text{Multiplying by the LCD}$$

$$50r + 750 = 80r \qquad \text{Simplifying}$$

$$750 = 30r$$

$$25 = r.$$

Graphical Solution

We graph the equations $y_1 = 50/x$ and $y_2 = 80/(x + 15)$, and look for a point of intersection. By adjusting the window dimensions and using the INTERSECT feature, we see that the graphs intersect at the point $(25, 2)$.

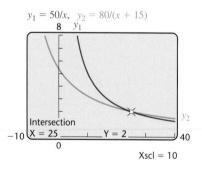

The solution is 25.

4. **Check.**   If our answer checks, the mountain bike is going 25 km/h and the racing bike is going $25 + 15 = 40$ km/h.

   Traveling 80 km at 40 km/h, the racer is riding for $\frac{80}{40} = 2$ hr. Traveling 50 km at 25 km/h, the person on the mountain bike is riding for $\frac{50}{25} = 2$ hr. Our answer checks since the two times are the same.

5. **State.**   The speed of the racer is 40 km/h, and the speed of the person on the mountain bike is 25 km/h.

In the following example, although the distance is the same in both directions, the key to the translation lies in an additional piece of given information.

EXAMPLE 4   A Hudson River tugboat goes 10 mph in still water. It travels 24 mi upstream and 24 mi back in a total time of 5 hr. What is the speed of the current?

Solution

1. **Familiarize.**   Let's guess that the speed of the current is 4 mph. The tugboat would then be moving $10 - 4 = 6$ mph upstream and $10 + 4 = 14$ mph downstream. The tugboat would require $\frac{24}{6} = 4$ hr to travel 24 mi upstream and $\frac{24}{14} = 1\frac{5}{7}$ hr to travel 24 mi downstream. Since the total time, $4 + 1\frac{5}{7} = 5\frac{5}{7}$ hr, is not the 5 hr mentioned in the problem, we know that our guess is wrong.

   Suppose that the current's speed $= c$ mph. The tugboat would then travel $10 - c$ mph when going upstream and $10 + c$ mph when going downstream.

   A sketch and table can help display the information.

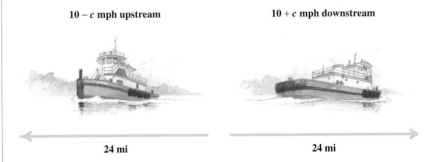

10 − c mph upstream   10 + c mph downstream

24 mi   24 mi

	Distance	Speed	Time
Upstream	24	$10 - c$	t_1
Downstream	24	$10 + c$	t_2

2. Translate.   From examining our guess, we see that the time traveled can be represented using the formula *Time = Distance/Rate*:

	Distance	Speed	Time
Upstream	24	$10 - c$	$24/(10 - c)$
Downstream	24	$10 + c$	$24/(10 + c)$

Since the total time upstream and back is 5 hr, we use the last column of the table to form an equation:

$$\frac{24}{10 - c} + \frac{24}{10 + c} = 5.$$

3. Carry out.   We solve the equation

$$\frac{24}{10 - c} + \frac{24}{10 + c} = 5$$

both algebraically and graphically.

Algebraic Solution

We have

$$\frac{24}{10 - c} + \frac{24}{10 + c} = 5$$

$$(10 - c)(10 + c)\left[\frac{24}{10 - c} + \frac{24}{10 + c}\right] = (10 - c)(10 + c)5$$

Multiplying by the LCD

$$24(10 + c) + 24(10 - c) = (100 - c^2)5$$
$$480 = 500 - 5c^2 \quad \text{Simplifying}$$
$$5c^2 - 20 = 0$$
$$5(c^2 - 4) = 0$$
$$5(c - 2)(c + 2) = 0$$
$$c = 2 \quad or \quad c = -2.$$

Graphical Solution

We graph

$$y_1 = 24/(10 - x) + 24/(10 + x) \quad \text{and}$$
$$y_2 = 5$$

and see that there are two points of intersection.

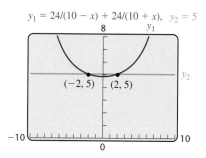

$y_1 = 24/(10 - x) + 24/(10 + x), \quad y_2 = 5$

The solutions are -2 and 2.

4. Check.   Since speed cannot be negative in this problem, -2 cannot be a solution. You should confirm that 2 checks in the original problem.

5. State.   The speed of the current is 2 mph.

6.5

Exercise Set

Solve.

1. The reciprocal of 3, plus the reciprocal of 6, is the reciprocal of what number?   2

2. The reciprocal of 5, plus the reciprocal of 7, is the reciprocal of what number? $\frac{35}{12}$

3. The sum of a number and 6 times its reciprocal is -5. Find the number. $-3, -2$

4. The sum of a number and 21 times its reciprocal is -10. Find the number. $-3, -7$

5. The reciprocal of the product of two consecutive integers is $\frac{1}{42}$. Find the two integers. -6 and -7, 7 and 6

6. The reciprocal of the product of two consecutive integers is $\frac{1}{72}$. Find the two integers. -9 and -8, 8 and 9

7. *Home Restoration.*   Cedric can refinish the floor of an apartment in 8 hr. Carolyn can refinish the floor in 6 hr. How long will it take them, working together, to refinish the floor? $3\frac{3}{7}$ hr

8. *Mail Order.*   Zoe, an experienced shipping clerk, can fill a certain order in 5 hr. Willy, a new clerk, needs 9 hr to complete the same job. Working together, how long will it take them to fill the order? $3\frac{3}{14}$ hr

9. *Filling a Tank.*   A community water tank can be filled in 18 hr by the town office well alone and in 22 hr by the high school well alone. How long will it take to fill the tank if both wells are working? $9\frac{9}{10}$ hr

10. *Filling a Pool.*   A swimming pool can be filled in 12 hr if water enters through a pipe alone or in 30 hr if water enters through a hose alone. If water is entering through both the pipe and the hose, how long will it take to fill the pool? $8\frac{4}{7}$ hr

11. *Hotel Management.*   The Honeywell HQ17 air cleaner can clean the air in a 12-ft by 14-ft conference room in 10 min. The HQ174 can clean the air in a room of the same size in 6 min. How long would it take the two machines together to clean the air in such a room? $3\frac{3}{4}$ min

12. *Printing.*   Pronto Press can print an order of booklets in 4.5 hr. Red Dot Printers can do the same job in 5.5 hr. How long will it take if both presses are used?   2.475 hr

13. *Cutting Firewood.*   Jake can cut and split a cord of firewood in 6 fewer hr than Skyler can. When they work together, it takes them 4 hr. How long would it take each of them to do the job alone?   Skyler: 12 hr; Jake: 6 hr

14. *Wood Cutting.*   Damon can clear a lot in 5.5 hr. His partner, Tyron, can complete the same job in 7.5 hr. How long will it take them to clear the lot working together? $3\frac{9}{52}$ hr

15. *Hotel Management.*   The Honeywell HQ17 air cleaner takes twice as long as the EV25 to clean the same volume of air. Together the two machines can clean the air in a 24-ft by 24-ft banquet room in 10 min. How long would it take each machine, working alone, to clean the air in the room?   HQ17: 30 min; EV25: 15 min

16. *Computer Printers.*   The HP Office Jet G85 works twice as fast as the Laser Jet II. When the machines work together, a university can produce all its staff manuals in 15 hr. Find the time it would take each machine, working alone, to complete the same job.   Office Jet G85: $22\frac{1}{2}$ hr; Laser Jet II: 45 hr

17. *Painting.*   Sara takes 3 hr longer to paint a floor than it takes Kate. When they work together, it takes them 2 hr. How long would each take to do the job alone?   Sara: 6 hr; Kate: 3 hr

18. *Painting.*   Claudia can paint a neighbor's house 4 times as fast as Jan can. The year they worked together it took them 8 days. How long would it take each to paint the house alone?   Claudia: 10 days; Jan: 40 days

19. *Waxing a Car.*   Rosita can wax her car in 2 hr. When she works together with Helga, they can wax the car in 45 min. How long would it take Helga, working by herself, to wax the car? $1\frac{1}{5}$ hr

20. *Newspaper Delivery.*   Zsuzanna can deliver papers three times as fast as Stan can. If they work together, it takes them 1 hr. How long would it take each to deliver the papers alone?   Zsuzanna: $\frac{4}{3}$ hr; Stan: 4 hr

21. *Sorting Recyclables.*   Together, it takes John and Deb 2 hr 55 min to sort recyclables. Alone, John would require 2 more hr than Deb. How long would it take Deb to do the job alone? (*Hint*: Convert minutes to hours or hours to minutes.)   300 min, or 5 hr

22. *Paving.*   Together, Larry and Mo require 4 hr 48 min to pave a driveway. Alone, Larry would require 4 hr more than Mo. How long would it take Mo to do the job alone? (*Hint*: Convert minutes to hours.)   8 hr

23. *Kayaking.*   The speed of the current in Catamount Creek is 3 mph. Zeno can kayak 4 mi upstream in the same time it takes him to kayak 10 mi downstream. What is the speed of Zeno's kayak in still water?   7 mph

24. *Boating.*   The current in the Lazy River moves at a rate of 4 mph. Monica's dinghy motors 6 mi upstream in the same time it takes to motor 12 mi downstream. What is the speed of the dinghy in still water?   12 mph

25. *Moving Sidewalks.*   The moving sidewalk at O'Hare Airport in Chicago moves 1.8 ft/sec. Walking on the moving sidewalk, Camille travels 105 ft forward in the time it takes to travel 51 ft in the opposite direction. How fast would Camille be walking on a nonmoving sidewalk?   5.2 ft/sec

26. *Moving Sidewalks.*   Newark Airport's moving sidewalk moves at a speed of 1.7 ft/sec. Walking on the moving sidewalk, Benny can travel 120 ft forward in the same time it takes to travel 52 ft in the opposite direction. How fast would Benny be walking on a nonmoving sidewalk?   4.3 ft/sec

27. *Train Speed.*   The speed of the A&M freight train is 14 mph less than the speed of the A&M passenger train. The passenger train travels 400 mi in the same time that the freight train travels 330 mi. Find the speed of each train.   Freight: 66 mph; passenger: 80 mph

28. *Walking.*   Rosanna walks 2 mph slower than Simone. In the time it takes Simone to walk 8 mi, Rosanna walks 5 mi. Find the speed of each person.   Rosanna: $3\frac{1}{3}$ mph; Simone: $5\frac{1}{3}$ mph

Aha! **29.** *Bus Travel.*   A local bus travels 7 mph slower than the express. The express travels 45 mi in the time it takes the local to travel 38 mi. Find the speed of each bus.   Express: 45 mph; local: 38 mph

30. *Train Speed.*   The A train goes 12 mph slower than the E train. The A train travels 230 mi in the same time that the E train travels 290 mi. Find the speed of each train.   A: 46 mph; E: 58 mph

31. *Boating.*   Audrey's paddleboat travels 2 km/h in still water. The boat is paddled 4 km downstream in the same time it takes to go 1 km upstream. What is the speed of the river? $1\frac{1}{5}$ km/h

32. *Boating.*   Laverne's Mercruiser travels 15 km/h in still water. She motors 140 km downstream in the same time it takes to travel 35 km upstream. What is the speed of the river?   9 km/h

33. *Shipping.*   A barge moves 7 km/h in still water. It travels 45 km upriver and 45 km downriver in a total time of 14 hr. What is the speed of the current?   2 km/h

34. *Moped Speed.*   Jaime's moped travels 8 km/h faster than Mara's. Jaime travels 69 km in the same time that Mara travels 45 km. Find the speed of each person's moped.   Jaime: 23 km/h; Mara: 15 km/h

35. *Aviation.*   A Citation II Jet travels 350 mph in still air and flies 487.5 mi into the wind and 487.5 mi with the wind in a total of 2.8 hr (*Source*: Eastern Air Charter). Find the wind speed.   25 mph

36. *Canoeing.*   Al paddles 55 m per minute in still water. He paddles 150 m upstream and 150 m downstream in a total time of 5.5 min. What is the speed of the current?   5 m per minute

37. *Train Travel.*   A freight train covered 120 mi at a certain speed. Had the train been able to travel 10 mph faster, the trip would have been 2 hr shorter. How fast did the train go?   20 mph

38. *Boating.*   Julia's Boston Whaler cruised 45 mi upstream and 45 mi back in a total of 8 hr. The speed of the river is 3 mph. Find the speed of the boat in still water.   12 mph

T.W. 39. Two steamrollers are paving a parking lot. Working together, will the two steamrollers take less than half as long as the slower steamroller would working alone? Why or why not?

T.W. 40. Two fuel lines are filling a freighter with oil. Will the faster fuel line take more or less than twice as long to fill the freighter by itself? Why?

Skill Maintenance

Simplify.

41. $\dfrac{35a^6b^8}{7a^2b^2}$ [1.4] $5a^4b^6$ **42.** $\dfrac{20x^9y^6}{4x^3y^2}$ [1.4] $5x^6y^4$

43. $\dfrac{36s^{15}t^{10}}{9s^5t^2}$ [1.4] $4s^{10}t^8$

$-2x^4 - 7x^2 + 11x$
44. $6x^4 - 3x^2 + 9x - (8x^4 + 4x^2 - 2x)$ [5.1]

45. $2(x^3 + 4x^2 - 5x + 7) - 5(2x^3 - 4x^2 + 3x - 1)$
 [5.1] $-8x^3 + 28x^2 - 25x + 19$

46. $9x^4 + 7x^3 + x^2 - 8 - (-2x^4 + 3x^2 + 4x + 2)$
 [5.1] $11x^4 + 7x^3 - 2x^2 - 4x - 10$

Synthesis

T.W. 47. Write a work problem for a classmate to solve. Devise the problem so that the solution is "Liane and Michele will take 4 hr to complete the job, working together."

T.W. 48. Write a work problem for a classmate to solve. Devise the problem so that the solution is "Jen takes 5 hr and Pablo takes 6 hr to complete the job alone."

49. *Filling a Bog.*   The Norwich cranberry bog can be filled in 9 hr and drained in 11 hr. How long will it take to fill the bog if the drainage gate is left open?
 $49\frac{1}{2}$ hr

50. *Filling a Tub.*   Justine's hot tub can be filled in 10 min and drained in 8 min. How long will it take to empty a full tub if the water is left on?   40 min

51. Refer to Exercise 24. How long will it take Monica to motor 3 mi downstream?   11.25 min

52. Refer to Exercise 23. How long will it take Zeno to kayak 5 mi downstream?   30 min

53. *Escalators.*   Together, a 100-cm-wide escalator and a 60-cm-wide escalator can empty a 1575-person auditorium in 14 min (*Source*: McGraw-Hill *Encyclopedia of Science and Technology*). The wider escalator moves twice as many people as the narrower one. How many people per hour does the 60-cm-wide escalator move?   2250 people per hour

54. *Aviation.*   A Coast Guard plane has enough fuel to fly for 6 hr, and its speed in still air is 240 mph. The plane departs with a 40-mph tailwind and returns to the same airport flying into the same wind. How far can the plane travel under these conditions?
 700 mi from the airport

55. *Boating.*   Shoreline Travel operates a 3-hr paddle-boat cruise on the Missouri River. If the speed of the boat in still water is 12 mph, how far upriver can the pilot travel against a 5-mph current before it is time to turn around? $14\frac{7}{8}$ mi

56. *Boating.*   The speed of a motor boat in still water is three times the speed of a river's current. A trip up the river and back takes 10 hr, and the total distance of the trip is 100 km. Find the speed of the current.
 $3\frac{3}{4}$ km/h

57. *Travel by Car.*   Melissa drives to work at 50 mph and arrives 1 min late. She drives to work at 60 mph and arrives 5 min early. How far does Melissa live from work?   30 mi

58. At what time after 4:00 will the minute hand and the hour hand of a clock first be in the same position?
 $21\frac{9}{11}$ min after 4:00

59. At what time after 10:30 will the hands of a clock first be perpendicular? $8\frac{2}{11}$ min after 10:30

Average speed is defined as total distance divided by total time.

60. Lenore drove 200 km. For the first 100 km of the trip, she drove at a speed of 40 km/h. For the second half of the trip, she traveled at a speed of 60 km/h. What was the average speed of the entire trip? (It was *not* 50 km/h.)   48 km/h

61. For the first 50 mi of a 100-mi trip, Chip drove 40 mph. What speed would he have to travel for the last half of the trip so that the average speed for the entire trip would be 45 mph? $51\frac{3}{7}$ mph

Collaborative Corner

Does the Model Hold Water?

Focus: Testing a mathematical model

Time: 20–30 minutes

Group size: 2–3

Materials: An empty 1-gal plastic jug, a kitchen or laboratory sink, a stopwatch or a watch capable of measuring seconds, an inexpensive pen or pair of scissors or a nail or knife for poking holes in plastic.

Problems like Exercises 49 and 50 can be solved algebraically and then checked at home or in a laboratory.

ACTIVITY

1. While one group member fills the empty jug with water, the other group member(s) should record how many seconds this takes.

2. After carefully poking a few holes in the bottom of the jug, record how many seconds it takes the full jug to empty.

3. Using the information found in parts (1) and (2) above, use algebra to predict how long it will take to fill the punctured jug.

4. Test your prediction by timing how long it takes for the pierced jug to be filled. Be sure to run the water at the same rate as in part (1).

5. How accurate was your prediction? How might your prediction have been made more accurate?

6.6

Divisor a Monomial ■ Divisor a Polynomial

Division of Polynomials

A rational expression indicates division. Division of polynomials, like division of real numbers, relies on our multiplication and subtraction skills.

Divisor a Monomial

To divide a monomial by a monomial, we can subtract exponents when bases are the same (see Section 1.4)—for example,

$$\frac{45x^{10}}{3x^4} = 15x^{10-4} = 15x^6, \qquad \frac{48a^2b^5}{-3ab^2} = \frac{48}{-3}a^{2-1}b^{5-2} = -16ab^3.$$

To divide a polynomial by a monomial, we regard the division as a sum of quotients of monomials. This uses the fact that since

$$\frac{A}{C} + \frac{B}{C} = \frac{A+B}{C}, \quad \text{we know that} \quad \frac{A+B}{C} = \frac{A}{C} + \frac{B}{C}.$$

TEACHING TIP

You may wish to review the terms dividend, divisor, quotient, and remainder.

EXAMPLE 1   Divide $12x^3 + 8x^2 + x + 4$ by $4x$.

Solution

$$(12x^3 + 8x^2 + x + 4) \div (4x) = \frac{12x^3 + 8x^2 + x + 4}{4x}$$

Writing a rational expression

$$= \frac{12x^3}{4x} + \frac{8x^2}{4x} + \frac{x}{4x} + \frac{4}{4x}$$

Writing as a sum of quotients

$$= 3x^2 + 2x + \frac{1}{4} + \frac{1}{x}$$

Performing the four indicated divisions

EXAMPLE 2   Divide: $(8x^4y^5 - 3x^3y^4 + 5x^2y^3) \div x^2y^3$.

Solution

$$\frac{8x^4y^5 - 3x^3y^4 + 5x^2y^3}{x^2y^3} = \frac{8x^4y^5}{x^2y^3} - \frac{3x^3y^4}{x^2y^3} + \frac{5x^2y^3}{x^2y^3}$$

Try to perform this step mentally.

$$= 8x^2y^2 - 3xy + 5$$

Division by a Monomial

To divide a polynomial by a monomial, divide each term of the polynomial by the monomial.

Divisor a Polynomial

When the divisor has more than one term, we use a procedure very similar to long division in arithmetic.

EXAMPLE 3   Divide $2x^2 - 7x - 15$ by $x - 5$.

Solution   We have

$$\begin{array}{r} 2x \\ x - 5 \overline{) 2x^2 - 7x - 15} \\ -(2x^2 - 10x) \\ \hline 3x \end{array}$$

Divide $2x^2$ by x: $2x^2/x = 2x$.
Multiply $x - 5$ by $2x$.
Subtract by mentally changing signs and adding: $-7x + 10x = 3x$.

We now "bring down" the other term in the dividend, -15.

$$\begin{array}{r} 2x + 3 \\ x - 5 \overline{) 2x^2 - 7x - 15} \\ 2x^2 - 10x \\ \hline 3x - 15 \\ -(3x - 15) \\ \hline 0 \end{array}$$

Divide $3x$ by x: $3x/x = 3$.
Multiply $x - 5$ by 3.
Subtract.

Check: $(x - 5)(2x + 3) = 2x^2 - 7x - 15$. The answer checks.

The quotient is $2x + 3$.

To understand why we perform long division as we do, note that Example 3 amounts to "filling in" an unknown polynomial:

$$(x - 5)(\ ? \ ) = 2x^2 - 7x - 15.$$

We see that $2x$ must be in the unknown polynomial if we are to get the first term, $2x^2$, from the multiplication. To see what else is needed, note that

$$(x - 5)(2x \quad) = 2x^2 - 10x \neq 2x^2 - 7x - 15.$$

The $2x$ can be regarded as a (poor) approximation of the quotient that we are after. To see how far off the approximation is, we subtract:

$$\begin{array}{r} 2x^2 - 7x - 15 \\ -(2x^2 - 10x) \\ \hline 3x - 15 \end{array}$$

Note where this appeared in the long division above.

To get the needed terms, $3x - 15$, we need another term in the unknown polynomial. We use 3 because $(x - 5) \cdot 3$ is $3x - 15$:

$$(x - 5)(2x + 3) = 2x^2 - 10x + 3x - 15$$
$$= 2x^2 - 7x - 15.$$

Now when we subtract the product $(x - 5)(2x + 3)$ from $2x^2 - 7x - 15$, the remainder is 0.

If a nonzero remainder occurs, when do we stop dividing? We continue until the degree of the remainder is less than the degree of the divisor.

EXAMPLE 4   Divide $x^2 + 5x + 8$ by $x + 3$.

Solution   We have

$$\begin{array}{r} x \\ x + 3 \overline{) x^2 + 5x + 8} \\ \underline{x^2 + 3x} \\ 2x \end{array}$$

Divide the first term of the dividend by the first term of the divisor: $x^2/x = x$.
Multiply x above by $x + 3$.
Subtract.

The subtraction above is $(x^2 + 5x) - (x^2 + 3x)$. Remember: To subtract, add the opposite (change the sign of every term, then add).

We now "bring down" the next term of the dividend—in this case, 8— and repeat the process:

$$\begin{array}{r} x \ + 2 \\ x + 3 \overline{) x^2 + 5x + 8} \\ \underline{x^2 + 3x} \\ 2x + 8 \\ \underline{2x + 6} \\ 2 \end{array}$$

Divide the first term by the first term: $2x/x = 2$.

The 8 has been "brought down."
Multiply 2 by $x + 3$.
Subtract: $(2x + 8) - (2x + 6)$.

The quotient is $x + 2$, with remainder 2. Note that the degree of the remainder is 0 and the degree of the divisor, $x + 3$, is 1. Since $0 < 1$, the process stops.

Check: $(x + 3)(x + 2) + 2 = x^2 + 5x + 6 + 2$ **Add the remainder to the product.**

$$= x^2 + 5x + 8$$

TEACHING TIP

You may want to introduce checking answers using an example from arithmetic such as $7\overline{)53}$.

We write our answer as $x + 2$, R2, or as

$$\text{Quotient} + \frac{\text{Remainder}}{\text{Divisor}}$$

$$x + 2 + \frac{2}{x + 3}.$$

This is how answers are listed at the back of the book.

The last answer in Example 4 can also be checked by multiplying:

$$(x + 3)\left[(x + 2) + \frac{2}{x + 3}\right] = (x + 3)(x + 2) + (x + 3)\frac{2}{x + 3}$$

Using the distributive law

$$= x^2 + 5x + 6 + 2$$
$$= x^2 + 5x + 8.$$

This was the dividend in Example 4.

An equivalent, but quicker, check is to multiply the divisor by the quotient and then add the remainder. This is precisely what we did in the check in Example 4.

You may have noticed that it is helpful to have all polynomials written in descending order.

Tips for Dividing Polynomials

1. Arrange polynomials in descending order.
2. If there are missing terms in the dividend, either write them with 0 coefficients or leave space for them.
3. Continue the long-division process until the degree of the remainder is less than the degree of the divisor.

EXAMPLE 5   Divide: $(9a^2 + a^3 - 5) \div (a^2 - 1)$.

Solution   We rewrite the problem in descending order:

$$(a^3 + 9a^2 - 5) \div (a^2 - 1).$$

Thus,

TEACHING TIP

It may be helpful to show students how to predict the degree of the quotient before dividing.

$$
\begin{array}{r}
a + 9 \\
a^2 - 1 \overline{)a^3 + 9a^2 + 0a - 5} \\
\underline{a^3 \quad\quad - a} \\
9a^2 + a - 5 \\
\underline{9a^2 \quad\quad - 9} \\
a + 4
\end{array}
$$

When there is a missing term, we can write it in, as in this example, or leave space, as in Example 6 below.

Subtracting:
$a^3 + 9a^2 - (a^3 - a) = 9a^2 + a$

The degree of the remainder is less than the degree of the divisor, so we are finished.

The answer is $a + 9 + \dfrac{a + 4}{a^2 - 1}$.

EXAMPLE 6 Let $f(x) = 125x^3 - 8$ and $g(x) = 5x - 2$. If $F(x) = (f/g)(x)$, find a simplified expression for $F(x)$.

Solution   Recall that $(f/g)(x) = f(x)/g(x)$. Thus,

$$F(x) = \frac{125x^3 - 8}{5x - 2}$$

and

$$
\begin{array}{r}
25x^2 + 10x + 4 \\
5x - 2 \overline{)125x^3 \qquad\qquad - 8} \\
\underline{125x^3 - 50x^2} \\
50x^2 \\
\underline{50x^2 - 20x} \\
20x - 8 \\
\underline{20x - 8} \\
0.
\end{array}
$$

Leaving space for the missing terms.
Subtracting:
$125x^3 - (125x^3 - 50x^2) = 50x^2$

Subtracting

Note that, because $F(x) = f(x)/g(x)$, $g(x)$ cannot be 0. Since $g(x)$ is 0 for $x = \frac{2}{5}$ (check this), we have

$$F(x) = 25x^2 + 10x + 4, \quad \text{provided } x \neq \frac{2}{5}.$$

6.6

Exercise Set

Divide and check.

1. $\dfrac{34x^6 + 18x^5 - 28x^2}{6x^2}$ $\dfrac{17}{3}x^4 + 3x^3 - \dfrac{14}{3}$

2. $\dfrac{30y^8 - 15y^6 + 40y^4}{5y^4}$ $6y^4 - 3y^2 + 8$

3. $\dfrac{21a^3 + 7a^2 - 3a - 14}{7a}$ $3a^2 + a - \dfrac{3}{7} - \dfrac{2}{a}$

4. $\dfrac{-25x^3 + 20x^2 - 3x + 7}{5x}$ $-5x^2 + 4x - \dfrac{3}{5} + \dfrac{7}{5x}$

5. $(14y^3 - 9y^2 - 8y) \div (2y^2)$ $7y - \dfrac{9}{2} - \dfrac{4}{y}$

6. $(6a^4 + 9a^2 - 8) \div (2a)$ $3a^3 + \dfrac{9a}{2} - \dfrac{4}{a}$

7. $(15x^7 - 21x^4 - 3x^2) \div (-3x^2)$ $-5x^5 + 7x^2 + 1$

8. $(36y^6 - 18y^4 - 12y^2) \div (-6y)$ $-6y^5 + 3y^3 + 2y$

9. $(a^2b - a^3b^3 - a^5b^5) \div (a^2b)$ $1 - ab^2 - a^3b^4$

10. $(x^3y^2 - x^3y^3 - x^4y^2) \div (x^2y^2)$ $x - xy - x^2$

11. $(6p^2q^2 - 9p^2q + 12pq^2) \div (-3pq)$ $-2pq + 3p - 4q$

12. $(16y^4z^2 - 8y^6z^4 + 12y^8z^3) \div (4y^4z)$ $4z - 2y^2z^3 + 3y^4z^2$

Aha! 13. $(x^2 + 10x + 21) \div (x + 7)$ $x + 3$

14. $(y^2 - 8y + 16) \div (y - 4)$ $y - 4$

15. $(a^2 - 8a - 16) \div (a + 4)$ $a - 12 + \dfrac{32}{a + 4}$

16. $(y^2 - 10y - 25) \div (y - 5)$ $y - 5 + \dfrac{-50}{y - 5}$

17. $(x^2 - 9x + 21) \div (x - 5)$ $x - 4 + \dfrac{1}{x - 5}$

18. $(x^2 - 11x + 23) \div (x - 7)$ $x - 4 + \dfrac{-5}{x - 7}$

19. $(y^2 - 25) \div (y + 5)$ $y - 5$

20. $(a^2 - 81) \div (a - 9)$ $a + 9$

21. $(y^3 - 4y^2 + 3y - 6) \div (y - 2)$ $y^2 - 2y - 1 + \dfrac{-8}{y - 2}$

22. $(x^3 - 5x^2 + 4x - 7) \div (x - 3)$ $x^2 - 2x - 2 + \dfrac{-13}{x - 3}$

23. $(2x^3 + 3x^2 - x - 3) \div (x + 2)$ $2x^2 - x + 1 + \dfrac{-5}{x + 2}$

24. $(3x^3 - 5x^2 - 3x - 2) \div (x - 2)$ $3x^2 + x - 1 + \dfrac{-4}{x - 2}$

25. $(a^3 - a + 10) \div (a - 4)$ $a^2 + 4a + 15 + \dfrac{70}{a - 4}$

26. $(x^3 - x + 6) \div (x + 2)$ $x^2 - 2x + 3 + \dfrac{8}{x + 2}$

27. $(10y^3 + 6y^2 - 9y + 10) \div (5y - 2)$ $2y^2 + 2y - 1 + \dfrac{8}{5y - 2}$

28. $(6x^3 - 11x^2 + 11x - 2) \div (2x - 3)$ $3x^2 - x + 4 + \dfrac{10}{2x - 3}$

29. $(2x^4 - x^3 - 5x^2 + x - 6) \div (x^2 + 2)$ $2x^2 - x - 9 + \dfrac{3x + 12}{x^2 + 2}$

30. $(3x^4 + 2x^3 - 11x^2 - 2x + 5) \div (x^2 - 2)$ $3x^2 + 2x - 5 + \dfrac{2x - 5}{x^2 - 2}$

For Exercises 31–38, $f(x)$ and $g(x)$ are as given. Find a simplified expression for $F(x)$ if $F(x) = (f/g)(x)$. (See Example 6.)

31. $f(x) = 8x^3 + 27, \ g(x) = 2x + 3$ $4x^2 - 6x + 9, x \neq -\frac{3}{2}$

32. $f(x) = 64x^3 - 8, \ g(x) = 4x - 2$ $16x^2 + 8x + 4, x \neq \frac{1}{2}$

33. $f(x) = 6x^2 - 11x - 10, \ g(x) = 3x + 2$ $2x - 5, x \neq -\frac{2}{3}$

34. $f(x) = 8x^2 - 22x - 21, \ g(x) = 2x - 7$ $4x + 3, x \neq \frac{7}{2}$

35. $f(x) = x^4 - 24x^2 - 25, \ g(x) = x^2 - 25$ $x^2 + 1, x \neq -5, x \neq 5$

36. $f(x) = x^4 - 3x^2 - 54, \ g(x) = x^2 - 9$ $x^2 + 6, x \neq -3, x \neq 3$

37. $f(x) = 8x^2 - 3x^4 - 2x^3 + 2x^5 - 5, \ g(x) = x^2 - 1$ $2x^3 - 3x^2 + 5, x \neq -1, x \neq 1$

38. $f(x) = 4x - x^3 - 10x^2 + 3x^4 - 8, \ g(x) = x^2 - 4$ $3x^2 - x + 2, x \neq -2, x \neq 2$

TW 39. Explain how factoring could be used to solve Example 6.

TW 40. Explain how to construct a polynomial of degree 4 that has a remainder of 3 when divided by $x + 1$.

Skill Maintenance

Solve.

41. $ab - cd = k$, for c [2.3] $c = \dfrac{ab - k}{d}$

42. $xy - wz = t$, for z [2.3] $z = \dfrac{xy - t}{w}$

43. Find three consecutive positive integers such that the product of the first and second integers is 26 less than the product of the second and third integers. [5.8]   12, 13, 14

44. If $f(x) = 2x^3$, find $f(-3a)$. [2.1] $-54a^3$

Solve.

45. $|2x - 3| > 7$ [4.3]   ▣   **46.** $|3x - 1| < 8$ [4.3]   ▣

Synthesis

TW 47. Explain how to construct a polynomial of degree 4 that has a remainder of 2 when divided by $x + c$.

TW 48. Do addition, subtraction, and multiplication of polynomials always result in a polynomial? Does division? Why or why not?

Divide.

49. $(4a^3b + 5a^2b^2 + a^4 + 2ab^3) \div (a^2 + 2b^2 + 3ab)$ $a^2 + ab$

50. $(x^4 - x^3y + x^2y^2 + 2x^2y - 2xy^2 + 2y^3) \div (x^2 - xy + y^2)$ $x^2 + 2y$

51. $(a^7 + b^7) \div (a + b)$ $a^6 - a^5b + a^4b^2 - a^3b^3 + a^2b^4 - ab^5 + b^6$

52. Find k such that when $x^3 - kx^2 + 3x + 7k$ is divided by $x + 2$, the remainder is 0. $\frac{14}{3}$

53. When $x^2 - 3x + 2k$ is divided by $x + 2$, the remainder is 7. Find k. $-\frac{3}{2}$

54. Let
$$f(x) = \frac{3x + 7}{x + 2}.$$
 $3 + \dfrac{1}{x + 2}$

a) Use division to find an expression equivalent to $f(x)$. Then graph f.

b) On the same set of axes, sketch both $g(x) = 1/(x + 2)$ and $h(x) = 1/x$.   ▣

c) How do the graphs of f, g, and h compare?   ▣

TW 55. Jamaladeen incorrectly states that
$$(x^3 + 9x^2 - 6) \div (x^2 - 1) = x + 9 + \frac{x + 4}{x^2 - 1}.$$
Without performing any long division, how could you show Jamaladeen that his division cannot possibly be correct?

▣ Answers to Exercises 45, 46, 54(b), and 54(c) can be found on p. A-63.

6.7   Streamlining Long Division ■ The Remainder Theorem

Synthetic Division

Streamlining Long Division

To divide a polynomial by a binomial of the type $x - a$, we can streamline the usual procedure to develop a process called *synthetic division*.

Compare the following. In each stage, we attempt to write a bit less than in the previous stage, while retaining enough essentials to solve the problem. At the end, we will return to the usual polynomial notation.

Stage 1

When a polynomial is written in descending order, the coefficients provide the essential information:

$$
\begin{array}{r}
4x^2 + 5x\ + 11 \\
x - 2\overline{)4x^3 - 3x^2 +\ \ \ x +\ 7} \\
\underline{4x^3 - 8x^2} \\
5x^2 +\ \ \ x \\
\underline{5x^2 - 10x} \\
11x +\ 7 \\
\underline{11x - 22} \\
29
\end{array}
\qquad
\begin{array}{r}
4 + 5 + 11 \\
1 - 2\overline{)4 - 3 +\ 1 +\ 7} \\
\underline{4 - 8} \\
5 +\ 1 \\
\underline{5 - 10} \\
11 +\ 7 \\
\underline{11 - 22} \\
29
\end{array}
$$

Because the leading coefficient in the divisor is 1, each time we multiply the divisor by a term in the answer, the leading coefficient of that product duplicates a coefficient in the answer. In the next stage, we don't bother to duplicate these numbers. We also show where -2 is used and drop the 1 from the divisor.

Stage 2

$$
\begin{array}{r}
4x^2 + 5x\ + 11 \\
x - 2\overline{)4x^3 - 3x^2 +\ \ \ x +\ 7} \\
\underline{4x^3 - 8x^2} \\
5x^2 +\ \ \ x \\
\underline{5x^2 - 10x} \\
11x +\ 7 \\
\underline{11x - 22} \\
29
\end{array}
$$

$$
\begin{array}{r}
4 + 5 + 11 \\
-2\overline{)4 - 3 +\ 1 +\ 7} \\
-\ 8 \qquad \\
5 +\ 1 \qquad \\
-\ 10 \quad \\
11 +\ 7 \\
-\ 22 \\
29
\end{array}
$$

Multiply: $-2 \cdot 4 = -8$.
Subtract: $-3 - (-8) = 5$.
Multiply: $-2 \cdot 5 = -10$.
Subtract: $1 - (-10) = 11$.
Multiply: $-2 \cdot 11 = -22$.
Subtract: $7 - (-22) = 29$.

To simplify further, we now reverse the sign of the -2 in the divisor and, in exchange, *add* at each step in the long division.

Stage 3

$$
\begin{array}{r}
4x^2 + 5x\ + 11 \\
x - 2\overline{)4x^3 - 3x^2 +\ \ \ x +\ 7} \\
\underline{4x^3 - 8x^2} \\
5x^2 +\ \ \ x \\
\underline{5x^2 - 10x} \\
11x +\ 7 \\
\underline{11x - 22} \\
29
\end{array}
$$

$$
\begin{array}{r}
4 + 5 + 11 \\
2\overline{)4 - 3 +\ 1 +\ 7} \\
8 \qquad \\
5 +\ 1 \qquad \\
10 \quad \\
11 +\ 7 \\
22 \\
29
\end{array}
$$

Replace the -2 with 2.
Multiply: $2 \cdot 4 = 8$.
Add: $-3 + 8 = 5$.
Multiply: $2 \cdot 5 = 10$.
Add: $1 + 10 = 11$.
Multiply: $2 \cdot 11 = 22$.
Add: $7 + 22 = 29$.

The blue numbers can be eliminated if we look at the red numbers instead.

Stage 4

$$
\begin{array}{r}
4x^2 + 5x + 11 \\
x - 2{\overline{\smash{\big)}\,4x^3 - 3x^2 + x + 7}} \\
\underline{4x^3 - 8x^2} \\
5x^2 + x \\
\underline{5x^2 - 10x} \\
11x + 7 \\
\underline{11x - 22} \\
29
\end{array}
$$

$$
\begin{array}{r}
4 \quad 5 \quad 11 \\
2\overline{\smash{\big)}\,4 \;\; -3 \quad 1 \quad 7} \\
8 \quad 10 \quad 22 \\
\overline{5 \quad 11 \quad 29}
\end{array}
$$

Don't lose sight of how the products 8, 10, and 22 are found. Also, note that the 5 and 11 preceding the remainder 29 coincide with the 5 and 11 following the 4 on the top line. By writing a 4 to the left of 5 on the bottom line, we can eliminate the top line in stage 4 and read our answer from the bottom line. This final stage is commonly called **synthetic division**.

Stage 5

$$
\begin{array}{r}
4 \quad 5 \quad 11 \\
2\overline{\smash{\big)}\,4 \;\; -3 \quad 1 \quad 7} \\
8 \quad 10 \quad 22 \\
\overline{5 \quad 11 \quad 29}
\end{array}
$$

$$
\begin{array}{r}
2 \rfloor 4 \;\; -3 \quad 1 \quad 27 \\
8 \quad 10 \quad 22 \\
\overline{4 \quad 5 \quad 11 \; | \; 29}
\end{array}
$$

⟵ This is the remainder.

This is the zero-degree coefficient.

This is the first-degree coefficient.

This is the second-degree coefficient.

The quotient is $4x^2 + 5x + 11$. The remainder is 29.

Remember that in order for this method to work, the divisor must be of the form $x - a$, that is, a variable minus a constant. The coefficient of the variable must be 1.

EXAMPLE 1   Use synthetic division to divide:

$$(x^3 + 6x^2 - x - 30) \div (x - 2).$$

Solution

$$
\begin{array}{r}
2 \rfloor 1 \quad 6 \quad -1 \quad -30 \\
\hline
1
\end{array}
$$

Write the 2 of $x - 2$ and the coefficients of the dividend.

Bring down the first coefficient.

$$
\begin{array}{r}
2 \rfloor 1 \quad 6 \quad -1 \quad -30 \\
2 \\
\hline
1 \quad 8
\end{array}
$$

Multiply 1 by 2 to get 2.

Add 6 and 2.

$$
\begin{array}{r}
2 \rfloor 1 \quad 6 \quad -1 \quad -30 \\
2 \quad 16 \\
\hline
1 \quad 8 \quad 15
\end{array}
$$

Multiply 8 by 2.

Add -1 and 16.

$$
\begin{array}{r|rrr}
2 & 1 & 6 & -1 & -30 \\
  &   & 2 & 16 & 30 \\
\hline
  & 1 & 8 & 15 & 0
\end{array}
$$

Multiply 15 by 2 and add.

Since the remainder is 0, we have

$$(x^3 + 6x^2 - x - 30) \div (x - 2) = x^2 + 8x + 15.$$

We can check this with a table, letting

$$y_1 = (x^3 + 6x^2 - x - 30) \div (x - 2) \quad \text{and} \quad y_2 = x^2 + 8x + 15.$$

The values of both expressions are the same except for $x = 2$, when y_1 is not defined.

X	Y₁	Y₂
-2	3	3
-1	8	8
0	15	15
1	24	24
2	ERROR	35
3	48	48
4	63	63

X = -2

The answer is $x^2 + 8x + 15$ with R0, or just $x^2 + 8x + 15$.

EXAMPLE 2 Use synthetic division to divide.

a) $(2x^3 + 7x^2 - 5) \div (x + 3)$
b) $(10x^2 - 13x + 3x^3 - 20) \div (4 + x)$

Solution

a) $(2x^3 + 7x^2 - 5) \div (x + 3)$

The dividend has no x-term, so we need to write 0 for its coefficient of x. Note that $x + 3 = x - (-3)$, so we write -3 inside the ⌋.

$$
\begin{array}{r|rrrr}
-3 & 2 & 7 & 0 & -5 \\
   &   & -6 & -3 & 9 \\
\hline
   & 2 & 1 & -3 & 4
\end{array}
$$

The answer is $2x^2 + x - 3$, with R4, or $2x^2 + x - 3 + \dfrac{4}{x + 3}$.

b) We first rewrite $(10x^2 - 13x + 3x^3 - 20) \div (4 + x)$ in descending order:

$$(3x^3 + 10x^2 - 13x - 20) \div (x + 4).$$

Next, we use synthetic division. Note that $x + 4 = x - (-4)$.

$$
\begin{array}{r|rrrr}
-4 & 3 & 10 & -13 & -20 \\
   &   & -12 & 8 & 20 \\
\hline
   & 3 & -2 & -5 & 0
\end{array}
$$

The answer is $3x^2 - 2x - 5$.

The Remainder Theorem

When a polynomial function $f(x)$ is divided by $x - a$, the remainder is related to the function value $f(a)$. Compare the following from Examples 1 and 2.

Polynomial Function	Divisor	Remainder	Function Value
Example 1: $f(x) = x^3 + 6x^2 - x - 30$	$x - 2$	0	$f(2) = 0$
Example 2(a): $g(x) = 2x^3 + 7x^2 - 5$	$x + 3$, or $x - (-3)$	4	$g(-3) = 4$
Example 2(b): $p(x) = 10x^2 - 13x + 3x^3 - 20$	$4 + x$, or $x - (-4)$	0	$p(-4) = 0$

When the remainder is 0, the divisor $x - a$ is a factor of the polynomial. Thus we can write $f(x) = x^3 + 6x^2 - x - 30$ in Example 1 as $f(x) = (x - 2)(x^2 + 8x + 15)$. Then

$$f(2) = (2 - 2)(2^2 + 8 \cdot 2 + 15)$$
$$= 0(2^2 + 8 \cdot 2 + 15) = 0.$$

Thus, when the remainder is 0 after division by $x - a$, the function value $f(a)$ is also 0. Remarkably, as Example 2(a) suggests, this pattern extends to nonzero remainders as well. The fact that the remainder and the function value coincide is predicted by the remainder theorem, which follows.

The Remainder Theorem

The remainder obtained by dividing $P(x)$ by $x - r$ is $P(r)$.

A proof of this result is outlined in Exercise 31.

EXAMPLE 3   Let $f(x) = 8x^5 - 6x^3 + x - 8$. Use synthetic division to find $f(2)$.

Solution   The remainder theorem tells us that $f(2)$ is the remainder when $f(x)$ is divided by $x - 2$. We use synthetic division to find that remainder:

```
2| 8   0   -6   0    1    -8
        16  32  52  104   210
   8   16  26  52  105 | 202
```

Although the bottom line can be used to find the quotient for the division $(8x^5 - 6x^3 + x - 8) \div (x - 2)$, what we are really interested in is the remainder. It tells us that $f(2) = 202$.

6.7

Exercise Set

Use synthetic division to divide.

1. $(x^3 - 2x^2 + 2x - 7) \div (x + 1)$ $\quad x^2 - 3x + 5 + \dfrac{-12}{x + 1}$

2. $(x^3 - 2x^2 + 2x - 7) \div (x - 1)$ $\quad x^2 - x + 1 + \dfrac{-6}{x - 1}$

3. $(a^2 + 8a + 11) \div (a + 3)$ $\quad a + 5 + \dfrac{-4}{a + 3}$

4. $(a^2 + 8a + 11) \div (a + 5)$ $\quad a + 3 + \dfrac{-4}{a + 5}$

5. $(x^3 - 7x^2 - 13x + 3) \div (x - 2)$ $\quad x^2 - 5x - 23 + \dfrac{-43}{x - 2}$

6. $(x^3 - 7x^2 - 13x + 3) \div (x + 2)$ $\quad x^2 - 9x + 5 + \dfrac{-7}{x + 2}$

7. $(3x^3 + 7x^2 - 4x + 3) \div (x + 3)$ ⊡

8. $(3x^3 + 7x^2 - 4x + 3) \div (x - 3)$ $\quad 3x^2 + 16x + 44 + \dfrac{135}{x - 3}$

9. $(y^3 - 3y + 10) \div (y - 2)$ $\quad y^2 + 2y + 1 + \dfrac{12}{y - 2}$

10. $(x^3 - 2x^2 + 8) \div (x + 2)$ $\quad x^2 - 4x + 8 + \dfrac{-8}{x + 2}$

11. $(x^5 - 32) \div (x - 2)$ $\quad x^4 + 2x^3 + 4x^2 + 8x + 16$

12. $(y^5 - 1) \div (y - 1)$ $\quad y^4 + y^3 + y^2 + y + 1$

13. $(3x^3 + 1 - x + 7x^2) \div \left(x + \frac{1}{3}\right)$ $\quad 3x^2 + 6x - 3 + \dfrac{2}{x + \frac{1}{3}}$

14. $(8x^3 - 1 + 7x - 6x^2) \div \left(x - \frac{1}{2}\right)$ $\quad 8x^2 - 2x + 6 + \dfrac{2}{x - \frac{1}{2}}$

Use synthetic division to find the indicated function value.

15. $f(x) = 5x^4 + 12x^3 + 28x + 9;\ f(-3)$ $\quad 6$

16. $g(x) = 3x^4 - 25x^2 - 18;\ g(3)$ $\quad 0$

17. $P(x) = 6x^4 - x^3 - 7x^2 + x + 2;\ P(-1)$ $\quad 1$

18. $F(x) = 3x^4 + 8x^3 + 2x^2 - 7x - 4;\ F(-2)$ $\quad 2$

19. $f(x) = x^4 - x^3 - 19x^2 + 49x - 30;\ f(4)$ $\quad 54$

20. $p(x) = x^4 + 7x^3 + 11x^2 - 7x - 12;\ p(2)$ $\quad 90$

ᴛᴡ 21. Why is it that we *add* when performing synthetic division, but *subtract* when performing long division?

ᴛᴡ 22. Explain how synthetic division could be useful when factoring a polynomial.

Skill Maintenance

Solve. [2.3]

23. $9 + cb = a - b$, for b $\quad b = \dfrac{a - 9}{c + 1}$

24. $8 + ac = bd + ab$, for a $\quad a = \dfrac{bd - 8}{c - b}$

Find the domain of f. [5.5]

25. $f(x) = \dfrac{5}{3x^2 - 75}$ $\quad \{x \mid x \text{ is a real number } and\ x \neq -5 \text{ } and\ x \neq 5\}$

$\{x \mid x \text{ is a real number } and\ x \neq -\frac{9}{2} \text{ } and\ x \neq 1\}$

26. $f(x) = \dfrac{7}{2x^2 + 7x - 9}$

Graph.

27. $y - 2 = \frac{3}{4}(x + 1)$ [2.6] ⊡

28. $y = -\frac{4}{3}x + 2$ [2.4] ⊡

Synthesis

ᴛᴡ 29. Let $Q(x)$ be a polynomial function with $p(x)$ a factor of $Q(x)$. If $p(3) = 0$, does it follow that $Q(3) = 0$? Why or why not? If $Q(3) = 0$, does it follow that $p(3) = 0$? Why or why not?

ᴛᴡ 30. What adjustments must be made if synthetic division is to be used to divide a polynomial by a binomial of the form $ax + b$, with $a > 1$?

31. To prove the remainder theorem, note that any polynomial $P(x)$ can be rewritten as $(x - r) \cdot Q(x) + R$, where $Q(x)$ is the quotient polynomial that arises when $P(x)$ is divided by $x - r$, and R is some constant (the remainder).

a) How do we know that R must be a constant? ⊡

b) Show that $P(r) = R$ (this says that $P(r)$ is the remainder when $P(x)$ is divided by $x - r$). ⊡

32. Let $f(x) = 4x^3 + 16x^2 - 3x - 45$. Find $f(-3)$ and then solve the equation $f(x) = 0$. $\quad 0; -3, -\frac{5}{2}, \frac{3}{2}$

33. Let $f(x) = 6x^3 - 13x^2 - 79x + 140$. Find $f(4)$ and then solve the equation $f(x) = 0$. $\quad 0; -\frac{7}{2}, \frac{5}{3}, 4$

Nested Evaluation. One way to evaluate a polynomial function like $P(x) = 3x^4 - 5x^3 + 4x^2 - 1$ is to successively factor out x as shown:

$$P(x) = x(x(x(3x - 5) + 4) + 0) - 1.$$

Computations are then performed using this "nested" form of $P(x)$.

34. Use nested evaluation to find $f(-3)$ in Exercise 32. Note the similarities to the calculations performed with synthetic division. $\quad 0$

35. Use nested evaluation to find $f(4)$ in Exercise 33. Note the similarities to the calculations performed with synthetic division. $\quad 0$

⊡ Answers to Exercises 7, 27, 28, and 31 can be found on p. A-63.

6.8

Formulas ■ Direct Variation ■ Inverse Variation ■ Joint and Combined Variation ■ Models

Formulas, Applications, and Variation

Formulas

Formulas occur frequently as mathematical models. Many formulas contain rational expressions, and to solve such formulas for a specified letter, we proceed as when solving rational equations.

EXAMPLE 1 Optics.   A camera's "f-stop" is a setting that determines how much light reaches the film. The formula $f = L/d$ tells how to calculate an f-stop. In this formula, f is the f-stop, L is the focal length (approximately the distance from the lens to the film), and d is the diameter of the lens. Solve for d.

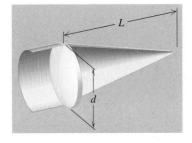

Solution   We solve this equation as we did those in Section 6.4:

$$f = \frac{L}{d}$$

$$d \cdot f = d \cdot \frac{L}{d} \qquad \textbf{Multiplying both sides by the LCD to clear fractions}$$

$$df = L$$

$$df \cdot \frac{1}{f} = L \cdot \frac{1}{f} \qquad \textbf{Multiplying both sides by } \frac{1}{f} \textbf{ or dividing by } f$$

$$d = \frac{L}{f}. \qquad \textbf{Simplifying and removing a factor equal to 1: } \frac{f}{f} = 1$$

The formula $d = L/f$ can now be used to determine the diameter of a lens if we know the focal length and the f-stop.

EXAMPLE 2 Astronomy.   The formula

$$\frac{V^2}{R^2} = \frac{2g}{R + h}$$

is used to find a satellite's *escape velocity* V, where R is a planet's radius, h is the satellite's height above the planet, and g is the planet's gravitational constant. Solve for h.

Solution We first clear fractions by multiplying by the LCD, which is $R^2(R + h)$:

$$\frac{V^2}{R^2} = \frac{2g}{R + h}$$

$$R^2(R + h)\frac{V^2}{R^2} = R^2(R + h)\frac{2g}{R + h}$$

$$\frac{R^2(R + h)V^2}{R^2} = \frac{R^2(R + h)2g}{R + h}$$

$$(R + h)V^2 = R^2 \cdot 2g. \quad \text{Removing factors equal to 1: } \frac{R^2}{R^2} = 1$$
$$\text{and } \frac{R + h}{R + h} = 1$$

Remember: We are solving for h. Although we *could* distribute V^2, since h appears only within the factor $R + h$, it is easier to divide both sides by V^2:

$$\frac{(R + h)V^2}{V^2} = \frac{2R^2g}{V^2} \quad \text{Dividing both sides by } V^2$$

$$R + h = \frac{2R^2g}{V^2} \quad \text{Removing a factor equal to 1: } \frac{V^2}{V^2} = 1$$

$$h = \frac{2R^2g}{V^2} - R. \quad \text{Subtracting } R \text{ from both sides}$$

The last equation can be used to determine the height of a satellite above a planet when the planet's radius and gravitational constant, along with the satellite's escape velocity, are known.

EXAMPLE 3 Acoustics (the Doppler Effect). The formula

$$f = \frac{sg}{s + v}$$

is used to determine the frequency f of a sound that is moving at velocity v toward a listener who hears the sound as frequency g. Here s is the speed of sound in a particular medium. Solve for s.

Solution We first clear fractions by multiplying by the LCD, $s + v$:

$$f \cdot (s + v) = \frac{sg}{s + v}(s + v)$$

$$fs + fv = sg. \quad \text{The variable for which we are solving appears on both sides, forcing us to distribute on the left side.}$$

Next, we must get all terms containing s on one side:

$$fv = sg - fs$$ **Subtracting fs from both sides**

$$fv = s(g - f)$$ **Factoring out s**

$$\frac{fv}{g - f} = s.$$ **Dividing both sides by $g - f$**

Since s is isolated on one side, we have solved for s. This last equation can be used to determine the speed of sound whenever f, v, and g are known.

To Solve a Rational Equation for a Specified Unknown

1. If necessary, multiply both sides by the LCD to clear fractions.
2. Multiply, as needed, to remove parentheses.
3. Get all terms with the specified unknown alone on one side.
4. Factor out the specified unknown if it is in more than one term.
5. Multiply or divide on both sides to isolate the specified unknown.

Variation

To extend our study of formulas and functions, we now examine three real-world situations: direct variation, inverse variation, and combined variation.

Direct Variation

A hair stylist earns \$18 per hour. In 1 hr, \$18 is earned. In 2 hr, \$36 is earned. In 3 hr, \$54 is earned, and so on. This gives rise to a set of ordered pairs:

$$(1, 18), (2, 36), (3, 54), (4, 72), \quad \text{and so on.}$$

Note that the ratio of earnings E to time t is $\frac{18}{1}$ in every case.

If a situation gives rise to pairs of numbers in which the ratio is constant, we say that there is **direct variation**. Here earnings *vary directly* as the time:

We have $\dfrac{E}{t} = 18,$ so $E = 18t$ or, using function notation,

$E(t) = 18t.$

Direct Variation

When a situation gives rise to a linear function of the form $f(x) = kx$, or $y = kx$, where k is a nonzero constant, we say that there is *direct variation*, that *y varies directly* as *x*, or that *y is proportional to x*. The number k is called the *variation constant*, or *constant of proportionality*.

Note that for $k > 0$, any equation of the form $y = kx$ indicates that as x increases, y increases as well.

EXAMPLE 4 Find the variation constant and an equation of variation if y varies directly as x, and $y = 32$ when $x = 2$.

Solution We know that $(2, 32)$ is a solution of $y = kx$. Therefore,

$$32 = k \cdot 2 \qquad \text{Substituting}$$
$$\frac{32}{2} = k, \quad \text{or} \quad k = 16. \qquad \text{Solving for } k$$

The variation constant is 16. The equation of variation is $y = 16x$. The notation $y(x) = 16x$ or $f(x) = 16x$ is also used.

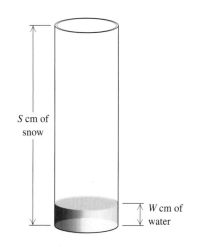

S cm of
snow

W cm of
water

EXAMPLE 5 **Water from Melting Snow.** The number of centimeters W of water produced from melting snow varies directly as the number of centimeters S of snow. Meteorologists know that under certain conditions, 150 cm of snow will melt to 16.8 cm of water. The average annual snowfall in Alta, Utah, is 500 in. Assuming the above conditions, how much water will replace the 500 in. of snow?

Solution

1. **Familiarize.** Because of the phrase "$W \ldots$ varies directly as $\ldots S$," we express the amount of water as a function of the amount of snow. Thus, $W(S) = kS$, where k is the variation constant. Knowing that 150 cm of snow becomes 16.8 cm of water, we have $W(150) = 16.8$. Because we are using ratios, it does not matter whether we work in inches or centimeters, provided the same units are used for W and S.

2. **Translate.** We find the variation constant using the data and then find the equation of variation:

$$W(S) = kS$$
$$W(150) = k \cdot 150 \qquad \text{Replacing } S \text{ with 150}$$
$$16.8 = k \cdot 150 \qquad \text{Replacing } W(150) \text{ with 16.8}$$
$$\frac{16.8}{150} = k \qquad \text{Solving for } k$$
$$0.112 = k. \qquad \text{This is the variation constant.}$$

The equation of variation is $W(S) = 0.112S$. This is the translation.

3. **Carry out.** To find how much water 500 in. of snow will become, we compute $W(500)$:

$$W(S) = 0.112S$$
$$W(500) = 0.112(500) \qquad \text{Substituting 500 for } S$$
$$W = 56.$$

4. **Check.** To check, we could reexamine all our calculations. Note that our answer seems reasonable since 500/56 and 150/16.8 are equal.

5. **State.** Alta's 500 in. of snow will be replaced with 56 in. of water.

Inverse Variation

To see what we mean by inverse variation, suppose a bus is traveling 20 mi. At 20 mph, the trip will take 1 hr. At 40 mph, it will take $\frac{1}{2}$ hr. At 60 mph, it will take $\frac{1}{3}$ hr, and so on. This gives rise to pairs of numbers, all having the same product:

$$(20, 1), \left(40, \tfrac{1}{2}\right), \left(60, \tfrac{1}{3}\right), \left(80, \tfrac{1}{4}\right), \quad \text{and so on.}$$

Note that the product of each pair of numbers is 20. Whenever a situation gives rise to pairs of numbers for which the product is constant, we say that there is **inverse variation**. Since $r \cdot t = 20$, the time t, in hours, required for the bus to travel 20 mi at r mph is given by

$$t = \frac{20}{r} \quad \text{or, using function notation,} \quad t(r) = \frac{20}{r}.$$

> ### Inverse Variation
>
> When a situation gives rise to a rational function of the form $f(x) = k/x$, or $y = k/x$, where k is a nonzero constant, we say that there is *inverse variation*, that *y varies inversely as x*, or that *y is inversely proportional to x*. The number k is called the *variation constant*, or *constant of proportionality*.

Note that for $k > 0$, any equation of the form $y = k/x$ indicates that as x increases, y decreases.

EXAMPLE 6 Find the variation constant and an equation of variation if y varies inversely as x, and $y = 32$ when $x = 0.2$.

Solution We know that $(0.2, 32)$ is a solution of

$$y = \frac{k}{x}.$$

Therefore,

$$32 = \frac{k}{0.2} \qquad \text{Substituting}$$
$$(0.2)32 = k$$
$$6.4 = k. \qquad \text{Solving for } k$$

The variation constant is 6.4. The equation of variation is

$$y = \frac{6.4}{x}.$$

There are many real-world problems that translate to an equation of inverse variation.

EXAMPLE 7   Commuter Travel.   For residents of suburban and rural areas, the number of trips made to a nearby city varies inversely as the distance of the residence from the city (*Source*: Kolars, John R., and John D. Nyusten, *Geography*. New York: McGraw-Hill, 1974). For a residential area 10 mi from a city, there will be, on average, 45 trips to the city per day per residential acre. How many daily trips per acre will be made for a residential area 25 mi from the city?

Solution

1. **Familiarize.**   Because of the phrase "... varies inversely as the distance," we express the number of daily trips t per residential acre as a function of the distance d from the city: $t(d) = k/d$.

2. **Translate.**   We use the given information to solve for k. Then we find the equation of variation.

$$t(d) = \frac{k}{d} \qquad \text{Using function notation}$$

$$t(10) = \frac{k}{10} \qquad \text{Replacing } d \text{ with 10}$$

$$45 = \frac{k}{10} \qquad \text{Replacing } t(10) \text{ with 45}$$

$$450 = k \qquad \text{Solving for } k, \text{ the variation constant}$$

The equation of variation is $t(d) = 450/d$. This is the translation.

3. **Carry out.**   To find the number of daily trips per acre for a residential area that is 25 mi from the city, we calculate $t(25)$:

$$t(25) = \frac{450}{25} = 18. \qquad t = 18 \text{ when } d = 25.$$

4. **Check.**   We could now recheck each step. Note that, as expected, as the distance from the city goes *up*, the number of daily trips goes *down*.

5. **State.**   For a residential area 25 mi from a city, there will be 18 daily trips to the city per residential acre.

Joint and Combined Variation

When a variable varies directly with more than one other variable, we say that there is *joint variation*. For example, in the formula for the volume of a right circular cylinder, $V = \pi r^2 h$, we say that V varies *jointly* as h and the square of r.

Joint Variation

y *varies jointly* as x and z if, for some nonzero constant k, $y = kxz$.

EXAMPLE 8   Find an equation of variation if y varies jointly as x and z, and $y = 30$ when $x = 2$ and $z = 3$.

Solution   We have

$$y = kxz,$$

so

$$30 = k \cdot 2 \cdot 3$$
$$k = 5. \qquad \text{The variation constant is 5.}$$

The equation of variation is $y = 5xz$.

Joint variation is one form of *combined variation*. In general, when a variable varies directly and/or inversely, at the same time, with more than one other variable, there is **combined variation**. Examples 8 and 9 are both examples of combined variation.

EXAMPLE 9   Find an equation of variation if y varies jointly as x and z and inversely as the square of w, and $y = 105$ when $x = 3$, $z = 20$, and $w = 2$.

Solution   The equation of variation is of the form

$$y = k \cdot \frac{xz}{w^2},$$

so, substituting, we have

$$105 = k \cdot \frac{3 \cdot 20}{2^2}$$
$$105 = k \cdot 15$$
$$k = 7.$$

Thus,

$$y = 7 \cdot \frac{xz}{w^2}.$$

Models

We may be able to recognize from a set of data whether two quantities vary directly or inversely. A graph like the one on the left below indicates that the quantities represented vary directly. The graph on the right below represents quantities that vary inversely. If we know the type of variation involved, we can choose one data point and calculate an equation of variation.

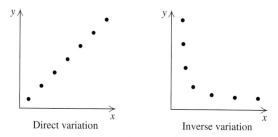

Direct variation   Inverse variation

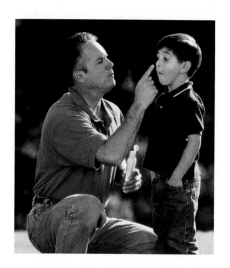

EXAMPLE 10   Ultraviolet Index.   The ultraviolet, or UV, index is a measure issued daily by the National Weather Service that indicates the strength of the sun's rays in a particular locale. The following table shows, for people whose skin is highly sensitive, how many minutes of exposure to the sun is safe (will not cause sunburn) on days with various UV ratings (*Source: Indianapolis Star*, 7/18/02).

UV Index	Safe Exposure Time (in minutes)
2	30
4	15
6	10
8	8
10	6

a) Determine whether the data indicate direct variation or inverse variation.

b) Find an equation of variation that approximates the data.

c) Use the equation to estimate the safe exposure time for a person with highly sensitive skin when the UV rating is 5.

Solution

a) We graph the data, letting x represent the UV rating and y the safe exposure time, in minutes. The points approximate the graph of a function of the type $f(x) = k/x$. The safe exposure time varies inversely as the UV index.

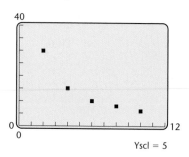

Yscl = 5

b) To find an equation of variation, we choose one point. We will use the point $(6, 10)$.

$$y = \frac{k}{x}$$ **An equation of inverse variation**

$$10 = \frac{k}{6}$$ **Substituting 6 for x and 10 for y**

$$60 = k$$ **Multiplying both sides by 6**

We now have an equation of inverse variation, $y = 60/x$. If we graph the equation along with the data, we can see that it does approximate the data.

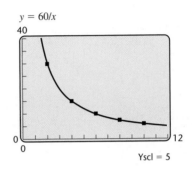

$y = 60/x$

c) To find the safe exposure time when the UV index is 5, we substitute 5 for x in the equation of variation and calculate y:

$$y = \frac{60}{x}$$ **The equation of variation**

$$y = \frac{60}{5}$$ **Substituting**

$$y = 12.$$

When the UV index is 5, the safe exposure time for a person with highly sensitive skin is 12 min.

6.8

Exercise Set

Solve the formula for the specified letter.

1. $\dfrac{W_1}{W_2} = \dfrac{d_1}{d_2};\ d_1$ $\quad d_1 = \dfrac{W_1 d_2}{W_2}$

2. $\dfrac{W_1}{W_2} = \dfrac{d_1}{d_2};\ W_1$ $\quad W_1 = \dfrac{W_2 d_1}{d_2}$

3. $s = \dfrac{(v_1 + v_2)t}{2};\ v_1$ ⊡

4. $s = \dfrac{(v_1 + v_2)t}{2};\ t$ ⊡

5. $\dfrac{1}{f} = \dfrac{1}{d_i} + \dfrac{1}{d_o};\ f$ ⊡

6. $\dfrac{1}{R} = \dfrac{1}{r_1} + \dfrac{1}{r_2};\ R$ ⊡

7. $I = \dfrac{2V}{R + 2r};\ R$ ⊡

8. $I = \dfrac{2V}{R + 2r};\ r$ ⊡

9. $R = \dfrac{gs}{g + s};\ g$ $\quad g = \dfrac{Rs}{s - R}$

10. $K = \dfrac{rt}{r - t};\ t$ $\quad t = \dfrac{Kr}{r + K}$

11. $I = \dfrac{nE}{R + nr};\ n$ $\quad n = \dfrac{IR}{E - Ir}$

12. $I = \dfrac{nE}{R + nr};\ r$ ⊡

13. $\dfrac{1}{p} + \dfrac{1}{q} = \dfrac{1}{f};\ q$ $\quad q = \dfrac{pf}{p - f}$

14. $\dfrac{1}{p} + \dfrac{1}{q} = \dfrac{1}{f};\ p$ $\quad p = \dfrac{qf}{q - f}$

15. $S = \dfrac{H}{m(t_1 - t_2)};\ t_1$ ⊡

16. $S = \dfrac{H}{m(t_1 - t_2)};\ H$ ⊡

17. $\dfrac{E}{e} = \dfrac{R + r}{r};\ r$ $\quad r = \dfrac{Re}{E - e}$

18. $\dfrac{E}{e} = \dfrac{R + r}{R};\ R$ $\quad R = \dfrac{er}{E - e}$

19. $S = \dfrac{a}{1 - r};\ r$ ⊡

20. $S = \dfrac{a - ar^n}{1 - r};\ a$ ⊡

Aha! **21.** $c = \dfrac{f}{(a + b)c};\ a + b$ ⊡

22. $d = \dfrac{g}{d(c + f)};\ c + f$ ⊡

23. *Taxable Interest.* The formula

$$I_t = \dfrac{I_f}{1 - T} \quad T = -\dfrac{I_f}{I_t} + 1,\text{ or } 1 - \dfrac{I_f}{I_t},\text{ or } \dfrac{I_t - I_f}{I_t}$$

gives the *taxable interest rate* I_t equivalent to the *tax-free interest rate* I_f for a person in the $(100 \cdot T)\%$ tax bracket. Solve for T.

24. *Interest.* The formula

$$P = \dfrac{A}{1 + r} \quad r = \dfrac{A}{P} - 1,\text{ or } \dfrac{A - P}{P}$$

is used to determine what principal P should be invested for one year at $(100 \cdot r)\%$ simple interest in order to have A dollars after a year. Solve for r.

25. *Electricity.* Electricians regularly use the formula

$$\dfrac{1}{R} = \dfrac{1}{r_1} + \dfrac{1}{r_2} \quad r_2 = \dfrac{Rr_1}{r_1 - R}$$

to determine the resistance R that corresponds to two resistors r_1 and r_2 connected in parallel. Solve for r_2.

26. *Work Rate.* The formula

$$\dfrac{1}{t} = \dfrac{1}{a} + \dfrac{1}{b} \quad t = \dfrac{ab}{b + a}$$

gives the total time t required for two workers to complete a job, if the workers' individual times are a and b. Solve for t.

27. *Average Acceleration.* The formula

$$a = \dfrac{v_2 - v_1}{t_2 - t_1} \quad t_1 = t_2 - \dfrac{v_2 - v_1}{a}$$

gives a vehicle's *average acceleration* when its velocity changes from v_1 at time t_1 to v_2 at time t_2. Solve for t_1.

28. *Average Speed.* The formula

$$v = \dfrac{d_2 - d_1}{t_2 - t_1} \quad t_2 = \dfrac{d_2 - d_1}{v} + t_1,\text{ or } \dfrac{d_2 - d_1 + t_1 v}{v}$$

gives an object's average speed v when that object has traveled d_1 miles in t_1 hours and d_2 miles in t_2 hours. Solve for t_2.

⊡ Answers to Exercises 3–8, 12, 15, 16 and 19–22 can be found on p. A-63.

29. *Semester Average.* The formula

$$A = \frac{2Tt + Qq}{2T + Q} \qquad Q = \frac{2Tt - 2AT}{A - q}$$

gives a student's average A after T tests and Q quizzes, where each test counts as 2 quizzes, t is the test average, and q is the quiz average. Solve for Q.

30. *Astronomy.* The formula

$$L = \frac{dR}{D - d}, \qquad D = \frac{dR}{L} + d, \text{ or } \frac{dR + dL}{L}$$

where D is the diameter of the sun, d is the diameter of the earth, R is the earth's distance from the sun, and L is some fixed distance, is used in calculating when lunar eclipses occur. Solve for D.

Find the variation constant and an equation of variation if y varies directly as x and the following conditions apply.

31. $y = 28$ when $x = 4$ $k = 7; y = 7x$

32. $y = 5$ when $x = 12$ $k = \frac{5}{12}; y = \frac{5}{12}x$

33. $y = 3.4$ when $x = 2$ $k = 1.7; y = 1.7x$

34. $y = 2$ when $x = 5$ $k = \frac{2}{5}; y = \frac{2}{5}x$

35. $y = 2$ when $x = \frac{1}{3}$ $k = 6; y = 6x$

36. $y = 0.9$ when $x = 0.5$ $k = 1.8; y = 1.8x$

37. *Hooke's Law.* Hooke's law states that the distance d that a spring is stretched by a hanging object varies directly as the mass m of the object. If the distance is 20 cm when the mass is 3 kg, what is the distance when the mass is 5 kg? $33\frac{1}{3}$ cm

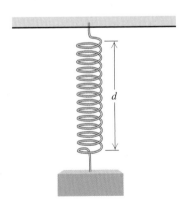

38. *Ohm's Law.* The electric current I, in amperes, in a circuit varies directly as the voltage V. When 15 volts are applied, the current is 5 amperes. What is the current when 18 volts are applied?   6 amperes

39. *Use of Aluminum Cans.* The number N of aluminum cans used each year varies directly as the number of people using the cans. If 250 people use 60,000 cans in one year, how many cans are used each year in Dallas, which has a population of 1,008,000?   241,920,000 cans

40. *Weekly Allowance.* According to Fidelity Investments *Investment Vision Magazine*, the average weekly allowance A of children varies directly as their grade level, G. In a recent year, the average allowance of a 9th-grade student was $9.66 per week. What was the average weekly allowance of a 4th-grade student? $4.29

Aha! **41.** *Mass of Water in a Human.* The number of kilograms W of water in a human body varies directly as the mass of the body. A 96-kg person contains 64 kg of water. How many kilograms of water are in a 48-kg person?   32 kg

42. *Weight on Mars.* The weight M of an object on Mars varies directly as its weight E on Earth. A person who weighs 95 lb on Earth weighs 38 lb on Mars. How much would a 100-lb person weigh on Mars?   40 lb

43. *Relative Aperture.* The relative aperture, or f-stop, of a 23.5-mm lens is directly proportional to the focal length F of the lens. If a lens with a 150-mm focal length has an f-stop of 6.3, find the f-stop of a 23.5-mm lens with a focal length of 80 mm.   3.36

44. *Lead Pollution.* The average U.S. community of population 12,500 released about 385 tons of lead into the environment in a recent year.* How many tons were released nationally? Use 250,000,000 as the U.S. population.   7,700,000 tons

Find the variation constant and an equation of variation in which y varies inversely as x, and the following conditions exist.

45. $y = 3$ when $x = 20$ $k = 60; y = \dfrac{60}{x}$

46. $y = 16$ when $x = 4$ $k = 64; y = \dfrac{64}{x}$

47. $y = 28$ when $x = 4$ $k = 112; y = \dfrac{112}{x}$

48. $y = 9$ when $x = 5$ $k = 45; y = \dfrac{45}{x}$

---

*Conservation Matters, Autumn 1995 issue. (Boston: Conservation Law Foundation), p. 30.

49. $y = 27$ when $x = \frac{1}{3}$ $k = 9; y = \dfrac{9}{x}$

50. $y = 81$ when $x = \frac{1}{9}$ $k = 9; y = \dfrac{9}{x}$

Solve.

51. *Ultraviolet Index.*   At an ultraviolet, or UV, rating of 4, those people who are moderately sensitive to the sun will burn in 70 min (*Source: Los Angeles Times,* 3/24/98). Given that the number of minutes it takes to burn, t, varies inversely with the UV rating, u, how long will it take moderately-sensitive people to burn when the UV rating is 14?   20 min

52. *Current and Resistance.*   The current I in an electrical conductor varies inversely as the resistance R of the conductor. If the current is $\frac{1}{2}$ ampere when the resistance is 240 ohms, what is the current when the resistance is 540 ohms? $\frac{2}{9}$ ampere

53. *Volume and Pressure.*   The volume V of a gas varies inversely as the pressure P upon it. The volume of a gas is 200 cm³ under a pressure of 32 kg/cm². What will be its volume under a pressure of 40 kg/cm²?
160 cm³

54. *Pumping Rate.*   The time t required to empty a tank varies inversely as the rate r of pumping. If a Briggs and Stratton pump can empty a tank in 45 min at the rate of 600 kL/min, how long will it take the pump to empty the tank at 1000 kL/min?   27 min

55. *Work Rate.*   The time T required to do a job varies inversely as the number of people P working. It takes 5 hr for 7 volunteers to pick up rubbish from 1 mi of roadway. How long would it take 10 volunteers to complete the job?   3.5 hr

56. *Wavelength and Frequency.*   The wavelength W of a radio wave varies inversely as its frequency F. A wave with a frequency of 1200 kilohertz has a length of 300 meters. What is the length of a wave with a frequency of 800 kilohertz?   450 m

Find an equation of variation in which:

57. y varies directly as the square of x, and $y = 6$ when $x = 3$. $y = \frac{2}{3}x^2$

58. y varies directly as the square of x, and $y = 0.15$ when $x = 0.1$. $y = 15x^2$

59. y varies inversely as the square of x, and $y = 6$ when $x = 3$. $y = \dfrac{54}{x^2}$

60. y varies inversely as the square of x, and $y = 0.15$ when $x = 0.1$. $y = \dfrac{0.0015}{x^2}$

61. y varies jointly as x and the square of z, and $y = 105$ when $x = 14$ and $z = 5$. $y = 0.3xz^2$

62. y varies jointly as x and z and inversely as w, and $y = \frac{3}{2}$ when $x = 2$, $z = 3$, and $w = 4$. $y = \dfrac{xz}{w}$

63. y varies jointly as w and the square of x and inversely as z, and $y = 49$ when $w = 3$, $x = 7$, and $z = 12$. $y = \dfrac{4wx^2}{z}$

64. y varies directly as x and inversely as w and the square of z, and $y = 4.5$ when $x = 15$, $w = 5$, and $z = 2$. $y = \dfrac{6x}{wz^2}$

Solve.

65. *Intensity of Light.*   The intensity I of light from a light bulb varies inversely as the square of the distance d from the bulb. Suppose I is 90 W/m² (watts per square meter) when the distance is 5 m. What would the intensity be 7.5 m from the bulb?
40 W/m²

66. *Stopping Distance of a Car.*   The stopping distance d of a car after the brakes have been applied varies directly as the square of the speed r. If a car traveling 60 mph can stop in 200 ft, what stopping distance corresponds to a speed of 36 mph?   72 ft

67. *Volume of a Gas.*   The volume V of a given mass of a gas varies directly as the temperature T and inversely as the pressure P. If $V = 231$ cm³ when $T = 42°$ and $P = 20$ kg/cm², what is the volume when $T = 30°$ and $P = 15$ kg/cm²?   220 cm³

68. *Intensity of a Signal.*   The intensity I of a television signal varies inversely as the square of the distance d from the transmitter. If the intensity is 25 W/m² at a distance of 2 km, what is the intensity 6.25 km from the transmitter?   2.56 W/m²

69. *Atmospheric Drag.*   Wind resistance, or atmospheric drag, tends to slow down moving objects. Atmospheric drag W varies jointly as an object's surface area A and velocity v. If a car traveling at a speed of 40 mph with a surface area of 37.8 ft² experiences a drag of 222 N (Newtons), how fast must a car with 51 ft² of surface area travel in order to experience a drag force of 430 N?   About 57.42 mph

70. *Drag Force.* The drag force F on a boat varies jointly as the wetted surface area A and the square of the velocity of the boat. If a boat going 6.5 mph experiences a drag force of 86 N when the wetted surface area is 41.2 ft^2, find the wetted surface area of a boat traveling 8.2 mph with a drag force of 94 N.   About 28.3 ft^2

71. *Ultraviolet Index.* The following table shows the safe exposure time for people with less sensitive skin (see Example 10).

UV Index	Safe Exposure Time (in minutes)
2	120
4	75
6	50
8	35
10	25

a) Determine whether the data indicate direct variation or inverse variation.   Inverse
b) Find an equation of variation that approximates the data. Use the data point (6, 50). $y = \dfrac{300}{x}$
c) Use the equation to predict the safe exposure time for people with less sensitive skin when the UV rating is 3.   100 min

72. *Perceived Height.* The following table shows the perceived height y of a pole when the observer is x feet from the pole.

Distance from Pole (in feet)	Perceived Height of Pole (in feet)
5	4
10	2
15	$1\frac{1}{3}$
20	1
40	0.5

a) Determine whether the data indicate direct variation or inverse variation.   Inverse
b) Find an equation of variation that approximates the data. Use the data point (10, 2). $y = \dfrac{20}{x}$
c) Use the equation to predict the perceived height of the pole when the observer is 2 ft from the pole.   10 ft

73. *Mail Order.* The following table shows the number of persons y ordering merchandise by telephone and the number of persons x ordering merchandise by mail for various age groups.

Age	Number Ordering by Mail (in millions)	Number Ordering by Telephone (in millions)
18–24	6.146	5.255
25–34	11.813	12.775
35–44	14.294	16.650
45–54	11.294	13.703
55–64	6.948	8.275
65+	10.541	9.358

Source: Simmons Market Research Bureau, *Study of Media and Markets*, New York, NY

a) Determine whether the number of people ordering by mail varies directly or inversely as the number ordering by telephone.   Directly
b) Find an equation of variation that approximates the data. Use the data point (11.813, 12.775). $y \approx 1.08x$
c) Use the equation to predict the number of persons ordering by telephone if 8 million order by mail.   8.64 million

74. *Motor Vehicle Registrations.* The following table shows the number of automobile registrations y and the number of drivers licensed x for various states.

State	Number of Driver's Licenses (in millions)	Number of Automobiles Registered (in millions)
Alabama	2.063	3.434
Colorado	1.843	2.946
Georgia	4.033	5.316
Idaho	0.502	0.863
Maryland	2.622	3.178
Texas	7.456	13.323
Virginia	3.774	4.787

a) Determine whether the number of automobiles registered varies directly or inversely as the number of driver's licenses.   Directly
b) Find an equation of variation that approximates the data. Use the data point (2.063, 3.434). $y \approx 1.66x$
c) Use the equation to estimate the number of automobiles registered in Hawaii, where there are 0.45 million licensed drivers.   0.747 million automobiles

TW **75.** If two quantities vary directly, as in Exercise 73, does this mean that one is "caused" by the other? Why or why not?

TW **76.** If y varies directly as x, does doubling x cause y to be doubled as well? Why or why not?

Skill Maintenance

Find the domain of f.

77. $f(x) = \dfrac{2x - 1}{x^2 + 1}$ [2.7] $\mathbb{R}$ **78.** $f(x) = |2x - 1|$ [2.7] $\mathbb{R}$

79. Graph on a plane: $6x - y < 6$. [4.4]

80. If $f(x) = x^3 - x$, find $f(2a)$. [2.1] $8a^3 - 2a$

81. Factor: $t^3 + 8b^3$. [5.7] $(t + 2b)(t^2 - 2bt + 4b^2)$

82. Solve: $6x^2 = 11x + 35$. [5.5] $-\frac{5}{3}, \frac{7}{2}$

Synthesis

TW **83.** Suppose that the number of customer complaints is inversely proportional to the number of employees hired. Will a firm reduce the number of complaints more by expanding from 5 to 10 employees, or from 20 to 25? Explain. Consider using a graph to help justify your answer.

TW **84.** Why do you think subscripts are used in Exercises 3 and 15 but not in Exercises 17 and 18?

85. *Escape Velocity.* A satellite's escape velocity is 6.5 mi/sec, the radius of the earth is 3960 mi, and the earth's gravitational constant is 32.2 ft/sec². How far is the satellite from the surface of the earth? (See Example 2.)   567 mi

86. The *harmonic mean* of two numbers a and b is a number M such that the reciprocal of M is the average of the reciprocals of a and b. Find a formula for the harmonic mean. $M = \dfrac{2ab}{b + a}$

87. *Health Care.* Young's rule for determining the size of a particular child's medicine dosage c is

$$c = \frac{a}{a + 12} \cdot d,$$

where a is the child's age and d is the typical adult dosage (*Source*: Olsen, June Looby, Leon J. Ablon, and Anthony Patrick Giangrasso, *Medical Dosage Calculations*, 6th ed.). If a child's age is doubled, the dosage increases. Find the ratio of the larger dosage to the smaller dosage. By what percent does the dosage increase?   ⊡

⊡ Answers to Exercises 87 and 89 can be found on p. A-63.

88. Solve for x:

$$x^2\left(1 - \frac{2pq}{x}\right) = \frac{2p^2q^3 - pq^2x}{-q}.$$ $x = pq$ or $x = 2pq$

89. *Average Acceleration.* The formula

$$a = \frac{\dfrac{d_4 - d_3}{t_4 - t_3} - \dfrac{d_2 - d_1}{t_2 - t_1}}{t_4 - t_2}$$

can be used to approximate average acceleration, where the d's are distances and the t's are the corresponding times. Solve for t_1.   ⊡

90. If y varies inversely as the cube of x and x is multiplied by 0.5, what is the effect on y? y is multiplied by 8.

Describe, in words, the variation given by the equation. Assume k is a constant.

91. $Q = \dfrac{kp^2}{q^3}$ Q varies directly as the square of p and inversely as the cube of q.

92. $W = \dfrac{km_1M_1}{d^2}$ W varies jointly as m_1 and M_1 and inversely as the square of d.

93. *Tension of a Musical String.* The tension T on a string in a musical instrument varies jointly as the string's mass per unit length m, the square of its length l, and the square of its fundamental frequency f. A 2-m–long string of mass 5 gm/m with a fundamental frequency of 80 has a tension of 100 N. How long should the same string be if its tension is going to be changed to 72 N?   About 1.697 m

94. *Volume and Cost.* A peanut butter jar in the shape of a right circular cylinder is 4 in. high and 3 in. in diameter and sells for $1.20. If we assume that cost is proportional to volume, how much should a jar 6 in. high and 6 in. in diameter cost? $7.20

95. *Golf Distance Finder.* A device used in golf to estimate the distance d to a hole measures the size s that the 7-ft pin *appears* to be in a viewfinder. The viewfinder uses the principle, diagrammed here, that s gets bigger when d gets smaller. If $s = 0.56$ in. when $d = 50$ yd, find an equation of variation that expresses d as a function of s. What is d when $s = 0.40$ in.? $d(s) = \dfrac{28}{s}$; 70 yd

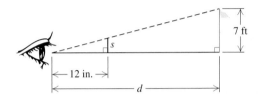

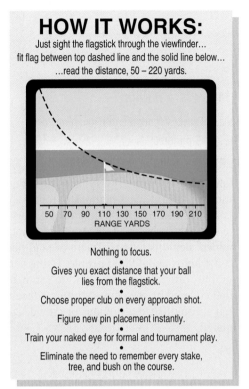

HOW IT WORKS:

Just sight the flagstick through the viewfinder...
fit flag between top dashed line and the solid line below...
...read the distance, 50 – 220 yards.

RANGE YARDS

Nothing to focus.

Gives you exact distance that your ball
lies from the flagstick.

Choose proper club on every approach shot.

Figure new pin placement instantly.

Train your naked eye for formal and tournament play.

Eliminate the need to remember every stake,
tree, and bush on the course.

Collaborative Corner

How Many Is a Million?

Focus: Direct variation and estimation

Time: 15 minutes

Group size: 2 or 3 and entire class

The National Park Service's estimates of crowd sizes for static (stationary) mass demonstrations vary directly as the area covered by the crowd. Park Service officials have found that at basic "shoulder-to-shoulder" demonstrations, 1 acre of land (about 45,000 ft^2) holds about 9000 people. Using aerial photographs, officials impose a grid to estimate the total area covered by the demonstrators. Once this has been accomplished, estimates of crowd size can be prepared.

ACTIVITY

1. In the grid imposed on the photograph below, each square represents 10,000 ft^2. Estimate the size of the crowd photographed. Then compare your group's estimate with those of other groups. What might explain discrepancies between estimates? List ways in which your group's estimate could be made more accurate.

2. Park Service officials use an "acceptable margin of error" of no more than 20%. Using all estimates from part (1) above and allowing for error, find a range of values within which you feel certain that the actual crowd size lies.

3. The Million Man March of 1995 was not a static demonstration because of a periodic turnover of people in attendance (many people stayed for only part of the day's festivities). How might you change your methodology to compensate for this complication?

6 Chapter Summary and Review

Key Terms

Rational expression, p. 398
Rational function, p. 398
Simplified, p. 403
Least common multiple, LCM, p. 413
Least common denominator, LCD, p. 414

Complex rational expression, p. 421
Rational equation, p. 432
Clear fractions, p. 432
Motion problem, p. 443
Synthetic division, p. 455
Direct variation, p. 463

Variation constant, p. 463
Constant of proportionality, p. 463
Inverse variation, p. 465
Joint variation, p. 467
Combined variation, p. 467

Important Properties and Formulas

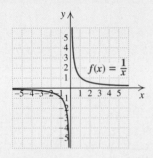

$f(x) = \dfrac{1}{x}$

Domain: $(-\infty, 0) \cup (0, \infty)$

Range: $(-\infty, 0) \cup (0, \infty)$

Addition: $\dfrac{A}{C} + \dfrac{B}{C} = \dfrac{A + B}{C}$

Subtraction: $\dfrac{A}{C} - \dfrac{B}{C} = \dfrac{A - B}{C}$

Multiplication: $\dfrac{A}{B} \cdot \dfrac{C}{D} = \dfrac{AC}{BD}$

Division: $\dfrac{A}{B} \div \dfrac{C}{D} = \dfrac{A}{B} \cdot \dfrac{D}{C}$

To find the least common multiple, LCM, use each factor the greatest number of times that it occurs in any one prime factorization.

Simplifying Complex Rational Expressions

I: By using multiplication by 1

1. Find the LCD of all rational expressions *within* the complex rational expression.

2. Multiply the complex rational expression by 1, writing 1 as the LCD divided by itself.

3. Distribute and simplify so that the numerator and the denominator of the complex rational expression are polynomials

4. Factor and, if possible, simplify.

II. By using division

1. Add or subtract, as necessary, to get one rational expression in the numerator.

2. Add or subtract, as necessary, to get one rational expression in the denominator.

3. Perform the indicated division (invert the divisor and multiply).

4. Simplify, if possible, by removing any factors equal to 1.

Modeling Work Problems

If

a = the time needed for A to complete the work alone,

b = the time needed for B to complete the work alone, and

t = the time needed for A and B to complete the work together,

then

$$\frac{t}{a} + \frac{t}{b} = 1 \quad \text{and} \quad \frac{1}{a} \cdot t + \frac{1}{b} \cdot t = 1$$

$$\text{and} \quad \frac{1}{a} + \frac{1}{b} = \frac{1}{t}.$$

Motion Formula

$$d = rt \quad \text{or} \quad r = d/t \quad \text{or} \quad t = d/r$$

The Remainder Theorem

The remainder obtained by dividing $P(x)$ by $x - r$ is $P(r)$.

Variation

y varies directly as x if there is some nonzero constant k such that $y = kx$.

y varies inversely as x if there is some nonzero constant k such that $y = k/x$.

y varies jointly as x and z if there is some nonzero constant k such that $y = kxz$.

Review Exercises

1. If

$$f(t) = \frac{t^2 - 3t + 2}{t^2 - 9},$$

find the following function values.

a) $f(0)$ [6.1] $-\frac{2}{9}$ **b)** $f(-1)$ [6.1] $-\frac{3}{4}$ **c)** $f(2)$ [6.1] 0

Find the LCD.

2. $\dfrac{7}{6x^3}, \quad \dfrac{y}{16x^2}$ [6.2] $48x^3$

3. $\dfrac{x + 8}{x^2 + x - 20}, \quad \dfrac{x}{x^2 + 3x - 10}$
[6.2] $(x + 5)(x - 2)(x - 4)$

Perform the indicated operations and, if possible, simplify.

4. $\dfrac{x^2}{x - 3} - \dfrac{9}{x - 3}$ [6.2] $x + 3$

5. $\dfrac{4x - 2}{x^2 - 5x + 4} - \dfrac{3x + 2}{x^2 - 5x + 4}$ [6.2] $\dfrac{1}{x - 1}$

6. $\dfrac{3a^2b^3}{5c^3d^2} \cdot \dfrac{15c^9d^4}{9a^7b}$ [6.1] $\dfrac{b^2c^6d^2}{a^5}$

7. $\dfrac{5}{6m^2n^3p} + \dfrac{7}{9mn^4p^2}$ [6.2] $\dfrac{15np + 14m}{18m^2n^4p^2}$

8. $\dfrac{y^2 - 64}{2y + 10} \cdot \dfrac{y + 5}{y + 8}$ [6.1] $\dfrac{y - 8}{2}$

9. $\dfrac{x^3 - 8}{x^2 - 25} \cdot \dfrac{x^2 + 10x + 25}{x^2 + 2x + 4}$ [6.1] $\dfrac{(x - 2)(x + 5)}{x - 5}$

10. $\dfrac{9a^2 - 1}{a^2 - 9} \div \dfrac{3a + 1}{a + 3}$ [6.1] $\dfrac{3a - 1}{a - 3}$

11. $\dfrac{x^3 - 64}{x^2 - 16} \div \dfrac{x^2 + 5x + 6}{x^2 - 3x - 18}$ [6.1] $\dfrac{(x^2 + 4x + 16)(x - 6)}{(x + 4)(x + 2)}$

12. $\dfrac{x}{x^2 + 5x + 6} - \dfrac{2}{x^2 + 3x + 2}$ [6.2] $\dfrac{x - 3}{(x + 1)(x + 3)}$

13. $\dfrac{-4xy}{x^2 - y^2} + \dfrac{x + y}{x - y}$ [6.2] $\dfrac{x - y}{x + y}$

14. $\dfrac{2x^2}{x - y} + \dfrac{2y^2}{y - x}$ [6.2] $2(x + y)$

15. $\dfrac{3}{y + 4} - \dfrac{y}{y - 1} + \dfrac{y^2 + 3}{y^2 + 3y - 4}$ [6.2] $\dfrac{-y}{(y + 4)(y - 1)}$

Simplify.

16. $\dfrac{\dfrac{5}{x} - 5}{\dfrac{7}{x} - 7}$ [6.3] $\dfrac{5}{7}$

17. $\dfrac{\dfrac{2}{a} + \dfrac{2}{b}}{\dfrac{4}{a^3} + \dfrac{4}{b^3}}$ [6.3] $\dfrac{a^2 b^2}{2(b^2 - ba + a^2)}$

18. $\dfrac{\dfrac{y^2 + 4y - 77}{y^2 - 10y + 25}}{\dfrac{y^2 - 5y - 14}{y^2 - 25}}$

19. $\dfrac{\dfrac{5}{x^2 - 9} - \dfrac{3}{x + 3}}{\dfrac{4}{x^2 + 6x + 9} + \dfrac{2}{x - 3}}$

Solve.

20. $\dfrac{6}{x} + \dfrac{4}{x} = 5$ [6.4] 2

21. $\dfrac{5}{3x + 2} = \dfrac{3}{2x}$ [6.4] 6

22. $\dfrac{4x}{x + 1} + \dfrac{4}{x} + 9 = \dfrac{4}{x^2 + x}$ [6.4] No solution

23. $\dfrac{x + 6}{x^2 + x - 6} + \dfrac{x}{x^2 + 4x + 3} = \dfrac{x + 2}{x^2 - x - 2}$ [6.4] 0

24. If
$$f(x) = \dfrac{2}{x - 1} + \dfrac{2}{x + 2},$$
find all a for which $f(a) = 1$.   [6.4] $-1, 4$

Solve.

25. Kim can set up for a banquet in 12 hr. Kelly can set up for the same banquet in 9 hr. How long would it take them, working together, to set up for the banquet?   [6.5] $5\frac{1}{7}$ hr

26. A research company uses personal computers to process data while the owner is not using the computer. A Pentium III 850 megahertz processor can process a megabyte of data in 15 sec less time than a Celeron 700 megahertz processor. Working together, the computers can process a megabyte of data in 18 sec. How long does it take each computer to process one megabyte of data?   [6.5] Celeron: 45 sec; Pentium III: 30 sec

27. The Gold River's current is 6 mph. A boat travels 50 mi downstream in the same time that it takes to travel 30 mi upstream. What is the speed of the boat in still water?   [6.5] 24 mph

28. A car and a motorcycle leave a rest area at the same time, with the car traveling 8 mph faster than the motorcycle. The car then travels 105 mi in the time it takes the motorcycle to travel 93 mi. Find the speed of each vehicle.   [6.5] Motorcycle: 62 mph; car 70 mph

Divide.

29. $(20r^2 s^3 + 15r^2 s^2 - 10r^3 s^3) \div (5r^2 s)$ [6.6] $4s^2 + 3s - 2rs^2$

30. $(y^3 + 125) \div (y + 5)$ [6.6] $y^2 - 5y + 25$

31. $(4x^3 + 3x^2 - 5x - 2) \div (x^2 + 1)$ [6.6] $4x + 3 + \dfrac{-9x - 5}{x^2 + 1}$

32. Divide using synthetic division:
$$(x^3 + 3x^2 + 2x - 6) \div (x - 3).$$

33. If $f(x) = 4x^3 - 6x^2 - 9$, use synthetic division to find $f(5)$.   [6.7] 341

Solve.

34. $R = \dfrac{gs}{g + s}$, for s [6.8] $s = \dfrac{Rg}{g - R}$

35. $S = \dfrac{H}{m(t_1 - t_2)}$, for m [6.8] $m = \dfrac{H}{S(t_1 - t_2)}$

36. $\dfrac{1}{ac} = \dfrac{2}{ab} - \dfrac{3}{bc}$, for c [6.8] $c = \dfrac{b + 3a}{2}$

37. $T = \dfrac{A}{v(t_2 - t_1)}$, for t_1 [6.8] $t_1 = \dfrac{-A}{vT} + t_2$, or $\dfrac{-A + vTt_2}{vT}$

38. The amount of waste generated by a restaurant varies directly as the number of customers served. A typical McDonalds that serves 2000 customers per day generates 238 lb of waste daily (*Source:* Environmental Defense Fund Study, November 1990). How many pounds of waste would be generated daily by a McDonalds that serves 1700 customers a day?   [6.8] About 202.3 lb

39. A warning dye is used by people in lifeboats to aid search planes. The volume V of the dye used varies directly as the square of the diameter d of the circular patch of water formed by the dye. If 4 L of dye is required for a 10-m wide circle, how much dye is needed for a 40-m wide circle?   [6.8] 64 L

40. Find an equation of variation in which y varies inversely as x, and $y = 3$ when $x = \frac{1}{4}$.   [6.8] $y = \dfrac{\frac{3}{4}}{x}$

---

18. [6.3] $\dfrac{(y + 11)(y + 5)}{(y - 5)(y + 2)}$ **19.** [6.3] $\dfrac{(14 - 3x)(x + 3)}{2x^2 + 16x + 6}$ **32.** [6.7] $x^2 + 6x + 20 + \dfrac{54}{x - 3}$

41. The following table shows the size y of one serving of breakfast cereal for the corresponding number of servings x obtained from a given box.

Number of Servings	Size of Serving (in ounces)
1	24
4	6
6	4
12	2
16	1.5

a) Determine whether the data indicate direct variation or inverse variation. [6.8] Inverse

b) Find an equation of variation that fits the data. Use the data point $(12, 2)$. [6.8] $y = \dfrac{24}{x}$

c) Use the equation to estimate the size of each serving when 8 servings are obtained from the box. [6.8] 3 oz

Synthesis

^{TW} **42.** Discuss at least three different uses of the LCD studied in this chapter.

^{TW} **43.** Explain the difference between a rational expression and a rational equation.

Solve.

44. $\dfrac{5}{x - 13} - \dfrac{5}{x} = \dfrac{65}{x^2 - 13x}$ [6.4] All real numbers except 0 and 13

45. $\dfrac{\dfrac{x}{x^2 - 25} + \dfrac{2}{x - 5}}{\dfrac{3}{x - 5} - \dfrac{4}{x^2 - 10x + 25}} = 1$ [6.3], [6.4] 45

46. A Pentium 4 1.5-gigahertz processor can process a megabyte of data in 20 sec. How long would it take a Pentium 4 working together with the Pentium III and Celeron processors (see Exercise 26) to process a megabyte of data? [6.5] $9\frac{9}{19}$ sec

Chapter Test 6

Simplify.

1. $\dfrac{t - 1}{t + 3} \cdot \dfrac{3t + 9}{4t^2 - 4}$ [6.1] $\dfrac{3}{4(t + 1)}$

2. $\dfrac{x^3 + 27}{x^2 - 16} \div \dfrac{x^2 + 8x + 15}{x^2 + x - 20}$ [6.1] $\dfrac{x^2 - 3x + 9}{x + 4}$

3. Find the LCD: [6.2] $(x - 3)(x + 11)(x - 9)$

$$\dfrac{3x}{x^2 + 8x - 33}, \quad \dfrac{x + 1}{x^2 - 12x + 27}.$$

Perform the indicated operation and simplify when possible.

4. $\dfrac{25x}{x + 5} + \dfrac{x^3}{x + 5}$ [6.2] $\dfrac{25x + x^3}{x + 5}$

5. $\dfrac{3a^2}{a - b} - \dfrac{3b^2 - 6ab}{b - a}$ [6.2] $3(a - b)$

6. $\dfrac{4ab}{a^2 - b^2} + \dfrac{a^2 + b^2}{a + b}$ [6.2] $\dfrac{a^3 - a^2b + 4ab + ab^2 - b^3}{(a - b)(a + b)}$

7. $\dfrac{6}{x^3 - 64} - \dfrac{4}{x^2 - 16}$ [6.2] $\dfrac{-2(2x^2 + 5x + 20)}{(x - 4)(x + 4)(x^2 + 4x + 16)}$

8. $\dfrac{4}{y + 3} - \dfrac{y}{y - 2} + \dfrac{y^2 + 4}{y^2 + y - 6}$ [6.2] $\dfrac{y - 4}{(y + 3)(y - 2)}$

Simplify.

9. $\dfrac{\dfrac{2}{a} + \dfrac{3}{b}}{\dfrac{5}{ab} + \dfrac{1}{a^2}}$ [6.3] $\dfrac{a(2b + 3a)}{5a + b}$

10. $\dfrac{\dfrac{x^2 - 5x - 36}{x^2 - 36}}{\dfrac{x^2 + x - 12}{x^2 - 12x + 36}}$ [6.3] $\dfrac{(x - 9)(x - 6)}{(x + 6)(x - 3)}$

11. $\dfrac{\dfrac{4}{x + 3} - \dfrac{2}{x^2 - 3x + 2}}{\dfrac{3}{x - 2} + \dfrac{1}{x^2 + 2x - 3}}$ [6.3] $\dfrac{4x^2 - 14x + 2}{3x^2 + 7x - 11}$

Solve.

12. $\dfrac{4}{2x-5} = \dfrac{6}{5x+3}$ [6.4] $-\dfrac{21}{4}$

13. $\dfrac{t+11}{t^2-t-12} + \dfrac{1}{t-4} = \dfrac{4}{t+3}$ [6.4] 15

For Exercises 14 and 15, let $f(x) = \dfrac{x+3}{x-1}$.

14. Find $f(2)$ and $f(-3)$.   [6.1] 5; 0

15. Find all a for which $f(a) = 7$.   [6.4] $\dfrac{5}{3}$

16. Kyla can lay vinyl in a kitchen in 3.5 hr. Brock can lay the same vinyl in 4.5 hr. How long will it take them, working together, to lay the vinyl?   [6.5] $1\frac{31}{32}$ hr

Divide.

17. $(16ab^3c - 10ab^2c^2 + 12a^2b^2c) \div (4a^2b)$

18. $(y^2 - 20y + 64) \div (y - 6)$ [6.6] $y - 14 + \dfrac{-20}{y-6}$

19. $(6x^4 + 3x^2 + 5x + 4) \div (x^2 + 2)$ [6.6] $6x^2 - 9 + \dfrac{5x+22}{x^2+2}$

20. Divide using synthetic division:
$$(x^3 + 5x^2 + 4x - 7) \div (x - 4).$$

21. If $f(x) = 3x^4 - 5x^3 + 2x - 7$, use synthetic division to find $f(4)$.   [6.7] 449

22. Solve $A = \dfrac{h(b_1 + b_2)}{2}$ for b_1.

23. The product of the reciprocals of two consecutive integers is $\frac{1}{30}$. Find the integers.   [6.5] 5 and 6; -6 and -5

24. Georgia bicycles 12 mph with no wind. Against the wind, she bikes 8 mi in the same time that it takes to bike 14 mi with the wind. What is the speed of the wind?   [6.5] $3\frac{3}{11}$ mph

25. The number of workers n needed to clean a stadium after a game varies inversely as the amount of time t allowed for the cleanup. If it takes 25 workers to clean the stadium when there are 6 hr allowed for the job, how many workers are needed if the stadium must be cleaned in 5 hr?   [6.8] 30 workers

26. The surface area of a balloon varies directly as the square of its radius. The area is 325 in^2 when the radius is 5 in. What is the area when the radius is 7 in.?   [6.8] 637 in^2

Synthesis

27. Let
$$f(x) = \dfrac{1}{x+3} + \dfrac{5}{x-2}.$$
Find all a for which $f(a) = f(a + 5)$.   [6.4] $-\dfrac{19}{3}$

28. Solve: $\dfrac{6}{x-15} - \dfrac{6}{x} = \dfrac{90}{x^2-15x}$.   [6.4] $\{x \,|\, x$ is a real number *and* $x \neq 0$ *and* $x \neq 15\}$

29. Find the x- and y-intercepts for the function given by   [6.3], [6.4] x-intercept: $(11, 0)$; y-intercept: $\left(0, -\dfrac{33}{5}\right)$
$$f(x) = \dfrac{\dfrac{5}{x+4} - \dfrac{3}{x-2}}{\dfrac{2}{x-3} + \dfrac{1}{x+4}}.$$

30. One summer, Hans mowed 4 lawns for every 3 lawns mowed by his brother Franz. Together, they mowed 98 lawns. How many lawns did each mow?   [6.5] Hans: 56 lawns; Franz: 42 lawns

---

17. [6.6] $\dfrac{4b^2c}{a} - \dfrac{5bc^2}{2a} + 3bc$ **20.** [6.7] $x^2 + 9x + 40 + \dfrac{153}{x-4}$ **22.** [6.8] $b_1 = \dfrac{2A}{h} - b_2$, or $\dfrac{2A - b_2h}{h}$

1-6   Cumulative Review

17. [4.2] $\left\{x \mid x < -\frac{4}{3} \ or \ x > 6\right\}$, or $\left(-\infty, -\frac{4}{3}\right) \cup (6, \infty)$

1. Evaluate

$$\frac{2x - y^2}{x + y}$$

for $x = 3$ and $y = -4$.   [1.1], [1.2] 10

2. Convert to scientific notation: 5,760,000,000.
[1.4] 5.76×10^9
3. Determine the slope and the y-intercept for the line
given by $7x - 4y = 12$.   [2.4] Slope: $\frac{7}{4}$; y-intercept:
$(0, -3)$
4. Find an equation for the line that passes through the
points $(-1, 7)$ and $(2, -3)$.   [2.6] $y = -\frac{10}{3}x + \frac{11}{3}$

5. Solve the system

$$5x - 2y = -23,$$
$$3x + 4y = 7.   [3.2] \ (-3, 4)$$

6. Solve the system

$$-3x + 4y + \ z = -5,$$
$$x - 3y - \ z = 6,$$
$$2x + 3y + 5z = -8.   [3.4] \ (-2, -3, 1)$$

7. Briar Creek Elementary School sold 45 pizzas for a
fundraiser. Small pizzas sold for $7.00 each and
large pizzas for $10.00 each. The total amount of
funds received from the sale was $402. How many
of each size pizza were sold?   [3.3] Small: 16; large: 29

8. The sum of three numbers is 20. The first number
is 3 less than twice the third number. The second
number minus the third number is -7. What are
the numbers?   [3.5] $12, \frac{1}{2}, 7\frac{1}{2}$

9. Trex Company makes decking material from waste
wood fibers and reclaimed polyethylene. Its sales
rose from $3.5 million in 1993 to $74.3 million in
1999 (*Source: Business Week*, May 29, 2000).
Calculate the rate at which sales were rising.
[2.4] $11.8 million per year
10. In 1989, the average length of a visit to a physician
in an HMO was 15.4 min; and in 1998, it was
17.9 min (*Sources:* Rutgers University Study;
National Center for Health Statistics). Let V represent the average length of a visit t years after 1989.

[2.6] $V(t) = \frac{5}{18}t + 15.4$
a) Find a linear function $V(t)$ that fits the data.
b) Use the function of part (a) to predict the average
length of a visit in 2005.   [2.6] About 19.8 min

11. If

$$f(x) = \frac{x - 2}{x - 5},$$

find **(a)** $f(3)$ and **(b)** the domain of f.   **(a)** [6.1] $-\frac{1}{2}$;
(b) [2.7] $\{x \mid x \text{ is a real number } and \ x \ne 5\}$
Solve.

12. $8x = 1 + 16x^2$ [5.6] $\frac{1}{4}$

13. $625 = 49y^2$ [5.6] $-\frac{25}{7}, \frac{25}{7}$

14. $20 > 2 - 6x$ [4.1] $\{x \mid x > -3\}$, or $(-3, \infty)$

15. $\frac{1}{3}x - \frac{1}{5} \ge \frac{1}{5}x - \frac{1}{3}$ [4.1] $\{x \mid x \ge -1\}$, or $[-1, \infty)$

16. $-8 < x + 2 < 15$
[4.2] $\{x \mid -10 < x < 13\}$, or $(-10, 13)$
17. $3x - 2 < -6 \ or \ x + 3 > 9$

18. $|x| > 6.4$ [4.3] $\{x \mid x < -6.4 \ or \ x > 6.4\}$, or
$(-\infty, -6.4) \cup (6.4, \infty)$
19. $|4x - 1| \le 14$ [4.3] $\left\{x \mid -\frac{13}{4} \le x \le \frac{15}{4}\right\}$, or $\left[-\frac{13}{4}, \frac{15}{4}\right]$

20. $\dfrac{2}{n} - \dfrac{7}{n} = 3$ [6.4] $-\frac{5}{3}$

21. $\dfrac{6}{x - 5} = \dfrac{2}{2x}$ [6.4] -1

22. $\dfrac{3x}{x - 2} - \dfrac{6}{x + 2} = \dfrac{24}{x^2 - 4}$ [6.4] No solution

23. $\dfrac{3x^2}{x + 2} + \dfrac{5x - 22}{x - 2} = \dfrac{-48}{x^2 - 4}$ [6.4] $\frac{1}{3}$

24. Let $f(x) = |3x - 5|$. Find all values of x for which
$f(x) = 2$.   [4.3] $1, \frac{7}{3}$

25. Write the domain of f using interval notation if
$f(x) = \sqrt{x - 7}$.   [4.2] $[7, \infty)$

Solve.

26. $5m - 3n = 4m + 12$, for n [2.3] $n = \dfrac{m - 12}{3}$

27. $P = \dfrac{3a}{a + b}$, for a [6.8] $a = \dfrac{Pb}{3 - P}$

Graph on a plane.

28. $4x \geq 5y + 20$ ⊡

29. $y = \frac{1}{3}x - 2$ ⊡

Perform the indicated operations and simplify.

30. $(2x^2 - 3x + 1) + (6x - 3x^3 + 7x^2 - 4)$
[5.1] $-3x^3 + 9x^2 + 3x - 3$

31. $(5x^3y^2)(-3xy^2)$ [5.2] $-15x^4y^4$

32. $(3a + b - 2c) - (-4b + 3c - 2a)$
[5.1] $5a + 5b - 5c$

33. $(5x^2 - 2x + 1)(3x^2 + x - 2)$
[5.2] $15x^4 - x^3 - 9x^2 + 5x - 2$

34. $(2x^2 - y)^2$ [5.2] $4x^4 - 4x^2y + y^2$

35. $(2x^2 - y)(2x^2 + y)$ [5.2] $4x^4 - y^2$

36. $(-5m^3n^2 - 3mn^3) +$ [5.1] $-m^3n^2 - m^2n^2 - 5mn^3$
$(-4m^2n^2 + 4m^3n^2) - (2mn^3 - 3m^2n^2)$

37. $\dfrac{y^2 - 36}{2y + 8} \cdot \dfrac{y + 4}{y + 6}$ [6.1] $\dfrac{y - 6}{2}$

38. $\dfrac{x^4 - 1}{x^2 - x - 2} \div \dfrac{x^2 + 1}{x - 2}$ [6.1] $x - 1$

39. $\dfrac{5ab}{a^2 - b^2} + \dfrac{a + b}{a - b}$ [6.2] $\dfrac{a^2 + 7ab + b^2}{(a - b)(a + b)}$

40. $\dfrac{2}{m + 1} + \dfrac{3}{m - 5} - \dfrac{m^2 - 1}{m^2 - 4m - 5}$
[6.2] $\dfrac{-m^2 + 5m - 6}{(m + 1)(m - 5)}$

41. $y - \dfrac{2}{3y}$ [6.2] $\dfrac{3y^2 - 2}{3y}$

42. Simplify: $\dfrac{\dfrac{1}{x} - \dfrac{1}{y}}{x + y}$. [6.3] $\dfrac{y - x}{xy(x + y)}$

43. Divide: $(9x^3 + 5x^2 + 2) \div (x + 2)$.
[6.6] $9x^2 - 13x + 26 + \dfrac{-50}{x + 2}$

Factor.

44. $4x^3 + 18x^2$ [5.3] $2x^2(2x + 9)$

45. $x^2 + 8x - 84$ [5.4] $(x - 6)(x + 14)$

46. $16y^2 - 81$ [5.6] $(4y - 9)(4y + 9)$

47. $64x^3 + 8$ [5.7] $8(2x + 1)(4x^2 - 2x + 1)$

48. $t^2 - 16t + 64$ [5.6] $(t - 8)^2$

49. $x^6 - x^2$ [5.6] $x^2(x - 1)(x + 1)(x^2 + 1)$

50. $0.027b^3 - 0.008c^3$ ⊡

51. $20x^2 + 7x - 3$ [5.5] $(4x - 1)(5x + 3)$

52. $3x^2 - 17x - 28$ [5.5] $(3x + 4)(x - 7)$

53. $x^5 - x^3y + x^2y - y^2$ [5.3] $(x^2 - y)(x^3 + y)$

54. If $f(x) = x^2 - 4$ and $g(x) = x^2 - 7x + 10$, find the domain of f/g. [2.7], [5.4] $\{x \mid x$ is a real number *and* $x \neq 2$ *and* $x \neq 5\}$

55. A digital data circuit can transmit a particular set of data in 4 sec. An analog phone circuit can transmit the same data in 20 sec. How long would it take, working together, for both circuits to transmit the data? [6.5] $3\frac{1}{3}$ sec

56. The floor area of a rental trailer is rectangular. The length is 3 ft more than the width. A rug of area 54 ft² exactly fills the floor of the trailer. Find the perimeter of the trailer. [5.8] 30 ft

57. The sum of the squares of three consecutive even integers is equal to 8 more than three times the square of the second number. Find the integers.
[5.8] All such sets of even integers satisfy this condition.

58. The volume of wood V in a tree trunk varies jointly as the height h and the square of the girth g (girth is distance around). If the volume is 35 ft³ when the height is 20 ft and the girth is 5 ft, what is the height when the volume is 85.75 ft³ and the girth is 7 ft?
[6.8] 25 ft

The following table shows the average tuition for public and private four-year colleges for various years. *

Year	Public College Tuition	Private College Tuition
1985	$1386	$ 6,843
1990	2035	10,348
1994	2820	13,874
1997	3323	16,552
1999	3644	18,237
2001	4021	20,145

59. Let x represent the number of years since 1985 and b represent the average public college tuition.

a) Use the data to draw a line graph. ⊡

b) Use the points $(5, 2035)$ and $(9, 2820)$ to find a linear function $b(x)$ that can be used to approximate public college tuition x years after 1985. [2.6] $b(x) = 196.25x + 1053.75$

c) Use the function found in part (b) to predict public college tuition in 2010. [2.6] $5960

Source: U.S. National Center for Education Statistics, *Digest of Education Statistics*, annual

⊡ Answers to Exercises 28, 29, 50, and 59(a) can be found on p. A-63.

60. Let b represent public college tuition and v represent private college tuition.

  a) Graph the data and determine whether private college tuition varies directly or inversely as private college tuition.   [6.8] Directly

  b) Use the point $(2820, 13,874)$ to find an equation of variation.   [6.8] $v \approx 4.92b$

  c) Use the equation of variation and the result of Exercise 59(c) to predict private college tuition in 2010.   [6.8] $29,323.20

61. Let x represent the number of years since 1985 and v represent the average private college tuition.

  a) Use linear regression to find a linear function $v(x)$ that can be used to approximate private college tuition x years after 1985.

  b) Use the function found in part (a) to predict private college tuition in 2010.

  c) Compare the result found in part (b) with that found in Exercise 60(c).

In Exercises 62–65, match each equation with one of the following graphs.

a)
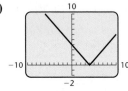

b)
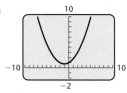

c)
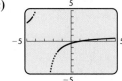

d)
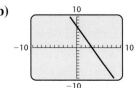

62. $y = -2x + 7$ [1.5] (b)   **63.** $y = |x - 4|$ [4.3] (a)

64. $y = x^2 + x + 1$ [5.1] (d)   **65.** $y = \dfrac{x - 1}{x + 3}$ [6.1] (c)

Find the solutions of each equation, inequality, or system from the given graph.

66. $10 - 5x = 3x + 2$ [2.2] 1

67. $f(x) \le g(x)$ [4.3] $\{x \mid x \ge 1\}$, or $[1, \infty)$

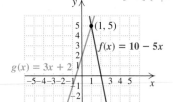

68. $y = x - 5$,
    $y = 1 - 2x$
    [3.1] $(2, -3)$

69. $x - 5 = 1 - 2x$
    [2.2] 2

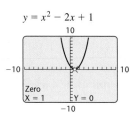

70. $x^2 - 2x + 1 = 0$ [5.3] 1

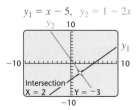

71. $|x - 3| \le 2$
[4.3] $\{x \mid 1 \le x \le 5\}$, or $[1, 5]$

72. $|x - 3| = 2$
[4.3] $\{1, 5\}$

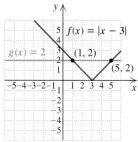

Synthesis

73. Multiply: $(x - 4)^3$.   [5.2] $x^3 - 12x^2 + 48x - 64$

74. Find all roots for $f(x) = x^4 - 34x^2 + 225$.
[5.6] $-3, 3, -5, 5$

Solve.

75. $4 \le |3 - x| \le 6$ [4.2], [4.3] $\{x \mid -3 \le x \le -1$
    or $7 \le x \le 9\}$, or $[-3, -1] \cup [7, 9]$

76. $\dfrac{18}{x - 9} + \dfrac{10}{x + 5} = \dfrac{28x}{x^2 - 4x - 45}$ [6.4] All real numbers except 9 and -5

77. $16x^3 = x$ [5.6] $-\frac{1}{4}, 0, \frac{1}{4}$

61. [2.6] **(a)** $v(x) = 837.605948x + 6515.511152$;
(b) $27,456; **(c)** The amount predicted in Exercise 60 was about $2000 more than the amount predicted in Exercise 61.

Exponents and Radical Functions

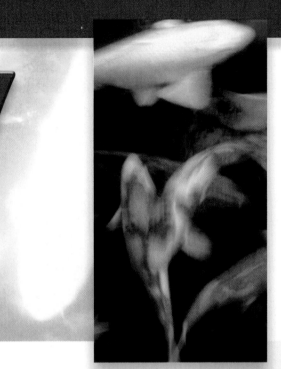

7

In this chapter, we learn about square roots, cube roots, fourth roots, and so on. These roots are studied in connection with the manipulation of radical expressions and the solution of real-world applications. Rational exponents are also studied and are used to ease some of our work with radicals. The chapter closes with an examination of the complex-number system.

APPLICATION

KOI GROWTH. Koi, a popular fish for backyard pools, grow from $\frac{1}{40}$ cm when newly hatched to an average length of 80 cm (*Source*: www.coloradokoi.com/koisize.htm). The following table shows the length y of a koi fish, in centimeters, after x months. Graph the data and determine whether a radical function can be used to model koi growth. The graph of the data indicates that a radical function could be used to model the data.

Age (in months)	Length (in centimeters)
1	2
2	5
4	10
13	30
16	32
30	46
36	50
48	59
60	64
72	68

Xscl = 10, Yscl = 10

This problem appears as Example 14 in Section 7.1.

7.1

Square Roots and Square Root Functions ■ Expressions of the Form $\sqrt{a^2}$ ■ Cube Roots ■ Odd and Even nth Roots ■ Radical Functions and Models

Radical Expressions, Functions, and Models

In this section, we consider roots, such as square roots and cube roots. We look at the symbolism that is used and ways in which symbols can be manipulated to get equivalent expressions. All of this will be important in problem solving.

Square Roots and Square Root Functions

When a number is raised to the second power, the number is squared. Often we need to know what number was squared in order to produce some value a. If such a number can be found, we call that number a *square root* of a.

> ### Square Root
> The number c is a *square root* of a if $c^2 = a$.

For example,

    9 has -3 and 3 as square roots because $(-3)^2 = 9$ and $3^2 = 9$.

    25 has -5 and 5 as square roots because $(-5)^2 = 25$ and $5^2 = 25$.

    -4 does not have a real-number square root because there is no real number c such that $c^2 = -4$.

Note that every positive number has two square roots, whereas 0 has only itself as a square root. Negative numbers do not have real-number square roots, although later in this chapter we will work with a number system in which such square roots do exist.

EXAMPLE 1   Find the two square roots of 64.

Solution   The square roots are 8 and -8, because $8^2 = 64$ and $(-8)^2 = 64$.

Whenever we refer to *the* square root of a number, we mean the nonnegative square root of that number. This is often referred to as the *principal square root* of the number.

> ### Principal Square Root
> The *principal square root* of a nonnegative number is its nonnegative square root. The symbol $\sqrt{}$ is called a *radical sign* and is used to indicate the principal square root of the number over which it appears.

EXAMPLE 2   Simplify each of the following.

a) $\sqrt{25}$ **b)** $\sqrt{\dfrac{25}{64}}$

c) $-\sqrt{64}$ **d)** $\sqrt{0.0049}$

Solution

a) $\sqrt{25} = 5$ $\sqrt{}$ indicates the principal square root. Note that $\sqrt{25} \neq -5$.

b) $\sqrt{\dfrac{25}{64}} = \dfrac{5}{8}$ Since $\left(\dfrac{5}{8}\right)^2 = \dfrac{25}{64}$

c) $-\sqrt{64} = -8$ Since $\sqrt{64} = 8, -\sqrt{64} = -8.$

d) $\sqrt{0.0049} = 0.07$ $(0.07)(0.07) = 0.0049$

In addition to being read as "the principal square root of a," $\sqrt{a}$ is also read as "the square root of a," or simply "root a." Any expression in which a radical sign appears is called a *radical expression*. The following are radical expressions:

$$\sqrt{5}, \qquad \sqrt{a}, \qquad -\sqrt{3x}, \qquad \sqrt{\dfrac{y^2 + 7}{y}}.$$

The expression under the radical sign is called the **radicand**. In the expressions above, the radicands are 5, a, $3x$, and $(y^2 + 7)/y$.

All but the most basic calculators give values for square roots. For example, to calculate $\sqrt{5}$ on most graphing calculators, we press $\boxed{\sqrt{}}$ $\boxed{5}$ $\boxed{)}$ $\boxed{\text{ENTER}}$. On some calculators, $\boxed{\sqrt{}}$ is pressed after 5 has been entered. A calculator will display an approximation like

2.23606798

for $\sqrt{5}$. The exact value of $\sqrt{5}$ is not given by any repeating or terminating decimal. The same is true for the square root of any whole number that is not a perfect square. We discussed such *irrational numbers* in Chapter 1.

The square-root function, given by

$$f(x) = \sqrt{x},$$

has the interval $[0, \infty)$ as its domain. We can draw its graph by selecting convenient values for x and calculating the corresponding outputs. Once these ordered pairs have been graphed, a smooth curve can be drawn.

As x increases, the output $\sqrt{x}$ increases. The range of $f(x)$ is $[0, \infty)$.

$f(x) = \sqrt{x}$

x	$\sqrt{x}$	$(x, f(x))$
0	0	$(0, 0)$
1	1	$(1, 1)$
4	2	$(4, 2)$
9	3	$(9, 3)$

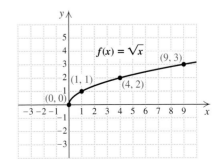

TEACHING TIP

This may be a good time to discuss the difference between the solution of $x^2 = 9$ and the symbol $\sqrt{9}$.

EXAMPLE 3 For each function, find the indicated function value.

a) $f(x) = \sqrt{3x - 2}$; $f(1)$

b) $g(z) = -\sqrt{6z + 4}$; $g(3)$

Solution

a) $f(1) = \sqrt{3 \cdot 1 - 2}$ Substituting

      $= \sqrt{1} = 1$ Simplifying

b) $g(3) = -\sqrt{6 \cdot 3 + 4}$ Substituting

      $= -\sqrt{22}$ Simplifying

      ≈ -4.69041576 Using a calculator to approximate $\sqrt{22}$

Expressions of the Form $\sqrt{a^2}$

It is tempting to write $\sqrt{a^2} = a$, but the next example shows that, as a rule, this is untrue.

EXAMPLE 4 Evaluate $\sqrt{x^2}$ for the following values: **(a)** 5; **(b)** 0; **(c)** -5.

Solution

a) $\sqrt{5^2} = \sqrt{25} = 5$ Same

b) $\sqrt{0^2} = \sqrt{0} = 0$ Same

c) $\sqrt{(-5)^2} = \sqrt{25} = 5$ Opposites   Note that $\sqrt{(-5)^2} \neq -5$.

You may have noticed that evaluating $\sqrt{a^2}$ is just like evaluating $|a|$.

Interactive Discovery

Use graphs or tables to determine which of the following are identities. Be sure to enclose the entire radicand in parentheses.

1. $\sqrt{x^2} = x$ Not an identity   **2.** $\sqrt{x^2} = -x$ Not an identity

3. $\sqrt{x^2} = |x|$ Identity   **4.** $\sqrt{(x + 3)^2} = x + 3$ Not an identity

5. $\sqrt{(x + 3)^2} = |x + 3|$ Identity   **6.** $\sqrt{x^8} = x^4$ Identity

In general, we cannot say that $\sqrt{x^2} = x$. However, we can simplify $\sqrt{x^2}$ using absolute value.

Simplifying $\sqrt{a^2}$

For any real number a,

$$\sqrt{a^2} = |a|.$$

(The principal square root of a^2 is the absolute value of a.)

When a radicand is the square of a variable expression, like $(x + 5)^2$ or $36t^2$, absolute-value signs are needed when simplifying. We use absolute-value signs unless we know that the expression being squared is nonnegative. This assures that our result is never negative.

EXAMPLE 5   Simplify each expression. Assume that the variable can represent any real number.

a) $\sqrt{(x + 1)^2}$ **b)** $\sqrt{x^2 - 8x + 16}$

c) $\sqrt{a^8}$ **d)** $\sqrt{t^6}$

Solution

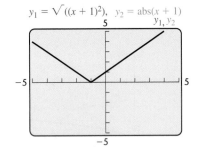

$y_1 = \sqrt{((x + 1)^2)}, \quad y_2 = \text{abs}(x + 1)$

a) $\sqrt{(x + 1)^2} = |x + 1|$ Since $x + 1$ might be negative (for example, if $x = -3$), absolute-value notation is necessary.

The graph at left confirms the identity.

b) $\sqrt{x^2 - 8x + 16} = \sqrt{(x - 4)^2} = |x - 4|$ Since $x - 4$ might be negative, absolute-value notation is necessary.

c) Note that $(a^4)^2 = a^8$ and that a^4 is never negative. Thus,

$$\sqrt{a^8} = a^4.$$ Absolute-value notation is unnecessary here.

d) Note that $(t^3)^2 = t^6$. Thus,

$$\sqrt{t^6} = |t^3|.$$ Since t^3 might be negative, absolute-value notation is necessary.

If we know that no radicands have been formed by raising negative quantities to even powers, we do not need absolute-value signs.

EXAMPLE 6   Simplify each expression. Assume that no radicands were formed by raising negative quantities to even powers.

a) $\sqrt{y^2}$ **b)** $\sqrt{a^{10}}$ **c)** $\sqrt{9x^2 - 6x + 1}$

Solution

a) $\sqrt{y^2} = y$ We are assuming that y is nonnegative, so no absolute-value notation is necessary. When y is negative, $\sqrt{y^2} \neq y$.

b) $\sqrt{a^{10}} = a^5$ Assuming that a^5 is nonnegative. Note that $(a^5)^2 = a^{10}$.

c) $\sqrt{9x^2 - 6x + 1} = \sqrt{(3x - 1)^2} = 3x - 1$ Assuming that $3x - 1$ is nonnegative

Cube Roots

We often need to know what number was cubed in order to produce a certain value. When such a number is found, we say that we have found a *cube root*. For example,

2 is the cube root of 8 because $2^3 = 2 \cdot 2 \cdot 2 = 8$;

-4 is the cube root of -64 because $(-4)^3 = (-4)(-4)(-4) = -64$.

> **Cube Root**
>
> The number c is the *cube root* of a if $c^3 = a$. In symbols, we write $\sqrt[3]{a}$ to denote the cube root of a.

The cube-root function, given by

$$f(x) = \sqrt[3]{x},$$

has $\mathbb{R}$ as its domain. We can draw its graph by selecting convenient values for x and calculating the corresponding outputs. Once these ordered pairs have been graphed, a smooth curve can be drawn. Note that the range of $f(x)$ is also $\mathbb{R}$.

$$f(x) = \sqrt[3]{x}$$

x	$\sqrt[3]{x}$	$(x, f(x))$
0	0	$(0, 0)$
1	1	$(1, 1)$
8	2	$(8, 2)$
-1	-1	$(-1, -1)$
-8	-2	$(-8, -2)$

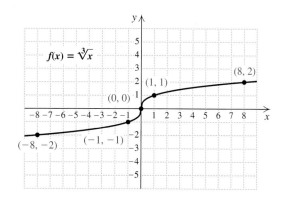

In the real-number system, every number has exactly one cube root. The cube root of a positive number is positive, the cube root of a negative number is negative, and the cube root of 0 is 0.

Interactive Discovery

Use graphs or tables to determine which of the following are identities. Be sure to enclose the entire radicand in parentheses. Note that on many calculators, the cube root option is located in the MATH MATH submenu.

1. $\sqrt[3]{x^3} = x$ Identity

2. $\sqrt[3]{x^3} = -x$ Not an identity

3. $\sqrt[3]{x^3} = |x|$ Not an identity

4. $\sqrt[3]{(-x)^3} = -x$ Identity

5. $\sqrt[3]{(x-1)^3} = |x-1|$
   Not an identity

6. $\sqrt[3]{(x-1)^3} = x - 1$ Identity

We see that although $\sqrt{x^2} = |x|$, $\sqrt[3]{x^3} \neq |x|$. In fact, $\sqrt[3]{x^3} = x$.

In simplifying expressions involving cube roots, we do not use absolute-value signs.

EXAMPLE 7 For each function, find the indicated function value.

a) $f(y) = \sqrt[3]{y}$; $f(125)$ **b)** $g(x) = \sqrt[3]{x - 3}$; $g(-24)$

Solution

a) $f(125) = \sqrt[3]{125} = 5$ Since $5 \cdot 5 \cdot 5 = 125$

b) $g(-24) = \sqrt[3]{-24 - 3}$
$= \sqrt[3]{-27}$
$= -3$ Since $(-3)(-3)(-3) = -27$

EXAMPLE 8 Simplify: $\sqrt[3]{-8y^3}$.

Solution

$\sqrt[3]{-8y^3} = -2y$ Since $(-2y)(-2y)(-2y) = -8y^3$

Odd and Even nth Roots

The fourth root of a number a is the number c for which $c^4 = a$. There are also 5th roots, 6th roots, and so on. We write $\sqrt[n]{a}$ for the nth root. The number n is called the **index** (plural, **indices**). When the index is 2, we do not write it.

EXAMPLE 9 Find each of the following.

a) $\sqrt[5]{32}$ **b)** $\sqrt[5]{-32}$
c) $-\sqrt[5]{32}$ **d)** $-\sqrt[5]{-32}$

Solution

a) $\sqrt[5]{32} = 2$ Since $2^5 = 32$
b) $\sqrt[5]{-32} = -2$ Since $(-2)^5 = -32$
c) $-\sqrt[5]{32} = -2$ Taking the opposite of $\sqrt[5]{32}$
d) $-\sqrt[5]{-32} = -(-2) = 2$ Taking the opposite of $\sqrt[5]{-32}$

Note that every number has just one real root when n is odd. Odd roots of positive numbers are positive and odd roots of negative numbers are negative. Absolute-value signs are not used when finding odd roots.

EXAMPLE 10 Find each of the following.

a) $\sqrt[7]{x^7}$ **b)** $\sqrt[9]{(x - 1)^9}$

Solution

a) $\sqrt[7]{x^7} = x$ **b)** $\sqrt[9]{(x - 1)^9} = x - 1$

When the index n is even, we say that we are taking an *even root*. Every positive real number has two real nth roots when n is even. One root is positive and one is negative. Negative numbers do not have real nth roots when n is even.

When n is even, the notation $\sqrt[n]{a}$ indicates the nonnegative nth root. Thus, when we are finding even nth roots, absolute-value signs are often necessary.

EXAMPLE 11 Simplify each expression, if possible. Assume that variables can represent any real number.

a) $\sqrt[4]{16}$ **b)** $-\sqrt[4]{16}$

c) $\sqrt[4]{-16}$ **d)** $\sqrt[4]{81x^4}$

e) $\sqrt[6]{(y + 7)^6}$

Solution

a) $\sqrt[4]{16} = 2$ Since $2^4 = 16$

b) $-\sqrt[4]{16} = -2$ Taking the opposite of $\sqrt[4]{16}$

c) $\sqrt[4]{-16}$ cannot be simplified. $\sqrt[4]{-16}$ is not a real number.

d) $\sqrt[4]{81x^4} = 3|x|$ Use absolute-value notation since x could represent a negative number.

e) $\sqrt[6]{(y + 7)^6} = |y + 7|$ Use absolute-value notation since $y + 7$ could be negative.

We summarize as follows.

Simplifying nth Roots

n	a	$\sqrt[n]{a}$	$\sqrt[n]{a^n}$
Even	Positive	Positive	$\|a\|$ (or a)
Even	Negative	Not a real number	$\|a\|$ (or $-a$)
Odd	Positive	Positive	a
Odd	Negative	Negative	a

Radical Functions and Models

A **radical function** is a function that can be described by a radical expression.

Radical Functions

When entering radical equations on a graphing calculator, care must be taken to place parentheses properly. For example, we enter

$$f(x) = \frac{3.5 + \sqrt{4.5 - 6x}}{5}$$

on the Y= screen as ⎡(⎤⎡3⎤⎡.⎤⎡5⎤⎡+⎤⎡2nd⎤⎡√⎤⎡4⎤⎡.⎤⎡5⎤⎡−⎤⎡6⎤⎡x⎤⎡)⎤⎡)⎤⎡÷⎤⎡5⎤. The outer parentheses enclose the numerator of the expression. The first right parenthesis) indicates the end of the radicand; the left parenthesis of the radicand is supplied by the calculator when ⎡√⎤ is pressed.

Some choices of window dimensions may not show a good representation of the graph of a radical function. If we can determine the domain of the function algebraically, we can use that information to choose a window that will show more of the function. For example, compare the following graphs of the function $f(x) = 2\sqrt{15 - x}$. Note that the domain of f is $\{x \mid x \leq 15\}$, or $(-\infty, 15]$.

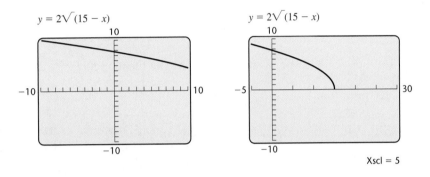

The graph on the left above uses the standard viewing window. The domain of the function is not clear from the graph, and the shape almost looks like a straight line. The viewing window for the graph on the right above is $[-5, 30, -10, 10]$. The part of the x-axis shown there includes the endpoint of the interval that gives the domain of the function. Knowing the domain of the function helps us choose Xmin and Xmax. For now, choosing appropriate values of Ymin and Ymax may require some trial and error.

If a function is given by a radical expression with an odd index, the domain is the set of all real numbers. If a function is given by a radical expression with an even index, the domain is the set of replacements for which the radicand is nonnegative.

EXAMPLE 12 Find the domain of the function given by each of the following equations. Check by graphing the function. Then, from the graph, estimate the range of the function.

a) $q(x) = \sqrt{-x}$

b) $t(x) = \sqrt{2x - 5} - 3$

c) $f(x) = \sqrt{x^2 + 1}$

Solution

a) We find all values of x for which the radicand is nonnegative:

$$-x \geq 0$$

$$x \leq 0. \qquad \text{Multiplying by } -1; \text{ reversing the direction of the inequality}$$

The domain is $\{x\,|\,x \le 0\}$, or $(-\infty, 0]$, as indicated on the graph by the shading on the x-axis. The range appears to be $[0, \infty)$, as indicated by the shading on the y-axis. This can also be seen by examining a table of values.

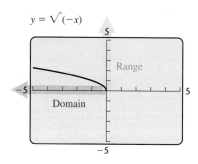

b) The function $t(x) = \sqrt{2x - 5} - 3$ is defined when $2x - 5 \ge 0$:

$$2x - 5 \ge 0$$
$$2x \ge 5 \qquad \textbf{Adding 5}$$
$$x \ge \tfrac{5}{2}. \qquad \textbf{Dividing by 2}$$

The domain is $\left\{x\,\middle|\,x \ge \tfrac{5}{2}\right\}$, or $\left[\tfrac{5}{2}, \infty\right)$, as indicated on the graph by the shading on the x-axis. The range appears to be $[-3, \infty)$, as indicated by the shading on the y-axis.

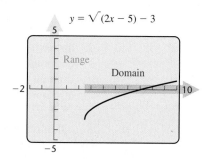

c) The radicand in $f(x) = \sqrt{x^2 + 1}$ is $x^2 + 1$. We must have

$$x^2 + 1 \ge 0$$
$$x^2 \ge -1.$$

Since x^2 is nonnegative for all real numbers x, the inequality is true for all real numbers. The domain is $(-\infty, \infty)$. The range appears to be $[1, \infty)$.

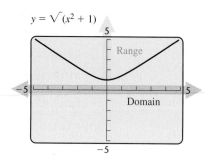

EXAMPLE 13   Determine the domain of the function given by

$$g(x) = \sqrt[6]{7 - 3x}.$$

Solution   Since the index is even, the radicand, $7 - 3x$, must be nonnegative. We solve the inequality:

$$7 - 3x \geq 0 \qquad \textbf{We cannot find the 6th root of a negative number.}$$
$$-3x \geq -7$$
$$x \leq \tfrac{7}{3}. \qquad \textbf{Multiplying both sides by } -\tfrac{1}{3} \textbf{ and}$$
$$\textbf{reversing the inequality}$$

Thus,

the domain of g is $\left\{x \,\middle|\, x \leq \tfrac{7}{3}\right\}$, or $\left(-\infty, \tfrac{7}{3}\right]$.

Some situations can be modeled using radical functions. The graphs of radical functions can have many different shapes. However, radical functions given by equations of the form

$$r(x) = \sqrt{ax + b}$$

will have the general shape of the graph of $f(x) = \sqrt{x}$.

We can determine whether a radical function might fit a set of data by plotting the points.

EXAMPLE 14   Koi, a popular fish for backyard pools, grow from $\tfrac{1}{40}$ cm when newly hatched to an average length of 80 cm. Generally, koi reach about 50% of their final adult length after 2 yr and 99% after 14 yr (*Source*: www. coloradokoi.com/koisize.htm).

The following table shows the length y of a koi fish, in centimeters, after x months. Graph the data and determine whether a radical function can be used to model koi growth.

Age (in months)	Length (in centimeters)
1	2
2	5
4	10
13	30
16	32
30	46
36	50
48	59
60	64
72	68

Solution   We graph the data, entering the ages in L₁ and the lengths in L₂. The data appear to follow the pattern of the graph of a radical function. We determine that a radical function could probably be used to model koi growth.

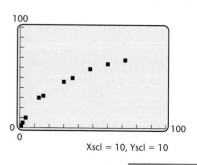

Most graphing calculators do not have the capability of fitting radical equations to data. The koi-growth model given in the following example was found using other methods.

EXAMPLE 15   Koi growth, as described in Example 14, can be modeled by the radical function given by

$$f(x) = 4.94 + \sqrt{59.52x - 129.59},$$

where $f(x)$ is the length, in centimeters, of a fish of age x months. Estimate the length of a koi at 8 months and at 20 months.

Solution   We enter the equation as

$$y = 4.94 + \sqrt{(59.52x - 129.59)}.$$

Graphing it along with the data from Example 14, as shown below, we can see that the model does fit the data. Using a Table with Indpt set to Ask to find $f(8)$ and $f(20)$, we determine that the length of a koi at 8 months is about 24 cm and the length at 20 months is about 38 cm.

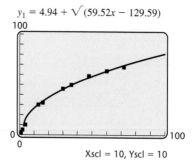

7.1

Exercise Set

For each number, find the square roots.

1. 16   4, −4

2. 49   7, −7

3. 144   12, −12

4. 9   3, −3

5. 81   9, −9

6. 400   20, −20

7. 900   30, −30

8. 225   15, −15

For each number, use a calculator to approximate the square roots to the nearest ten-thousandth.

9. 7   2.6458, −2.6458

10. 15   3.8730, −3.8730

11. 23.7   4.8683, −4.8683

12. 0.05   0.2236, −0.2236

13. $\frac{3}{4}$ 0.8660, −0.8660

14. $\frac{5}{16}$ 0.5590, −0.5590

Simplify.

15. $-\sqrt{\frac{49}{36}}$ $-\frac{7}{6}$

16. $-\sqrt{\frac{361}{9}}$ $-\frac{19}{3}$

17. $\sqrt{441}$ 21

18. $\sqrt{196}$ 14

19. $-\sqrt{\frac{16}{81}}$ $-\frac{4}{9}$

20. $-\sqrt{\frac{81}{144}}$ $-\frac{3}{4}$

21. $\sqrt{0.09}$ 0.3

22. $\sqrt{0.36}$ 0.6

23. $-\sqrt{0.0049}$ −0.07

24. $\sqrt{0.0144}$ 0.12

Identify the radicand and the index for each expression.

25. $5\sqrt{p^2 + 4}$ $p^2 + 4$; 2

26. $-7\sqrt{y^2 - 8}$ $y^2 - 8$; 2

27. $x^2 y^3 \sqrt[3]{\frac{x}{y + 4}}$ $\frac{x}{y + 4}$; 3

28. $a^2 b^3 \sqrt[3]{\frac{a}{a^2 - b}}$ $\frac{a}{a^2 - b}$; 3

For each function, find the specified function value, if it exists.

29. $f(t) = \sqrt{5t - 10}$; $f(6), f(2), f(1), f(-1)$ $\sqrt{20}$; 0; does not exist; does not exist

30. $g(x) = \sqrt{x^2 - 25}$; $g(-6), g(3), g(6), g(13)$ ☐

31. $t(x) = -\sqrt{2x + 1}$; $t(4), t(0), t(-1), t\left(-\frac{1}{2}\right)$ ☐

32. $p(z) = \sqrt{2z^2 - 20}$; $p(4), p(3), p(-5), p(0)$ ☐

33. $f(t) = \sqrt{t^2 + 1}$; $f(0), f(-1), f(-10)$ 1; $\sqrt{2}$; $\sqrt{101}$

34. $g(x) = -\sqrt{(x + 1)^2}$; $g(-3), g(4), g(-5)$ −2; −5; −4

35. $g(x) = \sqrt{x^3 + 9}$; $g(-2), g(-3), g(3)$ 1; does not exist; 6

36. $f(t) = \sqrt{t^3 - 10}$; $f(2), f(3), f(4)$ Does not exist; $\sqrt{17}$; $\sqrt{54}$

Simplify. Remember to use absolute-value notation when necessary.

37. $\sqrt{36x^2}$ $6|x|$

38. $\sqrt{25t^2}$ $5|t|$

39. $\sqrt{(-6b)^2}$ $6|b|$

40. $\sqrt{(-7c)^2}$ $7|c|$

41. $\sqrt{(7 - t)^2}$ $|7 - t|$

42. $\sqrt{(a + 1)^2}$ $|a + 1|$

43. $\sqrt{y^2 + 16y + 64}$ $|y + 8|$

44. $\sqrt{x^2 - 4x + 4}$ $|x - 2|$

45. $\sqrt{9x^2 - 30x + 25}$ $|3x - 5|$

46. $\sqrt{4x^2 + 28x + 49}$ $|2x + 7|$

47. $-\sqrt[4]{625}$ −5

48. $\sqrt[4]{256}$ 4

49. $-\sqrt[5]{3^5}$ −3

50. $\sqrt[5]{-1}$ −1

51. $\sqrt[5]{-\frac{1}{32}}$ $-\frac{1}{2}$

52. $\sqrt[5]{-\frac{32}{243}}$ $-\frac{2}{3}$

53. $\sqrt[8]{y^8}$ $|y|$

54. $\sqrt[6]{x^6}$ $|x|$

55. $\sqrt[4]{(7b)^4}$ $7|b|$

56. $\sqrt[4]{(5a)^4}$ $5|a|$

57. $\sqrt[12]{(-10)^{12}}$ 10

58. $\sqrt[10]{(-6)^{10}}$ 6

59. $\sqrt[1976]{(2a + b)^{1976}}$ $|2a + b|$

60. $\sqrt[414]{(a + b)^{414}}$ $|a + b|$

61. $\sqrt{x^{10}}$ $|x^5|$

62. $\sqrt{a^{22}}$ $|a^{11}|$

63. $\sqrt{a^{14}}$ $|a^7|$

64. $\sqrt{x^{16}}$ x^8

Simplify. Assume that no radicands were formed by raising negative quantities to even powers.

65. $\sqrt{25t^2}$ $5t$

66. $\sqrt{16x^2}$ $4x$

67. $\sqrt{(7c)^2}$ $7c$

68. $\sqrt{(6b)^2}$ $6b$

69. $\sqrt{(5 + b)^2}$ $5 + b$

70. $\sqrt{(a + 1)^2}$ $a + 1$

71. $\sqrt{9x^2 + 36x + 36}$ ☐

72. $\sqrt{4x^2 + 8x + 4}$ ☐

73. $\sqrt{25t^2 - 20t + 4}$ $5t - 2$

74. $\sqrt{9t^2 - 12t + 4}$ $3t - 2$

75. $-\sqrt[3]{64}$ −4

76. $\sqrt[3]{27}$ 3

77. $\sqrt[4]{81x^4}$ $3x$

78. $\sqrt[4]{16x^4}$ $2x$

79. $-\sqrt[5]{-100,000}$ 10

80. $\sqrt[3]{-216}$ −6

81. $-\sqrt[3]{-64x^3}$ $4x$

82. $-\sqrt[3]{-125y^3}$ $5y$

83. $\sqrt{a^{14}}$ a^7

84. $\sqrt{a^{22}}$ a^{11}

85. $\sqrt{(x + 3)^{10}}$ $(x + 3)^5$

86. $\sqrt{(x - 2)^8}$ $(x - 2)^4$

☐ Answers to Exercises 30–32, 71, and 72 can be found on p. A-63.

For each function, find the specified function value, if it exists.

87. $f(x) = \sqrt[3]{x + 1}$; $f(7)$, $f(26)$, $f(-9)$, $f(-65)$
2; 3; -2; -4

88. $g(x) = -\sqrt[3]{2x - 1}$; $g(0)$, $g(-62)$, $g(-13)$, $g(63)$
1; 5; 3; -5

89. $g(t) = \sqrt[4]{t - 3}$; $g(19)$, $g(-13)$, $g(1)$, $g(84)$
2; does not exist; does not exist; 3

90. $f(t) = \sqrt[4]{t + 1}$; $f(0)$, $f(15)$, $f(-82)$, $f(80)$
1; 2; does not exist; 3

Determine the domain of each function described.

91. $f(x) = \sqrt{x - 5}$ ⬚   **92.** $g(x) = \sqrt{x + 8}$ ⬚

93. $g(t) = \sqrt[4]{t + 3}$ ⬚   **94.** $f(x) = \sqrt[4]{x - 7}$ ⬚

95. $g(x) = \sqrt[4]{5 - x}$ ⬚   **96.** $g(t) = \sqrt[3]{2t - 5}$ ℝ

97. $f(t) = \sqrt[5]{2t + 9}$ ℝ   **98.** $f(t) = \sqrt[6]{2t + 5}$ ⬚

99. $h(z) = -\sqrt[6]{5z + 3}$ ⬚   **100.** $d(x) = -\sqrt[4]{7x - 5}$
$\{x | x \geq \frac{5}{7}\}$, or $[\frac{5}{7}, \infty)$

Aha! **101.** $f(t) = 7 + \sqrt[8]{t^8}$ ℝ

102. $g(t) = 9 + \sqrt[6]{t^6}$ ℝ

Determine algebraically the domain of each function described. Then use a graphing calculator to confirm your answer and to estimate the range.

103. $f(x) = \sqrt{5 - x}$ ⬚   **104.** $g(x) = \sqrt{2x + 1}$ ⬚

105. $f(t) = 1 - \sqrt{x + 1}$ ⬚

106. $g(t) = 2 + \sqrt{3x - 5}$ ⬚

107. $g(x) = 3 + \sqrt{x^2 + 4}$ ⬚

108. $f(x) = 5 - \sqrt{3x^2 + 1}$ ⬚

In Exercises 109–112, match each function with one of the following graphs without using a calculator.

a)
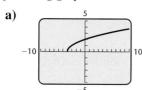

b)
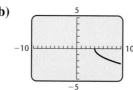

c)
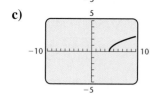

d)
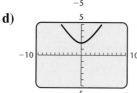

109. $f(x) = \sqrt{x - 4}$ (c)   **110.** $g(x) = \sqrt{x + 4}$ (a)

111. $h(x) = \sqrt{x^2 + 4}$ (d)   **112.** $f(x) = -\sqrt{x - 4}$ (b)

In Exercises 113–116, determine whether a radical function would be a good model of the given situation.

113. *Farm Size.* The average size of United States' farms increased during the last half of the 20th century. (See the table at the top of the next column.)
Yes

Year	Average Size of Farm (in acres)
1950	215
1960	302
1970	390
1980	426
2000	434

Source: U.S. Department of Agriculture

114. *Wind Chill.* When the wind is blowing, the air temperature feels lower than the actual temperature. This is referred to as the wind-chill temperature. The following table lists wind-chill temperatures for various wind speeds at a thermometer reading of 15°F.   Yes

Wind Speed (in miles per hour)	Wind-Chill Temperature (in degrees Fahrenheit)
5	7
10	3
15	0
20	-2
25	-4
30	-5
35	-7
40	-8

Source: National Weather Service

115. *Cable Television.* The number of households served by cable television increased from 1980 to 1998.   Yes

Year	Number of Households Served by Cable Television (in millions)
1980	17.7
1985	39.9
1990	54.9
1993	58.8
1994	60.5
1996	64.6
1998	67.4

Source: Nielsen Media Research

⬚ Answers to Exercises 91–95, 98, 99, and 103–108 can be found on p. A-63.

116. *Radio Stations.* The number of commercial FM radio stations in the United States increased from 1980 to 1998. No

Year	Number of Commercial FM Radio Stations
1980	3282
1985	3875
1990	4392
1993	4971
1994	5109
1996	5419
1998	5662

Source: Radio Advertising Bureau

117. *Cable Television.* The number of households h, in millions, served by cable television x years after 1970 can be modeled by the function

$$h(x) = 10.681 + \sqrt{177.971284x - 1744.994255}$$

(see Exercise 115). Use the function to estimate the number of households served by cable television in 1992 and in 2001. 57.3 million; 72.1 million

118. *Falling Object.* The number of seconds t that it takes for an object to fall x meters when thrown down at a velocity of 9.5 meters per second is given by the function

$$t(x) = -0.9694 + \sqrt{0.9397 + 0.2041x}.$$

A rock is thrown from the top of Turkey Bluff, 90 m above the Gasconade River in Missouri, at a velocity of 9.5 meters per second. After how many seconds will the rock hit the water? 3.4 sec

TW **119.** Explain how to write the negative square root of a number using radical notation.

TW **120.** Does the square root of a number's absolute value always exist? Why or why not?

Skill Maintenance   123. $\dfrac{a^6c^{12}}{8b^9}$ 124. $\dfrac{x^6y^2}{25z^4}$

Simplify. Do not use negative exponents in your answer. [1.4]

121. $(a^3b^2c^5)^3$ $a^9b^6c^{15}$ **122.** $(5a^7b^8)(2a^3b)$ $10a^{10}b^9$

123. $(2a^{-2}b^3c^{-4})^{-3}$ **124.** $(5x^{-3}y^{-1}z^2)^{-2}$

125. $\dfrac{8x^{-2}y^5}{4x^{-6}z^{-2}}$ $2x^4y^5z^2$ **126.** $\dfrac{10a^{-6}b^{-7}}{2a^{-2}c^{-3}}$ $\dfrac{5c^3}{a^4b^7}$

127. Use the points $(0, 3282)$ and $(16, 5419)$ to find a linear function that can be used to estimate the

number n of commercial FM radio stations t years after 1980 (see Exercise 116). [2.6] $n(t) = 133.5625t + 3282$

128. Use linear regression and the data in Exercise 116 to find an equation for a linear function that can be used to estimate the number f of commercial FM radio stations x years after 1980. [2.6] $f(x) = 134.4737456x + 3212.856476$

Synthesis

TW **129.** If the domain of $f = [1, \infty)$ and the range of $f = [2, \infty)$, find a possible expression for $f(x)$ and explain how such an expression is formulated.

TW **130.** Kelly obtains the following graph of

$$f(x) = \sqrt{x^2 - 4x - 12}$$

and concludes that the domain of f is $(-\infty, -2]$. Is she correct? If not, what mistake is she making?

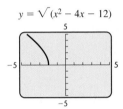

$y = \sqrt{(x^2 - 4x - 12)}$

TW **131.** Could the following situation possibly be modeled using a radical function? Why or why not? "For each year, the yield increases. The amount of increase is smaller each year."

TW **132.** Could the following situation possibly be modeled using a radical function? Why or why not? "For each year, the costs increase. The amount of increase is the same each year."

133. *Spaces in a Parking Lot.* A parking lot has attendants to park the cars. The number N of stalls needed for waiting cars before attendants can get to them is given by the formula $N = 2.5\sqrt{A}$, where A is the number of arrivals in peak hours. Find the number of spaces needed for the given number of arrivals in peak hours: **(a)** 25; **(b)** 36; **(c)** 49; **(d)** 64.   (a) 13; (b) 15; (c) 18; (d) 20

134. Find the domain of f if

$$f(x) = \frac{\sqrt{x + 3}}{\sqrt[4]{2 - x}}.$$ $\{x \mid -3 \le x < 2\}$, or $[-3, 2)$

135. Find the domain of g if

$$g(x) = \frac{\sqrt[4]{5 - x}}{\sqrt[6]{x + 4}}.$$ $\{x \mid -4 < x \le 5\}$, or $(-4, 5]$

7.2

Rational Exponents ■ Negative Rational Exponents ■ Laws of Exponents ■ Simplifying Radical Expressions

Rational Numbers as Exponents

In Section 1.1, we considered the natural numbers as exponents. Our discussion of exponents was expanded to include all integers in Section 1.4. In this section, we expand the study still further—to include all rational numbers. This will give meaning to expressions like $a^{1/3}$, $7^{-1/2}$, and $(3x)^{4/5}$. Such notation will help us simplify certain radical expressions.

Rational Exponents

Consider $a^{1/2} \cdot a^{1/2}$. If we still want to add exponents when multiplying, it must follow that $a^{1/2} \cdot a^{1/2} = a^{1/2+1/2}$, or a^1. This suggests that $a^{1/2}$ is a square root of a. Similarly, $a^{1/3} \cdot a^{1/3} \cdot a^{1/3} = a^{1/3+1/3+1/3}$, or a^1, so $a^{1/3}$ should mean $\sqrt[3]{a}$.

$$a^{1/n} = \sqrt[n]{a}$$

$a^{1/n}$ means $\sqrt[n]{a}$. When a is nonnegative, n can be any natural number greater than 1. When a is negative, n must be odd.

TEACHING TIP

Point out that when the index is 2, it is not written:
$x^{1/2} = \sqrt{x}$ and $\sqrt{3} = 3^{1/2}$.

Note that the denominator of the exponent becomes the index and the base becomes the radicand.

EXAMPLE 1   Write an equivalent expression using radical notation.

a) $x^{1/2}$ **b)** $(-8)^{1/3}$ **c)** $(abc)^{1/5}$

Solution

a) $x^{1/2} = \sqrt{x}$
b) $(-8)^{1/3} = \sqrt[3]{-8} = -2$ }   The denominator of the exponent becomes the index. The base becomes the radicand.
c) $(abc)^{1/5} = \sqrt[5]{abc}$

EXAMPLE 2   Write an equivalent expression using exponential notation.

a) $\sqrt[5]{9xy}$ **b)** $\sqrt[7]{\dfrac{x^3y}{4}}$ **c)** $\sqrt{5x}$

Solution   Parentheses are required to indicate the base.

a) $\sqrt[5]{9xy} = (9xy)^{1/5}$
b) $\sqrt[7]{\dfrac{x^3y}{4}} = \left(\dfrac{x^3y}{4}\right)^{1/7}$ }   The index becomes the denominator of the exponent. The radicand becomes the base.
c) $\sqrt{5x} = (5x)^{1/2}$ For square roots, the index 2 is understood without being written.

Radical Expressions and Rational Exponents

We can enter a radical expression in radical notation or by using rational exponents.

To use radical notation or a square root expression, use the $\boxed{\sqrt{}}$ key. Enter a cube root by using the $\sqrt[3]{}$ (option in the MATH MATH menu, as shown in the figure below. For a higher root, the $\sqrt[x]{}$ option in the same menu can be used. To use this notation, type the index, then choose $\sqrt[x]{}$, and then enter the radicand. Note that if this option is used, the left parenthesis is not supplied automatically.

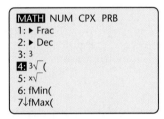

To use rational exponents, enter the radicand enclosed in parentheses, then press $\boxed{\wedge}$, and then enter the rational exponent, also enclosed in parentheses. If the rational exponent is written as a single decimal number, the parentheses around the exponent are unnecessary. If the decimal notation for a rational exponent must be rounded, the fraction notation should be used.

EXAMPLE 3   Graph: $f(x) = \sqrt[4]{2x - 7}$.

Solution   We can enter the equation in radical notation using $\sqrt[x]{}$. Alternatively, we can rewrite the radical expression using a rational exponent:

$$f(x) = (2x - 7)^{1/4}, \quad \text{or} \quad f(x) = (2x - 7)^{0.25}.$$

Then we let $y = (2x - 7)\wedge(1/4)$ or $y = (2x - 7)\wedge.25$. The screen on the left below shows all three forms of the equation; they are equivalent.

Knowing the domain of the function can help us determine an appropriate viewing window. Since the index is even, the domain is the set of all x for which the radicand is nonnegative, or $\left[\frac{7}{2}, \infty\right)$. We choose a viewing window of $[-1, 10, -1, 5]$. This will show the axes and the first quadrant. The graph is shown on the right below.

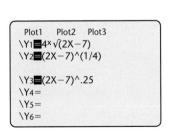

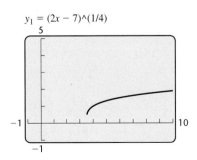

How shall we define $a^{2/3}$? If the property for multiplying exponents is to hold, we must have $a^{2/3} = (a^{1/3})^2$ and $a^{2/3} = (a^2)^{1/3}$. This would suggest that $a^{2/3} = (\sqrt[3]{a})^2$ and $a^{2/3} = \sqrt[3]{a^2}$. We make our definition accordingly.

Positive Rational Exponents

For any natural numbers m and n ($n \neq 1$) and any real number a for which $\sqrt[n]{a}$ exists,

$$a^{m/n} \quad \text{means} \quad (\sqrt[n]{a})^m, \quad \text{or} \quad \sqrt[n]{a^m}.$$

EXAMPLE 4   Write an equivalent expression using radical notation and simplify.

a) $27^{2/3}$

b) $25^{3/2}$

Solution

a) $27^{2/3} = \sqrt[3]{27^2}$, or $(\sqrt[3]{27})^2$ It is easier to simplify using $(\sqrt[3]{27})^2$.

$\qquad\quad = 3^2,$ or 9

b) $25^{3/2} = \sqrt[2]{25^3}$, or $(\sqrt[2]{25})^3$ We normally omit the index 2.

$\qquad\quad = 5^3,$ or 125 Taking the square root and cubing

---

EXAMPLE 5   Write an equivalent expression using exponential notation.

a) $\sqrt[3]{9^4}$

b) $(\sqrt[4]{7xy})^5$

Solution

a) $\sqrt[3]{9^4} = 9^{4/3}$

b) $(\sqrt[4]{7xy})^5 = (7xy)^{5/4}$

$\left.\vphantom{\begin{array}{c} a \\ b \end{array}}\right\}$ The index becomes the denominator of the fractional exponent.

Rational roots of numbers can be approximated on a calculator.

EXAMPLE 6   Approximate $\sqrt[5]{(-23)^3}$. Round to the nearest thousandth.

Solution   We first rewrite the expression using a rational exponent:

$$\sqrt[5]{(-23)^3} = (-23)^{3/5}.$$

Using a calculator, we have

$$(-23)\wedge(3/5) \approx -6.562.$$

Negative Rational Exponents

Recall that $x^{-2} = 1/x^2$. Negative rational exponents behave similarly.

> ### Negative Rational Exponents
>
> For any rational number m/n and any nonzero real number a for which $a^{m/n}$ exists,
>
> $$a^{-m/n} \quad \text{means} \quad \frac{1}{a^{m/n}}.$$

TEACHING TIP

You may wish to review negative exponents at this point.

> **CAUTION!**   A negative exponent does not indicate that the expression in which it appears is negative.

EXAMPLE 7   Write an equivalent expression with positive exponents and, if possible, simplify.

a) $9^{-1/2}$ **b)** $(5xy)^{-4/5}$ **c)** $64^{-2/3}$

d) $4x^{-2/3}y^{1/5}$ **e)** $\left(\dfrac{3r}{7s}\right)^{-5/2}$

Solution

a) $9^{-1/2} = \dfrac{1}{9^{1/2}}$ $9^{-1/2}$ **is the reciprocal of** $9^{1/2}$**.**

Since $9^{1/2} = \sqrt{9} = 3$, the answer simplifies to $\dfrac{1}{3}$.

b) $(5xy)^{-4/5} = \dfrac{1}{(5xy)^{4/5}}$ $(5xy)^{-4/5}$ **is the reciprocal of** $(5xy)^{4/5}$**.**

c) $64^{-2/3} = \dfrac{1}{64^{2/3}}$ $64^{-2/3}$ **is the reciprocal of** $64^{2/3}$**.**

Since $64^{2/3} = \left(\sqrt[3]{64}\right)^2 = 4^2 = 16$, the answer simplifies to $\dfrac{1}{16}$.

d) $4x^{-2/3}y^{1/5} = 4 \cdot \dfrac{1}{x^{2/3}} \cdot y^{1/5} = \dfrac{4y^{1/5}}{x^{2/3}}$

e) In Section 1.4, we found that $(a/b)^{-n} = (b/a)^n$. This property holds for *any* negative exponent:

$$\left(\frac{3r}{7s}\right)^{-5/2} = \left(\frac{7s}{3r}\right)^{5/2}. \qquad \begin{array}{l}\textbf{Writing the reciprocal of the base and}\\ \textbf{changing the sign of the exponent}\end{array}$$

Laws of Exponents

The same laws hold for rational exponents as for integer exponents.

Laws of Exponents

For any real numbers a and b and any rational exponents m and n for which a^m, a^n, and b^m are defined:

1. $a^m \cdot a^n = a^{m+n}$ In multiplying, add exponents if the bases are the same.

2. $\dfrac{a^m}{a^n} = a^{m-n}$ In dividing, subtract exponents if the bases are the same. (Assume $a \neq 0$.)

3. $(a^m)^n = a^{m \cdot n}$ To raise a power to a power, multiply the exponents.

4. $(ab)^m = a^m b^m$ To raise a product to a power, raise each factor to the power and multiply.

EXAMPLE 8   Use the laws of exponents to simplify.

a) $3^{1/5} \cdot 3^{3/5}$ **b)** $a^{1/4}/a^{1/2}$

c) $(7.2^{2/3})^{3/4}$ **d)** $(a^{-1/3}b^{2/5})^{1/2}$

Solution

a) $3^{1/5} \cdot 3^{3/5} = 3^{1/5+3/5} = 3^{4/5}$ **Adding exponents**

b) $\dfrac{a^{1/4}}{a^{1/2}} = a^{1/4 - 1/2} = a^{1/4 - 2/4}$ **Subtracting exponents after finding a common denominator**

$$= a^{-1/4}, \text{ or } \frac{1}{a^{1/4}}$$ $a^{-1/4}$ is the reciprocal of $a^{1/4}$.

c) $(7.2^{2/3})^{3/4} = 7.2^{2/3 \cdot 3/4} = 7.2^{6/12}$ **Multiplying exponents**

$$= 7.2^{1/2}$$ **Using arithmetic to simplify the exponent**

d) $(a^{-1/3}b^{2/5})^{1/2} = a^{-1/3 \cdot 1/2} \cdot b^{2/5 \cdot 1/2}$ **Raising a product to a power and multiplying exponents**

$$= a^{-1/6}b^{1/5}, \text{ or } \frac{b^{1/5}}{a^{1/6}}$$

Simplifying Radical Expressions

Many radical expressions can be simplified using rational exponents.

To Simplify Radical Expressions

1. Convert radical expressions to exponential expressions.

2. Use arithmetic and the laws of exponents to simplify.

3. Convert back to radical notation as needed.

EXAMPLE 9 Use rational exponents to simplify. Do not use fractional exponents in the final answer.

a) $\sqrt[6]{(5x)^3}$ **b)** $\sqrt[5]{t^{20}}$

c) $\left(\sqrt[3]{ab^2c}\right)^{12}$ **d)** $\sqrt{\sqrt[3]{x}}$

Solution

a) $\sqrt[6]{(5x)^3} = (5x)^{3/6}$ Converting to exponential notation

$\qquad\qquad = (5x)^{1/2}$ Simplifying the exponent

$\qquad\qquad = \sqrt{5x}$ Returning to radical notation

To check on a graphing calculator, we let $y_1 = (5x)\hat{\ }(3/6)$ and $y_2 = \sqrt{\ }(5x)$, and compare values in a table. If we scroll through the table (see the figure at left), we see that $y_1 = y_2$, so our simplification is probably correct.

b) $\sqrt[5]{t^{20}} = t^{20/5}$ Converting to exponential notation

$\qquad\quad = t^4$ Simplifying the exponent

c) $\left(\sqrt[3]{ab^2c}\right)^{12} = (ab^2c)^{12/3}$ Converting to exponential notation

$\qquad\qquad\quad = (ab^2c)^4$ Simplifying the exponent

$\qquad\qquad\quad = a^4b^8c^4$ Using the laws of exponents

d) $\sqrt{\sqrt[3]{x}} = \sqrt{x^{1/3}}$ Converting the radicand to exponential notation

$\qquad\quad = (x^{1/3})^{1/2}$ Try to go directly to this step.

$\qquad\quad = x^{1/6}$ Using the laws of exponents

$\qquad\quad = \sqrt[6]{x}$ Returning to radical notation

We can check by graphing $y_1 = \sqrt{\ }\left(\sqrt[3]{\ }(x)\right)$ and $y_2 = x\hat{\ }(1/6)$. The graphs coincide, as we also see by scrolling through the table of values shown at left.

$y_1 = (5x)\hat{\ }(3/6),\ \ y_2 = \sqrt{\ }(5x)$

X	Y1	Y2
0	0	0
1	2.2361	2.2361
2	3.1623	3.1623
3	3.873	3.873
4	4.4721	4.4721
5	5	5
6	5.4772	5.4772
X = 0		

$y_1 = \sqrt{\ }(\sqrt[3]{\ }(x)),\ \ y_2 = x\hat{\ }(1/6)$

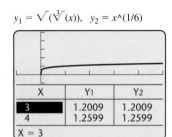

X	Y1	Y2
3	1.2009	1.2009
4	1.2599	1.2599
X = 3		

7.2

Exercise Set

FOR EXTRA HELP

Digital Video Tutor CD 7 Videotape 10 | Student's Solutions Manual | Tutor Center AW Math Tutor Center | InterAct Math | MathXL | MyMathLab

Note: Assume for all exercises that even roots are of non-negative quantities and that all denominators are nonzero.

Write an equivalent expression using radical notation and, if possible, simplify.

1. $x^{1/4}$ $\sqrt[4]{x}$

2. $y^{1/5}$ $\sqrt[5]{y}$

3. $16^{1/2}$ 4

4. $8^{1/3}$ 2

5. $81^{1/4}$ 3

6. $64^{1/6}$ 2

7. $9^{1/2}$ 3

8. $25^{1/2}$ 5

9. $(xyz)^{1/3}$ $\sqrt[3]{xyz}$

10. $(ab)^{1/4}$ $\sqrt[4]{ab}$

11. $(a^2b^2)^{1/5}$ $\sqrt[5]{a^2b^2}$

12. $(x^3y^3)^{1/4}$ $\sqrt[4]{x^3y^3}$

13. $a^{2/3}$ $\sqrt[3]{a^2}$

14. $b^{3/2}$ $\sqrt{b^3}$

15. $16^{3/4}$ 8

16. $4^{7/2}$ 128

17. $49^{3/2}$ 343

18. $27^{4/3}$ 81

19. $9^{5/2}$ 243

20. $81^{3/2}$ 729

21. $(81x)^{3/4}$

21. $\sqrt[4]{81^3x^3}$, or $27\sqrt[4]{x^3}$

22. $(125a)^{2/3}$ $25\sqrt[3]{a^2}$

23. $(25x^4)^{3/2}$ $125x^6$

24. $(9y^6)^{3/2}$ $27y^9$

Write an equivalent expression using exponential notation.

25. $\sqrt[3]{20}$ $20^{1/3}$

26. $\sqrt[3]{19}$ $19^{1/3}$

27. $\sqrt{17}$ $17^{1/2}$

28. $\sqrt{6}$ $6^{1/2}$

29. $\sqrt{x^3}$ $x^{3/2}$

30. $\sqrt{a^5}$ $a^{5/2}$

31. $\sqrt[5]{m^2}$ $m^{2/5}$ **32.** $\sqrt[5]{n^4}$ $n^{4/5}$ **33.** $\sqrt[4]{cd}$ $(cd)^{1/4}$

34. $\sqrt[5]{xy}$ $(xy)^{1/5}$ **35.** $\sqrt[5]{xy^2z}$ ⊡   **36.** $\sqrt[5]{x^3y^2z^2}$ ⊡

37. $\left(\sqrt{3mn}\right)^3$ ⊡   **38.** $\left(\sqrt[3]{7xy}\right)^4$ ⊡   **39.** $\left(\sqrt[5]{8x^2y}\right)^5$ Aha!

40. $\left(\sqrt[6]{2a^5b}\right)^7$ $(2a^5b)^{7/6}$ **41.** $\dfrac{2x}{\sqrt[3]{z^2}}$ $\dfrac{2x}{z^{2/3}}$ **42.** $\dfrac{3a}{\sqrt[5]{c^2}}$ $\dfrac{3a}{c^{2/5}}$

Write an equivalent expression with positive exponents and, if possible, simplify.

43. $x^{-1/3}$ $\dfrac{1}{x^{1/3}}$ **44.** $y^{-1/4}$ $\dfrac{1}{y^{1/4}}$

45. $(2rs)^{-3/4}$ $\dfrac{1}{(2rs)^{3/4}}$ **46.** $(5xy)^{-5/6}$ $\dfrac{1}{(5xy)^{5/6}}$

47. $\left(\dfrac{1}{8}\right)^{-2/3}$ 4 **48.** $\left(\dfrac{1}{16}\right)^{-3/4}$ 8

49. $\dfrac{1}{a^{-5/7}}$ $a^{5/7}$ **50.** $\dfrac{1}{a^{-3/5}}$ $a^{3/5}$

51. $2a^{3/4}b^{-1/2}c^{2/3}$ $\dfrac{2a^{3/4}c^{2/3}}{b^{1/2}}$ **52.** $5x^{-2/3}y^{4/5}z$ $\dfrac{5y^{4/5}z}{x^{2/3}}$

53. $2^{-1/3}x^4y^{-2/7}$ $\dfrac{x^4}{2^{1/3}y^{2/7}}$ **54.** $3^{-5/2}a^3b^{-7/3}$ $\dfrac{a^3}{3^{5/2}b^{7/3}}$

55. $\left(\dfrac{7x}{8yz}\right)^{-3/5}$ $\left(\dfrac{8yz}{7x}\right)^{3/5}$ **56.** $\left(\dfrac{2ab}{3c}\right)^{-5/6}$ $\left(\dfrac{3c}{2ab}\right)^{5/6}$

57. $\dfrac{7x}{\sqrt[3]{z}}$ $\dfrac{7x}{z^{1/3}}$ **58.** $\dfrac{6a}{\sqrt[4]{b}}$ $\dfrac{6a}{b^{1/4}}$

59. $\dfrac{5a}{3c^{-1/2}}$ $\dfrac{5ac^{1/2}}{3}$ **60.** $\dfrac{2z}{5x^{-1/3}}$ $\dfrac{2x^{1/3}z}{5}$

Graph using a graphing calculator.

61. $f(x) = \sqrt[4]{x+7}$ ⊡   **62.** $g(x) = \sqrt[5]{4-x}$ ⊡

63. $r(x) = \sqrt{3x-2}$ ⊡   **64.** $q(x) = \sqrt[6]{2x+3}$ ⊡

65. $f(x) = \sqrt[6]{x^3}$ ⊡   **66.** $g(x) = \sqrt[8]{x^2}$ ⊡

Approximate. Round to the nearest thousandth.

67. $\sqrt[5]{9}$ 1.552   **68.** $\sqrt[6]{13}$ 1.533

69. $\sqrt[4]{10}$ 1.778   **70.** $\sqrt[7]{-127}$ -1.998

71. $\sqrt[3]{(-3)^5}$ -6.240 **72.** $\sqrt[10]{(1.5)^6}$ 1.275

Use the laws of exponents to simplify. Do not use negative exponents in any answers.

73. $5^{3/4} \cdot 5^{1/8}$ $5^{7/8}$ **74.** $11^{2/3} \cdot 11^{1/2}$ $11^{7/6}$

75. $\dfrac{3^{5/8}}{3^{-1/8}}$ $3^{3/4}$ **76.** $\dfrac{8^{7/11}}{8^{-2/11}}$ $8^{9/11}$

77. $\dfrac{4.1^{-1/6}}{4.1^{-2/3}}$ $4.1^{1/2}$ **78.** $\dfrac{2.3^{-3/10}}{2.3^{-1/5}}$ $\dfrac{1}{2.3^{1/10}}$

79. $(10^{3/5})^{2/5}$ $10^{6/25}$ **80.** $(5^{5/4})^{3/7}$ $5^{15/28}$

81. $a^{2/3} \cdot a^{5/4}$ $a^{23/12}$ **82.** $x^{3/4} \cdot x^{2/3}$ $x^{17/12}$

83. $(64^{3/4})^{4/3}$ 64 **84.** $(27^{-2/3})^{3/2}$ $\dfrac{1}{27}$

85. $(m^{2/3}n^{-1/4})^{1/2}$ $\dfrac{m^{1/3}}{n^{1/8}}$ **86.** $(x^{-1/3}y^{2/5})^{1/4}$ $\dfrac{y^{1/10}}{x^{1/12}}$

Use rational exponents to simplify. Do not use fractional exponents in the final answer.

87. $\sqrt[6]{a^2}$ $\sqrt[3]{a}$ **88.** $\sqrt[6]{t^4}$ $\sqrt[3]{t^2}$

89. $\sqrt[3]{x^{15}}$ x^5 **90.** $\sqrt[4]{a^{12}}$ a^3

91. $\sqrt[6]{x^{18}}$ x^3 **92.** $\sqrt[5]{a^{10}}$ a^2

93. $\left(\sqrt[3]{ab}\right)^{15}$ a^5b^5 **94.** $\left(\sqrt[7]{xy}\right)^{14}$ x^2y^2

95. $\sqrt[8]{(3x)^2}$ $\sqrt[4]{3x}$ **96.** $\sqrt[4]{(7a)^2}$ $\sqrt{7a}$

97. $\left(\sqrt[10]{3a}\right)^5$ $\sqrt{3a}$ **98.** $\left(\sqrt[8]{2x}\right)^6$ $\sqrt[4]{8x^3}$

99. $\sqrt[4]{\sqrt{x}}$ $\sqrt[8]{x}$ **100.** $\sqrt[3]{\sqrt[6]{m}}$ $\sqrt[18]{m}$

101. $\sqrt{(ab)^6}$ a^3b^3 **102.** $\sqrt[4]{(xy)^{12}}$ x^3y^3

103. $\left(\sqrt[3]{x^2y^5}\right)^{12}$ x^8y^{20} **104.** $\left(\sqrt[5]{a^2b^4}\right)^{15}$ a^6b^{12}

105. $\sqrt[3]{\sqrt[4]{xy}}$ $\sqrt[12]{xy}$ **106.** $\sqrt[5]{\sqrt{2a}}$ $\sqrt[10]{2a}$

ᵀᵂ **107.** If $f(x) = (x+5)^{1/2}(x+7)^{-1/2}$, find the domain of f. Explain how you found your answer.

ᵀᵂ **108.** Explain why $\sqrt[3]{x^6} = x^2$ for any value of x, whereas $\sqrt{x^6} = x^3$ only when $x \geq 0$.

Skill Maintenance

Simplify. [5.2]

109. $3x(x^3 - 2x^2) + 4x^2(2x^2 + 5x)$ $11x^4 + 14x^3$

110. $5t^3(2t^2 - 4t) - 3t^4(t^2 - 6t)$ $-3t^6 + 28t^5 - 20t^4$

111. $(3a - 4b)(5a + 3b)$ $15a^2 - 11ab - 12b^2$

112. $(7x - y)^2$ $49x^2 - 14xy + y^2$

113. *Real Estate Taxes.* For homes under \$100,000, the real-estate transfer tax in Vermont is 0.5% of the selling price. Find the selling price of a home that had a transfer tax of \$467.50. [2.3]   \$93,500

114. What numbers are their own squares? [5.8]   0, 1

Synthesis

ᵀᵂ **115.** Let $f(x) = 5x^{-1/3}$. Under what condition will we have $f(x) > 0$? Why?

ᵀᵂ **116.** If $g(x) = x^{3/n}$, in what way does the domain of g depend on whether n is odd or even?

⊡ Answers to Exercises 35–39 and 61–66 can be found on pp. A-63 and A-64.

Use rational exponents to simplify.

117. $\sqrt[5]{x^2y\sqrt{xy}}$ $\sqrt[10]{x^5y^3}$

118. $\sqrt{x\sqrt[3]{x^2}}$ $\sqrt[6]{x^5}$

119. $\sqrt[4]{\sqrt[3]{8x^3y^6}}$ $\sqrt[4]{2xy^2}$

120. $\sqrt[12]{p^2 + 2pq + q^2}$ $\sqrt[6]{p + q}$

Music.   *The function* $f(x) = k2^{x/12}$ *can be used to determine the frequency, in cycles per second, of a musical note that is x half-steps above a note with frequency k.**

121. The frequency of middle C on a piano is 262 cycles per second. Find the frequency of the C that is one octave (12 half-steps) higher.   524 cycles per second

122. The frequency of concert A for a trumpet is 440 cycles per second. Find the frequency of the A that is two octaves (24 half-steps) above concert A (few trumpeters can reach this note).   1760 cycles per second

123. Show that the G that is 7 half-steps (a "perfect fifth") above middle C (see Exercise 121) has a frequency that is about 1.5 times that of middle C. $2^{7/12} \approx 1.498 \approx 1.5$

124. Show that the C sharp that is 4 half-steps (a "major third") above concert A (see Exercise 122) has a frequency that is about 25% greater than that of concert A. $2^{4/12} \approx 1.2599 \approx 1.25$, which is 25% greater than 1

125. *Road Pavement Messages.*   In a psychological study, it was determined that the proper length L of the letters of a word printed on pavement is given by

$$L = \frac{0.000169d^{2.27}}{h},$$

where d is the distance of a car from the lettering and h is the height of the eye above the surface of the road. All units are in meters. This formula says that if a person is h meters above the surface of the road and is to be able to recognize a message d meters away, that message will be the most

*This application was inspired by information provided by Dr. Homer B. Tilton of Pima Community College East.

recognizable if the length of the letters is L. Find L to the nearest tenth of a meter, given d and h.

a) $h = 1$ m, $d = 60$ m   1.8 m
b) $h = 0.9906$ m, $d = 75$ m   3.1 m
c) $h = 2.4$ m, $d = 80$ m   1.5 m
d) $h = 1.1$ m, $d = 100$ m   5.3 m

126. *Dating Fossils.*   The function $r(t) = 10^{-12}2^{-t/5700}$ expresses the ratio of carbon isotopes to carbon atoms in a fossil that is t years old. What ratio of carbon isotopes to carbon atoms would be present in a 1900-year-old bone?   About 7.937×10^{-13} to 1

127. *Physics.*   The equation $m = m_0(1 - v^2c^{-2})^{-1/2}$, developed by Albert Einstein, is used to determine the mass m of an object that is moving v meters per second and has mass m_0 before the motion begins. The constant c is the speed of light, approximately 3×10^8 m/sec. Suppose that a particle with mass 8 mg is accelerated to a speed of $\frac{9}{5} \times 10^8$ m/sec. Without using a calculator, find the new mass of the particle.   10 mg

128. Use a graphing calculator in the SIMULTANEOUS mode with the LABEL OFF format to graph

$$y_1 = x^{1/2}, \qquad y_2 = 3x^{2/5},$$
$$y_3 = x^{4/7}, \quad \text{and} \quad y_4 = \tfrac{1}{5}x^{3/4}.$$

Then, looking only at coordinates, match each graph with its equation.

Collaborative Corner

Are Equivalent Fractions Equivalent Exponents?

Focus: Functions and rational exponents

Time: 10–20 minutes

Group size: 3

Materials: Graph paper

In arithmetic, we have seen that $\frac{1}{3}$, $\frac{1}{6} \cdot 2$, and $2 \cdot \frac{1}{6}$ all represent the same number. Interestingly,

$$f(x) = x^{1/3},$$
$$g(x) = (x^{1/6})^2, \quad \text{and}$$
$$h(x) = (x^2)^{1/6}$$

represent three *different* functions.

ACTIVITY

1. Selecting a variety of values for x and using the definition of positive rational exponents, one group member should graph f, a second group member should graph g, and a third group member should graph h. Be sure to check whether negative x-values are in the domain of the function. All graphs should be done first by hand and then using a graphing calculator.

2. Compare the three graphs and check each other's work. How and why do the graphs differ?

3. Decide as a group which graph, if any, would best represent the graph of $k(x) = x^{2/6}$. Then be prepared to explain your reasoning to the entire class. (*Hint*: Study the definition of $a^{m/n}$ on p. 502 carefully.)

4. Graph $k(x) = x^{2/6}$ using a graphing calculator. Is the graph that is shown correct?

7.3

Multiplying Radical Expressions ■ Simplifying by Factoring ■ Multiplying and Simplifying

Multiplying Radical Expressions

Multiplying Radical Expressions

Note that $\sqrt{4}\,\sqrt{25} = 2 \cdot 5 = 10$. Also $\sqrt{4 \cdot 25} = \sqrt{100} = 10$. Likewise,

$$\sqrt[3]{27}\,\sqrt[3]{8} = 3 \cdot 2 = 6 \quad \text{and} \quad \sqrt[3]{27 \cdot 8} = \sqrt[3]{216} = 6.$$

These examples suggest the following.

TEACHING TIP

When they are simplifying radicals, it is useful for students to know these symbolic equivalencies (for $x \geq 0$):

(a) $\sqrt{x^2} = x$;
(b) $(\sqrt{x})^2 = x$;
(c) $\sqrt{x}\,\sqrt{x} = x$.

The Product Rule for Radicals

For any real numbers $\sqrt[n]{a}$ and $\sqrt[n]{b}$,

$$\sqrt[n]{a} \cdot \sqrt[n]{b} = \sqrt[n]{a \cdot b}.$$

(To multiply, when the indices match, multiply the radicands.)

Fractional exponents can be used to derive this rule:

$$\sqrt[n]{a} \cdot \sqrt[n]{b} = a^{1/n} \cdot b^{1/n} = (a \cdot b)^{1/n} = \sqrt[n]{a \cdot b}.$$

EXAMPLE 1   Multiply.

a) $\sqrt{3} \cdot \sqrt{5}$

b) $\sqrt{x + 3}\,\sqrt{x - 3}$

c) $\sqrt[3]{4} \cdot \sqrt[3]{5}$

d) $\sqrt[4]{\dfrac{y}{5}} \cdot \sqrt[4]{\dfrac{7}{x}}$

Solution

a) When no index is written, roots are understood to be square roots with an unwritten index of two. We apply the product rule:

$$\sqrt{3} \cdot \sqrt{5} = \sqrt{3 \cdot 5}$$
$$= \sqrt{15}.$$

b) $\sqrt{x + 3}\,\sqrt{x - 3} = \sqrt{(x + 3)(x - 3)}$ The product of two square roots is the square root of the product.

$$= \sqrt{x^2 - 9}$$

CAUTION!
$\sqrt{x^2 - 9} \neq \sqrt{x^2} - \sqrt{9}.$

c) Both $\sqrt[3]{4}$ and $\sqrt[3]{5}$ have indices of three, so to multiply we can use the product rule:

$$\sqrt[3]{4} \cdot \sqrt[3]{5} = \sqrt[3]{4 \cdot 5} = \sqrt[3]{20}.$$

d) $\sqrt[4]{\dfrac{y}{5}} \cdot \sqrt[4]{\dfrac{7}{x}} = \sqrt[4]{\dfrac{y}{5} \cdot \dfrac{7}{x}} = \sqrt[4]{\dfrac{7y}{5x}}$ In Section 7.4, we discuss other ways to write answers like this.

Important: The product rule for radicals applies only when radicals have the same index.

Connecting the Concepts

INTERPRETING GRAPHS: DOMAINS OF RADICAL FUNCTIONS

Although, as we saw in Example 1(b),

$$\sqrt{x + 3}\,\sqrt{x - 3} = \sqrt{x^2 - 9},$$

it is not true that

$$f(x) = \sqrt{x + 3}\,\sqrt{x - 3} \quad \text{and} \quad g(x) = \sqrt{x^2 - 9}$$

represent the same function. This is because the domains of the functions are not the same. As you can see from the graphs of the functions shown below, the domain of f is $[3, \infty)$, whereas the domain of g is $(-\infty, -3] \cup [3, \infty)$.

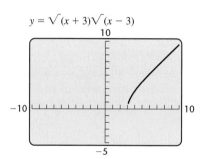

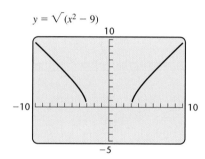

$y = \sqrt{(x + 3)}\sqrt{(x - 3)}$

$y = \sqrt{(x^2 - 9)}$

Note that the graphs do coincide where the domains coincide. We say that the expressions are equivalent because they represent the same number for all possible replacements for x in *both* expressions.

Simplifying by Factoring

An integer p is a *perfect square* if there exists a rational number q for which $q^2 = p$. We say that p is a *perfect cube* if $q^3 = p$ for some rational number q. In general, p is a *perfect nth power* if $q^n = p$ for some rational number q. The product rule allows us to simplify $\sqrt[n]{ab}$ when a or b is a perfect nth power.

Using the Product Rule to Simplify

$$\sqrt[n]{ab} = \sqrt[n]{a} \cdot \sqrt[n]{b}.$$

To illustrate, suppose we wish to simplify $\sqrt{20}$. Since this is a *square* root, we check to see if there is a factor of 20 that is a perfect square. There is one, 4, so we express 20 as $4 \cdot 5$ and use the product rule:

$$\sqrt{20} = \sqrt{4 \cdot 5} \qquad \text{Factoring the radicand (4 is a perfect square)}$$
$$= \sqrt{4} \cdot \sqrt{5} \qquad \text{Factoring into two radicals}$$
$$= 2\sqrt{5}. \qquad \text{Taking the square root of 4}$$

To Simplify a Radical Expression with Index *n* by Factoring

1. Express the radicand as a product in which one factor is the largest perfect *n*th power possible.
2. Take the *n*th root of each factor.
3. Simplification is complete when no radicand has a factor that is a perfect *n*th power.

EXAMPLE 2   Simplify by factoring: **(a)** $\sqrt{200}$; **(b)** $\sqrt[3]{32}$; **(c)** $\sqrt[4]{48}$; **(d)** $\sqrt{18x^2y}$.

Solution

a) $\sqrt{200} = \sqrt{100 \cdot 2} = \sqrt{100} \cdot \sqrt{2} = 10\sqrt{2}$ This is the largest perfect-square factor of 200.

b) $\sqrt[3]{32} = \sqrt[3]{8 \cdot 4} = \sqrt[3]{8} \cdot \sqrt[3]{4} = 2\sqrt[3]{4}$ This is the largest perfect-cube (third-power) factor of 32.

c) $\sqrt[4]{48} = \sqrt[4]{16 \cdot 3} = \sqrt[4]{16} \cdot \sqrt[4]{3} = 2\sqrt[4]{3}$ This is the largest fourth-power factor of 48.

d) $\sqrt{18x^2y} = \sqrt{9x^2 \cdot 2y}$ $9x^2$ is a perfect square.
$\qquad\quad\;\; = \sqrt{9x^2} \cdot \sqrt{2y}$ Factoring into two radicals
$\qquad\quad\;\; = |3x|\sqrt{2y}$, or $3|x|\sqrt{2y}$ Taking the square root of $9x^2$

EXAMPLE 3   If $f(x) = \sqrt{3x^2 - 6x + 3}$, find a simplified form for $f(x)$.

Solution

$$f(x) = \sqrt{3x^2 - 6x + 3}$$
$$\left. \begin{array}{l} = \sqrt{3(x^2 - 2x + 1)} \\ = \sqrt{(x-1)^2 \cdot 3} \end{array} \right\} \quad \begin{array}{l} \text{Factoring the radicand; } x^2 - 2x + 1 \\ \text{is a perfect square.} \end{array}$$
$$= \sqrt{(x-1)^2} \cdot \sqrt{3} \qquad \text{Factoring into two radicals}$$
$$= |x-1|\sqrt{3} \qquad \text{Taking the square root of } (x-1)^2$$

We can check Example 3 by graphing $y_1 = \sqrt{3x^2 - 6x + 3}$ and $y_2 = |x - 1|\sqrt{3}$, as shown in the graph on the left below.

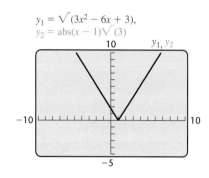

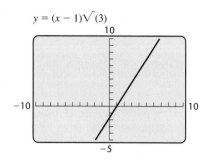

It appears that the graphs coincide, and scrolling through a table of values would show us that this is indeed the case. Note from the graph on the right above that the absolute-value sign is important since the graph of $y = (x - 1)\sqrt{3}$ is different from the graph of $y = \sqrt{3x^2 - 6x + 3}$.

In many situations that do not involve functions, it is safe to assume that no radicands were formed by raising negative quantities to even powers. We now make this assumption and thus discontinue the use of absolute-value notation when taking even roots unless functions are involved.

EXAMPLE 4 Simplify: **(a)** $\sqrt{x^7 y^{11} z^9}$; **(b)** $\sqrt[3]{16a^7 b^{14}}$.

Solution

a) There are many ways to factor $x^7 y^{11} z^9$. Because of the square root (index of 2), we identify the largest exponents that are multiples of 2:

$$\sqrt{x^7 y^{11} z^9} = \sqrt{x^6 \cdot x \cdot y^{10} \cdot y \cdot z^8 \cdot z} \qquad \text{Using the largest even powers of } x, y, \text{ and } z$$

$$= \sqrt{x^6} \sqrt{y^{10}} \sqrt{z^8} \sqrt{xyz} \qquad \text{Factoring into several radicals}$$

$$= x^{6/2} \; y^{10/2} \; z^{8/2} \sqrt{xyz} \qquad \text{Converting to fractional exponents}$$

$$= x^3 y^5 z^4 \sqrt{xyz}.$$

Check: $\left(x^3 y^5 z^4 \sqrt{xyz}\right)^2 = (x^3)^2 (y^5)^2 (z^4)^2 \left(\sqrt{xyz}\right)^2$
$$= x^6 \cdot y^{10} \cdot z^8 \cdot xyz = x^7 y^{11} z^9$$

Our check shows that $x^3 y^5 z^4 \sqrt{xyz}$ is the square root of $x^7 y^{11} z^9$.

TEACHING TIP

If students have difficulty simplifying, suggest they write the prime factorization of the radicand and group factors. For example, $\sqrt[3]{48x^7}$ can be written as

$$\sqrt[3]{2 \cdot 2 \cdot 2 \cdot 2 \cdot 3 \cdot x \cdot x \cdot x \cdot x \cdot x \cdot x \cdot x}.$$

Ask students to find groups of 3 identical factors (since the index is 3) and point out that each group represents a perfect-cube factor of the radicand.

b) There are many ways to factor $16a^7b^{14}$. Because of the cube root (index of 3), we identify factors with the largest exponents that are multiples of 3:

$$\sqrt[3]{16a^7b^{14}} = \sqrt[3]{8 \cdot 2 \cdot a^6 \cdot a \cdot b^{12} \cdot b^2}$$ Using the largest perfect-cube factors

$$= \sqrt[3]{8}\,\sqrt[3]{a^6}\,\sqrt[3]{b^{12}}\,\sqrt[3]{2ab^2}$$ Factoring into several radicals

$$= 2\quad a^{6/3}\quad b^{12/3}\sqrt[3]{2ab^2}$$ Converting to fractional exponents

$$= 2a^2b^4\sqrt[3]{2ab^2}$$

Check: $\left(2a^2b^4\sqrt[3]{2ab^2}\right)^3 = 2^3(a^2)^3(b^4)^3\left(\sqrt[3]{2ab^2}\right)^3$

$$= 8 \cdot a^6 \cdot b^{12} \cdot 2ab^2 = 16a^7b^{14}$$

We see that $2a^2b^4\sqrt[3]{2ab^2}$ is the cube root of $16a^7b^{14}$.

Example 4 demonstrates the following.

> To simplify an nth root, identify factors in the radicand with exponents that are multiples of n.

Multiplying and Simplifying

We have used the product rule for radicals to find products and also to simplify radical expressions. For some radical expressions, it is possible to do both: First find a product and then simplify.

EXAMPLE 5   Multiply and simplify.

a) $\sqrt{15}\,\sqrt{6}$ **b)** $3\sqrt[3]{25} \cdot 2\sqrt[3]{5}$ **c)** $\sqrt[4]{8x^3y^5}\,\sqrt[4]{4x^2y^3}$

Solution

a) $\sqrt{15}\,\sqrt{6} = \sqrt{15 \cdot 6}$ Multiplying radicands
$$= \sqrt{90} = \sqrt{9 \cdot 10}$$ 9 is a perfect square.
$$= 3\sqrt{10}$$

b) $3\sqrt[3]{25} \cdot 2\sqrt[3]{5} = 3 \cdot 2 \cdot \sqrt[3]{25 \cdot 5}$ Using a commutative law; multiplying radicands
$$= 6 \cdot \sqrt[3]{125}$$ 125 is a perfect cube.
$$= 6 \cdot 5, \text{ or } 30$$

c) $\sqrt[4]{8x^3y^5}\,\sqrt[4]{4x^2y^3} = \sqrt[4]{32x^5y^8}$ Multiplying radicands
$$= \sqrt[4]{16x^4y^8 \cdot 2x}$$ Identifying perfect fourth-power factors
$$= \sqrt[4]{16}\,\sqrt[4]{x^4}\,\sqrt[4]{y^8}\,\sqrt[4]{2x}$$ Factoring into radicals
$$= 2xy^2\sqrt[4]{2x}$$ Finding the fourth roots; assume $x \geq 0$.

The checks are left to the student.

7.3

Exercise Set

Multiply.

1. $\sqrt{10}\ \sqrt{7}$ $\sqrt{70}$

2. $\sqrt{5}\ \sqrt{7}$ $\sqrt{35}$

3. $\sqrt[3]{2}\ \sqrt[3]{5}$ $\sqrt[3]{10}$

4. $\sqrt[3]{7}\ \sqrt[3]{2}$ $\sqrt[3]{14}$

5. $\sqrt[4]{8}\ \sqrt[4]{9}$ $\sqrt[4]{72}$

6. $\sqrt[4]{6}\ \sqrt[4]{3}$ $\sqrt[4]{18}$

7. $\sqrt{5a}\ \sqrt{6b}$ $\sqrt{30ab}$

8. $\sqrt{2x}\ \sqrt{13y}$ $\sqrt{26xy}$

9. $\sqrt[5]{9t^2}\ \sqrt[5]{2t}$ $\sqrt[5]{18t^3}$

10. $\sqrt[5]{8y^3}\ \sqrt[5]{10y}$ $\sqrt[5]{80y^4}$

11. $\sqrt{x-a}\ \sqrt{x+a}$ $\sqrt{x^2-a^2}$

12. $\sqrt{y-b}\ \sqrt{y+b}$ $\sqrt{y^2-b^2}$

13. $\sqrt[3]{0.5x}\ \sqrt[3]{0.2x}$ $\sqrt[3]{0.1x^2}$

14. $\sqrt[3]{0.7y}\ \sqrt[3]{0.3y}$ $\sqrt[3]{0.21y^2}$ Aha!

15. $\sqrt[4]{x-1}\ \sqrt[4]{x^2+x+1}$ $\sqrt[4]{x^3-1}$

16. $\sqrt[5]{x-2}\ \sqrt[5]{(x-2)^2}$ ▫

17. $\sqrt{\dfrac{x}{6}}\ \sqrt{\dfrac{7}{y}}$ $\sqrt{\dfrac{7x}{6y}}$

18. $\sqrt{\dfrac{7}{t}}\ \sqrt{\dfrac{s}{11}}$ $\sqrt{\dfrac{7s}{11t}}$

19. $\sqrt[7]{\dfrac{x-3}{4}}\ \sqrt[7]{\dfrac{5}{x+2}}$ $\sqrt[7]{\dfrac{5x-15}{4x+8}}$

20. $\sqrt[6]{\dfrac{a}{b-2}}\ \sqrt[6]{\dfrac{3}{b+2}}$ $\sqrt[6]{\dfrac{3a}{b^2-4}}$

Simplify by factoring.

21. $\sqrt{50}$ $5\sqrt{2}$

22. $\sqrt{27}$ $3\sqrt{3}$

23. $\sqrt{28}$ $2\sqrt{7}$

24. $\sqrt{45}$ $3\sqrt{5}$

25. $\sqrt{8}$ $2\sqrt{2}$

26. $\sqrt{18}$ $3\sqrt{2}$

27. $\sqrt{198}$ $3\sqrt{22}$

28. $\sqrt{325}$ $5\sqrt{13}$

29. $\sqrt{36a^4b}$ $6a^2\sqrt{b}$

30. $\sqrt{175y^8}$ $5y^4\sqrt{7}$

31. $\sqrt[3]{8x^3y^2}$ $2x\sqrt[3]{y^2}$

32. $\sqrt[3]{27ab^6}$ $3b^2\sqrt[3]{a}$

33. $\sqrt[3]{-16x^6}$ $-2x^2\sqrt[3]{2}$

34. $\sqrt[3]{-32a^6}$ $-2a^2\sqrt[3]{4}$

Find a simplified form of $f(x)$. Assume that x can be any real number.

35. $f(x) = \sqrt[3]{125x^5}$ ▫

36. $f(x) = \sqrt[3]{16x^6}$ ▫

37. $f(x) = \sqrt{49(x-3)^2}$ ▫

38. $f(x) = \sqrt{81(x-1)^2}$ ▫

39. $f(x) = \sqrt{5x^2 - 10x + 5}$ $f(x) = |x-1|\sqrt{5}$

40. $f(x) = \sqrt{2x^2 + 8x + 8}$ $f(x) = |x+2|\sqrt{2}$

Simplify. Assume that no radicands were formed by raising negative numbers to even powers.

41. $\sqrt{a^3b^4}$ $ab^2\sqrt{a}$

42. $\sqrt{x^6y^9}$ $x^3y^4\sqrt{y}$

43. $\sqrt[3]{x^5y^6z^{10}}$ $xy^2z^3\sqrt[3]{x^2z}$

44. $\sqrt[3]{a^6b^7c^{13}}$ $a^2b^2c^4\sqrt[3]{bc}$

45. $\sqrt[5]{-32a^7b^{11}}$ $-2ab^2\sqrt[5]{a^2b}$

46. $\sqrt[4]{16x^5y^{11}}$ $2xy^2\sqrt[4]{xy^3}$

47. $\sqrt[5]{a^6b^8c^9}$ $abc\sqrt[5]{ab^3c^4}$

48. $\sqrt[5]{x^{13}y^8z^{17}}$ $x^2yz^3\sqrt[5]{x^3y^3z^2}$

49. $\sqrt[4]{810x^9}$ $3x^2\sqrt[4]{10x}$

50. $\sqrt[3]{-80a^{14}}$ $-2a^4\sqrt[3]{10a^2}$

Multiply and simplify.

51. $\sqrt{15}\ \sqrt{5}$ $5\sqrt{3}$

52. $\sqrt{6}\ \sqrt{3}$ $3\sqrt{2}$

53. $\sqrt{10}\ \sqrt{14}$ $2\sqrt{35}$

54. $\sqrt{15}\ \sqrt{21}$ $3\sqrt{35}$

55. $\sqrt[3]{2}\ \sqrt[3]{4}$ 2

56. $\sqrt[3]{9}\ \sqrt[3]{3}$ 3

57. $\sqrt{18a^3}\ \sqrt{18a^3}$ $18a^3$

58. $\sqrt{75x^7}\ \sqrt{75x^7}$ $75x^7$

59. $\sqrt[3]{5a^2}\ \sqrt[3]{2a}$ $a\sqrt[3]{10}$

60. $\sqrt[3]{7x}\ \sqrt[3]{3x^2}$ $x\sqrt[3]{21}$

61. $\sqrt{3x^5}\ \sqrt{15x^2}$ $3x^3\sqrt{5x}$

62. $\sqrt{5a^7}\ \sqrt{15a^3}$ $5a^5\sqrt{3}$

63. $\sqrt[3]{s^2t^4}\ \sqrt[3]{s^4t^6}$ $s^2t^3\sqrt[3]{t}$

64. $\sqrt[3]{x^2y^4}\ \sqrt[3]{x^2y^6}$ $xy^3\sqrt[3]{xy}$

65. $\sqrt[3]{(x+5)^2}\ \sqrt[3]{(x+5)^4}$

66. $\sqrt[3]{(a-b)^5}\ \sqrt[3]{(a-b)^7}$

67. $\sqrt[4]{12a^3b^7}\ \sqrt[4]{4a^2b^5}$ ▫

68. $\sqrt[4]{9x^7y^2}\ \sqrt[4]{9x^2y^9}$ ▫

69. $\sqrt[5]{x^3(y+z)^4}\ \sqrt[5]{x^3(y+z)^6}$ $x(y+z)^2\sqrt[5]{x}$

70. $\sqrt[5]{a^3(b-c)^4}\ \sqrt[5]{a^7(b-c)^4}$ $a^2(b-c)\sqrt[5]{(b-c)^3}$

TW **71.** Why do we need to know how to multiply radical expressions before learning how to simplify radical expressions?

TW **72.** Why is it incorrect to say that, in general, $\sqrt{x^2} = x$?

Skill Maintenance

Perform the indicated operation and, if possible, simplify. [6.2]

73. $\dfrac{3x}{16y} + \dfrac{5y}{64x}$ $\dfrac{12x^2 + 5y^2}{64xy}$

74. $\dfrac{2}{a^3b^4} + \dfrac{6}{a^4b}$ $\dfrac{2a + 6b^3}{a^4b^4}$

75. $\dfrac{4}{x^2-9} - \dfrac{7}{2x-6}$ $\dfrac{-7x-13}{2(x-3)(x+3)}$

76. $\dfrac{8}{x^2-25} - \dfrac{3}{2x-10}$ $\dfrac{-3x+1}{2(x-5)(x+5)}$

Simplify. [1.4]

77. $\dfrac{9a^4b^7}{3a^2b^5}$ $3a^2b^2$

78. $\dfrac{12a^2b^7}{4ab^2}$ $3ab^5$

▫ Answers to Exercises 16, 35–38, and 65–68 can be found on p. A-64.

Synthesis

TW **79.** Explain why it is true that
$$\sqrt[n]{ab} = \sqrt[n]{a} \cdot \sqrt[n]{b}.$$

TW **80.** Is the equation $\sqrt{(2x + 3)^8} = (2x + 3)^4$ always, sometimes, or never true? Why?

81. *Speed of a Skidding Car.*   Police can estimate the speed at which a car was traveling by measuring its skid marks. The function
$$r(L) = 2\sqrt{5L}$$

can be used, where L is the length of a skid mark, in feet, and $r(L)$ is the speed, in miles per hour. Find the exact speed and an estimate (to the nearest tenth mile per hour) for the speed of a car that left skid marks **(a)** 20 ft long; **(b)** 70 ft long; **(c)** 90 ft long.
(a) 20 mph; **(b)** 37.4 mph; **(c)** 42.4 mph

82. *Wind Chill Temperature.*   When the temperature is T degrees Fahrenheit and the wind speed is v meters per second, the *wind chill temperature*, T_W, is the temperature (with no wind) that it feels like. Here is a formula for finding wind chill temperature:
$$T_W = 91.4 - (91.4 - T)\left(0.478 + 0.301\sqrt{v} - 0.02v\right).$$

Estimate the wind chill temperature (to the nearest tenth of a degree) for the given actual temperatures and wind speeds. (Note that this formula can be used only when the wind speed is above 4 mph.)

a) $T = 30°F$, $v = 20$ mph   4.1°F
b) $T = 20°F$, $v = 20$ mph   −10.3°F
c) $T = 0°F$, $v = 35$ mph   −51.1°F
d) $T = −16°F$, $v = 40$ mph   −78.5°F

Simplify. Assume that all variables are nonnegative.

83. $\left(\sqrt{r^3t}\right)^7$ $r^{10}t^3\sqrt{rt}$

84. $\left(\sqrt[3]{25x^4}\right)^4$ $25x^5\sqrt[3]{25x}$

85. $\left(\sqrt[3]{a^2b^4}\right)^5$ $a^3b^6\sqrt[3]{ab^2}$

86. $\left(\sqrt{a^3b^5}\right)^7$ $a^{10}b^{17}\sqrt{ab}$

Draw and compare the graphs of each group of equations. Graph by hand or using a calculator.

87. $f(x) = \sqrt{x^2 - 2x + 1}$,
$g(x) = x - 1$,
$h(x) = |x - 1|$

88. $f(x) = \sqrt{x^2 + 2x + 1}$,
$g(x) = x + 1$,
$h(x) = |x + 1|$ $\{x \mid x \le -1 \text{ or } x \ge 4\}$, or $(-\infty, -1] \cup [4, \infty)$

89. If $f(x) = \sqrt{t^2 - 3t - 4}$, what is the domain of f?

90. What is the domain of g, if $g(x) = \sqrt{x^2 - 6x + 8}$?
$\{x \mid x \le 2 \text{ or } x \ge 4\}$, or $(-\infty, 2] \cup [4, \infty)$

Solve.

91. $\sqrt[3]{5x^{k+1}} \sqrt[3]{25x^k} = 5x^7$, for k 10

92. $\sqrt[5]{4a^{3k+2}} \sqrt[5]{8a^{6-k}} = 2a^4$, for k 6

TW **93.** Rony is puzzled. When he uses a graphing calculator to graph $y = \sqrt{x} \cdot \sqrt{x}$, he gets the following screen. Explain why Rony did not get the complete line $y = x$.

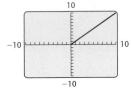

87. $f(x) = h(x); f(x) \ne g(x)$ **88.** $f(x) = h(x); f(x) \ne g(x)$

7.4 Dividing and Simplifying ▪ Rationalizing Denominators and Numerators

Dividing Radical Expressions

Study Tip

It is always best to study for a final exam over a period of at least two weeks. If you have only one or two days of study time, however, begin by studying the formulas, problems, properties, and procedures in each chapter Summary and Review. Then do the exercises in the Cumulative Reviews. Make sure to attend a review session if one is offered.

Dividing and Simplifying

Just as the root of a product can be expressed as the product of two roots, the root of a quotient can be expressed as the quotient of two roots. For example,

$$\sqrt[3]{\frac{27}{8}} = \frac{3}{2} \quad \text{and} \quad \frac{\sqrt[3]{27}}{\sqrt[3]{8}} = \frac{3}{2}.$$

This example suggests the following.

The Quotient Rule for Radicals

For any real numbers $\sqrt[n]{a}$ and $\sqrt[n]{b}$, $b \neq 0$,

$$\sqrt[n]{\frac{a}{b}} = \frac{\sqrt[n]{a}}{\sqrt[n]{b}}.$$

Remember that an nth root is simplified when its radicand has no factors that are perfect nth powers. Recall too that we assume that no radicands represent negative quantities raised to an even power.

EXAMPLE 1 Simplify by taking the roots of the numerator and the denominator.

a) $\sqrt[3]{\frac{27}{125}}$ **b)** $\sqrt{\frac{25}{y^2}}$

Solution

a) $\sqrt[3]{\frac{27}{125}} = \frac{\sqrt[3]{27}}{\sqrt[3]{125}} = \frac{3}{5}$ Taking the cube roots of the numerator and the denominator

b) $\sqrt{\frac{25}{y^2}} = \frac{\sqrt{25}}{\sqrt{y^2}} = \frac{5}{y}$ Taking the square roots of the numerator and the denominator. Assume $y > 0$.

TEACHING TIP
Explain that the student should use the form that makes the simplifying easiest.

As in Section 7.3, any radical expressions appearing in the answers should be simplified as much as possible.

EXAMPLE 2 Simplify: **(a)** $\sqrt{\frac{16x^3}{y^8}}$; **(b)** $\sqrt[3]{\frac{27y^{14}}{8x^3}}$.

Solution

a) $\sqrt{\dfrac{16x^3}{y^8}} = \dfrac{\sqrt{16x^3}}{\sqrt{y^8}}$

$\qquad = \dfrac{\sqrt{16x^2 \cdot x}}{\sqrt{y^8}} = \dfrac{4x\sqrt{x}}{y^4}$ **Simplifying the numerator and the denominator**

b) $\sqrt[3]{\dfrac{27y^{14}}{8x^3}} = \dfrac{\sqrt[3]{27y^{14}}}{\sqrt[3]{8x^3}}$

$\qquad = \dfrac{\sqrt[3]{27y^{12}y^2}}{\sqrt[3]{8x^3}} = \dfrac{\sqrt[3]{27y^{12}}\,\sqrt[3]{y^2}}{\sqrt[3]{8x^3}} = \dfrac{3y^4\sqrt[3]{y^2}}{2x}$ **Simplifying the numerator and the denominator**

If we read from right to left, the quotient rule tells us that to divide two radical expressions that have the same index, we can divide the radicands.

EXAMPLE 3   Divide and, if possible, simplify.

a) $\dfrac{\sqrt{80}}{\sqrt{5}}$ **b)** $\dfrac{5\sqrt[3]{32}}{\sqrt[3]{2}}$ **c)** $\dfrac{\sqrt{72xy}}{2\sqrt{2}}$ **d)** $\dfrac{\sqrt[4]{18a^9b^5}}{\sqrt[4]{3b}}$

Solution

a) $\dfrac{\sqrt{80}}{\sqrt{5}} = \sqrt{\dfrac{80}{5}} = \sqrt{16} = 4$ **Because the indices match, we can divide the radicands.**

b) $\dfrac{5\sqrt[3]{32}}{\sqrt[3]{2}} = 5\sqrt[3]{\dfrac{32}{2}} = 5\sqrt[3]{16}$

$\qquad = 5\sqrt[3]{8 \cdot 2}$

$\qquad = 5\sqrt[3]{8}\,\sqrt[3]{2} = 5 \cdot 2\sqrt[3]{2}$

$\qquad = 10\sqrt[3]{2}$

c) $\dfrac{\sqrt{72xy}}{2\sqrt{2}} = \dfrac{1}{2}\,\dfrac{\sqrt{72xy}}{\sqrt{2}} = \dfrac{1}{2}\sqrt{\dfrac{72xy}{2}} = \dfrac{1}{2}\sqrt{36xy}$ **Because the indices match, we can divide the radicands.**

$\qquad = \dfrac{1}{2}\sqrt{36}\,\sqrt{xy} = \dfrac{1}{2} \cdot 6\sqrt{xy}$

$\qquad = 3\sqrt{xy}$

d) $\dfrac{\sqrt[4]{18a^9b^5}}{\sqrt[4]{3b}} = \sqrt[4]{\dfrac{18a^9b^5}{3b}}$

$\qquad = \sqrt[4]{6a^9b^4} = \sqrt[4]{a^8b^4}\,\sqrt[4]{6a}$ **Note that 8 is the largest power less than 9 that is a multiple of the index 4.**

$\qquad = a^2b\sqrt[4]{6a}$ *Partial check:* $(a^2b)^4 = a^8b^4$

Rationalizing Denominators and Numerators

When a radical expression appears in a denominator, it can be useful to find an equivalent expression in which the denominator no longer contains a radical.* The procedure for finding such an expression is called **rationalizing the denominator**. We carry this out by multiplying by 1 in either of two ways.

One way is to multiply by 1 *under* the radical to make the denominator of the radicand a perfect power.

EXAMPLE 4   Rationalize each denominator.

a) $\sqrt{\dfrac{7}{3}}$

b) $\sqrt[3]{\dfrac{5}{16}}$

Solution

a) We multiply by 1 under the radical, using $\frac{3}{3}$. We do this so that the denominator of the radicand will be a perfect square:

$$\sqrt{\frac{7}{3}} = \sqrt{\frac{7}{3} \cdot \frac{3}{3}} \qquad \text{Multiplying by 1 under the radical}$$

$$= \sqrt{\frac{21}{9}} \qquad \text{The denominator, 9, is now a perfect square.}$$

$$= \frac{\sqrt{21}}{\sqrt{9}} \qquad \text{Using the quotient rule for radicals}$$

$$= \frac{\sqrt{21}}{3}.$$

b) Note that $16 = 4^2$. Thus, to make the denominator a perfect cube, we multiply under the radical by $\frac{4}{4}$:

$$\sqrt[3]{\frac{5}{16}} = \sqrt[3]{\frac{5}{4 \cdot 4} \cdot \frac{4}{4}} \qquad \begin{array}{l}\text{Since the index is 3, we need 3 identical} \\ \text{factors in the denominator.}\end{array}$$

$$= \sqrt[3]{\frac{20}{4^3}} \qquad \text{The denominator is now a perfect cube.}$$

$$= \frac{\sqrt[3]{20}}{\sqrt[3]{4^3}}$$

$$= \frac{\sqrt[3]{20}}{4}.$$

Another way to rationalize a denominator is to multiply by 1 *outside* the radical.

EXAMPLE 5   Rationalize each denominator.

a) $\sqrt{\dfrac{4}{5b}}$

b) $\dfrac{\sqrt[3]{a}}{\sqrt[3]{9x}}$

c) $\dfrac{3x}{\sqrt[5]{2x^2y^3}}$

*See Exercise 65 on p. 521.

Solution

a) We rewrite the expression as a quotient of two radicals. Then we simplify and multiply by 1:

$$\sqrt{\frac{4}{5b}} = \frac{\sqrt{4}}{\sqrt{5b}} = \frac{2}{\sqrt{5b}} \qquad \text{We assume } b > 0.$$

$$= \frac{2}{\sqrt{5b}} \cdot \frac{\sqrt{5b}}{\sqrt{5b}} \qquad \textbf{Multiplying by 1}$$

$$= \frac{2\sqrt{5b}}{(\sqrt{5b})^2} \qquad \textbf{Try to do this step mentally.}$$

$$= \frac{2\sqrt{5b}}{5b}.$$

b) To rationalize the denominator $\sqrt[3]{9x}$, note that $9x$ is $3 \cdot 3 \cdot x$. In order for this radicand to be a cube, we need another factor of 3 and two more factors of x. Thus we multiply by 1, using $\sqrt[3]{3x^2}/\sqrt[3]{3x^2}$:

$$\frac{\sqrt[3]{a}}{\sqrt[3]{9x}} = \frac{\sqrt[3]{a}}{\sqrt[3]{9x}} \cdot \frac{\sqrt[3]{3x^2}}{\sqrt[3]{3x^2}} \qquad \textbf{Multiplying by 1}$$

$$= \frac{\sqrt[3]{3ax^2}}{\sqrt[3]{27x^3}} \longleftarrow \quad \textbf{This radicand is now a perfect cube.}$$

$$= \frac{\sqrt[3]{3ax^2}}{3x}.$$

c) To change the radicand $2x^2y^3$ into a perfect fifth power, we need four more factors of 2, three more factors of x, and two more factors of y. Thus we multiply by 1, using $\sqrt[5]{2^4x^3y^2}/\sqrt[5]{2^4x^3y^2}$, or $\sqrt[5]{16x^3y^2}/\sqrt[5]{16x^3y^2}$:

$$\frac{3x}{\sqrt[5]{2x^2y^3}} = \frac{3x}{\sqrt[5]{2x^2y^3}} \cdot \frac{\sqrt[5]{16x^3y^2}}{\sqrt[5]{16x^3y^2}} \qquad \textbf{Multiplying by 1}$$

$$= \frac{3x\sqrt[5]{16x^3y^2}}{\sqrt[5]{32x^5y^5}} \longleftarrow \quad \textbf{This radicand is now a perfect fifth power.}$$

$$= \frac{3x\sqrt[5]{16x^3y^2}}{2xy} = \frac{3\sqrt[5]{16x^3y^2}}{2y}. \qquad \textbf{Always simplify if possible.}$$

Sometimes in calculus it is necessary to rationalize a numerator. To do so, we multiply by 1 to make the radicand in the *numerator* a perfect power.

EXAMPLE 6   Rationalize each numerator.

a) $\sqrt{\dfrac{7}{5}}$ 　　　　　　　　**b)** $\dfrac{\sqrt[3]{4a^2}}{\sqrt[3]{5b}}$

TEACHING TIP

Emphasize the importance of reading instructions carefully to see whether the numerator or the denominator is to be rationalized.

9. $\dfrac{3a\sqrt[3]{a}}{2b}$ **12.** $\dfrac{3x}{y^2 z}$ **13.** $\dfrac{ab^2}{c^2}\sqrt[4]{\dfrac{a}{c^2}}$

14. $\dfrac{x^2 y^3}{z}\sqrt[4]{\dfrac{x}{z^2}}$ **15.** $\dfrac{2x}{y^2}\sqrt[5]{\dfrac{x}{y}}$

16. $\dfrac{3a}{b^2}\sqrt[5]{\dfrac{a^4}{b^3}}$ **17.** $\dfrac{xy}{z^2}\sqrt[6]{\dfrac{y^2}{z^3}}$

18. $\dfrac{ab^2}{c^2}\sqrt[6]{\dfrac{a^3}{c}}$

Solution

a) $\sqrt{\dfrac{7}{5}} = \sqrt{\dfrac{7}{5}\cdot\dfrac{7}{7}}$ **Multiplying by 1 under the radical. We also could have multiplied by $\sqrt{7}/\sqrt{7}$ outside the radical.**

$= \sqrt{\dfrac{49}{35}}$ **The numerator is now a perfect square.**

$= \dfrac{\sqrt{49}}{\sqrt{35}}$ **Using the quotient rule for radicals**

$= \dfrac{7}{\sqrt{35}}$

b) $\dfrac{\sqrt[3]{4a^2}}{\sqrt[3]{5b}} = \dfrac{\sqrt[3]{4a^2}}{\sqrt[3]{5b}}\cdot\dfrac{\sqrt[3]{2a}}{\sqrt[3]{2a}}$ **Multiplying by 1**

$= \dfrac{\sqrt[3]{8a^3}}{\sqrt[3]{10ba}}$ ←——— **This radicand is now a perfect cube.**

$= \dfrac{2a}{\sqrt[3]{10ab}}$

In Section 7.5, we will discuss rationalizing denominators and numerators in which two terms appear.

7.4

Exercise Set

Simplify by taking the roots of the numerator and the denominator. Assume all variables represent positive numbers.

1. $\sqrt{\dfrac{25}{36}}$ $\dfrac{5}{6}$ **2.** $\sqrt{\dfrac{100}{81}}$ $\dfrac{10}{9}$ **3.** $\sqrt[3]{\dfrac{64}{27}}$ $\dfrac{4}{3}$

4. $\sqrt[3]{\dfrac{343}{1000}}$ $\dfrac{7}{10}$ **5.** $\sqrt{\dfrac{49}{y^2}}$ $\dfrac{7}{y}$ **6.** $\sqrt{\dfrac{121}{x^2}}$ $\dfrac{11}{x}$

7. $\sqrt{\dfrac{25y^3}{x^4}}$ $\dfrac{5y\sqrt{y}}{x^2}$ **8.** $\sqrt{\dfrac{36a^5}{b^6}}$ $\dfrac{6a^2\sqrt{a}}{b^3}$ **9.** $\sqrt[3]{\dfrac{27a^4}{8b^3}}$

10. $\sqrt[3]{\dfrac{64x^7}{216y^6}}$ $\dfrac{2x^2\sqrt[3]{x}}{3y^2}$ **11.** $\sqrt[4]{\dfrac{16a^4}{b^4c^8}}$ $\dfrac{2a}{bc^2}$ **12.** $\sqrt[4]{\dfrac{81x^4}{y^8z^4}}$

13. $\sqrt[4]{\dfrac{a^5b^8}{c^{10}}}$ **14.** $\sqrt[4]{\dfrac{x^9y^{12}}{z^6}}$ **15.** $\sqrt[5]{\dfrac{32x^6}{y^{11}}}$

16. $\sqrt[5]{\dfrac{243a^9}{b^{13}}}$ **17.** $\sqrt[6]{\dfrac{x^6y^8}{z^{15}}}$ **18.** $\sqrt[6]{\dfrac{a^9b^{12}}{c^{13}}}$

Divide and, if possible, simplify. Assume all variables represent positive numbers.

19. $\dfrac{\sqrt{35x}}{\sqrt{7x}}$ $\sqrt{5}$ **20.** $\dfrac{\sqrt{28y}}{\sqrt{4y}}$ $\sqrt{7}$ **21.** $\dfrac{\sqrt[3]{270}}{\sqrt[3]{10}}$ 3

22. $\dfrac{\sqrt[3]{40}}{\sqrt[3]{5}}$ 2 **23.** $\dfrac{\sqrt{40xy^3}}{\sqrt{8x}}$ $y\sqrt{5y}$ **24.** $\dfrac{\sqrt{56ab^3}}{\sqrt{7a}}$ $2b\sqrt{2b}$

25. $\dfrac{\sqrt[3]{96a^4b^2}}{\sqrt[3]{12a^2b}}$ **26.** $\dfrac{\sqrt[3]{189x^5y^7}}{\sqrt[3]{7x^2y^2}}$ **27.** $\dfrac{\sqrt{100ab}}{5\sqrt{2}}$ $\sqrt{2ab}$

28. $\dfrac{\sqrt{75ab}}{3\sqrt{3}}$ $\dfrac{5}{3}\sqrt{ab}$ **29.** $\dfrac{\sqrt[4]{48x^9y^{13}}}{\sqrt[4]{3xy^{-2}}}$ **30.** $\dfrac{\sqrt[5]{64a^{11}b^{28}}}{\sqrt[5]{2ab^{-2}}}$ $2a^2b^6$

31. $\dfrac{\sqrt[3]{x^3-y^3}}{\sqrt[3]{x-y}}$ $\sqrt[3]{x^2+xy+y^2}$ **32.** $\dfrac{\sqrt[3]{r^3+s^3}}{\sqrt[3]{r+s}}$ $\sqrt[3]{r^2-rs+s^2}$

Hint: Factor and then simplify.

Rationalize each denominator. Assume all variables represent positive numbers.

33. $\sqrt{\dfrac{5}{7}}$ $\dfrac{\sqrt{35}}{7}$

34. $\sqrt{\dfrac{11}{6}}$ $\dfrac{\sqrt{66}}{6}$

35. $\dfrac{6\sqrt{5}}{5\sqrt{3}}$ $\dfrac{2\sqrt{15}}{5}$

36. $\dfrac{4\sqrt{5}}{3\sqrt{2}}$ $\dfrac{2\sqrt{10}}{3}$

37. $\sqrt[3]{\dfrac{16}{9}}$ $\dfrac{2\sqrt[3]{6}}{3}$

38. $\sqrt[3]{\dfrac{2}{9}}$ $\dfrac{\sqrt[3]{6}}{3}$

39. $\dfrac{\sqrt[3]{3a}}{\sqrt[3]{5c}}$ $\dfrac{\sqrt[3]{75ac^2}}{5c}$

40. $\dfrac{\sqrt[3]{7x}}{\sqrt[3]{3y}}$ $\dfrac{\sqrt[3]{63xy^2}}{3y}$

41. $\dfrac{\sqrt[3]{5y^4}}{\sqrt[3]{6x^4}}$ ⊡

42. $\dfrac{\sqrt[3]{3a^4}}{\sqrt[3]{7b^2}}$ $\dfrac{a\sqrt[3]{147ab}}{7b}$

43. $\sqrt[3]{\dfrac{2}{x^2y}}$ $\dfrac{\sqrt[3]{2xy^2}}{xy}$

44. $\sqrt[3]{\dfrac{5}{ab^2}}$ $\dfrac{\sqrt[3]{5a^2b}}{ab}$

45. $\sqrt{\dfrac{7a}{18}}$ $\dfrac{\sqrt{14a}}{6}$

46. $\sqrt{\dfrac{3x}{10}}$ $\dfrac{\sqrt{30x}}{10}$

47. $\sqrt{\dfrac{9}{20x^2y}}$ ⊡

48. $\sqrt{\dfrac{7}{32a^2b}}$ ⊡ Aha!

49. $\sqrt{\dfrac{10ab^2}{72a^3b}}$ $\dfrac{\sqrt{5b}}{6a}$

50. $\sqrt{\dfrac{21x^2y}{75xy^5}}$ $\dfrac{\sqrt{7x}}{5y^2}$

Rationalize each numerator. Assume all variables represent positive numbers.

51. $\dfrac{\sqrt{5}}{\sqrt{7x}}$ $\dfrac{5}{\sqrt{35x}}$

52. $\dfrac{\sqrt{10}}{\sqrt{3x}}$ $\dfrac{10}{\sqrt{30x}}$

53. $\sqrt{\dfrac{14}{21}}$ $\dfrac{2}{\sqrt{6}}$

54. $\sqrt{\dfrac{12}{15}}$ $\dfrac{2}{\sqrt{5}}$

55. $\dfrac{4\sqrt{13}}{3\sqrt{7}}$ $\dfrac{52}{3\sqrt{91}}$

56. $\dfrac{5\sqrt{21}}{2\sqrt{5}}$ $\dfrac{105}{2\sqrt{105}}$

57. $\dfrac{\sqrt[3]{7}}{\sqrt[3]{2}}$ $\dfrac{7}{\sqrt[3]{98}}$

58. $\dfrac{\sqrt[3]{5}}{\sqrt[3]{4}}$ $\dfrac{5}{\sqrt[3]{100}}$

59. $\sqrt{\dfrac{7x}{3y}}$ $\dfrac{7x}{\sqrt{21xy}}$

60. $\sqrt{\dfrac{6a}{5b}}$ $\dfrac{6a}{\sqrt{30ab}}$

61. $\sqrt[3]{\dfrac{2a^5}{5b}}$ $\dfrac{2a^2}{\sqrt[3]{20ab}}$

62. $\sqrt[3]{\dfrac{2a^4}{7b}}$ $\dfrac{2a^2}{\sqrt[3]{28a^2b}}$

63. $\sqrt{\dfrac{x^3y}{2}}$ $\dfrac{x^2y}{\sqrt{2xy}}$

64. $\sqrt{\dfrac{ab^5}{3}}$ $\dfrac{ab^3}{\sqrt{3ab}}$

ᵀᵂ **65.** Explain why it is easier to approximate

$$\dfrac{\sqrt{2}}{2} \quad \text{than} \quad \dfrac{1}{\sqrt{2}}$$

if no calculator is available and $\sqrt{2} \approx 1.414213562$.

ᵀᵂ **66.** A student *incorrectly* claims that

$$\dfrac{5 + \sqrt{2}}{\sqrt{18}} = \dfrac{5 + \sqrt{1}}{\sqrt{9}} = \dfrac{5 + 1}{3}.$$

How could you convince the student that a mistake has been made? How would you explain the correct way of rationalizing the denominator?

Skill Maintenance

Multiply. [6.1]

67. $\dfrac{3}{x - 5} \cdot \dfrac{x - 1}{x + 5}$ $\dfrac{3(x - 1)}{(x - 5)(x + 5)}$

68. $\dfrac{7}{x + 4} \cdot \dfrac{x - 2}{x - 4}$ $\dfrac{7(x - 2)}{(x + 4)(x - 4)}$

Simplify.

69. $\dfrac{a^2 - 8a + 7}{a^2 - 49}$ [6.1] $\dfrac{a - 1}{a + 7}$

70. $\dfrac{t^2 + 9t - 22}{t^2 - 4}$ [6.1] $\dfrac{t + 11}{t + 2}$

71. $(5a^3b^4)^3$ [1.4] $125a^9b^{12}$

72. $(3x^4)^2(5xy^3)^2$ [1.4] $225x^{10}y^6$

Synthesis

ᵀᵂ **73.** Is it possible to understand how to rationalize a denominator without knowing how to multiply rational expressions? Why or why not?

ᵀᵂ **74.** Is the quotient of two irrational numbers always an irrational number? Why or why not?

75. *Pendulums.*   The *period* of a pendulum is the time it takes to complete one cycle, swinging to and fro. For a pendulum that is L centimeters long, the period T is given by the formula

$$T = 2\pi\sqrt{\dfrac{L}{980}},$$

where T is in seconds. Find, to the nearest hundredth of a second, the period of a pendulum of length **(a)** 65 cm; **(b)** 98 cm; **(c)** 120 cm. Use a calculator's $\boxed{\pi}$ key if possible.
(a) 1.62 sec; **(b)** 1.99 sec; **(c)** 2.20 sec

Perform the indicated operations.

76. $\dfrac{7\sqrt{a^2b}}{5\sqrt{a^{-4}b^{-1}}} \cdot \dfrac{\sqrt{25xy}}{\sqrt{49x^{-1}y^{-3}}}$ a^3bxy^2

77. $\dfrac{\left(\sqrt[3]{81mn^2}\right)^2}{\left(\sqrt[3]{mn}\right)^2}$ $9\sqrt[3]{9n^2}$

78. $\dfrac{\sqrt{44x^2y^9z} \cdot \sqrt{22y^9z^6}}{\left(\sqrt{11xy^8z^2}\right)^2}$ $2yz\sqrt{2z}$

79. $\sqrt{a^2 - 3} - \dfrac{a^2}{\sqrt{a^2 - 3}}$ $\dfrac{-3\sqrt{a^2 - 3}}{a^2 - 3}$, or $\dfrac{-3}{\sqrt{a^2 - 3}}$

⊡ Answers to Exercises 41, 47, and 48 can be found on p. A-64.

80. $5\sqrt{\dfrac{x}{y}} + 4\sqrt{\dfrac{y}{x}} - \dfrac{3}{\sqrt{xy}}$ $\dfrac{(5x + 4y - 3)\sqrt{xy}}{xy}$

81. Provide a reason for each step in the following derivation of the quotient rule:

$$\sqrt[n]{\dfrac{a}{b}} = \left(\dfrac{a}{b}\right)^{1/n} \underline{\hspace{4cm}}$$

$$= \dfrac{a^{1/n}}{b^{1/n}} \underline{\hspace{3cm}}$$

$$= \dfrac{\sqrt[n]{a}}{\sqrt[n]{b}} \underline{\hspace{3cm}}.$$

82. Show that $\dfrac{\sqrt[n]{a}}{\sqrt[n]{b}}$ is the nth root of $\dfrac{a}{b}$ by raising it to the nth power and simplifying.

83. Let $f(x) = \sqrt{18x^3}$ and $g(x) = \sqrt{2x}$. Find $(f/g)(x)$ and specify the domain of f/g. $(f/g)(x) = 3x$, where x is a real number and $x > 0$

84. Let $f(t) = \sqrt{2t}$ and $g(t) = \sqrt{50t^3}$. Find $(f/g)(t)$ and specify the domain of f/g.

85. Let $f(x) = \sqrt{x^2 - 9}$ and $g(x) = \sqrt{x - 3}$. Find $(f/g)(x)$ and specify the domain of f/g.
$(f/g)(x) = \sqrt{x + 3}$, where x is a real number and $x > 3$

7.5

Adding and Subtracting Radical Expressions ■ Products and Quotients of Two or More Radical Terms ■ Terms with Differing Indices

Expressions Containing Several Radical Terms

81. Step 1: $\sqrt[n]{a} = a^{1/n}$, by definition; Step 2: $\left(\dfrac{a}{b}\right)^n = \dfrac{a^n}{b^n}$, raising a quotient to a power; Step 3: $a^{1/n} = \sqrt[n]{a}$, by definition

82. A number c is the nth root of a/b if $c^n = a/b$. Let $c = \sqrt[n]{a}/\sqrt[n]{b}$.

$$c^n = \left(\dfrac{\sqrt[n]{a}}{\sqrt[n]{b}}\right)^n = \left(\dfrac{a^{1/n}}{b^{1/n}}\right)^n$$

$$= \dfrac{a^{(1/n)\cdot n}}{b^{(1/n)\cdot n}} = \dfrac{a}{b}$$

84. $(f/g)(t) = \dfrac{1}{5t}$, where t is a real number and $t > 0$

Radical expressions like $6\sqrt{7} + 4\sqrt{7}$ or $\left(\sqrt{a} + \sqrt{b}\right)\left(\sqrt{a} - \sqrt{b}\right)$ contain more than one *radical term* and can sometimes be simplified.

Adding and Subtracting Radical Expressions

When two radical expressions have the same indices and radicands, they are said to be **like radicals**. Like radicals can be combined (added or subtracted) in much the same way that we combined like terms earlier in this text.

EXAMPLE 1 Simplify by combining like radical terms.

a) $6\sqrt{7} + 4\sqrt{7}$ **b)** $\sqrt[3]{2} - 7x\sqrt[3]{2} + 5\sqrt[3]{2}$

c) $6\sqrt[5]{4x} + 3\sqrt[5]{4x} - \sqrt[3]{4x}$

Solution

a) $6\sqrt{7} + 4\sqrt{7} = (6 + 4)\sqrt{7}$ Using the distributive law (factoring out $\sqrt{7}$)

$= 10\sqrt{7}$ You can think: 6 square roots of 7 plus 4 square roots of 7 results in 10 square roots of 7.

b) $\sqrt[3]{2} - 7x\sqrt[3]{2} + 5\sqrt[3]{2} = (1 - 7x + 5)\sqrt[3]{2}$ Factoring out $\sqrt[3]{2}$

$= (6 - 7x)\sqrt[3]{2}$ These parentheses are important!

c) $6\sqrt[5]{4x} + 3\sqrt[5]{4x} - \sqrt[3]{4x} = (6 + 3)\sqrt[5]{4x} - \sqrt[3]{4x}$ Try to do this step mentally.

$= 9\sqrt[5]{4x} - \sqrt[3]{4x}$ Because the indices differ, we are done.

Our ability to simplify radical expressions can help us to find like radicals even when, at first, it may appear that none exists.

EXAMPLE 2   Simplify by combining like radical terms, if possible.

a) $3\sqrt{8} - 5\sqrt{2}$

b) $9\sqrt{5} - 4\sqrt{3}$

c) $\sqrt[3]{2x^6y^4} + 7\sqrt[3]{2y}$

Solution

a) $3\sqrt{8} - 5\sqrt{2} = 3\sqrt{4 \cdot 2} - 5\sqrt{2}$

$\phantom{3\sqrt{8} - 5\sqrt{2}} = 3\sqrt{4} \cdot \sqrt{2} - 5\sqrt{2}$ ⎫ Simplifying $\sqrt{8}$

$\phantom{3\sqrt{8} - 5\sqrt{2}} = 3 \cdot 2 \cdot \sqrt{2} - 5\sqrt{2}$ ⎭

$\phantom{3\sqrt{8} - 5\sqrt{2}} = 6\sqrt{2} - 5\sqrt{2}$

$\phantom{3\sqrt{8} - 5\sqrt{2}} = \sqrt{2}$ **Combining like radicals**

b) $9\sqrt{5} - 4\sqrt{3}$ cannot be simplified.

c) $\sqrt[3]{2x^6y^4} + 7\sqrt[3]{2y} = \sqrt[3]{x^6y^3 \cdot 2y} + 7\sqrt[3]{2y}$

$\phantom{\sqrt[3]{2x^6y^4} + 7\sqrt[3]{2y}} = \sqrt[3]{x^6y^3} \cdot \sqrt[3]{2y} + 7\sqrt[3]{2y}$ ⎫ Simplifying $\sqrt[3]{2x^6y^4}$

$\phantom{\sqrt[3]{2x^6y^4} + 7\sqrt[3]{2y}} = x^2y \cdot \sqrt[3]{2y} + 7\sqrt[3]{2y}$ ⎭

$\phantom{\sqrt[3]{2x^6y^4} + 7\sqrt[3]{2y}} = (x^2y + 7)\sqrt[3]{2y}$ **Factoring to combine like radical terms**

Products and Quotients of Two or More Radical Terms

Radical expressions often contain factors that have more than one term. The procedure for multiplying out such expressions is similar to finding products of polynomials. Some products will yield like radical terms, which we can now combine.

EXAMPLE 3   Multiply.

a) $\sqrt{3}(x - \sqrt{5})$

b) $\sqrt[3]{y}(\sqrt[3]{y^2} + \sqrt[3]{2})$

c) $(4\sqrt{3} + \sqrt{2})(\sqrt{3} - 5\sqrt{2})$

d) $(\sqrt{a} + \sqrt{b})(\sqrt{a} - \sqrt{b})$

Solution

a) $\sqrt{3}(x - \sqrt{5}) = \sqrt{3} \cdot x - \sqrt{3} \cdot \sqrt{5}$ **Using the distributive law**

$\phantom{\sqrt{3}(x - \sqrt{5})} = x\sqrt{3} - \sqrt{15}$ **Multiplying radicals**

b) $\sqrt[3]{y}(\sqrt[3]{y^2} + \sqrt[3]{2}) = \sqrt[3]{y} \cdot \sqrt[3]{y^2} + \sqrt[3]{y} \cdot \sqrt[3]{2}$ **Using the distributive law**

$\phantom{\sqrt[3]{y}(\sqrt[3]{y^2} + \sqrt[3]{2})} = \sqrt[3]{y^3} + \sqrt[3]{2y}$ **Multiplying radicals**

$\phantom{\sqrt[3]{y}(\sqrt[3]{y^2} + \sqrt[3]{2})} = y + \sqrt[3]{2y}$ **Simplifying $\sqrt[3]{y^3}$**

$$\textbf{c)}\ \left(4\sqrt{3} + \sqrt{2}\right)\left(\sqrt{3} - 5\sqrt{2}\right) = \overset{\text{F}}{4(\sqrt{3})^2} - \overset{\text{O}}{20\sqrt{3} \cdot \sqrt{2}} + \overset{\text{I}}{\sqrt{2} \cdot \sqrt{3}} - \overset{\text{L}}{5(\sqrt{2})^2}$$

$$= 4 \cdot 3 - 20\sqrt{6} + \sqrt{6} - 5 \cdot 2 \qquad \textbf{Multiplying radicals}$$

$$= 12 - 20\sqrt{6} + \sqrt{6} - 10$$

$$= 2 - 19\sqrt{6} \qquad \textbf{Combining like terms}$$

$$\textbf{d)}\ \left(\sqrt{a} + \sqrt{b}\right)\left(\sqrt{a} - \sqrt{b}\right) = (\sqrt{a})^2 - \sqrt{a}\,\sqrt{b} + \sqrt{a}\,\sqrt{b} - (\sqrt{b})^2$$

$$\textbf{Using FOIL}$$

$$= a - b \qquad \textbf{Combining like terms}$$

In Example 3(d) above, you may have noticed that since the outer and inner products in FOIL are opposites, the result, $a - b$, is not itself a radical expression. Pairs of radical terms, like $\sqrt{a} + \sqrt{b}$ and $\sqrt{a} - \sqrt{b}$, are called **conjugates**. The use of conjugates allows us to rationalize denominators or numerators with two terms.

EXAMPLE 4 Rationalize each denominator: **(a)** $\dfrac{4}{\sqrt{3} + x}$; **(b)** $\dfrac{4 + \sqrt{2}}{\sqrt{5} - \sqrt{2}}$.

Solution

a) $\dfrac{4}{\sqrt{3} + x} = \dfrac{4}{\sqrt{3} + x} \cdot \dfrac{\sqrt{3} - x}{\sqrt{3} - x}$ **Multiplying by 1, using the conjugate of $\sqrt{3} + x$, which is $\sqrt{3} - x$**

$$= \dfrac{4(\sqrt{3} - x)}{(\sqrt{3} + x)(\sqrt{3} - x)}$$ **Multiplying numerators and denominators**

$$= \dfrac{4(\sqrt{3} - x)}{(\sqrt{3})^2 - x^2}$$ **Using FOIL in the denominator**

$$= \dfrac{4\sqrt{3} - 4x}{3 - x^2}$$ **Simplifying**

b) $\dfrac{4 + \sqrt{2}}{\sqrt{5} - \sqrt{2}} = \dfrac{4 + \sqrt{2}}{\sqrt{5} - \sqrt{2}} \cdot \dfrac{\sqrt{5} + \sqrt{2}}{\sqrt{5} + \sqrt{2}}$ **Multiplying by 1, using the conjugate of $\sqrt{5} - \sqrt{2}$, which is $\sqrt{5} + \sqrt{2}$**

$$= \dfrac{(4 + \sqrt{2})(\sqrt{5} + \sqrt{2})}{(\sqrt{5} - \sqrt{2})(\sqrt{5} + \sqrt{2})}$$ **Multiplying numerators and denominators**

$$= \dfrac{4\sqrt{5} + 4\sqrt{2} + \sqrt{2}\,\sqrt{5} + (\sqrt{2})^2}{(\sqrt{5})^2 - (\sqrt{2})^2}$$ **Using FOIL**

$$= \dfrac{4\sqrt{5} + 4\sqrt{2} + \sqrt{10} + 2}{5 - 2}$$ **Squaring in the denominator and the numerator**

$$= \dfrac{4\sqrt{5} + 4\sqrt{2} + \sqrt{10} + 2}{3}$$

```
(4+√(2))/(√(5)−√
(2))
           6.587801273
(4√(5)+4√(2)+√(1
0)+2)/3
           6.587801273
```

We can check by approximating the value of the original expression and the value of the rationalized expression, as shown at left. Care must be taken to place the parentheses properly. The approximate values are the same, so our work is correct.

To rationalize a numerator with more than one term, we use the conjugate of the numerator.

EXAMPLE 5   Rationalize the numerator: $\dfrac{4 + \sqrt{2}}{\sqrt{5} - \sqrt{2}}$.

Solution   We have

$$\dfrac{4 + \sqrt{2}}{\sqrt{5} - \sqrt{2}} = \dfrac{4 + \sqrt{2}}{\sqrt{5} - \sqrt{2}} \cdot \dfrac{4 - \sqrt{2}}{4 - \sqrt{2}} \quad \begin{array}{l}\textbf{Multiplying by 1, using the}\\ \textbf{conjugate of } 4 + \sqrt{2}\textbf{, which}\\ \textbf{is } 4 - \sqrt{2}\end{array}$$

$$= \dfrac{16 - \left(\sqrt{2}\right)^2}{4\sqrt{5} - \sqrt{5}\,\sqrt{2} - 4\sqrt{2} + \left(\sqrt{2}\right)^2}$$

$$= \dfrac{14}{4\sqrt{5} - \sqrt{10} - 4\sqrt{2} + 2}.$$

We check by comparing the values of the original expression and the expression with a rationalized numerator, as shown at left.

(4+√(2))/(√(5)−√(2))

          6.587801273

14/(4√(5)−√(10)−4√(2)+2)

          6.587801273

Terms with Differing Indices

Sometimes it is necessary to determine products or quotients involving radical terms with indices that differ from each other. When this occurs, we can convert to exponential notation, use the rules for exponents, and then convert back to radical notation.

EXAMPLE 6   Divide and, if possible, simplify: $\dfrac{\sqrt[4]{(x + y)^3}}{\sqrt{x + y}}$.

Solution

$$\dfrac{\sqrt[4]{(x + y)^3}}{\sqrt{x + y}} = \dfrac{(x + y)^{3/4}}{(x + y)^{1/2}} \qquad \textbf{Converting to exponential notation}$$

$$= (x + y)^{3/4 - 1/2} \qquad \begin{array}{l}\textbf{Since the bases are identical, we can}\\ \textbf{subtract exponents: } \frac{3}{4} - \frac{1}{2} = \frac{3}{4} - \frac{2}{4} = \frac{1}{4}.\end{array}$$

$$\left.\begin{array}{l}= (x + y)^{1/4}\\ = \sqrt[4]{x + y}\end{array}\right\} \qquad \textbf{Converting back to radical notation}$$

TEACHING TIP

Point out that rational exponents are needed in Example 6 because the indices are different.

The steps used in Example 6 can be used in a variety of situations.

To Simplify Products or Quotients with Differing Indices

1. Convert all radical expressions to exponential notation.

2. When the bases are identical, subtract exponents to divide and add exponents to multiply. This may require finding a common denominator.

3. Convert back to radical notation and, if possible, simplify.

EXAMPLE 7 Multiply and simplify: $\sqrt{x^3}\ \sqrt[3]{x}$.

Solution

$$\sqrt{x^3}\ \sqrt[3]{x} = x^{3/2} \cdot x^{1/3} \qquad \text{Converting to exponential notation}$$
$$= x^{11/6} \qquad \text{Adding exponents: } \tfrac{3}{2} + \tfrac{1}{3} = \tfrac{9}{6} + \tfrac{2}{6}$$
$$= \sqrt[6]{x^{11}} \qquad \text{Converting back to radical notation}$$
$$\left.\begin{array}{l} = \sqrt[6]{x^6}\ \sqrt[6]{x^5} \\ = x\sqrt[6]{x^5} \end{array}\right\} \quad \text{Simplifying}$$

EXAMPLE 8 If $f(x) = \sqrt[3]{x^2}$ and $g(x) = \sqrt{x} + \sqrt[4]{x}$, find $(f \cdot g)(x)$.

Solution Recall from Section 2.7 that $(f \cdot g)(x) = f(x) \cdot g(x)$. Thus,

$$(f \cdot g)(x) = \sqrt[3]{x^2}\left(\sqrt{x} + \sqrt[4]{x}\right) \qquad x \text{ is assumed to be nonnegative.}$$
$$= x^{2/3}(x^{1/2} + x^{1/4}) \qquad \text{Converting to exponential notation}$$
$$= x^{2/3} \cdot x^{1/2} + x^{2/3} \cdot x^{1/4} \qquad \text{Using the distributive law}$$
$$= x^{2/3+1/2} + x^{2/3+1/4} \qquad \text{Adding exponents}$$
$$= x^{7/6} + x^{11/12} \qquad \tfrac{2}{3}+\tfrac{1}{2}=\tfrac{4}{6}+\tfrac{3}{6}; \tfrac{2}{3}+\tfrac{1}{4}=\tfrac{8}{12}+\tfrac{3}{12}$$
$$= \sqrt[6]{x^7} + \sqrt[12]{x^{11}} \qquad \text{Converting back to radical notation}$$
$$\left.\begin{array}{l} = \sqrt[6]{x^6}\ \sqrt[6]{x} + \sqrt[12]{x^{11}} \\ = x\sqrt[6]{x} + \sqrt[12]{x^{11}} \end{array}\right\} \quad \text{Simplifying}$$

If factors are raised to powers that share a common denominator, we can write the final result as a single radical expression.

EXAMPLE 9 Divide and, if possible, simplify: $\dfrac{\sqrt[3]{a^2b^4}}{\sqrt{ab}}$.

Solution

$$\frac{\sqrt[3]{a^2b^4}}{\sqrt{ab}} = \frac{(a^2b^4)^{1/3}}{(ab)^{1/2}} \qquad \text{Converting to exponential notation}$$
$$= \frac{a^{2/3}b^{4/3}}{a^{1/2}b^{1/2}} \qquad \text{Using the product and power rules}$$
$$= a^{2/3-1/2}b^{4/3-1/2} \qquad \text{Subtracting exponents}$$
$$= a^{1/6}b^{5/6}$$
$$= \sqrt[6]{a}\ \sqrt[6]{b^5} \qquad \text{Converting to radical notation}$$
$$= \sqrt[6]{ab^5} \qquad \text{Using the product rule for radicals}$$

7.5

Exercise Set

Add or subtract. Simplify by combining like radical terms, if possible. Assume that all variables and radicands represent nonnegative real numbers.

1. $3\sqrt{7} + 2\sqrt{7}$ $5\sqrt{7}$

2. $8\sqrt{5} + 9\sqrt{5}$ $17\sqrt{5}$

3. $9\sqrt[3]{5} - 6\sqrt[3]{5}$ $3\sqrt[3]{5}$

4. $14\sqrt[5]{2} - 6\sqrt[5]{2}$ $8\sqrt[5]{2}$

5. $4\sqrt[3]{y} + 9\sqrt[3]{y}$ $13\sqrt[3]{y}$

6. $9\sqrt[4]{t} - 3\sqrt[4]{t}$ $6\sqrt[4]{t}$

7. $8\sqrt{2} - 6\sqrt{2} + 5\sqrt{2}$ $7\sqrt{2}$

8. $2\sqrt{6} + 8\sqrt{6} - 3\sqrt{6}$ $7\sqrt{6}$

9. $9\sqrt[3]{7} - \sqrt{3} + 4\sqrt[3]{7} + 2\sqrt{3}$ $13\sqrt[3]{7} + \sqrt{3}$

10. $5\sqrt{7} - 8\sqrt[4]{11} + \sqrt{7} + 9\sqrt[4]{11}$ $6\sqrt{7} + \sqrt[4]{11}$

11. $8\sqrt{27} - 3\sqrt{3}$ $21\sqrt{3}$

12. $9\sqrt{50} - 4\sqrt{2}$ $41\sqrt{2}$

13. $3\sqrt{45} + 7\sqrt{20}$ $23\sqrt{5}$

14. $5\sqrt{12} + 16\sqrt{27}$ $58\sqrt{3}$

15. $3\sqrt[3]{16} + \sqrt[3]{54}$ $9\sqrt[3]{2}$

16. $\sqrt[3]{27} - 5\sqrt[3]{8}$ -7

17. $\sqrt{5a} + 2\sqrt{45a^3}$ $(1 + 6a)\sqrt{5a}$

18. $4\sqrt{3x^3} - \sqrt{12x}$ $(4x - 2)\sqrt{3x}$

19. $\sqrt[3]{6x^4} + \sqrt[3]{48x}$ $(x + 2)\sqrt[3]{6x}$

20. $\sqrt[3]{54x} - \sqrt[3]{2x^4}$ $(3 - x)\sqrt[3]{2x}$

21. $\sqrt{4a - 4} + \sqrt{a - 1}$ $3\sqrt{a - 1}$

22. $\sqrt{9y + 27} + \sqrt{y + 3}$ $4\sqrt{y + 3}$

23. $\sqrt{x^3 - x^2} + \sqrt{9x - 9}$ $(x + 3)\sqrt{x - 1}$

24. $\sqrt{4x - 4} - \sqrt{x^3 - x^2}$ $(2 - x)\sqrt{x - 1}$

Multiply. Assume all variables represent nonnegative real numbers.

25. $\sqrt{7}(3 - \sqrt{7})$ $3\sqrt{7} - 7$ **26.** $\sqrt{3}(4 + \sqrt{3})$ $4\sqrt{3} + 3$

27. $4\sqrt{2}(\sqrt{3} - \sqrt{5})$ ▣   **28.** $3\sqrt{5}(\sqrt{5} - \sqrt{2})$ $15 - 3\sqrt{10}$

29. $\sqrt{3}(2\sqrt{5} - 3\sqrt{4})$ ▣   **30.** $\sqrt{2}(3\sqrt{10} - 2\sqrt{2})$ ▣

31. $\sqrt[3]{2}(\sqrt[3]{4} - 2\sqrt[3]{32})$ -6 **32.** $\sqrt[3]{3}(\sqrt[3]{9} - 4\sqrt[3]{21})$ $3 - 4\sqrt[3]{63}$

33. $\sqrt[3]{a}(\sqrt[3]{a^2} + \sqrt[3]{24a^2})$ $a + 2a\sqrt[3]{3}$

34. $\sqrt[3]{x}(\sqrt[3]{3x^2} - \sqrt[3]{81x^2})$ $-2x\sqrt[3]{3}$

35. $(5 + \sqrt{6})(5 - \sqrt{6})$ 19

36. $(2 - \sqrt{5})(2 + \sqrt{5})$ -1

37. $(3 - 2\sqrt{7})(3 + 2\sqrt{7})$ -19

38. $(4 + 3\sqrt{2})(4 - 3\sqrt{2})$ -2

39. $(3 + \sqrt{5})^2$ $14 + 6\sqrt{5}$

40. $(7 + \sqrt{3})^2$ $52 + 14\sqrt{3}$

41. $(2\sqrt{7} - 4\sqrt{2})(3\sqrt{7} + 6\sqrt{2})$ -6

42. $(4\sqrt{5} + 3\sqrt{3})(3\sqrt{5} - 4\sqrt{3})$ $24 - 7\sqrt{15}$

43. $(2\sqrt[3]{3} - \sqrt[3]{2})(\sqrt[3]{3} + 2\sqrt[3]{2})$ $2\sqrt[3]{9} + 3\sqrt[3]{6} - 2\sqrt[3]{4}$

44. $(3\sqrt[4]{7} + \sqrt[4]{6})(2\sqrt[4]{9} - 3\sqrt[4]{6})$ $6\sqrt[4]{63} - 9\sqrt[4]{42} + 2\sqrt[4]{54} - 3\sqrt[4]{36}$

45. $(\sqrt{3x} + \sqrt{y})^2$ $3x + 2\sqrt{3xy} + y$

46. $(\sqrt{t} - \sqrt{2r})^2$ $t - 2\sqrt{2rt} + 2r$

Rationalize each denominator.

47. $\dfrac{2}{3 + \sqrt{5}}$ $\dfrac{3 - \sqrt{5}}{2}$ **48.** $\dfrac{3}{4 - \sqrt{7}}$ $\dfrac{4 + \sqrt{7}}{3}$

49. $\dfrac{2 + \sqrt{5}}{6 - \sqrt{3}}$ ▣   **50.** $\dfrac{1 + \sqrt{2}}{3 + \sqrt{5}}$ ▣

51. $\dfrac{\sqrt{a}}{\sqrt{a} + \sqrt{b}}$ $\dfrac{a - \sqrt{ab}}{a - b}$ **52.** $\dfrac{\sqrt{z}}{\sqrt{x} - \sqrt{z}}$ $\dfrac{z + \sqrt{xz}}{x - z}$

Aha!   **53.** $\dfrac{\sqrt{7} - \sqrt{3}}{\sqrt{3} - \sqrt{7}}$ -1 **54.** $\dfrac{\sqrt{7} + \sqrt{5}}{\sqrt{5} + \sqrt{2}}$ ▣

55. $\dfrac{3\sqrt{2} - \sqrt{7}}{4\sqrt{2} + \sqrt{5}}$ ▣   **56.** $\dfrac{5\sqrt{3} - \sqrt{11}}{2\sqrt{3} - 5\sqrt{2}}$ ▣

57. $\dfrac{5\sqrt{3} - 3\sqrt{2}}{3\sqrt{2} - 2\sqrt{3}}$ $\dfrac{3\sqrt{6} + 4}{2}$ **58.** $\dfrac{7\sqrt{2} + 4\sqrt{3}}{4\sqrt{3} - 3\sqrt{2}}$ $\dfrac{4\sqrt{6} + 9}{3}$

Rationalize each numerator.

59. $\dfrac{\sqrt{7} + 2}{5}$ $\dfrac{3}{5\sqrt{7} - 10}$ **60.** $\dfrac{\sqrt{3} + 1}{4}$ $\dfrac{1}{2\sqrt{3} - 2}$

▣ Answers to Exercises 27, 29, 30, 49, 50, and 54–56 can be found on p. A-64.

61. $\dfrac{\sqrt{6}-2}{\sqrt{3}+7}$ ⊡

62. $\dfrac{\sqrt{10}+4}{\sqrt{2}-3}$ ⊡

63. $\dfrac{\sqrt{x}-\sqrt{y}}{\sqrt{x}+\sqrt{y}}$ $\quad\dfrac{x-y}{x+2\sqrt{xy}+y}$

64. $\dfrac{\sqrt{a}+\sqrt{b}}{\sqrt{a}-\sqrt{b}}$ ⊡

Perform the indicated operation and simplify. Assume all variables represent nonnegative real numbers.

65. $\sqrt{a}\ \sqrt[4]{a^3}$ $\quad a\sqrt[4]{a}$

66. $\sqrt[3]{x^2}\ \sqrt[6]{x^5}$ $\quad x\sqrt{x}$

67. $\sqrt[5]{b^2}\ \sqrt{b^3}$ $\quad b\sqrt[10]{b^9}$

68. $\sqrt[4]{a^3}\ \sqrt[3]{a^2}$ $\quad a\sqrt[12]{a^5}$

69. $\sqrt{xy^3}\ \sqrt[3]{x^2y}$ $\quad xy\sqrt[6]{xy^5}$

70. $\sqrt[5]{a^3b}\ \sqrt{ab}$ $\quad a\sqrt[10]{ab^7}$

71. $\sqrt[4]{9ab^3}\ \sqrt{3a^4b}$ $\quad 3a^2b\sqrt[4]{ab}$

72. $\sqrt{2x^3y^3}\ \sqrt[3]{4xy^2}$ ⊡

73. $\sqrt[3]{xy^2z}\ \sqrt{x^3yz^2}$ ⊡

74. $\sqrt{a^4b^3c^4}\ \sqrt[3]{ab^2c}$ ⊡

75. $\dfrac{\sqrt[3]{x^2}}{\sqrt[5]{x}}$ $\quad\sqrt[15]{x^7}$

76. $\dfrac{\sqrt[3]{a^2}}{\sqrt[4]{a}}$ $\quad\sqrt[12]{a^5}$

77. $\dfrac{\sqrt[5]{a^4b}}{\sqrt[3]{ab}}$ $\quad\sqrt[15]{\dfrac{a^7}{b^2}}$

78. $\dfrac{\sqrt[4]{x^2y^3}}{\sqrt[3]{xy}}$ $\quad\sqrt[12]{x^2y^5}$

79. $\dfrac{\sqrt[5]{x^3y^4}}{\sqrt{xy}}$ $\quad\sqrt[10]{xy^3}$

80. $\dfrac{\sqrt{ab^3}}{\sqrt[5]{a^2b^3}}$ $\quad\sqrt[10]{ab^9}$

81. $\dfrac{\sqrt[3]{(2+5x)^2}}{\sqrt[4]{2+5x}}$ $\quad\sqrt[12]{(2+5x)^5}$

82. $\dfrac{\sqrt[4]{(3x-1)^3}}{\sqrt[5]{(3x-1)^3}}$ $\quad\sqrt[20]{(3x-1)^3}$

83. $\dfrac{\sqrt[4]{(5+3x)^3}}{\sqrt[3]{(5+3x)^2}}$ $\quad\sqrt[12]{5+3x}$

84. $\dfrac{\sqrt[3]{(2x+1)^2}}{\sqrt[5]{(2x+1)^2}}$ $\quad\sqrt[15]{(2x+1)^4}$

85. $\sqrt[3]{x^2y}\left(\sqrt{xy}-\sqrt[5]{xy^3}\right)$ $\quad x\sqrt[6]{xy^5}-\sqrt[15]{x^{13}y^{14}}$

86. $\sqrt[4]{a^2b}\left(\sqrt[3]{a^2b}-\sqrt[5]{a^2b^2}\right)$ $\quad a\sqrt[12]{a^2b^7}-\sqrt[20]{a^{18}b^{13}}$

87. $\left(m+\sqrt[3]{n^2}\right)\left(2m+\sqrt[4]{n}\right)$ $\quad 2m^2+m\sqrt[4]{n}+2m\sqrt[3]{n^2}+\sqrt[12]{n^{11}}$

88. $\left(r-\sqrt[4]{s^3}\right)\left(3r-\sqrt[5]{s}\right)$ $\quad 3r^2-r\sqrt[5]{s}-3r\sqrt[4]{s^3}+\sqrt[20]{s^{19}}$

In Exercises 89–92, $f(x)$ and $g(x)$ are as given. Find $(f\cdot g)(x)$. Assume all variables represent nonnegative real numbers.

89. $f(x)=\sqrt[4]{x}$, $g(x)=\sqrt[4]{2x}-\sqrt[4]{x^{11}}$ $\quad\sqrt[4]{2x^2}-x^3$

90. $f(x)=\sqrt[4]{x^7}+\sqrt[4]{3x^2}$, $g(x)=\sqrt[4]{x}$ $\quad x^2+\sqrt[4]{3x^3}$

91. $f(x)=x+\sqrt{7}$, $g(x)=x-\sqrt{7}$ $\quad x^2-7$

92. $f(x)=x-\sqrt{2}$, $g(x)=x+\sqrt{6}$ $\quad x^2+x\sqrt{6}-x\sqrt{2}-2\sqrt{3}$

Let $f(x)=x^2$. Find each of the following.

93. $f\left(5-\sqrt{2}\right)$ $\quad 27-10\sqrt{2}$

94. $f\left(7+\sqrt{3}\right)$ $\quad 52+14\sqrt{3}$

95. $f\left(\sqrt{3}+\sqrt{5}\right)$ $\quad 8+2\sqrt{15}$

96. $f\left(\sqrt{6}-\sqrt{3}\right)$ $\quad 9-6\sqrt{2}$

TW **97.** Why do we need to know how to multiply radical expressions before learning how to add them?

TW **98.** In what way(s) is combining like radical terms the same as combining like terms that are monomials?

⊡ Answers to Exercises 61, 62, 64, 72–74, and 111–113 can be found on p. A-64.

Skill Maintenance

Solve.

99. $\dfrac{12x}{x-4}-\dfrac{3x^2}{x+4}=\dfrac{384}{x^2-16}$ [6.4] 8

100. $\dfrac{2}{3}+\dfrac{1}{t}=\dfrac{4}{5}$ [6.4] $\dfrac{15}{2}$

101. The width of a rectangle is one-fourth the length. The area is twice the perimeter. Find the dimensions of the rectangle. [5.8] Length: 20 units; width: 5 units

102. The sum of a number and its square is 20. Find the number. [5.8] $-5,4$

103. $5x^2-6x+1=0$ [5.7] $\dfrac{1}{5},1$

104. $7t^2-8t+1=0$ [5.7] $\dfrac{1}{7},1$

Synthesis

TW **105.** Ramon *incorrectly* writes
$$\sqrt[5]{x^2}\cdot\sqrt{x^3}=x^{2/5}\cdot x^{3/2}=\sqrt[5]{x^3}.$$
What mistake do you suspect he is making?

TW **106.** After examining the expression $\sqrt[4]{25xy^3}\ \sqrt{5x^4y}$, Dyan (correctly) concludes that x and y are both nonnegative. Explain how she could reach this conclusion.

For Exercises 107–110, fill in the blanks by selecting from the following words:

radicands, indices, bases, denominators.

Words can be used more than once.

107. To add radical expressions, the _____ and the _____ must be the same. radicands, indices

108. To multiply radical expressions, the _____ must be the same. indices

109. To add rational expressions, the _____ must be the same. denominators

110. To find a product by adding exponents, the _____ must be the same. bases

Find a simplified form for $f(x)$. Assume $x\ge 0$.

111. $f(x)=\sqrt{20x^2+4x^3}-3x\sqrt{45+9x}+\dfrac{\boxed{\cdot}}{\sqrt{5x^2+x^3}}$

112. $f(x)=\sqrt{x^3-x^2}+\sqrt{9x^3-9x^2}-\sqrt{4x^3-4x^2}$

113. $f(x)=\sqrt[4]{x^5-x^4}+3\sqrt[4]{x^9-x^8}$ ⊡

114. $f(x)=\sqrt[4]{16x^4+16x^5}-2\sqrt[4]{x^8+x^9}$ $\quad f(x)=2x(1-x)\sqrt[4]{1+x}$

Simplify.

115. $\frac{1}{2}\sqrt{36a^5bc^4} - \frac{1}{2}\sqrt[3]{64a^4bc^6} + \frac{1}{6}\sqrt{144a^3bc^6}$

116. $7x\sqrt{(x+y)^3} - 5xy\sqrt{x+y} - 2y\sqrt{(x+y)^3}$

117. $\sqrt{27a^5(b+1)}\,\sqrt[3]{81a(b+1)^4}$

118. $\sqrt{8x(y+z)^5}\,\sqrt[3]{4x^2(y+z)^2}$ $4x(y+z)^3\sqrt[6]{2x(y+z)}$

119. $\dfrac{\dfrac{1}{\sqrt{w}} - \sqrt{w}}{\dfrac{\sqrt{w}+1}{\sqrt{w}}}$ $1 - \sqrt{w}$

120. $\dfrac{1}{4+\sqrt{3}} + \dfrac{1}{\sqrt{3}} + \dfrac{1}{\sqrt{3}-4}$ $\dfrac{7\sqrt{3}}{39}$

Express each of the following as the product of two radical expressions.

121. $x - 5$ **122.** $y - 7$ **123.** $x - a$

Multiply.

124. $\sqrt{9 + 3\sqrt{5}}\,\sqrt{9 - 3\sqrt{5}}$ 6

125. $\left(\sqrt{x+2} - \sqrt{x-2}\right)^2$ $2x - 2\sqrt{x^2 - 4}$

For Exercises 126–129, assume that all radicands are positive and that no denominator is 0.

Rationalize each denominator.

126. $\dfrac{a - \sqrt{a+b}}{\sqrt{a+b} - b}$ **127.** $\dfrac{b + \sqrt{b}}{1 + b + \sqrt{b}}$

$\dfrac{b^2 + \sqrt{b}}{1 + b + b^2}$

Rationalize each numerator.

128. $\dfrac{\sqrt{y+18} - \sqrt{y}}{18}$ **129.** $\dfrac{\sqrt{x+6} - 5}{\sqrt{x+6} + 5}$

115. $ac^2\left[(3a + 2c)\sqrt{ab} - 2\sqrt[3]{ab}\right]$ **116.** $(7x^2 - 2y^2)\sqrt{x+y}$ **117.** $9a^2(b+1)\sqrt[6]{243a^5(b+1)^5}$

121. $\left(\sqrt{x} + \sqrt{5}\right)\left(\sqrt{x} - \sqrt{5}\right)$ **122.** $\left(\sqrt{y} + \sqrt{7}\right)\left(\sqrt{y} - \sqrt{7}\right)$ **123.** $\left(\sqrt{x} + \sqrt{a}\right)\left(\sqrt{x} - \sqrt{a}\right)$

126. $\dfrac{ab + (a - b)\sqrt{a+b} - a - b}{a + b - b^2}$ **128.** $\dfrac{1}{\sqrt{y+18} + \sqrt{y}}$ **129.** $\dfrac{x - 19}{x + 10\sqrt{x+6} + 31}$

7.6

The Principle of Powers ■ Equations with Two Radical Terms

Solving Radical Equations

Connecting the Concepts

In Sections 7.1–7.5, we learned how to manipulate radical expressions as well as expressions containing rational exponents. We performed this work to find *equivalent expressions*.

Now that we know how to work with radicals and rational exponents, we can learn how to solve a new type of equation. As in our earlier work with equations, finding *equivalent equations* will be part of our strategy. What is different, however, is that now we will use a step that does not always produce equivalent equations. Checking solutions will therefore be more important than ever.

The Principle of Powers

A **radical equation** is an equation in which the variable appears in a radicand. Examples are

$$\sqrt[3]{2x} + 1 = 5,$$
$$\sqrt{a} + \sqrt{a - 2} = 7,$$
and $\quad 4 - \sqrt{3x + 1} = \sqrt{6 - x}.$

To solve such equations, we need a new principle. Suppose an equation $a = b$ is true. If we square both sides, we get another true equation: $a^2 = b^2$. This can be generalized.

The Principle of Powers

If $a = b$, then $a^n = b^n$ for any exponent n.

Note that the principle of powers is an "if–then" statement. The statement obtained by interchanging the two parts of the sentence—"if $a^n = b^n$ for some exponent n, then $a = b$"—is *not always true*. For example, $(3^2) = (-3)^2$ *is* true, but $3 = -3$ *is not* true. Thus if we raise both sides of an equation to the same power, we may not form an equivalent equation.

Interactive Discovery

Solve each pair of equations graphically, and compare the solution sets.

1. $x = 3;\ x^2 = 9$ {3}; {−3, 3}

2. $x = -2;\ x^2 = 4$ {−2}; {−2, 2}

3. $\sqrt{x} = 5;\ x = 25$ {25}; {25}

4. $\sqrt{x} = -3;\ x = 9$ ∅; {9}

We see that every solution of $x = a$ is a solution of $x^2 = a^2$, but not every solution of $x^2 = a^2$ is necessarily a solution of $x = a$. A similar statement can be made for any even exponent n; for example, the solution sets of $x^4 = 3^4$ and $x = 3$ are not the same. For this reason, when we raise both sides of an equation to an even power, it will be essential for us to check the answer in the original equation. We *will* find all the solutions; we *may* also find additional numbers that are not solutions of the original equation.

EXAMPLE 1   Solve: $\sqrt{x} - 3 = 4$.

Algebraic Solution

We have

$$\sqrt{x} - 3 = 4$$

$$\sqrt{x} = 7 \qquad \text{Adding 3 to both sides to isolate the radical}$$

$$\left(\sqrt{x}\right)^2 = 7^2 \qquad \text{Using the principle of powers}$$

$$x = 49.$$

Check:
$$\begin{array}{c|c} \sqrt{x} - 3 = 4 \\ \hline \sqrt{49} - 3 \ ? \ 4 \\ 7 - 3 \\ 4 & 4 \qquad \text{TRUE} \end{array}$$

The solution is 49.

Graphical Solution

We let $f(x) = \sqrt{x} - 3$ and $g(x) = 4$. Since the domain of f is $[0, \infty)$, we try a viewing window of $[-1, 10, -10, 10]$.

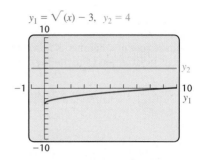

It appears as though the graphs may intersect at an x-value greater than 10. We adjust the viewing window to $[-10, 100, -5, 5]$ and find the point of intersection.

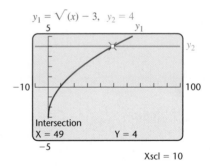

We have a solution of 49, which we can check by substituting into the original equation, as shown in the algebraic solution.

The solution is 49.

EXAMPLE 2   Solve: $\sqrt{x} - 5 = -7$.

Algebraic Solution

We have

$$\sqrt{x} - 5 = -7$$
$$\sqrt{x} = -2 \qquad \text{Adding 5 to both sides to isolate the radical}$$

The equation $\sqrt{x} = -2$ has no solution because the principal square root of a number is never negative. We continue as in Example 1 for comparison.

$$\left(\sqrt{x}\right)^2 = (-2)^2 \qquad \text{Using the principle of powers (squaring)}$$
$$x = 4.$$

Check:

$$\begin{array}{c|c} \sqrt{x} - 5 = -7 \\ \hline \sqrt{4} - 5 \;?\; -7 \\ 2 - 5 \\ -3 \;\big|\; -7 \qquad \text{FALSE} \end{array}$$

The number 4 does not check. There is no real-number solution.

Graphical Solution

We let $f(x) = \sqrt{x} - 5$ and $g(x) = -7$ and graph.

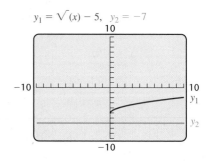

$y_1 = \sqrt{(x)} - 5, \quad y_2 = -7$

The graphs do not intersect. There is no real-number solution.

> **CAUTION!**   Raising both sides of an equation to an even power may not produce an equivalent equation. In this case, a check is essential.

Note in Example 2 that 4 is the solution of $x = 4$, but that $\sqrt{x} - 5 = -7$ has *no* solution. Thus the equations $x = 4$ and $\sqrt{x} - 5 = -7$ are *not* equivalent.

To Solve an Equation with a Radical Term

1. Isolate the radical term on one side of the equation.

2. Use the principle of powers and solve the resulting equation.

3. Check any possible solution in the original equation.

EXAMPLE 3   Solve: $x = \sqrt{x + 7} + 5$.

Algebraic Solution

We have

$$x = \sqrt{x + 7} + 5$$

$$x - 5 = \sqrt{x + 7}$$
Subtracting 5 from both sides. This isolates the radical term.

$$\left. \begin{array}{l} (x - 5)^2 = \left(\sqrt{x + 7}\right)^2 \\ x^2 - 10x + 25 = x + 7 \end{array} \right\}$$
Using the principle of powers; squaring both sides

$$x^2 - 11x + 18 = 0$$
Adding $-x - 7$ to both sides to write the quadratic equation in standard form

$$(x - 9)(x - 2) = 0$$
Factoring

$$x = 9 \quad or \quad x = 2.$$
Using the principle of zero products

The possible solutions are 9 and 2. Let's check.

Check: For 9:

$$\begin{array}{c|c} x = \sqrt{x + 7} + 5 \\ \hline 9 ? \sqrt{9 + 7} + 5 \\ 9 \mid 9 \end{array} \quad \text{TRUE}$$

For 2:

$$\begin{array}{c|c} x = \sqrt{x + 7} + 5 \\ \hline 2 ? \sqrt{2 + 7} + 5 \\ 2 \mid 8 \end{array} \quad \text{FALSE}$$

Since 9 checks but 2 does not, the solution is 9.

Graphical Solution

We graph $y_1 = x$ and $y_2 = \sqrt{(x + 7)} + 5$ and determine any points of intersection.

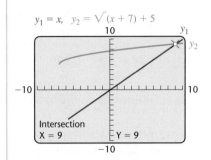

We obtain a solution of 9.

It is important to isolate a radical term before using the principle of powers. Suppose in Example 3 that both sides of the equation were squared *before* isolating the radical. We then would have had the expression $\left(\sqrt{x + 7} + 5\right)^2$ or $x + 7 + 10\sqrt{x + 7} + 25$ in the third step of the algebraic solution, and the radical would have remained in the problem.

EXAMPLE 4 Solve: $(2x + 1)^{1/3} + 5 = 0$.

Algebraic Solution

We need not use radical notation to solve:

$$(2x + 1)^{1/3} + 5 = 0$$

$(2x + 1)^{1/3} = -5$ Subtracting 5 from both sides

$[(2x + 1)^{1/3}]^3 = (-5)^3$ Cubing both sides

$(2x + 1)^1 = (-5)^3$ Multiplying exponents. Try to do this mentally.

$2x + 1 = -125$

$2x = -126$ Subtracting 1 from both sides

$x = -63.$

Because both sides were raised to an *odd* power, we need check only the accuracy of our work.

Graphical Solution

We let $f(x) = (2x + 1)^{1/3} + 5$ and determine any values of a for which $f(a) = 0$. We might start by graphing f using a standard viewing window.

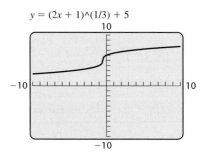

$y = (2x + 1)^\wedge(1/3) + 5$

It appears that if the graph of f does have an x-intercept, it will be at a value of x less than -10. Let's try a viewing window of $[-100, 100, -10, 10]$.

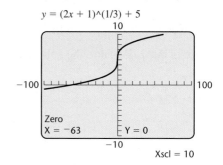

$y = (2x + 1)^\wedge(1/3) + 5$

Zero
X = −63   Y = 0

Xscl = 10

We find that $f(x) = 0$ when $x = -63$.

Since both approaches yield the same solution, our work is probably correct. As a final check, the student can show that -63 checks in the original equation and is the solution.

Equations with Two Radical Terms

A strategy for solving equations with two or more radical terms is as follows.

To Solve an Equation with Two or More Radical Terms

1. Isolate one of the radical terms.

2. Use the principle of powers.

3. If a radical remains, perform steps (1) and (2) again.

4. Solve the resulting equation.

5. Check possible solutions in the original equation.

EXAMPLE 5   Solve: $\sqrt{2x - 5} = 1 + \sqrt{x - 3}$.

Algebraic Solution

We have

$$\sqrt{2x - 5} = 1 + \sqrt{x - 3}$$
$$(\sqrt{2x - 5})^2 = (1 + \sqrt{x - 3})^2 \qquad \text{One radical is already isolated. We square both sides.}$$

> This is like squaring a binomial. We square 1 and then find twice the product of 1 and $\sqrt{x - 3}$ and then the square of $\sqrt{x - 3}$.

$$2x - 5 = 1 + 2\sqrt{x - 3} + (\sqrt{x - 3})^2$$
$$2x - 5 = 1 + 2\sqrt{x - 3} + (x - 3)$$
$$x - 3 = 2\sqrt{x - 3} \qquad \text{Isolating the remaining radical term}$$
$$(x - 3)^2 = (2\sqrt{x - 3})^2 \qquad \text{Squaring both sides}$$
$$x^2 - 6x + 9 = 4(x - 3) \qquad \text{Remember to square both the 2 and the } \sqrt{x - 3} \text{ on the right side.}$$

$$x^2 - 6x + 9 = 4x - 12$$
$$x^2 - 10x + 21 = 0$$
$$(x - 7)(x - 3) = 0 \qquad \text{Factoring}$$
$$x = 7 \quad or \quad x = 3. \qquad \text{Using the principle of zero products}$$

The possible solutions are 7 and 3.

Graphical Solution

We let $f(x) = \sqrt{2x - 5}$ and $g(x) = 1 + \sqrt{x - 3}$. The domain of f is $[\frac{5}{2}, \infty)$, and the domain of g is $[3, \infty)$. Since the intersection of their domains is $[3, \infty)$, any point of intersection will occur in that interval. We choose a viewing window of $[-1, 10, -1, 5]$.

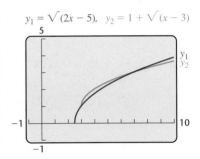

$y_1 = \sqrt{(2x - 5)}, \quad y_2 = 1 + \sqrt{(x - 3)}$

The graphs appear to intersect when x is approximately 3 and again when x is approximately 7. The points of intersection are $(3, 1)$ and $(7, 3)$. To confirm that there are no other points of intersection, we could examine the graphs using a larger viewing window.

We leave it to the student to show that 7 and 3 both check in the original equation and are the solutions.

> **CAUTION!**   A common error in solving equations like
>
> $$\sqrt{2x - 5} = 1 + \sqrt{x - 3}$$
>
> is to obtain $1 + (x - 3)$ as the square of the right side. This is wrong because $(A + B)^2 \neq A^2 + B^2$—for example,
>
> $$(1 + 2)^2 \neq 1^2 + 2^2$$
> $$3^2 \neq 1 + 4$$
> $$9 \neq 5.$$

EXAMPLE 6 Let $f(x) = \sqrt{x + 5} - \sqrt{x - 3}$. Find all x-values for which $f(x) = 2$.

Algebraic Solution

We must have $f(x) = 2$, or

$\sqrt{x + 5} - \sqrt{x - 3} = 2$. **Substituting for $f(x)$**

To solve, we isolate one radical term and square both sides:

$\sqrt{x + 5} = 2 + \sqrt{x - 3}$ **Adding $\sqrt{x - 3}$. This isolates one of the radical terms.**

$\left(\sqrt{x + 5}\right)^2 = \left(2 + \sqrt{x - 3}\right)^2$ **Using the principle of powers (squaring both sides)**

$x + 5 = 4 + 4\sqrt{x - 3} + (x - 3)$ **Using $(A + B)^2 = A^2 + 2AB + B^2$**

$5 = 1 + 4\sqrt{x - 3}$ **Adding $-x$ and combining like terms**

$\left.\begin{array}{l} 4 = 4\sqrt{x - 3} \\ 1 = \sqrt{x - 3} \end{array}\right\}$ **Isolating the remaining radical term**

$1^2 = \left(\sqrt{x - 3}\right)^2$ **Squaring both sides**

$1 = x - 3$

$4 = x$.

The solution is 4.

Graphical Solution

We graph $y_1 = \sqrt{(x + 5)} - \sqrt{(x - 3)}$ and $y_2 = 2$ and determine the coordinates of any points of intersection.

$y_1 = \sqrt{(x + 5)} - \sqrt{(x - 3)}, \quad y_2 = 2$

The solution is the x-coordinate of the point of intersection, which is 4.

Check: $f(4) = \sqrt{4 + 5} - \sqrt{4 - 3} = \sqrt{9} - \sqrt{1} = 3 - 1 = 2.$

We will have $f(x) = 2$ when $x = 4$.

7.6

Exercise Set

FOR EXTRA HELP

Digital Video Tutor CD 7 Videotape 10 | Student's Solutions Manual | Tutor Center AW Math Tutor Center | InterAct Math | MathXL | MyMathLab

Solve.

1. $\sqrt{x + 3} = 5$ 22

2. $\sqrt{5x + 1} = 8$ $\frac{63}{5}$

3. $\sqrt{2x - 1} = 2$ $\frac{9}{2}$

4. $\sqrt{3x + 1} = 6$ $\frac{25}{3}$

5. $\sqrt{x - 2} - 7 = -4$ 11

6. $\sqrt{y + 1} - 5 = 8$ 168

7. $\sqrt{y + 4} + 6 = 7$ -3

8. $\sqrt{x - 7} + 3 = 10$ 56

9. $\sqrt[3]{x - 2} = 3$ 29

10. $\sqrt[3]{x + 5} = 2$ 3

11. $\sqrt[4]{x + 3} = 2$ 13

12. $\sqrt[4]{y - 1} = 3$ 82

13. $8\sqrt{y} = y$ 0, 64

14. $3\sqrt{x} = x$ 0, 9

15. $3x^{1/2} + 12 = 9$
No solution

16. $2y^{1/2} - 7 = 9$ 64

17. $\sqrt[3]{y} = -4$ -64

18. $\sqrt[3]{x} = -3$ -27

19. $x^{1/4} - 2 = 1$ 81

20. $t^{1/3} - 2 = 3$ 125

Aha! **21.** $(y - 3)^{1/2} = -2$
No solution

22. $(x + 2)^{1/2} = -4$
No solution

23. $\sqrt[4]{3x + 1} - 4 = -1$ $\frac{80}{3}$

24. $\sqrt[4]{2x + 3} - 5 = -2$ 39

25. $(x + 7)^{1/3} = 4$ 57   **26.** $(y - 7)^{1/4} = 3$ 88

27. $\sqrt[3]{3y + 6} + 7 = 8$ $-\frac{5}{3}$ **28.** $\sqrt[3]{6x + 9} + 5 = 2$ -6

29. $\sqrt{3t + 4} = \sqrt{4t + 3}$ 1   **30.** $\sqrt{2t - 7} = \sqrt{3t - 12}$ 5

31. $3(4 - t)^{1/4} = 6^{1/4}$ $\frac{106}{27}$ **32.** $2(1 - x)^{1/3} = 4^{1/3}$ $\frac{1}{2}$

33. $3 + \sqrt{5 - x} = x$ 4   **34.** $x = \sqrt{x - 1} + 3$ 5

35. $\sqrt{4x - 3} = 2 + \sqrt{2x - 5}$ 3, 7

36. $3 + \sqrt{z - 6} = \sqrt{z + 9}$ 7

37. $\sqrt{20 - x} + 8 = \sqrt{9 - x} + 11$ $\frac{80}{9}$

38. $4 + \sqrt{10 - x} = 6 + \sqrt{4 - x}$ $\frac{15}{4}$

39. $\sqrt{x + 2} + \sqrt{3x + 4} = 2$ -1

40. $\sqrt{6x + 7} - \sqrt{3x + 3} = 1$ $-1, \frac{1}{3}$

41. If $f(x) = \sqrt{x} + \sqrt{x - 9}$, find x such that $f(x) = 1$.   No solution

42. If $g(x) = \sqrt{x} + \sqrt{x - 5}$, find x such that $g(x) = 5$.   9

43. If $f(x) = \sqrt{x - 2} - \sqrt{4x + 1}$, find a such that $f(a) = -3$.   2, 6

44. If $g(x) = \sqrt{2x + 7} - \sqrt{x + 15}$, find a such that $g(a) = -1$.   1

45. If $f(x) = \sqrt{2x - 3}$ and $g(x) = \sqrt{x + 7} - 2$, find x such that $f(x) = g(x)$.   2

46. If $f(x) = 2\sqrt{3x + 6}$ and $g(x) = 5 + \sqrt{4x + 9}$, find x such that $f(x) = g(x)$.   10

47. If $f(t) = 4 - \sqrt{t - 3}$ and $g(t) = (t + 5)^{1/2}$, find a such that $f(a) = g(a)$.   4

48. If $f(t) = 7 + \sqrt{2t - 5}$ and $g(t) = 3(t + 1)^{1/2}$, find a such that $f(a) = g(a)$.   15

^{TW} **49.** Explain in your own words why it is important to check your answers when using the principle of powers.

^{TW} **50.** Describe a procedure that could be used to write radical equations that have no solution.

Skill Maintenance

51. The base of a triangle is 2 in. longer than the height. The area is $31\frac{1}{2}$ in^2. Find the height and the base. [2.3]   Height: 7 in.; base: 9 in.

52. During a one-hour television show, there were 12 commercials. Some of the commercials were 30 sec long and the others were 60 sec long. If the number of 30-sec commercials was 6 less than the total number of minutes of commercial time during the show, how many 60-sec commercials were used? [3.3]   8 60-sec commercials

⊡ Answers to Exercises 55, 56, and 61 can be found on p. A-64.

53. Elaine can sew a quilt in 6 fewer hours than Gonzalo can. When they work together, it takes them 4 hr. How long would it take each of them alone to sew the quilt? [6.5]   Elaine: 6 hr; Gonzalo: 12 hr

54. Jackie can paint an apartment in 5.5 hr. Her partner, Grant, can paint the same apartment in 7.5 hr. How long would it take the two of them, working together, to paint the apartment? [6.5] $\frac{165}{52}$ hr, or about 3.2 hr

Graph.

55. $y > 3x + 5$ [4.4]   ⊡   **56.** $f(x) = \frac{2}{3}x - 7$ [2.4]   ⊡

Synthesis

^{TW} **57.** The principle of powers is an "if–then" statement that becomes false when the sentence parts are interchanged. Give an example of another such if–then statement.

^{TW} **58.** Is checking essential when the principle of powers is used with an odd power n? Why or why not?

Steel Manufacturing.   *In the production of steel and other metals, the temperature of the molten metal is so great that conventional thermometers melt. Instead, sound is transmitted across the surface of the metal to a receiver on the far side and the speed of the sound is measured. The formula*

$$S(t) = 1087.7 \sqrt{\frac{9t + 2617}{2457}}$$

gives the speed of sound $S(t)$, in feet per second, at a temperature of t degrees Celsius.

59. Find the temperature of a blast furnace in which sound travels 1502.3 ft/sec.   230°C

60. Find the temperature of a blast furnace in which sound travels 1880 ft/sec.   524.8°C

61. Solve the above equation for t.   ⊡

Automotive Repair. *For an engine with a displacement of 2.8 L, the function given by*

$$d(n) = 0.75\sqrt{2.8n}$$

can be used to determine the diameter size of the carburetor's opening, in millimeters. Here n is the number of rpm's at which the engine achieves peak performance. (Source: macdizzy.com)

62. If a carburetor's opening is 81 mm, for what number of rpm's will the engine produce peak power?
  About 4166 rpm

63. If a carburetor's opening is 84 mm, for what number of rpm's will the engine produce peak power?
  4480 rpm

Escape Velocity. *A formula for the escape velocity v of a satellite is*

$$v = \sqrt{2gr}\ \sqrt{\frac{h}{r+h}},$$

where g is the force of gravity, r is the planet or star's radius, and h is the height of the satellite above the planet or star's surface.

64. Solve for *h.* $h = \dfrac{v^2 r}{2gr - v^2}$ **65.** Solve for *r.* $r = \dfrac{v^2 h}{2gh - v^2}$

Sighting to the Horizon. *The function $D(h) = 1.2\sqrt{h}$ can be used to approximate the distance D, in miles, that a person can see to the horizon from a height h, in feet.*

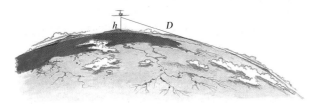

66. How far above sea level must a pilot fly in order to see a horizon that is 180 mi away? 22,500 ft

67. How high above sea level must a sailor climb in order to see 10.2 mi out to sea? 72.25 ft

Solve.

68. $\dfrac{x + \sqrt{x+1}}{x - \sqrt{x+1}} = \dfrac{5}{11}$ $-\frac{8}{9}$

69. $\left(\dfrac{z}{4} - 5\right)^{2/3} = \dfrac{1}{25}$ $\frac{2504}{125}, \frac{2496}{125}$

70. $(z^2 + 17)^{3/4} = 27$ $-8, 8$

71. $\sqrt{\sqrt{y} + 49} = 7$ 0

72. $x^2 - 5x - \sqrt{x^2 - 5x - 2} = 4$ $-1, 6$
  (*Hint:* Let $u = x^2 - 5x - 2$.)

73. $\sqrt{8 - b} = b\sqrt{8 - b}$ $1, 8$

Without graphing, determine the x-intercepts of the graphs given by each of the following.

74. $f(x) = \sqrt{x - 2} - \sqrt{x + 2} + 2$ $(2, 0)$

75. $g(x) = 6x^{1/2} + 6x^{-1/2} - 37$ $\left(\frac{1}{36}, 0\right), (36, 0)$

76. $f(x) = (x^2 + 30x)^{1/2} - x - (5x)^{1/2}$ $(0, 0), \left(\frac{125}{4}, 0\right)$

TW **77.** Saul is trying to solve Exercise 67 using a graphing calculator. Without resorting to trial and error, how can he determine a suitable viewing window for finding the solution?

Tailgater Alert

Focus: Radical equations and problem solving

Time: 15–25 minutes

Group size: 2–3

Materials: Calculators or square-root tables

The faster a car is traveling, the more distance it needs to stop. Thus it is important for drivers to allow sufficient space between their vehicle and the vehicle in front of them. Police recommend that for each 10 mph of speed, a driver allow 1 car length. Thus a driver going 30 mph should have at least 3 car lengths between his or her vehicle and the one in front.

In Exercise Set 7.3, the function $r(L) = 2\sqrt{5L}$ was used to find the speed, in miles per hour, that a car was traveling when it left skid marks L feet long.

ACTIVITY

1. Each group member should estimate the length of a car in which he or she frequently travels. (Each should use a different length, if possible.)
2. Using a calculator as needed, each group member should complete the table at right. Column 1

gives a car's speed s, column 2 lists the minimum amount of space between cars traveling s miles per hour, as recommended by police. Column 3 is the speed that a vehicle *could* travel were it forced to stop in the distance listed in column 2, using the above function.

Column 1 s (in miles per hour)	Column 2 L(s) (in feet)	Column 3 r(L) (in miles per hour)
20		
30		
40		
50		
60		
70		

3. Determine whether there are any speeds at which the "1 car length per 10 mph" guideline might not suffice. On what reasoning do you base your answer? Compare tables to determine how car length affects the results. What recommendations would your group make to a new driver?

7.7

Using the Pythagorean Theorem   ▪   Two Special Triangles

Geometric Applications

Using the Pythagorean Theorem

There are many kinds of problems that involve powers and roots. Many also involve right triangles and the Pythagorean theorem, which we studied in Section 5.8 and restate here.

The Pythagorean Theorem[*]

In any right triangle, if a and b are the lengths of the legs and c is the length of the hypotenuse, then

$$a^2 + b^2 = c^2.$$

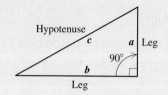

Hypotenuse

c a | Leg

90°

b

Leg

In using the Pythagorean theorem, we often make use of the following principle.

The Principle of Square Roots

For any nonnegative real number n,

If $x^2 = n$, then $x = \sqrt{n}$ or $x = -\sqrt{n}$.

When we apply the principle of square roots to find a length, we use only the positive root, since lengths are not negative.

EXAMPLE 1   Baseball.   A baseball diamond is actually a square 90 ft on a side. Suppose a catcher fields a ball along the third-base line 10 ft from home plate. How far would the catcher's throw to first base be? Give an exact answer and an approximation to three decimal places.

Solution   We first make a drawing and let $d =$ the distance, in feet, to first base. Note that a right triangle is formed in which the length of the leg from home to first base is 90 ft. The length of the leg from home to where the catcher fields the ball is 10 ft.

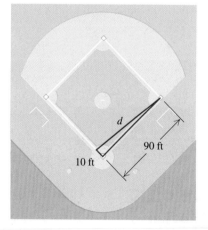

d

90 ft

10 ft

TEACHING TIP

You may wish to point out that $\sqrt{90^2 + 10^2} \ne 90 + 10$. Before calculating a square root, we must simplify the radicand.

---

[*]The converse of the Pythagorean theorem also holds. That is, if a, b, and c are the lengths of the sides of a triangle and $a^2 + b^2 = c^2$, then the triangle is a right triangle.

We substitute these values into the Pythagorean theorem to find d:

$$d^2 = 90^2 + 10^2$$
$$d^2 = 8100 + 100$$
$$d^2 = 8200.$$

We now use the principle of square roots. Since d is a length, it follows that d is the positive square root of 8200:

$d = \sqrt{8200}$ ft   **This is an exact answer.**

$d \approx 90.554$ ft.   **Using a calculator for an approximation**

EXAMPLE 2   Guy Wires.   The base of a 40-ft–long guy wire is located 15 ft from the telephone pole that it is anchoring. How high up the pole does the guy wire reach? Give an exact answer and an approximation to three decimal places.

Solution   We make a drawing and let $h =$ the height on the pole that the guy wire reaches. A right triangle is formed in which the length of one leg is 15 ft and the length of the hypotenuse is 40 ft. Using the Pythagorean theorem, we have

$$h^2 + 15^2 = 40^2$$
$$h^2 + 225 = 1600$$
$$h^2 = 1375$$
$$h = \sqrt{1375}.$$

Exact answer:
$$h = \sqrt{1375} \text{ ft}$$

Approximation:

$h \approx 37.081$ ft   Using a calculator

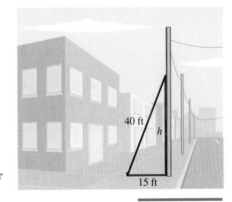

Two Special Triangles

When both legs of a right triangle are the same size, we call the triangle an *isosceles right triangle*, as shown at left. If one leg of an isosceles right triangle has length a, we can find a formula for the length of the hypotenuse as follows:

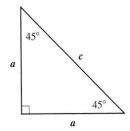

$c^2 = a^2 + b^2$

$c^2 = a^2 + a^2$ **Because the triangle is isosceles, both legs are the same size: $a = b$.**

$c^2 = 2a^2$.   **Combining like terms**

Next, we use the principle of square roots. Because a, b, and c are lengths, there is no need to consider negative square roots or absolute values. Thus,

$$c = \sqrt{2a^2} \quad \text{Using the principle of square roots}$$
$$c = \sqrt{a^2 \cdot 2} = a\sqrt{2}.$$

EXAMPLE 3 One leg of an isosceles right triangle measures 7 cm. Find the length of the hypotenuse. Give an exact answer and an approximation to three decimal places.

Solution   We substitute:

$$c = a\sqrt{2} \quad \text{This equation is worth memorizing.}$$
$$c = 7\sqrt{2}.$$

Exact answer: $c = 7\sqrt{2}$ cm

Approximation: $c \approx 9.899$ cm   **Using a calculator**

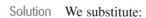

When the hypotenuse of an isosceles right triangle is known, the lengths of the legs can be found.

EXAMPLE 4 The hypotenuse of an isosceles right triangle is 5 ft long. Find the length of a leg. Give an exact answer and an approximation to three decimal places.

Solution   We replace c with 5 and solve for a:

$$5 = a\sqrt{2} \quad \text{Substituting 5 for } c \text{ in } c = a\sqrt{2}$$
$$\frac{5}{\sqrt{2}} = a \quad \text{Dividing both sides by } \sqrt{2}$$
$$\frac{5\sqrt{2}}{2} = a. \quad \text{Rationalize the denominator if desired.}$$

Exact answer: $a = \dfrac{5}{\sqrt{2}}$ ft, or $\dfrac{5\sqrt{2}}{2}$ ft

Approximation: $a \approx 3.536$ ft   **Using a calculator**

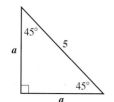

A second special triangle is known as a $30°-60°-90°$ right triangle, so named because of the measures of its angles. Note that in an equilateral triangle, all sides have the same length and all angles are $60°$. An altitude, drawn dashed in the figure, bisects, or splits in half, one angle and one side. Two $30°-60°-90°$ right triangles are thus formed.

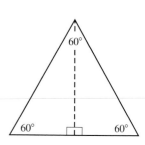

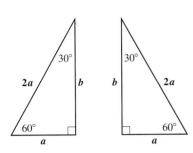

Because of the way in which the altitude is drawn, if a represents the length of the shorter leg in a $30°-60°-90°$ right triangle, then $2a$ represents the length of the hypotenuse. We have

$$a^2 + b^2 = (2a)^2 \qquad \text{Using the Pythagorean theorem}$$
$$a^2 + b^2 = 4a^2$$
$$b^2 = 3a^2 \qquad \text{Adding } -a^2 \text{ to both sides}$$
$$b = \sqrt{3a^2}$$
$$b = \sqrt{a^2 \cdot 3} = a\sqrt{3}.$$

EXAMPLE 5   The shorter leg of a $30°-60°-90°$ right triangle measures 8 in. Find the lengths of the other sides. Give exact answers and, where appropriate, an approximation to three decimal places.

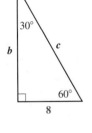

Solution   The hypotenuse is twice as long as the shorter leg, so we have

$$c = 2a \qquad \text{This relationship is worth memorizing.}$$
$$= 2 \cdot 8 = 16 \text{ in.}$$

The length of the longer leg is the length of the shorter leg times $\sqrt{3}$. This gives us

$$b = a\sqrt{3} \qquad \text{This is also worth memorizing.}$$
$$= 8\sqrt{3} \text{ in.}$$

Exact answer: $c = 16$ in., $b = 8\sqrt{3}$ in.

Approximation: $b \approx 13.856$ in.

EXAMPLE 6   The length of the longer leg of a $30°-60°-90°$ right triangle is 14 cm. Find the length of the hypotenuse. Give an exact answer and an approximation to three decimal places.

Solution   The length of the hypotenuse is twice the length of the shorter leg. We first find a, the length of the shorter leg, by using the length of the longer leg:

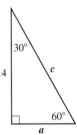

$$14 = a\sqrt{3} \qquad \text{Substituting 14 for } b \text{ in } b = a\sqrt{3}$$
$$\frac{14}{\sqrt{3}} = a. \qquad \text{Dividing by } \sqrt{3}$$

Since the hypotenuse is twice as long as the shorter leg, we have

$$c = 2 \cdot \frac{14}{\sqrt{3}} \qquad \text{Substituting } \frac{14}{\sqrt{3}} \text{ for } a \text{ in } c = 2a$$
$$= \frac{28}{\sqrt{3}} \text{ cm.}$$

Exact answer: $c = \dfrac{28}{\sqrt{3}}$ cm, or $\dfrac{28\sqrt{3}}{3}$ cm if the denominator is rationalized.

Approximation: $c \approx 16.166$ cm

Lengths Within Isosceles and 30°–60°–90° Right Triangles

The length of the hypotenuse in an isosceles right triangle is the length of a leg times $\sqrt{2}$.

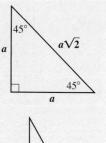

The length of the longer leg in a 30°–60°–90° right triangle is the length of the shorter leg times $\sqrt{3}$. The hypotenuse is twice as long as the shorter leg.

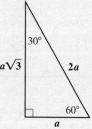

TEACHING TIP

Remind students that these are special types of right triangles. If they forget these relationships, they can use the Pythagorean theorem to derive them.

7.7

Exercise Set

FOR EXTRA HELP

Digital Video Tutor CD 7 Videotape 10 | Student's Solutions Manual | Tutor Center AW Math Tutor Center | InterAct Math | MathXL | MyMathLab

In a right triangle, find the length of the side not given. Give an exact answer and, where appropriate, an approximation to three decimal places.

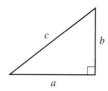

1. $a = 5, b = 3$ $\sqrt{34}$; 5.831   **2.** $a = 8, b = 10$ $\sqrt{164}$; 12.806

Aha! **3.** $a = 9, b = 9$ $9\sqrt{2}$; 12.728   **4.** $a = 10, b = 10$ $10\sqrt{2}$; 14.142

5. $b = 12, c = 13$ 5   **6.** $a = 5, c = 12$ $\sqrt{119}$; 10.909

7. $c = 6, a = \sqrt{5}$ $\sqrt{31}$; 5.568   **8.** $c = 8, a = 4\sqrt{3}$ 4

9. $b = 2, c = \sqrt{15}$ $\sqrt{11}$; 3.317   **10.** $a = 1, c = \sqrt{20}$ $\sqrt{19}$; 4.359

Aha! **11.** $a = 1, c = \sqrt{2}$ 1   **12.** $c = 2, a = 1$ $\sqrt{3}$; 1.732

In Exercises 13–20, give an exact answer and, where appropriate, an approximation to three decimal places.

13. *Guy Wire.* How long is a guy wire if it reaches from the top of a 15-ft pole to a point on the ground 10 ft from the pole? $\sqrt{325}$ ft; 18.028 ft

14. *Softball.* A slow-pitch softball diamond is actually a square 65 ft on a side. How far is it from home to second base? $\sqrt{8450}$, or $65\sqrt{2}$ ft; 91.924 ft

15. *Baseball.* Suppose the catcher in Example 1 makes a throw to second base from the same location. How far is that throw? $\sqrt{14{,}500}$ ft; 120.416 ft

16. *Television Sets.* What does it mean to refer to a 20-in. TV set or a 25-in. TV set? Such units refer to the diagonal of the screen. A 20-in. TV set has a width of 16 in. What is its height?   12 in.

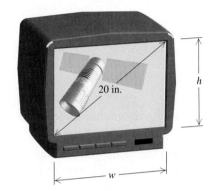

17. *Television Sets.* A 25-in. TV set has a screen with a height of 15 in. What is its width? (See Exercise 16.)   20 in.

18. *Speaker Placement.* A stereo receiver is in a corner of a 12-ft by 14-ft room. Speaker wire will run under a rug, diagonally, to a speaker in the far corner. If 4 ft of slack is required on each end, how long a piece of wire should be purchased? $\left(\sqrt{340} + 8\right)$ ft; 26.439 ft

19. *Distance over Water.* To determine the width of a pond, a surveyor locates two stakes at either end of the pond and uses instrumentation to place a third stake so that the distance across the pond is the length of a hypotenuse. If the third stake is 90 m from one stake and 70 m from the other, how wide is the pond? $\sqrt{13,000}$ m; 114.018 m

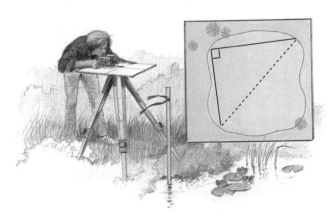

⊡ Answers to Exercises 25–28 can be found on p. A-64.

20. *Vegetable Garden.* Benito and Dominique are planting a 30-ft by 40-ft vegetable garden and are laying it out using string. They would like to know the length of a diagonal to make sure that right angles are formed. Find the length of a diagonal.   50 ft

For each triangle, find the missing length(s). Give an exact answer and, where appropriate, an approximation to three decimal places.

21.

$a = 5; c = 5\sqrt{2} \approx 7.071$

22.

$a = 14; c = 14\sqrt{2} \approx 19.799$

23.
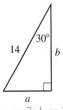
$a = 7; b = 7\sqrt{3} \approx 12.124$

24.
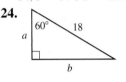
$a = 9; b = 9\sqrt{3} \approx 15.588$

25. ⊡
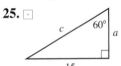

26. ⊡
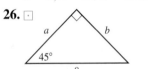

27. ⊡
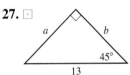

28. ⊡
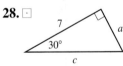

29.
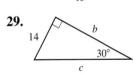
$b = 14\sqrt{3} \approx 24.249; c = 28$

30.
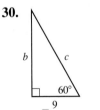
$b = 9\sqrt{3} \approx 15.588; c = 18$

31.
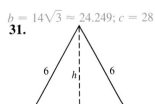
$3\sqrt{3} \approx 5.196$

32.
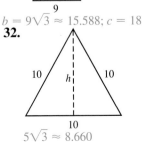
$5\sqrt{3} \approx 8.660$

33.

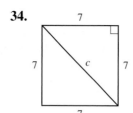

13

$13\sqrt{2} \approx 18.385$

34.

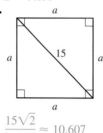

7

$7\sqrt{2} \approx 9.899$

35.

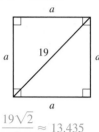

$\frac{19\sqrt{2}}{2} \approx 13.435$

36.

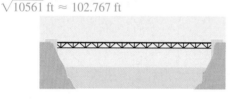

$\frac{15\sqrt{2}}{2} \approx 10.607$

In Exercises 37–42, give an exact answer and, where appropriate, an approximation to three decimal places.

37. *Bridge Expansion.* During the summer heat, a 2-mi bridge expands 2 ft in length. If we assume that the bulge occurs straight up the middle, how high is the bulge? (The answer may surprise you. Most bridges have expansion spaces to avoid such buckling.) $\sqrt{10561}$ ft ≈ 102.767 ft

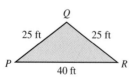

38. Triangle *ABC* has sides of lengths 25 ft, 25 ft, and 30 ft. Triangle *PQR* has sides of lengths 25 ft, 25 ft, and 40 ft. Which triangle has the greater area and by how much? Neither; they have the same area, 300 ft²

39. *Camping.* The entrance to a pup tent is the shape of an equilateral triangle. If the base of the tent is 4 ft wide, how tall is the tent? $h = 2\sqrt{3}$ ft ≈ 3.464 ft

40. Each side of a regular octagon has length *s*. Find a formula for the distance *d* between the parallel sides of the octagon. $d = s + s\sqrt{2}$

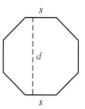

41. The diagonal of a square has length $8\sqrt{2}$ ft. Find the length of a side of the square. 8 ft

42. The length and the width of a rectangle are given by consecutive integers. The area of the rectangle is 90 cm². Find the length of a diagonal of the rectangle. $\sqrt{181}$ cm ≈ 13.454 cm

43. Find all points on the *y*-axis of a Cartesian coordinate system that are 5 units from the point (3, 0). (0, −4), (0, 4)

44. Find all points on the *x*-axis of a Cartesian coordinate system that are 5 units from the point (0, 4). (−3, 0), (3, 0)

TW 45. Write a problem for a classmate to solve in which the solution is: "The height of the tepee is $5\sqrt{3}$ yd."

TW 46. Write a problem for a classmate to solve in which the solution is: "The height of the window is $15\sqrt{3}$ ft."

Skill Maintenance

Simplify. [1.4]
47. $47(-1)^{19}$ −47
48. $(-5)(-1)^{13}$ 5

Factor. [5.3]
49. $x^3 - 9x$ $x(x-3)(x+3)$
50. $7a^3 - 28a$ $7a(a-2)(a+2)$

Solve. [4.3]
51. $|3x - 5| = 7$ $\{-\frac{2}{3}, 4\}$
52. $|2x - 3| = |x + 7|$ $\{-\frac{4}{3}, 10\}$

Synthesis

TW 53. Are there any right triangles, other than those with sides measuring 3, 4, and 5, that have consecutive numbers for the lengths of the sides? Why or why not?

TW 54. If a 30°–60°–90° triangle and an isosceles right triangle have the same perimeter, which will have the greater area? Why?

55. A cube measures 5 cm on each side. How long is the diagonal that connects two opposite corners of the cube? Give an exact answer. $\sqrt{75}$ cm

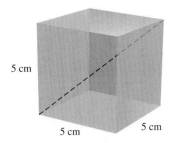

5 cm

5 cm   5 cm

56. *Roofing.*   Kit's cottage, which is 24 ft wide and 32 ft long, needs a new roof. By counting clapboards that are 4 in. apart, Kit determines that the peak of the roof is 6 ft higher than the sides. If one packet of shingles covers 100 square feet, how many packets will the job require?   9 packets

6 ft

10 ft

32 ft   24 ft

57. *Painting.*   (Refer to Exercise 56.) A gallon of paint covers about 275 square feet. If Kit's first floor is 10 ft high, how many gallons of paint should be bought to paint the house? What assumption(s) is made in your answer?   4 gal. The total area of the doors and windows is 164 ft² or more.

58. *Contracting.*   Oxford Builders has an extension cord on their generator that permits them to work, with electricity, anywhere in a circular area of 3850 ft². Find the dimensions of the largest square room they could work on without having to relocate the generator to reach each corner of the floor plan.   49.5 ft by 49.5 ft

59. *Contracting.*   Cerrelli Construction has an extension cord on their generator that permits them to work, with electricity, anywhere in a circular area of 6160 ft². Find the dimensions of the largest cube-shaped room they could work on without having to relocate the generator to reach the corners of the ceiling. Assume that the generator sits on the floor.   36.15 ft by 36.15 ft by 36.15 ft

---

7.8

The Complex Numbers

Imaginary and Complex Numbers ■ Addition and Subtraction ■ Multiplication ■ Conjugates and Division ■ Powers of *i*

Imaginary and Complex Numbers

Negative numbers do not have square roots in the real-number system. However, a larger number system that contains the real-number system is designed so that negative numbers *do* have square roots. That system is called the **complex-number system**, and it makes use of a number that is a square root of −1. We call this new number *i*.

The Number i

We define the number i such that $i = \sqrt{-1}$ and $i^2 = -1$.

To express roots of negative numbers in terms of i, we can use the fact that in the complex numbers, $\sqrt{-p} = \sqrt{-1}\,\sqrt{p} = i\sqrt{p}$ or $\sqrt{p}\,i$, for any positive number p.

EXAMPLE 1   Express in terms of i: **(a)** $\sqrt{-7}$; **(b)** $\sqrt{-16}$; **(c)** $-\sqrt{-13}$; **(d)** $-\sqrt{-50}$.

Solution

a) $\sqrt{-7} = \sqrt{-1 \cdot 7} = \sqrt{-1} \cdot \sqrt{7} = i\sqrt{7}$, or $\sqrt{7}\,i$ *i is **not** under the radical.*

b) $\sqrt{-16} = \sqrt{-1 \cdot 16} = \sqrt{-1} \cdot \sqrt{16} = i \cdot 4 = 4i$

c) $-\sqrt{-13} = -\sqrt{-1 \cdot 13} = -\sqrt{-1} \cdot \sqrt{13} = -i\sqrt{13}$, or $-\sqrt{13}\,i$

d) $-\sqrt{-50} = -\sqrt{-1} \cdot \sqrt{25} \cdot \sqrt{2} = -i \cdot 5 \cdot \sqrt{2} = -5i\sqrt{2}$, or $-5\sqrt{2}\,i$

Imaginary Numbers

An *imaginary number* is a number that can be written in the form $a + bi$, where a and b are real numbers and $b \neq 0$.

Don't let the name "imaginary" fool you. Imaginary numbers appear in fields such as engineering and the physical sciences. The following are examples of imaginary numbers:

$5 + 4i$,   Here $a = 5, b = 4.$

$\sqrt{5} - \pi i$,   Here $a = \sqrt{5}, b = -\pi.$

$17i$.   Here $a = 0, b = 17.$

When a and b are real numbers and b is allowed to be 0, the number $a + bi$ is said to be **complex**.

Complex Numbers

A *complex number* is any number that can be written in the form $a + bi$, where a and b are real numbers. (Note that a and b both can be 0.)

The following are examples of complex numbers:

$7 + 3i$ (here $a \neq 0, b \neq 0$); $4i$ (here $a = 0, b \neq 0$);

8 (here $a \neq 0, b = 0$); 0 (here $a = 0, b = 0$).

Complex numbers like $17i$ or $4i$, in which $a = 0$ and $b \neq 0$, are imaginary numbers with no real part. Such numbers are called *pure imaginary numbers*.

Note that when $b = 0$, we have $a + 0i = a$, so every real number is a complex number. The relationships among various real and complex numbers are shown below.

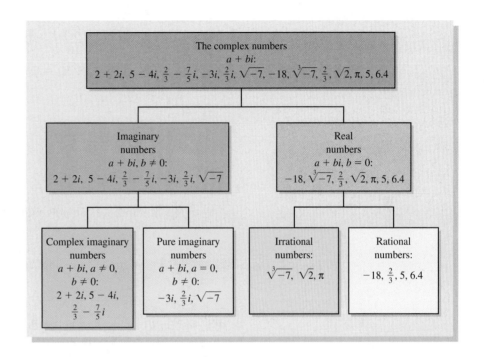

The complex numbers
$a + bi$:
$2 + 2i, \; 5 - 4i, \; \frac{2}{3} - \frac{7}{5}i, \; -3i, \; \frac{2}{3}i, \; \sqrt{-7}, \; -18, \; \sqrt[3]{-7}, \; \frac{2}{3}, \; \sqrt{2}, \; \pi, \; 5, \; 6.4$

Imaginary
numbers
$a + bi, b \neq 0$:
$2 + 2i, \; 5 - 4i, \; \frac{2}{3} - \frac{7}{5}i, \; -3i, \; \frac{2}{3}i, \; \sqrt{-7}$

Real
numbers
$a + bi, b = 0$:
$-18, \; \sqrt[3]{-7}, \; \frac{2}{3}, \; \sqrt{2}, \; \pi, \; 5, \; 6.4$

Complex imaginary
numbers
$a + bi, a \neq 0$,
$b \neq 0$:
$2 + 2i, \; 5 - 4i,$
$\frac{2}{3} - \frac{7}{5}i$

Pure imaginary
numbers
$a + bi, a = 0$,
$b \neq 0$:
$-3i, \; \frac{2}{3}i, \; \sqrt{-7}$

Irrational
numbers:
$\sqrt[3]{-7}, \; \sqrt{2}, \; \pi$

Rational
numbers:
$-18, \; \frac{2}{3}, \; 5, \; 6.4$

Note that although $\sqrt{-7}$ and $\sqrt[3]{-7}$ are both complex numbers, $\sqrt{-7}$ is imaginary whereas $\sqrt[3]{-7}$ is real.

Addition and Subtraction

The complex numbers obey the commutative, associative, and distributive laws. Thus we can add and subtract them as we do binomials.

EXAMPLE 2   Add or subtract and simplify.

a) $(8 + 6i) + (3 + 2i)$ **b)** $(4 + 5i) - (6 - 3i)$

Solution

a) $(8 + 6i) + (3 + 2i) = (8 + 3) + (6i + 2i)$ **Combining the real parts and the imaginary parts**

$$= 11 + (6 + 2)i = 11 + 8i$$

b) $(4 + 5i) - (6 - 3i) = (4 - 6) + [5i - (-3i)]$ **Note that the 6 and the $-3i$ are both being subtracted.**

$$= -2 + 8i$$

Multiplication

For complex numbers, the property $\sqrt{a} \ \sqrt{b} = \sqrt{ab}$ does *not* hold in general, but it does hold when $a = -1$ and b is nonnegative. To multiply square roots of negative real numbers, we first express them in terms of i. For example,

$$\sqrt{-2} \cdot \sqrt{-5} = \sqrt{-1} \cdot \sqrt{2} \cdot \sqrt{-1} \cdot \sqrt{5}$$
$$= i \cdot \sqrt{2} \cdot i \cdot \sqrt{5}$$
$$= i^2\sqrt{10}$$
$$= -1\sqrt{10} = -\sqrt{10} \text{ is correct!}$$

CAUTION! With complex numbers, simply multiplying radicands is *incorrect*: $\sqrt{-2} \cdot \sqrt{-5} \neq \sqrt{10}$.

With this in mind, we can now multiply complex numbers.

EXAMPLE 3 Multiply and simplify. When possible, write answers in the form $a + bi$.

a) $\sqrt{-16} \cdot \sqrt{-25}$ **b)** $\sqrt{-5} \cdot \sqrt{-7}$ **c)** $-3i \cdot 8i$

d) $-4i(3 - 5i)$ **e)** $(1 + 2i)(1 + 3i)$

Solution

a) $\sqrt{-16} \cdot \sqrt{-25} = \sqrt{-1} \cdot \sqrt{16} \cdot \sqrt{-1} \cdot \sqrt{25}$
$$= i \cdot 4 \cdot i \cdot 5$$
$$= i^2 \cdot 20$$
$$= -1 \cdot 20 \qquad i^2 = -1$$
$$= -20$$

b) $\sqrt{-5} \cdot \sqrt{-7} = \sqrt{-1} \cdot \sqrt{5} \cdot \sqrt{-1} \cdot \sqrt{7}$ **Try to do this step mentally.**
$$= i \cdot \sqrt{5} \cdot i \cdot \sqrt{7}$$
$$= i^2 \cdot \sqrt{35}$$
$$= -1 \cdot \sqrt{35} \qquad i^2 = -1$$
$$= -\sqrt{35}$$

c) $-3i \cdot 8i = -24 \cdot i^2$
$$= -24 \cdot (-1) \qquad i^2 = -1$$
$$= 24$$

d) $-4i(3 - 5i) = -4i \cdot 3 + (-4i)(-5i)$ **Using the distributive law**
$$= -12i + 20i^2$$
$$= -12i - 20 \qquad\qquad i^2 = -1$$
$$= -20 - 12i \qquad\qquad \textbf{Writing in the form } a + bi$$

e) $(1 + 2i)(1 + 3i) = 1 + 3i + 2i + 6i^2$ **Multiplying every term of one number by every term of the other (FOIL)**

$$= 1 + 3i + 2i - 6 \qquad i^2 = -1$$

$$= -5 + 5i \qquad \text{Combining like terms}$$

Conjugates and Division

Conjugates of complex numbers are defined as follows.

Conjugate of a Complex Number

The *conjugate* of a complex number $a + bi$ is $a - bi$, and the *conjugate* of $a - bi$ is $a + bi$.

EXAMPLE 4   Find the conjugate.

a) $-3 + 7i$

b) $14 - 5i$

c) $4i$

Solution

a) $-3 + 7i$ The conjugate is $-3 - 7i$.

b) $14 - 5i$ The conjugate is $14 + 5i$.

c) $4i$ The conjugate is $-4i$. Note that $4i = 0 + 4i$.

The product of a complex number and its conjugate is a real number.

EXAMPLE 5   Multiply: $(5 + 7i)(5 - 7i)$.

Solution

$$(5 + 7i)(5 - 7i) = 5^2 - (7i)^2 \qquad \text{Using } (A + B)(A - B) = A^2 - B^2$$

$$= 25 - 49i^2$$

$$= 25 - 49(-1) \qquad i^2 = -1$$

$$= 25 + 49 = 74$$

Conjugates are used when dividing complex numbers. The procedure is much like that used to rationalize denominators in Section 7.5.

EXAMPLE 6   Divide and simplify to the form $a + bi$.

a) $\dfrac{-5 + 9i}{1 - 2i}$

b) $\dfrac{7 + 3i}{5i}$

TEACHING TIP

Some students may need an explanation of how to write $\dfrac{-23 - i}{5}$ in the form $a + bi$.

Solution

a) To divide and simplify $(-5 + 9i)/(1 - 2i)$, we multiply by 1, using the conjugate of the denominator to form 1:

$$\frac{-5 + 9i}{1 - 2i} = \frac{-5 + 9i}{1 - 2i} \cdot \frac{1 + 2i}{1 + 2i}$$ **Multiplying by 1 using the conjugate of the denominator in the symbol for 1**

$$= \frac{(-5 + 9i)(1 + 2i)}{(1 - 2i)(1 + 2i)}$$ **Multiplying numerators; multiplying denominators**

$$= \frac{-5 - 10i + 9i + 18i^2}{1^2 - 4i^2}$$ **Using FOIL**

$$= \frac{-5 - i - 18}{1 - 4(-1)}$$ $i^2 = -1$

$$= \frac{-23 - i}{5}$$

$$= -\frac{23}{5} - \frac{1}{5}i$$ **Writing in the form $a + bi$**

b) When the denominator is a pure imaginary number, it is easiest if we multiply by i/i:

$$\frac{7 + 3i}{5i} = \frac{7 + 3i}{5i} \cdot \frac{i}{i}$$ **Multiplying by 1 using i/i. We can also use the conjugate of $5i$ to write $-5i/(-5i)$.**

$$= \frac{7i + 3i^2}{5i^2}$$ **Multiplying**

$$= \frac{7i + 3(-1)}{5(-1)}$$ $i^2 = -1$

$$= \frac{7i - 3}{-5}$$

$$= \frac{-3}{-5} + \frac{7}{-5}i, \text{ or } \frac{3}{5} - \frac{7}{5}i.$$

Powers of i

Answers to problems involving complex numbers are generally written in the form $a + bi$. In the following discussion, we show why there is no need to use powers of i (other than 1) when writing answers.

Recall that -1 raised to an *even* power is 1, and -1 raised to an *odd* power is -1. Simplifying powers of i can then be done by using the fact that $i^2 = -1$ and expressing the given power of i in terms of i^2. Consider the following:

$$i, \text{ or } \sqrt{-1},$$
$$i^2 = -1,$$
$$i^3 = i^2 \cdot i = (-1)i = -i,$$
$$i^4 = (i^2)^2 = (-1)^2 = 1,$$
$$i^5 = i^4 \cdot i = (i^2)^2 \cdot i = (-1)^2 \cdot i = i,$$
$$i^6 = (i^2)^3 = (-1)^3 = -1.$$

The pattern is now repeating.

Note that the powers of i cycle themselves through the values i, -1, $-i$, and 1 and that even powers of i are -1 or 1 whereas odd powers of i are i or $-i$.

EXAMPLE 7   Simplify: **(a)** i^{18}; **(b)** i^{24}; **(c)** i^{29}; **(d)** i^{75}.

Solution

a) $i^{18} = (i^2)^9$ Using the power rule

$\quad\quad = (-1)^9 = -1$ -1 to an odd power is -1

b) $i^{24} = (i^2)^{12}$ Using the power rule

$\quad\quad = (-1)^{12} = 1$ -1 to an even power is 1

c) $i^{29} = i^{28}i^1$ Using the product rule. This is a key step when i is raised to an odd power.

$\quad\quad = (i^2)^{14}i$ Using the power rule

$\quad\quad = (-1)^{14}i$

$\quad\quad = 1 \cdot i = i$

d) $i^{75} = i^{74}i^1$ Using the product rule

$\quad\quad = (i^2)^{37}i$ Using the power rule

$\quad\quad = (-1)^{37}i$

$\quad\quad = -1 \cdot i = -i$

Complex Numbers

Many graphing calculators can perform operations with complex numbers. To do so, first choose the a + bi setting in the MODE screen, as shown in the seventh line of the figure on the left below.

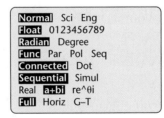

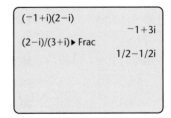

Now complex numbers can be entered as they are written. Press [2nd] [i] for the complex number i. ([i] is often the 2nd option associated with the [.] key.) For example, to calculate $(-1 + i)(2 - i)$, press [(] [(-)] [1] [+] [2nd] [i] [)] [(] [2] [−] [2nd] [i] [)] [ENTER]. As seen in the screen on the right above, the result is $-1 + 3i$. To calculate $\dfrac{2 - i}{3 + i}$ and write the result using fraction notation, press [(] [2] [−] [2nd] [i] [)] [÷] [(] [3] [+] [2nd] [i] [)] [MATH] [1] [ENTER]. Note the need for parentheses around the numerator and the denominator of the expression. The result appears in the form $a + bi$; i is not in the denominator of the fraction. The result is $\frac{1}{2} - \frac{1}{2}i$.

7.8

Exercise Set

16. $6 - 2i\sqrt{21}$, or $6 - 2\sqrt{21}i$ **18.** $\left(-2\sqrt{19} + 5\sqrt{5}\right)i$

Express in terms of i.

1. $\sqrt{-25}$ $5i$

2. $\sqrt{-36}$ $6i$

3. $\sqrt{-13}$ $i\sqrt{13}$, or $\sqrt{13}i$

4. $\sqrt{-19}$ $i\sqrt{19}$, or $\sqrt{19}i$

5. $\sqrt{-18}$ $3i\sqrt{2}$, or $3\sqrt{2}i$

6. $\sqrt{-98}$ $7i\sqrt{2}$, or $7\sqrt{2}i$

7. $\sqrt{-3}$ $i\sqrt{3}$, or $\sqrt{3}i$

8. $\sqrt{-4}$ $2i$

9. $\sqrt{-81}$ $9i$

10. $\sqrt{-27}$ $3i\sqrt{3}$, or $3\sqrt{3}i$

11. $\sqrt{-300}$ $10i\sqrt{3}$, or $10\sqrt{3}i$ **12.** $-\sqrt{-75}$ $-5i\sqrt{3}$, or $-5\sqrt{3}i$

13. $-\sqrt{-49}$ $-7i$

14. $-\sqrt{-125}$ $-5i\sqrt{5}$, or $-5\sqrt{5}i$

15. $4 - \sqrt{-60}$ $4 - 2\sqrt{15}i$ **16.** $6 - \sqrt{-84}$

17. $\sqrt{-4} + \sqrt{-12}$ $(2 + 2\sqrt{3})i$ **18.** $-\sqrt{-76} + \sqrt{-125}$

19. $\sqrt{-72} - \sqrt{-25}$ $(6\sqrt{2} - 5)i$ **20.** $\sqrt{-18} - \sqrt{-100}$ $(3\sqrt{2} - 10)i$

Perform the indicated operation and simplify. Write each answer in the form a + bi.

21. $(7 + 8i) + (5 + 3i)$ $12 + 11i$

22. $(4 - 5i) + (3 + 9i)$ $7 + 4i$

23. $(9 + 8i) - (5 + 3i)$ $4 + 5i$

24. $(9 + 7i) - (2 + 4i)$ $7 + 3i$

25. $(5 - 3i) - (9 + 2i)$ $-4 - 5i$

26. $(7 - 4i) - (5 - 3i)$ $2 - i$

27. $(-2 + 6i) - (-7 + i)$ $5 + 5i$

28. $(-5 - i) - (7 + 4i)$ $-12 - 5i$

29. $6i \cdot 9i$ -54

30. $7i \cdot 6i$ -42

31. $7i \cdot (-8i)$ 56

32. $(-4i)(-6i)$ -24

33. $\sqrt{-49}\sqrt{-25}$ -35

34. $\sqrt{-36}\sqrt{-9}$ -18

35. $\sqrt{-6}\sqrt{-7}$ $-\sqrt{42}$

36. $\sqrt{-5}\sqrt{-2}$ $-\sqrt{10}$

37. $\sqrt{-15}\sqrt{-10}$ $-5\sqrt{6}$

38. $\sqrt{-6}\sqrt{-21}$ $-3\sqrt{14}$

39. $2i(7 + 3i)$ $-6 + 14i$

40. $5i(2 + 6i)$ $-30 + 10i$

41. $-4i(6 - 5i)$ $-20 - 24i$

42. $-7i(3 - 4i)$ $-28 - 21i$

43. $(1 + 5i)(4 + 3i)$ $-11 + 23i$

44. $(1 + i)(3 + 2i)$ $1 + 5i$

45. $(5 - 6i)(2 + 5i)$ $40 + 13i$ **46.** $(6 - 5i)(3 + 4i)$ $38 + 9i$

47. $(-4 + 5i)(3 - 4i)$ $8 + 31i$

48. $(7 - 2i)(2 - 6i)$ $2 - 46i$

49. $(7 - 3i)(4 - 7i)$ $7 - 61i$ **50.** $(5 - 3i)(4 - 5i)$ $5 - 37i$

51. $(-3 + 6i)(-3 + 4i)$ $-15 - 30i$

52. $(-2 + 3i)(-2 + 5i)$ $-11 - 16i$

53. $(2 + 9i)(-3 - 5i)$ $39 - 37i$

54. $(-5 - 4i)(3 + 7i)$ $13 - 47i$

55. $(1 - 2i)^2$ $-3 - 4i$

56. $(4 - 2i)^2$ $12 - 16i$

57. $(3 + 2i)^2$ $5 + 12i$

58. $(2 + 3i)^2$ $-5 + 12i$

59. $(-5 - 2i)^2$ $21 + 20i$

60. $(-2 + 3i)^2$ $-5 - 12i$

61. $\dfrac{3}{2 - i}$ $\dfrac{6}{5} + \dfrac{3}{5}i$

62. $\dfrac{4}{3 + i}$ $\dfrac{6}{5} - \dfrac{2}{5}i$

63. $\dfrac{3i}{5 + 2i}$ $\dfrac{6}{29} + \dfrac{15}{29}i$

64. $\dfrac{4i}{5 - 3i}$ $-\dfrac{6}{17} + \dfrac{10}{17}i$

65. $\dfrac{7}{9i}$ $-\dfrac{7}{9}i$

66. $\dfrac{5}{8i}$ $-\dfrac{5}{8}i$

67. $\dfrac{5 - 3i}{4i}$ $-\dfrac{3}{4} - \dfrac{5}{4}i$

68. $\dfrac{2 + 7i}{5i}$ $\dfrac{7}{5} - \dfrac{2}{5}i$

Aha!   **69.** $\dfrac{7i + 14}{7i}$ $1 - 2i$

70. $\dfrac{6i + 3}{3i}$ $2 - i$

71. $\dfrac{4 + 5i}{3 - 7i}$ $-\dfrac{23}{58} + \dfrac{43}{58}i$

72. $\dfrac{5 + 3i}{7 - 4i}$ $\dfrac{23}{65} + \dfrac{41}{65}i$

73. $\dfrac{3 - 2i}{4 + 3i}$ $\dfrac{6}{25} - \dfrac{17}{25}i$

74. $\dfrac{5 - 2i}{3 + 6i}$ $\dfrac{1}{15} - \dfrac{4}{5}i$

Simplify.

75. i^7 $-i$

76. i^{11} $-i$

77. i^{24} 1

78. i^{35} $-i$

79. i^{42} -1

80. i^{64} 1

81. i^9 i

82. $(-i)^{71}$ i

83. $(-i)^6$ -1

84. $(-i)^4$ 1

85. $(5i)^3$ $-125i$

86. $(-3i)^5$ $-243i$

87. $i^2 + i^4$ 0

88. $5i^5 + 4i^3$ i

TW   **89.** Is the product of two imaginary numbers always an imaginary number? Why or why not?

TW   **90.** In what way(s) are conjugates of complex numbers similar to the conjugates used in Section 7.5?

Skill Maintenance

For Exercises 91–94, let

$$f(x) = x^2 - 3x \quad \text{and} \quad g(x) = 2x - 5.$$

91. Find $(f + g)(-2)$.   [2.7] 1

92. Find $(f - g)(4)$.   [2.7] 1

93. Find $(f \cdot g)(5)$. [2.7] 50

94. Find $(f/g)(3)$. [2.7] 0

Solve.

95. $28 = 3x^2 - 17x$ [5.7] $-\frac{4}{3}, 7$

96. $|3x + 7| < 22$ [4.3] $\left\{x \mid -\frac{29}{3} < x < 5\right\}$, or $\left(-\frac{29}{3}, 5\right)$

Synthesis

TW **97.** Is the set of real numbers a subset of the complex numbers? Why or why not?

TW **98.** Is the union of the set of imaginary numbers and the set of real numbers the set of complex numbers? Why or why not?

A function g is given by
$$g(z) = \frac{z^4 - z^2}{z - 1}.$$

99. Find $g(3i)$. $-9 - 27i$ **100.** Find $g(1 + i)$. $-2 + 4i$

101. Find $g(5i - 1)$. $50 - 120i$ **102.** Find $g(2 - 3i)$. $-51 - 21i$

103. Evaluate
$$\frac{1}{w - w^2} \quad \text{for} \quad w = \frac{1 - i}{10}. \quad \frac{250}{41} + \frac{200}{41}i$$

Simplify.

104. $\dfrac{i^5 + i^6 + i^7 + i^8}{(1 - i)^4}$ 0

105. $(1 - i)^3(1 + i)^3$ 8

106. $\dfrac{5 - \sqrt{5}i}{\sqrt{5}i}$ $-1 - \sqrt{5}i$

107. $\dfrac{6}{1 + \dfrac{3}{i}}$ $\dfrac{3}{5} + \dfrac{9}{5}i$

108. $\left(\dfrac{1}{2} - \dfrac{1}{3}i\right)^2 - \left(\dfrac{1}{2} + \dfrac{1}{3}i\right)^2$ $-\dfrac{2}{3}i$

109. $\dfrac{i - i^{38}}{1 + i}$ 1

---

7 # Chapter Summary and Review

Key Terms

Square root, p. 486	Even root, p. 492	30°–60°–90° right triangle,
Principal square root,	Radical function, p. 492	p. 542
p. 486	Rational exponent, p. 500	Complex-number system,
Radical sign, p. 486	Perfect square, p. 510	p. 547
Radical expression, p. 487	Perfect cube, p. 510	Imaginary number, p. 548
Radicand, p. 487	Perfect nth power, p. 510	Complex number, p. 548
Square-root function,	Rationalizing, p. 518	Pure imaginary number,
p. 487	Radical term, p. 522	p. 549
Cube root, p. 490	Like radicals, p. 522	Conjugate of a complex
nth root, p. 491	Conjugates, p. 524	number, p. 551
Index (plural, indices),	Radical equation, p. 530	
p. 491	Isosceles right triangle,	
Odd root, p. 491	p. 541	

Important Properties and Formulas

The number c is a square root of a if $c^2 = a$.

The number c is the cube root of a if $c^3 = a$.

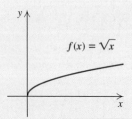

$$f(x) = \sqrt{x}$$

Domain of f: $[0, \infty)$

Range of f: $[0, \infty)$

For any real number a:

a) $\sqrt[n]{a^n} = |a|$ when n is even. Unless a is known to be nonnegative, absolute-value notation is needed when n is even.

b) $\sqrt[n]{a^n} = a$ when n is odd. Absolute-value notation is not used when n is odd.

$a^{1/n}$ means $\sqrt[n]{a}$. When a is nonnegative, n can be any natural number greater than 1. When a is negative, n must be odd.

For any natural numbers m and n ($n \neq 1$), and any real number a,

$$a^{m/n} \quad \text{means} \quad \left(\sqrt[n]{a}\right)^m \quad \text{or} \quad \sqrt[n]{a^m}.$$

When a is negative, n must be odd.

For any rational number m/n and any nonzero real number a for which $a^{m/n}$ exists,

$$a^{-m/n} \quad \text{means} \quad \frac{1}{a^{m/n}}.$$

For any real numbers a and b and any rational exponents m and n for which a^m, a^n, and b^m are defined:

1. $a^m \cdot a^n = a^{m+n}$ In multiplying, add exponents if the bases are the same.

2. $\dfrac{a^m}{a^n} = a^{m-n}$ In dividing, subtract exponents if the bases are the same. (Assume $a \neq 0$.)

3. $(a^m)^n = a^{m \cdot n}$ To raise a power to a power, multiply the exponents.

4. $(ab)^m = a^m b^m$ To raise a product to a power, raise each factor to the power and multiply.

The Product Rule for Radicals

For any real numbers $\sqrt[n]{a}$ and $\sqrt[n]{b}$,

$$\sqrt[n]{a}\, \sqrt[n]{b} = \sqrt[n]{a \cdot b}.$$

The Quotient Rule for Radicals

For any real numbers $\sqrt[n]{a}$ and $\sqrt[n]{b}$, $b \neq 0$,

$$\sqrt[n]{\frac{a}{b}} = \frac{\sqrt[n]{a}}{\sqrt[n]{b}}.$$

Some Ways to Simplify Radical Expressions

1. *Simplifying by factoring.* Factor the radicand and look for factors raised to powers that are divisible by the index.

   Example: $\sqrt[3]{a^6 b} = \sqrt[3]{a^6}\, \sqrt[3]{b} = a^2 \sqrt[3]{b}$

2. *Using rational exponents to simplify.* Convert to exponential notation and then use arithmetic and the laws of exponents to simplify the exponents. Then convert back to radical notation as needed.

   Example: $\sqrt[3]{p} \cdot \sqrt[4]{q^3} = p^{1/3} \cdot q^{3/4}$
   $$= p^{4/12} \cdot q^{9/12}$$
   $$= \sqrt[12]{p^4 q^9}$$

3. *Combining like radical terms.*

Example: $\sqrt{8} + 3\sqrt{2} = \sqrt{4} \cdot \sqrt{2} + 3\sqrt{2}$
$= 2\sqrt{2} + 3\sqrt{2} = 5\sqrt{2}$

The Principle of Powers

If $a = b$, then $a^n = b^n$ for any exponent n.

To solve an equation with a radical term:

1. Isolate the radical term on one side of the equation.

2. Use the principle of powers and solve the resulting equation.

3. Check any possible solution in the original equation.

To solve an equation with two or more radical terms:

1. Isolate one of the radical terms.

2. Use the principle of powers.

3. If a radical remains, repeat steps (1) and (2).

4. Solve the resulting equation.

5. Check possible solutions in the original equation.

The Pythagorean Theorem

$a^2 + b^2 = c^2$

The Principle of Square Roots

If $x^2 = n$, then $x = \sqrt{n}$ or $x = -\sqrt{n}$.

Special Triangles

The length of the hypotenuse in an isosceles right triangle is the length of a leg times $\sqrt{2}$.

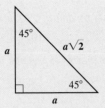

The length of the longer leg in a $30°-60°-90°$ right triangle is the length of the shorter leg times $\sqrt{3}$. The hypotenuse is twice as long as the shorter leg.

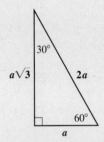

A complex number is any number that can be written in the form $a + bi$, where a and b are real numbers and $i = \sqrt{-1}$.

Review Exercises

Simplify.

1. $\sqrt{\dfrac{49}{36}}$ [7.1] $\dfrac{7}{6}$

2. $-\sqrt{0.25}$ [7.1] -0.5

Let $f(x) = \sqrt{2x - 7}$. Find the following [7.1] $\left\{x \middle| x \geq \frac{7}{2}\right\}$, or $\left[\frac{7}{2}, \infty\right)$

3. $f(16)$ [7.1] 5

4. The domain of f

Simplify. Assume that each variable can represent any real number.

5. $\sqrt{49a^2}$ [7.1] $7|a|$

6. $\sqrt{(c + 8)^2}$ [7.1] $|c + 8|$

7. $\sqrt{x^2 - 6x + 9}$ [7.1] $|x - 3|$

8. $\sqrt{4x^2 + 4x + 1}$ [7.1] $|2x + 1|$

9. $\sqrt[5]{-32}$ [7.1] -2

10. $\sqrt[3]{-\dfrac{64x^6}{27}}$ [7.4] $-\dfrac{4x^2}{3}$

11. $\sqrt[4]{x^{12}y^8}$ [7.3] $|x^3y^2|$, or $|x^3|y^2$ **12.** $\sqrt[6]{64x^{12}}$ [7.3] $2x^2$

13. Write an equivalent expression using exponential notation: $\left(\sqrt[3]{5ab}\right)^4$. [7.2] $(5ab)^{4/3}$

14. Write an equivalent expression using radical notation: $(16a^6)^{3/4}$. [7.2] $8a^4\sqrt{a}$

Use rational exponents to simplify. Assume x, y ≥ 0.

15. $\sqrt{x^6y^{10}}$ [7.2] x^3y^5 **16.** $\left(\sqrt[6]{x^2y}\right)^2$ [7.2] $\sqrt[3]{x^2y}$

Simplify. Do not use negative exponents in the answers.

17. $(x^{-2/3})^{3/5}$ [7.2] $\dfrac{1}{x^{2/5}}$ **18.** $\dfrac{7^{-1/3}}{7^{-1/2}}$ [7.2] $7^{1/6}$

19. If $f(x) = \sqrt{25(x-3)^2}$, find a simplified form for $f(x)$. [7.3] $f(x) = 5|x-3|$

Perform the indicated operation and, if possible, simplify. Write all answers using radical notation.

20. $\sqrt{5x}\ \sqrt{3y}$ [7.3] $\sqrt{15xy}$ **21.** $\sqrt[3]{a^5b}\ \sqrt[3]{27b}$ [7.3] $3a\sqrt[3]{a^2b^2}$

22. $\sqrt[3]{-24x^{10}y^8}\ \sqrt[3]{18x^7y^4}$ [7.3] $-6x^5y^4\sqrt[3]{2x^2}$ **23.** $\dfrac{\sqrt[3]{60xy^3}}{\sqrt[3]{10x}}$ [7.4] $y\sqrt[3]{6}$

24. $\dfrac{\sqrt{75x}}{2\sqrt{3}}$ [7.4] $\dfrac{5\sqrt{x}}{2}$ **25.** $\sqrt[4]{\dfrac{48a^{11}}{c^8}}$ [7.4] $\dfrac{2a^2\sqrt[4]{3a^3}}{c^2}$

26. $5\sqrt[3]{x} + 2\sqrt[3]{x}$ [7.5] $7\sqrt[3]{x}$ **27.** $2\sqrt{75} - 7\sqrt{3}$ [7.5] $3\sqrt{3}$

28. $\sqrt[3]{8x^4} + \sqrt[3]{xy^6}$ [7.5] $(2x + y^2)\sqrt[3]{x}$

29. $\sqrt{50} + 2\sqrt{18} + \sqrt{32}$ [7.5] $15\sqrt{2}$

30. $\left(\sqrt{5} - 3\sqrt{8}\right)\left(\sqrt{5} + 2\sqrt{8}\right)$ [7.5] $-43 - 2\sqrt{10}$

31. $\sqrt[4]{x}\ \sqrt{x}$ [7.5] $\sqrt[4]{x^3}$ **32.** $\dfrac{\sqrt[3]{x^2}}{\sqrt[4]{x}}$ [7.5] $\sqrt[12]{x^5}$

33. If $f(x) = x^2$, find $f\left(a - \sqrt{2}\right)$. [7.5] $a^2 - 2a\sqrt{2} + 2$

34. Rationalize the denominator:
$$\dfrac{2\sqrt{3}}{\sqrt{2} + \sqrt{3}}.$$ [7.5] $-2\sqrt{6} + 6$

35. Rationalize the numerator of the expression in Exercise 34. [7.5] $\dfrac{6}{3 + \sqrt{6}}$

Solve.

36. $\sqrt{y + 4} - 2 = 3$ [7.6] 21

37. $(x + 1)^{1/3} = -5$ [7.6] -126

38. $1 + \sqrt{x} = \sqrt{3x - 3}$ [7.6] 4

39. If $f(x) = \sqrt[4]{x + 2}$, find a such that $f(a) = 2$. [7.6] 14

Solve. Give an exact answer and, where appropriate, an approximation to three decimal places.

40. The diagonal of a square has length 10 cm. Find the length of a side of the square. [7.7] $5\sqrt{2}$ cm; 7.071 cm

41. A bookcase is 5 ft tall and has a 7-ft diagonal brace, as shown. How wide is the bookcase? [7.7] $\sqrt{24}$ ft; 4.899 ft

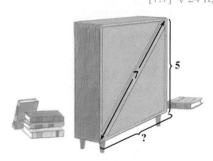

42. Find the missing lengths. Give exact answers and, where appropriate, an approximation to three decimal places. [7.7] $a = 10$; $b = 10\sqrt{3} \approx 17.321$

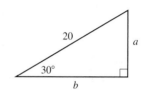

43. Express in terms of i and simplify: $-\sqrt{-8}$. [7.8] $-2i\sqrt{2}$, or $-2\sqrt{2}i$

44. Add: $(-4 + 3i) + (2 - 12i)$. [7.8] $-2 - 9i$

45. Subtract: $(4 - 7i) - (3 - 8i)$. [7.8] $1 + i$

Multiply.

46. $(2 + 5i)(2 - 5i)$ [7.8] 29

47. i^{13} [7.8] i

48. $(6 - 3i)(2 - i)$ [7.8] $9 - 12i$

49. Divide and simplify to the form $a + bi$:
$$\dfrac{7 - 2i}{3 + 4i}.$$ [7.8] $\dfrac{13}{25} - \dfrac{34}{25}i$

Synthesis

TW **50.** Explain why $\sqrt[n]{x^n} = |x|$ when n is even, but $\sqrt[n]{x^n} = x$ when n is odd.

TW **51.** What is the difference between real numbers and complex numbers?

52. Solve:
$$\sqrt{11x + \sqrt{6 + x}} = 6.$$ [7.6] 3

53. Simplify:
$$\dfrac{2}{1 - 3i} - \dfrac{3}{4 + 2i}.$$ [7.8] $-\dfrac{2}{5} + \dfrac{9}{10}i$

Chapter Test 7

Simplify. Assume that variables can represent any real number.

1. $\sqrt{75}$ [7.3] $5\sqrt{3}$

2. $\sqrt[3]{-\dfrac{8}{x^6}}$ [7.4] $-\dfrac{2}{x^2}$

3. $\sqrt{100a^2}$ [7.1] $10|a|$

4. $\sqrt{x^2 - 8x + 16}$ [7.1] $|x - 4|$

5. $\sqrt[5]{x^{12}y^8}$ [7.3] $x^2y\sqrt[5]{x^2y^3}$

6. $\sqrt{\dfrac{25x^2}{36y^4}}$ [7.4] $\left|\dfrac{5x}{6y^2}\right|$, or $\dfrac{5|x|}{6y^2}$

7. $\sqrt[3]{2x}\ \sqrt[3]{5y^2}$ [7.3] $\sqrt[3]{10xy^2}$

8. $\dfrac{\sqrt[5]{x^3y^4}}{\sqrt[5]{xy^2}}$ [7.4] $\sqrt[5]{x^2y^2}$

9. $\sqrt[4]{x^3y^2}\ \sqrt{xy}$ [7.5] $xy\sqrt[4]{x}$

10. $\dfrac{\sqrt[5]{a^2}}{\sqrt[4]{a}}$ [7.5] $\sqrt[20]{a^3}$

11. $7\sqrt{2} - 2\sqrt{2}$ [7.5] $5\sqrt{2}$

12. $\sqrt{x^4y} + \sqrt{9y^3}$ [7.5] $(x^2 + 3y)\sqrt{y}$

13. $\left(7 + \sqrt{x}\right)\left(2 - 3\sqrt{x}\right)$ [7.5] $14 - 19\sqrt{x} - 3x$

14. Write an equivalent expression using radical notation: $(2a^3b)^{5/6}$.   [7.2] $\sqrt[6]{(2a^3b)^5}$

15. Write an equivalent expression using exponential notation: $\sqrt{7xy}$.   [7.2] $(7xy)^{1/2}$

16. If $f(x) = \sqrt{8 - 4x}$, determine the domain of f.

17. If $f(x) = x^2$, find $f\left(5 + \sqrt{2}\right)$.   [7.5] $27 + 10\sqrt{2}$

18. Rationalize the denominator:

$$\frac{\sqrt{3}}{1 + \sqrt{2}}. \quad \text{[7.5] } \sqrt{6} - \sqrt{3}$$

Solve.

19. $x = \sqrt{2x - 5} + 4$ [7.6] 7

20. $\sqrt{x} = \sqrt{x + 1} - 5$ [7.6] No solution

16. [7.1] $\{x \mid x \le 2\}$, or $(-\infty, 2]$

Solve. Give exact answers and, where appropriate, approximations to three decimal places.

21. One leg of an isosceles right triangle is 7 cm long. Find the lengths of the other sides.   [7.7] Leg: 7 cm; hypotenuse: $7\sqrt{2}$ cm ≈ 9.899 cm

22. A referee jogs diagonally from one corner of a 50-ft by 90-ft basketball court to the far corner. How far does she jog? Give an exact answer and an approximation to three decimal places.   [7.7] $\sqrt{10,600}$ ft ≈ 102.956 ft

23. Express in terms of i and simplify: $\sqrt{-50}$.   [7.8] $5i\sqrt{2}$, or $5\sqrt{2}i$

24. Subtract: $(7 + 8i) - (-3 + 6i)$.   [7.8] $10 + 2i$

25. Multiply: $\sqrt{-16}\ \sqrt{-36}$.   [7.8] -24

26. Multiply. Write the answer in the form $a + bi$.

$$(4 - i)^2 \quad \text{[7.8] } 15 - 8i$$

27. Divide and simplify to the form $a + bi$:

$$\frac{-3 + i}{2 - 7i}. \quad \text{[7.8] } -\frac{13}{53} - \frac{19}{53}i$$

28. Simplify: i^{37}.   [7.8] i

Synthesis

29. Solve:

$$\sqrt{2x - 2} + \sqrt{7x + 4} = \sqrt{13x + 10}. \quad \text{[7.6] } 3$$

30. Simplify:

$$\frac{1 - 4i}{4i(1 + 4i)^{-1}}. \quad \text{[7.8] } -\frac{17}{4}i$$

Quadratic Functions and Equations

8

I n translating problem situations to mathematics, we often obtain a function or equation containing a second-degree polynomial in one variable. Such functions or equations are said to be *quadratic*. In this chapter, we will study a variety of equations, inequalities, and applications for which we will need to solve quadratic equations or graph quadratic functions.

APPLICATION

TEEN SMOKING. According to the Centers for Disease Control and Prevention, the percent of high school students who reported having smoked a cigarette in the preceding 30 days declined from 1997 to 2001, after rising in the first part of the 1990s, as shown in the following table. Find a quadratic function that fits the data. The graph of the data shows that a quadratic function can be used to model the situation.

Years After 1991	Percent of Students Who Smoked in the Preceding 30 Days
0	27.5
2	30.5
4	34.9
6	36.4
8	34.9
10	28.5

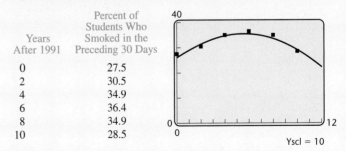

This problem appears as Example 4 in Section 8.8.

8.1

The Principle of Square Roots ■ Completing the Square ■ Problem Solving

Quadratic Equations

In Chapter 5, we solved *quadratic equations* like $x^2 = 10 + 3x$ by graphing and by factoring. One way to solve by graphing is to rewrite the equation above, for example, as $x^2 - 3x - 10 = 0$ and then graph the *quadratic function* given by $f(x) = x^2 - 3x - 10$. The solutions of the equation, -2 and 5, are the first coordinates of the x-intercepts of the graph of f.

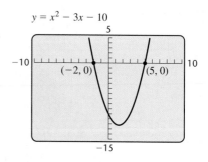

A quadratic equation will have no, one, or two real-number solutions. We can see this by examining the graphs of quadratic functions.

Interactive Discovery

Graph each function and determine the number of x-intercepts.

1. $f(x) = x^2$ 1

2. $g(x) = -x^2$ 1

3. $h(x) = (x - 2)^2$ 1

4. $p(x) = 2x^2 + 1$ 0

5. $f(x) = -1.5x^2 + x - 3$ 0

6. $g(x) = 4x^2 - 2x - 7$ 2

7. Describe the shape of the graph of a quadratic function.
A cup-shaped curve opening up or down

Although we will study graphs of quadratic functions in detail in Section 8.6, we can make some general observations here. The graph of a quadratic function is a cup-shaped curve called a *parabola*. It can open upward, like the graph of $f(x) = x^2$, or downward, like the graph of $g(x) = -x^2$.

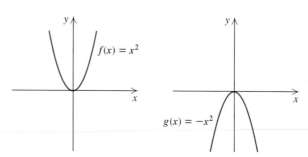

The graph of a quadratic function can have no, one, or two *x*-intercepts, as illustrated below. Thus a quadratic equation can have no, one, or two real-number roots.

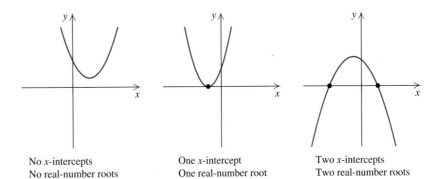

No *x*-intercepts
No real-number roots

One *x*-intercept
One real-number root

Two *x*-intercepts
Two real-number roots

To solve a quadratic equation by factoring, we write the equation in the *standard form* $ax^2 + bx + c = 0$, factor, and use the principle of zero products.

EXAMPLE 1   Solve: $3x^2 = 2 - x$.

Solution   We have

$$3x^2 = 2 - x$$
$$3x^2 + x - 2 = 0 \qquad \text{Adding } -2 + x \text{ to both sides to obtain standard form}$$
$$(3x - 2)(x + 1) = 0 \qquad \text{Factoring}$$
$$3x - 2 = 0 \quad or \quad x + 1 = 0 \qquad \text{Using the principle of zero products}$$
$$3x = 2 \quad or \qquad x = -1$$
$$x = \tfrac{2}{3} \quad or \qquad x = -1.$$

We check by substituting -1 and $\frac{2}{3}$ into the original equation. A graphical solution (see the figure at left) provides another check.

Check: For -1:

$$\begin{array}{c|c} \multicolumn{2}{c}{3x^2 = 2 - x} \\ \hline 3(-1)^2 \; ? \; 2 - (-1) \\ 3 \cdot 1 \;\big|\; 2 + 1 \\ 3 \;\big|\; 3 \end{array} \qquad \text{TRUE}$$

For $\frac{2}{3}$:

$$\begin{array}{c|c} \multicolumn{2}{c}{3x^2 = 2 - x} \\ \hline 3\left(\tfrac{2}{3}\right)^2 \; ? \; 2 - \tfrac{2}{3} \\ 3 \cdot \tfrac{4}{9} \;\big|\; \tfrac{6}{3} - \tfrac{2}{3} \\ \tfrac{4}{3} \;\big|\; \tfrac{4}{3} \end{array} \qquad \text{TRUE}$$

The solutions are -1 and $\frac{2}{3}$.

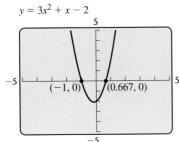

$y = 3x^2 + x - 2$
$(-1, 0)$ $(0.667, 0)$

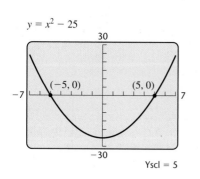

$y = x^2 - 25$

EXAMPLE 2   Solve: $x^2 = 25$.

Solution   We have

$$x^2 = 25$$
$$x^2 - 25 = 0 \qquad \text{Writing in standard form}$$
$$(x - 5)(x + 5) = 0 \qquad \text{Factoring}$$
$$x - 5 = 0 \quad or \quad x + 5 = 0 \qquad \text{Using the principle of zero products}$$
$$x = 5 \quad or \qquad x = -5.$$

The solutions are 5 and -5. The graph at left confirms the solutions.

The Principle of Square Roots

Consider the equation $x^2 = 25$ again. We know from Chapter 7 that the number 25 has two real-number square roots, namely, 5 and -5. Note that these are the solutions of the equation in Example 2. Thus square roots can provide a quick method for solving equations of the type $x^2 = k$.

> **The Principle of Square Roots**
> For any real number k, if $x^2 = k$, then $x = \sqrt{k}$ or $x = -\sqrt{k}$.

EXAMPLE 3   Solve: $3x^2 = 6$.

Solution   We have

$$3x^2 = 6$$
$$x^2 = 2 \qquad \text{Multiplying by } \tfrac{1}{3}$$
$$x = \sqrt{2} \quad or \quad x = -\sqrt{2}. \qquad \text{Using the principle of square roots}$$

We often use the symbol $\pm\sqrt{2}$ to represent the two numbers $\sqrt{2}$ and $-\sqrt{2}$. We check as follows.

Check: For $\sqrt{2}$:

$$\frac{3x^2 = 6}{3(\sqrt{2})^2 \ ? \ 6}$$
$$3 \cdot 2$$
$$6 \ | \ 6 \qquad \text{TRUE}$$

For $-\sqrt{2}$:

$$\frac{3x^2 = 6}{3(-\sqrt{2})^2 \ ? \ 6}$$
$$3 \cdot 2$$
$$6 \ | \ 6 \qquad \text{TRUE}$$

The solutions are $\sqrt{2}$ and $-\sqrt{2}$, or $\pm\sqrt{2}$.

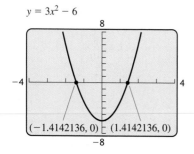

$y = 3x^2 - 6$

Note that a graphical solution of Example 3, shown at left, yields approximate solutions, since $\sqrt{2}$ is irrational. The solutions found graphically correspond to those found algebraically, since $\sqrt{2} \approx 1.4142136$.

Sometimes we rationalize denominators to simplify answers, although this is not as common as it once was.

EXAMPLE 4   Solve: $-5x^2 + 2 = 0$.

Solution   We have

$$-5x^2 + 2 = 0$$

$$x^2 = \frac{2}{5} \qquad \text{Isolating } x^2$$

$$x = \sqrt{\frac{2}{5}} \quad \text{or} \quad x = -\sqrt{\frac{2}{5}}. \qquad \text{Using the principle of square roots}$$

The solutions are $\sqrt{\dfrac{2}{5}}$ and $-\sqrt{\dfrac{2}{5}}$. This can also be written as $\pm\sqrt{\dfrac{2}{5}}$, or, if we rationalize the denominator, $\pm\dfrac{\sqrt{10}}{5}$. The checks are left to the student.

Sometimes we get solutions that are imaginary numbers.

EXAMPLE 5   Solve: $4x^2 + 9 = 0$.

Solution   We have

$$4x^2 + 9 = 0$$

$$x^2 = -\tfrac{9}{4} \qquad \text{Isolating } x^2$$

$$x = \sqrt{-\tfrac{9}{4}} \qquad \text{or} \quad x = -\sqrt{-\tfrac{9}{4}} \qquad \begin{array}{l}\text{Using the principle of} \\ \text{square roots}\end{array}$$

$$x = \sqrt{\tfrac{9}{4}}\sqrt{-1} \quad \text{or} \quad x = -\sqrt{\tfrac{9}{4}}\sqrt{-1}$$

$$x = \tfrac{3}{2}i \qquad\quad \text{or} \quad x = -\tfrac{3}{2}i.$$

Check: Since the solutions are opposites and the equation has an x^2-term and no x-term, we can check both solutions at once.

$$\begin{array}{c|c}
\multicolumn{2}{l}{4x^2 + 9 = 0} \\
\hline
4\left(\pm\tfrac{3}{2}i\right)^2 + 9 \; \stackrel{?}{|} \; 0 & \\
4 \cdot \tfrac{9}{4} \cdot i^2 + 9 & \\
9(-1) + 9 & \\
0 & 0 \qquad \text{TRUE}
\end{array}$$

The solutions are $\tfrac{3}{2}i$ and $-\tfrac{3}{2}i$, or $\pm\tfrac{3}{2}i$. The graph shown below indicates that there are no real-number solutions, because there are no x-intercepts.

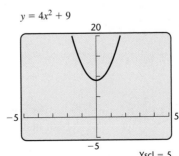

$y = 4x^2 + 9$

Yscl = 5

The principle of square roots can be restated in a more general form that pertains to more complicated algebraic expressions than just x.

The Principle of Square Roots (Generalized Form)

For any real number k and any algebraic expression X,

If $X^2 = k$, then $X = \sqrt{k}$ or $X = -\sqrt{k}$.

EXAMPLE 6 Let $f(x) = (x - 2)^2$. Find all x-values for which $f(x) = 7$.

Solution We are asked to find all x-values for which

$$f(x) = 7,$$

or

$$(x - 2)^2 = 7. \qquad \text{Substituting } (x - 2)^2 \text{ for } f(x)$$

We solve both algebraically and graphically.

Algebraic Solution

We use the generalized principle of square roots, replacing X with $x - 2$:

$$x - 2 = \sqrt{7} \quad or \quad x - 2 = -\sqrt{7}$$
$$x = 2 + \sqrt{7} \quad or \quad x = 2 - \sqrt{7}.$$

Thus the possible solutions are $2 + \sqrt{7}$ and $2 - \sqrt{7}$.

Graphical Solution

We graph the equations $y_1 = (x - 2)^2$ and $y_2 = 7$. If there are any real-number x-values for which $f(x) = 7$, they will be the x-coordinates of the points of intersection of the graphs.

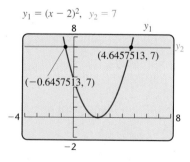

Rounded to the nearest thousandth, the solutions are approximately -0.646 and 4.646.

As one check, note that $2 + \sqrt{7} \approx 4.646$ and $2 - \sqrt{7} \approx -0.646$, so both methods give us the same solution. As a complete check, we can evaluate the function for both possible solutions. This can be done by hand or using a graphing calculator.

Algebraic Check

We have
$$f(2 + \sqrt{7}) = (2 + \sqrt{7} - 2)^2 = (\sqrt{7})^2 = 7.$$
Similarly,
$$f(2 - \sqrt{7}) = (2 - \sqrt{7} - 2)^2 = (-\sqrt{7})^2 = 7.$$
The numbers check.

Graphical Check

One way to check the solutions on a graphing calculator is to enter $y_1 = (x - 2)^2$ and then evaluate $y_1(2 + \sqrt{7})$ and $y_1(2 - \sqrt{7})$. The second calculation can be entered quickly by copying the first entry and editing it.

Y₁(2+√(7))	7
Y₁(2−√(7))	7

The solutions are $2 + \sqrt{7}$ and $2 - \sqrt{7}$, or $2 \pm \sqrt{7}$.

In Example 6, one side of the equation is the square of a binomial and the other side is a constant. Sometimes an equation must be factored in order to appear in this form.

EXAMPLE 7 Solve: $x^2 + 6x + 9 = 2$.

Solution   We have

$x^2 + 6x + 9 = 2$	The left side is the square of a binomial.
$(x + 3)^2 = 2$	Factoring
$x + 3 = \sqrt{2}$ or $x + 3 = -\sqrt{2}$	Using the principle of square roots
$x = -3 + \sqrt{2}$ or $x = -3 - \sqrt{2}.$	Adding −3 to both sides

The solutions are $-3 + \sqrt{2}$ and $-3 - \sqrt{2}$, or $-3 \pm \sqrt{2}$. The checks are left to the student.

Completing the Square

By using a method called *completing the square*, we can use the principle of square roots to solve *any* quadratic equation.

EXAMPLE 8 Solve: $x^2 + 6x + 4 = 0$.

Solution   We have

$x^2 + 6x + 4 = 0$	
$x^2 + 6x = -4$	Subtracting 4 from both sides
$x^2 + 6x + 9 = -4 + 9$	Adding 9 to both sides. We explain this shortly.
$(x + 3)^2 = 5$	Factoring the perfect-square trinomial
$x + 3 = \pm\sqrt{5}$	Using the principle of square roots. Remember that $\pm\sqrt{5}$ represents two numbers.
$x = -3 \pm \sqrt{5}.$	Adding −3 to both sides

TEACHING TIP

Review
$a^2 + 2ab + b^2 = (a + b)^2$
and
$a^2 - 2ab + b^2 = (a - b)^2$.

One way to check the solutions is to store $-3 + \sqrt{5}$ as x using the $\boxed{\text{STO} \rightarrow}$ key. To do this, we press $\boxed{(-)}$ $\boxed{3}$ $\boxed{+}$ $\boxed{\text{2nd}}$ $\boxed{\sqrt{}}$ $\boxed{5}$ $\boxed{)}$ $\boxed{\text{STO} \rightarrow}$ $\boxed{\text{X}}$. Then we evaluate $x^2 + 6x + 4$ for $x = -3 + \sqrt{5}$ by pressing $\boxed{\text{X}}$ $\boxed{x^2}$ $\boxed{+}$ $\boxed{6}$ $\boxed{\text{X}}$ $\boxed{+}$ $\boxed{4}$ $\boxed{\text{ENTER}}$. We repeat the process for $-3 - \sqrt{5}$ to check the second solution. Both solutions check. The solutions are $-3 + \sqrt{5}$ and $-3 - \sqrt{5}$.

⁻3+√(5)→X
-.7639320225
X2+6X+4
0

⁻3-√(5)→X
-5.236067977
X2+6X+4
0

Let's examine how the above solutions were found. The decision to add 9 to both sides in Example 8 was made because it creates a perfect-square trinomial on the left side. The 9 was determined by taking half of the coefficient of x and squaring it—that is,

$$\left(\tfrac{1}{2} \cdot 6\right)^2 = 3^2, \quad \text{or} \quad 9.$$

To help see why this procedure works, examine the following drawings.

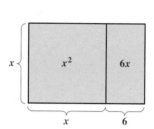 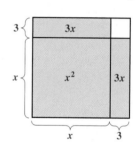

Note that the shaded areas in both figures represent the same area, $x^2 + 6x$. However, only the figure on the right, in which the $6x$ is halved, can be converted into a square with the addition of a constant term. The constant term, 9, can be interpreted as the "missing" piece of the diagram on the right. It *completes* the square.

To complete the square for $x^2 + bx$, we add $(b/2)^2$.

Example 9, which follows, is one of the few examples in this text for which we are neither solving an equation nor writing an equivalent expression. Instead we are simply gaining practice in finding numbers that complete the square. The trinomial that we create is *not* equivalent to the original binomial.

EXAMPLE 9   Complete the square. Then write the trinomial in factored form.

a) $x^2 + 14x$ **b)** $x^2 - 5x$ **c)** $x^2 + \frac{3}{4}x$

Solution

a) We take half of the coefficient of x and square it.

$$x^2 + 14x$$

$\longrightarrow$ Half of 14 is 7, and $7^2 = 49$. We add 49.

Thus, $x^2 + 14x + 49$ is a perfect-square trinomial. It is equivalent to $(x + 7)^2$. We must add 49 in order for $x^2 + 14x$ to become a perfect-square trinomial.

b) We take half of the coefficient of x and square it:

$$x^2 - 5x$$

$$\tfrac{1}{2} \cdot (-5) = -\tfrac{5}{2}, \quad \text{and} \quad \left(-\tfrac{5}{2}\right)^2 = \tfrac{25}{4}.$$

Thus, $x^2 - 5x + \tfrac{25}{4}$ is a perfect-square trinomial. It is equivalent to $\left(x - \tfrac{5}{2}\right)^2$. Note that for purposes of factoring, it is best *not* to convert $\tfrac{25}{4}$ to decimal notation.

c) We take half of the coefficient of x and square it:

$$x^2 + \tfrac{3}{4}x$$

$$\tfrac{1}{2} \cdot \tfrac{3}{4} = \tfrac{3}{8}, \quad \text{and} \quad \left(\tfrac{3}{8}\right)^2 = \tfrac{9}{64}.$$

Thus, $x^2 + \tfrac{3}{4}x + \tfrac{9}{64}$ is a perfect-square trinomial. It is equivalent to $\left(x + \tfrac{3}{8}\right)^2$.

We can now use the method of completing the square to solve equations similar to Example 8.

EXAMPLE 10   Solve: **(a)** $x^2 - 8x - 7 = 0$; **(b)** $x^2 + 5x - 3 = 0$.

Solution

a) $x^2 - 8x - 7 = 0$

$\quad x^2 - 8x \qquad = 7$ Adding 7 to both sides. We can now complete the square on the left side.

$\quad x^2 - 8x + 16 = 7 + 16$ Adding 16 to both sides to complete the square: $\tfrac{1}{2}(-8) = -4$, and $(-4)^2 = 16$

$\qquad (x - 4)^2 = 23$ Factoring

$\qquad x - 4 = \pm\sqrt{23}$ Using the principle of square roots

$\qquad x = 4 \pm \sqrt{23}$ Adding 4 to both sides

The solutions are $4 - \sqrt{23}$ and $4 + \sqrt{23}$, or $4 \pm \sqrt{23}$. The checks are left to the student.

> **CAUTION!**   Be sure to add 16 to *both sides* of the equation.

b) $x^2 + 5x - 3 = 0$

$\quad x^2 + 5x \qquad = 3$ Adding 3 to both sides

$\quad x^2 + 5x + \dfrac{25}{4} = 3 + \dfrac{25}{4}$ Completing the square: $\tfrac{1}{2} \cdot 5 = \tfrac{5}{2}$, and $\left(\tfrac{5}{2}\right)^2 = \tfrac{25}{4}$

$\qquad \left(x + \dfrac{5}{2}\right)^2 = \dfrac{37}{4}$ Factoring and simplifying

$\qquad x + \dfrac{5}{2} = \pm\dfrac{\sqrt{37}}{2}$ Using the principle of square roots and the quotient rule for radicals

$\qquad x = \dfrac{-5 \pm \sqrt{37}}{2}$ Adding $-\tfrac{5}{2}$ to both sides

TEACHING TIP

You may want to reinforce throughout the chapter that the symbol $\pm$ represents two numbers.

TEACHING TIP

You may want to point out that, unlike when we solve by factoring, we do not get 0 on one side. Instead, we isolate the variable terms before completing the square.

The checks are left to the student. The solutions are $\left(-5 - \sqrt{37}\right)/2$ and $\left(-5 + \sqrt{37}\right)/2$, or $\left(-5 \pm \sqrt{37}\right)/2$.

Before we can complete the square, the coefficient of x^2 must be 1. When it is not 1, we divide both sides of the equation by whatever that coefficient may be.

EXAMPLE 11 Solve: $3x^2 + 7x - 2 = 0$.

Solution We have

$$3x^2 + 7x - 2 = 0$$
$$3x^2 + 7x = 2 \qquad \text{Adding 2 to both sides}$$
$$x^2 + \frac{7}{3}x = \frac{2}{3} \qquad \text{Dividing both sides by 3}$$
$$x^2 + \frac{7}{3}x + \frac{49}{36} = \frac{2}{3} + \frac{49}{36} \qquad \text{Completing the square: } \left(\frac{1}{2} \cdot \frac{7}{3}\right)^2 = \frac{49}{36}$$
$$\left(x + \frac{7}{6}\right)^2 = \frac{73}{36} \qquad \text{Factoring and simplifying}$$
$$x + \frac{7}{6} = \pm \frac{\sqrt{73}}{6} \qquad \text{Using the principle of square roots and the quotient rule for radicals}$$
$$x = \frac{-7 \pm \sqrt{73}}{6}. \qquad \text{Adding } -\frac{7}{6} \text{ to both sides}$$

The solutions are

$$\frac{-7 - \sqrt{73}}{6} \quad \text{and} \quad \frac{-7 + \sqrt{73}}{6}, \quad \text{or} \quad \frac{-7 \pm \sqrt{73}}{6}.$$

The checks are left to the student.

The procedure used in Example 11 is important because it can be used to solve *any* quadratic equation.

To Solve a Quadratic Equation in x by Completing the Square

1. Isolate the terms with variables on one side of the equation, and arrange them in descending order.
2. Divide both sides by the coefficient of x^2 if that coefficient is not 1.
3. Complete the square by taking half of the coefficient of x and adding its square to both sides.
4. Express one side as the square of a binomial and simplify the other side.
5. Use the principle of square roots.
6. Solve for x by adding or subtracting on both sides.

Problem Solving

If you put money in a savings account, the bank will pay you interest. As interest is paid into your account, the bank will start paying you interest on both the original amount and the interest already earned. This is called **compounding interest**. If interest is paid yearly, we say that it is **compounded annually**.

The Compound-Interest Formula

If an amount of money P is invested at interest rate r, compounded annually, then in t years, it will grow to the amount A given by

$$A = P(1 + r)^t.$$

We can use quadratic equations to solve certain interest problems.

EXAMPLE 12   Investment Growth.   Rosa invested $4000 at interest rate r, compounded annually. In 2 yr, it grew to $4410. What was the interest rate?

Solution

1. **Familiarize.**   We are already familiar with the compound-interest formula. If we were not, we would need to consult an outside source.

2. **Translate.**   The translation consists of substituting into the formula:

$$A = P(1 + r)^t$$
$$4410 = 4000(1 + r)^2. \qquad \textbf{Substituting}$$

3. **Carry out.**   We solve for r both algebraically and graphically.

Algebraic Solution

We have

$$4410 = 4000(1 + r)^2$$

$\frac{4410}{4000} = (1 + r)^2$	**Dividing both sides by 4000**
$\frac{441}{400} = (1 + r)^2$	**Simplifying**
$\pm\sqrt{\frac{441}{400}} = 1 + r$	**Using the principle of square roots**
$\pm\frac{21}{20} = 1 + r$	**Simplifying**
$-\frac{20}{20} \pm \frac{21}{20} = r$	**Adding** -1, or $-\frac{20}{20}$, to both sides
$\frac{1}{20} = r \quad or \quad -\frac{41}{20} = r.$	

Since r represents an interest rate, we convert to decimal notation. We now have

$$r = 0.05 \quad or \quad r = -2.05.$$

Graphical Solution

We let $y_1 = 4410$ and $y_2 = 4000(1 + x)^2$. It may take several tries to find an appropriate viewing window. Since $y_1 = 4410$, the vertical axis should extend above 4500. Using the window $[-3, 1, 0, 5000]$, we see that the graphs intersect at two points.

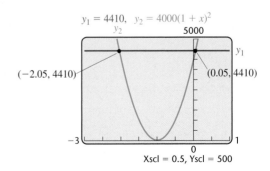

The first coordinates of the points of intersection are -2.05 and 0.05.

4. Check.   Since the interest rate cannot be negative, we need only check 0.05, or 5%. If $4000 were invested at 5% interest, compounded annually, then in 2 yr it would grow to $4000(1.05)^2$, or $4410. The number 0.05 checks.

5. State.   The interest rate was 5%.

EXAMPLE 13   Free-Falling Objects.   The formula $s = 16t^2$ is used to approximate the distance s, in feet, that an object falls freely from rest in t seconds. The RCA Building in New York City is 850 ft tall. How long will it take an object to fall from the top?

Solution

1. Familiarize.   We make a drawing to help visualize the problem.

2. Translate.   We substitute into the formula:

$$s = 16t^2$$
$$850 = 16t^2.$$

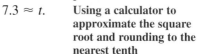

3. Carry out.   We solve for t:

$$850 = 16t^2$$
$$\frac{850}{16} = t^2$$
$$53.125 = t^2$$
$$\sqrt{53.125} = t$$ Using the principle of square roots; rejecting the negative square root since t cannot be negative in this problem

$$7.3 \approx t.$$ Using a calculator to approximate the square root and rounding to the nearest tenth

4. Check.   Since $16(7.3)^2 = 852.64 \approx 850$, our answer checks.

5. State.   It takes about 7.3 sec for an object to fall freely from the top of the RCA Building.

8.1
Exercise Set

Determine the number of real-number solutions of each equation from the given graph.

1. $x^2 + x - 12 = 0$ 2

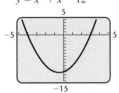

$y = x^2 + x - 12$

2. $-3x^2 - x - 7 = 0$ 0

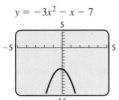

$y = -3x^2 - x - 7$

3. $4x^2 + 9 = 12x$ 1

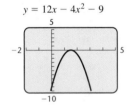

$y = 12x - 4x^2 - 9$

4. $2x^2 + 3 = 6x$ 2

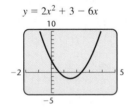

$y = 2x^2 + 3 - 6x$

5. $f(x) = 0$ 0

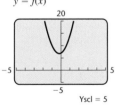

$y = f(x)$

Yscl = 5

6. $f(x) = 0$ 1

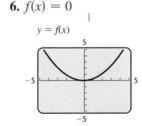

$y = f(x)$

Solve.

7. $7x^2 = 21$ $\pm\sqrt{3}$

8. $4x^2 = 20$ $\pm\sqrt{5}$

9. $25x^2 + 4 = 0$ $\pm\frac{2}{5}i$

10. $9x^2 + 16 = 0$ $\pm\frac{4}{3}i$

11. $3t^2 - 2 = 0$ ☉

12. $5t^2 - 7 = 0$ ☉

13. $(x + 2)^2 = 25$ $-7, 3$

14. $(x - 1)^2 = 49$ $-6, 8$

15. $(a + 5)^2 = 8$ $-5 \pm 2\sqrt{2}$

16. $(a - 13)^2 = 18$ $13 \pm 3\sqrt{2}$

17. $(x - 1)^2 = -49$ $1 \pm 7i$

18. $(x + 1)^2 = -9$ $-1 \pm 3i$

19. $\left(t + \frac{3}{2}\right)^2 = \frac{7}{2}$ $\dfrac{-3 \pm \sqrt{14}}{2}$

20. $\left(y + \frac{3}{4}\right)^2 = \frac{17}{16}$ $\dfrac{-3 \pm \sqrt{17}}{4}$

21. $x^2 - 6x + 9 = 100$ $-7, 13$

22. $x^2 - 10x + 25 = 64$ $-3, 13$

23. Let $f(x) = (x - 5)^2$. Find x such that $f(x) = 16$.   1, 9

24. Let $g(x) = (x - 2)^2$. Find x such that $g(x) = 25$. $-3, 7$

25. Let $F(t) = (t + 4)^2$. Find t such that $F(t) = 13$. $-4 \pm \sqrt{13}$

26. Let $f(t) = (t + 6)^2$. Find t such that $f(t) = 15$. $-6 \pm \sqrt{15}$

Aha! **27.** Let $g(x) = x^2 + 14x + 49$. Find x such that $g(x) = 49$. $-14, 0$

28. Let $F(x) = x^2 + 8x + 16$. Find x such that $F(x) = 9$. $-7, -1$

Complete the square. Then write the perfect-square trinomial in factored form.

29. $x^2 + 8x$
$x^2 + 8x + 16, (x + 4)^2$

30. $x^2 + 16x$
$x^2 + 16x + 64, (x + 8)^2$

31. $x^2 - 6x$
$x^2 - 6x + 9, (x - 3)^2$

32. $x^2 - 10x$
$x^2 - 10x + 25, (x - 5)^2$

33. $x^2 - 24x$
$x^2 - 24x + 144, (x - 12)^2$

34. $x^2 - 18x$
$x^2 - 18x + 81, (x - 9)^2$

35. $t^2 + 9t$
$t^2 + 9t + \frac{81}{4}, \left(t + \frac{9}{2}\right)^2$

36. $t^2 + 3t$
$t^2 + 3t + \frac{9}{4}, \left(t + \frac{3}{2}\right)^2$

37. $x^2 - 3x$
$x^2 - 3x + \frac{9}{4}, \left(x - \frac{3}{2}\right)^2$

38. $x^2 - 7x$
$x^2 - 7x + \frac{49}{4}, \left(x - \frac{7}{2}\right)^2$

39. $x^2 + \frac{2}{3}x$
$x^2 + \frac{2}{3}x + \frac{1}{9}, \left(x + \frac{1}{3}\right)^2$

40. $x^2 + \frac{2}{5}x$
$x^2 + \frac{2}{5}x + \frac{1}{25}, \left(x + \frac{1}{5}\right)^2$

41. $t^2 - \frac{5}{3}t$
$t^2 - \frac{5}{3}t + \frac{25}{36}, \left(t - \frac{5}{6}\right)^2$

42. $t^2 - \frac{5}{6}t$
$t^2 - \frac{5}{6}t + \frac{25}{144}, \left(t - \frac{5}{12}\right)^2$

43. $x^2 + \frac{9}{5}x$
$x^2 + \frac{9}{5}x + \frac{81}{100}, \left(x + \frac{9}{10}\right)^2$

44. $x^2 + \frac{9}{4}x$
$x^2 + \frac{9}{4}x + \frac{81}{64}, \left(x + \frac{9}{8}\right)^2$

Solve by completing the square. Show your work.

45. $x^2 + 6x = 7$ $-7, 1$

46. $x^2 + 8x = 9$ $-9, 1$

47. $x^2 - 10x = 22$ $5 \pm \sqrt{47}$

48. $x^2 - 4x = -9$ $2 \pm i\sqrt{5}$

49. $x^2 + 8x + 7 = 0$ $-7, -1$

50. $x^2 + 10x + 9 = 0$ $-9, -1$

51. $x^2 - 10x + 21 = 0$ 3, 7

52. $x^2 - 10x + 24 = 0$ 4, 6

53. $t^2 + 5t + 3 = 0$ ☉

54. $t^2 + 6t + 7 = 0$ $-3 \pm \sqrt{2}$

55. $x^2 + 10 = 6x$ $3 \pm i$

56. $x^2 + 23 = 10x$ $5 \pm \sqrt{2}$

57. $s^2 + 4s + 13 = 0$ $-2 \pm 3i$

58. $t^2 + 12t + 25 = 0$ $-6 \pm \sqrt{11}$

Solve by completing the square. Remember to first divide, as in Example 11, to make sure that the coefficient of x^2 is 1.

59. $2x^2 - 5x - 3 = 0$ $-\frac{1}{2}, 3$

60. $3x^2 + 5x - 2 = 0$ $-2, \frac{1}{3}$

61. $4x^2 + 8x + 3 = 0$ $-\frac{3}{2}, -\frac{1}{2}$

62. $9x^2 + 18x + 8 = 0$ $-\frac{4}{3}, -\frac{2}{3}$

☉ Answers to Exercises 11, 12, and 53 can be found on p. A-64.

$-\frac{3}{2}, \frac{5}{3}$ $-\frac{1}{2}, \frac{2}{3}$

63. $6x^2 - x = 15$ $\dfrac{-2 \pm \sqrt{2}}{2}$ **64.** $6x^2 - x = 2$ $-2, -\frac{1}{2}$

65. $2x^2 + 4x + 1 = 0$ **66.** $2x^2 + 5x + 2 = 0$

67. $3x^2 - 5x - 3 = 0$ **68.** $4x^2 - 6x - 1 = 0$

Interest. Use $A = P(1 + r)^t$ to find the interest rate in Exercises 69–74. Refer to Example 12.

69. $2000 grows to $2420 in 2 yr   10%

70. $2560 grows to $2890 in 2 yr   6.25%

71. $1280 grows to $1805 in 2 yr   18.75%

72. $1000 grows to $1440 in 2 yr   20%

73. $6250 grows to $6760 in 2 yr   4%

74. $6250 grows to $7290 in 2 yr   8%

Free-Falling Objects. Use $s = 16t^2$ for Exercises 75–78. Refer to Example 13.

75. The CN Tower in Toronto, at 1815 ft, is the world's tallest self-supporting tower (no guy wires) (*Source: The Guinness Book of Records*). How long would it take an object to fall freely from the top? About 10.7 sec

76. Reaching 745 ft above the water, the towers of California's Golden Gate Bridge are the world's tallest bridge towers (*Source: The Guinness Book of Records*). How long would it take an object to fall freely from the top?   About 6.8 sec

77. The Gateway Arch in St. Louis is 640 ft high. How long would it take an object to fall freely from the top?   About 6.3 sec

78. The Sears Tower in Chicago is 1454 ft tall. How long would it take an object to fall freely from the top?   About 9.5 sec

TW **79.** Explain in your own words a sequence of steps that can be used to solve any quadratic equation in the quickest way.

TW **80.** Describe how to write a quadratic equation that can be solved algebraically but not graphically.

Skill Maintenance

Evaluate. [1.2]

81. $at^2 - bt$, for $a = 3$, $b = 5$, and $t = 4$ 28

82. $mn^2 - mp$, for $m = -2$, $n = 7$, and $p = 3$ -92

Simplify. [7.3]

83. $\sqrt[3]{270}$ $3\sqrt[3]{10}$ **84.** $\sqrt{80}$ $4\sqrt{5}$

Let $f(x) = \sqrt{3x - 5}$.

85. Find $f(10)$. [7.1]   5 **86.** Find $f(18)$. [7.1]   7

Synthesis

TW **87.** What would be better: to receive 3% interest every 6 months or to receive 6% interest every 12 months? Why?

TW **88.** Example 12 can be solved with a graphing calculator by graphing each side of

$$4410 = 4000(1 + r)^2.$$

How could you determine, from a reading of the problem, a suitable viewing window?

Find b such that each trinomial is a square.

89. $x^2 + bx + 81$ ± 18 **90.** $x^2 + bx + 49$ ± 14

91. If $f(x) = 2x^5 - 9x^4 - 66x^3 + 45x^2 + 280x$ and $x^2 - 5$ is a factor of $f(x)$, find all a for which $f(a) = 0$. $-\frac{7}{2}, -\sqrt{5}, 0, \sqrt{5}, 8$

92. If

$$f(x) = \left(x - \tfrac{1}{3}\right)\left(x^2 + 6\right)$$

and

$$g(x) = \left(x - \tfrac{1}{3}\right)\left(x^2 - \tfrac{2}{3}\right),$$

find all a for which $(f + g)(a) = 0$.

93. *Boating.* A barge and a fishing boat leave a dock at the same time, traveling at a right angle to each other. The barge travels 7 km/h slower than the fishing boat. After 4 hr, the boats are 68 km apart. Find the speed of each boat.   Barge: 8 km/h; fishing boat: 15 km/h

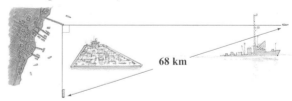

94. Find three consecutive integers such that the square of the first plus the product of the other two is 67.   5, 6, 7

67. $\dfrac{5 \pm \sqrt{61}}{6}$ **68.** $\dfrac{3 \pm \sqrt{13}}{4}$ **92.** $\dfrac{1}{3}, \pm \dfrac{2\sqrt{6}}{3}i$

8.2

The Quadratic Formula

There are at least two reasons for learning to complete the square. One is to enhance your ability to graph certain equations that appear later in this chapter. Another is to develop a general formula for solving quadratic equations.

Solving Using the Quadratic Formula

Each time you solve by completing the square, the procedure is the same. In mathematics, when a procedure is repeated many times, a formula is often developed to speed up our work.

We begin with a quadratic equation in standard form,

$$ax^2 + bx + c = 0,$$

with $a > 0$. For $a < 0$, a slightly different derivation is needed (see Exercise 52), but the result is the same. Let's solve by completing the square. As the steps are performed, compare them with Example 11 on p. 570.

$$ax^2 + bx = -c \qquad \text{Adding } -c \text{ to both sides}$$

$$x^2 + \frac{b}{a}x = -\frac{c}{a} \qquad \text{Dividing both sides by } a$$

Half of $\dfrac{b}{a}$ is $\dfrac{b}{2a}$ and $\left(\dfrac{b}{2a}\right)^2$ is $\dfrac{b^2}{4a^2}$. We add $\dfrac{b^2}{4a^2}$ to both sides:

$$x^2 + \frac{b}{a}x + \frac{b^2}{4a^2} = -\frac{c}{a} + \frac{b^2}{4a^2} \qquad \text{Adding } \frac{b^2}{4a^2} \text{ to complete the square}$$

$$\left(x + \frac{b}{2a}\right)^2 = -\frac{4ac}{4a^2} + \frac{b^2}{4a^2} \qquad \begin{array}{l}\textbf{Factoring on the left side;}\\ \textbf{finding a common denomi-}\\ \textbf{nator on the right side}\end{array}$$

$$\left(x + \frac{b}{2a}\right)^2 = \frac{b^2 - 4ac}{4a^2}$$

TEACHING TIP

You may want to remind students to write the fraction bar under the entire expression $-b \pm \sqrt{b^2 - 4ac}$.

$$x + \frac{b}{2a} = \pm\frac{\sqrt{b^2 - 4ac}}{2a} \qquad \begin{array}{l}\textbf{Using the principle of square}\\ \textbf{roots and the quotient rule}\\ \textbf{for radicals; since } a > 0,\\ \sqrt{4a^2} = 2a\end{array}$$

$$x = \frac{-b \pm \sqrt{b^2 - 4ac}}{2a}. \qquad \text{Adding } -\frac{b}{2a} \text{ to both sides}$$

It is important that you remember the quadratic formula and know how to use it.

The Quadratic Formula

The solutions of $ax^2 + bx + c = 0$, $a \neq 0$, are given by

$$x = \frac{-b \pm \sqrt{b^2 - 4ac}}{2a}.$$

EXAMPLE 1 Solve $5x^2 + 8x = -3$ using the quadratic formula.

Solution   We first find standard form and determine a, b, and c:

$$5x^2 + 8x + 3 = 0; \qquad \text{Adding 3 to both sides to get 0 on one side}$$
$$a = 5, \quad b = 8, \quad c = 3.$$

Next, we use the quadratic formula:

$$x = \frac{-b \pm \sqrt{b^2 - 4ac}}{2a}$$

$$x = \frac{-8 \pm \sqrt{8^2 - 4 \cdot 5 \cdot 3}}{2 \cdot 5} \qquad \text{Substituting}$$

$$x = \frac{-8 \pm \sqrt{64 - 60}}{10} \qquad \substack{\textbf{Be sure to write the fraction} \\ \textbf{bar all the way across.}}$$

$$x = \frac{-8 \pm \sqrt{4}}{10} = \frac{-8 \pm 2}{10}$$

$$x = \frac{-8 + 2}{10} \quad or \quad x = \frac{-8 - 2}{10}$$

$$x = \frac{-6}{10} \quad or \quad x = \frac{-10}{10}$$

$$x = -\frac{3}{5} \quad or \quad x = -1.$$

The solutions are $-\frac{3}{5}$ and -1. The checks are left to the student.

Because $5x^2 + 8x + 3$ can be factored as $(5x + 3)(x + 1)$, the quadratic formula may not have been the fastest way of solving Example 1. However, because the quadratic formula works for *any* quadratic equation, we need not spend too much time struggling to solve a quadratic equation by factoring.

> **To Solve a Quadratic Equation**
>
> **1.** If the equation can be easily written in the form $ax^2 = p$ or $(x + k)^2 = d$, use the principle of square roots as in Section 8.1.
>
> **2.** If step (1) does not apply, write the equation in the form $ax^2 + bx + c = 0$.
>
> **3.** Try factoring and using the principle of zero products.
>
> **4.** If factoring seems to be difficult or impossible, use the quadratic formula.
>
> The solutions of a quadratic equation can always be found using the quadratic formula. They cannot always be found by factoring.

Recall that a second-degree polynomial in one variable is said to be quadratic. Similarly, a second-degree polynomial function in one variable is said to be a **quadratic function**.

EXAMPLE 2   For the quadratic function given by $f(x) = 5x^2 - 8x - 3$, find all x for which $f(x) = 0$.

Solution   We substitute and solve for x:

$$f(x) = 0$$
$$5x^2 - 8x - 3 = 0 \qquad \textbf{Substituting. This cannot be solved by factoring.}$$
$$a = 5, \quad b = -8, \quad c = -3.$$

We then substitute into the quadratic formula:

$$x = \frac{-(-8) \pm \sqrt{(-8)^2 - 4 \cdot 5 \cdot (-3)}}{2 \cdot 5}$$

$$= \frac{8 \pm \sqrt{64 + 60}}{10}$$

$$= \frac{8 \pm \sqrt{124}}{10} \qquad \textbf{Note that 4 is a perfect-square factor of 124.}$$

$$= \frac{8 \pm \sqrt{4 \cdot 31}}{10} \qquad \textbf{124} = \textbf{4} \cdot \textbf{31}$$

$$= \frac{8 \pm 2\sqrt{31}}{10} \qquad \sqrt{\textbf{4}} = \textbf{2}$$

$$= \frac{2(4 \pm \sqrt{31})}{2 \cdot 5} = \frac{4 \pm \sqrt{31}}{5}. \qquad \textbf{Removing a factor equal to 1: } \tfrac{2}{2} = 1$$

> **CAUTION!**   To avoid a common error, *factor the numerator and the denominator* when removing a factor equal to 1.

To check using a calculator, we enter $y_1 = 5x^2 - 8x - 3$ and evaluate y_1 for both possible solutions. Care must be taken to place parentheses properly around the radicand as well as around the entire numerator.

```
Y₁((4+√ (31))/5)
                        0
Y₁((4−√ (31))/5)
                        0
```

Another way to check is by graphing $y_1 = 5x^2 - 8x - 3$, pressing TRACE, and entering $(4 + \sqrt{(31)})/5$. The y-value 0 appears along with a rational approximation for x.

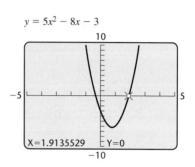

$y = 5x^2 - 8x - 3$

X=1.9135529   Y=0

We then check $(4 - \sqrt{(31)})/5$ in the same way. The solutions are

$$\frac{4 + \sqrt{31}}{5} \quad \text{and} \quad \frac{4 - \sqrt{31}}{5}.$$

Some quadratic equations have solutions that are imaginary numbers.

EXAMPLE 3   Solve: $x^2 + 2 = -x$.

Solution   We first find standard form:

$$x^2 + x + 2 = 0. \qquad \textbf{Adding } x \textbf{ to both sides}$$

Since we cannot factor $x^2 + x + 2$, we use the quadratic formula with $a = 1$, $b = 1$, and $c = 2$:

$$x = \frac{-1 \pm \sqrt{1^2 - 4 \cdot 1 \cdot 2}}{2 \cdot 1} \qquad \textbf{Substituting}$$

$$= \frac{-1 \pm \sqrt{1 - 8}}{2}$$

$$= \frac{-1 \pm \sqrt{-7}}{2}$$

$$= \frac{-1 \pm i\sqrt{7}}{2}, \text{ or } -\frac{1}{2} \pm \frac{\sqrt{7}}{2}i.$$

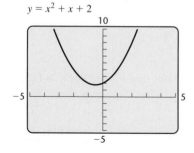

$y = x^2 + x + 2$

The solutions are $-\frac{1}{2} - \frac{\sqrt{7}}{2}i$ and $-\frac{1}{2} + \frac{\sqrt{7}}{2}i$. The checks are left to the student.

Note from the graph at left that there are no x-intercepts of the graph of $f(x) = x^2 + x + 2$, and thus no real-number solutions of the equation.

The quadratic formula is sometimes used to solve equations that do not originally appear to be quadratic.

EXAMPLE 4   If $f(x) = 2 + \dfrac{7}{x}$ and $g(x) = \dfrac{4}{x^2}$, find all x for which $f(x) = g(x)$.

Solution   We set $f(x)$ equal to $g(x)$ and solve:

$$f(x) = g(x)$$

$$2 + \frac{7}{x} = \frac{4}{x^2}. \qquad \textbf{Substituting. Note that } x \neq 0.$$

This is a rational equation similar to those in Section 6.4. To solve, we multiply both sides by the LCD, x^2:

$$x^2\left(2 + \frac{7}{x}\right) = x^2 \cdot \frac{4}{x^2}$$

$$2x^2 + 7x = 4 \qquad \textbf{Simplifying}$$

$$2x^2 + 7x - 4 = 0. \qquad \textbf{Subtracting 4 from both sides}$$

We have

$$a = 2, \quad b = 7, \quad \text{and} \quad c = -4.$$

Substituting then gives us

$$x = \frac{-7 \pm \sqrt{7^2 - 4 \cdot 2 \cdot (-4)}}{2 \cdot 2}$$

$$= \frac{-7 \pm \sqrt{49 + 32}}{4}$$

$$= \frac{-7 \pm \sqrt{81}}{4}$$

$$= \frac{-7 \pm 9}{4}$$

$$x = \frac{-7 + 9}{4} = \frac{1}{2} \quad or \quad x = \frac{-7 - 9}{4} = -4. \qquad \begin{array}{l}\textbf{Both answers should} \\ \textbf{check since } x \neq 0.\end{array}$$

You can confirm that $f\left(\tfrac{1}{2}\right) = g\left(\tfrac{1}{2}\right)$ and $f(-4) = g(-4)$. The solutions are $\tfrac{1}{2}$ and -4.

Interactive Discovery

Quadratic equations with real-number solutions can be solved algebraically or graphically. The equations in Examples 2 and 4 were solved algebraically using the quadratic formula.

1. Solve Example 2 graphically and compare with the algebraic solution. Which method is faster? Which method is more precise?

2. Solve Example 4 graphically and compare with the algebraic solution. Which method is faster? Which method is more precise?

The quadratic formula always yields exact solutions of a quadratic equation. Although solving by graphing may yield only approximate solutions, it may be faster.

Checking the solutions of Examples 2 and 3 can be cumbersome. Fortunately, when the quadratic formula is used to solve a quadratic equation, the results will always check in that equation provided the formula has been properly used. Thus checking for computational errors is usually sufficient.

Approximating Solutions

When the solution of an equation is irrational, a rational-number approximation is often useful. This is often the case in real-world applications similar to those found in Section 8.3.

EXAMPLE 5   Use a calculator to approximate the solutions of Example 2 to the nearest ten-thousandth.

Solution   The following sequence of keystrokes can be used to approximate $(4 + \sqrt{31})/5$:

$$\boxed{(}\ \boxed{4}\ \boxed{+}\ \boxed{\sqrt{}}\ \boxed{3}\ \boxed{1}\ \boxed{)}\ \boxed{)}\ \boxed{\div}\ \boxed{5}\ \boxed{\text{ENTER}}.$$

Similar keystrokes can be used to approximate $(4 - \sqrt{31})/5$.

The calculator will display results such as 1.913552873 and -0.3135528726, as shown at left. Note that these are *approximations*; the *exact* solutions are $(4 + \sqrt{31})/5$ and $(4 - \sqrt{31})/5$. Rounded to the nearest ten-thousandth, the solutions are approximately 1.9136 and -0.3136.

(4+√(31))/5
          1.913552873
(4−√(31))/5
          −.3135528726

8.2

Exercise Set

FOR EXTRA HELP

Digital Video Tutor CD 8 Videotape 11   Student's Solutions Manual   Tutor Center AW Math Tutor Center   InterAct Math   MathXL   MyMathLab

Solve.

1. $x^2 + 7x - 3 = 0$ $\dfrac{-7 \pm \sqrt{61}}{2}$

2. $x^2 - 7x + 4 = 0$ $\dfrac{7 \pm \sqrt{33}}{2}$

3. $3p^2 = 18p - 6$ $3 \pm \sqrt{7}$

4. $3u^2 = 8u - 5$ $1, \frac{5}{3}$

5. $x^2 + x + 2 = 0$ $-\frac{1}{2} \pm \frac{\sqrt{7}}{2}i$

6. $x^2 + x + 1 = 0$ $-\frac{1}{2} \pm \frac{\sqrt{3}}{2}i$

7. $x^2 + 13 = 4x$ $2 \pm 3i$

8. $x^2 + 13 = 6x$ $3 \pm 2i$

9. $h^2 + 4 = 6h$ $3 \pm \sqrt{5}$

10. $r^2 + 3r = 8$ $\dfrac{-3 \pm \sqrt{41}}{2}$

11. $3 + \dfrac{8}{x} = \dfrac{1}{x^2}$ $\dfrac{-4 \pm \sqrt{19}}{3}$

12. $2 + \dfrac{5}{x^2} = \dfrac{9}{x}$ $\dfrac{9 \pm \sqrt{41}}{4}$

13. $3x + x(x - 2) = 4$

14. $4x + x(x - 3) = 5$

15. $12t^2 + 9t = 1$ $\dfrac{-9 \pm \sqrt{129}}{24}$

16. $15t^2 + 7t = 2$ $-\frac{2}{3}, \frac{1}{5}$

17. $25x^2 - 20x + 4 = 0$ $\frac{2}{5}$

18. $36x^2 + 84x + 49 = 0$ $-\frac{7}{6}$

19. $7x(x + 2) + 5 = 3x(x + 1)$ $\dfrac{-11 \pm \sqrt{41}}{8}$

20. $5x(x - 1) - 7 = 4x(x - 2)$ $\dfrac{-3 \pm \sqrt{37}}{2}$

21. $14(x - 4) - (x + 2) = (x + 2)(x - 4)$ $5, 10$

22. $11(x - 2) + (x - 5) = (x + 2)(x - 6)$ $1, 15$

23. $5x^2 = 13x + 17$

24. $25x = 3x^2 + 28$ $\frac{4}{3}, 7$

25. $x^2 + 9 = 4x$ $2 \pm \sqrt{5}i$

26. $x^2 + 7 = 3x$ $\dfrac{3}{2} \pm \dfrac{\sqrt{19}}{2}i$

27. $x^3 - 8 = 0$ (*Hint:* Factor the difference of cubes. Then use the quadratic formula.) $2, -1 \pm \sqrt{3}i$

28. $x^3 + 1 = 0$ $-1, \dfrac{1}{2} \pm \dfrac{\sqrt{3}}{2}i$

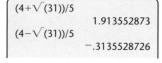

13. $\dfrac{-1 \pm \sqrt{17}}{2}$ **14.** $\dfrac{-1 \pm \sqrt{21}}{2}$

23. $\dfrac{13 \pm \sqrt{509}}{10}$

29. $\dfrac{5 \pm \sqrt{37}}{6}$ **30.** $\dfrac{1 \pm \sqrt{13}}{4}$

29. Let $f(x) = 3x^2 - 5x - 1$. Find x such that $f(x) = 0$.

30. Let $g(x) = 4x^2 - 2x - 3$. Find x such that $g(x) = 0$.

31. Let
$$f(x) = \frac{7}{x} + \frac{7}{x + 4}.$$
Find all x for which $f(x) = 1$. $5 \pm \sqrt{53}$

32. Let
$$g(x) = \frac{2}{x} + \frac{2}{x + 3}.$$
Find all x for which $g(x) = 1$. $-2, 3$

33. Let
$$F(x) = \frac{x + 3}{x} \quad \text{and} \quad G(x) = \frac{x - 4}{3}.$$
Find all x for which $F(x) = G(x)$. $\dfrac{7 \pm \sqrt{85}}{2}$

34. Let
$$f(x) = \frac{3 - x}{4} \quad \text{and} \quad g(x) = \frac{1}{4x}.$$
Find all x for which $f(x) = g(x)$. $\dfrac{3 \pm \sqrt{5}}{2}$

35. Let
$$f(x) = \frac{15 - 2x}{6} \quad \text{and} \quad g(x) = \frac{3}{x}.$$
Find all x for which $f(x) = g(x)$. $\frac{3}{2}, 6$

36. Let
$$f(x) = x + 5 \quad \text{and} \quad g(x) = \frac{3}{x - 5}.$$
Find all x for which $f(x) = g(x)$. $\pm 2\sqrt{7}$

Solve. Use a calculator to approximate solutions as rational numbers. Round to the nearest ten-thousandth.

37. $x^2 + 4x - 7 = 0$ $-5.3166, 1.3166$

38. $x^2 + 6x + 4 = 0$ $-5.2361, -0.7639$

39. $x^2 - 6x + 4 = 0$ $0.7639, 5.2361$

40. $x^2 - 4x + 1 = 0$ $0.2679, 3.7321$

41. $2x^2 - 3x - 7 = 0$ $-1.2656, 2.7656$

42. $3x^2 - 3x - 2 = 0$ $-0.4574, 1.4574$

TW 43. Are there any equations that can be solved by the quadratic formula but not by completing the square? Why or why not?

TW 44. The list on p. 577 does not mention completing the square as a method of solving quadratic equations. Why not?

Skill Maintenance

45. *Coffee Beans.* Twin Cities Roasters has Kenyan coffee for which they pay $6.75 a pound and Peruvian coffee for which they pay $11.25 a pound. How much of each kind should be mixed in order to obtain a 50-lb mixture that is worth $8.55 a pound? [3.3]   Kenyan: 30 lb; Peruvian: 20 lb

46. *Donuts.* South Street Bakers charges $1.10 for a cream-filled donut and 85¢ for a glazed donut. On a recent Sunday, a total of 90 glazed and cream-filled donuts were sold for $88.00. How many of each type were sold? [3.3]   Cream-filled: 46; glazed: 44

Simplify.

47. $\sqrt{27a^2b^5} \cdot \sqrt{6a^3b}$ [7.3] $9a^2b^3\sqrt{2a}$

48. $\sqrt{8a^3b} \cdot \sqrt{12ab^5}$ [7.3] $4a^2b^3\sqrt{6}$

49. $\dfrac{\dfrac{3}{x - 1}}{\dfrac{1}{x + 1} + \dfrac{2}{x - 1}}$ [6.3] $\dfrac{3(x + 1)}{3x + 1}$

50. $\dfrac{\dfrac{4}{a^2b}}{\dfrac{3}{a} - \dfrac{4}{b^2}}$ [6.3] $\dfrac{4b}{3ab^2 - 4a^2}$

Synthesis

TW 51. Suppose you had a large number of quadratic equations to solve and none of the equations had a constant term. Would you use factoring or the quadratic formula to solve these equations? Why?

TW 52. If $a < 0$ and $ax^2 + bx + c = 0$, then $-a$ is positive and the equivalent equation, $-ax^2 - bx - c = 0$, can be solved using the quadratic formula.

 a) Find this solution, replacing a, b, and c in the formula with $-a$, $-b$, and $-c$ from the equation.

 b) How does the result of part (a) indicate that the quadratic formula "works" regardless of the sign of a?

For Exercises 53–55, let
$$f(x) = \frac{x^2}{x - 2} + 1 \quad \text{and} \quad g(x) = \frac{4x - 2}{x - 2} + \frac{x + 4}{2}.$$

53. Find the x-intercepts of the graph of f. $(-2, 0), (1, 0)$

54. Find the x-intercepts of the graph of g. $\left(-5 - \sqrt{37}, 0\right), \left(-5 + \sqrt{37}, 0\right)$

55. Find all x for which $f(x) = g(x)$. $4 - 2\sqrt{2}, 4 + 2\sqrt{2}$

Solve using the quadratic formula.

56. $x^2 - 0.75x - 0.5 = 0$ $-0.4253905297, 1.17539053$

57. $z^2 + 0.84z - 0.4 = 0$ $-1.1792101, 0.3392101$

58. $\left(1 + \sqrt{3}\right)x^2 - \left(3 + 2\sqrt{3}\right)x + 3 = 0$ $\sqrt{3}, \dfrac{3 - \sqrt{3}}{2}$

59. $\sqrt{2}x^2 + 5x + \sqrt{2} = 0$ $\dfrac{-5\sqrt{2} \pm \sqrt{34}}{4}$

60. $ix^2 - 2x + 1 = 0$
$-i \pm i\sqrt{1 - i}$

61. One solution of $kx^2 + 3x - k = 0$ is -2. Find the other. $\frac{1}{2}$

TW **62.** Can a graph be used to solve *any* quadratic equation? Why or why not?

Solving Problems ▪ Solving Formulas

Applications Involving Quadratic Equations

Solving Problems

As we found in Section 6.5, some problems translate to rational equations. The solution of such rational equations can involve quadratic equations.

EXAMPLE 1 **Motorcycle Travel.** Makita rode her motorcycle 300 mi at a certain average speed. Had she averaged 10 mph more, the trip would have taken 1 hr less. Find the average speed of the motorcycle.

Solution

1. **Familiarize.** We make a drawing, labeling it with the known and unknown information. As in Section 6.5, we can organize the information in a table. We let r represent the rate, in miles per hour, and t the time, in hours, for Makita's trip.

300 miles	
Time t	Speed r
300 miles	
Time $t - 1$	Speed $r + 10$

Distance	Speed	Time
300	r	t	$\longrightarrow r = \dfrac{300}{t}$
300	$r + 10$	$t - 1$	$\longrightarrow r + 10 = \dfrac{300}{t - 1}$

Recall that the definition of speed, $r = d/t$, relates the three quantities.

2. Translate.   From the first two lines of the table, we obtain

$$r = \frac{300}{t} \quad \text{and} \quad r + 10 = \frac{300}{t-1}.$$

3. Carry out.   A system of equations has been formed. We substitute for r from the first equation into the second and solve the resulting equation:

$$\frac{300}{t} + 10 = \frac{300}{t-1} \qquad \text{Substituting } 300/t \text{ for } r$$

$$t(t-1) \cdot \left[\frac{300}{t} + 10 \right] = t(t-1) \cdot \frac{300}{t-1} \qquad \text{Multiplying by the LCD}$$

$$\cancel{t}(t-1) \cdot \frac{300}{\cancel{t}} + t(t-1) \cdot 10 = t\cancel{(t-1)} \cdot \frac{300}{\cancel{t-1}} \qquad \begin{array}{l}\text{Using the distributive} \\ \text{law and removing} \\ \text{factors that equal 1:} \\ \frac{t}{t} = 1; \frac{t-1}{t-1} = 1 \end{array}$$

$$\left. \begin{array}{l} 300(t-1) + 10(t^2 - t) = 300t \\ 300t - 300 + 10t^2 - 10t = 300t \\ 10t^2 - 10t - 300 = 0 \end{array} \right\} \qquad \begin{array}{l}\text{Rewriting in} \\ \text{standard form}\end{array}$$

$$t^2 - t - 30 = 0 \qquad \begin{array}{l}\text{Multiplying by } \frac{1}{10} \text{ or} \\ \text{dividing by 10}\end{array}$$

$$(t-6)(t+5) = 0 \qquad \text{Factoring}$$

$$t = 6 \quad or \quad t = -5. \qquad \begin{array}{l}\text{Principle of zero} \\ \text{products}\end{array}$$

4. Check.   As a partial check, we graph $y_1 = 300/x + 10$ and $y_2 = 300/(x - 1)$. The graph at left shows that the solutions are -5 and 6.

Note that we have solved for t, not r as required. Since negative time has no meaning here, we disregard the -5 and use 6 hr to find r:

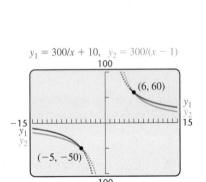

$y_1 = 300/x + 10, \quad y_2 = 300/(x - 1)$

(6, 60)

y_1
y_2

(−5, −50)

Yscl = 50

$$r = \frac{300 \text{ mi}}{6 \text{ hr}} = 50 \text{ mph}.$$

> **CAUTION!**   Always make sure that you find the quantity asked for in the problem.

To see if 50 mph checks, we increase the speed 10 mph to 60 mph and see how long the trip would have taken at that speed:

$$t = \frac{d}{r} = \frac{300 \text{ mi}}{60 \text{ mph}} = 5 \text{ hr.} \qquad \text{Note that mi/mph} = \text{mi} \div \frac{\text{mi}}{\text{hr}} =$$

$$\text{mi} \cdot \frac{\text{hr}}{\text{mi}} = \text{hr.}$$

This is 1 hr less than the trip actually took, so the answer checks.

5. State.   Makita's motorcycle traveled at an average speed of 50 mph.

Solving Formulas

Recall that to solve a formula for a certain letter, we use the principles for solving equations to get that letter alone on one side.

TEACHING TIP

We use the word "letter" instead of the word "variable" because not all letters are variables. Point out that π and g in Example 2 are both constants.

EXAMPLE 2 Period of a Pendulum.   The time T required for a pendulum of length l to swing back and forth (complete one period) is given by the formula $T = 2\pi\sqrt{l/g}$, where g is the earth's gravitational constant. Solve for l.

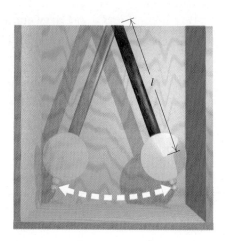

Solution   We have

$$T = 2\pi\sqrt{\dfrac{l}{g}} \qquad \textbf{This is a radical equation (see Section 7.6).}$$

$$T^2 = \left(2\pi\sqrt{\dfrac{l}{g}}\right)^2 \qquad \textbf{Principle of powers (squaring both sides)}$$

$$T^2 = 2^2\pi^2\dfrac{l}{g}$$

$$gT^2 = 4\pi^2 l \qquad \textbf{Multiplying both sides by } g \textbf{ to clear fractions}$$

$$\dfrac{gT^2}{4\pi^2} = l. \qquad \textbf{Dividing both sides by } 4\pi^2$$

We now have l alone on one side and l does not appear on the other side, so the formula is solved for l.

In most formulas, variables represent nonnegative numbers, so we do not need to use absolute-value signs when taking square roots.

EXAMPLE 3 Hang Time.*   An athlete's *hang time* is the amount of time that the athlete can remain airborne when jumping. A formula relating an athlete's vertical leap V, in inches, to hang time T, in seconds, is $V = 48T^2$. Solve for T.

---

*This formula is taken from an article by Peter Brancazio, "The Mechanics of a Slam Dunk," *Popular Mechanics*, November 1991. Courtesy of Professor Peter Brancazio, Brooklyn College.

Solution

$$48T^2 = V$$

$$T^2 = \frac{V}{48} \qquad \text{\textbf{Dividing by 48 to get } } T^2 \text{ \textbf{alone}}$$

$$T = \frac{\sqrt{V}}{\sqrt{48}} \qquad \text{\textbf{Using the principle of square roots and the quotient rule for radicals}}$$

$$= \frac{\sqrt{V}}{\sqrt{16}\sqrt{3}} = \frac{\sqrt{V}}{4\sqrt{3}}$$

$$= \frac{\sqrt{V}}{4\sqrt{3}} \cdot \frac{\sqrt{3}}{\sqrt{3}} = \frac{\sqrt{3V}}{12}. \qquad \Big\} \quad \text{\textbf{Rationalizing the denominator}}$$

EXAMPLE 4 Falling Distance. An object tossed downward with an initial speed (velocity) of v_0 will travel a distance of s meters, where $s = 4.9t^2 + v_0t$ and t is measured in seconds. Solve for t.

Solution   Since t is squared in one term and raised to the first power in the other term, the equation is quadratic in t.

$$4.9t^2 + v_0t = s$$

$$4.9t^2 + v_0t - s = 0 \qquad \text{\textbf{Writing standard form}}$$

$$a = 4.9, \quad b = v_0, \quad c = -s$$

$$t = \frac{-v_0 \pm \sqrt{v_0^2 - 4(4.9)(-s)}}{2(4.9)} \qquad \text{\textbf{Using the quadratic formula}}$$

Since the negative square root would yield a negative value for t, we use only the positive root:

$$t = \frac{-v_0 + \sqrt{v_0^2 + 19.6s}}{9.8}.$$

The following list of steps should help you when solving formulas for a given letter. Try to remember that when solving a formula, you use the same approach that you would to solve an equation.

> **To Solve a Formula for a Letter – Say, *b***
>
> **1.** Clear fractions and use the principle of powers, as needed. (In some cases, you may clear the fractions first, and in some cases, you may use the principle of powers first.) Perform these steps until radicals containing *b* are gone and *b* is not in any denominator.
>
> **2.** Combine all terms with b^2 in them. Also combine all terms with b in them.
>
> **3.** If b^2 does not appear, you can solve by using just the addition and multiplication principles as in Sections 2.3 and 6.8.
>
> **4.** If b^2 appears but b does not, solve the equation for b^2. Then use the principle of square roots to solve for *b*.
>
> **5.** If there are terms containing both b and b^2, put the equation in standard form and use the quadratic formula.

8.3

Exercise Set

Solve.

1. *Canoeing.* During the first part of a canoe trip, Tim covered 60 km at a certain speed. He then traveled 24 km at a speed that was 4 km/h slower. If the total time for the trip was 8 hr, what was the speed on each part of the trip?
First part: 12 km/h; second part: 8 km/h

2. *Car Trips.* During the first part of a trip, Meira's Honda traveled 120 mi at a certain speed. Meira then drove another 100 mi at a speed that was 10 mph slower. If Meira's total trip time was 4 hr, what was her speed on each part of the trip?
First part: 60 mph; second part: 50 mph

3. *Car Trips.* Sandi's Subaru travels 280 mi averaging a certain speed. If the car had gone 5 mph faster, the trip would have taken 1 hr less. Find Sandi's average speed.   35 mph

4. *Car Trips.* Petra's Plymouth travels 200 mi averaging a certain speed. If the car had gone 10 mph faster, the trip would have taken 1 hr less. Find Petra's average speed.   40 mph

5. *Air Travel.* A Cessna flies 600 mi at a certain speed. A Beechcraft flies 1000 mi at a speed that is 50 mph faster, but takes 1 hr longer. Find the speed of each plane.   Cessna: 150 mph, Beechcraft: 200 mph; or Cessna: 200 mph, Beechcraft: 250 mph

6. *Air Travel.* A turbo-jet flies 50 mph faster than a super-prop plane. If a turbo-jet goes 2000 mi in 3 hr less time than it takes the super-prop to go 2800 mi, find the speed of each plane.
Super-prop: 350 mph; turbo-jet: 400 mph

7. *Bicycling.* Naoki bikes the 40 mi to Hillsboro averaging a certain speed. The return trip is made at a speed that is 6 mph slower. Total time for the round trip is 14 hr. Find Naoki's average speed on each part of the trip.   To Hillsboro: 10 mph; return trip: 4 mph

8. *Car Speed.* On a sales trip, Gail drives the 600 mi to Richmond averaging a certain speed. The return trip is made at an average speed that is 10 mph slower. Total time for the round trip is 22 hr. Find Gail's average speed on each part of the trip.   Average speed to Richmond: 60 mph; average speed returning: 50 mph

9. *Navigation.* The current in a typical Mississippi River shipping route flows at a rate of 4 mph. In order for a barge to travel 24 mi upriver and then return in a total of 5 hr, approximately how fast must the barge be able to travel in still water?   About 11 mph

10. *Navigation.* The Hudson River flows at a rate of 3 mph. A patrol boat travels 60 mi upriver and returns in a total time of 9 hr. What is the speed of the boat in still water?   About 14 mph

11. *Filling a Pool.* Two wells are used to fill a swimming pool. Working together, they can fill the pool in 4 hr. One well, working alone, can fill the pool in 6 hr less time than the other. How long would the smaller one take, working alone, to fill the pool? 12 hr

12. *Filling a Tank.* Two pipes are connected to the same tank. Working together, they can fill the tank in 2 hr. The larger pipe, working alone, can fill the tank in 3 hr less time than the smaller one. How long would the smaller one take, working alone, to fill the tank? 6 hr

13. *Paddleboats.* Ellen paddles 1 mi upstream and 1 mi back in a total time of 1 hr. The speed of the river is 2 mph. Find the speed of Ellen's paddleboat in still water. About 3.24 mph

14. *Rowing.* Dan rows 10 km upstream and 10 km back in a total time of 3 hr. The speed of the river is 5 km/h. Find Dan's speed in still water. About 9.34 km/h

Solve each formula for the indicated letter. Assume that all variables represent nonnegative numbers.

15. $A = 4\pi r^2$, for r $\quad r = \dfrac{1}{2}\sqrt{\dfrac{A}{\pi}}$
(Surface area of a sphere)

16. $A = 6s^2$, for s $\quad s = \sqrt{\dfrac{A}{6}}$
(Surface area of a cube)

17. $A = 2\pi r^2 + 2\pi rh$, for r $\quad r = \dfrac{-\pi h + \sqrt{\pi^2 h^2 + 2\pi A}}{2\pi}$
(Surface area of a right cylindrical solid)

18. $F = \dfrac{Gm_1 m_2}{r^2}$, for r $\quad r = \sqrt{\dfrac{Gm_1 m_2}{F}}$
(Law of gravity)

19. $N = \dfrac{kQ_1 Q_2}{s^2}$, for s $\quad s = \sqrt{\dfrac{kQ_1 Q_2}{N}}$
(Number of phone calls between two cities)

20. $A = \pi r^2$, for r $\quad r = \sqrt{\dfrac{A}{\pi}}$
(Area of a circle)

21. $T = 2\pi\sqrt{\dfrac{l}{g}}$, for g $\quad g = \dfrac{4\pi^2 l}{T^2}$
(A pendulum formula)

22. $a^2 + b^2 = c^2$, for b $\quad b = \sqrt{c^2 - a^2}$
(Pythagorean formula in two dimensions)

23. $a^2 + b^2 + c^2 = d^2$, for c $\quad c = \sqrt{d^2 - a^2 - b^2}$
(Pythagorean formula in three dimensions)

24. $N = \dfrac{k^2 - 3k}{2}$, for k $\quad k = \dfrac{3 + \sqrt{9 + 8N}}{2}$
(Number of diagonals of a polygon)

25. $s = v_0 t + \dfrac{gt^2}{2}$, for t $\quad t = \dfrac{-v_0 + \sqrt{v_0^2 + 2gs}}{g}$
(A motion formula)

26. $A = \pi r^2 + \pi rs$, for r $\quad r = \dfrac{-\pi s + \sqrt{\pi^2 s^2 + 4\pi A}}{2\pi}$
(Surface area of a cone)

27. $N = \frac{1}{2}(n^2 - n)$, for n $\quad n = \dfrac{1 + \sqrt{1 + 8N}}{2}$
(Number of games if n teams play each other once)

28. $A = A_0(1 - r)^2$, for r $\quad r = 1 - \sqrt{\dfrac{A}{A_0}}$
(A business formula)

29. $V = 3.5\sqrt{h}$, for h $\quad h = \dfrac{V^2}{12.25}$
(Distance to horizon from a height)

30. $W = \sqrt{\dfrac{1}{LC}}$, for L $\quad L = \dfrac{1}{W^2 C}$
(An electricity formula)

Aha! **31.** $at^2 + bt + c = 0$, for t $\quad t = \dfrac{-b \pm \sqrt{b^2 - 4ac}}{2a}$
(An algebraic formula)

32. $A = P_1(1 + r)^2 + P_2(1 + r)$, for r $\quad r = -1 + \dfrac{-P_2 + \sqrt{P_2^2 + 4AP_1}}{2P_1}$
(An investment formula)

Solve. Refer to Exercises 15–32 and Examples 2–4 for the appropriate formula.

33. *Falling Distance.*

   a) An object is dropped 500 m from an airplane. How long does it take the object to reach the ground? 10.1 sec

   b) An object is thrown downward 500 m from the plane at an initial velocity of 30 m/sec. How long does it take the object to reach the ground? 7.49 sec

   c) How far will an object fall in 5 sec, when thrown downward at an initial velocity of 30 m/sec? 272.5 m

34. *Falling Distance.*

   a) An object is dropped 75 m from an airplane. How long does it take the object to reach the ground? 3.9 sec

   b) An object is thrown downward with an initial velocity of 30 m/sec from a plane 75 m above the ground. How long does it take the object to reach the ground? 1.9 sec

   c) How far will an object fall in 2 sec, if thrown downward at an initial velocity of 30 m/sec? 79.6 m

35. *Bungee Jumping.*   Jesse is tied to one end of a 40-m elasticized (bungee) cord. The other end of the cord is tied to the middle of a train trestle. If Jesse jumps off the bridge, for how long will he fall before the cord begins to stretch? (See Example 4 and let $v_0 = 0$.)   2.9 sec

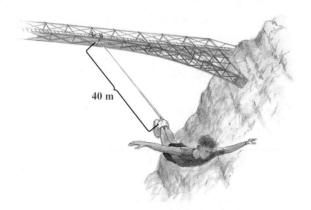

40 m

36. *Bungee Jumping.*   Sheila is tied to a bungee cord (see Exercise 35) and falls for 2.5 sec before her cord begins to stretch. How long is the bungee cord?
30.625 m

37. *Hang Time.*   The NBA's Vince Carter has a vertical leap of about 36 in. What is his hang time?   0.87 sec

38. *League Schedules.*   In a volleyball league, each team plays each of the other teams once. If a total of 66 games is played, how many teams are in the league?   12

39. *Downward Speed.*   An object thrown downward from a 100-m cliff travels 51.6 m in 3 sec. What was the initial velocity of the object?   2.5 m/sec

40. *Downward Speed.*   An object thrown downward from a 200-m cliff travels 91.2 m in 4 sec. What was the initial velocity of the object?   3.2 m/sec

41. *Compound Interest.*   A firm invests $3000 in a savings account for 2 yr. At the beginning of the second year, an additional $1700 is invested. If a total of $5253.70 is in the account at the end of the second year, what is the annual interest rate? (*Hint*: See Exercise 32.)   7%

42. *Compound Interest.*   A business invests $10,000 in a savings account for 2 yr. At the beginning of the second year, an additional $3500 is invested. If a total of $15,569.75 is in the account at the end of the second year, what is the annual interest rate? (*Hint*: See Exercise 32.)   8.5%

TW **43.** Marti is tied to a bungee cord that is twice as long as the cord tied to Pedro. Will Marti's fall take twice as long as Pedro's before their cords begin to stretch? Why or why not? (See Exercises 35 and 36.)

TW **44.** Under what circumstances would a negative value for t, time, have meaning?

Skill Maintenance

Evaluate.

45. $b^2 - 4ac$, for $a = 5$, $b = 6$, and $c = 7$ [1.2] -104

46. $\sqrt{b^2 - 4ac}$, for $a = 3$, $b = 4$, and $c = 5$ [7.8] $2i\sqrt{11}$

Simplify.

47. $\dfrac{x^2 + xy}{2x}$ [6.1] $\dfrac{x + y}{2}$ **48.** $\dfrac{a^3 - ab^2}{ab}$ [6.1] $\dfrac{a^2 - b^2}{b}$

49. $\dfrac{3 + \sqrt{45}}{6}$ [7.3] $\dfrac{1 + \sqrt{5}}{2}$ **50.** $\dfrac{2 - \sqrt{28}}{10}$ [7.3] $\dfrac{1 - \sqrt{7}}{5}$

Synthesis

TW **51.** In what ways do the motion problems of this section (like Example 1) differ from the motion problems in Chapter 6 (see p. 443)?

TW **52.** Write a problem for a classmate to solve. Devise the problem so that **(a)** the solution is found after solving a rational equation and **(b)** the solution is "The express train travels 90 mph."

53. *Biochemistry.*   The equation

$$A = 6.5 - \frac{20.4t}{t^2 + 36}$$

is used to calculate the acid level A in a person's blood t minutes after sugar is consumed. Solve for t.

54. *Special Relativity.*   Einstein found that an object of mass m_0, traveling velocity v, has its mass become

$$m = \frac{m_0}{\sqrt{1 - \dfrac{v^2}{c^2}}}, \qquad c = \frac{vm}{\sqrt{m^2 - m_0^2}}$$

where c is the speed of light. Solve the formula for c.

55. *The Golden Rectangle.*   For over 2000 yr, the proportions of a "golden" rectangle have been considered visually appealing. A rectangle of width w and length l is considered "golden" if

$$\frac{w}{l} = \frac{l}{w + l}.$$

Solve for l. $l = \dfrac{w + w\sqrt{5}}{2}$

56. *Diagonal of a Cube.*   Find a formula that expresses the length of the three-dimensional diagonal of a cube as a function of the cube's surface area.

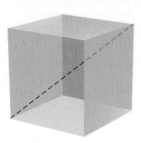

$$L(A) = \sqrt{\frac{A}{2}}$$

57. Find a number for which the reciprocal of 1 less than the number is the same as 1 more than the number. $\pm\sqrt{2}$

58. *Purchasing.*   A discount store bought a quantity of beach towels for \$250 and sold all but 15 at a profit of \$3.50 per towel. With the total amount received, the manager could buy 4 more than twice as many as were bought before. Find the cost per towel.   \$2.50

59. Solve for n:

$$mn^4 - r^2pm^3 - r^2n^2 + p = 0.$$

60. *Surface Area.*   Find a formula that expresses the diameter of a right cylindrical solid as a function of its surface area and its height.

61. A sphere is inscribed in a cube as shown in the figure below. Express the surface area of the sphere as a function of the surface area S of the cube.

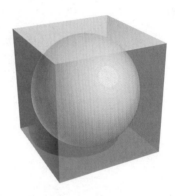

$$A(S) = \frac{\pi S}{6}$$

ᵀᵂ **62.** Explain how Exercises 1–14 can be solved without factoring, completing the square, or using the quadratic formula.

---

53. $t = \dfrac{-10.2 \pm 6\sqrt{-A^2 + 13A - 39.36}}{A - 6.5}$ **59.** $n = \pm\sqrt{\dfrac{r^2 \pm \sqrt{r^4 + 4m^4r^2p - 4mp}}{2m}}$ **60.** $d = \dfrac{-\pi h + \sqrt{\pi^2h^2 + 2\pi A}}{\pi}$

8.4

The Discriminant ■ Writing Equations from Solutions

Studying Solutions of Quadratic Equations

The Discriminant

Sometimes in mathematics it is enough to know what *type* of number a solution will be, without actually solving the equation. To illustrate, suppose we want to know if the equation $4x^2 + 7x - 15 = 0$ has rational solutions (and thus can be solved by factoring). Using the quadratic formula, we would have

$$x = \frac{-b \pm \sqrt{b^2 - 4ac}}{2a} = \frac{-7 \pm \sqrt{7^2 - 4 \cdot 4 \cdot (-15)}}{2 \cdot 4}.$$

Note that the radicand, $7^2 - 4 \cdot 4 \cdot (-15)$, determines what type of number the solutions will be. Since $7^2 - 4 \cdot 4 \cdot (-15) = 49 - 16(-15) = 289$, and since 289 is a perfect square $\left(\sqrt{289} = 17\right)$, we know that the solutions of the equation will be two rational numbers. This means that $4x^2 + 7x - 15 = 0$ *can* be solved by factoring.

It is the expression $b^2 - 4ac$, known as the **discriminant**, that determines what type of number the solutions of a quadratic equation will be:

- When $b^2 - 4ac$ simplifies to 0, it doesn't matter if we use $+\sqrt{b^2 - 4ac}$ or $-\sqrt{b^2 - 4ac}$; we get the same solution twice. Thus, when the discriminant is 0, there is one *repeated* solution and it will be rational.

- When the discriminant is positive, there are two different real-number solutions. As we saw above, when $b^2 - 4ac$ is a perfect square, these solutions are rational numbers. When $b^2 - 4ac$ is positive but not a perfect square, there are two irrational solutions and they will be conjugates of each other (see p. 524).

- When the discriminant is negative, there are two imaginary-number solutions and they will be complex conjugates of each other.

(see p. 524)

TEACHING TIP

You may wish to point out that this shows the types of solutions, not the solutions themselves.

Discriminant $b^2 - 4ac$	Nature of Solutions
0	One solution; a rational number

Positive
  Perfect square
  Not a perfect square
	Two different real-number solutions
	  Solutions are rational.
	  Solutions are irrational conjugates.

Negative	Two different imaginary-number solutions (complex conjugates)

EXAMPLE 1   For each equation, determine what type of number the solutions will be and how many solutions exist.

a) $9x^2 - 12x + 4 = 0$

b) $x^2 + 5x + 8 = 0$

c) $2x^2 + 7x - 3 = 0$

$y = 9x^2 - 12x + 4$

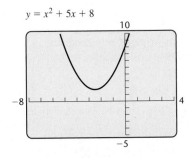

$y = x^2 + 5x + 8$

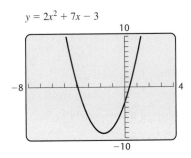

$y = 2x^2 + 7x - 3$

Solution

a) For $9x^2 - 12x + 4 = 0$, we have

$$a = 9, \quad b = -12, \quad c = 4.$$

We substitute and compute the discriminant:

$$b^2 - 4ac = (-12)^2 - 4 \cdot 9 \cdot 4$$
$$= 144 - 144 = 0.$$

There is just one solution, and it is rational. This tells us that $9x^2 - 12x + 4 = 0$ can be solved by factoring. The graph at left confirms that $9x^2 - 12x + 4 = 0$ has just one solution.

b) For $x^2 + 5x + 8 = 0$, we have

$$a = 1, \quad b = 5, \quad c = 8.$$

We substitute and compute the discriminant:

$$b^2 - 4ac = 5^2 - 4 \cdot 1 \cdot 8$$
$$= 25 - 32 = -7.$$

Since the discriminant is negative, there are two imaginary-number solutions that are complex conjugates of each other. As the graph at left shows, there are no x-intercepts of the graph of $y = x^2 + 5x + 8$, and thus no real solutions of the equation $x^2 + 5x + 8 = 0$.

c) For $2x^2 + 7x - 3 = 0$, we have

$$a = 2, \quad b = 7, \quad c = -3;$$
$$b^2 - 4ac = 7^2 - 4 \cdot 2(-3)$$
$$= 49 - (-24) = 73.$$

The discriminant is a positive number that is not a perfect square. Thus there are two irrational solutions that are conjugates of each other. From the graph at left, we see that there are two solutions of the equation. We cannot tell from simply observing the graph whether the solutions are rational or irrational.

As we saw in Example 1, discriminants can be used to determine the number of real-number solutions of $ax^2 + bx + c = 0$. This can be used as an aid in graphing.

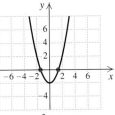

$y = ax^2 + bx + c$
$b^2 - 4ac > 0$
Two real solutions
of $ax^2 + bx + c = 0$
Two x-intercepts

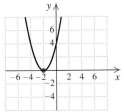

$y = ax^2 + bx + c$
$b^2 - 4ac = 0$
One real solution
of $ax^2 + bx + c = 0$
One x-intercept

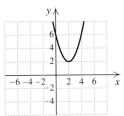

$y = ax^2 + bx + c$
$b^2 - 4ac < 0$
No real solutions
of $ax^2 + bx + c = 0$
No x-intercept

Writing Equations from Solutions

We know by the principle of zero products that $(x - 2)(x + 3) = 0$ has solutions 2 and -3. If we know the solutions of an equation, we can write an equation, using the principle in reverse.

EXAMPLE 2   Find an equation for which the given numbers are solutions.

a) 3 and $-\frac{2}{5}$ **b)** $2i$ and $-2i$

c) $5\sqrt{7}$ and $-5\sqrt{7}$ **d)** $-4, 0$, and 1

Solution

a)
$$x = 3 \quad or \quad x = -\tfrac{2}{5}$$
$$x - 3 = 0 \quad or \quad x + \tfrac{2}{5} = 0 \qquad \text{Getting 0's on one side}$$
$$(x - 3)\left(x + \tfrac{2}{5}\right) = 0 \qquad \text{Using the principle of zero products (multiplying)}$$
$$x^2 + \tfrac{2}{5}x - 3x - 3 \cdot \tfrac{2}{5} = 0 \qquad \text{Multiplying}$$
$$x^2 - \tfrac{13}{5}x - \tfrac{6}{5} = 0 \qquad \text{Combining like terms}$$
$$5x^2 - 13x - 6 = 0 \qquad \text{Multiplying both sides by 5 to clear fractions}$$

Note that $x^2 - \frac{13}{5}x - \frac{6}{5} = 0$ is an equation for which 3 and $-\frac{2}{5}$ are solutions. To find an equivalent equation with integer coefficients, we multiplied both sides by the LCD, 5, to clear the equation of fractions.
Had we preferred, we could have multiplied $x + \frac{2}{5} = 0$ by 5, thus clearing fractions *before* using the principle of zero products.

b)
$$x = 2i \quad or \quad x = -2i$$
$$x - 2i = 0 \quad or \quad x + 2i = 0 \qquad \text{Getting 0's on one side}$$
$$(x - 2i)(x + 2i) = 0 \qquad \text{Using the principle of zero products (multiplying)}$$
$$x^2 - (2i)^2 = 0 \qquad \text{Finding the product of a sum and difference}$$
$$x^2 - 4i^2 = 0$$
$$x^2 + 4 = 0 \qquad i^2 = -1$$

c)
$$x = 5\sqrt{7} \quad or \quad x = -5\sqrt{7}$$
$$x - 5\sqrt{7} = 0 \quad or \quad x + 5\sqrt{7} = 0 \qquad \text{Getting 0's on one side}$$
$$\left(x - 5\sqrt{7}\right)\left(x + 5\sqrt{7}\right) = 0 \qquad \text{Using the principle of zero products}$$
$$x^2 - \left(5\sqrt{7}\right)^2 = 0 \qquad \text{Finding the product of a sum and difference}$$
$$x^2 - 25 \cdot 7 = 0$$
$$x^2 - 175 = 0$$

d)
$$x = -4 \quad or \quad x = 0 \quad or \quad x = 1$$
$$x + 4 = 0 \quad or \quad x = 0 \quad or \quad x - 1 = 0 \qquad \text{Getting 0's on one side}$$
$$(x + 4)x(x - 1) = 0 \qquad \text{Using the principle of zero products}$$
$$x(x^2 + 3x - 4) = 0 \qquad \text{Multiplying}$$
$$x^3 + 3x^2 - 4x = 0$$

TEACHING TIP

Students tend to ignore the factor x. In Example 2(d), you may want to emphasize that if 0 is a solution, then x is a factor.

To check any of these equations, we can simply substitute one or more of the given solutions. For example, in Example 2(d) above,

$$(-4)^3 + 3(-4)^2 - 4(-4) = -64 + 3 \cdot 16 + 16$$
$$= -64 + 48 + 16 = 0.$$

The other checks are left to the student.

8.4

Exercise Set

FOR EXTRA HELP

Digital Video Tutor CD 8 Videotape 11 | Student's Solutions Manual | Tutor Center AW Math Tutor Center | InterAct Math | MathXL MathXL | MyMathLab

For each equation, determine what type of number the solutions are and how many solutions exist.

1. $x^2 - 5x + 3 = 0$ Two irrational
2. $x^2 - 7x + 5 = 0$ Two irrational
3. $x^2 + 5 = 0$ Two imaginary
4. $x^2 + 3 = 0$ Two imaginary
5. $x^2 - 3 = 0$ Two irrational
6. $x^2 - 5 = 0$ Two irrational
7. $4x^2 - 12x + 9 = 0$ One rational
8. $4x^2 + 8x - 5 = 0$ Two rational
9. $x^2 - 2x + 4 = 0$ Two imaginary
10. $x^2 + 4x + 6 = 0$ Two imaginary
11. $6t^2 - 19t - 20 = 0$ Two rational
12. $9t^2 - 48t + 64 = 0$ One rational
13. $6x^2 + 5x - 4 = 0$ Two rational
14. $10x^2 - x - 2 = 0$ Two rational
^Aha! **15.** $9t^2 - 3t = 0$ Two rational
16. $4m^2 + 7m = 0$ Two rational
17. $x^2 + 4x = 8$ Two irrational
18. $x^2 + 5x = 9$ Two irrational
19. $2a^2 - 3a = -5$ Two imaginary
20. $3a^2 + 5 = 7a$ Two imaginary
21. $y^2 + \frac{9}{4} = 4y$ Two irrational
22. $x^2 = \frac{1}{2}x - \frac{3}{5}$ Two imaginary

Write a quadratic equation having the given numbers as solutions.

23. $-7, 3$ $x^2 + 4x - 21 = 0$
24. $-6, 4$ $x^2 + 2x - 24 = 0$
25. 3, only solution (*Hint*: It must be a repeated solution.) $x^2 - 6x + 9 = 0$
26. -5, only solution $x^2 + 10x + 25 = 0$
27. $-2, -5$ $x^2 + 7x + 10 = 0$
28. $-1, -3$ $x^2 + 4x + 3 = 0$
29. $4, \frac{2}{3}$ $3x^2 - 14x + 8 = 0$
30. $5, \frac{3}{4}$ $4x^2 - 23x + 15 = 0$
31. $\frac{1}{2}, \frac{1}{3}$ $6x^2 - 5x + 1 = 0$
32. $-\frac{1}{4}, -\frac{1}{2}$ $8x^2 + 6x + 1 = 0$
33. $-0.6, 1.4$ $x^2 - 0.8x - 0.84 = 0$
34. $2.4, -0.4$ $x^2 - 2x - 0.96 = 0$
35. $-\sqrt{7}, \sqrt{7}$ $x^2 - 7 = 0$
36. $-\sqrt{3}, \sqrt{3}$ $x^2 - 3 = 0$
37. $3\sqrt{2}, -3\sqrt{2}$ $x^2 - 18 = 0$
38. $2\sqrt{5}, -2\sqrt{5}$ $x^2 - 20 = 0$
39. $3i, -3i$ $x^2 + 9 = 0$
40. $4i, -4i$ $x^2 + 16 = 0$
41. $5 - 2i, 5 + 2i$ $x^2 - 10x + 29 = 0$
42. $2 - 7i, 2 + 7i$ $x^2 - 4x + 53 = 0$
43. $2 - \sqrt{10}, 2 + \sqrt{10}$ $x^2 - 4x - 6 = 0$
44. $3 - \sqrt{14}, 3 + \sqrt{14}$ $x^2 - 6x - 5 = 0$

Write a third-degree equation having the given numbers as solutions.

45. $-2, 1, 5$ $x^3 - 4x^2 - 7x + 10 = 0$
46. $-5, 0, 2$ $x^3 + 3x^2 - 10x = 0$
47. $-1, 0, 3$ $x^3 - 2x^2 - 3x = 0$
48. $-2, 2, 3$ $x^3 - 3x^2 - 4x + 12 = 0$

ᵀᵂ **49.** While solving a quadratic equation of the form $ax^2 + bx + c = 0$ with a graphing calculator, Shawn-Marie gets the following screen.

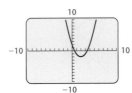

How could the sign of the discriminant help her check the graph?

ᵀᵂ **50.** Describe a procedure that could be used to write an equation having the first 7 natural numbers as solutions.

Skill Maintenance

Simplify. [1.4]

51. $(3a^2)^4$ $81a^8$
52. $(4x^3)^2$ $16x^6$

Find the x-intercepts of the graph of f. [5.4]

53. $f(x) = x^2 - 7x - 8$ $(-1, 0), (8, 0)$
54. $f(x) = x^2 - 6x + 8$ $(2, 0), (4, 0)$

55. During a one-hour television show, there were 12 commercials. Some of the commercials were 30 sec long and the others were 60 sec long. The amount of time for 30-sec commercials was 6 min less than the total number of minutes of commercial time during the show. How many 30-sec commercials were used? [3.3]   6 30-sec commercials

56. Graph by hand: $y = -\frac{3}{7}x + 4$. [2.4]   ⊡

Synthesis

TW **57.** If we assume that a quadratic equation has integers for coefficients, will the product of the solutions always be a real number? Why or why not?

TW **58.** Can a fourth-degree equation have three irrational solutions? Why or why not?

59. The graph of an equation of the form
$$y = ax^2 + bx + c \qquad a = 1, b = 2, c = -3$$
is a curve similar to the one shown below. Determine a, b, and c from the information given.

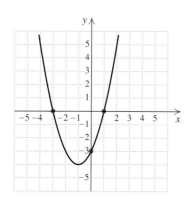

60. Show that the product of the solutions of $ax^2 + bx + c = 0$ is c/a.   ⊡

For each equation under the given condition, **(a)** *find k and* **(b)** *find the other solution.*

61. $kx^2 - 2x + k = 0$; one solution is -3 (a) $-\frac{3}{5}$; (b) $-\frac{1}{3}$

62. $x^2 - kx + 2 = 0$; one solution is $1 + i$
(a) 2; (b) $1 - i$

63. $x^2 - (6 + 3i)x + k = 0$; one solution is 3
(a) $9 + 9i$; (b) $3 + 3i$

64. Show that the sum of the solutions of $ax^2 + bx + c = 0$ is $-b/a$.   ⊡

65. Show that whenever there is just one solution of $ax^2 + bx + c = 0$, that solution is of the form $-b/2a$.   ⊡

66. Find h and k, where $3x^2 - hx + 4k = 0$, the sum of the solutions is -12, and the product of the solutions is 20. (*Hint*: See Exercises 60 and 64.)
$h = -36, k = 15$

67. Suppose that $f(x) = ax^2 + bx + c$, with $f(-3) = 0$, $f(\frac{1}{2}) = 0$, and $f(0) = -12$. Find a, b, and c.
$a = 8, b = 20, c = -12$

68. Find an equation for which $2 - \sqrt{3}, 2 + \sqrt{3}$, $5 - 2i$, and $5 + 2i$ are solutions.
$x^4 - 14x^3 + 70x^2 - 126x + 29 = 0$

Aha! **69.** Find an equation for which $1 - \sqrt{5}$ and $3 + 2i$ are two of the solutions. $x^4 - 8x^3 + 21x^2 - 2x - 52 = 0$

TW **70.** A discriminant that is a perfect square indicates that factoring can be used to solve the quadratic equation. Why?

⊡ Answers to Exercises 56, 60, 64, and 65 can be found on p. A-64.

8.5

Recognizing Equations in Quadratic Form ■ Radical and Rational Equations

Equations Reducible to Quadratic

Recognizing Equations in Quadratic Form

Certain equations that are not really quadratic can be thought of in such a way that they can be solved as quadratic. For example, because the square of x^2 is x^4, the equation $x^4 - 9x^2 + 8 = 0$ is said to be "quadratic in x^2":

$$x^4 - 9x^2 + 8 = 0$$

$$(x^2)^2 - 9(x^2) + 8 = 0 \qquad \text{Thinking of } x^4 \text{ as } (x^2)^2$$

$$u^2 - 9u + 8 = 0. \qquad \text{To make this clearer, write } u \text{ instead of } x^2.$$

TEACHING TIP

Many students have difficulty recognizing an equation in quadratic form. Point out that they should look for two terms with variables. The exponent of the first term is twice the exponent of the second term. It is the variable expression in the *second* term that is written as u.

We substitute u for x^2 in order to write the equation in a form with which we are familiar. The equation $u^2 - 9u + 8 = 0$ can be solved by factoring or by the quadratic formula. Then, remembering that $u = x^2$, we can solve for x. Equations that can be solved like this are said to be *reducible to quadratic*, or *in quadratic form*.

EXAMPLE 1   Solve: $x^4 - 9x^2 + 8 = 0$.

Solution   We solve both algebraically and graphically.

Algebraic Solution

Let $u = x^2$. We solve $x^4 - 9x^2 + 8 = 0$ by substituting u for x^2 and u^2 for x^4:

$$u^2 - 9u + 8 = 0$$
$$(u - 8)(u - 1) = 0 \qquad \text{Factoring}$$
$$u - 8 = 0 \quad or \quad u - 1 = 0 \qquad \text{Principle of zero products}$$
$$u = 8 \quad or \qquad u = 1.$$

We replace u with x^2 and solve these equations:

$$x^2 = 8 \qquad or \quad x^2 = 1$$
$$x = \pm\sqrt{8} \quad or \quad x = \pm 1$$
$$x = \pm 2\sqrt{2} \quad or \quad x = \pm 1.$$

To check, note that for $x = 2\sqrt{2}$, we have $x^2 = 8$ and $x^4 = 64$. Similarly, for $x = -2\sqrt{2}$, we have $x^2 = 8$ and $x^4 = 64$. When $x = 1$, we have $x^2 = 1$ and $x^4 = 1$, and when $x = -1$, we have $x^2 = 1$ and $x^4 = 1$. Thus instead of making four checks, we need make only two.

Check:
For $\pm 2\sqrt{2}$:

$$\frac{x^4 - 9x^2 + 8 = 0}{\left(\pm 2\sqrt{2}\right)^4 - 9\left(\pm 2\sqrt{2}\right)^2 + 8 \; ? \; 0}$$
$$64 - 9 \cdot 8 + 8 \;\Big|$$
$$0 \;\Big|\; 0 \qquad \text{TRUE}$$

For ± 1:

$$\frac{x^4 - 9x^2 + 8 = 0}{(\pm 1)^4 - 9(\pm 1)^2 + 8 \; ? \; 0}$$
$$1 - 9 + 8 \;\Big|$$
$$0 \;\Big|\; 0 \qquad \text{TRUE}$$

The solutions are 1, -1, $2\sqrt{2}$, and $-2\sqrt{2}$.

Graphical Solution

We let $f(x) = x^4 - 9x^2 + 8$ and determine the zeros of the function.

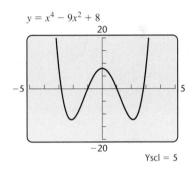

It appears as though the graph has 4 x-intercepts. Recall from Chapter 5 that a fourth-degree polynomial function has at most 4 zeros, so we know that we have not missed any zeros. Using ZERO four times, we obtain the solutions -2.8284271, -1, 1, and 2.8284271. The solutions are ± 1 and, approximately, ± 2.8284271.

Note that since $2\sqrt{2} \approx 2.8284271$, the solutions found using both methods agree.

Example 1 can be solved directly by factoring:

$$x^4 - 9x^2 + 8 = 0$$
$$(x^2 - 1)(x^2 - 8) = 0$$
$$x^2 - 1 = 0 \quad or \quad x^2 - 8 = 0$$
$$x^2 = 1 \quad or \quad x^2 = 8$$
$$x = \pm 1 \quad or \quad x = \pm 2\sqrt{2}.$$

There is nothing wrong with this approach. However, in the examples that follow, you will note that it becomes increasingly difficult to solve the equation without first making a substitution.

Radical and Rational Equations

Sometimes rational equations, radical equations, or equations containing fractional exponents are reducible to quadratic. It is especially important that answers to these equations be checked in the original equation.

EXAMPLE 2 Solve: $x - 3\sqrt{x} - 4 = 0$.

Solution   This radical equation could be solved using the method discussed in Section 7.6. However, if we note that the square of $\sqrt{x}$ is x, we can regard the equation as "quadratic in $\sqrt{x}$."

We let $u = \sqrt{x}$ and consequently $u^2 = x$:

$$x - 3\sqrt{x} - 4 = 0$$
$$u^2 - 3u - 4 = 0 \qquad \textbf{Substituting}$$
$$(u - 4)(u + 1) = 0$$
$$u = 4 \quad or \quad u = -1. \qquad \textbf{Using the principle of zero products}$$

CAUTION!   A common error is to solve for u and then forget to solve for x. Remember that you must find values for the *original* variable!

Next, we replace u with $\sqrt{x}$ and solve these equations:

$$\sqrt{x} = 4 \quad or \quad \sqrt{x} = -1.$$

Squaring gives us $x = 16$ or $x = 1$ and also makes checking essential.

Check:

For 16:

$$\frac{x - 3\sqrt{x} - 4 = 0}{16 - 3\sqrt{16} - 4 \; ? \; 0}$$
$$16 - 3 \cdot 4 - 4$$
$$0 \mid 0 \quad \text{TRUE}$$

For 1:

$$\frac{x - 3\sqrt{x} - 4 = 0}{1 - 3\sqrt{1} - 4 \; ? \; 0}$$
$$1 - 3 \cdot 1 - 4$$
$$-6 \mid 0 \quad \text{FALSE}$$

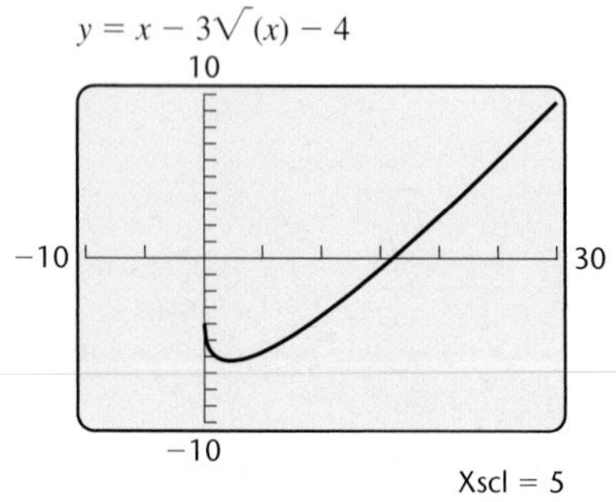

The number 16 checks, but 1 does not. Had we noticed that $\sqrt{x} = -1$ has no solution (since principal roots are never negative), we could have solved only the equation $\sqrt{x} = 4$. The graph at left provides further evidence that although 16 is a solution, -1 is not. The solution is 16.

TEACHING TIP

Students may recognize this more readily as quadratic in form if it is written with rational exponents:

$$x^1 - 3x^{1/2} - 4 = 0.$$

$y = x - 3\sqrt{(x)} - 4$

Xscl = 5

EXAMPLE 3   Find the x-intercepts of the graph of

$$f(x) = (x^2 - 1)^2 - (x^2 - 1) - 2.$$

Solution   The x-intercepts occur where $f(x) = 0$ so we must have

$$(x^2 - 1)^2 - (x^2 - 1) - 2 = 0.   \text{Setting } f(x) \text{ equal to } 0$$

The equation is quadratic in $x^2 - 1$, so we let $u = x^2 - 1$ and $u^2 = (x^2 - 1)^2$:

$$u^2 - u - 2 = 0   \text{Substituting in } (x^2 - 1)^2 - (x^2 - 1) - 2 = 0$$
$$(u - 2)(u + 1) = 0$$
$$u = 2 \quad or \quad u = -1.   \text{Using the principle of zero products}$$

Next, we replace u with $x^2 - 1$ and solve these equations:

$$
\begin{aligned}
x^2 - 1 = 2 & \quad or \quad x^2 - 1 = -1 \\
x^2 = 3 & \quad or \quad x^2 = 0 & \text{Adding 1 to both sides} \\
x = \pm\sqrt{3} & \quad or \quad x = 0. & \text{Using the principle of} \\
& & \text{square roots}
\end{aligned}
$$

The x-intercepts occur at $\left(-\sqrt{3}, 0\right)$, $(0, 0)$, and $\left(\sqrt{3}, 0\right)$. These solutions are confirmed by the graph below.

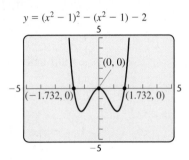

$$y = (x^2 - 1)^2 - (x^2 - 1) - 2$$

Sometimes great care must be taken in deciding what substitution to make.

EXAMPLE 4   Solve: $m^{-2} - 6m^{-1} + 4 = 0$.

Solution   Note that the square of m^{-1} is $(m^{-1})^2$, or m^{-2}. This allows us to regard the equation as quadratic in m^{-1}.
   We let $u = m^{-1}$ and $u^2 = m^{-2}$:

$$u^2 - 6u + 4 = 0   \text{Substituting}$$
$$u = \frac{-(-6) \pm \sqrt{(-6)^2 - 4 \cdot 1 \cdot 4}}{2 \cdot 1}   \text{Using the quadratic formula}$$
$$u = \frac{6 \pm \sqrt{20}}{2} = \frac{2 \cdot 3 \pm 2\sqrt{5}}{2}   \text{Simplifying}$$
$$u = 3 \pm \sqrt{5}.$$

Next, we replace u with m^{-1} and solve:

$$m^{-1} = 3 \pm \sqrt{5}$$

$$\frac{1}{m} = 3 \pm \sqrt{5} \qquad \text{Recall that } m^{-1} = \frac{1}{m}.$$

$$1 = m(3 \pm \sqrt{5}) \qquad \text{Multiplying both sides by } m$$

$$\frac{1}{3 \pm \sqrt{5}} = m. \qquad \text{Dividing both sides by } 3 \pm \sqrt{5}$$

We can check both solutions as follows.

Check:

For $1/(3 - \sqrt{5})$:

$$\begin{array}{c|c}
m^{-2} - 6m^{-1} + 4 = 0 & \\
\hline
\left(\dfrac{1}{3 - \sqrt{5}}\right)^{-2} - 6\left(\dfrac{1}{3 - \sqrt{5}}\right)^{-1} + 4 \; ? \; 0 & \\
(3 - \sqrt{5})^2 - 6(3 - \sqrt{5}) + 4 & \\
9 - 6\sqrt{5} + 5 - 18 + 6\sqrt{5} + 4 & \\
0 \;\bigm|\; 0 & \text{TRUE}
\end{array}$$

For $1/(3 + \sqrt{5})$:

$$\begin{array}{c|c}
m^{-2} - 6m^{-1} + 4 = 0 & \\
\hline
\left(\dfrac{1}{3 + \sqrt{5}}\right)^{-2} - 6\left(\dfrac{1}{3 + \sqrt{5}}\right)^{-1} + 4 \; ? \; 0 & \\
(3 + \sqrt{5})^2 - 6(3 + \sqrt{5}) + 4 & \\
9 + 6\sqrt{5} + 5 - 18 - 6\sqrt{5} + 4 & \\
0 \;\bigm|\; 0 & \text{TRUE}
\end{array}$$

We could also check the solutions using a calculator by entering $y_1 = x^{-2} - 6x^{-1} + 4$ and evaluating y_1 for both numbers, as shown on the left below.

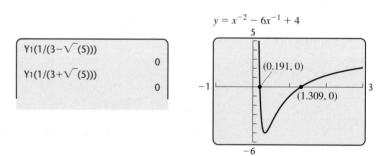

As another check, we solve the equation graphically. The first coordinates of the x-intercepts of the graph on the right above are approximate solutions.

Both numbers check. The solutions are $1/(3 - \sqrt{5})$ and $1/(3 + \sqrt{5})$, or approximately 1.309 and 0.191.

EXAMPLE 5 Solve: $t^{2/5} - t^{1/5} - 2 = 0$.

Solution Note that the square of $t^{1/5}$ is $(t^{1/5})^2$, or $t^{2/5}$. The equation is therefore quadratic in $t^{1/5}$, so we let $u = t^{1/5}$ and $u^2 = t^{2/5}$:

$$u^2 - u - 2 = 0 \qquad \text{Substituting}$$

$$(u - 2)(u + 1) = 0$$

$$u = 2 \quad or \quad u = -1. \qquad \text{Using the principle of zero products}$$

Now we replace u with $t^{1/5}$ and solve:

$$t^{1/5} = 2 \quad or \quad t^{1/5} = -1$$
$$t = 32 \quad or \quad t = -1.$$

Principle of powers; raising to the 5th power

Check:

For 32:

$$\frac{t^{2/5} - t^{1/5} - 2 = 0}{\begin{array}{c} 32^{2/5} - 32^{1/5} - 2 \;?\; 0 \\ (32^{1/5})^2 - 32^{1/5} - 2 \\ 2^2 - 2 - 2 \\ 0 \end{array}} \Bigg| \; 0 \quad \text{TRUE}$$

For -1:

$$\frac{t^{2/5} - t^{1/5} - 2 = 0}{\begin{array}{c} (-1)^{2/5} - (-1)^{1/5} - 2 \;?\; 0 \\ [(-1)^{1/5}]^2 - (-1)^{1/5} - 2 \\ (-1)^2 - (-1) - 2 \\ 0 \end{array}} \Bigg| \; 0 \quad \text{TRUE}$$

Both numbers check and are confirmed by the graph at left. The solutions are 32 and -1.

$y = x^{2/5} - x^{1/5} - 2$

$(-1, 0)$

$(32, 0)$

Xscl = 10

The following tips may prove useful.

To Solve an Equation That Is Reducible to Quadratic

1. The equation is quadratic in form if the variable factor in one term is the square of the variable factor in the other variable term.
2. Write down any substitutions that you are making.
3. Whenever you make a substitution, be sure to solve for the variable that is used in the original equation.
4. Check possible answers in the original equation.

8.5

Exercise Set

Solve.

1. $x^4 - 10x^2 + 9 = 0$ $\pm 1, \pm 3$

2. $x^4 - 5x^2 + 4 = 0$ $\pm 1, \pm 2$

3. $x^4 - 12x^2 + 27 = 0$ $\pm\sqrt{3}, \pm 3$

4. $x^4 - 9x^2 + 20 = 0$ $\pm\sqrt{5}, \pm 2$

5. $9x^4 - 14x^2 + 5 = 0$ $\pm\dfrac{\sqrt{5}}{3}, \pm 1$

6. $4x^4 - 19x^2 + 12 = 0$ $\pm\dfrac{\sqrt{3}}{2}, \pm 2$

7. $x - 4\sqrt{x} - 1 = 0$ $9 + 4\sqrt{5}$

8. $x - 2\sqrt{x} - 6 = 0$ $8 + 2\sqrt{7}$

9. $(x^2 - 7)^2 - 3(x^2 - 7) + 2 = 0$ $\pm 2\sqrt{2}, \pm 3$

10. $(x^2 - 1)^2 - 5(x^2 - 1) + 6 = 0$ $\pm\sqrt{3}, \pm 2$

11. $(1 + \sqrt{x})^2 + 5(1 + \sqrt{x}) + 6 = 0$ No solution

12. $(3 + \sqrt{x})^2 + 3(3 + \sqrt{x}) - 10 = 0$ No solution

13. $x^{-2} - x^{-1} - 6 = 0$ $-\frac{1}{2}, \frac{1}{3}$

14. $2x^{-2} - x^{-1} - 1 = 0$ $-2, 1$

15. $4x^{-2} + x^{-1} - 5 = 0$ $\quad -\frac{4}{5}, 1$

16. $m^{-2} + 9m^{-1} - 10 = 0$ $\quad -\frac{1}{10}, 1$

17. $t^{2/3} + t^{1/3} - 6 = 0$ $\quad -27, 8$

18. $w^{2/3} - 2w^{1/3} - 8 = 0$ $\quad -8, 64$

19. $y^{1/3} - y^{1/6} - 6 = 0$ $\quad 729$

20. $t^{1/2} + 3t^{1/4} + 2 = 0$ $\quad$ No solution

21. $t^{1/3} + 2t^{1/6} = 3$ $\quad 1$

22. $m^{1/2} + 6 = 5m^{1/4}$ $\quad 16, 81$

23. $(3 - \sqrt{x})^2 - 10(3 - \sqrt{x}) + 23 = 0$ $\quad$ No solution

24. $(5 + \sqrt{x})^2 - 12(5 + \sqrt{x}) + 33 = 0$ $\quad 4 + 2\sqrt{3}$

25. $16\left(\dfrac{x-1}{x-8}\right)^2 + 8\left(\dfrac{x-1}{x-8}\right) + 1 = 0$ $\quad \dfrac{12}{5}$

26. $9\left(\dfrac{x+2}{x+3}\right)^2 - 6\left(\dfrac{x+2}{x+3}\right) + 1 = 0$ $\quad -\dfrac{3}{2}$

Find all x-intercepts of the given function f. If none exists, state this.

27. $f(x) = 5x + 13\sqrt{x} - 6$ $\quad \left(\frac{4}{25}, 0\right)$

28. $f(x) = 3x + 10\sqrt{x} - 8$ $\quad \left(\frac{4}{9}, 0\right)$

29. $f(x) = (x^2 - 3x)^2 - 10(x^2 - 3x) + 24$

30. $f(x) = (x^2 - 6x)^2 - 2(x^2 - 6x) - 35$
$\quad (-1, 0), (1, 0), (5, 0), (7, 0)$

31. $f(x) = x^{2/5} + x^{1/5} - 6$ $\quad (-243, 0), (32, 0)$

32. $f(x) = x^{1/2} - x^{1/4} - 6$ $\quad (81, 0)$

Aha! **33.** $f(x) = \left(\dfrac{x^2 + 2}{x}\right)^4 + 7\left(\dfrac{x^2 + 2}{x}\right)^2 + 5$ $\quad$ No x-intercepts

34. $f(x) = \left(\dfrac{x^2 + 1}{x}\right)^4 + 4\left(\dfrac{x^2 + 1}{x}\right)^2 + 12$ $\quad$ No x-intercepts

TW **35.** To solve $25x^6 - 10x^3 + 1 = 0$, Don lets $u = 5x^3$ and Robin lets $u = x^3$. Can they both be correct? Why or why not?

TW **36.** While trying to solve $0.05x^4 - 0.8 = 0$ with a graphing calculator, Murray gets the following screen. Can Murray solve this equation with a graphing calculator? Why or why not?

$y_1 = .05x^4 - .8$

29. $\left(\dfrac{3 + \sqrt{33}}{2}, 0\right), \left(\dfrac{3 - \sqrt{33}}{2}, 0\right), (4, 0), (-1, 0)$

Skill Maintenance

Graph by hand.

37. $f(x) = \frac{3}{2}x$ [2.4]

38. $f(x) = -\frac{2}{3}x$ [2.4]

39. $f(x) = \dfrac{2}{x}$ [2.1]

40. $f(x) = \dfrac{3}{x}$ [2.1]

41. Solution A is 18% alcohol and solution B is 45% alcohol. How much of each should be mixed together to get 12 L of a solution that is 36% alcohol? [3.3]
$\quad$ A: 4 L; B: 8 L

42. If $g(x) = x^2 - x$, find $g(a + 1)$. [5.2] $\quad a^2 + a$

Synthesis

TW **43.** Describe a procedure that could be used to solve any equation of the form $ax^4 + bx^2 + c = 0$.

TW **44.** Describe a procedure that could be used to write an equation that is quadratic in $3x^2 - 1$. Then explain how the procedure could be adjusted to write equations that are quadratic in $3x^2 - 1$ and have no real-number solution.

Solve.

45. $5x^4 - 7x^2 + 1 = 0$ $\quad \pm\sqrt{\dfrac{7 \pm \sqrt{29}}{10}}$

46. $3x^4 + 5x^2 - 1 = 0$ $\quad \pm\sqrt{\dfrac{-5 \pm \sqrt{37}}{6}}$

47. $(x^2 - 4x - 2)^2 - 13(x^2 - 4x - 2) + 30 = 0$
$\quad -2, -1, 5, 6$

48. $(x^2 - 5x - 1)^2 - 18(x^2 - 5x - 1) + 65 = 0$
$\quad -2, -1, 6, 7$

49. $\dfrac{x}{x-1} - 6\sqrt{\dfrac{x}{x-1}} - 40 = 0$ $\quad \dfrac{100}{99}$

50. $\left(\sqrt{\dfrac{x}{x-3}}\right)^2 - 24 = 10\sqrt{\dfrac{x}{x-3}}$ $\quad \dfrac{432}{143}$

51. $a^5(a^2 - 25) + 13a^3(25 - a^2) + 36a(a^2 - 25) = 0$
$\quad -5, -3, -2, 0, 2, 3, 5$

52. $a^3 - 26a^{3/2} - 27 = 0$ $\quad 9$

53. $x^6 - 28x^3 + 27 = 0$ $\quad 1, 3$

54. $x^6 + 7x^3 - 8 = 0$ $\quad -2, 1$

37.

38.

39.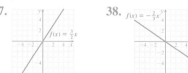

40.

8.6

The Graph of $f(x) = ax^2$ ■ The Graph of $f(x) = a(x - h)^2$ ■ The Graph of
$f(x) = a(x - h)^2 + k$

Quadratic Functions and Their Graphs

We have already used quadratic functions when we solved equations earlier in this chapter. In this section and the next, we learn to graph such functions.

The Graph of $f(x) = ax^2$

The most basic quadratic function is $f(x) = x^2$.

EXAMPLE 1   Graph: $f(x) = x^2$.

Solution   We choose some values for x and compute $f(x)$ for each. Then we plot the ordered pairs and connect them with a smooth curve.

x	$f(x) = x^2$	$(x, f(x))$
-3	9	$(-3, 9)$
-2	4	$(-2, 4)$
-1	1	$(-1, 1)$
0	0	$(0, 0)$
1	1	$(1, 1)$
2	4	$(2, 4)$
3	9	$(3, 9)$

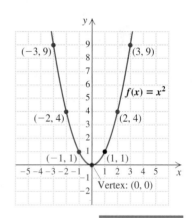

All quadratic functions have graphs similar to the one in Example 1. Such curves are called *parabolas*. They are cup-shaped and symmetric with respect to a vertical line known as the parabola's *axis of symmetry*. For the graph of $f(x) = x^2$, the y-axis (or the line $x = 0$) is the axis of symmetry. Were the paper folded on this line, the two halves of the curve would match. The point $(0, 0)$ is known as the *vertex* of this parabola.

Since x^2 is a polynomial, we know that the domain of $f(x) = x^2$ is the set of all real numbers, or $(-\infty, \infty)$. We see from the graph and the table in Example 1 that the range of f is $\{y \mid y \geq 0\}$, or $[0, \infty)$. The function has a *minimum value* of 0. This minimum value occurs when $x = 0$.

Now let's consider a quadratic function of the form $g(x) = ax^2$. How does the constant a affect the graph of the function?

Interactive Discovery

Graph each of the following equations along with $y_1 = x^2$, in a $[-5, 5, -10, 10]$ window. For each equation, answer the following questions.

a) What is the vertex of the graph of y_2?

b) What is the axis of symmetry of the graph of y_2?

c) Does the graph of y_2 open upward or downward?

d) Is the parabola narrower or wider than the parabola given by $y_1 = x^2$?

1. $y_2 = 3x^2$ ⊡   **2.** $y_2 = \frac{1}{3}x^2$ ⊡   **3.** $y_2 = 0.2x^2$ ⊡

4. $y_2 = -x^2$ ⊡   **5.** $y_2 = -4x^2$ ⊡   **6.** $y_2 = -\frac{2}{3}x^2$ ⊡

7. Describe the effect of multiplying x^2 by a when $a > 1$ and when $0 < a < 1$. ⊡

8. Describe the effect of multiplying x^2 by a when $a < -1$ and when $-1 < a < 0$. ⊡

⊡ Answers to Exercises 1–8 can be found on pp. A-64 and A-65.

You may have noticed that the constant a, when $|a| \ne 1$, made the graph wider or narrower than the graph of $f(x) = x^2$. When a was negative, the graph opened downward. We can generalize the results as follows.

Graphing $f(x) = ax^2$

The graph of $f(x) = ax^2$ is a parabola with $x = 0$ as its axis of symmetry. Its vertex is the origin. The domain of f is $(-\infty, \infty)$.

For $a > 0$, the parabola opens upward. The range of the function is $[0, \infty)$. A minimum function value of 0 occurs when $x = 0$.

For $a < 0$, the parabola opens downward. The range of the function is $(-\infty, 0]$. A maximum function value of 0 occurs when $x = 0$.

If $|a|$ is greater than 1, the parabola is narrower than $y = x^2$.

If $|a|$ is between 0 and 1, the parabola is wider than $y = x^2$.

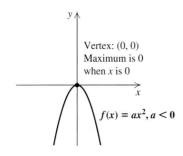

Vertex: $(0, 0)$
Minimum is 0
when x is 0

Vertex: $(0, 0)$
Maximum is 0
when x is 0

EXAMPLE 2   Graph: $g(x) = \frac{1}{2}x^2$.

Solution   The function is in the form $g(x) = ax^2$, where $a = \frac{1}{2}$. The vertex is $(0, 0)$ and the axis of symmetry is $x = 0$. Since $\frac{1}{2} > 0$, the parabola opens upward. The parabola is wider than $y = x^2$ since $\left|\frac{1}{2}\right|$ is between 0 and 1.

To graph the function, we calculate and plot points. Because we know the general shape of the graph, we can sketch the graph using only a few points.

x	$g(x) = \frac{1}{2}x^2$
-3	$\frac{9}{2}$
-2	2
-1	$\frac{1}{2}$
0	0
1	$\frac{1}{2}$
2	2
3	$\frac{9}{2}$

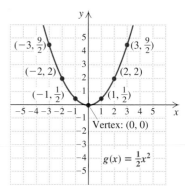

Note in Example 2 that since a parabola is symmetric, we can calculate function values for inputs to one side of the vertex and then plot the mirror images of those points on the other "half" of the graph.

The Graph of $f(x) = a(x - h)^2$

Why not now consider graphs of

$$f(x) = ax^2 + bx + c,$$

where b and c are not both 0? In effect, we will do that, but in a disguised form. It turns out to be convenient to first graph $f(x) = a(x - h)^2$, where h is some constant. This allows us to observe similarities to the graphs drawn above.

Interactive Discovery

Graph each of the following pairs of equations in a $[-5, 5, -10, 10]$ window. For each pair, answer the following questions.

a) What is the vertex of the graph of y_2?

b) What is the axis of symmetry of the graph of y_2?

c) Is the graph of y_2 narrower, wider, or the same shape as the graph of y_1?

1. $y_1 = x^2$; $y_2 = (x - 3)^2$

2. $y_1 = 2x^2$; $y_2 = 2(x + 1)^2$

3. $y_1 = -\frac{1}{2}x^2$; $y_2 = -\frac{1}{2}\left(x - \frac{3}{2}\right)^2$

4. $y_1 = -3x^2$; $y_2 = -3(x + 2)^2$

5. Describe the effect of h on the graph of $g(x) = a(x - h)^2$.

1. (a) $(3, 0)$; **(b)** $x = 3$; **(c)** same shape **2. (a)** $(-1, 0)$; **(b)** $x = -1$; **(c)** same shape **3. (a)** $\left(\frac{3}{2}, 0\right)$; **(b)** $x = \frac{3}{2}$; **(c)** same shape **4. (a)** $(-2, 0)$; **(b)** $x = -2$; **(c)** same shape **5.** The graph of $g(x) = a(x - h)^2$ looks like the graph of $f(x) = ax^2$, except that it is moved left or right.

The graph of $g(x) = a(x - h)^2$ looks just like the graph of $f(x) = ax^2$, except that it is moved, or *translated*, $|h|$ units horizontally.

Graphing $f(x) = a(x - h)^2$

The graph of $f(x) = a(x - h)^2$ has the same shape as the graph of $y = ax^2$. The domain of f is $(-\infty, \infty)$.

If h is positive, the graph of $y = ax^2$ is shifted h units to the right.

If h is negative, the graph of $y = ax^2$ is shifted $|h|$ units to the left.

The vertex is $(h, 0)$, and the axis of symmetry is $x = h$.

For $a > 0$, the range of f is $[0, \infty)$. A minimum function value of 0 occurs when $x = h$.

For $a < 0$, the range of f is $(-\infty, 0]$. A maximum function value of 0 occurs when $x = h$.

EXAMPLE 3 Graph $g(x) = -2(x + 3)^2$, and determine the maximum value of $g(x)$ and where it occurs.

1. (a) For any given x-value, the value of y_2 is 10 more than the value of y_1; **(b)** yes; **(c)** no
2. (a) For any given x-value, the value of y_2 is 5 less than the value of y_1; **(b)** yes; **(c)** no
3. (a) For any given x-value, the value of y_2 is 4 more than the value of y_1; **(b)** yes; **(c)** no
4. The graph of $g(x) = a(x - h)^2 + k$ looks like the graph of $f(x) = a(x - h)^2$, except that it is moved up or down.

Solution   We rewrite the equation as $g(x) = -2[x - (-3)]^2$. In this case, $a = -2$ and $h = -3$, so the graph looks like that of $y = 2x^2$ translated 3 units to the left and, since $-2 < 0$, it opens downward. The vertex is $(-3, 0)$, and the axis of symmetry is $x = -3$. Plotting points as needed, we obtain the graph shown below. We say that g is a *reflection* of the graph of $f(x) = 2(x + 3)^2$ across the x-axis.

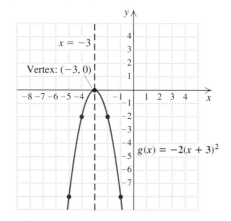

The maximum value of $g(x)$ is 0 when $x = -3$.

The Graph of $f(x) = a(x - h)^2 + k$

Given a graph of $f(x) = a(x - h)^2$, what happens if we add a constant k?

Interactive Discovery

For each pair of equations given below, create a table of values for both equations and graph both equations in the window $[-5, 5, -20, 20]$. Use the graph and the table to answer the following questions.

a) How do the values of y_1 and y_2 compare?
b) Do both graphs have the same axis of symmetry?
c) Do both graphs have the same vertex?

   1. $y_1 = 3(x - 2)^2$; $y_2 = 3(x - 2)^2 + 10$
   2. $y_1 = 3(x - 2)^2$; $y_2 = 3(x - 2)^2 - 5$
   3. $y_1 = -\frac{1}{2}(x + 1)^2$; $y_2 = -\frac{1}{2}(x + 1)^2 + 4$
   4. Describe the effect of k on the graph of $g(x) = a(x - h)^2 + k$.

     If we add a positive constant k to the graph of $f(x) = a(x - h)^2$, the graph is moved up. If we add a negative constant k, the graph is moved down. The axis of symmetry for the parabola remains at $x = h$, but the vertex will be at (h, k).

> **Graphing $f(x) = a(x - h)^2 + k$**
>
> The graph of $f(x) = a(x - h)^2 + k$ has the same shape as the graph of $y = a(x - h)^2$.
>
> If k is positive, the graph of $y = a(x - h)^2$ is shifted k units up.
>
> If k is negative, the graph of $y = a(x - h)^2$ is shifted $|k|$ units down.
>
> The vertex is (h, k), and the axis of symmetry is $x = h$.
>
> The domain of f is $(-\infty, \infty)$.
>
> For $a > 0$, the range of f is $[k, \infty)$. The minimum function value is k, which occurs when $x = h$.
>
> For $a < 0$, the range of f is $(-\infty, k]$. The maximum function value is k, which occurs when $x = h$.

TEACHING TIP

Point out that h and k change the vertex, and the sign of a determines whether the vertex gives a maximum or a minimum value.

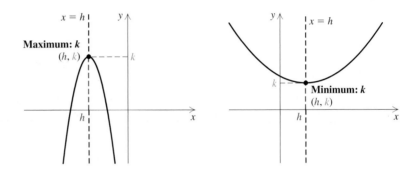

EXAMPLE 4 Graph $g(x) = (x - 3)^2 - 5$, and find the minimum function value and the range of f.

Solution The graph will look like that of $f(x) = (x - 3)^2$ but shifted 5 units down. You can confirm this by plotting some points.

The vertex is now $(3, -5)$, and the minimum function value is -5. The range of f is $[-5, \infty)$.

x	$g(x) = (x - 3)^2 - 5$	
0	4	
1	-1	
2	-4	
3	-5	← Vertex
4	-4	
5	-1	
6	4	

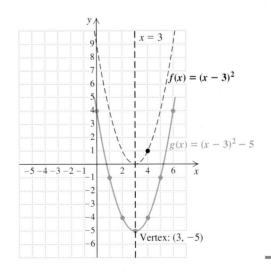

EXAMPLE 5 Graph $h(x) = \frac{1}{2}(x - 3)^2 + 5$, and find the minimum function value and the range of f.

Solution The graph looks just like that of $f(x) = \frac{1}{2}x^2$ but moved 3 units to the right and 5 units up. The vertex is $(3, 5)$, and the axis of symmetry is $x = 3$. We draw $f(x) = \frac{1}{2}x^2$ and then shift the curve over and up.

By plotting some points, we have a check. The minimum function value is 5, and the range is $[5, \infty)$.

x	$h(x) = \frac{1}{2}(x - 3)^2 + 5$	
0	$9\frac{1}{2}$	
1	7	
3	5	← Vertex
5	7	
6	$9\frac{1}{2}$	

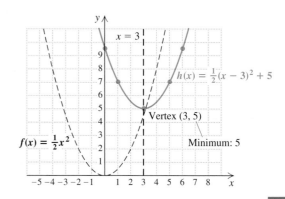

EXAMPLE 6 Graph $f(x) = -2(x + 3)^2 + 5$. Find the vertex, the axis of symmetry, the maximum or minimum function value, and the range of f.

Solution We first express the equation in the equivalent form

$$f(x) = -2[x - (-3)]^2 + 5.$$

The graph looks like that of $y = -2x^2$ translated 3 units to the left and 5 units up. The vertex is $(-3, 5)$, and the axis of symmetry is $x = -3$. Since -2 is negative, we know that 5, the second coordinate of the vertex, is the maximum function value. The range of f is $(-\infty, 5]$.

We compute a few points as needed, selecting convenient x-values on either side of the vertex. The graph is shown here.

x	$f(x) = -2(x + 3)^2 + 5$	
-4	3	
-3	5	← Vertex
-2	3	

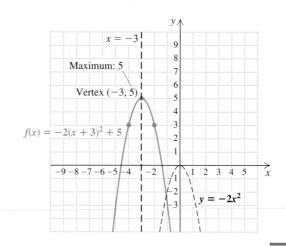

Because we can read the vertex and determine the orientation directly from the equation, it is actually faster to graph $y = a(x - h)^2 + k$ by hand than by using a graphing calculator.

8.6

Exercise Set

FOR EXTRA HELP

Digital Video Tutor CD 8 Videotape 12 · Student's Solutions Manual · Tutor Center · AW Math Tutor Center · InterAct Math · MathXL · MyMathLab

For each graph of a quadratic function
$$f(x) = a(x - h)^2 + k$$
in Exercises 1–6:

a) *Tell whether a is positive or negative.*
b) *Determine the vertex.*
c) *Determine the axis of symmetry.*
d) *Determine the range.*

1.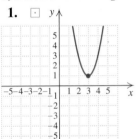

2.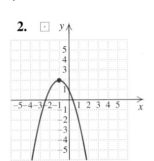

3.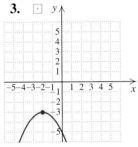

4.

5.

6.

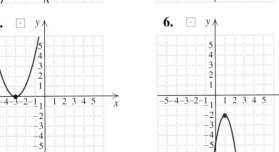

In Exercises 7–12, match each function with one of the following graphs.

a)   **b)**

c)   **d)**

e)   **f)**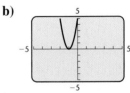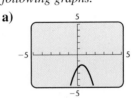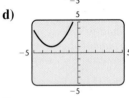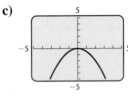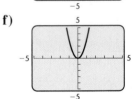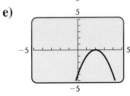

7. $f(x) = 3x^2$ (f)   **8.** $g(x) = -\frac{1}{2}x^2$ (c)

9. $h(x) = -(x - 2)^2$ (e)   **10.** $f(x) = 5(x + 1)^2$ (b)

11. $g(x) = \frac{2}{3}(x + 3)^2 + 1$ (d)   **12.** $h(x) = -2(x - \frac{1}{2})^2 - \frac{5}{3}$ (a)

Graph by hand.

13. $f(x) = x^2$ **14.** $f(x) = -x^2$

15. $f(x) = -2x^2$ **16.** $f(x) = -3x^2$

17. $g(x) = \frac{1}{3}x^2$ **18.** $g(x) = \frac{1}{4}x^2$

Aha! **19.** $h(x) = -\frac{1}{3}x^2$ **20.** $h(x) = -\frac{1}{4}x^2$

21. $f(x) = \frac{3}{2}x^2$ **22.** $f(x) = \frac{5}{2}x^2$

For each of the following, graph the function by hand, label the vertex, and draw the axis of symmetry.

23. $g(x) = (x + 1)^2$ **24.** $g(x) = (x + 4)^2$

25. $f(x) = (x - 2)^2$ **26.** $f(x) = (x - 1)^2$

Answers to Exercises 1–6 and 13–26 can be found on p. A-65.

27. $f(x) = -(x + 4)^2$ ☐   **28.** $f(x) = -(x - 2)^2$ ☐

29. $f(x) = 2(x + 1)^2$ ☐   **30.** $f(x) = 2(x + 4)^2$ ☐

31. $h(x) = -\frac{1}{2}(x - 3)^2$ ☐   **32.** $h(x) = -\frac{3}{2}(x - 2)^2$ ☐

33. $f(x) = \frac{1}{2}(x - 1)^2$ ☐   **34.** $f(x) = \frac{1}{3}(x + 2)^2$ ☐

For each of the following, graph the function by hand and find the vertex, the axis of symmetry, and the maximum value or the minimum value.

35. $f(x) = (x - 5)^2 + 1$ ☐

36. $f(x) = (x + 3)^2 - 2$ ☐

37. $f(x) = (x + 1)^2 - 2$ ☐

38. $g(x) = -(x - 2)^2 - 4$ ☐

39. $h(x) = -2(x - 1)^2 - 3$ ☐

40. $h(x) = -2(x + 1)^2 + 4$ ☐

41. $f(x) = 2(x + 4)^2 + 1$ ☐

42. $f(x) = 2(x - 5)^2 - 3$ ☐

43. $g(x) = -\frac{3}{2}(x - 1)^2 + 2$ ☐

44. $g(x) = \frac{3}{2}(x + 2)^2 - 1$ ☐

Without graphing, find the vertex, the axis of symmetry, the maximum value or the minimum value, and the range.

45. $f(x) = 8(x - 9)^2 + 5$ ☐

46. $f(x) = 10(x + 5)^2 - 8$ ☐

47. $h(x) = -\frac{2}{7}(x + 6)^2 + 11$ ☐

48. $h(x) = -\frac{3}{11}(x - 7)^2 - 9$ ☐

49. $f(x) = 5\left(x + \frac{1}{4}\right)^2 - 13$ ☐

50. $f(x) = 6\left(x - \frac{1}{4}\right)^2 + 19$ ☐

51. $f(x) = \sqrt{2}(x + 4.58)^2 + 65\pi$ ☐

52. $f(x) = 4\pi(x - 38.2)^2 - \sqrt{34}$ ☐

53. While trying to graph $y = -\frac{1}{2}x^2 + 3x + 1$, Omar gets the following screen.

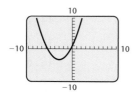

How can Omar tell at a glance that a mistake has been made?

54. Explain, without plotting points, why the graph of $y = (x + 2)^2$ looks like the graph of $y = x^2$ translated 2 units to the left.

Skill Maintenance

Graph using intercepts. [2.5]

55. $2x - 7y = 28$ ☐   **56.** $6x - 3y = 36$ ☐

Solve each system. [3.2]

57. $3x + 4y = -19$,   **58.** $5x + 7y = 9$,
    $7x - 6y = -29$ $3x - 4y = -11$
    $(-5, -1)$ $(-1, 2)$

Complete the square. [8.1]

59. $x^2 + 5x +$ _____   **60.** $x^2 - 9x +$ _____
    $x^2 + 5x + \frac{25}{4}$ $x^2 - 9x + \frac{81}{4}$

Synthesis

61. Before graphing a quadratic function, Sophie always plots five points. First, she calculates and plots the coordinates of the vertex. Then she plots *four* more points after calculating *two* more ordered pairs. How is this possible?

62. If the graphs of $f(x) = a_1(x - h_1)^2 + k_1$ and $g(x) = a_2(x - h_2)^2 + k_2$ have the same shape, what, if anything, can you conclude about the a's, the h's, and the k's? Why?

Write an equation for a function having a graph with the same shape as the graph of $f(x) = \frac{3}{5}x^2$, but with the given point as the vertex.

63. $(4, 1)$ ☐   **64.** $(2, 6)$ ☐   **65.** $(3, -1)$ ☐

66. $(5, -6)$ ☐   **67.** $(-2, -5)$ ☐   **68.** $(-4, -2)$ ☐

For each of the following, write the equation of the parabola that has the shape of $f(x) = 2x^2$ or $g(x) = -2x^2$ and has a maximum or minimum function value at the specified point.

69. Maximum: $(5, 0)$ **70.** Minimum: $(2, 0)$
    $g(x) = -2(x - 5)^2$ $f(x) = 2(x - 2)^2$

71. Minimum: $(-4, 0)$ **72.** Maximum: $(0, 3)$
    $f(x) = 2(x + 4)^2$ $g(x) = -2x^2 + 3$

73. Maximum: $(3, 8)$ **74.** Minimum: $(-2, 3)$
    $g(x) = -2(x - 3)^2 + 8$ $f(x) = 2(x + 2)^2 + 3$

Find an equation for a quadratic function F that satisfies the following conditions.

75. The graph of F is the same shape as the graph of f, where $f(x) = 3(x + 2)^2 + 7$, and $F(x)$ is a minimum at the same point that $g(x) = -2(x - 5)^2 + 1$ is a maximum. $F(x) = 3(x - 5)^2 + 1$

76. The graph of F is the same shape as the graph of f, where $f(x) = -\frac{1}{3}(x - 2)^2 + 7$, and $F(x)$ is a maximum at the same point that $g(x) = 2(x + 4)^2 - 6$ is a minimum. $F(x) = -\frac{1}{3}(x + 4)^2 - 6$

77. The graph of F has vertex $(2, -3)$ and passes through the point $(1, -5)$. $F(x) = -2(x - 2)^2 - 3$

☐ Answers to Exercises 27–52, 55, 56, and 63–68 can be found on pp. A-65 and A-66.

78. The maximum function value of F is -1 when x is 3, and $F(1) = -6$. $F(x) = -\frac{5}{4}(x - 3)^2 - 1$

Functions other than parabolas can be translated. When calculating $f(x)$, if we replace x with $x - h$, where h is a constant, the graph will be moved horizontally. If we replace $f(x)$ with $f(x) + k$, the graph will be moved vertically.

Use the graph below for Exercises 79–84.

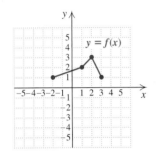

Draw a graph of each of the following.

79. $y = f(x - 1)$

80. $y = f(x + 2)$

81. $y = f(x) + 2$

82. $y = f(x) - 3$

83. $y = f(x + 3) - 2$

84. $y = f(x - 3) + 1$

79.

$y = f(x - 1)$

80.

$y = f(x + 2)$

81.

$y = f(x) + 2$

82.

$y = f(x) - 3$

83.
$y = f(x + 3) - 2$

84.

$y = f(x - 3) + 1$

Collaborative Corner

Match the Graph

Focus: Graphing quadratic functions

Time: 15–20 minutes

Group size: 6

Materials: Index cards

ACTIVITY

1. On each of six index cards, write one of the following equations:

   $y = \frac{1}{2}(x - 3)^2 + 1;$ $y = \frac{1}{2}(x - 1)^2 + 3;$
   $y = \frac{1}{2}(x + 1)^2 - 3;$ $y = \frac{1}{2}(x + 3)^2 + 1;$
   $y = \frac{1}{2}(x + 3)^2 - 1;$ $y = \frac{1}{2}(x + 1)^2 + 3.$

2. Fold each index card and mix up the six cards in a hat or bag. Then, one by one, each group member should select one of the equations. Do not let anyone see your equation.

3. Each group member should carefully graph the equation selected. Make the graph large enough so that when it is finished, it can be easily viewed by the rest of the group. Be sure to scale the axes and label the vertex, but **do not label the graph with the equation used**.

4. When all group members have drawn a graph, place the graphs in a pile. The group should then match and agree on the correct equation for each graph *with no help from the person who drew the graph*. If a mistake has been made and a graph has no match, determine what its equation *should* be.

5. Compare your group's labeled graphs with those of other groups to reach consensus within the class on the correct label for each graph.

8.7

Finding the Vertex   ■   Finding Intercepts

More About Graphing Quadratic Functions

Finding the Vertex

By *completing the square* (see Section 8.1), we can rewrite any polynomial $ax^2 + bx + c$ in the form $a(x - h)^2 + k$. Once that has been done, the procedures discussed in Section 8.6 will enable us to graph any quadratic function.

EXAMPLE 1   Graph: $g(x) = x^2 - 6x + 4$.

Solution   We have

$$g(x) = x^2 - 6x + 4$$
$$= (x^2 - 6x) + 4.$$

To complete the square inside the parentheses, we take half the x-coefficient, $\frac{1}{2} \cdot (-6) = -3$, and square it to get $(-3)^2 = 9$. Then we add $9 - 9$ inside the parentheses:

$$g(x) = (x^2 - 6x + 9 - 9) + 4$$ **The effect is of adding 0.**
$$= (x^2 - 6x + 9) + (-9 + 4)$$ **Using the associative law of addition to regroup**
$$= (x - 3)^2 - 5.$$ **Factoring and simplifying**

This equation was graphed in Example 4 of Section 8.6. The graph is that of $f(x) = x^2$ translated 3 units to the right and 5 units down. The vertex is $(3, -5)$.

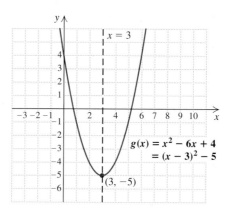

When the leading coefficient is not 1, we factor out that number from the first two terms. Then we complete the square.

EXAMPLE 2   Graph: $f(x) = 3x^2 + 12x + 13$.

Solution   Since the coefficient of x^2 is not 1, we need to factor out that number—in this case, 3—from the first two terms. Remember that we want the form $f(x) = a(x - h)^2 + k$:

$$f(x) = 3x^2 + 12x + 13$$
$$= 3(x^2 + 4x) + 13.$$

Now we complete the square as before. We take half of the x-coefficient, $\frac{1}{2} \cdot 4 = 2$, and square it: $2^2 = 4$. Then we add $4 - 4$ inside the parentheses:

$$f(x) = 3(x^2 + 4x + 4 - 4) + 13.$$ **Adding 4 − 4, or 0, inside the parentheses**

The distributive law allows us to separate the -4 from the perfect-square trinomial so long as it is multiplied by 3:

$$f(x) = 3(x^2 + 4x + 4) + 3(-4) + 13$$ **This leaves a perfect-square trinomial inside the parentheses.**

$$= 3(x + 2)^2 + 1.$$ **Factoring and simplifying**

The vertex is $(-2, 1)$, and the axis of symmetry is $x = -2$. The coefficient of x^2 is 3, so the graph is narrow and opens upward. We choose a few x-values on either side of the vertex, compute y-values, and then graph the parabola.

x	$f(x) = 3(x + 2)^2 + 1$	
-2	1	← Vertex
-3	4	
-1	4	

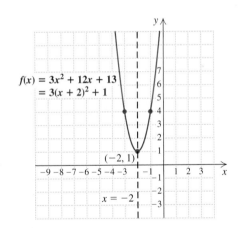

EXAMPLE 3   Graph: $f(x) = -2x^2 + 10x - 7$.

Solution   We first find the vertex by completing the square. To do so, we factor out -2 from the first two terms of the expression. This makes the coefficient of x^2 inside the parentheses 1:

$$f(x) = -2x^2 + 10x - 7$$
$$= -2(x^2 - 5x) - 7.$$

Now we complete the square as before. We take half of the x-coefficient and square it to get $\frac{25}{4}$. Then we add $\frac{25}{4} - \frac{25}{4}$ inside the parentheses:

$$f(x) = -2\left(x^2 - 5x + \tfrac{25}{4} - \tfrac{25}{4}\right) - 7$$

$$= -2\left(x^2 - 5x + \tfrac{25}{4}\right) + (-2)\left(-\tfrac{25}{4}\right) - 7 \qquad \text{Multiplying by } -2, \text{ using the distributive law, and regrouping}$$

$$= -2\left(x - \tfrac{5}{2}\right)^2 + \tfrac{11}{2}. \qquad \text{Factoring and simplifying}$$

The vertex is $\left(\frac{5}{2}, \frac{11}{2}\right)$, and the axis of symmetry is $x = \frac{5}{2}$. The coefficient of x^2, -2, is negative, so the graph opens downward. We plot a few points on either side of the vertex, including the y-intercept, $f(0)$, and graph the parabola.

x	$f(x)$	
$\frac{5}{2}$	$\frac{11}{2}$	← Vertex
0	-7	← y-intercept
1	1	
4	1	

$$f(x) = -2x^2 + 10x - 7$$
$$= -2(x - \tfrac{5}{2})^2 + \tfrac{11}{2}$$

The method used in Examples 1–3 can be generalized to find a formula for locating the vertex. We complete the square as follows:

$$f(x) = ax^2 + bx + c$$

$$= a\left(x^2 + \frac{b}{a}x\right) + c. \qquad \text{Factoring } a \text{ out of the first two terms. Check by multiplying.}$$

Half of the x-coefficient, $\frac{b}{a}$, is $\frac{b}{2a}$. We square it to get $\frac{b^2}{4a^2}$ and add $\frac{b^2}{4a^2} - \frac{b^2}{4a^2}$ inside the parentheses. Then we distribute the a and regroup terms:

$$f(x) = a\left(x^2 + \frac{b}{a}x + \frac{b^2}{4a^2} - \frac{b^2}{4a^2}\right) + c$$

$$= a\left(x^2 + \frac{b}{a}x + \frac{b^2}{4a^2}\right) + a\left(-\frac{b^2}{4a^2}\right) + c \qquad \text{Using the distributive law}$$

$$= a\left(x + \frac{b}{2a}\right)^2 + \frac{-b^2}{4a} + \frac{4ac}{4a} \qquad \text{Factoring and finding a common denominator}$$

$$= a\left[x - \left(-\frac{b}{2a}\right)\right]^2 + \frac{4ac - b^2}{4a}.$$

Thus we have the following.

The Vertex of a Parabola

The vertex of the parabola given by $f(x) = ax^2 + bx + c$ is

$$\left(-\frac{b}{2a}, f\left(-\frac{b}{2a}\right)\right), \quad \text{or} \quad \left(-\frac{b}{2a}, \frac{4ac - b^2}{4a}\right).$$

The x-coordinate of the vertex is $-b/(2a)$. The axis of symmetry is $x = -b/(2a)$. The second coordinate of the vertex is most commonly found by computing $f\left(-\frac{b}{2a}\right)$.

Let's reexamine Example 3 to see how we could have found the vertex directly. From the formula above,

$$\text{the } x\text{-coordinate of the vertex is } -\frac{b}{2a} = -\frac{10}{2(-2)} = \frac{5}{2}.$$

Substituting $\frac{5}{2}$ into $f(x) = -2x^2 + 10x - 7$, we find the second coordinate of the vertex:

$$\begin{aligned}
f\left(\tfrac{5}{2}\right) &= -2\left(\tfrac{5}{2}\right)^2 + 10\left(\tfrac{5}{2}\right) - 7 \\
&= -2\left(\tfrac{25}{4}\right) + 25 - 7 \\
&= -\tfrac{25}{2} + 18 \\
&= -\tfrac{25}{2} + \tfrac{36}{2} = \tfrac{11}{2}.
\end{aligned}$$

The vertex is $\left(\frac{5}{2}, \frac{11}{2}\right)$. The axis of symmetry is $x = \frac{5}{2}$.

Note that a quadratic function has a maximum or a minimum value at the vertex of its graph. Thus determining the maximum or minimum value of the function allows us to find the vertex of the graph.

Maximums and Minimums

On most graphing calculators, we can find a maximum or minimum function value for any given interval. This MAXIMUM or MINIMUM feature is often found in the CALC menu. (CALC is the second option associated with the TRACE key.)

To find a maximum or a minimum, enter and graph the function, choosing a viewing window that will show the vertex. Next, press [2nd] [CALC] and choose either the MAXIMUM or MINIMUM option in the menu. The graphing calculator will find the maximum or minimum function value over a specified interval, so the left and right endpoints, or bounds, of the interval must be entered as well as a guess near where the maximum or minimum occurs. The calculator will return the coordinates of the point for which the function value is a maximum or minimum within the interval.

EXAMPLE 4   Use a graphing calculator to determine the vertex of the graph of the function given by $f(x) = -2x^2 + 10x - 7$.

Solution   The coefficient of x^2 is negative, so the parabola opens downward and the function has a maximum value. We enter and graph the function, choosing a window that will show the vertex.

We choose the MAXIMUM option from the CALC menu. The calculator will prompt us first for a left bound. After visually locating the vertex, we either move the cursor to a point left of the vertex using the left and right arrow keys or type a value of x that is less than the x-value at the vertex. (See the graph on the top left below.) After pressing ENTER , we choose a right bound in a similar manner (as shown in the graph on the top right), and press ENTER again. Finally, we enter a guess that is close to the vertex (as in the graph on the lower left below), and press ENTER . (Many calculators will use the right bound as a guess if we simply press ENTER .) The calculator indicates that the coordinates of the vertex are $(2.5, 5.5)$. (See the graph on the lower right below.)

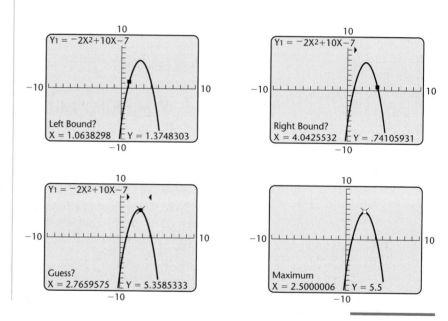

In Example 3, we found the coordinates of the vertex of the graph of $f(x) = -2x^2 + 10x - 7$, $\left(\frac{5}{2}, \frac{11}{2}\right)$, by completing the square. On p. 613, we also found these coordinates using the formula for the vertex of a parabola. The coordinates found using a calculator in Example 4 were in decimal notation. Because of the calculator's method used to find the maximum function value, the coordinates may not be exact and can indeed vary slightly for choices of windows. Since $2.5000006 \approx \frac{5}{2}$, and since $5.5 = \frac{11}{2}$, the coordinates check.

We have actually developed three methods for finding the vertex. One is by completing the square, the second is by using a formula, and the third is by using a graphing calculator. You should check with your instructor about which method to use.

TEACHING TIP

Lead students to realize that some quadratic equations of the form $y = ax^2 + bx + c$ may have no x-intercept, but all have a y-intercept.

Finding Intercepts

The points at which a graph crosses an axis are called intercepts. We saw in Chapter 2 and again in Example 3 that the y-intercept occurs at $f(0)$. For $f(x) = ax^2 + bx + c$, the y-intercept is simply $(0, c)$. To find x-intercepts, we look for points where $y = 0$ or $f(x) = 0$. To find the x-intercepts of the quadratic function given by $f(x) = ax^2 + bx + c$, we solve the equation

$$0 = ax^2 + bx + c.$$

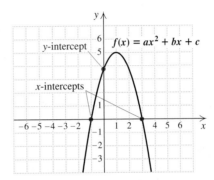

EXAMPLE 5   Find the x- and y-intercepts of the graph of

$$f(x) = x^2 - 2x - 2.$$

Solution   The y-intercept is simply $(0, f(0))$, or $(0, -2)$. To find the x-intercepts, we solve the equation

$$0 = x^2 - 2x - 2.$$

We are unable to factor $x^2 - 2x - 2$, so we use the quadratic formula and get $x = 1 \pm \sqrt{3}$. Thus the x-intercepts are $\left(1 - \sqrt{3}, 0\right)$ and $\left(1 + \sqrt{3}, 0\right)$.
   If graphing, we would approximate, to get $(-0.7, 0)$ and $(2.7, 0)$.

Connecting the Concepts

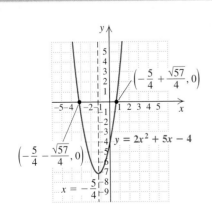

Because the graph of a quadratic equation is symmetric, the x-intercepts of the graph, if they exist, will be symmetric with respect to the axis of symmetry. This symmetry can be seen directly if the x-intercepts are found using the quadratic formula.
   For example, the x-intercepts of the graph of $y = 2x^2 + 5x - 4$ are

$$\left(-\frac{5}{4} + \frac{\sqrt{57}}{4}, 0\right) \quad \text{and} \quad \left(-\frac{5}{4} - \frac{\sqrt{57}}{4}, 0\right).$$

For this equation, the axis of symmetry is $x = -\dfrac{5}{4}$ and the x-intercepts are $\dfrac{\sqrt{57}}{4}$ units to the left and right of $-\dfrac{5}{4}$ on the x-axis.

The Graph of a Quadratic Function Given by $f(x) = ax^2 + bx + c$ or $f(x) = a(x - h)^2 + k$

The graph is a parabola.

The vertex is (h, k) or $\left(-\dfrac{b}{2a}, f\left(-\dfrac{b}{2a}\right)\right)$.

The axis of symmetry is $x = h$.

The y-intercept of the graph is $(0, c)$.

The x-intercepts can be found by solving $ax^2 + bx + c = 0$.
   If $b^2 - 4ac > 0$, there are two x-intercepts.
   If $b^2 - 4ac = 0$, there is one x-intercept.
   If $b^2 - 4ac < 0$, there are no x-intercepts.

The domain of the function is $(-\infty, \infty)$.

If a is positive: The graph opens upward.
   The function has a minimum value, given by k.
      This occurs when $x = h$.
   The range of the function is $[k, \infty)$.

If a is negative: The graph opens downward.
   The function has a maximum value, given by k.
      This occurs when $x = h$.
   The range of the function is $(-\infty, k]$.

---

1. (a) $f(x) = (x - 2)^2 + 1$; **(b)** vertex: $(2, 1)$; axis of symmetry: $x = 2$
2. (a) $f(x) = (x + 3)^2 + 4$; **(b)** vertex: $(-3, 4)$; axis of symmetry: $x = -3$
3. (a) $f(x) = -\left(x - \frac{3}{2}\right)^2 - \frac{31}{4}$; **(b)** vertex: $\left(\frac{3}{2}, -\frac{31}{4}\right)$; axis of symmetry: $x = \frac{3}{2}$ **4. (a)** $f(x) = \left(x + \frac{5}{2}\right)^2 - \frac{9}{4}$; **(b)** vertex: $\left(-\frac{5}{2}, -\frac{9}{4}\right)$; axis of symmetry: $x = -\frac{5}{2}$
5. (a) $f(x) = 2\left(x - \frac{7}{4}\right)^2 - \frac{41}{8}$; **(b)** vertex: $\left(\frac{7}{4}, -\frac{41}{8}\right)$; axis of symmetry: $x = \frac{7}{4}$
6. (a) $f(x) = -2\left(x - \frac{5}{4}\right)^2 + \frac{17}{8}$; **(b)** vertex: $\left(\frac{5}{4}, \frac{17}{8}\right)$; axis of symmetry: $x = \frac{5}{4}$

8.7

Exercise Set

*For each quadratic function, **(a)** write the function in the form $f(x) = a(x - h)^2 + k$ and **(b)** find the vertex and the axis of symmetry.*

1. $f(x) = x^2 - 4x + 5$

2. $f(x) = x^2 + 6x + 13$

3. $f(x) = -x^2 + 3x - 10$

4. $f(x) = x^2 + 5x + 4$

5. $f(x) = 2x^2 - 7x + 1$

6. $f(x) = -2x^2 + 5x - 1$

*For each quadratic function, **(a)** find the vertex and the axis of symmetry and **(b)** graph the function by hand.*

7. $f(x) = x^2 + 4x + 5$

8. $f(x) = x^2 + 2x - 5$

9. $f(x) = x^2 + 8x + 20$

10. $f(x) = x^2 - 10x + 21$

11. $h(x) = 2x^2 - 16x + 25$

12. $h(x) = 2x^2 + 16x + 23$

13. $f(x) = -x^2 + 2x + 5$

14. $f(x) = -x^2 - 2x + 7$

15. $g(x) = x^2 + 3x - 10$

16. $g(x) = x^2 + 5x + 4$

17. $h(x) = x^2 + 7x$

18. $h(x) = x^2 - 5x$

19. $f(x) = -2x^2 - 4x - 6$

20. $f(x) = -3x^2 + 6x + 2$

 Answers to Exercises 7–20 can be found on pp. A-66 and A-67.

21. $f(x) = -3x^2 + 5x - 2$ ☐

22. $f(x) = -3x^2 - 7x + 2$ ☐

23. $h(x) = \frac{1}{2}x^2 + 4x + \frac{19}{3}$ ☐

24. $h(x) = \frac{1}{2}x^2 - 3x + 2$ ☐

Use a graphing calculator to find the vertex of the graph of each function.

25. $f(x) = x^2 + x - 6$ $(-0.5, -6.25)$ **26.** $f(x) = x^2 + 2x - 5$ $(-1, -6)$

27. $f(x) = 5x^2 - x + 1$ $(0.1, 0.95)$

28. $f(x) = -4x^2 - 3x + 7$ $(-0.375, 7.5625)$

29. $f(x) = -0.2x^2 + 1.4x - 6.7$ $(3.5, -4.25)$

30. $f(x) = 0.5x^2 + 2.4x + 3.2$ $(-2.4, 0.32)$

Find the x- and y-intercepts. If no x-intercepts exist, state this.

31. $f(x) = x^2 - 6x + 3$ ☐   **32.** $f(x) = x^2 + 5x + 2$ ☐

33. $g(x) = -x^2 + 2x + 3$ **34.** $g(x) = x^2 - 6x + 9$
$(-1, 0), (3, 0); (0, 3)$ $(3, 0); (0, 9)$

Aha! **35.** $f(x) = x^2 - 9x$ **36.** $f(x) = x^2 - 7x$
$(0, 0), (9, 0); (0, 0)$ $(0, 0), (7, 0); (0, 0)$

37. $h(x) = -x^2 + 4x - 4$ **38.** $h(x) = 4x^2 - 12x + 3$ ☐
$(2, 0); (0, -4)$

39. $f(x) = 2x^2 - 4x + 6$ **40.** $f(x) = x^2 - x + 2$
No x-intercept; $(0, 6)$ No x-intercept; $(0, 2)$

For each quadratic function, find **(a)** *the maximum or minimum value and* **(b)** *the x- and y-intercepts. Round to the nearest hundredth.*

41. $f(x) = 2.31x^2 - 3.135x - 5.89$ ☐

42. $f(x) = -18.8x^2 + 7.92x + 6.18$ ☐

43. $g(x) = -1.25x^2 + 3.42x - 2.79$ ☐

44. $g(x) = 0.45x^2 - 1.72x + 12.92$ ☐

TW **45.** Does the graph of every quadratic function have a y-intercept? Why or why not?

TW **46.** Is it possible for the graph of a quadratic function to have only one x-intercept if the vertex is off the x-axis? Why or why not?

Skill Maintenance

Solve each system.

47. $5x - 3y = 16,$
$4x + 2y = 4$ [3.2] $(2, -2)$

48. $2x - 5y = 9,$
$5x - 15y = 20$ [3.2]
$(7, 1)$

49. $4a - 5b + c = 3,$
$3a - 4b + 2c = 3,$
$a + b - 7c = -2$
[3.4] $(3, 2, 1)$

50. $2a - 7b + c = 25,$
$a + 5b - 2c = -18,$
$3a - b + 4c = 14$
[3.4] $(1, -3, 2)$

Solve. [7.6]

51. $\sqrt{4x - 4} = \sqrt{x + 4} + 1$ 5

52. $\sqrt{5x - 4} + \sqrt{13 - x} = 7$ 4

Synthesis

TW **53.** If the graphs of two quadratic functions have the same x-intercepts, will they also have the same vertex? Why or why not?

TW **54.** Suppose that the graph of $f(x) = ax^2 + bx + c$ has $(x_1, 0)$ and $(x_2, 0)$ as x-intercepts. Explain why the graph of $g(x) = -ax^2 - bx - c$ will also have $(x_1, 0)$ and $(x_2, 0)$ as x-intercepts.

55. Graph the function

$$f(x) = x^2 - x - 6.$$

Then use the graph to approximate solutions to each of the following equations.

a) $x^2 - x - 6 = 2$ $-2.4, 3.4$

b) $x^2 - x - 6 = -3$ $-1.3, 2.3$

56. Graph the function

$$f(x) = \frac{x^2}{2} + x - \frac{3}{2}.$$

Then use the graph to approximate solutions to each of the following equations.

a) $\frac{x^2}{2} + x - \frac{3}{2} = 0$ $-3, 1$

b) $\frac{x^2}{2} + x - \frac{3}{2} = 1$ $-3.4, 1.4$

c) $\frac{x^2}{2} + x - \frac{3}{2} = 2$ $-3.8, 1.8$

Find an equivalent equation of the type
$$f(x) = a(x - h)^2 + k.$$

57. $f(x) = mx^2 - nx + p$ $f(x) = m\left(x - \dfrac{n}{2m}\right)^2 + \dfrac{4mp - n^2}{4m}$

58. $f(x) = 3x^2 + mx + m^2$ $f(x) = 3\left[x - \left(-\dfrac{m}{6}\right)\right]^2 + \dfrac{11m^2}{12}$

59. A quadratic function has $(-1, 0)$ as one of its intercepts and $(3, -5)$ as its vertex. Find an equation for the function.   ☐

60. A quadratic function has $(4, 0)$ as one of its intercepts and $(-1, 7)$ as its vertex. Find an equation for the function. $f(x) = -0.28x^2 - 0.56x + 6.72,$ or $f(x) = -\frac{7}{25}(x + 1)^2 + 7$

Graph.

61. $f(x) = |x^2 - 1|$ ☐

62. $f(x) = |x^2 - 3x - 4|$ ☐

63. $f(x) = |2(x - 3)^2 - 5|$ ☐

☐ Answers to Exercises 21–24, 31, 32, 38, 41–44, 59, and 61–63 can be found on p. A-67.

8.8

Maximum and Minimum Problems ■ Fitting Quadratic Functions to Data

Problem Solving and Quadratic Functions

Let's look now at some of the many situations in which quadratic functions are used for problem solving.

Maximum and Minimum Problems

We have seen that for any quadratic function f, the value of $f(x)$ at the vertex is either a maximum or a minimum. Thus problems in which a quantity must be maximized or minimized can often be solved by finding the coordinates of a vertex. This assumes that the problem can be modeled with a quadratic function.

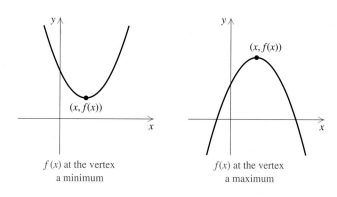

$f(x)$ at the vertex
a minimum

$f(x)$ at the vertex
a maximum

EXAMPLE 1 Newborn Calves. The number of pounds of milk per day recommended for a calf that is x weeks old can be approximated by $p(x)$, where $p(x) = -0.2x^2 + 1.3x + 6.2$ (*Source*: C. Chaloux, University of Vermont, 1998). When is a calf's milk consumption greatest and how much milk does it consume at that time?

Solution

1., 2. Familiarize and **Translate.**   We are given the function for milk consumption by a calf. Note that it is a quadratic function of x, the calf's age in weeks. The graph of the function (shown at left) indicates that the calf's consumption increases and then decreases.

3. Carry out.   We can either complete the square,

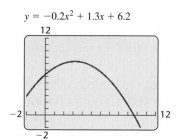

$y = -0.2x^2 + 1.3x + 6.2$

$$
\begin{aligned}
p(x) &= -0.2x^2 + 1.3x + 6.2 \\
&= -0.2(x^2 - 6.5x) + 6.2 \\
&= -0.2(x^2 - 6.5x + 3.25^2 - 3.25^2) + 6.2 \quad \textbf{Completing the} \\
&\hspace{8cm} \textbf{square; } \frac{6.5}{2} = \textbf{3.25} \\
&= -0.2(x^2 - 6.5x + 3.25^2) + (-0.2)(-3.25^2) + 6.2 \\
&= -0.2(x - 3.25)^2 + 8.3125, \quad \textbf{Factoring and simplifying}
\end{aligned}
$$

or we can use $-b/(2a) = -1.3/(-0.4) = 3.25$. Using a calculator, we find that

$$p(3.25) = -0.2(3.25)^2 + 1.3(3.25) + 6.2 = 8.3125.$$

4. Check. Both of the approaches in step (3) indicate that a maximum occurs when $x = 3.25$, or $3\frac{1}{4}$. We could also use the MAXIMUM feature on a graphing calculator as a check.

5. State. A calf's milk consumption is greatest when the calf is $3\frac{1}{4}$ weeks old. At that time, it drinks about 8.3 lb of milk per day.

EXAMPLE 2   Fenced-in Land. What are the dimensions of the largest rectangular pen that a farmer can enclose with 64 m of electric fence?

Solution

1. Familiarize. We make a drawing and label it. Recall these important formulas:

Perimeter: $2w + 2l$;

Area: $l \cdot w$.

To get a better feel for the problem, we can look at some possible dimensions for a rectangular pen that can be enclosed with 64 m of fence. All possibilities are chosen so that $2w + 2l = 64$.

We can make a table by hand or use a graphing calculator. To use a calculator, we let x represent the width of the pen. Solving $2w + 2l = 64$ for l, we have $l = (64 - 2w)/2$, or $l = 32 - w$. Thus, if $y_1 = 32 - x$, then the area is $y_2 = x \cdot y_1$.

X	Y1	Y2
22	10	220
21.5	10.5	225.75
21	11	231
20.5	11.5	235.75
20	12	240
19.5	12.5	243.75
19	13	247

X = 22

What choice of X will maximize Y2?

l	w	Perimeter	Area
22 m	10 m	64 m	220 m²
20 m	12 m	64 m	240 m²
18 m	14 m	64 m	252 m²
.	.		.
.	.		.
.	.		.

What choice of l and w will maximize A?

2. Translate. We have two equations: One guarantees that the perimeter is 64 m; the other expresses area in terms of length and width.

$$2w + 2l = 64$$

$$A = l \cdot w$$

3. Carry out. We solve the system of equations both algebraically and graphically.

Algebraic Solution

We need to express A as a function of l or w but not both. To do so, we solve for l in the first equation to obtain $l = 32 - w$. Substituting for l in the second equation, we get a quadratic function:

$$A = (32 - w)w \qquad \textbf{Substituting for } l$$
$$= -w^2 + 32w. \qquad \textbf{This is a parabola opening downward, so a maximum exists.}$$

Completing the square, we get

$$A = -(w^2 - 32w + 256 - 256)$$
$$= -(w - 16)^2 + 256.$$

The maximum function value, 256 m^2, occurs when $w = 16$ m and $l = 32 - 16$, or 16 m.

Graphical Solution

As in the *Familiarize* step, we let x represent the width of the pen. Then the length is given by $y_1 = 32 - x$ and the area is $y_2 = x \cdot y_1$, or $y = 32x - x^2$. The maximum area is the second coordinate of the vertex of the graph of $y = 32x - x^2$.

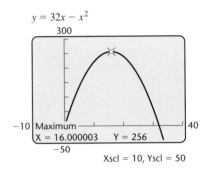

$$y = 32x - x^2$$

Maximum
X = 16.000003   Y = 256

Xscl = 10, Yscl = 50

The maximum is 256 m^2 when the width, x, is 16 m and the length, $32 - x$, is 16 m.

4. Check. Note that 256 m^2 is greater than any of the values for A found in the *Familiarize* step. To be more certain, we could check values other than those used in that step. For example, if $w = 15$ m, then $l = 32 - 15 = 17$ m, and $A = 15 \cdot 17 = 255 \text{ m}^2$. The same area results if $w = 17$ m and $l = 15$ m. Since 256 m^2 is greater than 255 m^2, it looks as though we have a maximum.

5. State. The largest rectangular pen that can be enclosed is 16 m by 16 m.

Fitting Quadratic Functions to Data

We can now model some real-world situations using quadratic functions. As always, before attempting to fit an equation to data, we should graph the data and compare the result to the shape of the graphs of different types of functions and ask what type of function seems appropriate.

Connecting the Concepts

The general shapes of the graphs of many functions that we have studied are shown below.

Linear function:
$$f(x) = mx + b$$

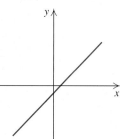

Absolute-value function:
$$f(x) = |x|$$

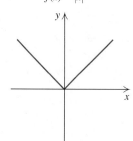

Rational function:
$$f(x) = \frac{1}{x}$$

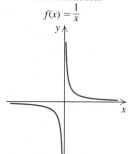

Radical function:
$$f(x) = \sqrt{x}$$

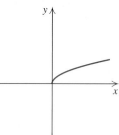

Quadratic function:
$$f(x) = ax^2 + bx + c, a > 0$$

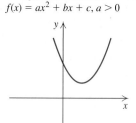

Quadratic function:
$$f(x) = ax^2 + bx + c, a < 0$$

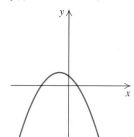

We see that in order for a quadratic function to fit a set of data, the graph of the data must approximate the shape of a parabola. Data that resemble one half of a parabola might also be modeled using a quadratic function with a restricted domain, as illustrated in the figure below.

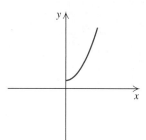

Quadratic function:
$$f(x) = ax^2 + bx + c,$$
$$a > 0, x \geq 0$$

Quadratic function:
$$f(x) = ax^2 + bx + c,$$
$$a < 0, x \geq 0$$

EXAMPLE 3 Determine whether a quadratic function can be used to model each of the following situations.

a) *Teen Smoking.* The percent of high school students who reported having smoked a cigarette in the past 30 days

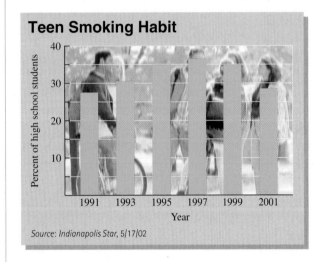

Teen Smoking Habit

Percent of high school students

Source: *Indianapolis Star, 5/17/02*

b) *Alternative-Fueled Vehicles.* The number of cars in the United States fueled by electricity

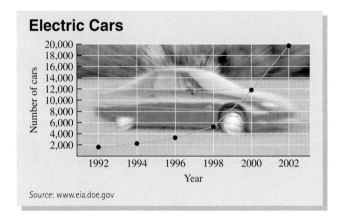

Electric Cars

Number of cars

Source: *www.eia.doe.gov*

c) *Prescriptions.* The number of prescriptions written in the United States for analeptic drugs, used to treat attention deficit disorder

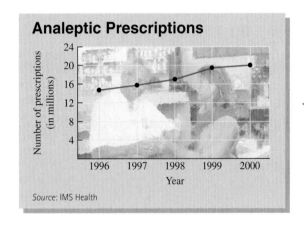

Analeptic Prescriptions

Number of prescriptions (in millions)

Source: *IMS Health*

d) *Driver Fatalities.* The number of driver fatalities by age

Driver Fatalities by Age

Number of driver deaths per 100,000

28   15   10   15   25

15–24 25–39 40–69 70–79 80+

Ages

Source: *National Highway Traffic Administration*

e) *Festival Revenues.*   The profit made by the Indianapolis 500 festival

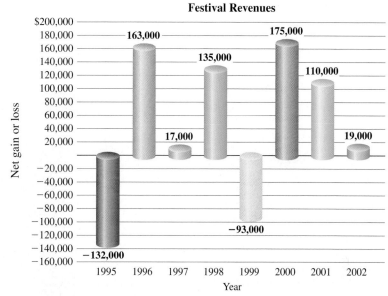

Festival Revenues

Source: 500 Festival

Solution

a) The data rise and then fall, resembling a parabola that opens downward. The situation could be modeled by a quadratic function $f(x) = ax^2 + bx + c, a < 0$.

b) The data resemble the right half of a parabola that opens upward. We
could use a quadratic function $f(x) = ax^2 + bx + c, a > 0, x \geq 0$, to
model the situation.

c) The data appear nearly linear. A linear function is a better model for this
situation than a quadratic function.

d) The data fall and then rise, resembling a parabola that opens upward. A
quadratic function $f(x) = ax^2 + bx + c, a > 0$, might be used as a model
for this situation.

e) The data do not resemble a parabola. A quadratic function would not be a
good model for this situation.

EXAMPLE 4 Teen Smoking. According to the Centers for Disease Control and Prevention, the percent of high school students who reported having smoked a cigarette in the preceding 30 days declined from 1997 to 2001, after rising in the first part of the 1990s. As we saw in Example 3(a), we can model the percent of high-schoolers who smoke as a quadratic function of the year. To do this, we let $x =$ the number of years after 1991.

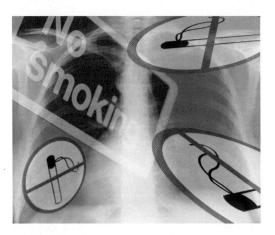

Years after 1991	Percent of High School Students Who Smoked a Cigarette in the Preceding 30 Days
0	27.5
2	30.5
4	34.9
6	36.4
8	34.9
10	28.5

Source: Indianapolis Star, 5/17/02

a) Use the data points $(0, 27.5)$, $(6, 36.4)$, and $(10, 28.5)$ to fit a quadratic equation $T(x)$ to the data.

b) Use the function from part (a) to estimate the percent of high school students in 2002 who smoked a cigarette in the preceding 30 days.

c) Use the REGRESSION feature of a graphing calculator to fit a quadratic function $H(x)$ to all the given data.

d) Use the function from part (c) to estimate the percent of high school students in 2002 who smoked a cigarette in the preceding 30 days, and compare the estimate with the estimate from part (b).

Solution

a) We are looking for a function of the form $T(x) = ax^2 + bx + c$, where $T(x)$ is the percent of high school students who smoked a cigarette in the preceding 30 days x years after 1991. We need values for a, b, and c. We substitute the given values of x and T from the three data points listed:

$27.5 = a \cdot 0^2 + b \cdot 0 + c,$ **Using the data point (0, 27.5)**

$36.4 = a \cdot 6^2 + b \cdot 6 + c,$ **Using the data point (6, 36.4)**

$28.5 = a \cdot 10^2 + b \cdot 10 + c.$ **Using the data point (10, 28.5)**

After simplifying, we see that we need to solve the system

$27.5 = c,$ **(1)**

$36.4 = 36a + 6b + c,$ **(2)**

$28.5 = 100a + 10b + c.$ **(3)**

We know from equation (1) that $c = 27.5$. Substituting that value into equations (2) and (3), we have

$36.4 = 36a + 6b + 27.5,$

$28.5 = 100a + 10b + 27.5.$

Simplifying, we obtain the system of equations

$8.9 = 36a + 6b,$ **(4)**

$1 = 100a + 10b.$ **(5)**

If we wish, we can clear decimals from the first equation by multiplying by 10:

$89 = 360a + 60b,$

$1 = 100a + 10b.$

To solve, we multiply equation (5) by -6 and add:

$$89 = 360a + 60b$$
$$\underline{-6 = -600a - 60b}$$
$$83 = -240a$$

$$-\frac{83}{240} = a.$$

Next, we solve for b, using equation (5) above:

$$1 = 100\left(-\frac{83}{240}\right) + 10b$$ **Substituting in equation (5)**

$$1 = -\frac{415}{12} + 10b$$

$$\frac{427}{12} = 10b$$

$$\frac{427}{120} = b.$$

Thus the function $T(x) = -\dfrac{83}{240}x^2 + \dfrac{427}{120}x + 27.5$ fits the three points given. As a partial check, we graph the function along with the data. The function goes through the three given points.

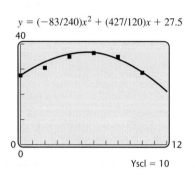

$y = (-83/240)x^2 + (427/120)x + 27.5$

Yscl = 10

b) To estimate the percent of high school students in 2002 who smoked a cigarette in the preceding 30 days, we evaluate $T(11)$, since 2002 is 11 years after 1991:

$$T(11) = -\dfrac{83}{240} \cdot 11^2 + \dfrac{427}{120} \cdot 11 + 27.5 \approx 24.8.$$

According to this model, in 2002, about 24.8% of high school students would have reported that they smoked a cigarette in the preceding 30 days.

c) To fit a quadratic function $H(x)$ to the data using regression, we enter the data into a graphing calculator and choose the QUADREG option in the STAT CALC menu.

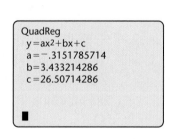

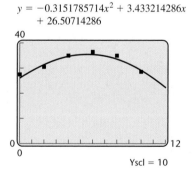

$y = -0.3151785714x^2 + 3.433214286x$
$+ 26.50714286$

Yscl = 10

From the screen on the left above, we obtain the quadratic function

$$H(x) = -0.3151785714x^2 + 3.433214286x + 26.50714286.$$

On the right above, we graph $H(x)$, along with the data points.

d) To estimate the percent of high school students in 2002 who smoked a cigarette in the preceding 30 days, we evaluate $H(11)$ using the VALUE option or a table, and obtain $H(11) \approx 26.1$. According to this model, in 2002, about 26.1% of high school students would have reported that they smoked a cigarette in the preceding 30 days. This estimate is 1.3% higher than the estimate given by $T(x)$ in part (b). The second model, which took into consideration all the data, is a better fit for the data. Without further information, it is difficult to tell which model more accurately predicts future trends.

Note in Example 4 that although a quadratic function fits the data, a quadratic model would not be a good predictor of long-range trends, since it would eventually predict negative percentages of high school smokers. This model may be valid only for the years 1991 through 2001. Another, more complex, model that takes into consideration other factors such as economic and educational changes may yield a more accurate prediction of future trends.

8.8

Exercise Set

FOR EXTRA HELP

| Digital Video Tutor CD 8 Videotape 12 | Student's Solutions Manual | Tutor Center AW Math Tutor Center | InterAct Math | MathXL | MyMathLab |

Solve.

1. *Stock Prices.*   The value of a share of R. P. Mugahti can be represented by $V(x) = x^2 - 6x + 13$, where x is the number of months after January 2001. What is the lowest value $V(x)$ will reach, and when will that occur? $4; 3 mos after January 2001

2. *Minimizing Cost.*   Aki's Bicycle Designs has determined that when x hundred bicycles are built, the average cost per bicycle is given by

$$C(x) = 0.1x^2 - 0.7x + 2.425,$$

where $C(x)$ is in hundreds of dollars. What is the minimum average cost per bicycle and how many bicycles should be built to achieve that minimum?

$120/bicycle; 350 bicycles

3. *Ticket Sales.*   The number of tickets sold each day for an upcoming performance of Handel's Messiah is given by

$$N(x) = -0.4x^2 + 9x + 11,$$

where x is the number of days since the concert was first announced. When will daily ticket sales peak and how many tickets will be sold that day?
11 days after the concert was announced; about 62 tickets

$P(x) = -x^2 + 980x - 3000; \$237,100$ at $x = 490$
4. *Maximizing Profit.*   Recall that total profit P is the difference between total revenue R and total cost C. Given $R(x) = 1000x - x^2$ and $C(x) = 3000 + 20x$, find the total profit, the maximum value of the total profit, and the value of x at which it occurs.

5. *Architecture.*   An architect is designing an atrium for a hotel. The atrium is to be rectangular with a perimeter of 720 ft of brass piping. What dimensions will maximize the area of the atrium?
180 ft by 180 ft

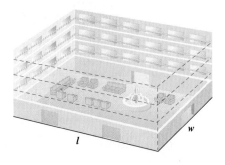

6. *Stained-Glass Window Design.* An artist is designing a rectangular stained-glass window with a perimeter of 84 in. What dimensions will yield the maximum area? 21 in. by 21 in.

7. *Garden Design.* A farmer decides to enclose a rectangular garden, using the side of a barn as one side of the rectangle. What is the maximum area that the farmer can enclose with 40 ft of fence? What should the dimensions of the garden be in order to yield this area? 200 ft²; 10 ft by 20 ft (The barn serves as a 20-ft side.)

8. *Patio Design.* A stone mason has enough stones to enclose a rectangular patio with 60 ft of perimeter, assuming that the attached house forms one side of the rectangle. What is the maximum area that the mason can enclose? What should the dimensions of the patio be in order to yield this area? 450 ft²; 15 ft by 30 ft (The house serves as the 30-ft side.)

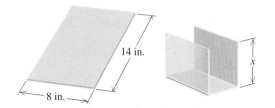

9. *Molding Plastics.* Economite Plastics plans to produce a one-compartment vertical file by bending the long side of an 8-in. by 14-in. sheet of plastic along two lines to form a U shape. How tall should the file be in order to maximize the volume that the file can hold? 3.5 in.

14 in.

8 in.

x

10. *Composting.* A rectangular compost container is to be formed in a corner of a fenced yard, with 8 ft of chicken wire completing the other two sides of the rectangle. If the chicken wire is 3 ft high, what dimensions of the base will maximize the container's volume? 4 ft by 4 ft

11. What is the maximum product of two numbers that add to 18? What numbers yield this product? 81; 9 and 9

12. What is the maximum product of two numbers that add to 26? What numbers yield this product? 169; 13 and 13

13. What is the minimum product of two numbers that differ by 8? What are the numbers? -16; 4 and -4

14. What is the minimum product of two numbers that differ by 7? What are the numbers? $-\frac{49}{4}$; $-\frac{7}{2}$ and $\frac{7}{2}$

Aha! **15.** What is the maximum product of two numbers that add to -10? What numbers yield this product? 25; -5 and -5

16. What is the maximum product of two numbers that add to -12? What numbers yield this product? 36; -6 and -6

For Exercises 17–26, state whether the graph appears to represent a quadratic function. Explain your reasoning.

17.

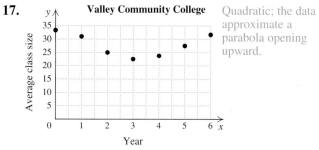

Quadratic; the data approximate a parabola opening upward.

18.

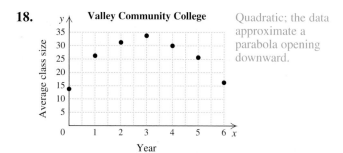

Quadratic; the data approximate a parabola opening downward.

19.

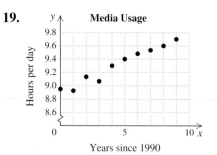

Not quadratic; the data do not approximate a parabola.

20.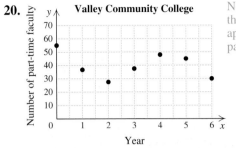
Valley Community College

Not quadratic; the data do not approximate a parabola.

23. Not quadratic; the data do not approximate a parabola.

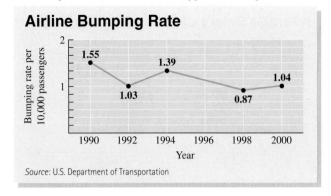

Airline Bumping Rate

Source: U.S. Department of Transportation

21. Not quadratic; the data do not approximate a parabola.

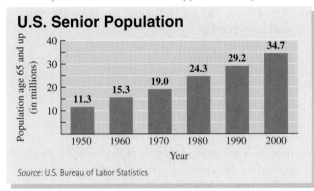

U.S. Senior Population

Source: U.S. Bureau of Labor Statistics

Quadratic; the data approximate half a parabola opening upward.
24.

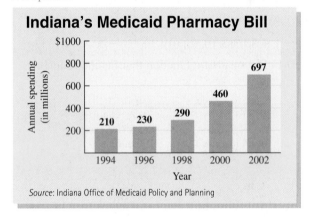

Indiana's Medicaid Pharmacy Bill

Source: Indiana Office of Medicaid Policy and Planning

22. Quadratic; the data approximate a parabola opening upward.

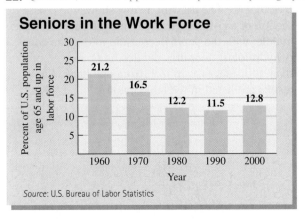

Seniors in the Work Force

Source: U.S. Bureau of Labor Statistics

Quadratic; the data approximate half a parabola opening upward.
25.

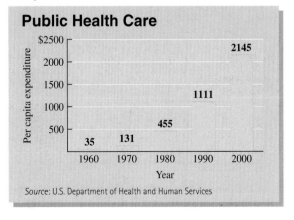

Public Health Care

Source: U.S. Department of Health and Human Services

26. Not quadratic; the data do not approximate a parabola.

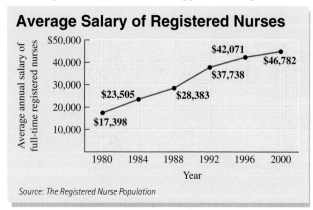

Average Salary of Registered Nurses

Average annual salary of full-time registered nurses

$50,000
40,000 — $42,071
$46,782
30,000 — $37,738
$23,505 — $28,383
20,000
$17,398
10,000

1980 1984 1988 1992 1996 2000

Year

Source: The Registered Nurse Population

Find a quadratic function that fits the set of data points.

27. $(1, 4), (-1, -2), (2, 13)$ $f(x) = 2x^2 + 3x - 1$

28. $(1, 4), (-1, 6), (-2, 16)$ $f(x) = 3x^2 - x + 2$

29. $(2, 0), (4, 3), (12, -5)$ $f(x) = -\frac{1}{4}x^2 + 3x - 5$

30. $(-3, -30), (3, 0), (6, 6)$ $f(x) = -\frac{1}{3}x^2 + 5x - 12$

31. a) Find a quadratic function that fits the following data. $A(s) = \frac{3}{16}s^2 - \frac{135}{4}s + 1750$

Travel Speed (in kilometers per hour)	Number of Nighttime Accidents (for every 200 million kilometers driven)
60	400
80	250
100	250

b) Use the function to estimate the number of night-time accidents that occur at 50 km/h.
About 531 accidents

32. a) Find a quadratic function that fits the following data. $A(s) = 0.05s^2 - 5.5s + 250$

Travel Speed (in kilometers per hour)	Number of Daytime Accidents (for every 200 million kilometers driven)
60	100
80	130
100	200

b) Use the function to estimate the number of day-time accidents that occur at 50 km/h.   100 accidents

33. *Archery.* The Olympic flame tower at the 1992 Summer Olympics was lit at a height of about 27 m by a flaming arrow that was launched about 63 m from the base of the tower. If the arrow landed about 63 m beyond the tower, find a quadratic function that expresses the height h of the arrow as a function of the distance d that it traveled horizontally.
$h(d) = -0.0068d^2 + 0.8571d$

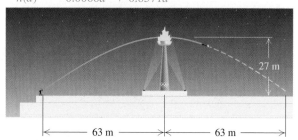

63 m   63 m

34. *Pizza Prices.* Pizza Unlimited has the following prices for pizzas.

Diameter	Price
8 in.	$ 6.00
12 in.	$ 8.50
16 in.	$11.50

Is price a quadratic function of diameter? It probably should be, because the price should be proportional to the area, and the area is a quadratic function of the diameter. (The area of a circular region is given by $A = \pi r^2$ or $(\pi/4) \cdot d^2$.)

a) Express price as a quadratic function of diameter using the data points $(8, 6), (12, 8.50)$, and $(16, 11.50)$. $P(d) = \frac{1}{64}d^2 + \frac{5}{16}d + \frac{5}{2}$

b) Use the function to find the price of a 14-in. pizza. $9.94

35. *Hydrology.* The drawing below shows the cross section of a river. Typically rivers are deepest in the middle, with the depth decreasing to 0 at the edges. A hydrologist measures the depths D, in feet, of a river at distances x, in feet, from one bank. The results are listed in the table below.

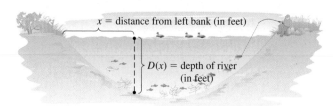

x = distance from left bank (in feet)

$D(x)$ = depth of river (in feet)

Distance x, from the Left Bank (in feet)	Depth, D, of the River (in feet)
0	0
15	10.2
25	17
50	20
90	7.2
100	0

$D(x) = -0.0082833093x^2 + 0.8242996891x + 0.2121786608$
a) Use regression to find a quadratic function that fits the data.

b) Use the function to estimate the depth of the river 70 ft from the left bank.   17.325 ft

36. *Work Force.*   The graph in Exercise 22 indicates that the percent of the U.S. population age 65 and over that is in the work force is increasing after several decades of decrease.

a) Use regression to find a quadratic function that can be used to estimate the percent p of the population 65 and over that is in the work force x years after 1960. $W(x) = 0.0111428571x^2 - 0.6637142857x + 21.42857143$

b) Use the function found in part (a) to predict the percent of those 65 and over who will be in the work force in 2010.   16.1%

37. *Alternative Fueled Vehicles.*   The number of cars fueled by electricity in the United States during several years is shown in the table below.

Year	Number of Cars
1992	1,607
1994	2,224
1996	3,280
1998	5,243
2000	11,834
2002	19,755

Source: www.eia.doe.gov

$c(x) = 261.875x^2 - 882.5642857x + 2134.571429$
As we saw in Example 3, a quadratic function can be used to model this data.

a) Use regression to find a quadratic function that can be used to estimate the number of cars c that are fueled by electricity x years after 1992.

b) Use the function found in part (a) to predict the number of cars fueled by electricity in 2004.   29,254 cars

(a) $h(x) = 1.577142857x^2 - 11.08571429x + 50.82857143$
38. *Public Health Care.*   The graph in Exercise 25 indicates that the per-capita expenditures for public health care have increased in a way that can be modeled by a quadratic function.

a) Use regression to find a quadratic function h that can be used to estimate the per-capita public health care expenditure x years after 1960.

b) Use the function found in part (a) to predict the per-capita public health care expenditure in 2010. $3439

ᵀᵂ 39. Does every nonlinear function have a minimum or maximum value? Why or why not?

ᵀᵂ 40. Explain how the leading coefficient of a quadratic function can be used to determine if a maximum or a minimum function value exists.

Skill Maintenance

Simplify.

41. $\dfrac{x}{x^2 + 17x + 72} - \dfrac{8}{x^2 + 15x + 56}$ [6.2] $\dfrac{x - 9}{(x + 9)(x + 7)}$

42. $\dfrac{x^2 - 9}{x^2 - 8x + 7} \div \dfrac{x^2 + 6x + 9}{x^2 - 1}$ [6.1] $\dfrac{(x - 3)(x + 1)}{(x - 7)(x + 3)}$

Solve. [4.1]

43. $5x - 9 < 31$ $\{x \mid x < 8\}$, or $(-\infty, 8)$

44. $3x - 8 \geq 22$ $\{x \mid x \geq 10\}$, or $[10, \infty)$

45. *Registered Nurses' Salary.*   The average annual salary of full-time registered nurses is shown in Exercise 26. Use the points $(4, 23{,}505)$ and $(20, 46{,}782)$ to find a linear function that can be used to estimate the average annual salary r of registered nurses x years after 1980. [2.6]
$r(x) = 1454.8125x + 17{,}685.75$

46. Use the data in Exercise 26 and regression to find a linear function that can be used to predict the average annual salary r of registered nurses x years after 1980. [2.6] $r(x) = 1514.092857x + 17{,}505.2381$

Synthesis

ᵀᵂ 47. Write a problem for a classmate to solve. Design the problem so that its solution requires finding a minimum or maximum function value.

ᵀᵂ 48. Explain what restrictions should be placed on the quadratic functions developed in Exercises 31 and 34 and why such restrictions are needed.

49. *Norman Window.*   A *Norman window* is a rectangle with a semicircle on top. Big Sky Windows is designing a Norman window that will require 24 ft of trim. What dimensions will allow the maximum amount of light to enter a house?

50. *Minimizing Area.*   A 36-in. piece of string is cut into two pieces. One piece is used to form a circle while the other is used to form a square. How should the string be cut so that the sum of the areas is a minimum?

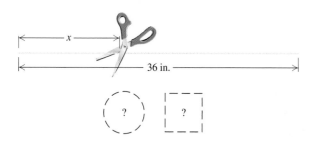

51. *Crop Yield.*   An orange grower finds that she gets an average yield of 40 bushels (bu) per tree when she plants 20 trees on an acre of ground. Each time she adds a tree to an acre, the yield per tree decreases by 1 bu, due to congestion. How many trees per acre should she plant for maximum yield?
30 trees per acre

52. *Cover Charges.*   When the owner of Sweet Sounds charges a $10 cover charge, an average of 80 people will attend a show. For each 25¢ increase in admission price, the average number attending decreases by 1. What should the owner charge in order to make the most money? $15

53. *Trajectory of a Launched Object.*   The height above the ground of a launched object is a quadratic function of the time that it is in the air. Suppose that a flare is launched from a cliff 64 ft above sea level. If 3 sec after being launched the flare is again level with the cliff, and if 2 sec after that it lands in the sea, what is the maximum height that the flare will reach?   78.4 ft

54. *Bridge Design.*   The cables supporting a straight-line suspension bridge are nearly parabolic in shape. Suppose that a suspension bridge is being designed with concrete supports 160 ft apart and with vertical cables 30 ft above road level at the midpoint of the bridge and 80 ft above road level at a point 50 ft from the midpoint of the bridge. How long are the longest vertical cables?   158 ft

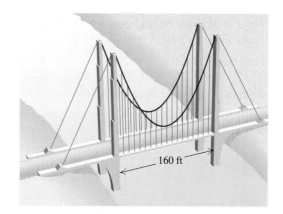

160 ft

---

49. The radius of the circular portion of the window and the height of the rectangular portion should each be $\dfrac{24}{\pi + 4}$ ft.

50. Length of piece used to form circle: $\dfrac{36\pi}{4 + \pi}$ in., length of piece used to form square: $\dfrac{144}{4 + \pi}$ in.

Quadratic Counter Settings

Focus: Modeling quadratic functions
Time: 20–30 minutes
Group size: 3 or 4
Materials: Graphing calculators are optional.

The Panasonic Portable Stereo System RX-DT680® has a counter for finding locations on an audio cassette. When a fully wound cassette with 45 min of music on a side begins to play, the counter is at 0. After 15 min of music has played, the counter reads 250, and after 35 min, it reads 487. When the 45-min side is finished playing, the counter reads 590.

ACTIVITY

1. The paragraph above describes four ordered pairs of the form (counter number, minutes played). Three pairs are enough to find a function of the form

$$T(n) = an^2 + bn + c,$$

where $T(n)$ is the time, in minutes, that the tape has run at counter reading n hundred. Each group member should select a different set of three points from the four given and then fit a quadratic function to the data.

2. Of the 3 or 4 functions found in part (1) above, which fits the data "best"? One way to answer this is to see how well each function predicts other pairs. The same counter used above reads 432 after a 45-min tape has played for 30 min. Which function comes closest to predicting this?

3. *Optional*: Use regression to fit a quadratic function to the data. How well does this function predict the reading after 30 min?

4. If a class member has access to a Panasonic System RX-DT680, see how well the functions developed above predict the counter readings for a tape that has played for 5 or 10 min.

8.9

Quadratic and Other Polynomial Inequalities ■ Rational Inequalities

Polynomial and Rational Inequalities

Quadratic and Other Polynomial Inequalities

Inequalities like the following are called *polynomial inequalities*:

$$x^3 - 5x > x^2 + 7, \qquad 4x - 3 < 9, \qquad 5x^2 - 3x + 2 \geq 0.$$

Second-degree polynomial inequalities in one variable are called *quadratic inequalities*. To solve polynomial inequalities, we often focus attention on where the outputs of a polynomial function are positive and where they are negative.

EXAMPLE 1   Solve: $x^2 + 3x - 10 > 0$.

Solution   Consider the "related" function $f(x) = x^2 + 3x - 10$ and its graph. Its graph opens upward since the leading coefficient is positive.

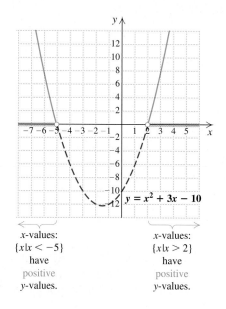

x-values:
$\{x | x < -5\}$
have
positive
y-values.

x-values:
$\{x | x > 2\}$
have
positive
y-values.

Values of y will be positive to the left and right of the x-intercepts, as shown. To find the intercepts, we set the polynomial equal to 0 and solve:

$$x^2 + 3x - 10 = 0$$
$$(x + 5)(x - 2) = 0$$
$$x + 5 = 0 \quad or \quad x - 2 = 0$$
$$x = -5 \quad or \quad x = 2.$$

Thus the solution set of the inequality is

$$\{x \,|\, x < -5 \,or\, x > 2\}, \quad or \quad (-\infty, -5) \cup (2, \infty).$$

Any inequality with 0 on one side can be solved by considering a graph of the related function and finding intercepts as in Example 1. Sometimes the quadratic formula is needed to find the intercepts.

EXAMPLE 2   Solve: $x^2 - 2x \le 2$.

Solution   We first find standard form with 0 on one side:

$$x^2 - 2x - 2 \le 0. \qquad \textbf{This is equivalent to the original inequality.}$$

The graph of $f(x) = x^2 - 2x - 2$ is a parabola opening upward. Values of $f(x)$ are negative for x-values between the x-intercepts. We find the x-intercepts by solving $f(x) = 0$:

$$x = \frac{-b \pm \sqrt{b^2 - 4ac}}{2a}$$

$$x = \frac{-(-2) \pm \sqrt{(-2)^2 - 4 \cdot 1(-2)}}{2 \cdot 1} \qquad \textbf{Substituting}$$

$$x = \frac{2 \pm \sqrt{12}}{2} = \frac{2 \pm 2\sqrt{3}}{2} = 1 \pm \sqrt{3}.$$

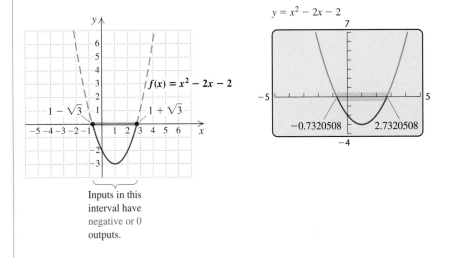

We can also find the x-intercepts using a graphing calculator. Because the numbers are irrational, this procedure will result in an approximation.

At the x-intercepts, $1 - \sqrt{3}$ and $1 + \sqrt{3}$, the value of $f(x)$ is 0. Thus the solution set of the inequality is

$$\left[1 - \sqrt{3}, 1 + \sqrt{3}\right], \quad \text{or} \quad \left\{x \mid 1 - \sqrt{3} \le x \le 1 + \sqrt{3}\right\}.$$

If the solutions are found graphically, the solution set is approximately $[-0.7320508, 2.7320508]$.

Note that in Example 2 the important information came from the location of the x-intercepts and the sign of $f(x)$ on each side of those intercepts.

In the next example, we solve a third-degree polynomial inequality, without graphing, by locating the x-intercepts, or *zeros*, of f and then using *test points* to determine the sign of $f(x)$ over each interval of the x-axis.

TEACHING TIP

You may want to point out that a polynomial function does not always change signs at a zero. Sketch some graphs as examples.

EXAMPLE 3   Solve: $5x^3 + 10x^2 - 15x > 0$.

Solution   We first solve the related equation:

$$5x^3 + 10x^2 - 15x = 0$$
$$5x(x^2 + 2x - 3) = 0$$
$$5x(x + 3)(x - 1) = 0$$
$$5x = 0 \quad or \quad x + 3 = 0 \quad or \quad x - 1 = 0$$
$$x = 0 \quad or \quad x = -3 \quad or \quad x = 1.$$

If $f(x) = 5x^3 + 10x^2 - 15x$, then the zeros of f are $-3, 0$, and 1. These zeros divide the number line, or x-axis, into four intervals: A, B, C, and D.

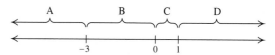

Next, using one convenient test value from each interval, we determine the sign of $f(x)$ for that interval. Within each interval, the sign of $f(x)$ cannot change. If it did, there would need to be another zero in that interval.

We choose -4 for a test value from interval A, -1 from interval B, 0.5 from interval C, and 2 from interval D. We enter $y_1 = 5x^3 + 10x^2 - 15x$ and use a table to evaluate the polynomial for each test value.

X	Y₁
−4	−100
−1	20
.5	−4.375
2	50

X = −4

We are interested only in the signs of each function value. From the table, we see that $f(-4)$ and $f(0.5)$ are negative and that $f(-1)$ and $f(2)$ are positive. We indicate on the number line the sign of $f(x)$ in each interval.

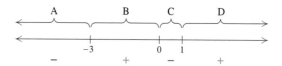

Recall that we are looking for all x for which $5x^3 + 10x^2 - 15x > 0$. The calculations above indicate that $f(x)$ is positive for any number in intervals B and D. The solution set of the original inequality is

$$(-3, 0) \cup (1, \infty), \quad or \quad \{x \mid -3 < x < 0 \: or \: x > 1\}.$$

The method of Example 3 works because polynomial function values can change signs only when the graph of the function crosses the x-axis. The graph of $p(x) = 5x^3 + 10x^2 - 15x$ illustrates the solution of Example 3. We can see from the graph at left that the function values are positive in the intervals $(-3, 0)$ and $(1, \infty)$.

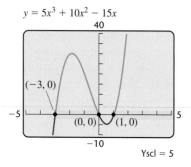

$y = 5x^3 + 10x^2 - 15x$
Yscl = 5

EXAMPLE 4 Solve: $x^4 + x^3 - 2x^2 \leq 0$.

Solution We first solve the related equation:

$$x^4 + x^3 - 2x^2 = 0$$
$$x^2(x^2 + x - 2) = 0$$
$$x^2(x + 2)(x - 1) = 0$$
$$x^2 = 0 \quad or \quad x + 2 = 0 \quad or \quad x - 1 = 0$$
$$x = 0 \quad or \quad x = -2 \quad or \quad x = 1.$$

The function $p(x) = x^4 + x^3 - 2x^2$ has zeros at -2, 0, and 1. We graph the function and determine the sign of $p(x)$ over each interval of the number line.

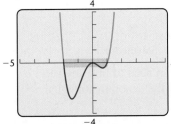 

We see from the graph that $p(x)$ is negative in the intervals $(-2, 0)$ and $(0, 1)$. Since the inequality is also true when the function is 0, we include the endpoints of the intervals. The solution is

$$[-2, 0] \cup [0, 1], \quad \text{or simply} \quad [-2, 1].$$

To Solve a Polynomial Inequality Using Factors

1. Get 0 on one side and solve the related polynomial equation $p(x) = 0$ by factoring.

2. Use the numbers found in step (1) to divide the number line into intervals.

3. Using a test value from each interval or the graph of the related function, determine the sign of $p(x)$ over each interval.

4. Select the interval(s) for which the inequality is satisfied and write set-builder notation or interval notation for the solution set. Include the endpoints of the intervals when $\leq$ or $\geq$ is used.

Note that if the polynomial cannot be factored, a graphing calculator can be used to at least approximate any x-intercepts of the graph of the function, as in Example 2.

Rational Inequalities

Inequalities involving rational expressions are called **rational inequalities**. Like polynomial inequalities, rational inequalities can be solved using test values. Unlike polynomials, however, rational expressions often have values for which the expression is undefined.

EXAMPLE 5 Solve: $\dfrac{x-3}{x+4} \geq 2$.

Solution We write the related equation by changing the $\geq$ symbol to $=$:

$$\dfrac{x-3}{x+4} = 2.$$

We show both algebraic and graphical solutions.

Algebraic Solution

We first solve the related equation:

$$(x+4) \cdot \dfrac{x-3}{x+4} = (x+4) \cdot 2 \qquad \text{\textbf{Multiplying both sides by the LCD, } } x+4$$

$$x - 3 = 2x + 8$$

$$-11 = x. \qquad \text{\textbf{Solving for } } x$$

In the case of rational inequalities, we also need to find any values that make the denominator 0. We set the denominator equal to 0 and solve:

$$\left. \begin{array}{l} x + 4 = 0 \\ x = -4. \end{array} \right\} \quad \begin{array}{l} \text{\textbf{This tells us that } } -4 \text{ \textbf{is not in the domain}} \\ \text{\textbf{of } } f \text{ \textbf{if } } f(x) = \dfrac{x-3}{x+4}. \end{array}$$

Now we use -11 and -4 to divide the number line into intervals:

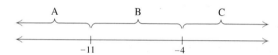

We test a number in each interval to see where the original inequality is satisfied:

$$\dfrac{x-3}{x+4} \geq 2.$$

If we enter $y_1 = (x-3)/(x+4)$ and $y_2 = 2$, the inequality is satisfied when $y_1 \geq y_2$.

X	Y₁	Y₂
−15	1.6364	2
−8	2.75	2
1	−.4	2

$y_1 < y_2$; -15 *is not* a solution.
$y_1 > y_2$; -8 *is* a solution.
$y_1 < y_2$; 1 *is not* a solution.

X =

The solution set includes the interval B. The endpoint -11 is included because the inequality symbol is $\geq$ and -11 is a solution of the related equation. The number -4 is *not* included because $(x-3)/(x+4)$ is undefined for $x = -4$. Thus the solution set of the original inequality is

$$[-11, -4), \quad \text{or} \quad \{x \mid -11 \leq x < -4\}.$$

Graphical Solution

We graph $y_1 = (x-3)/(x+4)$ and $y_2 = 2$ using DOT mode. The inequality is true for those values of x for which $y_1 \geq y_2$.

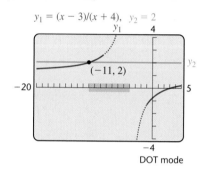

DOT mode

We see that $y_1 = y_2$ when $x = -11$, and $y_1 > y_2$ when $-11 < x < -4$. Thus the solution set is $[-11, -4)$.

To Solve a Rational Inequality

1. Change the inequality symbol to an equals sign and solve the related equation.

2. Find any replacements for which the rational expression is undefined.

3. Use the numbers found in steps (1) and (2) to divide the number line into intervals.

4. Substitute a test value from each interval into the inequality. If the number is a solution, then the interval to which it belongs is part of the solution set.

5. Select the interval(s) and any endpoints for which the inequality is satisfied and write set-builder or interval notation for the solution set. If the inequality symbol is ≤ or ≥, then the solutions from step (1) are also included in the solution set. Those numbers found in step (2) should be excluded from the solution set, even if they are solutions from step (1).

To Solve a Rational Inequality: Graphical Approach

1. Let y_1 = one side of the inequality and y_2 = the other side of the inequality. Examine the graph to determine the intervals that satisfy the inequality.

2. Select the intervals for which the inequality is satisfied. If the inequality symbol is ≤ or ≥, include any endpoints that also satisfy the inequality. Write set-builder or interval notation for the solution set.

8.9

Exercise Set

FOR EXTRA HELP

Digital Video Tutor CD 8 Videotape 12 | Student's Solutions Manual | Tutor Center AW Math Tutor Center | InterAct Math | Math XP MathXL | MyMathLab

Determine the solution set of each inequality from the given graph.

1. $p(x) \leq 0$

$\left[-4, \frac{3}{2}\right]$

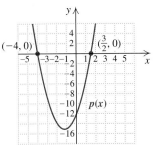

2. $p(x) < 0$

$\left(-4, -\frac{2}{3}\right)$

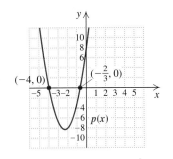

3. $x^4 + 12x > 3x^3 + 4x^2$ $(-\infty, -2) \cup (0, 2) \cup (3, \infty)$

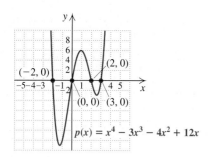

$p(x) = x^4 - 3x^3 - 4x^2 + 12x$

4. $x^4 + x^3 \geq 6x^2$ $(-\infty, -3] \cup \{0\} \cup [2, \infty)$

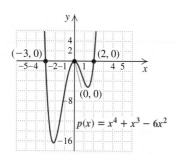

$p(x) = x^4 + x^3 - 6x^2$

5. $\dfrac{x - 1}{x + 2} < 3$ $\left(-\infty, -\frac{7}{2}\right) \cup (-2, \infty)$

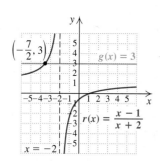

$r(x) = \dfrac{x - 1}{x + 2}$

6. $\dfrac{2x - 1}{x - 5} \geq 1$ $(-\infty, -4] \cup (5, \infty)$

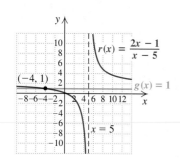

$r(x) = \dfrac{2x - 1}{x - 5}$

Solve.

7. $(x + 4)(x - 3) < 0$ $(-4, 3), \text{ or } \{x \mid -4 < x < 3\}$

8. $(x - 5)(x + 2) > 0$ $(-\infty, -2) \cup (5, \infty), \text{ or } \{x \mid x < -2 \text{ or } x > 5\}$

9. $(x + 7)(x - 2) \geq 0$ ▢

10. $(x - 1)(x + 4) \leq 0$ $[-4, 1], \text{ or } \{x \mid -4 \leq x \leq 1\}$

11. $x^2 - x - 2 < 0$ $(-1, 2), \text{ or } \{x \mid -1 < x < 2\}$

12. $x^2 + x - 2 < 0$ $(-2, 1), \text{ or } \{x \mid -2 < x < 1\}$

13. $25 - x^2 \geq 0$ $[-5, 5], \text{ or } \{x \mid -5 \leq x \leq 5\}$

14. $4 - x^2 \geq 0$ $[-2, 2], \text{ or } \{x \mid -2 \leq x \leq 2\}$

Aha! **15.** $x^2 + 4x + 4 < 0$ $\varnothing$

16. $x^2 + 6x + 9 < 0$ $\varnothing$

17. $x^2 - 4x < 12$ $(-2, 6), \text{ or } \{x \mid -2 < x < 6\}$

18. $x^2 + 6x > -8$ ▢

19. $3x(x + 2)(x - 2) < 0$ $(-\infty, -2) \cup (0, 2), \text{ or } \{x \mid x < -2 \text{ or } 0 < x < 2\}$

20. $5x(x + 1)(x - 1) > 0$ $(-1, 0) \cup (1, \infty), \text{ or } \{x \mid -1 < x < 0 \text{ or } x > 1\}$

21. $(x + 3)(x - 2)(x + 1) > 0$ $(-3, -1) \cup (2, \infty), \text{ or } \{x \mid -3 < x < -1 \text{ or } x > 2\}$

22. $(x - 1)(x + 2)(x - 4) < 0$ $(-\infty, -2) \cup (1, 4), \text{ or } \{x \mid x < -2 \text{ or } 1 < x < 4\}$

23. $(x + 3)(x + 2)(x - 1) < 0$ $(-\infty, -3) \cup (-2, 1), \text{ or } \{x \mid x < -3 \text{ or } -2 < x < 1\}$

24. $(x - 2)(x - 3)(x + 1) < 0$ $(-\infty, -1) \cup (2, 3), \text{ or } \{x \mid x < -1 \text{ or } 2 < x < 3\}$

25. $4.32x^2 - 3.54x - 5.34 \leq 0$ ▢

26. $7.34x^2 - 16.55x - 3.89 \geq 0$ ▢

27. $x^3 - 2x^2 - 5x + 6 < 0$ ▢

28. $\frac{1}{3}x^3 - x + \frac{2}{3} > 0$ ▢

29. $\dfrac{1}{x + 3} < 0$ $(-\infty, -3), \text{ or } \{x \mid x < -3\}$

30. $\dfrac{1}{x + 4} > 0$ $(-4, \infty), \text{ or } \{x \mid x > -4\}$

31. $\dfrac{x + 1}{x - 5} \geq 0$ ▢

32. $\dfrac{x - 2}{x + 5} \leq 0$ $(-5, 2], \text{ or } \{x \mid -5 < x \leq 2\}$

33. $\dfrac{3x + 2}{2x - 4} \leq 0$ ▢

34. $\dfrac{5 - 2x}{4x + 3} \leq 0$ ▢

35. $\dfrac{x + 1}{x + 6} > 1$ $(-\infty, -6), \text{ or } \{x \mid x < -6\}$

36. $\dfrac{x - 1}{x - 2} < 1$ $(-\infty, 2), \text{ or } \{x \mid x < 2\}$

37. $\dfrac{(x - 2)(x + 1)}{x - 5} \leq 0$ ▢

38. $\dfrac{(x + 4)(x - 1)}{x + 3} \geq 0$ ▢

39. $\dfrac{x}{x + 3} \geq 0$ ▢

40. $\dfrac{x - 2}{x} \leq 0$ $(0, 2], \text{ or } \{x \mid 0 < x \leq 2\}$

41. $\dfrac{x - 5}{x} < 1$ $(0, \infty), \text{ or } \{x \mid x > 0\}$

42. $\dfrac{x}{x - 1} > 2$ $(1, 2), \text{ or } \{x \mid 1 < x < 2\}$

43. $\dfrac{x - 1}{(x - 3)(x + 4)} \leq 0$ ▢

44. $\dfrac{x + 2}{(x - 2)(x + 7)} \geq 0$ ▢

45. $4 < \dfrac{1}{x}$ ▢

46. $\dfrac{1}{x} \leq 5$ ▢

▢ Answers to Exercises 9, 18, 25–28, 31, 33, 34, 37–39, and 43–46 can be found on p. A-67.

ᵀᵂ **47.** Explain how any quadratic inequality can be solved by examining a parabola.

ᵀᵂ **48.** Describe a method for creating a quadratic inequality for which there is no solution.

Skill Maintenance

Simplify. [1.4]

49. $(2a^3b^2c^4)^3$ $8a^9b^6c^{12}$ **50.** $(5a^4b^7)^2$ $25a^8b^{14}$

51. 2^{-5} $\frac{1}{32}$ **52.** 3^{-4} $\frac{1}{81}$

53. If $f(x) = 3x^2$, find $f(a + 1)$. [5.2] $3a^2 + 6a + 3$

54. If $g(x) = 5x - 3$, find $g(a + 2)$. [2.1] $5a + 7$

Synthesis

ᵀᵂ **55.** Step (5) on p. 639 states that even when the inequality symbol is ≤ or ≥, the solutions from step (1) are not always part of the solution set. Why?

ᵀᵂ **56.** Describe a method that could be used to create quadratic inequalities that have $(-\infty, a] \cup [b, \infty)$ as the solution set.

Find each solution set.

57. $x^2 + 2x < 5$ ☐

58. $x^4 + 2x^2 \geq 0$ $(-\infty, \infty)$, or $\mathbb{R}$

59. $x^4 + 3x^2 \leq 0$ {0}

60. $\left|\dfrac{x + 2}{x - 1}\right| \leq 3$ ☐

61. *Total Profit.*   Derex, Inc., determines that its total-profit function is given by

$$P(x) = -3x^2 + 630x - 6000.$$

a) Find all values of x for which Derex makes a profit. $(10, 200)$, or $\{x \mid 10 < x < 200\}$

b) Find all values of x for which Derex loses money. $[0, 10) \cup (200, \infty)$, or $\{x \mid 0 \leq x < 10 \text{ or } x > 200\}$

62. *Height of a Thrown Object.*   The function

$$S(t) = -16t^2 + 32t + 1920$$

gives the height S, in feet, of an object thrown from a cliff that is 1920 ft high. Here t is the time, in seconds, that the object is in the air.

a) For what times does the height exceed 1920 ft? $\{t \mid 0 \text{ sec} < t < 2 \text{ sec}\}$

b) For what times is the height less than 640 ft? $\{t \mid t > 10 \text{ sec}\}$

63. *Number of Handshakes.*   There are n people in a room. The number N of possible handshakes by the people is given by the function

$$N(n) = \frac{n(n - 1)}{2}.$$ $\{n \mid n \text{ is an integer and } 12 \leq n \leq 25\}$

For what number of people n is $66 \leq N \leq 300$?

64. *Number of Diagonals.*   A polygon with n sides has D diagonals, where D is given by the function

$$D(n) = \frac{n(n - 3)}{2}.$$

Find the number of sides n if

$27 \leq D \leq 230$. $\{n \mid n \text{ is an integer and } 9 \leq n \leq 23\}$

Use a graphing calculator to graph each function and find solutions of $f(x) = 0$. Then solve the inequalities $f(x) < 0$ and $f(x) > 0$.

65. $f(x) = x + \dfrac{1}{x}$ ☐

66. $f(x) = x - \sqrt{x}, x \geq 0$ ☐

67. $f(x) = \dfrac{x^3 - x^2 - 2x}{x^2 + x - 6}$ ☐

68. $f(x) = x^4 - 4x^3 - x^2 + 16x - 12$ ☐

69. *Welfare.*   The number of welfare cases in Indiana began to increase in 2001 after having decreased for several years. The following chart lists the caseload for several years.

Year	Number of Indiana Welfare Cases
1994	73,000
1997	44,000
2000	32,000
2003	53,000

Source: Indiana Family and Social Services Administration

$$w(x) = 1388.888889x^2 - 14,900x + 73,800$$

a) Use regression to find a quadratic function that can be used to estimate the number w of welfare cases in Indiana x years after 1994.

b) Use the function to predict the years after 1994 for which the number of welfare cases is greater than 50,000.   From 1994 to about 2 years after 1994, or 1994 to 1996, and after about 9 years after 1994, or after 2003

☐ Answers to Exercises 57, 60, and 65–68 can be found on p. A-67.

8 Chapter Summary and Review

Key Terms

Quadratic equation, p. 562
Quadratic function, p. 562
Parabola, p. 562
Principle of square roots, p. 564
Completing the square, p. 567
Compounding interest annually, p. 571

The quadratic formula, p. 576
Discriminant, p. 590
Reducible to quadratic, p. 595
Axis of symmetry, p. 601
Vertex, p. 601
Translated, p. 603
Minimum value, p. 603

Maximum value, p. 603
Reflection, p. 604
Polynomial inequality, p. 633
Quadratic inequality, p. 633
Zero, p. 635
Rational inequality, p. 637

Important Properties and Formulas

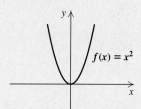

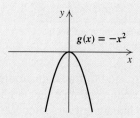

The Principle of Square Roots

For any real number k, if $x^2 = k$, then $x = \sqrt{k}$ or $x = -\sqrt{k}$.

For any real number k and any algebraic expression X, if $X^2 = k$, then $X = \sqrt{k}$ or $X = -\sqrt{k}$.

To complete the square for $x^2 + bx$, add $(b/2)^2$.

To solve a quadratic equation in x by completing the square:

1. Isolate the terms with variables on one side of the equation, and arrange them in descending order.
2. Divide both sides by the coefficient of x^2 if that coefficient is not 1.
3. Complete the square by taking half of the coefficient of x and adding its square to both sides.
4. Express one side as the square of a binomial and simplify the other side.
5. Use the principle of square roots.
6. Solve for x by adding or subtracting on both sides.

The Quadratic Formula

The solutions of $ax^2 + bx + c = 0$, $a \neq 0$, are given by

$$x = \frac{-b \pm \sqrt{b^2 - 4ac}}{2a}.$$

To solve a quadratic equation:

1. If the equation can be easily written in the form $ax^2 = p$ or $(x + k)^2 = d$, use the principle of square roots.

2. If step (1) does not apply, write the equation in $ax^2 + bx + c = 0$ form.

3. Try factoring and using the principle of zero products.

4. If factoring seems to be difficult or impossible, use the quadratic formula.

The solutions of a quadratic equation can always be found using the quadratic formula. They cannot always be found by factoring.

To solve a formula for a letter—say, b:

1. Clear fractions and use the principle of powers, as needed. (In some cases you may clear the fractions first, and in some cases you may use the principle of powers first.) Perform these steps until radicals containing b are gone and b is not in any denominator.

2. Combine all terms with b^2 in them. Also combine all terms with b in them.

3. If b^2 does not appear, you can solve by using just the addition and multiplication principles as in Sections 2.3 and 6.8.

4. If b^2 appears but b does not, solve the equation for b^2. Then use the principle of square roots to solve for b.

5. If there are terms containing both b and b^2, put the equation in standard form and use the quadratic formula.

Discriminant $b^2 - 4ac$	Nature of Solutions
0	One solution; a rational number
Positive	Two different real-number solutions
Perfect square	Solutions are rational.
Not a perfect square	Solutions are irrational conjugates.
Negative	Two different imaginary-number solutions (complex conjugates)

The graph of $g(x) = ax^2$ is a parabola with $x = 0$ as its axis of symmetry; its vertex is the origin.

For $a > 0$, the parabola opens upward. For $a < 0$, the parabola opens downward.

If $|a|$ is greater than 1, the parabola is narrower than $f(x) = x^2$.

If $|a|$ is between 0 and 1, the parabola is wider than $f(x) = x^2$.

The graph of $f(x) = a(x - h)^2$ has the same shape as the graph of $y = ax^2$.

If h is positive, the graph of $y = ax^2$ is shifted h units to the right.

If h is negative, the graph of $y = ax^2$ is shifted $|h|$ units to the left.

The vertex is $(h, 0)$, and the axis of symmetry is $x = h$.

The graph of $f(x) = a(x - h)^2 + k$ has the same shape as the graph of $y = a(x - h)^2$.

If k is positive, the graph of $y = a(x - h)^2$ is shifted k units up.

If k is negative, the graph of $y = a(x - h)^2$ is shifted $|k|$ units down.

The vertex is (h, k), and the axis of symmetry is $x = h$.

For $a > 0$, the range of f is $[k, \infty)$ and k is the minimum function value. For $a < 0$, the range of f is $(-\infty, k]$ and k is the maximum function value.

Formulas

Compound interest: $A = P(1 + r)^t$
Free-fall distance
(in feet): $s = 16t^2$

The vertex of the parabola given by $f(x) = ax^2 + bx + c$ is

$$\left(-\frac{b}{2a}, f\left(-\frac{b}{2a}\right)\right), \quad \text{or} \quad \left(-\frac{b}{2a}, \frac{4ac - b^2}{4a}\right).$$

The x-coordinate of the vertex is $-b/(2a)$. The axis of symmetry is $x = -b/(2a)$.

To solve a polynomial inequality using factors:

1. Get 0 on one side and solve the related polynomial equation $p(x) = 0$ by factoring.

2. Use the numbers found in step (1) to divide the number line into intervals.

3. Using a test value from each interval or the graph of the related function, determine the sign of each factor over each interval.

4. Select the interval(s) for which the inequality is satisfied and write set-builder or interval notation for the solution set. Include the endpoints of the intervals when $\le$ or $\ge$ is used.

To solve a rational inequality:

1. Change the inequality symbol to an equals sign and solve the related equation.

2. Find any replacements for which the rational expression is undefined.

3. Use the numbers found in steps (1) and (2) to divide the number line into intervals.

4. Substitute a test value from each interval into the inequality. If the number is a solution, then the interval to which it belongs is part of the solution set. To solve graphically, let $y_1 =$ one side of the inequality and $y_2 =$ the other side of the inequality, and examine the graph to determine what intervals satisfy the inequality.

5. Select the interval(s) and any endpoints for which the inequality is satisfied and write set-builder or interval notation for the solution set. If the inequality symbol is $\le$ or $\ge$, then the solutions to step (1) are also included in the solution set. Those numbers found in step (2) should be excluded from the solution set, even if they are solutions from step (1).

Review Exercises

1. Given the following graph of $f(x) = ax^2 + bx + c$.
   a) State the number of real-number solutions of $ax^2 + bx + c = 0$. [8.1] 2
   b) State whether a is positive or negative. [8.6] Positive
   c) Determine the minimum value of f. [8.6] -3

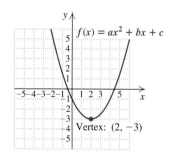

$f(x) = ax^2 + bx + c$

Vertex: $(2, -3)$

Solve.   [8.1] $0, -\frac{5}{14}$

2. $2x^2 - 7 = 0$

3. $14x^2 + 5x = 0$

4. $x^2 - 12x + 36 = 9$

5. $x^2 - 5x + 9 = 0$

[8.1] 3, 9

[8.2] $\dfrac{5 \pm i\sqrt{11}}{2}$

2. [8.1] $\pm\sqrt{\dfrac{7}{2}}$, or $\pm\dfrac{\sqrt{14}}{2}$

6. $x(3x + 4) = 4x(x - 1) + 15$ [8.2] 3, 5

7. $x^2 + 9x = 1$ [8.2] $\dfrac{-9 \pm \sqrt{85}}{2}$

8. $x^2 - 5x - 2 = 0$. Use a calculator to approximate the solutions with rational numbers to the nearest ten-thousandth. [8.2] $-0.3723, 5.3723$

9. Let $f(x) = 4x^2 - 3x - 1$. Find x such that $f(x) = 0$. [8.2] $-\frac{1}{4}$, 1

Complete the square. Then write the perfect-square trinomial in factored form. [8.1] $x^2 + \frac{3}{5}x + \frac{9}{100}$; $\left(x + \frac{3}{10}\right)^2$

10. $x^2 - 12x$ [8.1] $x^2 - 12x + 36$; $(x - 6)^2$

11. $x^2 + \frac{3}{5}x$

12. Solve by completing the square. Show your work.
   $$x^2 - 6x + 1 = 0$$ [8.1] $3 \pm 2\sqrt{2}$

13. \$2500 grows to \$3025 in 2 yr. Use the formula $A = P(1 + r)^t$ to find the interest rate. [8.1] 10%

14. The Peachtree Center Plaza in Atlanta, Georgia, is 723 ft tall. Use the formula $s = 16t^2$ to approximate how long it would take an object to fall from the top. [8.1] 6.7 sec

Solve.

15. A corporate pilot must fly from company headquarters to a manufacturing plant and back in 4 hr. The distance between headquarters and the plant is 300 mi. If there is a 20-mph headwind going and a 20-mph tailwind returning, how fast must the plane be able to travel in still air? [8.3] About 153 mph

16. Working together, Erica and Shawna can answer a day's worth of technical support questions in 4 hr. Working alone, Erica takes 6 hr longer than Shawna. How long would it take Shawna to answer the questions alone? [8.3] 6 hr

For each equation, determine what type of number the solutions will be.

17. $x^2 + 3x - 6 = 0$ [8.4] Two irrational

18. $x^2 + 2x + 5 = 0$ [8.4] Two imaginary

19. Write a quadratic equation having the solutions $\sqrt{5}$ and $-\sqrt{5}$. [8.4] $x^2 - 5 = 0$

20. Write a quadratic equation having -4 as its only solution. [8.4] $x^2 + 8x + 16 = 0$

21. Find all x-intercepts of the graph of
$$f(x) = x^4 - 13x^2 + 36.$$
[8.5] $(-3, 0), (-2, 0), (2, 0), (3, 0)$

Solve.

22. $15x^{-2} - 2x^{-1} - 1 = 0$ [8.5] $-5, 3$

23. $(x^2 - 4)^2 - (x^2 - 4) - 6 = 0$ [8.5] $\pm\sqrt{2}, \pm\sqrt{7}$

24. a) Graph: $f(x) = -3(x + 2)^2 + 4$. ▫
b) Label the vertex.
c) Draw the axis of symmetry.
d) Find the maximum or the minimum value.
Maximum: 4

25. For the function $f(x) = 2x^2 - 12x + 33$:
a) find the vertex and the axis of symmetry; ▫
b) graph the function. ▫

26. Find the x- and y-intercepts of
$$f(x) = x^2 - 9x + 14.$$ [8.7] $(2, 0), (7, 0); (0, 14)$

27. Solve $N = 3\pi\sqrt{1/p}$ for p. [8.3] $p = \dfrac{9\pi^2}{N^2}$

28. Solve $2A + T = 3T^2$ for T. [8.3] $T = \dfrac{1 \pm \sqrt{1 + 24A}}{6}$

▫ Answers to Exercises 24 and 25 can be found on p. A-67.

State whether the graph appears to represent a quadratic function. Explain your reasoning.

29. [8.8] Quadratic; the data approximate a parabola opening upward.

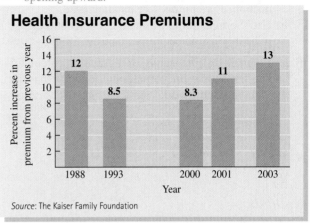

Source: The Kaiser Family Foundation

30.

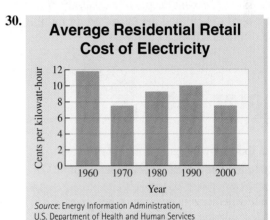

Source: Energy Information Administration, U.S. Department of Health and Human Services

[8.8] Not quadratic; the data do not approximate a parabola.

31. Eastgate Consignments wants to build an area for children to play in while their parents shop. They have 30 ft of low fencing. What is the maximum area they can enclose? What dimensions will yield this area? [8.8] 56.25 ft²; 7.5 ft by 7.5 ft

32. The following table lists the percent increase in health insurance premiums from the previous year. (See Exercise 29.)

Year	Percent Increase in Health Insurance Premiums
1988	12
1993	8.5
2000	8.3
2001	11
2003	13

$[8.8]\ h(x) = \frac{1}{8}x^2 - \frac{217}{120}x + 12$

a) Use the data points $(0, 12)$, $(12, 8.3)$, and $(15, 13)$ to find a quadratic function that can be used to estimate the percent increase h in health insurance premiums x years after 1988.

b) Use the function to estimate the percent increase in health insurance premiums from 1998 to 1999. About 7.2%

33. Use the REGRESSION feature of a graphing calculator and all the data to find a quadratic function that fits the data in Exercise 32. $[8.8]\ f(x) = 0.088311246x^2 - 1.286948848x + 12.19869334$, where x is the number of years after 1988

Solve.

34. $x^3 - 3x > 2x^2$
$[8.9]\ (-1, 0) \cup (3, \infty)$, or $\{x \mid -1 < x < 0\ or\ x > 3\}$

35. $\dfrac{x - 5}{x + 3} \le 0$
$[8.9]\ (-3, 5]$, or $\{x \mid -3 < x \le 5\}$

Synthesis

TW **36.** Explain how the x-intercepts of a quadratic function can be used to help find the maximum or minimum value of the function.

TW **37.** Suppose that the quadratic formula is used to solve a quadratic equation. If the discriminant is a perfect square, could factoring have been used to solve the equation? Why or why not?

TW **38.** What is the greatest number of solutions that an equation of the form $ax^4 + bx^2 + c = 0$ can have? Why?

TW **39.** Discuss two ways in which completing the square was used in this chapter.

40. A quadratic function has x-intercepts at -3 and 5. If the y-intercept is at -7, find an equation for the function. $[8.7]\ f(x) = \frac{7}{15}x^2 - \frac{14}{15}x - 7$

41. Find h and k if, for $3x^2 - hx + 4k = 0$, the sum of the solutions is 20 and the product is 80. $[8.4]\ h = 60,\ k = 60$

42. The average of two positive integers is 171. One of the numbers is the square root of the other. Find the integers. $[8.5]\ 18,\ 324$

Chapter Test 8

1. Given the following graph of $f(x) = ax^2 + bx + c$.

a) State the number of real-number solutions of $ax^2 + bx + c = 0$. $[8.1]\ 0$

b) State whether a is positive or negative. $[8.6]$ Negative

c) Determine the maximum function value of f. $[8.6]\ -1$

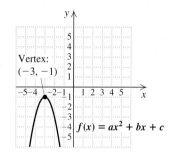

Vertex: $(-3, -1)$

$f(x) = ax^2 + bx + c$

Solve.

2. $3x^2 - 16 = 0$ $[8.1]\ \pm\dfrac{4\sqrt{3}}{3}$

3. $4x(x - 2) - 3x(x + 1) = -18$ $[8.2]\ 2, 9$

4. $x^2 + x + 1 = 0$ $[8.2]\ \dfrac{-1 \pm i\sqrt{3}}{2}$

5. $2x + 5 = x^2$ $[8.2]\ 1 \pm \sqrt{6}$

6. $x^{-2} - x^{-1} = \frac{3}{4}$ $[8.5]\ -2, \frac{2}{3}$

7. $x^2 + 3x = 5$. Use a calculator to approximate the solutions with rational numbers to the nearest ten-thousandth. $[8.2]\ -4.1926, 1.1926$

8. Let $f(x) = 12x^2 - 19x - 21$. Find x such that $f(x) = 0$. $[8.2]\ -\frac{3}{4}, \frac{7}{3}$

Complete the square. Then write the perfect-square trinomial in factored form. $[8.1]\ x^2 - \frac{2}{7}x + \frac{1}{49};\ \left(x - \frac{1}{7}\right)^2$

9. $x^2 + 14x$ $[8.1]\ x^2 + 14x + 49;\ (x + 7)^2$

10. $x^2 - \frac{2}{7}x$

11. Solve by completing the square. Show your work.
$x^2 + 10x + 15 = 0$ $[8.1]\ -5 \pm \sqrt{10}$

Solve.

12. The Connecticut River flows at a rate of 4 km/h for the length of a popular scenic route. In order for a cruiser to travel 60 km upriver and then return in a total of 8 hr, how fast must the boat be able to travel in still water? $[8.3]$ 16 km/h

13. Brock and Ian can assemble a swing set in $1\frac{1}{2}$ hr. Working alone, it takes Ian 4 hr longer than Brock to assemble the swing set. How long would it take Brock, working alone, to assemble the swing set? $[8.3]$ 2 hr

14. Determine the type of number that the solutions of $x^2 + 5x + 17 = 0$ will be.   [8.4] Two imaginary

15. Write a quadratic equation having solutions -2 and $\frac{1}{3}$.   [8.4] $3x^2 + 5x - 2 = 0$

16. Find all x-intercepts of the graph of
$$f(x) = (x^2 + 4x)^2 + 2(x^2 + 4x) - 3.\quad \square$$

17. a) Graph: $f(x) = 4(x - 3)^2 + 5$. $\square$
   b) Label the vertex.
   c) Draw the axis of symmetry.
   d) Find the maximum or the minimum function value.   Minimum: 5

18. For the function $f(x) = 2x^2 + 4x - 6$:
   a) find the vertex and the axis of symmetry; $\square$
   b) graph the function. $\square$

19. Find the x- and y-intercepts of
$$f(x) = x^2 - x - 6.\quad \text{[8.7] } (-2, 0), (3, 0); (0, -6)$$

20. Solve $V = \frac{1}{3}\pi(R^2 + r^2)$ for r.   [8.3] $r = \sqrt{\dfrac{3V}{\pi} - R^2}$

21. State whether the graph appears to represent a quadratic function. Explain your reasoning. $\square$

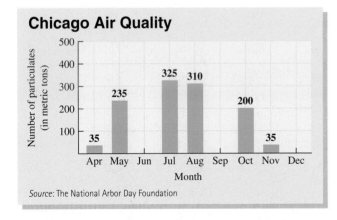

Chicago Air Quality

Source: The National Arbor Day Foundation

22. Jay's Custom Pickups has determined that when x hundred truck caps are built, the average cost per cap is given by
$$C(x) = 0.2x^2 - 1.3x + 3.4025,$$
where $C(x)$ is in hundreds of dollars. What is the minimum cost per truck cap and how many caps should be built to achieve that minimum?
[8.8] Minimum: \$129/cap when 325 caps are built

23. Trees improve air quality in part by retaining airborne particles, called particulates, from the air. The following table shows the number of metric tons of particulates retained by trees in Chicago during the spring, summer, and autumn months. (See Exercise 21.)

Month		Amount of Particulates Retained (in metric tons)
April	(0)	35
May	(1)	235
July	(3)	325
August	(4)	310
October	(6)	200
November	(7)	35

Source: The National Arbor Day Foundation

Use the data points $(0, 35)$, $(4, 310)$, and $(6, 200)$ to find a quadratic function that can be used to estimate the number of metric tons p of particulates retained x months after April.
[8.8] $p(x) = -\frac{165}{8}x^2 + \frac{605}{4}x + 35$

24. Use regression to find a quadratic function that fits the data in Exercise 23.   [8.8] $p(x) = -23.54166667x^2 + 162.2583333x + 57.61666667$

Solve.

25. $x^2 + 5x \le 6$ [8.9] $[-6, 1]$, or $\{x \mid -6 \le x \le 1\}$

26. $x - \dfrac{1}{x} > 0$ [8.9] $(-1, 0) \cup (1, \infty)$, or $\{x \mid -1 < x < 0 \text{ or } x > 1\}$

Synthesis

27. One solution of $kx^2 + 3x - k = 0$ is -2. Find the other solution. [8.4] $\frac{1}{2}$

28. Find a fourth-degree polynomial equation, with integer coefficients, for which $2 - \sqrt{3}$ and $5 - i$ are solutions.   [8.4] $x^4 - 14x^3 + 67x^2 - 114x + 26 = 0$; answers may vary

29. Find a polynomial equation, with integer coefficients, for which 5 is a repeated root and $\sqrt{2}$ and $\sqrt{3}$ are solutions.   [8.4] $x^6 - 10x^5 + 20x^4 + 50x^3 - 119x^2 - 60x + 150 = 0$; answers may vary

$\square$ Answers to Exercises 16–18 and 21 can be found on p. A-67.

Exponential and Logarithmic Functions

9

The functions that we consider in this chapter are interesting not only from a purely intellectual point of view, but also for their rich applications to many fields. We will look at applications such as compound interest and population growth, to name just two.

The basis of the theory centers on functions having variable exponents (*exponential functions*). Results follow from those functions and their properties.

APPLICATION

TULE ELK POPULATION.   In 1978, wildlife biologists introduced a herd of 10 Tule elk into the Point Reyes National Seashore near San Francisco. The size of the herd grew rapidly, as shown by the data in the following table. Predict the number of Tule elk in 2005. The data can be modeled by an *exponential function.*

Year	Tule Elk Population
1978	10
1982	24
1986	70
1996	200
2002	500

$y = 13.01608148(1.168547698)^x$

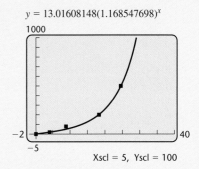

Xscl = 5, Yscl = 100

This problem appears as Example 9 in Section 9.7.

9.1

Composite Functions ■ Inverses and One-to-One Functions ■ Finding Formulas for Inverses ■ Graphing Functions and Their Inverses ■ Inverse Functions and Composition

Composite and Inverse Functions

Composite Functions

In the real world, functions frequently occur in which some quantity depends on a variable that, in turn, depends on another variable. For instance, a firm's profits may depend on the number of items the firm produces, which may in turn depend on the number of employees hired. Functions like this are called **composite functions**.

For example, the function g that gives a correspondence between women's shoe sizes in the United States and those in Italy is given by $g(x) = 2x + 24$, where x is the U.S. size and $g(x)$ is the Italian size. Thus a U.S. size 4 corresponds to a shoe size of $g(4) = 2 \cdot 4 + 24$, or 32, in Italy.

There is also a function that gives a correspondence between women's shoe sizes in Italy and those in Britain. This particular function is given by $f(x) = \frac{1}{2}x - 14$, where x is the Italian size and $f(x)$ is the corresponding British size. Thus an Italian size 32 corresponds to a British size $f(32) = \frac{1}{2} \cdot 32 - 14$, or 2.

It seems reasonable to conclude that a shoe size of 4 in the United States corresponds to a size of 2 in Britain and that some function h describes this correspondence. Can we find a formula for h? If we look at the following tables, we might guess that such a formula is $h(x) = x - 2$, and that is indeed correct. But, for more complicated formulas, we would need to use algebra.

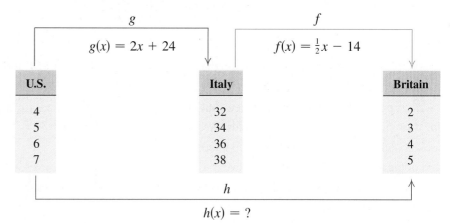

Size x shoes in the United States correspond to size $g(x)$ shoes in Italy, where

$$g(x) = 2x + 24.$$

Size n shoes in Italy correspond to size $f(n)$ shoes in Britain. Thus size $g(x)$ shoes in Italy correspond to size $f(g(x))$ shoes in Britain. Since the x in the expression $f(g(x))$ represents a U.S. shoe size, we can find the British shoe size that corresponds to a U.S. size x as follows:

$$f(g(x)) = f(2x + 24) = \tfrac{1}{2} \cdot (2x + 24) - 14 \qquad \text{Using } g(x) \text{ as an input}$$
$$= x + 12 - 14 = x - 2.$$

This gives a formula for h: $h(x) = x - 2$. Thus a shoe size of 4 in the United States corresponds to a shoe size of $h(4) = 4 - 2$, or 2, in Britain. The function h is called the *composition* of f and g and is denoted $f \circ g$ (read "the composition of f and g," "f composed with g," or "f circle g").

Composition of Functions

The *composite function* $f \circ g$, the *composition* of f and g, is defined as

$$(f \circ g)(x) = f(g(x)).$$

We can visualize the composition of functions as follows.

TEACHING TIP

You may wish to give students extra practice reading this notation and rewriting $(f \circ g)(x)$ as $f(g(x))$.

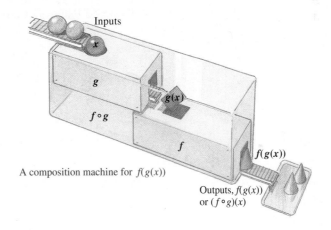

A composition machine for $f(g(x))$

Outputs, $f(g(x))$
or $(f \circ g)(x)$

EXAMPLE 1   Given $f(x) = 3x$ and $g(x) = 1 + x^2$:

a) Find $(f \circ g)(5)$ and $(g \circ f)(5)$.

b) Find $(f \circ g)(x)$ and $(g \circ f)(x)$.

Solution   Consider each function separately:

$$f(x) = 3x \qquad \textbf{This function multiplies each input by 3.}$$

and

$$g(x) = 1 + x^2. \qquad \textbf{This function adds 1 to the square of each input.}$$

a) To find $(f \circ g)(5)$, we first find $g(5)$ by substituting in the formula for g: Square 5 and add 1, to get 26. We then use 26 as an input for f:

$$(f \circ g)(5) = f(g(5)) = f(1 + 5^2) \qquad \textbf{Using } g(x) = 1 + x^2$$
$$= f(26) = 3 \cdot 26 = 78. \qquad \textbf{Using } f(x) = 3x$$

To find $(g \circ f)(5)$, we first find $f(5)$ by substituting into the formula for f: Multiply 5 by 3, to get 15. We then use 15 as an input for g:

$$(g \circ f)(5) = g(f(5)) = g(3 \cdot 5) \quad \text{Note that } f(5) = 3 \cdot 5 = 15.$$
$$= g(15) = 1 + 15^2 = 1 + 225 = 226.$$

b) We find $(f \circ g)(x)$ by substituting $g(x)$ for x in the equation for $f(x)$:

$$(f \circ g)(x) = f(g(x)) = f(1 + x^2) \quad \text{Using } g(x) = 1 + x^2$$
$$= 3(1 + x^2) = 3 + 3x^2. \quad \text{Using } f(x) = 3x. \text{ \textit{These} parentheses indicate multiplication.}$$

To find $(g \circ f)(x)$, we substitute $f(x)$ for x in the equation for $g(x)$:

$$(g \circ f)(x) = g(f(x)) = g(3x) \quad \text{Substituting } 3x \text{ for } f(x)$$
$$= 1 + (3x)^2 = 1 + 9x^2.$$

As a check, note that $(g \circ f)(5) = 1 + 9 \cdot 5^2 = 1 + 9 \cdot 25 = 226$, as expected from part (a) above.

Example 1 shows that, in general, $(f \circ g)(5) \neq (g \circ f)(5)$ and $(f \circ g)(x) \neq (g \circ f)(x)$.

EXAMPLE 2 Given $f(x) = \sqrt{x}$ and $g(x) = x - 1$, find $(f \circ g)(x)$ and $(g \circ f)(x)$.

Solution We have

$$(f \circ g)(x) = f(g(x)) = f(x - 1) = \sqrt{x - 1};$$
$$(g \circ f)(x) = g(f(x)) = g(\sqrt{x}) = \sqrt{x} - 1.$$

To check using a graphing calculator, let

$$y_1 = f(x) = \sqrt{\;}(x),$$
$$y_2 = g(x) = x - 1,$$
$$y_3 = \sqrt{\;}(x - 1),$$

and $\quad y_4 = y_1(y_2).$

If our work is correct, y_4 will be equivalent to y_3. We form a table of values, selecting only y_3 and y_4. The table on the left below indicates that the functions are probably equivalent.

X	Y3	Y4
1	0	0
1.5	.70711	.70711
2	1	1
2.5	1.2247	1.2247
3	1.4142	1.4142
3.5	1.5811	1.5811
4	1.7321	1.7321
X = 1		

X	Y5	Y6
1	0	0
1.5	.22474	.22474
2	.41421	.41421
2.5	.58114	.58114
3	.73205	.73205
3.5	.87083	.87083
4	1	1
X = 1		

Next, we let $y_5 = \sqrt{}(x) - 1$ and $y_6 = y_2(y_1)$ and select only y_5 and y_6. The table on the right at the bottom of p. 652 indicates that the functions are probably equivalent. Thus we conclude that

$$(f \circ g)(x) = \sqrt{x - 1} \quad \text{and} \quad (g \circ f)(x) = \sqrt{x} - 1.$$

Functions that are not defined by an equation can still be composed when the output of the first function is in the domain of the second function.

EXAMPLE 3   Use the following table to find each value, if possible.

a) $(y_2 \circ y_1)(3)$

b) $(y_1 \circ y_2)(3)$

X	Y1	Y2
0	-2	3
1	2	5
2	-1	7
3	1	9
4	0	11
5	0	13
6	-3	15

X = 0

Solution

a) Since $(y_2 \circ y_1)(3) = y_2(y_1(3))$, we first find $y_1(3)$. We locate 3 in the x-column and then move across to the y_1-column to find $y_1(3) = 1$.

X	Y1	Y2
0	-2	3
1	2	5
2	-1	7
3	1	9
4	0	11
5	0	13
6	-3	15

Y2 = 5

The output 1 now becomes an input for the function y_2:

$$y_2(y_1(3)) = y_2(1).$$

To find $y_2(1)$, we locate 1 in the x-column and then move across to the y_2-column to find $y_2(1) = 5$. Thus, $(y_2 \circ y_1)(3) = 5$.

b) Since $(y_1 \circ y_2)(3) = y_1(y_2(3))$, we find $y_2(3)$ by locating 3 in the x-column and moving across to the y_2-column. We have $y_2(3) = 9$, so

$$(y_1 \circ y_2)(3) = y_1(y_2(3)) = y_1(9).$$

However, y_1 is not defined for $x = 9$, so $(y_1 \circ y_2)(3)$ is undefined.

In fields ranging from chemistry to geology and economics, one needs to recognize how a function can be regarded as the composition of two "simpler" functions. This is sometimes called *de*composition.

> **EXAMPLE 4** If $h(x) = (7x + 3)^2$, find $f(x)$ and $g(x)$ such that $h(x) = (f \circ g)(x)$.
>
> Solution   To find $h(x)$, we can think of first forming $7x + 3$ and then squaring. This suggests that $g(x) = 7x + 3$ and $f(x) = x^2$. We check by forming the composition:
>
> $$h(x) = (f \circ g)(x) = f(g(x))$$
> $$= f(7x + 3) = (7x + 3)^2.$$
>
> This is probably the most "obvious" answer to the question. There can be other less obvious answers. For example, if
>
> $$f(x) = (x - 1)^2$$
>
> and
>
> $$g(x) = 7x + 4,$$
>
> then
>
> $$h(x) = (f \circ g)(x) = f(g(x)) = f(7x + 4)$$
> $$= (7x + 4 - 1)^2 = (7x + 3)^2.$$

Inverses and One-to-One Functions

Let's consider the following two functions. We think of them as relations, or correspondences.

Countries and Their Capitals

Domain (Set of Inputs)	Range (Set of Outputs)
Australia	Canberra
China	Beijing
Cyprus	Nicosia
Egypt	Cairo
Haiti	Port-au-Prince
Peru	Lima

U.S. Senators and Their States

Domain (Set of Inputs)	Range (Set of Outputs)
Clinton	New York
Schumer	
McCain	Arizona
Kyl	
Feinstein	California
Boxer	

Suppose we reverse the arrows. We obtain what is called the **inverse relation**. Are these inverse relations functions?

Countries and Their Capitals

Range (Set of Outputs)	Domain (Set of Inputs)
Australia ←	Canberra
China ←	Beijing
Cyprus ←	Nicosia
Egypt ←	Cairo
Haiti ←	Port-au-Prince
Peru ←	Lima

U.S. Senators and Their States

Range (Set of Outputs)	Domain (Set of Inputs)
Clinton ←	New York
Schumer ←	
McCain ←	Arizona
Kyl ←	
Feinstein ←	California
Boxer ←	

Recall that for each input, a function provides exactly one output. However, a function can have the same output for two or more different inputs. Thus it is possible for different inputs to correspond to the same output. Only when this possibility is *excluded* will the inverse be a function. For the functions listed above, this means the inverse of the "Capital" correspondence is a function, but the inverse of the "U.S. Senator" correspondence is not.

In the Capital function, different inputs have different outputs. It is an example of a **one-to-one function**. In the U.S. Senator function, *Clinton* and *Schumer* are both paired with *New York*. Thus the U.S. Senator function is not one-to-one.

One-To-One Function

A function f is *one-to-one* if different inputs have different outputs. That is, if for any $a \neq b$, we have $f(a) \neq f(b)$, the function f is one-to-one. If a function is one-to-one, then its inverse correspondence is also a function.

How can we tell graphically whether a function is one-to-one?

EXAMPLE 5 Shown here is the graph of a function similar to those we will study in Section 9.2. Determine whether the function is one-to-one and thus has an inverse that is a function.

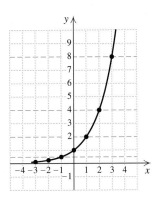

Solution   A function is one-to-one if different inputs have different out-puts—that is, if no two x-values have the same y-value. For this function, we cannot find two x-values that have the same y-value. Note that this means that no horizontal line can be drawn so that it crosses the graph more than once. The function is one-to-one so its inverse is a function.

    The graph of every function must pass the vertical-line test. In order for a function to have an inverse that is a function, it must pass the *horizontal-line test* as well.

The Horizontal-Line Test

A function is one-to-one, and thus has an inverse that is a function, if it is impossible to draw a horizontal line that intersects its graph more than once.

EXAMPLE 6   Determine whether the function $f(x) = x^2$ is one-to-one and thus has an inverse that is a function.

Solution   The graph of $f(x) = x^2$ is shown here. Many horizontal lines cross the graph more than once—in particular, the line $y = 4$. Note that where the line crosses, the first coordinates are -2 and 2. Although these are different inputs, they have the same output. That is, $-2 \neq 2$, but

$$f(-2) = (-2)^2 = 4 = 2^2 = f(2).$$

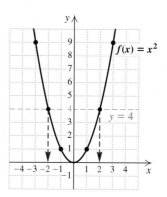

Thus the function is not one-to-one and no inverse function exists.

Finding Formulas for Inverses

When the inverse of f is also a function, it is denoted f^{-1} (read "f-inverse").

CAUTION!   The -1 in f^{-1} is *not* an exponent!

Suppose a function is described by a formula. If it has an inverse that is a function, how do we find a formula for the inverse? For any equation in two variables, if we interchange the variables, we obtain an equation of the inverse correspondence. If it is a function, we proceed as follows to find a formula for f^{-1}.

To Find a Formula for f^{-1}

First check a graph to make sure that f is one-to-one. Then:

1. Replace $f(x)$ with y.

2. Interchange x and y. (This gives the inverse function.)

3. Solve for y.

4. Replace y with $f^{-1}(x)$. (This is inverse function notation.)

EXAMPLE 7   Determine if each function is one-to-one and if it is, find a formula for $f^{-1}(x)$.

a) $f(x) = x + 2$ **b)** $f(x) = 2x - 3$

Solution

a) The graph of $f(x) = x + 2$ is shown below. It passes the horizontal-line test, so it is one-to-one. Thus its inverse is a function.

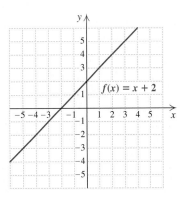

1. Replace $f(x)$ with y: $y = x + 2$.

2. Interchange x and y: $x = y + 2$.   **This gives the inverse function.**

3. Solve for y: $x - 2 = y$.

4. Replace y with $f^{-1}(x)$: $f^{-1}(x) = x - 2$.   **We also "reversed" the equation.**

In this case, the function f added 2 to all inputs. Thus, to "undo" f, the function f^{-1} must subtract 2 from its inputs.

b) The function $f(x) = 2x - 3$ is also linear. Any linear function that is not constant will pass the horizontal-line test. Thus, f is one-to-one.

1. Replace $f(x)$ with y: $y = 2x - 3$.

2. Interchange x and y: $x = 2y - 3$.

3. Solve for y: $x + 3 = 2y$

$$\frac{x + 3}{2} = y.$$

4. Replace y with $f^{-1}(x)$: $f^{-1}(x) = \dfrac{x + 3}{2}$.

Graphing Functions and Their Inverses

How do the graphs of a function and its inverse compare?

EXAMPLE 8   Graph $f(x) = 2x - 3$ and $f^{-1}(x) = (x + 3)/2$ on the same set of axes. Then compare.

Solution   The graph of each function follows. Note that the graph of f^{-1} can be drawn by reflecting the graph of f across the line $y = x$. That is, if we graph $f(x) = 2x - 3$ in wet ink and fold the paper along the line $y = x$, the graph of $f^{-1}(x) = (x + 3)/2$ will appear as the impression made by f.

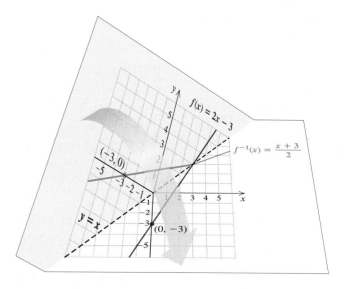

When x and y are interchanged to find a formula for the inverse, we are, in effect, reflecting or flipping the graph of $f(x) = 2x - 3$ across the line $y = x$. For example, when the coordinates of the y-intercept of the graph of f, $(0, -3)$, are reversed, we get the x-intercept of the graph of f^{-1}, $(-3, 0)$.

Visualizing Inverses

The graph of f^{-1} is a reflection of the graph of f across the line $y = x$.

EXAMPLE 9   Consider $g(x) = x^3 + 2$.

a) Determine whether the function is one-to-one.

b) If it is one-to-one, find a formula for its inverse.

c) Graph the inverse, if it exists.

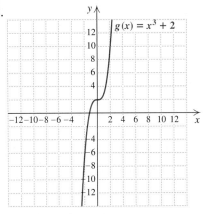

Solution

a) The graph of $g(x) = x^3 + 2$ is shown at right. It passes the horizontal-line test and thus has an inverse.

b) **1.** Replace $g(x)$ with y: $y = x^3 + 2.$ Using $g(x) = x^3 + 2$

   2. Interchange x and y: $x = y^3 + 2.$

   3. Solve for y: $x - 2 = y^3$

   $\sqrt[3]{x - 2} = y.$ **Since a number has only one cube root, we can solve for y.**

   4. Replace y with $g^{-1}(x)$: $g^{-1}(x) = \sqrt[3]{x - 2}.$

c) To find the graph, we reflect the graph of $g(x) = x^3 + 2$ across the line $y = x$, as we did in Example 8. We can also substitute into $g^{-1}(x) = \sqrt[3]{x - 2}$ and plot points. The graphs of g and g^{-1} are shown together below.

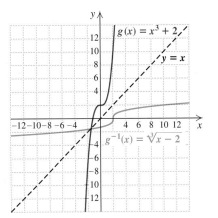

Inverse Functions and Composition

Let's consider inverses of functions in terms of a function machine. Suppose that a one-to-one function f is programmed into a machine. If the machine has a reverse switch, when the switch is thrown, the machine performs the inverse function f^{-1}. Inputs then enter at the opposite end, and the entire process is reversed.

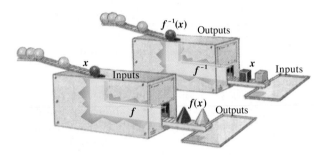

Consider $g(x) = x^3 + 2$ and $g^{-1}(x) = \sqrt[3]{x-2}$ from Example 9. For the input 3,

$$g(3) = 3^3 + 2 = 27 + 2 = 29.$$

The output is 29. Now we use 29 for the input in the inverse:

$$g^{-1}(29) = \sqrt[3]{29-2} = \sqrt[3]{27} = 3.$$

The function g takes 3 to 29. The inverse function g^{-1} takes the number 29 back to 3.

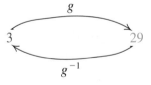

In general, for any output $f(x)$, the function f^{-1} takes that output back to x. Similarly, for any output $f^{-1}(x)$, the function f takes that output back to x.

Composition and Inverses

If a function f is one-to-one, then f^{-1} is the unique function for which

$$(f^{-1} \circ f)(x) = x \quad \text{and} \quad (f \circ f^{-1})(x) = x.$$

EXAMPLE 10   Let $f(x) = 2x + 1$. Show that

$$f^{-1}(x) = \frac{x - 1}{2}.$$

Solution   We find $(f^{-1} \circ f)(x)$ and $(f \circ f^{-1})(x)$ and check to see that each is x.

$$(f^{-1} \circ f)(x) = f^{-1}(f(x)) = f^{-1}(2x + 1)$$

$$= \frac{(2x + 1) - 1}{2}$$

$$= \frac{2x}{2} = x$$

$$(f \circ f^{-1})(x) = f(f^{-1}(x)) = f\left(\frac{x - 1}{2}\right)$$

$$= 2 \cdot \frac{x - 1}{2} + 1$$

$$= x - 1 + 1 = x$$

Inverse Functions: Graphing and Composition

In Example 9, we found that the inverse of $y_1 = x^3 + 2$ is $y_2 = \sqrt[3]{x - 2}$. We can check this visually using a graphing calculator by graphing both functions, along with the line $y = x$, on a "squared" set of axes. Since the graph of y_2 appears to be the reflection of the graph of y_1 across the line $y = x$, we conclude that y_1 and y_2 are inverses of each other.

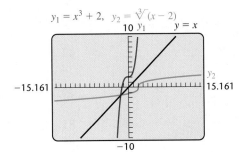

$y_1 = x^3 + 2, \quad y_2 = \sqrt[3]{(x - 2)}$

X	Y1	Y2
−3	−25	−3
−2	−6	−2
−1	1	−1
0	2	0
1	3	1
2	10	2
3	29	3

X = −3

For further verification, we can examine a table of values (on the right above) in which $y_1 = x^3 + 2$ and $y_2 = \sqrt[3]{y_1 - 2}$. Note that y_2 "undoes" what y_1 "does."

A final, visual check can be made by first graphing

$$y_1 = x^3 + 2.$$

We next draw the inverse of y_1 by going to the home screen and selecting the DRAWINV option from the DRAW menu and y_1 by using the VARS menu. We then graph $y_2 = \sqrt[3]{x - 2}$ using the path graph-style. The graph of y_2 should coincide with the graph of the inverse of y_1.

9.1

Exercise Set

Find $(f \circ g)(1)$, $(g \circ f)(1)$, $(f \circ g)(x)$, *and* $(g \circ f)(x)$.

1. $f(x) = x^2 + 3$; $g(x) = 2x + 1$ ⊡

2. $f(x) = 2x + 1$; $g(x) = x^2 - 5$ ⊡

3. $f(x) = 3x - 1$; $g(x) = 5x^2 + 2$ ⊡

4. $f(x) = 3x^2 + 4$; $g(x) = 4x - 1$ ⊡

5. $f(x) = x + 7$; $g(x) = 1/x^2$ ⊡

6. $f(x) = 1/x^2$; $g(x) = x + 2$ ⊡

Use the following table to find each value, if possible.

X	Y₁	Y₂
−3	−4	1
−2	−1	−2
−1	2	−3
0	5	−2
1	8	1
2	11	6
3	14	11

X =

7. $(y_1 \circ y_2)(-3)$ 8

8. $(y_2 \circ y_1)(-3)$ Not defined

9. $(y_1 \circ y_2)(-1)$ −4

10. $(y_2 \circ y_1)(-1)$ 6

11. $(y_2 \circ y_1)(1)$ Not defined

12. $(y_1 \circ y_2)(1)$ 8

Use the following table to find each value, if possible.

x	f(x)	g(x)
1	0	1
2	3	5
3	2	8
4	6	5
5	4	1

13. $(f \circ g)(2)$ 4

14. $(g \circ f)(4)$ Not defined

15. $f(g(3))$ Not defined

16. $g(f(5))$ 5

Find $f(x)$ *and* $g(x)$ *such that* $h(x) = (f \circ g)(x)$. *Answers may vary.*

17. $h(x) = (7 + 5x)^2$ ⊡

18. $h(x) = (3x - 1)^2$ ⊡

19. $h(x) = \sqrt{2x + 7}$
$f(x) = \sqrt{x}$; $g(x) = 2x + 7$

20. $h(x) = \sqrt{5x + 2}$
$f(x) = \sqrt{x}$; $g(x) = 5x + 2$

21. $h(x) = \dfrac{2}{x - 3}$ ⊡

22. $h(x) = \dfrac{3}{x} + 4$ ⊡

23. $h(x) = \dfrac{1}{\sqrt{7x + 2}}$ ⊡

24. $h(x) = \sqrt{x - 7} - 3$ ⊡

25. $h(x) = \dfrac{1}{\sqrt{3x}} + \sqrt{3x}$ ⊡

26. $h(x) = \dfrac{1}{\sqrt{2x}} - \sqrt{2x}$ ⊡

Determine whether each function is one-to-one.

27. $f(x) = x - 5$ Yes

28. $f(x) = 5 - 2x$ Yes

Aha! **29.** $f(x) = x^2 + 1$ No

30. $f(x) = 1 - x^2$ No

31. $g(x) = x^3$ Yes

32. $g(x) = \sqrt{x} + 1$ Yes

33. $g(x) = |x|$ No

34. $h(x) = |x| - 1$ No

For each function, **(a)** *determine whether it is one-to-one;* **(b)** *if it is one-to-one, find a formula for the inverse.*

35. $f(x) = x - 4$
(a) Yes; (b) $f^{-1}(x) = x + 4$

36. $f(x) = x - 2$
(a) Yes; (b) $f^{-1}(x) = x + 2$

37. $f(x) = 3 + x$
(a) Yes; (b) $f^{-1}(x) = x - 3$

38. $f(x) = 9 + x$
(a) Yes; (b) $f^{-1}(x) = x - 9$

39. $g(x) = x + 5$
(a) Yes; (b) $g^{-1}(x) = x - 5$

40. $g(x) = x + 8$
(a) Yes; (b) $g^{-1}(x) = x - 8$

41. $f(x) = 4x$
(a) Yes; (b) $f^{-1}(x) = \dfrac{x}{4}$

42. $f(x) = 7x$
(a) Yes; (b) $f^{-1}(x) = \dfrac{x}{7}$

43. $g(x) = 4x - 1$ ⊡

44. $g(x) = 4x - 6$ ⊡

45. $h(x) = 5$ (a) No

46. $h(x) = -2$ (a) No

Aha! **47.** $f(x) = \dfrac{1}{x}$ ⊡

48. $f(x) = \dfrac{3}{x}$ ⊡

49. $f(x) = \dfrac{2x + 1}{3}$ ⊡

50. $f(x) = \dfrac{3x + 2}{5}$ ⊡

51. $f(x) = x^3 - 5$ ⊡

52. $f(x) = x^3 + 2$ ⊡

53. $g(x) = (x - 2)^3$ ⊡

54. $g(x) = (x + 7)^3$ ⊡

55. $f(x) = \sqrt{x}$
(a) Yes; (b) $f^{-1}(x) = x^2$, $x \geq 0$

56. $f(x) = \sqrt{x - 1}$
(a) Yes; (b) $f^{-1}(x) = x^2 + 1$, $x \geq 0$

57. $f(x) = 2x^2 + 1$, $x \geq 0$ ⊡

58. $f(x) = 3x^2 - 2$, $x \geq 0$ (a) Yes; (b) $f^{-1}(x) = \sqrt{\dfrac{x + 2}{3}}$

Graph each function and its inverse using the same set of axes.

59. $f(x) = \frac{1}{3}x - 2$ ⊡

60. $g(x) = x + 4$ ⊡

61. $f(x) = x^3$ ⊡

62. $f(x) = x^3 - 1$ ⊡

⊡ Answers to Exercises 1–6, 17, 18, 21–26, 43, 44, 47–54, 57, and 59–62 can be found on p. A-68.

63. $g(x) = -2x + 3$ ☐   **64.** $g(x) = \sqrt{x}$ ☐

65. $F(x) = -\sqrt{x}$ ☐   **66.** $f(x) = -\frac{1}{2}x + 1$ ☐

67. $f(x) = 3 - x^2,\ x \geq 0$ ☐

68. $f(x) = x^2 - 1,\ x \leq 0$ ☐

69. Let $f(x) = \frac{4}{5}x$. Use composition to show that
$f^{-1}(x) = \frac{5}{4}x$.   ☐

70. Let $f(x) = (x + 7)/3$. Use composition to show that
$f^{-1}(x) = 3x - 7$.   ☐

71. Let $f(x) = (1 - x)/x$. Use composition to show that
$$f^{-1}(x) = \frac{1}{x + 1}.$$ ☐

72. Let $f(x) = x^3 - 5$. Use composition to show that
$f^{-1}(x) = \sqrt[3]{x + 5}$.   ☐

Use a graphing calculator to help determine whether or not the given functions are inverses of each other.

73. $f(x) = 0.75x^2 + 2;\ g(x) = \sqrt{\dfrac{4(x - 2)}{3}}$ No

74. $f(x) = 1.4x^3 + 3.2;\ g(x) = \sqrt[3]{\dfrac{x - 3.2}{1.4}}$ Yes

75. $f(x) = \sqrt{2.5x + 9.25}$;
$g(x) = 0.4x^2 - 3.7,\ x \geq 0$ Yes

76. $f(x) = 0.8x^{1/2} + 5.23$;
$g(x) = 1.25(x^2 - 5.23),\ x \geq 0$ No

In Exercises 77 and 78, match the graph of each function in Column A with the graph of its inverse in Column B.

77. This exercise includes the 4 graphs at the top of the next column.

Column A

(1)
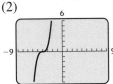

(2)
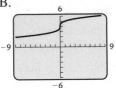

Column B

A.
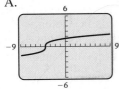

B.
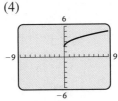

(3)
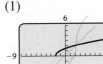

(4)
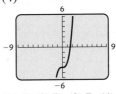

C.
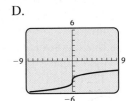

D.

(1) C; (2) D; (3) B; (4) A

78.

Column A

(1)
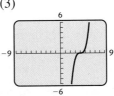

(2)
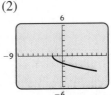

(3)
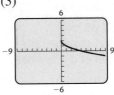

(4)
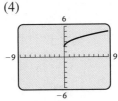

Column B

A.
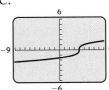

B.
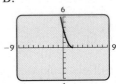

C.
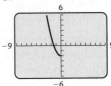

D.
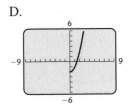

(1) D; (2) C; (3) B; (4) A

☐ Answers to Exercises 63–72 can be found on pp. A-68 and A-69.

79. *Dress Sizes in the United States and France.*   A size-6 dress in the United States is size 38 in France. A function that converts dress sizes in the United States to those in France is

$$f(x) = x + 32.$$

a) Find the dress sizes in France that correspond to sizes 8, 10, 14, and 18 in the United States.   40, 42, 46, 50

b) Determine whether this function has an inverse that is a function. If so, find a formula for the inverse. $f^{-1}(x) = x - 32$

c) Use the inverse function to find dress sizes in the United States that correspond to sizes 40, 42, 46, and 50 in France.   8, 10, 14, 18

80. *Dress Sizes in the United States and Italy.*   A size-6 dress in the United States is size 36 in Italy. A function that converts dress sizes in the United States to those in Italy is

$$f(x) = 2(x + 12).$$

a) Find the dress sizes in Italy that correspond to sizes 8, 10, 14, and 18 in the United States.   40, 44, 52, 60

b) Determine whether this function has an inverse that is a function. If so, find a formula for the inverse.   ⊡

c) Use the inverse function to find dress sizes in the United States that correspond to sizes 40, 44, 52, and 60 in Italy.   8, 10, 14, 18

TW **81.** Is there a one-to-one relationship between the numbers and letters on the keypad of a telephone? Why or why not?

TW **82.** Mathematicians usually try to select "logical" words when forming definitions. Does the term "one-to-one" seem logical? Why or why not?

Skill Maintenance

Simplify.

83. $(a^5b^4)^2(a^3b^5)$ [1.4] $a^{13}b^{13}$

84. $(x^3y^5)^2(x^4y^2)$ [1.4] $x^{10}y^{12}$

85. $27^{4/3}$ [7.2]   81

86. $25^{3/2}$ [7.2]   125

Solve. [2.3]

87. $x = \frac{2}{3}y - 7$, for y $y = \frac{3}{2}(x + 7)$

88. $x = 10 - 3y$, for y $y = \dfrac{10 - x}{3}$

Synthesis

TW **89.** The function $V(t) = 750(1.2)^t$ is used to predict the value, $V(t)$, of a certain rare stamp t years from 2001. Do not calculate $V^{-1}(t)$, but explain how V^{-1} could be used.

TW **90.** An organization determines that the cost per person of chartering a bus is given by the function

$$C(x) = \frac{100 + 5x}{x},$$

where x is the number of people in the group and $C(x)$ is in dollars. Determine $C^{-1}(x)$ and explain how this inverse function could be used.

For Exercises 91 and 92, graph the inverse of f.

91. ⊡
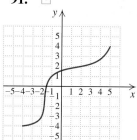

92. ⊡
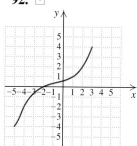

93. *Dress Sizes in France and Italy.*   Use the information in Exercises 79 and 80 to find a function for the French dress size that corresponds to a size x dress in Italy. $g(x) = \dfrac{x}{2} + 20$

94. *Dress Sizes in Italy and France.*   Use the information in Exercises 79 and 80 to find a function for the Italian dress size that corresponds to a size x dress in France. $h(x) = 2(x - 20)$

TW **95.** What relationship exists between the answers to Exercises 93 and 94? Explain how you determined this.

96. Show that function composition is associative by showing that $((f \circ g) \circ h)(x) = (f \circ (g \circ h))(x)$.   ⊡

97. Show that if $h(x) = (f \circ g)(x)$, then $h^{-1}(x) = (g^{-1} \circ f^{-1})(x)$. (*Hint*: Use Exercise 96.)   ⊡

⊡ Answers to Exercises 80(b), 91, 92, 96, and 97 can be found on p. A-69.

98. Match each function in Column A with its inverse from Column B.

Column A

(1) $y = 5x^3 + 10$ C

(2) $y = (5x + 10)^3$ A

(3) $y = 5(x + 10)^3$ B

(4) $y = (5x)^3 + 10$ D

Column B

A. $y = \dfrac{\sqrt[3]{x} - 10}{5}$

B. $y = \sqrt[3]{\dfrac{x}{5}} - 10$

C. $y = \sqrt[3]{\dfrac{x - 10}{5}}$

D. $y = \dfrac{\sqrt[3]{x - 10}}{5}$

TW **99.** Examine the following table. Does it appear that f and g could be inverses of each other? Why or why not?

x	$f(x)$	$g(x)$
6	6	6
7	6.5	8
8	7	10
9	7.5	12
10	8	14
11	8.5	16
12	9	18

100. Assume in Exercise 99 that f and g are both linear functions. Find equations for $f(x)$ and $g(x)$. Are f and g inverses of each other? $f(x) = \frac{1}{2}x + 3$; $g(x) = 2x - 6$; yes

101. Let $c(w)$ represent the cost of mailing a package that weighs w pounds. Let $f(n)$ represent the weight, in pounds, of n copies of a certain book. Explain what $(c \circ f)(n)$ represents.   The cost of mailing n copies of the book

102. Let $g(a)$ represent the number of gallons of sealant needed to seal a bamboo floor with area a. Let $c(s)$ represent the cost of s gallons of sealant. Which composition makes sense: $(c \circ g)(a)$ or $(g \circ c)(s)$? What does it represent? $(c \circ g)(a)$; It represents the cost of sealant for a bamboo floor with area a.

The following graphs show the rate of flow R, in liters per minute, of blood from the heart in a man who bicycles for 20 min, and the pressure p, in millimeters of mercury, in the artery leading to the lungs for a rate of blood flow R from the heart.*

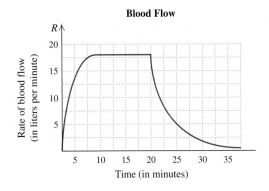

Blood Flow

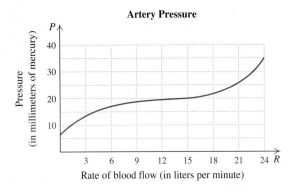

Artery Pressure

103. Estimate $P(R(10))$.   22 mm

104. Explain what $P(R(10))$ represents.   The pressure in the artery after 10 min of riding the bicycle

105. Estimate $P^{-1}(20)$.   15 L/min

106. Explain what $P^{-1}(20)$ represents.   The rate of blood flow from the heart when the pressure in the artery is 20 mm

---

*This problem was suggested by Kandace Kling, Portland Community College, Sylvania, Oregon.

9.2

Graphing Exponential Functions ■ Equations with *x* and *y* Interchanged ■
Applications of Exponential Functions

Exponential Functions

Composite and inverse functions, as shown in Section 9.1, are very useful in and of themselves. The reason they are included in this chapter, however, is because they are needed in order to understand the logarithmic functions that appear in Section 9.3. Here in Section 9.2, we make no reference to composite or inverse functions. Instead, we introduce a new type of function, the *exponential function*, so that we can study both it and its inverse in Sections 9.3–9.7.

Consider the graph below. The rapidly rising curve approximates the graph of an *exponential function*. We now consider such functions and some of their applications.

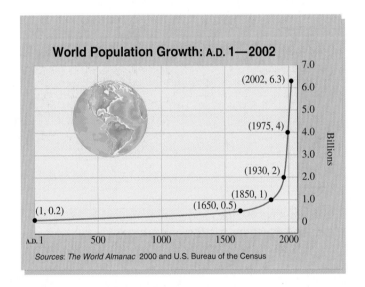

World Population Growth: A.D. 1—2002

Sources: *The World Almanac* 2000 and U.S. Bureau of the Census

Graphing Exponential Functions

In Chapter 7, we studied exponential expressions with rational-number exponents, such as $5^{1/4}$, $3^{-3/4}$, $7^{2.34}$, $5^{1.73}$. For example, $5^{1.73}$, or $5^{173/100}$, represents the 100th root of 5 raised to the 173rd power. What about expressions with irrational exponents, such as $5^{\sqrt{3}}$ or $7^{-\pi}$? To attach meaning to $5^{\sqrt{3}}$, consider a rational approximation, r, of $\sqrt{3}$. As r gets closer to $\sqrt{3}$, the value of 5^r gets closer to some real number p.

$\underbrace{r \text{ closes in on } \sqrt{3}.}$	$\underbrace{5^r \text{ closes in on some real number } p.}$
$1.7 < r < 1.8$	$15.426 \approx 5^{1.7} < p < 5^{1.8} \approx 18.119$
$1.73 < r < 1.74$	$16.189 \approx 5^{1.73} < p < 5^{1.74} \approx 16.452$
$1.732 < r < 1.733$	$16.241 \approx 5^{1.732} < p < 5^{1.733} \approx 16.267$

We define $5^{\sqrt{3}}$ to be the number p. To eight decimal places,

$$5^{\sqrt{3}} \approx 16.24245082.$$

Any positive irrational exponent can be defined in a similar way. Negative irrational exponents are then defined using reciprocals. Thus, so long as a is positive, a^x has meaning for *any* real number x. All of the laws of exponents still hold, but we will not prove that here. We now define an *exponential function*.

TEACHING TIP

You may wish to point out that the power in an exponential function need not be simply x.

> ## Exponential Function
>
> The function $f(x) = a^x$, where a is a positive constant, $a \neq 1$, is called the *exponential function*, base a.

We require the base a to be positive to avoid imaginary numbers that would result from taking even roots of negative numbers. The restriction $a \neq 1$ is made to exclude the constant function $f(x) = 1^x$, or $f(x) = 1$.

The following are examples of exponential functions:

$$f(x) = 2^x, \qquad f(x) = \left(\tfrac{1}{3}\right)^x, \qquad f(x) = 5^{-3x}. \qquad \text{Note that } 5^{-3x} = (5^{-3})^x.$$

Like polynomial functions, the domain of an exponential function is the set of all real numbers. In contrast to polynomial functions, exponential functions have a variable exponent. Because of this, graphs of exponential functions either rise or fall dramatically.

EXAMPLE 1   Graph the exponential function $y = f(x) = 2^x$.

Solution   We compute some function values, thinking of y as $f(x)$, and list the results in a table. It is a good idea to start by letting $x = 0$.

$$f(0) = 2^0 = 1; \qquad\qquad f(-1) = 2^{-1} = \frac{1}{2^1} = \frac{1}{2};$$
$$f(1) = 2^1 = 2;$$
$$f(2) = 2^2 = 4; \qquad\qquad f(-2) = 2^{-2} = \frac{1}{2^2} = \frac{1}{4};$$
$$f(3) = 2^3 = 8;$$
$$f(-3) = 2^{-3} = \frac{1}{2^3} = \frac{1}{8}$$

Next, we plot these points and connect them with a smooth curve.

x	y, or $f(x)$
0	1
1	2
2	4
3	8
-1	$\frac{1}{2}$
-2	$\frac{1}{4}$
-3	$\frac{1}{8}$

The curve comes very close to the x-axis, but does not touch or cross it.

$$y = f(x) = 2^x$$

Be sure to plot enough points to determine how steeply the curve rises.

Note that as x increases, the function values increase without bound. As x decreases, the function values decrease, getting very close to 0. The x-axis, or the line $y = 0$, is a horizontal *asymptote*, meaning that the curve gets closer and closer to this line the further we move to the left.

EXAMPLE 2   Graph the exponential function $y = f(x) = \left(\frac{1}{2}\right)^x$.

Solution   We compute some function values, thinking of y as $f(x)$, and list the results in a table. Before we do this, note that

$$y = f(x) = \left(\tfrac{1}{2}\right)^x = (2^{-1})^x = 2^{-x}.$$

Then we have

$$f(0) = 2^{-0} = 1;$$

$$f(1) = 2^{-1} = \frac{1}{2^1} = \frac{1}{2};$$

$$f(2) = 2^{-2} = \frac{1}{2^2} = \frac{1}{4};$$

$$f(3) = 2^{-3} = \frac{1}{2^3} = \frac{1}{8};$$

$$f(-1) = 2^{-(-1)} = 2^1 = 2;$$
$$f(-2) = 2^{-(-2)} = 2^2 = 4;$$
$$f(-3) = 2^{-(-3)} = 2^3 = 8.$$

x	y, or $f(x)$
0	1
1	$\frac{1}{2}$
2	$\frac{1}{4}$
3	$\frac{1}{8}$
−1	2
−2	4
−3	8

Next, we plot these points and connect them with a smooth curve. Note that this curve is a mirror image, or *reflection*, of the above graph of $y = 2^x$ across the y-axis. The line $y = 0$ is again an asymptote.

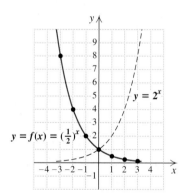

Interactive Discovery

If the values of $f(x)$ increase as x increases, we say that f is an *increasing function*. The function in Example 1, $f(x) = 2^x$, is increasing. If the values of $f(x)$ decrease as x increases, we say that f is a *decreasing function*. The function in Example 2, $f(x) = \left(\frac{1}{2}\right)^x$, is decreasing.

1. Graph each of the following functions, and determine whether the graph increases or decreases from left to right.

a) $f(x) = 3^x$ Increases
b) $g(x) = 4^x$ Increases
c) $h(x) = 1.5^x$ Increases
d) $r(x) = \left(\frac{1}{3}\right)^x$ Decreases
e) $t(x) = 0.75^x$ Decreases

2. How can you tell from the number a in $f(x) = a^x$ whether the graph of f
increases or decreases?   If $a > 1$, the graph increases; if $0 < a < 1$, the graph
decreases.

3. Compare the graphs of f, g, and h. Which curve is steepest? g

4. Compare the graphs of r and t. Which curve is steeper? r

We can make the following observations.

A. For $a > 1$, the graph of $f(x) = a^x$ increases from left to right. The greater
the value of a, the steeper the curve. (See the figure on the left below.)

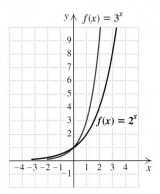
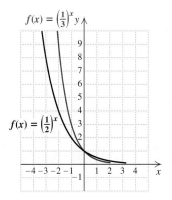

B. For $0 < a < 1$, the graph of $f(x) = a^x$ decreases from left to right. For
smaller values of a, the curve becomes steeper. (See the figure on the right
above.)

C. All graphs of $f(x) = a^x$ go through the y-intercept $(0, 1)$.

D. If $f(x) = a^x$, with $a > 0$, $a \neq 1$, the domain of f is all real numbers, and
the range of f is all positive real numbers.

E. For $a > 0$, $a \neq 1$, the function given by $f(x) = a^x$ is one-to-one. Its graph
passes the horizontal-line test.

EXAMPLE 3   Graph: $y = f(x) = 2^{x-2}$.

Solution   We construct a table of values. Then we plot the points and connect them with a smooth curve. Here $x - 2$ is the *exponent*.

$$f(0) = 2^{0-2} = 2^{-2} = \frac{1}{4};$$ $$f(-1) = 2^{-1-2} = 2^{-3} = \frac{1}{8};$$

$$f(1) = 2^{1-2} = 2^{-1} = \frac{1}{2};$$ $$f(-2) = 2^{-2-2} = 2^{-4} = \frac{1}{16}$$

$$f(2) = 2^{2-2} = 2^{0} = 1;$$

$$f(3) = 2^{3-2} = 2^{1} = 2;$$

$$f(4) = 2^{4-2} = 2^{2} = 4;$$

x	y, or $f(x)$
0	$\frac{1}{4}$
1	$\frac{1}{2}$
2	1
3	2
4	4
−1	$\frac{1}{8}$
−2	$\frac{1}{16}$

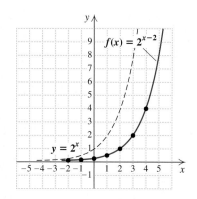

The graph looks just like the graph of $y = 2^x$, but it is translated 2 units to the right. The y-intercept of $y = 2^x$ is $(0, 1)$. The y-intercept of $y = 2^{x-2}$ is $\left(0, \frac{1}{4}\right)$. The line $y = 0$ is again the asymptote.

Connecting the Concepts

INTERPRETING GRAPHS: LINEAR AND EXPONENTIAL GROWTH

When the graph of a quantity over time approximates the graph of an exponential function, as in the World Population graph on p. 666, we say that there is *exponential* growth. This is different from *linear* growth, which we examined in Chapter 2.

Linear functions increase (or decrease) by a constant amount. The rate of change (slope) of a linear graph is constant. The rate of change of an exponential graph is not constant.

To illustrate the difference between linear and exponential functions, compare the salaries for two hypothetical jobs, shown in the following table and graph. Note that the starting salaries are the same, but the salary for job A increases by a constant amount (linearly) and the salary for job B increases by a constant percent (exponentially).

	Job A	**Job B**
Starting salary	$50,000	$50,000
Guaranteed raise	$5000 per year	8% per year
Salary function	$f(x) = 50{,}000 + 5000x$	$g(x) = 50{,}000(1.08)^x$

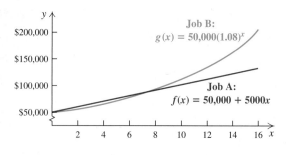

Note that the salary for job A is larger for the first few years only. The rate of change of the salary for job B increases each year, since the increase is based on a larger salary each year.

Equations with *x* and *y* Interchanged

It will be helpful in later work to be able to graph an equation in which the *x* and the *y* in $y = a^x$ are interchanged.

EXAMPLE 4   Graph: $x = 2^y$.

Solution   Note that *x* is alone on one side of the equation. To find ordered pairs that are solutions, we choose values for *y* and then compute values for *x*:

For $y = 0$, $x = 2^0 = 1$.
For $y = 1$, $x = 2^1 = 2$.
For $y = 2$, $x = 2^2 = 4$.
For $y = 3$, $x = 2^3 = 8$.
For $y = -1$, $x = 2^{-1} = \dfrac{1}{2}$.
For $y = -2$, $x = 2^{-2} = \dfrac{1}{4}$.
For $y = -3$, $x = 2^{-3} = \dfrac{1}{8}$.

x	y
1	0
2	1
4	2
8	3
$\frac{1}{2}$	−1
$\frac{1}{4}$	−2
$\frac{1}{8}$	−3

(1) Choose values for *y*.
(2) Compute values for *x*.

We plot the points and connect them with a smooth curve.

The curve does not touch or cross the *y*-axis, which serves as a vertical asymptote.

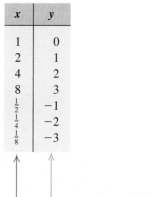

TEACHING TIP

Some students may recognize that $x = 2^y$ is the inverse of $y = 2^x$.

Note too that this curve looks just like the graph of $y = 2^x$, except that it is reflected across the line $y = x$, as shown here.

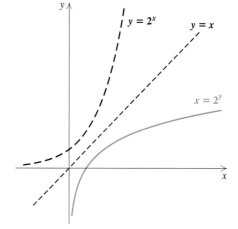

Applications of Exponential Functions

EXAMPLE 5 Interest Compounded Annually. The amount of money A that a principal P will be worth after t years at interest rate i, compounded annually, is given by the formula

$$A = P(1 + i)^t. \quad \textbf{You might review Example 12 in Section 8.1.}$$

Suppose that \$100,000 is invested at 8% interest, compounded annually.

a) Find a function for the amount in the account after t years.

b) Find the amount of money in the account at $t = 0$, $t = 4$, $t = 8$, and $t = 10$.

c) Graph the function.

Solution

a) If $P = \$100{,}000$ and $i = 8\% = 0.08$, we can substitute these values and form the following function:

$$A(t) = \$100{,}000(1 + 0.08)^t \quad \textbf{Using } A = P(1 + i)^t$$
$$= \$100{,}000(1.08)^t.$$

b) An efficient way to find the function values using a graphing calculator is to let $y_1 = 100{,}000(1.08)^{\wedge} x$ and use the TABLE feature with Indpnt set to Ask, as shown at left. We highlight a table entry if we want to view the function value to more decimal places than is shown in the table.

Alternatively, we can use $y_1(\ )$ notation to evaluate the function for each value. Using either procedure gives us

$$A(0) = \$100{,}000,$$
$$A(4) \approx \$136{,}048.90,$$
$$A(8) \approx \$185{,}093.02,$$

and $A(10) \approx \$215{,}892.50$.

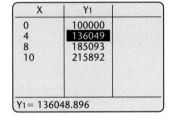

c) We can use the function values computed in part (b), and others if we wish, to draw the graph by hand. Whether we are graphing by hand or using a graphing calculator, the axes will be scaled differently because of the large function values.

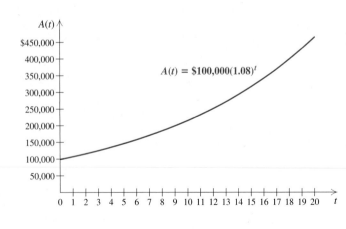

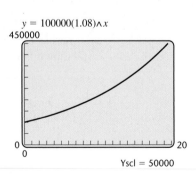

9.2

Exercise Set

In each of Exercises 1–4 is the graph of a function
$f(x) = a^x$. *Determine from the graph whether* $a > 1$
or $0 < a < 1$.

1.

$a > 1$

2.
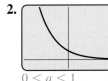
$0 < a < 1$

3.
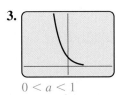
$0 < a < 1$

4.

$a > 1$

Graph.

5. $y = f(x) = 2^x$

6. $y = f(x) = 3^x$

7. $y = 5^x$

8. $y = 6^x$

9. $y = 2^x + 3$

10. $y = 2^x + 1$

11. $y = 3^x - 1$

12. $y = 3^x - 2$

13. $y = 2^{x-1}$

14. $y = 2^{x-2}$

15. $y = 2^{x+3}$

16. $y = 2^{x+1}$

17. $y = \left(\frac{1}{5}\right)^x$

18. $y = \left(\frac{1}{4}\right)^x$

19. $y = \left(\frac{1}{2}\right)^x$

20. $y = \left(\frac{1}{3}\right)^x$

21. $y = 2^{x-3} - 1$

22. $y = 2^{x+1} - 3$

23. $y = 1.7^x$

24. $y = 4.8^x$

25. $y = 0.15^x$

26. $y = 0.98^x$

27. $x = 3^y$

28. $x = 6^y$

29. $x = 2^{-y}$

30. $x = 3^{-y}$

31. $x = 5^y$

32. $x = 4^y$

33. $x = \left(\frac{3}{2}\right)^y$

34. $x = \left(\frac{4}{3}\right)^y$

Graph both equations using the same set of axes.

35. $y = 3^x, \quad x = 3^y$

36. $y = 2^x, \quad x = 2^y$

37. $y = \left(\frac{1}{2}\right)^x, \quad x = \left(\frac{1}{2}\right)^y$

38. $y = \left(\frac{1}{4}\right)^x, \quad x = \left(\frac{1}{4}\right)^y$

Aha! *In Exercises 39–44, match each equation with one of the following graphs.*

a)
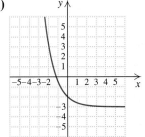

b)
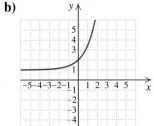

c)
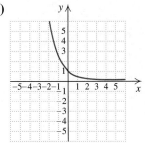

d)
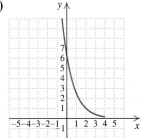

e)
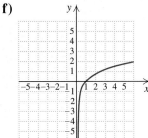

f)
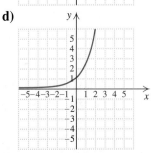

39. $y = \left(\frac{5}{2}\right)^x$ (d)

40. $y = \left(\frac{2}{5}\right)^x$ (e)

41. $x = \left(\frac{5}{2}\right)^y$ (f)

42. $y = \left(\frac{2}{5}\right)^x - 3$ (a)

43. $y = \left(\frac{2}{5}\right)^{x-2}$ (c)

44. $y = \left(\frac{5}{2}\right)^x + 1$ (b)

Answers to Exercises 5–38 can be found on p. A-69.

Solve.

45. *Population Growth.*   The world population $P(t)$, in billions, t years after 1975 can be approximated by

$$P(t) = 4(1.0164)^t$$

(*Sources*: Data from *The World Almanac* 2000 and U.S. Bureau of the Census).

a) What will the world population be in 2004? in 2008? in 2012? ⊡
b) Graph the function. ⊡

46. *Growth of Bacteria.*   The bacteria *Escherichi coli* can be found in the human bladder. Suppose that 3000 of the bacteria are present at time $t = 0$. Then t minutes later, the number of bacteria present will be

$$N(t) = 3000(2)^{t/20}.$$

a) How many bacteria will be present after 10 min? 20 min? 30 min? 40 min? 60 min? ⊡
b) Graph the function. ⊡

47. *Marine Biology.*   Due to excessive whaling prior to the mid 1970s, the humpback whale is considered an endangered species. The worldwide population of humpbacks, $P(t)$, in thousands, t years after 1900 ($t < 70$) can be approximated by*

$$P(t) = 150(0.960)^t.$$

a) How many humpback whales were alive in 1930? in 1960?   About 44,079 whales; about 12,953 whales
b) Graph the function. ⊡

---
*Based on information from the American Cetacean Society, 2001, and the ASK Archive, 1998

48. *Marine Biology.*   As a result of preservation efforts in most countries in which whaling was common, the humpback whale population has grown since the 1970s. The worldwide population of humpbacks, $P(t)$, in thousands, t years after 1982 can be approximated by*

$$P(t) = 5.5(1.047)^t.$$

a) How many humpback whales were alive in 1992? in 2001?   About 8706 whales; about 13,163 whales
b) Graph the function. ⊡

49. *Recycling Aluminum Cans.*   It is estimated that $\frac{2}{3}$ of all aluminum cans distributed will be recycled each year. A beverage company distributes 250,000 cans. The number still in use after time t, in years, is given by the exponential function

$$N(t) = 250,000\left(\tfrac{2}{3}\right)^t.$$

a) How many cans are still in use after 0 yr? 1 yr? 4 yr? 10 yr?   250,000; 166,667; 49,383; 4335
b) Graph the function. ⊡

50. *Salvage Value.*   A photocopier is purchased for $5200. Its value each year is about 80% of the value of the preceding year. Its value, in dollars, after t years is given by the exponential function

$$V(t) = 5200(0.8)^t.$$

a) Find the value of the machine after 0 yr, 1 yr, 2 yr, 5 yr, and 10 yr. ⊡
b) Graph the function. ⊡

51. *Cellular Phones.*   The number of cellular phones in use in the United States is increasing exponentially. The number N, in millions, in use is given by the exponential function

$$N(t) = 0.3(1.4477)^t,$$

where t is the number of years after 1985 (*Source*: Cellular Telecommunications and Internet Association).

a) Find the number of cellular phones in use in 1985, 1995, 2005, and 2010. ⊡
b) Graph the function. ⊡

⊡ Answers to Exercises 45, 46, 47(b), 48(b), 49(b), 50, and 51 can be found on pp. A-69 and A-70.

52. *Spread of Zebra Mussels.* Beginning in 1988, infestations of zebra mussels started spreading throughout North American waters.* These mussels spread with such speed that water treatment facilities, power plants, and entire ecosystems can become threatened. The function

$$A(t) = 10 \cdot 34^t$$

can be used to estimate the number of square centimeters of lake bottom that will be covered with mussels t years after an infestation covering 10 cm² first occurs.

a) How many square centimeters of lake bottom will be covered with mussels 5 years after an infestation covering 10 cm² first appears? 7 years after the infestation first appears?

b) Graph the function.

TW **53.** Without using a calculator, explain why 2^π must be greater than 8 but less than 16.

TW **54.** Suppose that \$1000 is invested for 5 yr at 7% interest, compounded annually. In what year will the most interest be earned? Why?

Skill Maintenance

Simplify.

55. 5^{-2} [1.4] $\frac{1}{25}$

56. 2^{-5} [1.4] $\frac{1}{32}$

57. $1000^{2/3}$ [7.2]   100

58. $25^{-3/2}$ [7.2] $\frac{1}{125}$

59. $\dfrac{10a^8b^7}{2a^2b^4}$ [1.4] $5a^6b^3$

60. $\dfrac{24x^6y^4}{4x^2y^3}$ [1.4] $6x^4y$

Synthesis

TW **61.** Examine Exercise 51. Do you believe that the equation for the number of cellular phones in use in the United States will be accurate 20 yr from now? Why or why not?

---

*Many thanks to Dr. Gerald Mackie of the Department of Zoology at the University of Guelph in Ontario for the background information for this exercise.

TW **62.** Why was it necessary to discuss irrational exponents before graphing exponential functions?

Determine which of the two numbers is larger. Do not use a calculator.

63. $\pi^{1.3}$ or $\pi^{2.4}$ $\pi^{2.4}$

64. $\sqrt{8^3}$ or $8^{\sqrt 3}$ $8^{\sqrt 3}$

Graph.

65. $y = 2^x + 2^{-x}$ ⊡

66. $y = \left|\left(\frac{1}{2}\right)^x - 1\right|$ ⊡

67. $y = |2^x - 2|$ ⊡

68. $y = 2^{-(x-1)^2}$ ⊡

69. $y = |2^{x^2} - 1|$ ⊡

70. $y = 3^x + 3^{-x}$ ⊡

Graph both equations using the same set of axes.

71. $y = 3^{-(x-1)}$, $x = 3^{-(y-1)}$ ⊡

72. $y = 1^x$, $x = 1^y$ ⊡

73. *Sales of DVD Players.* As prices of DVD players continue to drop, sales have grown from \$171 million in 1997 to \$421 million in 1998 and \$1099 million in 1999 (*Source: Statistical Abstract of the United States*, 2000). Use the REGRESSION feature in the STAT CALC menu to find an exponential function that models the total sales of DVD players t years after 1997. Then use that function to predict the total sales in 2005.

74. *Keyboarding Speed.* Ali is studying keyboarding. After he has studied for t hours, Ali's speed, in words per minute, is given by the exponential function

$$S(t) = 200[1 - (0.99)^t].$$

Use a graph and/or table of values to predict Ali's speed after studying for 10 hr, 40 hr, and 80 hr.
19 wpm, 66 wpm, 110 wpm

TW **75.** Consider any exponential function of the form $f(x) = a^x$ with $a > 1$. Will it always follow that $f(3) - f(2) > f(2) - f(1)$, and, in general, $f(n + 2) - f(n + 1) > f(n + 1) - f(n)$? Why or why not? (*Hint*: Think graphically.)

---

52. **(a)** 454,354,240 cm²; 525,233,501,400 cm²; **(b)**

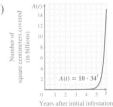

73. $A(t) = 169.3393318(2.535133248)^t$, where $A(t)$ is total sales, in millions of dollars; t years after 1997; \$288,911,061,500

⊡ Answers to Exercises 65–72 can be found on p. A-70.

Collaborative Corner

How Much Does a New Car Really Cost?

Focus: Car loans and exponential functions

Time: 30 minutes

Group size: 2

Materials: Calculators with exponentiation keys

The formula

$$M = \frac{Pr}{1 - (1 + r)^{-n}}$$

is used to determine the payment size, M, when a loan of P dollars is to be repaid in n equally sized monthly payments. Here r represents the monthly interest rate. Loans repaid in this fashion are said to be *amortized* (spread out equally) over a period of n months.

ACTIVITY

1. Suppose one group member is selling the other a car for $2600, financed at 1% interest per month for 24 months. What should be the size of each monthly payment?

2. Suppose both group members are shopping for the same model new car. To save time, each group member visits a different dealer. One dealer offers the car for $13,000 at 10.5% interest (0.00875 monthly interest) for 60 months (no down payment). The other dealer offers the same car for $12,000, but at 12% interest (0.01 monthly interest) for 48 months (no down payment).

   a) Determine the monthly payment size for each offer (remember to use the *monthly* interest rates). Then determine the total amount paid for the car under each offer. How much of each total is interest?

   b) Work together to find the annual interest rate for which the total cost of 60 monthly payments for the $13,000 car would equal the total amount paid for the $12,000 car (as found in part (a) above).

9.3

Graphs of Logarithmic Functions ■ Common Logarithms ■ Converting Exponential and Logarithmic Equations ■ Solving Certain Logarithmic Equations

Logarithmic Functions

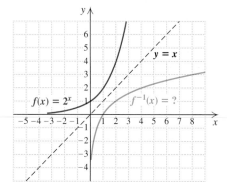

We are now ready to study inverses of exponential functions. These functions have many applications and are referred to as *logarithm*, or *logarithmic*, *functions*.

Graphs of Logarithmic Functions

Consider the exponential function $f(x) = 2^x$. Like all exponential functions, f is one-to-one. Can a formula for f^{-1} be found?

To answer this, we use the method of Section 9.1:

1. Replace $f(x)$ with y: $y = 2^x$.

2. Interchange x and y: $x = 2^y$.

3. Solve for y: $y =$ the power to which we raise 2 to get x.

4. Replace y with $f^{-1}(x)$: $f^{-1}(x) =$ the power to which we raise 2 to get x.

We now define a new symbol to replace the words "the power to which we raise 2 to get x":

$\log_2 x$, **read "the logarithm, base 2, of x," or "log, base 2, of x,"
means "the power to which we raise 2 to get x."**

Thus if $f(x) = 2^x$, then $f^{-1}(x) = \log_2 x$. Note that $f^{-1}(8) = \log_2 8 = 3$, because 3 is *the power to which we raise 2 to get 8*.

EXAMPLE 1   Simplify: **(a)** $\log_2 32$; **(b)** $\log_2 1$; **(c)** $\log_2 \frac{1}{8}$.

Solution

a) Think of the meaning of $\log_2 32$. It is the exponent to which we raise 2 to get 32. That exponent is 5. Therefore, $\log_2 32 = 5$.

b) We ask ourselves: "To what power do we raise 2 in order to get 1?" That power is 0 (recall that $2^0 = 1$). Thus, $\log_2 1 = 0$.

c) To what power do we raise 2 in order to get $\frac{1}{8}$? Since $2^{-3} = \frac{1}{8}$, we have $\log_2 \frac{1}{8} = -3$.

Although expressions like $\log_2 13$ can only be approximated, we must remember that $\log_2 13$ represents *the power to which we raise 2 to get* 13. That is, $2^{\log_2 13} = 13$. A calculator can be used to show that $\log_2 13 \approx 3.7$ and $2^{3.7} \approx 13$. Later in this chapter, we will discuss how a calculator can be used to find such approximations.

For any exponential function $f(x) = a^x$, the inverse is called a **logarithmic function, base a**. The graph of the inverse can, of course, be drawn by reflecting the graph of $f(x) = a^x$ across the line $y = x$. It will be helpful to remember that the inverse of $f(x) = a^x$ is given by $f^{-1}(x) = \log_a x$.

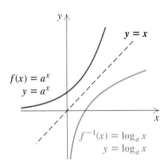

TEACHING TIP

This definition is crucial to later work with logarithms. It is essential that students understand that a logarithm is an exponent.

The Meaning of $\log_a x$

For $x > 0$ and a a positive constant other than 1, $\log_a x$ is the power to which a must be raised in order to get x. Thus,

$$a^{\log_a x} = x \qquad \text{or equivalently,} \qquad \text{if } y = \log_a x, \text{ then } a^y = x.$$

It is important to remember that *a logarithm is an exponent*. It might help to repeat several times: "The logarithm, base a, of a number x is the power to which a must be raised in order to get x."

EXAMPLE 2   Simplify: $7^{\log_7 85}$.

Solution   Remember that $\log_7 85$ is the power to which 7 is raised to get 85. Raising 7 to that power, we have

$$7^{\log_7 85} = 85.$$

Because logarithmic and exponential functions are inverses of each other, the result in Example 2 should come as no surprise: If $f(x) = \log_7 x$, then

for $f(x) = \log_7 x$, we have $f^{-1}(x) = 7^x$

and $f^{-1}(f(x)) = f^{-1}(\log_7 x) = 7^{\log_7 x} = x.$

Thus, $f^{-1}(f(85)) = 7^{\log_7 85} = 85.$

The following is a comparison of exponential and logarithmic functions.

Exponential Function	Logarithmic Function
$y = a^x$	$x = a^y$
$f(x) = a^x$	$f(x) = \log_a x$
$a > 0, a \neq 1$	$a > 0, a \neq 1$
The domain is $\mathbb{R}$.	The range is $\mathbb{R}$.
$y > 0$ (Outputs are positive.)	$x > 0$ (Inputs are positive.)
$f^{-1}(x) = \log_a x$	$f^{-1}(x) = a^x$
The graph of $f(x)$ contains the points $(0, 1)$ and $(1, a)$.	The graph of $f(x)$ contains the points $(1, 0)$ and $(a, 1)$.

EXAMPLE 3   Graph: $y = f(x) = \log_5 x$.

Solution   If $y = \log_5 x$, then $5^y = x$. We can find ordered pairs that are solutions by choosing values for y and computing the x-values.

For $y = 0, x = 5^0 = 1.$
For $y = 1, x = 5^1 = 5.$
For $y = 2, x = 5^2 = 25.$
For $y = -1, x = 5^{-1} = \frac{1}{5}.$
For $y = -2, x = 5^{-2} = \frac{1}{25}.$

(1) Select y.
(2) Compute x.

This table shows the following:

$\log_5 1 = 0;$
$\log_5 5 = 1;$
$\log_5 25 = 2;$
$\log_5 \frac{1}{5} = -1;$
$\log_5 \frac{1}{25} = -2.$

These can all be checked using the equations above.

x, or 5^y	y
1	0
5	1
25	2
$\frac{1}{5}$	-1
$\frac{1}{25}$	-2

We plot the set of ordered pairs and connect the points with a smooth curve. The graph of $y = 5^x$ is shown only for reference.

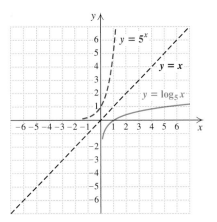

Common Logarithms

Any positive number other than 1 can serve as the base of a logarithmic function. However, some numbers are easier to use than others, and there are logarithmic bases that fit into certain applications more naturally than others.

Base-10 logarithms, called **common logarithms**, are useful because they have the same base as our "commonly" used decimal system. Before calculators became so widely available, common logarithms were helpful when performing tedious calculations. In fact, that is why logarithms were invented. Before the advent of calculators, tables were developed to list common logarithms. Today we find common logarithms using calculators.

Here, and in most books, the abbreviation **log**, with no base written, is understood to mean logarithm base 10, or a common logarithm. Thus,

$\log 17$ means $\log_{10} 17$. **It is important to remember this abbreviation.**

Common Logarithms

To find the common logarithm of a number on most graphing calculators, press $\boxed{\log}$, the number, and then $\boxed{\text{ENTER}}$. Some calculators automatically supply the left parenthesis when $\boxed{\log}$ is pressed. Of these, some require that a right parenthesis be entered and some assume one is present at the end of the expression. Even if a calculator does not require an ending parenthesis, it is a good idea to include one.

For example, if we are using a calculator that supplies the left parenthesis, log 5 can be found by pressing $\boxed{\log}$ $\boxed{5}$ $\boxed{)}$ $\boxed{\text{ENTER}}$. Note from the screen on the left below that log $5 \approx 0.6990$.

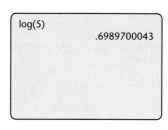

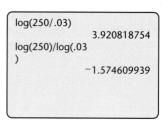

Parentheses must be placed carefully when evaluating expressions containing logarithms. For example, to evaluate

$$\log \frac{250}{0.03},$$

press $\boxed{\text{log}}$ $\boxed{2}$ $\boxed{5}$ $\boxed{0}$ $\boxed{\div}$ $\boxed{.}$ $\boxed{0}$ $\boxed{3}$ $\boxed{)}$ $\boxed{\text{ENTER}}$. To evaluate

$$\frac{\log 250}{\log 0.03},$$

press $\boxed{\text{log}}$ $\boxed{2}$ $\boxed{5}$ $\boxed{0}$ $\boxed{)}$ $\boxed{\div}$ $\boxed{\text{log}}$ $\boxed{.}$ $\boxed{0}$ $\boxed{3}$ $\boxed{)}$ $\boxed{\text{ENTER}}$. Note from the screen on the right above that $\log \dfrac{250}{0.03} \approx 3.9208$ and $\dfrac{\log 250}{\log 0.03} \approx -1.5746$.

The inverse of a logarithmic function is an exponential function. Because of this, on many calculators the $\boxed{\text{log}}$ key serves as the $\boxed{10^x}$ key after the $\boxed{\text{2nd}}$ key is pressed. As with logarithms, some calculators automatically supply a left parenthesis before the exponent. Using such a calculator, we find

$$10^{3.417}$$

by pressing $\boxed{\text{2nd}}$ $\boxed{10^x}$ $\boxed{3}$ $\boxed{.}$ $\boxed{4}$ $\boxed{1}$ $\boxed{7}$ $\boxed{)}$ $\boxed{\text{ENTER}}$. The result is approximately 2612.1614.

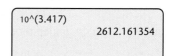

We can use a graphing calculator to show that $f(x) = \log x$ and $g(x) = 10^x$ are inverses of each other. We graph $y_1 = \log x$, $y_2 = 10^x$, and $y_3 = x$ using a squared window. The graphs of y_1 and y_2 on the left below appear to be reflections of each other across the line $y = x$.

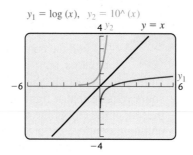

We can also let $y_3 = y_1(y_2)$ and $y_4 = y_2(y_1)$ to show, using a table as on the left at the bottom of p. 680, that $y_3 = y_4 = x$.

A graphing calculator can quickly draw graphs of logarithmic functions, base 10. It is important to remember to place parentheses properly when entering logarithmic expressions.

EXAMPLE 4   Graph: $f(x) = \log \dfrac{x}{5} + 1$.

Solution   We enter $y = \log (x/5) + 1$. Since logarithms of negative numbers are not defined, we set the window accordingly. The graph is shown below. Note that although on the calculator the graph appears to touch the y-axis, a table of values or TRACE would show that the y-axis is indeed an asymptote.

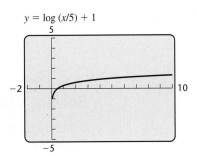

$y = \log (x/5) + 1$

Converting Exponential and Logarithmic Equations

We use the definition of logarithm to convert from *exponential equations* to *logarithmic equations*:

$$y = \log_a x \quad \text{is equivalent to} \quad a^y = x.$$

CAUTION!   **Do not forget this relationship!** It is probably the most important definition in the chapter. Many times this definition will be used to justify a property we are considering.

EXAMPLE 5   Convert each to a logarithmic equation.

a) $8 = 2^x$ **b)** $y^{-1} = 4$ **c)** $a^b = c$

Solution

a) $8 = 2^x$ is equivalent to $x = \log_2 8$ The exponent is the logarithm.

                                      The base remains the same.

b) $y^{-1} = 4$ is equivalent to $-1 = \log_y 4$

c) $a^b = c$ is equivalent to $b = \log_a c$

We also use the definition of logarithm to convert from logarithmic equations to exponential equations.

EXAMPLE 6   Convert each to an exponential equation.

a) $y = \log_3 5$ **b)** $-2 = \log_a 7$ **c)** $a = \log_b d$

Solution

a) $y = \log_3 5$ is equivalent to $3^y = 5$ The logarithm is the exponent.

The base remains the same.

b) $-2 = \log_a 7$ is equivalent to $a^{-2} = 7$
c) $a = \log_b d$ is equivalent to $b^a = d$

Solving Certain Logarithmic Equations

Some logarithmic equations can be solved by converting to exponential equations.

EXAMPLE 7   Solve: **(a)** $\log_2 x = -3$; **(b)** $\log_x 16 = 2$.

Solution

a) $\log_2 x = -3$

$\quad 2^{-3} = x$ Converting to an exponential equation

$\quad \frac{1}{8} = x$ Computing 2^{-3}

Check: $\log_2 \frac{1}{8}$ is the power to which 2 is raised to get $\frac{1}{8}$. Since that power is -3, we have a check. The solution is $\frac{1}{8}$.

b) $\log_x 16 = 2$

$\quad x^2 = 16$ Converting to an exponential equation

$\quad x = 4 \quad or \quad x = -4$ Principle of square roots

Check: $\log_4 16 = 2$ because $4^2 = 16$. Thus, 4 is a solution of $\log_x 16 = 2$. Because all logarithmic bases must be positive, -4 cannot be a solution. Logarithmic bases must be positive because logarithms are defined using exponential functions that require positive bases. The solution is 4.

One method for solving certain logarithmic and exponential equations relies on the following property, which results from the fact that exponential functions are one-to-one.

TEACHING TIP

You may want to give a few
examples illustrating the principle
of exponential equality.

The Principle of Exponential Equality
For any real number b, where $b \neq -1, 0$, or 1,

$$b^x = b^y \quad \text{is equivalent to} \quad x = y.$$

(Powers of the same base are equal if and only if the exponents are equal.)

EXAMPLE 8   Solve: **(a)** $\log_{10} 1000 = x$; **(b)** $\log_4 1 = t$.

Solution

a) We convert $\log_{10} 1000 = x$ to exponential form and solve:

$$10^x = 1000 \qquad \text{Converting to an exponential equation}$$
$$10^x = 10^3 \qquad \text{Writing 1000 as a power of 10}$$
$$x = 3. \qquad \text{Equating exponents}$$

Check: This equation can also be solved directly by determining the power to which we raise 10 in order to get 1000. In both cases we find that $\log_{10} 1000 = 3$, so we have a check. The solution is 3.

b) We convert $\log_4 1 = t$ to exponential form and solve:

$$4^t = 1 \qquad \text{Converting to an exponential equation}$$
$$4^t = 4^0 \qquad \text{Writing 1 as a power of 4. This can be done mentally.}$$
$$t = 0. \qquad \text{Equating exponents}$$

Check: As in part (a), this equation can be solved directly by determining the power to which we raise 4 in order to get 1. In both cases we find that $\log_4 1 = 0$, so we have a check. The solution is 0.

Example 8 illustrates an important property of logarithms.

$\log_a 1$
The logarithm, base a, of 1 is always 0: $\log_a 1 = 0$.

This follows from the fact that $a^0 = 1$ is equivalent to the logarithmic equation $\log_a 1 = 0$. Thus, $\log_{10} 1 = 0$, $\log_7 1 = 0$, and so on.

Another property results from the fact that $a^1 = a$. This is equivalent to the equation $\log_a a = 1$.

$\log_a a$
The logarithm, base a, of a is always 1: $\log_a a = 1$.

Thus, $\log_{10} 10 = 1$, $\log_8 8 = 1$, and so on.

9.3

Exercise Set

Simplify.

1. $\log_{10} 100$ 2

2. $\log_{10} 1000$ 3

3. $\log_2 8$ 3

4. $\log_2 16$ 4

5. $\log_3 81$ 4

6. $\log_3 27$ 3

7. $\log_4 \frac{1}{16}$ -2

8. $\log_4 \frac{1}{4}$ -1

9. $\log_7 \frac{1}{7}$ -1

10. $\log_7 \frac{1}{49}$ -2

11. $\log_5 625$ 4

12. $\log_5 125$ 3

13. $\log_6 6$ 1

14. $\log_7 1$ 0

15. $\log_8 1$ 0

16. $\log_8 8$ 1

Aha! **17.** $\log_9 9^7$ 7

18. $\log_9 9^{10}$ 10

19. $\log_{10} 0.1$ -1

20. $\log_{10} 0.01$ -2

21. $\log_9 3$ $\frac{1}{2}$

22. $\log_{16} 4$ $\frac{1}{2}$

23. $\log_9 27$ $\frac{3}{2}$

24. $\log_{16} 64$ $\frac{3}{2}$

25. $\log_{1000} 100$ $\frac{2}{3}$

26. $\log_{27} 9$ $\frac{2}{3}$

27. $5^{\log_5 7}$ 7

28. $6^{\log_6 13}$ 13

Graph by hand.

29. $y = \log_{10} x$

30. $y = \log_2 x$

31. $y = \log_3 x$

32. $y = \log_7 x$

33. $f(x) = \log_6 x$

34. $f(x) = \log_4 x$

35. $f(x) = \log_{2.5} x$

36. $f(x) = \log_{1/2} x$

Graph both functions using the same set of axes.

37. $f(x) = 3^x,\ f^{-1}(x) = \log_3 x$

38. $f(x) = 4^x,\ f^{-1}(x) = \log_4 x$

Use a calculator to find each of the following rounded to four decimal places.

39. $\log 4$ 0.6021

40. $\log 5$ 0.6990

41. $\log 13,400$ 4.1271

42. $\log 93,100$ 4.9689

43. $\log 0.527$ -0.2782

44. $\log 0.493$ -0.3072

Use a calculator to find each of the following rounded to four decimal places.

45. $10^{2.3}$ 199.5262

46. $10^{0.173}$ 1.4894

47. $10^{-2.9523}$ 0.0011

48. $10^{4.8982}$ 79,104.2833

49. $10^{0.0012}$ 1.0028

50. $10^{-3.89}$ 0.0001

Graph using a graphing calculator.

51. $\log (x + 2)$

52. $\log (x - 5)$

53. $\log (1 - 2x)$

54. $\log (3x + 2.7)$

55. $\log (x^2)$

56. $\log (x^2 + 1)$

Convert to logarithmic equations.

57. $10^2 = 100$ $2 = \log_{10} 100$

58. $10^4 = 10,000$ $4 = \log_{10} 10,000$

59. $4^{-5} = \frac{1}{1024}$ $-5 = \log_4 \frac{1}{1024}$

60. $5^{-3} = \frac{1}{125}$ $-3 = \log_5 \frac{1}{125}$

61. $16^{3/4} = 8$ $\frac{3}{4} = \log_{16} 8$

62. $8^{1/3} = 2$ $\frac{1}{3} = \log_8 2$

63. $10^{0.4771} = 3$ $0.4771 = \log_{10} 3$

64. $10^{0.3010} = 2$ $0.3010 = \log_{10} 2$

65. $p^k = 3$ $k = \log_p 3$

66. $m^n = r$ $n = \log_m r$

67. $p^m = V$ $m = \log_p V$

68. $Q^t = x$ $t = \log_Q x$

69. $e^3 = 20.0855$ $3 = \log_e 20.0855$

70. $e^2 = 7.3891$ $2 = \log_e 7.3891$

71. $e^{-4} = 0.0183$ $-4 = \log_e 0.0183$

72. $e^{-2} = 0.1353$ $-2 = \log_e 0.1353$

Convert to exponential equations.

73. $t = \log_3 8$ $3^t = 8$

74. $h = \log_7 10$ $7^h = 10$

75. $\log_5 25 = 2$ $5^2 = 25$

76. $\log_6 6 = 1$ $6^1 = 6$

77. $\log_{10} 0.1 = -1$ $10^{-1} = 0.1$

78. $\log_{10} 0.01 = -2$ $10^{-2} = 0.01$

79. $\log_{10} 7 = 0.845$ $10^{0.845} = 7$

80. $\log_{10} 3 = 0.4771$ $10^{0.4771} = 3$

81. $\log_c m = 8$ $c^8 = m$

82. $\log_b n = 23$ $b^{23} = n$

83. $\log_t Q = r$ $t^r = Q$

84. $\log_m P = a$ $m^a = P$

85. $\log_e 0.25 = -1.3863$ $e^{-1.3863} = 0.25$

86. $\log_e 0.989 = -0.0111$ $e^{-0.0111} = 0.989$

87. $\log_r T = -x$ $r^{-x} = T$

88. $\log_c M = -w$ $c^{-w} = M$

Solve.

89. $\log_3 x = 2$ 9

90. $\log_4 x = 3$ 64

91. $\log_x 64 = 3$ 4

92. $\log_x 125 = 3$ 5

93. $\log_5 25 = x$ 2

94. $\log_2 16 = x$ 4

95. $\log_4 16 = x$ 2

96. $\log_3 27 = x$ 3

97. $\log_x 7 = 1$ 7

98. $\log_x 8 = 1$ 8

99. $\log_9 x = 1$ 9

100. $\log_6 x = 0$ 1

 Answers to Exercises 29–38 and 51–56 can be found on p. A-70.

101. $\log_3 x = -2$ $\frac{1}{9}$ **102.** $\log_2 x = -1$ $\frac{1}{2}$

103. $\log_{32} x = \frac{2}{5}$ 4   **104.** $\log_8 x = \frac{2}{3}$ 4

TW **105.** Express in words what number is represented by $\log_b c$.

TW **106.** Is it true that $2 = b^{\log_b 2}$? Why or why not?

Skill Maintenance

Simplify.

107. $\dfrac{x^{12}}{x^4}$ [1.4] x^8 **108.** $\dfrac{a^{15}}{a^3}$ [1.4] a^{12}

109. $(a^4 b^6)(a^3 b^2)$ [1.4] $a^7 b^8$ **110.** $(x^3 y^5)(x^2 y^7)$ [1.4] $x^5 y^{12}$

111. $\dfrac{\dfrac{3}{x} - \dfrac{2}{xy}}{\dfrac{2}{x^2} + \dfrac{1}{xy}}$ [6.3] $\dfrac{x(3y-2)}{2y+x}$ **112.** $\dfrac{\dfrac{4+x}{x^2+2x+1}}{\dfrac{3}{x+1} - \dfrac{2}{x+2}}$ [6.3] $\dfrac{x+2}{x+1}$

Synthesis

TW **113.** Would a manufacturer be pleased or unhappy if sales of a product grew logarithmically? Why?

TW **114.** Explain why the number $\log_2 13$ must be between 3 and 4.

115.

$y = \left(\frac{3}{2}\right)^x$ $y = \log_{3/2} x$

116.
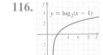
$y = \log_2(x-1)$

117.
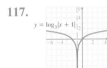
$y = \log_3 |x+1|$

115. Graph both equations using the same set of axes:
$$y = \left(\tfrac{3}{2}\right)^x, \qquad y = \log_{3/2} x.$$

Graph by hand.

116. $y = \log_2 (x - 1)$ **117.** $y = \log_3 |x + 1|$

Solve.

118. $|\log_3 x| = 2$ $\frac{1}{9}, 9$ **119.** $\log_{125} x = \frac{2}{3}$ 25

120. $\log_4 (3x - 2) = 2$ 6   **121.** $\log_8 (2x + 1) = -1$ $-\frac{7}{16}$

122. $\log_{10} (x^2 + 21x) = 2$ $-25, 4$

Simplify.

123. $\log_{1/4} \frac{1}{64}$ 3

124. $\log_{1/5} 25$ -2

125. $\log_{81} 3 \cdot \log_3 81$ 1

126. $\log_{10} (\log_4 (\log_3 81))$ 0

127. $\log_2 (\log_2 (\log_4 256))$ 1

128. Show that $b^x = b^y$ is *not* equivalent to $x = y$ for $b = 0$ or $b = 1$.

TW **129.** If $\log_b a = x$, does it follow that $\log_a b = 1/x$? Why or why not?

---

128. Let $b = 0$, and suppose that $x = 1$ and $y = 2$. Then $0^1 = 0^2$, but $1 \neq 2$. Then let $b = 1$, and suppose that $x = 1$ and $y = 2$. Then $1^1 = 1^2$, but $1 \neq 2$.

9.4

Logarithms of Products ■ Logarithms of Powers ■ Logarithms of Quotients ■ Using the Properties Together

Properties of Logarithmic Functions

Logarithmic functions are important in many applications and in more advanced mathematics. We now establish some basic properties that are useful in manipulating expressions involving logarithms.

Interactive Discovery

For each of the following, use a calculator to determine which is the equivalent logarithmic expression.

1. $\log (20 \cdot 5)$ (c)
    a) $\log (20) \cdot \log (5)$
    b) $5 \cdot \log (20)$
    c) $\log (20) + \log (5)$

2. $\log 20^5$ (b)
    a) $\log (20) \cdot \log (5)$
    b) $5 \cdot \log (20)$
    c) $\log (20) - \log (5)$

3. $\log \left(\dfrac{20}{5} \right)$ (c)

a) $\log (20) \cdot \log (5)$
b) $\log (20)/\log (5)$
c) $\log (20) - \log (5)$

4. $\log \dfrac{1}{20}$ (a)

a) $-\log (20)$
b) $\log (1)/\log (20)$
c) $\log (1) \cdot \log (20)$

In this section, we will state and prove the results you may have observed. As their proofs reveal, the properties of logarithms are related to the properties of exponents.

TEACHING TIP

You may want to give concrete examples for the properties introduced in this section—for example, $8 = 2^3$, $16 = 2^4$, and $8 \cdot 16 = 2^3 \cdot 2^4 = 2^{3+4} = 2^7$. Thus, $\log_2 8 \cdot 16 = \log_2 2^7 = 7$ and $\log_2 8 \cdot 16 = \log_2 8 + \log_2 16 = 3 + 4 = 7$.

Logarithms of Products

The first property we discuss is related to the product rule for exponents: $a^m \cdot a^n = a^{m+n}$.

The Product Rule for Logarithms

For any positive numbers M, N, and a ($a \neq 1$),

$$\log_a MN = \log_a M + \log_a N.$$

(The logarithm of a product is the sum of the logarithms of the factors.)

EXAMPLE 1   Express as a sum of logarithms: $\log_2 (4 \cdot 16)$.

Solution   We have

$$\log_2 (4 \cdot 16) = \log_2 4 + \log_2 16.$$ **Using the product rule for logarithms**

As a check, note that

$$\log_2 (4 \cdot 16) = \log_2 64 = 6 \qquad 2^6 = 64$$

and that

$$\log_2 4 + \log_2 16 = 2 + 4 = 6. \qquad 2^2 = 4 \text{ and } 2^4 = 16$$

EXAMPLE 2   Express as a single logarithm: $\log_b 7 + \log_b 5$.

Solution   We have

$$\log_b 7 + \log_b 5 = \log_b (7 \cdot 5) \qquad \textbf{Using the product rule for logarithms}$$
$$= \log_b 35.$$

The check is left to the student.

A PROOF OF THE PRODUCT RULE.   Let $\log_a M = x$ and $\log_a N = y$. Converting to exponential equations, we have $a^x = M$ and $a^y = N$.
    Now we multiply the last two equations, to obtain

$$MN = a^x \cdot a^y, \quad \text{or} \quad MN = a^{x+y}.$$

Converting back to a logarithmic equation, we get

$$\log_a MN = x + y.$$

Recalling what x and y represent, we get

$$\log_a MN = \log_a M + \log_a N.$$

Logarithms of Powers

The second basic property is related to the power rule for exponents: $(a^m)^n = a^{mn}$.

> ### The Power Rule for Logarithms
>
> For any positive numbers M and a ($a \neq 1$), and any real number p,
>
> $$\log_a M^p = p \cdot \log_a M.$$
>
> (The logarithm of a power of M is the exponent times the logarithm of M.)

To better understand the power rule, note that

$$\log_a M^3 = \log_a (M \cdot M \cdot M) = \log_a M + \log_a M + \log_a M$$
$$= 3 \log_a M.$$

EXAMPLE 3   Express as a product: **(a)** $\log_a 9^{-5}$; **(b)** $\log_7 \sqrt[3]{x}$.

Solution

a) $\log_a 9^{-5} = -5 \log_a 9$ **Using the power rule for logarithms**

b) $\log_7 \sqrt[3]{x} = \log_7 x^{1/3}$ **Writing exponential notation**

$\phantom{\log_7 \sqrt[3]{x}} = \frac{1}{3} \log_7 x$ **Using the power rule for logarithms**

A PROOF OF THE POWER RULE.   Let $x = \log_a M$. We then write the equivalent exponential equation, $a^x = M$. Raising both sides to the pth power, we get

$$(a^x)^p = M^p, \quad \text{or} \quad a^{xp} = M^p. \qquad \text{Multiplying exponents}$$

Converting back to a logarithmic equation gives us

$$\log_a M^p = xp.$$

But $x = \log_a M$, so substituting, we have

$$\log_a M^p = (\log_a M)p = p \cdot \log_a M.$$

Logarithms of Quotients

The third property that we study is similar to the quotient rule for exponents: $a^m/a^n = a^{m-n}$.

The Quotient Rule for Logarithms

For any positive numbers M, N, and a ($a \neq 1$),

$$\log_a \frac{M}{N} = \log_a M - \log_a N.$$

(The logarithm of a quotient is the logarithm of the dividend minus the logarithm of the divisor.)

To better understand the quotient rule, note that

$$\log_a \left(\frac{b^5}{b^3} \right) = \log_a b^2 = 2 \log_a b = 5 \log_a b - 3 \log_a b$$

$$= \log_a b^5 - \log_a b^3.$$

EXAMPLE 4 Express as a difference of logarithms: $\log_t (6/U)$.

Solution

$$\log_t \frac{6}{U} = \log_t 6 - \log_t U \qquad \text{Using the quotient rule for logarithms}$$

EXAMPLE 5 Express as a single logarithm: $\log_b 17 - \log_b 27$.

Solution

$$\log_b 17 - \log_b 27 = \log_b \frac{17}{27} \qquad \text{Using the quotient rule for logarithms "in reverse"}$$

A PROOF OF THE QUOTIENT RULE. Our proof uses both the product and power rules:

$$\log_a \frac{M}{N} = \log_a MN^{-1} \qquad \text{Rewriting } \frac{M}{N} \text{ with a negative exponent}$$

$$= \log_a M + \log_a N^{-1} \qquad \text{Using the product rule for logarithms}$$

$$= \log_a M + (-1)\log_a N \qquad \text{Using the power rule for logarithms}$$

$$= \log_a M - \log_a N.$$

Using the Properties Together

EXAMPLE 6 Express in terms of the individual logarithms of x, y, and z.

a) $\log_b \dfrac{x^3}{yz}$

b) $\log_a \sqrt[4]{\dfrac{xy}{z^3}}$

Solution

a) $\log_b \dfrac{x^3}{yz} = \log_b x^3 - \log_b yz$ **Using the quotient rule for logarithms**

$\qquad\qquad = 3 \log_b x - \log_b yz$ **Using the power rule for logarithms**

$\qquad\qquad = 3 \log_b x - (\log_b y + \log_b z)$ **Using the product rule for logarithms. Because of the subtraction, parentheses are essential.**

$\qquad\qquad = 3 \log_b x - \log_b y - \log_b z$ **Using the distributive law**

b) $\log_a \sqrt[4]{\dfrac{xy}{z^3}} = \log_a \left(\dfrac{xy}{z^3}\right)^{1/4}$ **Writing exponential notation**

$\qquad\qquad = \dfrac{1}{4} \cdot \log_a \dfrac{xy}{z^3}$ **Using the power rule for logarithms**

$\qquad\qquad = \dfrac{1}{4} \left(\log_a xy - \log_a z^3\right)$ **Using the quotient rule for logarithms. Parentheses are important.**

$\qquad\qquad = \dfrac{1}{4} \left(\log_a x + \log_a y - 3 \log_a z\right)$ **Using the product and power rules for logarithms**

CAUTION!   When subtraction or multiplication precedes use of the product or quotient rule, parentheses are needed, as in Example 6.

EXAMPLE 7   Express as a single logarithm.

a) $\dfrac{1}{2} \log_a x - 7 \log_a y + \log_a z$ **b)** $\log_a \dfrac{b}{\sqrt{x}} + \log_a \sqrt{bx}$

Solution

a) $\dfrac{1}{2} \log_a x - 7 \log_a y + \log_a z$

$\qquad = \log_a x^{1/2} - \log_a y^7 + \log_a z$ **Using the power rule for logarithms**

$\qquad = \left(\log_a \sqrt{x} - \log_a y^7\right) + \log_a z$ **Using parentheses to emphasize the order of operations; $x^{1/2} = \sqrt{x}$**

$\qquad = \log_a \dfrac{\sqrt{x}}{y^7} + \log_a z$ **Using the quotient rule for logarithms**

$\qquad = \log_a \dfrac{z\sqrt{x}}{y^7}$ **Using the product rule for logarithms**

b) $\log_a \dfrac{b}{\sqrt{x}} + \log_a \sqrt{bx} = \log_a \dfrac{b \cdot \sqrt{bx}}{\sqrt{x}}$ **Using the product rule for logarithms**

$\qquad\qquad = \log_a b\sqrt{b}$ **Removing a factor equal to 1: $\dfrac{\sqrt{x}}{\sqrt{x}} = 1$**

$\qquad\qquad = \log_a b^{3/2},\ \text{or}\ \dfrac{3}{2} \log_a b$ **Since $b\sqrt{b} = b^1 \cdot b^{1/2}$**

TEACHING TIP

You may want to explain what is meant by "in terms of individual logarithms" and "a single logarithm."

If we know the logarithms of two different numbers (to the same base), the properties allow us to calculate other logarithms.

EXAMPLE 8   Given $\log_a 2 = 0.431$ and $\log_a 3 = 0.683$, find each of the following.

a) $\log_a 6$ b) $\log_a \frac{2}{3}$ c) $\log_a 81$

d) $\log_a \frac{1}{3}$ e) $\log_a 2a$ f) $\log_a 5$

Solution

a) $\log_a 6 = \log_a (2 \cdot 3) = \log_a 2 + \log_a 3$ **Using the product rule for logarithms**

$$= 0.431 + 0.683 = 1.114$$

Check: $a^{1.114} = a^{0.431} \cdot a^{0.683} = 2 \cdot 3 = 6$

b) $\log_a \frac{2}{3} = \log_a 2 - \log_a 3$ **Using the quotient rule for logarithms**

$$= 0.431 - 0.683 = -0.252$$

c) $\log_a 81 = \log_a 3^4 = 4 \log_a 3$ **Using the power rule for logarithms**

$$= 4(0.683) = 2.732$$

d) $\log_a \frac{1}{3} = \log_a 1 - \log_a 3$ **Using the quotient rule for logarithms**

$$= 0 - 0.683 = -0.683$$

e) $\log_a 2a = \log_a 2 + \log_a a$ **Using the product rule for logarithms**

$$= 0.431 + 1 = 1.431$$

f) $\log_a 5$ *cannot be found using these properties.* $(\log_a 5 \neq \log_a 2 + \log_a 3)$

A final property follows from the product rule: Since $\log_a a^k = k \log_a a$, and $\log_a a = 1$, we have $\log_a a^k = k$.

The Logarithm of the Base to a Power

For any base a,

$$\log_a a^k = k.$$

(The logarithm, base a, of a to a power is the power.)

This property also follows from the definition of logarithm: k is the power to which you raise a in order to get a^k.

TEACHING TIP

You may want to make sure that students understand why $\log_a a^k = k$, and remind them to be alert for expressions like $\log_a a^k$ when simplifying logarithms.

EXAMPLE 9   Simplify: **(a)** $\log_3 3^7$; **(b)** $\log_{10} 10^{-5.2}$.

Solution

a) $\log_3 3^7 = 7$ **7 is the power to which you raise 3 in order to get 3^7.**

b) $\log_{10} 10^{-5.2} = -5.2$

We summarize the properties covered in this section as follows.

21. $\log_a \frac{15}{3}$, or $\log_a 5$ **22.** $\log_b \frac{42}{7}$, or $\log_b 6$ **23.** $\log_b \frac{36}{4}$, or $\log_b 9$
24. $\log_a \frac{26}{2}$, or $\log_a 13$
27. $5\log_a x + 7\log_a y + 6\log_a z$
28. $\log_a x + 4\log_a y + 3\log_a z$
29. $\log_b x + 2\log_b y - 3\log_b z$
30. $2\log_b x + 5\log_b y - 4\log_b w - 7\log_b z$
31. $4\log_a x - 3\log_a y - \log_a z$
32. $4\log_a x - \log_a y - 2\log_a z$
33. $\log_b x + 2\log_b y - \log_b w - 3\log_b z$
34. $2\log_b w + \log_b x - 3\log_b y - \log_b z$
35. $\frac{1}{2}(7\log_a x - 5\log_a y - 8\log_a z)$
36. $\frac{1}{3}(4\log_c x - 3\log_c y - 2\log_c z)$
37. $\frac{1}{3}(6\log_a x + 3\log_a y - 2 - 7\log_a z)$
38. $\frac{1}{4}(8\log_a x + 12\log_a y - 3 - 5\log_a z)$

> For any positive numbers M, N, and a $(a \neq 1)$:
> $$\log_a MN = \log_a M + \log_a N; \qquad \log_a M^p = p \cdot \log_a M;$$
> $$\log_a \frac{M}{N} = \log_a M - \log_a N; \qquad \log_a a^k = k.$$

CAUTION! Keep in mind that, in general,

$$\log_a (M + N) \neq \log_a M + \log_a N, \qquad \log_a MN \neq (\log_a M)(\log_a N),$$

$$\log_a (M - N) \neq \log_a M - \log_a N, \qquad \log_a \frac{M}{N} \neq \frac{\log_a M}{\log_a N}.$$

9.4

Exercise Set

Express as a sum of logarithms.

1. $\log_3 (81 \cdot 27)$
   $\log_3 81 + \log_3 27$
2. $\log_2 (16 \cdot 32)$
   $\log_2 16 + \log_2 32$
3. $\log_4 (64 \cdot 16)$
   $\log_4 64 + \log_4 16$
4. $\log_5 (25 \cdot 125)$
   $\log_5 25 + \log_5 125$
5. $\log_c rst$
   $\log_c r + \log_c s + \log_c t$
6. $\log_t 3ab$
   $\log_t 3 + \log_t a + \log_t b$

Express as a single logarithm.

7. $\log_a 5 + \log_a 14$
   $\log_a (5 \cdot 14)$, or $\log_a 70$
8. $\log_b 65 + \log_b 2$
   $\log_b (65 \cdot 2)$, or $\log_b 130$
9. $\log_c t + \log_c y$ $\log_c (t \cdot y)$
10. $\log_t H + \log_t M$
   $\log_t (H \cdot M)$

Express as a product.

11. $\log_a r^8$ $8\log_a r$
12. $\log_b t^5$ $5\log_b t$
13. $\log_c y^6$ $6\log_c y$
14. $\log_{10} y^7$ $7\log_{10} y$
15. $\log_b C^{-3}$ $-3\log_b C$
16. $\log_c M^{-5}$ $-5\log_c M$

Express as a difference of logarithms.

17. $\log_2 \frac{53}{17}$ $\log_2 53 - \log_2 17$
18. $\log_3 \frac{23}{9}$ $\log_3 23 - \log_3 9$
19. $\log_b \frac{m}{n}$ $\log_b m - \log_b n$
20. $\log_a \frac{y}{x}$ $\log_a y - \log_a x$

Express as a single logarithm.

21. $\log_a 15 - \log_a 3$
22. $\log_b 42 - \log_b 7$

23. $\log_b 36 - \log_b 4$
24. $\log_a 26 - \log_a 2$
25. $\log_a 7 - \log_a 18$ $\log_a \frac{7}{18}$
26. $\log_b 5 - \log_b 13$ $\log_b \frac{5}{13}$

Express in terms of the individual logarithms of w, x, y, and z.

27. $\log_a x^5 y^7 z^6$
28. $\log_a xy^4 z^3$
29. $\log_b \frac{xy^2}{z^3}$
30. $\log_b \frac{x^2 y^5}{w^4 z^7}$
31. $\log_a \frac{x^4}{y^3 z}$
32. $\log_a \frac{x^4}{yz^2}$
33. $\log_b \frac{xy^2}{wz^3}$
34. $\log_b \frac{w^2 x}{y^3 z}$
35. $\log_a \sqrt{\frac{x^7}{y^5 z^8}}$
36. $\log_c \sqrt[3]{\frac{x^4}{y^3 z^2}}$
37. $\log_a \sqrt[3]{\frac{x^6 y^3}{a^2 z^7}}$
38. $\log_a \sqrt[4]{\frac{x^8 y^{12}}{a^3 z^5}}$

Express as a single logarithm and, if possible, simplify.

39. $7\log_a x + 3\log_a z$ $\log_a x^7 z^3$

40. $2 \log_b m + \frac{1}{2} \log_b n$ $\log_b m^2 n^{1/2}$, or $\log_b m^2 \sqrt{n}$

41. $\log_a x^2 - 2 \log_a \sqrt{x}$ $\log_a x$

42. $\log_a \dfrac{a}{\sqrt{x}} - \log_a \sqrt{ax}$ $\log_a \dfrac{\sqrt{a}}{x}$

43. $\frac{1}{2} \log_a x + 5 \log_a y - 2 \log_a x$ $\log_a \dfrac{y^5}{x^{3/2}}$

44. $\log_a 2x + 3(\log_a x - \log_a y)$ $\log_a \dfrac{2x^4}{y^3}$

45. $\log_a (x^2 - 4) - \log_a (x + 2)$ $\log_a (x - 2)$

46. $\log_a (2x + 10) - \log_a (x^2 - 25)$ $\log_a \dfrac{2}{x - 5}$

Given $\log_b 3 = 0.792$ and $\log_b 5 = 1.161$. If possible, find each of the following.

47. $\log_b 15$ 1.953

48. $\log_b \frac{5}{3}$ 0.369

49. $\log_b \frac{3}{5}$ -0.369

50. $\log_b \frac{1}{3}$ -0.792

51. $\log_b \frac{1}{5}$ -1.161

52. $\log_b \sqrt{b}$ $\frac{1}{2}$

53. $\log_b \sqrt{b^3}$ $\frac{3}{2}$

54. $\log_b 3b$ 1.792

55. $\log_b 8$ Cannot be found

56. $\log_b 45$ 2.745

57. $\log_b 75$ 3.114

58. $\log_b 20$ Cannot be found

Simplify.

Aha! **59.** $\log_t t^9$ 9

60. $\log_p p^4$ 4

61. $\log_e e^m$ m

62. $\log_Q Q^{-2}$ -2

Find each of the following.

Aha! **63.** $\log_5 (125 \cdot 625)$ 7

64. $\log_3 (9 \cdot 81)$ 6

65. $\log_2 \left(\dfrac{128}{16} \right)$ 3

66. $\log_3 \left(\dfrac{243}{27} \right)$ 2

TW **67.** A student *incorrectly* reasons that

$$\log_b \frac{1}{x} = \log_b \frac{x}{xx}$$

$$= \log_b x - \log_b x + \log_b x = \log_b x.$$

What mistake has the student made?

TW **68.** How could you convince someone that

$$\log_a c \neq \log_c a?$$

Skill Maintenance

Graph by hand.

69. $f(x) = \sqrt{x} - 3$

70. $g(x) = \sqrt{x} + 2$

71. $g(x) = \sqrt[3]{x} + 1$

72. $f(x) = \sqrt[3]{x} - 1$

Simplify. [1.4]

73. $(a^3 b^2)^5 (a^2 b^7)$ $a^{17} b^{17}$

74. $(x^5 y^3 z^2)(x^2 y z^2)^3$ $x^{11} y^6 z^8$

Synthesis

TW **75.** Is it possible to express $\log_b \dfrac{x}{5}$ as a difference of two logarithms without using the quotient rule? Why or why not?

TW **76.** Is it true that $\log_a x + \log_b x = \log_{ab} x$? Why or why not?

Express as a single logarithm and, if possible, simplify.

77. $\log_a (x^8 - y^8) - \log_a (x^2 + y^2)$ $\log_a (x^6 - x^4 y^2 + x^2 y^4 - y^6)$

78. $\log_a (x + y) + \log_a (x^2 - xy + y^2)$ $\log_a (x^3 + y^3)$

Express as a sum or difference of logarithms.

79. $\log_a \sqrt{1 - s^2}$

80. $\log_a \dfrac{c - d}{\sqrt{c^2 - d^2}}$

81. If $\log_a x = 2$, $\log_a y = 3$, and $\log_a z = 4$, what is

$$\log_a \dfrac{\sqrt[3]{x^2 z}}{\sqrt[3]{y^2 z^{-2}}}?$$ $\dfrac{10}{3}$

82. If $\log_a x = 2$, what is $\log_a (1/x)$? -2

83. If $\log_a x = 2$, what is $\log_{1/a} x$? -2

Classify each of the following as true or false. Assume a, x, P, and $Q > 0$.

84. $\log_a \left(\dfrac{P}{Q} \right)^x = x \log_a P - \log_a Q$ False

85. $\log_a (Q + Q^2) = \log_a Q + \log_a (Q + 1)$ True

86. Use graphs to show that

$$\log x^2 \neq \log x \cdot \log x.$$

69.

70. $g(x) = \sqrt{x} + 2$

71. $g(x) = \sqrt[3]{x} + 1$

72. $f(x) = \sqrt[3]{x} - 1$

79. $\frac{1}{2} \log_a (1 - s) + \frac{1}{2} \log_a (1 + s)$ **80.** $\frac{1}{2} \log_a (c - d) - \frac{1}{2} \log_a (c + d)$ **86.** $y_1 = \log x^2, \ y_2 = \log x \cdot \log x$

9.5

The Base e and Natural Logarithms ■ Changing Logarithmic Bases ■ Graphs of Exponential and Logarithmic Functions, Base e

Natural Logarithms and Changing Bases

Any positive number other than 1 can serve as the base of a logarithmic function. However, there are logarithm bases that fit into certain applications more naturally than others. We have already looked at logarithms with one such base, base-10 logarithms, or common logarithms. Another logarithm base widely used today is an irrational number named e.

The Base e and Natural Logarithms

When interest is computed n times a year, the compound interest formula is

$$A = P\left(1 + \frac{r}{n}\right)^{nt},$$

where A is the amount that an initial investment P will be worth after t years at interest rate r. Suppose that \$1 is invested at 100% interest for 1 year (no bank would pay this). The preceding formula becomes a function A defined in terms of the number of compounding periods n:

$$A(n) = \left(1 + \frac{1}{n}\right)^{n}.$$

Interactive Discovery

1. What happens to the function values of $A(n) = \left(1 + \frac{1}{n}\right)^{n}$ as n gets larger?

To find out, use the TABLE feature with Indpnt set to Ask to fill in the following table. Round each entry to six decimal places.

n	$A(n) = \left(1 + \frac{1}{n}\right)^{n}$
1 (compounded annually)	$\left(1 + \frac{1}{1}\right)^{1}$, or \$2.00
2 (compounded semiannually)	$\left(1 + \frac{1}{2}\right)^{2}$, or \$⬚ \$2.25
3	\$2.370370
4 (compounded quarterly)	\$2.441406
12 (compounded monthly)	\$2.613035
52 (compounded weekly)	\$2.692597
365 (compounded daily)	\$2.714567
8760 (compounded hourly)	\$2.718127

2. Which of the following statements appears to be true?   (c)

  a) $A(n)$ gets very large as n gets very large.

  b) $A(n)$ gets very small as n gets very large.

  c) $A(n)$ approaches a certain number as n gets very large.

The numbers in the table approach a very important number in mathematics, called e. Because e is irrational, its decimal representation does not terminate or repeat.

The Number *e*

$e \approx 2.7182818284\ldots$

Logarithms base e are called **natural logarithms**, or **Napierian logarithms**, in honor of John Napier (1550–1617), who first "discovered" logarithms.

The abbreviation "ln" is generally used with natural logarithms. Thus,

  $\ln 53$ means $\log_e 53$.

Natural Logarithms

Natural logarithms are entered on a graphing calculator much like common logarithms are (see pp. 679–680). For example, if we are using a calculator that supplies the left parenthesis, ln 8 can be found by pressing $\boxed{\text{ln}}$ $\boxed{8}$ $\boxed{)}$ $\boxed{\text{ENTER}}$. Note from the screen below that $\ln 8 \approx 2.0794$.

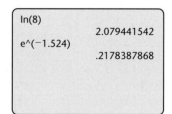

On many calculators, the $\boxed{\text{ln}}$ key serves as the $\boxed{e^x}$ key after the $\boxed{\text{2nd}}$ key is pressed. Some calculators automatically supply a left parenthesis before the exponent. If such a calculator is used,

  $e^{-1.524}$

is found by pressing $\boxed{\text{2nd}}$ $\boxed{e^x}$ $\boxed{(-)}$ $\boxed{1}$ $\boxed{.}$ $\boxed{5}$ $\boxed{2}$ $\boxed{4}$ $\boxed{)}$ $\boxed{\text{ENTER}}$. As shown on the screen above, $e^{-1.524} \approx 0.2178$.

Changing Logarithmic Bases

Most calculators can find both common logarithms and natural logarithms. To find a logarithm with some other base, a conversion formula is needed.

The Change-of-Base Formula

For any logarithmic bases a and b, and any positive number M,

$$\log_b M = \frac{\log_a M}{\log_a b}.$$

(To find the log, base b, of some number M, find the log of M using another base—usually 10 or e—and divide by the log of b to that same base.)

PROOF.   Let $x = \log_b M$. Then,

$$b^x = M \qquad \text{Rewriting } x = \log_b M \text{ in exponential form}$$
$$\log_a b^x = \log_a M \qquad \text{Taking the logarithm, base } a, \text{ on both sides}$$
$$x \log_a b = \log_a M \qquad \text{Using the power rule for logarithms}$$
$$x = \frac{\log_a M}{\log_a b}. \qquad \text{Dividing both sides by } \log_a b$$

But at the outset we stated that $x = \log_b M$. Thus, by substitution, we have

$$\log_b M = \frac{\log_a M}{\log_a b},$$

which is the change-of-base formula.

EXAMPLE 1   Find $\log_5 8$ using the change-of-base formula.

Solution   We use the change-of-base formula with $a = 10$, $b = 5$, and $M = 8$:

$$\log_5 8 = \frac{\log_{10} 8}{\log_{10} 5} \qquad \text{Substituting into } \log_b M = \frac{\log_a M}{\log_a b}$$

$$\approx \frac{0.903089987}{0.6989700043} \qquad \text{Using } \boxed{\log} \text{ twice}$$

$$\approx 1.2920. \qquad \text{When using a calculator, it is best not to round before dividing.}$$

The figure below shows the computation using a graphing calculator.

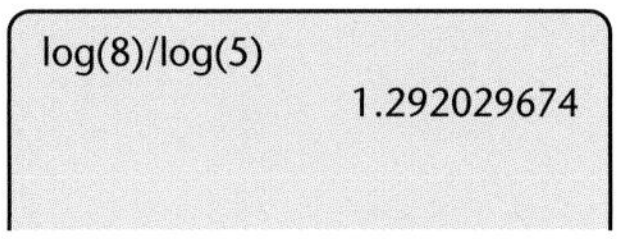

```
log(8)/log(5)
              1.292029674
```

To check, note that $\ln 8 / \ln 5 \approx 1.2920$. We can also use a calculator to verify that $5^{1.2920} \approx 8$.

TEACHING TIP

You may need to remind students that
$\frac{\log 8}{\log 5} \neq \log 8 - \log 5$.

EXAMPLE 2   Find $\log_4 31$.

Solution   As indicated in the check of Example 1, base e can also be used.

$$\log_4 31 = \frac{\log_e 31}{\log_e 4} \qquad \text{Substituting into } \log_b M = \frac{\log_a M}{\log_a b}$$

$$= \frac{\ln 31}{\ln 4} \qquad \text{Using } \boxed{\text{ln}} \text{ twice}$$

$$\approx 2.4771. \qquad \textit{Check: } 4^{2.4771} \approx 31$$

The calculation and check are shown below.

```
ln(31)/ln(4)
                    2.477098155
4^Ans
                             31
```

Graphs of Exponential and Logarithmic Functions, Base e

EXAMPLE 3   Graph $f(x) = e^x$ and $g(x) = e^{-x}$ and state the domain and the range of f and g.

Solution   We use a calculator with an $\boxed{e^x}$ key to find approximate values of e^x and e^{-x}. Using these values, we can graph the functions.

x	e^x	e^{-x}
0	1	1
1	2.7	0.4
2	7.4	0.1
−1	0.4	2.7
−2	0.1	7.4

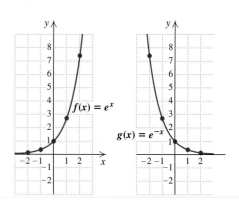

The domain of each function is $\mathbb{R}$ and the range of each function is $(0, \infty)$.

EXAMPLE 4   Graph $f(x) = e^{-0.5x} + 1$ and state the domain and the range of f.

Solution   To graph by hand, we find some solutions with a calculator, plot them, and then draw the graph. For example,

$$f(2) = e^{-0.5(2)} + 1 = e^{-1} + 1 \approx 1.4.$$

To graph using a graphing calculator, we enter

$$y = e\,^{\wedge}\,(-0.5x) + 1$$

and choose a viewing window that shows more of the y-axis than the x-axis, since the function is exponential.

x	$e^{-0.5x} + 1$
0	2
1	1.6
2	1.4
3	1.2
−1	2.6
−2	3.7
−3	5.5

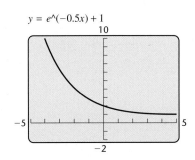

The domain of f is $\mathbb{R}$ and the range is $(1, \infty)$.

EXAMPLE 5   Graph and state the domain and the range of each function.

a) $g(x) = \ln x$ **b)** $f(x) = \ln (x + 3)$

Solution

a) To graph by hand, we find some solutions with a calculator and then draw the graph. As expected, the graph is a reflection across the line $y = x$ of the graph of $y = e^x$. We can also graph $y = \ln (x)$ using a graphing calculator. Since a logarithmic function is the inverse of an exponential function, we choose a viewing window that shows more of the x-axis than the y-axis.

x	$\ln x$
1	0
4	1.4
7	1.9
0.5	−0.7

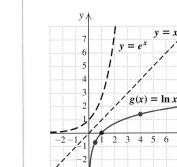

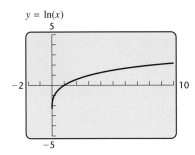

The domain of g is $(0, \infty)$ and the range is $\mathbb{R}$.

b) We find some solutions with a calculator, plot them, and draw the graph by hand. Note that the graph of $y = \ln(x + 3)$ is the graph of $y = \ln x$ translated 3 units to the left. To graph using a graphing calculator, we enter $y = \ln(x + 3)$.

x	$\ln(x + 3)$
0	1.1
1	1.4
2	1.6
3	1.8
4	1.9
−1	0.7
−2	0
−2.5	−0.7

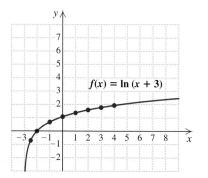

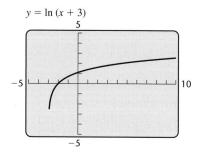

The domain of f is $(-3, \infty)$ and the range is $\mathbb{R}$.

Logarithmic functions with bases other than 10 or e can be drawn on a graphing calculator using the change-of-base formula.

EXAMPLE 6 Graph: $f(x) = \log_7 x + 2$.

Solution   We use the change-of-base formula with natural logarithms. (We would get the same graph if we used common logarithms.)

$$f(x) = \log_7 x + 2 \qquad \text{Note that this is not } \log_7(x + 2).$$

$$= \frac{\log_e x}{\log_e 7} + 2 \qquad \text{Using the change-of-base formula for } \log_7 x$$

$$= \frac{\ln x}{\ln 7} + 2$$

We graph $y = \ln(x)/\ln(7) + 2$.

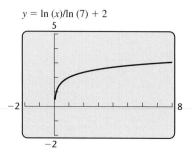

9.5

Exercise Set

Find each of the following. Round answers to the nearest ten-thousandth.

1. $\ln 5$ 1.6094   **2.** $\ln 2$ 0.6931   **3.** $\ln 52$
                                                                  3.9512
4. $\ln 30$ 3.4012   **5.** $\ln 0.0062$ **6.** $\ln 0.00073$
                                 −5.0832   −7.2225
7. $\dfrac{\ln 2300}{0.08}$ 96.7583 **8.** $\dfrac{\ln 1900}{0.07}$ **9.** $\ln \dfrac{34.2}{15.6}$
                                                     107.8516   0.7850
10. $\ln \dfrac{29}{43}$ −0.3939 **11.** $\dfrac{\ln 400}{\ln 75}$ 1.3877 **12.** $\dfrac{\ln 1236}{\ln 50}$
                                                                                        1.8199
13. $e^{2.71}$ 15.0293 **14.** $e^{3.06}$ 21.3276 **15.** $e^{-3.49}$
                                                                      0.0305
16. $e^{-2.64}$ 0.0714 **17.** $e^{4.7}$ 109.9472 **18.** $e^{1.23}$ 3.4212

Find each of the following logarithms using the change-of-base formula. Round answers to the nearest ten-thousandth.

19. $\log_6 92$ 2.5237 **20.** $\log_3 78$ 3.9656 **21.** $\log_2 100$
                                                                      6.6439
22. $\log_7 100$ **23.** $\log_7 65$ 2.1452 **24.** $\log_5 42$
          2.3666   2.3223
25. $\log_{0.5} 5$ **26.** $\log_{0.1} 3$ **27.** $\log_2 0.2$
          −2.3219   −0.4771   −2.3219
28. $\log_2 0.08$ **29.** $\log_\pi 58$ 3.5471 **30.** $\log_\pi 200$
          −3.6439   4.6284

Graph by hand or using a graphing calculator and state the domain and the range of each function.

31. $f(x) = e^x$ ⊡   **32.** $f(x) = e^{0.5x}$ ⊡
33. $f(x) = e^{-0.4x}$ ⊡   **34.** $f(x) = e^{-x}$ ⊡
35. $f(x) = e^x + 1$ ⊡   **36.** $f(x) = e^x + 2$ ⊡
37. $f(x) = e^x - 2$ ⊡   **38.** $f(x) = e^x - 3$ ⊡
39. $f(x) = 0.5e^x$ ⊡   **40.** $f(x) = 2e^x$ ⊡
41. $f(x) = 2e^{-0.5x}$ ⊡   **42.** $f(x) = 0.5e^{2x}$ ⊡
43. $f(x) = e^{x-2}$ ⊡   **44.** $f(x) = e^{x-3}$ ⊡
45. $f(x) = e^{x+3}$ ⊡   **46.** $f(x) = e^{x+2}$ ⊡
47. $g(x) = 2 \ln x$ ⊡   **48.** $g(x) = 3 \ln x$ ⊡
49. $g(x) = 0.5 \ln x$ ⊡   **50.** $g(x) = 0.4 \ln x$ ⊡
51. $g(x) = \ln x + 3$ ⊡   **52.** $g(x) = \ln x + 2$ ⊡
53. $g(x) = \ln x - 2$ ⊡   **54.** $g(x) = \ln x - 3$ ⊡
55. $g(x) = \ln (x + 1)$ ⊡   **56.** $g(x) = \ln (x + 2)$ ⊡
57. $g(x) = \ln (x - 3)$ ⊡   **58.** $g(x) = \ln (x - 1)$ ⊡

⊡ Answers to Exercises 31–64 can be found on pp. A-70–A-72.

Write an equivalent expression for the function that could be graphed using a graphing calculator. Then graph the function.

59. $f(x) = \log_5 x$ ⊡   **60.** $f(x) = \log_3 x$ ⊡
61. $f(x) = \log_2 (x - 5)$ ⊡ **62.** $f(x) = \log_5 (2x + 1)$ ⊡
63. $f(x) = \log_3 x + x$ ⊡   **64.** $f(x) = \log_2 x - x + 1$ ⊡

TW **65.** Using a calculator, Zeno *incorrectly* says that log 79 is between 4 and 5. How could you convince him, without using a calculator, that he is mistaken?

TW **66.** Examine Exercise 65. What mistake do you believe Zeno made?

Skill Maintenance

Solve. $0, \frac{7}{5}$
67. $4x^2 - 25 = 0$ [5.6] $-\frac{5}{2}, \frac{5}{2}$ **68.** $5x^2 - 7x = 0$ [5.3]
69. $17x - 15 = 0$ [2.2] $\frac{15}{17}$ **70.** $9\frac{9}{13} - 13x = 0$ [2.2]
71. $x^{1/2} - 6x^{1/4} + 8 = 0$ **72.** $2y - 7\sqrt{y} + 3 = 0$
[8.5] 16, 256   [8.5] $\frac{1}{4}$, 9

Synthesis

TW **73.** In an attempt to solve $\ln x = 1.5$, Emma gets the following graph.

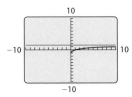

How can Emma tell at a glance that she has made a mistake?

TW **74.** Explain how the graph of $f(x) = \ln x$ could be used to graph the function given by $g(x) = e^{x-1}$.

Given that $\log 2 \approx 0.301$ and $\log 3 \approx 0.477$, find each of the following.
                                                              1.442
75. $\log_6 81$ 2.452 **76.** $\log_9 16$ 1.262 **77.** $\log_{12} 36$

78. Find a formula for converting common logarithms to natural logarithms. $\ln M = \dfrac{\log M}{\log e}$

79. Find a formula for converting natural logarithms to common logarithms. $\log M = \dfrac{\ln M}{\ln 10}$

Solve for x.

80. $\log (275x^2) = 38$ $\pm 6.0302 \times 10^{17}$

81. $\log (492x) = 5.728$ 1086.5129

82. $\dfrac{3.01}{\ln x} = \dfrac{28}{4.31}$ 1.5893

83. $\log 692 + \log x = \log 3450$ 4.9855

84. (a) Domain: $\{x \mid x > 0\}$, or $(0, \infty)$; range: $\mathbb{R}$; **(b)** $[-3, 10, -100, 1000]$, Yscl $= 100$; **(c)** $y = 7.4e^x \ln x$

Yscl = 100

For each function given below, **(a)** *determine the domain and the range,* **(b)** *set an appropriate window, and* **(c)** *draw the graph.*

84. $f(x) = 7.4e^x \ln x$

85. $f(x) = 3.4 \ln x - 0.25e^x$

86. $f(x) = x \ln (x - 2.1)$

87. $f(x) = 2x^3 \ln x$

85. (a) Domain: $\{x \mid x > 0\}$, or $(0, \infty)$; range: $\{y \mid y < 0.5135\}$, or $(-\infty, 0.5135)$; **(b)** $[-1, 5, -10, 5]$; **(c)** $y = 3.4 \ln x - 0.25e^x$

9.6

Solving Exponential Equations ▪ Solving Logarithmic Equations

Solving Exponential and Logarithmic Equations

86. (a) Domain: $\{x \mid x > 2.1\}$, or $(2.1, \infty)$; range: $\mathbb{R}$; **(b)** $[-1, 10, -10, 20]$, Yscl $= 5$; **(c)** $y = x \ln (x - 2.1)$

87. (a) Domain: $\{x \mid x > 0\}$, or $(0, \infty)$; range: $\{y \mid y > -0.2453\}$, or $(-0.2453, \infty)$; **(b)** $[-1, 5, -1, 10]$; **(c)** $y = 2x^3 \ln x$

Solving Exponential Equations

Equations with variables in exponents, such as $5^x = 12$ and $2^{7x} = 64$, are called **exponential equations**. In Section 9.3, we solved certain exponential equations by using the principle of exponential equality. We restate that principle below.

The Principle of Exponential Equality

For any real number b, where $b \neq -1, 0$, or 1,

$b^x = b^y$ is equivalent to $x = y$.

(Powers of the same base are equal if and only if the exponents are equal.)

EXAMPLE 1 Solve: $4^{3x-5} = 16$.

Solution Note that $16 = 4^2$. Thus we can write each side as a power of the same number:

$$4^{3x-5} = 4^2.$$

Since the base is the same, 4, the exponents must be the same. Thus,

$$3x - 5 = 2 \qquad \textbf{Equating exponents}$$
$$3x = 7$$
$$x = \tfrac{7}{3}.$$

We check both algebraically and graphically.

Check:

$$4^{3x-5} = 16$$

$$\begin{array}{c|c} 4^{3\cdot 7/3 - 5} \; ? \; 16 \\ 4^{7-5} \\ 4^2 \quad\Big|\quad 16 \qquad \text{TRUE} \end{array}$$

The solution is $\frac{7}{3}$.

As another check, we solve the equation graphically. We graph $y_1 = 4\char`^(3x - 5)$ and $y_2 = 16$ and determine the x-coordinates of any points of intersection. As the figure below indicates, the graphs intersect at the point $(2.3333333, 16)$. Since $\frac{7}{3} \approx 2.3333333$, the answer checks.

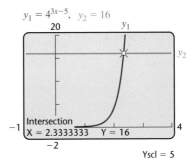

The solution is $\frac{7}{3}$.

When it does not seem possible to write both sides of an equation as powers of the same base, we can use the following principle along with the properties developed in Section 9.4.

The Principle of Logarithmic Equality

For any logarithmic base a, and for $x, y > 0$,

$x = y$ is equivalent to $\log_a x = \log_a y$.

(Two expressions are equal if and only if the logarithms of those expressions are equal.)

Because calculators can generally find only common or natural logarithms (without resorting to the change-of-base formula), we usually take the common or natural logarithm on both sides of the equation.

The principle of logarithmic equality is useful anytime a variable appears as an exponent.

EXAMPLE 2 Solve: $5^x = 12$.

Algebraic Solution

We have

$$5^x = 12$$

$$\log 5^x = \log 12$$ **Using the principle of logarithmic equality to take the common logarithm on both sides. Natural logarithms also would work.**

$$x \log 5 = \log 12$$ **Using the power rule for logarithms**

$$x = \frac{\log 12}{\log 5}$$ ← **CAUTION!**   This is *not* $\log 12 - \log 5$.

$$\approx 1.544.$$ **Using a calculator and rounding to three decimal places**

Graphical Solution

We graph $y_1 = 5^{\wedge}x$ and $y_2 = 12$.

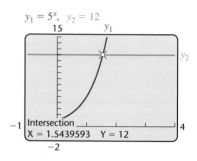

Rounded to three decimal places, the x-coordinate of the point of intersection is 1.544.

Since $5^{1.544} \approx 12$, we have a check. As another check, note that both approaches yield the same solution. The solution is $\log 12 / \log 5$, or approximately 1.544.

EXAMPLE 3 Solve: $e^{0.06t} = 1500$.

Algebraic Solution

Since one side is a power of e, we take the *natural logarithm* on both sides:

$$\ln e^{0.06t} = \ln 1500$$ **Taking the natural logarithm on both sides**

$$0.06t = \ln 1500$$ **Finding the logarithm of the base to a power: $\log_a a^k = k$**

$$t = \frac{\ln 1500}{0.06}$$ **Dividing both sides by 0.06**

$$\approx 121.887.$$ **Using a calculator and rounding to three decimal places**

Graphical Solution

We graph $y_1 = e^{\wedge}(0.06x)$ and $y_2 = 1500$. Since $y_2 = 1500$, we choose a value for Ymax that is greater than 1500. It may require trial and error to choose appropriate units for the x-axis.

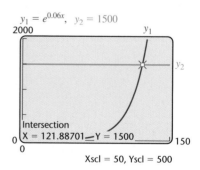

Rounded to three decimal places, the x-coordinate of the point of intersection is 121.887.

We check by substituting 121.887 for t in the original equation. The solution is approximately 121.887.

Some equations, like the one in Example 3, are readily solved algebraically. There are other exponential equations for which we do not have the tools to solve algebraically, but we can nevertheless solve graphically.

EXAMPLE 4   Solve: $xe^{3x-1} = 5$.

Solution   We graph $y_1 = xe^{\wedge}(3x - 1)$ and $y_2 = 5$ and determine the coordinates of any points of intersection.

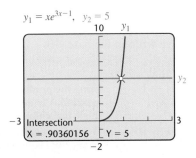

The x-coordinate of the point of intersection is approximately 0.90360156. Thus the solution is approximately 0.904.

Solving Logarithmic Equations

Equations containing logarithmic expressions are called **logarithmic equations**. We saw in Section 9.3 that certain logarithmic equations can be solved by writing an equivalent exponential equation.

EXAMPLE 5   Solve: $\log_4 (8x - 6) = 3$.

Solution   We write an equivalent exponential equation:

$$4^3 = 8x - 6 \qquad \textbf{Remember: } \log_a X = y \textbf{ is equivalent to } a^y = X.$$

$$64 = 8x - 6$$

$$70 = 8x \qquad \textbf{Adding 6 to both sides}$$

$$x = \tfrac{70}{8}, \text{ or } \tfrac{35}{4}.$$

The check is left to the student. The solution is $\frac{35}{4}$.

Note that to write an equivalent exponential equation, the variable must appear in just one logarithmic expression. Often the properties for logarithms are needed to first write an equivalent equation with one logarithmic expression. We then isolate that term and solve as in Example 5. This approach will yield all solutions to the original equation but may also yield numbers that are not solutions. Thus all possible solutions must be checked.

EXAMPLE 6   Solve: $\log x + \log (x - 3) = 1$.

Algebraic Solution

As an aid in solving, we write in the base, 10.

$\log_{10} x + \log_{10} (x - 3) = 1$

$\quad \log_{10} [x(x - 3)] = 1$ **Using the product rule for logarithms to obtain a single logarithm**

$\qquad\qquad x(x - 3) = 10^1$ **Writing an equivalent exponential equation**

$\qquad\qquad x^2 - 3x = 10$

$\qquad x^2 - 3x - 10 = 0$

$\quad (x + 2)(x - 5) = 0$ **Factoring**

$\quad x + 2 = 0 \quad or \quad x - 5 = 0$ **Using the principle of zero products**

$\qquad x = -2 \quad or \qquad x = 5$

The possible solutions are -2 and 5.

Graphical Solution

We graph $y_1 = \log (x) + \log (x + 3)$ and $y_2 = 1$ and determine the coordinates of any points of intersection.

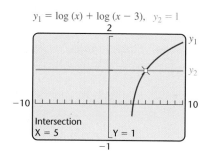

$y_1 = \log (x) + \log (x - 3), \ y_2 = 1$

There is one point of intersection, $(5, 1)$. The solution is the first coordinate of that point, or 5.

Note that the algebraic approach resulted in two possible solutions and the graphical approach in only one. The graphical solution indicates that -2 is not a solution of the equation. To verify this, we substitute both -2 and 5 in the original equation.

Check:

For -2:

$$\frac{\log x + \log (x - 3) = 1}{\log (-2) + \log (-2 - 3) \ ? \ 1} \qquad \text{FALSE}$$

For 5:

$$\frac{\log x + \log (x - 3) = 1}{\log 5 + \log (5 - 3) \ ? \ 1}$$
$$\log 5 + \log 2$$
$$\log 10$$
$$1 \ | \ 1 \qquad \text{TRUE}$$

The number -2 *does not check* because the logarithm of a negative number is undefined. The solution is 5.

| **EXAMPLE 7**   Solve: $\log_2 (x + 7) - \log_2 (x - 7) = 3$.

Algebraic Solution

We have

$$\log_2 (x + 7) - \log_2 (x - 7) = 3$$

$$\log_2 \frac{x + 7}{x - 7} = 3 \qquad \text{Using the quotient rule for logarithms to obtain a single logarithm}$$

$$\frac{x + 7}{x - 7} = 2^3 \qquad \text{Writing an equivalent exponential equation}$$

$$\frac{x + 7}{x - 7} = 8$$

$$x + 7 = 8(x - 7) \qquad \text{Multiplying by the LCD, } x - 7$$

$$x + 7 = 8x - 56 \qquad \text{Using the distributive law}$$

$$63 = 7x$$

$$9 = x. \qquad \text{Dividing by 7}$$

Graphical Solution

We first use the change-of-base formula to write the base-2 logarithms using common logarithms. Then we graph and determine the coordinates of any points of intersection. (We could use natural logarithms if we wish.)

$y_1 = \log (x + 7)/\log (2) - \log (x - 7)/\log (2),$
$y_2 = 3$

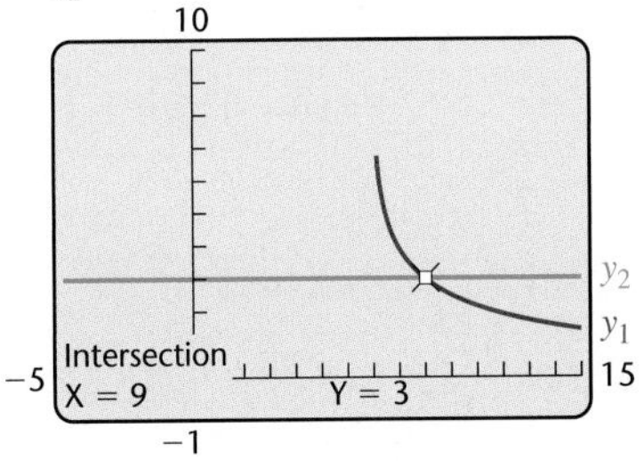

The graphs intersect at $(9, 3)$. The solution is 9.

Check:

$$\frac{\log_2 (x + 7) - \log_2 (x - 7) = 3}{\log_2 (9 + 7) - \log_2 (9 - 7) \; ? \; 3}$$

$$\log_2 16 - \log_2 2$$

$$4 - 1$$

$$3 \mid 3 \qquad \text{TRUE}$$

The solution is 9.

EXAMPLE 8 Solve: $\log_7 (x + 1) + \log_7 (x - 1) = \log_7 8$.

Algebraic Solution

We have

$$\log_7 (x + 1) + \log_7 (x - 1) = \log_7 8$$
$$\log_7 [(x + 1)(x - 1)] = \log_7 8 \qquad \text{Using the product rule for logarithms}$$
$$\log_7 (x^2 - 1) = \log_7 8 \qquad \text{Multiplying}$$
$$x^2 - 1 = 8 \qquad \text{Using the principle of logarithmic equality. Study this step carefully.}$$
$$x^2 - 9 = 0$$
$$(x - 3)(x + 3) = 0 \qquad \text{Solving the quadratic equation}$$
$$x = 3 \quad or \quad x = -3.$$

Graphical Solution

Using the change-of-base formula, we graph

$$y_1 = \log (x + 1)/\log (7)$$
$$+ \log (x - 1)/\log (7)$$

and

$$y_2 = \log (8)/\log (7).$$

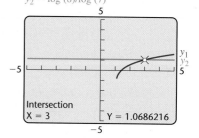

$y_1 = \log (x + 1)/\log (7) + \log (x - 1)/\log (7)$, $y_2 = \log (8)/\log (7)$

Intersection
X = 3   Y = 1.0686216

The graphs intersect at $(3, 1.0686216)$. We have a solution of 3.

The graphical approach indicates that 3 is a solution but -3 is not. The student should confirm that 3 checks in the equation, but -3 does not. The solution is 3.

9.6

Exercise Set

17. $\dfrac{\log 13}{\log 4} - 1 \approx 0.850$ 18. $\dfrac{\log 17}{\log 8} + 1 \approx 2.362$

Solve. Where appropriate, include approximations to the nearest thousandth.

1. $2^x = 16$ 4

2. $2^x = 8$ 3

3. $3^x = 27$ 3

4. $5^x = 125$ 3

5. $2^{x+3} = 32$ 2

6. $4^{x-2} = 64$ 5

7. $5^{3x} = 625$ $\frac{4}{3}$

8. $3^{2x} = 27$ $\frac{3}{2}$

Aha! **9.** $7^{4x} = 1$ 0

10. $8^{5x} = 1$ 0 $\frac{5}{2}$

11. $4^{2x-1} = 64$ 2

12. $5^{2x-3} = 25$

13. $3^{x^2} \cdot 3^{3x} = 81$ $-4, 1$

14. $3^{4x} \cdot 3^{x^2} = \frac{1}{27}$ $-3, -1$

15. $2^x = 15$ $\dfrac{\log 15}{\log 2} \approx 3.907$

16. $2^x = 19$ $\dfrac{\log 19}{\log 2} \approx 4.248$

17. $4^{x+1} = 13$

18. $8^{x-1} = 17$

19. $e^t = 100$ $\ln 100 \approx 4.605$

20. $e^t = 1000$ $\ln 1000 \approx 6.908$

21. $e^{-0.07t} + 3 = 3.08$ ⊡

22. $e^{0.03t} + 2 = 7$ ⊡

23. $2^x = 3^{x-1}$ ⊡

24. $5^x = 3^{x+1}$ ⊡

25. $7.2^x - 65 = 0$ ⊡

26. $4.9^x - 87 = 0$ ⊡

27. $e^{0.5x} - 7 = 2x + 6$
$-6.480, 6.519$

28. $e^{-x} - 3 = x^2$ -1.873

29. $\log_5 x = 3$ 125

30. $\log_3 x = 4$ 81

31. $\log_4 x = \frac{1}{2}$ 2

32. $\log_2 x = -3$ $\frac{1}{8}$

33. $\log x = 3$ 1000

34. $\log x = 1$ 10

35. $2 \log x = -8$ $\frac{1}{10,000}$

36. $4 \log x = -16$ $\frac{1}{10,000}$

Aha! **37.** $\ln x = 1$ $e \approx 2.718$

38. $\ln x = 2$ $e^2 \approx 7.389$

39. $5 \ln x = -15$ $e^{-3} \approx 0.050$

40. $3 \ln x = -3$
$e^{-1} \approx 0.368$

41. $\log_2 (8 - 6x) = 5$ -4

42. $\log_5 (2x - 7) = 3$ 66

43. $\log (x - 9) + \log x = 1$ 10

44. $\log (x + 9) + \log x = 1$ 1

45. $\log x - \log (x + 3) = 1$ No solution

46. $\log x - \log (x + 7) = -1$ $\frac{7}{9}$

47. $\log_4 (x + 3) - \log_4 (x - 5) = 2$ $\frac{83}{15}$

48. $\log_2 (x + 3) + \log_2 (x - 3) = 4$ 5

49. $\log_7 (x + 1) + \log_7 (x + 2) = \log_7 6$ 1

50. $\log_6 (x + 3) + \log_6 (x + 2) = \log_6 20$ 2

51. $\log_3 (x + 4) + \log_3 (x - 4) = 2$ 5

52. $\log_{14} (x + 3) + \log_{14} (x - 2) = 1$ 4

53. $\log_{12} (x + 5) - \log_{12} (x - 4) = \log_{12} 3$ $\frac{17}{2}$

54. $\log_6 (x + 7) - \log_6 (x - 2) = \log_6 5$ $\frac{17}{4}$

55. $\log_2 (x - 2) + \log_2 x = 3$ 4

56. $\log_4 (x + 6) - \log_4 x = 2$ $\frac{2}{5}$

57. $\ln (3x) = 3x - 8$ $0.000112, 3.445$

58. $\ln (x^2) = -x^2$ $-0.753, 0.753$

59. Find the value of x for which the natural logarithm is the same as the common logarithm. 1

60. Find all values of x for which the common logarithm of the square of x is the same as the square of the common logarithm of x. $1, 100$

ᴛᵂ **61.** Could Example 2 have been solved by taking the natural logarithm on both sides? Why or why not?

ᴛᵂ **62.** Christina finds that the solution of $\log_3 (x + 4) = 1$ is -1, but rejects -1 as an answer. What mistake is she making?

Skill Maintenance

Between 1991 and 2001, the waiting list for organ transplants grew much faster than the number of transplants performed, as shown on the following graph. Use the graph for Exercises 63–70.

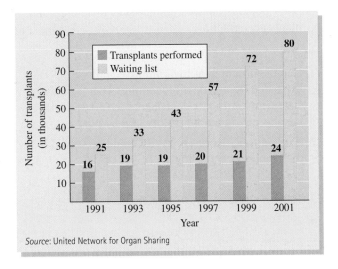

Source: United Network for Organ Sharing

63. Determine whether the number of transplants performed can be modeled better by a linear function or by a quadratic function. [2.6]   Linear

64. Determine whether the number of patients on the transplant waiting list can be modeled better by a linear function or by a quadratic function. [8.8]
Quadratic

65. Use the data from 1991 and from 2001 to find a linear function that can be used to estimate the number of transplants t performed x years after 1991. [2.6] $t(x) = \frac{4}{5}x + 16$

66. Use the data from 1991, 1995, and 2001 to find a quadratic function that can be used to estimate the number of patients p on the waiting list x years after 1991. [8.8] $p(x) = \frac{1}{6}x^2 + \frac{23}{6}x + 25$

67. Use the functions found in Exercises 65 and 66 to estimate the year in which there will be five times as many patients on the waiting list as there are transplants performed. [8.3]   2010

68. Use linear regression and the data for all the given years to find a linear function that can be used to estimate the number of transplants t performed x years after 1991. [2.6] $t(x) = 0.7571428571x + 15.71428571$

69. Use regression and the data for all the given years to find a quadratic function that can be used to estimate the number of patients p on the waiting list x years after 1991. [8.8] $p(x) = 0.0892857143x^2 + 4.907142857x + 23.85714286$

70. Use the functions found in Exercises 68 and 69 to estimate the year in which there will be five times as many patients on the waiting list as there are transplants performed. [8.3] 2010

Synthesis

TW **71.** Can the principle of logarithmic equality be expanded to include all functions? That is, is the statement "$m = n$ is equivalent to $f(m) = f(n)$" true for any function f? Why or why not?

TW **72.** Explain how Exercises 33–36 could be solved using the graph of $f(x) = \log x$.

Solve.

73. $100^{3x} = 1000^{2x+1}$ No solution

74. $27^x = 81^{2x-3}$ $\frac{12}{5}$

75. $8^x = 16^{3x+9}$ -4

76. $\log_x (\log_3 27) = 3$ $\sqrt[3]{3}$

77. $\log_6 (\log_2 x) = 0$ 2

78. $x \log \frac{1}{8} = \log 8$ -1

79. $\log_5 \sqrt{x^2 - 9} = 1$ $\pm\sqrt{34}$

80. $2^{x^2+4x} = \frac{1}{8}$ $-3, -1$

81. $\log (\log x) = 5$ $10^{100,000}$

82. $\log_5 |x| = 4$ $-625, 625$

83. $\log x^2 = (\log x)^2$ $1, 100$

84. $\log \sqrt{2x} = \sqrt{\log 2x}$ $\frac{1}{2}, 5000$

85. $\log x^{\log x} = 25$ $\frac{1}{100,000}, 100,000$

86. $3^{2x} - 8 \cdot 3^x + 15 = 0$ $1, \dfrac{\log 5}{\log 3} \approx 1.465$

87. $(81^{x-2})(27^{x+1}) = 9^{2x-3}$ $-\frac{1}{3}$

88. $3^{2x} - 3^{2x-1} = 18$ $\frac{3}{2}$

89. Given that $2^y = 16^{x-3}$ and $3^{y+2} = 27^x$, find the value of $x + y$. 38

90. If $x = (\log_{125} 5)^{\log_5 125}$, what is the value of $\log_3 x$? -3

91. *Supply and Demand.* The supply and demand for the sale of stereos by Sound Ideas are given by

$$S(x) = e^x \quad \text{and} \quad D(x) = 162{,}755e^{-x},$$

where $S(x)$ is the price at which the company is willing to supply x stereos and $D(x)$ is the demand price for a quantity of x stereos. Find the equilibrium point. (For reference, see Section 3.7.)
(6, $403)

9.7

Applications of Logarithmic Functions ■ Applications of Exponential Functions

Applications of Exponential and Logarithmic Functions

We now consider applications of exponential and logarithmic functions.

Applications of Logarithmic Functions

EXAMPLE 1 Sound Levels. To measure the volume, or "loudness," of a sound, the *decibel* scale is used. The loudness L, in decibels (dB), of a sound is given by

$$L = 10 \cdot \log \frac{I}{I_0},$$

where I is the intensity of the sound, in watts per square meter (W/m^2), and $I_0 = 10^{-12}$ W/m^2. (I_0 is approximately the intensity of the softest sound that can be heard by the human ear.)

a) It is common for the intensity of sound at live performances of rock music to reach 10^{-1} W/m² (even higher close to the stage). How loud, in decibels, is the sound level?

b) The Occupational Safety and Health Administration (OSHA) considers sound levels above 85 dB unsafe. What is the intensity of such sounds?

Solution

a) To find the loudness, in decibels, we use the above formula:

$$L = 10 \cdot \log \frac{I}{I_0}$$

$$= 10 \cdot \log \frac{10^{-1}}{10^{-12}} \qquad \textbf{Substituting}$$

$$= 10 \cdot \log 10^{11} \qquad \textbf{Subtracting exponents}$$

$$= 10 \cdot 11 \qquad \textbf{log } 10^a = a$$

$$= 110.$$

The volume of the music is 110 decibels.

b) We substitute and solve for I:

$$L = 10 \cdot \log \frac{I}{I_0}$$

$$85 = 10 \cdot \log \frac{I}{10^{-12}} \qquad \textbf{Substituting}$$

$$8.5 = \log \frac{I}{10^{-12}} \qquad \textbf{Dividing both sides by 10}$$

$$8.5 = \log I - \log 10^{-12} \qquad \textbf{Using the quotient rule for logarithms}$$

$$8.5 = \log I - (-12) \qquad \textbf{log } 10^a = a$$

$$-3.5 = \log I \qquad \textbf{Adding } -12 \textbf{ to both sides}$$

$$10^{-3.5} = I. \qquad \textbf{Converting to an exponential equation}$$

Earplugs would be recommended for sounds with intensities exceeding $10^{-3.5}$ W/m².

TEACHING TIP

You may need to remind students that $\log 10^{-12}$ means $\log_{10} 10^{-12}$. Thus, $\log_{10} 10^{-12} = -12$ since $\log_a a^k = k$.

EXAMPLE 2 Chemistry: pH of Liquids. In chemistry, the pH of a liquid is a measure of its acidity. We calculate pH as follows:

$$pH = -\log [H^+],$$

where $[H^+]$ is the hydrogen ion concentration in moles per liter.

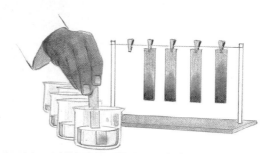

a) The hydrogen ion concentration of human blood is normally about 3.98×10^{-8} moles per liter. Find the pH.

b) The pH of seawater is about 8.3. Find the hydrogen ion concentration.

Solution

a) To find the pH of blood, we use the above formula:

$$
\begin{aligned}
pH &= -\log [H^+] \\
&= -\log [3.98 \times 10^{-8}] \\
&\approx -(-7.400117) \quad \text{**Using a calculator**} \\
&\approx 7.4.
\end{aligned}
$$

The pH of human blood is normally about 7.4.

b) We substitute and solve for $[H^+]$:

$$
\begin{aligned}
8.3 &= -\log [H^+] & &\text{**Using pH} = -\log [H^+]\text{**} \\
-8.3 &= \log [H^+] & &\text{**Dividing both sides by } -1\text{**} \\
10^{-8.3} &= [H^+] & &\text{**Converting to an exponential equation**} \\
5.01 \times 10^{-9} &\approx [H^+]. & &\text{**Using a calculator; writing scientific notation**}
\end{aligned}
$$

The hydrogen ion concentration of seawater is about 5.01×10^{-9} moles per liter.

Applications of Exponential Functions

EXAMPLE 3 Interest Compounded Annually. Suppose that $30,000 is invested at 8% interest, compounded annually. In t years, it will grow to the amount A given by the function

$$A(t) = 30,000(1.08)^t.$$

(See Example 5 in Section 9.2.)

a) How long will it take to accumulate $150,000 in the account?

b) Find the amount of time it takes for the $30,000 to double itself.

Solution

a) We set $A(t) = 150,000$ and solve for t:

$$150,000 = 30,000(1.08)^t$$

$$\frac{150,000}{30,000} = 1.08^t \qquad \text{Dividing both sides by 30,000}$$

$$5 = 1.08^t$$

$$\log 5 = \log 1.08^t \qquad \text{Taking the common logarithm on both sides}$$

$$\log 5 = t \log 1.08 \qquad \text{Using the power rule for logarithms}$$

$$\frac{\log 5}{\log 1.08} = t \qquad \text{Dividing both sides by log 1.08}$$

$$20.9 \approx t. \qquad \text{Using a calculator}$$

Remember that when doing a calculation like this on a calculator, it is best to wait until the end to round off. We check by solving graphically, as shown at left. At an interest rate of 8% per year, it will take about 20.9 yr for $30,000 to grow to $150,000.

b) To find the *doubling time*, we replace $A(t)$ with 60,000 and solve for t:

$$60,000 = 30,000(1.08)^t$$

$$2 = (1.08)^t \qquad \text{Dividing both sides by 30,000}$$

$$\log 2 = \log (1.08)^t \qquad \text{Taking the common logarithm on both sides}$$

$$\log 2 = t \log 1.08 \qquad \text{Using the power rule for logarithms}$$

$$t = \frac{\log 2}{\log 1.08} \approx 9.0. \qquad \text{Dividing both sides by log 1.08 and using a calculator}$$

At an interest rate of 8% per year, the doubling time is about 9 yr.

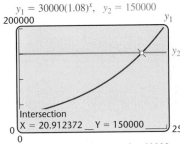

$y_1 = 30000(1.08)^x, \quad y_2 = 150000$

Intersection
X = 20.912372   Y = 150000

Xscl = 5, Yscl = 50000

Like investments, populations often grow exponentially.

TEACHING TIP

You may want to remind students that the *y*-axis must often be scaled differently than the *x*-axis when we are graphing exponential functions.

Exponential Growth

An **exponential growth model** is a function of the form

$$P(t) = P_0 e^{kt}, \quad k > 0,$$

where P_0 is the population at time 0, $P(t)$ is the population at time t, and k is the **exponential growth rate** for the situation. The **doubling time** is the amount of time necessary for the population to double in size.

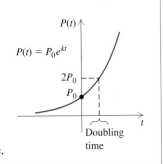

The exponential growth rate is the rate of growth of a population at any *instant* in time. Since the population is continually growing, the percent of total growth after one year will exceed the exponential growth rate.

EXAMPLE 4   Growth of Zebra Mussel Populations.   Zebra mussels, inadvertently imported from Europe, began fouling North American waters in 1988. These mussels are so prolific that lake and river bottoms, as well as water intake pipes, can become blanketed with them, altering an entire ecosystem. In 2000, a portion of the Hudson River contained an average of 10 zebra mussels per square mile. The exponential growth rate was 340% per year.

a) Find the exponential growth model.

b) Estimate the number of mussels per square mile in 2003.

Solution

a) In 2000, at $t = 0$, the population was $10/\text{mi}^2$. We substitute 10 for P_0 and 340%, or 3.4, for k. This gives the exponential growth function

$$P(t) = 10e^{3.4t}.$$

b) In 2003, we have $t = 3$ (since 3 yr have passed since 2000). To find the population in 2003, we compute $P(3)$:

$$P(3) = 10e^{3.4(3)} \qquad \textbf{Using } P(t) = 10e^{3.4t} \textbf{ from part (a)}$$
$$= 10e^{10.2}$$
$$\approx 269{,}000. \qquad \textbf{Using a calculator}$$

The population of zebra mussels in the specified portion of the Hudson River will reach approximately 269,000 per square mile in 2003.

EXAMPLE 5   Spread of a Computer Virus.   The number of computers infected by a virus t days after it first appears usually increases exponentially. In 2000, the "Love Bug" virus spread from 100 computers to about 1,000,000 computers in 2 hr (120 min).

a) Find the exponential growth rate and the exponential growth function.

b) Assuming exponential growth, estimate how long it took the Love Bug virus to infect 80,000 computers.

Solution

a) We use $N(t) = N_0 e^{kt}$, where t is the number of minutes since the first 100 computers were infected. Substituting 100 for N_0 gives

$$N(t) = 100e^{kt}.$$

To find the exponential growth rate, k, note that after 120 min, 1,000,000 computers were infected:

$$\left.\begin{array}{l} N(120) = 100e^{k \cdot 120} \\ 1{,}000{,}000 = 100e^{120k} \end{array}\right\} \quad \textbf{Substituting}$$

$$10{,}000 = e^{120k} \qquad \textbf{Dividing both sides by 100}$$

$$\ln 10{,}000 = \ln e^{120k} \qquad \textbf{Taking the natural logarithm on both sides}$$

$$\ln 10{,}000 = 120k \qquad \textbf{ln } e^a = a$$

$$\frac{\ln 10{,}000}{120} = k \qquad \textbf{Dividing both sides by 120}$$

$$0.077 \approx k. \qquad \textbf{Using a calculator and rounding}$$

The exponential growth function is given by $N(t) = 100e^{0.077t}$.

b) To estimate how long it took for 80,000 computers to be infected, we replace $N(t)$ with 80,000 and solve for t:

$$80{,}000 = 100e^{0.077t}$$

$$800 = e^{0.077t} \qquad \textbf{Dividing both sides by 100}$$

$$\ln 800 = \ln e^{0.077t} \qquad \textbf{Taking the natural logarithm on both sides}$$

$$\ln 800 = 0.077t \qquad \textbf{ln } e^a = a$$

$$\frac{\ln 800}{0.077} = t \qquad \textbf{Dividing both sides by 0.077}$$

$$86.8 \approx t. \qquad \textbf{Using a calculator}$$

Rounding up to 87, we see that, according to this model, it took about 87 min, or 1 hr 27 min, for 80,000 computers to be infected.

TEACHING TIP

Again, you may wish to remind students that $\ln e^{120k}$ means $\log_e e^{120k}$. Thus, $\ln e^{120k} = 120k$ since $\log_a a^k = k$.

EXAMPLE 6 Interest Compounded Continuously. Suppose that an amount of money P_0 is invested in a savings account at interest rate k, compounded continuously. That is, suppose that interest is computed every "instant" and added to the amount in the account. The balance $P(t)$, after t years, is given by the exponential growth model

$$P(t) = P_0 e^{kt}.$$

a) Suppose that $30,000 is invested and grows to $44,754.75 in 5 yr. Find the exponential growth function.

b) What is the doubling time?

Solution

a) We have $P(0) = 30,000$. Thus the exponential growth function is

$$P(t) = 30,000 e^{kt}, \quad \text{where } k \text{ must still be determined.}$$

Knowing that for $t = 5$ we have $P(5) = 44,754.75$, we can solve for k:

$$44,754.75 = 30,000 e^{k(5)} = 30,000 e^{5k}$$

$\dfrac{44,754.75}{30,000} = e^{5k}$	**Dividing both sides by 30,000**
$1.491825 = e^{5k}$	
$\ln 1.491825 = \ln e^{5k}$	**Taking the natural logarithm on both sides**
$\ln 1.491825 = 5k$	**$\ln e^a = a$**
$\dfrac{\ln 1.491825}{5} = k$	**Dividing both sides by 5**
$0.08 \approx k.$	**Using a calculator and rounding**

The interest rate is about 0.08, or 8%, compounded continuously. Note that since interest is being compounded continuously, the interest earned each year is more than 8%. The exponential growth function is

$$P(t) = 30,000 e^{0.08t}.$$

TEACHING TIP

You may want to do this example with other amounts to illustrate that the doubling time depends only on the interest rate.

b) To find the doubling time T, we replace $P(T)$ with 60,000 and solve for T:

$$60,000 = 30,000 e^{0.08T}$$

$2 = e^{0.08T}$	**Dividing both sides by 30,000**
$\ln 2 = \ln e^{0.08T}$	**Taking the natural logarithm on both sides**
$\ln 2 = 0.08T$	**$\ln e^a = a$**
$\dfrac{\ln 2}{0.08} = T$	**Dividing both sides by 0.08**
$8.7 \approx T.$	**Using a calculator and rounding**

Thus the original investment of $30,000 will double in about 8.7 yr.

As Examples 3(b) and 6(b) imply, for any specified interest rate, continuous compounding gives the highest yield and the shortest doubling time.

In some real-life situations, a quantity or population is *decreasing* or *decaying* exponentially.

Exponential Decay

An **exponential decay model** is a function of the form

$$P(t) = P_0 e^{-kt}, \quad k > 0,$$

where P_0 is the quantity present at time 0, $P(t)$ is the amount present at time t, and k is the **decay rate**. The **half-life** is the amount of time necessary for half of the quantity to decay.

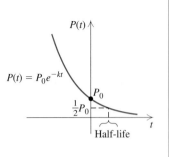

EXAMPLE 7   Carbon Dating.   The radioactive element carbon-14 has a half-life of 5750 yr. The percentage of carbon-14 present in the remains of organic matter can be used to determine the age of that organic matter. Recently, while digging in Chaco Canyon, New Mexico, archaeologists found corn pollen that had lost 38.1% of its carbon-14. The age of this corn pollen was evidence that Native Americans had been cultivating crops in the Southwest centuries earlier than scientists had thought. What was the age of the pollen? (*Source: American Anthropologist*)

Solution   We first find k. To do so, we use the concept of half-life. When $t = 5750$ (the half-life), $P(t)$ will be half of P_0. Then

$$0.5P_0 = P_0 e^{-k(5750)} \qquad \text{Substituting in } P(t) = P_0 e^{-kt}$$
$$0.5 = e^{-5750k} \qquad \text{Dividing both sides by } P_0$$
$$\ln 0.5 = \ln e^{-5750k} \qquad \text{Taking the natural logarithm on both sides}$$
$$\ln 0.5 = -5750k \qquad \ln e^a = a$$
$$\frac{\ln 0.5}{-5750} = k \qquad \text{Dividing}$$
$$0.00012 \approx k. \qquad \text{Using a calculator and rounding}$$

Now we have a function for the decay of carbon-14:

$$P(t) = P_0 e^{-0.00012t}. \qquad \textbf{This completes the first part of our solution.}$$

(*Note*: This equation can be used for any subsequent carbon-dating problem.) If the corn pollen has lost 38.1% of its carbon-14 from an initial amount P_0, then $100\% - 38.1\%$, or 61.9%, of P_0 is still present. To find the age t of the pollen, we solve this equation for t:

$$0.619P_0 = P_0 e^{-0.00012t} \qquad \text{We want to find } t \text{ for which } P(t) = 0.619P_0.$$
$$0.619 = e^{-0.00012t} \qquad \text{Dividing both sides by } P_0$$
$$\ln 0.619 = \ln e^{-0.00012t} \qquad \text{Taking the natural logarithm on both sides}$$
$$\ln 0.619 = -0.00012t \qquad \ln e^a = a$$
$$\frac{\ln 0.619}{-0.00012} = t \qquad \text{Dividing}$$
$$4000 \approx t. \qquad \text{Using a calculator}$$

The pollen is about 4000 yr old.

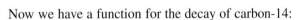

Connecting the Concepts

We can now add the exponential functions $f(t) = P_0 e^{kt}$ and $f(t) = P_0 e^{-kt}$, $k > 1$, and the logarithmic function $f(x) = \log_b x$, $b > 1$, to our library of functions.

By looking at the graph of a set of data, we can tell whether a population or other quantity is growing or decaying exponentially.

Linear function:
$f(x) = mx + b$

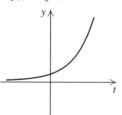

Absolute-value function:
$f(x) = |x|$

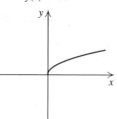

Rational function:
$f(x) = \dfrac{1}{x}$

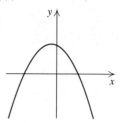

Radical function:
$f(x) = \sqrt{x}$

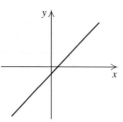

Quadratic function:
$f(x) = ax^2 + bx + c, a > 0$

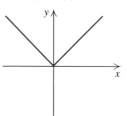

Quadratic function:
$f(x) = ax^2 + bx + c, a < 0$

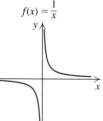

Exponential growth function:
$f(t) = P_0 e^{kt}, k > 0$

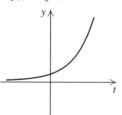

Exponential decay function:
$f(t) = P_0 e^{-kt}, k > 0$

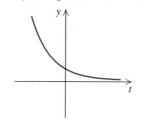

Logarithmic function:
$f(x) = \log_b x, b > 1$

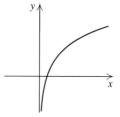

EXAMPLE 8   For each of the following graphs, determine whether an exponential function might fit the data.

a)

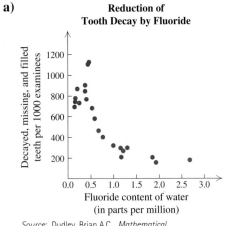

Source: Dudley, Brian A.C., *Mathematical and Biological Interrelations*

b)

c)

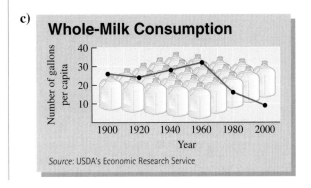

d)

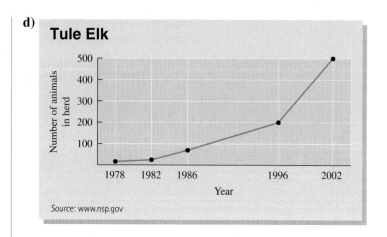

Tule Elk

Source: www.nsp.gov

Solution

a) As the fluoride content in the water increases, the amount of tooth decay decreases. The amount of decrease gets smaller as the amount of fluoride increases. It appears that an exponential decay function might fit the data.

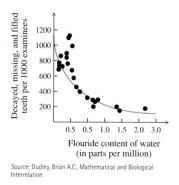

Source: Dudley, Brian A.C., Mathematical and Biological Interrelation

b) The number of salespersons increased between 1992 and 2000 at approximately the same rate each year. It does not appear that an exponential function models the data; instead, a linear function might be appropriate.

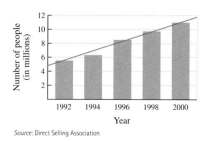

Source: Direct Selling Association

c) The number of gallons of whole milk consumed, per capita, first fell, then rose, then fell again. This does not fit an exponential model.

d) The elk population increased from 1978 to 2002, and the amount of yearly growth also increased during that time. This suggests that an exponential growth function could be used to model this situation.

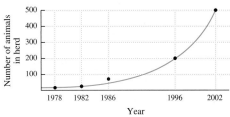

Source: www.nsp.gov

Exponential Regression

Most graphing calculators can use regression to fit an exponential function to a set of data. Often the feature will be an option labeled ExpReg in the STAT CALC menu. After entering the data, with the independent variable as L1 and the dependent variable as L2, choose the ExpReg option in the STAT CALC menu. For the data shown on the left below, the calculator will return a screen like that shown on the right below.

L1	L2	L3	1
1	12	------	
2	60		
3	300		
4	1500		
5	7500		
6	37500		
------	------		
L1(1)=1			

ExpReg
y=a*b^x
a=2.4
b=5

The exponential function found is of the form $f(x) = ab^x$, or, for this example, $f(x) = 2.4(5)^x$.

We can use the function in this form or convert it to an exponential function of the form $f(x) = ae^{kx}$. The number a remains unchanged. To find k, we solve the equation $5^x = e^{kx}$ for k:

$$5^x = e^{kx}$$
$$5^x = (e^k)^x \qquad \text{Writing both sides as a power of } x$$
$$5 = e^k \qquad \text{The bases must be equal.}$$
$$\ln 5 = \ln e^k \qquad \text{Taking the natural logarithm on both sides}$$
$$\ln 5 = k. \qquad \ln e^k = k$$

Thus to write $f(x) = 2.4(5)^x$ in the form $f(x) = ae^{kx}$, we can write $f(x) = 2.4e^{(\ln 5)x}$, or

$$f(x) \approx 2.4e^{1.609x}.$$

This process can be generalized as follows.

> **Any exponential function of the form $f(x) = ab^x$ can be written in the form $f(x) = ae^{kx}$, where $k = \ln b$.**

EXAMPLE 9 Tule Elk Population. In 1800, over 500,000 Tule elk inhabited the state of California. By the late 1800s, after the California Gold Rush, there were fewer than 50 elk remaining in the state. In 1978, wildlife biologists introduced a herd of 10 Tule elk into the Point Reyes National Seashore near San Francisco. By 1982, the herd had grown to 24 elk. There were 70 elk in 1986, 200 in 1996, and 500 in 2002. (This problem was suggested by Kevin Yokoyama of the College of the Redwoods in Eureka, California. *Source of data*: www.nsp.gov.)

a) Use regression to fit an exponential function to the data and graph the function.

b) Determine the exponential growth rate.

c) Use the exponential function to estimate the number of elk in Point Reyes National Seashore in 2005, if none of the herd is removed.

Solution

a) We let x represent the number of years since 1978, and enter the data, as shown on the left below. Then we use regression to find the function and copy it to the equation-editor screen as y_1. We have

$$f(x) = 13.01608148(1.168547698)^x.$$

The graph of the function, along with the data points, is shown on the right below.

Wildlife biologists monitor the size of the herd of Tule elk, described in Example 9, because Point Reyes National Seashore cannot support a large population of elk. In fact, as of 2002, the population was large enough that some elk were transferred to a new location.

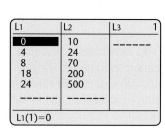

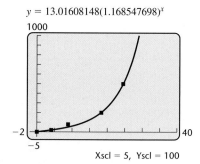

$y = 13.01608148(1.168547698)^x$

Xscl = 5, Yscl = 100

b) The exponential growth rate is the number k in the equation $f(x) = ae^{kx}$. To write $f(x) = 13.01608148(1.168547698)^x$ in the form $f(x) = ae^{kx}$, we find k:

$$k = \ln b$$
$$\approx \ln 1.168547698$$
$$\approx 0.155761694.$$

Thus the exponential growth rate is approximately 0.156, or 15.6%.

c) The year 2005 is 27 years after 1978, so to estimate the number of elk in Point Reyes National Seashore, we find $f(27)$ using a table of values, using TRACE, or by entering Y1(27):

$$f(27) \approx 873.$$

In 2005, there will be about 873 elk in the herd.

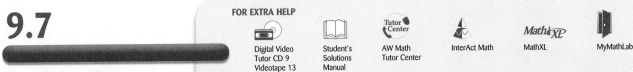

9.7

Exercise Set

Solve.

1. *Cellular Phones.*   The number of cellular phones in use in the United States t years after May 2002 can be predicted by

$$N(t) = 153(1.37)^t,$$

where $N(t)$ is the number of cellular phones in use, in millions.

a) Determine the year in which 200 million cellular phones would be in use.   2003

b) What is the doubling time for the number of cellular phones in use?   2.2 yr

2. *Compact Discs.*   The number of compact discs N purchased each year, in millions, can be approximated by

$$N(t) = 384(1.13)^t,$$

where t is the number of years after 1991.*

a) After what amount of time will two billion compact discs be sold in a year?   13.5 yr

b) What is the doubling time for the number of compact discs sold in one year?   5.7 yr

3. *Student Loan Repayment.*   A college loan of $29,000 is made at 8% interest, compounded annually. After t years, the amount due, A, is given by the function

$$A(t) = 29,000(1.08)^t.$$

a) After what amount of time will the amount due reach $40,000?   4.2 yr

b) Find the doubling time.   9.0 yr

4. *Spread of a Rumor.*   The number of people who have heard a rumor increases exponentially. If all those who hear a rumor repeat it to two people a day, and if 20 people start the rumor, the number of people N who have heard the rumor after t days is given by

$$N(t) = 20(3)^t.$$

a) After what amount of time will 1000 people have heard the rumor?   3.6 days

b) What is the doubling time for the number of people who have heard the rumor?   0.6 day

5. *Skateboarding.*   The number of skateboarders of age x (with $x \geq 16$), in thousands, can be approximated by

$$N(x) = 600(0.873)^{x-16}$$

(*Sources*: Based on figures from the U.S. Bureau of the Census and *Statistical Abstract of the United States*, 2000).   20,114

a) Estimate the number of 41-yr-old skateboarders.

b) At what age are there only 2000 skateboarders?   About 58

*Based on data from the Recording Industry Association, Washington, DC.

6. *Recycling Aluminum Cans.*   Approximately two thirds of all aluminum cans distributed will be recycled each year. A beverage company distributes 250,000 cans. The number still in use after t years is given by the function

$$N(t) = 250,000\left(\tfrac{2}{3}\right)^t.$$

a) After how many years will 60,000 cans still be in use?   3.5 yr
b) After what amount of time will only 1000 cans still be in use?   13.6 yr

Use the pH formula given in Example 2 for Exercises 7–10.

7. *Chemistry.*   The hydrogen ion concentration of fresh-brewed coffee is about 1.3×10^{-5} moles per liter. Find the pH.   4.9

8. *Chemistry.*   The hydrogen ion concentration of milk is about 1.6×10^{-7} moles per liter. Find the pH.   6.8

9. *Medicine.*   When the pH of a patient's blood drops below 7.4, a condition called *acidosis* sets in. Acidosis can be deadly when the patient's pH reaches 7.0. What would the hydrogen ion concentration of the patient's blood be at that point? 10^{-7} moles per liter

10. *Medicine.*   When the pH of a patient's blood rises above 7.4, a condition called *alkalosis* sets in. Alkalosis can be deadly when the patient's pH reaches 7.8. What would the hydrogen ion concentration of the patient's blood be at that point? 1.58×10^{-8} moles per liter

Use the decibel formula given in Example 1 for Exercises 11–14.

11. *Audiology.*   The intensity of sound in normal conversation is about 3.2×10^{-6} W/m². How loud in decibels is this sound level?   65 dB

12. *Audiology.*   The intensity of a riveter at work is about 3.2×10^{-3} W/m². How loud in decibels is this sound level?   95 dB

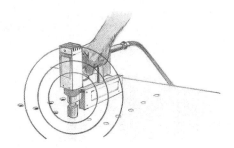

13. *Music.*   At a recent performance of the band U2, sound measurements of 105 dB were recorded. What is the intensity of such sounds? $10^{-1.5}$ W/m²

14. *Music.*   At a recent performance of the band Strange Folk in Burlington, VT, sound levels reached 111 dB (*Source*: Melissa Garrido, *Burlington Free Press*). What is the intensity of such sounds? $10^{-0.9}$ W/m²

Use the compound-interest formula in Example 6 for Exercises 15 and 16.

15. *Interest Compounded Continuously.*   Suppose that P_0 is invested in a savings account where interest is compounded continuously at 6% per year.

a) Express $P(t)$ in terms of P_0 and 0.06. $P(t) = P_0 e^{0.06t}$
b) Suppose that \$5000 is invested. What is the balance after 1 yr? after 2 yr?   \$5309.18, \$5637.48
c) When will an investment of \$5000 double itself?   11.6 yr

16. *Interest Compounded Continuously.*   Suppose that P_0 is invested in a savings account where interest is compounded continuously at 5% per year.

a) Express $P(t)$ in terms of P_0 and 0.05. $P(t) = P_0 e^{0.05t}$
b) Suppose that \$1000 is invested. What is the balance after 1 yr? after 2 yr?   \$1051.27, \$1105.17
c) When will an investment of \$1000 double itself?   13.9 yr

17. *Population Growth.*   In 2002, the population of the United States was 288.3 million and the exponential growth rate was 1.3% per year (*Source*: U.S. Bureau of the Census).

a) Find the exponential growth function.
b) Predict the U.S. population in 2005.   299.8 million
c) When will the U.S. population reach 325 million?   2011

17. (a) $P(t) = 288.3e^{0.013t}$, where t is the number of years after 2002 and $P(t)$ is in millions

18. (a) $P(t) = 6.3e^{0.014t}$, where t is the number of years after 2002 and $P(t)$ is in billions

18. *World Population Growth.* In 2002, the world population was 6.3 billion and the exponential growth rate was 1.4% per year (*Source:* U.S. Bureau of the Census).

a) Find the exponential growth function.

b) Predict the world population in 2005. 6.6 billion

c) When will the world population be 8.0 billion? 2019

19. *Growth of Bacteria.* The bacteria *Escherichi coli* are commonly found in the human bladder. Suppose that 3000 of the bacteria are present at time $t = 0$. Then t minutes later, the number of bacteria present is

$$N(t) = 3000(2)^{t/20}.$$

a) After what amount of time will there be 60,000 bacteria? 86.4 min

b) If 100,000,000 bacteria accumulate, a bladder infection can occur. What amount of time would have to pass in order for a possible infection to occur? 300.5 min

c) What is the doubling time? 20 min

20. *Population Growth.* The exponential growth rate of the population of Central America is 3.5% per year (one of the highest in the world). What is the doubling time? 19.8 yr

21. *Advertising.* A model for advertising response is given by

$$N(a) = 2000 + 500 \log a, \quad a \geq 1,$$

where $N(a)$ is the number of units sold and a is the amount spent on advertising, in thousands of dollars.

a) How many units were sold after spending $1000 ($a = 1$) on advertising? 2000

b) How many units were sold after spending $8000? 2452

c) Graph the function. ⊡

d) How much would have to be spent in order to sell 5000 units? $1,000,000 thousand, or $1,000,000,000

22. *Forgetting.* Students in an English class took a final exam. They took equivalent forms of the exam at monthly intervals thereafter. The average score $S(t)$, in percent, after t months was found to be given by

$$S(t) = 68 - 20 \log (t + 1), \quad t \geq 0.$$

a) What was the average score when they initially took the test, $t = 0$? 68%

b) What was the average score after 4 months? after 24 months? 54%, 40%

c) Graph the function. ⊡

d) After what time t was the average score 50? 6.9 months

⊡ Answers to Exercises 21(c), 22(c), and 27(a) can be found on p. A-72.

23. *Public Health.* In 1995, an outbreak of Herpes infected 17 people in a large community. By 1996, the number of those infected had grown to 29.

a) Find an exponential growth function that fits the data. $N(t) = 17e^{0.534t}$, where t is the number of years since 1995

b) Estimate the number of people infected in 2001. About 419

24. *Heart Transplants.* In 1967, Dr. Christiaan Barnard of South Africa stunned the world by performing the first heart transplant. There were 1418 heart transplants in 1987, and 2185 such transplants in 1999. $N(t) = 1418e^{0.036t}$, where t is the number of years since 1987

a) Find an exponential growth function that fits the data from 1987 and 1999.

b) Use the function to predict the number of heart transplants in 2012. About 3488 heart transplants

25. *Oil Demand.* The exponential growth rate of the demand for oil in the United States is 10% per year. In what year will the demand be double that of May 1998? 2005

26. *Coal Demand.* The exponential growth rate of the demand for coal in the world is 4% per year. When will the demand be double that of 1995? 2012

27. *Decline of Discarded Yard Waste.* The amount of discarded yard waste has declined considerably in recent years because of increased recycling and composting. In 1996, 17.5 million tons were discarded, but by 1998 the figure dropped to 14.5 million tons (*Source: Statistical Abstract of the United States*, 2000). Assume the amount of discarded yard waste is decreasing according to the exponential decay model.

a) Find the value k, and write an exponential function that describes the amount of yard waste discarded t years after 1996. ⊡

b) Predict the amount of discarded yard waste in 2006. 6.8 million tons

c) In what year (theoretically) will only 1 ton of yard waste be discarded? 2173

28. *Decline in Cases of Mumps.* The number of cases of mumps has dropped exponentially from 5300 in 1990 to 800 in 1996 (*Source: Statistical Abstract of the United States*, 2000). $k \approx 0.315$; $M(t) = 5300e^{-0.315t}$, where t is the number of years since 1990

a) Find the value k, and write an exponential function that can be used to estimate the number of cases t years after 1990.

b) Estimate the number of cases of mumps in 2004. 64 cases

c) In what year (theoretically) will there be only 1 case of mumps? 2017

29. *Archaeology.* When archaeologists found the Dead Sea scrolls, they determined that the linen wrapping had lost 22.3% of its carbon-14. How old is the linen wrapping? (See Example 7.) About 2103 yr

30. *Archaeology.* In 1996, researchers found an ivory tusk that had lost 18% of its carbon-14. How old was the tusk? (See Example 7.) 1654 yr

31. *Chemistry.* The exponential decay rate of iodine-131 is 9.6% per day. What is its half-life?
About 7.2 days

32. *Chemistry.* The decay rate of krypton-85 is 6.3% per year. What is its half-life? 11 yr

33. *Home Construction.* The chemical urea formaldehyde was found in some insulation used in houses built during the mid to late 60s. Unknown at the time was the fact that urea formaldehyde emitted toxic fumes as it decayed. The half-life of urea formaldehyde is 1 yr. What is its decay rate?
69.3% per year

34. *Plumbing.* Lead pipes and solder are often found in older buildings. Unfortunately, as lead decays, toxic chemicals can get in the water resting in the pipes. The half-life of lead is 22 yr. What is its decay rate?
3.15% per year

35. *Value of a Sports Card.* Legend has it that because he objected to young people smoking, and because his first baseball card was issued in cigarette packs, the great shortstop Honus Wagner halted production of his card before many were produced. One of these cards was sold in 1996 for $640,500 and again in 2000 for $1.1 million. For the following questions, assume that the card's value increases exponentially, as it has for many years.

WAGNER, PITTSBURG

$k \approx 0.135$;
$V(t) = 640,500e^{0.135t}$, where t is the number of years since 1996

a) Find the exponential growth rate k, and determine an exponential function V that can be used to estimate the dollar value, $V(t)$, of the card t years after 1996. About $2.47 million

b) Predict the value of the card in 2006.

c) What is the doubling time for the value of the card? 5.1 yr

d) In what year will the value of the card first exceed $2,000,000? 2004

36. *Portrait of Dr. Gachet.* As of May 2003, the most ever paid for a painting is $82.5 million, paid in 1990 for Vincent Van Gogh's *Portrait of Dr. Gachet.* The same painting sold for $58 million in 1987. Assume that the growth in the value V of the painting is exponential.

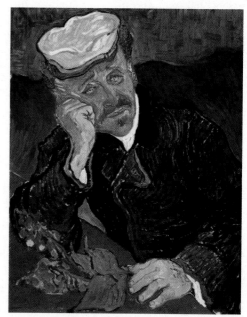

Van Gogh's Portrait of Dr. Gachet, *oil on canvas*
$k \approx 0.117$; $V(t) = 58e^{0.117t}$, where t is the number of years since 1987 and $V(t)$ is in millions of years

a) Find the exponential growth rate k, and determine the exponential growth function V, for which $V(t)$ is the painting's value, in millions of dollars, t years after 1987. About $602.1 million

b) Predict the value of the painting in 2007.

c) What is the doubling time for the value of the painting? 5.9 yr

d) How long after 1987 will the value of the painting be $1 billion? 24.3 yr

In Exercises 37–40, determine whether an exponential function might fit the data.

37.

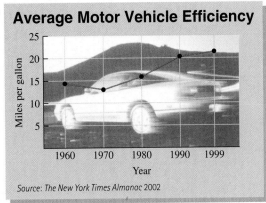

Average Motor Vehicle Efficiency

Source: The New York Times Almanac 2002

No

38. Yes

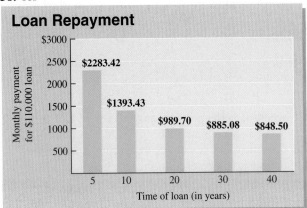

Loan Repayment

39. Boston Red Sox Infield Roof Box Seats

Year	Price
1980	$ 8
1991	16
1997	20
1998	25
1999	35
2000	45
2001	55
2002	60
2003	65

Sources: harvard-magazine.com and boston.com

Yes

40. World Automobile Production

Year	Number of Automobiles Produced (in millions)
1950	8
1960	13
1970	23
1980	29
1990	36
2000	41

Source: ibike.org

No

41. *Baseball.* The price of an infield roof box seat for Boston Red Sox fans has been growing exponentially. The table in Exercise 39 shows the price per game for season ticket holders for various years.

$$p(x) = 6.501242197(1.096109091)^x$$

a) Use regression to find an exponential function that can be used to estimate the price p of an infield roof box seat x years after 1980.   0.0918, or

b) Determine the exponential growth rate.   9.18%

c) Predict the price of an infield roof box seat in 2006. $71

42. *Salary.* In 1995, Tonya started a commercial cleaning business. Her profits have grown exponentially from $2000 in 1995 to $4300 in 1998, $9300 in 2001, and $15,500 in 2003.

$$p(x) = 1998.198072(1.291876847)^x$$

a) Use regression to find an exponential function that can be used to estimate the profit p of Tonya's company t years after 1995.

b) Determine the exponential growth rate.

c) Predict the profits in 2006.   0.256, or 25.6%
   $33,425

(a) $P(x) = 37.29665447(1.111922582)^x$

43. *Stadium Construction.* The amount spent on construction of stadiums and arenas grew from $50 million in 1965 to $100 million in 1975, $150 million in 1985, and $1.5 billion in 1995 (*Source*: harvard-magazine.com).

a) Use regression to find an exponential function of the form $P(x) = a \cdot b^x$ that can be used to estimate the amount spent P, in millions of dollars, on new stadium construction t years after 1965.

b) Predict the amount spent on construction of new stadiums in 2005.   About $2.6 billion

44. *World Population.*   The population of the world is growing exponentially, as shown by the graph on p. 666.

a) Use the four most recent data points and regression to fit an exponential function of the form $P(t) = P_0 e^{kt}$ to the data, where t is the year and P is in billions. $P(t) = 2.2624 \times 10^{-10} e^{0.0119579321t}$

b) Predict the world population in 2050.   10.02 billion

45. *Fluoride Water Content.*   The number of decayed, missing, and fitted teeth in dental patients decreases exponentially as the fluoride content of the water increases, as shown by these data from one study.

Fluoride Content of Water (in parts per million)	Number of Decayed, Missing, or Filled Teeth per 100 Participants in Study
0.2	805
0.4	410
0.7	360
1	350
1.7	270

Source: Dudley, Brian A. C., *Mathematical and Biological Interrelations.* Chichester: John Wiley & Sons, 1977

$f(x) = 647.6297124(0.5602992676)^x$

a) Use regression to fit an exponential function of the form $f(x) = ab^x$ to the data.

b) Estimate the number of decayed, missing, or filled teeth per 100 patients if the fluoride count of the water is 2.5 ppm.   152 teeth

46. *Cellular Phones.*   The number of U.S. subscribers of cellular phones has grown exponentially, as shown by the data in the following table.

Year	Number of Subscribers (in millions)
1985	0.3
1989	3.5
1994	24.1
1996	44
1998	69
1999	86
2001	117

$P(t) = 0.5434943782e^{0.369163743t}$

a) Use regression to fit an exponential function of the form $P(t) = P_0 e^{kt}$ to the data, where t is the number of years since 1985 and P is in millions.

b) Predict the number of cellular-phone subscribers in 2005.   874 million subscribers

☐ Answers to Exercises 49–52 can be found on p. A-72.

TW 47. Will the model used to predict the number of cellular phones in Exercise 1 still be realistic in 2015? Why or why not?

TW 48. Examine the restriction on t in Exercise 22.
a) What upper limit might be placed on t?
b) In practice, would this upper limit ever be enforced? Why or why not?

Skill Maintenance

Graph. [8.7]

49. $y = x^2 - 8x$ ☐

50. $y = x^2 - 5x - 6$ ☐

51. $f(x) = 3x^2 - 5x - 1$ ☐

52. $g(x) = 2x^2 - 6x + 3$ ☐

Solve by completing the square. [8.1]

53. $x^2 - 8x = 7$ $4 \pm \sqrt{23}$

54. $x^2 + 10x = 6$ $-5 \pm \sqrt{31}$

Synthesis

TW 55. *Atmospheric Pressure.*   Atmospheric pressure P at altitude a is given by
$$P = P_0 e^{-0.00005a},$$
where P_0 is the pressure at sea level ≈ 14.7 lb/in^2 (pounds per square inch). Explain how a barometer, or some other device for measuring atmospheric pressure, can be used to find the height of a skyscraper.

TW 56. Write a problem for a classmate to solve in which information is provided and the classmate is asked to find an exponential growth function. Make the problem as realistic as possible.

57. *Sports Salaries.*   In 2001, Derek Jeter of the New York Yankees signed a $189 million 10-yr contract that will pay him $21 million in 2010. How much would Yankee owner George Steinbrenner need to invest in 2001 at 5% interest compounded continuously, in order to have the $21 million for Jeter in 2010? (This is much like finding what $21 million in 2010 will be worth in 2001 dollars.) $13.4 million

58. *Nuclear Energy.*   Plutonium-239 (Pu-239) is used in nuclear energy plants. The half-life of Pu-239 is 24,360 yr (*Source*: *Microsoft Encarta 97 Encyclopedia*). How long will it take for a fuel rod of Pu-239 to lose 90% of its radioactivity?   About 80,922 yr, or with rounding of decay rate, about 80,792 yr

59. Use Exercises 1 and 17 to form a model for the percentage of U.S. residents owning a cellular phone t years after 2002. $P(t) = 53e^{0.302t}$, where t is the number of years after 2002 and $P(t)$ is a percent.

TW **60.** Use the model developed in Exercise 59 to predict the percentage of U.S. residents who will own cellular phones in 2010. Does your prediction seem plausible? Why or why not?

Logistic Curves. *Realistically, most quantities that are growing exponentially eventually level off. The quantity may continue to increase, but at a decreasing rate. This pattern of growth can be modeled by a logistic function*

$$f(x) = \frac{c}{1 + ae^{-bx}}.$$

The general shape of this family of functions is shown by the following graph.

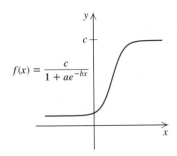

Many graphing calculators can fit a logistic function to a set of data.

61. *Telephones.* The percentage of telephones in U.S. households for various years is shown in the following table.

Year	Percentage of Households with Telephones
1920	35.0
1930	40.9
1950	61.8
1960	78.3
1970	90.5
1980	92.9
1990	94.8
1998	94.1

Sources: U.S. Bureau of the Census; Federal Communications Commission

(b) $f(x) = 35.0(1.0192)^x$

a) Graph the data from 1920 to 1970. Do the data appear to be growing exponentially?   Yes

b) Use the data from 1920 and 1970 to find an exponential function $f(x) = ab^x$ that could be used to estimate the percentage of U.S. households with telephones x years after 1920.

c) Use the function found in part (b) to predict the percentage of U.S. households with telephones in 2010. Does the estimate make sense?   194%; no

d) Using all the data in the table, find a logistic function

$$f(x) = \frac{c}{1 + ae^{-bx}}$$

that could be used to estimate the percentage of U.S. households with telephones x years after 1920.

e) Use the function found in part (d) to predict the percentage of U.S. households with telephones in 2010. Does the estimate make sense?   99.6%; yes

62. *Home Videos.*   The following table shows the total revenue from sales and rentals of home videos for various years.

Year	Total Revenue (in billions)
1985	$ 3.41
1990	9.81
1995	14.83
2000	18.18

Source: Paul Kagan Associates, Inc.

a) Find a logistic function

$$f(x) = \frac{c}{1 + ae^{-bx}}$$

that could be used to estimate the total revenue of the home video industry x years after 1985.

b) Use the function from part (a) to predict the total revenue from home videos in 2005. $18.82 billion

61. (d) $f(x) = \dfrac{102.4604343}{1 + 2.25595242e^{-0.0483573334x}}$

62. (a) $f(x) = \dfrac{19.11252764}{1 + 4.175198683e^{-0.2798470918x}}$

Collaborative Corner

Television Rights

Focus: Models
Time: 20 minutes
Group size: 3–6

Sometimes simply looking at the graph of a particular set of data will give a good indication of an appropriate model for the data. Often, more than one type of model is a possibility. One way to determine which is the best model is to calculate several possible functions, graph them along with the data, and determine which one seems to best fit the data. Another indication of a good model is its ability to predict another, known, data point not included in determining the model.

ACTIVITY

The following table shows the amount that television networks have either paid or agreed to pay for the television rights to the Summer Olympic Games.

Year	Amount Paid for Television Rights (in millions)
1968	$ 4.5
1972	7.5
1976	25
1980	87
1984	225
1988	300
1992	401
1996	456
2000	705
2004	793

Source: NBC Sports

1. Each group member should graph the data. The group should then agree on at least two possible models for the data from the following list: linear, quadratic, cubic, quartic, or exponential. (See the library of functions on the inside front cover.)

2. Form a model of each type chosen in part (1), and graph each model along with the data. Determine, as a group, which model appears to fit the data best.

3. NBC has agreed to pay $894 million for the television rights to the 2008 Summer Olympics. Determine, as a group, which model best predicts this amount.

9 Chapter Summary and Review

Key Terms

Composite functions, p. 650
Inverse relation, p. 654
One-to-one function, p. 655

Horizontal-line test, p. 656
Inverse function, p. 656
Exponential function, p. 667

Asymptote, p. 667
Logarithmic function, p. 667
Common logarithm, p. 679

Important Properties and Formulas

To Find a Formula for the Inverse of a Function

First check a graph to make sure that the function f is one-to-one. Then:

1. Replace $f(x)$ with y.
2. Interchange x and y.
3. Solve for y.
4. Replace y with $f^{-1}(x)$.

Composition of f and g:

$$(f \circ g)(x) = f(g(x))$$

Composition and inverses:

$$(f^{-1} \circ f)(x) = (f \circ f^{-1})(x) = x$$

Exponential function:

$$f(x) = a^x, \quad a > 0, \quad a \neq 1$$

Interest compounded annually:

$$A = P(1 + i)^t$$

For $x > 0$ and a a positive constant other than 1, the number $\log_a x$ is the power to which a must be raised in order to get x. Thus, $a^{\log_a x} = x$, or equivalently, if $y = \log_a x$, then $a^y = x$. *A logarithm is an exponent.*

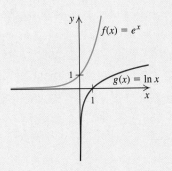

The Principle of Exponential Equality

For any real number b, $b \neq -1, 0$, or 1:

$$b^x = b^y \quad \text{is equivalent to} \quad x = y.$$

The Principle of Logarithmic Equality

For any logarithmic base a, and for $x, y > 0$:

$$x = y \quad \text{is equivalent to} \quad \log_a x = \log_a y.$$

Properties of Logarithms

$$\log_a MN = \log_a M + \log_a N,$$

$$\log_a \frac{M}{N} = \log_a M - \log_a N,$$

$$\log_a M^p = p \cdot \log_a M,$$

$$\log_a 1 = 0,$$

$$\log_a a = 1,$$

$$\log_a a^k = k,$$

$$\log M = \log_{10} M,$$

$$\ln M = \log_e M,$$

$$\log_b M = \frac{\log_a M}{\log_a b}$$

$$e \approx 2.7182818284\ldots$$

Loudness of sound: $L = 10 \cdot \log \dfrac{I}{I_0}$

pH: $\text{pH} = -\log [H^+]$

Exponential growth: $P(t) = P_0 e^{kt},\ k > 0$

Exponential decay: $P(t) = P_0 e^{-kt},\ k > 0$

Interest compounded continuously: $P(t) = P_0 e^{kt}$, where P_0 is the principal invested for t years at interest rate k

Carbon dating: $P(t) = P_0 e^{-0.00012t}$

Review Exercises

1. Find $(f \circ g)(x)$ and $(g \circ f)(x)$ if $f(x) = x^2 + 1$ and $g(x) = 2x - 3$. [9.1] $(f \circ g)(x) = 4x^2 - 12x + 10$; $(g \circ f)(x) = 2x^2 - 1$

2. If $h(x) = \sqrt{3 - x}$, find $f(x)$ and $g(x)$ such that $h(x) = (f \circ g)(x)$. Answers may vary. [9.1] $f(x) = \sqrt{x}$; $g(x) = 3 - x$

3. Determine whether $f(x) = 4 - x^2$ is one-to-one. [9.1] No

Find a formula for the inverse of each function.

4. $f(x) = x - 3$ [9.1] $f^{-1}(x) = x + 3$

5. $g(x) = \dfrac{3x + 1}{2}$ [9.1] $g^{-1}(x) = \dfrac{2x - 1}{3}$

6. $f(x) = 27x^3$ [9.1] $f^{-1}(x) = \dfrac{\sqrt[3]{x}}{3}$

Graph.

7. $f(x) = 3^x + 1$ ▣

8. $x = \left(\tfrac{1}{4}\right)^y$ ▣

9. $y = \log_5 x$ ▣

Simplify.

10. $\log_3 9$ [9.3] 2

11. $\log_{10} \tfrac{1}{10}$ [9.3] -1

12. $\log_{25} 5$ [9.3] $\tfrac{1}{2}$

13. $\log_3 3^{12}$ [9.3] 12

Convert to logarithmic equations.

14. $10^{-2} = \tfrac{1}{100}$ ▣

15. $25^{1/2} = 5$ [9.3] $\log_{25} 5 = \tfrac{1}{2}$

Convert to exponential equations.

16. $\log_4 16 = x$ [9.3] $16 = 4^x$

17. $\log_8 1 = 0$ [9.3] $1 = 8^0$

Express in terms of logarithms of x, y, and z.

18. $\log_a x^4 y^2 z^3$ [9.4] $4 \log_a x + 2 \log_a y + 3 \log_a z$

19. $\log_a \dfrac{x^3}{yz^2}$ [9.4] , or $3 \log_a x - \log_a y - 2 \log_a z$ $3 \log_a x - \log_a y - 2 \log_a z$

20. $\log \sqrt[4]{\dfrac{z^2}{x^3 y}}$ [9.4] $\tfrac{1}{4}(2 \log z - 3 \log x - \log y)$

Express as a single logarithm and, if possible, simplify.

21. $\log_a 8 + \log_a 15$ [9.4] $\log_a (8 \cdot 15)$, or $\log_a 120$

22. $\log_a 72 - \log_a 12$ [9.4] $\log_a \tfrac{72}{12}$, or $\log_a 6$

23. $\tfrac{1}{2} \log a - \log b - 2 \log c$ [9.4] $\log \dfrac{a^{1/2}}{bc^2}$

24. $\tfrac{1}{3}[\log_a x - 2 \log_a y]$ [9.4] $\log_a \sqrt[3]{\dfrac{x}{y^2}}$

Simplify.

25. $\log_m m$ [9.4] 1

26. $\log_m 1$ [9.4] 0

27. $\log_m m^{17}$ [9.4] 17

Given $\log_a 2 = 1.8301$ and $\log_a 7 = 5.0999$, find each of the following.

28. $\log_a 14$ [9.4] 6.93

29. $\log_a \tfrac{2}{7}$ [9.4] -3.2698

30. $\log_a 28$ [9.4] 8.7601

31. $\log_a 3.5$ [9.4] 3.2698

32. $\log_a \sqrt{7}$ [9.4] 2.54995

33. $\log_a \tfrac{1}{4}$ [9.4] -3.6602

Use a calculator to find each of the following to the nearest ten-thousandth.

34. $\log 82$ [9.3] 1.9138

35. $10^{1.789}$ [9.3] 61.5177

36. $\ln 0.05$ [9.5] -2.9957

37. $e^{-0.98}$ [9.5] 0.3753

Find each of the following logarithms using the change-of-base formula. Round answers to the nearest ten-thousandth.

38. $\log_5 2$ [9.5] 0.4307

39. $\log_{12} 70$ [9.5] 1.7097

Graph and state the domain and the range of each function.

40. $f(x) = e^x - 1$ ▣

41. $g(x) = 0.6 \ln x$ ▣

Solve. Where appropriate, include approximations to the nearest ten-thousandth.

42. $2^x = 32$ [9.6] 5

43. $3^x = \tfrac{1}{9}$ [9.6] -2

44. $\log_3 x = -2$ [9.6] $\tfrac{1}{9}$

45. $\log_x 32 = 5$ [9.6] 2

46. $\log x = -4$ [9.6] $\tfrac{1}{10,000}$

47. $3 \ln x = -6$ [9.6] $e^{-2} \approx 0.1353$

48. $4^{2x-5} = 16$ [9.6] $\tfrac{7}{2}$

49. $2^{x^2} \cdot 2^{4x} = 32$ [9.6] $-5, 1$

50. $4^x = 8.3$ ▣

51. $e^{-0.1t} = 0.03$ ▣

52. $2 \ln x = -6$ [9.6] $e^{-3} \approx 0.0498$

53. $\log_3 (2x - 5) = 1$ [9.6] 4

54. $\log_4 x + \log_4 (x - 6) = 2$ [9.6] 8

▣ Answers to Exercises 7–9, 14, 40, 41, 50, and 51 can be found on pp. A-72 and A-73.

55. $\log x + \log (x - 15) = 2$ [9.6] 20

56. $\log_3 (x - 4) = 3 - \log_3 (x + 4)$ [9.6] $\sqrt{43}$

57. In a business class, students were tested at the end of the course with a final exam. They were tested again after 6 months. The forgetting formula was determined to be

$$S(t) = 62 - 18 \log (t + 1),$$

where t is the time, in months, after taking the first test.

a) Determine the average score when they first took the test (when $t = 0$).   [9.7] 62

b) What was the average score after 6 months?   [9.7] 46.8

c) After what time was the average score 34?   [9.7] 35 months

58. A color photocopier is purchased for $5200. Its value each year is about 80% of its value in the pre-ceding year. Its value in dollars after t years is given by the exponential function

$$V(t) = 5200(0.8)^t.$$

a) After what amount of time will the salvage value be $1200?   [9.7] 6.6 yr

b) After what amount of time will the salvage value be half the original value?   [9.7] 3.1 yr

59. The number of customer service complaints against U.S. airlines grew exponentially from 667 in 1995 to 4535 in 2000 (*Source*: U.S. Department of Transportation, Office of Consumer Affairs).   [9.7] 0.383; $C(t) = 667e^{0.383t}$

a) Find the value k, and write an exponential func-tion that describes the number of customer service complaints t years after 1995.

b) Predict the number of complaints in 2005.   [9.7] About 30,724

c) In what year will there be 20,000 customer service complaints?   [9.7] 2004

60. The value of Jose's stock market portfolio doubled in 3 yr. What was the exponential growth rate?   [9.7] 23.105% per year

61. How long will it take $7600 to double itself if it is invested at 8.4%, compounded continuously?   [9.7] 8.25 yr

62. How old is a skeleton that has lost 34% of its carbon-14? (Use $P(t) = P_0 e^{-0.00012t}$.)   [9.7] 3463 yr

63. What is the pH of a substance if its hydrogen ion concentration is 2.3×10^{-7} moles per liter? (Use $\text{pH} = -\log [\text{H}^+]$.)   [9.7] 6.6

64. The intensity of the sound of water at the foot of the Niagara Falls is about 10^{-3} W/m^2.* How loud in decibels is this sound level?   [9.7] 90 dB

$$\left(\text{Use } L = 10 \cdot \log \frac{I}{10^{-12}}. \right)$$

Synthesis

TW **65.** Explain why negative numbers do not have logarithms.

TW **66.** Explain why taking the natural or common loga-rithm on each side of an equation produces an equivalent equation.

Solve.

67. $\ln (\ln x) = 3$ [9.6] e^{e^3}

68. $2^{x^2+4x} = \frac{1}{8}$ [9.6] $-3, -1$

69. $5^{x+y} = 25,$
    $2^{2x-y} = 64$ [9.6] $\left(\frac{8}{3}, -\frac{2}{3} \right)$

Sound and Hearing, Life Science Library. (New York: Time Incorporated, 1965), p. 173.

Chapter Test 9

1. Find $(f \circ g)(x)$ and $(g \circ f)(x)$ if $f(x) = x + x^2$ and $g(x) = 2x + 1$.   [9.1] $(f \circ g)(x) = 2 + 6x + 4x^2$;
$(g \circ f)(x) = 2x^2 + 2x + 1$

2. If

$$h(x) = \frac{1}{2x^2 + 1},$$

find $f(x)$ and $g(x)$ such that $h(x) = (f \circ g)(x)$. Answers may vary.   [9.1] $f(x) = \frac{1}{x}$; $g(x) = 2x^2 + 1$

3. Determine whether $f(x) = |x + 1|$ is one-to-one.   [9.1] No

Find a formula for the inverse of each function.

4. $f(x) = 4x - 3$ [9.1] $f^{-1}(x) = \dfrac{x + 3}{4}$

5. $g(x) = (x + 1)^3$ [9.1] $g^{-1}(x) = \sqrt[3]{x} - 1$

Graph.

6. $f(x) = 2^x - 3$

7. $g(x) = \log_7 x$

6. [9.2]

7. [9.2]
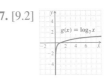

Simplify.

8. $\log_5 125$ [9.3] 3

9. $\log_{100} 10$ [9.3] $\frac{1}{2}$

10. $3^{\log_3 18}$ [9.3] 18

Convert to logarithmic equations.

11. $4^{-3} = \frac{1}{64}$
[9.3] $\log_4 \frac{1}{64} = -3$

12. $256^{1/2} = 16$
[9.3] $\log_{256} 16 = \frac{1}{2}$

Convert to exponential equations.

13. $m = \log_7 49$ [9.3] $49 = 7^m$

14. $\log_3 81 = 4$
[9.3] $81 = 3^4$

15. Express in terms of logarithms of a, b, and c:

$$\log \frac{a^3 b^{1/2}}{c^2}.$$ [9.4] $3 \log a + \frac{1}{2} \log b - 2 \log c$

16. Express as a single logarithm:

$\frac{1}{3} \log_a x + 2 \log_a z.$ [9.4] $\log_a \left(z^2 \sqrt[3]{x} \right)$

Simplify.

17. $\log_p p$ [9.4] 1

18. $\log_t t^{23}$ [9.4] 23

19. $\log_c 1$ [9.4] 0

Given $\log_a 2 = 0.301$, $\log_a 6 = 0.778$, and $\log_a 7 = 0.845$, find each of the following.

20. $\log_a \frac{2}{7}$
[9.4] -0.544

21. $\log_a 12$
[9.4] 1.079

22. $\log_a 16$
[9.4] 1.204

Use a calculator to find each of the following to the nearest ten-thousandth.

23. $\log 12.3$ [9.5] 1.0899

24. $10^{-0.8}$ [9.5] 0.1585

25. $\ln 0.035$ [9.5] -3.3524

26. $e^{4.8}$ [9.5] 121.5104

27. Find $\log_3 10$ using the change-of-base formula. Round to the nearest ten-thousandth. [9.5] 2.0959

Graph and state the domain and the range of each function.

28. $f(x) = e^x + 3$ ⊡

29. $g(x) = \ln(x - 4)$ ⊡

Solve. Where appropriate, include approximations to the nearest ten-thousandth.

30. $2^x = \frac{1}{32}$ [9.6] -5

31. $\log_x 25 = 2$ [9.6] 5

32. $\log_4 x = \frac{1}{2}$ [9.6] 2

33. $\log x = 4$ [9.6] 10,000

34. $5^{4-3x} = 125$ [9.6] $\frac{1}{3}$

35. $7^x = 1.2$ [9.6] $\frac{\log 1.2}{\log 7} \approx 0.0937$

36. $\ln x = \frac{1}{4}$ [9.6] $e^{1/4} \approx 1.2840$

37. $\log(x - 3) + \log(x + 1) = \log 5$ [9.6] 4

⊡ Answers to Exercises 28 and 29 can be found on p. A-73.

(a) [9.7] 2.46 ft/sec

38. The average walking speed R of people living in a city of population P, in thousands, is given by $R = 0.37 \ln P + 0.05$, where R is in feet per second.

a) The population of Albuquerque, New Mexico, is 679,000. Find the average walking speed.

b) A city has an average walking speed of 2.6 ft/sec. Find the population. [9.7] 984,262

39. The population of Kenya was 30 million in 2000, and the exponential growth rate was 1.5% per year.

a) Write an exponential function describing the population of Kenya.

b) What will the population be in 2003? in 2010?

c) When will the population be 50 million? [9.7] 2034

d) What is the doubling time? [9.7] 46.2 yr

40. The U.S. Consumer Price Index, a method of comparing prices of basic items, grew exponentially from 60.0 in 1920 to 511.5 in 2000 (*Sources*: U.S. Bureau of Labor Statistics and U.S. Department of Labor).

a) Find the value k, and write an exponential function that approximates the Consumer Price Index t years after 1920. [9.7] $k \approx 0.027$; $C(t) = 60e^{0.027t}$

b) Predict the Consumer Price Index in 2010. [9.7] 681.5

c) In what year will the Consumer Price Index be 1000? [9.7] 2024

41. An investment with interest compounded continuously doubled itself in 15 yr. What is the interest rate? [9.7] 4.6%

42. How old is an animal bone that has lost 43% of its carbon-14? (Use $P(t) = P_0 e^{-0.00012t}$.) [9.7] 4684 yr

43. The sound of traffic at a busy intersection averages 75 dB. What is the intensity of such a sound?

$\left(\text{Use } L = 10 \cdot \log \frac{I}{I_0}. \right)$ [9.7] $10^{-4.5}$ W/m²

44. The hydrogen ion concentration of water is 1.0×10^{-7} moles per liter. What is the pH? (Use pH $= -\log [H^+]$.) [9.7] 7.0

Synthesis

45. Solve: $\log_5 |2x - 7| = 4$. [9.6] $-309, 316$

46. If $\log_a x = 2$, $\log_a y = 3$, and $\log_a z = 4$, find

$$\log_a \frac{\sqrt[3]{x^2 z}}{\sqrt[3]{y^2 z^{-1}}}.$$ [9.4] 2

39. (a) [9.7] $P(t) = 30e^{0.015t}$, where $P(t)$ is in millions and t is the number of years after 2000; **(b)** [9.7] 31.4 million; 34.9 million

1-9 Cumulative Review

1. Evaluate $\dfrac{x^0 + y}{-z}$ for $x = 6$, $y = 9$, and $z = -5$.
[1.1], [1.4] 2

Simplify.

2. $\left| -\frac{5}{2} + \left(-\frac{7}{2} \right) \right|$ [1.2] 6

3. $(-2x^2y^{-3})^{-4}$ [1.4] $\dfrac{y^{12}}{16x^8}$

4. $(-5x^4y^{-3}z^2)(-4x^2y^2)$ [1.4] $\dfrac{20x^6z^2}{y}$

5. $\dfrac{3x^4y^6z^{-2}}{-9x^4y^2z^3}$ [1.4] $\dfrac{-y^4}{3z^5}$

6. $2x - 3 - 2[5 - 3(2 - x)]$ [1.3] $-4x - 1$

7. $3^3 + 2^2 - (32 \div 4 - 16 \div 8)$ [1.1] 25

Solve.

8. $8(2x - 3) = 6 - 4(2 - 3x)$ [2.2] $\frac{11}{2}$

9. $4x - 3y = 15,$
    $3x + 5y = 4$ [3.2] $(3, -1)$

10. $x + y - 3z = -1,$
    $2x - y + z = 4,$
    $-x - y + z = 1$ [3.4] $(1, -2, 0)$

11. $x(x - 3) = 10$ [5.4] $-2, 5$

12. $\dfrac{7}{x^2 - 5x} - \dfrac{2}{x - 5} = \dfrac{4}{x}$ [6.4] $\dfrac{9}{2}$

13. $\dfrac{8}{x + 1} + \dfrac{11}{x^2 - x + 1} = \dfrac{24}{x^3 + 1}$ [6.4], [8.2] $\dfrac{5}{8}$

14. $\sqrt{4 - 5x} = 2x - 1$ [7.6] $\frac{3}{4}$

15. $\sqrt[3]{2x} = 1$ [7.6] $\frac{1}{2}$

16. $3x^2 + 75 = 0$ [8.1] $\pm 5i$

17. $x - 8\sqrt{x} + 15 = 0$ [8.5] 9, 25

18. $x^4 - 13x^2 + 36 = 0$ [8.5] $\pm 2, \pm 3$

19. $\log_8 x = 1$ [9.3] 8

20. $\log_x 49 = 2$ [9.3] 7

21. $9^x = 27$ [9.6] $\frac{3}{2}$

22. $3^{5x} = 7$ [9.6] $\dfrac{\log 7}{5 \log 3} \approx 0.3542$

23. $\log x - \log (x - 8) = 1$ [9.6] $\frac{80}{9}$

24. $x^2 + 4x > 5$ [8.9] $(-\infty, -5) \cup (1, \infty)$ or
$\{x \mid x < -5 \text{ or } x > 1\}$

25. If $f(x) = x^2 + 6x$, find a such that $f(a) = 11$.
[8.2] $-3 \pm 2\sqrt{5}$

26. If $f(x) = |2x - 3|$, find all x for which $f(x) \geq 9$.
[4.3] $\{x \mid x \leq -3 \text{ or } x \geq 6\}$, or $(-\infty, -3] \cup [6, \infty)$

Solve.

27. $D = \dfrac{ab}{b + a}$, for a [6.8] $a = \dfrac{Db}{b - D}$

28. $\dfrac{1}{p} + \dfrac{1}{q} = \dfrac{1}{f}$, for q [6.8] $q = \dfrac{pf}{p - f}$

29. $M = \dfrac{2}{3}(A + B)$, for B
[2.3] $B = \dfrac{3M - 2A}{2}$, or $B = \dfrac{3}{2}M - A$

In Exercises 30–33, match each function with one of the following graphs.

a) 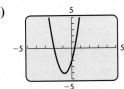   **b)**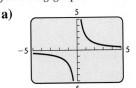

c) 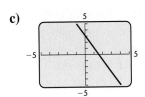   **d)**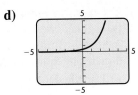

30. $f(x) = -2x + 3.1$
[2.4] (c)

31. $p(x) = 3x^2 + 5x - 2$
[8.7] (b)

32. $h(x) = \dfrac{2.3}{x}$ [6.7] (a)

33. $g(x) = e^{x-1}$ [9.5] (d)

34. Find the domain of the function f given by

$$f(x) = \dfrac{-4}{3x^2 - 5x - 2}.$$ [5.8] $\{x \mid x \text{ is a real number}$ and $x \neq -\frac{1}{3}$ and $x \neq 2\}$

Solve.

35. The number of Americans filing taxes on the Internet (e-filers) increased from 6.7 million in 2001 to 8.5 million in 2002 (*Source*: www.businessweek.com).

   a) At what rate was the number of e-filers increasing?   [2.4] 1.8 million e-filers per year

   b) Find a linear function $E(t)$ that fits the data. Let t represent the number of years since 2001.   ⊡

   c) Use the function of part (b) to predict the number of e-filers in 2005.   [2.4] 13.9 million

For each of the following sets of data, graph the data and determine whether a linear or exponential function would best model the situation.

36.   ⊡

Age	Monthly Premium for $250,000 Term Life Insurance, for Males
35	$14.85
40	18.53
45	21.60
50	33.05
55	47.25

37.   ⊡

Year	Factory Sales of Trucks and Buses (in millions)
1995	5.7
1996	5.8
1997	6.2
1998	6.4
1999	6.7
2000	7.0

Source: Ward's Communications, *Ward's Motor Vehicles Facts & Figures, 2000*.

38. The monthly premium for a term life insurance policy increases as the age of the insured person increases. Some premiums for a $250,000 policy for a male are shown in the table in Exercise 36.

   a) Use the data for ages 35 and 55 to find an exponential function that can be used to estimate the premium m for a male who is x years older than 35.   ⊡

   b) Use regression and all the data to find an exponential function that can be used to estimate the premium y for a male of age x.   ⊡

39. The sales of trucks and buses increased steadily from 1995 to 2000, as shown in the table in Exercise 37.

⊡ Answers to Exercises 35(b) and 36–39 can be found on p. A-73.

   a) Use the data for 1996 and 2000 to find a linear function that can be used to predict the factory sales f of trucks and buses x years after 1995.   ⊡

   b) Use regression and all the data to find a linear function that can be used to predict the factory sales y of trucks and buses x years after 1995.   ⊡

40. The perimeter of a rectangular garden is 112 m. The length is 16 m more than the width. Find the length and the width.   [2.3] Length: 36 m; width: 20 m

41. In triangle *ABC*, the measure of angle *B* is three times the measure of angle *A*. The measure of angle *C* is 105° greater than the measure of angle *A*. Find the angle measures.   [2.3] *A*: 15°; *B*: 45°; *C*: 120°

42. Good's Candies makes all their chocolates by hand. It takes Anne 10 min to coat a tray of candies in chocolate. It takes Clay 12 min to coat a tray of candies. How long would it take Anne and Clay, working together, to coat the candies?   [6.5] $5\frac{5}{11}$ min

43. Joe's Thick and Tasty salad dressing gets 45% of its calories from fat. The Light and Lean dressing gets 20% of its calories from fat. How many ounces of each should be mixed in order to get 15 oz of dressing that gets 30% of its calories from fat?   [3.3] Thick and Tasty: 6 oz; Light and Lean: 9 oz

44. A fishing boat with a trolling motor can move at a speed of 5 km/h in still water. The boat travels 42 km downstream in the same time that it takes to travel 12 km upstream. What is the speed of the stream?   [6.5] $2\frac{7}{9}$ km/h

45. What is the minimum product of two numbers whose difference is 14? What are the numbers that yield this product? [8.8] -49; -7 and 7

Students in a biology class just took a final exam. A formula for determining what the average exam grade will be t months later is

$$S(t) = 78 - 15 \log (t + 1).$$

46. The average score when the students first took the test occurs when $t = 0$. Find the students' average score on the final exam. [9.7] 78

47. What would the average score be on a retest after 4 months? [9.7] 67.5

The population of Mozambique was 19.4 million in 2001, and the exponential growth rate was 1.5% per year.

48. Write an exponential function describing the growth of the population of Mozambique. ⊡

49. Predict what the population will be in 2005 and in 2012. [9.7] 20.6 million, 22.9 million

50. What is the doubling time of the population? [9.7] 46.2 yr

51. y varies directly as the square of x and inversely as z, and $y = 2$ when $x = 5$ and $z = 100$. What is y when $x = 3$ and $z = 4$? [6.8] 18

Perform the indicated operations and simplify.

52. $(5p^2q^3 + 6pq - p^2 + p) + (2p^2q^3 + p^2 - 5pq - 9)$ [5.1] $7p^2q^3 + pq + p - 9$

53. $(11x^2 - 6x - 3) - (3x^2 + 5x - 2)$ [5.1] $8x^2 - 11x - 1$

54. $(3x^2 - 2y)^2$ [5.2] $9x^4 - 12x^2y + 4y^2$

55. $(5a + 3b)(2a - 3b)$ [5.2] $10a^2 - 9ab - 9b^2$

56. $\dfrac{x^2 + 8x + 16}{2x + 6} \div \dfrac{x^2 + 3x - 4}{x^2 - 9}$ [6.1] $\dfrac{(x+4)(x-3)}{2(x-1)}$

57. $\dfrac{1 + \dfrac{3}{x}}{x - 1 - \dfrac{12}{x}}$ [6.3] $\dfrac{1}{x - 4}$

58. $\dfrac{a^2 - a - 6}{a^3 - 27} \cdot \dfrac{a^2 + 3a + 9}{6}$ [6.1] $\dfrac{a + 2}{6}$

59. $\dfrac{3}{x + 6} - \dfrac{2}{x^2 - 36} + \dfrac{4}{x - 6}$ [6.2] $\dfrac{7x + 4}{(x+6)(x-6)}$

Factor.

60. $xy - 2xz + xw$ [5.3] $x(y - 2z + w)$

61. $1 - 125x^3$ [5.7] $(1 - 5x)(1 + 5x + 25x^2)$

62. $6x^2 + 8xy - 8y^2$ [5.5] $2(3x - 2y)(x + 2y)$

63. $x^4 - 4x^3 + 7x - 28$ [5.3] $(x^3 + 7)(x - 4)$

64. $2m^2 + 12mn + 18n^2$ [5.6] $2(m + 3n)^2$

65. $x^4 - 16y^4$ [5.6] $(x - 2y)(x + 2y)(x^2 + 4y^2)$

66. For the function described by

$$h(x) = -3x^2 + 4x + 8,$$

find $h(-2)$. [2.1] -12

67. Divide: $(x^4 - 5x^3 + 2x^2 - 6) \div (x - 3)$. ⊡

68. Multiply $(5.2 \times 10^4)(3.5 \times 10^{-6})$. Write scientific notation for the answer. [1.4] 1.82×10^{-1}

For the radical expressions that follow, assume that all variables represent positive numbers.

69. Divide and simplify:

$\dfrac{\sqrt[3]{40xy^8}}{\sqrt[3]{5xy}}$. [7.4] $2y^2\sqrt[3]{y}$

70. Multiply and simplify: $\sqrt{7xy^3} \cdot \sqrt{28x^2y}$. [7.4] $14xy^2\sqrt{x}$

71. Rewrite without rational exponents: $(27a^6b)^{4/3}$. [7.2] $81a^8b\sqrt[3]{b}$

72. Rationalize the denominator:

$\dfrac{3 - \sqrt{y}}{2 - \sqrt{y}}$. [7.5] $\dfrac{6 + \sqrt{y} - y}{4 - y}$

73. Divide and simplify:

$\dfrac{\sqrt{x + 5}}{\sqrt[5]{x + 5}}$. [7.4] $\sqrt[10]{(x + 5)^3}$

74. Multiply these complex numbers:

$\left(1 + i\sqrt{3}\right)\left(6 - 2i\sqrt{3}\right)$. [7.8] $12 + 4\sqrt{3}i$

75. Add: $(3 - 2i) + (5 + 3i)$. [7.8] $8 + i$

76. Find the inverse of f if $f(x) = 7 - 2x$. ⊡

77. Find a linear function with a graph that contains the points $(0, -3)$ and $(-1, 2)$. [2.6] $f(x) = -5x - 3$

78. Find an equation of the line whose graph has a y-intercept of $(0, 7)$ and is perpendicular to the line given by $2x + y = 6$. [2.5] $y = \frac{1}{2}x + 7$

Graph by hand.

79. $5x = 15 + 3y$ ⊡

80. $y = 2x^2 - 4x - 1$ ⊡

81. $y = \log_3 x$ ⊡

82. $y = 3^x$ ⊡

83. $-2x - 3y \le 6$ ⊡

84. Graph: $f(x) = 2(x + 3)^2 + 1$. ⊡

a) Label the vertex.

b) Draw the axis of symmetry.

c) Find the maximum or minimum value.

⊡ Answers to Exercises 48, 67, 76, and 79–84 can be found on p. A-73.

85. Graph $f(x) = 2e^x$ and determine the domain and the range. ☐

86. Express in terms of logarithms of a, b, and c:
$$\log\left(\frac{a^2c^3}{b}\right). \quad \text{[9.4]} \ 2\log a + 3\log c - \log b$$

87. Express as a single logarithm:
$$3\log x - \tfrac{1}{2}\log y - 2\log z. \quad \text{[9.4]} \log\left(\frac{x^3}{y^{1/2}z^2}\right)$$

88. Convert to an exponential equation: $\log_a 5 = x$.
[9.3] $a^x = 5$

89. Convert to a logarithmic equation: $x^3 = t$.
[9.3] $\log_x t = 3$

Find each of the following using a calculator. Round to the nearest ten-thousandth.

90. $\log 0.05566$ [9.3] -1.2545 **91.** $10^{2.89}$ [9.3] 776.2471

92. $\ln 12.78$ [9.5] 2.5479 **93.** $e^{-1.4}$ [9.5] 0.2466

Synthesis

Solve.

94. $\dfrac{5}{3x-3} + \dfrac{10}{3x+6} = \dfrac{5x}{x^2+x-2}$ [6.4] All real numbers except 1 and -2

95. $\log\sqrt{3x} = \sqrt{\log 3x}$ [9.6] $\tfrac{1}{3}, \tfrac{10{,}000}{3}$

96. A train travels 280 mi at a certain speed. If the speed had been increased by 5 mph, the trip could have been made in 1 hr less time. Find the actual speed. [8.3] 35 mph

85. [9.5]
Domain: $\mathbb{R}$; range: $(0,\infty)$

Sequences, Series, and the Binomial Theorem

10

The first three sections of this chapter are devoted to sequences and series. A sequence is simply an ordered list. For example, when a baseball coach writes a batting order, a sequence is being formed. When the members of a sequence are numbers, we can discuss their sum. Such a sum is called a series.

Section 10.4 presents the binomial theorem, which is used to expand expressions of the form $(a + b)^n$. Such an expression is itself a series.

APPLICATION

INTERNET ACCESS. The number of U.S. households with Internet access during various years is shown in the following table and graph. Write the number of households as an arithmetic sequence and find a formula for the general term.

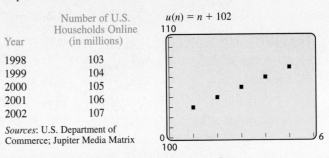

Year	Number of U.S. Households Online (in millions)
1998	103
1999	104
2000	105
2001	106
2002	107

Sources: U.S. Department of Commerce; Jupiter Media Matrix

This problem appears as Exercise 65 in Exercise Set 10.2.

10.1

Sequences ■ Finding the General Term ■ Sums and Series ■ Sigma Notation ■ Graphs of Sequences

Sequences and Series

Sequences

Suppose that $1000 is invested at 8%, compounded annually. The amounts to which the money grows after 1 year, 2 years, 3 years, and so on, are as follows:

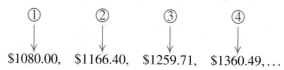

$$\$1080.00, \quad \$1166.40, \quad \$1259.71, \quad \$1360.49, \ldots$$

We can regard this as a function that pairs 1 with $1080.00, 2 with $1166.40, 3 with $1259.71, and so on. A **sequence** (or **progression**) is thus a function, where the domain is a set of consecutive positive integers beginning with 1, and the range varies from sequence to sequence.

If we continue computing the amounts in the account forever, we obtain an **infinite sequence**, with function values

$$\$1080.00, \ \$1166.40, \ \$1259.71, \ \$1360.49, \ \$1469.33, \ \$1586.87, \ldots$$

The three dots at the end indicate that the sequence goes on without stopping. If we stop after a certain number of years, we obtain a **finite sequence**:

$$\$1080.00, \ \$1166.40, \ \$1259.71, \ \$1360.49.$$

> ## Sequences
>
> An *infinite sequence* is a function having for its domain the set of natural numbers: $\{1, 2, 3, 4, 5, \ldots\}$.
>
> A *finite sequence* is a function having for its domain a set of natural numbers: $\{1, 2, 3, 4, 5, \ldots, n\}$, for some natural number n.

As another example, consider the sequence given by

$$a(n) = 2^n, \quad \text{or} \quad a_n = 2^n.$$

The notation a_n means the same as $a(n)$ but is used more commonly with sequences. Some function values (also called *terms* of the sequence) follow:

$$a_1 = 2^1 = 2,$$
$$a_2 = 2^2 = 4,$$
$$a_3 = 2^3 = 8,$$
$$a_6 = 2^6 = 64.$$

The first term of the sequence is a_1, the fifth term is a_5, and the nth term, or **general term**, is a_n. This sequence can also be denoted in the following ways:

$$2, 4, 8, \ldots;$$

or $\quad 2, 4, 8, \ldots, 2^n, \ldots.$ The 2^n emphasizes that the nth term of this sequence is found by raising 2 to the nth power.

Sequences

Sequences are entered into a graphing calculator and graphed much like functions. The difference is that the SEQUENCE MODE must be selected for some options. To do this, first press $\boxed{\text{MODE}}$ and then move the cursor to the line that reads Func Par Pol Seq, the fourth line in the screen shown on the left below. Next, move the cursor to Seq and press $\boxed{\text{ENTER}}$. The functions will be called $u(n)$ and $v(n)$ instead of y_1 and y_2, and the variable n is used instead of x; the variable is still entered using the $\boxed{\text{X,T,}\theta\text{,}n}$ key.

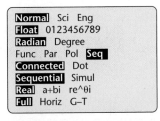

Most graphing calculators will write the terms of a sequence as a list without using the SEQUENCE mode. Usually found in the LIST OPS submenu, the SEQ option will list a finite sequence when the general term and the beginning and ending values of n are given. The CUMSUM option will list the cumulative sums of the elements of a sequence.

If the SEQ option is used, information must be supplied in the following order:

seq(general term, variable, value of n for the first term, value of n for the last term).

For example, to list the first 6 terms of the sequence given by

$$a_n = 2^n,$$

enter $\text{seq}(2^\wedge n, n, 1, 6)$. This is done by selecting SEQ under the LIST OPS menu, shown on the left below, and then pressing $\boxed{2}\ \boxed{\wedge}\ \boxed{\text{X,T,}\theta\text{,}n}\ \boxed{,}\ \boxed{\text{X,T,}\theta\text{,}n}\ \boxed{,}\ \boxed{1}\ \boxed{,}$ $\boxed{6}\ \boxed{)}\ \boxed{\text{ENTER}}$. Note that if this sequence of keystrokes is done in the FUNCTION mode, rather than the SEQUENCE mode, the variable used will be x, but the resulting sequence will be the same.

The result is shown in the screen on the right below. The first 6 terms of the sequence are 2, 4, 8, 16, 32, 64.

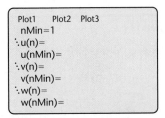

The CUMSUM (cumulative sums) option can be used in connection with the SEQ option. Instead of listing the sequence itself, CUMSUM lists the first term, then the sum of the first two terms, then the sum of the first three terms, and so on. The screen on the right above shows the cumulative sums of the sequence given by $a_n = 2^n$. The first term is 2, the sum of the first two terms is 6, the sum of the first three terms is 14, and so on. Pressing the right arrow key will show more cumulative sums of the sequence.

EXAMPLE 1   Find the first 4 terms and the 13th term of the sequence for which the general term is given by $a_n = (-1)^n n^2$.

Solution   We solve both algebraically and using a calculator.

By Hand

We have $a_n = (-1)^n n^2$, so

$$a_1 = (-1)^1 \cdot 1^2 = -1,$$
$$a_2 = (-1)^2 \cdot 2^2 = 4,$$
$$a_3 = (-1)^3 \cdot 3^2 = -9,$$
$$a_4 = (-1)^4 \cdot 4^2 = 16,$$
$$a_{13} = (-1)^{13} \cdot 13^2 = -169.$$

Using a Graphing Calculator

We let $u(n) = (-1)^\wedge n * n^2$.

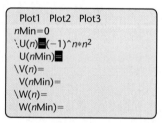

We set up a table with Indpnt set to Ask, and then supply 1, 2, 3, 4, and 13 as values for n.

n	$u(n)$	
1	−1	
2	4	
3	−9	
4	16	
13	−169	

$n =$

Note in Example 1 that the expression $(-1)^n$ causes the signs of the terms to alternate between positive and negative, depending on whether n is even or odd.

EXAMPLE 2   Use a graphing calculator to find the first 5 terms of the sequence for which the general term is given by $a_n = n/(n + 1)^2$.

Solution   We use the seq(feature, supplying the formula for the general term, the variable, and the values of n for the first and last terms that we wish to calculate.

```
seq(n/(n+1)²,n,1
,5)▶Frac
{1/4  2/9  3/16  4…
```

Here we used ▶Frac to write each term in the sequence in fraction notation.

The first three terms are listed on the screen; the remaining can be found by pressing ▶. We have

$$a_1 = \tfrac{1}{4}, \qquad a_2 = \tfrac{2}{9}, \qquad a_3 = \tfrac{3}{16}, \qquad a_4 = \tfrac{4}{25}, \quad \text{and} \quad a_5 = \tfrac{5}{36}.$$

Finding the General Term

When only the first few terms of a sequence are known, it is impossible to be certain what the general term is, but a prediction can be made by looking for a pattern.

EXAMPLE 3 For each sequence, predict the general term.

a) $1, 4, 9, 16, 25, \ldots$ **b)** $-1, 2, -4, 8, -16, \ldots$

c) $2, 4, 8, \ldots$

Solution

a) $1, 4, 9, 16, 25, \ldots$

These are squares of consecutive positive integers, so the general term could be n^2.

b) $-1, 2, -4, 8, -16, \ldots$

These are powers of 2 with alternating signs, so the general term may be $(-1)^n[2^{n-1}]$. To check, note that 8 is the fourth term, and

$$
\begin{aligned}
(-1)^4[2^{4-1}] &= 1 \cdot 2^3 \\
&= 8.
\end{aligned}
$$

c) $2, 4, 8, \ldots$

We regard the pattern as powers of 2, in which case 16 would be the next term and 2^n the general term. The sequence could then be written with more terms as

$$2, 4, 8, 16, 32, 64, 128, \ldots$$

In part (c) above, suppose that the second term is found by adding 2, the third term by adding 4, the next term by adding 6, and so on. In this case, 14 would be the next term and the sequence would be

$$2, 4, 8, 14, 22, 32, 44, 58, \ldots$$

This illustrates that the fewer terms we are given, the greater the uncertainty about the nth term.

TEACHING TIP

You may need to emphasize that when the signs of the terms alternate, -1 is a factor of the general term.

Sums and Series

Series

Given the infinite sequence

$$a_1, a_2, a_3, a_4, \ldots, a_n, \ldots,$$

the sum of the terms

$$a_1 + a_2 + a_3 + \cdots + a_n + \cdots$$

is called an *infinite series*. A *partial sum* is the sum of the first n terms:

$$a_1 + a_2 + a_3 + \cdots + a_n.$$

A partial sum is also called a *finite series* and is denoted S_n.

EXAMPLE 4 For the sequence $-2, 4, -6, 8, -10, 12, -14$, find: **(a)** S_2; **(b)** S_3; **(c)** S_7.

Solution

a) $S_2 = -2 + 4 = 2$ This is the sum of the first 2 terms.

b) $S_3 = -2 + 4 + (-6) = -4$ This is the sum of the first 3 terms.

c) $S_7 = -2 + 4 + (-6) + 8 + (-10) + 12 + (-14) = -8$

This is the sum of the first 7 terms.

We can use a graphing calculator to find partial sums of a sequence for which the general term is given by a formula.

EXAMPLE 5 Use a graphing calculator to find S_1, S_2, S_3, and S_4 for the sequence in which the general term is given by $a_n = (-1)^n/(n + 1)$.

Solution We use the CUMSUM option in the LIST OPS menu, along with the SEQ option.

```
cumSum(seq((−1)^
n/(n+1),n,1,4))▶
Frac
      {−1/2 −1/6 −5/1...
```

Here we used ▶Frac to write each sum in fraction notation.

The first two sums, S_1 and S_2, are listed on the screen; S_3 and S_4 can be seen by pressing ▶ . We have

$$S_1 = -\tfrac{1}{2}, \qquad S_2 = -\tfrac{1}{6}, \qquad S_3 = -\tfrac{5}{12}, \quad \text{and} \quad S_4 = -\tfrac{13}{60}.$$

Sigma Notation

When the general term of a sequence is known, the Greek letter Σ (capital sigma) can be used to write a series. For example, the sum of the first four terms of the sequence $3, 5, 7, 9, 11, \ldots, 2k + 1, \ldots$ can be named as follows, using *sigma notation*, or *summation notation*:

$$\sum_{k=1}^{4} (2k + 1). \qquad \begin{array}{l} \textbf{This represents} \\ \mathbf{(2 \cdot 1 + 1) + (2 \cdot 2 + 1) + (2 \cdot 3 + 1) + (2 \cdot 4 + 1).} \end{array}$$

This is read "the sum as k goes from 1 to 4 of $(2k + 1)$." The letter k is called the *index of summation*. The index of summation need not start at 1.

EXAMPLE 6 Write out and evaluate each sum.

a) $\displaystyle\sum_{k=1}^{5} k^2$ **b)** $\displaystyle\sum_{k=4}^{6} (-1)^k (2k)$ **c)** $\displaystyle\sum_{k=0}^{3} (2^k + 5)$

Solution

a) $\displaystyle\sum_{k=1}^{5} k^2 = 1^2 + 2^2 + 3^2 + 4^2 + 5^2 = 1 + 4 + 9 + 16 + 25 = 55$

> **Evaluate k^2 for all integers from 1 through 5. Then add.**

b) $\displaystyle\sum_{k=4}^{6} (-1)^k (2k) = (-1)^4 (2 \cdot 4) + (-1)^5 (2 \cdot 5) + (-1)^6 (2 \cdot 6)$

$$= 8 - 10 + 12 = 10$$

c) $\displaystyle\sum_{k=0}^{3} (2^k + 5) = (2^0 + 5) + (2^1 + 5) + (2^2 + 5) + (2^3 + 5)$

$$= 6 + 7 + 9 + 13 = 35$$

EXAMPLE 7 Write sigma notation for each sum.

a) $1 + 4 + 9 + 16 + 25$ **b)** $-1 + 3 - 5 + 7$

c) $3 + 9 + 27 + 81 + \cdots$

Solution

a) $1 + 4 + 9 + 16 + 25$

Note that this is a sum of squares, $1^2 + 2^2 + 3^2 + 4^2 + 5^2$, so the general term is k^2. Sigma notation is

$$\sum_{k=1}^{5} k^2. \qquad \textbf{The sum starts with } 1^2 \textbf{ and ends with } 5^2.$$

Answers may vary here. For example, another—perhaps less obvious—way of writing $1 + 4 + 9 + 16 + 25$ is

$$\sum_{k=2}^{6} (k - 1)^2.$$

TEACHING TIP

You may wish to point out that Examples 7(a) and 7(b) are finite series and that Example 7(c) is an infinite series.

TEACHING TIP

Consider asking students how the series in Example 7(b) would differ if the general term were $(-1)^{k+1}(2k - 1)$.

b) $-1 + 3 - 5 + 7$

Except for the alternating signs, this is the sum of the first four positive odd numbers. Note that $2k - 1$ is a formula for the kth positive odd number. It is also important to note that since $(-1)^k = 1$ when k is even and $(-1)^k = -1$ when k is odd, the factor $(-1)^k$ can be used to create the alternating signs. The general term is thus $(-1)^k(2k - 1)$, beginning with $k = 1$. Sigma notation is

$$\sum_{k=1}^{4} (-1)^k (2k - 1).$$

c) $3 + 9 + 27 + 81 + \cdots$

This is a sum of powers of 3, and it is also an infinite series. We use the symbol ∞ for infinity and write the series using sigma notation:

$$\sum_{k=1}^{\infty} 3^k.$$

Graphs of Sequences

Because the domain of a sequence is a set of integers, the graph of a sequence is a set of points that are not connected. We can use the DOT mode to graph a sequence when a formula for the general term is known.

EXAMPLE 8 Graph the sequence for which the general term is given by $a_n = (-1)^n/n$.

Solution   We let $u(n) = (-1)^{\wedge}n/n$ and set the graph mode to DOT. You may wish to examine a table of values to help set up the window dimensions for the graph. From the table on the left below, the terms appear to be between -1 and 1, so we let Ymin $= -1$ and Ymax $= 1$.

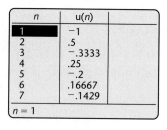

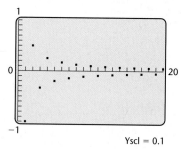

Yscl = 0.1

For the window settings, we must determine nMin and nMax, as well as Xmin, Xmax, Ymin, and Ymax. Here nMin is the smallest value of n for which we wish to evaluate the sequence, and nMax is the largest value. We let nMin $= 1$ and nMax $= 20$. If the graphing calculator also allows us to set PlotStart and PlotStep, we set each of those to 1.

We see from the graph on the right above that the absolute value of the terms gets smaller as n gets larger. The graph also illustrates that the signs of the terms alternate.

10.1

Exercise Set

In each of the following, the nth term of a sequence is given. In each case, find the first 4 terms; the 10th term, a_{10}; and the 15th term, a_{15}.

1. $a_n = 5n - 2$
  3, 8, 13, 18; 48; 73

2. $a_n = 2n + 3$
  5, 7, 9, 11; 23; 33

3. $a_n = \dfrac{n}{n+1}$ $\frac{1}{2}, \frac{2}{3}, \frac{3}{4}, \frac{4}{5}; \frac{10}{11}; \frac{15}{16}$

4. $a_n = n^2 + 2$
  3, 6, 11, 18; 102; 227

5. $a_n = n^2 - 2n$
  −1, 0, 3, 8; 80; 195

6. $a_n = \dfrac{n^2 - 1}{n^2 + 1}$ ▢

7. $a_n = n + \dfrac{1}{n}$ ▢

8. $a_n = \left(-\dfrac{1}{2}\right)^{n-1}$ ▢

9. $a_n = (-1)^n n^2$ −1, 4, −9, 16; 100; −225

10. $a_n = (-1)^n (n + 3)$ −4, 5, −6, 7; 13; −18

11. $a_n = (-1)^{n+1}(3n - 5)$ −2, −1, 4, −7; −25; 40

12. $a_n = (-1)^n (n^3 - 1)$ 0, 7, −26, 63; 999; −3374

Find the indicated term of each sequence.

13. $a_n = 2n - 5$; a_7 9

14. $a_n = 3n + 2$; a_8 26

15. $a_n = (3n + 1)(2n - 5)$; a_9 364

16. $a_n = (3n + 2)^2$; a_6 400

17. $a_n = (-1)^{n-1}(3.4n - 17.3)$; a_{12} −23.5

18. $a_n = (-2)^{n-2}(45.68 - 1.2n)$; a_{23} −37,916,508.16

19. $a_n = 3n^2(9n - 100)$; a_{11} −363

20. $a_n = 4n^2(2n - 39)$; a_{22} 9680

21. $a_n = \left(1 + \dfrac{1}{n}\right)^2$; a_{20} $\frac{441}{400}$

22. $a_n = \left(1 - \dfrac{1}{n}\right)^3$; a_{15} $\frac{2744}{3375}$

Look for a pattern and then predict the general term, or nth term, a_n, of each sequence. Answers may vary.

23. $1, 3, 5, 7, 9, \ldots$ $2n - 1$

24. $2, 4, 6, 8, \ldots$ $2n$

25. $1, -1, 1, -1, \ldots$ $(-1)^{n+1}$

26. $-1, 1, -1, 1, \ldots$ $(-1)^n$

27. $-1, 2, -3, 4, \ldots$ $(-1)^n \cdot n$

28. $1, -2, 3, -4, \ldots$ $(-1)^{n+1} \cdot n$

29. $-2, 6, -18, 54, \ldots$ $(-1)^n \cdot 2 \cdot (3)^{n-1}$

30. $-2, 3, 8, 13, 18, \ldots$ $5n - 7$

31. $\frac{1}{2}, \frac{2}{3}, \frac{3}{4}, \frac{4}{5}, \frac{5}{6}, \ldots$ $\dfrac{n}{n+1}$

32. $1 \cdot 2, 2 \cdot 3, 3 \cdot 4, 4 \cdot 5, \ldots$ $n(n + 1)$

33. $5, 25, 125, 625, \ldots$ 5^n

34. $4, 16, 64, 256, \ldots$ 4^n

35. $-1, 4, -9, 16, \ldots$ $(-1)^n \cdot n^2$

36. $1, -4, 9, -16, \ldots$ $(-1)^{n+1} \cdot n^2$

Find the indicated partial sum for each sequence.

37. $1, -2, 3, -4, 5, -6, \ldots$; S_7 4

38. $1, -3, 5, -7, 9, -11, \ldots$; S_8 −8

39. $2, 4, 6, 8, \ldots$; S_5 30

40. $1, \frac{1}{4}, \frac{1}{9}, \frac{1}{16}, \frac{1}{25}, \ldots$; S_5 $\frac{5269}{3600}$

Write out and evaluate each sum.

41. $\displaystyle\sum_{k=1}^{5} \dfrac{1}{2k}$ ▢

42. $\displaystyle\sum_{k=1}^{6} \dfrac{1}{2k - 1}$ ▢

43. $\displaystyle\sum_{k=0}^{4} 3^k$ ▢

44. $\displaystyle\sum_{k=4}^{7} \sqrt{2k + 1}$ ▢

45. $\displaystyle\sum_{k=1}^{8} \dfrac{k}{k + 1}$ ▢

46. $\displaystyle\sum_{k=1}^{4} \dfrac{k - 2}{k + 3}$ ▢

47. $\displaystyle\sum_{k=1}^{8} (-1)^{k+1} 2^k$ ▢

48. $\displaystyle\sum_{k=1}^{7} (-1)^k 4^{k+1}$ ▢

49. $\displaystyle\sum_{k=0}^{5} (k^2 - 2k + 3)$ ▢

50. $\displaystyle\sum_{k=0}^{5} (k^2 - 3k + 4)$ ▢

51. $\displaystyle\sum_{k=3}^{5} \dfrac{(-1)^k}{k(k + 1)}$ ▢

52. $\displaystyle\sum_{k=3}^{7} \dfrac{k}{2^k}$ ▢

▢ Answers to Exercises 6–8 and 41–52 can be found on p. A-73.

Rewrite each sum using sigma notation. Answers may vary.

53. $\dfrac{2}{3} + \dfrac{3}{4} + \dfrac{4}{5} + \dfrac{5}{6} + \dfrac{6}{7}$ $\displaystyle\sum_{k=1}^{5} \dfrac{k+1}{k+2}$

54. $3 + 6 + 9 + 12 + 15$ $\displaystyle\sum_{k=1}^{5} 3k$

55. $1 + 4 + 9 + 16 + 25 + 36$ $\displaystyle\sum_{k=1}^{6} k^2$

56. $\dfrac{1}{1^2} + \dfrac{1}{2^2} + \dfrac{1}{3^2} + \dfrac{1}{4^2} + \dfrac{1}{5^2}$ $\displaystyle\sum_{k=1}^{5} \dfrac{1}{k^2}$

57. $4 - 9 + 16 - 25 + \cdots + (-1)^n n^2$ $\displaystyle\sum_{k=2}^{n} (-1)^k k^2$

58. $9 - 16 + 25 - \cdots + (-1)^{n+1} n^2$ $\displaystyle\sum_{k=3}^{n} (-1)^{k+1} k^2$

59. $5 + 10 + 15 + 20 + 25 + \cdots$ $\displaystyle\sum_{k=1}^{\infty} 5k$

60. $7 + 14 + 21 + 28 + 35 + \cdots$ $\displaystyle\sum_{k=1}^{\infty} 7k$

61. $\dfrac{1}{1\cdot 2} + \dfrac{1}{2\cdot 3} + \dfrac{1}{3\cdot 4} + \dfrac{1}{4\cdot 5} + \cdots$ $\displaystyle\sum_{k=1}^{\infty} \dfrac{1}{k(k+1)}$

62. $\dfrac{1}{1\cdot 2^2} + \dfrac{1}{2\cdot 3^2} + \dfrac{1}{3\cdot 4^2} + \dfrac{1}{4\cdot 5^2} + \cdots$ $\displaystyle\sum_{k=1}^{\infty} \dfrac{1}{k(k+1)^2}$

TW 63. The sequence $1, 4, 9, 16, \ldots$ can be written as $f(x) = x^2$ with the domain the set of all positive integers. Explain how the graph of f would compare with the graph of $y = x^2$.

TW 64. Eric says he expects he will prefer sequences to functions because he dislikes fractions. Will his expectations prove correct? Why or why not?

Skill Maintenance

Evaluate. [1.1]

65. $\dfrac{7}{2}(a_1 + a_7)$, for $a_1 = 8$ and $a_7 = 14$ 77

66. $a_1 + (n-1)d$, for $a_1 = 3$, $n = 6$, and $d = 4$ 23

Multiply. [5.2]

67. $(x + y)^3$ $x^3 + 3x^2y + 3xy^2 + y^3$

68. $(a - b)^3$ $a^3 - 3a^2b + 3ab^2 - b^3$

69. $(2a - b)^3$ $8a^3 - 12a^2b + 6ab^2 - b^3$

70. $(2x + y)^3$ $8x^3 + 12x^2y + 6xy^2 + y^3$

Synthesis

TW 71. Explain why the equation
$$\sum_{k=1}^{n} (a_k + b_k) = \sum_{k=1}^{n} a_k + \sum_{k=1}^{n} b_k$$
is true for any positive integer n. What laws are used to justify this result?

TW 72. Consider the sums
$$\sum_{k=1}^{5} 3k^2 \quad \text{and} \quad 3\sum_{k=1}^{5} k^2.$$

a) Which is easier to evaluate and why?

b) Is it true that
$$\sum_{k=1}^{n} ca_k = c\sum_{k=1}^{n} a_k?$$
Why or why not?

Some sequences are given by a recursive *definition. The value of the first term, a_1, is given, and then we are told how to find any subsequent term from the term preceding it. Find the first six terms of each of the following recursively defined sequences.*

73. $a_1 = 1$, $a_{n+1} = 5a_n - 2$ 1, 3, 13, 63, 313, 1563

74. $a_1 = 0$, $a_{n+1} = a_n^2 + 3$ 0, 3, 12, 147, 21,612, 467,078,547

75. *Cell Biology.* A single cell of bacterium divides into two every 15 min. Suppose that the same rate of division is maintained for 4 hr. Give a sequence that lists the number of cells after successive 15-min periods.   1, 2, 4, 8, 16, 32, 64, 128, 256, 512, 1024, 2048, 4096, 8192, 16,384, 32,768, 65,536

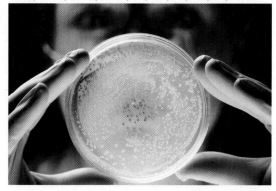

76. *Value of a Copier.* The value of a color photocopier is \$5200. Its scrap value each year is 75% of its value the year before. Give a sequence that lists the scrap value of the machine at the start of each year for a 10-yr period.   \$5200, \$3900, \$2925, \$2193.75, \$1645.31, \$1233.98, \$925.49, \$694.12, \$520.59, \$390.44

77. Find S_{100} and S_{101} for the sequence in which $a_n = (-1)^n$. $S_{100} = 0$; $S_{101} = -1$

Find the first five terms of each sequence; then find S_5.

78. $\dfrac{3}{2}, \dfrac{3}{2}, \dfrac{9}{8}, \dfrac{3}{4}, \dfrac{15}{32}; \dfrac{171}{32}$

78. $a_n = \dfrac{1}{2^n} \log 1000^n$

79. $a_n = i^n$, $i = \sqrt{-1}$ $i, -1, -i, 1, i; i$

80. Find all values for x that solve the following:
$$\sum_{k=1}^{x} i^k = -1.$$
$\{x \mid x = 4n - 1, \text{ where } n \text{ is a natural number}\}$

81. The nth term of a sequence is given by

$$a_n = n^5 - 14n^4 + 6n^3 + 416n^2 - 655n - 1050.$$

Use a graphing calculator with a TABLE feature to determine what term in the sequence is 6144.   11th term

82. To define a sequence recursively on a graphing calculator (see Exercises 73 and 74), the SEQ MODE is used. The general term U_n or V_n can often be expressed in terms of U_{n-1} or V_{n-1} by pressing $\boxed{\text{2nd}}$ $\boxed{7}$ or $\boxed{\text{2nd}}$ $\boxed{8}$. The starting values of U_n, V_n, and n are set as one of the WINDOW variables.

Use recursion to determine how many handshakes will occur if a group of 50 people shake hands with one another. To develop the recursion formula, begin with a group of 2 and determine how many additional handshakes occur with the arrival of each new group member. (See the Collaborative Corner following Exercise Set 5.1 on p. 322.)

1225 handshakes

10.2

Arithmetic Sequences ■ Sum of the First n Terms of an Arithmetic Sequence ■ Problem Solving

Arithmetic Sequences and Series

In this section, we concentrate on sequences and series that are said to be arithmetic (pronounced ar-ith-MET-ik).

Arithmetic Sequences

In an **arithmetic sequence** (or **progression**), any term (other than the first) can be found by adding the same number to its preceding term. For example, the sequence 2, 5, 8, 11, 14, 17, . . . is arithmetic because adding 3 to any term produces the next term.

> ### Arithmetic Sequence
>
> A sequence is *arithmetic* if there exists a number d, called the *common difference*, such that $a_{n+1} = a_n + d$ for any integer $n \geq 1$.

TEACHING TIP

Point out that when the terms are decreasing, the common difference must be negative. Encourage students to determine the sign of the common difference before calculating it.

EXAMPLE 1 For each arithmetic sequence, identify the first term, a_1, and the common difference, d.

a) 4, 9, 14, 19, 24, . . . **b)** 27, 20, 13, 6, -1, -8, . . .

Solution   To find a_1, we simply use the first term listed. To find d, we choose any term beyond the first and subtract the preceding term from it.

Sequence	First Term, a_1	Common Difference, d
a) 4, 9, 14, 19, 24, . . .	4	5 ⟵ 9 − 4 = 5
b) 27, 20, 13, 6, -1, -8, . . .	27	-7 ⟵ 20 − 27 = −7

To find the common difference, we subtracted a_1 from a_2. Had we subtracted a_2 from a_3 or a_3 from a_4, we would have found the same values for d.

Check: As a check, note that when d is added to each term, the result is the next term in the sequence.

a) $4 + 5 = 9$, $9 + 5 = 14$, $14 + 5 = 19$, $19 + 5 = 24$
b) $27 + (-7) = 20$, $20 + (-7) = 13$, $13 + (-7) = 6$,
$\quad 6 + (-7) = -1$, $-1 + (-7) = -8$

To find a formula for the general, or nth, term of any arithmetic sequence, we denote the common difference by d and write out the first few terms:

$a_1,$

$a_2 = a_1 + d,$

$a_3 = a_2 + d = (a_1 + d) + d = a_1 + 2d,$ Substituting $a_1 + d$ for a_2

$a_4 = a_3 + d = (a_1 + 2d) + d = a_1 + 3d.$ Substituting $a_1 + 2d$ for a_3

Note that the coefficient of d in each case is 1 less than the subscript.

Generalizing, we obtain the following formula.

To Find a_n for an Arithmetic Sequence

The nth term of an arithmetic sequence with common difference d is

$$a_n = a_1 + (n - 1)d, \quad \text{for any integer } n \geq 1.$$

EXAMPLE 2   Find the 14th term of the arithmetic sequence $6, 9, 12, 15, \ldots$

Solution   First we note that $a_1 = 6$, $d = 3$, and $n = 14$. Using the formula for the nth term of an arithmetic sequence, we have

$$a_n = a_1 + (n - 1)d$$
$$a_{14} = 6 + (14 - 1) \cdot 3 = 6 + 13 \cdot 3 = 6 + 39 = 45.$$

The 14th term is 45.

EXAMPLE 3   For the sequence in Example 2, which term is 300? That is, find n if $a_n = 300$.

Solution   We substitute into the formula for the nth term of an arithmetic sequence and solve for n:

$$a_n = a_1 + (n - 1)d$$
$$300 = 6 + (n - 1) \cdot 3$$
$$300 = 6 + 3n - 3$$
$$297 = 3n$$
$$99 = n.$$

The term 300 is the 99th term of the sequence.

Given two terms and their places in an arithmetic sequence, we can construct the sequence.

EXAMPLE 4   The 3rd term of an arithmetic sequence is 14, and the 16th term is 79. Find a_1 and d and construct the sequence.

Solution   We know that $a_3 = 14$ and $a_{16} = 79$. Thus we would have to add d 13 times to get from 14 to 79. That is,

$$14 + 13d = 79. \qquad a_3 \text{ and } a_{16} \text{ are 13 terms apart; } 16 - 3 = 13$$

Solving $14 + 13d = 79$, we obtain

$$13d = 65 \qquad \textbf{Subtracting 14 from both sides}$$
$$d = 5. \qquad \textbf{Dividing both sides by 13}$$

We subtract d twice from a_3 to get to a_1. Thus,

$$a_1 = 14 - 2 \cdot 5 = 4. \qquad a_1 \text{ and } a_3 \text{ are 2 terms apart; } 3 - 1 = 2$$

The sequence is $4, 9, 14, 19, \ldots$. Note that we could have subtracted d 15 times from a_{16} in order to find a_1.

In general, d should be subtracted $(n - 1)$ times from a_n in order to find a_1.

What will the graph of an arithmetic sequence look like?

Interactive Discovery

Graph each of the following arithmetic sequences. What pattern do you observe?

1. $a_n = -5 + (n - 1)3$ The points lie on a straight line with positive slope.

2. $a_n = \frac{1}{2} + (n - 1)\frac{3}{2}$ The points lie on a straight line with positive slope.

3. $a_n = 60 + (n - 1)(-5)$ The points lie on a straight line with negative slope.

4. $a_n = -1.7 + (n - 1)(-0.2)$ The points lie on a straight line with negative slope.

The pattern you may have observed above is true in general:

The graph of an arithmetic sequence is a set of points that lie on a straight line.

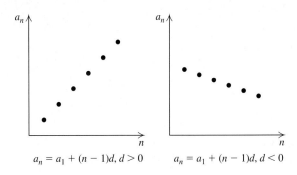

$a_n = a_1 + (n - 1)d, d > 0$ $\qquad$ $a_n = a_1 + (n - 1)d, d < 0$

Sum of the First n Terms of an Arithmetic Sequence

When the terms of an arithmetic sequence are added, an **arithmetic series** is formed. To find a formula for computing S_n when the series is arithmetic, we denote the first n terms as follows:

This is the next-to-last term. If you add d to this term, the result is a_n.

$$a_1, (a_1 + d), (a_1 + 2d), \ldots, \underbrace{(a_n - 2d)}, \overbrace{(a_n - d)}, a_n$$

This term is two terms back from the end. If you add d to this term, you get the next-to-last term, $a_n - d$.

Thus, S_n is given by

$$S_n = a_1 + (a_1 + d) + (a_1 + 2d) + \cdots + (a_n - 2d) + (a_n - d) + a_n.$$

Using a commutative law, we have a second equation:

$$S_n = a_n + (a_n - d) + (a_n - 2d) + \cdots + (a_1 + 2d) + (a_1 + d) + a_1.$$

Adding corresponding terms on each side of the above equations, we get

$$2S_n = [a_1 + a_n] + [(a_1 + d) + (a_n - d)] + [(a_1 + 2d) + (a_n - 2d)] \\ + \cdots + [(a_n - 2d) + (a_1 + 2d)] + [(a_n - d) + (a_1 + d)] \\ + [a_n + a_1].$$

This simplifies to

$$2S_n = [a_1 + a_n] + [a_1 + a_n] + [a_1 + a_n] \\ + \cdots + [a_n + a_1] + [a_n + a_1] + [a_n + a_1].$$

There are n bracketed sums.

Since $[a_1 + a_n]$ is being added n times, it follows that

$$2S_n = n[a_1 + a_n].$$

Dividing both sides by 2 leads to the following formula.

TEACHING TIP

Make sure students realize that they must verify that the sequence is arithmetic before using this formula.

To Find S_n for an Arithmetic Sequence

The sum of the first n terms of an arithmetic sequence is given by

$$S_n = \frac{n}{2}(a_1 + a_n).$$

EXAMPLE 5   Find the sum of the first 100 positive even numbers.

Solution   The sum is

$$2 + 4 + 6 + \cdots + 198 + 200.$$

This is the sum of the first 100 terms of the arithmetic sequence for which

$$a_1 = 2, \quad n = 100, \quad \text{and} \quad a_n = 200.$$

Substituting in the formula

$$S_n = \frac{n}{2}(a_1 + a_n),$$

we get

$$S_{100} = \frac{100}{2}(2 + 200)$$

$$= 50(202) = 10,100.$$

The above formula is useful when we know the first and last terms, a_1 and a_n. To find S_n when a_n is unknown, but a_1, n, and d are known, we can use the formula $a_n = a_1 + (n - 1)d$ to calculate a_n and then proceed as in Example 5.

EXAMPLE 6   Find the sum of the first 15 terms of the arithmetic sequence 4, 7, 10, 13, ...

Solution   Note that

$$a_1 = 4, \quad n = 15, \quad \text{and} \quad d = 3.$$

Before using the formula for S_n, we find a_{15}:

$$a_{15} = 4 + (15 - 1)3 \qquad \text{Substituting into the formula for } a_n$$

$$= 4 + 14 \cdot 3 = 46.$$

Thus, knowing that $a_{15} = 46$, we have

$$S_{15} = \tfrac{15}{2}(4 + 46) \qquad \text{Using the formula for } S_n$$

$$= \tfrac{15}{2}(50) = 375.$$

Problem Solving

For some problem-solving situations, the translation may involve sequences or series. In Examples 7 and 8, the calculations and translations can be done in a number of ways. There is often a variety of ways in which a problem can be solved. You should use the one that is best or easiest for you. In this chapter, however, we will try to emphasize sequences and series and their related formulas.

EXAMPLE 7 Hourly Wages. Chris accepts a job managing a CD shop, starting with an hourly wage of $14.25, and is promised a raise of 15¢ per hour every 2 months for 5 years. After 5 years of work, what will be Chris's hourly wage?

Solution

1. Familiarize. It helps to write down the hourly wage for several two-month time periods.

Beginning:	14.25,
After two months:	14.40,
After four months:	14.55,

and so on.

What appears is a sequence of numbers: 14.25, 14.40, 14.55,.... Since the same amount is added each time, the sequence is arithmetic.

We list what we know about arithmetic sequences. The pertinent formulas are

$$a_n = a_1 + (n - 1)d$$

and

$$S_n = \frac{n}{2}(a_1 + a_n).$$

In this case, we are not looking for a sum, so we use the first formula. We want to determine the last term in a sequence. To do so, we need to know a_1, n, and d. From our list above, we see that

$$a_1 = 14.25 \quad \text{and} \quad d = 0.15.$$

What is n? That is, how many terms are in the sequence? After 1 year, there have been 6 raises, since Chris gets a raise every 2 months. There are 5 years, so the total number of raises will be $5 \cdot 6$, or 30. Altogether, there will be 31 terms: the original wage and 30 increased rates.

2. Translate. We want to find a_n for the arithmetic sequence in which $a_1 = 14.25$, $n = 31$, and $d = 0.15$.

3. Carry out. Substituting in the formula for a_n gives us

$$a_{31} = 14.25 + (31 - 1) \cdot 0.15$$
$$= 18.75.$$

4. Check. We can check by redoing the calculations or we can calculate in a slightly different way for another check. For example, at the end of a year, there will be 6 raises, for a total raise of $0.90. At the end of 5 years, the total raise will be $5 \times \$0.90$, or $4.50. If we add that to the original wage of $14.25, we obtain $18.75. The answer checks.

5. State. After 5 years, Chris's hourly wage will be $18.75.

EXAMPLE 8   Telephone Pole Storage.   A stack of telephone poles has 30 poles in the bottom row, 29 poles in the second row, 28 in the next row, and so on. How many poles are in the stack if there are 5 poles in the top row?

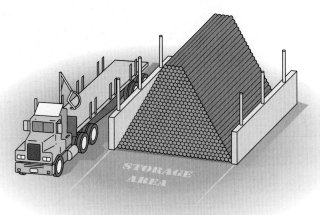

Solution

1. Familiarize.   The following figure shows the ends of the poles and the way in which they stack. There are 30 poles on the bottom, and we see that there will be one fewer in each succeeding row. How many rows will there be?

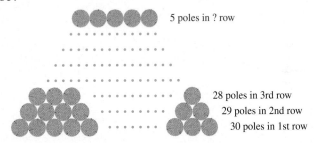

5 poles in ? row

28 poles in 3rd row
29 poles in 2nd row
30 poles in 1st row

Note that there are $30 - 1 = 29$ poles in the 2nd row, $30 - 2 = 28$ poles in the 3rd row, $30 - 3 = 27$ poles in the 4th row, and so on. The pattern leads to $30 - 25 = 5$ poles in the 26th row.

The situation is represented by the equation

$30 + 29 + 28 + \cdots + 5.$ **There are 26 terms in this series.**

Thus we have an arithmetic series. We recall the formula

$$S_n = \frac{n}{2}(a_1 + a_n).$$

2. Translate.   We want to find the sum of the first 26 terms of an arithmetic sequence in which $a_1 = 30$ and $a_{26} = 5$.

3. Carry out.   Substituting into the above formula gives us

$$S_{26} = \frac{26}{2}(30 + 5)$$
$$= 13 \cdot 35 = 455.$$

4. Check.   In this case, we can check the calculations by doing them again. A longer, harder way would be to do the entire addition:

$$30 + 29 + 28 + \cdots + 5.$$

5. State.   There are 455 poles in the stack.

10.2

Exercise Set

Find the first term and the common difference.

1. $2, 6, 10, 14, \ldots$ $a_1 = 2, d = 4$

2. $1.06, 1.12, 1.18, 1.24, \ldots$ $a_1 = 1.06, d = 0.06$

3. $6, 2, -2, -6, \ldots$ $a_1 = 6, d = -4$

4. $-9, -6, -3, 0, \ldots$ $a_1 = -9, d = 3$

5. $\frac{3}{2}, \frac{9}{4}, 3, \frac{15}{4}, \ldots$ $a_1 = \frac{3}{2}, d = \frac{3}{4}$

6. $\frac{3}{5}, \frac{1}{10}, -\frac{2}{5}, \ldots$ $a_1 = \frac{3}{5}, d = -\frac{1}{2}$

7. $\$5.12, \$5.24, \$5.36, \$5.48, \ldots$ $a_1 = \$5.12, d = \$0.12$

8. $\$214, \$211, \$208, \$205, \ldots$ $a_1 = \$214, d = -\$3$

9. Find the 12th term of the arithmetic sequence $3, 7, 11, \ldots$ 47

10. Find the 11th term of the arithmetic sequence $0.07, 0.12, 0.17, \ldots$ 0.57

11. Find the 17th term of the arithmetic sequence $7, 4, 1, \ldots$ -41

12. Find the 14th term of the arithmetic sequence $3, \frac{7}{3}, \frac{5}{3}, \ldots$ $-\frac{17}{3}$

13. Find the 13th term of the arithmetic sequence $\$1200, \$964.32, \$728.64, \ldots$ $-\$1628.16$

14. Find the 10th term of the arithmetic sequence $\$2345.78, \$2967.54, \$3589.30, \ldots$ $\$7941.62$

15. In the sequence of Exercise 9, what term is 107?   27th

16. In the sequence of Exercise 10, what term is 1.67?   33rd

17. In the sequence of Exercise 11, what term is -296?   102nd

18. In the sequence of Exercise 12, what term is -27?   46th

19. Find a_{17} when $a_1 = 2$ and $d = 5$.   82

20. Find a_{20} when $a_1 = 14$ and $d = -3$. -43

21. Find a_1 when $d = 4$ and $a_8 = 33$.   5

22. Find a_1 when $d = 8$ and $a_{11} = 26$. -54

23. Find n when $a_1 = 5$, $d = -3$, and $a_n = -76$.   28

24. Find n when $a_1 = 25$, $d = -14$, and $a_n = -507$.   39

25. For an arithmetic sequence in which $a_{17} = -40$ and $a_{28} = -73$, find a_1 and d. Write the first five terms of the sequence. $a_1 = 8; d = -3; 8, 5, 2, -1, -4$

26. In an arithmetic sequence, $a_{17} = \frac{25}{3}$ and $a_{32} = \frac{95}{6}$. Find a_1 and d. Write the first five terms of the sequence. $a_1 = \frac{1}{3}, d = \frac{1}{2}; \frac{1}{3}, \frac{5}{6}, \frac{4}{3}, \frac{11}{6}, \frac{7}{3}$

Aha! **27.** Find a_1 and d if $a_{13} = 13$ and $a_{54} = 54$. $a_1 = 1, d = 1$

28. Find a_1 and d if $a_{12} = 24$ and $a_{25} = 50$. $a_1 = 2, d = 2$

29. Find the sum of the first 20 terms of the arithmetic series $1 + 5 + 9 + 13 + \cdots$.   780

30. Find the sum of the first 14 terms of the arithmetic series $11 + 7 + 3 + \cdots$. -210

31. Find the sum of the first 250 natural numbers.   31,375

32. Find the sum of the first 400 natural numbers.   80,200

33. Find the sum of the even numbers from 2 to 100, inclusive.   2550

34. Find the sum of the odd numbers from 1 to 99, inclusive.   2500

35. Find the sum of all multiples of 6 from 6 to 102, inclusive.   918

36. Find the sum of all multiples of 4 that are between 15 and 521.   34,036

37. An arithmetic series has $a_1 = 4$ and $d = 5$. Find S_{20}.
1030

38. An arithmetic series has $a_1 = 9$ and $d = -3$.
Find S_{32}. -1200

Solve.

39. *Band Formations.*   The Duxbury marching band has 14 marchers in the front row, 16 in the second row, 18 in the third row, and so on, for 15 rows. How many marchers are in the last row? How many marchers are there altogether?   42; 420

40. *Gardening.*   A gardener is planting bulbs near an entrance to a college. She has 39 plants in the front row, 35 in the second row, 31 in the third row, and so on. If the pattern is consistent, how many plants will be in the last row? How many plants will there be altogether?   3; 210

41. *Telephone Pole Piles.*   How many poles will be in a pile of telephone poles if there are 50 in the first layer, 49 in the second, and so on, until there are 6 in the last layer?   1260

42. *Accumulated Savings.*   If 10¢ is saved on October 1, another 20¢ on October 2, another 30¢ on October 3, and so on, how much is saved during October? (October has 31 days.) $49.60

43. *Accumulated Savings.*   Renata saves money in an arithmetic sequence: $600 for the first year, another $700 the second, and so on, for 20 yr. How much does she save in all (disregarding interest)? $31,000

44. *Spending.*   Jacob spent $30 on August 1, $50 on August 2, $70 on August 3, and so on. How much did Jacob spend in August? (August has 31 days.)
$10,230

45. *Auditorium Design.*   Theaters are often built with more seats per row as the rows move toward the back. The Sanders Amphitheater has 20 seats in the first row, 22 in the second, 24 in the third, and so on, for 19 rows. How many seats are in the amphitheater?   722

46. *Accumulated Savings.*   Shirley sets up an investment such that it will return $5000 the first year, $6125 the second year, $7250 the third year, and so on, for 25 yr. How much in all is received from the investment? $462,500

TW **47.** It is said that as a young child, the mathematician Karl F. Gauss (1777–1855) was able to compute the sum $1 + 2 + 3 + \cdots + 100$ very quickly in his head. Explain how Gauss might have done this and present a formula for the sum of the first n natural numbers. (*Hint*: $1 + 99 = 100$.)

TW **48.** If every number in a sequence is doubled and then added, is the result the same as if the numbers were first added and the sum then doubled? Why or why not?

Skill Maintenance

Simplify. [6.2]

49. $\dfrac{3}{10x} + \dfrac{2}{15x}$ $\dfrac{13}{30x}$

50. $\dfrac{2}{9t} + \dfrac{5}{12t}$ $\dfrac{23}{36t}$

Convert to an exponential equation. [9.3]

51. $\log_a P = k$ $a^k = P$

52. $\ln t = a$ $e^a = t$

Solve. [9.6]

53. $4^{3x} = 8^{x+2}$ 2

54. $\log x + \log (x - 3) = 1$ 5

Synthesis

TW **55.** Write a problem for a classmate to solve. Devise the problem so that its solution requires computing S_{17} for an arithmetic sequence.

TW **56.** The sum of the first n terms of an arithmetic sequence is also given by

$$S_n = \frac{n}{2}[2a_1 + (n - 1)d].$$

Use the earlier formulas for a_n and S_n to explain how this equation was developed.

57. Find a formula for the sum of the first n consecutive odd numbers starting with 1:

$$1 + 3 + 5 + \cdots + (2n - 1).$$ $S_n = n^2$

58. Find three numbers in an arithmetic sequence for which the sum of the first and third is 10 and the product of the first and second is 15.   3, 5, 7

59. In an arithmetic sequence, $a_1 = \$8760$ and $d = -\$798.23$. Find the first 10 terms of the sequence. $\$8760, \$7961.77, \$7163.54, \$6365.31, \$5567.08, \$4768.85, \$3970.62, \$3172.39, \$2374.16, \$1575.93$

60. Find the sum of the first 10 terms of the sequence given in Exercise 59. $\$51,679.65$

61. Prove that if p, m, and q are consecutive terms in an arithmetic sequence, then $\quad$ Let d = the common difference. Since p, m, and q form an arithmetic sequence, $m = p + d$

$$m = \frac{p + q}{2}.$$
and $q = p + 2d$. Then $\frac{p+q}{2} = \frac{p+(p+2d)}{2} = p + d = m.$

62. *Straight-line Depreciation.* A company buys a color copier for $5200 on January 1 of a given year. The machine is expected to last for 8 yr, at the end of which time its *trade-in*, or *salvage*, *value* will be $1100. If the company figures the decline in value to be the same each year, then the trade-in values after t years, $0 \le t \le 8$, form an arithmetic sequence given by

$$a_t = C - t\left(\frac{C - S}{N}\right),$$

where C is the original cost of the item, N the years of expected life, and S the salvage value.

a) Find the formula for a_t for the straight-line depreciation of the copier. $a_t = \$5200 - \$512.50t$

b) Find the salvage value after 0 yr, 1 yr, 2 yr, 3 yr, 4 yr, 7 yr, and 8 yr. $\$5200, \$4687.50, \$4175, \$3662.50, \$3150, \$1612.50, \$1100$

c) Find a formula that expresses a_t recursively. $a_0 = \$5200, a_t = a_{t-1} - \$512.50$

63. Use your answer to Exercise 31 to find the sum of all integers from 501 through 750. $156,375$

In Exercises 64–67, graph the data in each table, and state whether or not the graph could be the graph of an arithmetic sequence. If it can, find a formula for the general term of the sequence.

64. *Aerobic Exercise.* Arithmetic; $a_n = -0.75n + 165$, where $n = 1$ corresponds to age 20, $n = 2$ corresponds to age 21, and so on

Age	Target Heart Rate (in beats per minute)
20	150
40	135
60	120
80	105

65. *Internet Access.* Arithmetic; $a_n = n + 102$, where $n = 1$ corresponds to 1998

Year	Number of U.S. Households Online (in millions)
1998	103
1999	104
2000	105
2001	106
2002	107

Sources: U.S. Department of Commerce; Jupiter Media Matrix

66. *High School Athletics.* Not arithmetic

Year	Number of Females Participating in High School Athletics (in millions)
1991–1992	1.9
1992–1993	2.0
1993–1994	2.1
1994–1995	2.2
1995–1996	2.4
1996–1997	2.5
1997–1998	2.6
1998–1999	2.7
1999–2000	2.7
2000–2001	2.7

Source: National Federation of State High School Associations, Indianapolis, IN

67. *House Sales.* Not arithmetic

Year	Number of Existing One-Family Houses Sold in the United States (in millions)
1997	4382
1998	4970
1999	5205
2000	5113

Source: National Association of Realtors

10.3

Geometric Sequences and Series

Geometric Sequences ■ Sum of the First n Terms of a Geometric Sequence ■ Infinite Geometric Series ■ Problem Solving

In an arithmetic sequence, a certain number is added to each term to get the next term. When each term in a sequence is *multiplied* by a certain number to get the next term, the sequence is **geometric**. In this section, we examine *geometric sequences* (or *progressions*) and *geometric series*.

Geometric Sequences

Consider the sequence

$$2, 6, 18, 54, 162, \ldots.$$

If we multiply each term by 3, we obtain the next term. The multiplier is called the *common ratio* because it is found by dividing any term by the preceding term.

> ### Geometric Sequence
>
> A sequence is *geometric* if there exists a number r, called the *common ratio*, for which
>
> $$\frac{a_{n+1}}{a_n} = r, \quad \text{or} \quad a_{n+1} = a_n \cdot r \quad \text{for any integer } n \geq 1.$$

EXAMPLE 1   For each geometric sequence, find the common ratio.

a) $3, 6, 12, 24, 48, \ldots$
b) $3, -6, 12, -24, 48, -96, \ldots$
c) $\$5200, \$3900, \$2925, \$2193.75, \ldots$

Solution

Sequence	Common Ratio	
a) $3, 6, 12, 24, 48, \ldots$	2	$\frac{6}{3} = 2, \frac{12}{6} = 2,$ **and so on**
b) $3, -6, 12, -24, 48, -96, \ldots$	-2	$\frac{-6}{3} = -2, \frac{12}{-6} = -2,$ **and so on**
c) $\$5200, \$3900, \$2925, \$2193.75, \ldots$	0.75	$\frac{\$3900}{\$5200} = 0.75, \frac{\$2925}{\$3900} = 0.75$

TEACHING TIP

Encourage students to determine the sign of the common ratio before calculating it.

To develop a formula for the general, or nth, term of a geometric sequence, let a_1 be the first term and let r be the common ratio. We write out the first few terms as follows:

$a_1,$

$a_2 = a_1 r,$

$a_3 = a_2 r = (a_1 r)r = a_1 r^2,$ Substituting $a_1 r$ for a_2

$a_4 = a_3 r = (a_1 r^2)r = a_1 r^3.$ Substituting $a_1 r^2$ for a_3

Note that the exponent is 1 less than the subscript.

Generalizing, we obtain the following.

To Find a_n for a Geometric Sequence

The nth term of a geometric sequence with common ratio r is given by

$$a_n = a_1 r^{n-1}, \quad \text{for any integer } n \geq 1.$$

EXAMPLE 2 Find the 7th term of the geometric sequence $4, 20, 100, \ldots$.

Solution   First we note that

$$a_1 = 4 \quad \text{and} \quad n = 7.$$

To find the common ratio, we can divide any term (other than the first) by the term preceding it. Since the second term is 20 and the first is 4,

$$r = \frac{20}{4}, \quad \text{or } 5.$$

The formula

$$a_n = a_1 r^{n-1}$$

gives us

$$a_7 = 4 \cdot 5^{7-1} = 4 \cdot 5^6 = 4 \cdot 15{,}625 = 62{,}500.$$

EXAMPLE 3 Find the 10th term of the geometric sequence

$$64, -32, 16, -8, \ldots.$$

Solution   First, we note that

$$a_1 = 64, \quad n = 10, \quad \text{and} \quad r = \frac{-32}{64} = -\frac{1}{2}.$$

Then, using the formula for the nth term of a geometric sequence, we have

$$a_{10} = 64 \cdot \left(-\frac{1}{2}\right)^{10-1} = 64 \cdot \left(-\frac{1}{2}\right)^9 = 2^6 \cdot \left(-\frac{1}{2^9}\right) = -\frac{1}{2^3} = -\frac{1}{8}.$$

The 10th term is $-\frac{1}{8}$.

What does the graph of a geometric series look like?

Interactive Discovery

Graph each of the following geometric series. What patterns, if any, do you observe?

1. $a_n = 5 \cdot 2^{n-1}$ The points lie on an exponential curve with $a > 1$.
2. $a_n = \frac{1}{2} \cdot \left(\frac{5}{4}\right)^{n-1}$ The points lie on an exponential curve with $a > 1$.
3. $a_n = 2.3 \cdot (0.75)^{n-1}$ The points lie on an exponential curve with $0 < a < 1$.
4. $a_n = 3\left(\frac{9}{10}\right)^{n-1}$ The points lie on an exponential curve with $0 < a < 1$.

The pattern you may have observed is true in general for $r > 0$.

The graph of a geometric series is a set of points that lie on the graph of an exponential function.

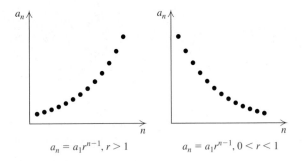

$a_n = a_1 r^{n-1}, r > 1$ $a_n = a_1 r^{n-1}, 0 < r < 1$

Sum of the First *n* Terms of a Geometric Sequence

We next develop a formula for S_n when a sequence is geometric:

$$a_1, a_1r, a_1r^2, a_1r^3, \ldots, a_1r^{n-1}, \ldots.$$

The **geometric series** S_n is given by

$$S_n = a_1 + a_1r + a_1r^2 + \cdots + a_1r^{n-2} + a_1r^{n-1}. \qquad \textbf{(1)}$$

Multiplying both sides by r gives us

$$rS_n = a_1r + a_1r^2 + a_1r^3 + \cdots + a_1r^{n-1} + a_1r^n. \qquad \textbf{(2)}$$

When we subtract corresponding sides of equation (2) from equation (1), the color terms drop out, leaving

$$S_n - rS_n = a_1 - a_1r^n$$
$$S_n(1 - r) = a_1(1 - r^n), \qquad \textbf{Factoring}$$

or

$$S_n = \frac{a_1(1 - r^n)}{1 - r}. \qquad \textbf{Dividing both sides by } 1 - r$$

> **To Find S_n for a Geometric Sequence**
>
> The sum of the first n terms of a geometric sequence with common ratio r is given by
>
> $$S_n = \frac{a_1(1 - r^n)}{1 - r}, \quad \text{for any } r \neq 1.$$

EXAMPLE 4 Find the sum of the first 7 terms of the geometric sequence $3, 15, 75, 375, \ldots$.

Solution First, we note that

$$a_1 = 3, \quad n = 7, \quad \text{and} \quad r = \frac{15}{3} = 5.$$

Then, substituting in the formula $S_n = \dfrac{a_1(1 - r^n)}{1 - r}$, we have

$$S_7 = \frac{3(1 - 5^7)}{1 - 5} = \frac{3(1 - 78{,}125)}{-4}$$

$$= \frac{3(-78{,}124)}{-4}$$

$$= 58{,}593.$$

Infinite Geometric Series

Suppose we consider the sum of the terms of an infinite geometric sequence, such as $2, 4, 8, 16, 32, \ldots$. We get what is called an **infinite geometric series**:

$$2 + 4 + 8 + 16 + 32 + \cdots.$$

Here, as n grows larger and larger, the sum of the first n terms, S_n, becomes larger and larger without bound. There are also infinite series that get closer and closer to some specific number. Here is an example:

$$\frac{1}{2} + \frac{1}{4} + \frac{1}{8} + \frac{1}{16} + \cdots + \frac{1}{2^n} + \cdots.$$

Let's consider S_n for the first four values of n:

$$S_1 = \tfrac{1}{2} \qquad\qquad\qquad = \tfrac{1}{2} = 0.5,$$
$$S_2 = \tfrac{1}{2} + \tfrac{1}{4} \qquad\qquad = \tfrac{3}{4} = 0.75,$$
$$S_3 = \tfrac{1}{2} + \tfrac{1}{4} + \tfrac{1}{8} \qquad = \tfrac{7}{8} = 0.875,$$
$$S_4 = \tfrac{1}{2} + \tfrac{1}{4} + \tfrac{1}{8} + \tfrac{1}{16} = \tfrac{15}{16} = 0.9375.$$

The denominator of the sum is 2^n, where n is the subscript of S. The numerator is $2^n - 1$.

Thus, for this particular series, we have

$$S_n = \frac{2^n - 1}{2^n} = \frac{2^n}{2^n} - \frac{1}{2^n} = 1 - \frac{1}{2^n}.$$

Note that the value of S_n is less than 1 for any value of n, but as n gets larger and larger, the values of $1/2^n$ get closer to 0 and the values of S_n get closer to 1. We can visualize S_n by considering the area of a square. For S_1, we shade half the square. For S_2, we shade half the square plus half the remaining part, or $\frac{1}{4}$. For S_3, we shade the parts shaded in S_2 plus half the remaining part. Again we see that the values of S_n will continue to get close to 1 (shading the complete square).

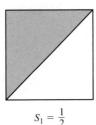 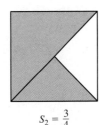 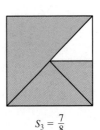 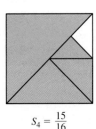

$S_1 = \frac{1}{2}$ $\qquad$ $S_2 = \frac{3}{4}$ $\qquad$ $S_3 = \frac{7}{8}$ $\qquad$ $S_4 = \frac{15}{16}$

We say that 1 is the *limit* of S_n and that 1 is the sum of this infinite geometric sequence. The sum of an infinite geometric sequence is an infinite geometric series and is written S_∞.

How can we tell if an infinite geometric series has a sum?

For each of the following geometric sequences, **(a)** determine the common ratio, **(b)** list the first 10 partial sums, S_1 through S_{10}, and **(c)** estimate, if possible, S_∞.

1. $a_n = \left(\frac{1}{3}\right)^{n-1}$ $\qquad\qquad$ **2.** $a_n = \left(-\frac{7}{2}\right)^{n-1}$

3. $a_n = 4(2)^{n-1}$ $\qquad\qquad$ **4.** $a_n = 1.3(-0.4)^{n-1}$

5. What relationship do you observe between the common ratio and the existence of S_∞? S_∞ does not exist if $|r| > 1$.

It can be shown (but we will not do it here) that the pattern you may have observed is true in general.

The sum of the terms of an infinite geometric sequence exists if and only if $|r| < 1$.

To find a formula for the sum of an infinite geometric sequence, we first consider the sum of the first n terms:

$$S_n = \frac{a_1(1 - r^n)}{1 - r} = \frac{a_1 - a_1 r^n}{1 - r}. \qquad \text{Using the distributive law}$$

For $|r| < 1$, it follows that values of r^n get closer to 0 as n gets larger. (Check this by selecting a number between -1 and 1 and finding larger and larger powers on a calculator.) As r^n gets closer to 0, so does $a_1 r^n$. Thus, S_n gets closer to $a_1/(1 - r)$.

The Limit of an Infinite Geometric Series

When $|r| < 1$, the limit of an infinite geometric series is given by

$$S_\infty = \frac{a_1}{1 - r}. \quad \text{(For } |r| \geq 1, \text{ no limit exists.)}$$

EXAMPLE 5 Determine whether each series has a limit. If a limit exists, find it.

a) $1 + 3 + 9 + 27 + \cdots$ **b)** $-2 + 1 - \frac{1}{2} + \frac{1}{4} - \frac{1}{8} + \cdots$

Solution

a) Here $r = 3$, so $|r| = |3| = 3$. Since $|r| \not< 1$, the series *does not* have a limit.

b) Here $r = -\frac{1}{2}$, so $|r| = \left|-\frac{1}{2}\right| = \frac{1}{2}$. Since $|r| < 1$, the series *does* have a limit. We find the limit by substituting into the formula for S_∞:

$$S_\infty = \frac{-2}{1 - \left(-\frac{1}{2}\right)} = \frac{-2}{\frac{3}{2}} = -2 \cdot \frac{2}{3} = -\frac{4}{3}.$$

EXAMPLE 6 Find fraction notation for $0.63636363\ldots$.

Solution We can express this as

$$0.63 + 0.0063 + 0.000063 + \cdots.$$

This is an infinite geometric series, where $a_1 = 0.63$ and $r = 0.01$. Since $|r| < 1$, this series has a limit:

$$S_\infty = \frac{a_1}{1 - r} = \frac{0.63}{1 - 0.01} = \frac{0.63}{0.99} = \frac{63}{99}.$$

Thus fraction notation for $0.63636363\ldots$ is $\frac{63}{99}$, or $\frac{7}{11}$.

Problem Solving

For some problem-solving situations, the translation may involve geometric sequences or series.

EXAMPLE 7 Daily Wages. Suppose someone offered you a job for the month of September (30 days) under the following conditions. You will be paid $0.01 for the first day, $0.02 for the second, $0.04 for the third, and so on, doubling your previous day's salary each day. How much would you earn? (Would you take the job? Make a guess before reading further.)

Solution

1. Familiarize. You earn $0.01 the first day, $0.01(2)$ the second day, $0.01(2)(2)$ the third day, and so on. Since each day's wages are a constant multiple of the previous day's wages, a geometric sequence is formed.

2. **Translate.**   The amount earned is the geometric series

$$\$0.01 + \$0.01(2) + \$0.01(2^2) + \$0.01(2^3) + \cdots + \$0.01(2^{29}),$$

where

$$a_1 = \$0.01, \qquad n = 30, \quad \text{and} \quad r = 2.$$

3. **Carry out.**   Using the formula

$$S_n = \frac{a_1(1 - r^n)}{1 - r},$$

we have

$$S_{30} = \frac{\$0.01(1 - 2^{30})}{1 - 2}$$

$$= \frac{\$0.01(-1,073,741,823)}{-1} \qquad \text{Using a calculator}$$

$$= \$10,737,418.23.$$

4. **Check.**   The calculations can be repeated as a check.

5. **State.**   The pay exceeds $10.7 million for the month. Most people would probably take the job!

EXAMPLE 8   Loan Repayment.   Francine's student loan is in the amount of $6000. Interest is to be 9% compounded annually, and the entire amount is to be paid after 10 yr. How much is to be paid back?

Solution

1. **Familiarize.**   Suppose we let P represent any principal amount. At the end of one year, the amount owed will be $P + 0.09P$, or $1.09P$. That amount will be the principal for the second year. The amount owed at the end of the second year will be $1.09 \times$ New principal $= 1.09(1.09P)$, or 1.09^2P. Thus the amount owed at the beginning of successive years is as follows:

$$①\qquad②\qquad③\qquad④$$
$$\downarrow\qquad\downarrow\qquad\quad\downarrow\qquad\quad\downarrow$$
$$P,\quad 1.09P,\quad 1.09^2P,\quad 1.09^3P,\quad \text{and so on.}$$

We have a geometric sequence. The amount owed at the beginning of the 11th year will be the amount owed at the end of the 10th year.

2. **Translate.**   We have a geometric sequence with $a_1 = 6000$, $r = 1.09$, and $n = 11$. The appropriate formula is

$$a_n = a_1 r^{n-1}.$$

3. **Carry out.**   We substitute and calculate:

$$a_{11} = \$6000(1.09)^{11-1} = \$6000(1.09)^{10}$$

$$\approx \$14,204.18. \qquad \text{Using a calculator and rounding to the nearest hundredth}$$

4. **Check.**   A check, by repeating the calculations, is left to the student.

5. **State.**   Francine will owe $14,204.18 at the end of 10 yr.

TEACHING TIP

You might want to ask students to explain why, in Example 8, n is 11 rather than 10.

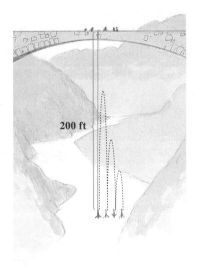

200 ft

TEACHING TIP

You might want to ask students why 200 is not included in the sequence in Example 9.

EXAMPLE 9   Bungee Jumping. A bungee jumper rebounds 60% of the height jumped. A bungee jump is made using a cord that stretches to 200 ft.

a) After jumping and then rebounding 9 times, how far has a bungee jumper traveled upward (the total rebound distance)?

b) Approximately how far will a jumper have traveled upward (bounced) before coming to rest?

Solution

1. Familiarize. Let's do some calculations and look for a pattern.

First fall:	200 ft
First rebound:	0.6×200, or 120 ft
Second fall:	120 ft, or 0.6×200
Second rebound:	0.6×120, or $0.6(0.6 \times 200)$, which is 72 ft
Third fall:	72 ft, or $0.6(0.6 \times 200)$
Third rebound:	0.6×72, or $0.6(0.6(0.6 \times 200))$, which is 43.2 ft

The rebound distances form a geometric sequence:

$$120, \quad 0.6 \times 120, \quad 0.6^2 \times 120, \quad 0.6^3 \times 120, \ldots.$$

2. Translate.

a) The total rebound distance after 9 bounces is the sum of a geometric sequence. The first term is 120 and the common ratio is 0.6. There will be 9 terms, so we can use the formula

$$S_n = \frac{a_1(1 - r^n)}{1 - r}.$$

b) Theoretically, the jumper will never stop bouncing. Realistically, the bouncing will eventually stop. To approximate the actual distance, we consider an infinite number of bounces and use the formula

$$S_\infty = \frac{a_1}{1 - r}. \qquad \text{Since } r = 0.6 \text{ and } |0.6| < 1, \text{ we know that } S_\infty \text{ exists.}$$

3. Carry out.

a) We substitute into the formula and calculate:

$$S_9 = \frac{120[1 - (0.6)^9]}{1 - 0.6} \approx 297. \qquad \text{Using a calculator}$$

b) We substitute and calculate:

$$S_\infty = \frac{120}{1 - 0.6} = 300.$$

4. Check. We can do the calculations again.

5. State.

a) In 9 bounces, the bungee jumper will have traveled upward a total distance of about 297 ft.

b) The jumper will have traveled upward a total of about 300 ft before coming to rest.

10.3

Exercise Set

Find the common ratio for each geometric sequence.

1. $7, 14, 28, 56, \ldots$ 2

2. $2, 6, 18, 54, \ldots$ 3

3. $5, -5, 5, -5, \ldots$ -1

4. $-5, -0.5, -0.05, -0.005, \ldots$ 0.1

5. $\frac{1}{2}, -\frac{1}{4}, \frac{1}{8}, -\frac{1}{16}, \ldots$ $-\frac{1}{2}$ **6.** $\frac{2}{3}, -\frac{4}{3}, \frac{8}{3}, -\frac{16}{3}, \ldots$ -2

7. $75, 15, 3, \frac{3}{5}, \ldots$ $\frac{1}{5}$ **8.** $12, -4, \frac{4}{3}, -\frac{4}{9}, \ldots$ $-\frac{1}{3}$

9. $\dfrac{1}{m}, \dfrac{3}{m^2}, \dfrac{9}{m^3}, \dfrac{27}{m^4}, \ldots$ $\dfrac{3}{m}$ **10.** $4, \dfrac{4m}{5}, \dfrac{4m^2}{25}, \dfrac{4m^3}{125}, \ldots$ $\dfrac{m}{5}$

Find the indicated term for each geometric sequence.

11. $3, 6, 12, \ldots$; the 7th term   192

12. $2, 8, 32, \ldots$; the 9th term   131,072

13. $5, 5\sqrt{2}, 10, \ldots$; the 9th term   80

14. $4, 4\sqrt{3}, 12, \ldots$; the 8th term $108\sqrt{3}$

15. $-\frac{8}{243}, \frac{8}{81}, -\frac{8}{27}, \ldots$; the 10th term   648

16. $\frac{7}{625}, \frac{-7}{125}, \frac{7}{25}, \ldots$; the 13th term   2,734,375

17. $\$1000, \$1080, \$1166.40, \ldots$; the 12th term $2331.64

18. $\$1000, \$1070, \$1144.90, \ldots$; the 11th term $1967.15

Find the nth, or general, term for each geometric sequence.

19. $1, 3, 9, \ldots$ $a_n = 3^{n-1}$ **20.** $25, 5, 1, \ldots$ $a_n = 5^{3-n}$

21. $1, -1, 1, -1, \ldots$ $a_n = (-1)^{n-1}$ **22.** $2, 4, 8, \ldots$ $a_n = 2^n$

23. $\dfrac{1}{x}, \dfrac{1}{x^2}, \dfrac{1}{x^3}, \ldots$ $a_n = \dfrac{1}{x^n}$ **24.** $5, \dfrac{5m}{2}, \dfrac{5m^2}{4}, \ldots$ $a_n = 5\left(\dfrac{m}{2}\right)^{n-1}$

For Exercises 25–32, use the formula for S_n to find the indicated sum.

25. S_7 for the geometric series $6 + 12 + 24 + \cdots$ 762

26. S_6 for the geometric series $16 - 8 + 4 - \cdots$ 10.5

27. S_7 for the geometric series $\frac{1}{18} - \frac{1}{6} + \frac{1}{2} - \cdots$ $\frac{547}{18}$

Aha! **28.** S_5 for the geometric series $7 + 0.7 + 0.07 + \cdots$ 7.7777

29. $\dfrac{1-x^8}{1-x}$, or $(1+x)(1+x^2)(1+x^4)$

29. S_8 for the series $1 + x + x^2 + x^3 + \cdots$

30. S_{10} for the series $1 + x^2 + x^4 + x^6 + \cdots$ $\dfrac{1-x^{20}}{1-x^2}$

31. S_{16} for the geometric sequence
$\$200, \$200(1.06), \$200(1.06)^2, \ldots$ $5134.51

32. S_{23} for the geometric sequence
$\$1000, \$1000(1.08), \$1000(1.08)^2, \ldots$ $60,893.30

Determine whether each infinite geometric series has a limit. If a limit exists, find it.

33. $16 + 4 + 1 + \cdots$ $\frac{64}{3}$ **34.** $8 + 4 + 2 + \cdots$ 16

35. $7 + 3 + \frac{9}{7} + \cdots$ $\frac{49}{4}$ **36.** $12 + 9 + \frac{27}{4} + \cdots$ 48

37. $3 + 15 + 75 + \cdots$ No   **38.** $2 + 3 + \frac{9}{2} + \cdots$ No

39. $4 - 6 + 9 - \frac{27}{2} + \cdots$ No

40. $-6 + 3 - \frac{3}{2} + \frac{3}{4} - \cdots$ -4

41. $0.43 + 0.0043 + 0.000043 + \cdots$ $\frac{43}{99}$

42. $0.37 + 0.0037 + 0.000037 + \cdots$ $\frac{37}{99}$

43. $\$500(1.02)^{-1} + \$500(1.02)^{-2} + \$500(1.02)^{-3} + \cdots$ $25,000

44. $\$1000(1.08)^{-1} + \$1000(1.08)^{-2} + \$1000(1.08)^{-3} + \cdots$ $12,500

Find fraction notation for each infinite sum. (These are geometric series.)

45. $0.7777\ldots$ $\frac{7}{9}$ **46.** $0.2222\ldots$ $\frac{2}{9}$

47. $8.3838\ldots$ $\frac{830}{99}$ **48.** $7.4747\ldots$ $\frac{740}{99}$

49. $0.15151515\ldots$ $\frac{5}{33}$ **50.** $0.12121212\ldots$ $\frac{4}{33}$

Solve. Use a calculator as needed for evaluating formulas.

51. *Population Growth.*   Yorktown has a current population of 100,000, and the population is increasing by 3% each year. What will the population be in 15 yr?   155,797

52. *Doubling Time.*   How long will it take for the population of Yorktown to double? (See Exercise 51.)   About 24 yr

53. *Rebound Distance.* A ping-pong ball is dropped from a height of 20 ft and always rebounds one fourth of the distance fallen. How high does it rebound the 6th time? $\frac{5}{1024}$ ft

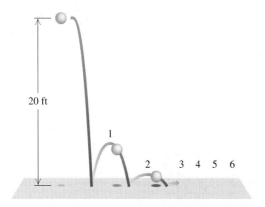

20 ft

1

2   3   4   5   6

54. *Rebound Distance.* Approximate the total of the rebound heights of the ball in Exercise 53. $6\frac{2}{3}$ ft

55. *Amount Owed.* Gilberto borrows $15,000. The loan is to be repaid in 13 yr at 8.5% interest, compounded annually. How much will be repaid at the end of 13 yr? $43,318.94

56. *Shrinking Population.* A population of 5000 fruit flies is dying off at a rate of 4% per minute. How many flies will be alive after 15 min? 2710 fruit flies

57. *Shrinking Population.* For the population of fruit flies in Exercise 56, how long will it take for only 1800 fruit flies to remain alive? (See Exercise 56 and use logarithms.) Round to the nearest minute. 25 min

58. *Investing.* Leslie is saving money in a retirement account. At the beginning of each year, she invests $1000 at 7%, compounded annually. How much will be in the retirement fund at the end of 40 yr? $213,609.57

59. *Rebound Distance.* A superball dropped from the top of the Washington Monument (556 ft high) rebounds three fourths of the distance fallen. How far (up and down) will the ball have traveled when it hits the ground for the 6th time? 3100.35 ft

60. *Rebound Distance.* Approximate the total distance that the ball of Exercise 59 will have traveled when it comes to rest. 3892 ft

61. *Stacking Paper.* Construction paper is about 0.02 in. thick. Beginning with just one piece, a stack is doubled again and again 10 times. Find the height of the final stack. 20.48 in.

62. *Monthly Earnings.* Suppose you accepted a job for the month of February (28 days) under the following conditions. You will be paid $0.01 the first day, $0.02 the second, $0.04 the third, and so on, doubling your previous day's salary each day. How much would you earn? $2,684,354.55

In Exercises 63–68, state whether the graph is that of an arithmetic sequence or a geometric sequence.

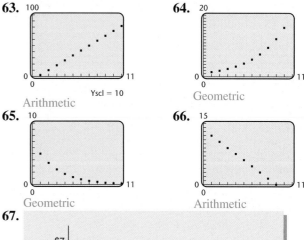

63.
Arithmetic   Yscl = 10

64.
Geometric

65.
Geometric

66.
Arithmetic

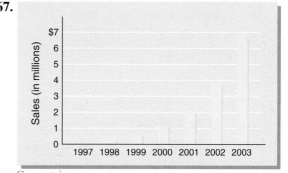

67.

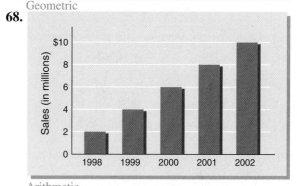

Geometric

68.

Arithmetic

69. Under what circumstances is it possible for the 5th term of a geometric sequence to be greater than the 4th term but less than the 7th term?

70. When r is negative, a series is said to be *alternating.* Why do you suppose this terminology is used?

Skill Maintenance

Multiply. [5.2]

71. $(x + y)(x^2 + 2xy + y^2)$ $x^3 + 3x^2y + 3xy^2 + y^3$

72. $(a - b)(a^2 - 2ab + b^2)$ $a^3 - 3a^2b + 3ab^2 - b^3$

Solve the system.

73. $5x - 2y = -3,$
$2x + 5y = -24$
[3.2] $\left(-\frac{63}{29}, -\frac{114}{29}\right)$

74. $x - 2y + 3z = 4,$
$2x - y + z = -1,$
$4x + y + z = 1$
[3.4] $(-1, 2, 3)$

Synthesis

TW **75.** Using Example 5 and Exercises 33–44, explain how the graph of a geometric sequence can be used to determine whether a geometric series has a limit.

TW **76.** The infinite series

$$S_\infty = 2 + \frac{1}{2} + \frac{1}{2 \cdot 3} + \frac{1}{2 \cdot 3 \cdot 4}$$
$$+ \frac{1}{2 \cdot 3 \cdot 4 \cdot 5} + \frac{1}{2 \cdot 3 \cdot 4 \cdot 5 \cdot 6} + \cdots$$

is not geometric, but it does have a sum. Using S_1, S_2, S_3, S_4, S_5, and S_6, make a conjecture about the value of S_∞ and explain your reasoning.

77. Find the sum of the first n terms of
$$x^2 - x^3 + x^4 - x^5 + \cdots.$$ $\dfrac{x^2[1 - (-x)^n]}{1 + x}$

78. Find the sum of the first n terms of
$$1 + x + x^2 + x^3 + \cdots.$$ $\dfrac{1 - x^n}{1 - x}$

79. The sides of a square are each 16 cm long. A second square is inscribed by joining the midpoints of the sides, successively. In the second square we repeat the process, inscribing a third square. If this process is continued indefinitely, what is the sum of all of the areas of all the squares? (*Hint:* Use an infinite geometric series.)   512 cm²

80. Show that $0.999\ldots$ is 1. $0.999\ldots$ can be expressed as $0.9 + 0.09 + 0.009 + \cdots.$ Let $a_1 = 0.9$ and $r = 0.1$. Then $S_\infty = \dfrac{0.9}{1 - 0.1} = \dfrac{0.9}{0.9} = 1.$

Collaborative Corner

Bargaining for a Used Car

Focus: Geometric series

Time: 30 minutes

Group size: 2

Materials: Graphing calculators are optional.

ACTIVITY*

1. One group member ("the seller") has a car for sale and is asking $3500. The second ("the buyer") offers $1500. The seller splits the difference ($2000 ÷ 2 = $1000) and lowers the price to $2500. The buyer then splits the difference again ($1000 ÷ 2 = $500) and counters with $2000. Continue in this manner and stop when you are able to agree on the car's selling price to the nearest penny.

2. What should the buyer's initial offer be in order to achieve a purchase price of $2000? (Check several guesses to find the appropriate initial offer.)

---

*This activity is based on the article, "Bargaining Theory, or Zeno's Used Cars," by James C. Kirby, *The College Mathematics Journal*, **27**(4), September 1996.

(*continued*)

3. The seller's price in the bargaining above can be
modeled recursively (see Exercises 73, 74, and
82 in Section 10.1) by the sequence

$$a_1 = 3500, \qquad a_n = a_{n-1} - \frac{d}{2^{2n-3}},$$

where d is the difference between the initial price
and the first offer. Use this recursively defined
sequence to solve parts (1) and (2) above either
manually or by using the SEQ MODE and the TABLE
feature of a graphing calculator.

4. The first four terms in the sequence in part (3)
can be written as

$$a_1, \quad a_1 - \frac{d}{2}, \quad a_1 - \frac{d}{2} - \frac{d}{8},$$

$$a_1 - \frac{d}{2} - \frac{d}{8} - \frac{d}{32}.$$

Use the formula for the limit of an infinite
geometric series to find a simple algebraic
formula for the eventual sale price, P, when the
bargaining process from above is followed.
Verify the formula by using it to solve parts (1)
and (2) above.

10.4

Binomial Expansion Using Pascal's Triangle ■ Binomial Expansion Using
Factorial Notation

The Binomial Theorem

Connecting the Concepts

Sequences and series occur in many settings, some
of which were mentioned in Sections 10.1–10.3.
Although you may not have viewed it this way
before, the expression $(x + y)^2$ can be regarded as
a series: $x^2 + 2xy + y^2$.

In Chapter 5, we found that the expansion of
$(x + y)^n$, for powers greater than 2, can be quite

time-consuming. The reason for this extends all the
way back to Chapter 1 and the rules for the order of
operations and the properties of exponents:
$(x + y)^n \neq x^n + y^n$. Since the terms in the expansion
of $(x + y)^n$ have many uses, we devote this section to
two methods that streamline the expansion of this
important algebraic expression.

Binomial Expansion Using Pascal's Triangle

TEACHING TIP

Some students may need a definition
of "expansion."

Consider the following expanded powers of $(a + b)^n$:

$$(a + b)^0 = 1$$
$$(a + b)^1 = a + b$$
$$(a + b)^2 = a^2 + 2a^1b^1 + b^2$$
$$(a + b)^3 = a^3 + 3a^2b^1 + 3a^1b^2 + b^3$$
$$(a + b)^4 = a^4 + 4a^3b^1 + 6a^2b^2 + 4a^1b^3 + b^4$$
$$(a + b)^5 = a^5 + 5a^4b^1 + 10a^3b^2 + 10a^2b^3 + 5a^1b^4 + b^5.$$

Each expansion is a polynomial. There are some patterns to be noted:

1. There is one more term than the power of the binomial, n. That is, there are $n + 1$ terms in the expansion of $(a + b)^n$.

2. In each term, the sum of the exponents is the power to which the binomial is raised.

3. The exponents of a start with n, the power of the binomial, and decrease to 0 (since $a^0 = 1$, the last term has no factor of a). The first term has no factor of b, so powers of b start with 0 and increase to n.

4. The coefficients start at 1, increase through certain values, and then decrease through these same values back to 1. Let's study the coefficients further.

Suppose we wish to expand $(a + b)^8$. The patterns we noticed above indicate 9 terms in the expansion:

$$a^8 + c_1a^7b + c_2a^6b^2 + c_3a^5b^3 + c_4a^4b^4 + c_5a^3b^5 + c_6a^2b^6 + c_7ab^7 + b^8.$$

How can we determine the values for the c's? One method seems very simple, but it has some drawbacks. It involves writing down the coefficients in a triangular array as follows. We form what is known as **Pascal's triangle**:

```
(a + b)⁰:                    1
(a + b)¹:                 1     1
(a + b)²:              1     2     1
(a + b)³:           1     3     3     1
(a + b)⁴:        1     4     6     4     1
(a + b)⁵:     1     5    10    10     5     1
```

There are many patterns in the triangle. Find as many as you can.

Perhaps you discovered a way to write the next row of numbers, given the numbers in the row above it. There are always 1's on the outside. Each remaining number is the sum of the two numbers above:

```
                    1
                 1     1
              1     2     1
           1     3     3     1
        1     4     6     4     1
     1     5    10    10     5     1
  1     6    15    20    15     6     1
```

We see that in the bottom (seventh) row

the 1st and last numbers are 1;

the 2nd number is $1 + 5$, or 6:

the 3rd number is $5 + 10$, or 15;

the 4th number is $10 + 10$, or 20;

the 5th number is $10 + 5$, or 15; and

the 6th number is $5 + 1$, or 6.

Thus the expansion of $(a + b)^6$ is

$$(a + b)^6 = 1a^6 + 6a^5b + 15a^4b^2 + 20a^3b^3 + 15a^2b^4 + 6ab^5 + 1b^6.$$

To expand $(a + b)^8$, we complete two more rows of Pascal's triangle:

$$
\begin{array}{ccccccccccccccccc}
 & & & & & & & & 1 & & & & & & & & \\
 & & & & & & & 1 & & 1 & & & & & & & \\
 & & & & & & 1 & & 2 & & 1 & & & & & & \\
 & & & & & 1 & & 3 & & 3 & & 1 & & & & & \\
 & & & & 1 & & 4 & & 6 & & 4 & & 1 & & & & \\
 & & & 1 & & 5 & & 10 & & 10 & & 5 & & 1 & & & \\
 & & 1 & & 6 & & 15 & & 20 & & 15 & & 6 & & 1 & & \\
 & 1 & & 7 & & 21 & & 35 & & 35 & & 21 & & 7 & & 1 & \\
1 & & 8 & & 28 & & 56 & & 70 & & 56 & & 28 & & 8 & & 1 \\
\end{array}
$$

Thus the expansion of $(a + b)^8$ has coefficients found in the 9th row above:

$$(a + b)^8 = 1a^8 + 8a^7b + 28a^6b^2 + 56a^5b^3 + 70a^4b^4 + 56a^3b^5 + 28a^2b^6 + 8ab^7 + 1b^8.$$

We can generalize our results as follows:

The Binomial Theorem (Form 1)

For any binomial $a + b$ and any natural number n,

$$(a + b)^n = c_0a^nb^0 + c_1a^{n-1}b^1 + c_2a^{n-2}b^2 + \cdots + c_{n-1}a^1b^{n-1} + c_na^0b^n,$$

where the numbers $c_0, c_1, c_2, \ldots, c_n$ are from the $(n + 1)$st row of Pascal's triangle.

EXAMPLE 1   Expand: $(u - v)^5$.

Solution   Using the binomial theorem, we have $a = u$, $b = -v$, and $n = 5$. We use the 6th row of Pascal's triangle: 1   5   10   10   5   1. Thus,

$$
\begin{aligned}
(u - v)^5 &= [u + (-v)]^5 \qquad \text{\textbf{Rewriting } } u - v \text{ \textbf{as a sum}} \\
&= 1(u)^5 + 5(u)^4(-v)^1 + 10(u)^3(-v)^2 + 10(u)^2(-v)^3 \\
&\quad + 5(u)^1(-v)^4 + 1(-v)^5 \\
&= u^5 - 5u^4v + 10u^3v^2 - 10u^2v^3 + 5uv^4 - v^5.
\end{aligned}
$$

Note that the signs of the terms alternate between $+$ and $-$. When $-v$ is raised to an odd power, the sign is $-$.

EXAMPLE 2   Expand: $\left(2t + \dfrac{3}{t}\right)^6$.

Solution   Note that $a = 2t$, $b = 3/t$, and $n = 6$. We use the 7th row of Pascal's triangle: 1   6   15   20   15   6   1. Thus,

$$\left(2t + \frac{3}{t}\right)^6 = 1(2t)^6 + 6(2t)^5\left(\frac{3}{t}\right)^1 + 15(2t)^4\left(\frac{3}{t}\right)^2 + 20(2t)^3\left(\frac{3}{t}\right)^3$$

$$+ 15(2t)^2\left(\frac{3}{t}\right)^4 + 6(2t)^1\left(\frac{3}{t}\right)^5 + 1\left(\frac{3}{t}\right)^6$$

$$= 64t^6 + 6(32t^5)\left(\frac{3}{t}\right) + 15(16t^4)\left(\frac{9}{t^2}\right) + 20(8t^3)\left(\frac{27}{t^3}\right)$$

$$+ 15(4t^2)\left(\frac{81}{t^4}\right) + 6(2t)\left(\frac{243}{t^5}\right) + \frac{729}{t^6}$$

$$= 64t^6 + 576t^4 + 2160t^2 + 4320 + 4860t^{-2} + 2916t^{-4}$$

$$+ 729t^{-6}.$$

Binomial Expansion Using Factorial Notation

The drawback to using Pascal's triangle is that we must compute all the preceding rows in the table to obtain the row needed for the expansion in which we are interested. The following method avoids this difficulty. It will also enable us to find a specific term—say, the 8th term—without computing all the other terms in the expansion. This method is useful in such courses as finite mathematics, calculus, and statistics.

To develop the method, we need some new notation. Products of successive natural numbers, such as $6 \cdot 5 \cdot 4 \cdot 3 \cdot 2 \cdot 1$ and $8 \cdot 7 \cdot 6 \cdot 5 \cdot 4 \cdot 3 \cdot 2 \cdot 1$, have a special notation. For the product $6 \cdot 5 \cdot 4 \cdot 3 \cdot 2 \cdot 1$, we write 6!, read "6 factorial."

Factorial Notation

For any natural number n,

$$n! = n(n - 1)(n - 2) \cdots (3)(2)(1).$$

Here are some examples:

$$6! = 6 \cdot 5 \cdot 4 \cdot 3 \cdot 2 \cdot 1 = 720,$$
$$5! = 5 \cdot 4 \cdot 3 \cdot 2 \cdot 1 = 120,$$
$$4! = 4 \cdot 3 \cdot 2 \cdot 1 = 24,$$
$$3! = 3 \cdot 2 \cdot 1 = 6,$$
$$2! = 2 \cdot 1 = 2,$$
$$1! = 1 = 1.$$

We also define 0! to be 1 for reasons explained shortly.

To simplify expressions like

$$\frac{8!}{5!\,3!},$$

note that

$$8! = 8 \cdot 7 \cdot 6 \cdot 5 \cdot 4 \cdot 3 \cdot 2 \cdot 1 = 8 \cdot 7! = 8 \cdot 7 \cdot 6! = 8 \cdot 7 \cdot 6 \cdot 5!,$$

and so on.

CAUTION! $\dfrac{6!}{3!} \neq 2!$ To see this, note that

$$\frac{6!}{3!} = \frac{6 \cdot 5 \cdot 4 \cdot \cancel{3} \cdot \cancel{2} \cdot \cancel{1}}{\cancel{3} \cdot \cancel{2} \cdot \cancel{1}} = 6 \cdot 5 \cdot 4.$$

EXAMPLE 3   Simplify: $\dfrac{8!}{5!\,3!}$.

Solution

$$\frac{8!}{5!\,3!} = \frac{8 \cdot 7 \cdot 6 \cdot 5!}{5! \cdot 3 \cdot 2 \cdot 1} = 8 \cdot 7$$

Removing a factor equal to 1:
$$\frac{6 \cdot 5!}{5! \cdot 3 \cdot 2} = 1$$

$$= 56.$$

The following notation is used in our second formulation of the binomial theorem.

$\dbinom{n}{r}$ **Notation**

For n, r nonnegative integers with $n \geq r$,

$$\binom{n}{r}, \quad \text{read "}n \text{ choose } r\text{,"} \quad \text{means} \quad \frac{n!}{(n-r)!\,r!}.^{*}$$

EXAMPLE 4   Simplify: **(a)** $\dbinom{7}{2}$; **(b)** $\dbinom{9}{6}$; **(c)** $\dbinom{6}{6}$.

Solution

a) $\dbinom{7}{2} = \dfrac{7!}{(7-2)!\,2!}$

$$= \frac{7!}{5!\,2!} = \frac{7 \cdot 6 \cdot 5!}{5! \cdot 2 \cdot 1} = \frac{7 \cdot 6}{2}$$

$$= 7 \cdot 3$$

$$= 21$$

---

*In many books and for many calculators, the notation $_nC_r$ is used instead of $\dbinom{n}{r}$.

b) $\displaystyle\binom{9}{6} = \frac{9!}{3!\,6!}$

$\displaystyle\qquad = \frac{9 \cdot 8 \cdot 7 \cdot 6!}{3 \cdot 2 \cdot 1 \cdot 6!} = \frac{9 \cdot 8 \cdot 7}{3 \cdot 2}$

$\displaystyle\qquad = 3 \cdot 4 \cdot 7$

$\displaystyle\qquad = 84$

c) $\displaystyle\binom{6}{6} = \frac{6!}{0!\,6!} = \frac{6!}{1 \cdot 6!}\qquad$ Since $0! = 1$

$\displaystyle\qquad = \frac{6!}{6!}$

$\displaystyle\qquad = 1$

Factorials and $\displaystyle\binom{n}{r}$

The PRB submenu of the MATH menu provides access to both factorial calculations and $\displaystyle\binom{n}{r}$, or $_nC_r$. In both cases, a number is entered on the home screen before the option is accessed.

To find 7!, press $\boxed{7}$, select option 4: ! from the MATH PRB menu, and press $\boxed{\text{ENTER}}$.

To find $\displaystyle\binom{9}{2}$, press $\boxed{9}$, select option 3: nCr from the MATH PRB menu, press $\boxed{2}$, and then press $\boxed{\text{ENTER}}$.

The screen on the left below shows the options listed in the MATH PRB menu. From the screen on the right below, note that $7! = 5040$ and $\displaystyle\binom{9}{2} = {_9C_2} = 36$.

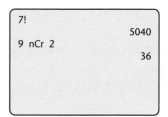

```
MATH  NUM  CPX  PRB
1: rand
2: nPr
3: nCr
4: !
5: randInt(
6: randNorm(
7: randBin(
```

```
7!
                    5040
9 nCr 2
                      36
```

Now we can restate the binomial theorem using our new notation.

The Binomial Theorem (Form 2)

For any binomial $a + b$ and any natural number n,

$$(a + b)^n = \binom{n}{0}a^n + \binom{n}{1}a^{n-1}b + \binom{n}{2}a^{n-2}b^2 + \cdots + \binom{n}{n}b^n.$$

EXAMPLE 5 Expand: $(3x + y)^4$.

Solution We use the binomial theorem (Form 2) with $a = 3x$, $b = y$, and $n = 4$:

$$(3x + y)^4 = \binom{4}{0}(3x)^4 + \binom{4}{1}(3x)^3y + \binom{4}{2}(3x)^2y^2 + \binom{4}{3}(3x)y^3 + \binom{4}{4}y^4$$

$$= \frac{4!}{4!\,0!}3^4x^4 + \frac{4!}{3!\,1!}3^3x^3y + \frac{4!}{2!\,2!}3^2x^2y^2 + \frac{4!}{1!\,3!}3xy^3 + \frac{4!}{0!\,4!}y^4$$

$$= 81x^4 + 108x^3y + 54x^2y^2 + 12xy^3 + y^4. \quad \textbf{Simplifying}$$

EXAMPLE 6 Expand: $(x^2 - 2y)^5$.

Solution In this case, $a = x^2$, $b = -2y$, and $n = 5$:

$$(x^2 - 2y)^5 = \binom{5}{0}(x^2)^5 + \binom{5}{1}(x^2)^4(-2y) + \binom{5}{2}(x^2)^3(-2y)^2$$

$$+ \binom{5}{3}(x^2)^2(-2y)^3 + \binom{5}{4}(x^2)(-2y)^4 + \binom{5}{5}(-2y)^5$$

$$= \frac{5!}{5!\,0!}x^{10} + \frac{5!}{4!\,1!}x^8(-2y) + \frac{5!}{3!\,2!}x^6(-2y)^2 + \frac{5!}{2!\,3!}x^4(-2y)^3$$

$$+ \frac{5!}{1!\,4!}x^2(-2y)^4 + \frac{5!}{0!\,5!}(-2y)^5$$

$$= x^{10} - 10x^8y + 40x^6y^2 - 80x^4y^3 + 80x^2y^4 - 32y^5.$$

Note that in the binomial theorem (Form 2), $\binom{n}{0}a^nb^0$ gives us the first term, $\binom{n}{1}a^{n-1}b^1$ gives us the second term, $\binom{n}{2}a^{n-2}b^2$ gives us the third term, and so on. This can be generalized to give a method for finding a specific term without writing the entire expansion.

Finding a Specific Term
The $(r + 1)$st term of $(a + b)^n$ is

$$\binom{n}{r}a^{n-r}b^r.$$

17. $m^5 + 5m^4n + 10m^3n^2 + 10m^2n^3 + 5mn^4 + n^5$

18. $a^4 - 4a^3b + 6a^2b^2 - 4ab^3 + b^4$

19. $x^6 - 6x^5y + 15x^4y^2 - 20x^3y^3 + 15x^2y^4 - 6xy^5 + y^6$

20. $p^7 + 7p^6q + 21p^5q^2 + 35p^4q^3 + 35p^3q^4 + 21p^2q^5 + 7pq^6 + q^7$

21. $x^{10} - 15x^8y + 90x^6y^2 - 270x^4y^3 + 405x^2y^4 - 243y^5$

22. $2187c^7 - 5103c^6d + 5103c^5d^2 - 2835c^4d^3 + 945c^3d^4 - 189c^2d^5 + 21cd^6 - d^7$

23. $729c^6 - 1458c^5d + 1215c^4d^2 - 540c^3d^3 + 135c^2d^4 - 18cd^5 + d^6$

24. $t^{-12} + 12t^{-10} + 60t^{-8} + 160t^{-6} + 240t^{-4} + 192t^{-2} + 64$

25. $x^3 - 3x^2y + 3xy^2 - y^3$

26. $x^5 - 5x^4y + 10x^3y^2 - 10x^2y^3 + 5xy^4 - y^5$

27. $x^9 + \dfrac{18x^8}{y} + \dfrac{144x^7}{y^2} + \dfrac{672x^6}{y^3} + \dfrac{2016x^5}{y^4} + \dfrac{4032x^4}{y^5} + \dfrac{5376x^3}{y^6} + \dfrac{4608x^2}{y^7} + \dfrac{2304x}{y^8} + \dfrac{512}{y^9}$

28. $19{,}683s^9 + \dfrac{59{,}049s^8}{t} + \dfrac{78{,}732s^7}{t^2} + \dfrac{61{,}236s^6}{t^3} + \dfrac{30{,}618s^5}{t^4} + \dfrac{10{,}206s^4}{t^5} + \dfrac{2268s^3}{t^6} + \dfrac{324s^2}{t^7} + \dfrac{27s}{t^8} + \dfrac{1}{t^9}$

29. $a^{10} - 5a^8b^3 + 10a^6b^6 - 10a^4b^9 + 5a^2b^{12} - b^{15}$

30. $x^{15} - 10x^{12}y + 40x^9y^2 - 80x^6y^3 + 80x^3y^4 - 32y^5$

31. $9 - 12\sqrt{3}t + 18t^2 - 4\sqrt{3}t^3 + t^4$

32. $125 + 150\sqrt{5}t + 375t^2 + 100\sqrt{5}t^3 + 75t^4 + 6\sqrt{5}t^5 + t^6$

EXAMPLE 7   Find the 5th term in the expansion of $(2x - 3y)^7$.

Solution   First, we note that $5 = 4 + 1$. Thus, $r = 4$, $a = 2x$, $b = -3y$, and $n = 7$. Then the 5th term of the expansion is

$$\binom{7}{4}(2x)^{7-4}(-3y)^4, \quad \text{or} \quad \frac{7!}{3!\,4!}(2x)^3(-3y)^4, \quad \text{or} \quad 22{,}680x^3y^4.$$

It is because of the binomial theorem that $\binom{n}{r}$ is called a *binomial coefficient*. We can now explain why 0! is defined to be 1. In the binomial expansion, we want $\binom{n}{0}$ to equal 1 and we also want the definition

$$\binom{n}{r} = \frac{n!}{(n-r)!\,r!}$$

to hold for all whole numbers n and r. Thus we must have

$$\binom{n}{0} = \frac{n!}{(n-0)!\,0!} = \frac{n!}{n!\,0!} = 1.$$

This is satisfied only if 0! is defined to be 1.

10.4

Exercise Set

FOR EXTRA HELP

Digital Video Tutor CD 10 Videotape 14   Student's Solutions Manual   AW Math Tutor Center   InterAct Math   MathXL   MyMathLab

Simplify.

1. $8!$ 40,320

2. $9!$ 362,880

3. $10!$ 3,628,800

4. $11!$ 39,916,800

5. $\dfrac{7!}{4!}$ 210

6. $\dfrac{8!}{6!}$ 56

7. $\dfrac{10!}{7!}$ 720

8. $\dfrac{9!}{5!}$ 3024

9. $\binom{8}{2}$ 28

10. $\binom{7}{4}$ 35

11. $\binom{10}{6}$ 210

12. $\binom{9}{5}$ 126

13. $\binom{20}{18}$ 190

14. $\binom{30}{3}$ 4060

15. $\binom{35}{2}$ 595

16. $\binom{40}{38}$ 780

Expand. Use both of the methods shown in this section.

17. $(m + n)^5$

18. $(a - b)^4$

19. $(x - y)^6$

20. $(p + q)^7$

21. $(x^2 - 3y)^5$

22. $(3c - d)^7$

23. $(3c - d)^6$

24. $(t^{-2} + 2)^6$

25. $(x - y)^3$

26. $(x - y)^5$

27. $\left(x + \dfrac{2}{y}\right)^9$

28. $\left(3s + \dfrac{1}{t}\right)^9$

29. $(a^2 - b^3)^5$

30. $(x^3 - 2y)^5$

31. $\left(\sqrt{3} - t\right)^4$

32. $\left(\sqrt{5} + t\right)^6$

33. $x^{-8} + 4x^{-4} + 6 + 4x^4 + x^8$ **34.** $x^{-3} - 6x^{-2} + 15x^{-1} - 20 + 15x - 6x^2 + x^3$

33. $(x^{-2} + x^2)^4$ **34.** $\left(\dfrac{1}{\sqrt{x}} - \sqrt{x}\right)^6$

Find the indicated term for each binomial expression.

35. 3rd, $(a + b)^6$ $15a^4b^2$ **36.** 6th, $(x + y)^7$ $21x^2y^5$

37. 12th, $(a - 3)^{14}$ **38.** 11th, $(x - 2)^{12}$
$-64,481,508a^3$ $67,584x^2$

39. 5th, $(2x^3 + \sqrt{y})^8$ **40.** 4th, $\left(\dfrac{1}{b^2} + c\right)^7$ $\dfrac{35c^3}{b^8}$
$1120x^{12}y^2$

41. Middle, $(2u - 3v^2)^{10}$ $-1,959,552u^5v^{10}$

42. Middle two, $(\sqrt{x} + \sqrt{3})^5$ $30x\sqrt{x}, 30x\sqrt{3}$

Aha! **43.** 9th, $(x - y)^8$ y^8 **44.** 10th, $(a - b)^9$ $-b^9$

TW **45.** Maya claims that she can calculate mentally the first two and the last two terms of the expansion of $(a + b)^n$ for any whole number n. How do you think she does this?

TW **46.** Without performing any calculations, how can you tell if the expansions of $(x - y)^8$ and $(y - x)^8$ are equal?

Skill Maintenance

Solve. [9.6]

47. $\log_2 x + \log_2(x - 2) = 3$ 4

48. $\log_3(x + 2) - \log_3(x - 2) = 2$ $\frac{5}{2}$

49. $e^t = 280$ 5.6348

50. $\log_5 x^2 = 2$ ± 5

Synthesis

TW **51.** Explain how someone can determine the x^2-term of the expansion of $\left(x - \frac{3}{x}\right)^{10}$ without calculating any other terms.

TW **52.** Devise two problems requiring the use of the binomial theorem. Design the problems so that one is solved more easily using Form 1 and the other is solved more easily using Form 2. Then explain what makes one form easier to use than the other in each case.

53. Show that there are exactly $\dbinom{5}{3}$ ways of forming a subset of size 3 from a set of 5 elements.   ·

54. *Baseball.* At one point in a recent season, Derek Jeter of the New York Yankees had a batting average of .325. At that time, if someone were to randomly

select 5 of his "at-bats," the probability of his getting exactly 3 hits would be the 3rd term of the binomial expansion of $(0.325 + 0.675)^5$. Find that term and use a calculator to estimate the probability. ·

55. *Widows or Divorcees.* The probability that a woman will be either widowed or divorced is 85%. If 8 women are randomly selected, the probability that exactly 5 of them will be either widowed or divorced is the 6th term of the binomial expansion of $(0.15 + 0.85)^8$. Use a calculator to estimate that probability. ·

56. *Baseball.* In reference to Exercise 54, the probability that Jeter will get *at most* 3 hits is found by adding the last 4 terms of the binomial expansion of $(0.325 + 0.675)^5$. Find these terms and use a calculator to estimate the probability. ·

57. *Widows or Divorcees.* In reference to Exercise 55, the probability that *at least* 6 of the women will be widowed or divorced is found by adding the last three terms of the binomial expansion of $(0.15 + 0.85)^8$. Find these terms and use a calculator to estimate the probability. ·

58. Prove that
$$\binom{n}{r} = \binom{n}{n - r}$$
for any whole numbers n and r. Assume $r \leq n$.   ·

59. Find the term of
$$\left(\frac{3x^2}{2} - \frac{1}{3x}\right)^{12} \quad \frac{55}{144}$$
that does not contain x.

60. Find the middle term of $(x^2 - 6y^{3/2})^6$. $-4320x^6y^{9/2}$

61. Find the ratio of the 4th term of
$$\left(p^2 - \frac{1}{2}p\sqrt[3]{q}\right)^5 \quad \frac{-\sqrt[3]{q}}{2p}$$
to the 3rd term.

62. Find the term containing $\dfrac{1}{x^{1/6}}$ of
$$\left(\sqrt[3]{x} - \frac{1}{\sqrt{x}}\right)^7. \quad -\frac{35}{x^{1/6}}$$

63. What is the degree of $(x^2 + 3)^4$? 8

Aha! **64.** Multiply: $(x^2 + 2xy + y^2)(x^2 + 2xy + y^2)^2(x + y)$.
$x^7 + 7x^6y + 21x^5y^2 + 35x^4y^3 + 35x^3y^4 + 21x^2y^5 + 7xy^6 + y^7$

· Answers to Exercises 53–58 can be found on pp. A-73 and A-74.

10 Chapter Summary and Review

Key Terms

Important Properties and Formulas

Arithmetic sequence: $a_{n+1} = a_n + d$

nth term of an arithmetic sequence: $a_n = a_1 + (n-1)d$

Sum of the first n terms of an arithmetic sequence: $S_n = \dfrac{n}{2}(a_1 + a_n)$

Geometric sequence: $a_{n+1} = a_n \cdot r$

nth term of a geometric sequence: $a_n = a_1 r^{n-1}$

Sum of the first n terms of a geometric sequence: $S_n = \dfrac{a_1(1 - r^n)}{1 - r}$

Limit of an infinite geometric series: $S_\infty = \dfrac{a_1}{1 - r}, \quad |r| < 1$

Factorial notation: $n! = n(n-1)(n-2)\cdots 3 \cdot 2 \cdot 1$

Binomial coefficient: $\dbinom{n}{r} = {}_nC_r = \dfrac{n!}{(n-r)!\,r!}$

Binomial theorem: $(a+b)^n = \dbinom{n}{0}a^n + \dbinom{n}{1}a^{n-1}b + \dbinom{n}{2}a^{n-2}b^2 + \cdots + \dbinom{n}{n}b^n$

$(r+1)$st term of $(a+b)^n$: $\dbinom{n}{r}a^{n-r}b^r$

Review Exercises

6. [10.1] $-3 + (-5) + (-7) + (-9) + (-11) + (-13) = -48$

Find the first four terms; the 8th term, a_8; and the 12th term, a_{12}.

[10.1] $0, \frac{1}{5}, \frac{1}{5}, \frac{3}{17}; \frac{7}{65}; \frac{11}{145}$

1. $a_n = 4n - 3$
[10.1] 1, 5, 9, 13; 29; 45

2. $a_n = \dfrac{n-1}{n^2+1}$

Predict the general term. Answers may vary.

3. $-2, -4, -6, -8, -10, \ldots$ [10.1] $a_n = -2n$

4. $-1, 3, -5, 7, -9, \ldots$ [10.1] $a_n = (-1)^n(2n-1)$

Write out and evaluate each sum.

5. $\displaystyle\sum_{k=1}^{5} (-2)^k$

6. $\displaystyle\sum_{k=2}^{7} (1 - 2k)$

[10.1] $-2 + 4 + (-8) + 16 + (-32) = -22$

Rewrite using sigma notation.

7. $4 + 8 + 12 + 16 + 20$ [10.1] $\displaystyle\sum_{k=1}^{5} 4k$

8. $\dfrac{-1}{2} + \dfrac{1}{4} + \dfrac{-1}{8} + \dfrac{1}{16} + \dfrac{-1}{32}$ [10.1] $\displaystyle\sum_{k=1}^{5} \dfrac{1}{(-2)^k}$

9. Find the 14th term of the arithmetic sequence $-6, 1, 8, \ldots$.   [10.2] 85

10. Find d when $a_1 = 11$ and $a_{10} = 35$. Assume an arithmetic sequence.   [10.2] $\frac{8}{3}$

11. Find a_1 and d when $a_{12} = 25$ and $a_{24} = 40$. Assume an arithmetic sequence.   [10.2] $d = 1.25, a_1 = 11.25$

12. Find the sum of the first 17 terms of the arithmetic series $-8 + (-11) + (-14) + \cdots$.   [10.2] -544

13. Find the sum of all the multiples of 6 from 12 to 318, inclusive.   [10.2] 8580

14. Find the 20th term of the geometric sequence $2, 2\sqrt{2}, 4, \ldots$.   [10.3] $1024\sqrt{2}$

15. Find the common ratio of the geometric sequence $2, \frac{4}{3}, \frac{8}{9}, \ldots$.   [10.3] $\frac{2}{3}$

16. Find the nth term of the geometric sequence $-2, 2, -2, \ldots$.   [10.3] $a_n = 2(-1)^n$

17. Find the nth term of the geometric sequence $3, \frac{3}{4}x, \frac{3}{16}x^2, \ldots$.   [10.3] $a_n = 3\left(\dfrac{x}{4}\right)^{n-1}$

18. Find S_6 for the geometric series $3 + 12 + 48 + \cdots$.   [10.3] 4095

19. Find S_{12} for the geometric series $3x - 6x + 12x - \cdots$.   [10.3] $-4095x$

Determine whether each infinite geometric series has a limit. If a limit exists, find it.

20. $6 + 3 + 1.5 + 0.75 + \cdots$ [10.3] 12

21. $7 - 4 + \frac{16}{7} - \cdots$ [10.3] $\frac{49}{11}$

22. $2 + (-2) + 2 + (-2) + \cdots$ [10.3] No

23. $0.04 + 0.08 + 0.16 + 0.32 + \cdots$ [10.3] No

24. $\$2000 + \$1900 + \$1805 + \$1714.75 + \cdots$ [10.3] $40,000

25. Find fraction notation for $0.555555\ldots$.   [10.3] $\frac{5}{9}$

26. Find fraction notation for $1.39393939\ldots$.   [10.3] $\frac{46}{33}$

Solve.

27. Adam took a telemarketing job, starting with an hourly wage of \$11.40. He was promised a raise of 20¢ per hour every 3 mos for 8 yr. At the end of 8 yr, what will be his hourly wage?   [10.2] \$17.80

28. A stack of poles has 42 poles in the bottom row. There are 41 poles in the second row, 40 poles in the third row, and so on, ending with 1 pole in the top row. How many poles are in the stack?   [10.2] 903

29. A student loan is in the amount of \$10,000. Interest is 7%, compounded annually, and the amount is to be paid off in 12 yr. How much is to be paid back?   [10.3] \$22,521.92

30. Find the total rebound distance of a ball, given that it is dropped from a height of 12 m and each rebound is one-third of the preceding one.   [10.3] 6 m

Simplify.

31. $7!$ [10.4] 5040

32. $\dbinom{8}{3}$ [10.4] 56

33. Find the 3rd term of $(a + b)^{20}$.   [10.4] $190a^{18}b^2$

34. Expand: $(x - 2y)^4$.   [10.4] $x^4 - 8x^3y + 24x^2y^2 - 32xy^3 + 16y^4$

Synthesis

TW **35.** What happens to the terms of a geometric sequence with $|r| < 1$ as n gets larger? Why?

TW **36.** Compare the two forms of the binomial theorem given in the text. Under what circumstances would one be more useful than the other?

37. Find the sum of the first n terms of the geometric series $1 - x + x^2 - x^3 + \cdots$.   [10.3] $\dfrac{1 - (-x)^n}{x + 1}$

38. Expand: $(x^{-3} + x^3)^5$.   [10.4] $x^{-15} + 5x^{-9} + 10x^{-3} + 10x^3 + 5x^9 + x^{15}$

Chapter Test 10

1. Find the first five terms and the 16th term of a sequence with general term $a_n = 6n - 5$.
[10.1] 1, 7, 13, 19, 25; 91

2. Predict the general term of the sequence
$$\frac{4}{3}, \frac{4}{9}, \frac{4}{27}, \ldots. \quad [10.1] \; a_n = 4\left(\frac{1}{3}\right)^n$$

3. Write out and evaluate:
$$\sum_{k=1}^{5} (3 - 2^k). \quad [10.1]$$
$$1 + (-1) + (-5) + (-13) + (-29) = -47$$

4. Rewrite using sigma notation:
$$1 + (-8) + 27 + (-64) + 125. \quad [10.1] \sum_{k=1}^{5} (-1)^{k+1} k^3$$

5. Find the 12th term, a_{12}, of the arithmetic sequence $9, 4, -1, \ldots.$ [10.2] -46

Assume arithmetic sequences for Questions 6 and 7.

6. Find the common difference d when $a_1 = 9$ and $a_7 = 11\frac{1}{4}$. [10.2] $\frac{3}{8}$

7. Find a_1 and d when $a_5 = 16$ and $a_{10} = -3$.
[10.2] $a_1 = 31.2$; $d = -3.8$

8. Find the sum of all the multiples of 12 from 24 to 240, inclusive. [10.2] 2508

9. Find the 6th term of the geometric sequence $72, 18, 4\frac{1}{2}, \ldots.$ [10.3] $\frac{9}{128}$

10. Find the common ratio of the geometric sequence $22\frac{1}{2}, 15, 10, \ldots.$ [10.3] $\frac{2}{3}$

11. Find the nth term of the geometric sequence $3, -9, 27, \ldots.$ [10.3] $(-1)^{n+1} 3^n$

12. Find the sum of the first nine terms of the geometric series.
[10.3] $511 + 511x$
$$(1 + x) + (2 + 2x) + (4 + 4x) + \cdots.$$

Determine whether each infinite geometric series has a limit. If a limit exists, find it.

13. $0.5 + 0.25 + 0.125 + \cdots$ [10.3] 1

14. $0.5 + 1 + 2 + 4 + \cdots$ [10.3] No

15. $\$1000 + \$80 + \$6.40 + \cdots$ [10.3] $\frac{\$25,000}{23} \approx \$1086.96$

16. Find fraction notation for $0.85858585\ldots.$ [10.3] $\frac{85}{99}$

17. An auditorium has 31 seats in the first row, 33 seats in the second row, 35 seats in the third row, and so on, for 18 rows. How many seats are in the 17th row? [10.2] 63

18. Lindsay's Uncle Ken gave her \$100 for her first birthday, \$200 for her second birthday, \$300 for her third birthday, and so on, until her eighteenth birthday. How much did he give her in all? [10.2] \$17,100

19. Each week the price of a \$15,000 boat will be reduced 5% of the previous week's price. If we assume that it is not sold, what will be the price after 10 weeks? [10.3] \$8981.05

20. Find the total rebound distance of a ball that is dropped from a height of 18 m, with each rebound two thirds of the preceding one. [10.3] 36 m

21. Simplify: $\begin{pmatrix} 13 \\ 11 \end{pmatrix}$. [10.4] 78

22. Expand: $(x^2 - 3y)^5$.
[10.4] $x^{10} - 15x^8 y + 90x^6 y^2 - 270x^4 y^3 + 405x^2 y^4 - 243y^5$

23. Find the 4th term in the expansion of $(a + x)^{12}$.
[10.4] $220a^9 x^3$

Synthesis

24. Find a formula for the sum of the first n even natural numbers:
$$2 + 4 + 6 + \cdots + 2n. \quad [10.2] \; n(n + 1)$$

25. Find the sum of the first n terms of
$$1 + \frac{1}{x} + \frac{1}{x^2} + \frac{1}{x^3} + \cdots.$$

$$[10.3] \; \frac{1 - \left(\frac{1}{x}\right)^n}{1 - \frac{1}{x}}, \text{ or } \frac{x^n - 1}{x^{n-1}(x - 1)}$$

1–10 Cumulative Review

Simplify.

1. $(-9x^2y^3)(5x^4y^{-7})$ [1.4] $-45x^6y^{-4}$, or $\dfrac{-45x^6}{y^4}$

2. $|-3.5 + 9.8|$ [1.2] 6.3

3. $2y - [3 - 4(5 - 2y) - 3y]$ [1.3] $-3y + 17$

4. $(10 \cdot 8 - 9 \cdot 7)^2 - 54 \div 9 - 3$ [1.1] 280

5. Evaluate

$$\frac{ab - ac}{bc}$$ [1.1], [1.2] $\dfrac{7}{6}$

for $a = -2$, $b = 3$, and $c = -4$.

Perform the indicated operations and simplify.

6. $(5a^2 - 3ab - 7b^2) - (2a^2 + 5ab + 8b^2)$ [5.1] $3a^2 - 8ab - 15b^2$

7. $(-3x^2 + 4x^3 - 5x - 1) + (9x^3 - 4x^2 + 7 - x)$ [5.1] $13x^3 - 7x^2 - 6x + 6$

8. $(2a - 1)(3a + 5)$ [5.2] $6a^2 + 7a - 5$

9. $(3a^2 - 5y)^2$ [5.2] $9a^4 - 30a^2y + 25y^2$

10. $\dfrac{1}{x - 2} - \dfrac{4}{x^2 - 4} + \dfrac{3}{x + 2}$ [6.2] $\dfrac{4}{x + 2}$

11. $\dfrac{x^2 - 6x + 8}{3x + 9} \cdot \dfrac{x + 3}{x^2 - 4}$ [6.1] $\dfrac{x - 4}{3(x + 2)}$

12. $\dfrac{3x + 3y}{5x - 5y} \div \dfrac{3x^2 + 3y^2}{5x^3 - 5y^3}$ [6.1] $\dfrac{(x + y)(x^2 + xy + y^2)}{x^2 + y^2}$

13. $\dfrac{x - \dfrac{a^2}{x}}{1 + \dfrac{a}{x}}$ [6.3] $x - a$

Factor.

14. $4x^2 - 12x + 9$ [5.6] $(2x - 3)^2$

15. $27a^3 - 8$ [5.7] $(3a - 2)(9a^2 + 6a + 4)$

16. $a^3 + 3a^2 - ab - 3b$ [5.3] $(a + 3)(a^2 - b)$

17. $15y^4 + 33y^2 - 36$ [5.5] $3(y^2 + 3)(5y^2 - 4)$

18. For the function described by
$$f(x) = 3x^2 - 4x,$$ [2.1] 20
find $f(-2)$.

19. Divide:
$$(7x^4 - 5x^3 + x^2 - 4) \div (x - 2).$$ [6.6] $7x^3 + 9x^2 + 19x + 38 + \dfrac{72}{x - 2}$

Solve.

20. $9(x - 1) - 3(x - 2) = 1$ [2.2] $\frac{2}{3}$

21. $\dfrac{6}{x} + \dfrac{6}{x + 2} = \dfrac{5}{2}$ [6.4] $-\dfrac{6}{5}, 4$

22. $2x + 1 > 5 \ or \ x - 7 \le 3$ [4.2] $\mathbb{R}$, or $(-\infty, \infty)$

23. $5x + 3y = 2,$
$3x + 5y = -2$ [3.2] $(1, -1)$

24. $x + y - z = 0,$
$3x + y + z = 6,$
$x - y + 2z = 5$ [3.4] $(2, -1, 1)$

25. $3\sqrt{x - 1} = 5 - x$ [7.6] 2

26. $x^4 - 29x^2 + 100 = 0$ [8.5] $\pm 2, \pm 5$

27. $\ln e^{4.67} = x$ [9.6] 4.67

28. $5^x = 8$ [9.6] $\dfrac{\ln 8}{\ln 5} \approx 1.2920$

29. $\log(x^2 - 25) - \log(x + 5) = 3$ [9.6] 1005

30. $\log_4 x = -2$ [9.6] $\frac{1}{16}$

31. $7^{2x+3} = 49$ [9.6] $-\frac{1}{2}$

32. $|2x - 1| \le 5$ [4.3] $\{x \mid -2 \le x \le 3\}$, or $[-2, 3]$

33. $7x^2 + 14 = 0$ [8.1] $\pm i\sqrt{2}$

34. $x^2 + 4x = 3$ [8.2] $-2 \pm \sqrt{7}$

35. $y^2 + 3y > 10$ [8.9] $\{y \mid y < -5 \ or \ y > 2\}$, or $(-\infty, -5) \cup (2, \infty)$

36. Let $f(x) = x^2 - 2x$. Find a such that $f(a) = 48$.   [5.4] $-6, 8$

37. If $f(x) = \sqrt{x + 1}$ and $g(x) = \sqrt{x - 2} + 3$, find a such that $f(a) = g(a)$.   [7.6] No solution

Solve.

38. Tomas sailed 12 mi south and then 8 mi east. How far is he from his starting point? Give an exact answer and an approximation to three decimal places.
[7.7] $\sqrt{208}$ mi, or approximately 14.422 mi

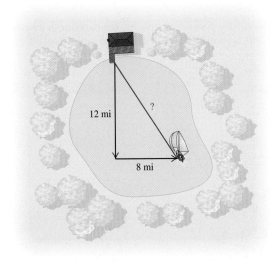

39. A music club offers two types of membership. Limited members pay a fee of $10 a year and can buy CDs for $10 each. Preferred members pay $20 a year and can buy CDs for $7.50 each. For what numbers of annual CD purchases would it be less expensive to be a preferred member? [4.1] More than 4

40. Find three consecutive integers whose sum is 198. [2.3] 65, 66, 67

41. A pentagon with all five sides the same size has a perimeter equal to that of an octagon in which all eight sides are the same size. One side of the pentagon is 2 less than three times one side of the octagon. What is the perimeter of each figure? [3.3] $11\frac{3}{7}$

42. Mark's Natural Foods mixes herbs that cost $2.68 an ounce with herbs that cost $4.60 an ounce to create a seasoning that costs $3.80 an ounce. How many ounces of each herb should be mixed together to make 24 oz of the seasoning? [3.3] $2.68 herb: 10 oz; $4.60 herb: 14 oz

43. An airplane can fly 190 mi with the wind in the same time it takes to fly 160 mi against the wind. The speed of the wind is 30 mph. How fast can the plane fly in still air? [6.5] 350 mph

44. Bianca can tap the sugar maple trees in Southway Park in 21 hr. Delia can tap the trees in 14 hr. How long would it take them, working together, to tap the trees? [6.5] $8\frac{2}{5}$ hr, or 8 hr 24 min

⊡ Answers to Exercises 47–51 can be found on p. A-74.

45. The centripetal force F of an object moving in a circle varies directly as the square of the velocity v and inversely as the radius r of the circle. If $F = 8$ when $v = 1$ and $r = 10$, what is F when $v = 2$ and $r = 16$? [6.8] 20

46. A farmer wants to fence in a rectangular area next to a river. (Note that no fence will be needed along the river.) What is the area of the largest region that can be fenced in with 100 ft of fencing? [8.9] 1250 ft^2

Graph by hand.

47. $3x - y = 6$ ⊡

48. $y = \log_2 x$ ⊡

49. $2x - 3y < -6$ ⊡

50. Graph $f(x) = \sqrt{4 - x}$ and use the graph to estimate the domain and the range of f. ⊡

51. Graph: $f(x) = -2(x - 3)^2 + 1$. ⊡
a) Label the vertex.
b) Draw the axis of symmetry.
c) Find the maximum or minimum value.

52. Solve $V = P - Prt$ for r. [2.3] $r = \dfrac{V - P}{-Pt}$, or $\dfrac{P - V}{Pt}$

53. Solve $I = \dfrac{R}{R + r}$ for R. [6.8] $R = \dfrac{Ir}{1 - I}$

54. Find a linear equation whose graph has a y-intercept of $(0, -3)$ and is parallel to the line whose equation is $3x - y = 6$. [2.5] $y = 3x - 3$

Find the domain of each function.

55. $f(x) = \sqrt{5 - 3x}$ [4.2] $\left\{x \middle| x \le \frac{5}{3}\right\}$, or $\left(-\infty, \frac{5}{3}\right]$

56. $g(x) = \dfrac{x - 4}{x^2 - 2x + 1}$ [5.4] $\{x \mid x$ is a real number *and* $x \ne 1\}$, or $(-\infty, 1) \cup (1, \infty)$

57. Multiply $(8.9 \times 10^{-17})(7.6 \times 10^4)$. Write scientific notation for the answer. [1.4] 6.764×10^{-12}

58. Multiply and simplify: $\sqrt{8x}\,\sqrt{8x^3y}$. [7.3] $8x^2\sqrt{y}$

59. Simplify: $(25x^{4/3}y^{1/2})^{3/2}$.   [7.2] $125x^2y^{3/4}$

60. Divide and simplify:

$$\frac{\sqrt[3]{15x}}{\sqrt[3]{3y^2}}.$$ [7.4] $\frac{\sqrt[3]{5xy}}{y}$

61. Rationalize the denominator:

$$\frac{1 - \sqrt{x}}{1 + \sqrt{x}}.$$ [7.5] $\frac{1 - 2\sqrt{x} + x}{1 - x}$

62. Multiply these complex numbers:

$$(3 + 2i)(4 - 7i).$$ [7.8] $26 - 13i$

63. Write a quadratic equation whose solutions are $5\sqrt{2}$ and $-5\sqrt{2}$.   [8.4] $x^2 - 50 = 0$

64. Simplify: $\log_t t^{37}$.   [9.4] 37

65. Express as a single logarithm:

$$\tfrac{2}{3} \log_a x - \tfrac{1}{2} \log_a y + 5 \log_a z.$$ [9.4] $\log_a \frac{\sqrt[3]{x^2} \cdot z^5}{\sqrt{y}}$

66. Convert to an exponential equation: $\log_a c = 5$.
   [9.3] $a^5 = c$

Find each of the following using a calculator.

67. $\log 5677.2$ [9.3] 3.7541

68. $10^{-3.587}$ [9.3] 0.0003

69. $\ln 5677.2$ [9.5] 8.6442

70. $e^{-3.587}$ [9.5] 0.0277

71. The number of personal computers in Mexico has grown exponentially from 0.12 million in 1985 to 6.0 million in 2000.   [9.6] $k \approx 0.261$; $C(t) = 0.12e^{0.261t}$

  a) Find the exponential growth rate k to three decimal places and write an exponential function describing the number of personal computers in Mexico t years after 1985.

  b) Predict the number of personal computers in Mexico in 2008.   About 48.6 million

72. Find the general term of the sequence   [10.1] $-n^2$
  $-1, -4, -9, -16, -25, \ldots$. Answers may vary.

73. Find the 21st term of the arithmetic sequence
  $19, 12, 5, \ldots$.   [10.2] -121

74. Find the sum of the first 25 terms of the arithmetic series $-1 + 2 + 5 + \cdots$.   [10.2] 875

75. Find the general term of the geometric sequence
  $16, 4, 1, \ldots$.   [10.3] $16\left(\tfrac{1}{4}\right)^{n-1}$

76. Find the 7th term of $(a - 2b)^{10}$.   [10.4] $13,440a^4b^6$

77. Find the sum of the first nine terms of the geometric series $x + 1.5x + 2.25x + \cdots$.   [10.3] $74.88671875x$

78. On Elyse's 9th birthday, her grandmother opened a savings account for her with $100. The account draws 6% interest, compounded annually. If Elyse neither adds to nor withdraws any money from the bank, how much will be in the account on her 18th birthday?   [10.3] $168.95

*For each of the following graphs, **(a)** determine whether the graph is that of a linear, quadratic, exponential, or logarithmic function and **(b)** use the graph to solve* $f(x) = 0$.

79.

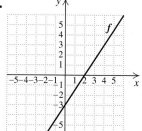

  (a) [2.4] Linear; **(b)** [2.3] 2

80.

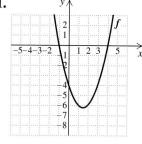

  (a) [9.3] Logarithmic; **(b)** [9.6] -1

81.

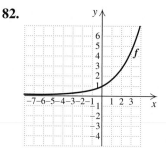

  (a) [8.7] Quadratic; **(b)** [8.1] $-1, 4$

82.

  (a) [9.2] Exponential; **(b)** [9.6] $\varnothing$

For Exercises 83–85, use the following graph showing the amount of money spent on lawn and landscaping services.

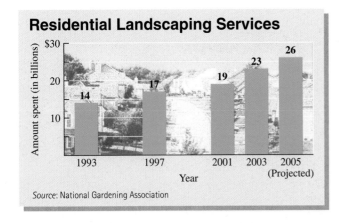

Residential Landscaping Services

Source: National Gardening Association

83. Determine whether a linear, quadratic, or exponential function would best fit the situation.
[2.6], [8.8], [9.7] Answers may vary.

84. Use the data from 1993 and 2005 to fit a linear function to the data. Let *x* represent the number of years since 1990.   [2.6] $f(x) = x + 11$, where *f* is in billions of dollars

85. Use linear regression to fit a function to the data. Let *x* represent the number of years since 1990.
[2.6] $f(x) = 0.9568965517x + 10.42241379$, where *f* is in billions of dollars

Synthesis

Solve.

86. $\dfrac{9}{x} - \dfrac{9}{x + 12} = \dfrac{108}{x^2 + 12x}$ [6.4] All real numbers except 0 and −12

87. $\log_2 (\log_3 x) = 2$ [9.6] 81

88. *y* varies directly as the cube of *x* and *x* is multiplied by 0.5. What is the effect on *y*?   [6.8] *y* gets divided by 8

89. Divide these complex numbers:
$$\dfrac{2\sqrt{6} + 4\sqrt{5}i}{2\sqrt{6} - 4\sqrt{5}i}.$$ [7.8] $-\dfrac{7}{13} + \dfrac{2\sqrt{30}}{13}i$

90. Diaphantos, a famous mathematician, spent $\frac{1}{6}$ of his life as a child, $\frac{1}{12}$ as an adolescent, and $\frac{1}{7}$ as a bachelor. Five years after he was married, he had a son who died 4 years before his father at half his father's final age. How long did Diaphantos live?
[3.5] 84 yr

Appendixes

Parabolas ■ The Distance and Midpoint Formulas ■ Circles

Conic Sections: Parabolas and Circles

Parabolas

When a cone is cut as shown in the figure at left, the *conic section* formed is a **parabola**. Parabolas have many applications in electricity, mechanics, and optics. A cross section of a contact lens or satellite dish is a parabola, and arches that support certain bridges are parabolas.

Equation of a Parabola

A parabola with a vertical axis of symmetry opens upward or downward and has an equation that can be written in the form

$$y = ax^2 + bx + c.$$

A parabola with a horizontal axis of symmetry opens to the right or left and has an equation that can be written in the form

$$x = ay^2 + by + c.$$

EXAMPLE 1   Graph: $y = x^2 - 4x + 9$.

Solution   To locate the vertex, we can use either of two approaches. One way is to complete the square:

$$y = (x^2 - 4x) + 9 \qquad \text{Note that half of } -4 \text{ is } -2, \text{ and } (-2)^2 = 4.$$
$$= (x^2 - 4x + 4 - 4) + 9 \qquad \text{Adding and subtracting 4}$$
$$= (x^2 - 4x + 4) + (-4 + 9) \qquad \text{Regrouping}$$
$$= (x - 2)^2 + 5. \qquad \text{Factoring and simplifying}$$

The vertex is $(2, 5)$.

A second way to find the vertex is to recall that the x-coordinate of the vertex of the parabola given by $y = ax^2 + bx + c$ is $-b/(2a)$:

$$x = -\frac{b}{2a} = -\frac{-4}{2(1)} = 2.$$

To find the y-coordinate of the vertex, we substitute 2 for x:

$$y = x^2 - 4x + 9 = 2^2 - 4(2) + 9 = 5.$$

Either way, the vertex is $(2, 5)$. Next, we calculate and plot some points on each side of the vertex. Since the x^2-coefficient, 1, is positive, the graph opens upward.

x	y	
2	5	← Vertex
0	9	← y-intercept
1	6	
3	6	
4	9	

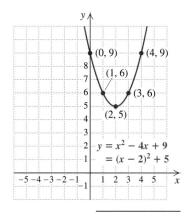

To Graph an Equation of the Form $y = ax^2 + bx + c$

1. Find the vertex (h, k) either by completing the square to find an equivalent equation $y = a(x - h)^2 + k$, or by using $-b/(2a)$ to find the x-coordinate and substituting to find the y-coordinate.

2. Choose other values for x on each side of the vertex, and compute the corresponding y-values.

3. The graph opens upward for $a > 0$ and downward for $a < 0$.

Equations of the form $x = ay^2 + by + c$ represent horizontal parabolas. These parabolas open to the right for $a > 0$, open to the left for $a < 0$, and have axes of symmetry parallel to the x-axis.

EXAMPLE 2   Graph: $x = y^2 - 4y + 9$.

Solution   This equation is like that in Example 1 except that x and y are interchanged. The vertex is $(5, 2)$ instead of $(2, 5)$. To find ordered pairs, we choose values for y on each side of the vertex. Then we compute values for x. Note that the x- and y-values of the table in Example 1 are now switched. You should confirm that, by completing the square, we get $x = (y - 2)^2 + 5$.

x	y	
5	2	← Vertex
9	0	← x-intercept
6	1	
6	3	
9	4	

(1) Choose these values for y.
(2) Compute these values for x.

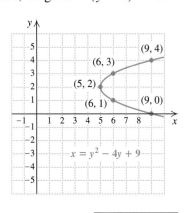

TEACHING TIP

Remind students that a parabola with a horizontal axis is not a function.

> **To Graph an Equation of the Form $x = ay^2 + by + c$**
>
> 1. Find the vertex (h, k) either by completing the square to find an equivalent equation
>
> $$x = a(y - k)^2 + h,$$
>
> or by using $-b/(2a)$ to find the y-coordinate and substituting to find the x-coordinate.
> 2. Choose other values for y that are above and below the vertex, and compute the corresponding x-values.
> 3. The graph opens to the right if $a > 0$ and to the left if $a < 0$.

EXAMPLE 3 Graph: $x = -2y^2 + 10y - 7$.

Solution We use the method of completing the square:

$$\begin{aligned} x &= -2y^2 + 10y - 7 \\ &= -2(y^2 - 5y \qquad) - 7 \\ &= -2\left(y^2 - 5y + \tfrac{25}{4}\right) - 7 - (-2)\tfrac{25}{4} && \textbf{$\tfrac{1}{2}(-5) = \tfrac{-5}{2}$; $\left(\tfrac{-5}{2}\right)^2 = \tfrac{25}{4}$; we add and subtract $(-2)\tfrac{25}{4}$.} \\ &= -2\left(y - \tfrac{5}{2}\right)^2 + \tfrac{11}{2}. && \textbf{Factoring and simplifying} \end{aligned}$$

The vertex is $\left(\frac{11}{2}, \frac{5}{2}\right)$.

For practice, we also find the vertex by first computing its y-coordinate, $-b/(2a)$, and then substituting to find the x-coordinate:

$$y = -\frac{b}{2a} = -\frac{10}{2(-2)} = \frac{5}{2}$$

$$x = -2y^2 + 10y - 7 = -2\left(\tfrac{5}{2}\right)^2 + 10\left(\tfrac{5}{2}\right) - 7 = \tfrac{11}{2}.$$

To find ordered pairs, we choose values for y on each side of the vertex and then compute values for x. A table is shown below, together with the graph. The graph opens to the left because the y^2-coefficient, -2, is negative.

x	y	
$\frac{11}{2}$	$\frac{5}{2}$	← **Vertex**
-7	0	← **x-intercept**
5	2	
5	3	
1	1	
1	4	
-7	5	

(1) Choose these values for y.
(2) Compute these values for x.

TEACHING TIP

You may want to point out that for $x = a(y - k)^2 + h$, the vertex is still (h, k), but the hyperbola is horizontal.

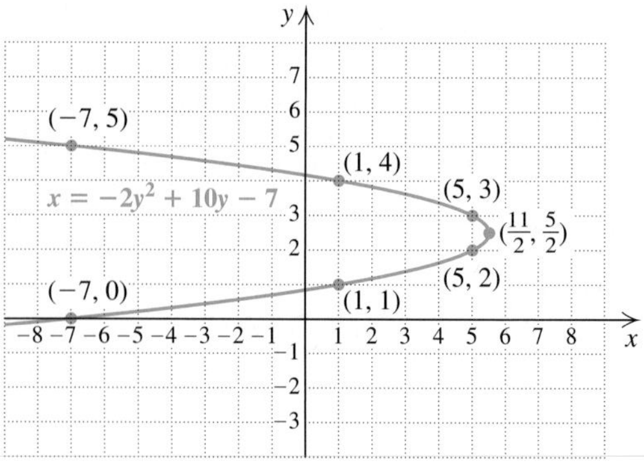

The Distance and Midpoint Formulas

Suppose that two points are on a horizontal line, and thus have the same second coordinate. We can find the distance between them by subtracting their first coordinates. This difference may be negative, depending on the order in which we subtract. So, to make sure we get a positive number, we take the absolute value of this difference. The distance between the points (x_1, y_1) and (x_2, y_1) on a horizontal line is thus $|x_2 - x_1|$. Similarly, the distance between the points (x_2, y_1) and (x_2, y_2) on a vertical line is $|y_2 - y_1|$.

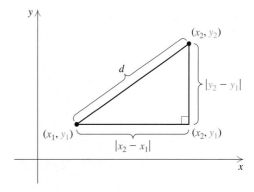

Now consider *any* two points (x_1, y_1) and (x_2, y_2). If $x_1 \neq x_2$ and $y_1 \neq y_2$, these points, along with the point (x_2, y_1), describe a right triangle. The lengths of the legs are $|x_2 - x_1|$ and $|y_2 - y_1|$. We find d, the length of the hypotenuse, by using the Pythagorean theorem:

$$d^2 = |x_2 - x_1|^2 + |y_2 - y_1|^2.$$

Since the square of a number is the same as the square of its opposite, we don't really need these absolute-value signs. Thus,

$$d^2 = (x_2 - x_1)^2 + (y_2 - y_1)^2.$$

Taking the principal square root, we obtain the distance between two points.

The Distance Formula

The distance d between any two points (x_1, y_1) and (x_2, y_2) is given by

$$d = \sqrt{(x_2 - x_1)^2 + (y_2 - y_1)^2}.$$

EXAMPLE 4 Find the distance between $(5, -1)$ and $(-4, 6)$. Find an exact answer and an approximation to three decimal places.

Solution   We substitute into the distance formula:

$$d = \sqrt{(-4 - 5)^2 + [6 - (-1)]^2} \qquad \text{Substituting}$$
$$= \sqrt{(-9)^2 + 7^2}$$
$$= \sqrt{130} \qquad\qquad \text{This is exact.}$$
$$\approx 11.402. \qquad\qquad \text{Using a calculator for an approximation}$$

The distance formula is needed to develop the formula for a circle, which follows, and to verify certain properties of conic sections. It is also needed to verify a formula for the coordinates of the *midpoint* of a segment connecting two points. We state the midpoint formula and leave its proof to the exercises. Note that although the distance formula involves both subtraction and addition, the midpoint formula uses only addition.

The Midpoint Formula

If the endpoints of a segment are (x_1, y_1) and (x_2, y_2), then the coordinates of the midpoint are

$$\left(\frac{x_1 + x_2}{2}, \frac{y_1 + y_2}{2} \right).$$

(To locate the midpoint, average the x-coordinates and average the y-coordinates.)

EXAMPLE 5 Find the midpoint of the segment with endpoints $(-2, 3)$ and $(4, -6)$.

Solution   Using the midpoint formula, we obtain

$$\left(\frac{-2 + 4}{2}, \frac{3 + (-6)}{2} \right), \quad \text{or} \quad \left(\frac{2}{2}, \frac{-3}{2} \right), \quad \text{or} \quad \left(1, -\frac{3}{2} \right).$$

Circles

One conic section, the **circle**, is a set of points in a plane that are a fixed distance r, called the **radius** (plural, **radii**), from a fixed point (h, k), called the **center**. Note that the word radius can mean either any segment connecting a point on a circle and the center of the circle or the length of such a segment.

If a point (x, y) is on the circle, then by the definition of a circle and the distance formula, it follows that

$$r = \sqrt{(x - h)^2 + (y - k)^2}.$$

Squaring both sides gives the equation of a circle in standard form.

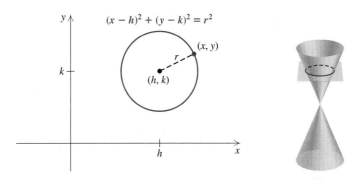

Equation of a Circle

The equation of a circle, centered at (h, k), with radius r, is given by

$$(x - h)^2 + (y - k)^2 = r^2.$$

Note that when $h = 0$ and $k = 0$, the circle is centered at the origin. Otherwise, the circle is translated $|h|$ units horizontally and $|k|$ units vertically.

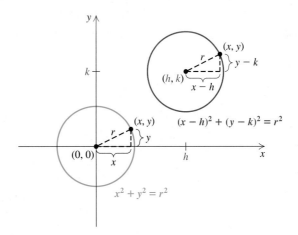

EXAMPLE 6   Find an equation of the circle having center $(4, 5)$ and radius 6.

Solution   Using the standard form, we obtain

$$(x - 4)^2 + (y - 5)^2 = 6^2, \qquad \text{Using } (x - h)^2 + (y - k)^2 = r^2$$

or

$$(x - 4)^2 + (y - 5)^2 = 36.$$

EXAMPLE 7 Find the center and the radius and then graph each circle.

a) $(x - 2)^2 + (y + 3)^2 = 4^2$
b) $x^2 + y^2 + 8x - 2y + 15 = 0$

Solution

a) We write standard form:

$$(x - 2)^2 + [y - (-3)]^2 = 4^2.$$

The center is $(2, -3)$ and the radius is 4. To graph, we can use a compass or plot the points $(2, 1)$, $(2, -7)$, $(-2, -3)$, and $(6, -3)$, which are 4 units above, below, left, and right of $(2, -3)$, respectively, and then sketch a circle by hand.

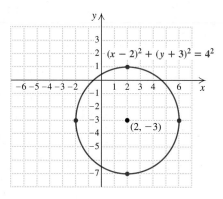

b) To write the equation $x^2 + y^2 + 8x - 2y + 15 = 0$ in standard form, we complete the square twice, once with $x^2 + 8x$ and once with $y^2 - 2y$:

$$x^2 + y^2 + 8x - 2y + 15 = 0$$

$$x^2 + 8x \qquad + y^2 - 2y \qquad = -15 \qquad \text{Grouping the } x\text{-terms and the } y\text{-terms; adding } -15 \text{ to both sides}$$

$$x^2 + 8x + 16 + y^2 - 2y + 1 = -15 + 16 + 1 \qquad \text{Adding } \left(\frac{8}{2}\right)^2, \text{ or } 16, \text{ and } \left(-\frac{2}{2}\right)^2, \text{ or } 1, \text{ to both sides}$$

$$(x + 4)^2 + (y - 1)^2 = 2 \qquad \text{Factoring}$$

$$[x - (-4)]^2 + (y - 1)^2 = \left(\sqrt{2}\right)^2. \qquad \text{Writing standard form}$$

The center is $(-4, 1)$ and the radius is $\sqrt{2}$.

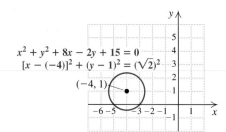

1.

2.

3.

4.

5.

6.

7.

8.

9.

10.

11.

12.

13.

14.

Graphing Circles

Because most graphing calculators can graph only functions, graphing the equation of a circle usually requires two steps:

1. Solve the equation for y. The result will include a $\pm$ sign in front of a radical.

2. Graph two functions, one for the $+$ sign and the other for the $-$ sign, on the same set of axes.

For example, to graph $(x - 3)^2 + (y + 1)^2 = 16$, solve for $y + 1$ and then y:

$$(y + 1)^2 = 16 - (x - 3)^2$$
$$y + 1 = \pm\sqrt{16 - (x - 3)^2},$$
$$y = -1 \pm \sqrt{16 - (x - 3)^2},$$

or
$$y_1 = -1 + \sqrt{16 - (x - 3)^2}$$
and
$$y_2 = -1 - \sqrt{16 - (x - 3)^2}.$$

When both functions are graphed (in a "squared" window to eliminate distortion), the result is as follows.

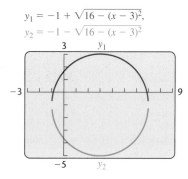

$$y_1 = -1 + \sqrt{16 - (x - 3)^2},$$
$$y_2 = -1 - \sqrt{16 - (x - 3)^2}$$

Circles can also be drawn using the CIRCLE option of the DRAW menu, but such graphs cannot be traced.

A
Exercise Set

FOR EXTRA HELP

MyMathLab

Graph. Be sure to label each vertex.

1. $y = -x^2$

2. $y = 2x^2$

3. $y = -x^2 + 4x - 5$

4. $x = 4 - 3y - y^2$

5. $x = y^2 - 4y + 1$

6. $y = x^2 + 2x + 3$

7. $x = y^2 + 1$

8. $x = 2y^2$

9. $x = -\frac{1}{2}y^2$

10. $x = y^2 - 1$

11. $x = -y^2 - 4y$

12. $x = y^2 + y - 6$

13. $x = 8 - y - y^2$

14. $y = x^2 + 2x + 1$

15. $y = x^2 - 2x + 1$ ⊡

16. $y = -\frac{1}{2}x^2$ ⊡

17. $x = -y^2 + 2y - 1$ ⊡

18. $x = -y^2 - 2y + 3$ ⊡

⊡ Answers to Exercises 15–18 can be found on p. A-74.

19. $x = -2y^2 - 4y + 1$ ▫   **20.** $x = 2y^2 + 4y - 1$ ▫

Find the distance between each pair of points. Where appropriate, find an approximation to three decimal places.

21. $(1, 6)$ and $(5, 9)$ 5

22. $(1, 10)$ and $(7, 2)$ 10

23. $(0, -7)$ and $(3, -4)$ $\sqrt{18} \approx 4.243$

24. $(6, 2)$ and $(6, -8)$ 10

25. $(-4, 4)$ and $(6, -6)$ $\sqrt{200} \approx 14.142$

26. $(5, 21)$ and $(-3, 1)$ $\sqrt{464} \approx 21.541$

27. $(8.6, -3.4)$ and $(-9.2, -3.4)$ 17.8

28. $(5.9, 2)$ and $(3.7, -7.7)$ $\sqrt{98.93} \approx 9.946$

29. $\left(\frac{5}{7}, \frac{1}{14}\right)$ and $\left(\frac{1}{7}, \frac{11}{14}\right)$ $\dfrac{\sqrt{41}}{7} \approx 0.915$

30. $\left(0, \sqrt{7}\right)$ and $\left(\sqrt{6}, 0\right)$ $\sqrt{13} \approx 3.606$

31. $\left(-\sqrt{6}, \sqrt{2}\right)$ and $(0, 0)$ $\sqrt{8} \approx 2.828$

32. $\left(\sqrt{5}, -\sqrt{3}\right)$ and $(0, 0)$ $\sqrt{8} \approx 2.828$

33. $\left(\sqrt{2}, -\sqrt{3}\right)$ and $\left(-\sqrt{7}, \sqrt{5}\right)$ $\sqrt{17 + 2\sqrt{14} + 2\sqrt{15}} \approx 5.677$

34. $\left(\sqrt{8}, \sqrt{3}\right)$ and $\left(-\sqrt{5}, -\sqrt{6}\right)$ $\sqrt{22 + 2\sqrt{40} + 2\sqrt{18}} \approx 6.568$

35. $(0, 0)$ and (s, t) $\sqrt{s^2 + t^2}$

36. (p, q) and $(0, 0)$ $\sqrt{p^2 + q^2}$

Find the midpoint of each segment with the given endpoints.

37. $(-7, 6)$ and $(9, 2)$ $(1, 4)$

38. $(6, 7)$ and $(7, -9)$ $\left(\frac{13}{2}, -1\right)$

39. $(2, -1)$ and $(5, 8)$ $\left(\frac{7}{2}, \frac{7}{2}\right)$

40. $(-1, 2)$ and $(1, -3)$ $\left(0, -\frac{1}{2}\right)$

41. $(-8, -5)$ and $(6, -1)$ $(-1, -3)$

42. $(8, -2)$ and $(-3, 4)$ $\left(\frac{5}{2}, 1\right)$

43. $(-3.4, 8.1)$ and $(2.9, -8.7)$ $(-0.25, -0.3)$

44. $(4.1, 6.9)$ and $(5.2, -6.9)$ $(4.65, 0)$

45. $\left(\frac{1}{6}, -\frac{3}{4}\right)$ and $\left(-\frac{1}{3}, \frac{5}{6}\right)$ $\left(-\frac{1}{12}, \frac{1}{24}\right)$

46. $\left(-\frac{4}{5}, -\frac{2}{3}\right)$ and $\left(\frac{1}{8}, \frac{3}{4}\right)$ $\left(-\frac{27}{80}, \frac{1}{24}\right)$

47. $\left(\sqrt{2}, -1\right)$ and $\left(\sqrt{3}, 4\right)$ $\left(\dfrac{\sqrt{2} + \sqrt{3}}{2}, \dfrac{3}{2}\right)$

48. $\left(9, 2\sqrt{3}\right)$ and $\left(-4, 5\sqrt{3}\right)$ $\left(\dfrac{5}{2}, \dfrac{7\sqrt{3}}{2}\right)$

Find an equation of the circle satisfying the given conditions.

49. Center $(0, 0)$, radius 6 $x^2 + y^2 = 36$

50. Center $(0, 0)$, radius 5 $x^2 + y^2 = 25$

51. Center $(7, 3)$, radius $\sqrt{5}$ $(x - 7)^2 + (y - 3)^2 = 5$

52. Center $(5, 6)$, radius $\sqrt{2}$ $(x - 5)^2 + (y - 6)^2 = 2$

53. Center $(-4, 3)$, radius $4\sqrt{3}$ $(x + 4)^2 + (y - 3)^2 = 48$

54. Center $(-2, 7)$, radius $2\sqrt{5}$ $(x + 2)^2 + (y - 7)^2 = 20$

55. Center $(-7, -2)$, radius $5\sqrt{2}$ $(x + 7)^2 + (y + 2)^2 = 50$

56. Center $(-5, -8)$, radius $3\sqrt{2}$ $(x + 5)^2 + (y + 8)^2 = 18$

57. Center $(0, 0)$, passing through $(-3, 4)$ $x^2 + y^2 = 25$

58. Center $(3, -2)$, passing through $(11, -2)$ $(x - 3)^2 + (y + 2)^2 = 64$

59. Center $(-4, 1)$, passing through $(-2, 5)$ $(x + 4)^2 + (y - 1)^2 = 20$

60. Center $(-1, -3)$, passing through $(-4, 2)$ $(x + 1)^2 + (y + 3)^2 = 34$

Find the center and the radius of each circle. Then graph the circle.

61. $x^2 + y^2 = 49$ ▫   **62.** $x^2 + y^2 = 36$ ▫

63. $(x + 1)^2 + (y + 3)^2 = 4$ ▫

64. $(x - 2)^2 + (y + 3)^2 = 1$ ▫

65. $(x - 4)^2 + (y + 3)^2 = 10$ ▫

66. $(x + 5)^2 + (y - 1)^2 = 15$ ▫

67. $x^2 + y^2 = 7$ ▫

68. $x^2 + y^2 = 8$ ▫

69. $(x - 5)^2 + y^2 = \frac{1}{4}$ ▫

70. $x^2 + (y - 1)^2 = \frac{1}{25}$ ▫

71. $x^2 + y^2 + 8x - 6y - 15 = 0$ ▫

72. $x^2 + y^2 + 6x - 4y - 15 = 0$ ▫

73. $x^2 + y^2 - 8x + 2y + 13 = 0$ ▫

74. $x^2 + y^2 + 6x + 4y + 12 = 0$ ▫

75. $x^2 + y^2 + 10y - 75 = 0$ ▫

76. $x^2 + y^2 - 8x - 84 = 0$ ▫

77. $x^2 + y^2 + 7x - 3y - 10 = 0$ ▫

78. $x^2 + y^2 - 21x - 33y + 17 = 0$ ▫

79. $36x^2 + 36y^2 = 1$ ▫

80. $4x^2 + 4y^2 = 1$ ▫

Graph using a graphing calculator.

81. $x^2 + y^2 - 16 = 0$ ▫

82. $4x^2 + 4y^2 = 100$ ▫

83. $x^2 + y^2 + 14x - 16y + 54 = 0$ ▫

84. $x^2 + y^2 - 10x - 11 = 0$ ▫

Synthesis

Find an equation of a circle satisfying the given conditions.

85. Center $(3, -5)$ and tangent to (touching at one point) the y-axis $(x - 3)^2 + (y + 5)^2 = 9$

▫ Answers to Exercises 19, 20, and 61–84 can be found on pp. A-74 and A-75.

86. Center $(-7, -4)$ and tangent to the x-axis
$(x + 7)^2 + (y + 4)^2 = 16$

87. The endpoints of a diameter are $(7, 3)$ and $(-1, -3)$.
$(x - 3)^2 + y^2 = 25$

88. Center $(-3, 5)$ with a circumference of 8π units
$(x + 3)^2 + (y - 5)^2 = 16$

89. Find the point on the y-axis that is equidistant from $(2, 10)$ and $(6, 2)$. $(0, 4)$

90. Find the point on the x-axis that is equidistant from $(-1, 3)$ and $(-8, -4)$. $(-5, 0)$

91. *Snowboarding.* Each side edge of the Burton® Twin 53 snowboard is an arc of a circle with a "running length" of 1150 mm and a "sidecut depth" of 19.5 mm (see the figure below).

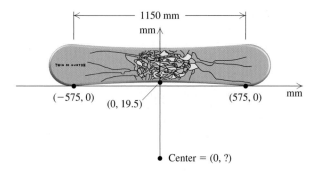

a) Using the coordinates shown, locate the center of the circle. (*Hint*: Equate distances.) $(0, -8467.8)$

b) What radius is used for the edge of the board?   8487.3 mm

92. *Snowboarding.* The Burton® Twin 44 snowboard has a running length of 1070 mm and a sidecut depth of 17.5 mm (see Exercise 91). What radius is used for the edge of this snowboard?   8186.6 mm

93. *Skiing.* The Rossignol® Cut 10.4 ski, when lying flat and viewed from above, has edges that are arcs of a circle. (Actually, each edge is made of two arcs of slightly different radii. The arc for the rear half of the ski edge has a slightly larger radius.)

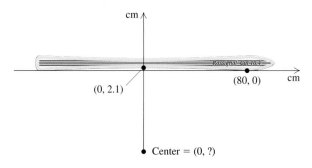

a) Using the coordinates shown, locate the center of the circle. (*Hint*: Equate distances.) $(0, -1522.8)$

b) What radius is used for the arc passing through $(0, 2.1)$ and $(80, 0)$?   1524.9 cm

94. *Doorway Construction.* Ace Carpentry needs to cut an arch for the top of an entranceway. The arch needs to be 8 ft wide and 2 ft high. To draw the arch, the carpenters will use a stretched string with chalk attached at an end as a compass.

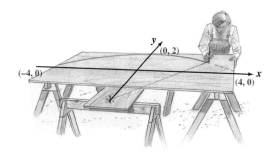

a) Using a coordinate system, locate the center of the circle. $(0, -3)$

b) What radius should the carpenters use to draw the arch?   5 ft

95. *Archaeology.* During an archaeological dig, Martina finds the bowl fragment shown below. What was the original diameter of the bowl?   29 cm

96. *Ferris Wheel Design.* A ferris wheel has a radius of 24.3 ft. Assuming that the center is 30.6 ft off the ground and that the origin is below the center, as in the following figure, find an equation of the circle.
$x^2 + (y - 30.6)^2 = 590.49$

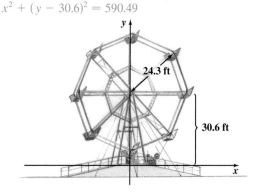

97. Use a graph of the equation $x = y^2 - y - 6$ to approximate to the nearest tenth the solutions of each of the following equations. $\quad$ $-2.4, 3.4$
a) $y^2 - y - 6 = 2$ (*Hint*: Graph $x = 2$ on the same set of axes as the graph of $x = y^2 - y - 6$.)
b) $y^2 - y - 6 = -3$ $\quad -1.3, 2.3$

98. *Power of a Motor.* The horsepower of a certain kind of engine is given by the formula

$$H = \frac{D^2 N}{2.5},$$

where N is the number of cylinders and D is the diameter, in inches, of each piston. Graph this equation, assuming that $N = 6$ (a six-cylinder engine). Let D run from 2.5 to 8.

99. Prove the midpoint formula by showing that
i) the distance from (x_1, y_1) to

$$\left(\frac{x_1 + x_2}{2}, \frac{y_1 + y_2}{2} \right)$$

equals the distance from (x_2, y_2) to

$$\left(\frac{x_1 + x_2}{2}, \frac{y_1 + y_2}{2} \right);$$

and
ii) the points

$$(x_1, y_1), \left(\frac{x_1 + x_2}{2}, \frac{y_1 + y_2}{2} \right),$$

and

$$(x_2, y_2)$$

lie on the same line. ⊡

⊡ Answer to Exercise 99 can be found on p. A-75.

| B | Ellipses ■ Hyperbolas ■ Hyperbolas (Nonstandard Form) ■ Classifying Graphs of Equations |

Conic Sections: Ellipses and Hyperbolas

When a cone is cut at an angle, as shown below, the conic section formed is an *ellipse*. To draw an ellipse, stick two tacks in a piece of cardboard. Then tie a string to the tacks, place a pencil as shown, and draw an oval by moving the pencil while keeping the string taut.

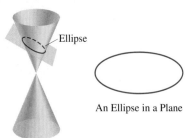

Ellipse

An Ellipse in a Plane

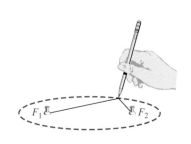

Ellipses

An **ellipse** is defined as the set of all points in a plane for which the *sum* of the distances from two fixed points F_1 and F_2 is constant. The points F_1 and F_2 are called **foci** (pronounced fō-sī), the plural of focus. In the figure above, the tacks are at the foci and the length of the string is the constant sum of the distances.

The midpoint of the segment F_1F_2 is the **center**. The equation of an ellipse is as follows. Its derivation is left to the exercises.

Equation of an Ellipse Centered at the Origin

The equation of an ellipse centered at the origin and symmetric with respect to both axes is

$$\frac{x^2}{a^2} + \frac{y^2}{b^2} = 1, \quad a, b > 0. \qquad \text{(Standard form)}$$

To graph an ellipse centered at the origin, it helps to first find the intercepts. If we replace x with 0, we can find the y-intercepts:

$$\frac{0^2}{a^2} + \frac{y^2}{b^2} = 1$$

$$\frac{y^2}{b^2} = 1$$

$$y^2 = b^2 \quad \text{or} \quad y = \pm b.$$

Thus the y-intercepts are $(0, b)$ and $(0, -b)$. Similarly, the x-intercepts are $(a, 0)$ and $(-a, 0)$. If $a > b$, the ellipse is said to be horizontal and $(-a, 0)$ and $(a, 0)$ are referred to as the **vertices** (singular, **vertex**). If $b > a$, the ellipse is said to be vertical and $(0, -b)$ and $(0, b)$ are then the vertices.

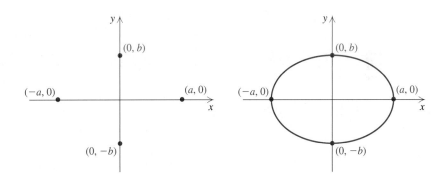

Plotting these four points and drawing an oval-shaped curve, we get a graph of the ellipse. If a more precise graph is desired, we can plot more points.

Using *a* and *b* to Graph an Ellipse

For the ellipse

$$\frac{x^2}{a^2} + \frac{y^2}{b^2} = 1,$$

the x-intercepts are $(-a, 0)$ and $(a, 0)$. The y-intercepts are $(0, -b)$ and $(0, b)$.

EXAMPLE 1 Graph the ellipse

$$\frac{x^2}{4} + \frac{y^2}{9} = 1.$$

Solution   Note that

$$\frac{x^2}{4} + \frac{y^2}{9} = \frac{x^2}{2^2} + \frac{y^2}{3^2}.$$ **Identifying a and b. Since $b > a$, the ellipse is vertical.**

Thus the x-intercepts are $(-2, 0)$ and $(2, 0)$, and the y-intercepts are $(0, -3)$ and $(0, 3)$. We plot these points and connect them with an oval-shaped curve. To plot some other points, we let $x = 1$ and solve for y:

$$\frac{1^2}{4} + \frac{y^2}{9} = 1$$

$$36\left(\frac{1}{4} + \frac{y^2}{9}\right) = 36 \cdot 1$$

$$36 \cdot \frac{1}{4} + 36 \cdot \frac{y^2}{9} = 36$$

$$9 + 4y^2 = 36$$

$$4y^2 = 27$$

$$y^2 = \frac{27}{4}$$

$$y = \pm\sqrt{\frac{27}{4}}$$

$$y \approx \pm 2.6.$$

Thus, $(1, 2.6)$ and $(1, -2.6)$ can also be used to draw the graph. Similarly, the points $(-1, 2.6)$ and $(-1, -2.6)$ can also be computed and plotted.

EXAMPLE 2   Graph: $4x^2 + 25y^2 = 100.$

Solution   To write the equation in standard form, we divide both sides by 100 to get 1 on the right side:

$$\frac{4x^2 + 25y^2}{100} = \frac{100}{100}$$ **Dividing by 100 to get 1 on the right side**

$$\frac{4x^2}{100} + \frac{25y^2}{100} = 1$$

$$\frac{x^2}{25} + \frac{y^2}{4} = 1$$ **Simplifying**

$$\frac{x^2}{5^2} + \frac{y^2}{2^2} = 1.$$ $a = 5, b = 2$

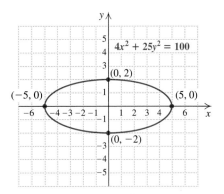

The x-intercepts are $(-5, 0)$ and $(5, 0)$, and the y-intercepts are $(0, -2)$ and $(0, 2)$. We plot the intercepts and connect them with an oval-shaped curve. Other points can also be computed and plotted.

Ellipses have many applications. Communications satellites move in elliptical orbits with the earth as a focus while the earth itself follows an elliptical path around the sun. A medical instrument, the lithotripter, uses shock waves originating at one focus to crush a kidney stone located at the other focus.

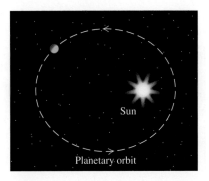

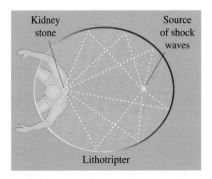

In some buildings, an ellipsoidal ceiling creates a "whispering gallery" in which a person at one focus can whisper and still be heard clearly at the other focus. This happens because sound waves coming from one focus are all reflected to the other focus. Similarly, light waves bouncing off an ellipsoidal mirror are used in a dentist's or surgeon's reflector light. The light source is located at one focus while the patient's mouth is at the other.

Hyperbolas

A **hyperbola** looks like a pair of parabolas, but the shapes are actually different. A hyperbola has two **vertices** and the line through the vertices is known as an **axis**. The point halfway between the vertices is called the **center**. The two curves that comprise a hyperbola are called **branches**.

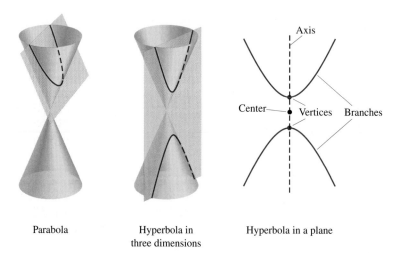

Parabola   Hyperbola in   Hyperbola in a plane
                  three dimensions

Equation of a Hyperbola Centered at the Origin

Hyperbolas with their centers at the origin have equations as follows:

$$\frac{x^2}{a^2} - \frac{y^2}{b^2} = 1 \qquad \text{(Axis horizontal)};$$

$$\frac{y^2}{b^2} - \frac{x^2}{a^2} = 1 \qquad \text{(Axis vertical)}.$$

Note that both equations have a 1 on the right-hand side and a subtraction symbol between the terms. For the discussion that follows, we assume a, $b > 0$.

To graph a hyperbola, it helps to begin by graphing two lines called **asymptotes**. Although the asymptotes themselves are not part of the graph, they serve as guidelines for an accurate sketch.

Asymptotes of a Hyperbola

For hyperbolas with equations as shown below, the asymptotes are the lines

$$y = \frac{b}{a}x \quad \text{and} \quad y = -\frac{b}{a}x.$$

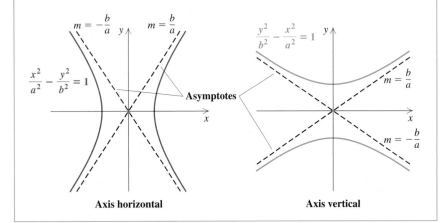

Axis horizontal   Axis vertical

As a hyperbola gets farther away from the origin, it gets closer and closer to its asymptotes. The larger $|x|$ gets, the closer the graph gets to an asymptote. The asymptotes act to "constrain" the graph of a hyperbola. Parabolas are *not* constrained by any asymptotes.

For hyperbolas, a and b can be used to determine the base and the height of a rectangle that can be used as an aid in sketching asymptotes and locating vertices. This is illustrated in the following example.

EXAMPLE 3   Graph: $\dfrac{x^2}{4} - \dfrac{y^2}{9} = 1$.

Solution   Note that

$$\frac{x^2}{4} - \frac{y^2}{9} = \frac{x^2}{2^2} - \frac{y^2}{3^2}, \qquad \text{Identifying } a \text{ and } b$$

so $a = 2$ and $b = 3$. The asymptotes are thus

$$y = \frac{3}{2}x \quad \text{and} \quad y = -\frac{3}{2}x.$$

To help us sketch asymptotes and locate vertices, we use a and b—in this case, 2 and 3—to form the pairs $(-2, 3)$, $(2, 3)$, $(2, -3)$, and $(-2, -3)$. We plot these pairs and lightly sketch a rectangle. The asymptotes pass through the corners and, since this is a horizontal hyperbola, the vertices are where the rectangle intersects the x-axis. Finally, we draw the hyperbola, as shown on the following page.

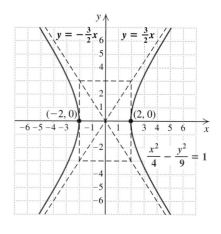

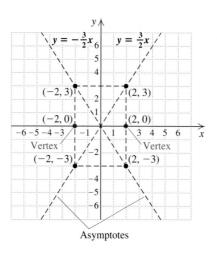

Asymptotes

EXAMPLE 4   Graph: $\dfrac{y^2}{36} - \dfrac{x^2}{4} = 1$.

Solution   Note that

$$\frac{y^2}{36} - \frac{x^2}{4} = \frac{y^2}{6^2} - \frac{x^2}{2^2} = 1.$$

> **Whether the hyperbola is horizontal or vertical is determined by the nonnegative term. Here there is a y in this term, so the hyperbola is vertical.**

Using ± 2 as x-coordinates and ± 6 as y-coordinates, we plot $(2, 6)$, $(2, -6)$, $(-2, 6)$, and $(-2, -6)$, and lightly sketch a rectangle through them. The asymptotes pass through the corners (see the figure on the left below). Since the hyperbola is vertical, its vertices are $(0, 6)$ and $(0, -6)$. Finally, we draw curves through the vertices toward the asymptotes, as shown below.

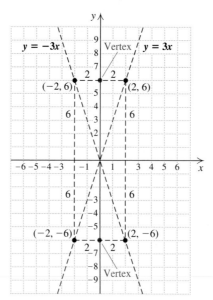

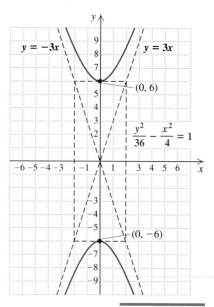

Hyperbolas (Nonstandard Form)

The equations for hyperbolas just examined are the standard ones, but there are other hyperbolas. We consider some of them.

Equation of a Hyperbola in Nonstandard Form

Hyperbolas having the x- and y-axes as asymptotes have equations as follows:

$$xy = c, \quad \text{where } c \text{ is a nonzero constant.}$$

EXAMPLE 5   Graph: $xy = -8$.

Solution   We first solve for y:

$$y = -\frac{8}{x}.$$ **Dividing both sides by x. Note that $x \neq 0$.**

Next, we find some solutions, keeping the results in a table. Note that x cannot be 0 and that for large values of $|x|$, y will be close to 0. Thus the x- and y-axes serve as asymptotes. We plot the points and draw two curves.

x	y
2	−4
−2	4
4	−2
−4	2
1	−8
−1	8
8	−1
−8	1

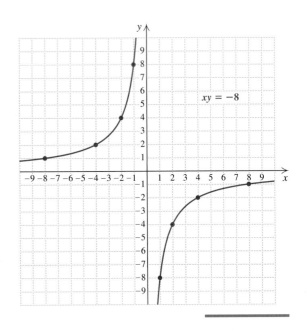

Hyperbolas have many applications. A jet breaking the sound barrier creates a sonic boom with a wave front the shape of a cone. The intersection of the cone with the ground is one branch of a hyperbola. Some comets travel in hyperbolic orbits, and a cross section of many lenses is hyperbolic in shape.

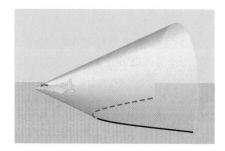

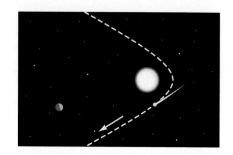

Graphing Ellipses and Hyperbolas

Graphing an ellipse or a hyperbola on a graphing calculator is much like graphing a circle.

To graph the ellipse given by the equation $4x^2 + 25y^2 = 100$, we first solve for y:

$$4x^2 + 25y^2 = 100$$
$$25y^2 = 100 - 4x^2$$
$$y^2 = 4 - \frac{4}{25}x^2$$
$$y = \pm\sqrt{4 - \frac{4}{25}x^2}.$$

Then, using a squared window, we graph.

$$y_1 = -\sqrt{4 - \frac{4}{25}x^2}, \quad y_2 = \sqrt{4 - \frac{4}{25}x^2}$$

To graph the hyperbola given by the equation

$$\frac{x^2}{25} - \frac{y^2}{49} = 1,$$

we again solve for y, which gives the equations

$$y_1 = \frac{\sqrt{49x^2 - 1225}}{5}$$
$$= \frac{7}{5}\sqrt{x^2 - 25}$$

and $y_2 = \dfrac{-\sqrt{49x^2 - 1225}}{5}$

$\qquad = -\dfrac{7}{5}\sqrt{x^2 - 25},$

or $y_2 = -y_1.$

When the two pieces are drawn on the same squared window, the result is as shown. Note the problem that the graphing calculator has at points where the graph is nearly vertical.

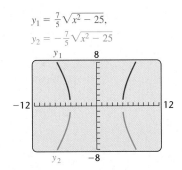

$y_1 = \frac{7}{5}\sqrt{x^2 - 25},$
$y_2 = -\frac{7}{5}\sqrt{x^2 - 25}$

Classifying Graphs of Equations

We summarize the equations and the graphs of the conic sections studied.

Parabola

$y = ax^2 + bx + c, \quad a > 0$
$\quad = a(x - h)^2 + k$

$y = ax^2 + bx + c, \quad a < 0$
$\quad = a(x - h)^2 + k$

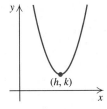

$x = ay^2 + by + c, \quad a > 0$
$\quad = a(y - k)^2 + h$

$x = ay^2 + by + c, \quad a < 0$
$\quad = a(y - k)^2 + h$

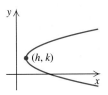

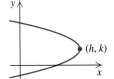

Circle

Center at the origin:

$$x^2 + y^2 = r^2$$

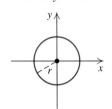

Center at (h, k):

$$(x - h)^2 + (y - k)^2 = r^2$$

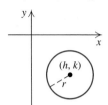

Ellipse

Center at the origin:

$$\frac{x^2}{a^2} + \frac{y^2}{b^2} = 1$$

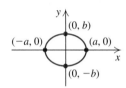

Hyperbola

Center at the origin:

$$\frac{x^2}{a^2} - \frac{y^2}{b^2} = 1$$

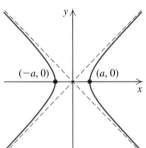

$$\frac{y^2}{b^2} - \frac{x^2}{a^2} = 1$$

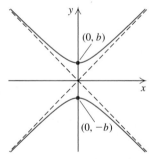

$$xy = c, \quad c > 0$$

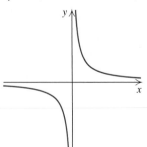

$$xy = c, \quad c < 0$$

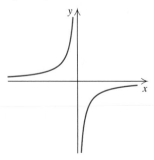

Algebraic manipulations may be needed to express an equation in one of the preceding forms.

EXAMPLE 6   Classify the graph of each equation as a circle, an ellipse, a parabola, or a hyperbola.

a) $5x^2 = 20 - 5y^2$ **b)** $x + 3 + 8y = y^2$

c) $x^2 = y^2 + 4$ **d)** $x^2 = 16 - 4y^2$

Solution

a) We get the terms with variables on one side by adding $5y^2$:

$$5x^2 + 5y^2 = 20.$$

Since x and y are *both* squared, we do not have a parabola. The fact that the squared terms are *added* tells us that we do not have a hyperbola. Do we have a circle? To find out, we need to get $x^2 + y^2$ by itself. We can do that by factoring the 5 out of both terms on the left and then dividing by 5:

$$5(x^2 + y^2) = 20 \qquad \text{Factoring out 5}$$
$$x^2 + y^2 = 4 \qquad \text{Dividing both sides by 5}$$
$$x^2 + y^2 = 2^2. \qquad \text{This is an equation for a circle.}$$

We can see that the graph is a circle with center at the origin and radius 2.

b) The equation $x + 3 + 8y = y^2$ has only one variable squared, so we solve for the other variable:

$$x = y^2 - 8y - 3. \qquad \text{This is an equation for a parabola.}$$

The graph is a horizontal parabola that opens to the right.

c) In $x^2 = y^2 + 4$, both variables are squared, so the graph is not a parabola. We subtract y^2 on both sides and divide by 4 to obtain

$$\frac{x^2}{2^2} - \frac{y^2}{2^2} = 1. \qquad \text{This is an equation for a hyperbola.}$$

The minus sign here indicates that the graph of this equation is a hyperbola. Because it is the x^2-term that is nonnegative, the hyperbola is horizontal.

d) In $x^2 = 16 - 4y^2$, both variables are squared, so the graph cannot be a parabola. We obtain the following equivalent equation:

$$x^2 + 4y^2 = 16.$$

If the coefficients of the terms were the same, we would have the graph of a circle, as in part (a), but they are not. Dividing both sides by 16 yields

$$\frac{x^2}{16} + \frac{y^2}{4} = 1. \qquad \text{This is an equation for an ellipse.}$$

The graph of this equation is a horizontal ellipse.

B

Exercise Set

Graph each of the following equations.

1. $\dfrac{x^2}{1} + \dfrac{y^2}{4} = 1$ ⊡

2. $\dfrac{x^2}{4} + \dfrac{y^2}{1} = 1$ ⊡

3. $\dfrac{x^2}{25} + \dfrac{y^2}{9} = 1$ ⊡

4. $\dfrac{x^2}{16} + \dfrac{y^2}{25} = 1$ ⊡

5. $4x^2 + 9y^2 = 36$ ⊡

6. $9x^2 + 4y^2 = 36$ ⊡

7. $16x^2 + 9y^2 = 144$ ⊡

8. $9x^2 + 16y^2 = 144$ ⊡

9. $2x^2 + 3y^2 = 6$ ⊡

10. $5x^2 + 7y^2 = 35$ ⊡

Aha! **11.** $5x^2 + 5y^2 = 125$ ⊡

12. $8x^2 + 5y^2 = 80$ ⊡

13. $3x^2 + 7y^2 - 63 = 0$ ⊡

14. $3x^2 + 8y^2 - 72 = 0$ ⊡

15. $8x^2 = 96 - 3y^2$ ⊡

16. $6y^2 = 24 - 8x^2$ ⊡

17. $16x^2 + 25y^2 = 1$ ⊡

18. $9x^2 + 4y^2 = 1$ ⊡

Graph each hyperbola. Label all vertices and sketch all asymptotes.

19. $\dfrac{y^2}{9} - \dfrac{x^2}{9} = 1$ ⊡

20. $\dfrac{x^2}{16} - \dfrac{y^2}{16} = 1$ ⊡

21. $\dfrac{x^2}{4} - \dfrac{y^2}{25} = 1$ ⊡

22. $\dfrac{y^2}{16} - \dfrac{x^2}{9} = 1$ ⊡

23. $\dfrac{y^2}{36} - \dfrac{x^2}{9} = 1$ ⊡

24. $\dfrac{x^2}{25} - \dfrac{y^2}{36} = 1$ ⊡

25. $y^2 - x^2 = 25$ ⊡

26. $x^2 - y^2 = 4$ ⊡

27. $25x^2 - 16y^2 = 400$ ⊡

28. $4y^2 - 9x^2 = 36$ ⊡

Graph.

29. $xy = -6$ ⊡

30. $xy = 6$ ⊡

31. $xy = 4$ ⊡

32. $xy = -9$ ⊡

33. $xy = -2$ ⊡

34. $xy = -1$ ⊡

35. $xy = 1$ ⊡

36. $xy = 2$ ⊡

Classify each of the following as the equation of a circle, an ellipse, a parabola, or a hyperbola.

37. $x^2 + y^2 - 10x + 8y - 40 = 0$ Circle

38. $y + 7 = 3x^2$ Parabola

39. $9x^2 + 4y^2 - 36 = 0$ Ellipse

40. $1 + 3y = 2y^2 - x$ Parabola

41. $4x^2 - 9y^2 - 72 = 0$ Hyperbola

42. $y^2 + x^2 = 8$ Circle

43. $x^2 + y^2 = 2x + 4y + 4$ Circle

44. $2y + 13 + x^2 = 8x - y^2$ Circle

45. $4x^2 = 64 - y^2$ Ellipse

46. $y = \dfrac{2}{x}$ Hyperbola

47. $x - \dfrac{3}{y} = 0$ Hyperbola

48. $x - 4 = y^2 - 3y$ Parabola

49. $y + 6x = x^2 + 5$ Parabola

50. $x^2 = 16 + y^2$ Hyperbola

51. $9y^2 = 36 + 4x^2$ Hyperbola

52. $3x^2 + 5y^2 + x^2 = y^2 + 49$ Circle

53. $3x^2 + y^2 - x = 2x^2 - 9x + 10y + 40$ Circle

54. $1 - 5x^2 - 3y^2 = 0$ Ellipse

55. $1 - 3x^2 = 4y$ Parabola

56. $1 - 8x^2 = 6y^2$ Hyperbola

Synthesis

Find an equation of an ellipse that contains the following points.

57. $(-9, 0), (9, 0), (0, -11),$ and $(0, 11)$ $\dfrac{x^2}{81} + \dfrac{y^2}{121} = 1$

58. $(-7, 0), (7, 0), (0, -5),$ and $(0, 5)$ $\dfrac{x^2}{49} + \dfrac{y^2}{25} = 1$

Find an equation of a hyperbola satisfying the given conditions.

59. Having intercepts $(0, 6)$ and $(0, -6)$ and asymptotes $y = 3x$ and $y = -3x$ $\dfrac{y^2}{36} - \dfrac{x^2}{4} = 1$

60. Having intercepts $(8, 0)$ and $(-8, 0)$ and asymptotes $y = 4x$ and $y = -4x$ $\dfrac{x^2}{64} - \dfrac{y^2}{1024} = 1$

⊡ Answers to Exercises 1–36 can be found on pp. A-75 and A-76.

61. *Astronomy.*   The maximum distance of the planet Mars from the sun is 2.48×10^8 mi. The minimum distance is 3.46×10^7 mi. The sun is at one focus of the elliptical orbit. Find the distance from the sun to the other focus. 2.134×10^8 mi

62. Let $(-c, 0)$ and $(c, 0)$ be the foci of an ellipse. Any point $P(x, y)$ is on the ellipse if the sum of the distances from the foci to P is some constant. Use $2a$ to represent this constant.

a) Show that an equation for the ellipse is given by

$$\frac{x^2}{a^2} + \frac{y^2}{a^2 - c^2} = 1.$$

b) Substitute b^2 for $a^2 - c^2$ to get standard form.

63. *President's Office.*   The Oval Office of the President of the United States is an ellipse 31 ft wide and 38 ft long. Show in a sketch precisely where the President and an adviser could sit to best hear each other using the room's acoustics. (*Hint:* See Exercise 62(b) and the discussion following Example 2.)

64. *Dentistry.*   The light source in a dental lamp shines against a reflector that is shaped like a portion of an ellipse in which the light source is one focus of the ellipse. Reflected light enters a patient's mouth at the other focus of the ellipse. If the ellipse from which the reflector was formed is 2 ft wide and 6 ft long, how far should the patient's mouth be from the light source? (*Hint:* See Exercise 62b.)   5.66 ft

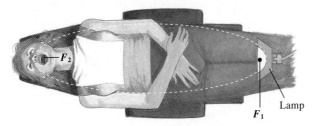

Lamp

62. (a) Let $F_1 = (-c, 0)$ and $F_2 = (c, 0)$. Then the sum of the distances from the foci to P is $2a$. By the distance formula,

$$\sqrt{(x + c)^2 + y^2} + \sqrt{(x - c)^2 + y^2} = 2a, \text{ or}$$
$$\sqrt{(x + c)^2 + y^2} = 2a - \sqrt{(x - c)^2 + y^2}.$$

Squaring, we get

$$(x + c)^2 + y^2 = 4a^2 - 4a\sqrt{(x - c)^2 + y^2}$$
$$+ (x - c)^2 + y^2,$$

or $x^2 + 2cx + c^2 + y^2$
$$= 4a^2 - 4a\sqrt{(x - c)^2 + y^2} + x^2 - 2cx + c^2 + y^2.$$

Thus $-4a^2 + 4cx = -4a\sqrt{(x - c)^2 + y^2}$
$$a^2 - cx = a\sqrt{(x - c)^2 + y^2}.$$

65. *Firefighting.*   The size and shape of certain forest fires can be approximated as the union of two "half-ellipses." For the blaze modeled below, the equation of the smaller ellipse—the part of the fire moving *into* the wind—is

$$\frac{x^2}{40,000} + \frac{y^2}{10,000} = 1.$$

The equation of the other ellipse—the part moving *with* the wind—is

$$\frac{x^2}{250,000} + \frac{y^2}{10,000} = 1.$$

(*Source for figure*: "Predicting Wind-Driven Wild Land Fire Size and Shape," Hal E. Anderson, Research Paper INT-305, U.S. Department of Agriculture, Forest Service, February 1983).

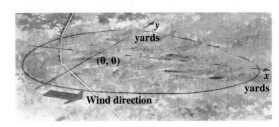

Determine the width and the length of the fire.
Length: 700 yd; width: 200 yd

63.

Squaring again, we get

$$a^4 - 2a^2cx + c^2x^2 = a^2(x^2 - 2cx + c^2 + y^2)$$
$$a^4 - 2a^2cx + c^2x^2 = a^2x^2 - 2a^2cx + a^2c^2 + a^2y^2,$$

or

$$x^2(a^2 - c^2) + a^2y^2 = a^2(a^2 - c^2)$$
$$\frac{x^2}{a^2} + \frac{y^2}{a^2 - c^2} = 1.$$

(b) When P is at $(0, b)$, it follows that $b^2 = a^2 - c^2$. Substituting, we have

$$\frac{x^2}{a^2} + \frac{y^2}{b^2} = 1.$$

Systems Involving One Nonlinear Equation ▪ Systems of Two Nonlinear Equations ▪ Problem Solving

Nonlinear Systems of Equations

The equations appearing in systems of two equations have thus far all been linear. We now consider systems of two equations in which at least one equation is nonlinear.

Systems Involving One Nonlinear Equation

Suppose that a system consists of an equation of a circle and an equation of a line. In what ways can the circle and the line intersect? The figures below represent three ways in which the situation can occur. We see that such a system will have 0, 1, or 2 real solutions.

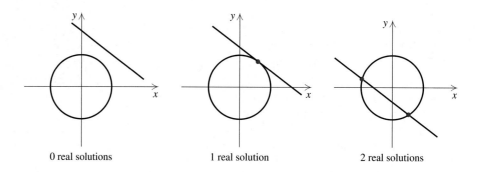

0 real solutions   1 real solution   2 real solutions

Recall that, in addition to graphing, we used both *elimination* and *substitution* to solve systems of linear equations. When solving systems in which one equation is of first degree and one is of second degree, it is preferable to use the *substitution* method.

TEACHING TIP

Encourage students to interpret the solution of the system geometrically and to describe verbally what the possible solutions are.

EXAMPLE 1   Solve the system

$$x^2 + y^2 = 25, \quad (1) \qquad \text{(The graph is a circle.)}$$
$$3x - 4y = 0. \quad (2) \qquad \text{(The graph is a line.)}$$

Solution   First, we solve the linear equation, (2), for x:

$$x = \tfrac{4}{3}y. \quad (3) \qquad \textbf{We could have solved for } y \textbf{ instead.}$$

Then we substitute $\tfrac{4}{3}y$ for x in equation (1) and solve for y:

$$\left(\tfrac{4}{3}y\right)^2 + y^2 = 25$$
$$\tfrac{16}{9}y^2 + y^2 = 25$$
$$\tfrac{25}{9}y^2 = 25$$
$$y^2 = 9 \qquad \textbf{Multiplying both sides by } \tfrac{9}{25}$$
$$y = \pm 3. \qquad \textbf{Using the principle of square roots}$$

Now we substitute these numbers for y in equation (3) and solve for x:

for $y = 3$, $\quad x = \frac{4}{3}(3) = 4$;

for $y = -3$, $\quad x = \frac{4}{3}(-3) = -4$.

Check: For $(4, 3)$:

$$\frac{x^2 + y^2 = 25}{4^2 + 3^2 \;?\; 25}$$
$$16 + 9 \quad\Big|$$
$$25 \;\Big|\; 25 \quad \text{TRUE}$$

$$\frac{3x - 4y = 0}{3(4) - 4(3) \;?\; 0}$$
$$12 - 12 \quad\Big|$$
$$0 \;\Big|\; 0 \quad \text{TRUE}$$

It is left to the student to confirm that $(-4, -3)$ also checks in both equations.

The pairs $(4, 3)$ and $(-4, -3)$ check, so they are solutions. We can see the solutions in the graph. Intersections occur at $(4, 3)$ and $(-4, -3)$.

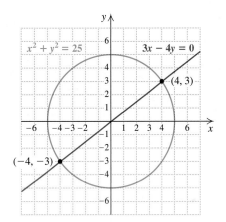

Although we may not know what the graph of each equation in a system looks like, the algebraic approach of Example 1 can still be used.

EXAMPLE 2   Solve the system

$$y + 3 = 2x, \qquad (1)$$
$$x^2 + 2xy = -1. \qquad (2)$$

Solution   First, we solve the linear equation (1) for y:

$$y = 2x - 3. \qquad (3)$$

Then we substitute $2x - 3$ for y in equation (2) and solve for x:

$$x^2 + 2x(2x - 3) = -1$$
$$x^2 + 4x^2 - 6x = -1$$
$$5x^2 - 6x + 1 = 0$$
$$(5x - 1)(x - 1) = 0 \qquad \text{\textbf{Factoring}}$$
$$5x - 1 = 0 \quad or \quad x - 1 = 0 \qquad \text{\textbf{Using the principle of zero}}$$
$$\text{\textbf{products}}$$
$$x = \tfrac{1}{5} \quad or \qquad x = 1.$$

Now we substitute these numbers for x in equation (3) and solve for y:

$$\text{for } x = \tfrac{1}{5}, \quad y = 2\left(\tfrac{1}{5}\right) - 3 = -\tfrac{13}{5};$$
$$\text{for } x = 1, \quad y = 2(1) - 3 = -1.$$

You can confirm that $\left(\tfrac{1}{5}, -\tfrac{13}{5}\right)$ and $(1, -1)$ check, so they are both solutions.

EXAMPLE 3 Solve the system

$$x + y = 5, \qquad (1) \qquad \text{(The graph is a line.)}$$
$$y = 3 - x^2. \qquad (2) \qquad \text{(The graph is a parabola.)}$$

Solution   We substitute $3 - x^2$ for y in the first equation:

$$x + 3 - x^2 = 5$$
$$-x^2 + x - 2 = 0 \qquad \textbf{Adding } -5 \textbf{ to both sides and rearranging}$$
$$x^2 - x + 2 = 0. \qquad \textbf{Multiplying both sides by } -1$$

Since $x^2 - x + 2$ does not factor, we need the quadratic formula:

$$x = \frac{-b \pm \sqrt{b^2 - 4ac}}{2a}$$

$$= \frac{-(-1) \pm \sqrt{(-1)^2 - 4 \cdot 1 \cdot 2}}{2(1)} \qquad \textbf{Substituting}$$

$$= \frac{1 \pm \sqrt{1 - 8}}{2} = \frac{1 \pm \sqrt{-7}}{2} = \frac{1}{2} \pm \frac{\sqrt{7}}{2}i.$$

Solving equation (1) for y gives us $y = 5 - x$. Substituting values for x gives

$$y = 5 - \left(\frac{1}{2} + \frac{\sqrt{7}}{2}i\right) = \frac{9}{2} - \frac{\sqrt{7}}{2}i \quad \text{and}$$

$$y = 5 - \left(\frac{1}{2} - \frac{\sqrt{7}}{2}i\right) = \frac{9}{2} + \frac{\sqrt{7}}{2}i.$$

The solutions are

$$\left(\frac{1}{2} + \frac{\sqrt{7}}{2}i, \frac{9}{2} - \frac{\sqrt{7}}{2}i\right) \quad \text{and} \quad \left(\frac{1}{2} - \frac{\sqrt{7}}{2}i, \frac{9}{2} + \frac{\sqrt{7}}{2}i\right).$$

There are no real-number solutions. Note in the figure at right that the graphs do not intersect. Getting only nonreal solutions tells us that the graphs do not intersect.

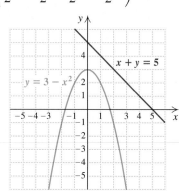

Systems of Two Nonlinear Equations

We now consider systems of two second-degree equations. Graphs of such systems can involve any two conic sections. The following figure shows some ways in which a circle and a hyperbola can intersect.

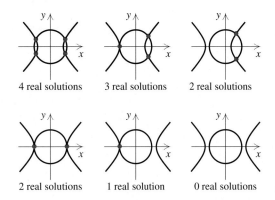

4 real solutions   3 real solutions   2 real solutions

2 real solutions   1 real solution   0 real solutions

To solve systems of two second-degree equations, we can use either substitution or elimination. The elimination method is generally better when both equations are of the form $Ax^2 + By^2 = C$. Then we can eliminate an x^2- or y^2-term in a manner similar to the procedure used in Chapter 3.

EXAMPLE 4   Solve the system

$$2x^2 + 5y^2 = 22, \qquad (1)$$
$$3x^2 - y^2 = -1. \qquad (2)$$

Solution   Here we multiply equation (2) by 5 and then add:

$$2x^2 + 5y^2 = 22$$
$$\underline{15x^2 - 5y^2 = -5} \qquad \text{**Multiplying both sides of equation (2) by 5**}$$
$$17x^2 \qquad = 17 \qquad \text{**Adding**}$$
$$x^2 = 1$$
$$x = \pm 1.$$

There is no x-term, and whether x is -1 or 1, we have $x^2 = 1$. Thus we can simultaneously substitute 1 and -1 for x in equation (2), and we have

$$\left. \begin{array}{r} 3 \cdot (\pm 1)^2 - y^2 = -1 \\ 3 - y^2 = -1 \\ -y^2 = -4 \end{array} \right\} \qquad \text{**Since $(-1)^2 = 1^2$, we can evaluate for $x = -1$ and $x = 1$ simultaneously.**}$$
$$y^2 = 4 \quad \text{or} \quad y = \pm 2.$$

Thus, if $x = 1$, then $y = 2$ or $y = -2$; and if $x = -1$, then $y = 2$ or $y = -2$. The four possible solutions are $(1, 2)$, $(1, -2)$, $(-1, 2)$, and $(-1, -2)$.

Check: Since $(2)^2 = (-2)^2$ and $(1)^2 = (-1)^2$, we can check all four pairs at once.

$$
\begin{array}{c}
\underline{2x^2 + 5y^2 = 22} \\
2(\pm 1)^2 + 5(\pm 2)^2 \;\overset{?}{=}\; 22 \\
2 + 20 \\
22 \;\Big|\; 22 \quad \text{TRUE}
\end{array}
\qquad
\begin{array}{c}
\underline{3x^2 - y^2 = -1} \\
3(\pm 1)^2 - (\pm 2)^2 \;\overset{?}{=}\; -1 \\
3 - 4 \\
-1 \;\Big|\; -1 \quad \text{TRUE}
\end{array}
$$

The solutions are $(1, 2)$, $(1, -2)$, $(-1, 2)$, and $(-1, -2)$.

When a product of variables is in one equation and the other equation is of the form $Ax^2 + By^2 = C$, we often solve for a variable in the equation with the product and then use substitution.

EXAMPLE 5 Solve the system

$$x^2 + 4y^2 = 20, \qquad (1)$$
$$xy = 4. \qquad (2)$$

Solution First, we solve equation (2) for y:

$$y = \frac{4}{x}. \qquad \textbf{Dividing both sides by } x. \textbf{ Note that } x \neq 0.$$

Then we substitute $4/x$ for y in equation (1) and solve for x:

$$x^2 + 4\left(\frac{4}{x}\right)^2 = 20$$

$$x^2 + \frac{64}{x^2} = 20$$

$$x^4 + 64 = 20x^2 \qquad \textbf{Multiplying by } x^2$$

$$x^4 - 20x^2 + 64 = 0 \qquad \begin{array}{l}\textbf{Obtaining standard form.}\\\textbf{This equation is reducible}\\\textbf{to quadratic.}\end{array}$$

$$(x^2 - 4)(x^2 - 16) = 0 \qquad \begin{array}{l}\textbf{Factoring. If you prefer,}\\\textbf{let } u = x^2 \textbf{ and substitute.}\end{array}$$

$$(x - 2)(x + 2)(x - 4)(x + 4) = 0 \qquad \textbf{Factoring}$$

$$x = 2 \quad or \quad x = -2 \quad or \quad x = 4 \quad or \quad x = -4. \qquad \begin{array}{l}\textbf{Using the}\\\textbf{principle of zero}\\\textbf{products}\end{array}$$

Since $y = 4/x$, for $x = 2$, we have $y = 4/2$, or 2. Thus, $(2, 2)$ is a solution. Similarly, $(-2, -2)$, $(4, 1)$, and $(-4, -1)$ are solutions. You can show that all four pairs check.

Problem Solving

We now consider applications that can be modeled by a system of equations in which at least one equation is not linear.

EXAMPLE 6   Architecture.   For a college gymnasium, an architect wants to lay out a rectangular piece of land that has a perimeter of 204 m and an area of 2565 m². Find the dimensions of the piece of land.

Solution

1. **Familiarize.**   We draw and label a sketch, letting $l =$ the length and $w =$ the width, both in meters.

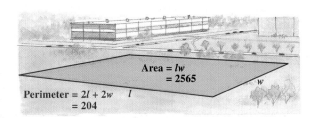

2. **Translate.**   We then have the following translation:

$$\text{Perimeter:}\quad 2w + 2l = 204;$$
$$\text{Area:}\qquad\quad lw = 2565.$$

3. **Carry out.**   We solve the system

$$2w + 2l = 204,$$
$$lw = 2565.$$

Solving the second equation for l gives us $l = 2565/w$. Then we substitute $2565/w$ for l in the first equation and solve for w:

$$2w + 2\left(\frac{2565}{w}\right) = 204$$

$$2w^2 + 2(2565) = 204w \qquad \textbf{Multiplying both sides by } w$$

$$2w^2 - 204w + 2(2565) = 0 \qquad \textbf{Standard form}$$

$$w^2 - 102w + 2565 = 0 \qquad \textbf{Multiplying by } \tfrac{1}{2}$$

> **Factoring could be used instead of the quadratic formula, but the numbers are quite large.**

$$w = \frac{-(-102) \pm \sqrt{(-102)^2 - 4 \cdot 1 \cdot 2565}}{2 \cdot 1}$$

$$w = \frac{102 \pm \sqrt{144}}{2} = \frac{102 \pm 12}{2}$$

$$w = 57 \quad or \quad w = 45.$$

If $w = 57$, then $l = 2565/w = 2565/57 = 45$. If $w = 45$, then $l = 2565/w = 2565/45 = 57$. Since length is usually considered to be longer than width, we have the solution $l = 57$ and $w = 45$, or $(57, 45)$.

4. **Check.**   If $l = 57$ and $w = 45$, the perimeter is $2 \cdot 57 + 2 \cdot 45$, or 204. The area is $57 \cdot 45$, or 2565. The numbers check.

5. **State.**   The length is 57 m and the width is 45 m.

EXAMPLE 7 HDTV Dimensions.   High-definition television (HDTV) offers greater clarity than conventional television. The Kaplans' new HDTV screen has an area of 1296 in^2 and has a $\sqrt{3033}$-in. (about 55-in.) diagonal screen. Find the width and the length of the screen.

Solution

1. **Familiarize.**   We make a drawing and label it. Note that there is a right triangle in the figure. We let l = the length and w = the width, both in inches.

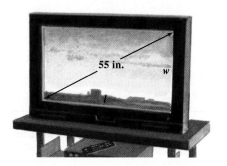

2. **Translate.**   We translate to a system of equations:

   $l^2 + w^2 = \sqrt{3033}^2$ **Using the Pythagorean theorem**

   $lw = 1296.$ **Using the formula for the area of a rectangle**

3. **Carry out.**   We solve the system

   $\left.\begin{array}{l} l^2 + w^2 = 3033, \\ lw = 1296 \end{array}\right\}$ **You should complete the solution of this system.**

   to get $(48, 27)$, $(27, 48)$, $(-48, -27)$, and $(-27, -48)$.

4. **Check.**   Measurements cannot be negative and length is usually greater than width, so we check only $(48, 27)$. In the right triangle, $48^2 + 27^2 = 2304 + 729 = 3033$ or $\sqrt{3033}^2$. The area is $48 \cdot 27 = 1296$, so our answer checks.

5. **State.**   The length is 48 in. and the width is 27 in.

C

FOR EXTRA HELP

MyMathLab

Exercise Set

Solve. Remember that graphs can be used to confirm all real solutions.

1. $x^2 + y^2 = 25$,
$y - x = 1$ $(-4, -3), (3, 4)$

2. $x^2 + y^2 = 100$,
$y - x = 2$
$(-8, -6), (6, 8)$

3. $9x^2 + 4y^2 = 36$,
$3x + 2y = 6$ $(2, 0), (0, 3)$

4. $4x^2 + 9y^2 = 36$,
$3y + 2x = 6$
$(0, 2), (3, 0)$

5. $y = x^2$, $(2, 4), (1, 1)$
$3x = y + 2$

6. $y^2 = x + 3$,
$2y = x + 4$ $(-2, 1)$

7. $2y^2 + xy + x^2 = 7$,
$x - 2y = 5$ $\left(\frac{11}{4}, -\frac{9}{8}\right), (1, -2)$

8. $x^2 - xy + 3y^2 = 27$,
$x - y = 2$ ⊡

9. $x^2 - y^2 = 16$,
$x - 2y = 1$ $\left(\frac{13}{3}, \frac{5}{3}\right), (-5, -3)$

10. $x^2 + 4y^2 = 25$,
$x + 2y = 7$ $\left(4, \frac{3}{2}\right), (3, 2)$

11. $m^2 + 3n^2 = 10$,
$m - n = 2$ ⊡

12. $x^2 - xy + 3y^2 = 5$,
$x - y = 2$ $\left(\frac{7}{3}, \frac{1}{3}\right), (1, -1)$

13. $2y^2 + xy = 5$,
$4y + x = 7$ $\left(-3, \frac{5}{2}\right), (3, 1)$

14. $3x + y = 7$, $\left(\frac{11}{4}, -\frac{5}{4}\right), (1, 4)$
$4x^2 + 5y = 24$

15. $p + q = -6$,
$pq = -7$ $(1, -7), (-7, 1)$

16. $a + b = 7$,
$ab = 4$ ⊡

17. $4x^2 + 9y^2 = 36$,
$x + 3y = 3$ $(3, 0), \left(-\frac{9}{5}, \frac{8}{5}\right)$

18. $2a + b = 1$,
$b = 4 - a^2$ $(3, -5), (-1, 3)$

19. $xy = 4$,
$x + y = 5$ $(1, 4), (4, 1)$

20. $a^2 + b^2 = 89$,
$a - b = 3$ $(-5, -8), (8, 5)$

Aha! **21.** $y = x^2$,
$x = y^2$ ⊡

22. $x^2 + y^2 = 25$,
$y^2 = x + 5$
$(-5, 0), (4, 3), (4, -3)$

23. $x^2 + y^2 = 9$,
$x^2 - y^2 = 9$
$(-3, 0), (3, 0)$

24. $y^2 - 4x^2 = 4$,
$4x^2 + y^2 = 4$
$(0, 2), (0, -2)$

25. $x^2 + y^2 = 25$,
$xy = 12$
$(-4, -3), (-3, -4), (3, 4), (4, 3)$

26. $x^2 - y^2 = 16$,
$x + y^2 = 4$
$(-5, 3), (-5, -3), (4, 0)$

27. $x^2 + y^2 = 4$,
$9x^2 + 16y^2 = 144$ ⊡

28. $x^2 + y^2 = 9$,
$25x^2 + 16y^2 = 400$ ⊡

29. $x^2 + y^2 = 16$,
$y^2 - 2x^2 = 10$ ⊡

30. $x^2 + y^2 = 14$,
$x^2 - y^2 = 4$ ⊡

31. $x^2 + y^2 = 5$, $(-2, -1), (-1, -2),$
$xy = 2$ $(1, 2), (2, 1)$

32. $x^2 + y^2 = 20$,
$xy = 8$ $(4, 2), (-4, -2), (2, 4), (-2, -4)$

33. $x^2 + y^2 = 13$, $(-3, -2), (-2, -3),$
$xy = 6$ $(2, 3), (3, 2)$

34. $x^2 + 4y^2 = 20$,
$xy = 4$ $(4, 1), (-4, -1), (2, 2), (-2, -2)$

$(2, 5), (-2, -5)$
35. $3xy + x^2 = 34$,
$2xy - 3x^2 = 8$

$(2, 1), (-2, -1)$
36. $2xy + 3y^2 = 7$,
$3xy - 2y^2 = 4$

37. $xy - y^2 = 2$, $(3, 2),$
$2xy - 3y^2 = 0$ $(-3, -2)$

38. $4a^2 - 25b^2 = 0$,
$2a^2 - 10b^2 = 3b + 4$ ⊡

39. $x^2 - y = 5$,
$x^2 + y^2 = 25$
$(-3, 4), (3, 4), (0, -5)$

40. $ab - b^2 = -4$,
$ab - 2b^2 = -6$
$\left(-\sqrt{2}, \sqrt{2}\right), \left(\sqrt{2}, -\sqrt{2}\right)$

Solve.

41. *Computer Parts.*   Dataport Electronics needs a rectangular memory board that has a perimeter of 28 cm and a diagonal of length 10 cm. What should the dimensions of the board be?   Length: 8 cm; width: 6 cm

42. *Geometry.*   A rectangle has an area of 2 yd² and a perimeter of 6 yd. Find its dimensions.
Length: 2 yd; width: 1 yd

43. *Geometry.*   A rectangle has an area of 20 in² and a perimeter of 18 in. Find its dimensions.
Length: 5 in.; width: 4 in.

44. *Tile Design.*   The New World tile company wants to make a new rectangular tile that has a perimeter of 6 in. and a diagonal of length $\sqrt{5}$ in. What should the dimensions of the tile be?   Length: 2 in.; width: 1 in.

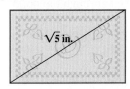

45. *Design of a Van.*   The cargo area of a delivery van must be 60 ft², and the length of a diagonal must accommodate a 13-ft board. Find the dimensions of the cargo area.   Length: 12 ft; width: 5 ft

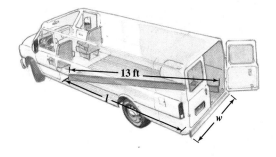

⊡ Answers to Exercises 8, 11, 16, 21, 27–30, and 38 can be found on p. A-76.

46. *Dimensions of a Rug.* The diagonal of a Persian rug is 25 ft. The area of the rug is 300 ft². Find the length and the width of the rug. Length: 20 ft; width: 15 ft

47. The product of the lengths of the legs of a right triangle is 156. The hypotenuse has length $\sqrt{313}$. Find the lengths of the legs. 13 and 12

48. The product of two numbers is 60. The sum of their squares is 136. Find the numbers. 6 and 10; −6 and −10

49. *Investments.* A certain amount of money saved for 1 yr at a certain interest rate yielded $225 in interest. If $750 more had been invested and the rate had been 1% less, the interest would have been the same. Find the principal and the rate. $3750, 6%

50. *Garden Design.* A garden contains two square peanut beds. Find the length of each bed if the sum of their areas is 832 ft² and the difference of their areas is 320 ft². 24 ft, 16 ft

51. The area of a rectangle is $\sqrt{3}$ m², and the length of a diagonal is 2 m. Find the dimensions. Length: $\sqrt{3}$ m; width: 1 m

52. The area of a rectangle is $\sqrt{2}$ m², and the length of a diagonal is $\sqrt{3}$ m. Find the dimensions. Length: $\sqrt{2}$ m; width: 1 m

Synthesis

53. Find the equation of a circle that passes through $(-2, 3)$ and $(-4, 1)$ and whose center is on the line $5x + 8y = -2$. $(x + 2)^2 + (y - 1)^2 = 4$

54. Find the equation of an ellipse centered at the origin that passes through the points $(2, -3)$ and $\left(1, \sqrt{13}\right)$. $4x^2 + 3y^2 = 43$

Solve.

55. $p^2 + q^2 = 13,$
$\dfrac{1}{pq} = -\dfrac{1}{6}$
$(-2, 3), (2, -3), (-3, 2), (3, -2)$

56. $a + b = \dfrac{5}{6},$ $\left(\dfrac{1}{3}, \dfrac{1}{2}\right),$
$\dfrac{a}{b} + \dfrac{b}{a} = \dfrac{13}{6}$ $\left(\dfrac{1}{2}, \dfrac{1}{3}\right)$

57. *Box Design.* Four squares with sides 5 in. long are cut from the corners of a rectangular metal sheet that has an area of 340 in². The edges are bent up to form an open box with a volume of 350 in³. Find the dimensions of the box. 10 in. by 7 in. by 5 in.

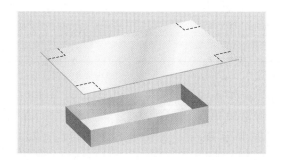

58. A piece of wire 100 cm long is to be cut into two pieces and those pieces are each to be bent to make a square. The area of one square is to be 144 cm² greater than that of the other. How should the wire be cut? 61.52 cm and 38.48 cm

Answers

Chapter 1

Exercise Set 1.1, pp. 11–12

1. 25   **3.** 16   **5.** 27   **7.** 0   **9.** 5   **11.** 25
13. 109   **15.** 17.5 sq ft   **17.** 11.2 sq m   **19. (a)** Yes;
(b) no; **(c)** no   **21. (a)** No; **(b)** yes; **(c)** yes   **23. (a)** Yes;
(b) no; **(c)** no   **25.** {a, e, i, o, u}, or {a, e, i, o, u, y}
27. {1, 3, 5, 7, ...}   **29.** {5, 10, 15, 20, ...}   **31.** {$x | x$ is an
odd number between 10 and 20}   **33.** {$x | x$ is a whole
number less than 5}   **35.** {$n | n$ is a multiple of 5 between 7
and 79}   **37.** True   **39.** True   **41.** False   **43.** True
45. True   **47.** True   **49.** 15   **51.** 41.4494
53. 25.125   **55.** 13,778   **57.** ᴛᵂ   **59.** ᴛᵂ   **61.** {0}
63. {5, 10, 15, 20, ...}   **65.** {1, 3, 5, 7, ...}
67.

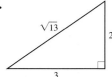

Interactive Discovery, p. 16

1. −3, −6   **2.** 5, 10

Exercise Set 1.2, pp. 20–22

1. 9   **3.** 6   **5.** 6.2   **7.** 0   **9.** $1\frac{7}{8}$ **11.** 4.21
13. True   **15.** False   **17.** True   **19.** True   **21.** True
23. False   **25.** 11   **27.** −11   **29.** −1.2   **31.** $-\frac{11}{35}$
33. −8.5   **35.** $\frac{5}{9}$ **37.** −4.5   **39.** 0   **41.** −6.4
43. −3.14   **45.** $4\frac{1}{3}$ **47.** 0   **49.** −7   **51.** 2.7
53. −1.79   **55.** 0   **57.** 7   **59.** −7   **61.** 4
63. −19   **65.** −3.1   **67.** $-\frac{11}{10}$ **69.** 0   **71.** 7.9

73. −30   **75.** 24   **77.** −21   **79.** $-\frac{3}{7}$ **81.** 0
83. 5.44   **85.** 5   **87.** −5   **89.** −73   **91.** 0
93. Undefined   **95.** $\frac{1}{4}$ **97.** $-\frac{1}{9}$ **99.** $\frac{3}{2}$ **101.** $-\frac{11}{3}$
103. $\frac{5}{6}$ **105.** $-\frac{6}{5}$ **107.** $\frac{1}{36}$ **109.** 1   **111.** 23
113. $-\frac{6}{11}$ **115.** Undefined   **117.** $\frac{11}{43}$ **119.** 28
121. −7   **123.** 10   **125.** −12.86   **127.** (a)
129. (d)   **131.** ᴛᵂ   **133.** 16; 16   **134.** 11; 11
135. ᴛᵂ   **137.** $(8 − 5)^3 + 9 = 36$
139. $5 \cdot 2^3 \div (3 − 4)^4 = 40$ **141.** −6.2

Interactive Discovery, p. 27

1. 0, 0; 185, 160; 33, 24; not equivalent
2. −30, −30; 70, 70; −90, −90; equivalent

Exercise Set 1.3, pp. 29–30

1. $7b + 4a$; $a4 + b7$; $4a + b7$ **3.** $y(7x)$; $(x7)y$ **5.** $3(xy)$
7. $(x + 2y) + 5$ **9.** $3a + 21$ **11.** $4x − 4y$
13. $−10a − 15b$ **15.** $9ab − 9ac + 9ad$ **17.** $5(x + 5)$
19. $3(p − 3)$ **21.** $7(x − 3y + 2z)$ **23.** $17(15 − 2b)$
25. $x(y + 1)$ **27.** $4x, −5y, 3$ **29.** $x^2, −6x, −7$
31. $10x$ **33.** $−2rt$ **35.** $10t^2$ **37.** $11a$ **39.** $−7n$
41. $10x$ **43.** $7x − 2x^2$ **45.** $5a + 11a^2$
47. $22x + 18$ **49.** $−5t^2 + 2t + 4t^3$ **51.** $5a − 5$
53. $−2m + 1$ **55.** $5d − 12$ **57.** $−7x + 14$
59. $−10x + 21$ **61.** $44a − 22$ **63.** $−100a − 90$
65. $−12y − 145$ **67.** First expression: −16.3, −12, −15;
second expression: 13.7, 18, 15; not equivalent   **69.** First
expression: 17.2, 0, 12; second expression: 17.2, 0, 12;
equivalent   **71.** ᴛᵂ   **73.** 3   **74.** 3.59   **75.** 35
76. $-\frac{1}{35}$ **77.** ᴛᵂ   **79.** $23a − 18b + 184$ **81.** $−4z$
83. $−x + 19$ **85.** ᴛᵂ

Interactive Discovery, p. 30

1. (b)   **2.** (c)   **3.** (a)   **4.** (d)

Interactive Discovery, p. 31

1. (d)   **2.** (a)   **3.** (b)   **4.** (c)

Exercise Set 1.4, pp. 41–42

1. 5^9 **3.** t^8 **5.** $18x^7$ **7.** $21m^{13}$ **9.** $x^{10}y^{10}$
11. a^6 **13.** $3t^5$ **15.** m^5n^4 **17.** $-4x^8y^6z^6$ **19.** -1
21. 1   **23.** 81   **25.** -81 **27.** $\frac{1}{16}$ **29.** $-\frac{1}{16}$
31. -1 **33.** $\dfrac{1}{n^6}$ **35.** 2^6, or 64   **37.** $\dfrac{2a^2}{b^5}$ **39.** $\dfrac{x^2y^4}{z^3}$
41. 3^{-4} **43.** $\dfrac{1}{x^{-5}}$ **45.** $\dfrac{6}{x^{-2}}$, or $\dfrac{1}{6^{-1}x^{-2}}$ **47.** $(5y)^{-3}$
49. 7^{-7}, or $\dfrac{1}{7^7}$ **51.** b^{-3}, or $\dfrac{1}{b^3}$ **53.** a^3 **55.** $10a^{-6}b^{-2}$,
or $\dfrac{10}{a^6b^2}$ **57.** 10^{-9}, or $\dfrac{1}{10^9}$ **59.** 2^{-2}, or $\dfrac{1}{2^2}$, or $\dfrac{1}{4}$
61. 8^{-5}, or $\dfrac{1}{8^5}$ **63.** $-\dfrac{1}{6}x^3y^{-2}z^{10}$, or $-\dfrac{x^3z^{10}}{6y^2}$ **65.** x^{12}
67. 9^{-12}, or $\dfrac{1}{9^{12}}$ **69.** 7^{40} **71.** $a^{12}b^4$ **73.** $5x^{-14}y^{-14}$, or
$\dfrac{5}{x^{14}y^{14}}$ **75.** $a^{-7}b^4$, or $\dfrac{b^4}{a^7}$ **77.** 1   **79.** $\dfrac{9}{2}x^8y^9$
81. $\dfrac{5^4y^{24}}{4^4x^{20}}$ **83.** 1   **85.** $\dfrac{5^2b^{12}}{6^2a^{10}}$ **87.** $\dfrac{8^3x^{24}}{6^3y^{12}}$, or $\dfrac{64x^{24}}{27y^{12}}$
89. 4.7×10^{10} **91.** 1.6×10^{-8} **93.** 4.07×10^{11}
95. 6.03×10^{-7} **97.** 0.0004   **99.** 673,000,000
101. 0.0000000008923   **103.** 90,300,000,000
105. 9.66×10^{-5} **107.** 1.3338×10^{-11}
109. 3.528448×10^3 **111.** 1.5×10^3 **113.** 3×10^{-5}
115. Approximately 1.79×10^{20} **117.** 1.1076×10^{11}
119. 8.3×10^{23} **121.** TW   **123.** $-\frac{1}{12}$ **124.** 30.96
125. $-8x + 12y$ **126.** $2(4x - 5)$ **127.** TW
129. TW   **131.** $-3x^{2a-1}$ **133.** 12^{6b-2ab}
135. $-5x^{2b}y^{-2a}$ **137.** 1.25×10^{22} **139.** 1

Exercise Set 1.5, pp. 52–53

1. $(5,3)$, $(-4,3)$, $(0,2)$, $(-2,-3)$, $(4,-2)$, and $(-5,0)$
3.

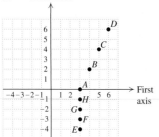

5.

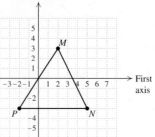

Triangle, 21 units^2

7. III   **9.** II   **11.** I   **13.** IV   **15.** Yes   **17.** No
19. Yes   **21.** Yes   **23.** Yes   **25.** Yes   **27.** Yes
29. No   **31.**

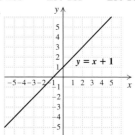

33.

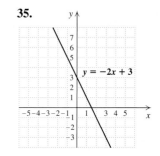

35.

37.

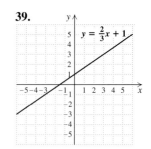

39.

41.

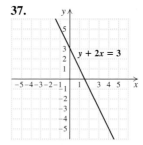

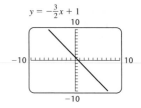

43.

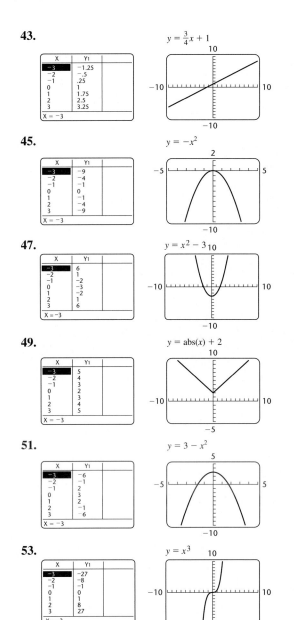

$y = \frac{3}{4}x + 1$

X	Y1
-3	-1.25
-2	-.5
-1	.25
0	1
1	1.75
2	2.5
3	3.25
X = -3	

45.

$y = -x^2$

X	Y1
-3	-9
-2	-4
-1	-1
0	0
1	-1
2	-4
3	-9
X = -3	

47.

$y = x^2 - 3$

X	Y1
-3	6
-2	1
-1	-2
0	-3
1	-2
2	1
3	6
X = -3	

49.

$y = \text{abs}(x) + 2$

X	Y1
-3	5
-2	4
-1	3
0	2
1	3
2	4
3	5
X = -3	

51.

$y = 3 - x^2$

X	Y1
-3	-6
-2	-1
-1	2
0	3
1	2
2	-1
3	-6
X = -3	

53.

$y = x^3$

X	Y1
-3	-27
-2	-8
-1	-1
0	0
1	1
2	8
3	27
X = -3	

55. (b)   **57.** (a)   **59.** (b)   **61.** Exercises 31–43 and 55 are linear.   **63.** TW   **65.** Yes   **66.** Yes   **67.** No
68. No   **69.** -2 **70.** 49   **71.** 0   **72.** -3
73. TW   **75.** TW   **77.** (a), (d)   **79.** $(-1, -2)$, $(-19, -2)$, or $(13, 10)$ **81.** Equations (a), (c), and (d) appear to be linear.

Exercise Set 1.6, pp. 67–71

1. Let n represent the number; $n - 3$ **3.** Let t represent the number; $12t$ **5.** Let x represent the number; $0.65x$, or $\frac{65}{100}x$
7. Let y represent the number; $2y + 10$ **9.** Let s represent the number; $0.1s + 8$, or $\frac{10}{100}s + 8$ **11.** Let m and n

represent the numbers; $m - n - 1$ **13.** $90 \div 4$, or $\frac{90}{4}$
15. Let x and $x + 7$ represent the numbers; $x + (x + 7) = 65$
17. Let x represent the number; $128 = 0.4x$ **19.** Let a represent the number; $\frac{a}{4} = 12.3$ **21.** Let w represent the rectangle's width; $21 = 2(2w) + 2w$ **23.** Let p represent the original price; $490 = p - 0.35p$ **25.** Let a represent the length of an American bullfrog; $a = 0.64(350)$ **27.** Let n represent the number of hours Rhonda rode; $25 = 15n$
29. Let x represent the measure of the smallest angle; $x + (x + 1) + (x + 2) = 180$ **31.** Let c represent the original cost of the order, in dollars; $c - 0.1c = 279$
33. Let t represent the number of minutes spent climbing; $3500t = 29{,}000 - 8000$ **35.** Let x represent the measure of the second angle, in degrees; $3x + x + (2x - 12) = 180$
37. Let x represent the first even number; $2x + 3(x + 2) = 76$
39. Let s represent the length, in centimeters, of a side of the smaller triangle; $3s + 3 \cdot 2s = 90$ **41.** Let x represent the score on the next test; $\dfrac{93 + 89 + 72 + 80 + 96 + x}{6} = 88$

43. 75 heart attacks per 10,000 men   **45.** 56%
47. 3.5 drinks

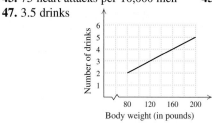

49.

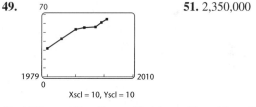

Xscl = 10, Yscl = 10

51. 2,350,000

53. 1993   **55.** TW   **57.** 16   **58.** -16 **59.** z^4
60. z^{-8}, or $\dfrac{1}{z^8}$ **61.** $2x^8y^{12}$ **62.** $16x^8y^{12}$ **63.** TW
65. (a) III; (b) II; (c) I; (d) IV

Review Exercises: Chapter 1, pp. 75–76

1. [1.1] 28   **2.** [1.1] 60.614425   **3.** [1.1] {2, 4, 6, 8, 10, 12}; $\{x \mid x$ is an even integer between 1 and 13$\}$
4. [1.1] 2100 cm^2 **5.** [1.1] (a) No; (b) yes
6. [1.1] (a) Yes; (b) yes   **7.** [1.2] 7.3   **8.** [1.2] 4.09
9. [1.2] 0   **10.** [1.2] -12.1 **11.** [1.2] $-\frac{23}{35}$
12. [1.2] $\frac{7}{15}$ **13.** [1.2] -11.5 **14.** [1.2] $-\frac{1}{6}$
15. [1.2] -5.4 **16.** [1.2] 15.3   **17.** [1.2] $-\frac{5}{12}$
18. [1.2] -9.1 **19.** [1.2] $-\frac{21}{4}$ **20.** [1.2] 4.01
21. [1.3] $a + 5$ **22.** [1.3] $y3$ **23.** [1.3] $y + 5x$, or $x5 + y$, or $y + x5$ **24.** [1.3] $4 + (a + b)$
25. [1.3] $x(y7)$ **26.** [1.3] $7m(n + 2)$

27. [1.3] $6x^3 - 8x^2 + 2$ **28.** [1.3] $47x - 60$
29. [1.4] $-10a^5b^8$ **30.** [1.4] $4xy^6$ **31.** [1.4] 1; 28.09;
-28.09 **32.** [1.4] 3^3, or 27   **33.** [1.4] $125a^6$
34. [1.4] $-\dfrac{a^9}{8b^6}$ **35.** [1.4] $\dfrac{z^8}{x^4y^6}$ **36.** [1.4] $\dfrac{b^{16}}{16a^{20}}$
37. [1.2] $-\frac{16}{7}$ **38.** [1.2] 0   **39.** [1.4] 1.03×10^{-7}
40. [1.4] 3.086×10^{13} **41.** [1.4] 3.741×10^7
42. [1.4] 8×10^{-6} **43.** [1.5] Yes   **44.** [1.5] No
45. [1.5] Yes   **46.** [1.5] No
47. [1.5]   **48.** [1.5]

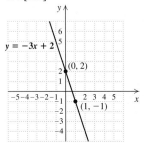   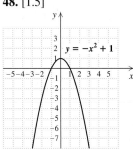

49. [1.5]   **50.** [1.5]

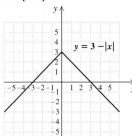   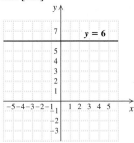

51. [1.5] 3, 1, -1, -3, -1, 1, 3

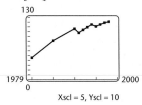

52. [1.6] Let n represent the number; $2n - 13 = 21$
53. [1.6] Let n represent the larger number;
$n + (n - 17) = 115$ **54.** [1.6] Let x represent the
measure of the second angle; $3x + x + 2x = 180$
55. [1.6] **(a)** About 15 unhealthy days; **(b)** 1995; **(c)** about 15
56. [1.6]

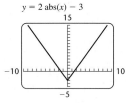

57. [1.1], [1.5] ᵀᵂ A solution of $y = 2x + 1$ is an ordered
pair; there are infinitely many such solutions. There is only

one solution of $3x + 5 = 2$; it is a number.
58. [1.1] ᵀᵂ Without a standard set of rules, an expression
could be interpreted to have more than one value.
59. [1.2], [1.4] $-\frac{23}{24}$ **60.** [1.4] $3^{-2a+2b-8ab}$
61. [1.3] $a2 + cb + cd + ad = ad + a2 + cb + cd = a(d + 2) + c(b + d)$ **62.** [1.2] 0.56556555655556...;
answers may vary

Test: Chapter 1, p. 77

1. [1.1] -47 **2.** [1.1] 3.75 cm^2 **3.** [1.1] **(a)** No;
(b) yes; **(c)** no   **4.** [1.2] -41 **5.** [1.2] -3.7
6. [1.2] -2.11 **7.** [1.2] -14.2 **8.** [1.2] -43.2
9. [1.2] -33.92 **10.** [1.2] $-\frac{19}{12}$ **11.** [1.2] $\frac{5}{49}$
12. [1.2] 6   **13.** [1.2] $-\frac{4}{3}$ **14.** [1.2] $-\frac{3}{2}$
15. [1.3] $y + 7x$, or $x7 + y$, or $y + x7$
16. [1.3] $-3y - 29$ **17.** [1.3] $3x + 8$ **18.** [1.4] $-\dfrac{72}{x^{10}y^6}$
19. [1.4] $-\frac{1}{9}$ **20.** [1.4] $\dfrac{y^8}{36x^4}$ **21.** [1.4] $\dfrac{x^6}{4y^8}$
22. [1.4] 1   **23.** [1.4] 2.0091×10^{-7}
24. [1.4] 2.0×10^{10} **25.** [1.4] 3.05×10^4
26. [1.5] Yes   **27.** [1.5] No
28. [1.5]   **29.** [1.5]

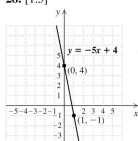   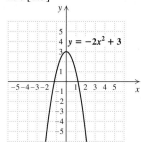

30. [1.5]

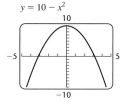

31. [1.6] Let m and n represent the numbers; $mn + 3$, or
$3 + mn$ **32.** [1.6] Let t represent the score on the sixth
test; $\dfrac{94 + 80 + 76 + 91 + 75 + t}{6} = 85$
33. [1.6] 45 mph   **34.** [1.6] 7 mpg
35. [1.4] $16^c x^{6ac} y^{2bc+2c}$ **36.** [1.4] $-9a^3$ **37.** [1.4] $\dfrac{4}{7y^2}$

Chapter 2

Exercise Set 2.1, pp. 89–93

1. No   **3.** Yes   **5.** Yes   **7.** Function   **9.** A relation
but not a function   **11.** Function   **13. (a)** -2;

(b) $\{x \mid -2 \le x \le 5\}$; **(c)** 4; **(d)** $\{y \mid -3 \le y \le 4\}$
15. (a) 3; **(b)** $\{x \mid -1 \le x \le 4\}$; **(c)** 3; **(d)** $\{y \mid 1 \le y \le 4\}$
17. (a) -2; **(b)** $\{x \mid -4 \le x \le 2\}$; **(c)** -2;
(d) $\{y \mid -3 \le y \le 3\}$ **19. (a)** 3; **(b)** $\{x \mid -4 \le x \le 3\}$;
(c) -3; **(d)** $\{y \mid -2 \le y \le 5\}$ **21. (a)** 1;
(b) $\{-3, -1, 1, 3, 5\}$; **(c)** 3; **(d)** $\{-1, 0, 1, 2, 3\}$
23. (a) 4; **(b)** $\{x \mid -3 \le x \le 4\}$; **(c)** $-1, 3$;
(d) $\{y \mid -4 \le y \le 5\}$ **25. (a)** 1; **(b)** $\{x \mid -4 < x \le 5\}$;
(c) $\{x \mid 2 < x \le 5\}$; **(d)** $\{-1, 1, 2\}$ **27.** Yes   **29.** Yes
31. No   **33.** No   **35. (a)** 3; **(b)** -5; **(c)** -11; **(d)** 19;
(e) $2a + 7$; **(f)** $2a + 5$ **37. (a)** 0; **(b)** 1; **(c)** 57;
(d) $5t^2 + 4t$; **(e)** $20a^2 + 8a$; **(f)** $10a^2 + 8a$ **39. (a)** $\frac{3}{5}$;
(b) $\frac{1}{3}$; **(c)** $\frac{4}{7}$; **(d)** 0; **(e)** $\dfrac{x-1}{2x-1}$ **41.** $4\sqrt{3}$ cm$^2 \approx 6.93$ cm^2
43. 36π in$^2 \approx 113.10$ in^2 **45.** $14°$F   **47.** 159.48 cm

49. About 24,000 cases

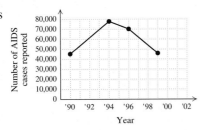

51. About 65,000

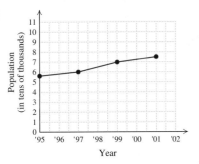

53. About $313,000

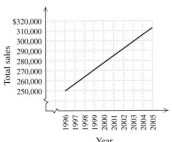

55. TW   **57.** $-6x + 10$ **58.** $-8x - 3$
59. $44x - 26$ **60.** $3y - 21$ **61.** $-10x + 71$
62. $-11x - 6$ **63.** $\frac{1}{3}$ **64.** -1 **65.** TW   **67.** 26;
99   **69.** About 2.29 cm   **71.** About 22 mm   **73.** TW

75.

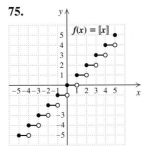

77. Bicycling 14 mph for 1 hr

Exercise Set 2.2, pp. 105–108

1. -2 **3.** -4 **5.** 8   **7.** 0   **9.** 5   **11.** 7
13. 3   **15.** 9   **17.** 1   **19.** 6   **21.** $1\frac{1}{3}$
23. 4 months   **25.** 2 hr 15 min   **27.** 13 lb
29. $4315.14   **31.** Yes   **33.** Yes   **35.** No   **37.** No
39. 16.3   **41.** 9   **43.** 18   **45.** 8   **47.** 7   **49.** 21
51. 2   **53.** 2   **55.** 3   **57.** 7   **59.** 5   **61.** 2
63. 2   **65.** $\frac{49}{9}$ **67.** -4.17619 **69.** $\frac{4}{5}$ **71.** $\frac{23}{8}$
73. $-\frac{1}{2}$ **75.** $\varnothing$; contradiction   **77.** $\{0\}$; conditional
79. $\varnothing$; contradiction   **81.** $\mathbb{R}$; identity   **83.** TW
85. $\{1, 2, 3, 4, 5, 6, 7, 8, 9\}$; $\{x \mid x$ is a positive integer less than
10$\}$ **86.** $\{-8, -7, -6, -5, -4, -3, -2, -1\}$; $\{x \mid x$ is a
negative integer greater than $-9\}$ **87.** Let m and n
represent the numbers; $mn - 3$ **88.** Let x and y represent
the numbers; $(x - y) + 10$, or $10 + (x - y)$ **89.** Let n
represent the number; $9 + 2n$, or $2n + 9$ **90.** Let t
represent the number; $0.42\left(\dfrac{t}{2}\right)$ **91.** TW   **93.** 8
95. $\frac{224}{29}$ **97.** 1   **99.** $-6, 2$ **101.** $-1, 2$
103.   **105.** TW

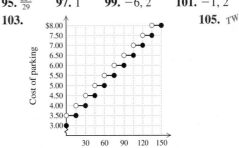

Exercise Set 2.3, pp. 115–119

1. 29, 36   **3.** $\frac{3}{2}$ **5.** Length: 5 ft; width: $\frac{5}{2}$ ft   **7.** $\frac{1}{2}$ hr
9. $\frac{100}{3}$ sec   **11.** $45°, 45°, 90°$ **13.** 4900 megawatts
15. $\frac{1}{5}$ lb   **17.** $2 million   **19.** 38 prints   **21.** $450
23. 6.79×10^8 km   **25.** 2.2×10^{-3} lb   **27.** 8 light years
29. 7.90×10^7 bacteria   **31.** 4.49×10^4 km/h
33. $r = \dfrac{d}{t}$ **35.** $a = \dfrac{F}{m}$ **37.** $I = \dfrac{W}{E}$ **39.** $h = \dfrac{V}{lw}$
41. $k = Ld^2$ **43.** $n = \dfrac{G - w}{150}$ **45.** $l = p - 2w - 2h$

47. $y = \dfrac{C - Ax}{B}$ **49.** $F = \frac{9}{5}C + 32$ **51.** $b_2 = \dfrac{2A}{h} - b_1$

53. $t = \dfrac{d_2 - d_1}{v}$ **55.** $d_1 = d_2 - vt$ **57.** $m = \dfrac{r}{1 + np}$

59. $a = \dfrac{y}{b - c^2}$ **61.** $y = \dfrac{3 - x}{2}$ **63.** $y = x + 7$

65. $y = -3x - 10$ **67.** $y = \dfrac{-x^2 + x + 1}{4}$ **69.** 12%

71. 6 cm   **73.** 8.5 cm   **75.** 9 ft   **77.** 1 year
79. 34 appointments   **81.** TW

83.
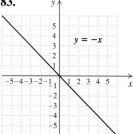

84.
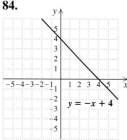

85.
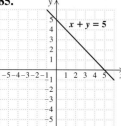

86.
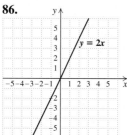

87. $y = \text{abs}(x + 1)$
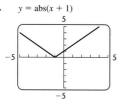

88. $y = -2x^2$
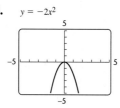

89. TW   **91.** About 10.9 g   **93.** About 7.4 cm
95. 10 points above the average   **97.** $110,000

99. $a = \dfrac{2s - 2v_1t}{t^2}$ **101.** $T_2 = \dfrac{T_1 P_2 V_2}{P_1 V_1}$ **103.** $b = \dfrac{ac}{1 + c}$

105. $a = \dfrac{bc}{1 - c}$ **107.** TW   **109.** TW

111. $y = -2x^2$

113. $y = \text{abs}(x + 2)/7$
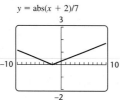

Interactive Discovery, p. 121

1. Yes   **2.** The origin, or $(0, 0)$ **3.** The origin, or $(0, 0)$

Interactive Discovery, p. 121

1. The value of y_2 is 5 more than that of y_1 for the same value of x. The value of y_3 is 7 less than that of y_1.   **2.** The graph of y_4 will look like the graph of y_1, shifted down 3.2 units.
3. The graph is shifted up or down, depending on the sign of b.

Interactive Discovery, pp. 122–123

1. Yes   **2.** No   **3.** No   **4.** Yes   **5.** m

Interactive Discovery, p. 125

1. y_1 and y_3 **2.** y_2 and y_4 **3.** The graph slants up if m is positive and down if m is negative.   **4.** The graph of $y = 7x + 1$ goes through $(0, 1)$ and is steeper than y_1, y_2, and y_3.   **5.** The larger m is, the steeper the graph.

Exercise Set 2.4, pp. 130–135

1.
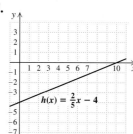

3.
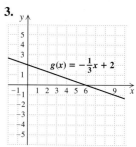

5.
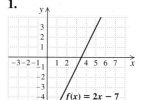

7. $(0, 7)$ **9.** $(0, -6)$

11. $(0, -4.5)$ **13.** $(0, -9)$ **15.** $(0, 204)$ **17.** 2
19. -2 **21.** $-\frac{1}{3}$ **23.** 0   **25.** (a) II; (b) IV; (c) III;
(d) I   **27.** Slope: $\frac{5}{2}$; y-intercept: $(0, 3)$

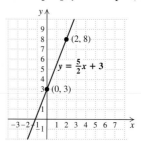

29. Slope: $-\frac{5}{2}$; y-intercept: $(0, 1)$

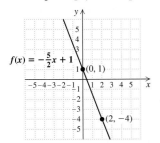

$f(x) = -\frac{5}{2}x + 1$　$(0, 1)$　$(2, -4)$

31. Slope: 2; y-intercept: $(0, -5)$

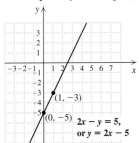

$(1, -3)$　$(0, -5)$　$2x - y = 5$, or $y = 2x - 5$

33. Slope: $\frac{1}{3}$; y-intercept: $(0, 2)$

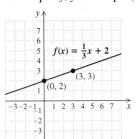

$f(x) = \frac{1}{3}x + 2$　$(3, 3)$　$(0, 2)$

35. Slope: $-\frac{2}{7}$; y-intercept: $(0, 1)$

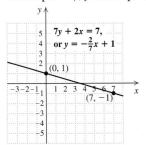

$7y + 2x = 7$, or $y = -\frac{2}{7}x + 1$　$(0, 1)$　$(7, -1)$

37. Slope: -0.25; y-intercept: $(0, 0)$

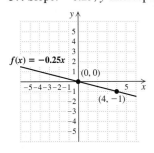

$f(x) = -0.25x$　$(0, 0)$　$(4, -1)$

39. Slope: $\frac{4}{5}$; y-intercept: $(0, -2)$

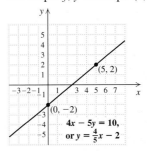

$(5, 2)$　$(0, -2)$　$4x - 5y = 10$, or $y = \frac{4}{5}x - 2$

41. Slope: $\frac{5}{4}$; y-intercept: $(0, -2)$

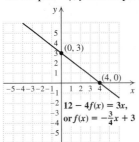

$(4, 3)$　$(0, -2)$　$f(x) = \frac{5}{4}x - 2$

43. Slope: $-\frac{3}{4}$; y-intercept: $(0, 3)$

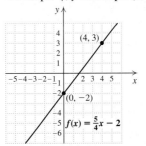

$(0, 3)$　$(4, 0)$　$12 - 4f(x) = 3x$, or $f(x) = -\frac{3}{4}x + 3$

45. Slope: 0; y-intercept: $(0, 2.5)$　　**47.** $f(x) = \frac{2}{3}x - 9$

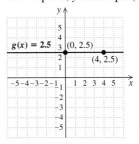

$g(x) = 2.5$　$(0, 2.5)$　$(4, 2.5)$

49. $f(x) = -5x + 2$　　**51.** $f(x) = -\frac{7}{9}x + 5$
53. $f(x) = 5x + \frac{1}{2}$　　**55.** The value is decreasing at a rate of $900 per year.　　**57.** The distance is increasing at a rate of 3 miles per hour.　　**59.** The number of pages read is increasing at a rate of 75 pages per day.　　**61.** The average SAT math score is increasing at a rate of 1 point per thousand dollars of family income.　　**63.** 12 km/h
65. 175,000 hits/yr　　**67.** 300 ft/min　　**69. (a)** II; **(b)** IV; **(c)** I; **(d)** III　　**71.** 25 signifies that the cost per person is $25; 75 signifies that the setup cost for the party is $75.

73. $\frac{1}{2}$ signifies that Tina's hair grows $\frac{1}{2}$ in. per month; 1 signifies that her hair is 1 in. long when cut.   **75.** $\frac{3}{20}$ signifies that the life expectancy of American women increases $\frac{3}{20}$ yr per year, for years after 1950; 72 signifies that the life expectancy in 1950 was 72 yr.   **77.** 2.6 signifies that sales increase \$2.6 billion per year, for years after 1975; 17.8 signifies that sales in 1975 were \$17.8 billion.
79. 0.75 signifies that the cost per mile of a taxi ride is \$0.75; 2 signifies that the minimum cost of a taxi ride is \$2.
81. (a) -5000 signifies a depreciation of \$5000 per year; 90,000 signifies the original value of the truck, \$90,000; **(b)** 18 yr; **(c)** $\{t \mid 0 \le t \le 18\}$ **83. (a)** -150 signifies that the depreciation is \$150 per winter of use; 900 signifies that the original value of the snowblower was \$900; **(b)** after 4 winters of use; **(c)** $\{0, 1, 2, 3, 4, 5, 6\}$ **85.** TW
87. $45x + 54$ **88.** $-125a^6b^9$ **89.** $-26m^7n^4$
90. $8x - 3y - 12$ **91.** $4x^2y^3$ **92.** $\dfrac{x^{10}y^{14}}{4}$ **93.** TW

95. TW   **97.** Slope: $-\dfrac{r}{p}$; y-intercept: $\left(0, \dfrac{s}{p}\right)$
99. Since (x_1, y_1) and (x_2, y_2) are two points on the graph of $y = mx + b$, then $y_1 = mx_1 + b$ and $y_2 = mx_2 + b$. Using the definition of slope, we have

$$\begin{aligned}\text{Slope} &= \frac{y_2 - y_1}{x_2 - x_1} \\ &= \frac{(mx_2 + b) - (mx_1 + b)}{x_2 - x_1} \\ &= \frac{m(x_2 - x_1)}{x_2 - x_1} \\ &= m.\end{aligned}$$

101. False   **103.** False   **105. (a)** III; **(b)** IV; **(c)** I; **(d)** II
107.

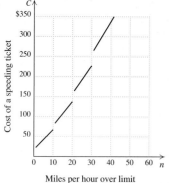

Interactive Discovery, p. 141

1. The lines appear to be parallel.   **2.** The lines appear to be parallel.   **3.** The lines in Exercise 1 still appear parallel; the lines in Exercise 2 are not parallel.

Exercise Set 2.5, pp. 145–146

1. 0   **3.** Undefined   **5.** 0   **7.** Undefined
9. Undefined   **11.** 0   **13.** Undefined   **15.** Undefined
17. $-\frac{2}{3}$
19.

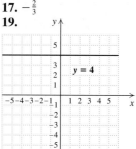

21.

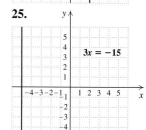

23.

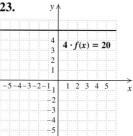

25.

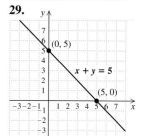

27.

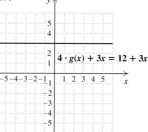

29.

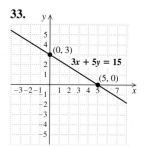

31.

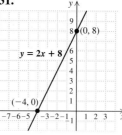

33.

35.

37.

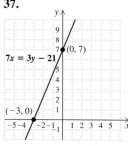

39.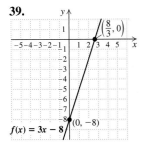

41.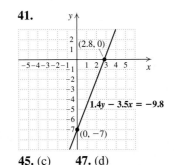

43.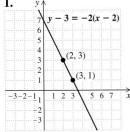

45. (c)   **47.** (d)

49. Yes   **51.** Yes   **53.** No   **55.** Yes   **57.** No
59. $f(x) = 3x + 9$ **61.** $f(x) = -2x - 5$
63. $f(x) = -\frac{2}{5}x - \frac{1}{3}$ **65.** $f(x) = -5$
67. $f(x) = -x + 4$ **69.** $f(x) = \frac{3}{2}x - 4$
71. $f(x) = -\frac{1}{5}x + \frac{1}{5}$ **73.** Linear; $\frac{5}{3}$ **75.** Linear; 0
77. Not linear   **79.** Linear; $\frac{14}{3}$ **81.** Not linear
83. Not linear   **85.** TW   **87.** -1 **88.** -1
89. $6x - 3y + 21$ **90.** $-2x - 10y + 2$ **91.** $-5x - 15$
92. $-2x - 8$ **93.** $\frac{2}{3}x - \frac{2}{3}$ **94.** $-\frac{3}{2}x - \frac{12}{5}$ **95.** TW
97. $4x - 5y = 20$ **99.** Linear   **101.** Linear
103. The slope of equation B is $\frac{1}{2}$ the slope of equation A.
105. $a = 7, b = -3$ **107.** (a) Yes; (b) no

Exercise Set 2.6, pp. 155–160

1.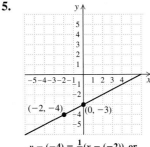

3.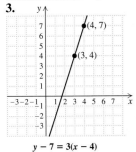

5.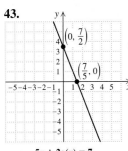

7.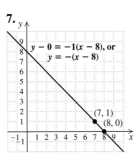

9.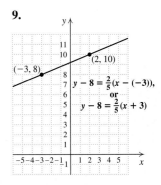

11. $\frac{2}{7}$; $(1, 4)$ **13.** -5; $(7, -2)$ **15.** $-\frac{5}{3}$; $(-2, 1)$
17. $\frac{4}{7}$; $(0, 0)$ **19.** $f(x) = 4x - 11$ **21.** $f(x) = -\frac{3}{5}x - \frac{23}{5}$
23. $f(x) = -0.6x - 5.8$ **25.** $f(x) = \frac{2}{7}x - 5$
27. $f(x) = \frac{1}{2}x + \frac{7}{2}$ **29.** $f(x) = 1.5x - 6.75$
31. $f(x) = 5x - 2$ **33.** $f(x) = \frac{3}{2}x$
35. (a) $R(t) = -0.075t + 46.8$; (b) 41.325 sec; 41.1 sec;
(c) 2021   **37.** (a) $A(t) = 8.0625t + 178.6$;
(b) \$291.475 million   **39.** (a) $N(t) = 3.68t + 43.8$;
(b) 87.96 million tons   **41.** (a) $E(t) = 0.1t + 78.8$;
(b) 80.6 yr   **43.** (a) $A(t) = 0.7t + 74.9$; (b) 83.3 million
acres   **45.** Linear   **47.** Not linear   **49.** Linear
51. (a) $S(t) = 1.25t + 3$; (b) 60,500 shopping centers
53. (a) $W(x) = 0.182x + 62.13333333$; (b) 81.8 yr; this
estimate is higher   **55.** (a) $C(x) = -5.027678571x +$
95.21892857; (b) \$14.78   **57.** TW   **59.** $3x^2 + 7x - 4$
60. $5t^2 - 6t - 3$ **61.** 0   **62.** 0   **63.** 11   **64.** -34
65. TW   **67.** \$1350   **69.** \$30   **71.** 82.8%
73. $y = -\frac{1}{2}x + \frac{17}{2}$ **75.** $y = \frac{1}{2}x + 4$
77. $\{p \mid 0 < p \le 10.6\}$ **79.** (a) $g(x) = x - 8$; (b) -10;
(c) 83

Interactive Discovery, p. 160

1. $f(x) + g(x)$ **2.** Add the y-values of y_1 and y_2 for each
x-value.   **3.** $y_3 = y_4$

Interactive Discovery, pp. 164–165

1. 0   **2.** -1 **3.** $-1, 0$ **4.** Yes   **5.** Yes   **6.** No;
3 is not in the domain of f/g.

Exercise Set 2.7, pp. 166–169

1. 1   **3.** -41 **5.** 12   **7.** $\frac{13}{18}$ **9.** 5
11. $x^2 - 3x + 3$ **13.** $x^2 - x + 3$ **15.** 23   **17.** 5
19. 56   **21.** $\frac{x^2 - 2}{5 - x}$, $x \ne 5$ **23.** $\frac{2}{7}$
25. $0.75 + 2.5 = 3.25$ **27.** Women under 30
29. About 50 million; the number of passengers using Newark
and LaGuardia in 1998   **31.** About 8 million; how many
more passengers used Kennedy than LaGuardia in 1994
33. About 89 million; the number of passengers using the
three airports in 1999   **35.** (a) $\{x \mid x$ is a real number *and*
$x \ne 3\}$; (b) $\{x \mid x$ is a real number *and* $x \ne 6\}$; (c) $\mathbb{R}$; (d) $\mathbb{R}$;

(e) $\{x \mid x$ is a real number *and* $x \neq \frac{5}{2}\}$; **(f)** $\mathbb{R}$ **37.** $\mathbb{R}$
39. $\{x \mid x$ is a real number *and* $x \neq 3\}$ **41.** $\{x \mid x$ is a real number *and* $x \neq 0\}$ **43.** $\{x \mid x$ is a real number *and* $x \neq 1\}$
45. $\{x \mid x$ is a real number *and* $x \neq 2$ *and* $x \neq 4\}$ **47.** $\{x \mid x$ is a real number *and* $x \neq 3\}$ **49.** $\{x \mid x$ is a real number *and* $x \neq 4\}$ **51.** $\{x \mid x$ is a real number *and* $x \neq 4$ *and* $x \neq 5\}$
53. $\{x \mid x$ is a real number *and* $x \neq -1$ *and* $x \neq -\frac{5}{2}\}$
55. 4; 3   **57.** 5; -1 **59.** $\{x \mid 0 \leq x \leq 9\}$;
$\{x \mid 3 \leq x \leq 10\}$; $\{x \mid 3 \leq x \leq 9\}$; $\{x \mid 3 \leq x \leq 9\}$
61.   **63.** ᵀᵂ   **65.** 270

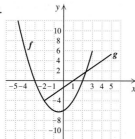

66. 16   **67.** 7.03×10^{-6} **68.** 4.506×10^{12}
69. 430,000,000   **70.** 0.0006037   **71.** Let x represent the first integer; $x + (x + 1) = 145$ **72.** Let n represent the number; $n - (-n) = 20$ **73.** ᵀᵂ   **75.** $\{x \mid x$ is a real number *and* $x \neq -\frac{5}{2}$ *and* $x \neq -3$ *and* $x \neq 1$ *and* $x \neq -1\}$
77. Answers may vary.

79. Domain of $f + g =$ Domain of $f - g =$ Domain of $f \cdot g = \{-2, -1, 0, 1\}$; Domain of $f/g = \{-2, 0, 1\}$
81. Answers may vary. $f(x) = \dfrac{1}{x + 2}$, $g(x) = \dfrac{1}{x - 5}$
83. The domain of $y_3 = \dfrac{2.5x + 1.5}{x - 3}$ is $\{x \mid x$ is a real number *and* $x \neq 3\}$. The CONNECTED mode graph contains the line $x = 3$, whereas the DOT mode graph contains no points having 3 as the first coordinate. Thus the DOT mode graph represents y_3 more accurately.   **85. (a)** IV; **(b)** I; **(c)** II; **(d)** III

Review Exercises: Chapter 2, pp. 171–173

1. [2.1] **(a)** 3; **(b)** $\{x \mid -2 \leq x \leq 4\}$; **(c)** -1;
(d) $\{y \mid 1 \leq y \leq 5\}$ **2.** [2.2] 2   **3.** [2.1] 10.53 yr
4. [2.2] 12   **5.** [2.2] 9   **6.** [2.2] 6.6   **7.** [2.2] $\frac{27}{2}$
8. [2.2] $-\frac{4}{11}$ **9.** [2.2] $\mathbb{R}$; identity   **10.** [2.2] $\varnothing$; contradiction   **11.** [2.2] 0.46170   **12.** [2.3] 49
13. [2.3] 55 in.   **14.** [2.3] 90°, 30°, 60°

15. [2.3] 1.422×10^4 mm³, or 1.422×10^{-5} m³
16. [2.3] $y = \dfrac{3 - 6x}{-5}$, or $\dfrac{6x - 3}{5}$ **17.** [2.3] $m = PS$
18. [2.3] $x = \dfrac{c}{m - r}$ **19.** [2.4] Slope: -4; y-intercept: $(0, -9)$ **20.** [2.4] Slope: $\frac{1}{3}$; y-intercept: $\left(0, -\frac{7}{6}\right)$
21. [2.1] About 510 cans per person   **22.** [2.1] About 670 cans per person   **23.** [2.4] $\frac{4}{7}$ **24.** [2.5] Undefined
25. [2.4] $2500 of personal income per year since high school
26. [2.4] 132,500 homes per month   **27.** [2.4] 645 signifies that tuition is increasing at a rate of $645 a year; 9800 represents tuition costs in 1997   **28.** [2.4] $f(x) = \frac{2}{7}x - 6$
29. [2.5]   **30.** [2.5]

31. [2.5]   **32.** [2.5]

33. [2.5] The window should show both intercepts: $(98, 0)$ and $(0, 14)$. One possibility is $[-10, 120, -5, 15]$, with Xscl $= 10$.
34. [2.5] Perpendicular   **35.** [2.5] Parallel
36. [2.5] Linear   **37.** [2.5] Linear   **38.** [2.5] Not linear
39. [2.5] Not linear   **40.** [2.6] $y - 4 = -2(x - (-3))$, or $y - 4 = -2(x + 3)$ **41.** [2.6] $f(x) = \frac{4}{3}x + \frac{7}{3}$
42. [2.5] 7 months   **43.** [2.6] **(a)** $W(t) = \frac{11}{105}t + 0.75$;
(b) $5.99   **44.** [2.6]

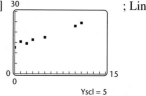

Yscl = 5

45. [2.6] $A(t) = 1.05t + 12.5$
46. [2.6] $A(x) = 0.9688442211x + 13.4281407$
47. [2.7] $\{x \mid x$ is a real number *and* $x \neq -3\}$ **48.** [2.7] $\mathbb{R}$
49. [2.1] -6 **50.** [2.1] 26   **51.** [2.7] 102
52. [2.7] -17 **53.** [2.7] $-\frac{9}{2}$ **54.** [2.1] $3a + 3b - 6$
55. [2.7] $\mathbb{R}$ **56.** [2.7] $\{x \mid x$ is a real number *and* $x \neq 2\}$
57. ᵀᵂ [2.1] For a function, every member of the domain

corresponds to *exactly one* member of the range. Thus, for any function, each member of the domain corresponds to *at least one* member of the range. Therefore, a function is a relation. In a relation, every member of the domain corresponds to *at least one,* but not necessarily *exactly one,* member of the range. Therefore, a relation may or may not be a function. **58.** TW [2.5] The slope of a line is the rise between two points on the line divided by the run between those points. For a vertical line, there is no run between any two points, and division by 0 is undefined; therefore, the slope is undefined. For a horizontal line, there is no rise between any two points, so the slope is 0/run, or 0. **59.** [1.4], [2.5] -9
60. [2.6] $-\frac{9}{2}$ **61.** [2.6] $f(x) = 3.09x + 3.75$
62. [2.4] **(a)** III; **(b)** IV; **(c)** I; **(d)** II

Test: Chapter 2, pp. 174–175

1. [2.1] **(a)** 1; **(b)** $\{x \mid -3 \le x \le 4\}$; **(c)** 3;
(d) $\{y \mid -1 \le y \le 2\}$ **2.** [2.1] **(a)** $40.6 billion;
(b) [2.4] 1.2 signifies that the rate of increase in U.S. book sales was $1.2 billion per year; 21.4 signifies that U.S. book sales were $21.4 billion in 1992 **3.** [2.1] 46 million
4. [2.2] 43 **5.** [2.2] -2 **6.** [2.2] $\mathbb{R}$; identity
7. [2.3] 18, 20, 22 **8.** [2.3] 20.175% increase
9. [2.3] About 4.2×10^8 mi **10.** [2.3] $l = \dfrac{S - 2wh}{2h + 2w}$

11. [2.3] $y = \dfrac{x - 7 - 3x^2}{2}$ **12.** [2.4] Slope: $-\frac{3}{5}$;
y-intercept: $(0, 12)$ **13.** [2.4] Slope: $-\frac{2}{5}$; y-intercept:
$\left(0, -\frac{7}{5}\right)$ **14.** [2.4] $\frac{5}{8}$ **15.** [2.4] 0 **16.** [2.4] 75 calories per 30 minutes, or 2.5 calories per minute
17. [2.4] $f(x) = -5x - 1$
18. [2.4] **19.** [2.5]

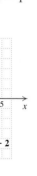
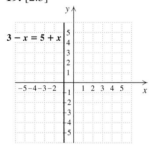

20. [2.5] **21.** [2.5] Yes

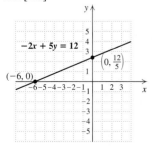

22. [2.5] Parallel **23.** [2.5] Perpendicular
24. [2.5] **(a)**, **(c)** **25.** [2.6] $y - (-4) = 4(x - (-2))$, or
$y + 4 = 4(x + 2)$ **26.** [2.6] $f(x) = -x + 2$
27. [2.6] **(a)** $C(m) = 0.3m + 25$; **(b)** $175
28. [2.6] **(a)** $A(x) = -0.554571187x + 90.14595817$;
(b) 29.1 accidental deaths **29.** [2.7] $\{x \mid x$ is a real number *and* $x \ne -\frac{1}{2}\}$ **30. (a)** [2.1] 5; **(b)** [2.7] -130;
(c) [2.7] $\{x \mid x$ is a real number *and* $x \ne -\frac{4}{3}\}$
31. [2.1], [2.4] **(a)** 30 mi; **(b)** 15 mph
32. [2.5], [2.6] $y = \frac{2}{5}x + \frac{16}{5}$ **33.** [2.5], [2.6] $y = -\frac{5}{2}x - \frac{11}{2}$

34. [2.6] $s = -\dfrac{3}{2}r + \dfrac{27}{2}$, or $\dfrac{27 - 3r}{2}$

35. [2.7] $h(x) = 7x - 2$

Chapter 3

Exercise Set 3.1, pp. 188–192

1. Yes **3.** No **5.** Yes **7.** Yes **9.** $(4, 1)$
11. $(2, -1)$ **13.** $(4, 3)$ **15.** $(-3, -2)$ **17.** $(-3, 2)$
19. Approximately $(1.53, 2.58)$ **21.** $(3, -7)$ **23.** $(7, 2)$
25. $(4, 0)$ **27.** Approximately $(2.23, 1.14)$ **29.** No
solution **31.** Approximately $(-0.73, 0)$, or $\left(-\frac{11}{15}, 0\right)$
33. $\{(x, y) \mid y = 3 - x\}$ **35.** No solution **37.** All but
Exercises 29 and 35 **39.** Exercise 33 **41.** Let x
represent the larger number and y the smaller number;
$x - y = 11, 3x + 2y = 123$ **43.** Let x represent the
number of $8.50 brushes sold and y the number of $9.75
brushes sold; $x + y = 45, 8.50x + 9.75y = 398.75$
45. Let x and y represent the angles; $x + y = 180$,
$x = 2y - 3$ **47.** Let x represent the number of two-point
shots and y the number of free throws; $x + y = 64$,
$2x + y = 100$ **49.** Let h represent the number of vials of
Humulin sold and n the number of vials of Novolin;
$h + n = 50, 23.97h + 34.39n = 1406.90$ **51.** Let l
represent the length, in feet, and w the width, in feet;
$2l + 2w = 228, w = l - 42$ **53.** Let w represent the
number of wins and t the number of ties; $2w + t = 60$,
$w = t + 9$ **55.** Let x represent the number of ounces of
lemon juice and y the number of ounces of linseed oil;
$y = 2x, x + y = 32$ **57.** Let x represent the number of
general-interest films rented and y the number of children's
films rented; $x + y = 77, 3x + 1.5y = 213$ **59.** Let y
represent the percent of the population in the work
force and x the number of years since 1970; $y = \frac{83}{180}x + \frac{422}{9}$,
$y = -\frac{5}{36}x + \frac{7091}{90}$; about 2023 **61.** Let y represent the
amount of paper, in millions of tons, and x the number of
years since 1990; $y = 0.825x + 77.5, y = 1.275x + 31.4$;
about 2092 **63.** Let y represent the per capita
consumption, in gallons, and x the number of years since
1950; $y = -\frac{7}{25}x + 38, y = \frac{43}{50}x + 10$; about 1975
65. Women: $y_1 = 0.5660768453x + 44.59989889$, men:

$y_2 = -0.1678968655x + 79.48407482$, where x is the number of years since 1970; about 2018
67. Waste generated: $y_1 = 1.41x + 72.68$, paper recycled: $y_2 = 1.81x + 28.86$, where x is the number of years since 1990; about 2100 **69.** TW **71.** 15 **72.** $\frac{19}{12}$ **73.** $\frac{9}{20}$
74. $\frac{13}{3}$ **75.** $y = -\frac{3}{4}x + \frac{7}{4}$ **76.** $y = \frac{2}{5}x - \frac{9}{5}$ **77.** TW
79. 1994 **81.** Answers may vary. **(a)** $x + y = 6$, $x - y = 4$; **(b)** $x + y = 1$, $2x + 2y = 3$; **(c)** $x + y = 1$, $2x + 2y = 2$ **83.** $A = -\frac{17}{4}, B = -\frac{12}{5}$ **85.** Let x and y represent the number of years that Lou and Juanita have taught at the university, respectively; $x + y = 46$, $x - 2 = 2.5(y - 2)$ **87.** Let s and v represent the number of ounces of baking soda and vinegar needed, respectively; $s = 4v, s + v = 16$ **89.** $(0,0), (1,1)$ **91.** (c)
93. (b)

Exercise Set 3.2, pp. 198–200

1. $(2, -3)$ **3.** $\left(\frac{21}{5}, \frac{12}{5}\right)$ **5.** $(2, -2)$
7. $\{(x,y) \mid 2x - 3 = y\}$ **9.** $(-2, 1)$ **11.** $\left(\frac{1}{2}, \frac{1}{2}\right)$
13. $\left(\frac{19}{8}, \frac{1}{8}\right)$ **15.** No solution **17.** $(1, 2)$ **19.** $(3, 0)$
21. $(-1, 2)$ **23.** $\left(\frac{128}{31}, -\frac{17}{31}\right)$ **25.** $(6, 2)$ **27.** No solution **29.** $\left(\frac{110}{19}, -\frac{12}{19}\right)$ **31.** $(3, -1)$
33. $\{(x,y) \mid -4x + 2y = 5\}$ **35.** $\left(\frac{140}{13}, -\frac{50}{13}\right)$
37. $(-2, -9)$ **39.** $(30, 6)$ **41.** $(x,y) \mid x = 2 + 3y\}$
43. No solution **45.** $(140, 60)$ **47.** $\left(\frac{1}{3}, -\frac{2}{3}\right)$ **49.** (d)
51. (b) **53.** TW **55.** 4 mi **56.** 86
57. $\$11\frac{2}{3}$ billion on bathrooms; $\$23\frac{1}{3}$ billion on kitchens
58. 30 m, 90 m, 360 m **59.** 450.5 mi **60.** 460.5 mi
61. TW **63.** $m = -\frac{1}{2}, b = \frac{5}{2}$ **65.** $a = 5, b = 2$
67. $\left(-\frac{32}{17}, \frac{38}{17}\right)$ **69.** $\left(-\frac{1}{5}, \frac{1}{10}\right)$

Exercise Set 3.3, pp. 212–214

1. 29, 18 **3.** $\$8.50$-brushes: 32; $\$9.75$-brushes: 13
5. $119°, 61°$ **7.** Two-point shots: 36; foul shots: 28
9. Humulin: 30; Novolin: 20 **11.** Width: 36 ft; length: 78 ft **13.** Wins: 23; ties: 14 **15.** Lemon juice: $10\frac{2}{3}$ oz; linseed oil: $21\frac{1}{3}$ oz **17.** General interest: 65; children's: 12
19. Boxes: 13; four-packs: 27 **21.** Kenyan: 8 lb; Sumatran: 12 lb **23.** 5 lb of each **25.** 25%-acid: 4 L; 50%-acid: 6 L **27.** $\$7500$ at 6%; $\$4500$ at 9%
29. Arctic Antifreeze: 12.5 L; Frost No-More: 7.5 L
31. Length: 76 m; width: 19 m **33.** Quarters: 17; fifty-cent pieces: 13 **35.** 375 km **37.** 14 km/h **39.** About 1489 mi **41.** TW **43.** 16 **44.** 11 **45.** -28
46. -10 **47.** $\frac{49}{12}$ **48.** $\frac{13}{10}$ **49.** TW **51.** Burl: 40; son: 20 **53.** Length: $\frac{288}{5}$ in.; width: $\frac{102}{5}$ in. **55.** $\frac{120}{7}$ lb
57. 4 km **59.** 82 **61.** Brown: 0.8 gal; neutral: 0.2 gal
63. 45 L **65.** $P(x) = \dfrac{0.1 + x}{1.5}$. (This expresses the percent as a decimal quantity.)

Exercise Set 3.4, pp. 221–223

1. No **3.** $(1, 2, 3)$ **5.** $(-1, 5, -2)$ **7.** $(3, 1, 2)$
9. $(-3, -4, 2)$ **11.** $(2, 4, 1)$ **13.** $(-3, 0, 4)$ **15.** The equations are dependent. **17.** $(3, -5, 8)$ **19.** $\left(\frac{3}{5}, \frac{2}{3}, -3\right)$
21. $\left(4, \frac{1}{2}, -\frac{1}{2}\right)$ **23.** $(17, 9, 79)$ **25.** $\left(\frac{1}{4}, -\frac{1}{2}, -\frac{1}{4}\right)$
27. $(20, 62, 100)$ **29.** No solution **31.** The equations are dependent. **33.** TW **35.** Let x and y represent the numbers; $x = 2y$ **36.** Let x and y represent the numbers; $x + y = 3x$ **37.** Let x represent the first number; $x + (x + 1) + (x + 2) = 45$ **38.** Let x and y represent the numbers; $x + 2y = 17$ **39.** Let $x, y,$ and z represent the numbers; $x + y = 5z$ **40.** Let x and y represent the numbers; $xy = 2(x + y)$ **41.** TW **43.** $(1, -1, 2)$
45. $(-3, -1, 0, 4)$ **47.** $\left(-\frac{1}{2}, -1, -\frac{1}{3}\right)$ **49.** 14
51. $z = 8 - 2x - 4y$

Exercise Set 3.5, pp. 227–229

1. 16, 19, 22 **3.** 8, 21, -3 **5.** $32°, 96°, 52°$
7. Automatic transmission: $\$865$; power door locks: $\$520$; air conditioning: $\$375$ **9.** Elrod: 20; Dot: 24; Wendy: 30
11. 10-oz cups: 11; 14-oz cups: 15; 20-oz cups: 8
13. First fund: $\$45,000$; second fund: $\$10,000$; third fund: $\$25,000$ **15.** Roast beef: 2; baked potato: 1; broccoli: 2
17. Man: 3.6; woman: 18.1; one-year-old child: 50
19. Two-point field goals: 32; three-point field goals: 5; foul shots: 13 **21.** TW **23.** -8 **24.** 33 **25.** -55
26. -71 **27.** $-14x + 21y - 35z$
28. $-24a - 42b + 54c$ **29.** $-5a$ **30.** $11x$ **31.** TW
33. 464 **35.** Adults: 5; students: 1; children: 94
37. $180°$

Exercise Set 3.6, pp. 236–237

1. $\left(-\frac{1}{3}, -4\right)$ **3.** $(-4, 3)$ **5.** $\left(\frac{3}{2}, \frac{5}{2}\right)$ **7.** $\left(2, \frac{1}{2}, -2\right)$
9. $(2, -2, 1)$ **11.** $\left(4, \frac{1}{2}, -\frac{1}{2}\right)$ **13.** $\left(\frac{93}{34}, -\frac{7}{34}, \frac{67}{34}\right)$
15. Approximately $(-0.53, -1.29, -0.62)$
17. $(1, -3, -2, -1)$ **19.** Dimes: 21; quarters: 22
21. $\$4.05$-granola: 5 lb; $\$2.70$-granola: 10 lb **23.** $\$400$ at 7%; $\$500$ at 8%; $\$1600$ at 9% **25.** TW **27.** 13
28. -22 **29.** 37 **30.** 422 **31.** TW **33.** 1324

Exercise Set 3.7, pp. 241–243

1. **(a)** $P(x) = 20x - 300,000$; **(b)** 15,000 units
3. **(a)** $P(x) = 50x - 120,000$; **(b)** 2400 units
5. **(a)** $P(x) = 45x - 22,500$; **(b)** 500 units
7. **(a)** $P(x) = 18x - 16,000$; **(b)** 889 units
9. **(a)** $P(x) = 50x - 100,000$; **(b)** 2000 units
11. $(\$70, 300)$ **13.** $(\$22, 474)$ **15.** $(\$50, 6250)$
17. $(\$10, 1070)$ **19.** **(a)** $C(x) = 125,300 + 450x$;
(b) $R(x) = 800x$; **(c)** $P(x) = 350x - 125,300$;
(d) $\$90,300$ loss, $\$14,700$ profit; **(e)** (358 computers, $\$286,400$) **21.** **(a)** $C(x) = 16,404 + 6x$; **(b)** $R(x) = 18x$;

(c) $P(x) = 12x - 16{,}404$; **(d)** \$19,596 profit, \$4404 loss;
(e) (1367 dozen caps, \$24,606)   **23. (a)** \$8.74;
(b) 793 units   **25.** TW   **27.** 12   **28.** 15   **29.** $\frac{8}{3}$
30. 4   **31.** $\frac{9}{2}$ **32.** $\frac{1}{3}$ **33.** TW   **35.** (\$5, 300 yo-yo's)
37. (a) $S(p) = 15.97p - 1.05$; **(b)** $D(p) = -11.26p + 41.16$;
(c) (\$1.55, 23.7 million jars)

Review Exercises: Chapter 3, pp. 246–247

1. [3.1] $(-2, 1)$ **2.** [3.1] $(3, 2)$ **3.** [3.2] $\left(-\frac{11}{15}, -\frac{43}{30}\right)$
4. [3.2] No solution   **5.** [3.2] $\left(-\frac{4}{5}, \frac{2}{5}\right)$ **6.** [3.2] $\left(\frac{37}{19}, \frac{53}{19}\right)$
7. [3.2] $\left(\frac{76}{17}, -\frac{2}{119}\right)$ **8.** [3.2] $(2, 2)$
9. [3.2] $\{(x, y) \mid 3x + 4y = 6\}$ **10.** [3.3] DVD: \$29;
videocassette: \$14   **11.** [3.3] 4 hr   **12.** [3.3] 8% juice:
10 L; 15% juice: 4 L   **13.** [3.4] $(4, -8, 10)$
14. [3.4] The equations are dependent.   **15.** [3.4] $(2, 0, 4)$
16. [3.2] No solution   **17.** [3.4] $\left(\frac{8}{9}, -\frac{2}{3}, \frac{10}{9}\right)$ **18.** [3.5] A:
90°; B: 67.5°; C: 22.5°   **19.** [3.5] 641   **20.** [3.5] \$20
bills; 7; \$5 bills: 3; \$1 bills: 4   **21.** [3.6] $\left(55, -\frac{89}{2}\right)$
22. [3.6] $(-1, 1, 3)$ **23.** [3.6] $\left(-\frac{32}{15}, -\frac{439}{45}, -\frac{421}{45}\right)$
24. [3.7] (\$3, 81)   **25.** [3.7] **(a)** $C(x) = 1.5x + 18{,}000$;
(b) $R(x) = 6x$; **(c)** $P(x) = 4.5x - 18{,}000$; **(d)** \$11,250 loss,
\$4500 profit; **(e)** (4000 pints of honey, \$24,000)   **26.** TW
27. TW   **28.** [3.7] 10,000 pints   **29.** [3.1] $(0, 2), (1, 3)$
30. [3.5] $a = -\frac{2}{3}$, $b = -\frac{4}{3}$, $c = 3$; $f(x) = -\frac{2}{3}x^2 - \frac{4}{3}x + 3$

Test: Chapter 3, pp. 247–248

1. [3.1] $(2, 4)$ **2.** [3.1] $(2, 1)$ **3.** [3.2] $\left(3, -\frac{11}{3}\right)$
4. [3.2] $\left(\frac{15}{7}, -\frac{18}{7}\right)$ **5.** [3.2] $\left(-\frac{3}{2}, -\frac{3}{2}\right)$ **6.** [3.2] No
solution   **7.** [3.3] Length: 30 units; width:
18 units   **8.** [3.5] Mortgage: \$74,000; car loan: \$600;
credit-card bill: \$700   **9.** [3.4] The equations are
dependent.   **10.** [3.4] $\left(2, -\frac{1}{2}, -1\right)$
11. [3.4] No solution   **12.** [3.4] $(0, 1, 0)$
13. [3.6] $\left(\frac{34}{107}, -\frac{104}{107}\right)$ **14.** [3.6] $(3, 1, -2)$
15. [3.6] $\left(\frac{202}{9}, \frac{281}{18}, \frac{490}{9}\right)$ **16.** [3.5] 3.5 hr   **17.** [3.7] (\$3, 55)
18. [3.7] **(a)** $C(x) = 25x + 40{,}000$; **(b)** $R(x) = 70x$;
(c) $P(x) = 45x - 40{,}000$; **(d)** \$26,500 loss, \$500 profit;
(e) (889 radios, \$62,230)   **19.** [2.4], [3.3] $m = 7$, $b = 10$
20. [3.5] Adult: 1346; senior citizen: 335; child: 1651

Cumulative Review: Chapters 1–3, pp. 248–250

1. [2.2] 14.87   **2.** [2.2] -22 **3.** [2.2] -42.9
4. [2.2] 20   **5.** [2.2] $-\frac{21}{4}$ **6.** [2.2] 6   **7.** [2.2] -5
8. [2.2] $\frac{10}{9}$ **9.** [2.2] $-\frac{32}{5}$ **10.** [2.2] $\frac{18}{17}$ **11.** [1.4] x^{11}
12. [1.4] $-\dfrac{40x}{y^5}$ **13.** [1.4] $-288x^4y^{18}$ **14.** [1.4] y^{10}
15. [1.4] $-\dfrac{2a^{11}}{5b^{33}}$ **16.** [1.4] $\dfrac{81x^{36}}{256y^8}$
17. [1.4] 1.11735×10^6 **18.** [1.4] 4×10^6

19. [2.3] $b = \dfrac{2A}{h} - t$, or $\dfrac{2A - ht}{h}$ **20.** [1.5] Yes

21. [2.4]
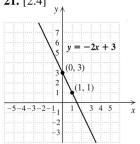

22. [2.1]
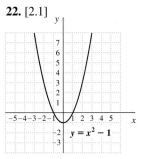

23. [2.5]
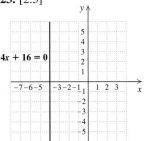

24. [2.5]
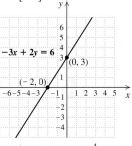

25. [2.4] Slope: $\frac{9}{4}$; y-intercept: $(0, -3)$ **26.** [2.4] $\frac{4}{3}$
27. [2.6] $y = -3x - 5$ **28.** [2.6] $y = -\frac{1}{10}x + \frac{12}{5}$
29. [2.5] Parallel   **30.** [2.5] $y = -\frac{3}{2}x - 3$
31. [2.1] $\{-5, -3, -1, 1, 3\}$; $\{-3, -2, 1, 4, 5\}$; -2; 3
32. [2.7] $\left\{x \mid x \text{ is a real number } and \ x \neq \frac{1}{2}\right\}$ **33.** [2.1] -31
34. [2.1] 3   **35.** [2.7] 7   **36.** [2.7] $8a^2 + 4a - 4$
37. [3.2] $(1, 1)$ **38.** [3.2] $(-2, 3)$ **39.** [3.2] $\left(-3, \frac{2}{5}\right)$
40. [3.4] $(-3, 2, -4)$ **41.** [3.4] $(0, -1, 2)$
42. [3.6] $\left(-\frac{33}{56}, -\frac{169}{224}\right)$ **43.** [3.6] $\left(\frac{11}{15}, \frac{2}{15}, \frac{16}{15}\right)$
44. [3.3] 8, 18   **45.** [2.4] 64.8 recovery plans per year
46. [2.4] 9 signifies that the number of U.S. pleasure trips is
increasing by 9 million each year; 616 signifies that there
were 616 million U.S. pleasure trips in 1994.
47. [2.6] **(a)** $A(t) = \frac{11}{60}t + 6.6$; **(b)** 10.45 million departures
48. [3.3] Chewies: 13 oz; Crunchies: 7 oz   **49.** [2.3] 3, 5, 7
50. [2.3] 90   **51.** [3.3] Length: 10 cm; width: 6 cm
52. [3.3] Nickels: 19; dimes: 15   **53.** [3.5] \$120
54. [3.5] Wins: 23; losses: 33; ties: 8   **55.** [3.5] Cookie: 90;
banana: 80; yogurt: 165
56. (a) [1.6]

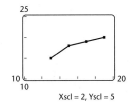

Xscl = 2, Yscl = 5

(b) [2.6] $f(x) = 0.8x + 5.2$; **(c)** [2.1] 18%
57. [1.4] $-12x^{2a}y^{b+y+3}$ **58.** [2.6] \$151,000
59. [2.6], [3.3] $m = -\frac{5}{9}$, $b = -\frac{2}{9}$

Chapter 4

Exercise Set 4.1, pp. 263–268

1. No, no, yes, yes **3.** No, yes, yes, no
5. $\{y \mid y < 6\}, (-\infty, 6)$
7. $\{x \mid x \geq -4\}, [-4, \infty)$
9. $\{t \mid t > -3\}, (-3, \infty)$
11. $\{x \mid x \leq -7\}, (-\infty, -7]$
13. $\{x \mid x > -5\}$, or $(-5, \infty)$
15. $\{a \mid a \leq -20\}$, or $(-\infty, -20]$
17. $\{x \mid x \leq 12\}$, or $(-\infty, 12]$
19. $\{y \mid y > -9\}$, or $(-9, \infty)$
21. $\{y \mid y \leq 14\}$, or $(-\infty, 14]$
23. $\{t \mid t < -9\}$, or $(-\infty, -9)$
25. $\{x \mid x < 50\}$, or $(-\infty, 50)$
27. $\{y \mid y \geq -0.4\}$, or $[-0.4, \infty)$
29. $\left\{y \mid y \geq \frac{9}{10}\right\}$, or $\left[\frac{9}{10}, \infty\right)$
31. $\{y \mid y > 3\}$, or $(3, \infty)$
33. $\{x \mid x \leq 4\}$, or $(-\infty, 4]$
35. $\left\{x \mid x < -\frac{2}{5}\right\}$, or $\left(-\infty, -\frac{2}{5}\right)$
37. $\{x \mid x \geq 11.25\}$, or $[11.25, \infty)$
39. $\{x \mid x \geq 2\}$, or $[2, \infty)$ **41.** $\{x \mid x < 3\}$, or $(-\infty, 3)$
43. (a) $\{x \mid x < 4\}$, or $(-\infty, 4)$; (b) $\{x \mid x \geq 2\}$, or $[2, \infty)$;
(c) $\{x \mid x \geq 3.2\}$, or $[3.2, \infty)$ **45.** $\left\{y \mid y \leq -\frac{53}{6}\right\}$, or $\left(-\infty, -\frac{53}{6}\right]$
47. $\left\{t \mid t < \frac{29}{5}\right\}$, or $\left(-\infty, \frac{29}{5}\right)$ **49.** $\left\{m \mid m > \frac{7}{3}\right\}$, or $\left(\frac{7}{3}, \infty\right)$
51. $\{x \mid x \geq 2\}$, or $[2, \infty)$ **53.** $\{y \mid y < 5\}$, or $(-\infty, 5)$
55. $\left\{x \mid x \leq \frac{4}{7}\right\}$, or $\left(-\infty, \frac{4}{7}\right]$ **57.** Mileages less than or equal
to 150 mi **59.** \$11,500 or more **61.** For 1175 min
or more **63.** More than 25 checks per month
65. Gross sales greater than \$7000 **67.** More than \$1850
69. At least 625 people **71.** (a) $\left\{x \mid x < 8181\frac{9}{11}\right\}$, or
$\{x \mid x \leq 8181\}$; (b) $\left\{x \mid x > 8181\frac{9}{11}\right\}$, or $\{x \mid x \geq 8182\}$
73. $t(x) = 8744.4x + 21,511.2$; years after 1998
75. Swimming: $f(x) = -1.1625x + 73.3125$; equipment:
$g(x) = 1.35x + 28.55$; years after 2002
77. $f(x) = 0.0056862962x + 1.891866531$; years after 2006
79. Swimming: $f(x) = -1.174729242x + 73.30288809$;
equipment: $g(x) = 1.273826715x + 31.20415162$; years after
2002 **81.** ᴛᴡ **83.** $\{x \mid x$ is a real number *and* $x \neq 2\}$
84. $\{x \mid x$ is a real number *and* $x \neq -3\}$ **85.** $\{x \mid x$ is a real
number *and* $x \neq \frac{7}{2}\}$ **86.** $\{x \mid x$ is a real number *and* $x \neq \frac{9}{4}\}$

87. $7x + 10$ **88.** $22x - 7$ **89.** ᴛᴡ
91. $\left\{x \mid x \leq \dfrac{2}{a-1}\right\}$ **93.** $\left\{y \mid y \geq \dfrac{2a+5b}{b(a-2)}\right\}$
95. $\left\{x \mid x > \dfrac{4m-2c}{d-(5c+2m)}\right\}$ **97.** False; $2 < 3$ and $4 < 5$,
but $2 - 4 = 3 - 5$. **99.** ᴛᴡ
101. $\mathbb{R}$
103. $\{x \mid x$ is a real number *and* $x \neq 0\}$

Interactive Discovery, p. 276

1. Domain of $f = (-\infty, 3]$; domain of $g = [-1, \infty)$
2. Domain of $f + g =$ domain of $f - g =$ domain of
$f \cdot g = [-1, 3]$ **3.** By finding the intersection of the
domains of f and g

Exercise Set 4.2, pp. 277–279

1. $\{9, 11\}$ **3.** $\{1, 5, 10, 15, 20\}$ **5.** $\{b, d, f\}$
7. $\{r, s, t, u, v\}$ **9.** $\varnothing$ **11.** $\{3, 5, 7\}$
13. $(3, 8)$
15. $[-6, -2]$
17. $(-\infty, -2) \cup (3, \infty)$
19. $(-\infty, -1] \cup (5, \infty)$
21. $(-2, 4]$
23. $(-2, 4)$
25. $(-\infty, 5) \cup (7, \infty)$
27. $(-\infty, -4] \cup [5, \infty)$
29. $[-6, 4)$
31. $[3, 7)$
33. $(-\infty, 5)$
35. $(-\infty, \infty)$
37. $(7, \infty)$
39. $\{t \mid -3 < t < 5\}$, or $(-3, 5)$
41. $\{x \mid -1 < x \leq 4\}$, or $(-1, 4]$
43. $\{a \mid -2 \leq a < 2\}$, or $[-2, 2)$
45. $\mathbb{R}$, or $(-\infty, \infty)$
47. $\{x \mid 1 \leq x \leq 3\}$, or $[1, 3]$
49. $\left\{x \mid -\frac{7}{2} < x \leq 7\right\}$, or $\left(-\frac{7}{2}, 7\right]$

51. $\{x \mid x \le 1 \ or \ x \ge 3\}$, or $(-\infty, 1] \cup [3, \infty)$

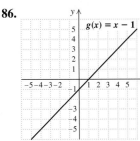

53. $\{x \mid x < 3 \ or \ x > 4\}$, or $(-\infty, 3) \cup (4, \infty)$

55. $\left\{a \mid a < \frac{7}{2}\right\}$, or $\left(-\infty, \frac{7}{2}\right)$

57. $\{a \mid a < -5\}$, or $(-\infty, -5)$

59. $\mathbb{R}$, or $(-\infty, \infty)$

61. $\{t \mid t \le 6\}$, or $(-\infty, 6]$

63. $(-1, 6)$ **65.** $(-\infty, -7) \cup (-7, \infty)$ **67.** $[6, \infty)$
69. $\left(-\infty, \frac{5}{2}\right) \cup \left(\frac{5}{2}, \infty\right)$ **71.** $[-4, \infty)$ **73.** $(-\infty, 4]$
75. $[-3, 4]$ **77.** $\left[\frac{3}{4}, \infty\right)$ **79.** $\left[-\frac{2}{3}, 5\right) \cup (5, \infty)$
81. TW   **83.**

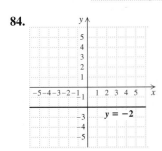

84.

85.

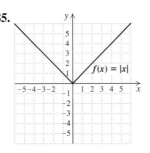

86.

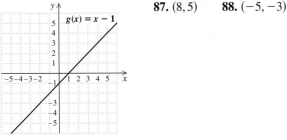

87. $(8, 5)$ **88.** $(-5, -3)$

89. TW   **91.** Between 12 and 240 trips   **93.** Sizes between 6 and 13   **95. (a)** $1945.4° \le F < 4820°$; **(b)** $1761.44° \le F < 3956°$
97. $\left\{a \mid -\frac{3}{2} \le a \le 1\right\}$, or $\left[-\frac{3}{2}, 1\right]$;

99. $\{x \mid -4 < x \le 1\}$, or $(-4, 1]$;

101. True   **103.** False   **105.** $\left[-\frac{5}{2}, 1\right) \cup (1, \infty)$
107. Left to the student

Interactive Discovery, p. 281
1. $\{-4, 4\}$ **2.** $-a, a$ **3.** $\{0\}$ **4.** $\varnothing$ **5.** No solution

Interactive Discovery, p. 285
1. $(-4, 4)$ **2.** $(-\infty, -4) \cup (4, \infty)$ **3.** $[-4, 4]$
4. $(-\infty, -4] \cup [4, \infty)$ **5.** $\mathbb{R}$ **6.** $\varnothing$

Exercise Set 4.3, pp. 288–289
1. $\{-5, 1\}$ **3.** $(-5, 1)$ **5.** $(-\infty, -5] \cup [1, \infty)$
7. $\{-4, 4\}$ **9.** $\varnothing$ **11.** $\{-7.3, 7.3\}$ **13.** $\{0\}$
15. $\left\{-\frac{9}{5}, 1\right\}$ **17.** $\varnothing$ **19.** $\{-5, 11\}$ **21.** $\{5, 7\}$
23. $\{-1, 9\}$ **25.** $\{-9, 9\}$ **27.** $\{-2, 2\}$ **29.** $\left\{-\frac{14}{5}, \frac{22}{5}\right\}$
31. $\{4, 10\}$ **33.** $\left\{\frac{3}{2}, \frac{7}{2}\right\}$ **35.** $\left\{-\frac{4}{3}, 4\right\}$ **37.** $\{-8.3, 8.3\}$
39. $\left\{-\frac{8}{3}, 4\right\}$ **41.** $\{1, 11\}$ **43.** $\left\{\frac{3}{2}\right\}$ **45.** $\left\{-4, -\frac{10}{9}\right\}$
47. $\mathbb{R}$ **49.** $\{1\}$ **51.** $\left\{32, \frac{8}{3}\right\}$
53. $\{a \mid -7 \le a \le 7\}$, or $[-7, 7]$
55. $\{x \mid x < -8 \ or \ x > 8\}$, or $(-\infty, -8) \cup (8, \infty)$
57. $\{t \mid t < 0 \ or \ t > 0\}$, or $(-\infty, 0) \cup (0, \infty)$
59. $\{x \mid -2 < x < 8\}$, or $(-2, 8)$
61. $\{x \mid -8 \le x \le 4\}$, or $[-8, 4]$
63. $\{x \mid x < -2 \ or \ x > 8\}$, or $(-\infty, -2) \cup (8, \infty)$
65. $\mathbb{R}$, or $(-\infty, \infty)$
67. $\left\{a \mid a \le -\frac{2}{3} \ or \ a \ge \frac{10}{3}\right\}$, or $\left(-\infty, -\frac{2}{3}\right] \cup \left[\frac{10}{3}, \infty\right)$
69. $\{y \mid -9 < y < 15\}$, or $(-9, 15)$;
71. $\{x \mid x \le -8 \ or \ x \ge 0\}$, or $(-\infty, -8] \cup [0, \infty)$
73. $\left\{y \mid y < -\frac{4}{3} \ or \ y > 4\right\}$, or $\left(-\infty, -\frac{4}{3}\right) \cup (4, \infty)$;
75. $\varnothing$
77. $\left\{x \mid x \le -\frac{2}{15} \ or \ x \ge \frac{14}{15}\right\}$, or $\left(-\infty, -\frac{2}{15}\right] \cup \left[\frac{14}{15}, \infty\right)$;
79. $\{m \mid -12 \le m \le 2\}$, or $[-12, 2]$;
81. $\{a \mid -6 < a < 0\}$, or $(-6, 0)$
83. $\left\{x \mid -\frac{1}{2} \le x \le \frac{7}{2}\right\}$, or $\left[-\frac{1}{2}, \frac{7}{2}\right]$

85. $\left\{x \mid x \le -\frac{7}{3} \text{ or } x \ge 5\right\}$, or $\left(-\infty, -\frac{7}{3}\right] \cup [5, \infty)$

87. $\{x \mid -4 < x < 5\}$, or $(-4, 5)$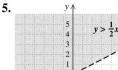

89. TW **91.** $\left(-\frac{16}{13}, -\frac{41}{13}\right)$ **92.** $(-2, -3)$
93. $(10, 4)$ **94.** $(-1, 7)$ **95.** $(1, 4)$ **96.** $(24, -41)$
97. TW **99.** $\left\{t \mid t \ge \frac{5}{3}\right\}$, or $\left[\frac{5}{3}, \infty\right)$
101. $\{x \mid -4 \le x \le -1 \text{ or } 3 \le x \le 6\}$, or $[-4, -1] \cup [3, 6]$
103. $\left\{x \mid x \le \frac{5}{2}\right\}$, or $\left(-\infty, \frac{5}{2}\right]$ **105.** $|y| \le 5$ **107.** $|x| > 4$
109. $|x + 2| < 3$ **111.** $|x - 5| < 1$, or $|5 - x| < 1$
113. $|x - 2| < 6$ **115.** $|x - 7| \le 5$

Exercise Set 4.4, pp. 299–301

1. Yes **3.** No **5.**
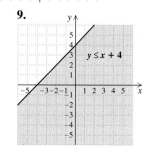

7.
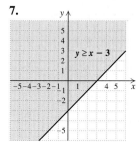

9.
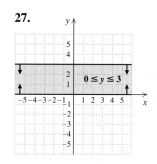

11.
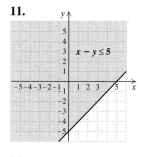

13.
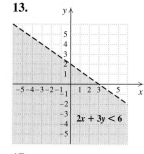

15.
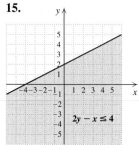

17.
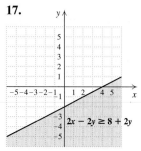

19.
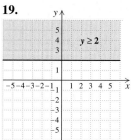

21.
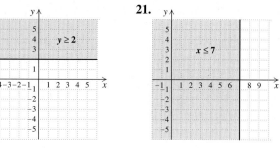

23.
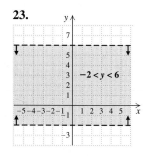

25.
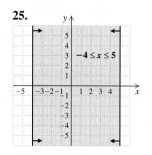

27.
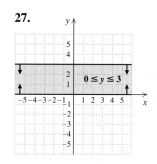

29. $y > x + 3.5$
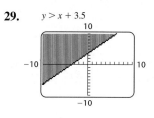

31. $8x - 2y < 11$
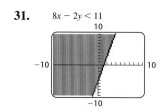

33.
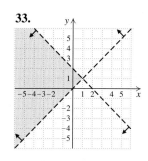

35.
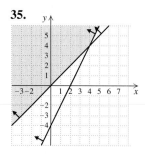

37.
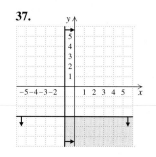

39.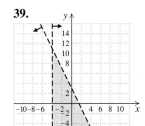

41.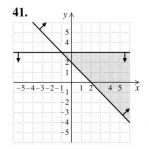

43.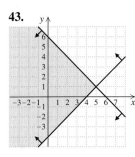

45.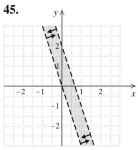

47.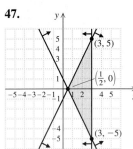

49.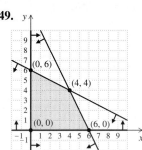

51.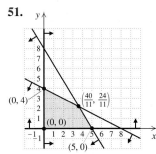

53.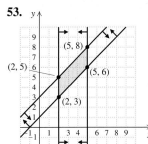

55. TW   **57.** Peanuts: $6\frac{2}{3}$ lb; fancy nuts: $3\frac{1}{3}$ lb
58. Hendersons: 10 bags; Savickis: 4 bags   **59.** Activity-

card holders: 128; noncard holders: 75   **60.** Students: 70;
adults: 130   **61.** 25 ft   **62.** 11%   **63.** TW

65.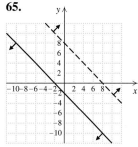

67.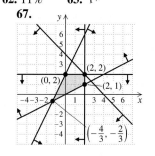

69. $0 < w \leq 62$,
$0 < h \leq 62$,
$62 + 2w + 2h \leq 108$,
or $w + h \leq 23$

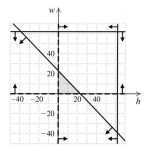

71. $35c + 75a > 1000$,
$c \geq 0$,
$a \geq 0$

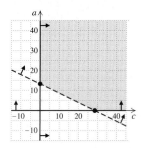

73. $y \leq x$,
$y \leq 2$

75. $y \leq x + 2$,
$y \leq -x + 4$,
$y \geq 0$

Review Exercises: Chapter 4, pp. 303–304

1. [4.1] $\{x \mid x \leq -2\}$, or $(-\infty, -2]$;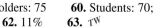

2. [4.1] $\{a \mid a \leq -21\}$, or $(-\infty, -21]$;

3. [4.1] $\{y \mid y \geq -7\}$, or $[-7, \infty)$;

4. [4.1] $\{y \mid y > -\frac{15}{4}\}$, or $\left(-\frac{15}{4}, \infty\right)$;

5. [4.1] $\{y \mid y > -30\}$, or $(-30, \infty)$;

6. [4.1] $\{x \mid x > -\frac{3}{2}\}$, or $\left(-\frac{3}{2}, \infty\right)$;

7. [4.1] $\{x \mid x < -3\}$, or $(-\infty, -3)$;

8. [4.1] $\left\{y \mid y > -\frac{220}{23}\right\}$, or $\left(-\frac{220}{23}, \infty\right)$;

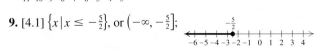

9. [4.1] $\left\{x \mid x \le -\frac{5}{2}\right\}$, or $\left(-\infty, -\frac{5}{2}\right]$;

10. [4.1] $\{x \mid x \le 4\}$, or $(-\infty, 4]$ **11.** [4.1] More than 125 hr
12. [4.1] \$1500   **13.** [4.2] $\{1, 5, 9\}$
14. [4.2] $\{1, 2, 3, 5, 6, 9\}$
15. [4.2] ; $(-5, 3]$

16. [4.2] ; $(-\infty, \infty)$

17. [4.2] $\{x \mid -7 < x \le 2\}$, or $(-7, 2]$

18. [4.2] $\left\{x \mid -\frac{5}{4} < x < \frac{5}{2}\right\}$, or $\left(-\frac{5}{4}, \frac{5}{2}\right)$

19. [4.2] $\{x \mid x < -3 \text{ or } x > 1\}$, or $(-\infty, -3) \cup (1, \infty)$

20. [4.2] $\{x \mid x < -11 \text{ or } x \ge -6\}$, or $(-\infty, -11) \cup [-6, \infty)$

21. [4.2] $\{x \mid x \le -6 \text{ or } x \ge 8\}$, or $(-\infty, -6] \cup [8, \infty)$

22. [4.2] $\left\{x \mid x < -\frac{2}{5} \text{ or } x > \frac{8}{5}\right\}$, or $\left(-\infty, -\frac{2}{5}\right) \cup \left(\frac{8}{5}, \infty\right)$

23. [4.2] $(-\infty, 3) \cup (3, \infty)$ **24.** [4.2] $[-3, \infty)$
25. [4.2] $\left(-\infty, \frac{8}{3}\right]$ **26.** [4.3] $\{-4, 4\}$
27. [4.3] $\{t \mid t \le -3.5 \text{ or } t \ge 3.5\}$, or $(-\infty, -3.5] \cup [3.5, \infty)$
28. [4.3] $\{-5, 9\}$ **29.** [4.3] $\left\{x \mid -\frac{17}{2} < x < \frac{7}{2}\right\}$, or $\left(-\frac{17}{2}, \frac{7}{2}\right)$
30. [4.3] $\left\{x \mid x \le -\frac{11}{3} \text{ or } x \ge \frac{19}{3}\right\}$, or $\left(-\infty, -\frac{11}{3}\right] \cup \left[\frac{19}{3}, \infty\right)$
31. [4.3] $\left\{-14, \frac{4}{3}\right\}$ **32.** [4.3] $\varnothing$
33. [4.3] $\{x \mid -12 \le x \le 4\}$, or $[-12, 4]$
34. [4.3] $\{x \mid x < 0 \text{ or } x > 10\}$, or $(-\infty, 0) \cup (10, \infty)$
35. [4.3] $\varnothing$ **36.** [4.4]

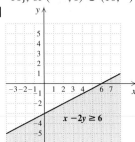

$x - 2y \ge 6$

37. [4.4]

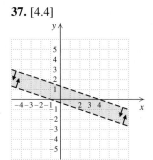

38. [4.4]

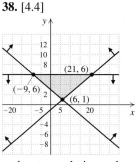

39. TW [4.3] The equation $|x| = p$ has two solutions when p is positive because x can be either p or $-p$. The same equation has no solution when p is negative because no number has a negative absolute value.   **40.** TW [4.4] The solution set of a system of inequalities is all ordered pairs that make *all* the individual inequalities true. This consists of ordered pairs that are common to all the individual solution sets, or the intersection of the graphs.   **41.** [4.3] $\left\{x \mid -\frac{8}{3} \le x \le -2\right\}$, or $\left[-\frac{8}{3}, -2\right]$ **42.** [4.1] False; $-4 < 3$ is true, but $(-4)^2 < 9$ is false.   **43.** [4.3] $|d - 1.1| \le 0.03$

Test: Chapter 4, p. 304

1. [4.1] $\{x \mid x < 12\}$, or $(-\infty, 12)$
2. [4.1] $\{y \mid y > -50\}$, or $(-50, \infty)$
3. [4.1] $\{y \mid y \le -2\}$, or $(-\infty, -2]$

4. [4.1] $\left\{a \mid a \le \frac{11}{5}\right\}$, or $\left(-\infty, \frac{11}{5}\right]$

5. [4.1] $\left\{x \mid x > \frac{5}{2}\right\}$, or $\left(\frac{5}{2}, \infty\right)$

6. [4.1] $\left\{x \mid x \le \frac{7}{4}\right\}$, or $\left(-\infty, \frac{7}{4}\right]$

7. [4.1] $\{x \mid x > 1\}$, or $(1, \infty)$ **8.** [4.1] More than $166\frac{2}{3}$ mi
9. [4.1] Less than or equal to 2.5 hr   **10.** [4.2] $\{3, 5\}$
11. [4.2] $\{1, 3, 5, 7, 9, 11, 13\}$ **12.** [4.2] $(-\infty, 7]$
13. [4.2] $\{x \mid -1 < x < 6\}$, or $(-1, 6)$

14. [4.2] $\left\{t \mid -\frac{2}{5} < t \le \frac{9}{5}\right\}$, or $\left(-\frac{2}{5}, \frac{9}{5}\right]$

15. [4.2] $\{x \mid x < 3 \text{ or } x > 6\}$, or $(-\infty, 3) \cup (6, \infty)$

16. [4.2] $\left\{x \mid x < -4 \text{ or } x > -\frac{5}{2}\right\}$, or $(-\infty, -4) \cup \left(-\frac{5}{2}, \infty\right)$

17. [4.2] $\left\{x \mid 4 \le x < \frac{15}{2}\right\}$, or $\left[4, \frac{15}{2}\right)$

18. [4.3] $\{-9, 9\}$

19. [4.3] $\{a\,|\,a < -3 \text{ or } a > 3\}$, or $(-\infty, -3) \cup (3, \infty)$

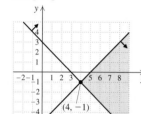

20. [4.3] $\{x\,|\,-\frac{7}{8} < x < \frac{11}{8}\}$, or $\left(-\frac{7}{8}, \frac{11}{8}\right)$

21. [4.3] $\{t\,|\,t \le -\frac{13}{5} \text{ or } t \ge \frac{7}{5}\}$, or $\left(-\infty, -\frac{13}{5}\right] \cup \left[\frac{7}{5}, \infty\right)$

22. [4.3] $\varnothing$
23. [4.2] $\{x\,|\,x < \frac{1}{2} \text{ or } x > \frac{7}{2}\}$, or $\left(-\infty, \frac{1}{2}\right) \cup \left(\frac{7}{2}, \infty\right)$

24. [4.3] $\{1\}$
25. [4.4]   **26.** [4.4]

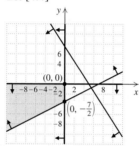

27. [4.3] $[-1, 0] \cup [4, 6]$ **28.** [4.2] $\left(\frac{1}{5}, \frac{4}{5}\right)$
29. [4.3] $|x + 13| \le 2$

Chapter 5

Interactive Discovery, p. 307

Answers may include: The graph of a polynomial function has no sharp corners; there are no holes or breaks; the domain is the set of all real numbers.

Interactive Discovery, pp. 314–315

1. The graphs of y_1 and y_2 should be the same.

$y_1 = (-3x^3 + 2x - 4) + (4x^3 + 3x^2 + 2),$
$y_2 = x^3 + 3x^2 + 2x - 2$

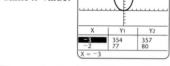

2. The graph of y_3 should be the x-axis.

$y_3 = y_2 - y_1$

3.

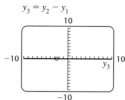

4. The graphs appear to coincide.

$y_1 = (-2x^4 + 3x^2 + 5) + (6x^4 - 2x - 8),$
$y_2 = 4x^4 + 3x^2 - 2x$

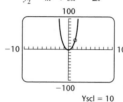

Yscl = 10

5. The table shows that the values of y_1 and y_2 are different for the same x-value.

6. The graph of y_3 is not the x-axis. It may be easier to see that the difference is not zero than that the graphs do not coincide.

7. No

Exercise Set 5.1, pp. 317–321

1. 5, 3, 2, 1, 0; 5   **3.** 3, 7, 6, 0; 7   **5.** 5, 6, 2, 1, 0; 6
7. $-4y^3 - 6y^2 + 7y + 19$; $-4y^3$; -4
9. $3x^7 + 5x^2 - x + 12$; $3x^7$; 3
11. $-a^7 + 8a^5 + 5a^3 - 19a^2 + a$; $-a^7$; -1
13. $-9 + 6x - 5x^2 + 3x^4$ **15.** $3xy^3 + x^2y^2 + 7x^3y - 5x^4$
17. $-7ab + 4ax - 7ax^2 + 4x^6$ **19.** 47; 7   **21.** -45;
$-8\frac{19}{27}$ **23.** -23 **25.** -12 **27.** 9   **29.** 6840
31. 144 ft   **33.** \$18,750   **35.** \$8375   **37.** 399
accidents   **39.** 5.2 million   **41.** 6.4 million
43. About 340 mg   **45.** $[0, 345]$ **47.** 56.5 in^2
49. $(-\infty, 3]$ **51.** $(-\infty, \infty)$ **53.** $[-4, \infty)$ **55.** $[-65, \infty)$
57. $[0, \infty)$ **59.** $(-\infty, 5]$ **61.** $(-\infty, \infty)$ **63.** $[-6.7, \infty)$
65. $2a^3 - a + 4$ **67.** $-6a^2b - 3b^2$
69. $10x^2 + 2xy + 15y^2$ **71.** $12a + 4b - c$

73. $-4a^2 - b^2 + 3c^2$ **75.** $-2x^2 + x - xy - 1$
77. $6x^2y - 4xy^2 + 5xy$ **79.** $9r^2 + 9r - 9$
81. $-\frac{5}{24}xy - \frac{27}{20}x^3y^2 + 1.4y^3$ **83.** $-(5x^3 - 7x^2 + 3x - 9)$,
$-5x^3 + 7x^2 - 3x + 9$ **85.** $-(-12y^5 + 4ay^4 - 7by^2)$,
$12y^5 - 4ay^4 + 7by^2$ **87.** $10x - 9$
89. $-4x^2 - 3x + 13$ **91.** $3a - 4b + 3c$
93. $-2x^2 + 6x$ **95.** $-4a^2 + 8ab - 5b^2$
97. $8a^2b + 16ab + 3ab^2$ **99.** $x^4 - x^2 - 1$ **101.** $9700
103. (a) and (c)   **105.** (b) and (d)   **107.** TW
109. $7x + 16$ **110.** $10a + 59$ **111.** $a^2 + 3a - 4$
112. $x^2 - x - 2$ **113.** x^9 **114.** a^8 **115.** TW
117. $68x^5 - 81x^4 - 22x^3 + 52x^2 + 2x + 250$
119. $45x^5 - 8x^4 + 208x^3 - 176x^2 + 116x - 25$
121. 494.55 cm^3 **123.** $5x^2 - 8x$ **125.** $8x^{2a} + 7x^a + 7$
127. $x^{5b} + 4x^{4b} + x^{3b} - 6x^{2b} - 9x^b$

Interactive Discovery, p. 326

1. Not an identity   **2.** Identity   **3.** Not an identity
4. Not an identity   **5.** Identity

Interactive Discovery, p. 328

1. Identity   **2.** Not an identity   **3.** Not an identity
4. Not an identity

Exercise Set 5.2, pp. 330–332

1. $32a^3$ **3.** $-20x^3y$ **5.** $-10x^5y^6$ **7.** $21x - 7x^2$
9. $20c^3d^2 - 25c^2d^3$ **11.** $6x^2 + 7x - 20$
13. $m^2 - mn - 6n^2$ **15.** $3y^2 - 13xy - 56x^2$
17. $a^4 - 5a^2b^2 + 6b^4$ **19.** $x^3 - 64$ **21.** $x^3 + y^3$
23. $a^4 + 5a^3 - 2a^2 - 9a + 5$
25. $3a^3b^2 - 8a^2b^2 + 3a^3b + 7ab^2 - 2a^2b + 3ab^3 - 6b^3$
27. $x^2 - \frac{3}{4}x + \frac{1}{8}$ **29.** $3x^2 - 1.5xy - 15y^2$
31. $12x^4 - 21x^3 - 17x^2 + 35x - 5$ **33.** $a^2 + 9a + 20$
35. $y^2 - 5y - 24$ **37.** $x^2 + 10x + 25$
39. $x^2 - 4xy + 4y^2$ **41.** $2x^2 + 13x + 18$
43. $100a^2 - 2.4ab + 0.0144b^2$ **45.** $4x^2 - 4xy - 3y^2$
47. $4x^6 - 12x^3y^2 + 9y^4$ **49.** $a^4b^4 + 2a^2b^2 + 1$
51. $16x^2 - 8x + 1$ **53.** $4x^2 - \frac{4}{3}x + \frac{1}{9}$ **55.** $c^2 - 4$
57. $16x^2 - 1$ **59.** $9m^2 - 4n^2$ **61.** $x^6 - y^2z^2$
63. $-m^2n^2 + m^4$, or $m^4 - m^2n^2$ **65.** $x^4 - 1$
67. $a^4 - 2a^2b^2 + b^4$ **69.** $a^2 + 2ab + b^2 - 1$
71. $4x^2 + 12xy + 9y^2 - 16$ **73.** $A = P + 2Pi + Pi^2$
75. (a) $t^2 - 2t + 6$; (b) $2ah + h^2$; (c) $2ah - h^2$ **77.** TW
79. $a = \dfrac{d}{b + c}$ **80.** $y = \dfrac{w}{x + z}$ **81.** $m = \dfrac{p}{n + 1}$
82. $s = \dfrac{t}{r + 1}$ **83.** Dimes: 5; nickels: 2; quarters: 6
84. 8 weekdays   **85.** TW   **87.** $x^{4a^2}y^{4ab}$
89. $\frac{1}{4}a^{7x}b^{2y+2}$ **91.** $y^{3n+3}z^{n+3} - 4y^4z^{3n}$
93. $a^2 + 2ac + c^2 - b^2 - 2bd - d^2$
95. $16x^4 + 4x^2y^2 + y^4$ **97.** $x^{4a} - y^{4b}$ **99.** $x^{a^2-b^2}$
101. 0   **103.** (b) and (c) are identities.

Interactive Discovery, p. 336

1. 1   **2.** 0   **3.** 2   **4.** 3   **5.** 2   **6.** 1
7. The number of real-number zeros is less than or equal to the degree.

Interactive Discovery, p. 337

1. $-4, 0$ **2.** $g(x)$: 0; $h(x)$: -4 **3.** The zeros of f are the zeros of g and h.

Exercise Set 5.3, pp. 343–346

1. $-3, 5$ **3.** $-2, 0$ **5.** $-3, 1$ **7.** $-4, 2$ **9.** 0, 5
11. 1, 3   **13.** 10, 15   **15.** 0, 1, 2   **17.** $-15, 6, 12$
19. $-0.42857, 0.33333$ **21.** $-5, 9$ **23.** $-0.5, 7$
25. $-1, 0, 3$ **27.** III   **29.** I   **31.** Expression
33. Equation   **35.** Equation   **37.** $2t(t + 4)$
39. $y^2(y + 9)$ **41.** $5x^2(3 - x^2)$ **43.** $4xy(x - 3y)$
45. $3(y^2 - y - 3)$ **47.** $2a(3b - 2d + 6c)$
49. $3x^2y^4z^2(3xy^2 - 4x^2z^2 + 5yz)$ **51.** $-5(x - 7)$
53. $-6(y + 12)$ **55.** $-2(x^2 - 2x + 6)$
57. $-3(-y + 8x)$, or $-3(8x - y)$ **59.** $-7(-s + 2t)$, or
$-7(2t - s)$ **61.** $-(x^2 - 5x + 9)$
63. $-a(a^3 - 2a^2 + 13)$ **65.** $(b - 5)(a + c)$
67. $(x + 7)(2x - 3)$ **69.** $(x - y)(a^2 - 5)$
71. $(c + d)(a + b)$ **73.** $(b - 1)(b^2 + 2)$
75. $(a - 3)(a^2 - 2)$ **77.** $12x(6x^2 - 3x + 2)$
79. $x^3(x - 1)(x^2 + 1)$ **81.** $(y^2 + 3)(2y^2 + 5)$
83. (a) $h(t) = -8t(2t - 9)$; (b) $h(1) = 56$ ft
85. $R(n) = n(n - 1)$ **87.** $P(x) = x(x - 3)$
89. $R(x) = 0.4x(700 - x)$ **91.** $N(x) = \frac{1}{6}(x^3 + 3x^2 + 2x)$
93. $H(n) = \frac{1}{2}n(n - 1)$ **95.** $-1, 0$ **97.** 0, 3
99. $-3, 0$ **101.** $-\frac{1}{3}, 0$ **103.** TW   **105.** -26
106. -1 **107.** -30 **108.** -19 **109.** 56, 58, 60
110. A: 75; B: 84; C: 63   **111.** TW
113. $(x + 4)(x - 2)$ **115.** $x^5y^4 + x^4y^6 = x^3y(x^2y^3 + xy^5)$
117. $(x^2 - x + 5)(r + s)$
119. $(x^4 + x^2 + 5)(a^4 + a^2 + 5)$ **121.** $x^{1/3}(1 - 7x)$
123. $x^{1/3}(1 - 5x^{1/6} + 3x^{5/12})$ **125.** $3a^n(a + 2 - 5a^2)$
127. $y^{a+b}(7y^a - 5 + 3y^b)$

Exercise Set 5.4, pp. 354–356

1. $(x + 2)(x + 6)$ **3.** $(t + 3)(t + 5)$
5. $(x - 9)(x + 3)$ **7.** $2(n - 5)(n - 5)$, or $2(n - 5)^2$
9. $a(a + 8)(a - 9)$ **11.** $(x + 9)(x + 5)$
13. $(x + 5)(x - 2)$ **15.** $3(x + 2)(x + 3)$
17. $-(x - 8)(x + 7)$, or $(-x + 8)(x + 7)$, or $(8 - x)(7 + x)$
19. $-y(y - 8)(y + 4)$, or $y(-y + 8)(y + 4)$, or
$y(8 - y)(4 + y)$ **21.** $x^2(x + 16)(x - 5)$ **23.** Prime
25. $(p - 8q)(p + 3q)$ **27.** $(y + 4z)(y + 4z)$, or
$(y + 4z)^2$ **29.** $p^2(p + 1)(p + 79)$ **31.** $-6, -2$
33. 5   **35.** $-8, 0, 9$ **37.** $-5, 1$ **39.** $-3, 2$

41. $-5, 9$ **43.** $-1, 0, 3$ **45.** $-9, 5$ **47.** $0, 9$
49. $-5, 0, 8$ **51.** $-4, 5$ **53.** $-7, 5$
55. $(x + 22)(x - 12)$ **57.** $(x + 24)(x + 16)$
59. $(x + 64)(x - 38)$ **61.** $f(x) = (x + 1)(x - 2)$, or
$f(x) = x^2 - x - 2$ **63.** $f(x) = (x + 7)(x + 10)$, or
$f(x) = x^2 + 17x + 70$ **65.** $f(x) = x(x - 1)(x - 2)$, or
$f(x) = x^3 - 3x^2 + 2x$ **67.** TW   **69.** $5x(2x^2 - 7x + 1)$
70. $4t(3t^2 - 10t - 2)$ **71.** $125a^{12}$ **72.** $-32x^{10}$
73. -24 **74.** 880 ft; 960 ft; 1024 ft; 624 ft; 240 ft
75. TW   **77.** $\{-1, 3\}$; $(-2, 4)$, or $\{x \mid -2 < x < 4\}$
79. $f(x) = 5x^3 - 20x^2 + 5x + 30$; answers may vary
81. 6.90   **83.** 3.48   **85.** $ab^2(2a^3b^4 - 3ab - 20)$
87. $\left(x + \frac{4}{5}\right)\left(x - \frac{1}{5}\right)$ **89.** $(y - 0.1)(y + 0.5)$
91. $(x + a)(x + b)$ **93.** $a^2(p^a + 2)(p^a - 1)$
95. $76, -76, 28, -28, 20, -20$ **97.** $x - 365$

Exercise Set 5.5, pp. 363–364

1. $(3x + 5)(2x - 5)$ **3.** $y(2y - 3)(5y + 4)$
5. $2(4a - 1)(3a - 1)$ **7.** $(7y + 4)(5y + 2)$
9. $2(5t - 3)(t + 1)$ **11.** $4(x - 4)(2x + 1)$
13. $x^2(2x - 3)(7x + 1)$ **15.** $4(3a - 4)(a + 1)$
17. $(3x + 4)(3x + 1)$ **19.** $(4x + 3)(x + 3)$
21. $-2(2t - 3)(2t + 5)$ **23.** $(4 + 3z)(2 - 3z)$
25. $xy(6y + 5)(3y - 2)$ **27.** $(x - 2)(24x + 1)$
29. $3x(7x + 3)(3x + 4)$ **31.** $2x^2(4x - 3)(6x + 5)$
33. $(4a - 3b)(3a - 2b)$ **35.** $(2x - 3y)(x + 2y)$
37. $(3x - 4y)(2x - 7y)$ **39.** $(3x - 5y)(3x - 5y)$, or
$(3x - 5y)^2$ **41.** $(9xy - 4)(xy + 1)$ **43.** $-\frac{5}{3}, \frac{5}{3}$
45. $-\frac{4}{3}, \frac{2}{3}$ **47.** $-\frac{4}{3}, -\frac{3}{7}, 0$ **49.** $\frac{2}{3}, 2$ **51.** $-2, -\frac{3}{4}, 0$
53. $-\frac{1}{3}, \frac{5}{2}$ **55.** $-\frac{7}{4}, \frac{4}{3}$ **57.** $-\frac{1}{2}, 7$ **59.** $-8, -4$
61. $-4, \frac{3}{2}$ **63.** $-9, -3$ **65.** $\{x \mid x$ is a real number *and*
$x \neq 5$ *and* $x \neq -1\}$ **67.** $\{x \mid x$ is a real number *and* $x \neq 0$
and $x \neq \frac{1}{2}\}$ **69.** $\{x \mid x$ is a real number *and* $x \neq 0$ *and*
$x \neq 2$ *and* $x \neq 5\}$ **71.** TW   **73.** 228 ft^2 **74.** -2
75. Slope: $\frac{4}{3}$; y-intercept: $\left(0, -\frac{8}{3}\right)$ **76.** $\{-27, 27\}$
77. $\{x \mid -\frac{33}{5} \leq x \leq 9\}$, or $\left[-\frac{33}{5}, 9\right]$
78. $\{x \mid x < -\frac{33}{5}$ *or* $x > 9\}$, or $\left(-\infty, -\frac{33}{5}\right) \cup (9, \infty)$
79. TW   **81.** $(2x + 15)(2x + 45)$
83. $3x(x + 68)(x - 18)$ **85.** $-\frac{11}{8}, -\frac{1}{4}, \frac{2}{3}$ **87.** 1
89. $(2a^2b^3 + 5)(a^2b^3 - 4)$ **91.** $(2x^a - 3)(2x^a + 1)$
93. Since $ax^2 + bx + c = (mx + r)(nx + s)$, from FOIL we
know that $a = mn$, $c = rs$, and $b = ms + rn$. If $P = ms$ and
$Q = rn$, then $b = P + Q$. Since $ac = mnrs = msrn$, we have
$ac = PQ$.

Exercise Set 5.6, pp. 371–372

1. $(x + 4)^2$ **3.** $(a + 8)^2$ **5.** $2(a + 2)^2$ **7.** $(y - 6)^2$
9. $a(a + 12)^2$ **11.** $2(4x + 3)^2$ **13.** $(5a - 8)^2$
15. $(0.5x + 0.3)^2$ **17.** $(p - q)^2$ **19.** $a(a - 5)^2$
21. $(5a - 3b)^2$ **23.** $4(t - r)^2$ **25.** $(x + 4)(x - 4)$
27. $(p + 7)(p - 7)$ **29.** $(ab + 9)(ab - 9)$

31. $6(x + y)(x - y)$ **33.** $7x(y^2 + z^2)(y + z)(y - z)$
35. $a(2a + 7)(2a - 7)$
37. $3(x^4 + y^4)(x^2 + y^2)(x + y)(x - y)$
39. $a^2(3a + 5b^2)(3a - 5b^2)$ **41.** $\left(\frac{1}{5} + x\right)\left(\frac{1}{5} - x\right)$
43. $(a + b + 3)(a + b - 3)$
45. $(x - 3 + y)(x - 3 - y)$
47. $(m - n + 5)(m - n - 5)$
49. $(6 + x + y)(6 - x - y)$
51. $(r - 1 + 2s)(r - 1 - 2s)$
53. $(4 + a + b)(4 - a - b)$
55. $(m - 7)(m + 2)(m - 2)$ **57.** $(a + b)(a - b)(a - 2)$
59. -4 **61.** $-4, 4$ **63.** 1   **65.** 6   **67.** $-3, 3$
69. $-\frac{1}{5}, \frac{1}{5}$ **71.** $-\frac{1}{2}, \frac{1}{2}$ **73.** $-1, 1, 3$ **75.** -1.541,
4.541 **77.** $-3.871, -0.129$ **79.** $-2.414, -1, 0.414$
81. 6   **83.** $-4, 4$ **85.** -1 **87.** $-1, 1, 2$ **89.** TW
91. $8a^{12}b^{15}$ **92.** $125x^6y^{12}$
93. $x^3 + 3x^2y + 3xy^2 + y^3$ **94.** $a^3 + 3a^2 + 3a + 1$
95. $(2, -1, 3)$ **96.** $\{x \mid x \leq -\frac{4}{7}$ *or* $x \geq 2\}$, or
$\left(-\infty, -\frac{4}{7}\right] \cup [2, \infty)$ **97.** $\{x \mid -\frac{4}{7} \leq x \leq 2\}$, or $\left[-\frac{4}{7}, 2\right]$
98. $\{x \mid x < \frac{14}{19}\}$, or $\left(-\infty, \frac{14}{19}\right)$ **99.** TW
101. $-\frac{1}{54}(4r + 3s)^2$ **103.** $(0.3x^4 + 0.8)^2$, or $\frac{1}{100}(3x^4 + 8)^2$
105. $(r + s + 1)(r - s - 9)$ **107.** $(x^{2a} + y^b)(x^{2a} - y^b)$
109. $(5y^a + x^b - 1)(5y^a - x^b + 1)$ **111.** $3(x + 3)^2$
113. $(3x^n - 1)^2$ **115.** $h(2a + h)$
117. **(a)** $\pi h(R + r)(R - r)$; **(b)** 3,014,400 cm^3

Exercise Set 5.7, pp. 376–377

1. $(t + 3)(t^2 - 3t + 9)$ **3.** $(x - 2)(x^2 + 2x + 4)$
5. $(m - 4)(m^2 + 4m + 16)$ **7.** $(2a + 1)(4a^2 - 2a + 1)$
9. $(3 - 2t)(9 + 6t + 4t^2)$ **11.** $(2x + 3)(4x^2 - 6x + 9)$
13. $(y - z)(y^2 + yz + z^2)$ **15.** $\left(x + \frac{1}{3}\right)\left(x^2 - \frac{1}{3}x + \frac{1}{9}\right)$
17. $2(y - 4)(y^2 + 4y + 16)$
19. $8(a + 5)(a^2 - 5a + 25)$
21. $r(s + 4)(s^2 - 4s + 16)$
23. $2(y - 3z)(y^2 + 3yz + 9z^2)$
25. $(y + 0.5)(y^2 - 0.5y + 0.25)$
27. $(5c^2 - 2d^2)(25c^4 + 10c^2d^2 + 4d^4)$
29. $3z^2(z - 1)(z^2 + z + 1)$ **31.** $(t^2 + 1)(t^4 - t^2 + 1)$
33. $(p + q)(p^2 - pq + q^2)(p - q)(p^2 + pq + q^2)$
35. $(a^3 + b^4c^5)(a^6 - a^3b^4c^5 + b^8c^{10})$ **37.** -1 **39.** $\frac{3}{2}$
41. 10   **43.** TW   **45.** 224 ft; 288 ft; 320 ft; 288 ft;
128 ft   **46.** $f(x) = -\frac{3}{4}x - 5$ **47.** $f(x) = 3x - 8$
48. $f(x) = 2x - 8$ **49.** $\frac{5}{3}$ **50.** $-\frac{7}{2}$ **51.** TW
53. $(x^{2a} - y^b)(x^{4a} + x^{2a}y^b + y^{2b})$ **55.** $2x(x^2 + 75)$
57. $5\left(xy^2 - \frac{1}{2}\right)\left(x^2y^4 + \frac{1}{2}xy^2 + \frac{1}{4}\right)$ **59.** $-(3x^{4a} + 3x^{2a} + 1)$
61. $(t - 8)(t - 1)(t^2 + t + 1)$
63. $h(2a + h)(a^2 + ah + h^2)(3a^2 + 3ah + h^2)$

Exercise Set 5.8, pp. 385–389

1. $-12, 11$ **3.** Length: 12 cm; width: 7 cm   **5.** 3 m
7. 3 cm   **9.** 10 ft   **11.** 16, 18, 20   **13.** Height: 6 ft;
base: 4 ft   **15.** Height: 16 m; base: 7 m   **17.** 41 ft

19. Length: 100 m; width: 75 m   **21.** 2 sets   **23.** 2 sec
25. 5 sec   **27. (a)** $E(x) = 0.0000256775x^4 -$
$0.0025507304x^3 + 0.0821803626x^2 - 0.9480449896x +$
$11.8202857;$ **(b)** 9.3¢ per kilowatt-hour; 8.2¢ per kilowatt-
hour; **(c)** 1964, 1977, 1992
29. (a) $d(x) = -105.9592768x^2 + 3621.572598x +$
$42307.79479;$ **(b)** 58,222 degrees; 39,263 degrees;
(c) 1976, 1998   **31. (a)** $b(x) = -1.143939394x^2 +$
$9.583333333x + 25.79545455;$ **(b)** 40%, 7%; **(c)** 1998
33. TW   **35.** $-\frac{5}{8}$ **36.** 2   **37.** Faster car: 54 mph;
slower car: 39 mph   **38.** Conventional: 36;
surround-sound: 42   **39.** $\{x \mid x > \frac{10}{3}\}$, or $\left(\frac{10}{3}, \infty\right)$
40. $(2, -2, -4)$ **41.** TW   **43.** Length: 28 cm;
width: 14 cm   **45.** Tugboat: 15 km/h; freighter: 8 km/h

Review Exercises: Chapter 5, pp. 392–394

1. [5.1] 7, 11, 3, 0; 11   **2.** [5.1] $-5x^3 + 2x^2 + 4x - 7$;
$-5x^3; -5$ **3.** [5.1] $-3x^2 + 2x^3 + 3x^6y - 7x^8y^3$
4. [5.1] 0; -6 **5.** [5.1] 4
6. [5.1] $-2a^3 + a^2 - 3a - 4$ **7.** [5.1] $-x^2y - 2xy^2$
8. [5.1] $-x^3 + 2x^2 + 5x + 2$
9. [5.1] $-3x^4 + 3x^3 - x + 16$
10. [5.1] $-5xy^2 - 2xy + x^2y$ **11.** [5.1] $9x - 7$
12. [5.1] $-2a + 6b + 7c$ **13.** [5.1] $6x^2 - 7xy + 3y^2$
14. [5.2] $-18x^3y^4$ **15.** [5.2] $x^8 - x^6 + 5x^2 - 3$
16. [5.2] $8a^2b^2 + 2abc - 3c^2$ **17.** [5.2] $4x^2 - 25y^2$
18. [5.2] $4x^2 - 20xy + 25y^2$ **19.** [5.2] $2x^2 + 5x - 3$
20. [5.2] $x^4 + 8x^2y^3 + 16y^6$ **21.** [5.2] $x^3 - 125$
22. [5.2] $x^2 - \frac{1}{2}x + \frac{1}{18}$ **23.** [5.3] $x(6x + 5)$
24. [5.3] $3y^2(3y^2 - 1)$ **25.** [5.3] $3x(5x^3 - 6x^2 + 7x - 3)$
26. [5.4] $(a - 9)(a - 3)$ **27.** [5.5] $(3m + 2)(m + 4)$
28. [5.6] $(5x + 2)^2$ **29.** [5.6] $4(y + 2)(y - 2)$
30. [5.4] $x(x - 2)(x + 7)$ **31.** [5.3] $(a + 2b)(x - y)$
32. [5.3] $(y + 2)(3y^2 - 5)$
33. [5.6] $(a^2 + 9)(a + 3)(a - 3)$
34. [5.3] $4(x^4 + x^2 + 5)$
35. [5.7] $(3x - 2)(9x^2 + 6x + 4)$
36. [5.7] $(0.4b - 0.5c)(0.16b^2 + 0.2bc + 0.25c^2)$
37. [5.3] $y(y^4 + 1)$ **38.** [5.3] $2z^6(z^2 - 8)$
39. [5.7] $2y(3x^2 - 1)(9x^4 + 3x^2 + 1)$
40. [5.6] $4(3x - 5)^2$ **41.** [5.5] $(3t + p)(2t + 5p)$
42. [5.6] $(x + 3)(x - 3)(x + 2)$
43. [5.6] $(a - b + 2t)(a - b - 2t)$
44. [5.3] $-2, 1, 5$ **45.** [5.4] 4, 7   **46.** [5.6] 10
47. [5.5] $\frac{2}{3}, \frac{3}{2}$ **48.** [5.3] $0, \frac{7}{4}$ **49.** [5.6] $-4, 4$
50. [5.4] $-3, 0, 7$ **51.** [5.4] $-1, 6$ **52.** [5.6] $-4, 4, 5$
53. [5.4] 12, 15   **54.** [5.4] $-4, 11$ **55.** [5.5] $\{x \mid x$ is a
real number *and* $x \neq -7$ *and* $x \neq \frac{2}{3}\}$ **56.** [5.8] 5
57. [5.8] 3, 5, 7; $-7, -5, -3$ **58.** [5.8] Length: 8 in.;
width: 5 in.   **59.** [5.8] 17 ft
60. [5.8] **(a)** $0.0273088783x^3 - 1.977521075x^2 +$
$37.02830561x + 38.79266348;$ **(b)** 166 billion kilowatt-hours;
(c) 1962, 1991, 2000   **61.** TW [5.3] The roots of a

polynomial function are the x-coordinates of the points at
which the graph of the function crosses the x-axis.
62. TW [5.3] The principle of zero products states that if a
product is equal to 0, at least one of the factors must be 0. If a
product is nonzero, we cannot conclude that any one of the
factors is a particular value.
63. [5.7] $2(2x - y)(4x^2 + 2xy + y^2)(2x + y)(4x^2 - 2xy + y^2)$
64. [5.7] $-2(3x^2 + 1)$
65. [5.2], [5.7] $a^3 - b^3 + 3b^2 - 3b + 1$ **66.** [5.2] z^{5n^5}
67. [5.6] $-1, -\frac{1}{2}$

Test: Chapter 5, pp. 394–395

1. [5.1] 9   **2.** [5.1] $5x^5y^4 - 2x^4y - 4x^2y + 3xy^3$
3. [5.1] $-4a^3$ **4.** [5.1] 4; 2   **5.** [5.2] $2ah + h^2 - 5h$
6. [5.1] $3xy + 3xy^2$ **7.** [5.1] $-3x^3 + 3x^2 - 6y - 7y^2$
8. [5.1] $7m^3 + 2m^2n + 3mn^2 - 7n^3$ **9.** [5.1] $6a - 8b$
10. [5.1] $2y^2 + 5y + y^3$ **11.** [5.2] $64x^3y^3$
12. [5.2] $12a^2 - 4ab - 5b^2$ **13.** [5.2] $x^3 - 2x^2y + y^3$
14. [5.2] $4x^6 + 20x^3 + 25$ **15.** [5.2] $16y^2 - 72y + 81$
16. [5.2] $x^2 - 4y^2$ **17.** [5.3] $5x^2(3 - x^2)$
18. [5.6] $(y + 5)(y + 2)(y - 2)$
19. [5.4] $(p - 14)(p + 2)$ **20.** [5.5] $(6m + 1)(2m + 3)$
21. [5.6] $(3y + 5)(3y - 5)$
22. [5.7] $3(r - 1)(r^2 + r + 1)$ **23.** [5.6] $(3x - 5)^2$
24. [5.6] $(x^4 + y^4)(x^2 + y^2)(x + y)(x - y)$
25. [5.6] $(y + 4 + 10t)(y + 4 - 10t)$
26. [5.6] $5(2a - b)(2a + b)$ **27.** [5.5] $2(4x - 1)(3x - 5)$
28. [5.7] $2ab(2a^2 + 3b^2)(4a^4 - 6a^2b^2 + 9b^4)$
29. [5.3] $4xy(y^3 + 9x + 2x^2y - 4)$ **30.** [5.3] $-3, -1, 2, 4$
31. [5.5] $-\frac{5}{2}, 8$ **32.** [5.4] $-3, 6$ **33.** [5.6] $-5, 5$
34. [5.5] $-7, -\frac{3}{2}$ **35.** [5.3] $-\frac{1}{3}, 0$ **36.** [5.6] 9
37. [5.4] $-3.372, 0, 2.372$ **38.** [5.5] 0, 5
39. [5.5] $\{x \mid x$ is a real number *and* $x \neq -1\}$
40. [5.8] Length: 8 cm; width: 5 cm   **41.** [5.8] $4\frac{1}{2}$ sec
42. [5.8] **(a)** $0.03135826x^3 - 3.21956840x^2 +$
$101.17516232x - 886.92962439;$ **(b)** 100; **(c)** 20, 30
43. (a) [5.2] $x^5 + x + 1;$
(b) [5.2], [5.7] $(x^2 + x + 1)(x^3 - x^2 + 1)$
44. [5.5] $(3x^n + 4)(2x^n - 5)$

Chapter 6

Interactive Discovery, p. 399

1. 3   **2.** The function values get very small.   **3.** The
function values get very large.   **4.** A table and Y1(3) return
error messages. The VALUE option of the CALC menu returns
nothing for Y when X = 3.

Interactive Discovery, p. 404

1. One vertical asymptote; $x = -\frac{1}{2}$ **2.** $x = -\frac{1}{2}$

Interactive Discovery, p. 406

1. No   **2.** No   **3.** No

Exercise Set 6.1, pp. 408–411

1. $\frac{40}{13}$ hr, or $3\frac{1}{13}$ hr   **3.** $\frac{2}{3}$; 28; $\frac{163}{10}$ **5.** $-\frac{9}{4}$; does not exist;

$-\frac{11}{9}$ **7.** $\frac{4x(x-3)}{4x(x+2)}$ **9.** $\frac{(t-2)(-1)}{(t+3)(-1)}$ **11.** $\frac{3}{x}$

13. $\frac{2}{3t^4}$ **15.** $a-5$ **17.** $\frac{3}{5a-6}$ **19.** $\frac{x-4}{x+5}$

21. $\frac{5}{x}$ **23.** $-\frac{1}{2}$ **25.** $\frac{8}{t+2}$ **27.** $-\frac{1}{1+2t}$ **29.** $-\frac{6}{5}$

31. $\frac{a-5}{a+5}$ **33.** $\frac{x+8}{x-4}$ **35.** $\frac{4+t}{4-t}$ **37.** $\frac{x^2+x+1}{x+1}$

39. $3(y+2)$ **41.** $x=-5$ **43.** No vertical asymptotes

45. $x=-3$ **47.** $x=2, x=4$ **49.** (b)   **51.** (f)

53. (a)   **55.** $\frac{7b^2}{6a^4}$ **57.** $\frac{8x^2}{25}$ **59.** $\frac{y+4}{4}$

61. $\frac{(x+4)(x-4)}{x(x+3)}$ **63.** $-\frac{a+1}{2+a}$ **65.** 1

67. $\frac{(x+5)(2x+3)}{7x}$ **69.** $c(c-2)$ **71.** $\frac{a^2+ab+b^2}{3(a+2b)}$

73. $\frac{1}{2x+3y}$ **75.** $6x^4y^7$ **77.** $\frac{5}{x^5}$ **79.** $\frac{(x+2)(x+4)}{x^7}$

81. $-\frac{5x+2}{x-3}$ **83.** $-\frac{1}{y^3}$ **85.** $\frac{(x+4)(x+2)}{3(x-5)}$

87. $\frac{y(y^2+3)}{(y+3)(y-2)}$ **89.** $\frac{x^2+4x+16}{(x+4)^2}$ **91.** $\frac{(2a+b)^2}{2(a+b)}$

93. TW   **95.** $-\frac{7}{30}$ **96.** $-\frac{13}{40}$ **97.** $\frac{5}{14}$ **98.** $\frac{1}{35}$

99. $4x^3-7x^2+9x-5$ **100.** $-2t^4+11t^3-t^2+10t-3$

101. TW   **103.** (a) $\frac{2x+2h+3}{4x+4h-1}$; (b) $\frac{2x+3}{8x-9}$;

(c) $\frac{x+5}{4x-1}$ **105.** $\frac{2s}{r+2s}$ **107.** $\frac{6t^2-26t+30}{8t^2-15t-21}$

109. $\frac{m-t}{m+t+1}$ **111.** $\frac{x^2+xy+y^2+x+y}{x-y}$

113. $-\frac{2x}{x-1}$

115. Domain: $(-\infty,-2)\cup(-2,1)\cup(1,\infty)$;
range: $(-\infty,2)\cup(2,3)\cup(3,\infty)$

117. Domain: $(-\infty,-1)\cup(-1,1)\cup(1,\infty)$;
range: $(-\infty,-1]\cup(0,\infty)$

Exercise Set 6.2, pp. 418–421

1. $\frac{4}{a}$ **3.** $-\frac{1}{a^2b}$ **5.** 2   **7.** $\frac{3y+5}{y-2}$ **9.** $\frac{1}{x-4}$

11. $\frac{-1}{a+5}$ **13.** $a+b$ **15.** $\frac{15}{x}$ **17.** $\frac{2}{x+4}$

19. $\frac{1}{t^2+4}$ **21.** $\frac{1}{m^2+mn+n^2}$ **23.** $\frac{2a^2-a+14}{(a-4)(a+3)}$

25. $\frac{5x+1}{x+1}$ **27.** $\frac{x+y}{x-y}$ **29.** $\frac{3x^2+7x+14}{(2x-5)(x-1)(x+2)}$

31. $\frac{8x+1}{(x+1)(x-1)}$ **33.** $\frac{-x+34}{20(x+2)}$

35. $\frac{-a^2+7ab-b^2}{(a-b)(a+b)}$ **37.** $\frac{x-5}{(x+5)(x+3)}$

39. $\frac{y}{(y-2)(y-3)}$ **41.** $\frac{7x+1}{x-y}$

43. $\frac{3y^2-3y-29}{(y-3)(y+8)(y-4)}$ **45.** $\frac{-2a^2-7a-10}{(a+4)(a-4)(a+6)}$

47. $\frac{10a-16}{(a-4)(a-1)(a-2)(a+2)}$ **49.** $\frac{2(3t-7)}{t-2}$

51. $\frac{-y}{(y+3)(y-1)}$ **53.** $-\frac{2y}{2y+1}$

55. $\frac{-6x+42}{(x-3)(x-2)(x+1)(x+3)}$ **57.** TW   **59.** $\frac{3y^6z^7}{7x^5}$

60. $\frac{7b^{11}c^7}{9a^2}$ **61.** $\frac{17s^{12}r^{40}}{5t^{60}}$ **62.** $y=\frac{5}{4}x+3$

63. Dimes: 2 rolls; nickels: 5 rolls; quarters: 5 rolls

64. 30-min tapes: 4; 60-min tapes: 8   **65.** TW

67. 420 days   **69.** 12 parts

71. $x^4(x^2+1)(x+1)(x-1)(x^2+x+1)(x^2-x+1)$

73. $8a^4, 8a^4b, 8a^4b^2, 8a^4b^3, 8a^4b^4, 8a^4b^5, 8a^4b^6, 8a^4b^7$

75. $\frac{x^4+6x^3+2x^2}{(x+2)(x-2)(x+5)}$ **77.** $\frac{x^5}{(x^2-4)(x^2+3x-10)}$

79. $\{x\,|\,x$ is a real number *and* $x\neq-5$ *and* $x\neq-2$ *and* $x\neq2\}$

81. $\frac{9x^2+28x+15}{(x-3)(x+3)^2}$ **83.** $\frac{1}{2x(x-5)}$ **85.** $-4t^4$

87. Domain: $(-\infty,-1)\cup(-1,\infty)$; range: $(-\infty,3)\cup(3,\infty)$

89. Domain: $(-\infty,0)\cup(0,1)\cup(1,\infty)$; range: $(0,\infty)$

Exercise Set 6.3, pp. 428–431

1. $\frac{7a+1}{1-3a}$ **3.** $\frac{x^2-1}{x^2+1}$ **5.** $\frac{6y+7x}{7y-6x}$ **7.** $\frac{x+y}{x}$

9. $\frac{x^2(3-y)}{y^2(2x-1)}$ **11.** $\frac{1}{a-b}$ **13.** 8   **15.** $-\frac{1}{x(x+h)}$

17. $\frac{(x-4)(x-7)}{(x-5)(x+6)}$ **19.** $\frac{4x-7}{7x-9}$ **21.** $\frac{a^2-3a-6}{a^2-2a-3}$

23. $\frac{x+2}{x+3}$ **25.** $\frac{a+1}{2a+5}$ **27.** $\frac{-1-3x}{8-2x}$, or $\frac{3x+1}{2x-8}$

29. $\frac{1}{y+5}$ **31.** $-y$ **33.** $\frac{2(5a^2+4a+12)}{5(a^2+6a+18)}$

35. $\frac{c^2-2c+4}{c}$ **37.** $\frac{xy}{x-y}$ **39.** $\frac{(2x+1)(x+2)}{2x(x-1)}$

41. $\frac{(2a-3)(a+5)}{2(a-3)(a+2)}$ **43.** -1 **45.** TW   **47.** 1

48. $\frac{11}{4}$ **49.** $y=\frac{t-rs}{r}$ **50.** $\{-2,5\}$ **51.** First frame:

15 cm; second frame: 29 cm   **52.** \$34   **53.** TW

55. $\dfrac{5(y+x)}{3(y-x)}$ **57.** $\dfrac{8c}{17}$ **59.** $\dfrac{-3}{x(x+h)}$

61. $\dfrac{2}{(1+x+h)(1+x)}$ **63.** $\{x\,|\,x$ is a real number *and*

$x \neq \pm1$ *and* $x \neq \pm4$ *and* $x \neq \pm5\}$ **65.** $\dfrac{2+a}{3+a}$ **67.** $\dfrac{x^4}{81}$;

$\{x\,|\,x$ is a real number *and* $x \neq 3\}$, $(-\infty, 3) \cup (3, \infty)$

Exercise Set 6.4, pp. 436–438

1. $\frac{51}{5}$ **3.** 144   **5.** -2 **7.** 7   **9.** No solution
11. $-\frac{1}{2}$ **13.** No solution   **15.** $-\frac{10}{3}$ **17.** -1
19. 2, 3   **21.** -145 **23.** -1 **25.** $-\frac{3}{2}$, 5   **27.** 14
29. $\frac{3}{4}$ **31.** 4   **33.** 3   **35.** No solution   **37.** $-\frac{7}{3}$
39. $\frac{1}{7}$ **41.** TW   **43.** Multiple-choice: 25; true–false: 30;
fill-in: 15   **44.** Child's: 118; adult's: 132   **45.** 24,640 m²
46. 16 and 18   **47.** (a) Inconsistent; (b) consistent
48. $\{x\,|\,x < -1$ *or* $x > 5\}$, or $(-\infty, -1) \cup (5, \infty)$
49. TW   **51.** $\frac{1}{5}$ **53.** $-\frac{7}{2}$ **55.** 0.0854697   **57.** Yes

Exercise Set 6.5, pp. 447–449

1. 2   **3.** $-3, -2$ **5.** -6 and -7, 7 and 6   **7.** $3\frac{3}{7}$ hr
9. $9\frac{9}{10}$ hr   **11.** $3\frac{3}{4}$ min   **13.** Skyler: 12 hr; Jake: 6 hr
15. HQ17: 30 min; EV25: 15 min   **17.** Sara: 6 hr;
Kate: 3 hr   **19.** $1\frac{1}{5}$ hr   **21.** 300 min, or 5 hr
23. 7 mph   **25.** 5.2 ft/sec   **27.** Freight: 66 mph;
passenger: 80 mph   **29.** Express: 45 mph; local: 38 mph
31. $1\frac{1}{5}$ km/h   **33.** 2 km/h   **35.** 25 mph   **37.** 20 mph
39. TW   **41.** $5a^4b^6$ **42.** $5x^6y^4$ **43.** $4s^{10}t^8$
44. $-2x^4 - 7x^2 + 11x$ **45.** $-8x^3 + 28x^2 - 25x + 19$
46. $11x^4 + 7x^3 - 2x^2 - 4x - 10$ **47.** TW   **49.** $49\frac{1}{2}$ hr
51. 11.25 min   **53.** 2250 people per hour   **55.** $14\frac{7}{8}$ mi
57. 30 mi   **59.** $8\frac{2}{11}$ min after 10:30   **61.** $51\frac{3}{7}$ mph

Exercise Set 6.6, pp. 454–455

1. $\dfrac{17}{3}x^4 + 3x^3 - \dfrac{14}{3}$ **3.** $3a^2 + a - \dfrac{3}{7} - \dfrac{2}{a}$

5. $7y - \dfrac{9}{2} - \dfrac{4}{y}$ **7.** $-5x^5 + 7x^2 + 1$

9. $1 - ab^2 - a^3b^4$ **11.** $-2pq + 3p - 4q$ **13.** $x + 3$

15. $a - 12 + \dfrac{32}{a+4}$ **17.** $x - 4 + \dfrac{1}{x-5}$ **19.** $y - 5$

21. $y^2 - 2y - 1 + \dfrac{-8}{y-2}$ **23.** $2x^2 - x + 1 + \dfrac{-5}{x+2}$

25. $a^2 + 4a + 15 + \dfrac{70}{a-4}$ **27.** $2y^2 + 2y - 1 + \dfrac{8}{5y-2}$

29. $2x^2 - x - 9 + \dfrac{3x+12}{x^2+2}$ **31.** $4x^2 - 6x + 9$, $x \neq -\frac{3}{2}$

33. $2x - 5$, $x \neq -\frac{2}{3}$ **35.** $x^2 + 1$, $x \neq -5$, $x \neq 5$
37. $2x^3 - 3x^2 + 5$, $x \neq -1$, $x \neq 1$ **39.** TW

41. $c = \dfrac{ab-k}{d}$ **42.** $z = \dfrac{xy-t}{w}$ **43.** 12, 13, 14
44. $-54a^3$ **45.** $\{x\,|\,x < -2$ *or* $x > 5\}$, or
$(-\infty, -2) \cup (5, \infty)$ **46.** $\left\{x\,\middle|\,-\frac{7}{3} < x < 3\right\}$, or $\left(-\frac{7}{3}, 3\right)$
47. TW   **49.** $a^2 + ab$ **51.** $a^6 - a^5b + a^4b^2 - a^3b^3 + a^2b^4 - ab^5 + b^6$ **53.** $-\frac{3}{2}$ **55.** TW

Exercise Set 6.7, p. 460

1. $x^2 - 3x + 5 + \dfrac{-12}{x+1}$ **3.** $a + 5 + \dfrac{-4}{a+3}$

5. $x^2 - 5x - 23 + \dfrac{-43}{x-2}$ **7.** $3x^2 - 2x + 2 + \dfrac{-3}{x+3}$

9. $y^2 + 2y + 1 + \dfrac{12}{y-2}$ **11.** $x^4 + 2x^3 + 4x^2 + 8x + 16$

13. $3x^2 + 6x - 3 + \dfrac{2}{x+\frac{1}{3}}$ **15.** 6   **17.** 1   **19.** 54

21. TW   **23.** $b = \dfrac{a-9}{c+1}$ **24.** $a = \dfrac{bd-8}{c-b}$
25. $\{x\,|\,x$ is a real number *and* $x \neq -5$ *and* $x \neq 5\}$
26. $\left\{x\,\middle|\,x$ is a real number *and* $x \neq -\frac{9}{2}$ *and* $x \neq 1\right\}$
27.
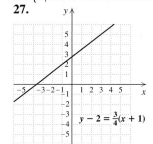
28.
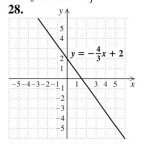

29. TW   **31.** (a) The degree of R must be less than 1, the
degree of $x - r$; (b) Let $x = r$. Then

$$P(r) = (r-r) \cdot Q(r) + R$$
$$= 0 \cdot Q(r) + R$$
$$= R.$$

33. 0; $-\frac{7}{2}, \frac{5}{3}, 4$ **35.** 0

Exercise Set 6.8, pp. 470–475

1. $d_1 = \dfrac{W_1d_2}{W_2}$ **3.** $v_1 = \dfrac{2s}{t} - v_2$, or $\dfrac{2s-tv_2}{t}$

5. $f = \dfrac{d_id_o}{d_o+d_i}$ **7.** $R = \dfrac{2V}{I} - 2r$, or $\dfrac{2V-2Ir}{I}$

9. $g = \dfrac{Rs}{s-R}$ **11.** $n = \dfrac{IR}{E-Ir}$ **13.** $q = \dfrac{pf}{p-f}$

15. $t_1 = \dfrac{H}{Sm} + t_2$, or $\dfrac{H+Smt_2}{Sm}$ **17.** $r = \dfrac{Re}{E-e}$

19. $r = 1 - \dfrac{a}{S}$, or $\dfrac{S-a}{S}$ **21.** $a + b = \dfrac{f}{c^2}$

23. $T = -\dfrac{I_f}{I_t} + 1$, or $1 - \dfrac{I_f}{I_t}$, or $\dfrac{I_t - I_f}{I_t}$ **25.** $r_2 = \dfrac{Rr_1}{r_1 - R}$

27. $t_1 = t_2 - \dfrac{v_2 - v_1}{a}$ **29.** $Q = \dfrac{2Tt - 2AT}{A - q}$ **31.** $k = 7$;
$y = 7x$ **33.** $k = 1.7$; $y = 1.7x$ **35.** $k = 6$; $y = 6x$
37. $33\frac{1}{3}$ cm   **39.** 241,920,000 cans   **41.** 32 kg

43. 3.36   **45.** $k = 60$; $y = \dfrac{60}{x}$ **47.** $k = 112$; $y = \dfrac{112}{x}$

49. $k = 9$; $y = \dfrac{9}{x}$ **51.** 20 min   **53.** 160 cm^3

55. 3.5 hr   **57.** $y = \frac{2}{3}x^2$ **59.** $y = \dfrac{54}{x^2}$ **61.** $y = 0.3xz^2$

63. $y = \dfrac{4wx^2}{z}$ **65.** 40 W/m^2 **67.** 220 cm^3

69. About 57.42 mph   **71. (a)** Inverse; **(b)** $y = \dfrac{300}{x}$;
(c) 100 min   **73. (a)** Directly; **(b)** $y \approx 1.08x$;
(c) 8.64 million   **75.** TW   **77.** $\mathbb{R}$ **78.** $\mathbb{R}$
79.   **80.** $8a^3 - 2a$

81. $(t + 2b)(t^2 - 2bt + 4b^2)$ **82.** $-\frac{5}{3}, \frac{7}{2}$ **83.** TW

85. 567 mi   **87.** Ratio is $\dfrac{a + 12}{a + 6}$; percent increase is

$\dfrac{6}{a + 6} \cdot 100\%$, or $\dfrac{600}{a + 6}\%$

89. $t_1 = t_2 + \dfrac{(d_2 - d_1)(t_4 - t_3)}{a(t_4 - t_2)(t_4 - t_3) + d_3 - d_4}$ **91.** Q varies
directly as the square of p and inversely as the cube of q.

93. About 1.697 m   **95.** $d(s) = \dfrac{28}{s}$; 70 yd

Review Exercises: Chapter 6, pp. 478–480

1. [6.1] **(a)** $-\frac{2}{9}$; **(b)** $-\frac{3}{4}$; **(c)** 0   **2.** [6.2] $48x^3$
3. [6.2] $(x + 5)(x - 2)(x - 4)$ **4.** [6.2] $x + 3$
5. [6.2] $\dfrac{1}{x - 1}$ **6.** [6.1] $\dfrac{b^2c^6d^2}{a^5}$ **7.** [6.2] $\dfrac{15np + 14m}{18m^2n^4p^2}$

8. [6.1] $\dfrac{y - 8}{2}$ **9.** [6.1] $\dfrac{(x - 2)(x + 5)}{x - 5}$

10. [6.1] $\dfrac{3a - 1}{a - 3}$ **11.** [6.1] $\dfrac{(x^2 + 4x + 16)(x - 6)}{(x + 4)(x + 2)}$

12. [6.2] $\dfrac{x - 3}{(x + 1)(x + 3)}$ **13.** [6.2] $\dfrac{x - y}{x + y}$

14. [6.2] $2(x + y)$ **15.** [6.2] $\dfrac{-y}{(y + 4)(y - 1)}$

16. [6.3] $\frac{5}{7}$ **17.** [6.3] $\dfrac{a^2b^2}{2(b^2 - ba + a^2)}$

18. [6.3] $\dfrac{(y + 11)(y + 5)}{(y - 5)(y + 2)}$ **19.** [6.3] $\dfrac{(14 - 3x)(x + 3)}{2x^2 + 16x + 6}$
20. [6.4] 2   **21.** [6.4] 6   **22.** [6.4] No solution
23. [6.4] 0   **24.** [6.4] $-1, 4$ **25.** [6.5] $5\frac{1}{7}$ hr
26. [6.5] Celeron: 45 sec; Pentium III: 30 sec
27. [6.5] 24 mph   **28.** [6.5] Motorcycle: 62 mph;
car: 70 mph   **29.** [6.6] $4s^2 + 3s - 2rs^2$
30. [6.6] $y^2 - 5y + 25$ **31.** [6.6] $4x + 3 + \dfrac{-9x - 5}{x^2 + 1}$

32. [6.7] $x^2 + 6x + 20 + \dfrac{54}{x - 3}$ **33.** [6.7] 341

34. [6.8] $s = \dfrac{Rg}{g - R}$ **35.** [6.8] $m = \dfrac{H}{S(t_1 - t_2)}$

36. [6.8] $c = \dfrac{b + 3a}{2}$ **37.** [6.8] $t_1 = \dfrac{-A}{vT} + t_2$, or

$\dfrac{-A + vTt_2}{vT}$ **38.** [6.8] About 202.3 lb   **39.** [6.8] 64 L

40. [6.8] $y = \dfrac{\frac{3}{4}}{x}$ **41.** [6.8] **(a)** Inverse; **(b)** $y = \dfrac{24}{x}$;

(c) 3 oz   **42.** [6.2], [6.3], [6.4] The least common
denominator was used to add and subtract rational
expressions, to simplify complex rational expressions, and to
solve rational equations.   **43.** [6.1], [6.4] A rational
expression is a quotient of two polynomials. Expressions can
be simplified, multiplied, or added, but they cannot be solved
for a variable. A rational *equation* is an equation containing
rational expressions. In a rational equation, we often can solve
for a variable.   **44.** [6.4] All real numbers except 0 and 13
45. [6.3], [6.4] 45   **46.** [6.5] $9\frac{9}{19}$ sec

Test: Chapter 6, pp. 480–481

1. [6.1] $\dfrac{3}{4(t + 1)}$ **2.** [6.1] $\dfrac{x^2 - 3x + 9}{x + 4}$

3. [6.2] $(x - 3)(x + 11)(x - 9)$ **4.** [6.2] $\dfrac{25x + x^3}{x + 5}$

5. [6.2] $3(a - b)$ **6.** [6.2] $\dfrac{a^3 - a^2b + 4ab + ab^2 - b^3}{(a - b)(a + b)}$

7. [6.2] $\dfrac{-2(2x^2 + 5x + 20)}{(x - 4)(x + 4)(x^2 + 4x + 16)}$

8. [6.2] $\dfrac{y - 4}{(y + 3)(y - 2)}$ **9.** [6.3] $\dfrac{a(2b + 3a)}{5a + b}$

10. [6.3] $\dfrac{(x - 9)(x - 6)}{(x + 6)(x - 3)}$ **11.** [6.3] $\dfrac{4x^2 - 14x + 2}{3x^2 + 7x - 11}$
12. [6.4] $-\frac{21}{4}$ **13.** [6.4] 15   **14.** [6.1] 5; 0
15. [6.4] $\frac{5}{3}$ **16.** [6.5] $1\frac{31}{32}$ hr

17. [6.6] $\dfrac{4b^2c}{a} - \dfrac{5bc^2}{2a} + 3bc$ **18.** [6.6] $y - 14 + \dfrac{-20}{y-6}$

19. [6.6] $6x^2 - 9 + \dfrac{5x+22}{x^2+2}$

20. [6.7] $x^2 + 9x + 40 + \dfrac{153}{x-4}$ **21.** [6.7] 449

22. [6.8] $b_1 = \dfrac{2A}{h} - b_2$, or $\dfrac{2A - b_2h}{h}$ **23.** [6.5] 5 and 6;

-6 and -5 **24.** [6.5] $3\frac{3}{11}$ mph **25.** [6.8] 30 workers
26. [6.8] 637 in^2 **27.** [6.4] $-\frac{19}{3}$
28. [6.4] $\{x \mid x$ is a real number *and* $x \neq 0$ *and* $x \neq 15\}$
29. [6.3], [6.4] x-intercept: $(11, 0)$; y-intercept: $\left(0, -\frac{33}{5}\right)$
30. [6.5] Hans: 56 lawns; Franz: 42 lawns

Cumulative Review: Chapters 1–6, pp. 482–484

1. [1.1], [1.2] 10 **2.** [1.4] 5.76×10^9 **3.** [2.4] Slope: $\frac{7}{4}$;
y-intercept: $(0, -3)$ **4.** [2.6] $y = -\frac{10}{3}x + \frac{11}{3}$
5. [3.2] $(-3, 4)$ **6.** [3.4] $(-2, -3, 1)$ **7.** [3.3] Small: 16;
large: 29 **8.** [3.5] 12, $\frac{1}{2}$, $7\frac{1}{2}$ **9.** [2.4] \$11.8 million per
year **10.** [2.6] **(a)** $V(t) = \frac{5}{18}t + 15.4$; **(b)** about 19.8 min
11. (a) [6.1] $-\frac{1}{2}$; **(b)** [2.7] $\{x \mid x$ is a real number *and* $x \neq 5\}$
12. [5.6] $\frac{1}{4}$ **13.** [5.6] $-\frac{25}{7}, \frac{25}{7}$ **14.** [4.1] $\{x \mid x > -3\}$, or
$(-3, \infty)$ **15.** [4.1] $\{x \mid x \geq -1\}$, or $[-1, \infty)$
16. [4.2] $\{x \mid -10 < x < 13\}$, or $(-10, 13)$
17. [4.2] $\{x \mid x < -\frac{4}{3}$ *or* $x > 6\}$, or $\left(-\infty, -\frac{4}{3}\right) \cup (6, \infty)$
18. [4.3] $\{x \mid x < -6.4$ *or* $x > 6.4\}$, or $(-\infty, -6.4) \cup (6.4, \infty)$
19. [4.3] $\{x \mid -\frac{13}{4} \leq x \leq \frac{15}{4}\}$, or $\left[-\frac{13}{4}, \frac{15}{4}\right]$ **20.** [6.4] $-\frac{5}{3}$
21. [6.4] -1 **22.** [6.4] No solution **23.** [6.4] $\frac{1}{3}$

24. [4.3] $1, \frac{7}{3}$ **25.** [4.2] $[7, \infty)$ **26.** [2.3] $n = \dfrac{m-12}{3}$

27. [6.8] $a = \dfrac{Pb}{3-P}$

28. [4.4]

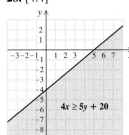

$4x \geq 5y + 20$

29. [2.4]

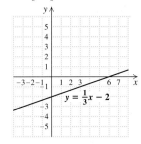

$y = \frac{1}{3}x - 2$

30. [5.1] $-3x^3 + 9x^2 + 3x - 3$ **31.** [5.2] $-15x^4y^4$
32. [5.1] $5a + 5b - 5c$
33. [5.2] $15x^4 - x^3 - 9x^2 + 5x - 2$
34. [5.2] $4x^4 - 4x^2y + y^2$ **35.** [5.2] $4x^4 - y^2$

36. [5.1] $-m^3n^2 - m^2n^2 - 5mn^3$ **37.** [6.1] $\dfrac{y-6}{2}$

38. [6.1] $x - 1$ **39.** [6.2] $\dfrac{a^2 + 7ab + b^2}{(a-b)(a+b)}$

40. [6.2] $\dfrac{-m^2 + 5m - 6}{(m+1)(m-5)}$ **41.** [6.2] $\dfrac{3y^2 - 2}{3y}$

42. [6.3] $\dfrac{y-x}{xy(x+y)}$ **43.** [6.6] $9x^2 - 13x + 26 + \dfrac{-50}{x+2}$
44. [5.3] $2x^2(2x+9)$ **45.** [5.4] $(x-6)(x+14)$
46. [5.6] $(4y-9)(4y+9)$
47. [5.7] $8(2x+1)(4x^2-2x+1)$ **48.** [5.6] $(t-8)^2$
49. [5.6] $x^2(x-1)(x+1)(x^2+1)$
50. [5.7] $(0.3b - 0.2c)(0.09b^2 + 0.06bc + 0.04c^2)$
51. [5.5] $(4x-1)(5x+3)$ **52.** [5.5] $(3x+4)(x-7)$
53. [5.3] $(x^2-y)(x^3+y)$ **54.** [2.7], [5.8] $\{x \mid x$ is a real
number *and* $x \neq 2$ *and* $x \neq 5\}$ **55.** [6.5] $3\frac{1}{3}$ sec
56. [5.8] 30 ft **57.** [5.8] All such sets of even integers
satisfy this condition. **58.** [6.8] 25 ft
59. (a) [1.5]

(b) [2.6] $b(x) = 196.25x + 1053.75$; **(c)** [2.6] \$5960
60. [6.8] **(a)** Directly; **(b)** $v \approx 4.92b$; **(c)** \$29,323.20
61. [2.6] **(a)** $v(x) = 837.605948x + 6515.511152$;
(b) \$27,456; **(c)** The amount predicted in Exercise 60 was
about \$2000 more than the amount predicted in Exercise 61.
62. [1.5] (b) **63.** [4.3] (a) **64.** [5.1] (d)
65. [6.1] (c) **66.** [2.2] 1 **67.** [4.3] $\{x \mid x \geq 1\}$, or $[1, \infty)$
68. [3.1] $(2, -3)$ **69.** [2.2] 2 **70.** [5.3] 1
71. [4.3] $\{x \mid 1 \leq x \leq 5\}$, or $[1, 5]$ **72.** [4.3] $\{1, 5\}$
73. [5.2] $x^3 - 12x^2 + 48x - 64$ **74.** [5.6] $-3, 3, -5, 5$
75. [4.2], [4.3] $\{x \mid -3 \leq x \leq -1$ *or* $7 \leq x \leq 9\}$, or
$[-3, -1] \cup [7, 9]$ **76.** [6.4] All real numbers except 9
and -5 **77.** [5.6] $-\frac{1}{4}, 0, \frac{1}{4}$

Chapter 7

Interactive Discovery, p. 488

1. Not an identity **2.** Not an identity **3.** Identity
4. Not an identity **5.** Identity **6.** Identity

Interactive Discovery, p. 490

1. Identity **2.** Not an identity **3.** Not an identity
4. Identity **5.** Not an identity **6.** Identity

Exercise Set 7.1, pp. 497–499

1. $4, -4$ **3.** $12, -12$ **5.** $9, -9$ **7.** $30, -30$
9. $2.6458, -2.6458$ **11.** $4.8683, -4.8683$
13. $0.8660, -0.8660$ **15.** $-\frac{7}{6}$ **17.** 21 **19.** $-\frac{4}{9}$

21. 0.3 **23.** -0.07 **25.** $p^2 + 4$; 2 **27.** $\dfrac{x}{y+4}$; 3

29. $\sqrt{20}$; 0; does not exist; does not exist
31. -3; -1; does not exist; 0 **33.** 1; $\sqrt{2}$; $\sqrt{101}$
35. 1; does not exist; 6 **37.** $6|x|$ **39.** $6|b|$
41. $|7 - t|$ **43.** $|y + 8|$ **45.** $|3x - 5|$ **47.** -5

49. -3 **51.** $-\frac{1}{2}$ **53.** $|y|$ **55.** $7|b|$ **57.** 10
59. $|2a + b|$ **61.** $|x^5|$ **63.** $|a^7|$ **65.** $5t$ **67.** $7c$
69. $5 + b$ **71.** $3(x + 2)$, or $3x + 6$ **73.** $5t - 2$
75. -4 **77.** $3x$ **79.** 10 **81.** $4x$ **83.** a^7
85. $(x + 3)^5$ **87.** $2; 3; -2; -4$ **89.** 2; does not exist;
does not exist; 3 **91.** $\{x \,|\, x \geq 5\}$, or $[5, \infty)$
93. $\{t \,|\, t \geq -3\}$, or $[-3, \infty)$ **95.** $\{x \,|\, x \leq 5\}$, or $(-\infty, 5]$
97. $\mathbb{R}$ **99.** $\{z \,|\, z \geq -\frac{3}{5}\}$, or $[-\frac{3}{5}, \infty)$ **101.** $\mathbb{R}$
103. Domain: $\{x \,|\, x \leq 5\}$, or $(-\infty, 5]$;
range: $\{y \,|\, y \geq 0\}$, or $[0, \infty)$
105. Domain: $\{x \,|\, x \geq -1\}$, or $[-1, \infty)$;
range: $\{y \,|\, y \leq 1\}$, or $(-\infty, 1]$
107. Domain: $\mathbb{R}$; range: $\{y \,|\, y \geq 5\}$, or $[5, \infty)$
109. (c)   **111.** (d)   **113.** Yes   **115.** Yes
117. 57.3 million; 72.1 million   **119.** TW   **121.** $a^9b^6c^{15}$
122. $10a^{10}b^9$ **123.** $\dfrac{a^6c^{12}}{8b^9}$ **124.** $\dfrac{x^6y^2}{25z^4}$ **125.** $2x^4y^5z^2$
126. $\dfrac{5c^3}{a^4b^7}$ **127.** $n(t) = 133.5625t + 3282$
128. $f(x) = 134.4737456x + 3212.856476$ **129.** TW
131. TW   **133.** (a) 13; (b) 15; (c) 18; (d) 20
135. $\{x \,|\, -4 < x \leq 5\}$, or $(-4, 5]$

Exercise Set 7.2, pp. 505–507

1. $\sqrt[4]{x}$ **3.** 4   **5.** 3   **7.** 3   **9.** $\sqrt[3]{xyz}$ **11.** $\sqrt[5]{a^2b^2}$
13. $\sqrt[3]{a^2}$ **15.** 8   **17.** 343   **19.** 243
21. $\sqrt[4]{81^3x^3}$, or $27\sqrt[4]{x^3}$ **23.** $125x^6$ **25.** $20^{1/3}$
27. $17^{1/2}$ **29.** $x^{3/2}$ **31.** $m^{2/5}$ **33.** $(cd)^{1/4}$
35. $(xy^2z)^{1/5}$ **37.** $(3mn)^{3/2}$ **39.** $(8x^2y)^{5/7}$ **41.** $\dfrac{2x}{z^{2/3}}$
43. $\dfrac{1}{x^{1/3}}$ **45.** $\dfrac{1}{(2rs)^{3/4}}$ **47.** 4   **49.** $a^{5/7}$ **51.** $\dfrac{2a^{3/4}c^{2/3}}{b^{1/2}}$
53. $\dfrac{x^4}{2^{1/3}y^{2/7}}$ **55.** $\left(\dfrac{8yz}{7x}\right)^{3/5}$ **57.** $\dfrac{7x}{z^{1/3}}$ **59.** $\dfrac{5ac^{1/2}}{3}$
61. $y = (x + 7)\wedge(1/4)$ **63.** $y = (3x - 2)\wedge(1/7)$

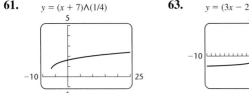

65. $y = x\wedge(3/6)$ **67.** 1.552   **69.** 1.778

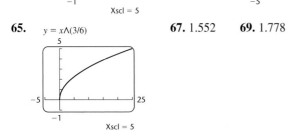

71. -6.240 **73.** $5^{7/8}$ **75.** $3^{3/4}$ **77.** $4.1^{1/2}$
79. $10^{6/25}$ **81.** $a^{23/12}$ **83.** 64   **85.** $\dfrac{m^{1/3}}{n^{1/8}}$ **87.** $\sqrt[3]{a}$

89. x^5 **91.** x^3 **93.** a^5b^5 **95.** $\sqrt[4]{3x}$ **97.** $\sqrt{3a}$
99. $\sqrt[8]{x}$ **101.** a^3b^3 **103.** x^8y^{20} **105.** $\sqrt[12]{xy}$
107. TW   **109.** $11x^4 + 14x^3$ **110.** $-3t^6 + 28t^5 - 20t^4$
111. $15a^2 - 11ab - 12b^2$ **112.** $49x^2 - 14xy + y^2$
113. \$93,500   **114.** 0, 1   **115.** TW   **117.** $\sqrt[10]{x^5y^3}$
119. $\sqrt[4]{2xy^2}$ **121.** 524 cycles per second
123. $2^{7/12} \approx 1.498 \approx 1.5$ **125.** (a) 1.8 m; (b) 3.1 m;
(c) 1.5 m; (d) 5.3 m   **127.** 10 mg

Exercise Set 7.3, pp. 514–515

1. $\sqrt{70}$ **3.** $\sqrt[3]{10}$ **5.** $\sqrt[4]{72}$ **7.** $\sqrt{30ab}$ **9.** $\sqrt[5]{18t^3}$
11. $\sqrt{x^2 - a^2}$ **13.** $\sqrt[3]{0.1x^2}$ **15.** $\sqrt[4]{x^3 - 1}$ **17.** $\sqrt{\dfrac{7x}{6y}}$
19. $\sqrt[7]{\dfrac{5x - 15}{4x + 8}}$ **21.** $5\sqrt{2}$ **23.** $2\sqrt{7}$ **25.** $2\sqrt{2}$
27. $3\sqrt{22}$ **29.** $6a^2\sqrt{b}$ **31.** $2x\sqrt[3]{y^2}$ **33.** $-2x^2\sqrt[3]{2}$
35. $f(x) = 5x\sqrt[3]{x^2}$ **37.** $f(x) = |7(x - 3)|$, or $7|x - 3|$
39. $f(x) = |x - 1|\sqrt{5}$ **41.** $ab^2\sqrt{a}$ **43.** $xy^2z^3\sqrt[3]{x^2z}$
45. $-2ab^2\sqrt[5]{a^2b}$ **47.** $abc\sqrt[5]{ab^3c^4}$ **49.** $3x^2\sqrt[4]{10x}$
51. $5\sqrt{3}$ **53.** $2\sqrt{35}$ **55.** 2   **57.** $18a^3$ **59.** $a\sqrt[3]{10}$
61. $3x^3\sqrt{5x}$ **63.** $s^2t^3\sqrt[3]{t}$ **65.** $(x + 5)^2$ **67.** $2ab^3\sqrt[4]{3a}$
69. $x(y + z)^2\sqrt[5]{x}$ **71.** TW   **73.** $\dfrac{12x^2 + 5y^2}{64xy}$
74. $\dfrac{2a + 6b^3}{a^4b^4}$ **75.** $\dfrac{-7x - 13}{2(x - 3)(x + 3)}$
76. $\dfrac{-3x + 1}{2(x - 5)(x + 5)}$ **77.** $3a^2b^2$ **78.** $3ab^5$ **79.** TW
81. (a) 20 mph; (b) 37.4 mph; (c) 42.4 mph   **83.** $r^{10}t^3\sqrt{rt}$
85. $a^3b^6\sqrt[3]{ab^2}$
87. $f(x) = h(x); f(x) \neq g(x)$

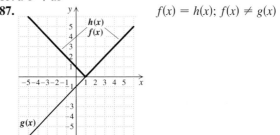

89. $\{x \,|\, x \leq -1 \ or \ x \geq 4\}$, or $(-\infty, -1] \cup [4, \infty)$ **91.** 10
93. TW

Exercise Set 7.4, pp. 520–522

1. $\dfrac{5}{6}$ **3.** $\dfrac{4}{3}$ **5.** $\dfrac{7}{y}$ **7.** $\dfrac{5y\sqrt{y}}{x^2}$ **9.** $\dfrac{3a\sqrt[3]{a}}{2b}$
11. $\dfrac{2a}{bc^2}$ **13.** $\dfrac{ab^2}{c^2}\sqrt[4]{\dfrac{a}{c^2}}$ **15.** $\dfrac{2x}{y^2}\sqrt[5]{\dfrac{x}{y}}$ **17.** $\dfrac{xy}{z^2}\sqrt[6]{\dfrac{y^2}{z^3}}$
19. $\sqrt{5}$ **21.** 3   **23.** $y\sqrt{5y}$ **25.** $2\sqrt[3]{a^2b}$
27. $\sqrt{2ab}$ **29.** $2x^2y^3\sqrt[4]{y^3}$ **31.** $\sqrt[3]{x^2 + xy + y^2}$
33. $\dfrac{\sqrt{35}}{7}$ **35.** $\dfrac{2\sqrt{15}}{5}$ **37.** $\dfrac{2\sqrt[3]{6}}{3}$ **39.** $\dfrac{\sqrt[3]{75ac^2}}{5c}$

41. $\dfrac{y\sqrt[3]{180x^2y}}{6x^2}$　**43.** $\dfrac{\sqrt[3]{2xy^2}}{xy}$　**45.** $\dfrac{\sqrt{14a}}{6}$　**47.** $\dfrac{3\sqrt{5y}}{10xy}$

49. $\dfrac{\sqrt{5b}}{6a}$　**51.** $\dfrac{5}{\sqrt{35x}}$　**53.** $\dfrac{2}{\sqrt{6}}$　**55.** $\dfrac{52}{3\sqrt{91}}$

57. $\dfrac{7}{\sqrt[3]{98}}$　**59.** $\dfrac{7x}{\sqrt{21xy}}$　**61.** $\dfrac{2a^2}{\sqrt[3]{20ab}}$　**63.** $\dfrac{x^2y}{\sqrt{2xy}}$

65. TW　**67.** $\dfrac{3(x-1)}{(x-5)(x+5)}$　**68.** $\dfrac{7(x-2)}{(x+4)(x-4)}$

69. $\dfrac{a-1}{a+7}$　**70.** $\dfrac{t+11}{t+2}$　**71.** $125a^9b^{12}$　**72.** $225x^{10}y^6$

73. TW　**75.** (a) 1.62 sec; (b) 1.99 sec; (c) 2.20 sec

77. $9\sqrt[3]{9n^2}$　**79.** $\dfrac{-3\sqrt{a^2-3}}{a^2-3}$, or $\dfrac{-3}{\sqrt{a^2-3}}$

81. Step 1: $\sqrt[n]{a}=a^{1/n}$, by definition; Step 2: $\left(\dfrac{a}{b}\right)^n=\dfrac{a^n}{b^n}$,

raising a quotient to a power; Step 3: $a^{1/n}=\sqrt[n]{a}$, by definition
83. $(f/g)(x)=3x$, where x is a real number and $x>0$
85. $(f/g)(x)=\sqrt{x+3}$, where x is a real number and $x>3$

Exercise Set 7.5, pp. 527–529

1. $5\sqrt{7}$　**3.** $3\sqrt[3]{5}$　**5.** $13\sqrt[3]{y}$　**7.** $7\sqrt{2}$

9. $13\sqrt[3]{7}+\sqrt{3}$　**11.** $21\sqrt{3}$　**13.** $23\sqrt{5}$　**15.** $9\sqrt[3]{2}$

17. $(1+6a)\sqrt{5a}$　**19.** $(x+2)\sqrt[3]{6x}$　**21.** $3\sqrt{a-1}$

23. $(x+3)\sqrt{x-1}$　**25.** $3\sqrt{7}-7$　**27.** $4\sqrt{6}-4\sqrt{10}$

29. $2\sqrt{15}-6\sqrt{3}$　**31.** -6　**33.** $a+2a\sqrt[3]{3}$　**35.** 19

37. -19　**39.** $14+6\sqrt{5}$　**41.** -6

43. $2\sqrt[3]{9}+3\sqrt[3]{6}-2\sqrt[3]{4}$　**45.** $3x+2\sqrt{3xy}+y$

47. $\dfrac{3-\sqrt{5}}{2}$　**49.** $\dfrac{12+2\sqrt{3}+6\sqrt{5}+\sqrt{15}}{33}$

51. $\dfrac{a-\sqrt{ab}}{a-b}$　**53.** -1

55. $\dfrac{24-3\sqrt{10}-4\sqrt{14}+\sqrt{35}}{27}$　**57.** $\dfrac{3\sqrt{6}+4}{2}$

59. $\dfrac{3}{5\sqrt{7}-10}$　**61.** $\dfrac{2}{14+2\sqrt{3}+3\sqrt{2}+7\sqrt{6}}$

63. $\dfrac{x-y}{x+2\sqrt{xy}+y}$　**65.** $a\sqrt[4]{a}$　**67.** $b\sqrt[10]{b^9}$

69. $xy\sqrt[6]{xy^5}$　**71.** $3a^2b\sqrt[4]{ab}$　**73.** $xyz\sqrt[6]{x^5yz^2}$

75. $\sqrt[15]{x^7}$　**77.** $\sqrt[15]{\dfrac{a^7}{b^2}}$　**79.** $\sqrt[10]{xy^3}$　**81.** $\sqrt[12]{(2+5x)^5}$

83. $\sqrt[12]{5+3x}$　**85.** $x\sqrt[6]{xy^5}-\sqrt[15]{x^{13}y^{14}}$

87. $2m^2+m\sqrt[4]{n}+2m\sqrt[3]{n^2}+\sqrt[12]{n^{11}}$　**89.** $\sqrt[4]{2x^2}-x^3$

91. x^2-7　**93.** $27-10\sqrt{2}$　**95.** $8+2\sqrt{15}$

97. TW　**99.** 8　**100.** $\dfrac{15}{2}$

101. Length: 20 units; width: 5 units　**102.** $-5, 4$

103. $\frac{1}{5}, 1$　**104.** $\frac{1}{7}, 1$　**105.** TW

107. Radicands, indices　**109.** Denominators

111. $f(x)=-6x\sqrt{5+x}$　**113.** $f(x)=(x+3x^2)\sqrt[4]{x-1}$

115. $ac^2\left[(3a+2c)\sqrt{ab}-2\sqrt[3]{ab}\right]$
117. $9a^2(b+1)\sqrt[6]{243a^5(b+1)^5}$　**119.** $1-\sqrt{w}$
121. $\left(\sqrt{x}+\sqrt{5}\right)\left(\sqrt{x}-\sqrt{5}\right)$
123. $\left(\sqrt{x}+\sqrt{a}\right)\left(\sqrt{x}-\sqrt{a}\right)$　**125.** $2x-2\sqrt{x^2-4}$
127. $\dfrac{b^2+\sqrt{b}}{1+b+b^2}$　**129.** $\dfrac{x-19}{x+10\sqrt{x+6}+31}$

Interactive Discovery, p. 530

1. {3}; {−3, 3}　**2.** {−2}; {−2, 2}　**3.** {25}; {25}
4. ∅; {9}

Exercise Set 7.6, pp. 536–538

1. 22　**3.** $\frac{9}{2}$　**5.** 11　**7.** −3　**9.** 29　**11.** 13
13. 0, 64　**15.** No solution　**17.** −64　**19.** 81
21. No solution　**23.** $\frac{80}{3}$　**25.** 57　**27.** $-\frac{5}{3}$　**29.** 1
31. $\frac{106}{27}$　**33.** 4　**35.** 3, 7　**37.** $\frac{80}{9}$　**39.** −1
41. No solution　**43.** 2, 6　**45.** 2　**47.** 4　**49.** TW
51. Height: 7 in.; base: 9 in.　**52.** 8 60-sec commercials
53. Elaine: 6 hr; Gonzalo: 12 hr　**54.** $\frac{165}{52}$ hr, or about 3.2 hr
55.

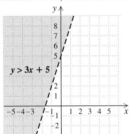

$y > 3x + 5$

56.

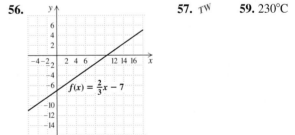

$f(x)=\frac{2}{3}x-7$

57. TW　**59.** 230°C

61. $t=\dfrac{1}{9}\left(\sqrt{\dfrac{S^2\cdot 2457}{1087.7^2}}-2617\right)$　**63.** 4480 rpm

65. $r=\dfrac{v^2h}{2gh-v^2}$　**67.** 72.25 ft　**69.** $\frac{2504}{125}, \frac{2496}{125}$　**71.** 0

73. 1, 8　**75.** $\left(\frac{1}{36}, 0\right), (36, 0)$　**77.** TW

Exercise Set 7.7, pp. 544–547

1. $\sqrt{34}$; 5.831　**3.** $9\sqrt{2}$; 12.728　**5.** 5
7. $\sqrt{31}$; 5.568　**9.** $\sqrt{11}$; 3.317　**11.** 1
13. $\sqrt{325}$ ft; 18.028 ft　**15.** $\sqrt{14,500}$ ft; 120.416 ft
17. 20 in.　**19.** $\sqrt{13,000}$ m; 114.018 m　**21.** $a=5$; $c=5\sqrt{2}\approx 7.071$　**23.** $a=7$; $b=7\sqrt{3}\approx 12.124$

25. $a = 5\sqrt{3} \approx 8.660$; $c = 10\sqrt{3} \approx 17.321$
27. $a = \dfrac{13\sqrt{2}}{2} \approx 9.192$; $b = \dfrac{13\sqrt{2}}{2} \approx 9.192$
29. $b = 14\sqrt{3} \approx 24.249$; $c = 28$ **31.** $3\sqrt{3} \approx 5.196$
33. $13\sqrt{2} \approx 18.385$ **35.** $\dfrac{19\sqrt{2}}{2} \approx 13.435$
37. $\sqrt{10561}$ ft ≈ 102.767 ft   **39.** $h = 2\sqrt{3}$ ft ≈ 3.464 ft
41. 8 ft   **43.** $(0, -4), (0, 4)$ **45.** TW   **47.** -47
48. 5   **49.** $x(x-3)(x+3)$ **50.** $7a(a-2)(a+2)$
51. $\left\{-\frac{2}{3}, 4\right\}$ **52.** $\left\{-\frac{4}{3}, 10\right\}$ **53.** TW   **55.** $\sqrt{75}$ cm
57. 4 gal. The total area of the doors and windows is 164 ft^2 or more.   **59.** 36.15 ft by 36.15 ft by 36.15 ft

Exercise Set 7.8, pp. 554–555

1. $5i$ **3.** $i\sqrt{13}$, or $\sqrt{13}i$ **5.** $3i\sqrt{2}$, or $3\sqrt{2}i$
7. $i\sqrt{3}$, or $\sqrt{3}i$ **9.** $9i$ **11.** $10i\sqrt{3}$, or $10\sqrt{3}i$
13. $-7i$ **15.** $4 - 2\sqrt{15}i$ **17.** $\left(2 + 2\sqrt{3}\right)i$
19. $\left(6\sqrt{2} - 5\right)i$ **21.** $12 + 11i$ **23.** $4 + 5i$
25. $-4 - 5i$ **27.** $5 + 5i$ **29.** -54 **31.** 56
33. -35 **35.** $-\sqrt{42}$ **37.** $-5\sqrt{6}$ **39.** $-6 + 14i$
41. $-20 - 24i$ **43.** $-11 + 23i$ **45.** $40 + 13i$
47. $8 + 31i$ **49.** $7 - 61i$ **51.** $-15 - 30i$
53. $39 - 37i$ **55.** $-3 - 4i$ **57.** $5 + 12i$
59. $21 + 20i$ **61.** $\frac{6}{5} + \frac{3}{5}i$ **63.** $\frac{6}{29} + \frac{15}{29}i$ **65.** $-\frac{7}{9}i$
67. $-\frac{3}{4} - \frac{5}{4}i$ **69.** $1 - 2i$ **71.** $-\frac{23}{58} + \frac{43}{58}i$
73. $\frac{6}{25} - \frac{17}{25}i$ **75.** $-i$ **77.** 1 **79.** -1 **81.** i
83. -1 **85.** $-125i$ **87.** 0 **89.** TW   **91.** 1
92. 1 **93.** 50 **94.** 0 **95.** $-\frac{4}{3}, 7$
96. $\left\{x \mid -\frac{29}{3} < x < 5\right\}$, or $\left(-\frac{29}{3}, 5\right)$ **97.** TW
99. $-9 - 27i$ **101.** $50 - 120i$ **103.** $\frac{250}{41} + \frac{200}{41}i$
105. 8 **107.** $\frac{3}{5} + \frac{9}{5}i$ **109.** 1

Review Exercises: Chapter 7, pp. 557–558

1. [7.1] $\frac{7}{6}$ **2.** [7.1] -0.5 **3.** [7.1] 5
4. [7.1] $\left\{x \mid x \geq \frac{7}{2}\right\}$, or $\left[\frac{7}{2}, \infty\right)$ **5.** [7.1] $7|a|$
6. [7.1] $|c + 8|$ **7.** [7.1] $|x - 3|$ **8.** [7.1] $|2x + 1|$
9. [7.1] -2 **10.** [7.4] $-\dfrac{4x^2}{3}$ **11.** [7.3] $|x^3y^2|$, or $|x^3|y^2$
12. [7.3] $2x^2$ **13.** [7.2] $(5ab)^{4/3}$ **14.** [7.2] $8a^4\sqrt{a}$
15. [7.2] x^3y^5 **16.** [7.2] $\sqrt[3]{x^2y}$ **17.** [7.2] $\dfrac{1}{x^{2/5}}$
18. [7.2] $7^{1/6}$ **19.** [7.3] $f(x) = 5|x - 3|$
20. [7.3] $\sqrt{15xy}$ **21.** [7.3] $3a\sqrt[3]{a^2b^2}$
22. [7.3] $-6x^5y^4\sqrt[3]{2x^2}$ **23.** [7.4] $y\sqrt[3]{6}$ **24.** [7.4] $\dfrac{5\sqrt{x}}{2}$
25. [7.4] $\dfrac{2a^2\sqrt[4]{3a^3}}{c^2}$ **26.** [7.5] $7\sqrt[3]{x}$ **27.** [7.5] $3\sqrt{3}$
28. [7.5] $(2x + y^2)\sqrt[3]{x}$ **29.** [7.5] $15\sqrt{2}$
30. [7.5] $-43 - 2\sqrt{10}$ **31.** [7.5] $\sqrt[4]{x^3}$ **32.** [7.5] $\sqrt[12]{x^5}$

33. [7.5] $a^2 - 2a\sqrt{2} + 2$ **34.** [7.5] $-2\sqrt{6} + 6$
35. [7.5] $\dfrac{6}{3 + \sqrt{6}}$ **36.** [7.6] 21 **37.** [7.6] -126
38. [7.6] 4 **39.** [7.6] 14 **40.** [7.7] $5\sqrt{2}$ cm; 7.071 cm
41. [7.7] $\sqrt{24}$ ft; 4.899 ft   **42.** [7.7] $a = 10$;
$b = 10\sqrt{3} \approx 17.321$ **43.** [7.8] $-2i\sqrt{2}$, or $-2\sqrt{2}i$
44. [7.8] $-2 - 9i$ **45.** [7.8] $1 + i$ **46.** [7.8] 29
47. [7.8] i **48.** [7.8] $9 - 12i$ **49.** [7.8] $\frac{13}{25} - \frac{34}{25}i$
50. TW [7.1] An absolute-value sign must be used to simplify $\sqrt[n]{x^n}$ when n is even, since x may be negative. If x is negative while n is even, the radical expression cannot be simplified to x, since $\sqrt[n]{x^n}$ represents the principal, or positive, root. When n is odd, there is only one root, and it will be positive or negative depending on the sign of x. Thus there is no absolute-value sign when n is odd.   **51.** TW [7.8] Every real number is a complex number, but there are complex numbers that are not real. A complex number $a + bi$ is not real if $b \neq 0$.   **52.** [7.6] 3 **53.** [7.8] $-\frac{2}{5} + \frac{9}{10}i$

Test: Chapter 7, p. 559

1. [7.3] $5\sqrt{3}$ **2.** [7.4] $-\dfrac{2}{x^2}$ **3.** [7.1] $10|a|$
4. [7.1] $|x - 4|$ **5.** [7.3] $x^2y\sqrt[5]{x^2y^3}$
6. [7.4] $\left|\dfrac{5x}{6y^2}\right|$, or $\dfrac{5|x|}{6y^2}$ **7.** [7.3] $\sqrt[3]{10xy^2}$ **8.** [7.4] $\sqrt[5]{x^2y^2}$
9. [7.5] $xy\sqrt[4]{x}$ **10.** [7.5] $\sqrt[20]{a^3}$ **11.** [7.5] $5\sqrt{2}$
12. [7.5] $(x^2 + 3y)\sqrt{y}$ **13.** [7.5] $14 - 19\sqrt{x} - 3x$
14. [7.2] $\sqrt[6]{(2a^3b)^5}$ **15.** [7.2] $(7xy)^{1/2}$
16. [7.1] $\{x \mid x \leq 2\}$, or $(-\infty, 2]$ **17.** [7.5] $27 + 10\sqrt{2}$
18. [7.5] $\sqrt{6} - \sqrt{3}$ **19.** [7.6] 7 **20.** [7.6] No solution
21. [7.7] Leg: 7 cm; hypotenuse: $7\sqrt{2}$ cm ≈ 9.899 cm
22. [7.7] $\sqrt{10,600}$ ft ≈ 102.956 ft   **23.** [7.8] $5i\sqrt{2}$, or
$5\sqrt{2}i$ **24.** [7.8] $10 + 2i$ **25.** [7.8] -24
26. [7.8] $15 - 8i$ **27.** [7.8] $-\frac{13}{53} - \frac{19}{53}i$ **28.** [7.8] i
29. [7.6] 3 **30.** [7.8] $-\frac{17}{4}i$

Chapter 8

Interactive Discovery, p. 562

1. 1   **2.** 1   **3.** 1   **4.** 0   **5.** 0   **6.** 2   **7.** A cup-shaped curve opening up or down

Exercise Set 8.1, pp. 573–574

1. 2   **3.** 1   **5.** 0   **7.** $\pm\sqrt{3}$ **9.** $\pm\dfrac{2}{5}i$
11. $\pm\sqrt{\dfrac{2}{3}}$, or $\pm\dfrac{\sqrt{6}}{3}$ **13.** $-7, 3$ **15.** $-5 \pm 2\sqrt{2}$

17. $1 \pm 7i$ **19.** $\dfrac{-3 \pm \sqrt{14}}{2}$ **21.** $-7, 13$ **23.** $1, 9$

25. $-4 \pm \sqrt{13}$ **27.** $-14, 0$ **29.** $x^2 + 8x + 16$, $(x + 4)^2$ **31.** $x^2 - 6x + 9$, $(x - 3)^2$

33. $x^2 - 24x + 144$, $(x - 12)^2$ **35.** $t^2 + 9t + \dfrac{81}{4}$,

$\left(t + \dfrac{9}{2}\right)^2$ **37.** $x^2 - 3x + \dfrac{9}{4}$, $\left(x - \dfrac{3}{2}\right)^2$

39. $x^2 + \dfrac{2}{3}x + \dfrac{1}{9}$, $\left(x + \dfrac{1}{3}\right)^2$ **41.** $t^2 - \dfrac{5}{3}t + \dfrac{25}{36}$,

$\left(t - \dfrac{5}{6}\right)^2$ **43.** $x^2 + \dfrac{9}{5}x + \dfrac{81}{100}$, $\left(x + \dfrac{9}{10}\right)^2$

45. $-7, 1$ **47.** $5 \pm \sqrt{47}$ **49.** $-7, -1$ **51.** $3, 7$

53. $\dfrac{-5 \pm \sqrt{13}}{2}$ **55.** $3 \pm i$ **57.** $-2 \pm 3i$

59. $-\dfrac{1}{2}, 3$ **61.** $-\dfrac{3}{2}, -\dfrac{1}{2}$ **63.** $-\dfrac{3}{2}, \dfrac{5}{3}$

65. $\dfrac{-2 \pm \sqrt{2}}{2}$ **67.** $\dfrac{5 \pm \sqrt{61}}{6}$ **69.** 10% **71.** 18.75%

73. 4% **75.** About 10.7 sec   **77.** About 6.3 sec
79. TW   **81.** 28 **82.** -92 **83.** $3\sqrt[3]{10}$ **84.** $4\sqrt{5}$
85. 5 **86.** 7 **87.** TW   **89.** ± 18 **91.** $-\frac{7}{2}, -\sqrt{5},$
$0, \sqrt{5}, 8$ **93.** Barge: 8 km/h; fishing boat: 15 km/h

Interactive Discovery, p. 579

1. $-0.3135529, 1.9135529$; answers may vary as to which method is faster; the algebraic method is more precise, since it yields an exact answer.   **2.** $-4, 0.5$; answers may vary as to which method is faster; both are equally precise.

Exercise Set 8.2, pp. 580–582

1. $\dfrac{-7 \pm \sqrt{61}}{2}$ **3.** $3 \pm \sqrt{7}$ **5.** $-\dfrac{1}{2} \pm \dfrac{\sqrt{7}}{2}i$

7. $2 \pm 3i$ **9.** $3 \pm \sqrt{5}$ **11.** $\dfrac{-4 \pm \sqrt{19}}{3}$

13. $\dfrac{-1 \pm \sqrt{17}}{2}$ **15.** $\dfrac{-9 \pm \sqrt{129}}{24}$ **17.** $\dfrac{2}{5}$

19. $\dfrac{-11 \pm \sqrt{41}}{8}$ **21.** $5, 10$ **23.** $\dfrac{13 \pm \sqrt{509}}{10}$

25. $2 \pm \sqrt{5}i$ **27.** $2, -1 \pm \sqrt{3}i$ **29.** $\dfrac{5 \pm \sqrt{37}}{6}$

31. $5 \pm \sqrt{53}$ **33.** $\dfrac{7 \pm \sqrt{85}}{2}$ **35.** $\dfrac{3}{2}, 6$

37. $-5.3166, 1.3166$ **39.** $0.7639, 5.2361$
41. $-1.2656, 2.7656$ **43.** TW   **45.** Kenyan: 30 lb;
Peruvian: 20 lb   **46.** Cream-filled: 46; glazed: 44

47. $9a^2b^3\sqrt{2a}$ **48.** $4a^2b^3\sqrt[3]{6}$ **49.** $\dfrac{3(x + 1)}{3x + 1}$

50. $\dfrac{4b}{3ab^2 - 4a^2}$ **51.** TW   **53.** $(-2, 0), (1, 0)$

55. $4 - 2\sqrt{2}, 4 + 2\sqrt{2}$ **57.** $-1.1792101, 0.3392101$

59. $\dfrac{-5\sqrt{2} \pm \sqrt{34}}{4}$ **61.** $\dfrac{1}{2}$

Exercise Set 8.3, pp. 586–589

1. First part: 12 km/h; second part: 8 km/h   **3.** 35 mph
5. Cessna: 150 mph, Beechcraft: 200 mph; or Cessna:
200 mph, Beechcraft: 250 mph   **7.** To Hillsboro: 10 mph;
return trip: 4 mph   **9.** About 11 mph   **11.** 12 hr

13. About 3.24 mph   **15.** $r = \dfrac{1}{2}\sqrt{\dfrac{A}{\pi}}$

17. $r = \dfrac{-\pi h + \sqrt{\pi^2 h^2 + 2\pi A}}{2\pi}$ **19.** $s = \sqrt{\dfrac{kQ_1 Q_2}{N}}$

21. $g = \dfrac{4\pi^2 l}{T^2}$ **23.** $c = \sqrt{d^2 - a^2 - b^2}$

25. $t = \dfrac{-v_0 + \sqrt{v_0^2 + 2gs}}{g}$ **27.** $n = \dfrac{1 + \sqrt{1 + 8N}}{2}$

29. $h = \dfrac{V^2}{12.25}$ **31.** $t = \dfrac{-b \pm \sqrt{b^2 - 4ac}}{2a}$

33. (a) 10.1 sec; (b) 7.49 sec; (c) 272.5 m   **35.** 2.9 sec
37. 0.87 sec   **39.** 2.5 m/sec   **41.** 7% **43.** TW

45. -104 **46.** $2i\sqrt{11}$ **47.** $\dfrac{x + y}{2}$ **48.** $\dfrac{a^2 - b^2}{b}$

49. $\dfrac{1 + \sqrt{5}}{2}$ **50.** $\dfrac{1 - \sqrt{7}}{5}$ **51.** TW

53. $t = \dfrac{-10.2 \pm 6\sqrt{-A^2 + 13A - 39.36}}{A - 6.5}$

55. $l = \dfrac{w + w\sqrt{5}}{2}$ **57.** $\pm\sqrt{2}$

59. $n = \pm\sqrt{\dfrac{r^2 \pm \sqrt{r^4 + 4m^4r^2p - 4mp}}{2m}}$ **61.** $A(S) = \dfrac{\pi S}{6}$

Exercise Set 8.4, pp. 593–594

1. Two irrational   **3.** Two imaginary   **5.** Two irrational
7. One rational   **9.** Two imaginary   **11.** Two rational
13. Two rational   **15.** Two rational   **17.** Two irrational
19. Two imaginary   **21.** Two irrational
23. $x^2 + 4x - 21 = 0$ **25.** $x^2 - 6x + 9 = 0$
27. $x^2 + 7x + 10 = 0$ **29.** $3x^2 - 14x + 8 = 0$
31. $6x^2 - 5x + 1 = 0$ **33.** $x^2 - 0.8x - 0.84 = 0$
35. $x^2 - 7 = 0$ **37.** $x^2 - 18 = 0$ **39.** $x^2 + 9 = 0$
41. $x^2 - 10x + 29 = 0$ **43.** $x^2 - 4x - 6 = 0$
45. $x^3 - 4x^2 - 7x + 10 = 0$ **47.** $x^3 - 2x^2 - 3x = 0$
49. TW   **51.** $81a^8$ **52.** $16x^6$ **53.** $(-1, 0), (8, 0)$
54. $(2, 0), (4, 0)$ **55.** 6 30-sec commercials

56.

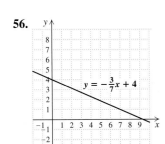

$y = -\frac{3}{7}x + 4$

57. TW

59. $a = 1, b = 2, c = -3$ **61.** (a) $-\frac{3}{5}$; (b) $-\frac{1}{3}$
63. (a) $9 + 9i$; (b) $3 + 3i$ **65.** The solutions of
$ax^2 + bx + c = 0$ are $x = \dfrac{-b \pm \sqrt{b^2 - 4ac}}{2a}$. When there is

just one solution, $b^2 - 4ac$ must be 0, so $x = \dfrac{-b \pm 0}{2a} = \dfrac{-b}{2a}$.

67. $a = 8, b = 20, c = -12$
69. $x^4 - 8x^3 + 21x^2 - 2x - 52 = 0$

Exercise Set 8.5, pp. 599–600

1. $\pm 1, \pm 3$ **3.** $\pm\sqrt{3}, \pm 3$ **5.** $\pm\dfrac{\sqrt{5}}{3}, \pm 1$

7. $9 + 4\sqrt{5}$ **9.** $\pm 2\sqrt{2}, \pm 3$ **11.** No solution
13. $-\frac{1}{2}, \frac{1}{3}$ **15.** $-\frac{4}{5}, 1$ **17.** $-27, 8$ **19.** 729   **21.** 1
23. No solution   **25.** $\frac{12}{5}$ **27.** $\left(\frac{4}{25}, 0\right)$

29. $\left(\dfrac{3 + \sqrt{33}}{2}, 0\right), \left(\dfrac{3 - \sqrt{33}}{2}, 0\right), (4, 0), (-1, 0)$

31. $(-243, 0), (32, 0)$ **33.** No x-intercepts   **35.** TW
37.

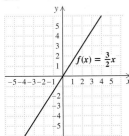

$f(x) = \frac{3}{2}x$

38.

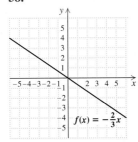

$f(x) = -\frac{2}{3}x$

39.

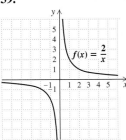

$f(x) = \frac{2}{x}$

40.

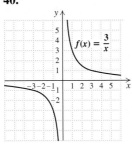

$f(x) = \frac{3}{x}$

41. A: 4 L; B: 8 L   **42.** $a^2 + a$ **43.** TW

45. $\pm\sqrt{\dfrac{7 \pm \sqrt{29}}{10}}$ **47.** $-2, -1, 5, 6$ **49.** $\frac{100}{99}$
51. $-5, -3, -2, 0, 2, 3, 5$ **53.** 1, 3

Interactive Discovery, p. 602

1. (a) $(0, 0)$; (b) $x = 0$; (c) upward; (d) narrower
2. (a) $(0, 0)$; (b) $x = 0$; (c) upward; (d) wider   **3.** (a) $(0, 0)$;
(b) $x = 0$; (c) upward; (d) wider   **4.** (a) $(0, 0)$; (b) $x = 0$;
(c) downward; (d) neither narrower nor wider   **5.** (a) $(0, 0)$;
(b) $x = 0$; (c) downward; (d) narrower   **6.** (a) $(0, 0)$;
(b) $x = 0$; (c) downward; (d) wider   **7.** When $a > 1$, the
graph of $y = ax^2$ is narrower than the graph of $y = x^2$. When
$0 < a < 1$, the graph of $y = ax^2$ is wider than the graph of
$y = x^2$.   **8.** When $a < -1$, the graph of $y = ax^2$ is
narrower than the graph of $y = x^2$ and the graph opens
downward. When $-1 < a < 0$, the graph of $y = ax^2$ is wider
than the graph of $y = x^2$ and the graph opens downward.

Interactive Discovery, p. 603

1. (a) $(3, 0)$; (b) $x = 3$; (c) same shape   **2.** (a) $(-1, 0)$;
(b) $x = -1$; (c) same shape   **3.** (a) $\left(\frac{3}{2}, 0\right)$; (b) $x = \frac{3}{2}$;
(c) same shape   **4.** (a) $(-2, 0)$; (b) $x = -2$; (c) same
shape   **5.** The graph of $g(x) = a(x - h)^2$ looks like the
graph of $f(x) = ax^2$, except that it is moved left or right.

Interactive Discovery, p. 604

1. (a) For any given x-value, the value of y_2 is 10 more than
the value of y_1; (b) yes; (c) no   **2.** (a) For any given
x-value, the value of y_2 is 5 less than the value of y_1; (b) yes;
(c) no   **3.** (a) For any given x-value, the value of y_2 is 4
more than the value of y_1; (b) yes; (c) no   **4.** The graph
of $g(x) = a(x - h)^2 + k$ looks like the graph of
$f(x) = a(x - h)^2$, except that it is moved up or down.

Exercise Set 8.6, pp. 607–609

1. (a) Positive; (b) $(3, 1)$; (c) $x = 3$; (d) $[1, \infty)$
3. (a) Negative; (b) $(-2, -3)$; (c) $x = -2$; (d) $(-\infty, -3]$
5. (a) Positive; (b) $(-3, 0)$; (c) $x = -3$; (d) $[0, \infty)$ **7.** (f)
9. (e)   **11.** (d)   **13.**

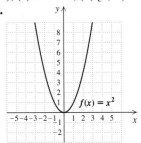

$f(x) = x^2$

15.

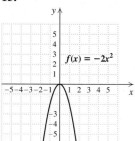

$f(x) = -2x^2$

17.

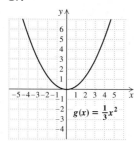

$g(x) = \frac{1}{3}x^2$

31. Vertex: $(3, 0)$; axis of symmetry: $x = 3$

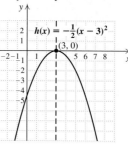

$h(x) = -\frac{1}{2}(x - 3)^2$

$(3, 0)$

33. Vertex: $(1, 0)$; axis of symmetry: $x = 1$

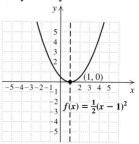

$(1, 0)$

$f(x) = \frac{1}{2}(x - 1)^2$

19.

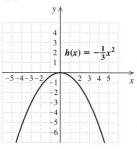

$h(x) = -\frac{1}{3}x^2$

21.

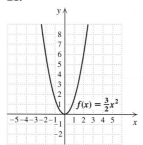

$f(x) = \frac{3}{2}x^2$

35. Vertex: $(5, 1)$; axis of symmetry: $x = 5$; minimum: 1

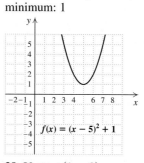

$f(x) = (x - 5)^2 + 1$

37. Vertex: $(-1, -2)$; axis of symmetry: $x = -1$; minimum: -2

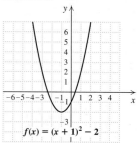

$f(x) = (x + 1)^2 - 2$

23. Vertex: $(-1, 0)$; axis of symmetry: $x = -1$

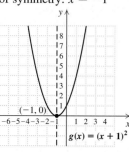

$(-1, 0)$

$g(x) = (x + 1)^2$

25. Vertex: $(2, 0)$; axis of symmetry: $x = 2$

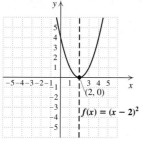

$(2, 0)$

$f(x) = (x - 2)^2$

39. Vertex: $(1, -3)$; axis of symmetry: $x = 1$; maximum: -3

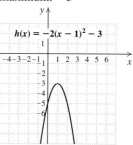

$h(x) = -2(x - 1)^2 - 3$

41. Vertex: $(-4, 1)$; axis of symmetry: $x = -4$; minimum: 1

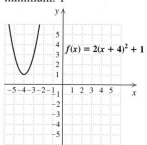

$f(x) = 2(x + 4)^2 + 1$

27. Vertex: $(-4, 0)$; axis of symmetry: $x = -4$

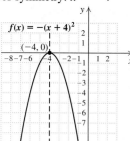

$f(x) = -(x + 4)^2$

$(-4, 0)$

29. Vertex: $(-1, 0)$; axis of symmetry: $x = -1$

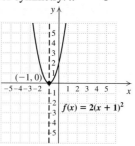

$(-1, 0)$

$f(x) = 2(x + 1)^2$

43. Vertex: $(1, 2)$; axis of symmetry: $x = 1$; maximum: 2

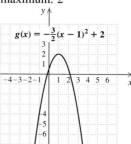

$g(x) = -\frac{3}{2}(x - 1)^2 + 2$

45. Vertex: $(9, 5)$; axis of symmetry: $x = 9$; minimum: 5; range: $[5, \infty)$

47. Vertex: $(-6, 11)$; axis of symmetry: $x = -6$; maximum: 11; range: $(-\infty, 11]$ **49.** Vertex: $\left(-\frac{1}{4}, -13\right)$; axis of symmetry: $x = -\frac{1}{4}$; minimum: -13; range: $[-13, \infty)$
51. Vertex: $(-4.58, 65\pi)$; axis of symmetry: $x = -4.58$; minimum: 65π; range: $[65\pi, \infty)$ **53.** TW
55.

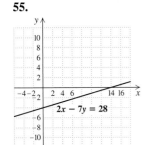

56.

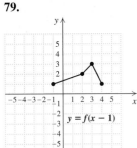

57. $(-5, -1)$ **58.** $(-1, 2)$ **59.** $x^2 + 5x + \frac{25}{4}$
60. $x^2 - 9x + \frac{81}{4}$ **61.** TW   **63.** $f(x) = \frac{3}{5}(x - 4)^2 + 1$
65. $f(x) = \frac{3}{5}(x - 3)^2 - 1$ **67.** $f(x) = \frac{3}{5}(x + 2)^2 - 5$
69. $g(x) = -2(x - 5)^2$ **71.** $f(x) = 2(x + 4)^2$
73. $g(x) = -2(x - 3)^2 + 8$ **75.** $F(x) = 3(x - 5)^2 + 1$
77. $F(x) = -2(x - 2)^2 - 3$
79.

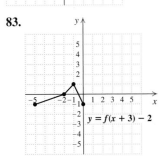

81.

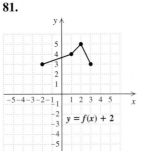

83.

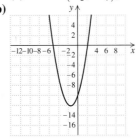

Exercise Set 8.7, pp. 616–617

1. (a) $f(x) = (x - 2)^2 + 1$; **(b)** vertex: $(2, 1)$; axis of symmetry: $x = 2$ **3. (a)** $f(x) = -\left(x - \frac{3}{2}\right)^2 - \frac{31}{4}$; **(b)** vertex: $\left(\frac{3}{2}, -\frac{31}{4}\right)$; axis of symmetry: $x = \frac{3}{2}$
5. (a) $f(x) = 2\left(x - \frac{7}{4}\right)^2 - \frac{41}{8}$; **(b)** vertex: $\left(\frac{7}{4}, -\frac{41}{8}\right)$; axis of symmetry: $x = \frac{7}{4}$

7. (a) Vertex: $(-2, 1)$; axis of symmetry: $x = -2$;   **(b)**

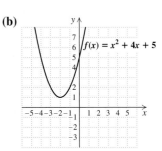

9. (a) Vertex: $(-4, 4)$; axis of symmetry: $x = -4$;   **(b)**

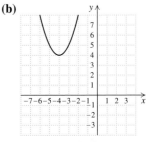

$f(x) = x^2 + 8x + 20$

11. (a) Vertex: $(4, -7)$; axis of symmetry: $x = 4$;
(b)

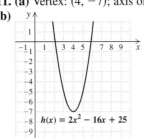

13. (a) Vertex: $(1, 6)$; axis of symmetry: $x = 1$;
(b)

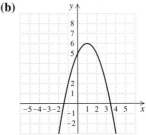

$f(x) = -x^2 + 2x + 5$

15. (a) Vertex: $\left(-\frac{3}{2}, -\frac{49}{4}\right)$; axis of symmetry: $x = -\frac{3}{2}$;
(b)

$g(x) = x^2 + 3x - 10$

17. (a) Vertex: $\left(-\frac{7}{2}, -\frac{49}{4}\right)$; axis of symmetry: $x = -\frac{7}{2}$;
(b)

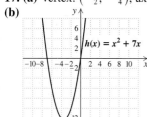

$h(x) = x^2 + 7x$

19. (a) Vertex: $(-1, -4)$; axis of symmetry: $x = -1$;
(b)

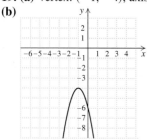

$f(x) = -2x^2 - 4x - 6$

21. (a) Vertex: $\left(\frac{5}{6}, \frac{1}{12}\right)$; axis of symmetry: $x = \frac{5}{6}$;
(b)

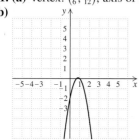

$f(x) = -3x^2 + 5x - 2$

23. (a) Vertex: $\left(-4, -\frac{5}{3}\right)$; axis of symmetry: $x = -4$;
(b)

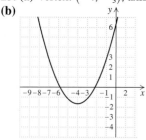

$h(x) = \frac{1}{2}x^2 + 4x + \frac{19}{3}$

25. $(-0.5, -6.25)$ **27.** $(0.1, 0.95)$ **29.** $(3.5, -4.25)$
31. $\left(3 - \sqrt{6}, 0\right), \left(3 + \sqrt{6}, 0\right)$; $(0, 3)$ **33.** $(-1, 0), (3, 0)$;
$(0, 3)$ **35.** $(0, 0), (9, 0)$; $(0, 0)$ **37.** $(2, 0)$; $(0, -4)$
39. No x-intercept; $(0, 6)$ **41. (a)** Minimum: -6.95;
(b) $(-1.06, 0), (2.41, 0)$; $(0, -5.89)$
43. (a) Maximum: -0.45; **(b)** no x-intercept; $(0, -2.79)$

45. TW **47.** $(2, -2)$ **48.** $(7, 1)$ **49.** $(3, 2, 1)$
50. $(1, -3, 2)$ **51.** 5 **52.** 4 **53.** TW
55. (a) $-2.4, 3.4$; **(b)** $-1.3, 2.3$
57. $f(x) = m\left(x - \dfrac{n}{2m}\right)^2 + \dfrac{4mp - n^2}{4m}$
59. $f(x) = \frac{5}{16}x^2 - \frac{15}{8}x - \frac{35}{16}$, or $f(x) = \frac{5}{16}(x - 3)^2 - 5$
61. **63.**

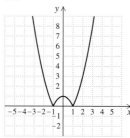 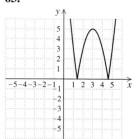

$f(x) = |x^2 - 1|$ $f(x) = |2(x - 3)^2 - 5|$

Exercise Set 8.8, pp. 627–632

1. \$4; 3 mos after January 2001 **3.** 11 days after the concert
was announced; about 62 tickets **5.** 180 ft by 180 ft
7. 200 ft²; 10 ft by 20 ft (The barn serves as a 20-ft side.)
9. 3.5 in. **11.** 81; 9 and 9 **13.** -16; 4 and -4
15. 25; -5 and -5 **17.** Quadratic; the data approximate a
parabola opening upward. **19.** Not quadratic; the data do
not approximate a parabola. **21.** Not quadratic; the data do
not approximate a parabola. **23.** Not quadratic; the data do
not approximate a parabola. **25.** Quadratic; the data
approximate half a parabola opening upward.
27. $f(x) = 2x^2 + 3x - 1$ **29.** $f(x) = -\frac{1}{4}x^2 + 3x - 5$
31. (a) $A(s) = \frac{3}{16}s^2 - \frac{135}{4}s + 1750$; **(b)** about 531 accidents
33. $h(d) = -0.0068d^2 + 0.8571d$
35. (a) $D(x) = -0.0082833093x^2 + 0.8242996891x + 0.2121786608$; **(b)** 17.325 ft
37. (a) $c(x) = 261.875x^2 - 882.5642857x + 2134.571429$;
(b) 29,254 cars **39.** TW **41.** $\dfrac{x - 9}{(x + 9)(x + 7)}$
42. $\dfrac{(x - 3)(x + 1)}{(x - 7)(x + 3)}$ **43.** $\{x \mid x < 8\}$, or $(-\infty, 8)$
44. $\{x \mid x \geq 10\}$, or $[10, \infty)$
45. $r(x) = 1454.8125x + 17,685.75$
46. $r(x) = 1514.092857x + 17,505.2381$
47. TW **49.** The radius of the circular portion of the
window and the height of the rectangular portion should each
be $\dfrac{24}{\pi + 4}$ ft. **51.** 30 trees per acre **53.** 78.4 ft

Exercise Set 8.9, pp. 639–641

1. $\left[-4, \frac{3}{2}\right]$ **3.** $(-\infty, -2) \cup (0, 2) \cup (3, \infty)$

5. $\left(-\infty, -\frac{7}{2}\right) \cup (-2, \infty)$ **7.** $(-4, 3)$, or $\{x \mid -4 < x < 3\}$
9. $(-\infty, -7] \cup [2, \infty)$, or $\{x \mid x \leq -7 \text{ or } x \geq 2\}$
11. $(-1, 2)$, or $\{x \mid -1 < x < 2\}$ **13.** $[-5, 5]$, or
$\{x \mid -5 \leq x \leq 5\}$ **15.** $\varnothing$ **17.** $(-2, 6)$, or
$\{x \mid -2 < x < 6\}$ **19.** $(-\infty, -2) \cup (0, 2)$, or
$\{x \mid x < -2 \text{ or } 0 < x < 2\}$ **21.** $(-3, -1) \cup (2, \infty)$, or
$\{x \mid -3 < x < -1 \text{ or } x > 2\}$ **23.** $(-\infty, -3) \cup (-2, 1)$, or
$\{x \mid x < -3 \text{ or } -2 < x < 1\}$ **25.** $[-0.78, 1.59]$, or
$\{x \mid -0.78 \leq x \leq 1.59\}$ **27.** $(-\infty, -2) \cup (1, 3)$, or
$\{x \mid x < -2 \text{ or } 1 < x < 3\}$ **29.** $(-\infty, -3)$, or $\{x \mid x < -3\}$
31. $(-\infty, -1] \cup (5, \infty)$, or $\{x \mid x \leq -1 \text{ or } x > 5\}$
33. $\left[-\frac{2}{3}, 2\right)$, or $\left\{x \mid -\frac{2}{3} \leq x < 2\right\}$ **35.** $(-\infty, -6)$, or
$\{x \mid x < -6\}$ **37.** $(-\infty, -1] \cup [2, 5)$, or
$\{x \mid x \leq -1 \text{ or } 2 \leq x < 5\}$ **39.** $(-\infty, -3) \cup [0, \infty)$, or
$\{x \mid x < -3 \text{ or } x \geq 0\}$ **41.** $(0, \infty)$, or $\{x \mid x > 0\}$
43. $(-\infty, -4) \cup [1, 3)$, or $\{x \mid x < -4 \text{ or } 1 \leq x < 3\}$
45. $\left(0, \frac{1}{4}\right)$, or $\left\{x \mid 0 < x < \frac{1}{4}\right\}$ **47.** ᴛᴡ   **49.** $8a^9b^6c^{12}$
50. $25a^8b^{14}$ **51.** $\frac{1}{32}$ **52.** $\frac{1}{81}$ **53.** $3a^2 + 6a + 3$
54. $5a + 7$ **55.** ᴛᴡ   **57.** $\left(-1 - \sqrt{6}, -1 + \sqrt{6}\right)$, or
$\left\{x \mid -1 - \sqrt{6} < x < -1 + \sqrt{6}\right\}$ **59.** $\{0\}$
61. **(a)** $(10, 200)$, or $\{x \mid 10 < x < 200\}$;
(b) $[0, 10) \cup (200, \infty)$, or $\{x \mid 0 \leq x < 10 \text{ or } x > 200\}$
63. $\{n \mid n \text{ is an integer } and \ 12 \leq n \leq 25\}$ **65.** $f(x)$ has no
zeros; $f(x) < 0$ for $(-\infty, 0)$, or $\{x \mid x < 0\}$; $f(x) > 0$ for $(0, \infty)$,
or $\{x \mid x > 0\}$ **67.** $f(x) = 0$ for $x = -1, 0$; $f(x) < 0$ for
$(-\infty, -3) \cup (-1, 0)$, or $\{x \mid x < -3 \text{ or } -1 < x < 0\}$;
$f(x) > 0$ for $(-3, -1) \cup (0, 2) \cup (2, \infty)$, or
$\{x \mid -3 < x < -1 \text{ or } 0 < x < 2 \text{ or } x > 2\}$
69. **(a)** $w(x) = 1388.888889x^2 - 14{,}900x + 73{,}800$; **(b)** from
1994 to about 2 years after 1994, or 1994 to 1996, and after
about 9 years after 1994, or after 2003

Review Exercises: Chapter 8, pp. 644–646

1. **(a)** [8.1] 2; **(b)** [8.6] positive; **(c)** [8.6] −3
2. [8.1] $\pm\sqrt{\dfrac{7}{2}}$, or $\pm\dfrac{\sqrt{14}}{2}$ **3.** [8.1] $0, -\frac{5}{14}$
4. [8.1] 3, 9   **5.** [8.2] $\dfrac{5 \pm i\sqrt{11}}{2}$ **6.** [8.2] 3, 5
7. [8.2] $\dfrac{-9 \pm \sqrt{85}}{2}$ **8.** [8.2] −0.3723, 5.3723
9. [8.2] $-\frac{1}{4}, 1$ **10.** [8.1] $x^2 - 12x + 36$; $(x - 6)^2$
11. [8.1] $x^2 + \frac{3}{5}x + \frac{9}{100}$; $\left(x + \frac{3}{10}\right)^2$ **12.** [8.1] $3 \pm 2\sqrt{2}$
13. [8.1] 10%   **14.** [8.1] 6.7 sec
15. [8.3] About 153 mph   **16.** [8.3] 6 hr
17. [8.4] Two irrational   **18.** [8.4] Two imaginary
19. [8.4] $x^2 - 5 = 0$ **20.** [8.4] $x^2 + 8x + 16 = 0$
21. [8.5] $(-3, 0), (-2, 0), (2, 0), (3, 0)$ **22.** [8.5] −5, 3
23. [8.5] $\pm\sqrt{2}, \pm\sqrt{7}$

24. [8.6]

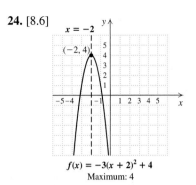

$f(x) = -3(x + 2)^2 + 4$
Maximum: 4

25. [8.7] **(a)** Vertex: $(3, 5)$; axis of symmetry: $x = 3$;
(b)

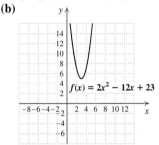

$f(x) = 2x^2 - 12x + 23$

26. [8.7] $(2, 0), (7, 0)$; $(0, 14)$ **27.** [8.3] $p = \dfrac{9\pi^2}{N^2}$
28. [8.3] $T = \dfrac{1 \pm \sqrt{1 + 24A}}{6}$ **29.** [8.8] Quadratic; the
data approximate a parabola opening upward.
30. [8.8] Not quadratic; the data do not approximate a
parabola.   **31.** [8.8] 56.25 ft²; 7.5 ft by 7.5 ft
32. [8.8] **(a)** $h(x) = \frac{1}{8}x^2 - \frac{217}{120}x + 12$; **(b)** about 7.2%
33. [8.8] $f(x) = 0.088311246x^2 - 1.286948848x + 12.19869334$, where x is the number of years after 1988
34. [8.9] $(-1, 0) \cup (3, \infty)$, or $\{x \mid -1 < x < 0 \text{ or } x > 3\}$
35. [8.9] $(-3, 5]$, or $\{x \mid -3 < x \leq 5\}$
36. ᴛᴡ [8.7], [8.8] The x-coordinate of the maximum or
minimum point lies halfway between the x-coordinates of the
x-intercepts.   **37.** ᴛᴡ [8.2], [8.4] Yes; if the discriminant is
a perfect square, then the solutions are rational numbers, p/q
and r/s. (Note that if the discriminant is 0, then $p/q = r/s$.)
Then the equation can be written in factored form,
$(qx - p)(sx - r) = 0$.   **38.** ᴛᴡ [8.5] Four; let $u = x^2$.
Then $au^2 + bu + c = 0$ has at most two solutions, $u = m$
and $u = n$. Now substitute x^2 for u and obtain $x^2 = m$ or
$x^2 = n$. These equations yield the solutions $x = \pm\sqrt{m}$ and
$x = \pm\sqrt{n}$. When $m \neq n$, the maximum number of solutions,
four, occurs.   **39.** ᴛᴡ [8.1], [8.2], [8.7] Completing the
square was used to solve quadratic equations and to graph
quadratic functions by rewriting the function in the form
$f(x) = a(x - h)^2 + k$.   **40.** [8.7] $f(x) = \frac{7}{15}x^2 - \frac{14}{15}x - 7$
41. [8.4] $h = 60, k = 60$ **42.** [8.5] 18, 324

Test: Chapter 8, pp. 646–647

1. (a) [8.1] 0; **(b)** [8.6] negative; **(c)** [8.6] −1
2. [8.1] $\pm\dfrac{4\sqrt{3}}{3}$ **3.** [8.2] 2, 9   **4.** [8.2] $\dfrac{-1 \pm i\sqrt{3}}{2}$
5. [8.2] $1 \pm \sqrt{6}$ **6.** [8.5] $-2, \frac{2}{3}$ **7.** [8.2] −4.1926,
1.1926   **8.** [8.2] $-\frac{3}{4}, \frac{7}{3}$ **9.** [8.1] $x^2 + 14x + 49$;
$(x + 7)^2$ **10.** [8.1] $x^2 - \frac{2}{7}x + \frac{1}{49}; \left(x - \frac{1}{7}\right)^2$
11. [8.1] $-5 \pm \sqrt{10}$ **12.** [8.3] 16 km/h   **13.** [8.3] 2 hr
14. [8.4] Two imaginary   **15.** [8.4] $3x^2 + 5x - 2 = 0$
16. [8.5] $(-3, 0), (-1, 0), \left(-2 - \sqrt{5}, 0\right), \left(-2 + \sqrt{5}, 0\right)$
17. [8.6]

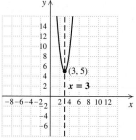

$f(x) = 4(x - 3)^2 + 5$
Minimum: 5

18. [8.7] **(a)** $(-1, -8), x = -1$; **(b)**

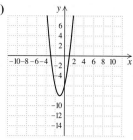

$f(x) = 2x^2 + 4x - 6$

19. [8.7] $(-2, 0), (3, 0); (0, -6)$
20. [8.3] $r = \sqrt{\dfrac{3V}{\pi} - R^2}$ **21.** [8.8] Quadratic; the data
approximate a parabola opening downward.
22. [8.8] Minimum $129/cap when 325 caps are built
23. [8.8] $p(x) = -\frac{165}{8}x^2 + \frac{605}{4}x + 35$
24. [8.8] $p(x) = -23.54166667x^2 + 162.2583333x +$
57.61666667   **25.** [8.9] $[-6, 1]$, or $\{x | -6 \le x \le 1\}$
26. [8.9] $(-1, 0) \cup (1, \infty)$, or $\{x | -1 < x < 0\ or\ x > 1\}$
27. [8.4] $\frac{1}{2}$ **28.** [8.4] $x^4 - 14x^3 + 67x^2 - 114x + 26 = 0$;
answers may vary.   **29.** [8.4] $x^6 - 10x^5 + 20x^4 + 50x^3 - 119x^2 - 60x + 150 = 0$; answers may vary

Chapter 9

Exercise Set 9.1, pp. 662–665

1. $(f \circ g)(1) = 12; (g \circ f)(1) = 9; (f \circ g)(x) = 4x^2 + 4x + 4;$

$(g \circ f)(x) = 2x^2 + 7$ **3.** $(f \circ g)(1) = 20; (g \circ f)(1) = 22;$
$(f \circ g)(x) = 15x^2 + 5; (g \circ f)(x) = 45x^2 - 30x + 7$
5. $(f \circ g)(1) = 8; (g \circ f)(1) = \frac{1}{64}; (f \circ g)(x) = \dfrac{1}{x^2} + 7;$
$(g \circ f)(x) = \dfrac{1}{(x + 7)^2}$ **7.** 8   **9.** −4   **11.** Not defined
13. 4   **15.** Not defined   **17.** $f(x) = x^2; g(x) = 7 + 5x$
19. $f(x) = \sqrt{x}; g(x) = 2x + 7$ **21.** $f(x) = \dfrac{2}{x};$
$g(x) = x - 3$ **23.** $f(x) = \dfrac{1}{\sqrt{x}}; g(x) = 7x + 2$
25. $f(x) = \dfrac{1}{x} + x; g(x) = \sqrt{3x}$ **27.** Yes   **29.** No
31. Yes   **33.** No   **35. (a)** Yes; **(b)** $f^{-1}(x) = x + 4$
37. (a) Yes; **(b)** $f^{-1}(x) = x - 3$ **39. (a)** Yes;
(b) $g^{-1}(x) = x - 5$ **41. (a)** Yes; **(b)** $f^{-1}(x) = \dfrac{x}{4}$
43. (a) Yes; **(b)** $g^{-1}(x) = \dfrac{x + 1}{4}$ **45. (a)** No
47. (a) Yes; **(b)** $f^{-1}(x) = \dfrac{1}{x}$ **49. (a)** Yes;
(b) $f^{-1}(x) = \dfrac{3x - 1}{2}$ **51. (a)** Yes; **(b)** $f^{-1}(x) = \sqrt[3]{x + 5}$
53. (a) Yes; **(b)** $g^{-1}(x) = \sqrt[3]{x} + 2$ **55. (a)** Yes;
(b) $f^{-1}(x) = x^2, x \ge 0$ **57. (a)** Yes;
(b) $f^{-1}(x) = \sqrt{\dfrac{x - 1}{2}}$ **59.**

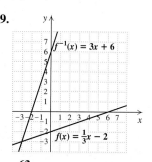

61.

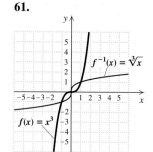

63.

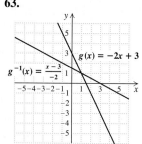

65.

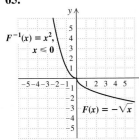

67.

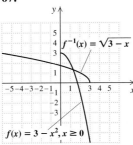

$I(x) = x$, $(f \circ I)(x) = f(I(x)) = f(x)$ for any function f.
(i) $((g^{-1} \circ f^{-1}) \circ h)(x) = ((g^{-1} \circ f^{-1}) \circ (f \circ g))(x)$
$= ((g^{-1} \circ (f^{-1} \circ f)) \circ g)(x)$
$= ((g^{-1} \circ I) \circ g)(x)$
$= (g^{-1} \circ g)(x) = x$
(ii) $(h \circ (g^{-1} \circ f^{-1}))(x) = ((f \circ g) \circ (g^{-1} \circ f^{-1}))(x)$
$= ((f \circ (g \circ g^{-1})) \circ f^{-1})(x)$
$= ((f \circ I) \circ f^{-1})(x)$
$= (f \circ f^{-1})(x) = x.$
Therefore, $(g^{-1} \circ f^{-1})(x) = h^{-1}(x)$.
99. TW   **101.** The cost of mailing n copies of the book
103. 22 mm   **105.** 15 L/min

69. (1) $(f^{-1} \circ f)(x) = f^{-1}(f(x)) = f^{-1}\left(\frac{4}{5}x\right) = \frac{5}{4}\left(\frac{4}{5}x\right) = x$;
(2) $(f \circ f^{-1})(x) = f(f^{-1}(x)) = f\left(\frac{5}{4}x\right) = \frac{4}{5}\left(\frac{5}{4}x\right) = x$

71. (1) $(f^{-1} \circ f)(x) = f^{-1}(f(x)) = f^{-1}\left(\dfrac{1-x}{x}\right)$

$= \dfrac{1}{\left(\dfrac{1-x}{x}\right) + 1}$

$= \dfrac{1}{\dfrac{1-x+x}{x}}$

$= x$;

(2) $(f \circ f^{-1})(x) = f(f^{-1}(x)) = f\left(\dfrac{1}{x+1}\right)$

$= \dfrac{1 - \left(\dfrac{1}{x+1}\right)}{\left(\dfrac{1}{x+1}\right)}$

$= \dfrac{\dfrac{x+1-1}{x+1}}{\dfrac{1}{x+1}} = x$

73. No   **75.** Yes   **77. (1)** C; **(2)** D; **(3)** B; **(4)** A
79. (a) 40, 42, 46, 50; **(b)** $f^{-1}(x) = x - 32$; **(c)** 8, 10, 14, 18
81. TW   **83.** $a^{13}b^{13}$ **84.** $x^{10}y^{12}$ **85.** 81   **86.** 125
87. $y = \frac{3}{2}(x + 7)$ **88.** $y = \dfrac{10 - x}{3}$ **89.** TW

91.

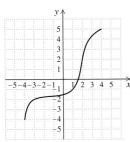

93. $g(x) = \dfrac{x}{2} + 20$

95. TW
97. Suppose that $h(x) = (f \circ g)(x)$. First, note that for

Interactive Discovery, pp. 668–669

1. (a) Increases; **(b)** increases; **(c)** increases; **(d)** decreases;
(e) decreases   **2.** If $a > 1$, the graph increases; if
$0 < a < 1$, the graph decreases.   **3.** (g)   **4.** (r)

Exercise Set 9.2, pp. 673–675

1. $a > 1$ **3.** $0 < a < 1$ **5.**

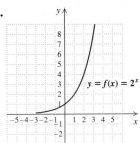

7.

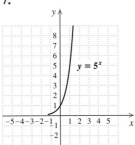

9.

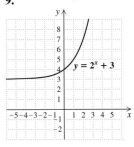

11.

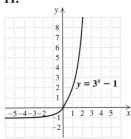

13.

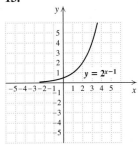

15.
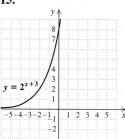

17.
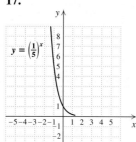

35.
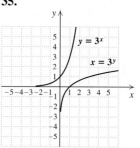

37.
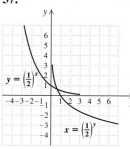

39. (d) **41.** (f) **43.** (c)
45. (a) About 6.4 billion; about 6.8 billion; about 7.3 billion;
(b)
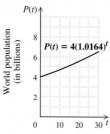

19.
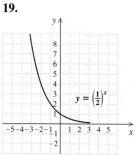

21.
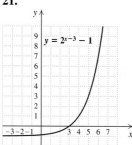

23.
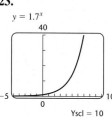

25.
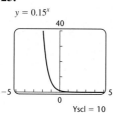

47. (a) About 44,079 whales; about 12,953 whales;
(b)
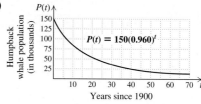

49. (a) 250,000; 166,667; 49,383; 4335;
(b)
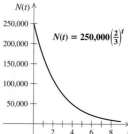

27.
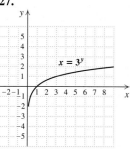

29.
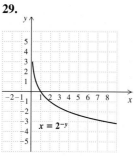

31.
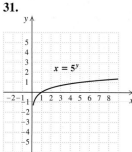

33.
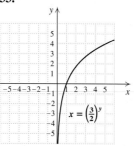

51. (a) 0.3 million, or 300,000; 12.1 million; 490.6 million;
3119.5 million; **(b)**
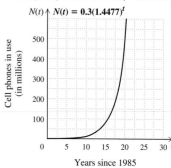

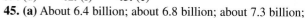

53. TW   **55.** $\frac{1}{25}$ **56.** $\frac{1}{32}$ **57.** 100   **58.** $\frac{1}{125}$
59. $5a^6b^3$ **60.** $6x^4y$ **61.** TW   **63.** $\pi^{2.4}$
65.

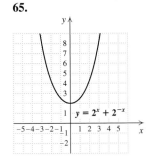
$y = 2^x + 2^{-x}$

67.

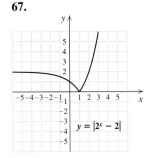
$y = |2^x - 2|$

69.

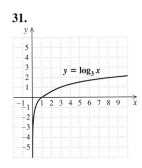
$y = |2^{x^2} - 1|$

71.

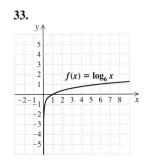
$y = 3^{-(x-1)}$
$x = 3^{-(y-1)}$

73. $A(t) = 169.3393318(2.535133248)^t$, where $A(t)$ is total sales, in millions of dollars, t years after 1997; $288,911,061,500   **75.** TW

Exercise Set 9.3, pp. 684–685

1. 2   **3.** 3   **5.** 4   **7.** −2   **9.** −1   **11.** 4
13. 1   **15.** 0   **17.** 7   **19.** −1   **21.** $\frac{1}{2}$ **23.** $\frac{3}{2}$
25. $\frac{2}{3}$ **27.** 7   **29.**

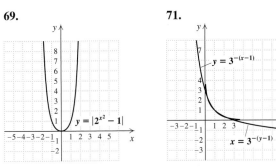
$y = \log_{10} x$

31.

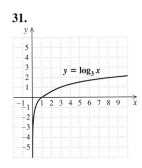
$y = \log_3 x$

33.

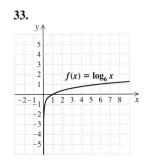
$f(x) = \log_6 x$

35.

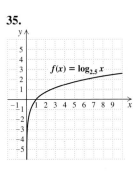
$f(x) = \log_{2.5} x$

37.

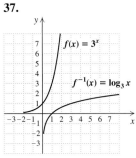
$f(x) = 3^x$
$f^{-1}(x) = \log_3 x$

39. 0.6021   **41.** 4.1271   **43.** −0.2782   **45.** 199.5262
47. 0.0011   **49.** 1.0028   **51.**

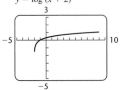
$y = \log(x + 2)$

53.

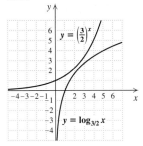
$y = \log(1 - 2x)$

55.

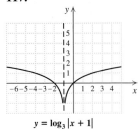
$y = \log(x^2)$

57. $2 = \log_{10} 100$ **59.** $-5 = \log_4 \frac{1}{1024}$ **61.** $\frac{3}{4} = \log_{16} 8$
63. $0.4771 = \log_{10} 3$ **65.** $k = \log_p 3$ **67.** $m = \log_p V$
69. $3 = \log_e 20.0855$ **71.** $-4 = \log_e 0.0183$
73. $3^t = 8$ **75.** $5^2 = 25$ **77.** $10^{-1} = 0.1$
79. $10^{0.845} = 7$ **81.** $c^8 = m$ **83.** $t^r = Q$
85. $e^{-1.3863} = 0.25$ **87.** $r^{-x} = T$ **89.** 9   **91.** 4
93. 2   **95.** 2   **97.** 7   **99.** 9   **101.** $\frac{1}{9}$ **103.** 4
105. TW   **107.** x^8 **108.** a^{12} **109.** a^7b^8
110. x^5y^{12} **111.** $\dfrac{x(3y - 2)}{2y + x}$ **112.** $\dfrac{x + 2}{x + 1}$ **113.** TW
115.

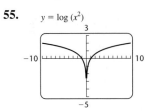
$y = \left(\frac{3}{2}\right)^x$
$y = \log_{3/2} x$

117.

$y = \log_3 |x + 1|$

119. 25   **121.** $-\frac{7}{16}$ **123.** 3   **125.** 1   **127.** 1
129. TW

Interactive Discovery, pp. 685–686

1. (c)   **2.** (b)   **3.** (c)   **4.** (a)

Exercise Set 9.4, pp. 691–692

1. $\log_3 81 + \log_3 27$ **3.** $\log_4 64 + \log_4 16$
5. $\log_c r + \log_c s + \log_c t$ **7.** $\log_a (5 \cdot 14)$, or $\log_a 70$
9. $\log_c (t \cdot y)$ **11.** $8 \log_a r$ **13.** $6 \log_c y$
15. $-3 \log_b C$ **17.** $\log_2 53 - \log_2 17$
19. $\log_b m - \log_b n$ **21.** $\log_a \frac{15}{3}$, or $\log_a 5$
23. $\log_b \frac{36}{4}$, or $\log_b 9$ **25.** $\log_a \frac{7}{18}$
27. $5 \log_a x + 7 \log_a y + 6 \log_a z$
29. $\log_b x + 2 \log_b y - 3 \log_b z$
31. $4 \log_a x - 3 \log_a y - \log_a z$
33. $\log_b x + 2 \log_b y - \log_b w - 3 \log_b z$
35. $\frac{1}{2}(7 \log_a x - 5 \log_a y - 8 \log_a z)$
37. $\frac{1}{3}(6 \log_a x + 3 \log_a y - 2 - 7 \log_a z)$ **39.** $\log_a x^7 z^3$

41. $\log_a x$ **43.** $\log_a \dfrac{y^5}{x^{3/2}}$ **45.** $\log_a (x - 2)$ **47.** 1.953

49. -0.369 **51.** -1.161 **53.** $\frac{3}{2}$ **55.** Cannot be found
57. 3.114 **59.** 9 **61.** m **63.** 7 **65.** 3 **67.** TW
69. **70.**

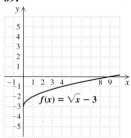

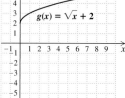

71. **72.**

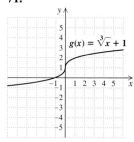

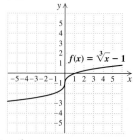

73. $a^{17}b^{17}$ **74.** $x^{11}y^6z^8$ **75.** TW
77. $\log_a (x^6 - x^4y^2 + x^2y^4 - y^6)$
79. $\frac{1}{2} \log_a (1 - s) + \frac{1}{2} \log_a (1 + s)$ **81.** $\frac{10}{3}$ **83.** -2
85. True

Interactive Discovery, pp. 693–694

1. \$2.25; \$2.370370; \$2.441406; \$2.613035; \$2.692597;
\$2.714567; \$2.718127 **2.** (c)

Exercise Set 9.5, pp. 699–700

1. 1.6094 **3.** 3.9512 **5.** -5.0832 **7.** 96.7583
9. 0.7850 **11.** 1.3877 **13.** 15.0293 **15.** 0.0305

17. 109.9472 **19.** 2.5237 **21.** 6.6439 **23.** 2.1452
25. -2.3219 **27.** -2.3219 **29.** 3.5471
31. Domain: $\mathbb{R}$; range: $(0, \infty)$

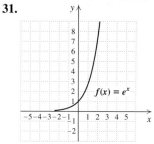

33. Domain: $\mathbb{R}$; range: $(0, \infty)$

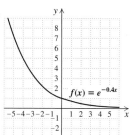

35. Domain: $\mathbb{R}$; range: $(1, \infty)$

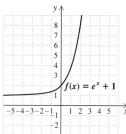

37. Domain: $\mathbb{R}$; range: $(-2, \infty)$

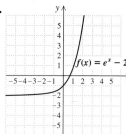

39. Domain: $\mathbb{R}$; range: $(0, \infty)$

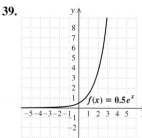

41. 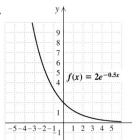 Domain: $\mathbb{R}$; range: $(0, \infty)$

$f(x) = 2e^{-0.5x}$

43. 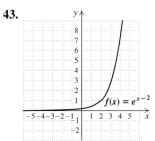 Domain: $\mathbb{R}$; range: $(0, \infty)$

$f(x) = e^{x-2}$

45. 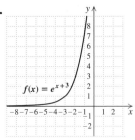 Domain: $\mathbb{R}$; range: $(0, \infty)$

$f(x) = e^{x+3}$

47. 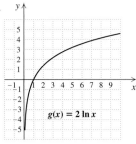 Domain: $(0, \infty)$; range: $\mathbb{R}$

$g(x) = 2 \ln x$

49. 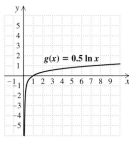 Domain: $(0, \infty)$; range: $\mathbb{R}$

$g(x) = 0.5 \ln x$

51. 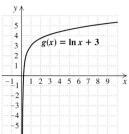 Domain: $(0, \infty)$; range: $\mathbb{R}$

$g(x) = \ln x + 3$

53. 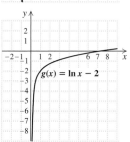 Domain: $(0, \infty)$; range: $\mathbb{R}$

$g(x) = \ln x - 2$

55. 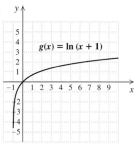 Domain: $(-1, \infty)$; range: $\mathbb{R}$

$g(x) = \ln (x + 1)$

57. 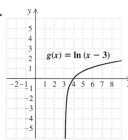 Domain: $(3, \infty)$; range: $\mathbb{R}$

$g(x) = \ln (x - 3)$

59. $f(x) = \log (x)/\log (5)$, or $f(x) = \ln (x)/\ln (5)$

$y = \log (x)/\log (5)$, or $y = \ln (x)/\ln (5)$

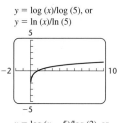

61. $f(x) = \log (x - 5)/\log (2)$, or $f(x) = \ln (x - 5)/\ln (2)$

$y = \log (x - 5)/\log (2)$, or $y = \ln (x - 5)/\ln (2)$

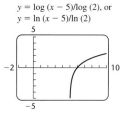

63. $f(x) = \log(x)/\log(3) + x$, or $f(x) = \ln(x)/\ln(3) + x$

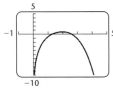

$y = \log(x)/\log(3) + x$, or $y = \ln(x)/\ln(3) + x$

65. TW   **67.** $-\frac{5}{2}, \frac{5}{2}$ **68.** $0, \frac{7}{5}$ **69.** $\frac{15}{17}$ **70.** $\frac{9}{13}$

71. 16, 256   **72.** $\frac{1}{4}, 9$ **73.** TW   **75.** 2.452

77. 1.442   **79.** $\log M = \dfrac{\ln M}{\ln 10}$ **81.** 1086.5129

83. 4.9855

85. **(a)** Domain: $\{x \mid x > 0\}$, or $(0, \infty)$; range: $\{y \mid y < 0.5135\}$, or $(-\infty, 0.5135)$; **(b)** $[-1, 5, -10, 5]$;

(c) $y = 3.4 \ln x - 0.25 e^x$

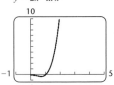

87. **(a)** Domain: $\{x \mid x > 0\}$, or $(0, \infty)$; range: $\{y \mid y > -0.2453\}$, or $(-0.2453, \infty)$; **(b)** $[-1, 5, -1, 10]$;

(c) $y = 2x^3 \ln x$

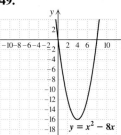

Exercise Set 9.6, pp. 706–708

1. 4   **3.** 3   **5.** 2   **7.** $\frac{4}{3}$ **9.** 0   **11.** 2   **13.** $-4, 1$

15. $\dfrac{\log 15}{\log 2} \approx 3.907$ **17.** $\dfrac{\log 13}{\log 4} - 1 \approx 0.850$

19. $\ln 100 \approx 4.605$ **21.** $\dfrac{\ln 0.08}{-0.07} \approx 36.082$

23. $\dfrac{\log 3}{\log 3 - \log 2} \approx 2.710$ **25.** $\dfrac{\log 65}{\log 7.2} \approx 2.115$

27. $-6.480, 6.519$ **29.** 125   **31.** 2   **33.** 1000
35. $\frac{1}{10,000}$ **37.** $e \approx 2.718$ **39.** $e^{-3} \approx 0.050$ **41.** -4
43. 10   **45.** No solution   **47.** $\frac{83}{15}$ **49.** 1   **51.** 5
53. $\frac{17}{2}$ **55.** 4   **57.** 0.000112, 3.445   **59.** 1
61. TW   **63.** Linear   **64.** Quadratic
65. $t(x) = \frac{4}{5}x + 16$ **66.** $p(x) = \frac{1}{6}x^2 + \frac{23}{6}x + 25$
67. 2010   **68.** $t(x) = 0.7571428571x + 15.71428571$
69. $p(x) = 0.0892857143x^2 + 4.907142857x + 23.85714286$
70. 2010   **71.** TW   **73.** No solution   **75.** -4
77. 2   **79.** $\pm\sqrt{34}$ **81.** $10^{100,000}$ **83.** 1, 100
85. $\frac{1}{100,000}$, 100,000   **87.** $-\frac{1}{3}$ **89.** 38   **91.** (6, $403)

Exercise Set 9.7, pp. 721–727

1. **(a)** 2003; **(b)** 2.2 yr   **3.** **(a)** 4.2 yr; **(b)** 9.0 yr
5. **(a)** 20,114; **(b)** about 58   **7.** 4.9
9. 10^{-7} moles per liter   **11.** 65 dB
13. $10^{-1.5}$ or about 3.2×10^{-2} W/m^2
15. **(a)** $P(t) = P_0 e^{0.06t}$; **(b)** \$5309.18, \$5637.48; **(c)** 11.6 yr
17. **(a)** $P(t) = 288.3 e^{0.013t}$, where t is the number of years after 2002 and $P(t)$ is in millions; **(b)** 299.8 million; **(c)** 2011
19. **(a)** 86.4 min; **(b)** 300.5 min; **(c)** 20 min
21. **(a)** 2000; **(b)** 2452; **(c)** $N(a)$

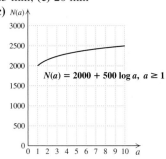

$N(a) = 2000 + 500 \log a, \ a \geq 1$

(d) \$1,000,000 thousand, or \$1,000,000,000
23. **(a)** $N(t) = 17 e^{0.534t}$, where t is the number of years since 1995; **(b)** about 419   **25.** 2005   **27.** **(a)** $k \approx 0.094$; $W(t) = 17.5 e^{-0.094t}$, where t is the number of years since 1996 and $W(t)$ is in millions of tons; **(b)** 6.8 million tons; **(c)** 2173
29. About 2103 yr   **31.** About 7.2 days   **33.** 69.3% per year   **35.** **(a)** $k \approx 0.135$; $V(t) = 640,500 e^{0.135t}$, where t is the number of years since 1996; **(b)** about \$2.47 million; **(c)** 5.1 yr; **(d)** 2004   **37.** No   **39.** Yes
41. **(a)** $p(x) = 6.501242197(1.096109091)^x$;
(b) 0.0918, or 9.18%; **(c)** \$71
43. **(a)** $P(x) = 37.29665447(1.111922582)^x$; **(b)** about \$2.6 billion   **45.** **(a)** $f(x) = 647.6297124(0.5602992676)^x$; **(b)** 152 teeth   **47.** TW
49.

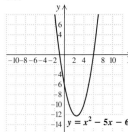

$y = x^2 - 8x$

50.

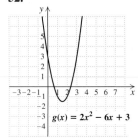

$y = x^2 - 5x - 6$

51.

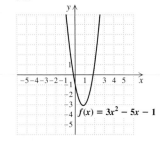

$f(x) = 3x^2 - 5x - 1$

52.

$g(x) = 2x^2 - 6x + 3$

53. $4 \pm \sqrt{23}$ **54.** $-5 \pm \sqrt{31}$ **55.** TW
57. $13.4 million   **59.** $P(t) = 53e^{0.302t}$, where t is the number of years after 2002 and $P(t)$ is a percent
61. (a) Yes; **(b)** $f(x) = 35.0(1.0192)^x$; **(c)** 194%; no;
(d) $f(x) = \dfrac{102.4604343}{1 + 2.25595242e^{-0.0483573334x}}$; **(e)** 99.6%; yes

Review Exercises: Chapter 9, pp. 730–732

1. [9.1] $(f \circ g)(x) = 4x^2 - 12x + 10$; $(g \circ f)(x) = 2x^2 - 1$
2. [9.1] $f(x) = \sqrt{x}$; $g(x) = 3 - x$ **3.** [9.1] No
4. [9.1] $f^{-1}(x) = x + 3$ **5.** [9.1] $g^{-1}(x) = \dfrac{2x - 1}{3}$
6. [9.1] $f^{-1}(x) = \dfrac{\sqrt[3]{x}}{3}$ **7.** [9.2]

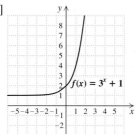

8. [9.2]

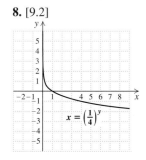

9. [9.3]

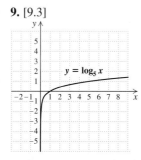

10. [9.3] 2   **11.** [9.3] -1 **12.** [9.3] $\frac{1}{2}$ **13.** [9.3] 12
14. [9.3] $\log_{10} \frac{1}{100} = -2$ **15.** [9.3] $\log_{25} 5 = \frac{1}{2}$
16. [9.3] $16 = 4^x$ **17.** [9.3] $1 = 8^0$
18. [9.4] $4 \log_a x + 2 \log_a y + 3 \log_a z$
19. [9.4] $3 \log_a x - (\log_a y + 2 \log_a z)$, or
$3 \log_a x - \log_a y - 2 \log_a z$
20. [9.4] $\frac{1}{4}(2 \log z - 3 \log x - \log y)$
21. [9.4] $\log_a(8 \cdot 15)$, or $\log_a 120$ **22.** [9.4] $\log_a \frac{72}{12}$, or
$\log_a 6$ **23.** [9.4] $\log \dfrac{a^{1/2}}{bc^2}$ **24.** [9.4] $\log_a \sqrt[3]{\dfrac{x}{y^2}}$
25. [9.4] 1   **26.** [9.4] 0   **27.** [9.4] 17   **28.** [9.4] 6.93
29. [9.4] -3.2698 **30.** [9.4] 8.7601   **31.** [9.4] 3.2698
32. [9.4] 2.54995   **33.** [9.4] -3.6602 **34.** [9.3] 1.9138
35. [9.3] 61.5177   **36.** [9.5] -2.9957 **37.** [9.5] 0.3753
38. [9.5] 0.4307   **39.** [9.5] 1.7097

40. [9.5]

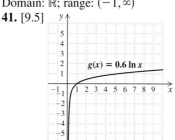

Domain: $\mathbb{R}$; range: $(-1, \infty)$
41. [9.5]

Domain: $(0, \infty)$; range: $\mathbb{R}$
42. [9.6] 5   **43.** [9.6] -2 **44.** [9.6] $\frac{1}{9}$ **45.** [9.6] 2
46. [9.6] $\frac{1}{10,000}$ **47.** [9.6] $e^{-2} \approx 0.1353$ **48.** [9.6] $\frac{7}{2}$
49. [9.6] $-5, 1$ **50.** [9.6] $\dfrac{\log 8.3}{\log 4} \approx 1.5266$
51. [9.6] $\dfrac{\ln 0.03}{-0.1} \approx 35.0656$ **52.** [9.6] $e^{-3} \approx 0.0498$
53. [9.6] 4   **54.** [9.6] 8   **55.** [9.6] 20   **56.** [9.6] $\sqrt{43}$
57. [9.7] **(a)** 62; **(b)** 46.8; **(c)** 35 months
58. [9.7] **(a)** 6.6 yr; **(b)** 3.1 yr   **59.** [9.7] **(a)** 0.383;
$C(t) = 667e^{0.383t}$; **(b)** about 30,724; **(c)** 2004
60. [9.7] 23.105% per yr   **61.** [9.7] 8.25 yr
62. [9.7] 3463 yr   **63.** [9.7] 6.6   **64.** [9.7] 90 dB
65. TW [9.3] Negative numbers do not have logarithms because logarithm bases are positive, and there is no power to which a positive number can be raised to yield a negative number.   **66.** TW [9.6] Taking the logarithm on each side of an equation produces an equivalent equation because the logarithm function is one-to-one. If two quantities are equal, their logarithms must be equal, and if the logarithms of two quantities are equal, the quantities must be the same.
67. [9.6] e^{e^3} **68.** [9.6] $-3, -1$ **69.** [9.6] $\left(\frac{8}{3}, -\frac{2}{3}\right)$

Test: Chapter 9, pp. 732–733

1. [9.1] $(f \circ g)(x) = 2 + 6x + 4x^2$;
$(g \circ f)(x) = 2x^2 + 2x + 1$ **2.** [9.1] $f(x) = \dfrac{1}{x}$;
$g(x) = 2x^2 + 1$ **3.** [9.1] No   **4.** [9.1] $f^{-1}(x) = \dfrac{x + 3}{4}$
5. [9.1] $g^{-1}(x) = \sqrt[3]{x} - 1$

6. [9.2] 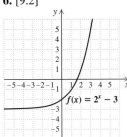   **7.** [9.2]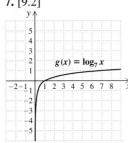

8. [9.3] 3   **9.** [9.3] $\frac{1}{2}$ **10.** [9.3] 18

11. [9.3] $\log_4 \frac{1}{64} = -3$ **12.** [9.3] $\log_{256} 16 = \frac{1}{2}$

13. [9.3] $49 = 7^m$ **14.** [9.3] $81 = 3^4$

15. [9.4] $3 \log a + \frac{1}{2} \log b - 2 \log c$

16. [9.4] $\log_a\left(z^2 \sqrt[3]{x}\right)$ **17.** [9.4] 1   **18.** [9.4] 23

19. [9.4] 0   **20.** [9.4] -0.544 **21.** [9.4] 1.079

22. [9.4] 1.204   **23.** [9.5] 1.0899   **24.** [9.5] 0.1585

25. [9.5] -3.3524 **26.** [9.5] 121.5104   **27.** [9.5] 2.0959

28. [9.5] 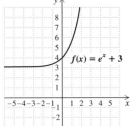   Domain: $\mathbb{R}$; range: $(3, \infty)$

29. [9.5] 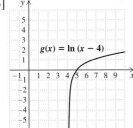   Domain: $(4, \infty)$; range: $\mathbb{R}$

30. [9.6] -5 **31.** [9.6] 5   **32.** [9.6] 2

33. [9.6] 10,000   **34.** [9.6] $\frac{1}{3}$ **35.** [9.6] $\frac{\log 1.2}{\log 7} \approx 0.0937$

36. [9.6] $e^{1/4} \approx 1.2840$ **37.** [9.6] 4

38. [9.7] **(a)** 2.46 ft/sec; **(b)** 984,262

39. [9.7] **(a)** $P(t) = 30e^{0.015t}$, where $P(t)$ is in millions and t is the number of years after 2000; **(b)** 31.4 million; 34.9 million; **(c)** 2034; **(d)** 46.2 yr

40. [9.7] **(a)** $k \approx 0.027$; $C(t) = 60e^{0.027t}$; **(b)** 681.5; **(c)** 2024

41. [9.7] 4.6%   **42.** [9.7] 4684 yr

43. [9.7] $10^{-4.5}$ W/m²   **44.** [9.7] 7.0

45. [9.6] $-309, 316$ **46.** [9.4] 2

Cumulative Review: Chapters 1–9, pp. 733–736

1. [1.1], [1.4] 2   **2.** [1.2] 6   **3.** [1.4] $\frac{y^{12}}{16x^8}$

4. [1.4] $\frac{20x^6z^2}{y}$ **5.** [1.4] $\frac{-y^4}{3z^5}$ **6.** [1.3] $-4x - 1$

7. [1.1] 25   **8.** [2.2] $\frac{11}{2}$ **9.** [3.2] $(3, -1)$

10. [3.4] $(1, -2, 0)$ **11.** [5.4] $-2, 5$ **12.** [6.4] $\frac{9}{2}$

13. [6.4], [8.2] $\frac{5}{8}$ **14.** [7.6] $\frac{3}{4}$ **15.** [7.6] $\frac{1}{2}$

16. [8.1] $\pm 5i$ **17.** [8.5] 9, 25   **18.** [8.5] $\pm 2, \pm 3$

19. [9.3] 8   **20.** [9.3] 7   **21.** [9.6] $\frac{3}{2}$

22. [9.6] $\frac{\log 7}{5 \log 3} \approx 0.3542$ **23.** [9.6] $\frac{80}{9}$

24. [8.9] $(-\infty, -5) \cup (1, \infty)$, or $\{x \mid x < -5 \text{ or } x > 1\}$

25. [8.2] $-3 \pm 2\sqrt{5}$ **26.** [4.3] $\{x \mid x \le -3 \text{ or } x \ge 6\}$, or $(-\infty, -3] \cup [6, \infty)$ **27.** [6.8] $a = \dfrac{Db}{b - D}$

28. [6.8] $q = \dfrac{pf}{p - f}$ **29.** [2.3] $B = \dfrac{3M - 2A}{2}$, or $B = \frac{3}{2}M - A$ **30.** [2.4] (c)   **31.** [8.7] (b)

32. [6.7] (a)   **33.** [9.5] (d)

34. [5.8] $\{x \mid x \text{ is a real number } and\ x \ne -\frac{1}{3} \text{ and } x \ne 2\}$

35. (a) [2.4] 1.8 million e-filers per year;
(b) [2.6] $E(t) = 1.8t + 6.7$, where E is in millions;
(c) [2.4] 13.9 million

36. [1.5], [9.7] 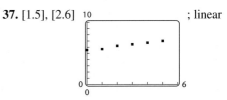 ; exponential

Xscl = 5, Yscl = 10

37. [1.5], [2.6] 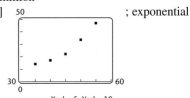 ; linear

38. [9.7] **(a)** $m(x) = 14.85e^{0.058x}$;
(b) $y(x) = 1.830390249(1.059578007)^x$

39. [2.6] **(a)** $f(x) = \frac{3}{10}x + \frac{11}{2}$, where f is in millions;
(b) $y(x) = 0.2685714286x + 5.628571429$

40. [2.3] Length: 36 m; width: 20 m   **41.** [2.3] A: 15°; B: 45°; C: 120°   **42.** [6.5] $5\frac{5}{11}$ min   **43.** [3.3] Thick and Tasty: 6 oz; Light and Lean: 9 oz   **44.** [6.5] $2\frac{7}{9}$ km/h

45. [8.8] $-49; -7$ and 7   **46.** [9.7] 78   **47.** [9.7] 67.5

48. [9.7] $P(t) = 19.4e^{0.015t}$, where t is the number of years since 2001 and P is in millions   **49.** [9.7] 20.6 million, 22.9 million   **50.** [9.7] 46.2 yr   **51.** [6.8] 18

52. [5.1] $7p^2q^3 + pq + p - 9$ **53.** [5.1] $8x^2 - 11x - 1$

54. [5.2] $9x^4 - 12x^2y + 4y^2$ **55.** [5.2] $10a^2 - 9ab - 9b^2$

56. [6.1] $\dfrac{(x + 4)(x - 3)}{2(x - 1)}$ **57.** [6.3] $\dfrac{1}{x - 4}$

58. [6.1] $\dfrac{a + 2}{6}$ **59.** [6.2] $\dfrac{7x + 4}{(x + 6)(x - 6)}$

60. [5.3] $x(y - 2z + w)$
61. [5.7] $(1 - 5x)(1 + 5x + 25x^2)$
62. [5.5] $2(3x - 2y)(x + 2y)$
63. [5.3] $(x^3 + 7)(x - 4)$ **64.** [5.6] $2(m + 3n)^2$
65. [5.6] $(x - 2y)(x + 2y)(x^2 + 4y^2)$ **66.** [2.1] -12
67. [6.6] $x^3 - 2x^2 - 4x - 12 + \dfrac{-42}{x - 3}$
68. [1.4] 1.8×10^{-1} **69.** [7.4] $2y^2\sqrt[3]{y}$
70. [7.4] $14xy^2\sqrt{x}$ **71.** [7.2] $81a^8b\sqrt[3]{b}$
72. [7.5] $\dfrac{6 + \sqrt{y} - y}{4 - y}$ **73.** [7.4] $\sqrt[10]{(x + 5)^3}$
74. [7.8] $12 + 4\sqrt{3}i$ **75.** [7.8] $8 + i$
76. [9.1] $f^{-1}(x) = \dfrac{x - 7}{-2}$, or $f^{-1}(x) = \dfrac{7 - x}{2}$
77. [2.6] $f(x) = -5x - 3$ **78.** [2.5] $y = \frac{1}{2}x + 7$
79. [2.5]   **80.** [8.7]

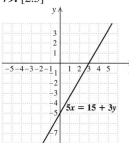

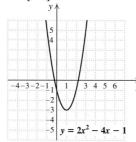

81. [9.3]   **82.** [9.1]

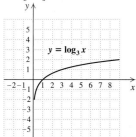

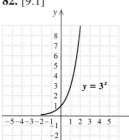

83. [4.4]   **84.** [8.7]

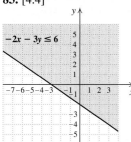

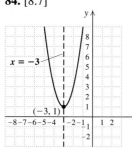

$f(x) = 2(x + 3)^2 + 1$
Minimum: 1

85. [9.5]

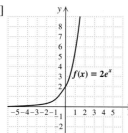

Domain: $\mathbb{R}$;
range: $(0, \infty)$

86. [9.4] $2 \log a + 3 \log c - \log b$ **87.** [9.4] $\log\left(\dfrac{x^3}{y^{1/2}z^2}\right)$
88. [9.3] $a^x = 5$ **89.** [9.3] $\log_x t = 3$
90. [9.3] -1.2545 **91.** [9.3] 776.2471
92. [9.5] 2.5479 **93.** [9.5] 0.2466 **94.** [6.4] All real numbers except 1 and -2 **95.** [9.6] $\frac{1}{3}, \frac{10,000}{3}$
96. [8.3] 35 mph

Chapter 10

Exercise Set 10.1, pp. 745–747

1. 3, 8, 13, 18; 48; 73   **3.** $\frac{1}{2}, \frac{2}{3}, \frac{3}{4}, \frac{4}{5}; \frac{10}{11}; \frac{15}{16}$
5. $-1, 0, 3, 8; 80; 195$ **7.** $2, 2\frac{1}{2}, 3\frac{1}{3}, 4\frac{1}{4}; 10\frac{1}{10}; 15\frac{1}{15}$
9. $-1, 4, -9, 16; 100; -225$ **11.** $-2, -1, 4, -7; -25; 40$
13. 9   **15.** 364   **17.** -23.5 **19.** -363 **21.** $\frac{441}{400}$
23. $2n - 1$ **25.** $(-1)^{n+1}$ **27.** $(-1)^n \cdot n$
29. $(-1)^n \cdot 2 \cdot (3)^{n-1}$ **31.** $\dfrac{n}{n + 1}$ **33.** 5^n
35. $(-1)^n \cdot n^2$ **37.** 4   **39.** 30
41. $\frac{1}{2} + \frac{1}{4} + \frac{1}{6} + \frac{1}{8} + \frac{1}{10} = \frac{137}{120}$
43. $3^0 + 3^1 + 3^2 + 3^3 + 3^4 = 121$
45. $\frac{1}{2} + \frac{2}{3} + \frac{3}{4} + \frac{4}{5} + \frac{5}{6} + \frac{6}{7} + \frac{7}{8} + \frac{8}{9} = \frac{15,551}{2520}$
47. $(-1)^2 2^1 + (-1)^3 2^2 + (-1)^4 2^3 + (-1)^5 2^4 + (-1)^6 2^5 + (-1)^7 2^6 + (-1)^8 2^7 + (-1)^9 2^8 = -170$
49. $(0^2 - 2 \cdot 0 + 3) + (1^2 - 2 \cdot 1 + 3) + (2^2 - 2 \cdot 2 + 3) + (3^2 - 2 \cdot 3 + 3) + (4^2 - 2 \cdot 4 + 3) + (5^2 - 2 \cdot 5 + 3) = 43$
51. $\dfrac{(-1)^3}{3 \cdot 4} + \dfrac{(-1)^4}{4 \cdot 5} + \dfrac{(-1)^5}{5 \cdot 6} = -\dfrac{1}{15}$ **53.** $\displaystyle\sum_{k=1}^{5} \dfrac{k + 1}{k + 2}$
55. $\displaystyle\sum_{k=1}^{6} k^2$ **57.** $\displaystyle\sum_{k=2}^{n} (-1)^k k^2$ **59.** $\displaystyle\sum_{k=1}^{\infty} 5k$
61. $\displaystyle\sum_{k=1}^{\infty} \dfrac{1}{k(k + 1)}$ **63.** TW   **65.** 77   **66.** 23
67. $x^3 + 3x^2y + 3xy^2 + y^3$ **68.** $a^3 - 3a^2b + 3ab^2 - b^3$
69. $8a^3 - 12a^2b + 6ab^2 - b^3$
70. $8x^3 + 12x^2y + 6xy^2 + y^3$ **71.** TW
73. 1, 3, 13, 63, 313, 1563   **75.** 1, 2, 4, 8, 16, 32, 64, 128, 256, 512, 1024, 2048, 4096, 8192, 16,384, 32,768, 65,536
77. $S_{100} = 0$; $S_{101} = -1$ **79.** $i, -1, -i, 1, i; i$
81. 11th term

Interactive Discovery, p. 749

1. The points lie on a straight line with positive slope.
2. The points lie on a straight line with positive slope.
3. The points lie on a straight line with negative slope.
4. The points lie on a straight line with negative slope.

Exercise Set 10.2, pp. 754–756

1. $a_1 = 2, d = 4$　　**3.** $a_1 = 6, d = -4$　　**5.** $a_1 = \frac{3}{2}, d = \frac{3}{4}$
7. $a_1 = \$5.12, d = \$0.12$　　**9.** 47　　**11.** -41
13. $-\$1628.16$　　**15.** 27th　　**17.** 102nd　　**19.** 82
21. 5　　**23.** 28　　**25.** $a_1 = 8; d = -3; 8, 5, 2, -1, -4$
27. $a_1 = 1; d = 1$　　**29.** 780　　**31.** 31,375　　**33.** 2550
35. 918　　**37.** 1030　　**39.** 42; 420　　**41.** 1260

43. \$31,000　　**45.** 722　　**47.** TW　　**49.** $\dfrac{13}{30x}$　　**50.** $\dfrac{23}{36t}$

51. $a^k = P$　　**52.** $e^a = t$　　**53.** 2　　**54.** 5　　**55.** TW
57. $S_n = n^2$　　**59.** \$8760, \$7961.77, \$7163.54, \$6365.31,
\$5567.08; \$4768.85, \$3970.62, \$3172.39, \$2374.16, \$1575.93
61. Let $d =$ the common difference. Since p, m, and q form
an arithmetic sequence, $m = p + d$ and $q = p + 2d$. Then
$\dfrac{p + q}{2} = \dfrac{p + (p + 2d)}{2} = p + d = m.$　　**63.** 156,735
65. Arithmetic; $a_n = n + 102$, where $n = 1$ corresponds to
1998　　**67.** Not arithmetic

Interactive Discovery, p. 759

1. The points lie on an exponential curve with $a > 1$.
2. The points lie on an exponential curve with $a > 1$.
3. The points lie on an exponential curve with $0 < a < 1$.
4. The points lie on an exponential curve with $0 < a < 1$.

Interactive Discovery, p. 761

1. (a) $\frac{1}{3}$; (b) 1, 1.33333, 1.44444, 1.48148, 1.49383, 1.49794,
1.49931, 1.49977, 1.49992, 1.49997 (sums are rounded to
5 decimal places); (c) 1.5
2. (a) $-\frac{7}{2}$; (b) 1, -2.5, 9.75, -33.125, 116.9375,
-408.28125, 1429.984375, -5003.945313, 17514.80859,
-61300.83008; (c) does not exist
3. (a) 2; (b) 4, 12, 28, 60, 124, 252, 508, 1020, 2044, 4092;
(c) does not exist
4. (a) -0.4; (b) 1.3, 0.78, 0.988, 0.9048, 0.93808, 0.92477,
0.93009, 0.92796, 0.92881, 0.92847 (sums are rounded to
5 decimal places); (c) 0.928　　**5.** S_∞ does not exist if $|r| > 1$.

Exercise Set 10.3, pp. 765–767

1. 2　　**3.** -1　　**5.** $-\frac{1}{2}$　　**7.** $\frac{1}{5}$　　**9.** $\dfrac{3}{m}$　　**11.** 192
13. 80　　**15.** 648　　**17.** \$2331.64　　**19.** $a_n = 3^{n-1}$

21. $a_n = (-1)^{n-1}$　　**23.** $a_n = \dfrac{1}{x^n}$　　**25.** 762　　**27.** $\frac{547}{18}$
29. $\dfrac{1 - x^8}{1 - x}$, or $(1 + x)(1 + x^2)(1 + x^4)$　　**31.** \$5134.51
33. $\frac{64}{3}$　　**35.** $\frac{49}{4}$　　**37.** No　　**39.** No　　**41.** $\frac{43}{99}$
43. \$25,000　　**45.** $\frac{7}{9}$　　**47.** $\frac{830}{99}$　　**49.** $\frac{5}{33}$　　**51.** 155,797
53. $\frac{5}{1024}$ ft　　**55.** \$43,318.94　　**57.** 25 min
59. 3100.35 ft　　**61.** 20.48 in.　　**63.** Arithmetic
65. Geometric　　**67.** Geometric　　**69.** TW
71. $x^3 + 3x^2y + 3xy^2 + y^3$
72. $a^3 - 3a^2b + 3ab^2 - b^3$　　**73.** $\left(-\frac{63}{29}, -\frac{114}{29}\right)$

74. $(-1, 2, 3)$　　**75.** TW　　**77.** $\dfrac{x^2[1 - (-x)^n]}{1 + x}$

79. 512 cm^2

Exercise Set 10.4, pp. 775–776

1. 40,320　　**3.** 3,628,800　　**5.** 210　　**7.** 720　　**9.** 28
11. 210　　**13.** 190　　**15.** 595
17. $m^5 + 5m^4n + 10m^3n^2 + 10m^2n^3 + 5mn^4 + n^5$
19. $x^6 - 6x^5y + 15x^4y^2 - 20x^3y^3 + 15x^2y^4 - 6xy^5 + y^6$
21. $x^{10} - 15x^8y + 90x^6y^2 - 270x^4y^3 + 405x^2y^4 - 243y^5$
23. $729c^6 - 1458c^5d + 1215c^4d^2 - 540c^3d^3 + 135c^2d^4 - 18cd^5 + d^6$　　**25.** $x^3 - 3x^2y + 3xy^2 - y^3$
27. $x^9 + \dfrac{18x^8}{y} + \dfrac{144x^7}{y^2} + \dfrac{672x^6}{y^3} + \dfrac{2016x^5}{y^4} + \dfrac{4032x^4}{y^5} +$
$\dfrac{5376x^3}{y^6} + \dfrac{4608x^2}{y^7} + \dfrac{2304x}{y^8} + \dfrac{512}{y^9}$
29. $a^{10} - 5a^8b^3 + 10a^6b^6 - 10a^4b^9 + 5a^2b^{12} - b^{15}$
31. $9 - 12\sqrt{3}t + 18t^2 - 4\sqrt{3}t^3 + t^4$
33. $x^{-8} + 4x^{-4} + 6 + 4x^4 + x^8$　　**35.** $15a^4b^2$
37. $-64,481,508a^3$　　**39.** $1120x^{12}y^2$
41. $-1,959,552u^5v^{10}$　　**43.** y^8　　**45.** TW　　**47.** 4
48. $\frac{5}{2}$　　**49.** 5.6348　　**50.** ± 5　　**51.** TW
53. Consider a set of 5 elements, $\{A, B, C, D, E\}$. List all the
subsets of size 3:

$\{A, B, C\}, \{A, B, D\}, \{A, B, E\}, \{A, C, D\}, \{A, C, E\},$
$\{A, D, E\}, \{B, C, D\}, \{B, C, E\}, \{B, D, E\}, \{C, D, E\}.$

There are exactly 10 subsets of size 3 and $\dbinom{5}{3} = 10$, so

there are exactly $\dbinom{5}{3}$ ways of forming a subset of size 3 from

a set of 5 elements.

55. $\dbinom{8}{5}(0.15)^3(0.85)^5 \approx 0.084$

57. $\dbinom{8}{6}(0.15)^2(0.85)^6 + \dbinom{8}{7}(0.15)(0.85)^7 +$

$\dbinom{8}{8}(0.85)^8 \approx 0.89$　　**59.** $\frac{55}{144}$　　**61.** $\dfrac{-\sqrt[3]{q}}{2p}$　　**63.** 8

Review Exercises: Chapter 10, p. 778

1. [10.1] 1, 5, 9, 13; 29; 45   **2.** [10.1] $0, \frac{1}{5}, \frac{1}{5}, \frac{3}{17}; \frac{7}{65}; \frac{11}{145}$
3. [10.1] $a_n = -2n$ **4.** [10.1] $a_n = (-1)^n(2n - 1)$
5. [10.1] $-2 + 4 + (-8) + 16 + (-32) = -22$
6. [10.1] $-3 + (-5) + (-7) + (-9) + (-11) +$
$(-13) = -48$ **7.** [10.1] $\sum\limits_{k=1}^{5} 4k$ **8.** [10.1] $\sum\limits_{k=1}^{5} \frac{1}{(-2)^k}$
9. [10.2] 85   **10.** [10.2] $\frac{8}{3}$
11. [10.2] $d = 1.25, a_1 = 11.25$ **12.** [10.2] -544
13. [10.2] 8580   **14.** [10.3] $1024\sqrt{2}$ **15.** [10.3] $\frac{2}{3}$
16. [10.3] $a_n = 2(-1)^n$ **17.** [10.3] $a_n = 3\left(\frac{x}{4}\right)^{n-1}$
18. [10.3] 4095   **19.** [10.3] $-4095x$ **20.** [10.3] 12
21. [10.3] $\frac{49}{11}$ **22.** [10.3] No   **23.** [10.3] No
24. [10.3] \$40,000   **25.** [10.3] $\frac{5}{9}$ **26.** [10.3] $\frac{46}{33}$
27. [10.2] \$17.80   **28.** [10.2] 903   **29.** [10.3] \$22,521.92
30. [10.3] 6 m   **31.** [10.4] 5040   **32.** [10.4] 56
33. [10.4] $190a^{18}b^2$
34. [10.4] $x^4 - 8x^3y + 24x^2y^2 - 32xy^3 + 16y^4$
35. ᴛᴡ [10.3] For a geometric sequence with $|r| < 1$, as n
gets larger, the absolute value of the terms gets smaller, since
$|r^n|$ gets smaller.   **36.** ᴛᴡ [10.4] The first form of the
binomial theorem draws the coefficients from Pascal's
triangle; the second form uses factorial notation. The second
form avoids the need to compute all preceding rows of
Pascal's triangle, and is generally easier to use when only one
term of an expression is needed. When several terms of an
expansion are needed and n is not larger (say, $n \le 8$), it is
often easier to use Pascal's triangle.   **37.** [10.3] $\dfrac{1 - (-x)^n}{x + 1}$
38. [10.4] $x^{-15} + 5x^{-9} + 10x^{-3} + 10x^3 + 5x^9 + x^{15}$

Test: Chapter 10, p. 779

1. [10.1] 1, 7, 13, 19, 25; 91   **2.** [10.1] $a_n = 4\left(\frac{1}{3}\right)^n$
3. [10.1] $1 + (-1) + (-5) + (-13) + (-29) = -47$
4. [10.1] $\sum\limits_{k=1}^{5} (-1)^{k+1}k^3$ **5.** [10.2] -46 **6.** [10.2] $\frac{3}{8}$
7. [10.2] $a_1 = 31.2; d = -3.8$ **8.** [10.2] 2508
9. [10.3] $\frac{9}{128}$ **10.** [10.3] $\frac{2}{3}$ **11.** [10.3] $(-1)^{n+1}3^n$
12. [10.3] $511 + 511x$ **13.** [10.3] 1   **14.** [10.3] No
15. [10.3] $\frac{\$25,000}{23} \approx \$1086.96$ **16.** [10.3] $\frac{85}{99}$
17. [10.2] 63   **18.** [10.2] \$17,100   **19.** [10.3] \$8981.05
20. [10.3] 36 m   **21.** [10.4] 78
22. [10.4] $x^{10} - 15x^8y + 90x^6y^2 - 270x^4y^3 +$
$405x^2y^4 - 243y^5$ **23.** [10.4] $220a^9x^3$
24. [10.2] $n(n + 1)$
25. [10.3] $\dfrac{1 - \left(\dfrac{1}{x}\right)^n}{1 - \dfrac{1}{x}}$, or $\dfrac{x^n - 1}{x^{n-1}(x - 1)}$

Cumulative Review: Chapters 1–10, pp. 780–783

1. [1.4] $-45x^6y^{-4}$, or $\dfrac{-45x^6}{y^4}$ **2.** [1.2] 6.3
3. [1.3] $-3y + 17$ **4.** [1.1] 280   **5.** [1.1], [1.2] $\frac{7}{6}$
6. [5.1] $3a^2 - 8ab - 15b^2$
7. [5.1] $13x^3 - 7x^2 - 6x + 6$ **8.** [5.2] $6a^2 + 7a - 5$
9. [5.2] $9a^4 - 30a^2y + 25y^2$ **10.** [6.2] $\dfrac{4}{x + 2}$
11. [6.1] $\dfrac{x - 4}{3(x + 2)}$ **12.** [6.1] $\dfrac{(x + y)(x^2 + xy + y^2)}{x^2 + y^2}$
13. [6.3] $x - a$ **14.** [5.6] $(2x - 3)^2$
15. [5.7] $(3a - 2)(9a^2 + 6a + 4)$
16. [5.3] $(a^2 - b)(a + 3)$ **17.** [5.5] $3(y^2 + 3)(5y^2 - 4)$
18. [2.1] 20   **19.** [6.6] $7x^3 + 9x^2 + 19x + 38 + \dfrac{72}{x - 2}$
20. [2.2] $\frac{2}{3}$ **21.** [6.4] $-\frac{6}{5}, 4$ **22.** [4.2] $\mathbb{R}$, or $(-\infty, \infty)$
23. [3.2] $(1, -1)$ **24.** [3.4] $(2, -1, 1)$ **25.** [7.6] 2
26. [8.5] $\pm 2, \pm 5$ **27.** [9.6] 4.67
28. [9.6] $\dfrac{\ln 8}{\ln 5} \approx 1.2920$ **29.** [9.6] 1005   **30.** [9.6] $\frac{1}{16}$
31. [9.6] $-\frac{1}{2}$ **32.** [4.3] $\{x \,|\, -2 \le x \le 3\}$, or $[-2, 3]$
33. [8.1] $\pm i\sqrt{2}$ **34.** [8.2] $-2 \pm \sqrt{7}$
35. [8.9] $\{y \,|\, y < -5 \text{ or } y > 2\}$, or $(-\infty, -5) \cup (2, \infty)$
36. [5.4] $-6, 8$ **37.** [7.6] No solution
38. [7.7] $\sqrt{208}$ mi, or approximately 14.422 mi
39. [4.1] More than 4   **40.** [2.3] 65, 66, 67
41. [3.3] $11\frac{3}{7}$ **42.** [3.3] \$2.68 herb: 10 oz;
\$4.60 herb: 14 oz   **43.** [6.5] 350 mph   **44.** [6.5] $8\frac{2}{5}$ hr or
8 hr, 24 min   **45.** [6.8] 20   **46.** [8.9] 1250 ft^2
47. [2.5]   **48.** [9.3]

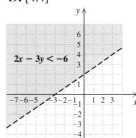

$3x - y = 6$

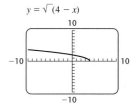

$y = \log_2 x$

49. [4.4]   **50.** [4.2]

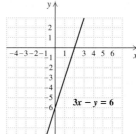

$2x - 3y < -6$

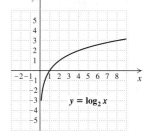

$y = \sqrt{(4 - x)}$

Domain: $(-\infty, 4]$;
range: $[0, \infty)$

51. [8.7]

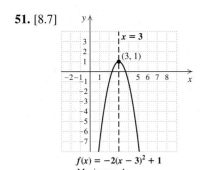

$f(x) = -2(x - 3)^2 + 1$
Maximum: 1

52. [2.3] $r = \dfrac{V - P}{-Pt}$, or $\dfrac{P - V}{Pt}$ **53.** [6.8] $R = \dfrac{Ir}{1 - I}$

54. [2.5] $y = 3x - 3$ **55.** [4.2] $\left\{x \mid x \le \frac{5}{3}\right\}$, or $\left(-\infty, \frac{5}{3}\right]$

56. [5.4] $\{x \mid x$ is a real number *and* $x \ne 1\}$, or $(-\infty, 1) \cup (1, \infty)$

57. [1.4] 6.764×10^{-12} **58.** [7.3] $8x^2\sqrt{y}$

59. [7.2] $125x^2y^{3/4}$ **60.** [7.4] $\dfrac{\sqrt[3]{5xy}}{y}$

61. [7.5] $\dfrac{1 - 2\sqrt{x} + x}{1 - x}$ **62.** [7.8] $26 - 13i$

63. [8.4] $x^2 - 50 = 0$ **64.** [9.4] 37

65. [9.4] $\log_a \dfrac{\sqrt[3]{x^2} \cdot z^5}{\sqrt{y}}$ **66.** [9.3] $a^5 = c$

67. [9.3] 3.7541   **68.** [9.3] 0.0003   **69.** [9.5] 8.6442

70. [9.5] 0.0277   **71.** [9.6] **(a)** $k \approx 0.261$;

$C(t) = 0.12e^{0.261t}$; **(b)** about 48.6 million   **72.** [10.1] $-n^2$

73. [10.2] -121 **74.** [10.2] 875   **75.** [10.3] $16\left(\frac{1}{4}\right)^{n-1}$

76. [10.4] $13{,}440a^4b^6$ **77.** [10.3] $74.88671875x$

78. [10.3] \$168.95   **79.** **(a)** [2.4] Linear; **(b)** [2.3] 2

80. **(a)** [9.3] Logarithmic; **(b)** [9.6] -1

81. **(a)** [8.7] Quadratic; **(b)** [8.1] $-1, 4$

82. **(a)** [9.2] Exponential; **(b)** [9.6] $\varnothing$

83. [2.6], [8.8], [9.7] Answers may vary.

84. [2.6] $f(x) = x + 11$, where f is in billions of dollars

85. [2.6] $f(x) = 0.9568965517x + 10.42241379$, where f is in

billions of dollars   **86.** [6.4] All real numbers except 0

and -12 **87.** [9.6] 81   **88.** [6.8] y gets divided by 8

89. [7.8] $-\dfrac{7}{13} + \dfrac{2\sqrt{30}}{13}i$ **90.** [3.5] 84 yr

Appendixes

Exercise Set A, pp. 793–796

1.

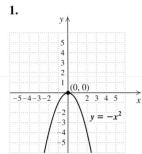

$y = -x^2$

3.

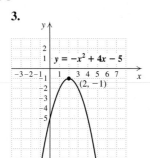

$y = -x^2 + 4x - 5$

5.

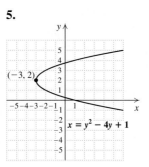

$x = y^2 - 4y + 1$

7.

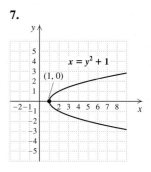

$x = y^2 + 1$

9.

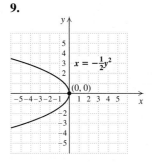

$x = -\frac{1}{2}y^2$

11.

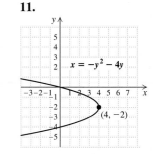

$x = -y^2 - 4y$

13.

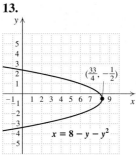

$x = 8 - y - y^2$

15.

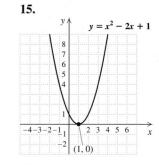

$y = x^2 - 2x + 1$

17.

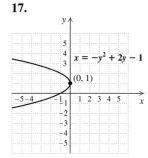

$x = -y^2 + 2y - 1$

19.

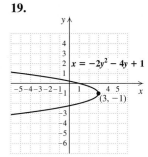

$x = -2y^2 - 4y + 1$

21. 5   **23.** $\sqrt{18} \approx 4.243$ **25.** $\sqrt{200} \approx 14.142$

27. 17.8   **29.** $\dfrac{\sqrt{41}}{7} \approx 0.915$ **31.** $\sqrt{8} \approx 2.828$

33. $\sqrt{17 + 2\sqrt{14} + 2\sqrt{15}} \approx 5.677$ **35.** $\sqrt{s^2 + t^2}$

37. $(1, 4)$ **39.** $\left(\frac{7}{2}, \frac{7}{2}\right)$ **41.** $(-1, -3)$

43. $(-0.25, -0.3)$ **45.** $\left(-\frac{1}{12}, \frac{1}{24}\right)$ **47.** $\left(\dfrac{\sqrt{2} + \sqrt{3}}{2}, \dfrac{3}{2}\right)$

49. $x^2 + y^2 = 36$ **51.** $(x - 7)^2 + (y - 3)^2 = 5$

53. $(x + 4)^2 + (y - 3)^2 = 48$

55. $(x + 7)^2 + (y + 2)^2 = 50$ **57.** $x^2 + y^2 = 25$
59. $(x + 4)^2 + (y - 1)^2 = 20$
61. $(0, 0)$; 7 **63.** $(-1, -3)$; 2

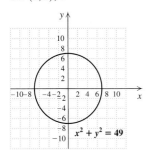

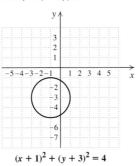

$x^2 + y^2 = 49$

$(x + 1)^2 + (y + 3)^2 = 4$

65. $(4, -3)$; $\sqrt{10}$ **67.** $(0, 0)$; $\sqrt{7}$

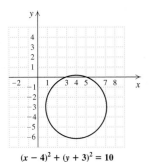

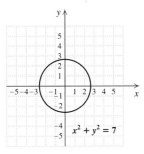

$(x - 4)^2 + (y + 3)^2 = 10$

$x^2 + y^2 = 7$

69. $(5, 0)$; $\frac{1}{2}$ **71.** $(-4, 3)$; $\sqrt{40}$, or $2\sqrt{10}$

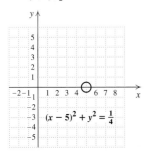

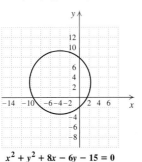

$(x - 5)^2 + y^2 = \frac{1}{4}$

$x^2 + y^2 + 8x - 6y - 15 = 0$

73. $(4, -1)$; 2 **75.** $(0, -5)$; 10

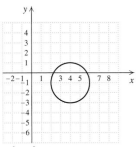

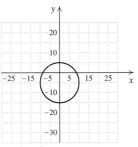

$x^2 + y^2 - 8x + 2y + 13 = 0$

$x^2 + y^2 + 10y - 75 = 0$

77. $\left(-\frac{7}{2}, \frac{3}{2}\right)$; $\sqrt{\frac{98}{4}}$, or $\frac{7\sqrt{2}}{2}$

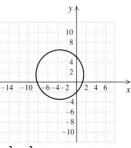

$x^2 + y^2 + 7x - 3y - 10 = 0$

79. $(0, 0)$; $\frac{1}{6}$

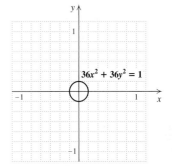

$36x^2 + 36y^2 = 1$

81. **83.**

$x^2 + y^2 - 16 = 0$ $x^2 + y^2 + 14x - 16y + 54 = 0$

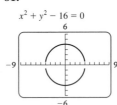

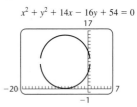

85. $(x - 3)^2 + (y + 5)^2 = 9$ **87.** $(x - 3)^2 + y^2 = 25$
89. $(0, 4)$ **91. (a)** $(0, -8467.8)$; **(b)** 8487.3 mm
93. (a) $(0, -1522.8)$; **(b)** 1524.9 cm **95.** 29 cm
97. (a) $-2.4, 3.4$; **(b)** $-1.3, 2.3$ **99.** Let $P_1 = (x_1, y_1)$,
$P_2 = (x_2, y_2)$, and $M = \left(\frac{x_1 + x_2}{2}, \frac{y_1 + y_2}{2}\right)$. Let $d(AB)$ denote
the distance from point A to point B.

(i) $d(P_1M) = \sqrt{\left(\frac{x_1 + x_2}{2} - x_1\right)^2 + \left(\frac{y_1 + y_2}{2} - y_1\right)^2}$
$= \frac{1}{2}\sqrt{(x_2 - x_1)^2 + (y_2 - y_1)^2}$;
$d(P_2M) = \sqrt{\left(\frac{x_1 + x_2}{2} - x_2\right)^2 + \left(\frac{y_1 + y_2}{2} - y_2\right)^2}$
$= \frac{1}{2}\sqrt{(x_1 - x_2)^2 + (y_1 - y_2)^2}$
$= \frac{1}{2}\sqrt{(x_2 - x_1)^2 + (y_2 - y_1)^2} = d(P_1M).$

(ii) $d(P_1M) + d(P_2M) = \dfrac{1}{2}\sqrt{(x_2-x_1)^2 + (y_2-y_1)^2}$

$+ \dfrac{1}{2}\sqrt{(x_2-x_1)^2 + (y_2-y_1)^2}$

$= \sqrt{(x_2-x_1)^2 + (y_2-y_1)^2}$
$= d(P_1P_2).$

Exercise Set B, pp. 808–809

1.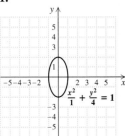

$\dfrac{x^2}{1} + \dfrac{y^2}{4} = 1$

3.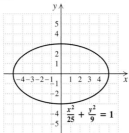

$\dfrac{x^2}{25} + \dfrac{y^2}{9} = 1$

5.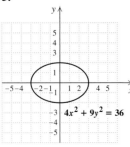

$4x^2 + 9y^2 = 36$

7.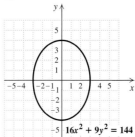

$16x^2 + 9y^2 = 144$

9.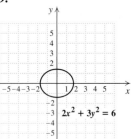

$2x^2 + 3y^2 = 6$

11.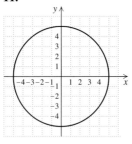

$5x^2 + 5y^2 = 125$

13.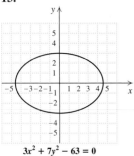

$3x^2 + 7y^2 - 63 = 0$

15.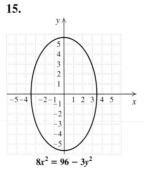

$8x^2 = 96 - 3y^2$

17.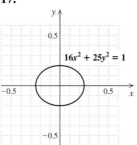

$16x^2 + 25y^2 = 1$

19.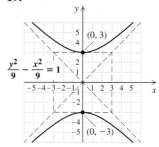

$\dfrac{y^2}{9} - \dfrac{x^2}{9} = 1$

21.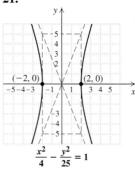

$\dfrac{x^2}{4} - \dfrac{y^2}{25} = 1$

23.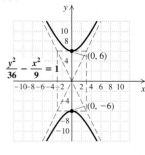

$\dfrac{y^2}{36} - \dfrac{x^2}{9} = 1$

25.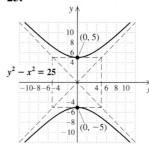

$y^2 - x^2 = 25$

27.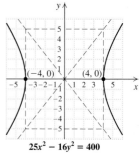

$25x^2 - 16y^2 = 400$

29.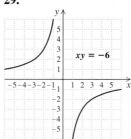

$xy = -6$

31.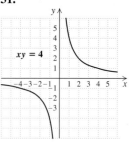

$xy = 4$

33.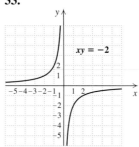

$xy = -2$

35.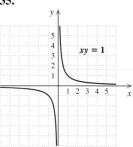

$xy = 1$

37. Circle **39.** Ellipse **41.** Hyperbola **43.** Circle
45. Ellipse **47.** Hyperbola **49.** Parabola
51. Hyperbola **53.** Circle **55.** Parabola
57. $\dfrac{x^2}{81} + \dfrac{y^2}{121} = 1$ **59.** $\dfrac{y^2}{36} - \dfrac{x^2}{4} = 1$
61. 2.134×10^8 mi **63.**

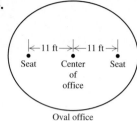

Oval office

65. Length: 700 yd; width: 200 yd

Exercise Set C, pp. 817–818

1. $(-4, -3), (3, 4)$ **3.** $(2, 0), (0, 3)$ **5.** $(2, 4), (1, 1)$
7. $\left(\frac{11}{4}, -\frac{9}{8}\right), (1, -2)$ **9.** $\left(\frac{13}{3}, \frac{5}{3}\right), (-5, -3)$
11. $\left(\dfrac{3 + \sqrt{7}}{2}, \dfrac{-1 + \sqrt{7}}{2}\right), \left(\dfrac{3 - \sqrt{7}}{2}, \dfrac{-1 - \sqrt{7}}{2}\right)$

13. $\left(-3, \frac{5}{2}\right), (3, 1)$ **15.** $(1, -7), (-7, 1)$
17. $(3, 0), \left(-\frac{9}{5}, \frac{8}{5}\right)$ **19.** $(1, 4), (4, 1)$ **21.** $(0, 0), (1, 1),$
$\left(-\dfrac{1}{2} + \dfrac{\sqrt{3}}{2}i, -\dfrac{1}{2} - \dfrac{\sqrt{3}}{2}i\right), \left(-\dfrac{1}{2} - \dfrac{\sqrt{3}}{2}i, -\dfrac{1}{2} + \dfrac{\sqrt{3}}{2}i\right)$
23. $(-3, 0), (3, 0)$ **25.** $(-4, -3), (-3, -4), (3, 4), (4, 3)$
27. $\left(\dfrac{4i\sqrt{35}}{7}, \dfrac{6\sqrt{21}}{7}\right), \left(-\dfrac{4i\sqrt{35}}{7}, \dfrac{6\sqrt{21}}{7}\right),$
$\left(\dfrac{4i\sqrt{35}}{7}, -\dfrac{6\sqrt{21}}{7}\right), \left(-\dfrac{4i\sqrt{35}}{7}, -\dfrac{6\sqrt{21}}{7}\right)$
29. $\left(-\sqrt{2}, -\sqrt{14}\right), \left(-\sqrt{2}, \sqrt{14}\right), \left(\sqrt{2}, -\sqrt{14}\right),$
$\left(\sqrt{2}, \sqrt{14}\right)$ **31.** $(-2, -1), (-1, -2), (1, 2), (2, 1)$
33. $(-3, -2), (-2, -3), (2, 3), (3, 2)$ **35.** $(2, 5), (-2, -5)$
37. $(3, 2), (-3, -2)$ **39.** $(-3, 4), (3, 4), (0, -5)$
41. Length: 8 cm; width: 6 cm **43.** Length: 5 in.;
width: 4 in. **45.** Length: 12 ft; width: 5 ft
47. 13 and 12 **49.** \$3750, 6% **51.** Length: $\sqrt{3}$ m;
width: 1 m **53.** $(x + 2)^2 + (y - 1)^2 = 4$
55. $(-2, 3), (2, -3), (-3, 2), (3, -2)$
57. 10 in. by 7 in. by 5 in.

Photo Credits

Additional Instructor's Answers

Chapter 1

Exercise Set 1.4, pp. 41–42

55. $10a^{-6}b^{-2}$, or $\dfrac{10}{a^6b^2}$ **56.** $6a^{-4}b^{-9}$, or $\dfrac{6}{a^4b^9}$

63. $-\dfrac{1}{6}x^3y^{-2}z^{10}$, or $-\dfrac{x^3z^{10}}{6y^2}$ **64.** $\dfrac{1}{3}a^{10}b^{-9}c^{-2}$, or $\dfrac{a^{10}}{3b^9c^2}$

73. $5x^{-14}y^{-14}$, or $\dfrac{5}{x^{14}y^{14}}$ **74.** $7a^{-15}b^{-20}$, or $\dfrac{7}{a^{15}b^{20}}$

94. 3.09×10^{12} **101.** 0.0000000008923
115. Approximately 1.79×10^{20} **116.** Approximately
3.02×10^{-13} **136.** $8 \cdot 10^{-90}$ is larger by $7.1 \cdot 10^{-90}$.

Exercise Set 1.5, pp. 52–53

3.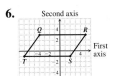

4.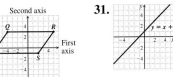

5.

Triangle, 21 units²

6.

Parallelogram, 36 units²

31.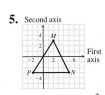

32.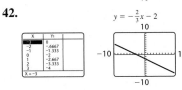

33.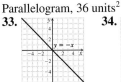

34.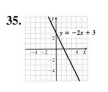

35.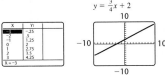

36.

37.

38.

39.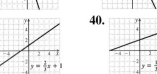

40.

41. $y = -\dfrac{3}{2}x + 1$

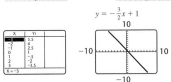

42. $y = -\dfrac{2}{3}x - 2$

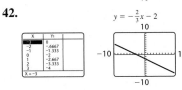

43. $y = \dfrac{3}{4}x + 1$

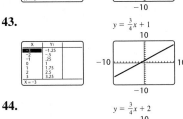

44. $y = \dfrac{3}{4}x + 2$

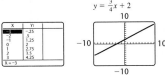

45.

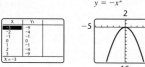

$y = -x^2$

46.

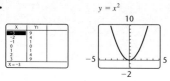

$y = x^2$

47.

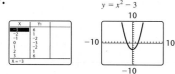

$y = x^2 - 3$

48.

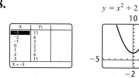

$y = x^2 + 2$

49.

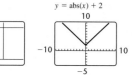

$y = \text{abs}(x) + 2$

50.

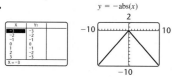

$y = -\text{abs}(x)$

51.

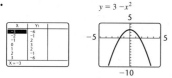

$y = 3 - x^2$

52.

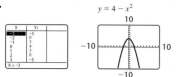

$y = 4 - x^2$

53.

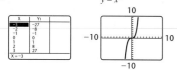

$y = x^3$

54.

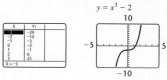

$y = x^3 - 2$

Exercise Set 1.6, pp. 67–71

40. Let x represent the length of one piece, in centimeters;
$$\left(\frac{x}{4}\right)^2 = \left(\frac{100 - x}{4}\right)^2 + 144$$
41. Let x represent the
score on the next test; $\dfrac{93 + 89 + 72 + 80 + 96 + x}{6} = 88$

42. Let x represent the price of the least expensive set, in dollars; $(x + 20) + (x + 6 \cdot 20) = x + 12 \cdot 20$

Review Exercises: Chapter 1, pp. 75–76

47. [1.5]

48. [1.5]

49. [1.5]

50. [1.5]

Test: Chapter 1, p. 77

28. [1.5]

29. [1.5]

30. [1.5]

$y = 10 - x^2$

32. [1.6] Let t represent the score on the sixth test;
$$\frac{94 + 80 + 76 + 91 + 75 + t}{6} = 85$$

Chapter 2

Exercise Set 2.1, pp. 89–93

13. (a) -2; **(b)** $\{x \mid -2 \le x \le 5\}$;
(c) 4; **(d)** $\{y \mid -3 \le y \le 4\}$

14. (a) −1; (b) $\{x\,|\,-4 \le x \le 3\}$;
(c) −3; (d) $\{y\,|\,-2 \le y \le 5\}$
15. (a) 3; (b) $\{x\,|\,-1 \le x \le 4\}$;
(c) 3; (d) $\{y\,|\,1 \le y \le 4\}$
16. (a) 1; (b) $\{x\,|\,-3 \le x \le 5\}$;
(c) −1; (d) $\{y\,|\,0 \le y \le 4\}$
17. (a) −2; (b) $\{x\,|\,-4 \le x \le 2\}$;
(c) −2; (d) $\{y\,|\,-3 \le y \le 3\}$
18. (a) 3; (b) $\{x\,|\,-4 \le x \le 3\}$;
(c) 0; (d) $\{y\,|\,-5 \le y \le 4\}$
19. (a) 3; (b) $\{x\,|\,-4 \le x \le 3\}$;
(c) −3; (d) $\{y\,|\,-2 \le y \le 5\}$
20. (a) 4; (b) $\{x\,|\,-3 \le x \le 4\}$;
(c) −1; (d) $\{y\,|\,0 \le y \le 5\}$
21. (a) 1; (b) $\{-3, -1, 1, 3, 5\}$;
(c) 3; (d) $\{-1, 0, 1, 2, 3\}$
22. (a) 3; (b) $\{-4, -3, -2, -1, 0, 1, 2\}$;
(c) −2, 0; (d) $\{1, 2, 3, 4\}$
23. (a) 4; (b) $\{x\,|\,-3 \le x \le 4\}$;
(c) −1, 3; (d) $\{y\,|\,-4 \le y \le 5\}$
24. (a) 2; (b) $\{x\,|\,-5 \le x \le 2\}$;
(c) −5, 1; (d) $\{y\,|\,-3 \le y \le 5\}$
25. (a) 1; (b) $\{x\,|\,-4 < x \le 5\}$,
(c) $\{x\,|\,2 < x \le 5\}$; (d) $\{-1, 1, 2\}$
26. (a) 2; (b) $\{x\,|\,-4 \le x \le 4\}$;
(c) $\{x\,|\,0 < x \le 2\}$; (d) $\{1, 2, 3, 4\}$
49. About 24,000 cases

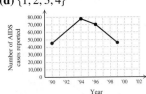

53. About $313,000

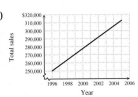

Exercise Set 2.3, pp. 115–119

44. $t = \dfrac{P - b}{0.5}$ **47.** $y = \dfrac{C - Ax}{B}$
48. $l = \dfrac{P - 2w}{2}$, or $\dfrac{P}{2} - w$ **49.** $F = \frac{9}{5}C + 32$
50. $I = \frac{10}{3}T + 12,000$

83.   **84.**

85.   **86.**

87.   **88.**

Exercise Set 2.4, pp. 130–135

27. Slope: $\frac{5}{2}$; y-intercept: $(0, 3)$

28. Slope: $\frac{2}{5}$; y-intercept: $(0, 4)$

29. Slope: $-\frac{5}{2}$; y-intercept: $(0, 1)$

30. Slope: $-\frac{2}{5}$; y-intercept: $(0, 3)$

31. Slope: 2; y-intercept: $(0, -5)$

32. Slope: −2; y-intercept: $(0, 4)$

33. Slope: $\frac{1}{3}$; y-intercept: $(0, 2)$

34. Slope: -3; y-intercept: $(0, 6)$

$f(x) = -3x + 6$

35. Slope: $-\frac{2}{7}$; y-intercept: $(0, 1)$

$7y + 2x = 7$, or $y = -\frac{2}{7}x + 1$

36. Slope: $\frac{1}{4}$; y-intercept: $(0, -5)$

$4y + 20 = x$, or $y = \frac{1}{4}x - 5$

37. Slope: -0.25; y-intercept: $(0, 0)$

$f(x) = -0.25x$

38. Slope: 1.5; y-intercept: $(0, -3)$

$f(x) = 1.5x - 3$

39. Slope: $\frac{4}{5}$; y-intercept: $(0, -2)$

$4x - 5y = 10$, or $y = \frac{4}{5}x - 2$

40. Slope: $-\frac{5}{4}$; y-intercept: $(0, 1)$

$5x + 4y = 4$, or $y = -\frac{5}{4}x + 1$

41. Slope: $\frac{5}{4}$; y-intercept: $(0, -2)$

$f(x) = \frac{5}{4}x - 2$

42. Slope: $\frac{4}{3}$; y-intercept: $(0, 2)$

$f(x) = \frac{4}{3}x + 2$

43. Slope: $-\frac{3}{4}$; y-intercept: $(0, 3)$

$12 - 4f(x) = 3x$,
or $f(x) = -\frac{3}{4}x + 3$

44. Slope: $-\frac{2}{5}$; y-intercept: $(0, -3)$

$15 + 5f(x) = -2x$,
or $f(x) = -\frac{2}{5}x - 3$

45. Slope: 0; y-intercept: $(0, 2.5)$

46. Slope: $\frac{3}{4}$; y-intercept: $(0, 0)$

$g(x) = \frac{3}{4}x$

57. The distance is increasing at a rate of 3 miles per hour.
58. The weight is increasing at a rate of 0.5 pound per bag.
59. The number of pages read is increasing at a rate of 75 pages per day.   **60.** The distance from home is increasing at a rate of 0.25 kilometer per minute.   **61.** The average SAT math score is increasing at a rate of 1 point per thousand dollars of family income.   **62.** The average SAT verbal score is increasing at a rate of 1 point per thousand dollars of family income.   **72.** 0.05 signifies that a salesperson earns 5% commission on sales; 200 signifies that a salesperson earns a base salary of $200 per week.
74. $\frac{1}{8}$ signifies that the grass grows $\frac{1}{8}$ in. per day; 2 signifies that the grass is 2 in. long when cut.   **75.** $\frac{3}{20}$ signifies that the life expectancy of American women increases $\frac{3}{20}$ yr per year, for years after 1950; 72 signifies that the life expectancy in 1950 was 72 yr.   **76.** $\frac{1}{5}$ signifies that the demand increases $\frac{1}{5}$ quadrillion joules per year, for years after 1960; 20 signifies that the demand was 20 quadrillion joules in 1960.   **77.** 2.6 signifies that sales increase $2.6 billion per year, for years after 1975; 17.8 signifies that sales in 1975 were $17.8 billion.   **78.** 0.1522 signifies that the price increases $0.1522 per year, for years since 1990; 4.29 signifies that the average cost of a movie ticket in 1990 was $4.29.   **79.** 0.75 signifies that the cost per mile of a taxi ride is $0.75; 2 signifies that the minimum cost of a taxi ride is $2.   **80.** 0.3 signifies that the cost per mile of renting the truck is $0.30; 20 signifies that the minimum cost is $20.
81. (a) -5000 signifies a depreciation of $5000 per year; 90,000 signifies the original value of the truck, $90,000
82. (a) -2000 signifies that the depreciation is $2000 per year; 15,000 signifies that the original value of the machine was $15,000   **83. (a)** -150 signifies that the depreciation

is $150 per winter of use; 900 signifies that the original value of the snowblower was $900   **84. (a)** -300 signifies that the depreciation is $300 per summer of use; 2400 signifies that the original value of the mower was $2400; **(b)** after 4 summers of use

98. Slope: $-\dfrac{r}{r+p}$; y-intercept: $\left(0, \dfrac{s}{r+p}\right)$

Exercise Set 2.5, pp. 145–146

19.

20.

21.

22.

23.

24.

$4 \cdot f(x) = 20$

$6 \cdot g(x) = 12$

25.

$3x = -15$

26.

$2x = 10$

27.

$4 \cdot g(x) + 3x = 12 + 3x$

28.

$3 - f(x) = 2$

29.

$x + y = 5$

30.

$x + y = 4$

31.

$y = 2x + 8$

32.

$y = 3x + 9$

33.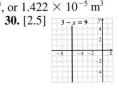

$3x + 5y = 15$

34.

$5x - 4y = 20$

35.

$2x - 3y = 18$

36.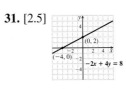

$3x + 2y = 12$

37.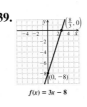

$7x = 3y - 21$

38. (graph)

$5y = -15 + 3x$

39. (graph)

$f(x) = 3x - 8$

40.

$g(x) = 2x - 9$

41.

$1.4y - 3.5x = -9.8$

42.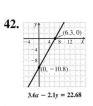

$3.6x - 2.1y = 22.68$

43.

$5x + 2g(x) = 7$

44.

$3x - 4f(x) = 11$

Exercise Set 2.7, pp. 166–169

29. About 50 million; the number of passengers using Newark and LaGuardia in 1998   **30.** About 41 million; the number of passengers using Kennedy and LaGuardia in 1998   **31.** About 8 million; how many more passengers used Kennedy than LaGuardia in 1994   **32.** About 1 million; how many more passengers used Kennedy than Newark in 1994   **33.** About 89 million; the number of passengers using the three airports in 1999   **34.** About 69 million; the number of passengers using the three airports in 1998   **35. (a)** $\{x \mid x$ is a real number *and* $x \neq 3\}$; **(b)** $\{x \mid x$ is a real number *and* $x \neq 6\}$; **(c)** $\mathbb{R}$; **(d)** $\mathbb{R}$; **(e)** $\{x \mid x$ is a real number *and* $x \neq \frac{5}{2}\}$; **(f)** $\mathbb{R}$ **36. (a)** $\{x \mid x$ is a real number *and* $x \neq 1\}$; **(b)** $\mathbb{R}$; **(c)** $\{x \mid x$ is a real number *and* $x \neq -3\}$; **(d)** $\{x \mid x$ is a real number *and* $x \neq -\frac{4}{3}\}$; **(e)** $\mathbb{R}$; **(f)** $\mathbb{R}$ **61.** 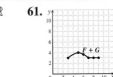   **62.**

79. Domain of $f + g =$ Domain of $f - g =$ Domain of $f \cdot g = \{-2, -1, 0, 1\}$; Domain of $f/g = \{-2, 0, 1\}$
81. Answers may vary. $f(x) = \dfrac{1}{x+2}, g(x) = \dfrac{1}{x-5}$

Review Exercises: Chapter 2, pp. 171–173

15. [2.3] 1.422×10^4 mm^3, or 1.422×10^{-5} m^3
29. [2.5] 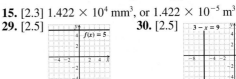 $f(x) = 5$

30. [2.5] 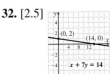 $3 - x = 9$

31. [2.5] (graph) $-2x + 4y = 8$

32. [2.5] (graph) $x + 7y = 14$

33. [2.5] The window should show both intercepts: $(98, 0)$ and $(0, 14)$. One possibility is $[-10, 120, -5, 15]$, with Xscl $= 10$.
40. [2.6] $y - 4 = -2(x - (-3))$, or $y - 4 = -2(x + 3)$

Test: Chapter 2, pp. 174–175

2. [2.1] **(b)** 1.2 signifies that the rate of increase in U.S. book sales was \$1.2 billion per year; 21.4 signifies that U.S. book sales were \$21.4 billion in 1992   **18.** [2.4]

19. [2.5]

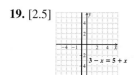

20. [2.5]

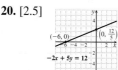

Chapter 3

Exercise Set 3.1, pp. 188–192

43. Let x represent the number of \$8.50 brushes sold and y the number of \$9.75 brushes sold; $x + y = 45$, $8.50x + 9.75y = 398.75$ **44.** Let p represent the number of polar fleece neckwarmers sold and w the number of wool neckwarmers sold; $p + w = 40$, $9.90p + 12.75w = 421.65$
47. Let x represent the number of two-point shots and y the number of free throws; $x + y = 64$, $2x + y = 100$
48. Let x represent the number of children's plates and y the number of adults' plates served; $x + y = 250$, $3.5x + 7y = 1347.5$ **49.** Let h represent the number of vials of Humulin sold and n the number of vials of Novolin; $h + n = 50$, $23.97h + 34.39n = 1406.90$ **50.** Let l represent the length, in feet, and w the width, in feet; $2l + 2w = 288$, $l = w + 44$ **51.** Let l represent the length, in feet, and w the width, in feet; $2l + 2w = 228$, $w = l - 42$ **52.** Let x represent the number of 2-pointers scored and y the number of 3-pointers scored; $x + y = 40$, $2x + 3y = 89$ **53.** Let w represent the number of wins and t the number of ties; $2w + t = 60$, $w = t + 9$
54. Let x represent the number of 30-sec commercials and y the number of 60-sec commercials; $x + y = 12$, $30x + 60y = 600$ **55.** Let x represent the number of ounces of lemon juice and y the number of ounces of linseed oil; $y = 2x$, $x + y = 32$ **56.** Let l represent the number of pallets of lumber produced and p the number of pallets of plywood produced; $l + p = 42$, $25l + 40p = 1245$
57. Let x represent the number of general-interest films rented and y the number of children's films rented; $x + y = 77$, $3x + 1.5y = 213$ **58.** Let c represent the number of coach-class seats and f the number of first-class seats; $c + f = 152$, $c = 5 + 6f$ **59.** Let y represent the percent of the population in the work force and x the number of years

since 1970; $y = \frac{83}{180}x + \frac{422}{9}$, $y = -\frac{5}{36}x + \frac{7091}{90}$; about 2023
60. Let y represent vehicle production, in thousands, and x the number of years since 1980; $y = -18.3125x + 6376$, $y = 255.125x + 1634$; about 1997   **61.** Let y represent the amount of paper, in millions of tons, and x the number of years since 1990; $y = 0.825x + 77.5$, $y = 1.275x + 31.4$; about 2092   **62.** Let y represent the number of meals eaten and x the number of years since 1990; $y = -\frac{2}{3}x + 64$, $y = \frac{4}{3}x + 55$; about 1995   **63.** Let y represent the per capita consumption, in gallons, and x the number of years since 1950; $y = -\frac{7}{25}x + 38$, $y = \frac{43}{50}x + 10$; about 1975
64. Let y represent the per capita consumption, in pounds, and x the number of years since 1980; $y = -\frac{1}{20}x + 17.5$, $y = \frac{1}{15}x + 7.1$; about 2069
65. Women: $y_1 = 0.5660768453x + 44.59989889$, men: $y_2 = -0.1678968655x + 79.48407482$, where x is the number of years since 1970; about 2018
66. Passenger cars: $y_1 = -8.073170732x + 6256.121951$, trucks: $y_2 = 266.6081301x + 1423.764228$, where x is the number of years since 1980; about 1998
67. Waste generated: $y_1 = 1.41x + 72.68$, paper recycled: $y_2 = 1.81x + 28.86$, where x is the number of years since 1990; about 2100
68. Regular: $y_1 = -0.0971857411x + 17.67298311$, lowfat: $y_2 = 0.0628517824x + 6.896622889$, where x is the number of years since 1980; about 2047   **85.** Let x and y represent the number of years that Lou and Juanita have taught at the university, respectively; $x + y = 46$, $x - 2 = 2.5(y - 2)$ **86.** Let l represent the original length, in inches, and w the original width, in inches; $2l + 2w = 156$, $l = 4(w - 6)$ **87.** Let s and v represent the number of ounces of baking soda and vinegar needed, respectively; $s = 4v$, $s + v = 16$

Exercise Set 3.3, pp. 212–214

1. 29, 18   **2.** 5, -47 **3.** \$8.50-brushes: 32; \$9.75-brushes: 13   **4.** Polarfleece: 31; wool: 9
5. 119°, 61°   **6.** 38°, 52°   **7.** Two-point shots: 36; foul shots: 28   **8.** Children's plates: 115; adults' plates: 135
9. Humulin: 30; Novolin: 20   **10.** Width: 50 ft; length: 94 ft   **11.** Width: 36 ft; length: 78 ft   **12.** Two-pointers: 31; three-pointers: 9   **13.** Wins: 23; ties: 14
14. 30-sec: 4; 60-sec: 8   **15.** Lemon juice: $10\frac{2}{3}$ oz; linseed oil: $21\frac{1}{3}$ oz   **16.** Pallets of lumber: 29; pallets of plywood: 13   **17.** General interest: 65; children's: 12
18. Coach: 131; first-class: 21   **51.** Burl: 40; son: 20
52. Lou: 32 yr; Juanita: 14 yr   **53.** Length: $\frac{288}{5}$ in.; width: $\frac{102}{5}$ in.   **54.** Baking soda: $\frac{64}{5}$ oz; vinegar: $\frac{16}{5}$ oz

Exercise Set 3.7, pp. 241–243

19. (d) \$90,300 loss; \$14,700 profit   **20. (b)** $R(x) = 85x$; **(d)** \$112,500 profit; \$4500 loss
21. (a) $C(x) = 16,404 + 6x$; **(d)** \$19,596 profit; \$4404 loss
22. (b) $R(x) = 80x$

Cumulative Review: Chapters 1–3, pp. 248–250

21. [2.4]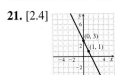

22. [2.1]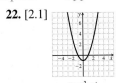

23. [2.5]   **24.** [2.5]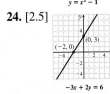

46. [2.4] 9 signifies that the number of U.S. pleasure trips is increasing by 9 million each year; 616 signifies that there were 616 million U.S. pleasure trips in 1994.
47. [2.6] **(a)** $A(t) = \frac{11}{60}t + 6.6$; **(b)** 10.45 million departures

Chapter 4

Exercise Set 4.1, pp. 263–268

13. $\{x \mid x > -5\}$, or $(-5, \infty)$
14. $\{x \mid x > -3\}$, or $(-3, \infty)$
15. $\{a \mid a \le -20\}$, or $(-\infty, -20]$
16. $\{a \mid a \le -21\}$, or $(-\infty, -21]$
17. $\{x \mid x \le 12\}$, or $(-\infty, 12]$
18. $\{t \mid t \ge -5\}$, or $[-5, \infty)$
19. $\{y \mid y > -9\}$, or $(-9, \infty)$
20. $\{y \mid y > -6\}$, or $(-6, \infty)$
21. $\{y \mid y \le 14\}$, or $(-\infty, 14]$
22. $\{x \mid x \le 9\}$, or $(-\infty, 9]$
23. $\{t \mid t < -9\}$, or $(-\infty, -9)$
24. $\{x \mid x \ge 3\}$, or $[3, \infty)$
25. $\{x \mid x < 50\}$, or $(-\infty, 50)$
26. $\{x \mid x < -60\}$, or $(-\infty, -60)$
27. $\{y \mid y \ge -0.4\}$, or $[-0.4, \infty)$
28. $\{x \mid x \le 0.9\}$, or $(-\infty, 0.9]$
29. $\{y \mid y \ge \frac{9}{10}\}$, or $[\frac{9}{10}, \infty)$
30. $\{x \mid x \le \frac{5}{6}\}$, or $(-\infty, \frac{5}{6}]$
31. $\{y \mid y > 3\}$, or $(3, \infty)$
32. $\{x \mid x < 6\}$, or $(-\infty, 6)$
33. $\{x \mid x \le 4\}$, or $(-\infty, 4]$

34. $\{y \mid y \le -3\}$, or $(-\infty, -3]$
35. $\{x \mid x < -\frac{2}{5}\}$, or $(-\infty, -\frac{2}{5})$
36. $\{x \mid x > \frac{2}{3}\}$, or $(\frac{2}{3}, \infty)$
37. $\{x \mid x \ge 11.25\}$, or $[11.25, \infty)$
38. $\{x \mid x \ge \frac{1}{2}\}$, or $[\frac{1}{2}, \infty)$
55. $\{x \mid x \le \frac{4}{7}\}$, or $(-\infty, \frac{4}{7}]$ **56.** $\{t \mid t \ge -\frac{27}{19}\}$, or $[-\frac{27}{19}, \infty)$
79. Swimming: $f(x) = -1.174729242x + 73.30288809$; equipment: $g(x) = 1.273826715x + 31.20415162$; years after 2002   **80.** Public: $f(x) = 0.0922664165x + 0.4818424015$; private: $g(x) = 0.0588968105x + 1.402990619$; years after 2007
93. $\left\{ y \mid y \ge \dfrac{2a + 5b}{b(a - 2)} \right\}$
94. $\left\{ x \mid x < \dfrac{4c + 3d}{6c - 2d} \right\}$
95. $\left\{ x \mid x > \dfrac{4m - 2c}{d - (5c + 2m)} \right\}$
96. $\left\{ x \mid x < \dfrac{-3a + 2d}{c - (4a + 5d)} \right\}$

Exercise Set 4.2, pp. 277–279

13. $(3, 8)$ **14.** $[0, 4]$
15. $[-6, -2]$ **16.** $[-9, -5)$
17. $(-\infty, -2) \cup (3, \infty)$
18. $(-\infty, -5) \cup (1, \infty)$
19. $(-\infty, -1] \cup (5, \infty)$
20. $(-\infty, -5] \cup (2, \infty)$
21. $(-2, 4]$ **22.** $(-7, -2)$
23. $(-2, 4)$ **24.** $(-3, 1]$
25. $(-\infty, 5) \cup (7, \infty)$
26. $(-\infty, -3) \cup [2, \infty)$
27. $(-\infty, -4] \cup [5, \infty)$
28. $(-\infty, -6) \cup (-3, \infty)$
29. $[-6, 4)$ **30.** $(-6, 0]$
31. $[3, 7)$ **32.** $[-3, 3)$
33. $(-\infty, 5)$ **34.** $(-1, \infty)$
35. $(-\infty, \infty)$ **36.** $(-\infty, \infty)$
37. $(7, \infty)$ **38.** $(-\infty, -4]$
39. $\{t \mid -3 < t < 5\}$, or $(-3, 5)$
40. $\{t \mid -4 < t \le 4\}$, or $(-4, 4]$

41. $\{x \mid -1 < x \le 4\}$, or $(-1, 4]$

42. $\{x \mid -3 < x < 7\}$, or $(-3, 7)$

43. $\{a \mid -2 \le a < 2\}$, or $[-2, 2)$

44. $\{n \mid -2 \le n \le 4\}$, or $[-2, 4]$

45. $\mathbb{R}$, or $(-\infty, \infty)$

46. $\{x \mid x < -8 \text{ or } x > -1\}$, or $(-\infty, -8) \cup [-1, \infty)$

47. $\{x \mid 1 \le x \le 3\}$, or $[1, 3]$

48. $\{x \mid 1 \le x \le 4\}$, or $[1, 4]$

49. $\{x \mid -\frac{7}{2} < x \le 7\}$, or $\left(-\frac{7}{2}, 7\right]$

50. $\left\{t \mid -4 < t \le -\frac{10}{3}\right\}$, or $\left(-4, -\frac{10}{3}\right]$

51. $\{x \mid x \le 1 \text{ or } x \ge 3\}$, or $(-\infty, 1] \cup [3, \infty)$

52. $\{x \mid x \le 1 \text{ or } x \ge 5\}$, or $(-\infty, 1] \cup [5, \infty)$

53. $\{x \mid x < 3 \text{ or } x > 4\}$, or $(-\infty, 3) \cup (4, \infty)$

54. $\left\{x \mid x < -4 \text{ or } x > \frac{2}{3}\right\}$, or $(-\infty, -4) \cup \left(\frac{2}{3}, \infty\right)$

55. $\left\{a \mid a < \frac{7}{2}\right\}$, or $\left(-\infty, \frac{7}{2}\right)$

56. $\mathbb{R}$, or $(-\infty, \infty)$

57. $\{a \mid a < -5\}$, or $(-\infty, -5)$

58. $\{a \mid a > 4\}$, or $(4, \infty)$

59. $\mathbb{R}$, or $(-\infty, \infty)$

60. $\{x \mid x \le -2 \text{ or } x > 3\}$, or $(-\infty, -2] \cup (3, \infty)$

61. $\{t \mid t \le 6\}$, or $(-\infty, 6]$

62. $\{a \mid a \ge -1\}$, or $[-1, \infty)$

83. $y = 5$

84. $y = -2$

85. $f(x) = |x|$

86. $g(x) = x - 1$

Exercise Set 4.3, pp. 288–289

45. $\left\{-4, -\frac{10}{9}\right\}$ **46.** $\left\{-\frac{3}{5}, 5\right\}$

53. $\{a \mid -7 \le a \le 7\}$, or $[-7, 7]$

54. $\{x \mid -2 < x < 2\}$, or $(-2, 2)$

55. $\{x \mid x < -8 \text{ or } x > 8\}$, or $(-\infty, -8) \cup (8, \infty)$

56. $\{a \mid a \le -3 \text{ or } a \ge 3\}$, or $(-\infty, -3] \cup [3, \infty)$

57. $\{t \mid t < 0 \text{ or } t > 0\}$, or $(-\infty, 0) \cup (0, \infty)$

58. $\{t \mid t \le -1.7 \text{ or } t \ge 1.7\}$, or $(-\infty, -1.7] \cup [1.7, \infty)$

59. $\{x \mid -2 < x < 8\}$, or $(-2, 8)$

60. $\{x \mid -2 < x < 4\}$, or $(-2, 4)$

61. $\{x \mid -8 \le x \le 4\}$, or $[-8, 4]$

62. $\{x \mid -5 \le x \le -3\}$, or $[-5, -3]$

63. $\{x \mid x < -2 \text{ or } x > 8\}$, or $(-\infty, -2) \cup (8, \infty)$

64. $\mathbb{R}$, or $(-\infty, \infty)$

65. $\mathbb{R}$, or $(-\infty, \infty)$

66. $\left\{y \mid y < -\frac{4}{3} \text{ or } y > 4\right\}$, or $\left(-\infty, -\frac{4}{3}\right) \cup (4, \infty)$

67. $\left\{a \mid a \le -\frac{2}{3} \text{ or } a \ge \frac{10}{3}\right\}$, or $\left(-\infty, -\frac{2}{3}\right] \cup \left[\frac{10}{3}, \infty\right)$

68. $\left\{a \mid a \le -\frac{3}{2} \text{ or } a \ge \frac{13}{2}\right\}$, or $\left(-\infty, -\frac{3}{2}\right] \cup \left[\frac{13}{2}, \infty\right)$

69. $\{y \mid -9 < y < 15\}$, or $(-9, 15)$;

70. $\{p \mid -1 < p < 5\}$, or $(-1, 5)$

71. $\{x \mid x \le -8 \text{ or } x \ge 0\}$, or $(-\infty, -8] \cup [0, \infty)$

72. $\{x \mid x \le 2 \text{ or } x \ge 8\}$, or $(-\infty, 2] \cup [8, \infty)$

73. $\left\{y \mid y < -\frac{4}{3} \text{ or } y > 4\right\}$, or $\left(-\infty, -\frac{4}{3}\right) \cup (4, \infty)$;

76. $\left\{a \mid -\frac{7}{2} \le a \le 6\right\}$, or $\left[-\frac{7}{2}, 6\right]$

77. $\left\{x \mid x \le -\frac{2}{15} \text{ or } x \ge \frac{14}{15}\right\}$, or $\left(-\infty, -\frac{2}{15}\right] \cup \left[\frac{14}{15}, \infty\right)$;

78. $\left\{x \mid x < -\frac{43}{24} \text{ or } x > \frac{9}{8}\right\}$, or $\left(-\infty, -\frac{43}{24}\right) \cup \left(\frac{9}{8}, \infty\right)$

79. $\{m \mid -12 \le m \le 2\}$, or $[-12, 2]$;

80. $\{t \mid t \le 6 \text{ or } t \ge 8\}$, or $(-\infty, 6] \cup [8, \infty)$

81. $\{a \mid -6 < a < 0\}$, or $(-6, 0)$

82. $\left\{a \mid -\frac{13}{2} < a < \frac{5}{2}\right\}$, or $\left(-\frac{13}{2}, \frac{5}{2}\right)$

83. $\left\{x \mid -\frac{1}{2} \le x \le \frac{7}{2}\right\}$, or $\left[-\frac{1}{2}, \frac{7}{2}\right]$

84. $\left\{x \mid -1 \le x \le \frac{1}{5}\right\}$, or $\left[-1, \frac{1}{5}\right]$

85. $\left\{x \mid x \le -\frac{7}{3} \text{ or } x \ge 5\right\}$, or $\left(-\infty, -\frac{7}{3}\right] \cup [5, \infty)$

86. $\left\{x \mid x \le -\frac{23}{9} \text{ or } x \ge 3\right\}$, or $\left(-\infty, -\frac{23}{9}\right] \cup [3, \infty)$

Exercise Set 4.4, pp. 299–301

5. $y > \frac{1}{2}x$

6. $y > 2x$

7. $y \geq x - 3$

8. $y < x + 3$

9. $y \leq x + 4$

10. $y > x - 2$

11. $x - y \leq 5$

12. $x + y < 4$

13. $2x + 3y < 6$

14. $3x + 4y \leq 12$

15. $2y - x \leq 4$

16. $2y - 3x > 6$

17. $2x - 2y \geq 8 + 2y$

18. $3x - 2 \leq 5x + y$

19. $y \geq 2$

20. $x < -5$

21. $x \leq 7$

22. $y > -3$

23. $-2 < y < 6$

24. $-4 < y < -1$

25. $-4 \leq x \leq 5$

26. $-3 \leq y \leq 4$

27. $0 \leq y \leq 3$

28. 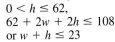 $0 \leq x \leq 6$

29. $y > x + 3.5$

30. $7y \leq 2x + 5$

31. $8x - 2y < 11$

32. $11x + 13y + 4 \geq 0$

33.

34.

35.

36.

37.

38.

39.

40.

41.

42.

43.

44.

45.

46.

47. $(3, 5)$ $\left(\frac{1}{2}, 0\right)$ $(3, -5)$

48. $(2, 2)$ $(1, -1)$ $(-4, -1)$

49. $(0, 6)$ $(4, 4)$ $(0, 0)$ $(6, 0)$

50. $(0, 4)$ $(6, 4)$ $(4, 2)$

51. $(0, 4)$ $\left(\frac{40}{11}, \frac{24}{11}\right)$ $(0, 0)$ $(5, 0)$

52. $(-12, 0)$ $(0, 0)$ $(-4, -6)$ $(0, -3)$

53. $(2, 5)$ $(5, 8)$ $(5, 6)$ $(2, 3)$

54. $\left(1, \frac{25}{6}\right)$ $\left(3, \frac{5}{2}\right)$ $\left(1, \frac{9}{4}\right)$ $\left(3, \frac{3}{4}\right)$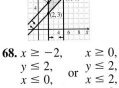

68. $x \geq -2$, $x \geq 0$, $x \geq 0$, $x \geq -2$,
$y \leq 2$, $y \leq 2$, $y \leq 0$, $y \leq 0$,
$x \leq 0$, or $x \leq 2$, or $x \leq 2$, or $x \leq 0$,
$y \geq 0$; $y \geq 0$; $y \geq -2$; $y \geq -2$

69. $0 < w \leq 62$,
$0 < h \leq 62$,
$62 + 2w + 2h \leq 108$
or $w + h \leq 23$

70. $2w + t \geq 60$,
$w \geq 0$,
$t \geq 0$

71. $35c + 75a > 1000,$
$c \geq 0,$
$a \geq 0$

72. $0 < L \leq 94,$

Review Exercises: Chapter 4, pp. 303–304

1. [4.1] $\{x \mid x \leq -2\}$, or $(-\infty, -2]$;

2. [4.1] $\{a \mid a \leq -21\}$, or $(-\infty, -21]$;

3. [4.1] $\{y \mid y \geq -7\}$, or $[-7, \infty)$;

4. [4.1] $\{y \mid y > -\frac{15}{4}\}$, or $(-\frac{15}{4}, \infty)$;

5. [4.1] $\{y \mid y > -30\}$, or $(-30, \infty)$;

6. [4.1] $\{x \mid x > -\frac{3}{2}\}$, or $(-\frac{3}{2}, \infty)$;

7. [4.1] $\{x \mid x < -3\}$, or $(-\infty, -3)$;

8. [4.1] $\{y \mid y > -\frac{220}{23}\}$, or $(-\frac{220}{23}, \infty)$;

9. [4.1] $\{x \mid x \leq -\frac{5}{2}\}$, or $(-\infty, -\frac{5}{2}]$;

17. [4.2] $\{x \mid -7 < x \leq 2\}$, or $(-7, 2]$

18. [4.2] $\{x \mid -\frac{5}{4} < x < \frac{5}{2}\}$, or $(-\frac{5}{4}, \frac{5}{2})$

19. [4.2] $\{x \mid x < -3 \text{ or } x > 1\}$, or $(-\infty, -3) \cup (1, \infty)$

20. [4.2] $\{x \mid x < -11 \text{ or } x \geq -6\}$, or $(-\infty, -11) \cup [-6, \infty)$

21. [4.2] $\{x \mid x \leq -6 \text{ or } x \geq 8\}$, or $(-\infty, -6] \cup [8, \infty)$

22. [4.2] $\{x \mid x < -\frac{2}{5} \text{ or } x > \frac{8}{5}\}$, or $(-\infty, -\frac{2}{5}) \cup (\frac{8}{5}, \infty)$

27. [4.3] $\{t \mid t \leq -3.5 \text{ or } t \geq 3.5\}$, or $(-\infty, -3.5] \cup [3.5, \infty)$

29. [4.3] $\{x \mid -\frac{17}{2} < x < \frac{7}{2}\}$, or $(-\frac{17}{2}, \frac{7}{2})$

30. [4.3] $\{x \mid x \leq -\frac{11}{3} \text{ or } x \geq \frac{19}{3}\}$, or $(-\infty, -\frac{11}{3}] \cup [\frac{19}{3}, \infty)$

31. [4.3] $\{-14, \frac{4}{3}\}$

Test: Chapter 4, p. 304

13. [4.2] $\{x \mid -1 < x < 6\}$, or $(-1, 6)$

14. [4.2] $\{t \mid -\frac{2}{5} < t \leq \frac{9}{5}\}$, or $(-\frac{2}{5}, \frac{9}{5}]$

16. [4.2] $\{x \mid x < -4 \text{ or } x > -\frac{5}{2}\}$, or $(-\infty, -4) \cup (-\frac{5}{2}, \infty)$

17. [4.2] $\{x \mid 4 \leq x < \frac{15}{2}\}$, or $[4, \frac{15}{2})$

18. [4.3] $\{-9, 9\}$

19. [4.3] $\{a \mid a < -3 \text{ or } a > 3\}$, or $(-\infty, -3) \cup (3, \infty)$

20. [4.3] $\{x \mid -\frac{7}{8} < x < \frac{11}{8}\}$, or $(-\frac{7}{8}, \frac{11}{8})$

21. [4.3] $\{t \mid t \leq -\frac{13}{5} \text{ or } t \geq \frac{7}{5}\}$, or $(-\infty, -\frac{13}{5}] \cup [\frac{7}{5}, \infty)$

23. [4.2] $\{x \mid x < \frac{1}{2} \text{ or } x > \frac{7}{2}\}$, or $(-\infty, \frac{1}{2}) \cup (\frac{7}{2}, \infty)$

Chapter 5

Exercise Set 5.2, pp. 330–332

102.

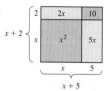

Exercise Set 5.4, pp. 354–356

61. $f(x) = (x + 1)(x - 2)$, or $f(x) = x^2 - x - 2$
62. $f(x) = (x - 2)(x - 5)$, or $f(x) = x^2 - 7x + 10$
63. $f(x) = (x + 7)(x + 10)$, or $f(x) = x^2 + 17x + 70$
64. $f(x) = (x - 8)(x + 3)$, or $f(x) = x^2 - 5x - 24$
65. $f(x) = x(x - 1)(x - 2)$, or $f(x) = x^3 - 3x^2 + 2x$
66. $f(x) = x(x + 3)(x - 5)$, or $f(x) = x^3 - 2x^2 - 15x$
77. $\{-1, 3\}$; $(-2, 4)$, or $\{x \mid -2 < x < 4\}$

Exercise Set 5.5, pp. 363–364

65. $\{x \mid x \text{ is a real number } and \ x \neq 5 \ and \ x \neq -1\}$
66. $\{x \mid x \text{ is a real number } and \ x \neq 6 \ and \ x \neq 1\}$
67. $\{x \mid x \text{ is a real number } and \ x \neq 0 \ and \ x \neq \frac{1}{2}\}$
68. $\{x \mid x \text{ is a real number } and \ x \neq 0 \ and \ x \neq \frac{1}{5}\}$

Exercise Set 5.8, pp. 385–389

27. (a) $E(x) = 0.0000256775x^4 - 0.0025507304x^3 + 0.0821803626x^2 - 0.9480449896x + 11.8202857;$
(b) 9.3¢ per kilowatt-hour; 8.2¢ per kilowatt-hour
28. (a) $f(x) = 0.0000001957x^4 - 0.0000570879x^3 + 0.0042982745x^2 - 0.0159423087x + 1.45236808;$
(b) 1.986 million farms; 2.6 million farms
29. (a) $d(x) = -105.9592768x^2 + 3621.572598x + 42307.79479;$ **(b)** 58,222 degrees; 39,263 degrees
30. (a) $r(x) = -0.0085518407x^3 + 0.390736534x^2 - 4.819375298x + 69.4855836$
31. (a) $b(x) = -1.143939394x^2 + 9.583333333x + 25.79545455$
32. (a) $n(x) = 0.0113125x^4 - 0.9444166667x^3 + 24.93375x^2 - 222.4083333x + 8174$

Review Exercises, pp. 392–394

36. [5.7] $(0.4b - 0.5c)(0.16b^2 + 0.2bc + 0.25c^2)$
55. [5.5] $\{x \mid x \text{ is a real number } and\ x \neq -7 \text{ and } x \neq \frac{2}{3}\}$
60. [5.8] **(a)** $0.0273088783x^3 - 1.977521075x^2 + 37.02830561x + 38.79266348$

Chapter 6

Exercise Set 6.2, pp. 418–421

18. $-\dfrac{1}{y + 5}$ **20.** $\dfrac{1}{y^2 + 9}$ **21.** $\dfrac{1}{m^2 + mn + n^2}$
22. $\dfrac{1}{r^2 + rs + s^2}$ **23.** $\dfrac{2a^2 - a + 14}{(a - 4)(a + 3)}$
24. $\dfrac{2a^2 + 22}{(a - 5)(a + 4)}$ **28.** $\dfrac{a^2 + 7ab + b^2}{(a + b)(a - b)}$
55. $\dfrac{-6x + 42}{(x - 3)(x - 2)(x + 1)(x + 3)}$

Exercise Set 6.3, pp. 428–431

27. $\dfrac{-1 - 3x}{8 - 2x}$, or $\dfrac{3x + 1}{2x - 8}$ **28.** $\dfrac{7 - 3x}{3 - 2x}$, or $\dfrac{3x - 7}{2x - 3}$

Exercise Set 6.6, pp. 454–455

45. $\{x \mid x < -2 \text{ or } x > 5\}$, or $(-\infty, -2) \cup (5, \infty)$
46. $\{x \mid -\frac{7}{3} < x < 3\}$, or $\left(-\frac{7}{3}, 3\right)$ **54. (b)**

(c) The graph of g is the graph of h shifted left 2 units. The graph of f is the graph of g shifted up 3 units.

Exercise Set 6.7, p. 460

7. $3x^2 - 2x + 2 + \dfrac{-3}{x + 3}$
27.
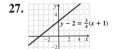
28.

31. (a) The degree of R must be less than 1, the degree of $x - r$; **(b)** Let $x = r$. Then
$$P(r) = (r - r) \cdot Q(r) + R$$
$$= 0 \cdot Q(r) + R$$
$$= R.$$

Exercise Set 6.8, pp. 470–475

3. $v_1 = \dfrac{2s}{t} - v_2$, or $\dfrac{2s - tv_2}{t}$ **4.** $t = \dfrac{2s}{v_1 + v_2}$

5. $f = \dfrac{d_i d_o}{d_o + d_i}$ **6.** $R = \dfrac{r_1 r_2}{r_2 + r_1}$ **7.** $R = \dfrac{2V}{I} - 2r$, or $\dfrac{2V - 2Ir}{I}$ **8.** $r = \dfrac{2V - IR}{2I}$ **12.** $r = \dfrac{nE - IR}{In}$
15. $t_1 = \dfrac{H}{Sm} + t_2$, or $\dfrac{H + Smt_2}{Sm}$ **16.** $H = m(t_1 - t_2)S$
19. $r = 1 - \dfrac{a}{S}$, or $\dfrac{S - a}{S}$ **20.** $a = \dfrac{S - Sr}{1 - r^n}$
21. $a + b = \dfrac{f}{c^2}$ **22.** $c + f = \dfrac{g}{d^2}$ **87.** Ratio is $\dfrac{a + 12}{a + 6}$;
percent increase is $\dfrac{6}{a + 6} \cdot 100\%$, or $\dfrac{600}{a + 6}\%$
89. $t_1 = t_2 + \dfrac{(d_2 - d_1)(t_4 - t_3)}{a(t_4 - t_2)(t_4 - t_3) + d_3 - d_4}$

Cumulative Review: Chapters 1–6, pp. 482–484

28. [4.4]   **29.** [2.4]
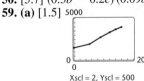

50. [5.7] $(0.3b - 0.2c)(0.09b^2 + 0.06bc + 0.04c^2)$
59. (a) [1.5]

Chapter 7

Exercise Set 7.1, pp. 497–499

30. $\sqrt{11}$; does not exist; $\sqrt{11}$; 12   **31.** -3; -1; does not exist; 0   **32.** $\sqrt{12}$; does not exist; $\sqrt{30}$; does not exist
71. $3(x + 2)$, or $3x + 6$ **72.** $2(x + 1)$, or $2x + 2$
91. $\{x \mid x \geq 5\}$, or $[5, \infty)$ **92.** $\{x \mid x \geq -8\}$, or $[-8, \infty)$
93. $\{t \mid t \geq -3\}$, or $[-3, \infty)$ **94.** $\{x \mid x \geq 7\}$, or $[7, \infty)$
95. $\{x \mid x \leq 5\}$, or $(-\infty, 5]$ **98.** $\{t \mid t \geq -\frac{5}{2}\}$, or $\left[-\frac{5}{2}, \infty\right)$
99. $\{z \mid z \geq -\frac{3}{5}\}$, or $\left[-\frac{3}{5}, \infty\right)$
103. Domain: $\{x \mid x \leq 5\}$, or $(-\infty, 5]$; range: $\{y \mid y \geq 0\}$, or $[0, \infty)$
104. Domain: $\{x \mid x \geq -\frac{1}{2}\}$, or $\left[-\frac{1}{2}, \infty\right)$; range: $\{y \mid y \geq 0\}$, or $[0, \infty)$
105. Domain: $\{x \mid x \geq -1\}$, or $[-1, \infty)$; range: $\{y \mid y \leq 1\}$, or $(-\infty, 1]$
106. Domain: $\{x \mid x \geq \frac{5}{3}\}$, or $\left[\frac{5}{3}, \infty\right)$; range: $\{y \mid y \geq 2\}$, or $[2, \infty)$
107. Domain: $\mathbb{R}$; range: $\{y \mid y \geq 5\}$, or $[5, \infty)$
108. Domain: $\mathbb{R}$; range: $\{y \mid y \leq 4\}$, or $(-\infty, 4]$

Exercise Set 7.2, pp. 505–507

35. $(xy^2z)^{1/5}$ **36.** $(x^3y^2z^2)^{1/7}$ **37.** $(3mn)^{3/2}$
38. $(7xy)^{4/3}$ **39.** $(8x^2y)^{5/7}$

61.

$y = (x + 7)\wedge(1/4)$

62.

$y = (4 - x)\wedge(1/5)$

63.

$y = (3x - 2)\wedge(1/7)$

64.

$y = (2x + 3)\wedge(1/6)$

65.
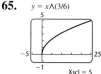
$y = x\wedge(3/6)$

66.

$y = x\wedge(2/8)$

Exercise Set 7.3, pp. 514–515

16. $\sqrt[5]{(x - 2)^3}$ **35.** $f(x) = 5x\sqrt[3]{x^2}$ **36.** $f(x) = 2x^2\sqrt[3]{2}$
37. $f(x) = |7(x - 3)|$, or $7|x - 3|$ **38.** $f(x) = |9(x - 1)|$,
or $9|x - 1|$ **65.** $(x + 5)^2$ **66.** $(a - b)^4$
67. $2ab^3\sqrt[4]{3a}$ **68.** $3x^2y^2\sqrt[4]{xy^3}$

Exercise Set 7.4, pp. 520–522

41. $\dfrac{y\sqrt[3]{180x^2y}}{6x^2}$ **47.** $\dfrac{3\sqrt{5y}}{10xy}$ **48.** $\dfrac{\sqrt{14b}}{8ab}$

Exercise Set 7.5, pp. 527–529

27. $4\sqrt{6} - 4\sqrt{10}$ **29.** $2\sqrt{15} - 6\sqrt{3}$ **30.** $6\sqrt{5} - 4$
49. $\dfrac{12 + 2\sqrt{3} + 6\sqrt{5} + \sqrt{15}}{33}$
50. $\dfrac{3 + 3\sqrt{2} - \sqrt{5} - \sqrt{10}}{4}$
54. $\dfrac{5 - \sqrt{10} - \sqrt{14} + \sqrt{35}}{3}$
55. $\dfrac{24 - 3\sqrt{10} - 4\sqrt{14} + \sqrt{35}}{27}$
56. $\dfrac{-30 - 25\sqrt{6} + 5\sqrt{22} + 2\sqrt{33}}{38}$
61. $\dfrac{2}{14 + 2\sqrt{3} + 3\sqrt{2} + 7\sqrt{6}}$
62. $\dfrac{6}{-12 + 4\sqrt{2} + 3\sqrt{10} - 2\sqrt{5}}$ **64.** $\dfrac{a - b}{a - 2\sqrt{ab} + b}$
72. $2xy^2\sqrt[6]{2x^5y}$ **73.** $xyz\sqrt[6]{x^5yz^2}$ **74.** $a^2b^2c^2\sqrt[6]{a^2bc^2}$
111. $f(x) = -6x\sqrt{5 + x}$ **112.** $f(x) = 2x\sqrt{x - 1}$
113. $f(x) = (x + 3x^2)\sqrt[4]{x - 1}$

Exercise Set 7.6, pp. 536–538

55.

$y > 3x + 5$

56.

$f(x) = \frac{2}{3}x - 7$

61. $t = \dfrac{1}{9}\left(\dfrac{S^2 \cdot 2457}{1087.7^2} - 2617\right)$

Exercise Set 7.7, pp. 544–547

25. $a = 5\sqrt{3} \approx 8.660$; $c = 10\sqrt{3} \approx 17.321$
26. $a = 4\sqrt{2} \approx 5.657$; $b = 4\sqrt{2} \approx 5.657$
27. $a = \dfrac{13\sqrt{2}}{2} \approx 9.192$; $b = \dfrac{13\sqrt{2}}{2} \approx 9.192$
28. $a = \dfrac{7\sqrt{3}}{3} \approx 4.041$; $c = \dfrac{14\sqrt{3}}{3} \approx 8.083$

Chapter 8

Exercise Set 8.1, pp. 573–574

11. $\pm\sqrt{\dfrac{2}{3}}$, or $\pm\dfrac{\sqrt{6}}{3}$ **12.** $\pm\sqrt{\dfrac{7}{5}}$, or $\pm\dfrac{\sqrt{35}}{5}$
53. $\dfrac{-5 \pm \sqrt{13}}{2}$

Exercise Set 8.4, pp. 593–594

56.

$y = -\frac{3}{7}x + 4$

60. $\left(\dfrac{-b + \sqrt{b^2 - 4ac}}{2a}\right)\left(\dfrac{-b - \sqrt{b^2 - 4ac}}{2a}\right) =$
$\dfrac{b^2 - (b^2 - 4ac)}{4a^2} = \dfrac{4ac}{4a^2} = \dfrac{c}{a}$
64. $\dfrac{-b + \sqrt{b^2 - 4ac}}{2a} + \dfrac{-b - \sqrt{b^2 - 4ac}}{2a} = \dfrac{-2b}{2a} = -\dfrac{b}{a}$
65. The solutions of $ax^2 + bx + c = 0$ are
$x = \dfrac{-b \pm \sqrt{b^2 - 4ac}}{2a}$. When there is just one solution,
$b^2 - 4ac$ must be 0, so $x = \dfrac{-b \pm 0}{2a} = \dfrac{-b}{2a}$.

Interactive Discovery, p. 602

1. (a) $(0, 0)$; **(b)** $x = 0$; **(c)** upward; **(d)** narrower
2. (a) $(0, 0)$; **(b)** $x = 0$; **(c)** upward; **(d)** wider
3. (a) $(0, 0)$; **(b)** $x = 0$; **(c)** upward; **(d)** wider

4. (a) (0, 0); **(b)** $x = 0$; **(c)** downward; **(d)** neither narrower nor wider
5. (a) (0, 0); **(b)** $x = 0$; **(c)** downward; **(d)** narrower
6. (a) (0, 0); **(b)** $x = 0$; **(c)** downward; **(d)** wider
7. When $a > 1$, the graph of $y = ax^2$ is narrower than the graph of $y = x^2$. When $0 < a < 1$, the graph of $y = ax^2$ is wider than the graph of $y = x^2$.
8. When $a < -1$, the graph of $y = ax^2$ is narrower than the graph of $y = x^2$ and the graph opens downward. When $-1 < a < 0$, the graph of $y = ax^2$ is wider than the graph of $y = x^2$ and the graph opens downward.

Exercise Set 8.6, pp. 607–609

1. (a) Positive; **(b)** (3, 1); **(c)** $x = 3$; **(d)** $[1, \infty)$
2. (a) Negative; **(b)** (−1, 2); **(c)** $x = -1$; **(d)** $(-\infty, 2]$
3. (a) Negative; **(b)** (−2, −3); **(c)** $x = -2$; **(d)** $(-\infty, -3]$
4. (a) Positive; **(b)** (2, 0); **(c)** $x = 2$; **(d)** $[0, \infty)$
5. (a) Positive; **(b)** (−3, 0); **(c)** $x = -3$; **(d)** $[0, \infty)$
6. (a) Negative; **(b)** (1, −2); **(c)** $x = 1$; **(d)** $(-\infty, -2]$

13.   **14.**

15.   **16.**

17.   **18.**

19.   **20.**

21.   **22.**

23. Vertex: (−1, 0);
axis of symmetry: $x = -1$

24. Vertex: (−4, 0);
axis of symmetry: $x = -4$

25. Vertex: (2, 0);
axis of symmetry: $x = 2$

26. Vertex: (1, 0);
axis of symmetry: $x = 1$

27. Vertex: (−4, 0);
axis of symmetry: $x = -4$

28. Vertex: (2, 0);
axis of symmetry: $x = 2$

29. Vertex: (−1, 0);
axis of symmetry: $x = -1$

30. Vertex: (−4, 0);
axis of symmetry: $x = -4$

31. Vertex: (3, 0);
axis of symmetry: $x = 3$

32. Vertex: (2, 0);
axis of symmetry: $x = 2$

33. Vertex: (1, 0);
axis of symmetry: $x = 1$

34. Vertex: (−2, 0);
axis of symmetry: $x = -2$

35. Vertex: (5, 1);
axis of symmetry: $x = 5$;
minimum: 1

36. Vertex: (−3, −2);
axis of symmetry: $x = -3$;
minimum: −2

37. Vertex: $(-1, -2)$;
axis of symmetry: $x = -1$;
minimum: -2

$f(x) = (x + 1)^2 - 2$

38. Vertex: $(2, -4)$;
axis of symmetry: $x = 2$;
maximum: -4

$g(x) = -(x - 2)^2 - 4$

39. Vertex: $(1, -3)$;
axis of symmetry: $x = 1$;
maximum: -3

$h(x) = -2(x - 1)^2 - 3$

40. Vertex: $(-1, 4)$;
axis of symmetry: $x = -1$;
maximum: 4

$h(x) = -2(x + 1)^2 + 4$

41. Vertex: $(-4, 1)$;
axis of symmetry: $x = -4$;
minimum: 1

$f(x) = 2(x + 4)^2 + 1$

42. Vertex: $(5, -3)$;
axis of symmetry: $x = 5$;
minimum: -3

$f(x) = 2(x - 5)^2 - 3$

43. Vertex: $(1, 2)$;
axis of symmetry: $x = 1$;
maximum: 2

$g(x) = -\frac{3}{2}(x - 1)^2 + 2$

44. Vertex: $(-2, -1)$;
axis of symmetry: $x = -2$;
minimum: -1

$g(x) = \frac{3}{2}(x + 2)^2 - 1$

45. Vertex: $(9, 5)$; axis of symmetry: $x = 9$; minimum: 5; range: $[5, \infty)$
46. Vertex: $(-5, -8)$; axis of symmetry: $x = -5$; minimum: -8; range: $[-8, \infty)$
47. Vertex: $(-6, 11)$; axis of symmetry: $x = -6$; maximum: 11; range: $(-\infty, 11]$
48. Vertex: $(7, -9)$; axis of symmetry: $x = 7$; maximum: -9; range: $(-\infty, -9]$
49. Vertex: $\left(-\frac{1}{4}, -13\right)$; axis of symmetry: $x = -\frac{1}{4}$; minimum: -13; range: $[-13, \infty)$
50. Vertex: $\left(\frac{1}{4}, 19\right)$; axis of symmetry: $x = \frac{1}{4}$; minimum: 19; range: $[19, \infty)$
51. Vertex: $(-4.58, 65\pi)$; axis of symmetry: $x = -4.58$; minimum: 65π; range: $[65\pi, \infty)$
52. Vertex: $\left(38.2, -\sqrt{34}\right)$; axis of symmetry: $x = 38.2$; minimum: $-\sqrt{34}$; range: $\left[-\sqrt{34}, \infty\right)$

55.

$2x - 7y = 28$

56.

$6x - 3y = 36$

63. $f(x) = \frac{3}{5}(x - 4)^2 + 1$
64. $f(x) = \frac{3}{5}(x - 2)^2 + 6$
65. $f(x) = \frac{3}{5}(x - 3)^2 - 1$
66. $f(x) = \frac{3}{5}(x - 5)^2 - 6$
67. $f(x) = \frac{3}{5}(x + 2)^2 - 5$
68. $f(x) = \frac{3}{5}(x + 4)^2 - 2$

Exercise Set 8.7, pp. 616–617

7. (a) Vertex: $(-2, 1)$;
axis of symmetry: $x = -2$;
(b)

$f(x) = x^2 + 4x + 5$

8. (a) Vertex: $(-1, -6)$;
axis of symmetry: $x = -1$;
(b)

$f(x) = x^2 + 2x - 5$

9. (a) Vertex: $(-4, 4)$;
axis of symmetry: $x = -4$;
(b)

$f(x) = x^2 + 8x + 20$

10. (a) Vertex: $(5, -4)$;
axis of symmetry: $x = 5$;
(b)

$f(x) = x^2 - 10x + 21$

11. (a) Vertex: $(4, -7)$;
axis of symmetry: $x = 4$;
(b)

$h(x) = 2x^2 - 16x + 25$

12. (a) Vertex: $(-4, -9)$;
axis of symmetry: $x = -4$;
(b)

$h(x) = 2x^2 + 16x + 23$

13. (a) Vertex: $(1, 6)$;
axis of symmetry: $x = 1$;
(b)

$f(x) = -x^2 + 2x + 5$

14. (a) Vertex: $(-1, 8)$;
axis of symmetry: $x = -1$;
(b)

$f(x) = -x^2 - 2x + 7$

15. (a) Vertex: $\left(-\frac{3}{2}, -\frac{49}{4}\right)$;
axis of symmetry: $x = -\frac{3}{2}$;
(b)

$g(x) = x^2 + 3x - 10$

16. (a) Vertex: $\left(-\frac{5}{2}, -\frac{9}{4}\right)$;
axis of symmetry: $x = -\frac{5}{2}$;
(b)

$g(x) = x^2 + 5x + 4$

17. (a) Vertex: $\left(-\frac{7}{2}, -\frac{49}{4}\right)$;
axis of symmetry: $x = -\frac{7}{2}$;
(b)

$h(x) = x^2 + 7x$

18. (a) Vertex: $\left(\frac{5}{2}, -\frac{25}{4}\right)$;
axis of symmetry: $x = \frac{5}{2}$;
(b)

$h(x) = x^2 - 5x$

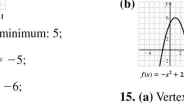

19. (a) Vertex: $(-1, -4)$;
axis of symmetry: $x = -1$;
(b)

$f(x) = -2x^2 - 4x - 6$

20. (a) Vertex: $(1, 5)$;
axis of symmetry: $x = 1$;
(b)

$f(x) = -3x^2 + 6x + 2$

21. (a) Vertex: $\left(\frac{5}{6}, \frac{1}{12}\right)$;
axis of symmetry: $x = \frac{5}{6}$;
(b)

$f(x) = -3x^2 + 5x - 2$

22. (a) Vertex: $\left(-\frac{7}{6}, \frac{73}{12}\right)$;
axis of symmetry: $x = -\frac{7}{6}$;
(b)

$f(x) = -3x^2 - 7x + 2$

23. (a) Vertex: $\left(-4, -\frac{5}{3}\right)$;
axis of symmetry: $x = -4$;
(b)

$h(x) = \frac{1}{3}x^2 + 4x + \frac{19}{3}$

24. (a) Vertex: $\left(3, -\frac{5}{2}\right)$;
axis of symmetry: $x = 3$;
(b)

$h(x) = \frac{1}{2}x^2 - 3x + 2$

31. $\left(3 - \sqrt{6}, 0\right), \left(3 + \sqrt{6}, 0\right)$; $(0, 3)$

32. $\left(\frac{-5 - \sqrt{17}}{2}, 0\right), \left(\frac{-5 + \sqrt{17}}{2}, 0\right)$; $(0, 2)$

38. $\left(\frac{3 - \sqrt{6}}{2}, 0\right), \left(\frac{3 + \sqrt{6}}{2}, 0\right)$; $(0, 3)$

41. (a) Minimum: -6.95; **(b)** $(-1.06, 0), (2.41, 0)$; $(0, -5.89)$
42. (a) Maximum: 7.01; **(b)** $(-0.40, 0), (0.82, 0)$; $(0, 6.18)$
43. (a) Maximum: -0.45; **(b)** no x-intercept; $(0, -2.79)$
44. (a) Minimum: 11.28; **(b)** no x-intercept; $(0, 12.92)$
59. $f(x) = \frac{5}{16}x^2 - \frac{15}{8}x - \frac{35}{16}$, or $f(x) = \frac{5}{16}(x - 3)^2 - 5$
61. **62.** **63.**

$f(x) = |x^2 - 1|$ $f(x) = |x^2 - 3x - 4|$ $f(x) = |2(x - 3)^2 - 5|$

Exercise Set 8.9, pp. 639–641

9. $(-\infty, -7] \cup [2, \infty)$, or $\{x \mid x \le -7 \text{ or } x \ge 2\}$
18. $(-\infty, -4) \cup (-2, \infty)$, or $\{x \mid x < -4 \text{ or } x > -2\}$
25. $[-0.78, 1.59]$, or $\{x \mid -0.78 \le x \le 1.59\}$
26. $(-\infty, -0.21] \cup [2.47, \infty)$, or $\{x \mid x \le -0.21 \text{ or } x \ge 2.47\}$
27. $(-\infty, -2) \cup (1, 3)$, or $\{x \mid x < -2 \text{ or } 1 < x < 3\}$
28. $(-2, 1) \cup (1, \infty)$, or $\{x \mid -2 < x < 1 \text{ or } x > 1\}$
31. $(-\infty, -1] \cup (5, \infty)$, or $\{x \mid x \le -1 \text{ or } x > 5\}$
33. $\left[-\frac{2}{3}, 2\right)$, or $\{x \mid -\frac{2}{3} \le x < 2\}$
34. $\left(-\infty, -\frac{3}{4}\right) \cup \left[\frac{5}{2}, \infty\right)$, or $\{x \mid x < -\frac{3}{4} \text{ or } x \ge \frac{5}{2}\}$
37. $(-\infty, -1] \cup [2, 5)$, or $\{x \mid x \le -1 \text{ or } 2 \le x < 5\}$
38. $[-4, -3) \cup [1, \infty)$, or $\{x \mid -4 \le x < -3 \text{ or } x \ge 1\}$
39. $(-\infty, -3) \cup [0, \infty)$, or $\{x \mid x < -3 \text{ or } x \ge 0\}$

43. $(-\infty, -4) \cup [1, 3)$, or $\{x \mid x < -4 \text{ or } 1 \le x < 3\}$
44. $(-7, -2] \cup (2, \infty)$, or $\{x \mid -7 < x \le -2 \text{ or } x > 2\}$
45. $\left(0, \frac{1}{4}\right)$, or $\{x \mid 0 < x < \frac{1}{4}\}$
46. $(-\infty, 0) \cup \left[\frac{1}{5}, \infty\right)$, or $\{x \mid x < 0 \text{ or } x \ge \frac{1}{5}\}$
57. $\left(-1 - \sqrt{6}, -1 + \sqrt{6}\right)$, or $\{x \mid -1 - \sqrt{6} < x < -1 + \sqrt{6}\}$
60. $\left(-\infty, \frac{1}{4}\right] \cup \left[\frac{5}{2}, \infty\right)$, or $\{x \mid x \le \frac{1}{4} \text{ or } x \ge \frac{5}{2}\}$
65. $f(x)$ has no zeros; $f(x) < 0$ for $(-\infty, 0)$, or $\{x \mid x < 0\}$; $f(x) > 0$ for $(0, \infty)$, or $\{x \mid x > 0\}$
66. $f(x) = 0$ for $x = 0, 1$; $f(x) < 0$ for $(0, 1)$, or $\{x \mid 0 < x < 1\}$; $f(x) > 0$ for $(1, \infty)$, or $\{x \mid x > 1\}$
67. $f(x) = 0$ for $x = -1, 0$; $f(x) < 0$ for $(-\infty, -3) \cup (-1, 0)$, or $\{x \mid x < -3 \text{ or } -1 < x < 0\}$; $f(x) > 0$ for $(-3, -1) \cup (0, 2) \cup (2, \infty)$, or $\{x \mid -3 < x < -1 \text{ or } 0 < x < 2 \text{ or } x > 2\}$
68. $f(x) = 0$ for $x = -2, 1, 2, 3$; $f(x) < 0$ for $(-2, 1) \cup (2, 3)$, or $\{x \mid -2 < x < 1 \text{ or } 2 < x < 3\}$; $f(x) > 0$ for $(-\infty, -2) \cup (1, 2) \cup (3, \infty)$, or $\{x \mid x < -2 \text{ or } 1 < x < 2 \text{ or } x > 3\}$

Review Exercises: Chapter 8, pp. 644–646

24. [8.6]
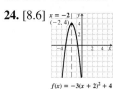
$f(x) = -3(x + 2)^2 + 4$
Maximum: 4

25. [8.7] **(a)** Vertex: $(3, 5)$; axis of symmetry: $x = 3$;
(b)

$f(x) = 2x^2 - 12x + 23$

Test: Chapter 8, pp. 646–647

16. [8.5] $(-3, 0), (-1, 0), \left(-2 - \sqrt{5}, 0\right), \left(-2 + \sqrt{5}, 0\right)$
17. [8.6]
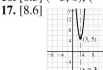
$f(x) = 4(x - 3)^2 + 5$
Minimum: 5

18. [8.7] **(a)** $(-1, -8)$, $x = -1$;
$f(x) = 2x^2 + 4x - 6$

21. [8.8] Quadratic; the data approximate a parabola opening downward.

Chapter 9

Exercise Set 9.1, pp. 662–665

1. $(f \circ g)(1) = 12$; $(g \circ f)(1) = 9$;
$(f \circ g)(x) = 4x^2 + 4x + 4$; $(g \circ f)(x) = 2x^2 + 7$
2. $(f \circ g)(1) = -7$; $(g \circ f)(1) = 4$; $(f \circ g)(x) = 2x^2 - 9$;
$(g \circ f)(x) = 4x^2 + 4x - 4$
3. $(f \circ g)(1) = 20$; $(g \circ f)(1) = 22$; $(f \circ g)(x) = 15x^2 + 5$;
$(g \circ f)(x) = 45x^2 - 30x + 7$
4. $(f \circ g)(1) = 31$; $(g \circ f)(1) = 27$;
$(f \circ g)(x) = 48x^2 - 24x + 7$; $(g \circ f)(x) = 12x^2 + 15$
5. $(f \circ g)(1) = 8$; $(g \circ f)(1) = \frac{1}{64}$; $(f \circ g)(x) = \frac{1}{x^2} + 7$;

$(g \circ f)(x) = \frac{1}{(x + 7)^2}$

6. $(f \circ g)(1) = \frac{1}{9}$; $(g \circ f)(1) = 3$; $(f \circ g)(x) = \frac{1}{(x + 2)^2}$;

$(g \circ f)(x) = \frac{1}{x^2} + 2$

17. $f(x) = x^2$; $g(x) = 7 + 5x$
18. $f(x) = x^2$; $g(x) = 3x - 1$
21. $f(x) = \frac{2}{x}$; $g(x) = x - 3$

22. $f(x) = x + 4$; $g(x) = \frac{3}{x}$

23. $f(x) = \frac{1}{\sqrt{x}}$; $g(x) = 7x + 2$

24. $f(x) = \sqrt{x} - 3$; $g(x) = x - 7$
25. $f(x) = \frac{1}{x} + x$; $g(x) = \sqrt{3x}$

26. $f(x) = \frac{1}{x} - x$; $g(x) = \sqrt{2x}$

43. (a) Yes; **(b)** $g^{-1}(x) = \frac{x + 1}{4}$

44. (a) Yes; **(b)** $g^{-1}(x) = \frac{x + 6}{4}$

47. (a) Yes; **(b)** $f^{-1}(x) = \frac{1}{x}$

48. (a) Yes; **(b)** $f^{-1}(x) = \frac{3}{x}$

49. (a) Yes; **(b)** $f^{-1}(x) = \frac{3x - 1}{2}$

50. (a) Yes; **(b)** $f^{-1}(x) = \frac{5x - 2}{3}$

51. (a) Yes; **(b)** $f^{-1}(x) = \sqrt[3]{x + 5}$
52. (a) Yes; **(b)** $f^{-1}(x) = \sqrt[3]{x - 2}$
53. (a) Yes; **(b)** $g^{-1}(x) = \sqrt[3]{x + 2}$
54. (a) Yes; **(b)** $g^{-1}(x) = \sqrt[3]{x - 7}$

57. (a) Yes; **(b)** $f^{-1}(x) = \sqrt{\dfrac{x - 1}{2}}$

59.

60.

61.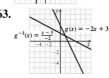

62.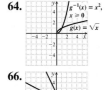

63.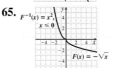

64.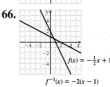

65.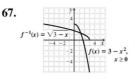

66.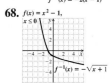

67.

68. $f(x) = x^2 - 1$,

69. (1) $(f^{-1} \circ f)(x) = f^{-1}(f(x)) = f^{-1}\left(\frac{4}{5}x\right) = \frac{5}{4}\left(\frac{4}{5}x\right) = x$;
(2) $(f \circ f^{-1})(x) = f(f^{-1}(x)) = f\left(\frac{5}{4}x\right) = \frac{4}{5}\left(\frac{5}{4}x\right) = x$

70. (1) $(f^{-1} \circ f)(x) = 3\left(\dfrac{x + 7}{3}\right) - 7 = x + 7 - 7 = x$;

(2) $(f \circ f^{-1})(x) = \dfrac{(3x - 7) + 7}{3} = \dfrac{3x}{3} = x$

71. (1) $(f^{-1} \circ f)(x) = f^{-1}(f(x)) = f^{-1}\left(\dfrac{1 - x}{x}\right)$

$= \dfrac{1}{\left(\dfrac{1 - x}{x}\right) + 1}$

$= \dfrac{1}{\dfrac{1 - x + x}{x}}$

$= x$;

(2) $(f \circ f^{-1})(x) = f(f^{-1}(x)) = f\left(\dfrac{1}{x + 1}\right)$

$= \dfrac{1 - \left(\dfrac{1}{x + 1}\right)}{\left(\dfrac{1}{x + 1}\right)}$

$= \dfrac{\dfrac{x + 1 - 1}{x + 1}}{\dfrac{1}{x + 1}} = x$

72. (1) $(f^{-1} \circ f)(x) = \sqrt[3]{x^3 - 5 + 5} = \sqrt[3]{x^3} = x;$
(2) $(f \circ f^{-1})(x) = (\sqrt[3]{x + 5})^3 - 5 = x + 5 - 5 = x$

80. (b) $f^{-1}(x) = \dfrac{x - 24}{2}$, or $\dfrac{x}{2} - 12$

91.   **92.**

96. $((f \circ g) \circ h)(x) = (f \circ g)(h(x))$
$\qquad = f(g(h(x))) = f((g \circ h)(x))$
$\qquad = (f \circ (g \circ h))(x)$

97. Suppose that $h(x) = (f \circ g)(x)$. First, note that for
$I(x) = x, (f \circ I)(x) = f(I(x)) = f(x)$ for any function f.
(i) $((g^{-1} \circ f^{-1}) \circ h)(x) = ((g^{-1} \circ f^{-1}) \circ (f \circ g))(x)$
$\qquad = ((g^{-1} \circ (f^{-1} \circ f)) \circ g)(x)$
$\qquad = ((g^{-1} \circ I) \circ g)(x)$
$\qquad = (g^{-1} \circ g)(x) = x$
(ii) $(h \circ (g^{-1} \circ f^{-1}))(x) = ((f \circ g) \circ (g^{-1} \circ f^{-1}))(x)$
$\qquad = ((f \circ (g \circ g^{-1})) \circ f^{-1})(x)$
$\qquad = ((f \circ I) \circ f^{-1})(x)$
$\qquad = (f \circ f^{-1})(x) = x.$
Therefore, $(g^{-1} \circ f^{-1})(x) = h^{-1}(x).$

Exercise Set 9.2, pp. 673–675

5.   **6.**   **7.**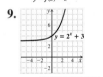

$y = f(x) = 2^x$ $y = f(x) = 3^x$

8.   **9.**   **10.**

11.   **12.**

13.   **14.**

15.   **16.**   **17.**

18.   **19.** **20.**

21.   **22.**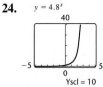

23. $y = 1.7^x$ **24.** $y = 4.8^x$
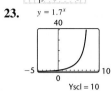 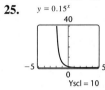
Yscl = 10 Yscl = 10

25. $y = 0.15^x$ **26.** $y = 0.98^x$
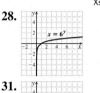 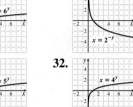
Yscl = 10 Xscl = 10

27.   **28.**   **29.**

30. 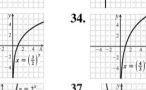 **31.** **32.**

33. 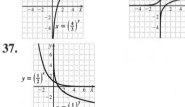 **34.** **35.**

36. **37.**

38.

45. (a) About 6.4 billion; about 6.8 billion; about 7.3 billion;
(b)

$P(t) = 4(1.0164)^t$
World population (in billions)
Years since 1975

46. (a) 4243; 6000; 8485; 12,000; 24,000;
(b)

$N(t) = 3000(2)^{t/20}$
Number of bacteria
Minutes

47. (b)
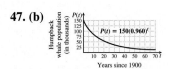

48. (b)
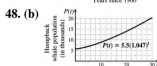

49. (b)
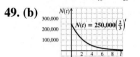

50. (a) $5200; $4160; $3328; $1703.94; $558.35;
(b)
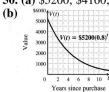

51. (a) 0.3 million, or 300,000; 12.1 million; 490.6 million; 3119.5 million; **(b)**

65.
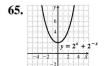

66.

67.
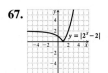

68.

69.

70.

71.

72.

Exercise Set 9.3, pp. 684–685

29.
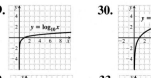

30.

31.

32.

33.

34.

35.

36.

37.
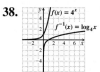

38.

51. $y = \log(x + 2)$ **52.** $y = \log(x - 5)$
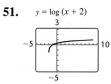

53. $y = \log(1 - 2x)$ **54.** $y = \log(3x + 2.7)$
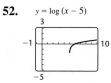
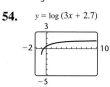

55. $y = \log(x^2)$ **56.** $y = \log(x^2 + 1)$
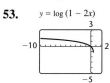
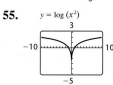

Exercise Set 9.5, pp. 699–700

31.   Domain: $\mathbb{R}$; range: $(0, \infty)$
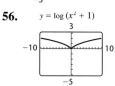

32.   Domain: $\mathbb{R}$; range: $(0, \infty)$
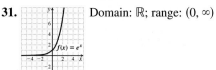

33.   Domain: $\mathbb{R}$; range: $(0, \infty)$
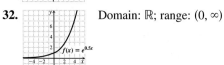
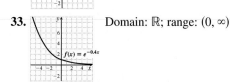
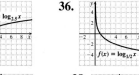

34. Domain: $\mathbb{R}$; range: $(0, \infty)$

35.  Domain: $\mathbb{R}$; range: $(1, \infty)$

36.  Domain: $\mathbb{R}$; range: $(2, \infty)$

37. Domain: $\mathbb{R}$; range: $(-2, \infty)$

38. Domain: $\mathbb{R}$; range: $(-3, \infty)$

39. Domain: $\mathbb{R}$; range: $(0, \infty)$

40. Domain: $\mathbb{R}$; range: $(0, \infty)$

41. 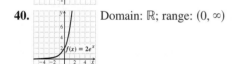 Domain: $\mathbb{R}$; range: $(0, \infty)$

42. Domain: $\mathbb{R}$; range: $(0, \infty)$

43. 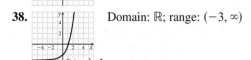 Domain: $\mathbb{R}$; range: $(0, \infty)$

44. $f(x) = e^{x-3}$ Domain: $\mathbb{R}$; range: $(0, \infty)$
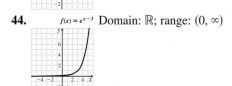

45. 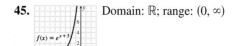 Domain: $\mathbb{R}$; range: $(0, \infty)$

46. Domain: $\mathbb{R}$; range: $(0, \infty)$

47. 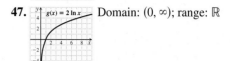 Domain: $(0, \infty)$; range: $\mathbb{R}$

48. 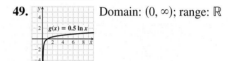 Domain: $(0, \infty)$; range: $\mathbb{R}$

49. Domain: $(0, \infty)$; range: $\mathbb{R}$

50. Domain: $(0, \infty)$; range: $\mathbb{R}$

51. 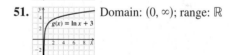 Domain: $(0, \infty)$; range: $\mathbb{R}$

52. Domain: $(0, \infty)$; range: $\mathbb{R}$

53. 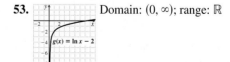 Domain: $(0, \infty)$; range: $\mathbb{R}$

54. Domain: $(0, \infty)$; range: $\mathbb{R}$

55. 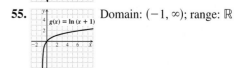 Domain: $(-1, \infty)$; range: $\mathbb{R}$

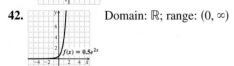

56.   Domain: $(-2, \infty)$; range: $\mathbb{R}$

57.   Domain: $(3, \infty)$; range: $\mathbb{R}$

58.   Domain: $(1, \infty)$; range: $\mathbb{R}$

59. $f(x) = \log(x)/\log(5)$, or $f(x) = \ln(x)/\ln(5)$
$y = \log(x)/\log(5)$, or
$y = \ln(x)/\ln(5)$

60. $f(x) = \log(x)/\log(3)$, or $f(x) = \ln(x)/\ln(3)$
$y = \log(x)/\log(3)$, or
$y = \ln(x)/\ln(3)$

61. $f(x) = \log(x-5)/\log(2)$, or $f(x) = \ln(x-5)/\ln(2)$
$y = \log(x-5)/\log(2)$, or
$y = \ln(x-5)/\ln(2)$

62. $f(x) = \log(2x+1)/\log(5)$, or $f(x) = \ln(2x+1)/\ln(5)$
$y = \log(2x+1)/\log(5)$, or
$y = \ln(2x+1)/\ln(5)$

63. $f(x) = \log(x)/\log(3) + x$, or $f(x) = \ln(x)/\ln(3) + x$
$y = \log(x)/\log(3) + x$, or
$y = \ln(x)/\ln(3) + x$
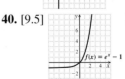

64. $f(x) = \log(x)/\log(2) - x + 1$, or
$f(x) = \ln(x)/\ln(2) - x + 1$
$y = \log(x)/\log(2) - x + 1$, or
$y = \ln(x)/\ln(2) - x + 1$

Exercise Set 9.6, pp. 706–708

21. $\dfrac{\ln 0.08}{-0.07} \approx 36.082$ **22.** $\dfrac{\ln 5}{0.03} \approx 53.648$

23. $\dfrac{\log 3}{\log 3 - \log 2} \approx 2.710$ **24.** $\dfrac{\log 3}{\log 5 - \log 3} \approx 2.151$

25. $\dfrac{\log 65}{\log 7.2} \approx 2.115$ **26.** $\dfrac{\log 87}{\log 4.9} \approx 2.810$

Exercise Set 9.7, pp. 721–727

21. (c)   **22. (c)**

27. (a) $k \approx 0.094$; $W(t) = 17.5e^{-0.094t}$, where t is the number of years since 1996 and $W(t)$ is in millions of tons

49. 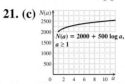   **50.**

51. 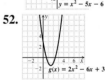   **52.**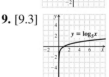

Review Exercises: Chapter 9, pp. 730–732

7. [9.2] 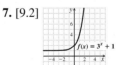   **8.** [9.2]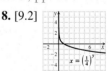

9. [9.3]   **14.** [9.3] $\log_{10} \frac{1}{100} = -2$

40. [9.5]   Domain: $\mathbb{R}$; range: $(-1, \infty)$

41. [9.5] 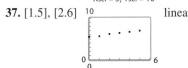 Domain: $(0, \infty)$; range: $\mathbb{R}$

50. [9.6] $\dfrac{\log 8.3}{\log 4} \approx 1.5266$ **51.** [9.6] $\dfrac{\ln 0.03}{-0.1} \approx 35.0656$

Test: Chapter 9, pp. 732–733

28. [9.5] Domain: $\mathbb{R}$; range: $(3, \infty)$

29. [9.5] Domain: $(4, \infty)$; range: $\mathbb{R}$

Cumulative Review: Chapters 1–9, pp. 733–736

35. (b) [2.6] $E(t) = 1.8t + 6.7$, where E is in millions
36. [1.5], [9.7]   exponential

Xscl = 5, Yscl = 10

37. [1.5], [2.6]   linear

38. [9.7] **(a)** $m(x) = 14.85e^{0.058x}$;
(b) $y(x) = 1.830390249(1.059578007)^x$
39. [2.6] **(a)** $f(x) = \frac{3}{10}x + \frac{11}{2}$, where f is in millions;
(b) $y(x) = 0.2685714286x + 5.628571429$
48. [9.7] $P(t) = 19.4e^{0.015t}$, where t is the number of years since 2001 and P is in millions

67. [6.6] $x^3 - 2x^2 - 4x - 12 + \dfrac{-42}{x-3}$

76. [9.1] $f^{-1}(x) = \dfrac{x-7}{-2}$, or $f^{-1}(x) = \dfrac{7-x}{2}$

79. [2.5]   **80.** [8.7]

$5x = 15 + 3y$ $y = 2x^2 - 4x - 1$

81. [9.3]   **82.** [9.1]

$y = \log_3 x$ $y = 3^x$

83. [4.4] $-2x - 3y \le 6$ **84.** [8.7]

$x = -3$
$f(x) = 2(x + 3)^2 + 1$
Minimum: 1

(−3, 1)

Chapter 10

Exercise Set 10.1, pp. 745–747

6. $0, \frac{3}{5}, \frac{4}{5}, \frac{15}{17}, \frac{99}{101}; \frac{112}{113}$ **7.** $2, 2\frac{1}{2}, 3\frac{1}{3}, 4\frac{1}{4}; 10\frac{1}{10}; 15\frac{1}{15}$
8. $1, -\frac{1}{2}, \frac{1}{4}, -\frac{1}{8}; -\frac{1}{512}; \frac{1}{16,384}$ **41.** $\frac{1}{2} + \frac{1}{4} + \frac{1}{6} + \frac{1}{8} + \frac{1}{10} = \frac{137}{120}$
42. $1 + \frac{1}{3} + \frac{1}{5} + \frac{1}{7} + \frac{1}{9} + \frac{1}{11} = \frac{6508}{3465}$
43. $3^0 + 3^1 + 3^2 + 3^3 + 3^4 = 121$
44. $\sqrt{9} + \sqrt{11} + \sqrt{13} + \sqrt{15} \approx 13.7952$
45. $\frac{1}{2} + \frac{2}{3} + \frac{3}{4} + \frac{4}{5} + \frac{5}{6} + \frac{6}{7} + \frac{7}{8} + \frac{8}{9} = \frac{15,551}{2520}$
46. $-\frac{1}{4} + 0 + \frac{1}{6} + \frac{2}{7} = \frac{17}{84}$
47. $(-1)^2 2^1 + (-1)^3 2^2 + (-1)^4 2^3 + (-1)^5 2^4 + (-1)^6 2^5 +$
$(-1)^7 2^6 + (-1)^8 2^7 + (-1)^9 2^8 = -170$
48. $-4^2 + 4^3 - 4^4 + 4^5 - 4^6 + 4^7 - 4^8 = -52,432$
49. $(0^2 - 2 \cdot 0 + 3) + (1^2 - 2 \cdot 1 + 3) +$
$(2^2 - 2 \cdot 2 + 3) + (3^2 - 2 \cdot 3 + 3) + (4^2 - 2 \cdot 4 + 3) +$
$(5^2 - 2 \cdot 5 + 3) = 43$
50. $4 + 2 + 2 + 4 + 8 + 14 = 34$
51. $\dfrac{(-1)^3}{3 \cdot 4} + \dfrac{(-1)^4}{4 \cdot 5} + \dfrac{(-1)^5}{5 \cdot 6} = -\dfrac{1}{15}$
52. $\frac{3}{8} + \frac{4}{16} + \frac{5}{32} + \frac{6}{64} + \frac{7}{128} = \frac{119}{128}$

Exercise Set 10.4, pp. 775–776

53. Consider a set of 5 elements, $\{A, B, C, D, E\}$. List all the subsets of size 3:

$\{A, B, C\}, \{A, B, D\}, \{A, B, E\}, \{A, C, D\}, \{A, C, E\},$
$\{A, D, E\}, \{B, C, D\}, \{B, C, E\}, \{B, D, E\}, \{C, D, E\}$

There are exactly 10 subsets of size 3 and $\binom{5}{3} = 10$, so there are exactly $\binom{5}{3}$ ways of forming a subset of size 3 from a set of 5 elements.
54. $\binom{5}{2}(0.325)^3(0.675)^2 \approx 0.156$
55. $\binom{8}{5}(0.15)^3(0.85)^5 \approx 0.084$
56. $\binom{5}{2}(0.325)^3(0.675)^2 + \binom{5}{3}(0.325)^2(0.675)^3 +$
$\binom{5}{4}(0.325)(0.675)^4 + \binom{5}{5}(0.675)^5 \approx 0.959$
57. $\binom{8}{6}(0.15)^2(0.85)^6 + \binom{8}{7}(0.15)(0.85)^7 +$
$\binom{8}{8}(0.85)^8 \approx 0.89$

58. $\dbinom{n}{n-r} = \dfrac{n!}{[n-(n-r)]!\,(n-r)!}$

$= \dfrac{n!}{r!\,(n-r)!} = \dbinom{n}{r}$

Cumulative Review: Chapters 1–10, pp. 780–783

47. [2.5]

$3x - y = 6$

48. [9.3]

$y = \log_2 x$

49. [4.4]

$2x - 3y < -6$

50. [4.2] $y = \sqrt{(4-x)}$ Domain: $(-\infty, 4]$; range: $[0, \infty)$

51. [8.7]
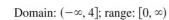
$f(x) = -2(x-3)^2 + 1$
Maximum: 1

Appendixes

Exercise Set A, pp. 793–796

15.

$y = x^2 - 2x + 1$ $(1, 0)$

16.

$y = -\frac{1}{2}x^2$ $(0, 0)$

17.

$x = -y^2 + 2y - 1$

18.

$x = -y^2 - 2y + 3$

19.

$x = -2y^2 - 4y + 1$

20.

$x = 2y^2 + 4y - 1$

61. $(0, 0)$; 7

$x^2 + y^2 = 49$

62. $(0, 0)$; 6

$x^2 + y^2 = 36$

63. $(-1, -3)$; 2

$(x+1)^2 + (y+3)^2 = 4$

64. $(2, -3)$; 1

$(x-2)^2 + (y+3)^2 = 1$

65. $(4, -3)$; $\sqrt{10}$

$(x-4)^2 + (y+3)^2 = 10$

66. $(-5, 1)$; $\sqrt{15}$
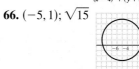
$(x+5)^2 + (y-1)^2 = 15$

67. $(0, 0)$; $\sqrt{7}$

$x^2 + y^2 = 7$

68. $(0, 0)$; $\sqrt{8}$, or $2\sqrt{2}$

$x^2 + y^2 = 8$

69. $(5, 0)$; $\frac{1}{2}$

$(x-5)^2 + y^2 = \frac{1}{4}$

70. $(0, 1)$; $\frac{1}{5}$

$x^2 + (y-1)^2 = \frac{1}{25}$

71. $(-4, 3)$; $\sqrt{40}$, or $2\sqrt{10}$

$x^2 + y^2 + 8x - 6y - 15 = 0$

72. $(-3, 2)$; $\sqrt{28}$, or $2\sqrt{7}$

$x^2 + y^2 + 6x - 4y - 15 = 0$

73. $(4, -1)$; 2

$x^2 + y^2 - 8x + 2y + 13 = 0$

74. $(-3, -2)$; 1

$x^2 + y^2 + 6x + 4y + 12 = 0$

75. $(0, -5)$; 10

$x^2 + y^2 + 10y - 75 = 0$

76. $(4, 0)$; 10

$x^2 + y^2 - 8x - 84 = 0$

77. $\left(-\frac{7}{2}, \frac{3}{2}\right)$; $\sqrt{\frac{98}{4}}$, or $\frac{7\sqrt{2}}{2}$

$x^2 + y^2 + 7x - 3y - 10 = 0$

78. $\left(\frac{21}{2}, \frac{33}{2}\right)$; $\frac{\sqrt{1462}}{2}$

$x^2 + y^2 - 21x - 33y + 17 = 0$

79. $(0, 0)$; $\frac{1}{6}$

$36x^2 + 36y^2 = 1$

80. $(0, 0)$; $\frac{1}{2}$

$4x^2 + 4y^2 = 1$

81. $x^2 + y^2 - 16 = 0$

82. $4x^2 + 4y^2 = 100$

83. $x^2 + y^2 + 14x - 16y + 54 = 0$

84. $x^2 + y^2 - 10x - 11 = 0$

99. Let $P_1 = (x_1, y_1)$, $P_2 = (x_2, y_2)$, and $M = \left(\frac{x_1 + x_2}{2}, \frac{y_1 + y_2}{2}\right)$. Let $d(AB)$ denote the distance from point A to point B.

(i) $d(P_1M) = \sqrt{\left(\frac{x_1 + x_2}{2} - x_1\right)^2 + \left(\frac{y_1 + y_2}{2} - y_1\right)^2}$
$= \frac{1}{2}\sqrt{(x_2 - x_1)^2 + (y_2 - y_1)^2};$

$d(P_2M) = \sqrt{\left(\frac{x_1 + x_2}{2} - x_2\right)^2 + \left(\frac{y_1 + y_2}{2} - y_2\right)^2}$
$= \frac{1}{2}\sqrt{(x_1 - x_2)^2 + (y_1 - y_2)^2}$
$= \frac{1}{2}\sqrt{(x_2 - x_1)^2 + (y_2 - y_1)^2} = d(P_1M).$

(ii) $d(P_1M) + d(P_2M) = \frac{1}{2}\sqrt{(x_2 - x_1)^2 + (y_2 - y_1)^2}$
$+ \frac{1}{2}\sqrt{(x_2 - x_1)^2 + (y_2 - y_1)^2}$
$= \sqrt{(x_2 - x_1)^2 + (y_2 - y_1)^2}$
$= d(P_1P_2).$

Exercise Set B, pp. 808–809

1.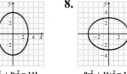
$\frac{x^2}{1} + \frac{y^2}{4} = 1$

2.
$\frac{x^2}{4} + \frac{y^2}{1} = 1$

3.
$\frac{x^2}{25} + \frac{y^2}{9} = 1$

4.
$\frac{x^2}{16} + \frac{y^2}{25} = 1$

5.
$4x^2 + 9y^2 = 36$

6.
$9x^2 + 4y^2 = 36$

7.
$16x^2 + 9y^2 = 144$

8.
$9x^2 + 16y^2 = 144$

9.
$2x^2 + 3y^2 = 6$

10.
$5x^2 + 7y^2 = 35$

11.
$5x^2 + 5y^2 = 125$

12.
$8x^2 + 5y^2 = 80$

13.
$3x^2 + 7y^2 - 63 = 0$

14.
$3x^2 + 8y^2 - 72 = 0$

15.
$8x^2 = 96 - 3y^2$

16.
$6y^2 = 24 - 8x^2$

17.
$16x^2 + 25y^2 = 1$

18.
$9x^2 + 4y^2 = 1$

19.
$$\frac{y^2}{9} - \frac{x^2}{9} = 1$$

20.
$$\frac{x^2}{16} - \frac{y^2}{16} = 1$$

21.
$$\frac{x^2}{4} - \frac{y^2}{25} = 1$$

22.
$$\frac{y^2}{16} - \frac{x^2}{9} = 1$$

23.
$$\frac{y^2}{36} - \frac{x^2}{9} = 1$$

24.
$$\frac{x^2}{25} - \frac{y^2}{36} = 1$$

25.
$$y^2 - x^2 = 25$$

26.
$$x^2 - y^2 = 4$$

27.
$$25x^2 - 16y^2 = 400$$

28.
$$4y^2 - 9x^2 = 36$$

29.
$$xy = -6$$

30.
$$xy = 6$$

31.
$$xy = 4$$

32.
$$xy = -9$$

33.
$$xy = -2$$

34.
$$xy = -1$$

35.
$$xy = 1$$

36.
$$xy = 2$$

Exercise Set C, pp. 817–818

8. $\left(\dfrac{5 + \sqrt{70}}{3}, \dfrac{-1 + \sqrt{70}}{3}\right), \left(\dfrac{5 - \sqrt{70}}{3}, \dfrac{-1 - \sqrt{70}}{3}\right)$

11. $\left(\dfrac{3 + \sqrt{7}}{2}, \dfrac{-1 + \sqrt{7}}{2}\right), \left(\dfrac{3 - \sqrt{7}}{2}, \dfrac{-1 - \sqrt{7}}{2}\right)$

16. $\left(\dfrac{7 - \sqrt{33}}{2}, \dfrac{7 + \sqrt{33}}{2}\right), \left(\dfrac{7 + \sqrt{33}}{2}, \dfrac{7 - \sqrt{33}}{2}\right)$

21. $(0,0), (1,1), \left(-\dfrac{1}{2} + \dfrac{\sqrt{3}}{2}i, -\dfrac{1}{2} - \dfrac{\sqrt{3}}{2}i\right),$
$\left(-\dfrac{1}{2} - \dfrac{\sqrt{3}}{2}i, -\dfrac{1}{2} + \dfrac{\sqrt{3}}{2}i\right)$

27. $\left(\dfrac{4i\sqrt{35}}{7}, \dfrac{6\sqrt{21}}{7}\right), \left(-\dfrac{4i\sqrt{35}}{7}, \dfrac{6\sqrt{21}}{7}\right),$
$\left(\dfrac{4i\sqrt{35}}{7}, -\dfrac{6\sqrt{21}}{7}\right), \left(-\dfrac{4i\sqrt{35}}{7}, -\dfrac{6\sqrt{21}}{7}\right)$

28. $\left(\dfrac{16}{3}, \dfrac{5\sqrt{7}}{3}i\right), \left(\dfrac{16}{3}, -\dfrac{5\sqrt{7}}{3}i\right), \left(-\dfrac{16}{3}, \dfrac{5\sqrt{7}}{3}i\right),$
$\left(-\dfrac{16}{3}, -\dfrac{5\sqrt{7}}{3}i\right)$

29. $\left(-\sqrt{2}, -\sqrt{14}\right), \left(-\sqrt{2}, \sqrt{14}\right), \left(\sqrt{2}, -\sqrt{14}\right),$
$\left(\sqrt{2}, \sqrt{14}\right)$

30. $\left(-3, -\sqrt{5}\right), \left(-3, \sqrt{5}\right), \left(3, -\sqrt{5}\right), \left(3, \sqrt{5}\right)$

38. $\left(2, -\frac{4}{5}\right), \left(-2, -\frac{4}{5}\right), (5, 2), (-5, 2)$

Index

Index of Applications

Index of Calculator References

Videotape and CD Index

Test/Video/ CD Section	Exercise Numbers	Example Numbers
1.1	5, 9, 23, 53	4, 9
1.2	25, 29, 37, 51, 53, 65, 69, 77, 83, 89, 103	3, 7c, 9b, 13
1.3	5, 9, 11, 25, 43, 51	1, 2, 4, 5, 7, 10
1.4	7, 90, 92, 116	1a, 2a, 3a, 3b, 3c, 4b, 6a, 9b, 13
1.5	43	—
1.6	15, 35	8
2.1	7, 9	5, 6, 8
2.2	19, 53	3
2.3	23, 47	8
2.4	9, 19, 53, 79	1, 5, 8, 9
2.5	—	1, 2, 7a, 10
2.6	5, 39	2, 5, 6
2.7	25, 38, 53, 59	1, 2
3.1	5, 9	1, 4b, 4c
3.2	13	1, 3
3.3	37	—
3.4	1, 29, 31	—
3.5	5	1
3.6	1, 9	3
3.7	9	1a, 1b, 1c, 1d, 2
4.1	—	1, 2, 5b
4.2	11, 23	1, 2, 4, 5, 6
4.3	—	5
4.4	13, 49	—
5.1	3, 37, 89	3, 9
5.2	23, 45, 59	1a, 2b, 3, 7a, 8a, 9a, 11a
5.3	75, 99	1, 2, 5, 12
5.4	7, 49, 59	1, 11
5.5	13, 41, 59, 67	1, 3, 4
5.6	25, 33	2b, 3a, 8
5.7	17, 31	1, 2a, 2d, 3
5.8	—	3
6.1	63	2, 5b
6.2	37	3, 4a, 5

Test/Video/ CD Section	Exercise Numbers	Example Numbers
6.3	7, 23, 39	1, 3
6.4	33	1, 3
6.5	10, 21, 24	—
6.6	36	2, 5
6.7	7	1
6.8	1, 9, 31, 43, 61, 63	6, 7, 10
7.1	39, 43, 75, 81, 95	2a, 2c, 3, 9b, 11d
7.2	11, 23, 27, 39, 49, 55, 77, 97	1a, 3a, 4a, 7c, 8d, 9c
7.3	1, 9, 21, 39, 43, 51, 67	2b, 4a
7.4	25, 31, 61	1a, 2a, 4a, 5a, 5c
7.5	25, 37, 71, 81	1a, 1c, 3c
7.6	34	1, 5
7.7	17, 27	1, 5
7.8	13, 17, 21, 31, 45, 49, 55, 71, 73, 79	1a, 2b, 3a, 5, 6a, 7c
8.1	63	2, 7
8.2	9	2, 3
8.3	11, 25	—
8.4	7, 43	1b, 2b
8.5	3, 15	2, 3
8.6	25, 41, 43	1
8.7	11	3
8.8	8	—
8.9	29	1, 5
9.1	27, 29	—
9.2	27, 45	1, 2
9.3	37, 62, 70, 79, 85, 96, 98, 100	—
9.4	7	1, 3b
9.5	15, 63	1, 3
9.6	41	1, 6c
9.7	7, 11, 33	3, 6, 9
10.1	7, 23, 39	3b, 5a
10.2	17, 43	1b, 5
10.3	5, 37, 53, 59	—
10.4	39	4a, 6

Frequently Used Symbols and Formulas

SYMBOLS

$=$	Is equal to
$\approx$	Is approximately equal to
$>$	Is greater than
$<$	Is less than
$\geq$	Is greater than or equal to
$\leq$	Is less than or equal to
$\in$	Is an element of
$\subseteq$	Is a subset of
$\lvert x \rvert$	The absolute value of x
$\{x \mid x...\}$	The set of all x such that $x...$
∞	Infinity
$-x$	The opposite of x
$\sqrt{x}$	The square root of x
$\sqrt[n]{x}$	The nth root of x
LCM	Least Common Multiple
LCD	Least Common Denominator
π	Pi, approximately 3.14
i	$\sqrt{-1}$
$f(x)$	f of x, or f at x
$f^{-1}(x)$	f inverse of x
$(f \circ g)(x)$	$f(g(x))$
e	Approximately 2.7
Σ	Summation
$n!$	Factorial notation

FORMULAS

$m = \dfrac{y_2 - y_1}{x_2 - x_1}$	Slope of a line
$y = mx + b$	Slope–intercept form of a linear equation
$y - y_1 = m(x - x_1)$	Point–slope form of a linear equation
$(A + B)(A - B) = A^2 - B^2$	Product of the sum and difference of the same two terms
$\left.\begin{array}{l}(A + B)^2 = A^2 + 2AB + B^2, \\ (A - B)^2 = A^2 - 2AB + B^2\end{array}\right\}$	Square of a binomial
$d = rt$	Formula for distance traveled
$\dfrac{1}{a} \cdot t + \dfrac{1}{b} \cdot t = 1$	Work principle
$s = 16t^2$	Free-fall distance
$y = kx$	Direct variation
$y = \dfrac{k}{x}$	Inverse variation
$x = \dfrac{-b \pm \sqrt{b^2 - 4ac}}{2a}$	Quadratic formula
$P(t) = P_0 e^{kt}, k > 0$	Exponential growth
$P(t) = P_0 e^{-kt}, k > 0$	Exponential decay
$d = \sqrt{(x_2 - x_1)^2 + (y_2 - y_1)^2}$	Distance formula
$\dbinom{n}{r} = \dfrac{n!}{(n - r)!\, r!}$	$\dbinom{n}{r}$ notation